7/27/94

# GEOLOGY OF
# ENGLAND
# AND WALES

Tucking Mill Cottage, near Bath, Somerset, bears a plaque in memory of William Smith, "Father of English Geology", who lived here for 20 years. Constructed of Bath Stone, the building stands adjacent to the course of the Somerset Coal Canal for which Smith, at the age of 24, surveyed the proposed route. Just off picture to the right the track of a narrow, horse-drawn railway leads to a quarry, whose financial losses led to his bankruptcy and reluctant sale of the property.

R. H. Roberts

Cover design by John McWilliam; View along Crib Goch ridge to Snowdon (left) and Crib y Ddysgyl (right) and the Seven Sisters from Seaford Head, East Sussex.

# GEOLOGY OF
# ENGLAND
## AND
# WALES

Edited by

## P. McL. D. DUFF
*Department of Geology and Geophysics,*
*University of Edinburgh*

and

## A. J. SMITH
*Royal Holloway and Bedford New College,*
*University of London*

1992

Published by

The Geological Society

London

# THE GEOLOGICAL SOCIETY

The Society was founded in 1807 as the Geological Society of London and is thus the oldest geological society in the world. It received its Royal Charter in 1825 for the purpose of 'investigating the mineral structure of the Earth'. The Society is Britain's national learned society for geology with a Fellowship exceeding 6,000. It has countrywide coverage and approximately one quarter of its membership resides overseas. The Society is responsible for promoting all aspects of the geological sciences and will also embrace professional matters on the completion of the reunification with the Institution of Geologists. The Society has its own publishing house to produce its international journals, books and maps, and is the European distributor for materials published by the American Association of Petroleum Geologists.

Fellowship is open to those holding a recognised honours degree in geology or a cognate subject and who have at least two years' relevant postgraduate experience, or have not less than six years' relevant experience in geology or a cognate subject. A fellow who has not less than five years' relevant postgraduate experience in the practice of Geology may apply for validation and subject to approval will be able to use the designatory letters C. Geol (Chartered Geologist). Further information about the Society is available from the Membership Manager, Geological Society, Burlington House, Piccadilly, London W1V 0JU.

Published by The Geological Society from:
The Geological Society Publishing House
Unit 7
Brassmill Enterprise Centre
Brassmill Lane
Bath
Avon BA1 3JN
UK
(*Orders*: Tel: 0225 445046)

First published 1992

**Distributors**

USA
    AAPG Bookstore
    PO Box 979
    Tulsa
    Oklahoma 74101–0979
    USA
(*Orders*: Tel: (918)584–2555)

Australia
    Australian Mineral Foundation
    63 Conyngham Street
    Glenside
    South Australia 5065
    Australia
(*Orders*: Tel: (08)379–0444)

**British Library Cataloguing in Publication Data**
A catalogue record for this book is available from the British Library.

ISBN 0–903317–70–2 (hardback)
ISBN 0–903317–71–0 (paperback)

Printed in Great Britain at the Alden Press, Oxford

# CONTENTS

# CONTRIBUTING AUTHORS

| | |
|---|---|
| D. V. Ager | University College, Swansea |
| M. Audley-Charles | University of London (University College) |
| M. G. Bassett | National Museum of Wales, Cardiff |
| G. S. Boulton | University of Edinburgh |
| J. D. Collinson | University of Bergen |
| J. W. Cowie | University of Bristol |
| D. Curry | formerly University of London (University College) |
| D. L. Dineley | University of Bristol |
| P. McL. D. Duff | University of Edinburgh |
| F. W. Dunning | formerly Geological Museum, London |
| E. H. Francis | University of Leeds |
| D. H. Griffiths | University of Birmingham |
| A. Hallam | University of Birmingham |
| A. L. Harris | University of Liverpool |
| C. H. Holland | Trinity College, Dublin |
| G. Kelling | University of Keele |
| M. R. Leeder | University of Leeds |
| P. F. Rawson | University of London (University College) |
| A. J. Smith | University of London (R. H. B. New College) |
| D. B. Smith | Geoperm & University of Durham |
| G. K. Westbrook | University of Birmingham |
| A. D. Wright | Queen's University, Belfast |

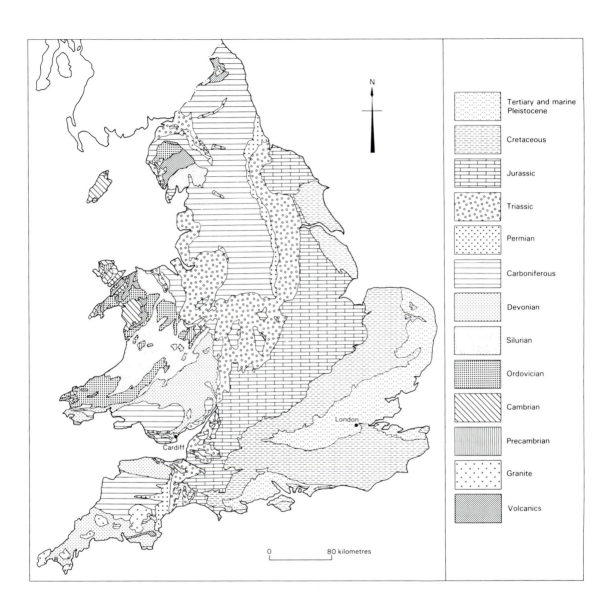

N

| | Tertiary and marine Pleistocene |
| | Cretaceous |
| | Jurassic |
| | Triassic |
| | Permian |
| | Carboniferous |
| | Devonian |
| | Silurian |
| | Ordovician |
| | Cambrian |
| | Precambrian |
| | Granite |
| | Volcanics |

London

Cardiff

0        80 kilometres

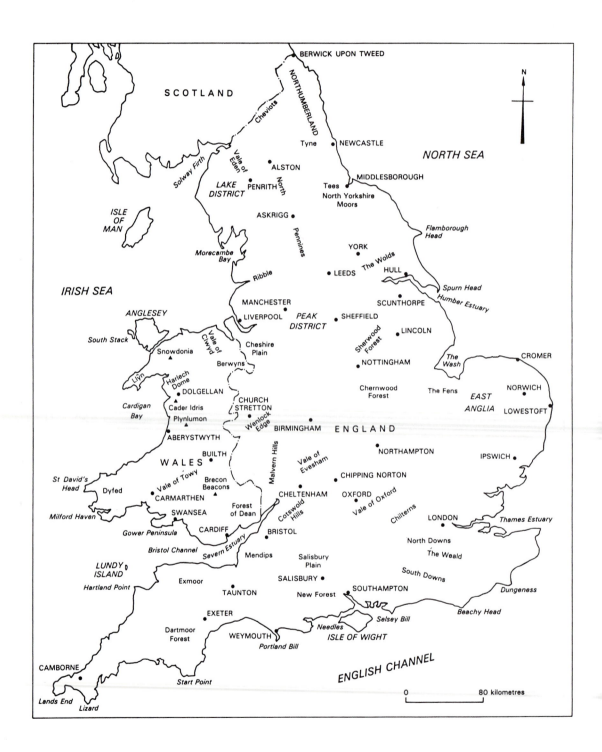

'Old' Administrative Boundaries

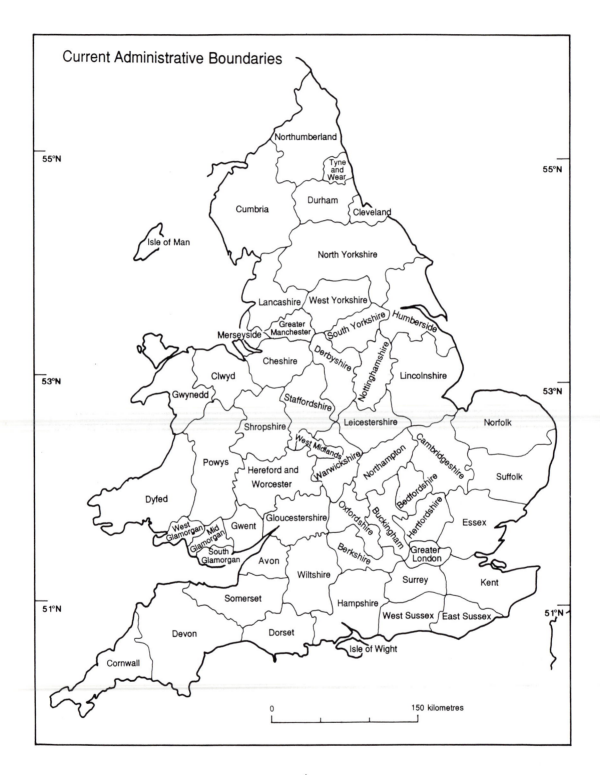

Current Administrative Boundaries

# PREFACE

Since the comprehensive *Handbook of the Geology of Great Britain*, edited by Evans and Stubblefield in 1929, no detailed summary of the geology of England has appeared. Wales has fared better, with two companion volumes being published (Wood (ed.) 1969 and Owen (ed.) in 1974).

Excellent accounts are available, of course, of specific areas and regions of both England and Wales by the British Geological Survey (and its predecessors) and by academics, as are student textbooks on stratigraphy (e.g. Anderton *et al.* 1979 and Rayner 1981). With the publication of the *Geology of Scotland* (1968, 1983) and the *Geology of Ireland* (1981), however, – both of which have new editions in preparation – there was an obvious need for similar comprehensive summaries of England and Wales.

This volume, therefore, represents an attempt to fill a gap, to provide a reference book to the vast literature and to give the most recent available summaries of the complex geology of England and Wales. The two countries are treated together because they are inextricably linked geologically and historically, particularly in the history of the development of geology. The Systems, for example, which now comprise the Palaeozoic were all established in England and Wales: the Carboniferous by William Coneybeare and William Phillips in 1822; the Cambrian by Adam Sedgwick in 1835; the Silurian by Roderick Impey Murchison also in 1835; the Devonian by Sedgwick and Murchison in 1839; the Ordovician, somewhat later, in 1879, by Charles Lapworth. Some of the background to this and the often acrimonious debates which resulted from the difficulties of defining systems are mentioned in Chapter 1. (Nowadays debates are neither so personalised nor so heated but problems still abound with international committees attempting to achieve agreed standards and definitions whilst seeking to satisfy national pride). Many of the contributors to this volume have been deeply involved in such debates and, it should also be added, during the period of the production of this volume. Most contributors have also been embroiled in fundamental (Government imposed), often quite radical changes which have affected the Geological Survey, the country's museums and the geology departments of the universities of Britain over the immediate past decade which it has taken to compile this volume. In addition there has been an exponential increase in geological meetings, symposia and consequent publications during the 1970s and 1980s which have added to the authors' difficulties in successive rewrites. Indeed, the editors are convinced that there can have been few similar works which have undergone so many delays, rewrites and revised deadlines because of events beyond their and their authors' control. Style, fashion and genuine progress in the science are all involved and affect interpretations of the geology of England and Wales. The use and modification of long-established terms give cause for thought: one example concerns the trio *Armorican, Hercynian* and *Variscan*. Faced with some confusion the editors elected only to use the term Armorican with reference to that particular region of NW France and events there. The tectonic events which affected Carboniferous and earlier rocks are generally referred to as the Hercynian, though views, even amongst our collaborators have differed strongly. Variscan is an equivalent form to Hercynian but is generally used in the British context when referring to the northern limit or 'front' of Hercynian compression – the Variscan Front. There has been no attempt to force contributors into a set vocabulary: indeed the very differences in usage may assist the reader in relating contemporary usage to traditional usage.

Variation in the use of proper names for geological features is particularly true for the offshore areas where the very speed of progress has led to the introduction of different names for the same areas and the replacement of names long familiar in maritime circles. The Manx–Furness Basin, sometimes called the Irish Sea Basin and the East Irish Sea Basin, is one such example. Again the editors have tended to allow the authors freedom of action in the belief that the reader will find the synonyms of value.

Even land localities and features are not immune from such variation. Administrative boundaries were changed in England and Wales in 1974 and the 'old' county names, which had been so long established and which had become so much part of geological literature were replaced by new names (see p. xvi). Nowhere was the impact greater than in Wales where all the old counties with the exception of Glamorgan disappeared: gone were Pembrokeshire, Merionethshire, Cardiganshire, Caernarvonshire and the rest to be replaced by Powys, Clwyd, Gwynedd, Dyfed and Gwent — even Glamorgan was divided into West, Mid- and South Glamorgan. Parts of England were similarly renamed. Further complications arise with the introduction of Welsh rather than anglicised spelling for many places and features in Wales: Twyi for Towy is one such example. Both languages appear in the book with the anglicised version being used for geological features (as was the case in the original sources) and the Welsh names being used for localities.

At one stage during the preparation of this book there had been an effort to give grid-references for localities. The editors, however, decided that this interrupted the prose and except for some very specific cases have omitted the grid-references in the text.

Prominent geographical features and towns can generally be located on the map on p. xiv, and many specific localities are shown in the figures illustrating the different chapters. Duplication of almost identical figures does occur in different chapters simply to avoid the inconvenience to the reader of consulting subject matter many pages apart.

The published literature on the geology of England and Wales is enormous and we wonder if any area of similar size in the world has had such detailed treatment! The reader wishing to consult original sources therefore will need to use the libraries of the universities, museums (particularly the National History Museum), the learned Societies and the British Geological Survey. The last named will also be able to assist by informing enquirers about Geological Survey maps and other publications available for purchase.

## Photographs

Those with B.G.S. numbers after the captions are reproduced from British Geological Survey transparencies by kind permission of the Director, British Geological Survey.

Figures 5.2 and 5.3 are reproduced by kind permission of the Curator of Aerial Photography, University of Cambridge, Figures 4.6 and 4.8 by the National Museum of Wales.

Figures 19.4 (a and b) and 19.11 were kindly supplied by Taylor Woodrow Services Ltd. and the British Petroleum Company plc, respectively.

# Bibliography

ANDERTON, R., BRIDGES, P. H., LEEDER, M. R. & SELLWOOD, B. W. 1979. *A dynamic stratigraphy of the British Isles*. George Allen & Unwin, London, 301 pp.

CRAIG, G. Y. (ed.). 1983. *The Geology of Scotland* (Second Edition). Scottish Academic Press, Edinburgh, 472 pp.

EVANS, J. W. & STUBBLEFIELD, C. J. (eds.). 1929. *Handbook of the Geology of Great Britain*. Murby, London, 556 pp.

HOLLAND, C. H. (ed.). 1981. *A Geology of Ireland*. Scottish Academic Press, Edinburgh, 335 pp.

OWEN, T. R. (ed.). 1974. *The Upper Palaeozoic and Post-Palaeozoic Rocks of Wales*. Univ. Wales Press, Cardiff, 426 pp.

RAYNER, D. 1981. *The stratigraphy of the British Isles* (Second Edition). Cambridge University Press, Cambridge, 460 pp.

WOOD, A. (ed.). 1969. *The Pre-Cambrian and Lower Palaeozoic Rocks of Wales*. Univ. Wales Press, Cardiff, 461 pp.

# Acknowledgements

Many thanks are due to Julie Brown and Monica Hayward for their excellent secretarial work and to (in alphabetical order) D. A. Ardus, J. Brooks, Yvonne Cooper, J. M. Dean, R. A. Downing, M. J. Gallagher, Sheila Jones, R. P. McIntosh, M. McGarr, W. L. G. Nash, R. Roberts, Lesley Russell, R. Stoneley, I. Strachan. A. Stride, and C. D. Will, for a variety of ready assistance, and to others acknowledged at the end of specific chapters.

BP Exploration (through the good offices of Drs A. J. Martin and J. P. B. Lovell) kindly paid in advance for copies of the three volumes on the British Isles to provide prize sets for selected University students.

Particular thanks are due to Jean Duff and Anita Kenyon-Smith whose help and encouragement sustained us through many crises, to Dr Douglas Grant of Scottish Academic Press, whose legendary patience and good humour must have been sorely tried by the unconscionable delays and broken deadlines throughout the years this volume was being compiled, to the Officers and Council of the Geological Society and to Mike Collins of the Geological Society Publishing House.

P. McL. D. DUFF                                                    A. J. SMITH
Edinburgh                                                              Egham

1992

## Editors' Note

We would like to acknowledge the forbearance of our contributors and the publisher over the untoward delay that has occurred in the production of this book. This is primarily due to the problems with the Ordovician chapter which was originally intended to be written by two contributors. One, without warning, dropped out leaving the task to Dr M. A. Bassett. Because of the many problems associated with the reorganisation of the museums of this country he found it impossible to complete the chapter and Professor A. D. Wright generously offered assistance. In the event he, too, found his contribution had to be limited, mainly to Northern England, and Dr Bassett once again agreed to complete the chapter. He, however, was hopelessly overcommitted and consequently unable to keep to any time-table or deadline. In the end we were forced, albeit inadequately, to fill in the gaps. *A symposium volume can be as complete as the chapters received by the editor: what purports to be a complete stratigraphic compilation, however, cannot have a whole system or part of a system omitted!*

As previously mentioned, many authors, as chairmen of Departments of Geology, have been (and still are) heavily involved in the unprecedented reorganisation and time-consuming exercise of changing the face of University geology in Britain. Inevitably, some have been unable, due to circumstances beyond their control, to update their chapters as they would have liked. Delay in publication was compounded by the fact that both editors changed employment and places of residence during the production of the volume. To all our authors, therefore, our apologies. We have taken the liberty in some cases where thematic volumes have been published since authors' page proofs were completed of inserting reference to specific volumes in italics in their chapters.

# I

# INTRODUCTION

# THE GROWTH AND STRUCTURE OF ENGLAND AND WALES

# D. V. Ager

## The geological setting of England and Wales

It is, perhaps, appropriate for an Englishman living in Wales to write the introduction to this book, even if it originated – paradoxically – with the Scottish Academic Press.

England and Wales have, together, a very real geological unity and form a critical part of the earth's surface both because of the nature of their rocks and because of their place in the history of our science.

It is not for me to try to summarise the contributions of so many distinguished specialist authors. But there are some generalities about England and Wales which may be considered at this stage.

In the first place, whatever the late General de Gaulle may have thought, England and Wales are clearly part of Europe. Their geological history and structures are closely related to those of the European continent (Fig. 1.1). Admittedly they show some influence of the Atlantic to the west, but this is not so obvious as it is in Scotland and Ireland.

Our Caledonian fold belt continues into Scandinavia, our Variscan fold belt continues into France and on into central Europe (Fig. 1.1). In the very south of our island we can almost smell the alpine meadows on our folded Mesozoic and Palaeogene rocks. Our stratigraphy cannot be studied in isolation but relates directly to that of the neighbouring continent. We are therefore not only an island but, as John Donne said, 'a peece of the Continent, a part of the maine'.

What is more, within this island the border between England and Scotland is surprisingly well placed geologically and gives this book a scientific as well as a

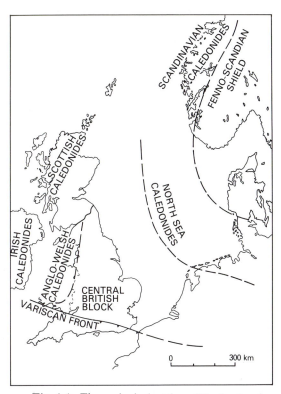

**Fig. 1.1.** The geological setting of England and Wales in relation to the continent of Europe.

1

political logic. The southern edge of the Lower Palaeozoic trough of the Southern Uplands against the Lower Carboniferous rocks of northernmost England coincides remarkably closely with the old frontier that was formerly torn by cattle-raids and controversy. The Cheviot Granite and its associated lava flows, though entirely in England, forms a seal on this ancient suture. It is interesting to note that the first unconformity ever described (by James Hutton) was just over the border where an extension of the flat-lying English Upper Palaeozoic rocks rests on the largely Scottish folded Lower Palaeozoic. Palaeogeographically the border marks the edge of the Proto-Atlantic ocean that some people like to call Iapetus and which later became a line of Caledonian folding.

The border between England and Wales is far less natural. This remark may anger my patriotic Welsh friends, but to console them I may suggest that a more meaningful boundary would be farther to the east. It would be more logical geologically to hand over to Wales all the older rocks at the surface as far as the line of the Malvern Hills. Then the border would lie along the west side of the Worcester Graben which opened and largely filled in Triassic times.

However, whatever may be said to justify or dispute the political linking of England and Wales more closely than the other constituent parts of the United Kingdom, it is more important for us that the geology clearly argues for their being linked in this book.

One of the projects of the International Geological Correlation Programme concerns itself with the problem of the south-western margin of the Russian Platform. The activities of that project are wholly concentrated way over in eastern Europe. It can be argued, however (indeed I have argued it), that the south-western margin of the Russian Platform lies roughly along the borders of England and Wales where the flat-lying Early Palaeozoic sediments of central England are suddenly replaced by the folded rocks of Wales and the Welsh Borderland.

Conventionally one would start any account of the geology of England and Wales with the ancient rocks of Anglesey. Indeed that is the procedure followed by the specialists in the main chapters of this book. But I prefer to think of England and Wales as being centred on the seemingly placid rocks of what I have called the Central British Block. This is delimited to the west by the Anglo-Welsh Caledonides; it is delimited to the south by the frontal structures of the Variscan folding of Late Palaeozoic times and it is delimited to the east by a much less clear but evidently metamorphic Caledonian belt running down the North Sea, only recognised in a few boreholes.

The Lower Palaeozoic rocks of the Central British Block are little disturbed. One thinks of the gentle dip slopes and escarpments such as Wenlock Edge in Shropshire. One thinks of the gentle domes of the Wren's Nest and Dudley near Birmingham. All these seem to rest on Late Precambrian volcanic rocks, as seen in Charnwood Forest near Leicester, and on a comparatively low density continental crust. There was no Caledonian orogeny here and one can see this Central British Block as a separate block of the stable platform of eastern Europe just as the Colorado Plateau in the south-western United States is separated from the Central Stable Region of North America with an orogenic belt going down either side of it. So, looked at in this way, one can say that the Russian Platform extended all the way to the Severn Estuary.

England and Wales are therefore delimited by Caledonian fold belts to the north and east; whilst their western part itself constitutes a paratectonic Caledonian fold belt and its southern part represents the northern ranges of the Variscan fold belt to the south. Indeed the Channel Islands (considered briefly herein) are obviously part of French Armorica and the orthotectonic Variscides.

The geological unity of England and Wales is clearly indicated by a traverse from north-west to south-east across them. If one follows a line south-east from Anglesey to London with a couple of careful slight swerves (north of the Harlech Dome and south of the Longmynd) one achieves an almost perfect and complete ascent from the metamorphic Precambrian to unconsolidated Tertiary in a gentle south-easterly dipping succession. This was roughly the section drawn by William Smith in the early days of geology (Fig. 1.2) when the older rocks had hardly been studied at all.

## The geological diversity of England and Wales

The geological glory of England and Wales is the diversity of their rocks together with the distinguished succession of geologists who unravelled them, though only a very few of the latter may be mentioned here.

We may leave to Scotland its fragment of the Fenno-Scandian Shield in the western highlands and islands but for the rest the southern part of Great Britain is remarkably fortunate in the variety of its rocks, its structures and its natural resources.

Even the basement is not completely hidden for we have exposed in Anglesey and thereabouts a complex and controversial display of very ancient metamorphic rocks associated with Proterozoic sediments and volcanics. Glimpses of some of these are seen again at various places in South Wales, in the Welsh Borderland and in the English Midlands.

We can claim in the rocks of England and Wales to have recognised the main outlines of earth history through Phanerozoic times. It was Celtic tribes living

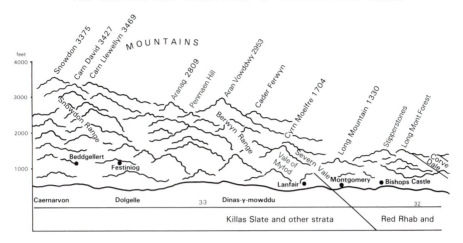

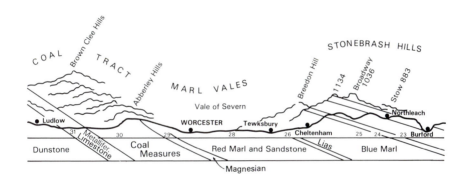

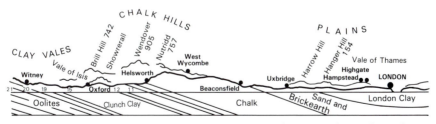

**Fig. 1.2.** William Smith's 1817 cross section from Snowdon to London, redrawn from Fitton, W. H. F. 1832, Notes on the History of English Geology, *Philosophical Magazine*, **1**, Plate 2, Fig. 4. (Described by Fitton, 1833, *op. cit.* **2**, p. 54, as "one of the most perfect sections of any portion of the globe so complex, which ever had been produced" (Eds.).

in what is now England and Wales that gave their names to the Cambrian, Ordovician and Silurian Systems. The beautiful south-western county of Devon gave its name to the Devonian, which was so nearly called the Erian after Lake Erie.

Lower in the stratigraphical scale we have many international stages (and so-called stages) named after English and Welsh localities, from the Thanetian to the Tremadocian, from the Wolstonian to the Wenlockian and from the Callovian to the Caradocian. I am particularly fond of the last, which was named after Caradoc or Caractacus, the Celtic chief who made his last heroic stand against the Roman invaders at so many different places in the Welsh Borderland. That confusing point in itself illustrates the lack of the obvious 'stratotypes' which are so beloved by the stratigraphers of the European continent. Then there is a whole great heap of trivial names in the Quaternary nomenclature derived from ephemeral and inadequate sections in East Anglia. As a genuine Angle, ancestrally originating from that part of the world, I can justifiably reject the chauvinism that seems to go with claiming the 'real', the only true Ruritanian or whatever it is. I have no enthusiasm for these divisions of our geological yesterday, though many will certainly leap up to defend the names of their own favourite backyards. Geology is in every sense an international subject and it is ironic to remember that all those familiar stages of the Jurassic, such as Oxfordian and Kimmeridgian, which remind one of English towns and villages, were named by a Frenchman (Alcide d'Orbigny) who never visited England. It is also both ironic and paradoxical (and the English love paradoxes) that the system that is probably the best displayed in Britain (and where the science of stratigraphy started) is the Jurassic, which was then named after the beautiful mountains of France and Switzerland, whilst the places after which d'Orbigny named his Jurassic stages are almost all lacking in complete or even useful sections. But if we look to priority then most of the names and many of the studies started here and we have an international responsibility to provide all the information we can and to welcome visiting geologists from abroad.

Nevertheless, if we were truly international in our attitudes, we would probably agree that it would be better sometimes to look elsewhere for standards than in those overgrown ditches and abandoned quarries that provide so much of our inland geology in southern Britain.

However, it should also be said that we do have a great advantage in being an island, not only from the point of view of unity and security, but also in that we have so many magnificent coast sections from Kent to Cornwall and from Dyfed to Durham. And our mountains and hills, though minor by the standards of most countries, provide us with splendid views of what used to be called the 'Mountain Limestone' of Early Carboniferous age, of the Ordovician volcanics of Snowdonia and the Lake District and the Permian granites of Devon and Cornwall. All these also happen to be the locations of our most valuable metalliferous mineralisation, though again it is minor by international standards.

Our cliffs and hills display rocks belonging to every division of geological time with the single, rather minor, exception of the Miocene. That too, we now know, is well developed within the thick Tertiary sediments of the central North Sea. For we must also remember that the geology of England and Wales has, in a sense, expanded considerably in recent years. This is not through imperialistic conquest, but from our increasing knowledge of the geology of the sea-floor around our island.

Up to a point our study of the continental shelf – though it has yielded great economic benefits – has provided no great geological surprises. With the benefit of hindsight we may say that there is very little there we should not have expected from the onshore evidence. Thus the English Channel simply forms a direct connecting link between England and France with the geomorphology perhaps more surprising than the geology. The North Sea has revealed metamorphosed Lower Palaeozoic rocks, marine Devonian quite far to the north and unexpected volcanics in the Jurassic and Cretaceous. There were, however, good on-shore reasons for suspecting all these things. Even the hydrocarbon-filled grabens should not have taken us entirely by surprise from the evidence on the east coast of Scotland and Yorkshire and that in the Netherlands.

Apart from the completeness of our record and the serendipitous juxtaposition of fold belts of different ages in England and Wales, we are also fortunate in other ways. Thus we have had a fascinating record of plate migration with changing climates. Our island has been covered from time to time by coral seas, tropical forests, arid deserts and great ice sheets. For long periods we seem to have straddled the northern limits of coral reef formation and evaporite deposition. During much of our history we have welcomed Boreal immigrants from the north and Tethyan immigrants from the south.

Our fossil record is remarkably complete, from Precambrian jelly-fish to Pleistocene elephants. We have most of the major groups of organisms represented here for most of the periods.

We have all kinds of igneous rocks, intruded and extruded in oceanic and continental settings. England and Wales are a little thin, admittedly, on metamorphic rocks and on mineralisation, but at least we had enough of the latter to start the Industrial Revolution.

## Anglo-Welsh geologists

We seem to enjoy intellectual self-flagellation in Britain and positively boast of the many things which we discovered or invented, but which were then developed and exploited by others. So it is with the geological record. We tell our students yet again the story of Adam Sedgwick, the cleric, climbing eastwards up his Cambrian System in Wales, and Roderick Murchison, the old soldier, climbing westwards down his Silurian System from England. We hear again how they overlapped and quarrelled whilst Charles Lapworth waited until they were both safely dead and then named the Ordovician System for the disputed strata. Then there was Lonsdale and his difficult Devonian System founded on those confused rocks down in south-west England. It is recorded in the *History of the Geological Society of London* that when Robert Etheridge was sent to north Devon to sort out the local rocks, he sat down at the end of his first day in the field and 'wept at the immensity of his task'.

So we can climb up the Anglo-Welsh succession following the footsteps of more and more geological pioneers.

We always call William Smith, the canal engineer and mine surveyor, born and bred on the beautiful Jurassic rocks of the Cotswolds, the 'Father of English Geology'. (We may remember that he remained a bachelor all his life and draw our own conclusions!) His demonstration in 1799 to his friends on Dundry Hill near Bath of the succession of local rocks and their fossils was perhaps the first geological excursion.

There had been others before him who recognised certain distinctive distributions of rocks and minerals, but as a German writer has said, William Smith 'opened the book of earth history' and his work was truly the beginning of historical geology as we know it today.

Then there was Henry De la Beche (English in spite of his gallicised name) who founded what is now the **British Geological Survey in 1835, the** first in the world. There was Gideon Mantell, the Sussex doctor, who recognised the first dinosaurs, and young Mary Anning, down in Dorset, finding the first giant marine reptiles.

We must also remember the wealthy English lawyer and Scottish landowner, Charles Lyell, who – without being particularly interested in strata as such – laid the foundations of modern geology on the basis of theorising about processes going on at the present day. Charles Darwin had a copy of Lyell's great book *Principles of Geology* on his voyage around the world in the H.M.S. *Beagle* as did the young Lenin in his study bedroom at Simbirsk on the Volga. Lyell is always given credit for the doctrine of uniformitarianism as the basis for all modern geological thought and though we have, in recent years, gone back a little towards the so-called 'catastrophism' which he rejected, there is no doubt that geology would not have reached its present form without the deep thought and persuasive pen of Charles Lyell. Indeed plate tectonics itself is another application of the Lyellian principle of studying the world as it is today and then using that knowledge in the interpretation of the past. It was Lyell too who first divided Cainozoic sediments into so-called systems on the basis of the percentage of still-living species which they contain.

There are many other great English and Welsh geologists I should mention, people such as W. J. Arkell who did so much for the Jurassic both of Britain and the world, great palaeontologists such as the Sowerbys, father and son, whose names are appended to so many fossil names. There was that remarkable man H. C. Sorby who introduced the study of rocks in thin section, and Charles Darwin himself figures in the history of our Geological Society, mentioned above, as a geologist not as a biologist. But the list could be almost endless and it would still be invidious.

## The voyage of England and Wales through space and time

The chapters that follow will consider the rocks of England and Wales as we see them at the moment and their record of the past. It is useful at this point, however, to consider how these rocks have come to be where they are today and the changing geographies through which they have passed.

The earliest rocks are certainly closely related to those of eastern North America and cannot be considered separately. Palaeomagnetic and other evidence is not really good enough to say where exactly England and Wales were situated on the globe at the beginning of our record. In fact I would regard such data as highly suspect for all of Precambrian and much of Early Palaeozoic time.

Red fluviatile deposits, such as the Wentnor Group of the Late Proterozoic Longmyndian (probably the youngest of our Precambrian rocks), lend support, perhaps, to the reconstructions that place England and Wales right on the equator at the beginning of Phanerozoic times. The complex of late Precambrian sediments, metamorphics and intrusives in north-west Wales (including pillow-lavas and tiny exposures of serpentinite) have been interpreted as all that is left of a subduction zone along the margin of some long-departed ocean. It is tempting to postulate that this was some very early version of the Atlantic that was still to live a life of repeated opening and closing before it came to be what it is today. Certainly these Precambrian igneous rocks are now marginal to the Caledonian paratectonic belt which crosses our country from south-west to north-east. On the other hand the more acidic Proterozoic igneous rocks of Shropshire were presumably derived from continental crust. So the Precambrian rocks can be related to the Caledonian story which followed.

In Early Palaeozoic times there developed a great trough of muddy sediments running from south-west to north-east across Wales and on into the English Lake District. This was long called a geosyncline following O. T. Jones's classic survey (1938) and later a eugeosyncline. Though these terms are no longer fashionable, the rocks remain the same and there is little doubt that there were deeper water deposits that passed eastwards into the shallow water sandstones, shales and limestones of the English Midlands. Another trough probably extended down the east side of England and on into the Ardennes.

In global terms England and Wales are usually placed in the tropics at this time, albeit in the southern hemisphere. However, one notes the absence of much in the way of red beds and certainly of evaporites such as are developed so thickly in the Silurian of the Great Lakes region. We must picture England and Wales in Early Palaeozoic times as lying on the south-eastern margin of an ocean which I will call the Proto-Atlantic, but others call the Iapetus. On the other shore of this ocean lay what is now north-west Scotland and most of North America. This ocean was to close in Silurian times with the resultant collison of the continents producing the Caledonian mountains or Caledonides.

The Anglo-Welsh Caledonides are all paratectonic in modern jargon. That is to say they were not affected by the intense tectonism and metamorphism that made the so-called orthotectonic Caledonides of the Scottish highlands and Ireland so much less interesting to my kind of geologist. In spite of the great depth of burial of the oldest Phanerozoic sediments in North Wales they have barely reached the lowest grade of metamorphism.

The intense volcanicity which started in the Cambrian and reached its climax in the Ordovician, with the craggy mountains of North Wales and the Lake District, is nowadays related to the subduction zone as that early version of the Atlantic Ocean was consumed beneath the European continent. However, it is reassuring to find that the oceanic influence is less marked in the volcanics as one moves eastwards until they disappear altogether under the rigid block of central England.

The Caledonian orogeny brought the Early Palaeozoic Anglo-Welsh trough to an end. The volcanicity died away in the Silurian and the new mountains rose parallel to the old depositional axis.

Our interpretation of the Caledonian structure of Britain has been considerably modified by the coming of the concept of 'displaced terranes'. These may be defined as large units along active plate margins limited by major lateral faults and with distinctive geological histories. The adjective 'suspect' has been added to those terranes whose origins could not be proved and it is felt by some of us that we are here departing from the fine old British legal tradition of 'innocent until proved guilty'. When guilt has been fully proved and the unit is thought to have arrived from outside the area concerned, then the adjective 'exotic' may be substituted.

Undoubtedly the terrane hypothesis applies equally up and down the stratigraphical record and indeed it might be said that it has always been acceptable when one considers some of the anomalies of structure and of allochthonous faunas and floras. Terrane movements may well have been responsible for the main deformations of the orogenic belts, as I have suggested for the equivalent microplates in the African system (Ager 1980).

Devonian times saw the differentiation of a completely new palaeogeography over the area that is now England and Wales. Evidently our home-to-be was sailing slowly northwards through the tropics. In the south a new trough developed in which were deposited the coral and brachiopod limestones of south Devon, notably around Torquay and Plymouth. Reefs developed towards the end of Devonian times, but on an insignificant scale compared with those of the same age in other parts of the world as far distant as Alberta and Western Australia. There are also marine shales and pillow lavas and even a mid Palaeozoic complex, including serpentinites, in the Lizard, which have made some people think of ocean floors. In fact these have been used as evidence of a supposed lost ocean heading eastwards into central Europe in Devonian and Carboniferous times. However, this ocean has been postulated in several different places on dubious grounds and must remain highly suspect, at least to me.

Marine deposits extend northwards from Cornwall and south Devon (even quite a long way up the North Sea) but before one reaches north Devon there are intertidal deposits and the first interdigitations of what may be called the Old Red Sandstone. That famous group of rocks (which extends in effect from Canada to the USSR and even to northern India) represents the rivers and lakes of a great northern continent. In Britain these were the detritus of the new Caledonian mountains laid down as sands, muds and conglomerates in inter-montane basins over large parts of Wales and the Welsh Borderland.

However, a land-mass seems to have risen where is now the Bristol Channel, cutting off my view of my English homeland. That land-mass seems to have lingered on into Carboniferous times pouring sediments into Wales from the south and not from the north, whence they were previously presumed to have come. So often in our history what is now down (like the North Sea) was formerly up and vice versa, following the biblical promise that 'Every valley shall be exalted, and every mountain and hill shall be made low'. The old Caledonian mountain range still extended across northern England and was probably further uplifed in mid-Devonian times. Its worn and folded roots were not buried until Carboniferous times. Northwards again the Old Red Sandstone reappears around and between the Caledonian mountains of Scotland, but they are beyond our concern.

There was some volcanicity in this northern continent, though of a different type from that in the marine realm to the south. Mention has already been made of the Cheviot Granite and its accompanying lava flows, of early Devonian age, which forms the northern watch-tower of our twin countries. This must have formed within continental crust, well away from the already departed Proto-Atlantic Ocean.

By the end of Devonian times the Caledonian mountains were worn down and the country must have been largely a flat red plain. There is no particular evidence of its being an arid plain, for there are no wind-blown sand-dunes or evaporites, but it was certainly hot. Fish abounded locally in ephemeral pools and the land was being rapidly colonised by plants. Some of the earliest known land plants have been found, for example, below the Old Red Sandstone escarpment that forms the northern frontier of South Wales.

Across this plain spread the Early Carboniferous (Dinantian) sea coming from the trough in the south in which there was no real break between the Devonian and the Carboniferous sediments.

The dominant deposits of that sea were of calcium carbonate. The resultant Dinantian limestones used to be called the 'Mountain Limestone' because they form so many of the upland areas of England and Wales – the Pennines and the Peak District of northern England, the Mendips of the west and the escarpments of Wales such as Eglwyseg above Llangollen. Between the shallow water limestone platforms were faulted basins filled with deeper water deposits. The limestones are packed with fossils – for example along the sea-cliffs of South Wales – and one pictures a shallow tropical sea full of colonies of corals, thickets of crinoids and huge brachiopods. There was also volcanism on a minor scale. Somewhat later, the limestones became host to the products of mineralisation, notably lead-zinc deposits which have been worked sporadically since at least Roman times.

Events to the north then caused great sandy deltas to spread southwards across northern England and to a lesser extent across Wales. The marine trough was still there to the south and its northern margin oscillated to and fro so that black shales full of goniatites and other molluscs interdigitate with the sandstones of the deltas. Thus whilst hard rocks of Namurian age form the heights of Derbyshire such as Kinderscout, softer sediments of the same age form the bays of the Gower Peninsula in South Wales between the cliffs of Dinantian limestone.

In later Carboniferous (Silesian) times the southern trough filled up with shales, thin sandstones and poor coals whilst northwards the great Coal Measure swamps spread northwards over England and Wales. At first there were marine intercalations in these too, but later the sea was excluded from the whole of our country as a new mountain range rose in the south.

A ridge, called the Mercian Highlands, used to be postulated across the centre of England separating the south from the Midlands and the north. However, deep boreholes have shown that this was not as important as was formerly supposed, since there are coal basins cutting across it.

The Silesian coal swamps had a complex history with dense, tropical forests growing periodically on their upper surfaces. Latitudinally England and Wales were situated right on the equator and were presumably receiving heavy tropical rainfall from the ocean to the south. The vegetation accumulated to form peat and thence coal in an episodic manner, controlled by rising ridges and graben-type faulting. Over and over again deltaic sands built up to support dense plant growth and then to be buried in black muds. Sometimes the sea swept in briefly from the south and west.

We think of the Carboniferous in Britain chiefly in terms of coal. It was, however, the proximity of coal and ironstone in our Late Carboniferous sediments that made our island the birth-place of the Industrial Revolution. Ironstone had earlier been worked in the Cretaceous of SE England and was until recently worked in the Jurassic of the Midlands and elsewhere. Coal, however, still battles on, in an atmosphere of continuing acrimony following the ill-conceived 1984–85 strike.

By the end of Carboniferous times the new Variscan mountains had reared up from west to east across southernmost Britain and the rest of our country was cut off from the seas to the south. England and Wales had also moved out of the humid tropics into the desert belt of the northern hemisphere with its prevailing trade winds always blowing from the east.

This time there was no doubt about the aridity, which began in Late Carboniferous times with 'barren' (i.e. coal-less) red beds. The so-called 'New Red Sandstone' of Permian and Triassic age contrasts with the 'Old Red Sandstone' of Devonian age in its abundant evaporite deposits, blown sands and other evidence of the environments of a hot desert. England and Wales were now in the equivalent of the modern Sahara.

One of the biggest problems in British stratigraphy is drawing the boundary between the Permian and the Triassic, between the Palaeozoic and the Mesozoic. Only up in the north of England is there a brief record of a late Permian sea coming across from the European continent.

Though named by an Englishman – Murchison again – he would hardly have recognised the Permian System at all from the evidence available in this country. He had to go to the other end of Europe – to the foothills of the Urals – to recognise this major division of Palaeozoic time on the basis of the thick carbonates laid down on the eastern edge of the East European Platform.

Our marine record is largely confined to the oddly named 'Magnesian Limestone' which is little more than slices of carbonate in a complex hamburger of evaporites.

It is generally accepted that we have across southern Britain the frontal structures of the Late Palaeozoic orogenic movements, though it is not generally accepted exactly where this Variscan Front lies. Naturally I prefer it to pass under the campus of the University College of Swansea. Northwards we have the more gentle structures of the South Wales coalfield, southwards we have the more intense structures of Gower, Pembrokeshire and south-west England. This then is the boundary between Palaeo-Europa and Meso-Europa in Stille's terminology. We are doubly fortunate in this small island therefore in having both the Caledonian Front and the Variscan Front clearly displayed.

We have in southern Britain an unrivalled record of the Variscan mountain belt (which I refuse to call Hercynian or Armorican). Much of this relates directly to the Brittany peninsula of France – the little Britain which gives Great Britain its seemingly conceited title.

The intensely folded rocks of SW England, which disappear eastwards under younger sediments, together with the more gently folded Late Palaeozoic rocks farther north, clearly indicate a new mountain chain. The associated Late Palaeozoic pillow lavas and serpentinite encourage some geologists to make the southern coast of SW England the margin of a mid-European ocean which disappeared down a subduction zone with continental collisions at this time. If so perhaps the English Channel is the last vestige of that ocean just as the Mediterranean is the last vestige of the later Tethys. But the evidence is extremely shaky.

Certainly the Permian granites of SW England were derived from good continental crust and are perhaps all part of a single great batholith. Associated with these granites is our other main source of metallic minerals in England, including the tin of Cornwall which first brought England into history with the Phoenician traders.

The Triassic followed the Permian in England and Wales with a continuation of its story of arid deserts.

The Triassic was also a time of tension in England and Wales. Grabens formed and filled with continental sediments, though not with the fissure eruptions and intrusions that we associate with similar phenomena across on the other side of the Atlantic and farther south in Europe and north-west Africa.

In the Anglo-Welsh Triassic there is almost no marine record at all until the peculiar but distinctive Rhaetian transgression at the end of the period. Our island was by now north of the tropical zone, but still very much in the rain shadow area of the new Variscan mountains and cut off from the wide waters of the Tethys to the south.

The deposits of the Anglo-Welsh Triassic were dominantly red, indeed redness has become the index fossil of the system. Widespread conglomerates at the base suggest renewed uplift and these are followed by sandstones and then, as the uplands wore down, by the mudstones that are called (quite wrongly) the 'Keuper Marl'.

Though it has been suggested that there was a sea out there to the west, there is not really much evidence of it in the British successions and we cannot yet speak of even an embryonic Atlantic. The only hints of marine conditions in the pre-Rhaetian Trias can be interpreted as the outermost manifestations of the late Muschelkalk Sea that spread across northern Europe in Mid-Triassic times from the Tethys to the south.

Towards the end of the period the whole island was worn down to a red monotonous plain across which spread the Rhaetian sea with its sulphurous black muds and undersized, low diversity fauna.

As the North Atlantic began to open, Britain hesitated. It was as if our island home could not make up its mind whether to go west with brash young America or to stay with wrinkled old Europe. Early in that opening Wales nearly separated from England in a tentative act of devolution long before the red dragon, the Welsh language or the game of rugby. The north-south Worcester graben formed and, as it formed, filled with Triassic and Jurassic sediments. Westwards, Rockall headed for the sunset and further graben opened and filled with the sediments in which we have high hopes for future hydrocarbon discoveries.

Eastwards, in the North Sea, there began to form the graben that were later to contain hydrocarbon-producing and hydrocarbon-containing sediments. So the first hints of Atlantic opening nearly separated the two countries of our book and at the same time serendipitously produced the oily graben which are so valuable to us today.

In Jurassic times England and Wales were still just in the tropics. Coral reefs were established from time to time, though only briefly, and there is evidence that the corals were sometimes killed off by falls of volcanic ash. The latter presumably came from the cracks opening in the North Sea. Starting with the Rhaetian transgression at the end of the Triassic, a shallow sea spread across England and the whole classic Jurassic succession is essentially one great marine cycle of oscillating deposition – limestones and shales and sandstones – ending with the Purbeckian regression at the end of Jurassic times. Islands stood up in places, notably in the Mendips and southernmost Wales. Conventionally a land-mass is postulated over most of Wales (unkindly called St George's Land after the patron saint of England) but recent evidence, notably the Mochras borehole in North Wales, suggests that Wales, like England, was largely covered by at least the early Jurassic sea.

Apart from the corals, many other fossils, such as giant ammonites, suggest southern affinities and there are some which clearly indicate a Tethyan connection. On the other hand there are others, especially in the Upper Jurassic deposits, which are clearly of Boreal affinities and come from the north. Some indeed are close to forms known otherwise only from Greenland.

The classic Jurassic succession of England includes repeated bituminous shaly clays, diachronous sandstones and some notable limestones including the golden oolites of central England where William Smith began his work, and the cliffs of Portland Limestone at the top of the system, much used for ecclesiastical and bureaucratic building. England and Wales were situated in the latitude of about 30° N and the water was still quite warm!

At the end of Jurassic times southernmost England was a lowland area covered with fresh-water lakes and meandering rivers. It was hot, for our island was still on the edge of the tropics. The episodically dry areas were covered with dense vegetation of ferns and cycad-like plants and evaporites were being deposited in places, notably in Sussex. Farther north in England and all over Wales there was no deposition and, one must presume, erosion, though no very marked topography. England and Wales were together a green and pleasant land though not a very exciting one apart from the tramp of the occasional passing dinosaur.

The fresh- and clear-water lakes of the latest Jurassic were followed by the muddy morasses of the earliest Cretaceous. The Wealden deposits of SE England have been variously interpreted as lakes and rivers, deltas and swamps. Presumably more mud was coming from the lands to the north and periodically there were spreads of fine-grained sands.

The plants were beginning to show growth-rings but everything continued to look sub-tropical rather than temperate and seasonal. A sea spread into NE England from a northern ocean that extended to Russia. Its outermost waves may have reached as far as the Dorset coast in the south-west, where a thin bed of oysters may announce the beginning of the Cretaceous Period. It took a long time, however, for the southern sea, which had withdrawn as far as the Alps, to come back to our shores. The earliest Cretaceous deposits of southern England are almost all non-marine in nature and chiefly famous for their plants and vertebrates, including the first dinosaur ever discovered. Everything seems to have been flat and placid, awaiting the gentle swamping of lowland Britain by another warm sea abounding in animal life, especially cephalopods.

All this time things had been happening out in the Atlantic to the west. The tensions which first expressed themselves in Triassic times with the development of the Worcester Graben and those in the North Sea had continued until by Cretaceous times there was something clearly recognisable as an ocean between Britain and North America. Mediterranean animals were already moving northwards, bypassing Britain and reaching as far north as Greenland, presumably carried on a newly formed North Atlantic Drift or 'Gulf Stream'. By late Cretaceous times England and Wales had moved almost to their present latitudes and were right out of the tropical belt. Away to the south were the clear carbonate seas of the Mediterranean and Middle East, with their abundant large rudists and giant forams.

Sandy and then muddy seas spread across southern England bringing some marine shells from the south and then the mud slowly cleared to give the clear blue waters of the Chalk sea which extended westwards to the Mississippi and eastwards to the Soviet Union. Perhaps the best-known rock in all Britain is the Senonian chalk of the white cliffs of Dover, looking out defiantly at both Napoleon and Hitler and repeated in the many white cliffs of southern and eastern England and in the rolling inland downs and wolds. Whether the Chalk sea ever covered Wales is an old argument, supported by a few flint gravels and by the Chalk of northern Ireland.

Chalk deposition terminated rather abruptly and awaited the transgression of the first Palaeogene sea. There is evidence that some of the so-called 'Alpine' structures were already beginning to form. We have no record of the very latest Mesozoic or the earliest Caenozoic.

It may be said (though it may also be argued with) that much of the outline of England and Wales was already shaped out by the end of Mesozoic times. Though Caenozoic folding and uplift obviously had fundamental later effects and the Quaternary glaciations carried out further modifications, the main pattern is of pre-Mesozoic structures and Mesozoic basins. Thus nearly every arcuate stretch of our coastline now seems to be the edge of a Mesozoic basin, every major downwarp seems to have been already forming in that era.

Though we are so proud of our Palaeozoic and Mesozoic successions, it is worth remembering that the tumbling Palaeogene cliffs of Whitecliff Bay on the east side of the Isle of Wight have been called 'the finest Tertiary section in Europe'. We see in southern England the cyclic deposits of a sea, centred between us and the continent, which periodically expanded and contracted in Eocene and Oligocene times. At the same time it episodically flooded the Paris and Belgian basins and one is tempted to correlate this with the displacement of water caused by the periodic rise and fall of the Mid-Atlantic Ridge. It is also fascinating to see at times colder water forms coming in from the north (as in the London Clay) and other suggestions of the warm south, as in the nummulites of the Bracklesham Beds. The Oligocene part of the story has less in it of the sea and more of continental deposits. That led to a major break in our record with the absence of much in the way of Neogene deposits on the land areas of England and Wales.

Undoubtedly, the main opening of the North Atlantic happened in Tertiary times and still continues. Associated with this tension was another great phase of British volcanism, though this is chiefly seen in the peninsulas and islands to the west of Scotland. Almost all that we have in England and Wales of this Tertiary activity are a few dykes, the little granitic island of Lundy in the Bristol Channel and the submarine basic rocks now known to be associated with it. Tertiary tuffs and other pyroclastic material in the North Sea suggest continuing tensions there.

The one major European orogeny to which the English and the Welsh can lay little claim is that of the Alpine chains. Nevertheless we have the steeply dipping or vertical Mesozoic and early Tertiary sediments of the Hampshire Basin where these sediments were folded into steep north-facing monoclines. We always show visitors the Lulworth crumple on the Dorset coast as the furthermost manifestation of the Alpine movements and I always call a fault-related crumple in the Vale of Glamorgan the 'Welsh Alps'. In fact it is surprising how many little disturbances there are – for example along the Bristol Channel – that relate to very late movements. In SW England we have clear evidence of intra-Mesozoic movements, albeit on a small scale, and in the North Sea the Cimmerian movements are an important factor in hydrocarbon exploration. It has sometimes been wrongly thought that their name comes from Kimmeridge Bay in Dorset with its late Jurassic shales and little oil rig. In fact it comes from the far-away Crimea.

As has already been said, we cannot boast of much of a Neogene record in Britain. The Miocene is only known out in the centre of the North Sea and in the Western Approaches to the English Channel, and we only see the Pliocene on shore in the form of a few scraps of shelly sands in eastern England and some pockets of clay in the west. A better record perhaps is provided by the erosion surfaces which give us the monotonous, flat inland geology of Cornwall and Pembrokeshire (Dyfed if you must).

So England and Wales continued to move north, and east too, perhaps, as Europe separated more and more from North America. In early Eocene times we still had hot, humid, sub-tropical forests in southern England, but from then on the climate deteriorated.

By Quaternary times we had reached our present geographical position and it was the arctic climatic belt that came down to us, rather than the other way about. But even with the Quaternary record we are fortunate, for the Pleistocene ice-front lay across southern England and Wales and we have a diversity of glacial and periglacial deposits. Whether or not we have the four main glaciations of continental Europe preserved for us in East Anglia is another matter. These were clearly more in the subconscious expectations of Quaternary geologists than in the clays and erratics of the Norfolk cliffs. Certainly the ice came and went at least twice and oscillated each time.

Since the glaciations we have a splendid record in pollen and snails, bringing the record of climatic, vegetational and topographical change right up to the present day. We can even see the record of early man cutting down the forests for his agriculture. And though the notorious Piltdown Man may no longer be called 'the first Englishman' and though the 'red lady of Paviland' in Wales was neither red nor a lady, we have in Swanscombe Man in the Thames Valley one of the first true examples of *Homo sapiens*.

*Acknowledgments*

Thanks are due to Mrs J. Nuttall for typing the MS and to J. H. Edwards for drawing Fig. 1.1.

## REFERENCES

More specific references are given after each of the chapters which follow, but attention should be drawn here to the following correlation tables of British strata (with accompanying notes) produced by the Geological Society of London and published as Special Reports: Precambrian (No. 6), Cambrian (No. 2), Ordovician (No. 3), Silurian (No. 1), Devonian (No. 8), Dinantian (No. 7), Silesian (No. 10), Permian (No. 5), Triassic (No. 13), Jurassic (two parts: Nos. 14 and 15), Cretaceous (No. 9), Tertiary (No. 12), Quaternary (No. 4).

There are also the 'British Regional Geology Books' produced by the British Geological Survey as follows:

Northern England (1971), Pennines & Adjacent Areas (1954), Eastern England (1980), North Wales (1961), South Wales (1970), the Welsh Borderland (1971), Central England (1969), East Anglia and Adjoining Areas (1961), Bristol & Gloucester District (1948), Hampshire Basin and Adjoining Areas (1982), Wealden District (1965).

In addition, attention may be drawn to the following recent general texts:

AGER, D. V. — 1980 — *The Geology of Europe.* McGraw-Hill (UK), London, 535 pp.

ANDERTON, R., BRIDGES, P. H., LEEDER, M. R. & SELWOOD — 1979 — *A Dynamic Stratigraphy of the British Isles.* Allen & Unwin, London, 301 pp.

BASSETT, M. G. (edit.) — 1984 — Focus on Wales. *Proc. geol. Assoc. London,* **95,** 289–398.

DEWEY, J. F. — 1982 — Plate tectonics and the evolution of the British Isles. *J. geol. Soc. London,* **139,** 371–412.

JOHNSON, M. R. W. & STEWART, F. H. (Eds.) — 1983 — *The British Caledonides.* Oliver & Boyd, Edinburgh, 280 pp.

JONES, O. T. — 1938 — On the evolution of a geosyncline. *Q. J. geol. Soc. London,* **94,** 60–110.

OWEN, T. R. (Ed.) — 1980 — *United Kingdom. Introduction to general geology.* XXVI Internat. geol. Congr., Paris, 177 pp.

OWEN, T. R. (Ed.) — 1974 — *The Upper Palaeozoic and Post-Palaeozoic Rocks of Wales.* Univ. Wales Press, Cardiff, 426 pp.

OWEN, T. R. — 1976 — *The Geological Evolution of the British Isles.* Pergamon, London, 161 pp.

RAYNER, D. H. — 1981 — *The Stratigraphy of the British Isles* (2nd edition). Cambridge University Press, Cambridge, 460 pp.

SOPER, N. J., GIBBONS, W. & MCKERROW, W. S. — 1989 — Displaced terranes in Britain and Ireland. *J. geol. Soc. London,* **146,** 365–367 (and subsequent papers in volume).

TRUEMAN, A. E. — 1972 — *Geology and Scenery in England and Wales.* Pelican, London, 349 pp.

WOOD, A. (Ed.) — 1969 — *The Pre-Cambrian and Lower Palaeozoic Rocks of Wales.* Univ. Wales Press, Cardiff, 461 pp.

# 2

# PRECAMBRIAN

# A. L. Harris

## Introduction

The Precambrian rocks of England and Wales form part of the Avalonian Southeastern Marginal Zone of the Caledonide/Appalachian belt (Kennedy 1979, Rast *et al.*, 1976, 1988). They provide insight into the early crustal evolution of England and Wales and thus lend to the extremely sparse and fragmented outcrop of the Precambrian rocks an importance that is disproportionate to the area that they underlie. Moreover, the Precambrian rocks and the deeply penetrating and long-lived structures which largely control their present-day outcrop influenced the behaviour of the Caledonian and later rocks. Thus, the Malvern line (Ch. 17) appears not only to separate important Precambrian units but to control, to some extent, later deposition and Hercynian structures, while the Church Stretton/Pontesford Linley fault zone (Ch. 17), probably initiated in the Precambrian, was of major importance in determining subsequent depositional facies and Caledonian structures in the Welsh Borderland (Woodcock 1984).

Much of England and Wales is comparable with the Avalon Platform of Newfoundland while the rocks of Anglesey may find their counterpart in the Gander zone of Newfoundland. Of the Precambrian rocks reviewed here the Pentevrian of the Channel Islands consists of ancient (Archaean–early Proterozoic) gneisses, possibly of comparable age and origin to the Rosslare gneisses of SE Ireland. Elsewhere in England and Wales incontrovertibly ancient continental basement gneisses are lacking, although Greenly (supported by his Geological Survey colleague, C. T. Clough) regarded the Anglesey gneisses as similar to the Lewisian of NW Scotland and thought that they formed a basement to the 'Bedded Succession'.

Current uncertainty about the status of some, if not all, of the gneisses of Anglesey and the Llŷn taken together with the geophysical evidence of the Lithosphere Seismic Profile in Britain (LISPB – see Chapter 18) (Bamford *et al.* 1976, 1977) which indicates lower crustal seismic velocities of <6·3 km sec$^{-1}$ (Bamford 1979) points to the absence of continuous dense continental basement below Southern Britain. Rather the petrology of the rocks and the geochemical evidence of Thorpe (1979, 1981) indicate that the rocks reviewed here are largely continental material formed in the late Precambrian as the products of island arc magmatism and related sedimentation and metamorphism. Thus, much of the geology outlined in this chapter relates to events leading up to and probably overlapping with the Caledonian history of Britain as they can be traced on the SE side of the British Caledonian belt. Lynas (1985) drawing on evidence for major strike-slip faults in N Wales and the Welsh Borderland discussed the possibility that the Precambrian of Wales and perhaps that of the English Midland Platform is a collage of accreted terranes. Pauley (1986) considered that the Longmyndian was involved in strike-slip deformation of probable late Precambrian/Cambrian age.

## Regional descriptions

Although well displayed in some localities (e.g. coastal Anglesey and in Dyfed) exposure of the Precambrian rocks in many inland areas is indifferent but local sequences, both stratigraphical and structural, can be determined with some confidence. Nevertheless relationships *between* areas inevitably are difficult to establish and in spite of increasing geochemical knowledge (e.g. Thorpe 1979, 1981) must remain tentative. Figure 2.1 shows the main areas of Precambrian outcrop, while Table 17.1 shows the possible time-relationships of these outcrops.

13

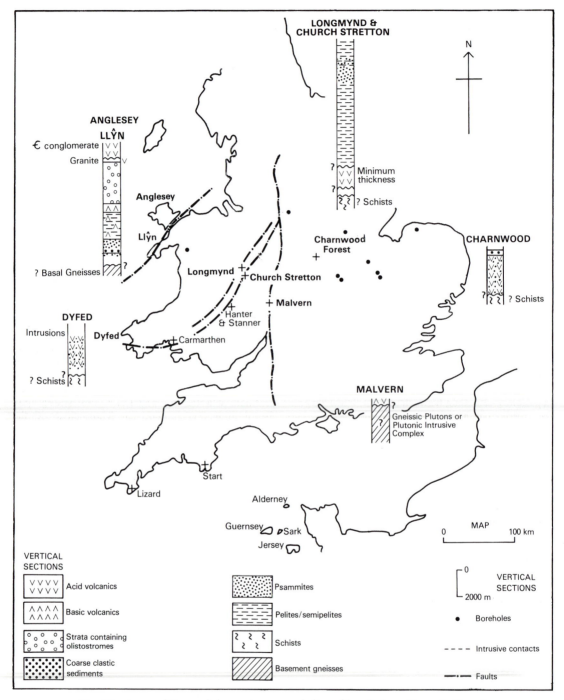

**Fig. 2.1.** Areas of Precambrian outcrops, and sites of boreholes with probable sub-crops.

Because the structure of England and Wales is described by F. W. Dunning (Ch. 17) more emphasis is given here to stratigraphy and lithology than to the detail of structure.

## North Wales

Precambrian rocks in North Wales crop out on the Llŷn Peninsula, on Anglesey and in the immediately adjacent mainland (Figs. 2.2, 2.3) whilst a borehole (Fig. 2.1) sunk through the oldest Cambrian rocks of the Harlech dome encountered *c.* 160 m of calc–alkaline volcanic rocks and sediments which may be Precambrian (Allen & Jackson 1978). Greenly (1919) distinguished two divisions in the Precambrian of Anglesey – the Mona Complex and the Arvonian (volcanic) Series. The former consists of gneisses, now of uncertain status, and the Bedded Succession (Monian Supergroup of Shackleton 1975), comprising metasedimentary and predominantly basic metavolcanic rocks of greenschist to amphibolite facies. Both the Mona Complex and the Arvonian volcanics were intruded by late Precambrian/early Cambrian granite.

## The Monian Supergroup

The formations and groups (Fig. 2.3) within the Monian Supergroup were formally established by Shackleton (1975, table 5). Greenly, having recognised a gradational contact between each stratigraphic unit, considered that the succession must be continuous and 6,000 m thick. It is still accepted as continuous, but Shackleton (1954, 1969) using evidence of sedimentary 'way-up' within and below the New Harbour Group demonstrated that Greenly's succession must be revised to read as shown in Figure 2.3a. He subsequently concluded (1975) that, although the Gwna Group is younger than the Skerries Group and older than the Fydlyn Felsitic Formation, the sequence within the Gwna Group itself is uncertain. The succession shown by Shackleton (1975, table 5) is *c.* 7,000 m. Only the upper part of the sequence proposed by Shackleton is present in the Llŷn Peninsula.

Structural-facing techniques (*sensu* Shackleton 1958) confirm the succession and show that the upright-to-overturned major folds of Holy Island (Fig. 2.4) consistently face upwards and to the south-east. Thus the recumbent folds proposed by Greenly cannot exist and the succession, regarded by Shackleton as an upward-shallowing sequence, largely NW-derived, is regionally right-way-up.

Since 1975, however, the stratigraphical and structural interpretation of Anglesey has become the subject of controversy and the view of Shackleton was challenged, principally by Barber & Max (1979). Probably the conflicting stratigraphical opinions outlined later and summarised in Figure 2.3, will not be resolved until the structural interpretations have been properly tested by a detailed study of the whole island. Nevertheless certain features of the succession seem to be generally agreed, and are set out below.

Neither the base nor the top of the Monian Supergroup is seen but the oldest rocks are NW-derived turbidites (South Stack and Rhoscolyn Formations) separated by a laterally bifurcating and thinning submarine sand-flow deposit (Holyhead Quartzite Formation) which may, in part, be the lateral equivalent of the Rhoscolyn Formation (Shackleton 1969, fig. 2). These turbidites, which in many respects resemble the upper parts of the Dalradian succession with which they may be broadly coeval, are consistent with submarine-fan deposition, perhaps building out from a continental slope. The pelites and semipelites (thin turbidites?) with basic volcanic rocks, including pillow lavas, of the overlying New Harbour Group are probably ocean-floor deposits, especially as they incorporate serpentinite and gabbro (Maltman 1975, Thorpe 1978). The remarkable Gwna Melange contains blocks ranging from millimetres to kilometres across. Thought by Greenly to be a tectonic breccia, the melange is regarded as an olistostrome (Shackleton 1969, pp. 9–10) largely on the grounds that it is of regional extent, has an unstratified matrix, has sharp concordant contacts with undisturbed formations above and below and, although normally chaotic, in places has a ghost stratigraphy based on the clast content. This is consistent with the opinion of Wood (1974) who interpreted the association of olistostrome melange, ophiolites (serpentine, gabbro and pillow basalts) and deep-water sediments together with the glaucophane (crossite)-schist facies metamorphism of SE Anglesey as indicating deposition and deformation in a subduction-related trench.

More controversially Barber & Max (1979) elaborated on the idea of subduction-related sedimentation and structure. They proposed an elaborate explanation of the Monian Supergroup succession and cast doubts on its validity as an entirely stratigraphical sequence. They subdivided the Bedded Succession into three stratigraphical/structural units – the South Stack, New Harbour and Cemlyn units (Barber & Max 1979, fig. 4). In their model, it was proposed that the New Harbour Unit is the distal (oceanward) equivalent of the South Stack Unit. Although agreeing that the former regionally overlies the latter they suggested that it was thrust into this position after undergoing intense deformation, arguing that this early deformation was caused by SE-directed subduction. To this subduction is related a zone of mylonite traced intermittently NE–SW across northern Anglesey (Fig. 2.2) as well as metamorphism

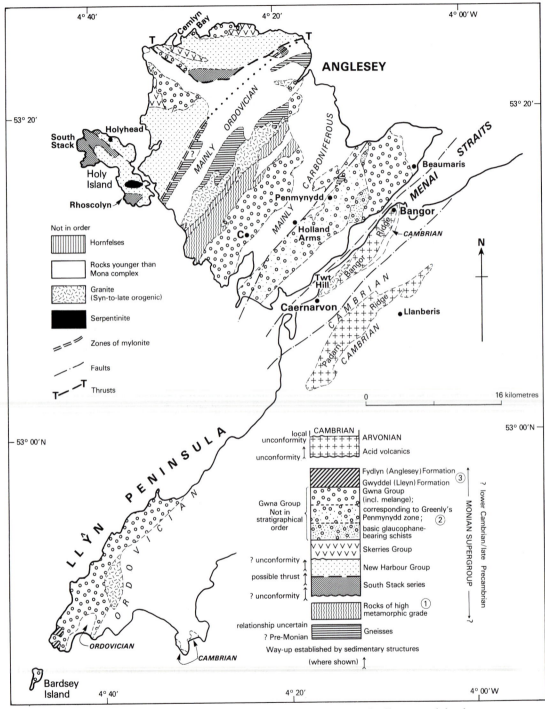

**Fig. 2.2.** Precambrian of Llŷn Peninsula and Anglesey and adjacent mainland.

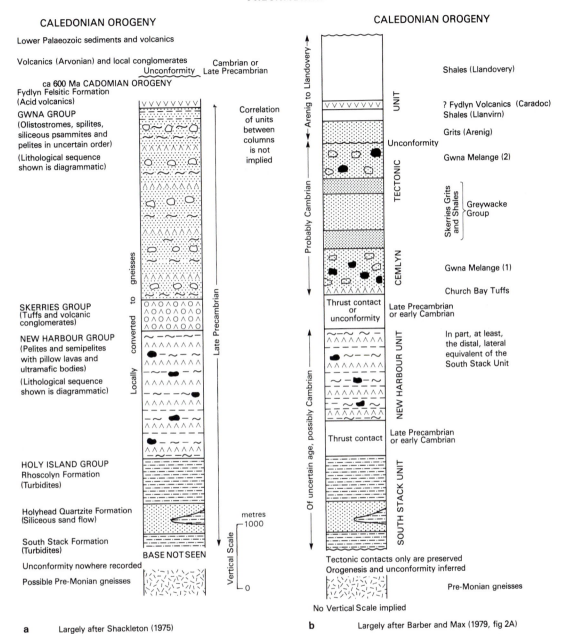

**CALEDONIAN OROGENY**

Lower Palaeozoic sediments and volcanics

Volcanics (Arvonian) and local conglomerates   Cambrian or
              Unconformity   Late Precambrian
     ca 600 Ma CADOMIAN OROGENY
Fydlyn Felsitic Formation
(Acid volcanics)
GWNA GROUP
(Olistostromes, spilites,                         Correlation
siliceous psammites and                           of units
pelites in uncertain order)                       between
                                                  columns
(Lithological sequence                            is not
shown is diagrammatic)                            implied

SKERRIES GROUP
(Tuffs and volcanic
conglomerates)

NEW HARBOUR GROUP
(Pelites and semipelites
with pillow lavas and
ultramafic bodies)

(Lithological sequence
shown is diagrammatic)

HOLY ISLAND GROUP
Rhoscolyn Formation
(Turbidites)

Holyhead Quartzite Formation
(Siliceous sand flow)                          metres
                                              ┌ 1000
South Stack Formation
(Turbidites)
                              BASE NOT SEEN
Unconformity nowhere recorded
Possible Pre-Monian gneisses
                                              └ 0

Locally converted to gneisses

Late Precambrian

Vertical Scale

**CALEDONIAN OROGENY**

Shales (Llandovery)

? Fydlyn Volcanics (Caradoc)
Shales (Llanvirn)

Grits (Arenig)
                        Unconformity
Gwna Melange (2)

Skerries Grits
and Shales          } Greywacke Group

Gwna Melange (1)

Church Bay Tuffs

Thrust contact     Late Precambrian
or                 or early Cambrian
unconformity

In part, at least,
the distal, lateral
equivalent of the
South Stack Unit

Thrust contact     Late Precambrian
                   or early Cambrian

Tectonic contacts only are preserved
Orogenesis and unconformity inferred

Pre-Monian gneisses

No Vertical Scale implied

Arenig to Llandovery

Probably Cambrian

Of uncertain age, possibly Cambrian

TECTONIC UNIT

CEMLYN

NEW HARBOUR UNIT

SOUTH STACK UNIT

a    Largely after Shackleton (1975)      b      Largely after Barber and Max (1979, fig 2A)

**Fig. 2.3.** Stratigraphic/structural classifications of Precambrian of North Wales.

(including blue-schist metamorphism), in SE Anglesey. According to Barber and Max the New Harbour Unit, comprising much of the New Harbour Group of fig. 3a, is overlain either unconformably or tectonically by the Cemlyn Unit; the latter, like the South Stack Unit, lacks the early intense deformation which is a feature of the New Harbour Unit. Coward & Siddans (1979) agreed that the South Stack and New Harbour Units may have an entirely tectonic contact, but suggested that all the units have a common structural history which includes Precambrian NW-directed structures overprinted by Caledonian major

**Fig. 2.4.** Folding in South Stack 'Formation', Mona Complex, Holy Island, Anglesey. (B.G.S. NW2257) (R. H. Roberts)

folds and cleavage. They believed that the contrast in strain across the South Stack/New Harbour contact, and the concentration of strain there, resulted from competence differences.

An important feature of the Barber and Max hypothesis is their recognition of melange facies at two stratigraphical levels (Fig. 2.3), (Barber & Max 1979, fig. 4) within the Gwna Group, although this conclusion may be an oversimplification (see Wood in discussion). Wood elaborated on his own (1974) model, pointing out (Barber & Max 1979, p. 426) that the Gwna Group is 'a *series* of easterly directed olistostromes and interslide members which become progressively younger towards the NW'. Stratigraphic units such as the Church Bay Tuffs and the Skerries Group are locally absent because they have been reworked and redeposited in olistostromes.

On the basis of palaeontological evidence from the Gwna Group (Downie 1975, Muir *et al.* 1979) Barber and Max regarded the deposition of the Cemlyn Unit as having been entirely Cambrian, and they interpreted the break between the lower Ordovician rocks and the melange in northern Anglesey as only slight. They found that a single important (Caledonian) cleavage is common to both Ordovician rocks and the Mona complex, and that neither the South Stack nor Cemlyn Units suffered significant pre-Caledonian deformation. Thus, in terms of their hypothesis, the Precambrian elements of the Mona complex are the gneisses (discussed below), the sediments and ocean-floor igneous rocks of the South Stack and New Harbour Units and the deformation and metamorphism (including blue-schist facies) related to Precambrian subduction. This interpretation requires that, far from the Mona Complex of Anglesey and Llŷn being an area of positive relief during the Cambrian (cf. Bennison & Wright 1969, figs. 4–5), it was the site of Cemlyn Unit deposition. This view must take account of the sedimentary patterns and provenance of adjacent rocks of lower Cambrian (and younger) age. In this respect it is significant that Crimes (1970 and pers. comm. 1979) showed that the sedimentary structures in lower Cambrian rocks of North Wales demonstrate clear evidence of transport from a northerly direction and that much of the clast content in these rocks is consistent with an origin in the rocks of the Mona complex.

The present writer believes that the Monian

forms a largely unbroken succession of late Pre-cambrian (possibly including lowest Cambrian) rocks deposited partly on an ocean floor and in an ocean trench. Having undergone late Precambrian (possibly early Cambrian) deformation and metamorphism which was related to SE-directed subduction they were subsequently intruded by granites. The Mona Com-plex thereafter formed the basement for the Arvonian acid volcanics (discussed below) and the basement and volcanics together became an important source of clastic material by late lower Cambrian times. Gibbons (1983a) rejected subduction-related hypoth-eses as 'oversimplistic' and argued for an explanation of the Mona Complex as a tectonic *collage* assembled by early transcurrent movements and moved laterally into position during the Cambrian.

## The Gneisses

Gneisses in Anglesey and the Llŷn Peninsula include both orthogneisses and paragneisses (Shack-leton 1975, p. 76). The nature of the relationship between them and the less severely metamorphosed Monian Supergroup remains a matter of discussion (Fig. 2.3).

Greenly (1919) regarded the Gneisses in Anglesey as forming a basement on which his Bedded Suc-cession was laid down unconformably though the unconformable contact was and is nowhere exposed and (*in* Matley 1928, p. 452, 479) interpreted the Llŷn gneisses also as forming pre-Monian basement. Baker (1969) considered that the 'gradational' con-tacts between the Gneisses and other Monian meta-morphic rocks in Llŷn are tectonic and that the Gneisses form a pre-Monian basement comparable with the Rosslare complex of SE Ireland. Barber and Max also supported these views and reported that in Anglesey and at least at one Llŷn locality exposed contacts between the Gneisses and members of the Bedded Succession are tectonically modified, some being marked by mylonite zones. They concluded that nowhere is there a *progressive* metamorphic transition from Bedded Succession through schists into gneisses and that the migmatitic gneisses, in fact, are inliers of continental basement.

Shackleton (1969) concluded that the Gneisses of Anglesey and Llŷn are probably migmatitic de-rivatives of the Monian Supergroup. He based this conclusion on his observations that in Anglesey there is everywhere a rapid increase in metamorphic grade from Bedded Succession into Gneisses. He further pointed out (1969, p. 14) that the Gneisses are not adjacent to the *oldest* parts of the Bedded Succession and that the distribution of the highest grades of metamorphism is independent of stratigraphic level.

Certain lithological associations such as Greenly's (1919, p. 220) Triple Group Formation (limestone, graphitic pelite and orthoquartzite) are found in Gneisses as well as in low-grade areas. Shackleton (1969) claimed that in the Lleyn there is a gradational change from low-grade Gwna Group rocks into the schists of the Penmynydd Zone described below and that these in turn prograde into gneisses which pass by migmatisation into the Sarn Granite. The differing views on the status of the gneisses is discussed below.

The change from low- to high-grade metamorphism takes place on Anglesey across the Penmynydd Zone of Metamorphism. First recognised by Greenly (1919, pp. 110-128) he characterised it as a zone of fine-grained mica schists carrying lenses of schistose quartzite, foliated limestone and graphitic schist. Greenly drew attention to the similarity of the fine-grained mica schists to rocks occurring above the Moine Thrust plane in Scotland, but did not call them mylonites. He did, however, find evidence that the Penmynydd Zone schists were derived from the Gwna Group, the Fydlyn Felsites (acid volcanics) and, significantly, from granite. Matley (1928) working on the Llŷn Peninsula recognised a unit similar to the Penmynydd Zone of Anglesey, separating the Gwna group of Llŷn from the Sarn tonalitic and gneissic complex.

Shackleton's (1969) view of a rapid but progressive increase in grade from low-grade Gwna Group into Gneisses has been questioned by a number of workers (Baker 1969, Barber & Max 1979, Gibbons 1983b). All of them invoked considerable shearing in the zone in Llŷn. Baker (1969) thought of the zone as one into which the low-grade Gwna Group prograded while the gneisses retrogressed with accompanying cataclasis. Gibbons (1983b) regarded the zone as mylonitic with contributions from both the Sarn Complex and low-grade Gwna Group, a view con-sistent with Greenly's (1919) observation of the zone in Anglesey.

Rb–Sr whole-rock isochrons from ortho- and paragneisses collected at Gaerwen Quarry [SH480728] indicate a metamorphic event at $595 \pm 12$ Ma, while a low initial $^{87}Sr/^{86}Sr$ ratio shows that a long crustal history for these rocks is improbable (Beckinsale & Thorpe 1979). This makes unlikely a basement-cover relationship between Gwna Group and Gneisses. Given the lithostratigraphic similarities of the high- and low-grade rocks, it seems possible that Shackle-ton's progressive metamorphic sequence has been disrupted and foreshortened by subsequent shearing, while the juxtaposition of high-grade rocks against low-grade would, during thermal relaxation, account for blastomylonitic characteristics of the transitional zone.

## Arvonian

The Arvonian Series consists largely of acid volcanic rocks of calc-alkaline character shown by Dearnley (1966) to be mainly rhyolitic ignimbrites, with andesites and dacites. Thorpe (1979) regarded them as having formed in an island-arc/young-continental-margin setting. Wood (1969) regarded the small outliers of acid volcanic rocks near Holland Arms and Beaumaris (Fig. 2.2) as belonging to the Arvonian Series and as being unconformable upon the Monian. Although the contact has been tectonically modified, the overlying volcanics are clearly less deformed than the underlying Monian metamorphic rocks since they contain unmodified vitroclastic textures. The Mona complex, by inference, therefore underwent orogenesis before deposition of the Arvonian volcanics.

At Llanberis (Fig. 2.2) the Arvonian passes up with little or no break into lower Cambrian conglomerates, volcanics and slates (Wood 1969). The Twt Hill granite cuts the Arvonian and had been supposed to supply debris to Cambrian conglomerates indicating a time-break between the Arvonian volcanics and the Cambrian conglomerate. However, the granite has been isotopically dated at $498 \pm 7$ Ma (Gale et al. 1983), which seems to invalidate this conclusion. The base of the Cambrian in North Wales is traditionally but somewhat arbitrarily taken at the base of the conglomerates. At Bangor (Fig. 2.2) conglomerates rest with slight discordance on the Arvonian, but even so it is not impossible that all the Arvonian volcanics are lower Cambrian. Certainly the supposed Cambrian conglomerate (Greenly 1919), Wood 1969) (C on Fig. 2.2 c. 8 km W of Holland Arms) which carries both Arvonian and Monian clasts lies with strong unconformity on the Mona complex and it appears that the break between the Cambrian/Arvonian and Mona complex is far greater than that between Arvonian rocks and the supposed lowest Cambrian strata.

## Granites

The Mona complex was intruded by two late Precambrian/early Cambrian granites the relative age of which is unknown. The Coedana granite of Anglesey was isotopically dated at between 580 and 610 Ma (Moorbath & Shackleton 1966) and by Rb–Sr whole-rock isochron at $603 \pm 34$ Ma by Beckinsale & Thorpe (1979). This suggests that the emplacement of the granite and the gneiss-forming metamorphism of the country rocks ($595 \pm 12$ Ma) are, as suggested by Shackleton (1969), broadly contemporaneous. Relationships between the Coedana granite and its country rocks are obscure. Greenly's (1919) observation that it intrudes and merges with the gneisses which crop out to the NW implies contemporaneity. On the other hand the postulated shallow depth of emplacement (3 to 7 km – Thorpe 1982) suggests that the gneisses must have formed before emplacement of the granite. By contrast Shackleton (1969, p. 15) reported sediments in contact with and hornfelsed by the granite which were apparently not schistose before its emplacement.

The emplacement of the Sarn granite on the Llŷn Peninsula is regarded by Shackleton (1969) as closely related to the formation of gneisses and schists. New geochronological studies by Beckinsale et al. (1984) indicate a metamorphic event in the Parwyd gneisses at $542 \pm 17$ Ma approximately coinciding with the intrusion of the Sarn complex at $549 \pm 19$ Ma. These Rb–Sr ages are somewhat younger than ages obtained from comparable rocks of the Mona Complex of Anglesey. The Llŷn Gneisses and granite are not distinguished from one another on Figure 2.2.

## Central England, Welsh borderland and South Wales

Figure 2.1 shows that a large part of central and eastern England is thought to be underlain by upper Proterozoic volcanic rocks, encountered in the few boreholes which have passed through the Phanerozoic cover, and in small, structurally controlled outcrops such as at Charnwood Forest near Leicester. More diverse Precambrian lithological units occur sporadically at the surface along a structurally controlled zone which has an arcuate outcrop through the Welsh borderland into Dyfed (Pembrokeshire). Much of this zone has a very extended history of displacement and probably marked the SE margin of the Welsh Lower Palaeozoic basin. Relationships preserved in the Welsh borderland between the Precambrian units and the Lower Palaeozoic rocks certainly suggest that Precambrian rocks were at the contemporary surface in this zone during several Lower Palaeozoic episodes. Another major Precambrian outcrop, that forming the Malvern Hills, lies on the western side of a major N–S fault of prolonged history.

The Ingletonian rocks (Ch. 4), formerly believed to be Precambrian are now regarded as Lower Palaeozoic, but the occurrence of metamorphic clasts in the Ingletonian (Rastall 1906) perhaps indicates a not-too-distant source of basement metamorphic rocks. This source may be part of the volcanic/metamorphic basement area envisaged by Bott (1967) and shown also by Wills (1975, pl. 1). This basement area is regarded on gravity and magnetic evidence to lie at no great depth below part of the Askrigg Block (Fig. 18.3) and to extend south to the Welsh borders and southeast to the Wash. Much of this area was thought by Wills (1975) to be composed of Precambrian rocks which were exposed during much of the Devonian.

Original relationships between the Precambrian rocks exposed in scattered outcrops have not been satisfactorily determined. It is possible, however, that the Malvernian and related rocks are the oldest and the Longmyndian the youngest.

## Malvernian and related rocks

It is likely that the Malvernian consists almost entirely of plutonic rocks in which dioritic types predominate (Callaway 1893, Lambert & Holland 1971); these range in composition from mafic diorite to leucocratic tonalite. Other rock types include syenite (in veins), and granite, which is variously massive and foliated and which is pegmatitic in veins and segregations. Brammall's umpublished opinion, reported in Dunning (1975), was that the Malvernian originated as a series of sediments and intrusions which suffered such high-grade metamorphism that they came to resemble plutonic rocks; these were subsequently retrograded and intruded by basic dykes. Original relationships, mineralogy and textures are so obscured by the later dislocation and retrogressive metamorphism that conflicting opinions are difficult to test conclusively. It is recognised that foliation in intrusive plutonic rocks may closely resemble relict gneissic foliation in high-grade metamorphic rocks, while rocks which are almost indistinguishable from metasedimentary schists may result from the cataclastic reworking of acid igneous rocks. Therefore although it is possible that some of the schistose rocks are the remains of a sedimentary sequence into which plutonic bodies were emplaced, it is much more likely that the whole of the Malvernian is a somewhat gneissic plutonic complex of variable composition, possibly regionally metamorphosed at high grade (Fitch et al. 1969), but heterogeneously reworked by largely brittle deformation.

This brittle deformation (Fitch et al. 1969, p. 44), which also affected the post Malvernian but pre-Cambrian Warren House volcanic rocks, may coincide with the last major isotopic event ($590 \pm 20$ Ma) to affect the Malvernian (Lambert & Rex 1966). The date of this event falls within the range of 580–640 Ma whole-rock and mineral ages reported from the Malverns and from Anglesey and it also coincides very closely with Malvern mineral ages at c. 595 Ma (Fitch et al. 1969).

Dunning & Max (1975) represented the Malvernian as intruding older basement rocks as a plutonic complex; this relationship would not, however, necessarily invalidate the conclusion of Fitch et al. (1969) that the complex itself underwent high grade metamorphism and had a protracted Precambrian history. If the latter conclusion is correct the Malvernian (and its country rocks) may be a fragment of the continental basement. The results of the LISPB experiment (Ch. 18) make this possibility less likely, for although they strongly support the presence of continental basement in England and Wales they suggest that it is not high-grade gneiss or granulite but rather low-grade metasediments. Present evidence (Thorpe 1979, Beckinsale et al. 1981) suggests that it is most likely that the Malvernian is a plutonic complex closely related to the calc–alkaline late-Precambrian volcanics. Its age $681 \pm 53$ Ma, based on an Rb–Sr whole-rock isochron (Beckinsale et al. 1981), and its low initial $^{87}Sr/^{86}Sr$ ratio suggesting a short crustal history are consistent with this suggestion. The diorite, hornblende-schists and biotite-gneisses of the Primrose Hill inlier (SE of the Wrekin some 25 m NE of Church Stretton, Fig. 2.1) seem likely to be correlatives of the Malvernian; they occur within the Wrekin inlier in close association with Uriconian calc–alkaline volcanics. Farther south, the Hanter and Stanner rocks (Fig. 2.1) comprise an intrusive complex within which early concordant bodies of dolerite and gabbro are cross-cut by acid and felsic dykes. Felsic bodies on Stanner Hill have yielded a $702 \pm 8$ Ma whole rock Rb–Sr isochron to Patchett et al. (1980). Some aspects of the secondary changes in these rocks, involving their alteration to hornblende- and biotite-bearing assemblages, are similar to retrograde changes in the Malvernian (Baker 1971), and like the Malvernian it is most likely that they are genetically related to late Precambrian effusive rocks, possibly being part of a subvolcanic complex.

The Rushton Schists (Fig. 2.5) were thought at one time to have been related to the Malvernian. They consist of garnetiferous and epidote-bearing quartz-mica schist and seem to be rather low-grade (greenschist-facies) regionally metamorphosed sediments faulted against Uriconian volcanic rocks and unconformably overlain by Lower Cambrian quartzites. Hence they are unlikely to be directly related to Malvernian, although their age of $667 \pm 20$ Ma reported in Thorpe et al. (1984, p. 525) is similar.

## Uriconian, Pebidian and Charnian volcanic units

Figure 2.1 shows the supposed distribution of upper Proterozoic volcanic rocks in England and Wales. The main present-day outcrops occur: (1) in the Welsh borderland, where the Uriconian is spatially associated with the Pontesford–Linley/Church Stretton fault system (Fig. 2.5); (2) in Dyfed where the Pebidian and Bentonian have structurally controlled outcrops; and (3) in Leicestershire where the Charnian occurs. Boreholes in England marked on Figure 2.1 all penetrated volcanic rocks of supposed Precambrian age. Details of the rocks encountered in the boreholes are given by Dunning (1975, pp. 89–90).

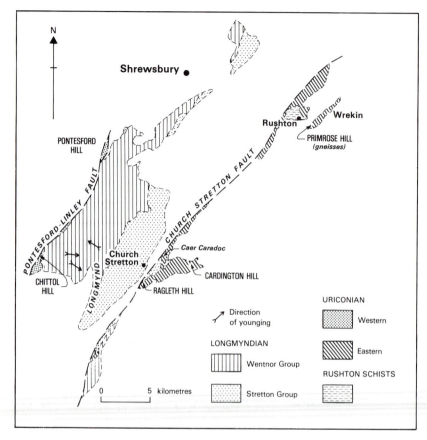

**Fig. 2.5.** Precambrian of Welsh Borderland.

The *Uriconian* of the Welsh borderland is divided into two – the Eastern and Western (Fig. 2.5). The Western Uriconian consists of rhyolites, rhyolite tuffs and basalts intruded by acid, intermediate and basic rocks. It occurs as a series of fault-bounded, lenticular outcrops closely related to the Pontesford–Linley fault-zone (Fig. 2.5). James (1956) proposed that the upper part of the Longmyndian succession (p. 25) unconformably overlay the Western Uriconian on Chittol Hill (Fig. 2.5) in Shropshire, the original unconformity having been subsequently overturned. Wright (1969, p. 101) regarded this interpretation as unlikely and believed that the 'unconformity' is either a thrust fault or that the sequence is right-way-up and the 'Uriconian' here is a volcanic sequence younger than the Longmyndian. Pauley (1986) interpreted the Western Uriconian/Westnor Group junction in this area as a thrust and all the Longmyndian/Western Uriconian contacts as faults rather than uncon-formities. The Eastern Uriconian occurs on and adjacent to the Wrekin and on Caer Caradoc, near Church Stretton (Fig. 2.5). Relationships to the younger, but still Precambrian, Longmyndian and to

Cambrian rocks are clearly shown on the published Geological Survey (1:25,000) Sheet (SO 49) and in Earp & Hains (1971, figs. 7, 8). Because the outcrop of the Eastern Uriconian is broken into blocks by faults the stratigraphical succession for the unit as a whole cannot be given. Almost 1,300 m of acid, intermediate and basic lavas and tuffs are exposed on Caer Caradoc and over 1,100 m on Cardington Hill (Fig. 2.5 and Table 2.1), while thinner local volcanic successions have been established for other fault-bounded blocks. The presence of ash-flow tuffs (Dearnley 1966) sug-gests that some of the succession is subaerial. Litho-stratigraphical correlation across the faults is not feasible. (For the detail and petrography of local successions see Greig *et al.* (1968).)

Basic and acid quartz–porphyry intrusions are associated with both Eastern and Western Uriconian. Evidence from the published Geological Survey map of Caer Caradoc shows that the Eastern Uriconian, at least, had been folded or tilted and thrust before deposition of the Longmyndian. However, Pauley (1986) maintained that the deformation of the Uricon-ian was essentially coeval with the deformation of the

Longmyndian and disputed the previously accepted unconformity between the Longmyndian and the Eastern Uriconian. This conclusion follows from the inclusion of the Ragleth Tuffs within the Longmyndian Supergroup rather than the Uriconian Volcanic complex. The Ragleth Tuff Formation has faulted contacts with the Uriconian Volcanic Complex. Dearnley (*in* Greig *et al*. 1968, pp. 30–32) and Thorpe (1972, 1979) showed that the Uriconian rocks have calc–alkaline affinities while Dearnley also advanced evidence of the nature of the pre-Uriconian metamorphic basement; fragments of mica schist and quartz-mica schist comparable with Mona complex rocks are found as fragments in volcanics. Fitch *et al*. (1969) recorded whole-rock K–Ar dates of $632 \pm 32$ Ma and $677 \pm 72$ Ma for rhyolites of The Wrekin and Pontesford Hill respectively while a basalt dyke from the Eastern Uriconian yielded a date of $638 \pm 81$ Ma. Dates more recently obtained for the Wrekin Uriconian volcanics give an Rb–Sr isochron of $558 \pm 16$ Ma (Patchett *et al*. 1980) while the subvolcanic Ercall granophyre of the Wrekin yielded $533 \pm 13$ Ma (Cope

---

**Table 2.1.** Successions of Uriconian rocks

CAER CARADOC

Top not seen: unconformity below Longmyndian

| | |
|---|---:|
| †Ragleth Tuffs – acid tuffs with thin rhyolite flows | 460 m |
| Cwms Rhyolites – flows | 105 m |
| Caer Caradoc Andesites – flows | 120 m |
| Caer Caradoc Rhyolites – flows | 90 m |
| ———— Probable Thrust ———— | |
| Little Caradoc Basalts – flows | 460 m |
| Little Caradoc Tuffs – bedded acid tuffs | 15 m |
| Comley Andesites – flows | 30 + m |

(Base not seen)

CARDINGTON HILL

Top not seen

| | |
|---|---:|
| Hope Batch Dacites ? flows | 30 + m |
| Woodgate tuffs – acid | 270 m |
| Woodgate Batch Dacites and Andesites | 210 m |
| Middle Hill Andesites and Dacites | 360 m |
| North Hill Dacites | 270 m |
| Stoneacton Tuffs and Andesites | ? |

(Base not seen)

---

† The Ragleth Tuffs and the Helmeth Grit (Table 2.2) were lithostratigraphically correlated on provenance grounds by Pauley (1986), the Ragleth Tuff Formation having a faulted contact with Uriconian rocks of Caer Caradoc.

& Gibbons 1987). The latter date has been questioned (Dr A. E. Wright, *pers. comm*.) and implications for the age of overlying L. Cambrian sediments may not be valid.

Farther south the *Warren House Group* consists of Precambrian keratophyric and rhyolitic effusive rocks as well as spilites (Platt 1933). A Precambrian age is implied because it yielded clasts to the adjacent lower Cambrian Malvern Quartzite (Earp & Hains 1971). Formerly regarded as unconformable on the Malvernian, Thorpe *et al*. (1984) suggested that it is a tectonically emplaced slice of ocean floor. Thorpe (1972, 1979) interpreted the Warren House rocks as differentiated tholeiite and for this reason distinguished this group from calc–alkaline Uriconian.

In SW Wales Precambrian volcanic rocks, probably of similar age to the Uriconian rocks of the Welsh borders, crop out in the core of ENE-trending, faulted 'Caledonian' anticlines (see p. 535) of St Davids, Haycastle and Roch–Trefgarn (Fig. 2.6). The predominantly pyroclastic rocks are known as the *Pebidian complex* (Shackleton 1975); early work on their field occurrence and petrography was carried out by Green (1908) (St Davids) and Thomas & Jones (1912) (Haycastle) and continued by Williams (1934). Volcanic rocks, mainly flows, correlated with the Pebidian also occur to the south of Haverfordwest (Fig. 2.6) in an important Hercynian structure – the Johnston–Benton thrust zone. *Benton Volcanic Group* rocks yielded K–Ar dates of 613 and 625 Ma (Moorbath *in* Shackleton 1975, p. 82).

In the St Davids anticline about 1,400 m of Pebidian acid-to-intermediate pyroclastic rocks crop out. These have been divided into four groups by Shackleton (1975, table 6). Repeated to the east at Haycastle only the two upper groups are recognised with a much diminished thickness. Because, however, there is no positive lithostratigraphical correlation between groups at different localities and the base of the succession is nowhere seen, this should not be taken to imply that the Pebidian sequence is thinning towards the east.

The nature of the basement on which the Pebidian complex (and Benton volcanics) was deposited is suggested by the presence of pebbles of siliceous schist in Arenig conglomerates and the occurrence of the quartzose, metasedimentary Dutch Gin schists which occur in association (probably as an enclave) within the dioritic gneisses of the Johnston Series. The status of the Dutch Gin schists is uncertain. Claxton (1963) interpreted the 100 $m^2$ of foliated rocks which form a headland at the southern end of St Brides Bay (Fig. 2.6) as greenschist–facies metasediments. This interpretation has been challenged by Baker *et al*. (1968) who regarded the 'schists' as foliated and mylonitized rocks from the Johnston Series itself.

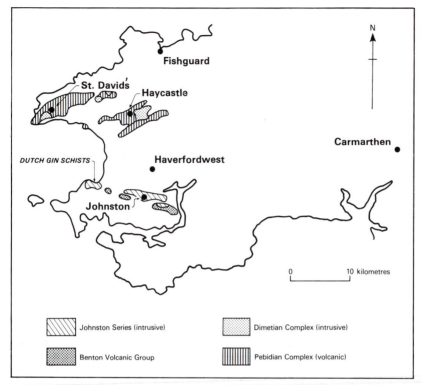

**Fig. 2.6.** Precambrian of South Wales.

The *Johnston Series* like the *Dimetian complex* probably were intrusive into, though broadly contemporary with, Benton and Pebidian volcanics respectively. The Dimetian consists of granitic and quartz-bearing dioritic rocks and clearly intrudes Pebidian volcanics in the St Davids area, Dimetian granophyres being cut by basic dykes many of which may themselves be Precambrian. Wright (1969) suggested that the acid-to-basic gneisses of the Johnston Series together with metasediments may be part of a high-grade crystalline basement on which the Pebidian volcanic group were deposited and he suggested that the Johnston gneisses may be correlated with elements of the Malvernian and of the Rosslare complex of SE Ireland. If they are part of a basement complex the Johnston rocks must be substantially older than the Dimetian complex, which is probably much the same age as the Pebidian volcanics. The substantial time gap between Johnston and Dimetian complexes implied by this view may not be borne out by the U–Pb ages ($643 \pm _{28}^{5}$ Ma and $587 \pm _{14}^{25}$ Ma respectively) (Patchett & Jocelyn 1979). It is much more likely that the Pebidian and Benton volcanics were fed and intruded by various units of the Dimetian and Johnston complexes during a period of unknown duration taking place about 600 Ma ago. Sr-isotope ratios (Patchett & Jocelyn 1979) suggest that

Dimetian and Johnston complexes are mantle-derived. The volcanics were possibly deposited on a low-grade metasedimentary schist terrain which was pierced by the Dimetian intrusions. Much of the ambiguity in the relationships as seen today is the result of vigorous Caledonian and Hercynian tectonic reworking.

The very careful and well-documented work of Thomas & Jones (1912) makes it clear that the Caerfai Group basal conglomerates (placed in the lower part of the lower Cambrian by Cowie *et al.* (1972)) post-date all the intrusive rocks grouped with the Dimetian and are younger than the penetrative deformation which affected the Pebidian rocks. Structures and bedding in the Pebidian commonly strike with strong obliquity to those in the Lower Palaeozoic cover. It is not clear from published sources if this deformation pre- or post-dated emplacement of the Dimetian, but considerable time must have elapsed between the deposition of the Pebidian complex and the Caerfai Group and Green (1908, p. 379) presented clear evidence for the pre-Cambrian age of the Dimetian granophyre of the St David's inlier. He showed that, in a temporary exposure, the granophyre is overlain by basal Cambrian conglomerate. The implied lapse of time between Pebidian and Cambrian is in marked contrast to the Arvonian–lower Cambrian relationships of North Wales, and the inferred interval makes it difficult to correlate Pebidian and Arvonian. It is largely on this basis that other workers (e.g. Shackleton 1975) concluded that the Pebidian (and the Benton Volcanic Group) should probably be correlated with the Uriconian rather than with the Arvonian.

Rhyolitic volcanics to the south of Carmarthen (Fig. 2.6) formerly thought to be Lower Palaeozoic age should similarly be correlated with the Uriconian; Cope (1977) recorded an *Ediacara* fauna regarded as late Precambrian (Glaessner 1971) in rocks which appear to rest on these rhyolites.

The *Charnian* consists of about 2,600 m of Precambrian volcanics and sediments; in its type locality – Charnwood Forest, NW of Leicester – the Charnian is overlain by Triassic sediments and occurs in the core of an anticline. Rocks having close affinities with the Charnian occur at Barnt Green (Lapworth *et al.* 1898, pp. 328–330) at the southern end of the South Staffordshire coalfield, and at Nuneaton (Allen 1968). At both these localities (Fig. 2.1) volcanic and volcaniclastic rocks with intrusions are overlain by Cambrian quartzites. Porphyritic igneous rocks some of which may have been extrusive have been dated at *c.* 694 ± 29 Ma by Meneisey & Miller (1963) and these were emplaced before the overlying Cambrian rocks which contain boulders of the diorite (540 ± 57 Ma: Cribb 1975) which is intrusive into the volcanic rocks. Further evidence of the Precambrian age of the Charnian and of Precambrian deformation comes from the penetrative cleavage which is a feature of the Charnian but which is absent from adjacent Cambrian strata.

The Charnian succession of the type locality was described in detail by Dunning (1975, p. 88). It starts with the tuffaceous Blackbrook Formation (900 m) (base unseen); this is overlain by the Maplewell Group (1,400 m) consisting largely of acid-to-intermediate pyroclastic rocks, varying from fine lithic to coarse crystal tuffs and coarse agglomerates. The strata contain evidence of contemporary slumping and of rapid facies variation from fine to coarse agglomerates. Towards the top of the Maplewell Group rocks contain sedimentary structures and supposed algae or coelenteratal algae (*Charnia masoni* and *Charnodiscus concentricus* (Ford 1958, 1968)) which strongly suggest a Vendian (Late Proterozoic) age (Downie 1975). The Brand Group forms the topmost 300 m of the Charnian and consists of an upward-fining sequence from coarse conglomerates through gritty sands to silts (now the Switherland Slates). The clastic grains in the coarser units are to a great extent derived from volcanic rocks, but the presence of metamorphic rocks below or adjacent to the Charnian volcanoes is suggested by the clasts of foliated quartzite in the Woodhouse and Bradgate Formation at the top of the Maplewell Group.

## Longmyndian

The Longmyndian consists of a sequence of sedimentary rocks within which some units are tuffaceous. Pauley (1986) proposed a thickness of 6500 m for this sequence. On Ragleth Hill in the Church Stretton fault zone the lowest member of the Stretton Group has been shown on the Geological Survey Sheet 166 (1967) as unconformable on Uriconian tuffs (see also Greig *et al.* 1968, p. 37). Although Cobbold & Whittard (1935) believed that this basal (Helmeth Grit) member (Table 2.2) is a coarse volcaniclastic deposit of Uriconian affinity and possibly a unit transitional from Uriconian to Longmyndian, the strong disparity of dip between Uriconian and Longmyndian suggested an unconformity. Pauley (1990), however, included the Ragleth Tuffs with the Longmyndian and regarded the contact between the tuffs and the Uriconian as a fault. He believed that the Uriconian deformation was of the same age as that in the Longmyndian. Rocks believed to belong to the Wentnor Group are overlain by basal Cambrian in the Church Stretton Fault zone, Cobbold (1927) describing the contact as a fault and inferring an unconformity. The main evidence for the Precambrian deformation of the Longmyndian derives from its strong, large-scale folding which is completely lacking in the adjacent Cambrian. Bath (1974) using Rb–Sr whole-rock data suggested that 600 Ma is a reasonable date for deposition of the Longmyndian, but Dunning (1975) argued that the date coincides with similar K–Ar ages in the Precambrian elsewhere in England and Wales and hence could be an age of deformation. Thus the folding of the Longmyndian may very well coincide with the brittle retrogressive reworking of, for example, the Malvernian rocks. More recently, Naeser *et al.* (1982) obtained dates of *c.* 530 Ma from fission-track studies which were interpreted as uplift ages. The Rb–Sr and fission-track data strongly suggest a deformational and uplift episode at *c.* 530 Ma (Pauley 1986).

The sequence shown in Table 2.2 is disposed in a major isoclinal syncline which is overturned and faces up towards the east. Younging evidence in the Bridges Formation which forms the core of this syncline is explicit and was demonstrated by Greig *et al.* (1968, fig. 8). A full range of the sedimentary structures identified in the Longmyndian, which include current- and graded bedding, ripple marks, groove casts, mud cracks and rain drop imprints are described in detail in Greig *et al.* (1968, pp. 68, 73). Dearnley (*in* Greig *et al.* 1968, pp. 73–74) has indicated that, in addition to the adjacent and underlying Uriconian, an important source of detritus for the Longmyndian was a

**Table 2.2.** Succession of Longmyndian rocks

WENTNOR GROUP (3,600 m)

| | |
|---|---|
| Bridges Fm (600–1,200 m) | Dominantly silts with subordinate sands. Flaggy |
| Bayston–Oakswood Fm (1,200–2,400) | Massive coarse sands with subordinate conglomerates |

STRETTON GROUP (4,300 m)

| | |
|---|---|
| Portway Fm (200–1,100 m) | Purple and green shaley silts with 20 m basal conglomerate |
| Lightspout Fm (520–820 m) | Silts with sandstones decreasing upwards |
| Synalds Fm (500–850 m) | Muds with subordinate sands; bands of tuff (Batch Volcanics) |
| Burway Fm (600 m) | Laminated silts with sandy layers; tuff at base. Cardingmill Grit at top (< 30 m) |
| Stretton Fm† (900 m) | Mudstones with coarse tuffaceous basal member (Helmeth Grit)† |

? Unconformity on Uriconian

† The Ragleth Tuffs (Table 2.1) and the Helmeth Grit have been lithostratigraphically correlated on provenance grounds by Pauley (1986), the Ragleth Tuff Formation having a faulted contact with Uriconian rocks of Caer Caradoc.

metamorphic terrain of greenschist facies, comparable with the Mona complex. This metamorphic detritus could, however, have been derived from a sheared igneous complex, similar to the Malvernian, in the 'Midland Platform' (Pauley 1986). Wright (*in* Greig *et al.* 1968, pp. 75–76) suggested that the Longmyndian was deposited as essentially non-marine fluviatile sediments in a graben, possibly bounded by the Church Stretton and Pontesford–Linley faults. Certainly some sensitive, continuously responding mechanism is required to balance fluviatile sedimentation with subsidence through some 8,000 m of deposition, and fault-bounded blocks could well provide such a mechanism.

G. Newall and E. G. Anderson (pers. comm. 1979) proposed a modification of this interpretation. Although regarding much of the succession as essentially fluviatile they suggested that deep-water Stretton Shale Formation muds are succeeded by a turbidite sequence (Burway Formation) which becomes increasingly proximal in character. The cross-bedded Cardingmill Grit member (30 m) at the top of the Burway Formation is interpreted as a near-shoreline (possibly shallow marine bar) deposit, while all the younger formations of the Stretton Group have the characteristics of a fluviatile sequence largely starved of coarse detritus. The unconformity between the Stretton and Wentnor groups is not regarded by Newall (op. cit.) as of regional importance, but rather as a series of disconformities – a result of the increased vigour of fluviatile transport and deposition when the hinterland was temporarily rejuvenated. Thus the Bayston–Oakswood/Bridges sequence is regarded as an upward-fining fluviatile sequence which followed this rejuvenation.

Pauley (1986) interpreted the Cardingmill Grit as a fluvial distributary channel deposit and the top of the Burway Formation as subaqueous deltaic deposits. He interpreted the main facies in a similar manner to Newall & Anderson, and regarded the Bayston–Oakswood Formation as an alluvial braidplain deposit. Pauley recognised no major unconformity at the base of the Wentnor Group and disputed the proposed unconformity between the Wentnor Group and the Western Uriconian (James 1956). The source for the Longmyndian Supergroup was shown to be a contemporary magmatic arc which lay to the south and south-east (Pauley 1986).

### Channel Islands and South-West England

Precambrian orthogneisses, metasediments and plutonic rocks outcrop in the Channel Islands while rocks of more doubtful age and status, but which have been thought of as Precambrian, occur in the Lizard and at Start Point (see Ch. 17). The Channel Island occurrences are at first sight more closely related to the French rather than to British Precambrian. Although the Channel Island rocks do not appear to have been significantly reworked during the Hercynian orogeny they occur within the Hercynides. For detailed correlation of Precambrian events recorded within the Channel Islands, the reader is referred to Bishop *et al.* (1975, fig. 17). The oldest rocks are part of the Pentevrian complex, the type locality of which is in Brittany. Of the rocks in the Channel Islands which have been referred to the Pentevrian, the oldest are the orthogneisses of Guernsey which suffered the Icartian event ($2,620 \pm 50$ Ma) (Roach *et al.* 1972). Younger quartz diorite and granodiorite gneisses of Alderney ($2,220 \pm 120$ Ma) and Guernsey also form part of the Pentevrian complex, while according to Bishop *et al.* (1975, p. 105) migmatitic metasediments and metabasic rocks on Sark 'have a Pentevrian aspect'. Foliated granitic rocks occurring in the Sark gneisses were dated by Adams (1967a) as $650 \pm 90$ Ma and Bishop *et al.* (1975) concluded that this implies some post Pentevrian (Cadomian) elements in the Sark rocks. K–Ar mineral ages from the Pentevrian basement strongly reflect Cadomian reworking (Adams 1967b) and range from 590 to 530 Ma.

Post-Pentevrian metasediments and volcanic rocks occur in Britanny and Normandy. These are regarded as a late Precambrian (possibly late Riphean/Vendian) succession which underwent Cadomian orogenesis. Minor occurrences of rocks referred to the Brioverian occur in the Channel Islands – the Jersey 'shale' Formation and the Pleinmont Formation of Guernsey. The Jersey 'shale' formation, a sequence of turbidite sediments, is cut by gabbro and diorite. Diorites of this suite, regarded as Cadomian in age, were described by Bishop & Key (1983). Trace fossils in the Jersey rocks suggested a Vendian age to Downie (*in* Bishop *et al.* 1975, p. 104). Orogenic reworking of the Pentevrian, giving radiometric dates of 900 Ma and 1,200 Ma may set a lower limit for Brioverian deposition, though they cannot be younger than the l'Erée adamellite of Guernsey in which Pleinmont sediments occur as a raft (Roach 1965). The adamellite was dated at $660 \pm 25$ Ma by Adams (*in* Bishop *et al.* 1975, p. 105), who also reported an Rb–Sr whole-rock age of $565 \pm 40$ Ma for a post metamorphic granite. This latter gives a minimum age for Cadomian deformation and metamorphism. Other elements of the Cadomian igneous complex were emplaced later and ages for these range from $570 \pm 15$ Ma to $490 \pm 15$ Ma.

The Lizard rocks, formerly thought to be Precambrian because of their intensely deformed and metamorphosed appearance, are now believed to be a tectonically emplaced ophiolite complex (Doody & Brooks 1986). They were interpreted by Davies (1984) as having formed and been obducted about 375 Ma ago. (For further discussion see Chapters 16 and 17.)

## Discussion and summary

Evidence of oceanic crust of Precambrian age is forthcoming from Anglesey where the New Harbour Group contains not only pillow basalts but serpentinite derived from a variety of ultramafic rocks (Maltman 1975, Thorpe 1978). Nevertheless, England and Wales is almost wholly underlain by continental basement (Bamford *et al.* 1976) the nature of which is obscure. The LISPB experiment (Ch. 18) suggested that much of the English basement is less dense than that of Scotland where the seismic velocities are greater. Thus it may not consist of gneisses and granulites but low-grade metasediments; fragments of such lithologies do occur as xenoliths and clasts in Precambrian and older Phanerozoic rocks.

Direct evidence of the nature of the low-grade metasediments envisaged is lacking. Borehole evidence suggests that much of the basement area indicated by Bott (1967) consists of volcanic rocks, exposed as the Uriconian, Charnian and Pebidian complexes. The only extensive area of undoubted metamorphic rocks forms the Mona complex and although this is older than the Eocambrian Arvonian acid volcanics there is no reason to believe that the Mona rocks are significantly older than the volcanic rocks which cover much of the English Midland platform; indeed many of the plate-tectonics reviewers think of the Mona complex and these volcanics as being essentially contemporary. Thus the only evidence of metamorphic rocks forming part of the basement are the Malvernian, Primrose Hill and Anglesey gneisses which are themselves the subject of controversy while the Rushton and Dutch Gin schists are of quite uncertain significance. Only Pentevrian gneisses are incontrovertibly basement rocks of great age and because of their position within the Hercynides they may not be related in any way to the basement of most of England and Wales (see Chapter 18 for discussion).

The Precambrian volcanic and plutonic rocks of England and Wales have been the subject of detailed chemical study by Thorpe. In a number of papers (e.g. 1978, 1979) he advanced cogent evidence to show that almost all volcanic rocks are the produce of broadly contemporaneous igneous activity at about $600 \pm 50$ Ma[1] and that this activity was dominantly calc–alkaline and characteristic of island-arc or continental margin. To this volcanicity he linked subvolcanic complexes such as the Malvernian and the Dimetian while Sr ratios indicate that the late Precambrian igneous activity (Ch. 16) largely added to the continental crust and was not derived from the recycling and mobilisation of existing continental crust. Locally, conditions during and possibly after volcanicity produced sequences of sedimentary rock. Of these the Longmyndian is by far the thickest but does bear some witness, in the form of tuffaceous bands, to distant contemporary volcanism. Although purely local conditions may have been responsible, the late Precambrian volcanic (Uriconian, Pebidian and Charnian) and sedimentary sequences may be distinguished from the Arvonian rocks which pass up more

---

[1] *Tucker & Pharaoh 1991* published new U–Pb zircon age dates from volcanic and intrusive rocks from southern Britain. These add considerably greater precision to this discussion but do not materially affect the conclusions (Eds.).

or less conformably into the Cambrian, while resting on the Precambrian metamorphic rocks of the Mona complex.

Elsewhere there is clear evidence of a break between late Precambrian sequences and the Cambrian, and that tectonic fabrics had been imposed on the Pebidian, Charnian, Uriconian and Longmyndian before deposition of adjacent Cambrian strata. This deformation was almost certainly a wholly, albeit late, Precambrian event and was very widespread. The deformation and metamorphism of the Mona complex, folding and cleaving of the volcanic and sedimentary sequences and the late-stage retrograde features of the Malvernian have been linked to late Precambrian radiometric ages. They seem, on these grounds, to be broadly coeval, and may well be linked as suggested by a number of reviewers, to the Cadomian orogenesis which deformed the Brioverian sequence of Northern France and imposed late Precambrian mineral ages on the Pentevrian gneisses.

The coincidence of deformation, blue-schist facies rocks (both in Anglesey and in Southern Brittany) together with calc–alkaline volcanicity and plutons has led to a variety of plate-tectonic reconstructions (Fig. 17.4). Baker (1973) and Virdi (1976) envisaged NW-directed subduction of oceanic crust below Anglesey, but most workers including Barber & Max (1979) preferred variations on Dewey's (1969) theme of SE-directed subduction in the Anglesey area to account for blue schist, ophiolite and olistostromes in the Mona complex. Wright (1976) linked the calc–alkaline volcanics of Dyfed, the Welsh borders and English Midlands to the SE-directed subduction, postulating an island-arc or continental-margin setting, with the deformation of the Mona complex, at least, being related to continental collision. Barber & Max (1979) indeed recognised the gneissic basement of both continents on Anglesey and traced zones of disruption of these gneisses and a gneiss-derived mylonite belt interpreted as the result of continental collision. Rast et al. (1976, fig. 4) thought of the main subduction zone as lying to the south-east of the Avalon Platform, of which the English Midland Platform forms a part basing their views on knowledge of eastern North America as well as Europe. They suggested that subduction was directed northwards in a southern ocean to yield the late Precambrian calc–alkaline volcanics and intrusions which are a widespread feature of the *whole* Avalon Platform. They regarded the Mona complex as being deposited in a small back-arc basin between the Avalon volcanic arcs underlain by continental crust and the main North American/North European continent to the north and explained the small scale of the Monian orogenesis in terms of the limited extent of this basin.

It could be argued strongly that almost all of the depositional history and orogenic activity described in this section are closely related to the Caledonian orogeny, although taking place in Precambrian time. Many authors believe that the events ascribed to the Precambrian of England and Wales took place at the SE margin of, and adjacent to, the Iapetus Ocean; on and beyond the other margin, deposition of much of the Dalradian Supergroup and the younger Moine rocks was accomplished before the Cambrian. In pointing out the contrast between the ancient continental crust of Scotland and the comparatively young basement of England and Wales, Thorpe et al. (1984) concluded that the latter is the result of the continuous accretion of island arcs, associated accretionary prisms and forearc-basin sediments between 900 Ma and 400 Ma. It is with the early part of this history that we have been concerned in this section.

*Acknowledgments*

The writer is greatly indebted to many colleagues who have helped in the compilation of this review and he would like to thank Professor Aguirre and Drs Barber, Max, Thorpe, Crimes, Newall and Brenchley in particular for advice and permission to quote prepublication manuscripts. Dr J. C. Pauley provided valuable comments based on his unpublished Ph.D. thesis on the Precambrian of the Welsh Borders; his help in updating this chapter at proof stage is especially acknowledged.

## REFERENCES

ADAMS, C. J. D.    1967    A geochronological and related isotopic study of rocks from northwestern France and the Channel Islands (United Kingdom). *Unpublished D.Phil. Thesis, Univ. Oxford.*

1967b    K–Ar ages from the basement complex of the Channel Islands (United Kingdom and the adjacent French mainland). *Earth planet. Sci. Lett.* **2**, 52–56.

ALLEN, J. R. L.     1968     Precambrian rocks. C. The Nuneaton District. *In* Sylvester-Bradley, P. C. & Ford, T. D. (Eds.) *The geology of the East Midlands*. Leicester University Press.

ALLEN, P. M. &     1978     Bryn-teg Borehole, North Wales. *Bull. geol. Surv. G.B.*, **61**, JACKSON, A. A.     51 pp.

BAKER, J. W.     1969     Correlation problems of unmetamorphosed Pre-Cambrian rocks in Wales and Southeast Ireland. *Geol. Mag.*, **106**, 246–259.

    1971     The Proterozoic history of Southern Britain. *Proc. Geol. Assoc. London*, **82**, 249–266.

    1973     A marginal late Proterozoic ocean basin in the Welsh region. *Geol. Mag.*, **110**, 447–455.

BAKER, J. W., LEMON, G. G.,     1968     The Dutch Gin Schists. *Geol. Mag.*, **105**, 493–494. GAYER, R. A. & MARSHMAN, R. R.

BAMFORD, D., &     1976     A lithosphere seismic profile in Britain – I. Preliminary EIGHT OTHERS     results. *Geophys. J. R. astron. Soc.*, **44**, 145–160.

BAMFORD, D., NUNN, K.,     1977     Upper crustal structure of northern Britain. *J. geol. Soc.* PRODEHL, C. &     *London*, **133**, 481–488. JACOBS, B.

BARBER, A. J. &     1979     A new look at the Mona Complex (Anglesey, North MAX, M. D.     Wales). *J. geol. Soc. London*, **136**, 407–432.

BATH, A. H.     1974     New isotopic data on rocks from the Long Mynd, Shropshire. *J. geol. Soc. London*, **130**, 567–574.

BECKINSALE, R. D.,     1984     Rb–Sr whole-rock ages, $\delta^{18}O$ values and geochemical data EVANS, J. A.,     for the Sarn Igneous Complex and the Parwyd gneisses of THORPE, R. S.,     the Mona Complex of Llŷn, N. Wales. *J. geol. Soc. London*, GIBBONS, W. &     **141**, 701–709. HARMON, R. S.

BECKINSALE, R. D. &     1979     Rubidium-strontium whole-rock isochron evidence for the THORPE, R. S.     age of metamorphism and magmatism in the Mona Complex of Anglesey. *J. geol. Soc. London*, **136**, 433–439.

BECKINSALE, R. D.,     1981     Rb–Sr whole-rock isochron evidence for the age of the THORPE, R. S.,     Malvern Hills complex. *J. geol. Soc. London*, **138**, 69–73. PANKHURST, R. J., & EVANS, J. A.

BENNISON, G. M. &     1969     *The geological history of the British Isles*. Edward Arnold, WRIGHT, A. E.     London.

BISHOP, A. C. &     1983     Nature and origin of layering in the diorites of SE Jersey, KEY, C. H.     Channel Islands. *J. geol. Soc. London*, **140**, 921–937.

BISHOP, A. C.,     1975     Precambrian rocks within the Hercynides. *In* Harris, A. L. ROACH, R. A. &     *et al.* (eds) A correlation of Precambrian rocks in the British ADAMS, C. J. D.     Isles. *Spec. Rep. geol. Soc. London*, **6**, 102–107.

BOTT, M. H. P.     1967     Geophysical investigations of the northern Pennine basement rocks. *Proc. Yorkshire geol. Soc.*, **36**, 139–168.

CALLAWAY, C.     1893     On the origin of the crystalline schists of the Malvern Hills. *Q. J. geol. Soc. London*, **49**, 398–425.

CLAXTON, C. W.     1963     An occurrence of regionally metamorphosed Pre-Cambrian schists in South-West Pembrokeshire. *Geol. Mag.*, **100**, 219–223.

COBBOLD, E. S. &     1935     The Helmeth Grits of the Caradoc Range, Church WHITTARD, W. F.     Stretton; their Bearing on Part of the Pre-Cambrian Succession of Shropshire. *Proc. Geol. Assoc. London*, **46**, 348–359.

COPE, J. C. W.                    1977    An Ediacara-type fauna from South Wales. *Nature, London,* **268**, 624.

COPE, J. C. W. &                  1987    New evidence for the relative age of the Ercall Granophyre
GIBBONS, W.                                and its bearing on the Precambrian–Cambrian Boundary in southern Britain. *Geol. J.,* **21**, 425–431.

COWARD, M. P. &                   1979    The tectonic evolution of the Welsh Caledonides. *In* Harris,
SIDDANS, A. W. B.                          A. L. *et al.* (eds) The Caledonides of the British Isles – reviewed. *Spec. Publ. geol. Soc. London,* **8**, 187–198.

COWIE, J. W.,                     1972    A correlation of Cambrian rocks in the British Isles. *Spec.*
RUSHTON, A. W. A. &                        *Rep. geol. Soc. London,* **1**, 42 pp.
STUBBLEFIELD, C. J.

CRIBB, S. J.                      1975    Rubidium–strontium ages and strontium isotope ratios from the igneous rocks of Leicestershire. *J. geol. Soc. London,* **131**, 203–212.

CRIMES, T. P.                     1970    A facies analysis of the Cambrian of Wales. *Palaeogeography Palaeoclimatol., Palaeoecol.,* **7**, 113–170.

DAVIES, G. R.                     1984    Isotopic evolution of the Lizard Complex. *J. geol. Soc. London,* **141**, 3–14.

DEARNLEY, R.                      1966    Ignimbrites from the Uriconian and Arvonian. *Bull. geol. Surv. G. B.,* **24**, 1–6.

DEWEY, J. F.                      1969    Evolution of the Appalachian/Caledonian orogen. *Nature, London,* **222**, 124–129.

DOODY, J. J. &                    1986    Seismic refraction investigation of the structural setting of
BROOKS, M.                                 the Lizard and Start complexes, SW England. *J. geol. Soc. London,* **143**, 135–139.

DOWNIE, C.                        1975    Precambrian of the British Isles – Palaeontology. *In* Harris, A. L. *et al.* (Eds.) A correlation of Precambrian rocks in the British Isles. *Spec. Rep. geol. Soc. London,* **6**, 113–115.

DUNNING, F. W.                    1975    Precambrian craton of central England and the Welsh Borders. *In* Harris, A. L. *et al.* (Eds.) A correlation of Precambrian rocks in the British Isles. *Spec. Rep. geol. Soc. London,* **6**, 83–95.

DUNNING, F. W. &                  1975    Explanatory notes to the geological map (fig. 4) of the
MAX, M. D.                                 exposed and concealed Precambrian basement of the British Isles. *In* Harris, A. L. *et al.* (Eds.) A correlation of Precambrian rocks in the British Isles. *Spec. Rep. geol. Soc. London,* **6**, 11–14.

EARP, J. R. &                     1971    *The Welsh Borderland. Br. reg. Geol.* (3rd Edit.).
HAINS, B. A.

FITCH, F. J.                      1969    *In discussion* of Fitch *et al.* 1969, p. 44. (*q.v.*).

FITCH, F. J.,                     1969    Isotopic age determinations on rocks from Wales and the
MILLER, J. A.,                             Welsh Borders. *In* Wood, A. (Ed.) *The Precambrian and*
EVANS, A. L.,                              *Lower Palaeozoic rocks of Wales.* University of Wales
GRASTY, R. L. &                            Press, Cardiff, pp. 23–45.
MENEISY. M. Y.

FORD, T. D.                       1958    Precambrian fossils from Charnwood Forest. *Proc. Yorkshire geol. Soc.,* **31**, 211–217.

                                  1968    Precambrian rocks. B. The Precambrian palaeontology of Charnwood Forest. *In* Sylvester-Bradley, P. C. & Ford, T. D. (Eds.) *The geology of the East Midlands.* Leicester University Press, 12–14.

GALE, N. H. &                     1983    Comments on the paper 'Fission-track dating of British
BECKINSALE, R. D.                          Ordovician and Silurian stratotypes' by R. J. Ross and others. *Geol. Mag.,* **120**, 295–302.

| | | |
|---|---|---|
| GIBBONS, W. | 1983a | Stratigraphy, subduction and strike-slip faulting in the Mona Complex of North Wales – a review. *Proc. Geol. Assoc. London*, **94**, 147–163. |
| | 1983b | The Monian 'Penmynydd Zone of Metamorphism in Llŷn, North Wales. *Geol. J.*, **18**, 21–41. |
| GLAESSNER, M. F. | 1971 | Geographic distribution and time range of the Ediacara Precambrian fauna. *Bull. geol. Soc. Am.*, **82**, 509–513. |
| GREEN, J. F. N. | 1908 | The geological structure of the St. David's area. *Q. J. geol. Soc. London*, **64**, 363–383. |
| GREENLY, E. | 1919 | The geology of Anglesey. *Mem. geol. Surv. G.B.* 2 Vols. |
| GREIG, D. C., WRIGHT, J. E., HAINS, B. A. & MITCHELL, G. H. | 1968 | Geology of the country around Church Stretton, Craven Arms, Wenlock Edge and Brown Clee. *Mem. geol. Surv. G.B.* |
| *Geological Survey Sheet* 166. | 1974 | Church Stretton. *Inst. Geol. Sci. 1 : 50 000.* |
| JAMES, J. H. | 1956 | The structure and stratigraphy of part of the Precambrian outcrop between Church Stretton and Linley, Shropshire. *Q. J. geol. Soc. London*, **112**, 315–337. |
| KENNEDY, M. J. | 1979 | The continuation of the Canadian Appalachians into the Caledonides of Britain and Ireland. *In* Harris, A. L. *et al.* (Eds.) The Caledonides of the British Isles – reviewed. *Spec. Publ. geol. Soc. London*, **8**, 33–64. |
| LAMBERT, R. ST. J. & HOLLAND, J. G. | 1971 | The petrography and chemistry of the igneous complex of the Malvern Hills, England. *Proc. Geol. Assoc. London*, **82**, 323–351. |
| LAMBERT, R. ST. J. & REX, D. C. | 1966 | Isotopic ages of minerals from the Precambrian complex of the Malverns. *Nature. London*, **209**, 605–606. |
| LAPWORTH, C., WATTS, W. S. & HARRISON, J. | 1898 | Sketch of the geology of the Birmingham district. *Proc. Geol. Assoc. London*, **15**, 313–416. |
| LYNAS, B. | 1985 | Discussion on the Pontesford Lineament, Welsh Borderland. *J. geol. Soc. London*, **142**, 935–937. |
| MALTMAN, A. | 1975 | Ultramafic rocks in Anglesey – their non-tectonic emplacement. *J. geol. Soc. London*, **131**, 593–606. |
| MATLEY, C. A. | 1928 | The Precambrian complex and associated rocks of southwestern Lleyn. *Q. J. geol. Soc. London*, **84**, 440–504. |
| MENEISEY, M. Y. & MILLER, J. A. | 1963 | A geochronological study of the crystalline rocks of Charnwood Forest, England. *Geol. Mag.*, **100**, 507–523. |
| MOORBATH, S. & SHACKLETON, R. M. | 1966 | Isotopic ages from the Precambrian Mona Complex of Anglesey, North Wales (Great Britain) *Earth planet. Sci. Lett.*, **1**, 113–117. |
| MUIR, M. D., BLISS, G. M., GRANT, P. R. & FISHER, M. J. | 1979 | Palaeontological evidence for the age of some supposedly Precambrian rocks in Anglesey, North Wales. *J. geol. Soc. London*, **136**, 61–64. |
| NAESER, C. W., TOGHILL, P. & ROSS, R. J. | 1982 | Fission-track ages from the Precambrian of Shropshire. *Geol. Mag.* **119**, 213–214. |
| PATCHETT, P. J., GALE, N. H., GOODWIN, R. & HUMM, M. J. | 1980 | Rb–Sr whole-rock isochron ages of late Precambrian to Cambrian igneous rocks from southern Britain. *J. geol. Soc. London*, **137**, 649–656. |
| PATCHETT, P. J. & JOCELYN, J. | 1979 | U-Pb zircon ages for late Precambrian igneous rocks in South Wales. *J. geol. Soc. London*, **136**, 13–19. |

PAULEY, J. C.                    1986    The Longmyndian Supergroup: facies, stratigraphy and
                                         structure. Unpublished Ph.D. thesis. University of
                                         Liverpool. Two volumes.
                                 1990    The Longmyndian Supergroup of the Welsh Borderland
                                         (U.K.). *In* D'Lemos *et al.* (Eds.), The Cadomian Orogeny.
                                         *Spec. Publ. geol. Soc. London*, **51**, 341–351.

PLATT, J. I.                     1933    The petrology of the Warren House Series, Malvern. *Geol.
                                         Mag.*, **70**, 423–429.

RAST, N.,                        1976    Relationships between Precambrian and Lower Palaeozoic
  O'BRIEN, B. H. &                       rocks of the 'Avalon Platform' in New Brunswick, the
  WARDLE, R. J.                          Northeast Appalachians and the British Isles. *Tectono-
                                         physics*, **30**, 315–338.

RAST, N.,                        1988    Early deformation in the Caledonian–Appalachian
  STURT, B. A. &                         orogen. *In* Harris, A. L., & Fettes, D. J. (Eds.) The
  HARRIS, A. L.                          Caledonian–Appalachian Orogen. *Spec. Publ. geol. Soc.
                                         London*, **38**, 111–122.

RASTALL, R. H.                   1906    The Ingletonian Series: West Yorkshire. *Proc. Yorkshire
                                         geol. Soc.*, **16**, 87–100.

ROACH, R. A.                     1965    Outline and guide to the geology of Guernsey. *Rep. Trans.
                                         Soc. Guernes.*, **17**, 751–755.

ROACH, R. A.,                    1972    The Precambrian stratigraphy of the Armorican Massif,
  ADAMS, C. J. D.,                       N.W. France. *Int. geol. Congr.*, **1**, 246–252.
  BROWN, M.,
  POWER, G. &
  RYAN, P.

SHACKLETON, R. M.               1954    The structure and succession of Anglesey and the Lleyn
                                         Peninsula. *Advanc. Sci. Lond. XI (41)*, 106–108.
                                 1958    Downward-facing structures of the Highland Border. *Q. J.
                                         geol. Soc. London*, **113** (for 1957), 361–392.
                                 1969    The Precambrian of North Wales. *In* Wood, A. (Ed.) *The
                                         Precambrian and Lower Palaeozoic rocks of Wales.*
                                         University of Wales Press, Cardiff, pp. 1–22.
                                 1975    Precambrian rocks of Wales. *In* Harris, A. L. *et al.* (Eds.) A
                                         correlation of Precambrian rocks in the British Isles. *Spec.
                                         Rep. geol. Soc. London*, **6**, 76–82.

THOMAS, H. H. &                  1912    The Precambrian and Cambrian rocks of Pembrokeshire.
  JONES, O. T.                           *Q. J. geol. Soc. London*, **68**, 374–400.

THORPE, R. S.                    1972    The geochemistry and correlation of the Warren House,
                                         the Uriconian and the Charnian volcanic rocks from the
                                         English Precambrian. *Proc. Geol. Assoc. London*, **83**, 269–
                                         285.
                                 1978    Tectonic emplacement of ophiolite rocks in the Pre-
                                         cambrian Mona Complex of Anglesey *Nature, London*,
                                         **275**, 57–58.
                                 1979    Late Precambrian igneous activity in southern Britain. *In*
                                         Harris, A. L., Holland, C. H., & Leake, B. E. (Eds.) The
                                         Caledonides of the British Isles – reviewed. *Spec. Publ. geol.
                                         Soc. London*, **8**, 579–584.
                                 1982    Precambrian igneous rocks of England, Wales and south-
                                         east Ireland. *In* Sutherland, D. S. (Ed.) *Igneous rocks of the
                                         British Isles*, 19–35.

THORPE, R. S.,                   1984    Crustal growth and late Precambrian–early Palaeozoic
  BECKINSALE, R. D.,                     plate tectonic evolution of England and Wales. *J. geol. Soc.
  PATCHETT, P. J.,                       London*, **141**, 521–536.
  PIPER, J. D. A.,
  DAVIES, G. R. &
  EVANS, J. A.

TUCKER, R. D. &   1991   U–Pb zircon ages for late Precambrian igneous rocks in
  PHARAOH, T. C.            southern Britain. *J. geol. Soc. London*, **148**, 435–443.

VIRDI, N. S.      1978    The Mona Complex of Anglesey, North Wales and Plate
                          Tectonics – a reappraisal. *Bull. Ind. Geol. Ass.*, **11**, 11–23.

WILLIAMS, T. G.   1934    The Precambrian and Lower Palaeozoic rocks of the
                          eastern end of the St. David's Precambrian area, Pem-
                          brokeshire. *Q. J. geol. Soc. London*, **90**, 32–75.

WILLS, L. J.      1975 &  A palaeogeological map of the Lower Palaeozoic floor
                  1978    below the cover of Upper Devonian, Carboniferous and
                          later formations. *Mem. geol. Soc. London*, **8**, 36 pp.
                          (containing Plate 1 dated 1975).

WOOD, D. S.       1969    The base and correlation of the Cambrian rocks of North
                          Wales. *In* Wood, A. (Ed.) *The Precambrian and Lower
                          Palaeozoic rocks of Wales*. University of Wales Press,
                          Cardiff, pp. 47–66.

                  1974    Ophiolites, melanges, blueschists and ignimbrites: early
                          Caledonian subduction in Wales? *In* Dott, R. H. & Shaver,
                          R. H. (Eds.) Modern and ancient geosynclinal sedimenta-
                          tion. *Spec. Publ. Soc. econ. Paleontol. Mineral*, **12**, 334–344.

WOODCOCK, N. H.   1984    The Pontesford Lineament, Welsh Borderland. *J. geol. Soc.
                          London*, **141**, 1001–1014.

WRIGHT, A. E.     1969    Precambrian rocks of England, Wales and Southeast
                          Ireland. *In* Kay, M. (Ed.) North Atlantic – geology and
                          continental drift. *Mem. Am. Assoc. Petrol. Geol.*, **12**,
                          93–109.

                  1976    Alternating subduction direction and the evolution of the
                          Atlantic Caledonides. *Nature, London*, **264**, 156–160.

# 3

# CAMBRIAN

# J. W. Cowie

## Introduction

In late Precambrian and early Cambrian times, the Welsh, English and other European areas were probably on the south-east margin of the Iapetus Ocean (Ch. 2) and were affected by subsidence, volcanism, plutonism and deformation which may have been due to the effects of destructive plate-margin subduction. This probably ended in early Cambrian times with a major transgression which in some parts covered the previous continental margin at different times during the Cambrian period. An 'Atlantic' faunal province is identified and was probably then situated on the south-east side of the Iapetus Ocean in contrast to a 'Pacific' faunal province, identified in the British area in the Cambrian of NW Scotland which at this time probably lay on the opposite side of the ocean (Cowie 1971).

From late Precambrian to and through Lower Palaeozoic times the region of Wales, the Welsh Borderland and Central England was apparently related to a marginal trough or basin along the south-eastern flank of the Iapetus Ocean. This depositional area had its axis through Central Wales although this may not have been so simply true during the Cambrian Period according to Wright (1981, p. 289) who stated 'The Iapetus Ocean thus only existed as a two-sided closing ocean for probably the last part of the Ordovician and Silurian times, at least in the Britain–Appalachia segment of the Caledonides'.

This marginal trough (the Welsh Basin) is thought to have had a source area to the west and north-west, in the vicinity of a line running from what is now Anglesey to SE Ireland. This area has been interpreted in various ways (which are not mutually exclusive) as for example a submarine-rise; a landmass; a fault-bounded horst; a destructive plate-margin subduction zone. Great thicknesses of terrigenous material were

brought into the Welsh Basin, particularly in North Wales, from this source, the 'Irish Sea Landmass' or 'Horst', throughout Cambrian times. The southern margin of the Welsh Basin extended from South-West Wales (Dyfed) in an arc to Eastern Wales (Powys). East and South-East of the basin lay a shallow shelf covering the western English Midlands, the Welsh Borderlands and South Wales (Glamorgan–Gwent) and perhaps part of the Bristol region of SW England.

Three main Lower Palaeozoic structural facies belts are recognised:

1. Harlech Dome–Bangor–Anglesey (in Gwynedd, North Wales).
2. Central Wales Basin.
3. South Wales–Welsh Borderland–Midlands of England.

At this time the Iapetus Ocean, of unknown width, occupied the area between what are now the Welsh Borders and NW Scotland but there may well have been islands which were a source for supplying subsiding sea-floor areas with mud and sand. The Welsh Borderland was then a stable platform shelf area on one side of this ocean and received sediment, mainly of relatively pure sand or lime-rich muds or sands, in meagre amounts. It did not seem to have been associated with a tectonic plate margin; although there is thought to have been a destructive plate margin to the west of the present North Wales area in Proterozoic times but this may not have persisted into the Cambrian.

Although the Cambrian System has its historical type-area in Wales, and has a considerable thickness of basinal sedimentary rocks there, the areal extent of outcrop is small by world standards and in some, mainly the older, parts of the succession the biostratigraphy and aspects of the chronostratigraphy are difficult to establish because of lack of fossils. The earliest strata at present assigned to the Cambrian

System often appear to be underlain by an unconformity but the rocks below this unconformity may be only speculatively assigned to Precambrian times. The evidence for a stratigraphical break of any magnitude in North Wales is equivocal and so is any selected position for the Precambrian–Cambrian Boundary on chronostratigraphic criteria. There and elsewhere disconformities or non-sequences may exist but their duration could have been small. The situation differs in South Wales, but in England there are many non-sequences and angular unconformities in rocks ranging from late Precambrian to undoubted Lower Cambrian strata. A summary of Lower Cambrian transgressions and related factors has been published by Brasier (1980). Geochronometry has so far been unsuccessful in calibrating Cambrian chronostratigraphy with isotopic dates: attempts so far are inadequate for England and Wales and also globally for the Cambrian period and its boundaries. An estimate from the contradictory evidence globally available in early 1987 is 530/600 Ma for the Precambrian–Cambrian Boundary and 495/515 Ma for the Cambrian–Ordovician Boundary (Cowie & Johnson 1985). In accordance with recommendations of the Working Group on the Cambrian–Ordovician Boundary and the Subcommission on Ordovician Stratigraphy (I.U.G.S. bodies) the Tremadoc Series is included in the Ordovician system. The candidate global stratotype section and point of the Precambrian–Cambrian Boundary is located in Yunnan Province, China at Meishucun and lies between Zones I and II of the Meishucun stage on present recommendation of the I.U.G.S. Working Group on the Precambrian–Cambrian Boundary (Table 3.1). (See Cowie & Brasier 1989.)

## Biostratigraphy and chronostratigraphy

The international subdivision of the Cambrian period and definition of its boundaries are currently under study by the International Commission on Stratigraphy (I.U.G.S.) through its constituent working groups. Historically, and colloquially at present, the primary divisions are often taken as Lower, Middle and Upper. In an endeavour to up-date the British stratigraphical nomenclature Cowie *et al.* (1972) suggested local series names but could not at that time propose British stages. Mainly Scandinavian trilobite zones were used which are chiefly assemblage and local-range zones. The validity of all Lower Cambrian zones is still debatable especially the non-trilobite or pre-Olenellid zone in Britain.

Within England and Wales lithological correlation of mappable formations is often good. However, correlation with other areas using a standard zonal scheme, such as that given in Table 3.1 is far from satisfactory due to inadequate biostratigraphic controls. Consequently Table 3.1 must be regarded as a scheme for discussion – owing much to the the work of Jan Bergström and co-workers, see Cowie *et al.* (1972), Rushton (1974) and Thomas, Owens & Rushton (1984, fig. 1) – only temporarily suggesting a possible correlation applying both to Britain and to other parts of the world.

## Distribution of Cambrian outcrops

Outcrops of rocks of accepted Cambrian age in Wales and England (Fig. 3.1) occur as follows:

WALES
    Anglesey and the 'Caernarvonshire' Slate Belt, Gwynedd (previously in Caernarvonshire)
    Harlech region, Gwynedd (previously in Merionethshire)
    St Tudwal's Peninsula, Gwynedd (previously in Caernarvonshire)
    St David's region, Dyfed (previously in Pembrokeshire)
    Llangynog area, SW of Carmarthen, Dyfed.

ENGLAND
    Shropshire (Salop)
    Malvern Hills (SW Midlands)
    Nuneaton area and Lickey Hills (Central Midlands).

While there are many Cambrian successions known from boreholes (Cowie *et al.* 1972) these and offshore occurrences are not dealt with in this account. In addition to the rocks of accepted Cambrian age there are outcrops where the age is still the subject of controversy. These include (a) *The Ingleton Slates*, Horton-in-Ribblesdale, West Yorkshire, which may be Precambrian, Cambrian or Ordovician (O'Nions *et al.* 1973); (b) *The Manx Slates* of the Isle of Man, possibly Arenig and no longer generally regarded as putative Cambrian strata but, at the oldest, of early Ordovician (Tremadoc) age (p. 123); (c) *The Dodman Phyllites* of South Cornwall, now referred to the Devonian period (cf. *Gramsatho Beds*, p. 197); (d) *Anglesey*. Since Greenly's pioneering work (1919) the Mona Complex, originally considered entirely Precambrian, has been divided into a ?Pre-Monian suite of gneiss and a Monian Supergroup (see Ch. 2). The three groups of the Monian Supergroup have been the subject of much work because of the complexity of their structural, metamorphic and sedimentary relationships. Shackleton (1975), for example, inverted Greenly's succession. Barber & Max (1979) proposed the scheme shown in Table 3.2 which accepts that the Cemlyn Unit is entirely Cambrian, its base being marked by what may be a thrust or unconformity.

**Table 3.1**

| Period | Series | Wales and England | Scandinavia | International Stages | International Biozones |
|---|---|---|---|---|---|
| Late Cambrian | MERIONETH SERIES | Acerocare | Acerocare | Acerocare | Acerocare |
| | | Peltura etc. | Peltura etc. | Shidertinian | Peltura scarabaeoides |
| | | | Peltura etc. | Shidertinian | Peltura minor |
| | | | Peltura etc. | Shidertinian | Protopeltura praecursor |
| | | Leptoplastus | Leptoplastus | Shidertinian | Leptoplastus |
| | | Parabolina spinulosa | Parabolina spinulosa | Shidertinian | Parabolina spinulosa |
| | | Olenus | Olenus | Tuorian | Olenus and Homagnostus obesus |
| | | Agnostus pisiformis | Agnostus pisiformis | Tuorian | Agnostus pisiformis |
| Medial Cambrian | ST. DAVID'S SERIES | Lejopyge laevigata | Lejopyge laevigata | 'Paradoxides forchhammeri' | Lejopyge laevigata |
| | | Solenopleura brachymetopa | Solenopleura brachymetopa | 'Paradoxides forchhammeri' | Solenopleura brachymetopa |
| | | Goniagnostus nathorsti | Ptychagnostus lundgreni and Goniagnostus nathorsti | 'Paradoxides forchhammeri' | Ptychagnostus lundgreni and Goniagnostus nathorsti |
| | | Ptychagnostus punctuosus | Ptychagnostus punctuosus | 'Paradoxides paradoxissimus' | Ptychagnostus punctuosus |
| | | Hypagnostus parvifrons | Hypagnostus parvifrons | 'Paradoxides paradoxissimus' | Hypagnostus parvifrons |
| | | Tomagnostus fissus | Ptychagnostus atavus and Tomagnostus fissus | 'Paradoxides paradoxissimus' | Tomagnostus fissus and Ptychagnostus atavus |
| | | Ptychagnostus gibbus (present?) | Ptychagnostus gibbus | 'Paradoxides paradoxissimus' | Ptychagnostus gibbus |
| | | Eccaparadoxides oelandicus | Eccaparadoxides pinus | 'Eccaparadoxides oelandicus' | Eccaparadoxides oelandicus pinus |
| | | (not recognized) | Eccaparadoxides insularis | 'Eccaparadoxides oelandicus' | Eccaparadoxides insularis |
| Early Cambrian | COMLEY SERIES | Protolenid-Strenuellid | (not represented) | Toyonian | Anabaraspis |
| | | Protolenid-Strenuellid | Strenuella linnarssoni | Toyonian | Lermontovia dzevanovskii and Paramicmacca |
| | | Protolenid-Strenuellid | | Botomian | Bergeroniellus expansus |
| | | | | Botomian | Bergeroniellus micmacciformis |
| | | Olenellid | Holmia kjerulfi | Atdabanian | Judomia and Dipharus atteborensis |
| | | Olenellid | | Atdabanian | Judomia (and Fallotaspis) |
| | | non-trilobite | Mobergella holsti | Tommotian / Meishucun | Meishucun |

CAMBRIAN

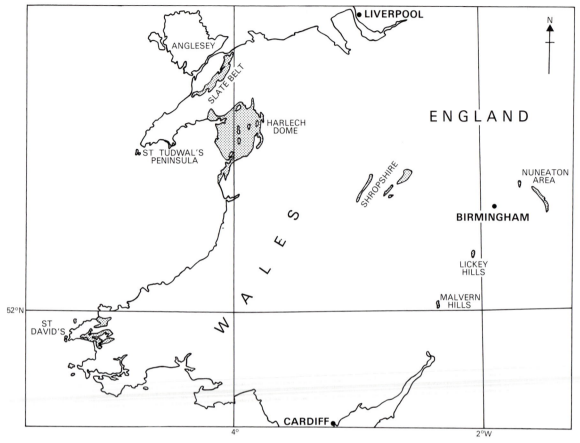

**Fig. 3.1.** Outcrops of the Cambrian System in Wales and England (Comley Series to Merioneth Series, excluding the Ordovician Tremadoc Series).

Vendian–Cambrian stromatolites have been described from limestone blocks in the melange of the Gwna Group and graphitic material from its matrix yielded palynomorphs (acritarchs) of probable Lower Cambrian age (Downie 1975, Muir *et al.* 1979). Microfossils of Lower Cambrian age have also been reported from the Gwna Group by Bliss (1977 – unpublished thesis).

### The 'Caernarvonshire' (Gwynedd) Slate Belt, North Wales

The Gwynedd Slate Belt lies between the Ordovician rocks of the mountains of Snowdonia and the Precambrian–Cambrian massif of Anglesey. The Cambrian rocks are strongly fractured, cleaved and metamorphosed but here the relationship with the Arvonian volcanic rocks is seen clearly in outcrop.

From Anglesey to the mainland of North Wales

there is a considerable lack of correlation, with differences in pre-Carboniferous geology, which Nutt & Smith (1981) explained by Devonian transcurrent faulting with a large dextral movement deduced from geological compatibility between the English Lake District in Cumbria and mainland North Wales. From evidence in SW Llŷn Peninsula this is denied by Tegerdine *et al.* (1981) who claim Lake District and North Wales compatibility is equivocal.

The Slate Belt succession (Table 3.3) was probably laid down between the subsiding Welsh Basin to the south and the uplifting 'Irish Sea landmass/platform' to the north-west (see Figs. 3.1, 3.2). The ignimbrites of the Arvonian Volcanic Series were transgressed by submarine conglomerates and grits. These were in turn succeeded by the muds which now form the Llanberis Slates. Non-deposition and possible erosion followed before the Bronllwyd Grits commenced with turbidites which gave way ultimately to shallow-water sandstones. There was a pause before fine-grained turbidites of the Maentwrog Slates and the shallow-water sandstones of the Festiniog Beds were laid down. The Dolgelly Beds are only 6 m thick and these are overlain unconformably by either the Arenig or Llanvirn Series so that the Tremadoc Series are not represented in Northern Gwynedd.

The Cambrian has been taken traditionally to begin with the first conglomerate at the top of the Arvonian volcanic succession at Llyn Padarn (Llanberis) and Bangor. The correlation of the 'Caernarvonshire' Slate Belt is still poorly established and great reliance is attached to the *Pseudatops viola* horizon in correlation with beds at Comley, Shropshire (pp. 51–52).

The *Tryfan Grits* are known best around Nantlle near Moel Tryfan in the tunnel driven through the mountain and may represent the earliest local Cambrian marine submergence. The formation rests with uncertain relationship on the Arvonian Clogwyn Volcanic Group. Coarse cross-bedded feldspathic sandstones are succeeded by cleaved feldspathic tuffs which in turn give way to 200 m of siltstones.

In the Llanberis–Bethesda district the basal Cambrian rests conformably on Arvonian ignimbrites (Woods 1969) and the first indication of marine transgression seen suggests a shallow sea. In contrast, in the Bangor area an erosional level at the top of the

**Table 3.2.** The Anglesey Succession

| Tectonic Units | Stratigraphic Units |
|---|---|
| CALEDONIAN OROGENY | |
| | Ordovician and Silurian sedimentaries |
| ————————————————unconformity———————————— | |
| Cemlyn Unit | Gwna Melange (2) Greywacke Group Gwna Group (inc. Melange) (1)   CAMBRIAN SYSTEM Church Bay Tuffs |
| ————discordance————thrust or unconformity———— | |
| New Harbour Unit | New Harbour Group   ? |
| ————structural discordance————thrust———— | |
| South Stack Unit   Rhoscolyn Fm. Holyhead Quartzite Fm.    Holy Island Group South Stack Fm. OROGENY | |
| ————————fault contact usually———————— | |
| GNEISSES | |

**Table 3.3.** The Cambrian succession of the Slate Belt of Gwynedd

| | |
|---|---|
| Ordovician: Arenig *or* Llanvirn Series | |
| Lingula Flags  { Dolgelly Beds 6 m   } Festinoig Beds 500 m   Merioneth Series   Maentwrog Beds 300 m } | |
| Cambrian    Cymffrych (200 m) *or* Bronllwyd Grits (500 m) | |
|    Llanberis Slates    ∼1000 m Glog Grits    600 m?    Comley Series Cilgwyn Conglomerate   150 m–400 m Tryfan Grits    300 m | |
| ————————possible conformable contact———————— | |

(Cambrian *or* Precambrian) Arvonian Volcanic Series e.g. Clogwyn Volcanic Group

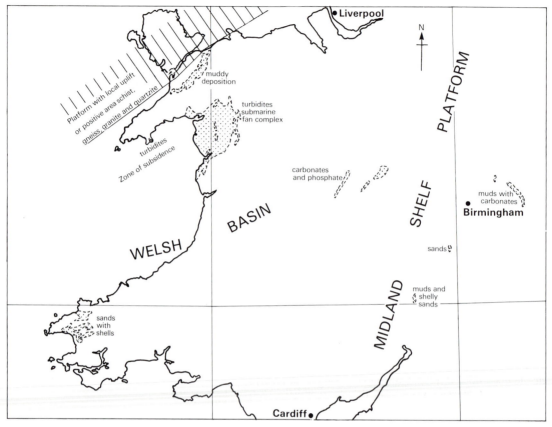

**Fig. 3.2.** Cambrian palaeogeography (excluding Ordovician Tremadoc Series) of Wales and England based on outcrops and concealed rocks.

Arvonian occurs and tuffs overly the basal conglomerates. In the Tregarth area (Fig. 3.3) the base of the Cambrian sequence is not known. The succession consists of what can be interpreted as contemporaneous water-laid tuffs and marine sediments.

The shallow-water *Cilgwyn Conglomerate* (max. 150 m) around Nantlle lies on an eroded surface of the Tryfan Grits with slight disconformity. The rounded pebbles are mainly of Arvonian ignimbrites and are up to 15 cm in diameter while pebbles of jasper and quartzite can be matched with the Monian Gwna Group.

The *Glog Grits* can only be given a tentative thickness (?600 m) due to strike faulting and consist of shallow sandstones, quartzites, siltstones and occasional slate beds.

The highly cleaved *Llanberis Slates* (Fig. 3.4) famed for their roofing qualities are thought to be about 1,000 m thick and exhibit some variations in colour

and grain size. The occasional greywackes probably represent proximal, and the siltstones distal, turbidites. Trilobites such as *Pseudatops viola*, *Protolenus*? sp., *Strenuella*? sp., *Serrodiscus*? sp., together with hyolithids and a crustacean, establish a Comley Series (Lower Cambrian) age near the base of the *Protolenus* Zone for the upper part of this slate sequence.

The *Bronllwyd, Cymffrych*, and *Dinas Grits* are comprised of massive sandstone beds and greywackes. Post-Comley Series in age they have an unclear locally conformable/locally disconformable relationship with the underlying slates. The Bronllwyd Grit (500 m) includes graded greywackes and manganiferous shales in its lower part while higher in the succession it becomes conglomeratic with 20 mm diameter pebbles and 100 mm slate fragments in places. The nature of the contact with the succeeding Maentwrog Beds of the Lingula Flags is unclear. (Precise correlation with the Grits and associated mudstones which underlie the Lingula Flags of the Harlech Dome (Table 3.4) is, in the absence of fossils, inconclusive.)

## Lingula Flags Group

This group is usually divided into three formations (see Table 3.3) in the Slate Belt region but it is not as well developed there as in the Harlech Dome. Typical trilobites are absent except in the topmost formation, the Dolgelly Beds. Rushton (1974, p. 82) summarised the regional details and variations.

Worthy of special mention are the Carnedd Filiast Grits of the Festiniog Beds – coarse, massive pebbly quartzites and feldspathic sandstones deposited in shallow waters from the north-east (Crimes 1966, 1970b). They show current bedding, ripples and *Cruziana* trace fossils.

The *Dolgelly Beds*, only 6 m thick, consist of dark laminated shales with siltstone layers up to 10 mm thick with a good fauna of trilobites (*Briscoia* sp., *Lakella* sp., *Parabolina* sp., and an agnostid) and the brachiopod *Orusia*. These beds are correlated with the *Parabolina spinulosa* Zone (Table 3.1). The small thickness of the Dolgelly Beds and the Tremadoc Series in the Slate Belt is considered to be due primarily to overstep by later Ordovician beds and to faulting. In the north-east, coarse sandstones, conglomerates and greywackes constitute much of the succession. *Lingulella* occurs in the interbedded

mudstones and siltstones. Trace fossils include *Cruziana*, *Rusophycus*, *Phycodes*, *Diplichnites*, *Dimorphichnus*, *Planolites* and near the top *Scolithus*. The sedimentary and biogenic structures suggest an accumulation in the infratidal zone.

## Harlech Dome Region, Gwynedd

By far the most impressive and scenic outcrops of Cambrian rocks in Wales and England occur in this relatively mountainous moorland region south of the Slate Belt (Fig. 3.5). Wooded valleys dissect the area and the massive coarse sandstones are interbedded with more shaly formations causing varied relief. A generalised succession is given in Table 3.4.

A remarkable feature of this succession is that the 3,000 m or more of strata give no evidence of unconformities or stratigraphical breaks of importance. All the rocks of the Harlech Region are somewhat recrystallised and many are cleaved. Metamorphism mainly reaches chlorite grade but in places micas were formed in the pelites.

The lower formations can be interpreted as submarine fan complexes with turbidity currents differently directed at various times and related to varying depths of deposition and changing submarine topography. Sedimentation and subsidence throughout more or less kept pace; only occasionally did shallow water deposition occur.

**Fig. 3.3.** Cambrian outcrops of the 'Caernarvonshire' Slate Belt (Gwynedd), north Wales (after Rushton 1974).

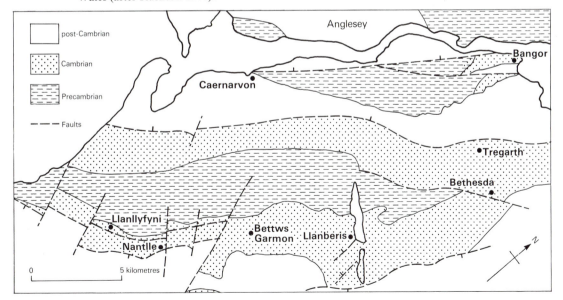

**Fig. 3.4.** Llanberis Slates (Comley Series) near Llanberis, Gwynedd (F. W. Dunning).

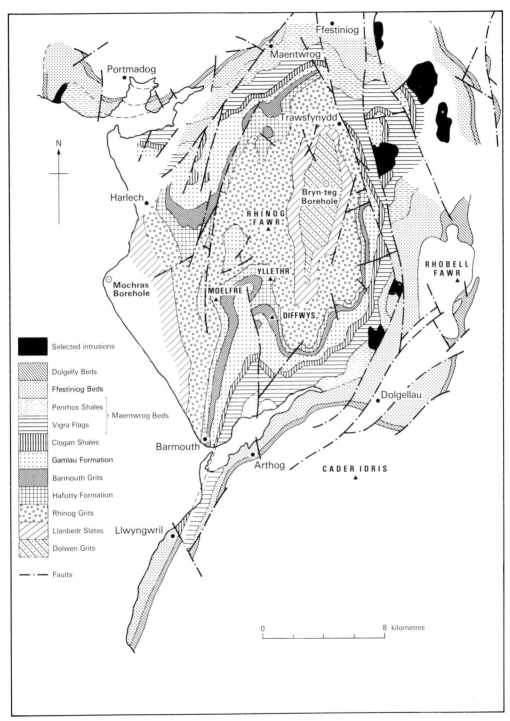

**Fig. 3.5.** Cambrian outcrops, Harlech Region, Gwynedd, north Wales (after Matley & Wilson 1946).

**Fig. 3.6.** Rhinog Grits (Harlech Grits Group) above Roman Steps, Gwynned (F. W. Dunning).

**Table 3.4.** The Cambrian Succession of the Harlech Dome Region

| | | | | |
|---|---|---|---|---|
| Lingula Flags Group | { Dolgelly Beds | | 60–180 m | |
| | Festiniog Beds | | 520–800 m | Merioneth |
| | Maentwrog Beds | Penrhos Shales | 340 m | Series |
| | | Vigra Flags | 310 m | |
| Harlech Grits Group | Clogau Shales | | 75–100 m | |
| | Cefn Coch Grits | | 0–15 m | St David's |
| | Gamlan Formation | | 230–300 m | Series |
| | Barmouth Grits | | 100–200 m | |
| | Hafotty (Manganese) Formation | | 250 m | |
| | Rhinog Grits | | 400–800 m | |
| | Llanbedr Slates | | 90–200 m | Comley |
| | Dolwen Grits* | | 150 + m | Series |

(base not seen at surface outcrop)

Bryn-teg Volcanic Formation* (in borehole only)

(Cambrian *or* Precambrian)

*Note\** The base of the Dolwen Grits is not seen in outcrop but a borehole (Fig. 3.5) (Allen & Jackson 1978) near the centre of the Dome (a true structural dome with dips generally away from a central region) penetrated 283 metres of Dolwen Grits lithology overlying 160 metres of volcanic rocks (Bryn-teg Volcanic Formation) correlatable with some doubt with the Arvonian or north-west Wales.

## Harlech Grits Group

Details of the successions are available in Cowie *et al.* (1972) and Rushton (1974), therefore only brief comments are given here.

The group (Table 3.4) consists of greywacke formations alternating with the beds of slate. A probable source was the region which included the Mona Complex of Anglesey (Fig. 3.2).

*Dolwen Grits.* These poorly exposed grits occur in the centre of the dome and are feldspathic pebbly-greywackes which are often lenticular, cross-bedded and of possible shallow water origin. The top of the succession includes arenaceous slates and a sandstone bed.

*Llanbedr Slates.* These are bluish slates which in places are intruded by sills. Inarticulate brachiopods have been recorded from near Harlech (Matley & Wilson 1946).

*Rhinog Grits* (Fig. 3.6). Greywackes predominate with accompanying shales, fine-grained flaggy sandstones, and persistent pebbly bands. The beds are massive and the formation forms high rugged hills with steep slopes and fine scenery. Sole markings and current-bedding orientation suggested to Knill (1958) and Crimes (1970b) that transport was from the north-east and the latter author suggested a deep-water proximal turbidite origin.

*Hafotty (Manganese) Formation.* This unit is placed at the base of the Middle Cambrian St David's Series and includes a distinctive manganese-rich bed which makes a useful mapping horizon. The Middle Cambrian age is based partly on debated correlation with similar deposits in Newfoundland and a Comley Series (Lower Cambrian) age cannot be ruled out. Mainly consisting of cleaved mudstones, the manganese Ore-Bed, however, is a hard, uncleaved flinty rock of very fine grain composed of manganese carbonate and manganiferous garnet minerals with some quartz. Textural features and geochemical studies suggest deposition of the ore by chemical precipitation in very quiet water in a partially enclosed, possibly fresh water, basin.

*Barmouth Grits.* These beds resemble the Rhinog Grits but coarse rocks are more common. Current-bedding, slumps and flow-markings were reported in Knill (1958) and Crimes (1970b) who interpreted deposition as being from north-flowing proximal turbidity currents.

*Gamlan Formation.* Greywacke grits, flags and shales comprise this formation but there is local variation with the thickest succession in the north-east where massive grits predominate while in the south-west grits are few and thin. Graded beds, flow markings, load-casts, convoluted bedding and groove casts indicate turbidity currents from the south. Worm tubes or burrows are plentiful but only one trilobite fragment and a few inarticulate brachiopods are found.

*Cefn Coch Grit.* These greywackes form a mappable unit (up to 15 m thick) which locally passes laterally into argillaceous beds.

The Lower Cambrian age of the Harlech Grit Group was until recently based on correlation with lithologically similar strata in St Tudwal's Peninsula. The *Pseudatops viola* horizon known from the 'Caernarvonshire' Slate Belt has not been found in the Harlech Dome.

*Platysolenites antiquissimus* Eichwald, (? foraminiferid or worm of Early Cambrian age (Føyn & Glaessner 1979)) was found at about 11 m above the Dolwen Formation base in the Bryn Teg borehole (Rushton 1972, 1978) (see p. 44). Palynomorphs at other stratigraphic levels proved unidentifiable; otherwise only a sinuous burrow trace fossil was found at a depth of about 72 m from the surface. The volcanic sediments, tuffites, lavas (andesite and dacite), tuffs and basic intrusives in the underlying Bryn-teg Volcanic Formation represent a calc–alkaline succession which can be likened to an island-arc succession correlatable with the Avalonian rocks in Newfoundland (Rast & Skehan 1981). They do not compare well on petrological grounds with other pre-Ordovician volcanic formations in North Wales, including the Arvonian.

*Clogau Shales.* This formation consists of dark blue-grey laminated mudstones with much pyrite; it probably accumulated slowly in quiet reducing waters. Fossil species indicate a Middle Cambrian age: trilobites are *Eodiscus, Paradoxides, Anopolenus, Meneviella*; agnostids including *Ptychagnostus*, while inarticulate brachiopods also occur.

## Lingula Flags Group

The Merioneth Series in North Wales is often given the collective term of *Lingula* Flags, a term used here for its informal usefulness. Rushton (1974, p. 63) discussed the various aspects of this usage. The Flags are a conformable succession divisible into three parts (Table 3.4).

*Maentwrog Beds.* This formation is divided into the Vigra Flags below and the Penrhos Shales above. The base of the Vigra Flags consists of tough silty arenaceous beds ('ringers' – ringing under hammer blows) resting conformably on the Clogau Shales. The flags are fine-grained ungraded sandy or silty greywackes showing convoluted bedding, ripple bedding and cross-lamination with intervening shales. Crimes (1970b) interpreted the unit as turbidites, becoming more distal higher in the succession.

The *Penrhos Shales* lack 'ringers' and have a

gradational junction with the *Vigra Flags*. The formation is replaced above by Festiniog Beds where 'ringers' reappear. Pyrite occurs and olenid and agnostid trilobites are abundant but distorted badly by the cleavage; recognisable are species of *Olenus* and *Homagnostus* suggesting correlation with a high subzone of the *Olenus* Zone.

*Festiniog Beds*. The Lower Festiniog Beds include alternations of 'ringers' with slaty beds. Upper Festiniog Beds include micaceous flags and grey silts and, again, 'ringers'. Crimes (1970a, b) considered deposition took place under the influence of currents from the south in a shallowing basin. The occurrence of the trace-fossils *Cruziana* and *Rusophycus* and *Skolithos* burrows, support the shallow-water hypothesis. The trilobite *Parabolinoides bucephalus* occurs in several widespread localities and supports an *Olenus scanicus* Subzone and/or *Parabolina brevispina* Subzone age.

*Dolgelly Beds*. The laminated mudstones and slates of the Lower Dolgelly Beds (*c*. 90 m) grade up from the Festiniog Beds and contain pyrites. The trilobite *Parabolina spinulosa* is found indicating its Subzone. A species of the trilobite *Briscoia* and the brachiopod *Orusia* is also found. Shallow sea conditions appear to have prevailed.

The Upper Dolgelly beds (*c*. 90 m) are laminated mudstones containing free carbon, pyrite and, locally, radioactive minerals. A condensed deposit is inferred apparently accumulating slowly under reducing conditions. Rich fossiliferous beds are common with trilobite species of *Leptoplastus*, *Eurycare*, *Paradoxides*, *Ctenopyge*, *Sphaerophthalmus*, *Parabolinites*, *Lakella*, *Lotagnostus*, the brachiopods *Orusia* and horny brachiopods. The *Leptoplastus* Zone and part of the *Peltura scarabaeoides* Zone are indicated. The uppermost Cambrian zone of *Acerocare* is not identified.

The overlying basal Tremadoc Series is represented by beds of the zone of *Dictyonema flabelliforme* with a gradual upward change in lithology.

## Mawddach Group

Allen *et al.* in 1981 gave the name Mawddach Group to rocks ranging from late St David's to Tremadoc age in the Harlech Dome which had previously been called Menevian Beds, Lingula Flags and Tremadoc Slates (Table 3.5).

## St Tudwal's Peninsula, Gwynedd

The small Cambrian outcrop (Fig. 3.7) is important because of its correlation with the main North Wales Harlech Grit Group (Comley Series) outcrop and its fauna supports the construction of a generalised succession in this part of the stratigraphical record. Because of its physical separation from the 'Caernarvonshire' Slate Belt and the Harlech Dome Region it is put in a separate section here. The Cambrian rocks have been slightly affected by cleavage; intrusions are not seen and may be absent. Descriptions have been given for the area first by Nicholas (1915, 1916) and more recently by Matley *et al.* (1939), Bassett & Walton (1960), Crimes (1970b) and Bassett *et al.* (1976) (Table 3.6).

*Hell's Mouth Grits*. Bassett & Walton (1960) recognised 11 members in this formation, though the base can not be seen. The main thickness is of coarse, crudely graded, massive greywackes with minor mudstones. The upper beds show cross-lamination and rarer convoluted bedding; also noted ripples, groove-casts, flute-casts, gouging and load-casts. Turbidity currents apparently came from the northeast. The lithology is closely comparable with the uppermost Rhinog Grits and near the top mudstone beds show enrichment in phosphates.

In most of the formation only trace fossils are present such as burrow casts and sponge spicules but in the upper part finds of trilobites (e.g. *Hamatolenus* (*Myopsolenus*) *douglasi*, *Kerberodiscus succinctus*, and *Serrodiscus ctenoa*?) associated with hexactinellid sponge spicules, a single inarticulate brachiopod and trace fossils are of considerable stratigraphic importance. The fauna is typical of the Comley Series and indicates the upper part of the Protolenid–Strenuellid Zone (Table 3.1). However, this horizon is younger than the *Pseudatops viola* horizon in the Llanberis Slates. Additional evidence as to biostratigraphic correlation is cited by Bassett *et al.* (1976) from acritarchs in six horizons of the Hell's Mouth Grits indicative of a level high in the Lower Cambrian close to the Lower–Middle Cambrian boundary.

*Mulfran Beds*. Manganese-rich shales, mudstones and greywackes (the Manganese Beds) appear to be in part equivalent to the Ore-Bed Shales of the Harlech Region's Hafotty Formation. The lowest 15 m are sandy mudstones with calcareous layers, as with the Ore-Bed the manganese enrichment causes weathering to a shiny purple-black surface. The upper 120 m of the Mulfran Beds are made up equally of arenaceous beds and manganiferous mudstones. The upper boundary of the Mulfran Beds is probably of a different age to that of the top of the Hafotty Formation (Nicholas 1915). Acritarchs have been reported from the base and, more sparsely, higher up.

*Cilan Grits*. They consist of massive graded greywackes with thin mudstones. Some of the coarse beds

**Table 3.5.** Tremadoc Rhobell Volcanic Group or Arenig sedimentary and volcanic rocks

unconformity – – – – – – – – – – – – – – – – – – – – – – – – – – – – – – – – – – – – – –

| | | |
|---|---|---|
| Mawddach Group (grey argillaceous rocks interbedded with quartzose sandstone and siltstone) | Cwmhesgen Formation Festiniog Flags* Formation Maentwrog Formation Clogau Formation | Dolgelly Member* Dol-cyn-afon Member |

――――――――――――――――conformable――――――――――――――

Harlech Grits Group

\* Eponymous terms: Dolgelly = Dolgellau   Festiniog = Ffestiniog

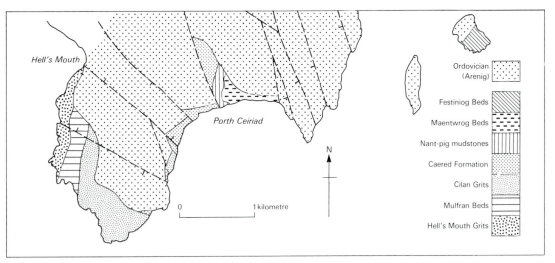

**Fig. 3.7.** Cambrian outcrops, St. Tudwal's Peninsula, Gwynedd, north Wales (after Nicholas 1915 and Rushton 1974).

**Table 3.6.** The St Tudwal's Succession

Ordovician Arenig Series

– – – – – – – – – – – – – – – – – – – – – – – – – – – – – – – – – – – – – – – – – – – –

| | | | |
|---|---|---|---|
| Festiniog Beds | c. 100 m seen | | Merioneth Series |
| unexposed interval | | | |
| Maentwrog Beds unknown thickness | | | |
| Nant-pig Mudstones | > 66 m | | |
| Caered Mudstones and Flags | Upper Caered Mudstones | > 84 m | St David's Series |
| | Caered Flags | > 35 m | |
| | Lower Caered Mudstones | > 34 m | |
| Cilan Grits | 300 m | | |
| Mulfran Beds | 135 m | | |
| Hell's Mouth Grits (base not seen) | > 176 m | | Comley Series |

contain pebbles comparable with the rocks of the Mona Complex. Though resembling the Barmouth Grits lithologically, the current direction appears to have been from the north-east and may have been axial to the trough. Crimes (1970b) reported *Planolites*, bifurcating burrows, *Sinusites*, intertwining burrows and burrow network (cf. *Paleodictyon*) from the upper part of the formation.

*Caered Mudstones and Flags.* An incomplete faulted succession of mudstones, commencing with a porcellanous vitric tuff at the base, is separated into two parts by flaggy sandstones in the middle. In the Upper Caered Mudstones trilobites are found: *Tomagnostus fissus* with species of *Paradoxides, Peronopsis, Ptychagnostus* and *Eodiscus*.

These indicate the Zone of *Tomagnostus fissus* and the formation as a whole is loosely correlated with the Gamlan Formation; depositing currents came from the south according to Crimes (1970b). Middle Cambrian acritarchs are reported to be present (Bassett *et al.* 1976) in this formation and in the underlying Cilan Grits.

*Nant-pig Mudstones.* Laminated pyritous mudstones occur with thin silty and calcareous, including black limestone, layers. In the lower part of this formation the zone of *Tomagnostus fissus* is indicated by *T. fissus* and *Parasolenopleura applanata*. The upper Nant-pig Mudstones give *Ptychagnostus, Pleuroctenium, Peronopsis, Eodiscus* and *Meneviella* referable to the zone of *Hypagnostus parvifrons*. A conglomerate at the top of the formation yields *Linguagnostus, Dorypyge* and *Bailiaspis* which suggest a *Paradoxides forchhammeri* (St David's Series) age.

The Nant-pig Mudstones environment of deposition seems to have been of calm water of moderate depth with shallowing and erosion at the top of the formation; there is strong bioturbation in places.

*Maentwrog Beds.* Greywackes and siltstones commencing with a calcareous conglomerate, alternate with shales with 'ringers' which seem equivalent to Vigra Flags (Table 3.4). There are non-sequences which reduce the thickness and no fossils indicating the *Agnostus pisiformis* Zone are found although the *Olenus* Zone is indicated by *Olenus* sp. and *Homagnostus obesus*. Flute, groove and load casts, other sole markings and 'flame' structures occur; the tops of many beds are ripple-marked.

*Festiniog Beds.* On a small island east of the peninsula, siltstones and mudstones (~100 m) thick yield *Lingulella davisi* which near the top of the succession constitute the '*Lingulella* Band'. Overlying beds belong to the Ordovician Arenig Tudwal Sandstone so that probably much of the Festiniog Formation is absent, as are Dolgelly and Tremadoc strata. The formation is rich in trace-fossils with *Cruziana* and *Skolithus*.

## St David's region, Dyfed, South-west Wales

During Cambrian times mainly shallow marine deposits accumulated on the shelf-margin which flanked up-faulted-blocks of Precambrian igneous rocks (Fig. 3.8) (Table 3.7).

The Precambrian basement was substantially eroded before the local transgression of Comley Series times. A shallow subsiding basin with overall fining-upwards trends experienced some pauses in deposition and possibly some erosion of sediments. The St David's Series started with shallow high energy environments followed by quieter conditions while the Lingula Flags exhibit an abruptly changed situation with fine grained turbidites laid down in a more rapidly subsiding shallow-water environment. The Cambrian rocks are exposed steeply dipping off the flanks of two anticlines showing Precambrian cores at Hayscastle and St David's (Fig. 3.8).

### Caerfai Group

The main outcrops occur in two areas – the St David's Anticline and the Hayscastle Anticline (Welsh Hook Beds) – and in both places show a transgressive sequence deposited in an open current-swept shallow sea.

*Basal Conglomerate.* Composed of well-rounded pebbles and boulders of quartzite and acid igneous rocks set in an arenaceous and argillaceous matrix, it is, in places, represented by a coarse orthoquartzite. It rests unconformably on the Precambrian rock suites of Pebidian tuffs and Dimetian granophyre and contains fragments of these Precambrian rocks, abundantly in the Hayscastle Anticline but sparsely elsewhere.

*St Non's Sandstone.* This includes bioturbated sandstone with some small-scale current bedding; *Skolithos* worm burrows indicate that shallow water prevailed.

*Caerfai Bay Shales.* Consist of thin but uniformly developed red shales (useful as markers) with pale silty layers, possibly tuffs, which exhibit small-scale current bedding and probable convoluted bedding. Fossils include inarticulate brachiopods, a bradoriid crustacean and worm burrows. The presence of trilobites has not been substantiated. A tentative age is Comley Series (Lower Cambrian) but this is not provable.

*Caerbwdy Sandstone* (150 m). In the St David's area it consists of an unfossiliferous fine-grained purple micaceous–feldspathic sandstone. Near the top coarser layers contain pebbles resembling the Dimetian granophyre. The Solva Group succeeds more or less abruptly but conformably.

In the Hayscastle Anticline area the purple sandstones are thinner and are overlain by coarser greenish quartzitic and micaceous sandstones more than 30 m thick which may be basal Solva beds or the uppermost part of the Caerfai Group.

## Solva Group

Well displayed at Solva, around St David's and the Hayscastle Anticline, it may rest unconformably on eroded Caerbwdy Sandstone at Caerfai Bay (Jones 1940). Stead & Williams (1971) however, maintain it is conformable there as elsewhere.

*Lower Solva Beds.* At the base are coarse greenish sandstones with *Skolithos* worm burrows, trilobites with species of *Paradoxides* referable to the *P. pinus* Zone and *Bailiella, Metadiscus, Condylopyge,* hyolithids and sponge spicules. Solva beds near Newgale yielded acritarchs of possible Middle Cambrian age (*P. oelandicus* zones) (Davies & Downie 1964). The formation was divided into three members by Stead & Williams (1971).

**Table 3.7.** Cambrian succession of southwest Wales

| Ordovician Arenig Series | | | |
|---|---|---|---|
| ————————————— unconformity ————————————— | | | |
| Lingula Flags | | over 600 m | Merioneth Series |
| Menevian Group | | about 230 m | St David's Series |
| Solva Group | | about 500 m | |
| Caerfai Group up to 350 m | Caerbwdy Sandstones | 80 up to 150 m | Comley Series |
| | Caerfai Bay (Red) Shales | 15 m | |
| | St Non's (Green) Sandstone | 140 m | |
| | basal conglomerate | 10/50 m | |
| ————————————— unconformity ————————————— | | | |
| Precambrian volcanic rocks (Pebidian & Dimetian) | | | |

**Fig. 3.8.** Cambrian outcrops of St David's region, Dyfed, south Wales (after Rushton 1974).

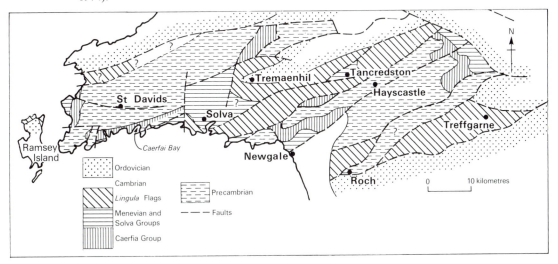

*Middle Solva Beds.* A thick sequence, with ill-defined boundaries, of purplish and greenish mudstones and sandstones which are reduced in thickness in places by strike faulting. Fossils include species of *Paradoxides*, *Ctenocephalus* and *Parasolenopleura* and inarticulate brachiopods; trace-fossils suggest shallow water.

*Upper Solva Beds.* These consist of greenish sandstones and grey flaggy mudstones with worm-burrows and concretions. Rare fossils are the trilobites *Paradoxides* and *Bailiaspis* which indicate an early *Paradoxides paradoxissimus* age (? Zone of *Ptychagnostus gibbus*).

## Menevian Group

This group occurs at Porth-y-rhaw (its type section), Solva, Tancredston and Hayscastle. In common with most Cambrian strata of the St David's area, there has been no recent published revision of the stratigraphy but correlation with the Clogau Shales is widely accepted.

*Lower Menevian Beds.* Generally more massive than the Upper Solva Beds, its shaly mudstones contain beds which yield the zonal index fossil *Tomagnostus fissus* (Linnarsson) and *Paradoxides*, *Hartshillia*, *Parasolenopleura*, *Eodiscus*, *Peronopsis* and *Ptychagnostus*.

*Middle Menevian Beds.* Dark shales or mudstones with silty and sandy lenticular beds and thin beds of ash suggest deposition in a calm sea poor in oxygen. A large fauna, however, is known and indicates the *Hypagnostus parvifrons* Zone and the *Ptychagnostus punctuosus* Zone and includes *Paradoxides*, *Anopolenus*, *Clarella*, *Holocephalina Meneviella*, *Solenopleuropsis?*, *Eodiscus*, *Cotalagnostus*, *Peronopsis*, *Pleuroctenium*, *Ptychagnostus* and also hyolithids and brachiopods.

*Upper Menevian Beds.* Consists of coarse massive sandstones with interbedded shales, succeeded by turbidites which continue into the *Lingula* Flags. Yielding *Billingsella* and undescribed species of *Paradoxides* and '*Conocoryphe*' it may be that these rocks partly correspond with the *Paradoxides forchhammeri* zonal group of Scandinavia.

## 'Lingula Flags' Group

Micaceous flaggy shales alternating with sandy mudstones and siliceous sandstone probably represent turbidite conditions. It is correlated with the Lingula Flags Group of North Wales and outcrops in the St David's and Hayscastle Anticlines and elsewhere. The base, where visible, is conformable upon the Menevian but the upper limit is generally faulted against Arenig rocks though in places there are indications of an angular unconformity.

The lithology is similar to that of North Wales with trace fossils and sedimentary structures indicating shallow-water currents. *Lingulella davisi* is abundant at several localities making a *coquina* in places while the trilobites *Homagnostus* and *Olenus* are recorded. The St David's 'Lingula Flags' probably span from the St David's Series to the top of the *Olenus* Zone of the early Merioneth Series but Festiniog beds and Dolgelly Beds appear to be absent, possibly removed by erosion before the deposition of the Arenig rocks. No close correlation at this level has been established with sequences elsewhere.

The St David's region is complex because of Caledonian deformation and post Cambrian igneous activity; there are fine cliff exposures but outcrops are sparse inland.

## Llangynog area, south-west of Carmarthen, Dyfed

South-west of Carmarthen exposures of Precambrian and Cambrian rocks were only identified and described recently by Cope (1977, 1979).

The oldest rocks are rhyolitic and andesitic lavas and tuffs with intrusions of dolerite. Beds interbedded with these igneous rocks have recently yielded probably medusoid impressions of Ediacarian (Vendian Sinian) type and on the basis of current practice (Cowie 1978) the age is considered to be Precambrian. The apparent succession is shown in Table 3.8.

It seems clear that Precambrian metamorphic basement and younger igneous rocks lie at shallow depths in the vicinity of Carmarthen Bay. This view is supported by the nature of the derived material in the local Lower Palaeozoic rocks with igneous rock pebbles in the basal Cambrian and the large amount of detrital mica in the Lingula Flags.

Cope (1979) suggested that the outcrops imply a position near the margins of the Welsh Lower Palaeozoic depositional basin associated with the fault-bounded margin of a Precambrian landmass. In late Precambrian times there existed a shallow-shelf sea in which locally derived ashes entombed medusoids. Subsequent movements were followed by a Cambrian marine transgression. A non-sequence means that Middle Cambrian rocks are unrepresented but renewed subsidence in the Upper Cambrian brought in conglomerates and the metamorphic rocks of the landmass provided mica-rich sands in which lingulids thrived. Further subsidence produced black shales which have yielded olenid trilobite remains. Shallower water conditions during Ordovician times are indicated by the presence of detrital mica derived from Precambrian mica schists.

**Table 3.8.** Succession in the Llangynog area of South Wales

| | |
|---|---|
| Ordovician | Tremadoc Beds (*Clonograptus tenellus* and *Shumardia pusilla* Zones and younger) |
| Upper Cambrian Lingula Flags (Merioneth Series) | Shales, conglomerates and very highly micaceous siltstones with *Lingulella davisii* and trilobites, about 400 m |
| ? Lower Cambrian | Green feldspathic siltstones |
| ? Lower Cambrian | Conglomerate 2 m thick |
| ----------------------------------------- unconformity | |
| Precambrian | Rhyolites, andesitic lavas and tuffs with intrusions of dolerite. Interbedded sediments with probable Ediacaran type medusoid impressions. |

An important fracture line in the Llangynog region is the Llandyfaelog Disturbance and several authors (Jones 1938; Dunning 1966; Owen 1967, 1971) have linked it north-eastwards with the Church Stretton Fault (see Ch. 17). The Llandyfaelog Disturbance shows Variscan down-faulting of Carboniferous Limestone but this may be a rejuvenated episode of movement along a fundamentally Caledonian fracture.

## Shropshire (Salop) and Herefordshire (Hereford & Worcester)

For nearly 60 years during the 19th century the Longmyndian sedimentary sequence was assumed to be of Cambrian age. It is thick and largely unfossiliferous, save for some possible trace fossils, and cryptarch microfossils (Peat 1984) which may be Precambrian or Cambrian in age. Later research by Callaway (1878), Lapworth (1882, 1886) and Cobbold (1919, 1927) between 1870 and 1936 established the conclusion that the Wrekin Quartzite was the oldest Cambrian formation and that the Longmyndian was Precambrian. This conclusion was based on grounds which may not be acceptable under current stratigraphic practice but is still widely held by active workers. It is, however, a distinct speculative possibility that part, at least, of the Longmyndian sequence belongs to the Cambrian system in terms of the current international proposal for the chronostratigraphic level of the Precambrian–Cambrian Boundary established by a global stratotype section and point.

The Church Stretton Fault-line (Figs. 3.9, 3.10) is of great significance, its movements possibly in Pre-

cambrian times and certainly in Phanerozoic times, were complex and occurred in different episodes widely separated in time. Possible wrench movements could have brought into juxtaposition rock units which were once widely separated. Conditions of sedimentation were affected by earth movements during much of Cambrian times as shown by the numerous interruptions in deposition. Locally derived shallow-water sandstones, sometimes glauconitic, phosphatic and/or calcareous at many levels and localities, are considered to have been deposited on a slowly subsiding surface. Frequent non-sequences in the Comley Limestones (Table 3.9) demonstrate that sedimentation and erosion alternated, though each limestone bed is only a few centimetres thick. The whole Comley Limestone unit (p. 52), with its time hiatuses, may represent a considerable length of time.

The unconformity between the folded and eroded Comley Series and the shallow water sandstones of the St David's Series marks the local Lower-Middle Cambrian Boundary. Instability continued during the deposition of the Merioneth Series as shown by non-sequences and erosion which on occasions cut into the Upper Comley Sandstone. However most of the Merioneth Series was deposited in calm seas with free faunal movement and sedimentation progressed to a more stable basinal character with the deposition of the Ordovician Tremadoc Series.

## Comley Series

The *Wrekin Quartzite* (Table 3.9) rests unconformably on the Uriconian volcanics, e.g. at Wrekin and Caer Caradoc, and on the Rushton Schist at Rushton (Fig. 3.10). Pebbly beds at the base of the Quartzite in-

clude Uriconian volcanic fragments. The main lithology is an orthoquartzite with rounded quartz grains,
some feldspar and glauconite. The upper beds coarsen
upwards through irregular quartzitic and conglomeratic beds with shaly partings into the Lower Comley
Sandstone. Worm traces including ? *Diplocraterion*
traces are taken to indicate a Lower Cambrian age
(Brasier *et al.* 1981, p. 31); a recent find of horny
brachiopod fragmentary material supports this (A. W.
A. Rushton, *pers. comm.*) in a similar way to the
brachiopods in the Lower Cambrian Malvern
Quartzite (Fig. 3.1).

The *Lower Comley Sandstone* outcropping near
Comley, Rushton, Hill End (diggings some 4 km SE of
Comley) and Lilleshall consists of a medium to fine-
grained glauconitic sandstone, often flaggy and
micaceous with calcareous developments. Using
contained fossils, Cobbold (1927) subdivided this unit
into his $Ab_1$ to $Ab_4$ and $Ac_1$. $Ab_1$ to $Ab_4$ include species
of *Paterina, Obolella, Walcottina, Hyolithes*, and
*Mobergella* and *bradoriid* crustacea. The base of the
Lower Comley Sandstone is reported to yield the
problematical phosphatic disc, *Mobergella radiolata*
Bengtson (Brasier *et al.* 1978, p. 34). Near the middle
of the formation a single olenellid cephalon was
originally referred to *Kjerulfia?* but it may be a species
of *Fallotaspis* (Hupe 1953) and may be faunistically
associated with the overlying Lower Comley
Limestones which contain species of *Callavia, Kjerulfia?, Hebediscus*, horny brachiopods, and *Wanneria?*
or *Judomia*.

The *Lower Comley Limestones* outcrop at Comley,
the Cwms (excavation), at Rushton (mainly excavation) and in a borehole at Lilleshall (12 km NE of the
Wrekin) but was not found in the author's excavation
at Hill End. This is the most interesting of the
Shropshire formations from a faunal and international correlative viewpoint. It is less than 2 m thick
but has five members with distinct lithologies and
faunas each separated from its neighbour by disconformities. The formation displays a trend in

lithological change from arenaceous to phosphatic,
the whole being deposited in shallow marine
conditions.

Cobbold's (1927) subdivisions are $Ac_2$ to $Ac_5$ and
Ad:

$Ac_2$.  '*Olenellus* Limestone' – nodular
calcareous sandstone, max. 0·75 m, the
fauna is, as already noted, similar to $Ac_1$
with species of the trilobites *Callavia,
Nevadia, Hebediscus, Angusteva,
Micmacca, Strenuella* accompanied by
horny brachiopods, bradoriid crustaceans
and conoidal shells of controversial
affinities.

$Ac_3$.  '*Eodiscus' bellimarginatus* Limestone –
pale grey-pink limestone *c.* 0·5 m, fauna:
species of trilobites *Callavia, Pseudatops,
Strenuella* and *Serrodiscus* together with
horny brachiopods and conoidal shells.

$Ac_4$.  *Strenuella* Limestone – reddish-purple to
grey arenaceous limestone, *c.* 0·2–0·4 m.
Fauna: *Callavia* sp., *Strenuella* spp.,
*Calodiscus* sp. and *Serrodiscus* sp.
represent the trilobites along with
brachiopods, hyolithids and other
conoidal fossils.

$Ac_5$.  *Protolenus* Limestone – dark to pale grey
or brownish phosphatic limestone 0·15 m.
Rich fauna: echinoderm plates, trilobites
spp. – *Protolenus, Mohicana, Strettonia,
Calodiscus, Serrodiscus, Cobboldites,
Runcinodiscus*, with brachiopods,
hyolithids, tubes and cones of uncertain
affinity and spicules of the sponge
*Chancelloria*.

Ad.  *Lapworthella* Limestone – black
(occasionally white or pink) with
phosphatic granules some of which have a
concentric structure and are possibly algal
in origin. The presence of manganese and

**Table 3.9.**  The Cambrian strata of Shropshire

| | | | |
|---|---|---|---|
| Ordovician Tremadoc Series | | Shineton Shales | |
| | Bentleyford Shales | 4 m | Merioneth Series |
| | *Orusia* Shales | 150 m | |
| Cambrian System | Upper Comley Sandstone | max. 200 m | St. David's Series |
| | Comley Limestones | 1·8 m | |
| | Lower Comley Sandstones | 150 m | Comley Series |
| | Wrekin Quartzite | max. 40 m | |

? Precambrian sedimentary, volcanic and metamorphic rocks (Longmyndian and
Uriconian)

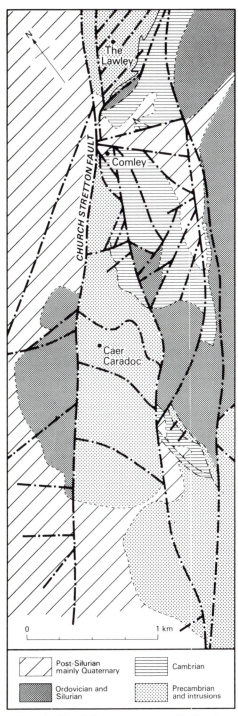

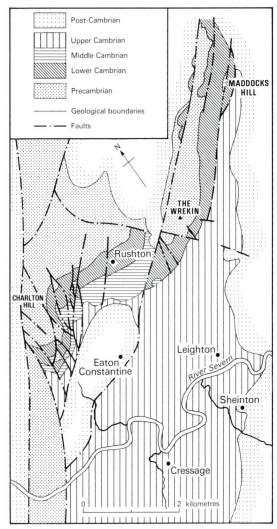

carbon add to the geochemical–sedimentological interest. The fauna shows changes with new incomers: species of genera with varying affinities and only a few doubtful fragments of trilobites. It is by no means certain that it belongs to the Comley Series (Lower Cambrian) but the St David's Series (Middle Cambrian) Upper Comley Sandstones overlie this unit unconformably so in the classic manner it is put with the other Comley Limestones as part of Comley Series strata.

**Fig. 3.9.** Cambrian outcrops of Comley area, near Church Stretton, Shropshire (Salop), England.

**Fig. 3.10.** Cambrian outcrops of the Wrekin area, near Wellington, Shropshire (Salop), England.

## St David's Series

The *Upper Comley Sandstone* is known from Comley and Rushton and can be divided into several units of grits, sandstones, flags and shales. A hiatus, marked by uplift, slight folding and local erosion with overlap and overstep expressing a transgression, followed the deposition of the Lower Comley Limestone (which terminated with the *Lapworthella* Limestone).

The detailed stratigraphy and faunal horizons have been given by Rushton (1974, pp. 97–100). Species of *Paradoxides, Kootenia, Bailiella, Corynexochus, Cobboldites, Bailiaspis, Parasolenopleura, Eodiscus, Ptychagnostus, Agraulos, Peronopsis, Agaso, Dorypyge, Cotalagnostus* and *Hypagnostus* have been found. Brachiopods, conoidal genera and bradoriids also occur.

## Merioneth Series

The *Orusia Shales* outcrop near Comley, Hill End and Bentleyford Brook (a tributary of the River Severn south of the Wrekin and about 4 km NE of Comley) and probably near Rushton. Grey silty shales with bands of sandstone yield the trilobites *Parabolina, Parabolinites* and *Parabolinoides* and the brachiopod *Orusia*.

The *Bentleyford Shales* in Bentleyford Brook is a thin formation mainly of fine-grained sandstone but also contains radioactive black concretions of bituminous limestone which yield olenid trilobites ascribed to *Ctenopyge* from the zone of *Peltura minor* (Table 3.1). Loose blocks of similar limestone from the Rushton area yield the trilobites *Leptoplastus, Eurycare, Ctenopyge* and *Sphaerophthalmus*? which Henningsmoen (1957, p. 48) considered to represent the *Leptoplastus* Zone and the *Protopeltura praecursor* Zone. Similar faunas have been found in a borehole at Lilleshall (Rushton 1972).

A number of small inliers of Cambrian rocks in the south-west Midlands of England are associated with Precambrian rocks and overlain conformably by Tremadoc Series strata which are in turn followed unconformably by rock of Silurian age; post-Tremadoc rocks are present only as intrusives. Only the largest inlier (at the southern end of the tectonically significant axis of the Malvern Hills) is described here. The rocks are much folded and faulted and the succession is incomplete and poorly known. Early work by Phillips (1848) and Holl (1865) was consolidated into the present stratigraphic scheme by Groom culminating in his papers of 1902 and 1910. The Comley Series arenaceous beds suggest a shallow high-energy marine environment changing over their considerable thickness from clean sands to glauconitic beds in a sequence similar to that of Shropshire.

Absence of St David's Series strata may be due to non-deposition during emergence. Renewed deposition in the form of the Merioneth Series shales suggests that they were laid down in a low-energy environment. There may even be more non-sequences present (Table 3.9).

The Cambrian and associated Tremadoc rocks which here form a unit are extensively intruded by basic igneous rocks which form bosses, sills and dykes. Truncation by the unconformity at the base of the Silurian suggests that these intrusives are part of an Ordovician igneous episode (Chs. 4 and 16).

## Malvern Hills, southwest Midlands, England

*Malvern Quartzite* (Table 3.10) is only found in small faulted slices in contact with underlying Precambrian Malvernian rocks. Apart from two outcrops in the northern Malverns and near Martley (14 km N of Gt Malvern) the slices are in the southern Malverns. In addition to faulting between Cambrian and Precambrian, an unconformable contact has been described (Jones *et al.* 1970). The formation is often false-bedded and locally the quartzite passes into a conglomerate with rounded pebbles (50 mm dia.) some of which appear to be derived from the Malvernian. Glauconite is present and cementation is by secondary silica. Fossils are important because they prove a Comley Series (Lower Cambrian) age which by lithological correlation embraces the Wrekin and Lickey Quartzites, and also the Lower Comley Sandstone fauna; they are brachiopod species of *Paterina, Obolella* and *Kutorgina* together with hyolithids. These quartzites from three regions are probably diachronous.

The *Hollybush Sandstone* was subdivided by Groom (1902) as follows:

3. Sandstones and some conglomerates ~200 m.
2. Sandstones with quartzites and some conglomerate ~100 m.
1. Sandstones with thin shaly interbeds and thin impure limestones apparently passing down into Malvern Quartzite ~20 m.

(There must, however, be some doubt in the sequence in this highly faulted region which also displays overturned strata.)

Heavy mineral analysis has suggested a comparison of the Hollybush Sandstone with the Comley Sandstone of Shropshire and the similarity in the field is striking. The sparse fossils could be of Comley Series age: the brachiopod *Paterina* and hyolithids.

## Merioneth Series

The *White-Leaved Oak Shales* are everywhere faulted against the Hollybush Sandstone. The lowest beds are black shales with thin bands of coarse dark quartzitic sandstones with glauconite which yield the bradoriid ostracod-like crustacean *Cyclotron* and horny brachiopods referable to the *Olenus* Zone.

The main part of the succession is dark grey to black soft shales rich in pyrites and organic material. In places they have been hardened and bleached by intrusions. Fossils include the brachiopods *Broeggeria* and *Orusia* and the trilobites *Lotagnostus, Pseudagnostus, Peltura, Sphaerophthalmus, Ctenopyge* and *Lakella*. More than one zonal horizon may be present but the stratigraphical provenance and sequence of the fossils is inadequately known. Most of the species fall naturally, however, into the Zone of *Peltura scarabaeoides* in this main part of the succession.

The formation passes conformably and with little marked lithological change into Bronsil Shales (Tremadoc Series) of the Ordovician system.

## The central Midlands of England

Cambrian rocks outcrop at the margins of two exposed coalfields: near Nuneaton and Atherstone on the eastern flank and at Dosthill (south of Tamworth) on the north-western margin of the Warwickshire Coalfield outcrops; in the Lickey Hills near Barnt Green (south of Birmingham) at the southern end of the South Staffordshire Coalfield (Fig. 3.1).

It is known, however, that Cambrian rocks are widely present beneath the Upper Palaeozoic and Mesozoic cover and they have been found in borings in many places (Rushton 1974, pp. 105–106). Upfaulted blocks bring Precambrian and Lower Palaeozoic rocks to outcrop.

Lapworth (1882, 1886, 1898) described the central Midlands Cambrian and was followed by other workers showing the succession to be sedimentary quartzites followed by shales. These represent shallow water clean quartz sands and micaceous laminated muds followed by muds of shallow but open marine origin, conditions which permitted fauna from elsewhere in the Atlantic (Acado–Baltic) Province to invade. Apart from two considerable breaks, sedimentation was slow and only interrupted for short periods. The fullest sequence of the Merioneth Series is found in the Nuneaton district.

## Nuneaton area, Leicestershire

The general succession in the Nuneaton inlier is given in Table 3.11.

## Hartshill Formation – Comley Series

This forms a prominent feature and is predominantly orthoquartzite; its hardness is caused by secondary cementation by silica. Occasionally some beds contain igneous grains and pebbles from conglomeratic horizons and suggest derivation from underlying volcanic rocks in the vicinity of Nuneaton which have usually been ascribed to the Precambrian without containing faunal, floral or radiometric evidence. These, the *Caldcote Volcanic Formation*, can, however, be equated on petrological similarity with Precambrian *Charnia*-bearing tuffs some 20 km to the north-east (Brasier 1984). More fissile partings in the main quartzite sequence are of micaceous shales or sandstones. It is these partings in more conspicuous development which are the basis of five members noted in the table above: Park Hill, Tuttle Hill, Jee's, Home Farm and Woodlands which represent, together with the lower part of the Purley Shales, the *Comley Series*. Shallow-water origin of the Hartshill Formation is evinced by conglomerates, burrowing, current-bedding, ripple marks and possible desiccation cracking.

The basal layers of the *Park Hill Member* overly the Caldecote volcanic rocks with angular unconformity and contain boulders and fragments of these volcanics. In one area the conglomerate rests on a syenitic intrusion which had already been deeply subaerially

**Table 3.10.** Cambrian succession in Malvern Hills area

| | | |
|---|---|---|
| Ordovician | Bronsil Shales | Tremadoc Series |
| Cambrian | White-Leaved-Oak Shales ~ 150 m | Merioneth Series |
| | ————————time gap———————— | (St David's Series missing) |
| Cambrian | Hollybush Sandstone ? 300 + m | Comley Series |
| | Malvern Quartzite ? 100 + m | |
| | ————————fault and probable unconformity———————— | |
| Precambrian | Malvernian igneous and metamorphic rocks | |

**Table 3.11.** Cambrian succession in Nuneaton area

|  |  |  |  |
|---|---|---|---|
|  | Merevale Shales |  | Tremadoc Series (Ordovician) |
| Stockingford Shales | Monks Park Shales | 80 m |  |
|  | Moor Wood Flags and Shales | 15 m | Merioneth Series |
|  | Outwood Shales | 300 m |  |
|  | Manchester Grits and Shales | 50 m | St David's Series |
|  | Abbey Shales | 15–40 m |  |
|  | Purley Shales | 210 m | Comley & St David's Series |
| Hartshill Quartzite Formation | Woodlands Member | ? m |  |
|  | Home Farm Member | 2 m |  |
|  | Jee's Member | 6 m | Comley Series |
|  | Tuttle Hill Member | 160–190 m |  |
|  | Park Hill Member | 30–60 m |  |
|  | ———————— angular unconformity ———————— |  |  |
| Charnian | Brand Group | Precambrian |  |
|  | Maplewell Group |  |  |

weathered, probably in early Cambrian times, before being covered by the transgressing Cambrian sea.

The Park Hill Member contains a suite of trace fossils – *Arenicolites, Planolites, Didymaulichnus, Monocraterion, Neonereites* and ? *Bergaueria* (Brasier 1984) which may prove to be stratigraphically diagnostic: it is probably early Cambrian and may be correlated with the Atdabanian or Tommotian Stages of Siberia but the more abundant trace fossils of South China – Meishucun or Qiongzhusi Stages – may give better correlations with that region.

The *Tuttle Hill Member* contains *Arenicolites, Planolites, Didymaulichnus, Neonereites* and *Cordia*.

The *Jee's Member* contains *Arenicolites, Planolites, Didymaulichnus* and the arthropod trace *Isopodichnus*.

Trace fossil research on late Precambrian and early Cambrian strata is proceeding apace and may soon produce biostratigraphic frameworks and zonal schemes which will expedite correlations at this level in the stratigraphic column (T. P. Crimes *pers. comm.*)

The *Woodlands, Home Farm* and *Jee's Members* include horizons of glauconitic sandstones and conglomerates rich in a matrix with up to 14 per cent of calcium phosphate (Lapworth 1898) and a highly calcareous sandstone with a fauna of five species of *Hyolithes s.l.*, and according to Cobbold (1919), species of *Coleoloides, Paterina, Helcionella, Stenotheca* and *Pelagiella* have been recorded. Missarzhevsky (Cowie *et al.* 1972) also noted *Hyolithellus, Torellella, Camenella, Chancelloria* and *Helenia*.

The coastal transgressive deposits early in the formation contain the meandering trails: *Planolites* and *Arenicolites*. 'Cruziana-type facies' in laminated sands and muds, shelly and stromatolitic limestone facies contain an early Cambrian fauna. The first trilobite fragments are seen in this region in monotonous muds in the overlying Purley Shales indicating deeper water sediment deposition.

The *Hartshill* (Quartzite) *Formation* of the Nuneaton region in the Midlands of England is of importance because of its significance in biostratigraphic correlations with other parts of the world (Brasier 1979, 1980; Brasier *et al.* 1978, 1979, 1981). In 1978 Brasier *et al.* correlated the lower part of the Hartshill Formation with the top of the Tommotian Stage of the East Siberian Cambrian but Matthews & Cowie (1979) cast doubt on this correlation. In 1981, however, Brasier revised his estimate of the age of the lower part of the Hartshill Quartzite on the basis of further palaeontological and palaeoecological studies. There now seems to be evidence only for a correlation with certainty to the Atdabanian Stage with little chance of late Tommotian and less or none of Vendian. Both the low-diversity trace fossil assemblage and the body fossil assemblage may be condensed and/or transported and/or reworked and are probably diachronous facies faunas.

Brasier (*in* Cowie 1981) stated that the 'earliest English shelly faunas may be approximately of *Schmidtiellus mickwitzi* or *Fallotaspis* age': these trilobites indicate an Atdabanian age.

The full fauna of Assemblages I–VI the *Home Farm Member* (Brasier 1984) includes species of *Micromitra, Torellella, Hyolithellus, Tommotia, Camenella, Sunnaginia, Amphigeisina, Coleoloides, Glauderia, Allatheca, ? Fordilla, ? Bija, Chancelloria, Eccentrotheca, Halkeria, Hertzina, Burithes, Spinulith-*

*eca, Tuojdachithes?, Teichichnus, Doliutus, Gracilith-eca, Hyperammina?, Igorella, Paterina, Prosinuites, Randomia, Tuojdachites* and *Yanischevskyites.* Brasier (1984) gave descriptions and illustrations of microfossils and small shelly fossils from the Home Farm Member: 32 species (20 described and illustrated from Nuneaton for the first time and 8 are new records). The varied fauna includes phosphatic and calcareous microproblematica (especially *Colcoloides typicalis*), protoconodonts, agglutinated foraminifera, cap-shaped molluscs, primitive bivalves, hyoliths, inarticulate brachiopods and sponges. The fauna has affinity with fossils from the top Tommotian to lower Atdabanian rocks from Siberia and China, from the Baltic area and the belt from SE Newfoundland to Massachusetts.

The lithological and faunal similarity between the Charnwood–Nuneaton succession and the SE Newfoundland succession is considerable (Brasier 1984) and both were probably part of an Avalon Platform in early Cambrian times.

The *Purley Shales* a lithologically uniform formation, succeeds the Hartshill Formation with an abrupt, though apparently conformable, change of lithology. Mainly blocky, poorly bedded fine-grained mudstone, these are interbedded with occasionally ripple-marked, sandstone. Just above the Hartshill Formation the lower beds contain calcareous nodules which sometimes contain fossils, including a dubious *?Callavia.* The upper 30 m have been assigned to the St David's Series leaving the lower 180 m in the Comley Series. Horny brachiopods and *Coleoloides* occur in thin fossiliferous seams at various levels in the succession; 70 m above the base a fauna of eodiscid and strenuellid trilobites includes *Serrodiscus bellimarginatus* (Shaler & Foerste) which may indicate a high horizon in the Olenellid Zone. At 140 m above the base are found species of *Acidiscus, Chelediscus, Serrodiscus, Tannudiscus, Condylopyge* and *Ellipsostrenua* suggesting an horizon high in the Comley Series (Rushton 1966).

A *St David's Series* fauna is found near the top of the Purley Shales with indications of *Paradoxides oelandicus* age (according to Rushton probably referable to the Zone of *P. pinus*) including species of *Condylopyge, Ptychagnostus, Eodiscus, Paradoxides, Bailiella* and *Bailiaspis.*

The *Abbey Shales* succeed transitionally the Purley Shales and are a well-laminated dark blue-grey variegated succession of shales. Highly siliceous brittle layers occur along with darker unlaminated mudstones containing limonite and dark calcareous, slightly phosphatic and manganiferous, concretions. There can be considerable variation in thickness and composition along the strike with fine-grained sandstone beds and laminae, also on occasion containing calcareous, phosphatic and glauconitic concretions. There is a stratigraphic break high in the St David's Series. Erosion of the formation before the highest beds were deposited is suggested by variation in thickness from NW to SE.

The large and diverse fauna includes inarticulate brachiopods (e.g. *Linnarsonia* sp.) hyolithids, *Stenotheca,* bradoriids, sponge spicules and many species of trilobites from more than twenty horizons ranging from the *Paradoxides aurora* fauna up to the Upper *Paradoxides davidis* fauna.

The *Mancetter Grits and Shales* succeed the Abbey Shales with slight unconformity. A calcareous conglomerate at the base, 20–250 mm thick, contains pebbles of acid igneous rocks along with grains of quartz, glauconite and fragments of the underlying shales, set in a calcareous matrix. The main part of the formation is comprised of evenly bedded micaceous shales with thin beds of mudstone and characteristic thin (200 mm) hard sandstone bands which contain coarse quartz and glauconite set in a muddy matrix. The few fossils indicate the *Lejopyge laevigata* Zone at the top of the St David's Series and include species of *Hypagnostus, Svealuta* and *Grandagnostus.*

The *Outwoods Shales* grade up from the previous formation and are chiefly of rapid alternations of burrowed mudstones and laminated pyritous shales with many traces of algae. Silty micaceous layers are common, particularly in the upper beds. Near the base, the lowest Upper Cambrian zone of *Agnostus pisiformis* is clearly indicated in the fossiliferous beds with species of the trilobites *Grandagnostus, Homagnostus, Olenus, Glyptagnostus, Proceratopyge, Sulcatagnostus* and *Irvingella,* indicating subzones. The uppermost beds yield *Orusia lenticularis* and *Protopeltura aciculta* indicating the base of the overlying zone of *Parabolina spinulosa* (Taylor & Rushton 1972).

The *Moor Wood Flags Shales* consist of hard fine-grained pyritous sandstone displaying convoluted bedding alternating with thin seams of grey shale and have yielded *Orusia lenticularis* and are referred to the *Parabolina spinulosa* Zone by Taylor & Rushton (1972).

The *Monks Park Shales* closely resemble the Dolgelly Beds of North Wales in thickness, lithology and fauna. In the lower part they yield the brachiopod *Orusia lenticularis* and the trilobites *Parabolina brevispina* and *P. spinulosa.*

The upper part of the formation consists of very dark grey and black laminated mudstones which are highly pyritous and carbonaceous and contain nodules of dolomite and barytes. The fauna includes horny brachiopods and species of the trilobites *Leptoplastus, Eurycare, Ctenopyge* and *Sphaerophthalmus* indicating several subzones of the *Leptoplastus* and *Peltura* Zones (Taylor & Rushton 1972).

These dark Monks Park Shales grade into the overlying greenish-grey Tremadocian Merevale Shales with no apparent break but zones may be missing.

## Lickey Hills

The *Lickey Quartzite* in the south-east part of the South Staffordshire Coalfield is found in a narrow outcrop and has an unknown thickness of quartzites with few thin shaly seams. At the southern end of the inlier the quartzites are associated with pyroclastics near Barnt Green. These pyroclastics are speculatively given a Precambrian age. At the northern end the Quartzite is overlain unconformably by the Llandovery Rubery Sandstone. Although without body fossils and yielding only trace fossils, the Quartzite is correlated on lithological grounds with the Nuneaton Inlier quartzites.

## Underground distribution of Cambrian rocks in the English Midlands

Quartzite similar to the Lickey Quartzite is known beneath Carboniferous strata north of Birmingham. In the Warwickshire Coalfield (Taylor & Rushton 1972) a Stockingford Shales lithology is found, but considered to be Tremadoc Series in age, while between Coventry and Nuneaton similar shales are correlatable with the Merioneth Series. Elsewhere in the Midlands Tremadoc strata have been recorded underground though Cambrian shales probably also occur in association with these Ordovician Tremadoc beds.

## REFERENCES

ALLEN, P. M. & JACKSON, A. A. 1978 Bryn-teg Borehole, North Wales. *Bull. Geol. Surv. G.B.*, **61**, 1–52.

ALLEN, P. M., JACKSON, A. A. & RUSHTON, A. W. A. 1981 Stratigraphy of the Mawddach Group in the Cambrian succession of North Wales. *Proc. Yorkshire Geol. Soc.*, **43** (3): 295–329, pls. 16–17.

BARBER, A. J. & MAX, M. D. 1979 A new look at the Mona Complex (Anglesey, North Wales). *J. geol. Soc. London*, **136**, 407–432.

BASSETT, D. A. & WALTON, E. K. 1960 The Hell's Mouth Grits: Cambrian greywackes in St. Tudwal's Peninsula, North Wales. *Q. J. geol. Soc. London*, **116**, 85–95.

BASSETT, M. G., OWENS, R. M. & RUSHTON, A. W. A. 1976 Lower Cambrian fossils from the Hell's Mouth Grits, St. Tudwal's Peninsula, North Wales. *J. geol. Soc. London*, **132**, 623–644.

BLISS, G. M. 1977 *The micropalaeontology of the Dalradian.* Unpubl. Ph.D. thesis, Univ. of London.

BRASIER, M. D. 1979 The Cambrian radiation event. *In* House, M. R. (Ed.) The Origin of Major Invertebrate Groups. *Systemics Assoc. Spec. Publ.*, **12**. Academic Press, London, 103–159.

1980 The Lower Cambrian transgression and glauconite-phosphate facies in western Europe. *J. geol. Soc. London*, **137**, 695–703.

1984 Microfossils and small shelly fossils from the Lower Cambrian *Hyolithes* Limestone at Nuneaton English Midlands. *Geol. Mag.*, **121**, 229–253, figs., 3 pls.

BRASIER, M. D., HEWITT, R. A. & BRASIER, C. J. 1978 On the late Precambrian-early Cambrian Hartshill Formation of Warwickshire. *Geol. Mag.*, **115**, 21–36.

BRASIER, M. D. & HEWITT, R. A. 1979 Environmental setting of fossiliferous rocks from the Uppermost Proterozoic-Lower Cambrian of central England. *Palaeogeogr. Palaeoclimatol. Palaeoecol.*, **27**, 35–57.

BRASIER, M. D. & HEWITT, R. A. 1981 Faunal sequence within the Lower Cambrian 'Non-Trilobite' Zone (S.L.) of Central England and correlated regions. *In* Taylor, M. E. (Ed.) *Short papers for the second International Symposium on the Cambrian System 1981.* U.S.G.S. Open File Report 81–743, 29–33.

CALLAWAY, C. 1878 On the quartzites of Shropshire. *Q. J. geol. Soc. London*, **34**, 754–763.

COBBOLD, E. S. 1919 Cambrian Hyolithidae, etc., from Hartshill in the Nuneaton District, Warwickshire. *Geol. Mag.*, **6**, 149–158.

1927 The stratigraphy and geological structure of the Cambrian area of Comley (Shropshire). *Q. J. geol. Soc. London*, **83**, 551–573.

COPE, J. C. W. 1977 An Ediacara-type fauna from South Wales. *Nature, London*, **278**, 624.

1979 Early history of the southern margin of the Tywi anticline in the Carmarthen area, South Wales. *In* Harris, A. L. *et al.* (Eds.) The Caledonides of the British Isles – reviewed. *Spec. Publ. geol. Soc. London*, **8**, 527–532.

COWIE, J. W. 1971 Lower Cambrian faunal provinces. *In* Middlemiss, F. A., Rawson, P. F. & Newall, G. (Eds.) *Faunal Provinces in Space and Time.* Seel House Press, Liverpool, 31–44.

1978 IUGS/IGCP Project 29 Precambrian-Cambrian Boundary Working Group in Cambridge, 1978. *Geol. Mag.*, **115**, 151–152.

1981 The Proterozoic-Phanerozoic transition and the Precambrian-Cambrian boundary. *Precambrian Reserch*, **15**, 187–190.

COWIE, J. W. & BRASIER, M. D. (Eds.) 1989 *The Precambrian–Cambrian Boundary.* Oxford Science Publications, Clarendon Press, Oxford, 213 pp.

COWIE, J. W. & JOHNSON, M. R. W. 1985 Late Precambrian and Cambrian geological time-scale. *In* Snelling, N. J. (Ed.) The Chronology of the Geological Record. *Mem. geol. Soc. London*, **10**, 47–64.

COWIE, J. W., RUSHTON, A. W. A. & STUBBLEFIELD, C. J. 1972 A correlation of Cambrian rocks in the British Isles. *Spec. Rep. geol. Soc. London*, **2**, 1–4.

CRIMES, T. P. 1966 Palaeocurrent directions in the Upper Cambrian of North Wales. *Nature, London*, **210**, 1246.

1970a Trilobite tracks and other trace fossils from the Upper Cambrian of North Wales. *Geol. J.*, **7**, 47–68.

1970b A facies analysis of the Cambrian of Wales. *Palaeogeogr. Palaeoclimatol. Palaeoecol.*, **7**, 113–170.

DAVIES, H. G. & DOWNIE, C. 1964 Age of the Newgale Beds. *Nature, London*, **203**, 71.

DOWNIE, C. 1975 Precambrian of the British Isles. *In* Harris, A. L. *et al.* (Eds.) A correlation of the Precambrian rocks of the British Isles. *Spec. Rep. Geol. Soc. London*, **6**, 113–115.

DUNNING, F. W. 1966 *Tectonic map of Great Britain and Northern Ireland.* 1:1,584,000. Institute of Geological Sciences, London.

FØYN, S. & GLAESSNER, M. F. 1979 *Platysolenites*, other animal fossils, and the Precambrian-Cambrian transition in Norway. *Norsk. geol. Tidsskr.*, **59**, 25–46.

GREENLY, E. 1919 The geology of Anglesey. *Mem. geol. Surv. G.B.* 2 Vols.

GROOM, T. T. 1902 The sequence of Cambrian and associated beds of the Malvern Hills. *Q. J. geol. Soc. London*, **58**, 89.

1910 The Malvern and Abberley Hills, and the Ledbury District. *Geology in the field*, **4**, 698. *Geol. Assoc.. London*.

HENNINGSMOEN, G.    1957    The trilobite family Olenidae with description of Norwegian material and remarks on the Olenid and Tremadocian Series. *Skr. norske Vidensk Akad.*, **1**. Mat. natur. Kl., 1957, no. 1. 303 pp.

HOLL, H. B.    1865    On the geological structure of the Malvern Hills and adjacent districts. *Q. J. geol. Soc. London*, **21**, 72.

HUPÉ, P.    1953 (dated 1952)    Contribution à l'etude du Cambrien inférieur et du Precambrien III de l'Anti-Atlas Marocain. Notes Mém. Serv. Mines Cartes geol. Maroc **103**, 102, 127.

JONES, O. T.    1938    On the evolution of a geosyncline. *Q. J. geol. Soc. London*, **94**, lx–cx.

1940    Some Lower Palaeozoic contacts in Pembrokeshire. *Geol. Mag.*, **77**, 405–409.

JONES, R. K., BROOKS, M., BASSETT, M. G., AUSTIN, R. L. & ALDRIDGE, R. J.    1970    An Upper Llandovery limestone overlying Hollybush Sandstone (Cambrian) in Hollybush Quarry, Malvern Hills. *Geol. Mag.*, **106**, 457–469.

KNILL, J. L.    1958    Axial and marginal sedimentation in geosynclinal basins. *J. Sediment. Petrol.*, **29**, 317–325.

LAPWORTH, C.    1882    On the discovery of Cambrian rocks in the neighbourhood of Birmingham. *Geol. Mag.*, **9**, 563–566.

1886    On the sequence and systematic position of the Cambrian rocks of Nuneaton. *Geol. Mag.*, **3**, 319–322.

LAPWORTH, C., WATTS, W. W. & HARRISON, W. J.    1898    A sketch of the Geology of the Birmingham District. *Proc. Geol. Assoc. London*, **15**, 313–316.

MATLEY, C. A., NICHOLAS, T. C. & HEARD, A.    1939    Summer field meeting to western part of the Lleyn Peninsula. *Proc. Geol. Assoc. London*, **50**, 83–100.

MATLEY, C. A. & WILSON, T. S.    1946    The Harlech Dome, north of the Barmouth Estuary. *Q. J. geol. Soc. London*, **102**, 1–40.

MATTHEWS, S. C. & COWIE, J. W.    1979    Early Cambrian transgression. *J. geol. Soc. London*, **136**, 133–135.

MUIR, M. D., BLISS, G. M., GRANT, P. R. & FISHER, M. J.    1979    Palaeontological evidence for the age of some supposedly pre-Cambrian rocks in Anglesey, North Wales. *J. geol. Soc. London*, **136**, 61–64.

NICHOLAS, T. C.    1915    The geology of the St. Tudwal's Peninsula (Carnarvonshire). *Q. J. geol. Soc. London*, **71**, 83–143.

1916    Notes on the trilobite fauna of the Middle Cambrian of the St. Tudwal's Peninsula (Carnarvonshire). *Q. J. geol. Soc. London*, **71** (for 1915), 451–472.

NUTT, M. J. C. & SMITH, E. G.    1981    Transcurrent faulting and the anomalous position of pre-Carboniferous Anglesey. *Nature, London*, **290**, 492–495.

OWEN, T. R.    1967    'From the south'. A discussion. *Proc. Geol. Assoc. London*, **78**, 595–601.

1971    The relationship of Carboniferous sedimentation to structure in South Wales. *C.r. 6e Congres. Intern. Strat. Geol. Carbonif. Sheffield 1967*, 1305–1316.

PEAT, C. J.    1984    Precambrian microfossils from the Longmyndian of Shropshire. *Proc. Geol. Assoc. London*, **95**, 17–22.

PHILLIPS, J.    1848    The Malvern Hills, compared with the Palaeozoic districts of Abberley, Woolhope, May Hill, Tortworth and Usk. *Mem. Geol. Surv. G.B.*, **2** (1).

RAST, N. & SKEHAN, J. W.    1981    Possible correlation of Precambrian rocks of Newport, Rhode Island, with those of Anglesey, Wales. *Geology*, **9**, 596–601.

RUSHTON, A. W. A.    1966    The Cambrian trilobites from the Purley Shales of Warwickshire. *Monogr. palaeontologr. Soc. London.* 1–55, 6 pls.

1972    In *Ann. Rep. Inst. geol. Sci.* for 1971: 93. H.M.S.O., London.

1974    The Cambrian of Wales and England. *In* Holland, C. H. (Ed.) *Cambrian of the British Isles, Norden and Spitzbergen.* John Wiley, 43–120.

1978    Description of the macrofossils from the Dolwen Formation. *In* Allen, P. M. & Jackson, A. A. (Eds.) Bryn-teg borehole, North Wales. *Bull. Geol. Surv. G.B.* Appendix 3D, 46–48.

SHACKLETON, R. M.    1975    Precambrian rocks of Wales. *In* Harris, M. L., Holland, C. H. & Leake, B. E. (Eds.) A correlation of Precambrian rocks in the British Isles. *Spec. Rep. geol. Soc. London,* **6**, 76–82.

STEAD, J. T. G. &    1971    The Cambrian rocks of north Pembrokeshire. *In* Bassett,
WILLIAMS, B. P. J.        D. & Bassett, M. G. (Eds.) *Geological excursions in South Wales and the Forest of Dean,* Cardiff.

TAYLOR, K. &    1972    The pre-Westphalian geology of the Warwickshire Coal-
RUSHTON, A. W. A.        field, with a description of three boreholes in the Merevale Area. *Bull. Geol. Surv. G.B.,* **35**.

TEGERDINE, G. D.,    1981    Transcurrent faulting and pre-Carboniferous Anglesey.
CAMPBELL, S. D. G. &      *Nature, London,* **293**, 760–762.
WOODCOCK, N. H.

THOMAS, A. T.,    1984    Trilobites in British stratigraphy. *Spec. Rep. geol. Soc.*
OWENS, R. M. &       *London,* **16**, 78 pp. and 29 figs.
RUSHTON, A. W. A.

WOOD, A. (Editor)    1969    *The Pre-Cambrian and Lower Palaeozoic rocks of Wales.* Univ. of Wales Press, Cardiff, 461 pp.

WOODS, D. S.    1969    The base and correlation of the Cambrian rocks of North Wales. *In* Wood, A. (Ed.) *The Precambrian and Lower Palaeozoic rocks of Wales.* Univ. of Wales Press, Cardiff, 47–66.

WRIGHT, A. E.    1981    Lower Palaeozoic oceans and Wilson cycles. *Nature, London,* **294**, 289.

# 4

# ORDOVICIAN

# M. G. Bassett, A. D. Wright
# and P. McL. D. Duff

## Introduction

Pronounced geographical differentiation and progressive taxonomic diversification characterise Ordovician faunas on a global scale (Jaanusson 1979, 1984a). In combination with these changing biotic relationships, the wide variety of both sedimentary and igneous rock types developed across England and Wales during the Period produced a facies mosaic that is probably more heterogeneous than in any other geological System. The major factors responsible are considered to be the positions of the Lake District and Wales relative to the southern margin of the closing Iapetus Ocean; the influence of underlying subduction movements and accompanying magmatism; contemporaneous tectonic activity.

Marine sedimentation predominated in both areas with, in the Lake District, some sedimentary facies variations and structures now being thought due to volcano–tectonic faulting and caldera collapse during ensialic volcanic-arc activity (Branney & Soper 1988). In North Wales sedimentation from late Tremadoc to Caradoc times, was affected by ensialic destructive plate-margin volcanism, which in turn changed locale due to changing degrees of extension, through time, across N–S graben-like structures (Kokelaar 1988).

Vast quantities of ashes, lavas and volcaniclastic debris were shed into the troughs and across neighbouring platform areas, modifying and amplifying sedimentation both locally and regionally. At the same time, persistent subsidence and instability heightened the contrasts between the environments in the troughs and those of the relatively stable, shallow-marine platform region that lay mainly to the east and south. Complementary biofacies contrasts are reflected in a preponderance of benthic organisms on the platform and around the volcanic centres, with planktonic biotas dominated by graptolites being more common in the offshore and deeper areas of the troughs.

The net result is that stratigraphy commonly differs markedly both within and between the preserved areas of Ordovician rocks (Figs. 4.1, 4.2). However, there is generally sufficient interdigitation of lithological units and/or faunas to allow correlations and to relate the diverse successions within a broader chronostratigraphical framework applicable to the region as a whole (Fig. 4.3). Williams et al. (1972) presented a detailed account of the correlation of most of the Ordovician rocks in the British Isles (excluding the Tremadoc Series), and their paper thus forms an essential background to this chapter; similarly, as the Tremadoc Series is included here in the Ordovician System (see below), the review of correlation by Cowie et al. (1972) is invaluable.

## Stratigraphical limits and subdivision of the Ordovician System

Charles Lapworth's original definition (1879, p. 14) of the Ordovician System as the 'Strata included between the base of the Lower Llandovery formation and that of the Lower Arenig' would appear at first sight to be an unequivocal statement from which no misconceptions could arise, particularly within Britain where both these units have their historical type areas. Yet the imprecisions of geological correlation have confused the definition almost since it was first published, and internationally fixed and agreed upper and lower limits are still the subject of considerable debate.

The question of the inclusion of rocks of Tremadoc age within the Ordovician System has been discussed by many authors and need not be repeated here (e.g. see Whittard 1960; Whittington & Williams 1964; Skevington 1966, 1969; Williams 1969, *in* Williams *et al.* 1972; Henningsmoen 1973; Whittington *et al.* 1984). Global reconstructions of earth history must be set within an internationally agreed time scale and

**Fig. 4.1.** Outcrop of Ordovician rocks in England and Wales, with boreholes that penetrate proved or probable Ordovician deposits; the density of boreholes in central England is too great to plot them all at this scale, but the most important are shown to indicate known subsurface limits. See Fig. 4.2 for detail of localities in Wales and the Welsh Borderland. Borehole data partly from Wills (1978) and Bulman & Rushton (1973).

while a formal decision has not yet been made, there is a growing acceptance that the lower boundary of the Ordovician should be defined in relation to an horizon at or close to the base of the Tremadoc Series (M. G. Bassett & Dean 1982, p. 4); this practice is adopted here, with the base of the Tremadoc taken at a level coincident with the first appearance in North Wales of the planktic dendroid graptolite *Rhabdinopora* [*Dictyonema*] *flabelliformis sensu lato*, following Skevington (1966, 1969) and Rushton (1982).

Defining the base of the Ordovician System involves the magnitude of a stratigraphical Series; the problem

**Fig. 4.2.** Distribution of Ordovician rocks in Wales and the Welsh Borderland, with principal structures that affect facies patterns.

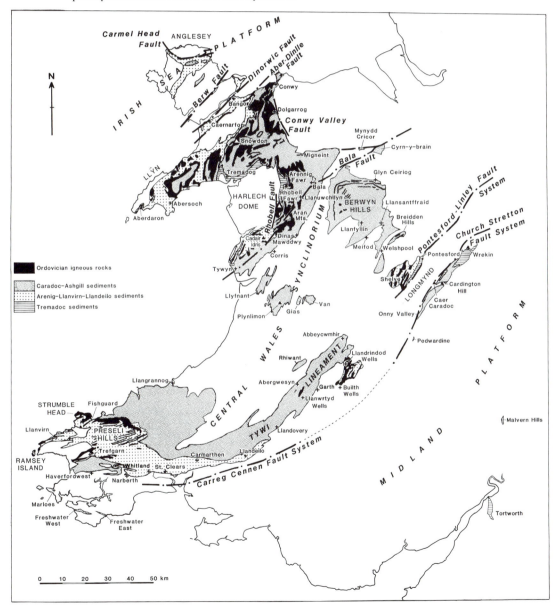

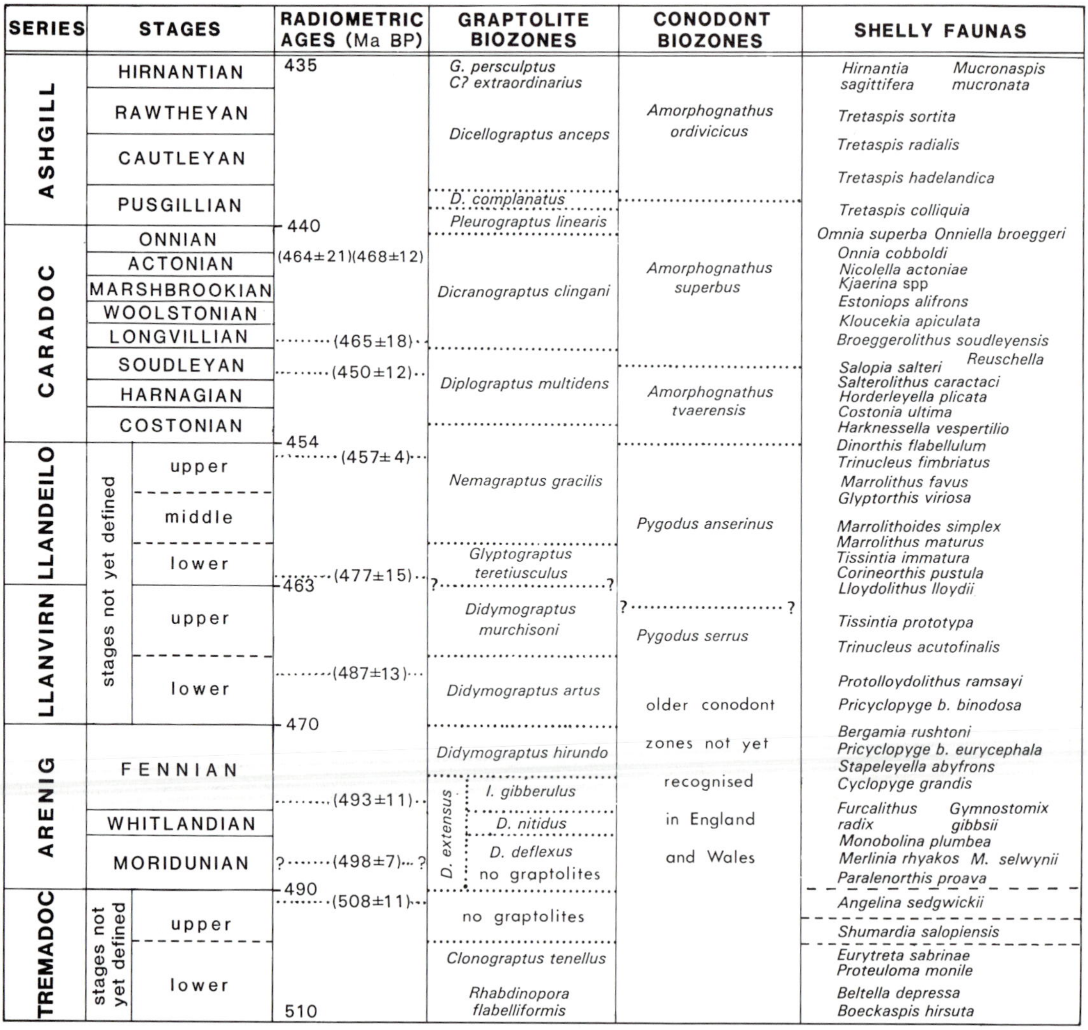

| SERIES | STAGES | | RADIOMETRIC AGES (Ma BP) | GRAPHTOLITE BIOZONES | CONODONT BIOZONES | SHELLY FAUNAS |
|---|---|---|---|---|---|---|
| ASHGILL | HIRNANTIAN | | 435 | G. persculptus C? extraordinarius | | Hirnantia Mucronaspis sagittifera mucronata |
| ASHGILL | RAWTHEYAN | | | Dicellograptus anceps | Amorphognathus ordivicicus | Tretaspis sortita |
| ASHGILL | CAUTLEYAN | | | Dicellograptus anceps | Amorphognathus ordivicicus | Tretaspis radialis |
| ASHGILL | PUSGILLIAN | | 440 | D. complanatus | | Tretaspis hadelandica |
| ASHGILL | PUSGILLIAN | | 440 | Pleurograptus linearis | | Tretaspis colliquia |
| CARADOC | ONNIAN | | (464±21)(468±12) | Dicranograptus clingani | Amorphognathus superbus | Omnia superba Onniella broeggeri |
| CARADOC | ACTONIAN | | (464±21)(468±12) | Dicranograptus clingani | Amorphognathus superbus | Onnia cobboldi Nicolella actoniae |
| CARADOC | MARSHBROOKIAN | | | Dicranograptus clingani | Amorphognathus superbus | Kjaerina spp |
| CARADOC | WOOLSTONIAN | | | Dicranograptus clingani | Amorphognathus superbus | Estoniops alifrons |
| CARADOC | LONGVILLIAN | | (465±18) | Dicranograptus clingani | Amorphognathus superbus | Kloucekia apiculata |
| CARADOC | SOUDLEYAN | | (450±12) | Diplograptus multidens | | Broeggerolithus soudleyensis Reuschella |
| CARADOC | HARNAGIAN | | | Diplograptus multidens | Amorphognathus tvaerensis | Salopia salteri Salterolithus caractaci Horderleyella plicata |
| CARADOC | COSTONIAN | | 454 | Diplograptus multidens | Amorphognathus tvaerensis | Costonia ultima Harknessella vespertilio |
| LLANDEILO | stages not yet defined | upper | (457±4) | Nemagraptus gracilis | | Dinorthis flabellulum Trinucleus fimbriatus Marrolithus favus Glyptorthis viriosa |
| LLANDEILO | stages not yet defined | middle | | Nemagraptus gracilis | Pygodus anserinus | Marrolithoides simplex Marrolithus maturus |
| LLANDEILO | stages not yet defined | lower | 463 (477±15) ? | Glyptograptus teretiusculus ? | | Tissintia immatura Corineorthis pustula Lloydolithus lloydii |
| LLANVIRN | | upper | | Didymograptus murchisoni | ? ? | Tissintia prototypa |
| LLANVIRN | | upper | | Didymograptus murchisoni | Pygodus serrus | Trinucleus acutofinalis |
| LLANVIRN | | lower | (487±13) | Didymograptus artus | older conodont | Protolloydolithus ramsayi Pricyclopyge b. binodosa |
| LLANVIRN | | lower | 470 | Didymograptus hirundo | zones not yet | Bergamia rushtoni Pricyclopyge b. eurycephala Stapeleyella abyfrons |
| ARENIG | FENNIAN | | (493±11) | Didymograptus hirundo | recognised | Cyclopyge grandis |
| ARENIG | FENNIAN | | (493±11) | I. gibberulus D. extensus | in England | Furcalithus Gymnostomix radix gibbsii |
| ARENIG | WHITLANDIAN | | | D. nitidus | and Wales | Monobolina plumbea |
| ARENIG | MORIDUNIAN | | ? (498±7) ? | D. deflexus no graptolites | and Wales | Merlinia rhyakos M. selwynii Paralenorthis proava |
| TREMADOC | stages not yet defined | upper | 490 (508±11) | no graptolites | | Angelina sedgwickii |
| TREMADOC | stages not yet defined | upper | | no graptolites | | Shumardia salopiensis |
| TREMADOC | stages not yet defined | lower | | Clonograptus tenellus | | Eurytreta sabrinae Proteuloma monile |
| TREMADOC | stages not yet defined | lower | 510 | Rhabdinopora flabelliformis | | Beltella depressa Boeckaspis hirsuta |

**Fig. 4.3.** Summary of chronostratigraphical classification of Ordovician rocks in England and Wales, with data for geochronological calibration and biostratigraphical bases for correlation. Radiometric ages for the base of each Series are interpolations using data from many sources worldwide as summarised by Gale & Beckinsale (1983) and McKerrow *et al.* (1985); ages in roman type on the left of this column are restricted to samples from England and Wales whose stratigraphical limits are closely defined and whose $2\sigma$ errors are 2% or less; bracketed ages also refer to stratigraphically precise samples but whose analytical errors are considered to be outside an acceptable range; these latter ages are included here to indicate the order of magnitude of all stratigraphically accurate samples from the region. Correlation of the graptolite and conodont biozones incorporates data from outside England and Wales, although only pre-upper Llanvirn conodonts are not represented directly in the region. Apart from in the upper Tremadoc, the shelly faunas (trilobites and brachiopods) do not constitute biozones but are examples of successive assemblages that can be used for reasonably diagnostic correlation within varying limits of accuracy.

of the upper limit (defined technically by the base of the Silurian System) has centred on a single graptolite biozone. Lapworth's (1879) strict definition in relation to the Llandovery rocks in their historical type area is difficult to apply because of the absence of fossils in the lowest beds but within Britain and elsewhere it has long been accepted that this level approximates fairly closely to the base of the *Glyptograptus persculptus* Biozone in graptolitic facies; for all practical purposes the latter datum has been taken as the top of the Ordovician (e.g. see Whittard 1960, Whittington & Williams 1964, Skevington 1969, Cocks *et al.* 1970, Lawson 1971, Rickards 1976). However, as summarised by Williams (*in* Williams *et al.* 1972), problems have arisen in some regions where *persculptus* faunas are in association with shelly fossils considered on other grounds to be of latest Ordovician age. The matter has been resolved by international agreement (Holland *et al.* 1984, M. G. Bassett 1985, Cocks 1985) and the Ordovician–Silurian boundary is now defined in relation to the base of the *Parakidograptus acuminatus* Biozone, leaving the underlying *persculptus* Biozone and its correlatives within the uppermost Ordovician System; such correlatives as can be identified in England and Wales are therefore described in this chapter.

The primary subdivision of the Ordovician of England and Wales into six Series (Fig. 4.3) is thus well founded (Whittington *et al.* 1984) although precise relationships between them remain tentative at some levels because of the geographical separation and facies differences of the type sections (see Williams 1969 for summary). Chronostratigraphical subdivision into Stages has not yet been made for the Tremadoc, Llanvirn and Llandeilo Series, but informal relative terms (Fig. 4.3) are useful until such procedures are carried out. The Stages of the Arenig, Caradoc and Ashgill Series are based on sections in essentially shelly facies so are of restricted application outside the Anglo–Welsh platform areas. For background and definition of the Series of the Ordovician and of the Caradoc–Ashgill Stages see Whittard (1960), Dean (1958), Williams (1969), Ingham and Wright (1970), Hurst (1979c), and Whittington *et al.* (1984). Stage divisions for the Arenig Series based on sections in South Wales are proposed by Fortey and Owens (1987).

Jaanusson (1960) and Whittington and Williams (1964) summarised the confusions arising from the use of the terms 'Lower', 'Middle' and 'Upper' Ordovician in different senses; in this chapter they have no place in formal classification.

## Geochronology

Most calculated radiometric dates from rocks of known stratigraphical limits suggest a time span for the Ordovician of *c*. 75 million years (Ma) (Fig. 4.3; see Gale & Beckinsale 1983 and McKerrow *et al.* 1985 for summaries). Estimates for the beginning and end of the Period still vary by up to 20 Ma, but there is fairly good agreement from most data that the Tremadoc and Arenig together represent close to 50 per cent of Ordovician time (both approximately 20 Ma), with the Llanvirn (7 Ma), Llandeilo (9 Ma), Caradoc (14 Ma) and Ashgill (5 Ma) spanning shorter intervals. McKerrow *et al.* (1985) estimated that the figures from mid-Arenig rocks upwards are probably in error by no more than 7 Ma. Assuming the reliability of these dates, Fig. 4.3 emphasises the considerably different times represented by individual stages and biozones recognised within each Series of the Ordovician System.

# Regional stratigraphy and facies

## North Wales

In naming the Ordovician System after the *Ordovices*, an ancient tribe that inhabited the northern parts of Wales, Lapworth (1879, p. 13) clearly saw this region as the type area. Much of the rugged, glacially modified mountain scenery, centred on Snowdonia, is carved in Ordovician volcanic rocks that originated from a number of centres both subaqueously and subaerially. Piecemeal mapping until fairly recently rendered correlation from area to area difficult and resulted in a plethora of localised (and to the non-Welsh speaker unpronounceable!) names for lithostratigraphic units. During the past two decades, however, the British Geological Survey and a number of university and museum workers have contributed greatly to the better understanding of the complex stratigraphy.

### North-west Wales
#### Anglesey

Over most of their outcrop area on Anglesey (Fig. 4.2), Ordovician rocks (entirely of post-Tremadoc age), sit with profound unconformity on Precambrian to ?early Cambrian basement of the Mona Complex (Greenly 1919; Bates 1972, 1974; Woods 1974) (see Chapter 2). Breaks within the succession (Fig. 4.4), together with rapid facies changes that include the introduction at some levels of slide conglomerates and slump breccias, suggest contemporaneous movement on basement faults (Bates 1972, p. 32; Beckly 1987).

Basal Arenig deposits are everywhere transgressive across a surface of some relief; there is a general coarsening of these basal facies towards the north indicating a source in that direction. North of the Carmel Head Thrust (Fig. 4.2), spectacular con-

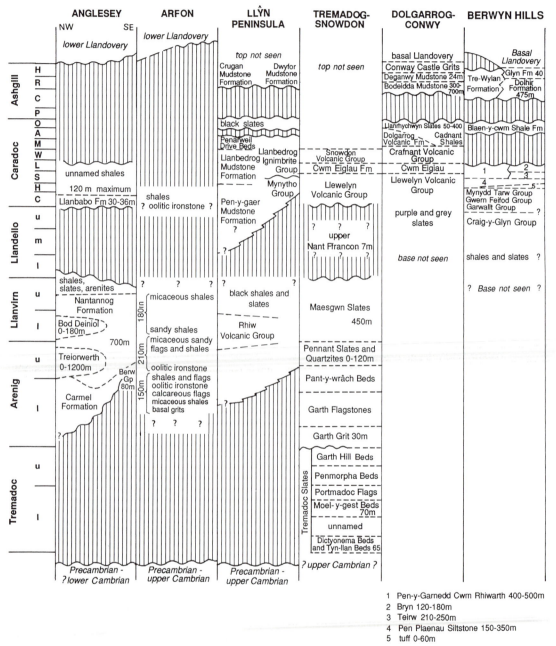

**Fig. 4.4.** Stratigraphy and correlation of representative Ordovician sequences in North Wales; notations used for the subdivision of Series are initial letters of the Stages and other units as shown in Fig. 4.3; thicknesses in metres.

glomerates and breccias containing cobbles and boulders are beach deposits incorporating material from the underlying quartzose, jasper and quartzite basement, set in a coarse sandy and muddy matrix. Higher rudites and arenites of this Porth Wen Group (Bates 1968, 1972) overlap the basal beds and contain brachiopods dominated by the genus *Paralenorthis* typical of shallow inshore environments. Across central Anglesey, basal Arenig sediments (e.g. Carmel and Foel formations, Fig. 4.4) are mostly finer grained than the Porth Wen Group, comprising coarse, ferrous, cross-bedded and rippled sandstones, although lenticular pebbly grits and conglomerates are developed sporadically. Orthid brachiopods are again present together with rare trilobites including *Neseuretus*, typical of inshore associations in relatively high energy environments (Fortey & Owens 1978). By comparison with faunas elsewhere in Wales it is likely that most of the various basal beds are no older than early Fennian or possibly latest Whitlandian (Beckly 1987, fig. 3; Fortey & Owens 1987, p. 97, fig. 10). South of the Berw Fault (Fig. 4.2), the Berw Group includes basal arenites succeeded by turbiditic siltstones and shales, again with a trilobite/brachiopod fauna but also including the graptolite *Didymograptus* (*Expansograptus*) *hirundo*, suggestive of a correlation within that biozone in the late Arenig Fennian Stage. The basal transgressive deposits may therefore be diachronous (from Whitlandian–early Fennian to late Fennian) southwards and south-eastwards across Anglesey. Along the south-easternmost coast of the island, however, the presence of *gibberulus* Biozone graptolites (Greenly 1919, p. 432; Bates 1972, p. 37) indicates *pre-hirundo* beds occur there.

In northernmost Anglesey, higher Arenig and Llanvirn rocks are absent, but in the central region the Treiorwerth, Nantannog, and Bod Deiniol formations contain distinctive shelly and graptolite faunas spanning these ages. All three units comprise suites of conglomerates, breccias and sandstones, with blocks of schist, jasper, quartzite, quartz and phyllite set in a finer matrix of the same materials, all of which are readily matched in the Mona Complex. The sequences appear to have been derived from a short distance to the west, possibly from a group of small islands, and were deposited on a sea floor of variable depth and configuration (Neuman & Bates 1978, p. 573). An origin as debris flows generated in response to faulting is likely for some units, with up to 650 m of rudaceous deposits banked against an easterly dipping fault scarp in places (Beckly 1987, p. 28). The Treiorwerth Formation sits directly on the Carmel Formation (Fig. 4.4), but also wedges laterally into and over Nantannog facies. Greenly's record (1919, p. 442) of *Tetragraptus* in Treiorwerth sandstones suggests an Arenig age, whilst throughout the more extensive Nantannog

beds, *extensus, hirundo, artus* and *murchisoni* graptolite assemblages occur sporadically. The Bod Deiniol Formation is a recurrence of Treiorwerth rudaceous and arenaceous lithologies, again occurring as a wedge into the Nantannog facies, which are distinguished principally by the presence throughout of a more muddy matrix. An early Llanvirn age for the Bod Deiniol beds is suggested by their position above the lowest occurrence in the sequence of *Didymograptus* '*bifidus*' [*artus*] (Bates 1968, p. 134; Neuman & Bates 1978, p. 572).

Rich assemblages of brachiopods with less common trilobites occur through the Treiorwerth, Nantannog and Bod Deiniol deposits, mostly as abraded and disarticulated specimens in lenses formed as lag concentrates (Bates 1968, Neuman & Bates 1978). As in the older Arenig beds, robust, thick-shelled or coarsely ribbed brachiopods are dominant, such as *Paralenorthis, Monorthis, Ffynnonia, Tritoechia* and *Rugostrophia*. A number of taxa are considered to be endemic and related to development in shallow water around islands (Neuman & Bates 1978, Neuman 1984). Pelmatozoan and bryozoan debris is common throughout.

Laterally away from the Nantannog Formation and its westerly derived debris wedges, contemporaneous beds are mostly micaceous shales with thin sandstones and siltstones. The arenites are graded, convoluted, and cross-laminated in places, and small-scale slumping is common; these beds represent turbiditic sands distal to the inshore coarse clastic deposits. *Didymograptus* (*Didymograptus*) *artus* and *D.* (*D.*) *murchisoni* occur in the shales, together with trinucleid trilobites (e.g. Dulas Formation and correlatives; Bates 1972, fig. 2). Successions at these levels to the south-east of the Berw Fault generally lack arenaceous beds and deposition was entirely as muds in a distal environment. Similar shales with slates and only thin arenites occur locally above the Nantannog Formation (Fig. 4.4) and are probably entirely of late Llanvirn age. Oolitic ironstones occur locally at this level, suggestive of shoaling within the offshore areas (e.g. Trythall *et al.* 1987).

There is no firm evidence for the presence of Llandeilo rocks on Anglesey. As noted by Bates (1972, p. 44), previous records of *Glyptograptus teretiusculus* Biozone faunas from some areas can equally be interpreted as indicating an early Caradoc age. In the north-west, the Garn Formation and its correlatives are intertonguing massive breccias, shales, coarse quartz-conglomerates, and feldspathic sandstones and pebbly grits that sit directly on the Precambrian; this was probably an erosive source area throughout Ordovician times, with repeated formation of debris beds along the fault scarp structures (Bates 1972, p. 57). Blocks within these chaotic olistostromal deposits are up to 3 metres long, incorporating both local Precambrian and Ordovician materials. Shales in the Garn Formation contain graptolites of the *Nemagraptus gracilis* Biozone; limestone blocks have yielded brachiopods, trilobites, bryozoans, crinoids and conodonts of the same age (Bates 1972, p. 51; Bergström 1981), indicating penecontemporaneous movement and incorporation into the debris beds. Grits and conglomerates lateral to the Garn Formation are rippled and cross-bedded, and are possibly more distal parts of debris flows.

Elsewhere on Anglesey, Caradoc rocks sit disconformably or unconformably on the older Ordovician beds. Much of the Caradoc succession is characterised by the presence of oolitic and pisolitic ironstones (Pulfrey 1933), interpreted as marking regional coastal onlap related either to the early *gracilis* eustatic transgression (Trythall *et al.* 1987, p. 41) or to minor regressive events (Hallam & Bradshaw 1979). Generally, however, the basal Caradoc grits and sandstones with alternating shales suggest continued dominant transgression across the older Ordovician, although regional influxes of terrigenous material may have initiated temporary regressive cycles. In the main outcrop area, conglomerates and grits of the Llanbabo Formation (Fig. 4.4) have yielded Costonian brachiopod faunas, including species of *Salopia, Bilobia, Leptestiina, Platystrophia, Dolerorthis, Nicolella* and *Dinorthis* (Bates 1972, p. 53). Graptolites in shales from the same level are from the *gracilis* Biozone.

At Parys Mountain in north-east Anglesey, a volcanic suite of rhyolites, autobreccias, ignimbrites, tuffs and shales (Hawkins 1966) rests on Llanvirn sediments and is probably of Caradoc age. The volcanics are overlain directly by Silurian shales, but in parts of Anglesey the basal Caradoc sediments are succeeded by poorly fossiliferous shales with rare graptolites that suggest an extension as high as the *clingani* Biozone; these youngest Ordovician deposits seen on the island were deposited entirely as muds in low energy environments and indicate continued transgressive deepening through middle Caradoc times.

## Arfon

On the North Wales mainland between Bangor and Caernarfon (Fig. 4.2), a succession of un-named grits, ironstones and calcareous and micaceous sandy shales resembles the more distal Ordovician facies preserved on Anglesey (Fig. 4.4). These rocks were deposited in the Arfon Basin between the Dinorwic and Aber–Dinlle faults, which controlled patterns of sedimentation in the region from Cambrian times onwards (Reedman *et al.* 1983, 1984). The lowest Ordovician beds are transgressive, locally micaceous arenites that

step across Cambrian deposits onto Precambrian basement, although contacts are mostly faulted (Greenly 1944). These basal beds are assignable to an inshore *Neseuretus* sandstone facies (Beckly 1987, p. 21, fig. 3), and the presence of *Azygograptus eivionicus* in flaggy sandstones at a slightly higher level suggests that the oldest Arenig horizons in the Arfon tract are no older than Whitlandian. Graptolite faunas in still higher calcareous flags and shales include *Isograptus gibberulus, Didymograptus (Expansograptus) extensus* and other rare didymograptids, indicating horizons beginning probably low in the upper Arenig (Fennian) *gibberulus* Biozone (Elles 1904, pp. 200–204; Greenly 1944, p. 78; Jenkins 1982, p. 222); the presence of the trilobite *Pricyclopyge binodosa eurycephala* supports such a correlation by comparison with the South Wales successions (Beckly 1987, fig. 7; Fortey & Owens 1987, table 1). Younger beds in continuous sequence contain fairly rich *hirundo* and lower Llanvirn *artus* Biozone faunas in similar offshore shelf facies, and Greenly (1944, p. 77) considered that still higher unfossiliferous shales might extend up to the *murchisoni* Biozone. The uppermost Ordovician rocks preserved in the Arfon tract contain *Glyptograptus teretiusculus*, but as on Anglesey it is possible that they represent part of the *gracilis* Biozone, with much or all of the Llandeilo missing. Apart from graptolites, the complete sequence contains rare shelly faunas, mostly trilobites such as *Cyclopyge*, but also with gastropods, inarticulate brachiopods and ostracodes. Also as on Anglesey, the main oolitic ironstones occur within the *teretiusculus* beds, but they are present throughout the succession and reflect local regressive and shoaling episodes that interrupt the overall gradual, transgressive deepening (Trythall *et al.* 1987).

## Llŷn Peninsula

The lower Ordovician succession of south-westernmost Llŷn (Figs. 4.2, 4.4) records a depositional[1] history across a topography that was laterally continuous from Anglesey and Arfon; similar transgressive facies reflect the sedimentary contiguity, albeit with faulting exercising control in different local basins (Beckly 1987). Tremadoc rocks are again absent, with various levels of an Arenig littoral to shallow sublittoral sequence sitting on gneissose Mona Complex basement. Junctions are mostly faulted, but the few preserved stratigraphical relationships confirm the scale of regional unconformity below the Arenig (Matley 1932, Shackleton 1954, Hawkins 1983). Eastwards across the peninsula the sub-Arenig surface becomes progressively younger (Fig. 4.5), rising to the upper Cambrian at St Tudwal's (Nicholas 1915, pl. 13). Records of the occurrence of the asaphid trilobite *Merlinia selwynii* together with *A. eivionicus* and didymograptid graptolites of the *deflexus* group (e.g. Beckly 1987, fig. 3; Nicholas 1915, p. 111) suggest that locally the base of the Ordovician on Llŷn may be high in the Moridunian Stage, but more generally the available faunas indicate Fennian *gibberulus* and *hirundo* Biozone ages (Elles 1904, Jenkins 1982, Beckly 1987) and approximate contemporaneity with the basal successions of Arfon and Anglesey.

**Fig. 4.5.** Schematic representation of the sub-Arenig surface and magnitude of easterly Arenig transgression and overstep across Anglesey and the Llŷn Peninsula. Partly after Shackleton (1954) and George (1963). On the map the distributional limits of successively overstepped beds are zero isopachs reconstructed in part to eliminate the effects of later faulting; in the section the assumption of a pre-Arenig dip of approximately 10° follows Shackleton (p. 268).

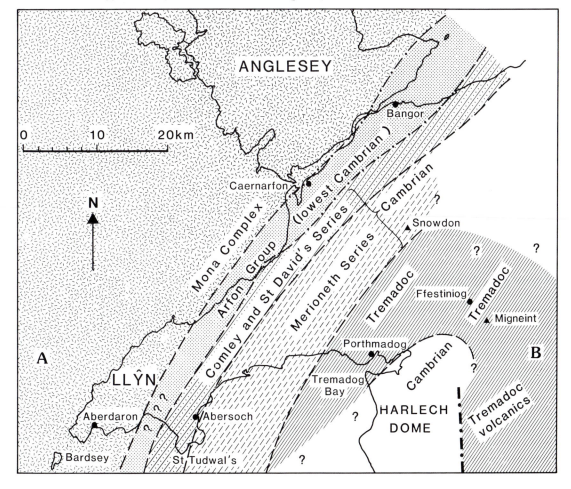

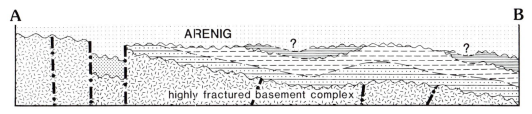

Crimes (1970b) described a complex of eight facies through the Arenig of Llŷn, deposited in transgressive–regressive pulses from within the intertidal zone to just below wave base. The lithostratigraphy (Fig. 4.4) is summarised by Matley (1928, fig. 7), Nicholas (1915, fig. 5) and Roberts (1979, fig. 5). Basal, locally derived conglomeratic and coarse sandy facies of the Parwyd Grit and lower Tudwal Sandstones contain specimens of '*Bolopora undosa*' (e.g. Matley *et al.* 1939), once thought to be a bryozoan diagnostic of an early Arenig age, but now known to be a phosphatic, oncolitic accretionary structure of chemogenic and/or bacteriogenic origin (Hofmann 1975). The basal sediments pass up rapidly through medium-grained sandstones and siltstones with dominant *Cruziana* ichnofacies into finer arenites and shales with fodinichinia ichnofacies, reflecting an offshore transgressive shift from the littoral zone; these facies re-occur repeatedly through the succession in response to the shifting energy levels and sediment supply. In the extreme south-west, the silty mudstone facies of the Meudwy Valley Slates, containing graptolites and *Merlinia selwynii*, suggest an early extension to sub-wave base sedimentation, but followed upwards by a return to arenite-dominated facies in the Porth Meudwy beds; coarse conglomerates with slump balls and clasts up to boulder grade reappear in this unit, indicating re-activation of the local sediment source, probably in response to faulting. Associated with some conglomerates are sandstones with a dominant *Skolithos* ichnofacies, again suggesting a littoral to sublittoral environment; similar sandstone facies also occur discretely through the succession (Crimes 1970b, fig. 4). *Skolithos, Planolites* and surface trails and burrows are distinctive trace fossils. Differences between the *Skolithos* and conglomeratic facies are probably a function of varying sediment supply and local water energy rather than variations in depth. The better sorted *Cruziana* sandstone facies are symmetrically rippled on a large scale, with tabular and trough cross-bedding resulting from reworking by wave ripples. Trilobite resting impressions (*Rusophycus*) are most common in this facies, although body fossils themselves are scarce.

Feeding burrows (*Phycodes* and *Teichichnus*) are abundant in the fodinichinia siltstones and fine sandstones (Crimes 1969, 1970b). The Bodwrdda Slates of south-west Llŷn, and the laterally equivalent Llanengan Mudstone Formation of St Tudwal's are developed dominantly in mixed to mud-dominated heterolithic facies of an offshore platform environment. These beds contain graptolites including *Azygograptus, Phyllograptus* and *Didymograptus* species, suggesting extension up into the Fennian *hirundo* Biozone (Nicholas 1915, Matley 1928), together with a more varied suite of trilobites and

inarticulate brachiopods than is found in the underlying arenitic facies. At the base of the Bodwrdda Slates is a distinctive horizon of spiculitic cherts and cherty mudstones (Daron Cherts); the spicules are mainly tetractinellid sponges, but no other fossils occur with them (Matley 1928, pp. 490–491; Crimes 1970b, p. 229, fig. 4). In slightly older Porth Meudwy beds there is an horizon of ferruginous ooliths within the mud facies as further evidence of interruption in the pulsatory Arenig transgression.

The sedimentary structures in the Arenig of Llŷn record dominantly north-easterly and south-westerly trends, with some subsidiary vectors indicating easterly flow (Crimes 1970b, fig. 3). In conjunction with the regional facies relationships, the data point to a north-east to south-west aligned shoreline close to western Llŷn, with the main currents flowing parallel to the coast, whose position and alignment may have been fault controlled (Crimes 1970b).

In central western Llŷn, thin tuffaceous seams occur within the Arenig mudrocks, first associated with the Daron Cherts and heralding the effusion of the Rhiw Volcanic Group (Matley 1932; P. M. Allen 1982, p. 87) through late Arenig to early Llanvirn times (*hirundo* to *bifidus* biozones, Fig. 4.4). Graptolites, trilobites and inarticulate brachiopods occur in sediments interbedded with the volcanics, whose base is taken at a distinctive manganese ore containing rhodocrosite, rhodonite and pyrolusite. Matley (1932, p. 260) interpreted this and a slightly lower oolitic and pisolitic chalybitic ironstone as metasomitised tuffs. The Rhiw volcanics, which thin out northwards, comprise spilitic pillow lavas, pillowed andesites, and crystal tuffs; the occurrence in some tuffs of rhyolite and spilite, with crystals of feldspar and quartz, may indicate contemporaneous acid and basic sources. Some of the tuffs are ignimbrites. Although not extensive, the volcanic centres were presumably emergent intermittently during this interval.

Grey and brown shales and siltstones associated with the Rhiw volcanics pass laterally and upwards across central Llŷn into darker argillites which contain graptolites of the *murchisoni* Biozone (e.g. Matley 1938, p. 563; Roberts 1979, p. 19). These black shales and slates were deposited as offshore muds and confirm the palaeogradient eastwards and south-eastwards as transgressive deepening continued through upper Llanvirn times.

As on Anglesey and elsewhere throughout north-west Wales, there is no unequivocal evidence for a continuation of deposition on Llŷn from the upper Llanvirn to the lower Llandeilo. Nicholas (1915) and Matley (1932) reported *Glyptograptus teretiusculus* and associated graptolites at a number of localities, but the faunas could equally indicate the *gracilis* Biozone in the later Llandeilo or earliest Caradoc.

Regionally there is a stratigraphical or tectonic break at this level (Shackleton 1959, Roberts 1979). In the north of St Tudwal's, the Hendy-Capel Ironstone is a complex of massively bedded pisolitic ores within tuffaceous and micaceous siltstones and black mudstones (Pen-y-gaer Mudstone) bearing a *gracilis* fauna (Nicholas 1915, p. 123); although the contact with the underlying beds is a crush zone these ferruginous facies appear to confirm a renewal of shallower water deposition following a break in sedimentation.

Across central and eastern Llŷn the break is even greater (Fig. 4.4), as the Llanbedrog grits and mudstones and equivalent sediments around Llanystumdwy (Matley 1938, Harper 1956) contain rich Soudleyan–Longvillian–Woolstonian shelly faunas. The basal, quartzo–feldspathic grits represent shallow marine transgressive sands derived probably from a re-activated Precambrian source; rapid upward transition into the mudstones reflects the rate of shift to an offshore regime as Caradoc transgression proceeded.

Intertonguing with and succeeding the Llanbedrog and related sediments are two major volcanic suites (Fig. 4.4) which crop out extensively through the area as, presumably, lateral equivalents of the Snowdon volcanics to the north-east. The lower, Mynytho Volcanic Group comprises mainly basaltic andesites, rhyodacites and dacite flows erupted mostly as submarine lavas, with interbedded sediments, ash-flow tuffs, air-fall tuffs, agglomerates and water-deposited tuffs (Matley 1938, Fitch 1967, Tremlett 1969, P. M. Allen 1982); at least two eruptive centres were present, and there was local uplift and erosion within this interval. The unconformably overlying Llanbedrog Ignimbrite Group is entirely rhyolitic (Fitch 1967). In easternmost Llŷn the major Llwyd Mawr ignimbrite sheet is part of this later episode, and was probably emplaced in Soudleyan–Longvillian times (Roberts 1967, 1969); its basal portion is a non-welded pumice lapilli-tuff, but throughout most of its thickness it is welded and strongly eutaxitic.

Ordovician igneous intrusions are comparatively rare in Llŷn, but small, high level rhyolitic sills and vent intrusions cut the Arenig–Llanvirn sediments in places (Matley 1938, Fitch 1967) and appear to be related to the mid-Caradoc volcanicity.

Conformably above the volcanics, the Penarwel Drive beds (Matley 1938, Fitch 1967) were deposited as offshore, pyritic black muds, with abundant graptolites including *Dicranograptus clingani* and other species of that biozone. Harper (1956) described the faunal sequence in detail around Llanystwmdwy in the east. In turn these beds are overlain unconformably by black slates with a locally conglomeratic base and containing a high *clingani* to *linearis* Biozone fauna; these dark beds, probably reflecting original restricted circulation, may represent the distinctive Nod Glas

facies seen widely across North Wales (Cave 1965).

The succession on Llŷn is then terminated upwards by further richly fossiliferous mudstones, assigned to the Crugan Mudstone Formation in central areas, and the Dwyfor Mudstone Formation in the east (Matley 1938, Harper 1956, Price 1981). These beds are of Rawtheyan age, and thus indicate a further break above the Nod Glas slates. The distinctive brachiopod *Foliomena folium* occurs in both formations, generally interpreted as an indication of offshore, relatively deep, low energy environments (e.g. Sheehan 1973). Trilobite associations in the Crugan mudstones, with genera such as *Tretaspis, Nankinolithus, Encrinuroides* and *Ceraurinella* pass laterally into assemblages dominated by *Opsimasaphus* and *Nankinolithus*, which Price (1981) interpreted as indicating a slight downslope environmental transect from west to east at some distance from a platform margin.

## Snowdonia

Snowdonia National Park (Fig. 4.6) includes Conwy Valley, Snowdon, Harlech, Bala and Cadair Idris Parks; for the purposes of this account Snowdonia is a suitable term to use when describing the extensive belt of Ordovician rocks surrounding the Harlech Dome (Fig. 4.2). It covers a region where the complex interplay of tectonics and igneous activity (from a number of different centres in space and time) and their effect on sedimentation is exemplified (see e.g. Kokelaar *et al.* 1984, Kokelaar 1988). The descriptions that follow are given in a clockwise direction from north to south around the Harlech Dome.

## Tremadog-Snowdon-Dolgarrog-Conwy

This includes the mountainous area to the north and north-east of the Harlech Dome. In general terms the end of the Cambrian Period saw uplift, shallowing of the sea with erosion and re-working of earlier sediments during the early Ordovician. At Tremadog there is a conformable transition from the underlying Cambrian Merioneth Series into the basal Tremadoc Series. In most of the area, however, Tremadoc beds are absent and a pronounced unconformity lies below the Arenig Series (see e.g. Reedman *et al.* 1983, Howells *et al.* 1983) with successively lower horizons of the Cambrian being overstepped north-westwards from Tremadog, while in many outcrops the Cambrian–Ordovician boundary is obscured by faulting.

At Tremadog the Tremadoc Series (Fig. 4.4) consists essentially of cleaved marine mudstones. The basal Tyn-llan Beds, some 65 m in thickness, contain trilobites and arthropods and are followed upwards by blue-grey mudstones yielding many *'Dictyonema'* sp.

(the well-known Dictyonema Band). The succeeding Moel-y-gest grey slates and mudstones have rare brachiopods and gastropods, and the overlying Portmadoc Flags trilobites. The fossiliferous mudstone and slate of the Penmorpha and Garth Hill Beds are notable for the presence of *Shumardia* and *Angelina sedgwickii* respectively.

The base of the Arenig Series is frequently marked by an impersistent bioturbated shallow-water sandstone – the Garth Grit in the south of the area (Williams *et al*. 1972), the Graianog Sandstone farther north (Howells *et al*. 1983) – which is notable for the frequent presence of *'Bolopora undosa'* (see p. 72). Trace fossils include *Phycodes circinatum* and *Teichichnus* sp. Near Blaenau Ffestiniog, some 12 km ENE of Tremadog, the Tremadoc beds are represented by the sandstones of the Glanypwll Formation; the Garth Grit marks the base of an Arenig age sequence of sandstones and siltstones forming the Cwmorthin Formation (Table 4.1). In the Moelwyn and Manod mountains a 500 m sequence of volcaniclastics accompanied by rhyolitic extrusives and intrusives have their lowermost 150 m assigned to the Moelwyn Formation, the overlying remainder forming, in this area, the basal part of the Nant Ffrancon Formation (Howells *et al*. 1983), which in places totals 1,800 m in thickness. This consists of a monotonous series of dark marine siltstones and mudstones with the occa-

**Fig. 4.6.** Typical glacially modified scenery in Ordovician igneous rocks of Snowdonia. View to the WSW along Crib Goch ridge to Snowdon in the left background and Crib y Ddysgyl in the right background. The rugged crags in the foreground showing prominent bedding and cleavage planes are formed of intrusive rhyolites. Most of Snowdon and the cwm to the right are in Bedded Pyroclastic Formation, with thin rhyolites near the summit. Slopes leading up to Crib y Ddysgyl are in Bedded Pyroclastic Formation, then dolerites (forming darker shadows) and Upper Rhyolitic Tuff Formation with subsidiary rhyolites up to the summit, which is formed by a further dolerite sill.

**Table 4.1.** Generalised section of Ordovician rocks near Blaenau Ffestiniog

|  |  | Thickness (metres) |
|---|---|---|
| Black Slates of Dolwyddelan |  | — |
| Upper Crafnant Volcanic Fm. |  | 120 |
| Bedded Pyroclastic Fm. | (equivalent to the Snowdon Volcanic Group) | 130 |
| Lower Crafnant Volcanic Fm. |  | 30 |
| Cwm Eigiau Fm. | (marine siltstones and mudstones) | 170 |
| Capel Curig Volcanic Fm. | (distal, isolated occurrences of Llewelyn Volcanic Group) |  |
| Nant Ffrancon Fm. | (mainly argillites) | c. 1,800 |
| Moelwyn Fm. | (acid intrusives and extrusives and volcaniclastics) | c. 150 |
| Cwmorthin Fm. | (siltstones and sandstones with Garth Grit marking base of Arenig) | c. 150 |
| Glanypwll Fm. | (sandstones of Tremadoc age) | c. 340 |

(Compiled from Br. Geol. Surv. 1:25,000 sheet, Bangor, 1989.)

sional impersistent sandstone, localised names being given to the beds from place to place (see Fig. 4.4). The littoral environment of the lower beds changes to a bathyal one up the succession perhaps suggestive of a eustatic transgression (Legget 1980), though intermittent shallowing is indicated by the occurrence of oolitic and pisolitic ironstones (Trythall *et al.* 1987).

Mudstone breccias and small-scale slump-bedding in the highermost beds reflect the tectonic activity that preceded later volcanicity.

Evidence for the age of the Nant Ffrancon Formation is scant but Howells *et al.* (1985, p. 8), for example, list evidence from their work in the northwest of the B.G.S. 1:50,000 Bangor Sheet indicating the formation extends from the Arenig to lower Caradoc.

Two major volcanic sequences, each more than 1 km in thickness (see Ch. 16), and separated by marine sediments of the Cwm Eigiau Formation, are recognised above the Nant Ffrancon Formation, the Llewelyn and Snowdon Volcanic Groups (Fig. 4.4) (Howells *et al.* 1981a, Howells *et al.* 1983, Kokelaar *et al.* 1984) with the Snowdon Volcanic Group, the higher, being regarded as the equivalent of the Crafnant Volcanic Group of the Dolgarrog–Conwy area (see Tables 4.1 and 4.2).

The Llewelyn Volcanic Group (180–1,700 m thick) encompasses a number of distinctive volcanic centres which erupted acid to basic lavas and pyroclastic rocks. The Snowdon and Crafnant Volcanic Groups are typified by the eruption of mainly acid pyroclastics (with classic welded tuffs). Within all three Groups, however, there were periods when mainly marine sedimentation continued during periods of quiescence and enough fossil evidence is present to indicate that the volcanism occurred within the *Diplograptus multidens* Biozone (and is part Soudleyan – part Longvillian).

The Llewelyn Volcanic Group can be divided into formations, the highest of which, the Capel Curig Volcanic Formation (up to 400 m thick) represents the most important episode of Caradocian volcanism in north-east Snowdonia, with the first major eruptions of acid ash-flows and debris-flows, both subaerial and subaqueous from three distinct centres (see Howells & Leveridge 1980, Howells *et al.* 1980). Traceable from Dolwydellan in the south to Conwy in the north, the Capel Curig Volcanic Formation is spectacularly exposed at Tryfan to the west of Capel Curig. In places

**Table 4.2.** Generalised section of Ordovician rocks in Dolgarrog area

|  | Thickness (metres) |
|---|---|
| Grinllwm Slates | 180 to 320 |
| Trefriw Tuff/Mudstone | up to 80 |
| Llanrhychwyn Slates | 50 to 400 |
| Dolgarrog Volcanic Fm. | 50 to 400 |
| Crafnant Volcanic Group |  |
| Middle and Upper Crafnant Volcanic Fms. | up to 1,000 |
| Lower Crafnant Volcanic Fm. | 360 to 500 |
| Cwm Eigiau Fm. |  |
| Mudstones and siltstones | up to 280 |
| Tuffs and sandstones | up to 365 |
| Mudstones and siltstones | 0 to 680 |
| Llewelyn Volcanic Group |  |
| Capel Curig Volcanic Fm. | 300 |
| Sandstones, siltstones, tuffs and tuffites | 350 to 400 |
| Conwy Rhyolite Fm. and | 920 |
| Foel Fras Volcanic Complex | 100 |

(From Howells *et al.* 1981b.)

the volcanics become intercalated with near-shore sediments and yield a rich fauna of brachiopods, molluscs and trilobites. Below the Capel Curig Volcanic Formation the volcanics have been divided into the Conwy Rhyolite Formation (and Foel Fras Volcanic Complex), the Foel Grach Basalt Formation and the basal Braich-tu-du Volcanic Formation though in the core of the Tryfan Anticline there is a considerable thickness of shallow water marine sandstones and siltstones with a brachiopod fauna of 'Soudleyan aspect' (Howells et al. 1985).

In the north of the area the occurrence of coarse clastics, interpreted as being subaerial alluvial fans, are thought to indicate fault-block uplift. A north-east–south-west coastline, with marine conditions existing down the palaeoslope to the south-east, is considered to be due to flexure from the subsidence of a basement fault downthrowing to the south-east. Continued movement of this flexure affected the depositional environments of the Cwm Eigiau Formation which overlies the Llewelyn Volcanic Group. Alluvial and fluvial sandstones (some 450 m in thickness) occur to the north of the hinge line in the Tal-y-Fan area whereas to the east (e.g. at Capel Curig) up to 1,300 m of deeper water mudstones, siltstones and sandstones occur, the last named yielding brachiopods and trilobites of late Soudleyan to Longvillian age. Associated with shallow-water fossiliferous sandstones, between the Llanberis and Nant Ffrancon Passes, the well-known Pitt's Head Tuff occurs, near the top of the formation. It is a welded ash-flow tuff which can be traced at outcrop for about 20 km northwards from Moel Hebog (south of Snowdon) and can reach 700 m in thickness, and is sub-aqueous in places (see Fig. 16.6 and p. 498).

Locally basalt flows and basalt breccias are present below the Lower Rhyolitic Tuff (L.R.T.) Formation which marks the base of the Snowdon Volcanic Group. The L.R.T. Formation (see Chapter 16), which reaches perhaps 600 m in thickness on the south side of the Snowdon massif, consists there essentially of sub-aqueous rhyolitic ash-flow tuffs but commonly has accompanying sedimentary, volcaniclastic and pyroclastic breccias, including the megabreccias such as the Llyn Dinas Breccia of Beavon (1963, 1980).

Northwards and eastwards the formation thins and intercalated siltstones and mudstones confirm a deepened marine environment. Overlying the Formation in central Snowdonia the Bedded Pyroclastic Formation consists of basic lavas (often pillowed) and basic pyroclastics. The latter exhibit sedimentary structures indicating a shallow tidal environment as do the contained Longvillian fossils such as *Chasmops cambrensis* and *Nicolella actoniae obesa*.

At Capel Curig, about 11 km east-north-east of Snowdon the Snowdon Volcanic Group passes laterally into the Crafnant Volcanic Group (Table 4.2), the Lower Crafnant Volcanic Formation of which consist of three ash-flow tuffs (the basal one containing brachiopod shells) interbedded with marine sediments. The Middle Crafnant Volcanic Formation, restricted in area, is a well-bedded alternation of dark graptolitic mudstone, occasional sandstones (turbidites) and acid tuffs (with *Amplexograptus arctus, A. fallax* and climacograptids, glyptograptids and orthograptids. The Upper Crafnant Volcanic Formation (Howells et al. 1978) is a massive tuffite thought to have accumulated as a high-density gravity flow.

Farther north the Middle and Upper Crafnant Volcanic Formations cannot be differentiated but their thickness at Dolgarrog (Table 4.2) must be of the order of 1,000 m of acidic tuffs. The contrast between the basaltic Bedded Pyroclastic Formation (and its interbedded shallow-water sediments) of Central Snowdonia with the deep-water acid volcanics of the eastern area may be due to the presence of a physical barrier between two environments (Howells et al. 1978).

Above the volcanic formations lie mainly graptolitic shales, though the basic Tal-y-Fan Volcanic Formation and the submarine basaltic Dolgarrog Volcanic Formation locally intervene. In the north-east the Llanrhychwyn Slates (or Cadnant Shales (Howells et al. 1985)) are considered to be within the *Diplograptus multidens* zone though the stratigraphic equivalent black slates of Dolwyddelan farther south are in part *clingani* Biozone. Overlying the Llanrhychwyn Slates the Grinllwm Slates (correlated with the Bodeidda Mudstone and part of the Deganwy (or Conwy) Mudstone are considered to be of Ashgill age. South of the Crafnant Valley the intervention of siltstones and sandstones which pass laterally northwards into the basaltic Trefriw Tuff below the Grinllwm Slates indicates shallowing, perhaps due to uplift, in an otherwise continuing deep-water environment.

South-eastwards from Betws-y-coed, towards Bala, Campbell (1984) established detailed correlations in the Caradoc and Ashgill successions, emphasising facies variations in the shallow marine sediments, the nature of the Caradoc–Ashgill unconformity and the absence of Ashgill volcanicity.

## Migneint–Arennig

It is within this area that Sedgwick (1838, 1845, 1852) defined the Arenig and Bala groups and that later Lapworth (1897) erected the type section of his new Ordovician System. The succession, which now includes the Tremadoc Series, is of the order of 3,800 m in thickness. Lynas (1973) described the

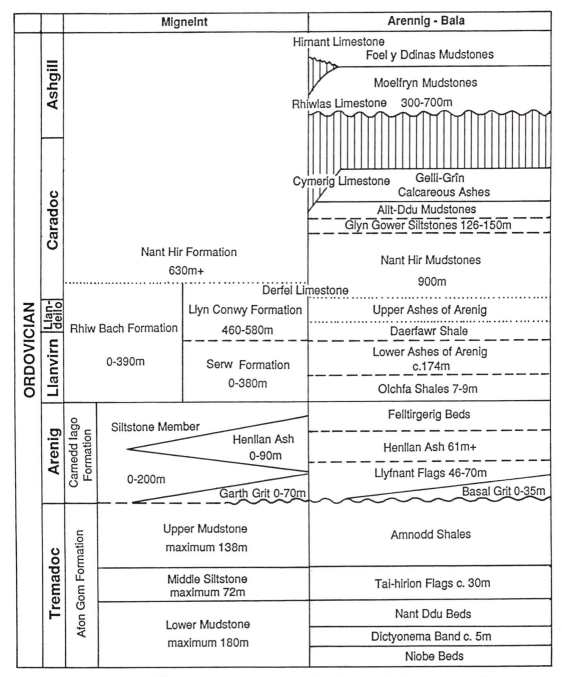

|  |  |  | Migneint | Arennig - Bala |
|---|---|---|---|---|
| ORDOVICIAN | Ashgill | | | Hirnant Limestone |
| | | | | Foel y Ddinas Mudstones |
| | | | | Moelfryn Mudstones |
| | | | | Rhiwlas Limestone 300-700m |
| | Caradoc | | | Cymerig Limestone / Gelli-Grîn Calcareous Ashes |
| | | | | Allt-Ddu Mudstones |
| | | | | Glyn Gower Siltstones 126-150m |
| | | | Nant Hir Formation 630m+ | Nant Hir Mudstones 900m |
| | Llan-deilo | | Derfel Limestone | |
| | | | Llyn Conwy Formation 460-580m | Upper Ashes of Arenig |
| | Llanvirn | | Rhiw Bach Formation | Daerfawr Shale |
| | | | 0-390m | Lower Ashes of Arenig c.174m |
| | | | Serw Formation 0-380m | Olchfa Shales 7-9m |
| | Arenig | Carnedd Iago Formation | Siltstone Member / Henllan Ash 0-90m | Felltirgerig Beds |
| | | | 0-200m | Henllan Ash 61m+ |
| | | | | Llyfnant Flags 46-70m |
| | | | Garth Grit 0-70m | Basal Grit 0-35m |
| | Tremadoc | Afon Gom Formation | Upper Mudstone maximum 138m | Amnodd Shales |
| | | | Middle Siltstone maximum 72m | Tai-hirion Flags c. 30m |
| | | | Lower Mudstone maximum 180m | Nant Ddu Beds |
| | | | | Dictyonema Band c. 5m |
| | | | | Niobe Beds |

**Fig. 4.7.** Stratigraphic nomenclature in Migneint–Arennig–Bala area (after Whittington & Whitworth 1974).

Cambrian and the Ordovician rocks of the Migneint area in detail (Fig. 4.7) emphasising the importance of the N–S-trending faults of the area, movement of which in Lower and Middle Ordovician times produced pre-Arenig and pre-*gracilis* unconformities and many disconformities. Concurrent with the fault movements was igneous activity, with basic to acid intrusions and extrusives in the form of feldspathic tuffs, occasional laharic breccias and agglomerates.

Whittington & Whitworth (1974, p. 19) described the sequence of the area in general as being ". . . made up largely of alternations of siltstones and mudstones with considerable thicknesses of acid to basic extrusive and intrusive rocks in the lower part, and with ash beds and thin limestone bands in the upper (the latter having long been key beds in mapping).

"The bulk of the rocks are unfossiliferous, brachiopod–trilobite faunas being only abundant in the Gelligrin Calcareous Ashes (Caradoc) and in the Rhiwlas Limestone (Ashgill). Elsewhere, fossils occur at restricted horizons and localities, but the occurrences are sufficient to enable reasonably accurate correlations to be made with other areas.

"The Arenig strata apparently succeed those of the Tremadoc conformably in the west of the area but unconformably farther east. The basal Garth Grit contains only *Bolopora undosa* Lewis. The *extensus*, *hirundo* and *bifidus* Zone graptolite faunas have been recognised in the Migneint and Arenig areas and one shelly fauna in the Henllan Ash of the Lower Arenig (Whittington 1966). The Upper Ashes of Arenig, which form the higher slopes of Arenig Fawr, have not yielded any fossils, the underlying Daerfawr Shales being doubtfully as young as Upper Llanvirn. On the east side of Arenig Fawr the Upper Ashes are succeeded by the early Caradoc Derfel Limestone, and a similar relationship obtains on the north side of Migneint. No faunas of Llandeilo age are known in the Arenig–Bala area."

Williams *et al.* (1972) outlined the palaeontological evidence for their division of the succession in the Arenig–Bala area and emphasised, as noted above, the correlative importance of the Cymerig, Rhiwlas and Hirnant limestones. Zalasiewicz (1984), however, after detailed study considered the Arenig type-section to be incomplete, only the *extensus* Biozone being identifiable with certainty with possible unconformities being present at both top and bottom of the type section (see also Whittington *et al.* 1984 and p. 65).

As pointed out on p. 76 Campbell (1984) noted differences between the Bala and Snowdon Upper Ordovician successions. With regard to the Caradoc–Ashgill unconformity, he suggested it might mark the 'erosion of an uplifted dome with major uplift control along NE–SW to N–S faults' and linked the dis-tribution of faunas to the water depth above the inferred eroded-dome topography.

## Bala-Cadair Idris-Corris

Ridgway (1971), having studied some 27 sections from 9 areas within this arcuate belt of alternating volcanics and sediments, published in 1975 and 1976 an important synthesis of the stratigraphy (omitting Rhobell Fawr (see p. 61), then classified with the Cambrian System). He defined a lithostratigraphic unit, the Aran Volcanic Group (comprising nine formations (Table 4.3) which ranged in age through Arenig to Caradoc, emphasising that he recognised six disconformities within the Group.

Allen *et al.* (1981) in the course of mapping part of the 1:50,000 Harlech Sheet (135) between Bala and Dolgellau used the term Mawddach Group to encompass the beds east of the Harlech Dome (previously described as the (Cambrian) Menevian Beds), Lingula Flags and Tremadoc Slates. Their topmost division, the Cwmhesgen Formation, comprised two members – the upper, the Dol-cyn-afon Member of Tremadoc age now being classified as basal Ordovician. Conformably lying on the Cambrian beds below, it consists mainly of dark mudstones with occasional units containing sand-sized chloritic grains, heterolithic 'mudstones' with phosphatic nodules, and tuffaceous siltstones. *Rhabdinopora flabelliformis socialis* (Salter) and *R. flabelliformis flabelliformis* (Eichwald *sensu* Bulman) mark the base of the member.

Allen & Jackson (1985) then proposed a new stratigraphic classification of the Aran Volcanic Group (Table 4.3) for the Harlech Sheet comparing it with previous classifications and discussing the correlation difficulties in a sedimentary/volcanic regime where lateral variability is endemic (see e.g. Allen & Jackson *op. cit.*, fig. 7 and pp. 31–32).

Sedimentation commenced in an upper shelf marine environment with the mudstones and silts of the Dol-cyn-afon Member of the Cwmhesgen Formation. Sporadic volcanic activity is seen in the occasional occurrence of tuffaceous beds and lenses. The Rhobell Volcanic Group (outcropping in the central western part of the arcuate belt) lies unconformably on Cambro–Ordovician beds of the Cwmhesgen Formation – late Tremadoc folding and faulting being attributed to the commencement of the extensive volcanic activity that was to follow.

The Aran Volcanic Group thereafter follows unconformably on either the Rhobell Volcanic Group or the Mawddach Group. Its basal Allt Lŵyd Formation consists of shallow-water, mainly clastic beds. The lowest, the Garth Grit Member (with its familiar '*Bolopora undosa*'), was probably derived from erosion of the Harlech Grits to the west but most of the clastics in the higher members are of probable contemporaneous volcanic derivation (e.g. the alluvial Aran Boulder Bed). Succeeding formations display an alternation of basic and acid extrusives, with marine black silty muds being deposited during periods of volcanic quiescence. Basic rocks include spilites and tuffs and are taken to represent local submarine eruptions, while acid ash-flow tuffs are thicker and extend over considerable areas and probably originated from a single volcanic centre to the south-east (Allen & Jackson 1985).

In the southern part of the belt the Aran Volcanic Group includes, in ascending order, the Offrwm, Brithion and Melau formations between the Allt Lŵyd Formation and the Benglog Formation (see Table 4.3 and Allen & Jackson op. cit., fig. 7). The Offrwm Formation records the first major acidic eruptions in the area while the Melau Formation represents the first of the many basaltic eruptives to come. The Brithion Formation, which is mainly developed in the eastern part of the arcuate belt, is, however, acidic. In the Cadair Idris area (Davies 1959, Ridgway 1975, Dunkley 1979) the alternating sequences of acidic and

**Table 4.3.** Aran Volcanic Group

| Harlech Sheet | | Ridgway (1971, 1975) | |
|---|---|---|---|
| Formation | Lithology | Formation | Lithology |
| Aran Fawddwy | acid tuff | Craig-y-Llam | Ignimbrite |
| Pistyllion | mudflow breccias with intercalated tuffs and lavas | Craig-y-Bwlch | Mudflow |
| Craig-y-Ffynnon | acid tuff | | Ignimbrite |
| Benglog Volcanic | intermediate crystal tuff with lavas hyaloclastites and minor siltstone | Pen-y-Gader | Mudflow |
| | | | Basic tuffs lavas and siltstones |
| AVG | siltstone and tuffaceous siltstone with unnamed basic tuff horizon | Nant Ffridd Fawr | Ignimbrite |
| | | Llyn-y-Gadr | Mudflow |
| Brithion | acid tuff | Cefn-hir | Mudflow |
| AVG | siltstone and tuffaceous siltstone | Gwynant | Mudstone |
| Allt Lŵyd | Aran Boulder Bed | Pared-yr-Ychain | Boulder Bed |
| | Volcanic Sandstone | | Siltstone, sandstone and feldspathic wackes |
| | Interbedded siltstone and sandstone | | |

Comparison of formation names used by Ridgway (1975) (for section about 7 km east of Dolgellau) with those used by Allen & Jackson (1985, table 6).
    In other areas covered by the Harlech Sheet the Brithion Fm. can be underlain by acid ash-flow tuffs of the Offrwm Fm. and overlain by the silty mudstones with sporadic interbedded agglomeratic tuffs and tuffites of the Melau Fm. (AVG = Aran Volcanic Group beds not named by B.G.S.).

**Fig. 4.8.** The Bala Fault, one of the most distinctive structural features cutting through Ordovician rocks in Wales, accentuated by Pleistocene glaciation. Aerial view northeastwards along the fault a few miles inland of the coast at Tywyn (Fig. 4.2); the village of Dolgoch is at lower centre, with Abergynolwyn and Tal-y-llyn Lake beyond. Slopes to the right are mostly in Caradoc-Ashgill mudstones and turbidites, with volcanics of the Aran Mountains in the far distance. Rugged crags to the left are in the Aran Volcanic group at the southern edge of the Cadair Idris range. (Crown Copyright Reserved.)

basic extrusive rocks (with intercalations of sediments) can reach a thickness of 2 km and are considered to be mainly subaqueous in origin. The large granophyre intrusion of Cadair Idris (Fig. 16.5) was considered by Davies (1959) to represent part of the magma feeding the acid extrusions of this area. South-west of Cadair Idris (Fig. 4.8), the succession is much thinner, with some sedimentary and volcanic units being missing (Dunkley *op. cit.*)

Beds of the Allt Lwyd Formation can be correlated with beds north of the Harlech Dome which are considered to be within the *extensus* Biozone. Several localities within the area of the Harlech Sheet contain shelly beds with fossils comparable with the 'Henlian Ash' fauna of the Arennig area (Fig. 4.7) and thus support an *extensus* Biozone age. A Llanvirnian age, however, is indicated (by the presence of *D. artus* Biozone fossils) for the shales below and above the Brithion Formation. Wells (1925) recorded graptolites 'no older than the *Nemagraptus gracilis* Zone' in the Afon Eiddon (Allen & Jackson *op. cit.*, p. 44) indicating a condensed Llanvirn–Llandeilo succession. The Aran Fawddwy Formation (Craig-y-Llam of Ridgeway – see Table 4.3) is considered probably to be of Caradoc age on the basis of its unconformable junction with the underlying rocks and it (or its correlatives) being overlain by the Derfel Limestone (Fig. 4.9). This limestone, at the base of the Nant Hir Mudstone (Ceiswyn Mudstone), yields fauna which are attributed to the Costonian stage of the Caradoc (Williams 1963, Ridgway 1975, Dunkley 1980, Lockley 1980).

The succession above the Aran Volcanic Group between Bala and Corris is summarised in Fig. 4.9, again displaying great lateral variation, but demonstrating the gradual dying out of the volcanism, which is represented by the occasional occurrence of tuffs in Caradoc times. Ashgill beds, mainly mudstones, lie unconformably above.

## Rhobell Fawr

The geology of the south-eastern flank of the Harlech Dome (Fig. 4.2) is dominated by the Rhobell Volcanic Group (Wells 1925; Kokelaar 1980, 1986). It oversteps unconformably Tremadoc and upper Cambrian strata, and in turn is overlain unconformably by Arenig beds. Sections through the Tremadoc to the north of Rhobell Fawr are important stratigraphically in demonstrating the only detailed evidence in Wales outside the type Tremadog area of a conformable upward transition from the uppermost Cambrian Merioneth Series (P. M. Allen *et al.* 1981, Rushton 1982). Within the Cwmhesgen Formation at the top of the Mawddach Group, dark carbonaceous mudstones of the Dolgellau Member pass up into more silty mudstones with rare tuffaceous horizons and fine–medium arenites of the Dol-cyn-afon Member; the occurrence of *R. flabelliformis* just above the base of this latter member marks the level of the base of the Tremadoc. Lower beds containing *R. flabelliformis socialis* are succeeded by beds with *R. flabelliformis flabelliformis*, suggesting a subzonal division useful for wider correlation. The graptolites are accompanied by fairly diverse trilobite and inarticulate brachiopod faunas, together with sponge spicules, hyoliths and fragmented echinoderms. Of the brachiopods, *Eurytreta sabrinae* occurs here only above the base of the Tremadoc, whilst the trilobite species *Boeckaspis hirsuta* and *B. mobergi* occur in sequence as conformation of the subzonal correlation based on graptolites (Rushton 1982, p. 45).

Some silty and sandy mudstones of the Dol-cyn-afon Member are cross laminated and convoluted, with other mixed mud/sand heterolithic facies containing chloritic and muddy clasts and phosphatic nodules; deposition was in a muddy shelf environment subjected to periodic storms (P. M. Allen *et al.* 1981, p. 322). There is no evidence for post *flabelliformis* Biozone horizons in the Dol-cyn-afon Member; fine tuffaceous bands reflect the onset of distal volcanism, followed by uplift, emergence and deep erosion throughout the later part of the early Tremadoc.

Within the Rhobell Volcanic Group (see also Chapter 16), successive lava flows overlap and overstep eastwards and south-eastwards across a highly irregular surface, in some cases involving ponding against ridges approximately 50 m high (Kokelaar 1980, p. 591). The pile comprises basaltic lava flows with associated autobreccias and locally developed epiclastic deposits. No pyroclastic rocks are present, suggesting relatively quiet effusion; some of the 'explosion tuffs' described by Wells (1925, p. 476) were reinterpreted by Kokelaar (1980) as brecciated, heterolithic wet mass-flow accumulations, which pass laterally into poorly sorted feldspathic arenites representing flash flood deposits in small alluvial fans developed around the peripheries of the volcanic centres.

The Rhobell basalts were erupted subaerially from a series of fault controlled fissures aligned N–S, to the west of the present main outcrop (Kokelaar 1980, fig. 2). A suite of hypabyssal, subvolcanic intrusions marks the sites of these fissures, preserved as a number of dyke complexes, minor sills and a laccolith; the intrusions comprise dolerites, microdiorites and microtonalites.

Kokelaar (*op. cit.*) regarded the igneous activity as indicative of a fractionation continuum of tholeiitic arc-like volcanism (see also Bevins *et al.* 1984).

A K–Ar age of $508 \pm 11$ Ma derived from a metabasalt in the Rhobell Volcanic Group provides an

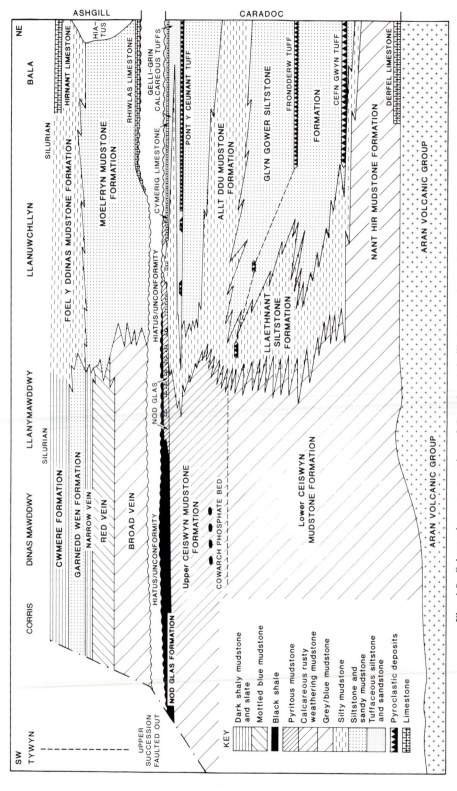

**Fig. 4.9.** Schematic reconstruction of Caradoc-Ashgill stratigraphy and facies variation south-westwards along the Ordovician outcrop south of the Bala Fault from Bala to the coast at Tywyn. Partly after Lockley (1980), not to scale.

acceptable calibration point for the late Tremadoc (Fig. 4.3; Kokelaar *et al.* 1982).

At the end of Tremadoc times, the Rhobell Volcanic Group was uplifted along a N–S axis and folded and eroded deeply prior to renewed deposition in Arenig times (e.g. Wells 1925, fig. 6; Kokelaar 1980, 1988; Kokelaar and others 1984a, b). This unconformably overlying sequence of sediments and volcanics comprises part of the Aran Volcanic Group (p. 78) (Ridgway 1975, 1976).

## Berwyn Hills

Although referred to commonly as the 'Berwyn Anticline' or the 'Berwyn Dome', the extensive tract of Ordovician rocks in the northern part of east Wales (Figs. 4.2, 4.4, 4.10) is deceptively complicated in structure. Separated from the Snowdonia–Harlech–Bala fold systems by the north-easterly extension of the Central Wales Synclinorium, and truncated in the north-west by the Bala Fault complex, the Berwyn Hills are folded into a succession of sub-parallel to en-echelon arcuate flexures in which two major periclines are disposed about a more gentle syncline, and in all of which the magnitude of folding is greater on the southerly limbs. In places the southern components are vertical to overturned and sheared, passing downward probably into high-angle reverse faults (e.g. Wedd *et al.* 1929, pp. 15–17; Shackleton 1954, p. 283; Coward & Siddans 1980, fig. 3).

The greater part of the region comprises over 1,500 m of Caradoc–Ashgill mudstones and siltstones with subordinate sandstones, limestones, black shales and various pyroclastic deposits, including ignimbrites (Brenchley 1969b). In addition, three small inliers in the central Berwyns to the north of Llanrhaeadr ym Mochnant include beds of Llandeilo age, the only palaeontologically proved Llandeilo succession in northern Wales (Wedd *et al.* 1929, p. 28; MacGregor 1961, 1962). In places, basal pebbly calcareous tuffs of this Craig-y-glyn Group rest with apparent conformity on yet older blue shales and slates, which Wedd *et al.* (1929, pp. 28–33) regarded questionably as being of Llanvirn age. As yet no diagnostic faunas have been found. The upward succession of tuffs, sandy and muddy limestones, some rippled and cross-bedded, and finally, calcareous shales and flags of the Craig-y-glyn Group in contrast contains fairly rich faunas of brachiopods and trilobites. Typical Llandeilo taxa include *Corineorthis biconvexa*, *Macrocoelia llandeiloensis*, *Ogygiocarella debuchii*, *Basilicus tyrranus* and *Marrolithus favus* (MacGregor 1961, 1962). The latter species occur through the upper two units suggesting an upper Llandeilo age; it remains possible that the lower beds may extend down into the middle Llandeilo.

The overlying Garwallt, Gwern Feifod and Mynydd Tarw groups lack the calcareous development of the older beds, reflecting greater clastic and volcaniclastic input, but they are otherwise similar in lithology and presumably environment (i.e. muddy shelf). Fossils are extremely rare or absent but the record of *Pseudoclimacograptus scharenbergi* from the Mynydd Tarw Group suggests an early Caradoc age.

Above the Mynydd Tarw Group the Cwm Clwyd Tuff Formation forms a distinctive marker horizon, except in the extreme south where it is cut out by faulting and in the south-west where it may have been removed by penecontemporaneous erosion (Brenchley 1972, 1978). The tuff is coarsest and thickest in the eastern outcrops, where lithic lapilli tuffs and agglomerates form graded beds locally in steep channels. Elsewhere the beds are mainly flaggy or cross-bedded pumice and crystal tuffs. Much of the ash sedimentation in the east was sub-aerial and probably occurred low on the flanks of a volcanic cone just to the east of the Berwyns (Brenchley 1964, 1972). To the south and south-west sedimentation was in a marine environment, where the tuffs can be traced progressively into re-worked volcaniclastic sandstones.

In the north and north-west Berwyns the remaining middle Caradoc (essentially Soudleyan–Longvillian) succession (Figs. 4.4, 4.10) comprises repetitive grey mudstones, siltstones and fine sandstones with variably developed pyroclastics. Two further major tuff formations (Swch Gorge Tuff and Pandy Tuff) permit subdivision of the monotonous terrigenous sequence, in which there is a high volcaniclastic matrix component throughout (e.g. Brenchley 1969a). More localised tuffs also occur as distinctive markers, such as the Pen Bwlch Tuff Member and the Pen-y-graig Tuff Member (Fig. 4.10). Both the Swch Gorge and Pandy tuffs are ignimbrites with associated volcaniclastic sediments (Brenchley 1978). They may have been deposited sub-aerially in part,. although overlying and underlying sediments are of shallow marine origin, and some show evidence of soft-sediment deformation following deposition of the ashes. Derivation from the north-west is inferred from general thickness and textural evidence.

The oldest fossiliferous sediments in the Caradoc sequence are in the Pen-plaenau Siltstone, which contains sparse trilobites such as *Brongniartella* and brachiopods including *Dinorthis*, *Sowerbyella* and lingulaceans; trace fossils and bioturbation are ubiquitous, and the beds are commonly rippled, cross-laminated and channelled. The Swch Gorge Tuff carries a rich shelly fauna in places, including *Broeggerolithus broeggeri*, *Flexicalymene limba*, *Dinorthis berwynensis* and *Reuschella oblonga*, indicating an early Soudleyan age by comparison with the Shropshire succession.

The laminated siltstones and mudstones of the Teirw Siltstone Formation are bioturbated in places, with assemblages of trace fossils including *Skolithos*, *Planolites* and *Dimorphichnus*. Massive, lenticular sandstones occupy broad channel fills, and there are intraformational mudstone conglomerates, desiccation cracks, ripples and large-scale cross-bedding. Within this sequence the Pen Bwlch Tuff is

**Fig. 4.10.** Reconstruction of Caradoc-Ashgill stratigraphy and facies variation in the Berwyn Hills. Partly after Brenchley (1978), Brenchley & Pickerill (1980), and Hiller (1981).

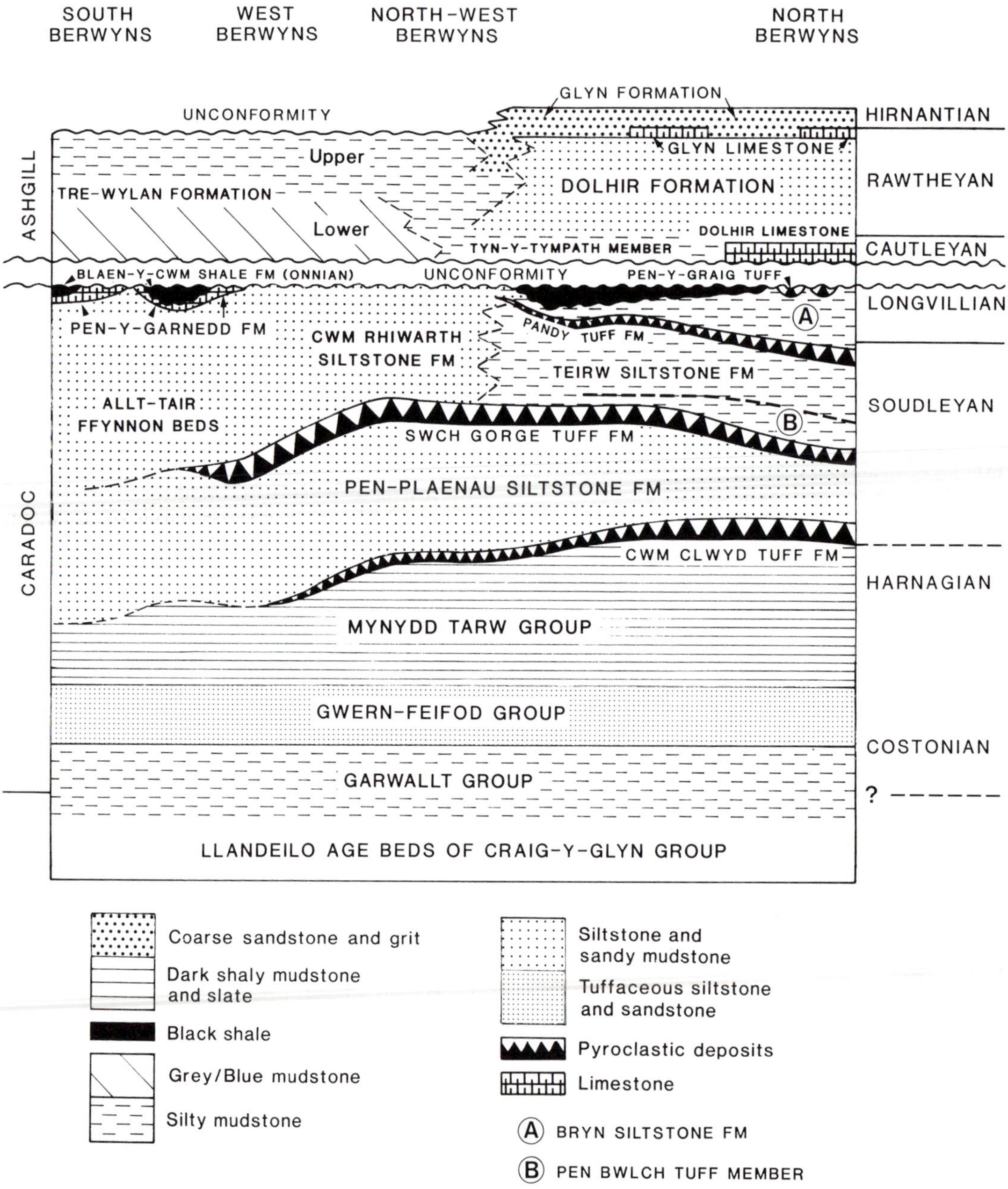

petrologically identical to the Frondderw Tuff of the Bala area (Schiener 1970, Brenchley 1978, p. 145) and their similar stratigraphical level suggests that the two are part of a single contiguous unit. Three successive faunas of some correlative value occur through this part of the sequence; the lowest is typified by the trilobite *Broeggerolithus broeggeri*, the second some 30 m higher by the brachiopod *Heterorthis retrorsistria*, and the upper one by a group of brachiopods including *Onniella* cf. *soudleyensis* together with *Dinorthis, Salopia, Sowerbyella* and *Howellites* species. Each assemblage contains a variety of shelly fossils permitting correlation both with Soudleyan faunas in Shropshire and those of the upper Glyn Gower through Allt Ddu formations of the Bala district (Fig. 4.9).

In the lower part of the Bryn Siltstone Formation (Fig. 4.4) faunas of early Longvillian age first appear, e.g. *Broegerolithus soudleyensis, Kloucekia apiculata, Dalmanella horderleyensis* and *Brongniartella ascripta*. Higher beds carry similar assemblages, together with *Nicolella actoniae*, suggesting that the Bryn sequence extends into the Woolstonian.

In the western Berwyns, the Teirw, Pandy and Lower Bryn formations pass laterally into a unit of grey siltstones, sandstones and mudstones known as the Cwm Rhiwarth Siltstone Formation. To the south, shales and mudstones become more dominant at the expense of sandstones, and are known as the Allt-fair-Ffynnon Formation (Wedd *et al.* 1929), mostly embracing what have also been named locally as the Llanyblodwel Beds (e.g. Whittington 1938). In the southern Berwyns, above the level of the Cwm Clwyd Tuff, other tuffs and tuffaceous sediments are rare implying dominantly northerly and northwesterly derivation. The Cwm Rhiwarth and equivalent strata are commonly rippled and channelled, with occasional mudstone conglomerates at the base of the channels. Trace fossil and body fossil assemblages generally match those of the northern and north-western Berwyns.

The succeeding Longvillian Pen-y-garnedd Formation comprises a heterolithic facies sequence of tuffaceous siltstones and sandstones, pebbly sandstones, dark grey mudstones and thin limestones (Brenchley 1978, p. 151). Conodonts from the limestones contain *Amorphognathus superbus* Biozone faunas (Savage & Bassett 1985), and there are fairly abundant mixed brachiopod–trilobite assemblages.

Brenchley and Pickerill (1980) interpreted the whole of the Soudleyan marine sequence as being of shallow subtidal origin, deposited in water depths of less than 25 m. Regular bedded facies were formed in relatively low energy environments affected periodically by storm turbulence which introduced rapidly deposited sheet sands. Irregularly layered, mottled, and homogeneous lithofacies are interbedded with these storm affected sediments repeatedly and randomly. The complete suite suggests that the sediments were deposited on the lower shore face or inner shelf of the shallow sea that covered northern Wales, with longshore currents flowing approximately along the east–west striking regional palaeoslope. The sediment was supplied almost entirely from intrabasinal sources. Similar conditions continued into Longvillian times, when current-affected sediments of the Bryn Siltstone Formation passed offshore into lower energy bioclastic and carbonate environments represented by the Pen-y-garnedd Formation. Here assemblages of the large lingulide brachiopod *Lingulasma tenuigranulata* are commonly found in life position (King 1928, Pickerill 1973).

At other levels the Caradoc faunas tend to occur as four regular associations, reflecting depth, turbulence, and rates of sedimentation (Pickerill & Brenchley 1979). Trace fossils are also common throughout (Pickerill 1977). Thus the *Dinorthis* Community lived in relatively high energy, inshore volcaniclastic sands, with the laterally contiguous *Macrococlia* sub-community existing on a preferred substrate of coarse silt and laminated fine sand. *Howellites, Dalmanella,* and *Nicolella* communities occupied increasingly offshore sites, with the latter two exhibiting slightly higher taxonomic diversities because of their more stable environmental settings (see Pickerill & Brenchley 1979, fig. 7).

Above the Longvillian successions in the northern, western and southern Berwyn Hills the Caradoc is capped by black or dark grey, sparsely graptolitic shales along with some bioclastic limestones containing phosphatic nodules and ooids; this Blaen-y-cwm Formation (Wills & Smith 1922, Brenchley 1978), is cut out across the north-western Berwyns (Fig. 4.10). Poorly preserved graptolites indicate a *Pleurograptus linearis* Biozone age (King 1923; Wedd *et al.* 1929) and these beds thus represent a part of the widespread Onnian Nod Glas facies (Cave 1965). They rest disconformably on the underlying Longvillian rocks, with much of the intervening upper Caradoc missing. As elsewhere across North Wales a reducing environment and sediment starvation is inferred for the Nod Glas.

Similar onshore–offshore facies patterns to those of the Caradoc are preserved in the Ashgill sediments across the Berwyns, which as elsewhere across North Wales are markedly unconformable on the older strata. The marked contrast is an absence of tuffs and volcaniclastic sediments throughout the Ashgill formations. The succession is best known in the Glyn Ceiriog area in the north-east, where the Glyn Valley Group spans the late Cautleyan, Rawtheyan and Hirnantian in mixed developments of mudstones and siltstones with subordinate sandstones and locally developed limestone members (Groom & Lake 1908, Wedd *et al*. 1927, Wills & Smith 1922, Hiller 1980, 1981). The Dolhir Limestone Member at the base of the Dolhir Formation in the east is a fine-grained muddy biosparite which passes westwards into the highly cleaved fissile mudstone of the Tyn-y-twmpath Member. Both grade upwards into terrigenous units with numerous thin sandstones containing rich shelly faunas in transgressive shelf facies. The thin arenites appear to represent storm events (Hiller 1981, p. 195). Southwards and south-westwards the Dolhir Formation passes into finer grained sediments of the Tre-Wylan Formation, containing the *Phillipsinella parabola–Staurocephalus clavifrons* trilobite fauna in the lower beds (e.g. King 1923, 1928, Whittington 1938), indicative of a Cautleyan age by comparison with the type Ashgill faunas of northern England (Price 1973, p. 539).

Upward passage from the Dolhir Formation into the Glyn Formation involves the incoming of coarse sandstones, grits and locally developed bioclastic limestones representing a regressive regime associated with the transition from open shelf to shoreline facies (Hiller 1981). These beds contain *ordovicicus* Biozone conodont faunas (Savage & Bassett 1985) together with rich coquinoid assemblages of brachiopods and trilobites of the *Hirnantia* fauna, including *Hirnantia sagittifera, Eostropheodonta hirnantensis*, and *Dalmanella testudinaria*. The brachiopod *Hindella* is a typical component of the sand facies in the Hirnantian of this part of the Berwyns, where deposition took place within wide channels developed near the shoreface. These associations passed laterally to the west and south-west into outer shelf facies in which *Hirnantia* itself became more common in mixed heterolithic silt and mud facies (Brenchley & Cullen 1984).

Throughout much of the north Berwyn Hills the uppermost Ordovician sediments are overlain with apparent conformity by lowest Silurian transgressive mud deposits containing *acuminatus* Biozone faunas (e.g. Wedd *et al*. 1927, p. 52; Hiller 1981, p. 189), although elsewhere Silurian sandstones sit unconformably across the Ashgill (e.g. King 1923, 1928; Whittington 1938).

## Central Wales Inliers

In the centre of the Welsh trough, uppermost Ordovician sediments occur in a series of inliers within elongated *en echelon* domes on both flanks of the axial line of the Central Wales Synclinorium (Fig. 4.2). The four main areas are referred to as the Llyfnant (or Machynlleth), Plynlimon, Gias and Van (or Clywedog) domes, each bounded by conformably succeeding Silurian strata, as are the smaller inliers

adjacent to the principal structures. The exposed Ordovician formations together constitute the Plynlimon Group of O. T. Jones (1909; a synonym of his later Van Formation – O. T. Jones 1922).

Interbedded flaggy mudstones (including hemipelagites) and coarse siltstones with local sandstones make up the Nant-y-Môch Formation at the base of the sequence, cropping out mainly in the Plynlimon [Pumlumon] inlier (O. T. Jones 1909, James 1971a) but also in a restricted belt along the faulted axial zone of the Llyfnant inlier (Cave & Hains 1986, fig. 3; James 1987, fig. 3). Graptolites from low in the formation include *Dicellograptus anceps* and *Orthograptus truncatus*, indicating a late Ashgill age for these oldest beds. A central, mudstone-dominated facies of about 12 m separates lower and upper facies of rhythmically alternating arenites and mudstones, with the percentage of sands and silts being greater in the lower unit. The facies assemblages are taken to indicate pulsatory incursions by turbidites from outer margins of submarine fans into the anoxic mud environments of the area, with only limited bottom scouring and tractional reworking. In the lower and middle facies the currents came dominantly from the SE, whilst in the upper unit the flow directions suggest a NNE source (James 1971a, fig. 1, p. 181; Cave 1980, 1984). Within the otherwise virtually unfossiliferous Nant-y-Môch Formation, burrows and trails of *Nereites* ichnofacies support the interpretation of deposition in relatively deep water (e.g. see Crimes 1970a).

Above the Nant-y-Môch turbidites there is a succession of chloritic and sericitic to feldspathic mudstones with a number of prominent sandstone units up to 10 m thick; rip-up clasts from underlying beds are common (Cave & Hains 1986, p. 10). These beds form the Drosgol Formation, which also occupies the core of the Gias and Van inliers (O. T. Jones 1909, W. D. V. Jones 1945, James 1983b); the higher greywackes and siltstones of the Drosgol Formation form the Pencerrigtewion Member, in which penecontemporaneously disturbed beds are common (Cave & Hains 1986). In each of these areas the succeeding Bryn Glâs Formation comprises mainly thick-bedded, dark mudstones with thin, sporadic rafts of distorted arenites suggestive of density flow within unlithified sediments. To the NW in the Llyfnant inlier, the lateral equivalents of both the Drosgol and Bryn Glâs formations are entirely in arenite and rudite dominated facies with interbedded siltstones and mudstones, assigned to a single stratigraphical unit, the Garnedd-Wen Formation (see James & James 1970, James 1972 for summaries).

The Drosgol arenites appear to represent a complex of relatively discrete fans, those at Plynlimon derived mainly from the south and those at Gias and Van from the east and ESE; the tops of some fans were reworked by distal turbidites represented by ripple-laminated siltstones, suggesting a more northerly flow in the east and a swing from north to north-westerly directions along the length of the Plynlimon dome. Inner-fan channel deposits locally preserve traces of levees.

Rudites and coarse arenites become increasingly common westwards from Plynlimon and into the Llyfnant inlier. James (1971b, 1972, 1983a) described a complex of nine main facies with a variety of bed forms and structures, for this area. Episodes of rapid deposition at least partly as intrabasinally generated fluxo-turbidites and mass-flows in slope-base environments were inferred (James 1987). Interbedded finer arenites, siltstones and mudstones represent turbidites from more distal sources. Palaeocurrent data suggest two principal transport components, towards the NNE and the WNW, the latter apparently associated with local slopes within the otherwise gentle gradients of the floor of the trough.

Apart from the *anceps* Biozone graptolites of the lower Nant-y-Môch Formation, no body fossils have been reported from other units of the Plynlimon Group, but throughout the area, within a metre or so above the top of the Bryn Glâs and Garnedd-Wen formations, there is a remarkably persistent mudstones horizon containing *Glyptograptus persculptus* and other graptolites of this biozone (O. T. Jones 1909; O. T. Jones & Pugh 1916, 1935; W. D. V. Jones 1945). The junction between the Plynlimon Group and the overlying Eisteddfa Formation (synonym Cwmere Formation) of the lower Ponterwyd Group is sharp but conformable, involving the abrupt replacement of rapidly deposited arenitic facies by mottled (bioturbated) and finely laminated mudstones; the latter represent distal muddy turbidites. Phosphatic concretionary layers are developed at some levels, and the intensity of bioturbation in the muds appears to indicate long pauses between turbidity flows (Cave 1980, p. 521); the burrows are referred to *Chondrites* (Cave & Hains 1986, p. 42). The upper part of the Eisteddfa Formation grades into Silurian grey and pyritous mudstones and shales of the *acuminatus* Biozone, with *P. acuminatus* itself a common component of the faunas. The dark, pyritous *persculptus* mudstones probably reflect deposition under anoxic conditions, following which the mottled beds represent early flushes of distal turbiditic muds heralding the main phases of early Silurian transgression.

## Southern Wales

The almost unbroken, arcuate belt of Ordovician rocks running from the south-west tip of Wales to Builth Wells (Figs. 4.1, 4.2) closely defines the southern and south-eastern marginal zone of the Welsh trough. Changes from platform to trough sedimentary facies through at least part of this zone were controlled by incipient folding and probable faulting along the Tywi lineament in later Ordovician times (D. A. Bassett 1969, Woodcock 1984b), whilst intense volcanism concentrated at both extremities of the outcrop testifies to persistent tectonic turbulence throughout the Period.

## Ramsey-Fishguard-Llangrannog

Upper Arenig to Llanvirn volcanic, volcaniclastic and high-level intrusive rocks dominate the succession exposed along much of the north Pembrokeshire coast from Ramsey Island to Strumble Head near Fishguard (Fig. 4.2). Regional background sedimentation to the volcanic activity was largely in black mud, low energy environments, punctuated by repeated turbidite emplacement. Volcanism, dominant in the early Llanvirn, was concentrated on two major centres, around Fishguard and on Ramsey, with a further minor centre near Abereiddi (Kokelaar *et al.* 1984a, b), although tuffaceous sediments at various horizons probably represent the feather edges of yet other accumulations that are no longer preserved.

On Ramsey, there are markedly different successions to the east and west of the N–S-trending Ramsey Fault. To the west there are no Ordovician rocks of pre-Llanvirn age, but to the east there is a thick basal Arenig sequence sitting unconformably on upper Cambrian sediments recording a transgressive upward-deepening episode. Pebbly sandstones with '*Bolopora undosa*' in the Ogof Hên Formation, grade up into laminated sandstones and cross laminated siltstones containing burrows and trilobite trails. Succeeding mudstones and siltstones contain a rich brachiopod and trilobite fauna of early Moridunian age including *Merlinia murchisoniae*, *Neseuretus ramseyensis*, *Monorthis menapiensis* and *Paralenorthis alata*, together with the distinctive early crinoid *Ramseyocrinus cambriensis* (Bates 1969, Kokelaar *et al.* 1985). The Ogof Hên Formation fauna found eastwards on the mainland as far as the Carmarthen area (Fortey & Owens 1987) is comparable; the continued presence of the *Neseuretus–Paralenorthis* biofacies in coarse arenites similarly defines the progression of early Arenig transgression.

Mid-Arenig (Whitlandian) rocks are absent on Ramsey, probably cut out erosively by later, Llanvirn age wet-sediment sliding (Kokelaar *et al.* 1985), but Fennian deposits are well developed. Strongly cleaved black mudstones of the Road Uchaf Formation represent the establishment of anoxic offshore conditions which were to persist generally through the subsequent episodes of volcanism, with only local, volcanotectonically induced shallowing and emergence. Rich graptolite faunas of the *gibberulus*

Biozone occur in these beds (Jenkins 1982). Upward transition into the Aber Mawr Formation takes place within the *gibberulus–hirundo* interval (late Fennian), with the dark argillites giving way to successions of rhyolitic turbiditic tuffs and cleaved black mudstones, often showing evidence of penecontemporaneous wet-sediment slides. The vitric and crystal tuffs can contain mudstone intraclasts and flute casts produced by turbidity currents; some are laminated and convoluted, while others are finely laminated. The tuffaceous deposits are presumably distal products of a late Arenig to early Llanvirn episode of explosive volcanism elsewhere in the region. Pencil slates within the Aber Mawr Formation contain abundant extensiform graptolites (*Expansograptus*) together with rare pendent *Didymograptus* and trilobites of the *Bergamia rushtoni* Biozone (Kokelaar *et al*. 1985, Fortey & Owens 1987). Dark mudstones at the top of this formation contain rare graptolites of the early Llanvirn *artus* Biozone.

Major, proximally emplaced, rhyolitic volcanic and volcanigenic successions related to tectonism dominate the remainder of the Llanvirn succession exposed on Ramsey Island. As with the underlying Ordovician, the Porth Llauog Formation is confined to the east of the Ramsey Fault, where it comprises volcanic debris-flows interbedded with mudstones, minor siltstones, and distal turbiditic tuffs. The succession is interpreted (Kokelaar *et al*. 1985) as a product of synsedimentary acid volcanotectonism in which activation of the Ramsey Fault, and probable similar faults to the east, led to large-scale sliding and emplacement of cohesive flows from uplifted western areas. Rhyolitic pebbles in the flows are well rounded as a result of shoreline erosion of a nearby emergent rhyolite block.

Uplift along the Ramsey Fault probably exceeded 1 km, leading to emergence and catastrophic erosion of the rhyolite volcanic island, followed closely by sub-mergence to depths of up to 500 m in which the Carn Llundain Formation then accumulated both in the east and west as an unconformably overlying succession of subaqueous rhyolitic ashes, etc., rhyolite lavas and a wide variety of debris-flow deposits. It appears that the Ramsey Fault acted as a channelway for magma. Intrusions (mainly acidic), were the direct cause of further slumping and sliding in overlying volcaniclastic sediments; some show evidence of emplacement into the fluidisation of unconsolidated wet sediment (Kokelaar 1982, *et al*. 1985). Black mudstones within the Porth Llauog succession contain trilobites and graptolites of the *artus* Biozone and the Carn Llundain volcanics were probably emplaced entirely during early Llanvirn times.

In the Abercastle–Abereiddi area of the mainland coast adjacent to Ramsey Island the Ordovician succession again begins with arenites of the lower Arenig

Ogof Hên succession resting with transgressive unconformity on upper Cambrian sediments; the full stratigraphical extent of Moridunian deposits in this area remains uncertain in the absence of diagnostic faunas (Fortey & Owens 1987). Overlying beds of the Abercastle Formation, which include the well-known Porth Gain Beds of Cox (1916), comprise sandy and micaceous mudstones with feldspathic turbiditic grits near the top and contain the Whitlandian trilobite *Ogyginus hybridus*. The succession indicates deepening upwards through the Penmaen Dewi and Aber Mawr Shale formations, mostly silty shales with dark calcareous nodules and thin tuffaceous horizons. The former contains a trilobite fauna in its upper part, belonging to the *Gymnostomix gibbsii* Biozone (Fortey & Owens 1987), whilst in the latter there are graptolites of the *artus* Biozone (Hughes *et al*. 1982), indicating an Arenig–Llanvirn transition across this interval. The remaining Llanvirn succession here (the historical type area for the Series (Hicks 1881)) is then dominated by the thick Llanrian Volcanic Formation (Cox 1916), mostly of *artus* Biozone age but whose upper member (the Abereiddi Tuff = Murchisoni Ash of Cox) occurs in the late Llanvirn *murchisoni* Biozone. The lower member of the Llanrian volcanics comprises mainly thick rhyolitic tuffs and tuff-aceous volcaniclastic deposits; crystal–vitric tuffs and volcaniclastic sandstones and siltstones predominate, deposited apparently from sediment gravity-flows (M. G. Bassett *et al*. 1986).

Intervening black graptolitic shales separate the lower Llanrian volcanics from the Abereiddi Tuff Member, which records the youngest Ordovician volcanic episode preserved in south-west Wales. By contrast with the underlying successions these tuffs are basaltic in composition. The coarser, more proximal deposits are debris-flows, but more distally and upwards they are succeeded by turbidites; they represent slumping from the unstable flanks of a tephra pile during eruption. The pile was eventually mantled by pelagic muds which also slumped periodically to form mixed tephra/mud accumulations (M. G. Bassett *et al*. 1986, Kokelaar *et al*. 1985).

Above the Llanrian volcanics dark, the tuffaceous lower Caerhys Shale Formation yields a rich graptolite fauna of upper Llanvirn *murchisoni* Biozone (Hughes *et al*. 1982). The shales become silty upwards (as distal turbidites) where there is a transition into the *teretiusculus* Biozone and there are shelly faunas of brachiopods, bivalves and trilobites such as *Protolloydolithus*, *Geragnostus*, *Ogygiocarella* and *Platycalymene*. Near the base of the Castell Limestone there is a fauna of gastropods, trinucleid trilobites and graptolites (*Dicellograptus*, *Pseudoclimacograptus*, *Orthograptus*, *Isograptus*), which may indicate the presence of the *gracilis* Biozone, although the first proof of this level is in shales immediately above the Limestone where *Nemagraptus* and *Diplograptus foliaceous* occur together with *Trinucleus* cf. *fimbriatus*. Within the micritic and silty carbonates of the Castell Limestone itself trinucleids are abundant, including both *T. fimbriatus* and *Talaeomarrolithus* cf. *intermedius*, which are also suggestive of a *gracilis* Biozone age (Black *et al.* 1971, Hughes *et al.* 1982). Low diversity conodont faunas from this level include *Amorphognathus tvaerensis* together with other elements taken to correlate with the lower part of that biozone and with the upper part of the Llandeilo Series in the Llandeilo area to the east (Fig. 4.3; Bergström 1964, Bergström *et al.* 1987).

The undifferentiated dark silty and pyritous shales above the Castell Limestone (Dicranograptus Shales of authors) probably extend well up into the Caradoc Series (Hughes *et al.* 1982). They mark a continuation of low energy mud deposition and transgressive deepening following the cessation of Llanvirn volcanicity.

Repetition of the volcanic-dominated successions takes place north-eastwards, where the Fishguard Volcanic Group represents the third of the eruptive centres in this region of south-west Wales, centred around Strumble Head and Fishguard itself and extending eastwards into the Preseli [Prescelly] Hills (Fig. 4.2). Shales within the Fishguard volcanics contain graptolites dated as early upper Llanvirn, including *Didymograptus murchisoni* and *D. artus* together with species of *Diplograptus* and *Glyptograptus* (Lowman & Bloxam 1981).

The entire Fishguard complex was emplaced subaqueously as a succession of acid to basic lava flows and volcaniclastic deposits, in which it is possible to recognise a tripartite lithostratigraphical subdivision (Thomas & Thomas 1956; see also Bevins 1982; Bevins & Roach 1980, 1982; Bevins *et al.* 1984; Kokelaar *et al.* 1984a; M. G. Bassett *et al.* 1986). The oldest, Porth Maen Melyn Formation comprises principally rhyolitic lavas, autobreccias, debris-flow breccias and bedded and massive tuffs, although rhyodacitic massive and pillow lavas also occur.

The overlying Strumble Head Volcanic Formation comprises mostly basaltic pillowed (Fig. 4.11) and massive lavas with only thin hyaloclastic and rare

**Fig. 4.11.** Basaltic pillow lavas in the Strumble Head Volcanic Formation, Pencaer, west of Fishguard, Dyfed; long axes of the pillows average 75 cm.

rhyolitic tuffaceous horizons. Vesicular and graded pumice also occurs. Eruption of these basic magmas took place relatively quietly on the sea floor, possibly along faults defining a graben structure (Kokelaar et al. 1984b). Upwards there is interdigitation of the Strumble Head lavas with the predominantly rhyolitic volcanics of the overlying Goodwick Formation, which includes rhyolitic lavas and associated autobreccias, tuffs, etc., bedded basaltic tuffs, and in the highest units, basalts that were intruded at a high level into wet silicic tuffs and tuffites, with accompanying contemporaneous sediment deformation.

Although the successions in the Fishguard volcanics to the west of Fishguard show a broadly repetitive acid–basic alternation, both magma compositions were generally available simultaneously (Kokelaar et al. 1984a). However, eastwards from Fishguard and into the Preseli Hills (Evans 1945, Lowman & Bloxam 1981) there are differences in the nature of the laterally equivalent volcanic complexes, with little evidence of basic flows; thick rhyolitic flows and domes are now dominant throughout, with an increase of mass flow deposits including subaqueous, often welded pyroclastic flows of a type not seen farther west.

Numerous high-level doleritic and gabbroic intrusions occur and were clearly coeval with the volcanic activity (Bevins 1982, Bevins & Roach 1980, Bevins et al. 1989). The intrusions, which also include diorites, microgranites and microtonalites, are especially common and thicker eastwards (Lowman & Bloxam 1981). Similar intrusions also extend well to the west beyond the outcrop of the Fishguard Volcanic Group including the gabbroic complexes of Carn Llidi and St David's Head (Roach 1969, Bevins & Roach 1982) and those of the Bishops and Clerks Islands beyond Ramsey (Bates et al. 1969).

Bevins (1982), Bevins et al. (1989) concluded that the variety of rock types within the Fishguard volcanics can be accounted for primarily by low pressure fractional crystallisation of a parental tholeiitic magma, derived by partial melting of mantle material and with a minor subduction-derived component. The complete complex developed in an extensional setting, influenced by contemporaneous graben or half-graben fault structures.

In sharp contrast to the successions south-west of Fishguard, those exposed in further magnificent coastal sections north-eastwards along Cardigan Bay as far as Llangrannog (Fig. 4.2) almost completely lack volcanic rocks apart from irregularly occurring distal tuffs; inland the exposure is poor. Turbiditic arenites and mudrocks make up the complete sequence.

The Fishguard Volcanic Group is probably entirely of Llanvirn age, though palaeontological evidence is lacking in the above arenaceous and calcareous flags

(Castle Point Beds of Cox 1930). The latter are succeeded by shales (Dicranograptus or Hendre Shales of authors) which yield graptolites including Nemagraptus gracilis indicative of that Biozone (Lowman & Bloxam 1981). A (? early) Llandeilo age is likely for the whole of this part of the sequence, as somewhat younger shales at Newport also contain a restricted graptolite fauna at about the teretiusculus–gracilis boundary in the middle-upper Llandeilo (Myers 1950). The latter fauna is important in confirming that the conformably continuous succession upwards onto Dinas Head is of Ordovician and not early Silurian age as interpreted by Reed (1895, plate 5).

Only limited faunal data are available from many of the coastal sections because of inaccessibility, but the sequence clearly passes up into the Caradoc towards Cardigan (Keeping 1881, 1882), accompanied by a fairly abrupt transition from Dicranograptus Shale facies below to thick, arenite-dominated associations above. At Poppit Sands, James (1975) described a sedimentary complex interpreted as showing a transition from mass-flow deposits to immature turbidites high on a middle fan environment. Two levels in the sequence are notable for the rare occurrence of dish structures (J. R. L. Allen 1981) Channel orientations within the turbidites complex are variable although more limited evidence from flute moulds suggests derivation from the south-west (James 1975).

Similar turbidites (Fig. 4.12) and mass-flow deposits also continue north-eastwards above the Caradoc and through the Ashgill Series into the Silurian, where there is a conformable transition in the sections around Llangrannog (Hendriks 1926, Anketell 1987). There the mudstones and siltstones of the Tresaith Formation contain pyritised graptolites in the upper beds including Orthograptus abbreviatus and Climacograptus miserabilis, indicating an age no younger than the anceps Biozone. The overlying Llangranog Formation displays rapid facies and thickness variations in sandstones, siltstones, mudstones and gritty mudstones, with numerous sedimentary structures resulting from turbidite, slump, and mass-flow deposition. Anketell (1987) described five members through the Formation, none of which has yielded diagnostic faunas, although rich assemblages of deep water trace fossils occur locally. In contrast, the succeeding Gaerglwyd Formation contains abundant, well-preserved graptolite faunas, with Glyptograptus persculptus occurring in dark pyritic mudstones close to the base and proving a late Hirnantian age; orthoconic nautiloids also occur in this part of the sequence (Hendriks 1926). Higher beds comprise mottled mudstones, finely laminated dark mudstones which weather to a distinctive yellow-orange colour, rare sandstones, and bands of calcareous nodules. Faunas from these units include *Parakidograptus acuminatus* with associated graptolites, confirming an earliest Silurian age and indicate a conformable Ordovician–Silurian boundary succession within the Gaerglwyd Formation (Anketell 1987).

## Haverfordwest-Narberth-Whitland

An extensive inland tract of Ordovician rocks occupies much of south-western Wales, running northwards into the Preseli Hills and beyond (Fig. 4.2), though exposure is generally poor. Nevertheless, a composite succession of dominantly heterolithic facies is known to occupy almost the complete extent of the Series and, by comparison with the Ramsey–Fishguard district volcanics are thinner.

At Trefgarn [Treffgarne] to the north of Haverfordwest, the oldest dated Ordovician sediments are underlain by basaltic andesites, andesitic lavas and volcaniclastic sediments of the Trefgarn Volcanic Formation, whose age remains equivocal. Thomas & Cox (1924) and Bevins *et al.* (1984) noted the petrographic similarities with the Rhobell Volcanic Group in North Wales, of Tremadoc age (see p. 81), and a comparable age and genesis are probable, contrasting with the extensional origin of the thick Arenig–Llanvirn volcanics of the coastal region. Epiclastic turbidites and debris-flow deposits directly overlying the Trefgarn volcanics are unfossiliferous and the contact may not be conformable (although a transitional succession was described by Thomas & Cox 1924), but then these in turn are succeeded by blue-grey tuffaceous, cross-laminated arenites and micaceous shales containing dendroid and extensiform graptolites. These beds pass laterally eastwards into beds yielding trilobites such as *Ogyginus hybridus* of the early middle Arenig (Whitlandian) Blaencediw Formation (Fortey & Owens 1987). Lava fragments of Trefgarn type as well as clasts from the Precambrian basement occur in the Blaencediw facies, which represent immature transgressive tidal sheets and muds that were generated in part by storm-triggered flows and then modified subsequently by wave activity (Traynor 1988). The Blaencediw beds extend well to the east beyond Whitland, where they pass laterally into the finer turbidites and shales of the Afon Ffinnant

**Fig. 4.12.** Slump folds in proximal Ashgill turbidites, Llangrannog, Dyfed.

Formation (see below; Fortey & Owens 1987). Early Arenig tectonism in the Haverfordwest area probably led to the emergence of one or more fault-bounded basement and volcanic blocks which were transgressed and eroded rapidly to provide the source of much of this material across south-west Wales.

By later Whitlandian times, arenite deposition was replaced by upward-deepening mud and silt-dominated regimes, now forming the beds of the Colomendy Formation. *O. hybridus* continues to be common in the lower beds, together with articulate brachiopods and other diagnostic trilobites such as *Gymnostomix gibbsi* and *Bohemopyge scutatrix*, but faunas are rarer in the upper mud-dominated facies. In the succeeding Cwmfelin Boeth Formation (the base of the Fennian Stage) coarse, well-graded turbidites re-appear, with rounded clasts of rhyolite, tuffs and mudstones, separated by unfossiliferous black shales. The turbidites reflect tectonic reactivation of a southern source area (Traynor 1988), but within the Fennian there was a rapid return to offshore, low energy, hemipelagic mud deposition as seen in the dark shales and blocky mudstones of the Pontyfenni Formation. This facies represents the most uniform and widespread open-sea conditions across south-west Wales in Arenig times, with a particularly distinctive association of trilobites that include large-eyed free swimming pelagic species (e.g. cyclopygids) and blind or nearly blind genera including *Ormathops, Illaenopsis, Dindymene, Ampyx* and *Bergamia*. This latter, benthic group, constitutes the atheloptic assemblage of Fortey & Owens (1987, fig. 13), adapted specifically for dwelling and burrowing in soft muds at water depths of *c.* 200 metres. Evidence for open-sea environments is supported further by graptolites such as *Pseudotrigonograptus* and *Isograptus*, which do not occur in onshore facies (Jenkins 1982, Fortey 1984, Fortey & Owens 1987). The trilobites *Stapeleyella abyfrons* and *Segmentagnostus whitlandensis* in the basal Pontyfenni Formation indicate the presence of the *abyfrons* Biozone, with the higher beds assigned to the *Bergamia rushtoni* Biozone; the graptolites indicate that the Fennian Stage spans at least part of the *Isograptus gibberulus* and all of the *Didymograptus hirundo* biozones.

The overlying Llanfallteg Formation is also of Fennian age at its base, but within the succession of light grey shales and mudstones there is then an abrupt incursion of pendent didymograptid graptolites of the *Didymograptus artus* Biozone marking the base of the Llanvirn Series. The basal, Arenig portion of the overlying mudstones and shales of the Llanfallteg Formation comprises the *Dionide levigena* trilobite Biozone. The lithological and faunal changes at this level within the uppermost Arenig are attributed to falling sea level as a result of eustatic regression

(Fortey 1984, Fortey & Owens 1987). An abrupt incursion of graptolites of the *Didymograptus artus* Biozone marks the base of the Llanvirn Series. Slightly higher beds contain numerous tuffaceous chinastones, probably derived in part from the thick volcanics to the west, and then through the remainder of the Llanvirn renewed transgressive deepening produced dark mud environments over most of the region (the 'Bifidus' and Murchisoni Shales of authors). Westwards from Whitland–Narberth, however, the upper beds of this Llanvirn succession, mainly mudstones with occasional sandstones and tuffs, are apparently cut out beneath overstepping Llandeilo sediments, so that no Murchisoni Shales crop out around Haverfordwest (Strahan *et al.* 1914).

To the north and east of Whitland, the upper Llanvirn–Llandeilo facies link directly with those of the contiguous St Clears–Carmarthen area (Evans 1906. Strahan *et al.* 1909, see below, p. 94), with the Asaphus Ash to Hendre Shale succession reflecting persistent offshore deposition. Along the southern outcrops and to the west, however, as in the Narberth district (Fig. 4.2), thinner carbonate facies are prominent in the Llandeilo sequence, representing more onshore platform environments and inviting closer comparison with the type Llandeilo area beyond Carmarthen. The lowest Llandeilo is generally poorly exposed or cut out by faulting, but the upper beds comprise argillaceous flaggy bedded limestones and thicker blue-grey calcarenites that together contain significant assemblages of trilobites, brachiopods and conodonts. In the basal units of this Narberth Group (Addison *in* Williams *et al.* 1972, p. 36; *in* D. A. Bassett *et al.* 1974; Bergström *et al* 1987, fig. 18.7) the presence of the trinucleid trilobite (*Marrolithus favus*) denotes a close correlation within the upper Llandeilo of the type area (Williams 1953, Wilcox & Lockley 1981). Accompanying conodonts indicate either the higher part of the *Pygodus anserinus* Biozone or the lower *Amorphognathus tvaerensis* Biozone (Fig. 4.3). The latter age is confirmed for the succeeding limestones which contain *A. tvaerensis* together with *Baltoniodus variabilis* and *Eoplacognathus elongatus*. Similar conodonts continue through the highest limestones of the Narberth Group, but are accompanied by the trilobite *Costonia elegans* and other silicified trilobites and brachiopods indicating an early Caradoc (Costonain) age by direct comparison with the faunal succession in the Welsh Borderland. The fact that the lower part of the type Caradoc sequence also contains a *tvaerensis* Biozone conodont fauna (Savage & Bassett 1986) is a clear indication that the Costonian faunal succession is in part coeval with and overlaps that of the supposed 'Llandeilo limestones' in the Narberth Group. Largely unfossiliferous, dark flags above the Narberth limestones then pass up conformably into black Mydrim [Dicranograptus] Shale facies containing *Nemagraptus gracilis* and other graptolites of that biozone, accompanied by rare inarticulate brachiopods (Strahan *et al.* 1914); these latter beds reflect a shift of offshore and deposition resulting from the widespread *gracilis* transgressive deepening.

Unlike the adjacent St Clears–Carmarthen district to the east, there is no evidence in the Narberth–Haverfordwest area that the Mydrim Shales extend above the level of the *multidens* Biozone. Some of the higher beds contain a restricted brachiopod–trilobite–mollusc fauna in pyritous mudstones and thin limestones, with '*Dalmanella' argentea* being typical at some levels, but for the most part graptolites remain the dominant fauna. Overlying strata are of Ashgill age, sitting unconformably on and diachronously across the Mydrim Shales (Strahan *et al.* 1909, 1914; Price 1973).

The non-depositional/erosional break extended at least from the mid to upper Caradoc into the lower Ashgill (Cautleyan), when renewed sedimentation then took place as carbonate mud and sand facies along the southern tracts of the present outcrop area. These facies of the Robeston Wathen and Sholeshook limestones mark a reversion to relatively onshore platform deposition by comparison with the Mydrim Shales, essentially repeating the environments represented by the Llandeilo limestones. The Sholeshook Limestone Formation (Marr & Roberts 1885, Price 1973) comprises variegated dark-bedded to nodular limestones with interbedded shales, siltstones and calcareous mudstones, carrying a fairly rich shelly fauna dominated by trilobites, brachiopods, corals and bryozoans. Species of *Tretaspis* confirm a Cautleyan age for these beds (Price 1973, fig. 6). The locally developed Robeston Wathen Limestone, also of Cautleyan age, directly underlies a thin development of Sholeshook mudstone and siltstone facies, from which it differs lithologically in comprising alternating limestones (with halysitid corals) and thinner shales.

The presence of *Tretaspis* cf. *granulata* and associated trilobites in both the lower Sholeshook Limestone and in the lower Slade and Redhill Mudstone Formation (Price 1973), to the north suggests a progressive transition to offshore mud facies from south to north. Southerly progradation of the mud facies in response to upward deepening is then indicated by the Slade and Redhill beds succeeding the Sholeshook carbonates throughout their outcrop, with this diachronous junction dated everywhere within the upper Cautleyan. In the depositional interval of the Slade and Redhill Mudstone Formation, lithologies in southerly outcrops are dominantly mudstones with micaceous and highly calcareous sandstones, con-

trasting with the northerly developments of uniform mudrocks (Strahan *et al.* 1914). Trilobites and brachiopods indicate that the top of the formation is of early to mid Rawtheyan age (Cocks & Price 1975).

In the overlying Portfield Formation, a basal member of dark shales probably is then followed fairly abruptly by sandstones, micaceous siltstones and shales, with feldspathic grits and felsitic conglomerates containing pebbles and cobbles of vein quartz (Strahan *et al.* 1914). Although fossils are lacking, a later Rawtheyan–early Hirnantian age is inferred for this succession (Cocks & Price 1975, table 1). The presumed source of the clasts is the northern margin of the southerly landmass known as Pretannia (Cope & Bassett 1987) which was exhumed by the late Ashgill glacio–eustatic fall in sea level following the global high stands of Rawtheyan times (Brenchley & Newall 1984). Similar emergent areas were widespread at this time around the platformal margin to the Welsh trough (see also below, p. 94). The Portfield Formation arenites and rudites were probably transported outwards and reworked across the platform in shallow marine environments as sea level fell. Overlying beds reflect the re-introduction of mud deposition and a new regime of mid-Hirnantian post-glacial transgressive deepening.

Above the Portfield Formation, olive-green bioturbated mudstones, shales and siltstones of the lowest Haverford Mudstone Formation (Cocks & Price 1975) contain a typical *Hirnantia* fauna of late Ashgill age. Subtidal platform environments are indicated across the whole area. Dating of the middle part of the Haverford Mudstone Formation is imprecise but in the upper mudstones and sandstones there is a rich fauna of early Llandovery age, indicating a conformable Ordovician–Silurian boundary transition in this lithostratigraphical unit.

Outside the main Ordovician core extending eastwards from Haverfordwest there are also relatively poorly exposed to poorly known successions running northwards to the Preseli Hills and in isolated coastal outcrops to the south. In the former case the sequence above the Trefgarn volcanics includes the Sealyham Group (Thomas & Cox 1924, W. D. Evans 1945) in which the upper, volcanic formation comprises rhyolitic and trachytic lava flows the tuffs of early Llanvirn age, approximately coeval with and related genetically to the volcanics on Ramsey Island; otherwise the succession forms a link between that of the Fishguard district and that running eastwards towards Carmarthen. By contrast, on the coast to the south, the outcrops are widely separated in fault blocks and are of restricted stratigraphical extent. South of Milford Haven there is only a thin succession of dark, finely micaceous Llanvirn shales with subsidiary sandstones preserved in fold cores between

Freshwater West and Freshwater East (Fig. 4.2); exposure is poor, but pyritic beds with *Didymograptus artus* are most extensive, with only limited evidence for development of succeeding beds of the *murchisoni* Biozone (Dixon 1921, M. G. Bassett 1982). The base of this succession is not known, and the beds are overlain conformably by sandstones of Wenlock (Silurian) age. On the north side of Milford Haven, grey tuffaceous flags, quartzites, dark mudstones, shales and impure limestones of early Llandeilo age crop out in a number of discrete localities near Marloes (Fig. 4.2; Cantril *et al.* 1916). Trilobites such as *Basilicus tyrannus*, ogygiids and trinucleids are not uncommon, accompanied by rare dalmanellid brachiopods; as with the Llanvirn argillites to the south, offshore low energy subtidal environments are indicated.

## Carmarthen–St. Clears

Fortey and Owens (1978) confirmed the conclusion of Smith and Stubblefield (1933, p. 374) that along the southern margin of the Tywi lineament in the Carmarthen district (Fig. 4.2), a considerable thickness of sediments long regarded as being of Tremadoc age in fact lies well above the base of the Arenig Series. However, true Tremadoc strata are now known to occur in the region (Cope *et al.* 1978, Cope 1980, Owens *et al.* 1982) and have also been shown to extend offshore under Carmarthen Bay (Brooks & James 1975, Tappin & Downie 1978). In the limited outcrop areas the thickness is unknown, but blue-grey shales appear to be overlain by slightly coarser olive-buff micaceous silty shales. *Rhabdinopora flabelliformis* occurs in the oldest beds exposed, with *Clonograptus tenellus* and *Adelograptus hunnebergensis* among the graptolites in higher shales, suggesting that the complete sequence is confined to these lower Tremadoc biozones. Accompanying faunas include trilobites, inarticulate brachiopods and sponges.

Unfossiliferous transgressive arenites with coarse lag-conglomerates form the basal member of the Arenig Ogof Hên Formation, which formed in sublittoral to very shallow subtidal environments (Fortey & Owens 1978). The igneous, metamorphic and vein-quartz clasts were presumably derived from an adjacent basement source immediately to the south (Cope & Bassett 1988, p. 318), and the considerable overstep by the basal units may imply the existence of a major pre-Arenig fault (Cope 1980, p. 529). Upward grading through fine sandstones and siltstones into silty mudstones and shales of the upper Ogof Hên beds is accompanied by the appearance of the coarsely ribbed orthacean brachiopod *Paralenorthis* and then a *Neseuretus–Merlinia* trilobite assemblage characteristic of shallow inshore

environment (Fortey & Owens 1978, p. 238). These are the earliest Moridunian faunas in the area, correlating closely with those from the same lithostratigraphical level in the coastal sections to the west (Fortey & Owens 1987, p. 96).

Transgressive deepening into offshore, dominantly low-energy mud environments continued through the remainder of Arenig times as reflected in the micaceous shales and mudstones of the Carmarthen Formation and Afon Ffinnant Formation. Thick, graded quartzitic and feldspathic pebbly turbidites occur in the middle of the Carmarthen Formation and there are similar beds in the low-middle Afon Ffinnant Formation, all lacking evidence of strong bottom-scouring but indicating the emplacement of fans across low palaeoslopes; derivation from a Precambrian source to the south is again inferred. As described by Fortey and Owens (1978, p. 238), the background muds of the lower Carmarthen Formation lack the hard-bottom benthos of underlying beds, but are characterised instead by infaunal bivalves and by Raphiophorid Community trilobites adapted for living on soft sediment surfaces. In higher mudstones these associations are replaced by Olenid Community trilobites believed to indicate yet deeper, poorly oxygenated environments.

Dendroid graptolites occur through the upper part of the Carmarthen Formation, and from that level upwards the remainder of the Ordovician in this area is in predominantly graptolitic or in mixed graptolitic/shelly facies. The incoming of the trilobite *Furcalithus radix* indicates that the base of the Whitlandian Stage is just above the base of the Afon Ffinnant Formation, with *Ogyginus hybridus* and then *Gymnostomix gibbsi* occurring as typical components of the Stage through the remainder of this unit. Equivalents of the Colomendy Formation and Pontyfenni Formation farther to the west contain late Whitlandian and Fennian trilobites in scattered outcrops (Fortey & Owens 1987, p. 97). The Llanvirn Artus and Murchisoni Shales are predominantly mud or fine silt-grade sediments with tuffs present throughout; faunas are mainly didymograptid graptolites but the silty and tuffaceous bands contain rich trilobite faunas. In the west of the area a thick tuff up to 12 m thick at the base of the Murchisoni Shales represents the same event as that displayed more fully in the coastal sections around Abereiddi (see above; Strahan *et al.* 1909, p. 36).

Above the Murchisoni Shales, the distinctive Asaphus Ash was long regarded as marking the local base of the Llandeilo Series (e.g. Strahan *et al.* 1909, p. 40; Toghill 1970, 1971), but revised correlation of the latter datum with the graptolite succession (Addison *in* Williams *et al.* 1972, p. 35; *in* D. A. Bassett *et al.* 1974, p. 11) now suggests that this sequence of silicified tuffs falls within the uppermost Llanvirn. Rich shelly faunas, in which asaphid trilobites are dominant, occur in shales and siltstones within the tuffs. Similar flaggy bedded argillaceous and silty units with thin tuffs form the local basal Llandeilo Flags, containing a mixed trilobite/graptolite fauna in which *Glyptograptus teretiusculus* first appears. Tuffs are conspicuously absent through most of the Hendre Shales in which trilobites and graptolites of early to late Llandeilo age are again abundant. Addison's important discovery (*in* Williams *et al.* 1972) of *Nemagraptus gracilis* in the Hendre Shales in association with the trinucleid trilobite *Lloydolithus lloydi* shows that the base of the *gracilis* Biozone begins just below the top of the lower Llandeilo (Fig. 4.3). Deposition through these levels was entirely in quiet offshore environments.

Reduction in the supply of terrigenous muds allowed impure calcareous muds to develop in early Caradoc times, now represented by the Mydrim Limestone which again contains *N. gracilis* together with other graptolites and trilobites. The overlying black pyritous shales and mudstones of the Mydrim Shales contain *gracilis, multidens, clingani* and possibly low *linearis* Biozone faunas (Strahan *et al.* 1909, p. 46; Toghill 1970). Above this level there appears to be a stratigraphical break corresponding to that in the Narberth–Haverfordwest area, with the succeeding Sholeshook Limestone and Slade and Redhill Mudstone facies indicating similar conditions throughout the remainder of the Ordovician.

## Llandeilo

Richly fossiliferous calcareous flags around Llandeilo and in adjacent areas of the Tywi valley (Fig. 4.2) led Murchison (1839) to define there what is now the Llandeilo Series. The succession ranges from the Llanvirn to the Ashgill, although south-westwards Arenig strata are present in facies linked directly to those of the Carmarthen district. Williams's (1953) revision of the stratigraphy emphasised the rapid lateral and vertical changes in lithofacies in response to volcanigenic input and tectonism, and the relationship of faunal patterns to these changes was demonstrated more recently in a number of studies (Lockley & Williams 1981, Williams *et al.* 1981, Wilcox & Lockley 1981, Lockley 1983).

The lower Llanvirn shales with *Didymograptus (Didymograptus) artus* (the 'Bifidus' Shales of earlier authors) accumulated as offshore shelf muds supporting an impoverished fauna of trilobites and rare brachiopods, with didymograptid and glyptograptid graptolites now preserved as the most common biota. Lower and upper shale divisions contain numerous water-lain vitric to crystal rhyolitic tuffs, the upper beds grading upwards into basal grits of the Ffairfach Group.

The five arenite-dominated and tuff formations of the upper Llanvirn Ffairfach Group document three successive regressive cycles in sublittoral to intertidal environments (Williams 1953; Williams *et al.* 1981, p. 673). Virtually unfossiliferous, winnowed, arkosic, basal grits succeed the earlier Llanvirn shales. They reflect shallow conditions and thicken sourcewards towards the Builth volcanic complex (see Goldring 1966). Thin shale beds towards the top of the Ffairfach Grits and finer, bioturbated and occasionally rippled sands with argillaceous laminae in the overlying Pebbly Sands Formation suggest a reduced rate of deposition, probably on and between subtidal or intertidal lenticular sand bodies. Shelly faunas become more common and lithologies coarsen upwards through the Pebbly Sands. Thin volcanic clays occur both in the higher beds and in the following finer siltstones and shales of the basal Flags and Grits Formation. These shales mark the end of the first depositional cycle, followed by renewed increase in arenite deposition culminating in conglomeratic grits whose composition and structure is consistent with derivation by the reworking of various lavas, pyroclastics, and banks and bars of rhyolitic ash (William *et al.* 1981, p. 673). Calcareous, fossiliferous siltstones are present throughout the Flags and Grits, which terminate as limestone beds.

The Ashes and Lavas Formation is formed mainly of crystal tuffs with arenaceous, argillaceous and agglomeratic interbeds. The sharply succeeding Rhyolitic Conglomerate Formation comprises conglomeratic, locally bioturbated sandstones with fine argillaceous and shelly calcareous partings. As throughout the complete Ffairfach Group, the argillaceous units probably do not indicate deepening but protected, low energy environments among ashy banks and bars. Eleven distinctive shelly faunal assocations were described by Williams *et al.* (1981) from the type Ffairfach sequence, two of them recurring five times in response to the changing substrate and prevalent sedimentary regime.

Fine sandstones at the base of the type Llandeilo flags mark the initiation of a major, fining upward transgressive cycle following the cessation of Llanvirn volcanism (Williams 1953, p. 202; Williams *et al.* 1981, p. 673). Repetitive alternations of flags and impure limestones, with some thin shales and sandstones, contain prolific, almost exclusively shelly faunas, of which Williams (1953) used rapidly evolving stocks of trinucleid trilobites to create fine biostratigraphical divisions within the lower, middle and upper Llandeilo lithofacies units. The progressive reduction in clastic input and the corresponding increase in carbonate deposition through the succession indicates a stable, level-bottom, subtidal to intertidal environment with an overall high to low energy gradient upward through the succession.

Wilcox and Lockley (1981) described this upward transition in terms of four main sedimentary facies from inshore/shoreface coarse sands through intertidal/subtidal sands and silts then silts and muds to offshore shallow shelf muds that accumulated below storm wave base. Complementary faunal changes accompanied the shifting lithofacies patterns, and they recognised persistent biotic associations grouped into three palaeocommunities which characterised successively offshore areas of the platform transect.

Important conodont data from the uppermost beds of the Ffaifach Group and the lowest Llandeilo Flags indicate a correlation within the upper part of the *Pygodus anserinus* Biozone, whilst younger type Llandeilo conodonts suggest an uppermost age limit for the Llandeilo Series within the lower part of the *Amorphognathus tvaerensis* Biozone (Bergström 1981, Bergström *et al.* 1987).

The Llandeilo transgression continued into Caradoc times with the deposition of dark graptolitic muds containing *N. gracilis*. Above this level, faulted lenticles of shelly, phosphatic, nodular and crystalline limestone (Crûg and Birdshill) reflect an eventual return to higher energy, inshore carbonate deposition, although evidence of the transition is lacking. Williams (1953, p. 195) assigned a Longvillian–Marshbrookian age to the Crûg Limestone on the basis of brachiopods and trilobites, and this was supported by Lindström's (1959) and Bergström's (1981, Fig. 7) assessment of conodont faunas. However, Owens (1973, p. 48) suggested possible equivalence of the Crûg Limestone with the younger Birdshill Limestone whose shelly faunas are of probable early Ashgill age (Pusgillian to earliest Cautleyan; Price 1973, p. 245); re-study of the Crûg conodonts (Savage & Bassett 1986) supported this latter correlation.

Uppermost Ordovician mudstones with sandstone and impure nodular limestone bands crop out in a broad belt north of Llandeilo (Williams 1953) and have yielded two shelly faunas of Cautleyan age; in facies development they link with the Slade and Redhill Mudstones to the west.

## Llandovery–Llanwrtyd–Garth

North-eastwards from Llandeilo through the Llandovery district and into east-central Wales, effects of the incipient late Ordovician growth of the Tywi lineament were particularly marked. The facies variations reflect the reparation of platform facies to the south-east from deeper water slope and trough environments to the north-west of the structural belt (Fig. 4.2).

On the south-eastern flank, the oldest shelf sequences around Llandovery are dark mudstones with intercalated winnowed shelly sandstones, possibly storm-generated, of the Tridwr Formation, yielding a Rawtheyan shelly fauna including the brachiopods *Sampo ruralis, Christiania tenuicincta* and *Chonetoidea papillosa*, plus graptolites such as *Orthograptus truncatus, Climacograptus angustus* and *C? supernus* indicative of the *anceps* Biozone (Jones 1925, 1949; Cocks *et al.* 1984).

The conformably overlying Hirnantian Scrach Formation contains a typically restricted *Hirnantia* fauna, including the eponymous brachiopod *Hirnantia sagittifera* plus *Eostropheodonta hirnantensis, Plectothyrella crassicostis* and *Dalmanella testudinaria*. In part these beds represent what Jones (1925) took to be basal Llandovery (A₁) sandstones but their correlation within the late Ordovician is now in no doubt. They comprise four main arenitic facies (Woodcock & Smallwood 1987) in a regressive sequence attributable to glacio-eustatic falling sea level. Interbedded shallow-water lenticular sandstones and burrowed mudstones are dominant. There is evidence of NW-directed tidal and storm currents. Sporadic sand sheets, represent storm events, whilst lenticular thicker bedded sandstone units formed as barrier facies in an area of high clastic input (Woodstock & Smallwood 1987). The fourth facies is characterised by lensoid beds of conglomerates and sandstones passing laterally into thicker bedded and lenticular sandstones which tend to thicken and coarsen upwards. Deposition took place as megaripples migrating in a discontinuous sheet across a subtidal flat. Derivation was dominantly from the south-west, involving transport from a nearby foreshore and barrier sands. The coastline, which formed part of the migrating northern margin of Pretannia (Cope & Bassett 1987) was defined partly by active NE-striking faults throughout late Ordovician–early Silurian times.

An abrupt top to the Scrach Formation marks a rapid rise in sea level with the beds passing up into mudstones of the Bronydd Formation. Discrete micaceous, storm generated shelly sandstones increase upwards through the mud-dominated marine shelf deposits which form part of a prograding (?prodeltaic) sedimentation system. The presence of *Climacograptus normalis* low in the Bronydd Formation suggests a *persculptus* or low *acuminatus* Biozone datum. There is thus an apparently conformable passage into the Silurian in this region, with post-glacial transgression and deepening beginning within the late Hirnantian (Cocks *et al.* 1984, Woodcock & Smallwood 1987).

Older rocks crop out in the core of the Tywi lineament to the north of Llanwrtyd Wells (Fig. 4.2), where dark Caradoc shales, mudstones and thin grits with a *clingani* Biozone graptolite fauna almost completely enclose an underlying succession of tuffs, volcanic breccias and sediments (Stamp & Wooldridge 1923; K. A. Davies 1926, 1933; Jones 1949). The lowest beds exposed are agglomeratic tuffs, succeeded by rhyolitic breccias, sandy mudstones, pyroclastic spilites and spilitic breccias, further variegated tuffaceous and calcareous mudstones, and then upper banded tuffs that contain small volcanic bombs, with interbedded black shales. Poorly documented shelly faunas, together with graptolites from the shales among the upper tuffs suggest a probable Costonian age (D. A. Bassett *in* Williams *et al.* 1972). The calcareous beds may represent a lateral equivalence of the Mydrim Limestone beyond Carmarthen to the south-west; see above) and thus support this correlation, and the conclusion that the Llanwrtyd volcanic events post-dated those of Llandeilo age occurring along the strike of the Ordovician outcrop at Builth to the north-east (Stamp & Wooldridge 1923).

Between Llanwrtyd and Builth, upper Ordovician sediments occur in lateral extension of the south-eastern Tywi flank facies. They are particularly well displayed to the north and north-east of Garth (Fig. 4.2) (see Andrew 1925 and Williams & Wright 1981). The oldest beds comprise bioturbated sandy and muddy siltstones of Rawtheyan age, initially containing a sparse shelly fauna of deep water aspect, with small brachiopods and micromorphic snails present, but then passing up into more diverse assemblages dominated by the trilobite *Sphaero-coryphe thomsoni*. This fauna also includes *Foliomena* and associated brachiopods of the offshore *Foliomena* fauna, and is then succeeded by a late Rawtheyan association containing *Mucronaspis* as the most abundant trilobite in a rich shelf assemblage dominated by gastropods and bivalves.

North of the Garth area, the basal laminated and bioturbated member of the overlying Wenallt Formation is again of Rawtheyan age, but the succeeding Ooid and Speckly Sandstone members contain typical elements of a diverse Hirnantian fauna, including *H. sagitiffera, E. Hirnantensis, D. testudinaria* and *Hindella incipiens* (Williams & Wright 1981). Beginning in the late Rawtheyan, the sedimentary sequence indicates shallowing from a distal shelf environment to near shore zones in which carbonate shoals developed briefly and in which storms periodically re-worked the sands and silts. The oolitic siltstones in the middle of the succession also contain mud clasts and abundant shell debris and pass northwards into finer sandstones and mudstones. The Speckly sandstones are also rich in skeletal fragments, chiefly pelmatozoans, bryozoans and brachiopods,

and similarly become finer grained and cross-bedded northwards.

The presence of *Eostropheodonta hirnantensis* in the overlying Cwm Clŷd Formation confirms a Hirnantian age. Rippled blue-grey sandstones with pebbles of chert, vein quartz and rhyolite become increasingly coarser and conglomeratic southwards, and more silty and muddy northwards. These sediments were deposited as shoreface and foreshore facies, including intertidal environments, longshore bars, and sand/silt wedges in depths of about 2 to 6 metres (Williams & Wright 1981). The coarse arenites and rudites are mostly clasts of older Ordovician sediments, probably reworked and transported outwards across the platform towards the Welsh trough as a result of extensive erosion following the late Rawtheyan–Hirnantian fall in sea level (Cope & Bassett 1987). The progressive southerly overstep and stratigraphical break at the base of the Cwm Clŷd Formation attributed by Williams and Wright (1981) to differential uplift linked with the shallowing was questioned by Woodcock and Smallwood (1987), who suggested instead that lateral facies changes towards the Llandovery area might account for the observed successions.

Upper beds of the Cwm Clŷd Formation contain increasing amounts of mudstone and siltstone, followed conformably by similar sediments of the unfossiliferous Garth Bank Formation, which also includes units rich in mica, bioturbated horizons, and trace fossils. Deepening is inferred in relation to renewed transgression through a shoreface zone and into shelf environments. In the absence of faunas it is difficult to locate the top of the Ordovician accurately here, but as in the Llandovery area to the south there is an apparently conformable transition into the Silurian, with evidence again that post-glacial deepening began within late Hirnantian times (Williams & Wright 1981).

## Builth–Llandrindod–Abbey Cwmhir

In the south-east of this area the Ordovician inlier extending from Builth to Llandrindod is composed of sedimentary and igneous rocks with ages ranging from Lower Llanvirn (*D. bifidus* Biozone) to the late Llandeilo (*N. gracilis* Biozone), and has been of long continuing interest (e.g. Murchison 1834, 1839; Elles 1940, Jones & Pugh 1941, 1949; Earp 1977). The volcanic rocks, which show considerable variation in thickness, include massive and pillowed basic lavas, hyaloclastites and intermediate and acid lavas and tuffs (see Chapter 16).

The Lower *Didymograptus bifidus* Beds are a 500 m sequence of grey shales containing *Ogyginus corn-densis* as well as the zone fossil and other graptolites;

the Upper *D. bifidus* Beds, 250 m thick, are again graptolite shales, more fossiliferous and containing other forms such as *D. speciosus, D. geminus* and *Bettonia chamberlaini*. This upper sequence is distinguished lithologically by interbedded tuff layers, commonly thin but occasionally attaining about 50 m as in a coarse rhyolitic tuff near the base.

The bulk of the volcanic rocks are of Upper Llanvirn age, with intercalated shales in the volcanic sequence yielding *D. murchisoni* and other fossils. Rhyolitic tuffs at the base are frequently attenuated by the overlying main tuff group (Earp 1977) with coarse, red-weathering basal agglomeratic tuffs that pass up into finer tuffs with abundantly fossiliferous shales; these then give way to a widespread pebbly tuff. Volcanics above this horizon are essentially confined to the south, where 90 m of a spilite boulder tuff passes laterally into about 150 m of spilitic lavas that are succeeded by about 45 m of keratophyres.

In the south these volcanics are overlain by very coarse sandstones at the base of which are boulder beds containing well-rounded boulders of igneous rocks and interpreted as coastal debris around a volcanic pile. The shore-line environment is supported by the distribution of the sands, suggestive of partially exhumed cliffs and sea stacks (Jones & Pugh 1949). To the north, the main tuff group is overlain by flinty tuffaceous beds, with the succeeding poorly exposed Upper *D. murchisoni* Shales here probably approaching 150 m.

The overlying *Glyptograptus teretiusculus* Shales, of Lower Llandeilo age (Addison *in* Williams *et al.* 1972), are estimated as up to 600 m thick and composed of shales with occasional thin tuff or limestone bands; again exposure is poor. In addition to the zone fossil and other graptolites including *Amplexograptus perexcavatus* and *Climacograptus antiquus*, trilobites are common and include *Ogygiocarella debuchii, Cnemidopyge nuda* and *Whittardolithus instabilis*. The sediments grade up into the overlying *Nemagraptus gracilis* Shales, again dominantly dark grey shales and mudstones, which are of late Llandeilo age (Hughes *in* Williams *et al.* 1972, p. 30). *Trinucleus fimbriatus, Ogygiocarella debuchii* and *Platycalymene duplicata* are the most common trilobites in a rich assemblage dominated by trilobites, graptolites and inarticulate brachiopods.

At Abbey Cwmhir, Caradoc rocks of graptolitic facies are brought up along the western side of the Carmel Fault. Here the Carmel Group of Roberts (1929) consists of up to about 300 m of coal-black shales and mudstones with laterally restricted bands of turbiditic arenites. The rather poorly preserved graptolite assemblages indicate the presence of the *D. multidens* and at least part of the *D. clingani* Biozones. A stratigraphic break separates the Carmel Group

from the overlying Camlo Hill Group, shown by Roberts (1925) to have developed differently the west and east of the Carmel Fault (although his estimates of the substantial thicknesses are unconvincing (James 1983a)). The western sequence he subdivided into three argillaceous units with conspicuous coarse arenites in the lower and upper divisions. One record of graptolites and some sparse transported shelly assemblages indicate a probable Rawtheyan age. The only diagnostic fossil occurs in the massive grits and sandy mudstones of the upper division – *Eostropheodonta hirnantensis*, indicative of a Hirnantian age. Shales with *G. persculptus* overlie these beds in the west. In the development to the east of the Carmel Fault coarse clastics are absent, and although shelly faunas are present they are in general poorly preserved and fragmentary apart from a brachiopod–gastropod assemblage (Roberts 1929) that would appear to be a Rawtheyan muddy-shelf assemblage. The Hirnantian is absent here, probably reflecting the intra-Hirnantian sea-level fall (James 1983a).

## Welsh Borderland and eastern Wales

In contrast to both North and South Wales, Ordovician outcrops along the eastern margin of the Welsh trough are widely scattered and discontinuous (Figs. 4.1, 4.2). Nevertheless, this region again emphasises the role of contemporaneous tectonism in influencing sedimentation, in dictating the site of volcanism close to the trough/platform margin, and in defining that boundary as a zone of separation between the subsiding region to the west and the relatively stable area to the east and south-east. The major controlling structures here comprise the Pontesford–Linley fault system (Fig. 4.2) and its continuing lineament to the south-west (Woodcock 1984a, b). Associated parallel intra-platform faults such as those of the Church Stretton system were responsible for facies changes and breaks in sedimentation within the shallow marine environments.

## South Shropshire

Within the Church Stretton fault complex along the east side of the Longmynd horst, Ordovician sediments are preserved in different stratigraphical settings (Fig. 4.13) to the north and south of the Precambrian mass of Cardington Hill. Pre-Caradoc strata crop out only to the north, as a thick sequence of Tremadoc shales with subsidiary sandstones and siltstones referred to collectively as the Shineton Shales. They are faulted against or are unconformable on Precambrian–upper Cambrian beds in three outcrop areas. The Tremadoc succession is most complete in the Wrekin district (Fig. 4.2), where six mixed

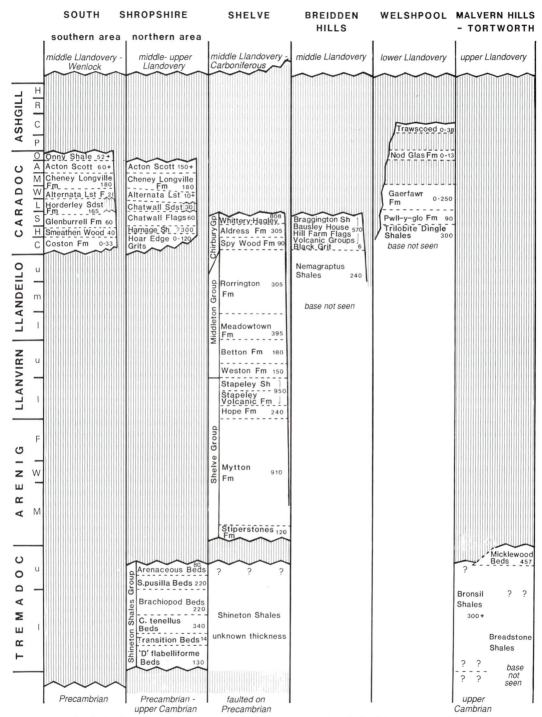

**Fig. 4.13.** Stratigraphy and correlation of representative Ordovician sequences in the Welsh Borderland and eastern Wales; notations used for the subdivisions of Series are initial letters of the Stages and other units as shown in Fig. 4.3; thicknesses in metres.

biostratigraphical/lithostratigraphical units were described by Stubblefield and Bulman (1927). Lowermost Tremadoc correlatives may not be represented at the base of the Shineton Shales Group, but the abundance of *R. flabelliformis socialis* in the 'Dictyonema' Beds clearly assigns them to the lower part of that biozone. Mottled blue-green and reddish shales and mudstones reflect low energy deposition and contain common trilobites such as *Proteuloma monile* and *Macropyge chermi*, plus acrotretid and lingulellid brachiopods (*Eurytreta sabrinae* and *Lingulella nicholsoni*), gastropods, hyoliths and the sponge *Protospongia*. Essentially similar, uniform mud environments persisted throughout the area from early to mid-late Tremadoc times; calcareous cone-in-cone concretions are typical lithological features at some levels. The Transition Beds contain alternating *R. flabelliformis/ C. tenellus* assemblages over a narrow interval, with a more diverse soft-bottom shelly fauna than in the underlying shales, and are then succeeded by Clonograptus Beds with *C. tenellus* and *Bryograptus* cf. *hunnebergensis* indicative of the *tenellus* Biozone proper; similar shelly faunas are again present. The Brachiopod Beds contain only the inarticulate brachiopods *Palaeobolus quadratus*, *Eurytreta sabrinae* and *Lingulella nicholsoni*, all three of which occur in the underlying *tenellus* Beds and suggest that this unit also belongs within the higher part of the lower Tremadoc (Fig. 4.13); the occurrence of many large, unbroken specimens of brachiopods in life position reinforces the interpretation of a quiet water depositional environment (Stubblefield & Bulman 1927, p. 111). The Shumardia 'pusilla' Beds become micaceous upwards and contain the zonal species (i.e. *S. salopiensis*) as a common component together with other trilobites, brachiopods and ostracodes. *Niobella homfrayi* is a distinctive addition to the trilobite fauna at this level in Shropshire, although in North Wales it extends throughout the Tremadoc and the underlying uppermost Cambrian (Rushton 1982, p. 45). Highest Tremadoc beds in the area, the Arenaceous Beds, comprise strongly bioturbated, micaceous, fine–medium sandstones interbedded gradationally with micaceous concretionary shales; sedimentary features reflect changing inshore energy conditions during a regressive phase. Faunas are restricted mostly to inarticulate brachiopods. There is no evidence that the Arenaceous Beds are significantly younger than the underlying S. 'pusilla' Beds and it is unlikely that deposits of uppermost Tremadoc age are now represented in the area (Fig. 4.13). Rich assemblages of acritarchs occur throughout the succession in the Wrekin area, arranged in a detailed succession of zones as a basis for wider correlations (Rasul & Downie 1974, Rasul 1979).

In the Lawley and Cardington Tremadoc outcrops (Stubblefield & Bulman 1927, pl. 5, pp. 115, 116; Stubblefield 1930), facies and faunas are the same as those described above. Only the 'Dictyonema' Beds are preserved in the former area, with both these and higher Tremadoc beds in the latter.

Post-Tremadoc Ordovician deposits in South Shropshire are confined to a relatively thin succession of the Caradoc Series (Fig. 4.13), for which this is the historical type area (Murchison 1839) (Figs. 4.1, 4.2). Unconformable on the Tremadoc in the northern area, the basal Caradoc deposits step southwards with increasing intensity onto Precambrian rocks to the east of the southern Longmynd. Rapid lateral facies changes and minor local, non-depositional breaks and diachronism are characteristic, but the rich assemblages of shelly faunas provide a sound basis for both correlation and environmental analysis (e.g. Bancroft 1933, 1945, 1949; Dean 1958, 1964; Hurst 1979a, b, c). Fine–medium terrigenous sands and silts are the dominant sediments, mostly with carbonate cements, but gravels, bioclastic accumulations, and muds are significant components at some levels. There is considerable evidence of local derivation and reworking of sediments; well preserved, reworked acritarchs throughout the succession suggest that Tremadoc, Arenig and Llanvirn strata formed part of the source, probably from a formerly extensive cover across adjacent areas of the Midland Platform (Turner 1982).

The lower formations of the type Caradoc record a fairly rapid but pulsatory marine encroachment from the west, with the shoreline aligned close to the present outcrop areas; minor differences in stratigraphical nomenclature from south to north (Fig. 4.13) are an expression of the changing lateral facies and grain size in the near-shore environments. Quartz conglomerates, feldspathic grits and coarse–medium sandstones of the basal Coston–Hoar Edge formations represent littoral to sub-littoral gravels and sands derived largely from the underlying exhumed Precambrian surface. Wind-faceted pebbles occur in some beds (Pocock *et al.* 1938, p. 83). The bed forms and lensoid accumulations of pebbles indicate the presence of a succession of transgressed foreshore and shoreface ridges and barriers with washover accumulations and gravel/shell lags in shallow tidal channels. Calcareous sandstones and lenticular shelly limestones become increasingly common upwards. Apart from in the basal conglomerates, fairly rich shell associations occur throughout the Costonian strata, typified by species of the brachiopods *Dinorthis*, *Harknessella* and *Horderleyella* and by the trilobite *Costonia*. The occurrence of the graptolite *Nemagraptus gracilis* in shales within the low–middle Hoar Edge Grits is evidence of the extension of the *gracilis* Biozone into the Costonian (Fig. 4.3; Pocock *et al.* 1938, p. 84; Dean 1958, p. 227).

Deposition of high-energy sands continued into Harnagian times in the northern district, as the upper Hoar Edge Grits contain diagnostic trilobites and brachiopods of this age, notably *Salterolithus* species and *Salopia salteri* (Dean 1958). Upward and lateral transition into the Harnage Shales and Smeathen Wood Formations reflects decreasing water energy and reduction of coarse clastic supply related to an off-shore shift as a result of continued transgressive deepening; fine sand, silt and mud-grade micaceous sediments are characteristic, with only occasional thin pebbly grits as residual lag deposits. Shelly faunas are again common, particularly *S. salteri* and the trinucleid trilobites *Reuscholithus reuschi* and *Salterolithus caractaci*, together with gastropods, conulariids and ostracodes. A thin porphyritic basalt flow in the Smeathen Wood Formation is the only sign of Caradoc volcanism within South Shropshire (Whittard 1952, p. 162; Dean 1964, p. 275; Greig *et al.* 1968, pp. 108, 121).

Mixed silt–mud heterolithic offshore facies continued through Soudleyan times in southern districts (Glenburrell Formation), passing laterally (inshore) into tabular fine sands of the Chatwall Flags. The basal Glenburrell beds contain remnant Harnagian faunas, replaced upwards by associations in which the trinucleid *Broeggerolithus broeggeri* is particularly characteristic. Higher Soudleyan beds throughout the outcrop contain abundant fragments of the crinoid *Balacrinus basalis*.

The Horderley–Chatwall Sandstone formations mark a major spread of sands across the shallow platform, beginning in late Soudleyan times. The middle Horderley Sandstone contains thick, hummocky cross-stratified units (Figs. 4.14) mantled by finer silts and muds which were deposited from suspension following storm activity. As interpreted by Brenchley and Newall (1982), these sand bodies formed inner shelf lobes brought offshore from the shoreface by storm surge ebb-currents. Coquinas of dalmanellid and sowerbyellid brachiopods occur in erosional hollows of the sand waves. The Chatwall Sandstone sequence lateral to the Horderley Sandstone is notably coarser grained (Dean 1960; Hurst 1979a, fig. 5), incorporating conglomerates and pebbly sandstones that probably represent part of the

**Fig. 4.14.** Hummocky cross stratification in the lower Horderley Sandstone Formation, northeast side of A489 road, Onny Valley, Shropshire.

foreshore/barrier system. Uplift and erosion related to fault movements in adjacent platform areas may have caused the rapid and substantial influx of sands during this interval. Minor stratigraphical breaks are reported within the Horderley–Chatwall sandstones themselves (Fig. 4.15), although they may be a result partly of erosive scour and ecological gaps rather than original non-deposition (cf. Dean 1960, Hurst 1979a).

In the overlying Woolstonian to Onnian succession, Hurst (1979a, b, c) described an array of seven heterolithic lithofacies groups containing nine related faunal associations. In general the sediments reflect a resumption and continuation of transgressive deepening but again with significant local interruptions and influxes of sands under higher energy conditions. The Alternata Limestone Formation is dominated by coquinoid layers and lags of the eponymous brachiopod *Heterorthis alternata*, with shell-rich laminated beds of sandstone and siltstone. In central and southern areas the presence of phosphatic nodule horizons through the lower Alternata Limestone suggests transgression over a previously non-depositional surface.

Another storm-generated, hummocky, offshore sand/mud facies occurs in the Cheney Longville Formation (Brenchley & Newall 1982). Faunas are mostly dalmanellid and strophomenid brachiopods together with trilobites such as *Broeggerolithus* and current oriented associations of tentaculitids. The Acton Scott Formation is mainly comprised of mudstones and calcareous siltstones, marking a continued shift to offshore low-energy deposition; the change in lithofacies is accompanied by a replacement of cryptolithid trilobite associations by faunas of *Platylichas, Chasmops* and *Calyptaulax* accompanied by bivalves and dalmanellid brachiopods such as *Onniella, Cryptothyris* and *Reuschella* (Dean 1964). There is at least one thick bentonite in the middle of the sequence as evidence of distant volcanism; a fission-track age of $466 \pm 15$ Ma derived from this bentonite is plotted in Fig. 4.3 (Ross *et al.* 1982). Hard calcareous sandstones form much of the Acton Scott succession towards the centre of the outcrop area, with ripple laminated and bioturbated beds containing coquinoid lenses of brachiopods, bryozoans and tentaculitids together with less common trilobites; the beds formed as sand sheets, probably on a local topographic high.

Bioturbated mudstones and laminated shales of the Onny Shale Formation at the top of the type Caradoc sequence were deposited entirely in a low energy, outer shelf environment; these beds are not preserved in the northern area (Fig. 4.13). Faunas are dominated by soft-bottom associations of the cryptolithid trilobite *Onnia* and the brachiopods *Onniella* and *Sericoidea*, together with burrowing bivalves, gastropods and brachiopods (Hurst 1979b, p. 227). The Onny Shale Formation is overstepped by Silurian strata (Fig. 4.13).

Comparatively rare graptolites from throughout the type Caradoc area suggest that the sequence does not extend above the *clingani* Biozone (e.g. Dean 1958; *in* Williams *et al.* 1972, p. 40). Conodonts from calcareous lenses through the same sequence confirm a correlation within the *tvaerensis* and *superbus* biozones, although nowhere are the zonal boundaries sharply defined (Savage & Bassett 1985).

## Shelve

The area of Tremadoc to lower Caradoc (Soudleyan) rocks occupying some 120 sq km of west Shropshire and north Powys (Fig. 4.2) forms what is probably the best-developed Ordovician succession in Britain. Separated sharply from the Precambrian mass of the Longmynd to the east by the Pontesford–Linley fault system, the Shelve sequence of over 4,500 m is largely argillaceous, but silty, sandy and volcanic lithologies are dominant at a number of levels. Intimately mixed trilobite, brachiopod and graptolite faunas through much of the sequence (Whittard 1955–67, Williams 1974, Strachan 1981, 1986) provide some of the fullest data available for the integration and correlation of stratigraphical schemes based on these different fossil groups, enhancing the value of the area as a geographical link between the trough environments to the west and those of the surrounding platform regions. Whittard's (1979) detailed account of the Shelve post-Tremadoc succession emphasises its almost unbroken nature (Figs. 4.13, 4.15), in contrast to the type Caradoc district only some 12 km to the east across the Pontesford–Linley and Church Stretton faults.

At least 150 m of Tremadoc beds of typical Shineton Shales lithologies with sparse graptolites, trilobites and inarticulate brachiopods crop out at the base of the Shelve sequence along the eastern margin of the area. Relatively rare occurrences of *Rhabdinopora flabelliformis, Clonograptus tenellus* and *Shumardia* (*Conophrys*) *salopiensis* confirm the presence of those biozones (Stubblefield & Bulman 1927, p. 116; Whittard 1931a, p. 324; 1931b, p. 344), but uppermost Tremadoc strata are apparently absent. Upward coarsening into fine arenaceous facies within the *salopiensis* Biozone suggests incursions of micaceous sands approximately contemporaneous with those in the Wrekin area to the east.

The Shelve, Middleton and Chirbury groups of the Arenig–Caradoc succession (Fig. 4.13) embrace three broad, upward-fining sedimentary cycles, each interrupted periodically by the deposition of volcanic

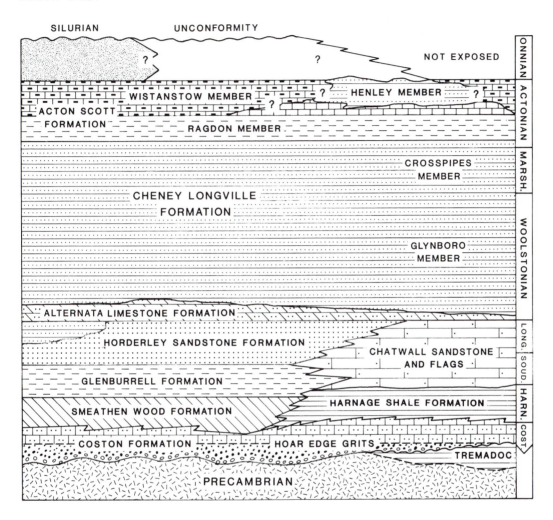

SOUTH-WEST                                                      NORTH-EAST

Fig. 4.15. Schematic reconstruction of Caradoc stratigraphy and facies variation in South Shropshire. Partly after Hurst (1979a), not to scale.

ash complexes. A basal polymict conglomerate (7·5 cm) of the Stiperstones Formation is disconformable on the Tremadoc, and may not have been derived entirely from nearby Precambrian positive sources but is perhaps the product of long-shore drift from the SW (Whittard 1931a, p. 324; 1979, p. 14). Above is a series of thick-bedded, rippled quartz arenites, formed as fairly clean, sublittoral transgressive sands; shaly and pebbly bands occur throughout, the rudites probably representing lags reworked by the migrating swash zone.

*Skolithos–Cruziana–Rusophycus* ichnofacies support the shallow marine origin of the sands (Crimes 1970a, fig. 8; Whittard 1979, p. 16), as does the presence of rare *Neseuretus ramseyensis* (Fortey & Owens 1987, p. 98); this latter species suggests a Moridunian age for the Stiperstones Formation.

Upward transition into sandy and gritty micaceous flags and then shaly flags and siltstones of the Mytton Formation reflect sedimentation farther offshore in response to rising sea level. Arenig shelly and graptolite faunas occur through all but the lowest Mytton beds, with associations of inarticulate brachiopods and dendroid graptolites particularly abundant in some of the finer grained lithologies, which in places contain volcanic dust (Whittard 1931a, p. 325). Low to middle horizons also contain other graptolites such as *Didymograptus* (*Expansograptus*) *extensus* and *D.* (*Corymbograptus*) *deflexus* indicative of a position high in the Moridunian Stage, whilst trilobites such as *Cylopyge grandis grandis* in higher beds suggest levels within the Whitlandian to early Fennian (Fortey & Owens 1987, p. 98); the uppermost 30 m or so of the Mytton Formation (Tankerville Flags facies) contain the Fennian zonal species *D.* (*Expansograptus*) *hirundo* (Strachan 1981) and a varied trilobite assemblage that contains species common to the South Wales sequences; Fortey and Owens (1987, p. 98) suggested that there may be gaps within the Shelve Arenig onshore succession, and that the uppermost Fennian in particular may be unrepresented.

Dark shales of the Hope and Stapeley Shale formations indicate a persistence of offshore mud environments across Shelve through early Llanvirn times, interrupted by the first major volcanic episode (the Stapeley Volcanic Formation); all three units contain dominantly cyclopygid trilobites, with soft-bottom, inarticulate brachiopod associations and graptolites. Pendent didymograptids are the most common, including *Didymograptus artus* suggesting correlation of this part of the sequence within the *artus* Biozone (Strachan 1981). Fine andesitic dust tuffs first occur in the middle and upper Hope shales, and then coarsen into crystal and lithic scoriaceous tuffs and agglomerates of the Stapeley Volcanic Formation (Blyth 1938, Lynas 1983). The tuffs and associated lavas were probably extruded from two local volcanic island centres (Lynas *op. cit.*), around which occasional debris flows and boulder beach deposits formed before the islands were eventually submerged later in early Llanvirn times.

Grits, calcareous sandstones and siltstones of the Weston Formation and their associated sedimentary features (Whittard 1979) are taken to reflect inner shelf deposition. The return to a higher energy regime, probably caused by encroachment of inshore environments was accompanied by a biofacies change from trilobite/graptolite dominance to infaunal bivalve, gastropod and lingulid brachiopod associations – typical of recent environmental analogues (Belderson & Stride 1966); the dalmanellid brachiopod *Tissintia* also became an important component of the Weston faunas (Williams 1974, p. 19; Lockley 1983a, b). The Weston sediments are in part volcanigenic, derived from the underlying Stapeley islands (Lynas 1983, p. 541). Renewed transgression and a return to offshore mud environments is reflected in the dark micaceous shales and flags with thin tuffs of the Betton Formation; infaunal brachiopods and *Tissintia* remain common, accompanied again by dominant trilobites of which mud-dwelling forms such as *Ogyginus* and *Trinucleus* are the most typical. Both the Weston and Betton formations contain upper Llanvirn *Didymograptus murchisoni* Biozone graptolite faunas (Strachan 1981).

Rare diplograptids, dicellograptids and dicranograptids in the Meadowtown Formation, accompanied by *Nemagraptus* and *Leptograptus* in the Rorrington Formation indicate a Llandeilo age for these beds, spanning the *teretiusculus*–lower *gracilis* biozonal interval. Low energy sedimentation persisted from the Llanvirn as dark calcareous silts and muds though the Meadowtown Formation silts and sands were probably sand sheets brought offshore in periodic storms. Thin calcarenites in this part of the sequence are an indication of the generally low clastic input. Mud-dwelling ogyginid/trinucleid and inarticulate/dalmanellid brachiopod associations continued as the dominant biota, but with the brachiopods in particular becoming increasingly diverse (Williams 1974) to include sowerbyellids and strophomenids that were also capable of living in a recumbent attitude on fine grained soft-bottoms (Jaanusson 1984b, p. 131). Upper Rorrington beds become increasingly fine grained and sparsely fossiliferous, suggesting a gradual offshore environmental shift associated with transgressive deepening, but then followed by rapid transition into arenaceous beds of the Spy Wood Formation.

The reintroduction of arenites at this level marks the beginning of the third sedimentary cycle at Shelve. The *gracilis* transgression reflected in the underlying beds had by now extended onto the Precambrian and older Palaeozoic surfaces to the east, bringing fresh supplies of calcareous sands offshore to Shelve as sheets directly equivalent laterally to basal Caradoc littoral and inshore sands of South Shropshire; correlation is confirmed by the presence in the Spy Wood beds of the Costonian zonal trilobite *Costonia ultima* (Whittard 1979, p. 50). Articulate brachiopods, particularly orthids and dalmanellids, form the main biota (Williams 1974, p. 21), although ostracodes and echinoderms become dominant locally. Grey/green micaceous shales within the Spy Wood arenites contain graptolites, of which *Orthograptus uplandicus* suggests a correlation within the *gracilis* Biozone close to the junction with the overlying *multidens* Biozone (Strachan 1981, p. 22).

Graptolites through the remainder of the preserved Shelve succession are diagnostic of the *multidens* Biozone. Above the Spy Wood Formation there is a fairly rapid reversion to fining-upward, mud dominated lithologies formed as offshore shelf facies. The Aldress, Hagley Shale, and Whittery Shale formations are all micaceous and tuffaceous shales, fine siltstones and mudstones with an increasingly diverse shelly fauna; the tuffaceous sediments tend to support significantly different associations from those of the indigenous muds (e.g. Williams 1974, p. 21). Interjected between the Aldress and Whittery shales are the Hagley Volcanic Formation and the Whittery Volcanic Formation, both comprising andesitic and rhyolitic, crystal and lithic tuffs and agglomerates; both units are rippled and interbedded with thin shales. The youngest Shelve Ordovician beds (Whittery Shale) contain the trilobites *Salterolithus caractaci* and *Broeggerolithus broeggeri*, characteristic of the lower Soudleyan Glenburrell Formation of South Shropshire. Unlike their equivalents to the east, these Whittery mudstones and shales contain evidence of continuing volcanic activity as tuffs with mud balls which may have been deposited by lahars (Williams 1974, p. 21).

Intrusive igneous rocks cut the Shelve sediments at a number of levels, most commonly through the Mytton, Hope, Stapeley Shale, Weston and Aldress Shale formations (Whittard 1979, geological map). The intrusions are mostly small dykes and high level sills of picrite, andesite and dolerite (Blyth 1944, P. M. Allen 1982). None of the intrusions penetrates the unconformable envelope of Upper Llandovery (Silurian) rocks; they are assumed to be of late Ordovician age. Variably reset isotopic ages from some of the Shelve igneous rocks (Lynas *et al.* 1985) attest to the low grade metamorphism associated with this episode. The possibility that the intrusive episode was partly of early Llandovery age cannot be discounted.

## Pontesford

Two small but significant areas of Caradoc micaceous shales and mudstones crop out in a fault zone only 1·2 km NE of the Shelve district but to the E of the Pontesford–Linley fault (Figs. 4.1, 4.2; Whitehead *in* Pocock *et al.* 1938, p. 90; Dean & Dineley 1961; Whittard 1979, p. 61). At the base of the sequence, a poorly sorted polymict conglomerate (3–4 m) is unconformable on the Precambrian and passes up through grits and sandstones into the argillites that comprise the bulk (? up to 300 m) of these Pontesford Shales. The lower shales contain abundant graptolites of the *multidens* Biozone, including the zonal species for which this is the type area. Lithologically the beds closely resemble the Harnage Shales of south Shropshire. Higher mudstones contain a Soudleyan shelly fauna typical of the Glenburrell Beds and thus also emphasise relationships with areas to the east rather than with the nearby Shelve outcrops (Dean *in* Williams *et al.* 1972, p. 40; Whittard 1979, p. 61). These relationships reflect the role of the Pontesford–Linley fault line in controlling different depositional histories on either side of the platform marginal zone.

## Pedwardine

Some 25–30 km S of the Shelve and south Shropshire outcrops, Tremadoc strata referable to the Shineton Shales are preserved at Pedwardine (Fig. 4.1) in a narrow, fault-bounded strip parallel to but a few km east of the main Church Stretton fault zone (Cox 1912; Stubblefield & Bulman 1927). The olive-green to grey shales, of uncertain thickness, are probably entirely of early Tremadoc age and contain *R. flabelliformis socialis* together with horny inarticulate brachiopods and hyoliths. Overlying the shales with gross unconformity in a very small outcrop area are a metre or so of quartz-pebble and quartzite conglomerates and grits of the Letton Formation. These arenites contain only poorly preserved orthacean brachiopods whose age has never been clearly assessed, but Cox (1912) demonstrated their lithological similarity to basal Caradoc beds in south Shropshire and they are here regarded as evidence of the same Costonian transgressive event.

## Breidden Hills

In the poorly known Ordovician tract of the Breidden Hills (Fig. 4.2), both the sedimentary and volcanic facies show close affinities with the higher beds of the Shelve district only some 14 km to the SSE (e.g. Watts 1885, 1925, p. 343; Wedd 1932; Dean *in* Williams *et al.* 1972, p. 41; Whittard 1979, p. 60). Dark micaceous graptolitic shales in the north of the area are the oldest beds seen, the presence of *Nemagraptus* suggesting a probable correlation within the *gracilis* Biozone. Fossils are unknown from the succeeding black, flaggy-bedded sandstones and grits, but they are considered contemporaneous with the basal Caradoc arenites of Shelve and south Shropshire. Interbedded micaceous shales, siltstones, thin limestones and sandstones, flags, tuffs and volcaniclastic conglomerates make up the remainder of the sequence (Fig. 4.13), which contains graptolite faunas at some levels indicative of the *multidens* Biozone throughout. Two major volcaniclastic formations are present, probably lateral equivalents of the Hagley–Whittery volcanics of Shelve. Shelly faunas support these correlations, particularly the typical Soudleyan trilobite *Broeggerolithus broeggeri*. A few dolerite dykes cutting the sequence were probably emplaced in the same intrusive episode as that seen at Shelve, and add further similarity to the geological history of the two areas.

## Welshpool

Following Murchison's (1839, p. 290) recognition of his 'Caradoc sandstone' and associated strata in and around Welshpool, this area of eastern Wales (Fig. 4.2) has long been linked descriptively with the Welsh Borderland. Geologically, however, the succession (Fig. 4.13) has equal ties with nearby areas of north Wales, and particularly the adjacent Berwyn Hills. On the one hand, the absence of volcanigenic formations in the Welshpool Caradoc rocks and the overstep of the local Silurian across the Ordovician find parallels in south Shropshire; on the other hand, the presence of the distinctive Onnian phosphatic Nod Glas facies (Cave 1965) and of Ashgill strata suggest a depositional history related in part to North Wales.

The lowest exposures are fine grey shales and nodular mudstones of the Trilobite Dingle Formation (Wade 1911), which contain the Harnagian trilobite *Salterolithus caractaci* (Cave 1957). The deposits also contain other shelly faunas, in which burrowing bivalves and gastropods occur, plus fairly rich graptolite associations. Thin, regularly alternating, parallel-bedded sandstones and micaceous grits within the flaggy shales and siltstones of the overlying Pwll-y-glo Formation probably represent suspension-deposited sheet sands brought offshore by storm surge ebb currents from a prograding Soudleyan inner shelf. Graptolites are sparse but the shelly faunas, particularly trilobites, are more diverse than in older beds, and molluscs continue to be significant. Longvillian–Woolstonian shelly faunas dominated by brachiopods (Wade 1911, p. 427) characterise the Gaerfawr Formation, in which bioturbated arenites increase in thickness and regularity at the expense of siltstones and shales. Both the upward change in faunal emphasis

and the increase of sheet sand deposits reflect further increase of water energy and encroachment of inner platform facies across the area with time.

The absence of Marshbrookian–Actonian beds above the Gaerfawr Formation suggests an abrupt cessation of clastic input and a period of non-deposition, followed by a quite different sedimentary regime represented by the Nod Glas Formation. Dark phosphatic limestones succeeded by pyritous black shales and mudstones with phosphatic nodules suggest slow deposition with restricted circulation, possibly on a topographic high that developed in mid Caradoc times (Cave 1965, p. 293). The Nod Glas sediments contain a mixed graptolite-shelly fauna, in which the trilobite *Onnia gracilis* provides a good tie with the mid Onnian of south Shropshire. The uppermost beds of this formation contain coarser siltstones and mudstones, heralding a return to more open platform environments. Following a further non-sequence in which lowest Ashgill strata are unrepresented, the bioturbated muds and shelly faunas of the Cautleyan Trawscoed Mudstone Formation indicate a continuing trend back to conditions that prevailed across the area in the earlier part of the Caradoc (Cave & Price 1978, p. 192). Younger Ashgill horizons are not present.

## Malvern Hills

Upper Cambrian shales along the western flank of the southern Malvern Hills (Fig. 4.1) are overlain by the Bronsil Shales of Tremadoc age, but the precise age relationships are uncertain (Fig. 4.13). The grey-green to blue, silty and micaceous Bronsil Shales are similar to the Shineton Shales of south Shropshire and Shelve, and the succession begins in the lower Tremadoc as indicated by *R. flabelliformis socialis* and ?*C. tenellus callavei* (Groom 1902, Stubblefield and Bulman 1927, table facing p. 118). Trilobites, horny inarticulate brachiopods, and hyoliths are also present. Definitive faunal evidence for upper Tremadoc horizons is lacking, although equivalents of the *salopiensis* [*pusilla*] Biozone may be present below the unconformably overlying lower Silurian (Bulman & Rushton 1973, p. 6). Thin doleritic and andesitic dykes and sills intrude the Bronsil Shales and older strata but not the Silurian of the area, and are thus probably of Ordovician age (Groom 1901, 1902; Blyth 1935).

## Tortworth

Tremadoc sediments are preserved in three small inliers immediately south of the Severn Estuary (Fig. 4.1) close to the village of Tortworth (Smith & Stubblefield 1933, Curtis 1968). Exposure is poor and older rocks are not seen so that the total Tremadoc thickness is uncertain, but it is probably over 1,000 m. The lower unit, the Breadstone Shales, contains *R. flabelliformis* in more northerly sections, followed southwards by beds with *Adelograptus* sp. and *C.* cf. *tenellus callavei*, indicating the presence of both the *flabelliformis* and *tenellus* biozones; the co-occurrence of *R. flabelliformis* with the trilobite *Beltella depressa* close to the base of the Tortworth sequence confirms a correlation with strata at or near the base of the Series in the type Tremadoc area (Tyn-llan Beds). Thinly bedded grey, blue-grey, and greenish micaceous shales make up the bulk of these strata, which besides graptolites contain trilobites and horny inarticulate brachiopods with rare gastropods, ostracodes, hyoliths and worms. Sandy partings and thin, intermittent, lensoid beds of sandstone within the shales represent incursions of fine sands into the low energy mud environments. In places the sandstones contain a *Cruziana* ichnofacies with rare *Rusophycus* traces (Crimes 1970a).

Upward transition from the Breadstone Shales to the Micklewood Beds is nowhere exposed, but the latter may be close to 460 m thick (Curtis 1968) and are characterised by an increased proportion of sandy beds. Soft to slightly siliceous, micaceous grey shales (containing fragmentary lingulellid brachiopods) are dominant. The whitish-grey sandstone units may thicken up to about 25 cm or rarely 1·5 m, but they generally wedge out rapidly laterally. Increased current activity in the Micklewood Beds sands is reflected in a variety of typical sedimentary structures. Rare desiccation cracks suggest that shallowing related to the sand incursions occasionally culminated in emersion.

Apart from the brachiopods, the only body fossils reported from the Micklewood Beds are the trilobites *Angelina sedgwickii* and *Peltura olenoides*, both suggesting correlation with the uppermost Tremadoc Garth Hill Beds of North Wales. The underlying *salopiensis* Biozone has not been recognised, probably because of the sparse exposure and rarity of fossils rather than because of breaks in the succession. No post-Tremadoc Ordovician rocks are present in the area.

## English Midlands
### Nuneaton–Dosthill

East of the Welsh Borderland shelf sequences, the only Ordovician rocks preserved at the surface are Tremadoc strata confined to two small strips along the eastern and north-western side of the Warwickshire Coalfield, near Nuneaton (Merevale) and at Dosthill, respectively (Fig. 4.1; see Taylor & Rushton 1971, p. 34 for summary). All the exposures are referred to the Merevale Shales, which are mostly grey-green silty and micaceous shales and mudstones similar to parts of the Shineton Shales to the west. Beds more than 90 m above the base of the formation at Merevale yield *R. flabelliformis* cf. *socialis*, possibly from a high level in the *flabelliformis* Biozone (Bulman & Rushton 1973, p. 10).

### Central–eastern England boreholes

Following the first record at Calvert by Davies and Pringle (1913), numerous boreholes have proved the presence of Tremadoc rocks across central England (Fig. 4.1) directly below an Upper Palaeozoic to Mesozoic cover. The borehole records are summarised by Taylor and Rushton (1971, table 4), Bulman and Rushton (1973, pp. 10–15) and Cowie *et al.* (1972, table 2). Fine sandstones are recorded from some cores, but otherwise lithologies compare directly with those of the Merevale Shales in outcrop, suggesting low energy mud and silt deposition everywhere. Graptolite, trilobite, inarticulate brachiopod, hyolith and trace fossil faunas are well represented through the sequences, with direct evidence of the *flabelliformis, tenellus* and *salopiensis* biozones at different localities. Both in the Merevale cores and in adjacent outcrops the Tremadoc appears to follow conformably on the upper Cambrian (Taylor & Rushton 1971, p. 35).

Post-Tremadoc Ordovician beds are proved only from three boreholes, at Great Paxton, Huntingdon, and Bobbing (Fig. 4.1; Stubblefield 1967; Strachan *in* Williams *et al.* 1972, p. 42). At Great Paxton, some 27 m of grey silty mudstones contain fine sandy, silty and micaceous interbeds showing grading, cross lamination, contortion, channelling and load structures. The sequence contains rich Llanvirn graptolite, trilobite, ostracode and chitinozoan faunas, with less common brachiopods, bivalves and nautiloids (Rushton & Hughes 1981). Skevington (1973) correlated the graptolites with those of the upper Llanvirn *murchisoni* Biozone, but Jenkins (1983) suggested that an early Llanvirn age is more likely. The Raphiophorid Community trilobites and mud sediments suggest a low energy offshore environment, with sands possibly brought offshore by storm-induced currents. The lithology and faunas at Huntingdon are similar to those at Great Paxton, indicating a generally similar age and environment. Pre-Llanvirn beds were not penetrated in these boreholes. Further boreholes at Culford and off the coast of eastern England (Fig. 4.1) provide tentative evidence that the same Llanvirn sediments extend eastwards and then northwards from Huntingdon and Great Paxton in an arcuate subcrop (Wills 1978, p. 30, pls. 1, 2), but fossil data for this age dating are as yet unpublished.

The Bobbing borehole provides evidence of the most south-easterly known occurrence of Ordovician rocks in England (Lister *et al.* 1970). Unconformably below Jurassic strata, 6 m of grey micaceous siltstones and sandstones contain poorly preserved shelly faunas and graptolites suggestive of a Caradoc or Ashgill age; associated chitinozoans and acritarchs support the former. The extent here of any older Ordovician is unknown.

## South Cornwall

Between Gorran Haven and Veryan Bay on the south coast of Cornwall (Fig. 4.1), relatively small, discontinuous outcrops of Ordovician quartzites form tectonic slices or phacoids within a slaty mélange of Devonian rocks along the line of the Lizard–Dodman–Start disturbance zone. Referred to collectively as the Gorran Quartzites, the Ordovician sediments are massive to thickly bedded and in composition are quartz arenites to feldspathic quartz wackes. Fossils are generally uncommon, but some blocks have yielded brachiopods and trilobites indicative of a Llandeilo age (Sadler 1974, M. G. Bassett 1981); crinoidal debris forms impersistent lenses at some localities. In association with the arenaceous sediments, the presence of coarsely ribbed orthacean brachiopods and the trilobite *Neseuretus* as the most common faunal elements invites comparison with the shallow-water, inshore *Neseuretus* palaeocommunity of Fortey and Owens (1978). The Gorran Quartzites are of unknown original thickness, but appear to represent the feather edge of one of the sheets of quartz sands that spread northwards from the Armorican area of NW France on a number of occasions during early and mid Ordovician times (e.g. see Renouf 1975, Henry 1980, M. G. Bassett 1981).

## Lake District

By far the most extensive and impressive area of Ordovician outcrop in the north of England is that of the Lake District of which, along with Silurian rocks to the south, it forms the core. Immediately to the north, the Solway line suggests a position for the Iapetus suture defining the northern limit of the plate underlying the Anglo–Welsh area, and along which the crust of the Protoatlantic (Wilson 1966) or Iapetus (Harland & Gayer 1972) Ocean subducted south-eastwards beneath what is now the Lake District (Dewey 1969, Fitton & Hughes 1970). Modern studies of terrane relationships are leading to a reappraisal of this model of the suture and its south-east subduction (Allen 1987, Soper 1987), although the suggestion (Allen 1987) that the Solway Line does not mark the position of the suture has been strongly refuted (McKerrow & Soper 1989, Fortey *et al.* 1989). No

rocks of Precambrian or Cambrian age are known from the Lake District, the oldest deposits being sediments of the Skiddaw Group in which the earliest dated fauna is of early Tremadoc age (Cooper & Molyneux 1990). These earliest Ordovician sediments are regarded as having been deposited directly on the relatively young and probably thin margin of the southern continental plate (O'Brien *et al.* 1985), the crustal basement of which is envisaged as being low-grade metamorphic, igneous and sedimentary rocks of a very much younger age (see Chapter 18) than the ancient basement of the continent to the north of the suture (Thorpe *et al.* 1984).

The traditional tripartite subdivision of the Ordovician of the Lake District in particular and the north of England in general into the Skiddaw (Slate) Group, the Borrowdale Volcanic Group and the Coniston Limestone Group was modified when Downie and Soper (1972) recognised the Eycott Volcanic Group as a fourth major lithostratigraphical unit, both older and chemically distinct from the Borrowdale Volcanic Group. More recently Moseley (1984) has proposed a 'Windermere Group' to subsume the Coniston Limestone Group in with the overlying much thicker sequence of Silurian sediments. The latter certainly need to be encompassed within a major stratigraphical division and, as the term 'Windermere Rocks' has already been applied to a substantial part of the succession above the Coniston Limestone (Sharpe 1842), the name seems appropriate. But the Coniston Limestone Group is a distinctive association of rocks with the top clearly defined by a major lithological change to the black shales of the Skelgill Beds, a change that is reflected also topographically by a distinctive depression and in palaeogeographical terms by flooding and deepening following the major Hirnantian glacial event. Moreover in northern England generally, although there is no angular discordance between the Skelgill Beds and the underlying Ashgill Shales, there is most likely a non-sequence for it is only in the western part of the Lake District (Rickards 1978) and at Keisley in the Cross Fell Inlier (Wright 1985) that there is a clear sequential passage upwards through the *Glyptograptus persculptus*, *Parakidograptus acuminatus* and *Atavograptus atavus* Biozones. Accordingly the long-established usage of Coniston Limestone Group is retained as a valuable collective term for these upper Ordovician sediments.

Although the areal extent of the Lake District is large compared with the Pennine inliers of Cross Fell, Cautley and Dent, and Craven (Fig. 4.16), it is convenient to treat the region on the basis of its lithostratigraphical grouping rather than to subdivide it areally. The igneous history is discussed both here and in Chapter 16.

## Skiddaw Group

The study of the lower Ordovician Skiddaw Group has had a chequered history since the term was used by Sedgwick (1832) for the suite of mudstones, siltstones and sandstones underlying the main volcanic rocks of the Lake District. This is due largely to structural complexities that have made the establishment of any lithostratigraphical sequence controversial, and to the general paucity of graptolite faunas which has inhibited any effective biostratigraphical control. Many workers have contributed to an understanding of these rocks and Jackson (1978), in his review of the Group, tabulated the principal changes in nomenclature and stratigraphical sequence (Table 4.4).

Controversy continues over what constitutes the Skiddaw Group but, despite problems of dating and field relationships, it primarily concerns stratigraphical nomenclature.

The Eycott Volcanic Group was designated by Downie and Soper (1972), on the basis of microfossils from interbedded mudstones which demonstrated that it was the lateral equivalent of the upper part of the Skiddaw Slate Group and thus older than the Borrowdale Volcanic Group. Marr (1894), it may be noted, had remarked that 'The term Skiddaw Slates has been used for all sedimentary *and contemporaneous volcanic deposits* of the Lake District and its neighbourhood *which lie below* the great volcanic group known as the Borrowdale Series' (our italics), a view endorsed by Elles (1898). However, Wadge (1978) restricted his use of the term Skiddaw Group to those sediments, wholly of Arenig age, below the Eycott Volcanics. Those *D.*

**Fig. 4.16.** Distribution of Ordovician lithostratigraphic units in the main outcrop and inliers of northern England. Subdivision of upper and lower Borrowdale Volcanic Group after Moseley & Millward (1982). Major intrusions included for reference.

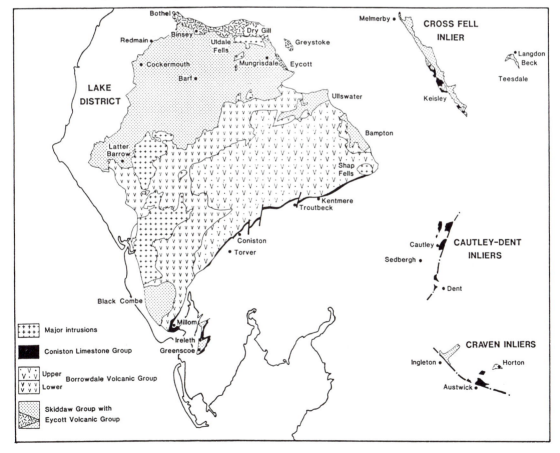

*artus* Biozone mudstones interbedded with the lavas, and the overlying mudstones including those of the *D. murchisoni* Biozone from the Tarn Moor tunnel (Wadge *et al.* 1972), he incorporated in his Eycott Group. Moseley (1984) pointed out that this approach effectively uses chronostratigraphic definitions for lithostratigraphic units, and proposed that the Skiddaw Group be retained for the full sequence of turbiditic sediments and that the Eycott Volcanic Group be applied to the partly contemporaneous outpourings of lavas and tuffs of the one complex volcano. This proposal is followed here.

As regards the age of the Skiddaw Slates, the early claims by Ward (1876, 1879) that the lower part included strata equivalent to the Tremadoc Slates was accepted by Marr (1894) and Elles (1898) on the basis of the occurrence of *Bryograptus* and, with less certainty, *Clonograptus tenellus* at the classic locality of Barf (Fig. 4.16). Subsequent doubts about a pre-Arenig age were confirmed when revisions of these graptolites (Bulman 1941, 1971) showed that the

forms present were not typical Tremadoc species, and when Jackson (1962) demonstrated that the other graptolites from the supposed Tremadoc localities were of Arenig age. Molyneux and Rushton (1984) described a late Tremadoc age biota from the River Calder to the west of Latter Barrow Hill (Fig. 4.16) consisting of acritarchs and sparse macrofossils including *Peltocare olenoides* as the most common element; the assemblage indicates the *A. sedgwickii* Biozone (Fig. 4.3). Further evidence of pre-Arenig age comes from the Uldale Fells to the north, where Jackson (1979) obtained a graptolite fauna with *Didymograptus protobalticus*, thought to be slightly older than the *D. deflexus* Biozone. Subsequently Rushton (1985) obtained *Dictyonema pulchellum* (inviting comparison with the middle Lancefieldian of the Australian graptolite succession). Again a pre-Arenig age is likely, although this form cannot be correlated precisely with faunas in North Wales. Cooper and Molyneux (1990), however, have recorded an early Tremadoc age, probably of the *Clono-*

**Table. 4.4.** Changes in stratigraphical nomenclature and ordering of formations in the Skiddaw Group in the Skiddaw Inlier during the last hundred years (after Jackson 1978).

| Ward (1876) Whitehaven to Mungrisdale | Dixon (1925) Ennerdale to Loweswater | Eastwood et al. (1931) North and South of Ennerdale | Eastwood et al. (1968) Cockermouth area |
|---|---|---|---|
| Black slates of Skiddaw | Mosser striped Slates | Latterbarrow Sandstone | Kirk Stile Slates |
| Gritty bed of Gatesgarth, Latterbarrow, Tongue Beck, Watch Hill and Gt Cockup | Loweswater Flags | Mosser Slates with Kirkstile Slates and Watch Hill Grits | Mosser Slates |
| | Kirkstile Slates | | Skiddaw Grits |
| Dark slates | Blake Fell Mudstones | Loweswater Flags | |
| Sandstones of Grasmoor and Whiteside | | Blakefell Mudstones | |
| Dark slates of Kirk Stile | | | |

| Rose (1954) Keswick–Buttermere | Jackson (1961) Lorton–Mungrisdale | Simpson (1967) Egremont–Bassenthwaite | Jackson (1978) Skiddaw Inlier |
|---|---|---|---|
| Mosser–Kirkstile Slates | Mosser–Kirkstile Slates | Latterbarrow Sandstone | Latterbarrow Sandstone |
| Loweswater Flags | Loweswater Flags | Sunderland Slates | Kirk Stile Slates |
| | Hope Beck Slates | Watch Hill Grits | Loweswater Flags |
| | | Mosser Slates | Hope Beck Slates |
| | | Loweswater Flags | |
| | | Kirkstile Slates | |
| | | Blakefell Mudstones | |
| | | Buttermere Flags | |
| | | Buttermere Slates | |

*graptus tenellus* Biozone, from within the Buttermere Formation olistostrome (based on an acritarch assemblage, S. G. Molyneux, *pers. comm.*).

The highest biozones in the Skiddaw Group are those of Llanvirn age in the Ullswater and Bampton inliers on the eastern edge of the Lake District. The *D. bifidus* [i.e. *D. artus*] Biozone was recorded by Elles (1898) from Aik Beck and Thornship Beck, but is best known from the grey-black shales alternating with tuffaceous horizons in the Tailbert–Lanshaw tunnel of the Bampton Inlier (Skevington 1970). *D. artus* is present in an extensive fauna dominated by *Cryptograptus tricornis* and *Glyptograptus dentatus*. The youngest Llanvirn faunas are those from the Tarn Moor tunnel, which contain graptolites of the *D. murchisoni* Biozone, described by Skevington (*in* Wadge *et al.* 1972); the enclosing sandy and silty black mudstones (the Tarn Moor Mudstones) have faulted contacts in the tunnel and are not known from surface outcrop.

Leaving aside the Eycott Volcanic Group, the Skiddaw Group is composed of mudstones with siltstones and sandstones representing turbidite deposits laid down in a relatively deep-water slope environment at the edge of the Anglo–Gondwanan Continent (Cocks & Fortey 1982) bordering the Iapetus Ocean. The oldest part of the sequence is formed by those sediments with the Tremadoc fossils noted above in the River Calder. This outcrop is close to an outcrop of the Latterbarrow Sandstone which, within 1 km, rests on Skiddaw Slates with an Arenig fauna, so that the Latterbarrow Sandstone is clearly unconformable on the Skiddaw Group (Molyneux & Rushton 1984). The pre-Arenig sediments from the Uldale Fells are also older than the earliest named subdivision of the Skiddaw Group in Jackson's (1978) revised sequence, the Hope Beck Slates (Fig. 4.17). Jackson did not obtain fossils from this striped silty mudstone formation, but Molyneux and Rushton (1984) reported a graptolite suggesting the *D. deflexus* Biozone.

This biozone spans the lower half of the overlying Loweswater Flags Formation, a coarser rhythmic sequence of grey flaggy sandstones and occasional fine conglomerates alternating with mudstones. The arenites display plentiful sedimentary structures of turbidite origin with the bottom currents flowing NNE (Jackson 1961) down the slope into the Iapetus ocean deeps to the north. Near the base of the formation a brief volcanic intercalation occurs in the form of the Watch Hill Lava, a 10 m thick felsite flow. The appearance of *Didymograptus nitidus* about half way up the formation is taken as the base of the *nitidus* Biozone, which persists to the top of the sequence. Here the sandstones become thinner as the formation

grades upwards into the dark, striped silty mudstones of the Kirk Stile Slates.

The Kirk Stile Slates form a substantial thickness (1,000 m) of argillaceous sediment (Jackson 1978). Near the top, rare coarser deposits occur in the form of mudstone breccias interpreted by Jackson (1978) as fluxoturbidites, rocks regarded as being diagnostic of slope deposits. Palaeontologically, the lower half of the Kirk Stile Slates is characterised by *Phyllograptus*, with *Isograptus gibberulus* as the zone fossil. The appearance of *D. hirundo* heralds the *hirundo* Biozone in the upper part of the sequence, where it is associated with abundant specimens of *Glyptograptus* sp., *Tetragraptus bigsbyi*, *Cryptograptus* and *Tristichograptus*. A basal *D. artus* Biozone fauna is recorded by Bulman (1968) from a locality 1 km south of Mungrisdale. This is the only Llanvirn record from the main outcrop of the Skiddaw Group accepted as valid by Jackson (1978); the other *artus* faunas and also the single record of the upper Llanvirn *D. murchisoni* Biozone in the highest beds of the Skiddaw Group are all confined to the eastern inliers of Ullswater and Bampton as noted above.

Above the Kirk Stile Slates, the Latterbarrow Sandstone is of uncertain stratigraphical placement as it has yet to yield fossils. The unit was considered to be part of the Skiddaw Group by Eastwood *et al.* (1931), Trotter *et al.* (1937), Jackson (1961) and Wadge (1978), whilst Simpson (1967) contended that the Latterbarrow Sandstone rests with unconformity on the Skiddaw Group and forms an impersistent non-marine basal unit to the Borrowdale Volcanic Group. Although the base is nowhere exposed, the presence of an underlying unconformity is supported by the faunas (Molyneux & Rushton 1984), which indicate that the sandstone rests on different horizons of the Skiddaw Group.

The Latterbarrow Sandstone occurs in the area around Latter Barrow, about 6 km east of Egremont. Lithologically it is a greenish- or purplish-grey medium-grained quartzose sandstone with some pebbly lenses (Eastwood *et al.* 1931). P. M. Allen and Cooper (1986) estimated a thickness of about 400 m, with the sedimentary structures implying an estuarine environment, contrasting with the turbidite units of the Skiddaw Group. Their work also supported the view that the Latterbarrow Sandstone is related to the Borrowdale Volcanic Group. Further, they considered some anomalous petrographic features of the upper part of the sandstone formation are best explained by a penecontemporaneous volcanic input. A short period of emergence is envisaged prior to the main Borrowdale volcanic episode, which resulted in the sandstones being succeeded by coarse volcanic debris flows, except on Boat How where they are directly overlain by andesitic lava. Allen and Cooper

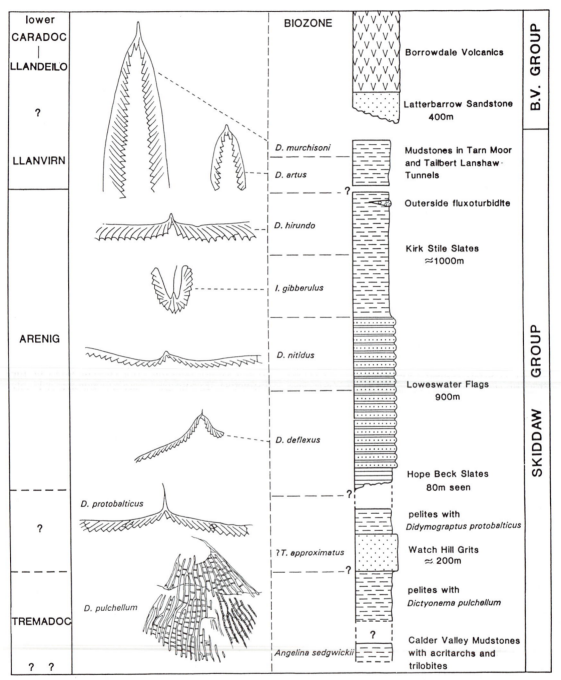

**Fig. 4.17.** Summary of the stratigraphy of the Skiddaw Group (based on Jackson 1978, Molyneux & Rushton 1984, 1988; Cooper & Molyneux 1990). The presence of the *Tetragraptus approximatus* Biozone has yet to be confirmed from fossil evidence.

(*op. cit.*) regarded the Latterbarrow Sandstone as being deposited in late Llanvirn or early Llandeilo times. The record of Eastwood *et al.* (1968) of quartzite clasts 'indistinguishable from the Latterbarrow Sandstone' near the base of the Eycott Volcanic Group is not accepted by Allen and Cooper (1986) on the basis of re-examination of the single thin section. Genuine Latterbarrow Sandstone here would be critical as this part of the sequence yielded the microfauna that Downie and Soper (1972) interpreted as being of earliest Llanvirn age.

An isolated outcrop of sandstone which crops out from beneath the Carboniferous cover north north-east of Cockermouth is usually regarded as the Latterbarrow Sandstone. The unit occurs in Tommy Gill, Redmain, and was termed the Redmain Formation by Allen and Cooper (1986) who differentiated it, by the lack of contained evidence of contemporaneous volcanicity, from the Latterbarrow Sandstone some 20 km to the south.

In the south-west, the Skiddaw Group crops out in the inliers of Black Combe and Greenscoe (Fig. 4.16). At Black Combe the sequence was subdivided by Helm (1970) into three formations, although the stratigraphical succession is somewhat uncertain in this structurally complex group. The formations comprise the Whicham Blue Slates, banded blue siltstones which pass into blue mudstones; the Townend Olive Slates, olive mudstones with only a fine striping effect from the siltier horizons; and the Fellside Mudstones, less fissile and predominantly greenish-grey mudstones of mottled appearance. The lower Ordovician fossils recorded by Smith (1912) and Helm (1970) from the Whicham Blue Slates were not zonally diagnostic but Rushton and Molyneux (1989) have recovered a *D. hirundo* Biozone association. Immediately to the south mudstones associated with tuffs and volcanics in the Whicham Valley have yielded mid-Arenig acritarchs (Turner & Wadge 1979). This record is important for demonstrating that local volcanic activity in the Lake District started much earlier than previously suspected. Across the Duddon Estuary at Greenscoe an assemblage of high *D. hirundo* to low *D. bifidus* Biozones was recorded from Skiddaw Group mudstones by Knipe and Grieve (Soper 1970). A rich microfauna was interpreted as probably of *D. bifidus* Biozone age, but the *D. murchisoni* Biozone was not ruled out by Downie (Rose & Dunham 1977).

Recent mapping of the Skiddaw Group by the British Geological Survey has shed new light on this group through the recognition of a major lineament coinciding with the Causey Pike Fault. This cuts across the Skiddaw Inlier along an ENE trend and can further be traced into the Cross Fell Inlier (Cooper & Molyneux 1990). The structural style differs on either side of the fault (Webb & Cooper 1988), and two distinct turbidite sequences are now recognised. To the north, a regularly bedded sequence shows slump folding overturned to the E and SE. To the south, the Buttermere Formation (Webb & Cooper 1988) additionally contains sedimentary breccias in the clastic sequence, with slump folds overturned to the N and NW. This Formation is interpreted as an olistostrome, probably emplaced in Llanvirn times. The bedding is disrupted and includes units from early Tremadoc (*Clonograptus tenellus* Biozone?) possibly to basal *D. artus* Biozone. Overlying the Buttermere Formation are the younger Llanvirn mudstones noted above.

*Eycott Volcanic Group*

These rocks (Downie & Soper 1972) are exposed abutting against the Carboniferous along the northern edge of the Ordovician outcrop in a narrow continuous east–west strip from Bothel to Linewath, at Eycott Hill on the north-east and as an inlier in the Carboniferous at Greystoke Park (Fig. 4.16). As already noted, the Group is contemporaneous with the upper part of the Skiddaw Group, an early Llanvirn fauna being obtained close to the base of the sequence south of Binsey. The succession, about 2,500 m, was divided by Eastwood and co-workers (1968) into the Binsey Formation and an overlying High Ireby Formation; both contain flows of the distinctive 'Eycott-type' basaltic andesite (a highly feldsparphyric rock with large (3–4 cm) labradorite phenocrysts).

The Binsey Formation, about 1,200 m on Binsey, starts with three of these flows (Lower Eycott Lava) with interbedded Skiddaw Slate marine sediments. They are followed by non-porphyritic andesites with tuff and agglomerates, with the upper half of the succession dominated by microporphyritic andesites with feldspar and pyroxene phenocrysts. The Formation thins eastwards, with the Lower Eycott Lavas absent from Eycott Hill where the 120 m present is mainly composed of microporphyritic andesites overlying andesitic tuffs with intercalated tuffaceous mudstones and sandstones near the base.

The succeeding High Ireby Formation, attaining about 1,300 m, is marked at the base by the Middle Eycott Lavas, poorly exposed north of Binsey, with the thickest sequence (300 m) recorded from five flows at Linewath, reducing to only two flows (75 m) at Eycott Hill. At High Ireby these lavas are succeeded by aphyric and microporphyritic andesites in poorly exposed ground which shows in addition a pink banded rhyolite and an Eycott-type lava (Upper Eycott Lava). At Greystoke Park, a flow banded rhyolite about 100 m thick occurs in a sequence of 800 m of andesite and tuffs; a thin rhyolite (12 m) is also present on Eycott Hill where the succession is dominated by andesites above the Middle Eycott Lavas.

The evidence for the source of these volcanics is discussed by Eastwood *et al.* (1968). For the Binsey Formation, thickening of the lavas points to an eruptive centre lying to the north or north-west of the present outcrops, with the source of the Eycott lavas not far from Binsey where they occur at three well-separated horizons. The number as well as the thickness of the Middle Eycott flows also decreases to the south and east, again pointing to a north to north-west centre. The same source area continued to be active after the Middle Eycott Lava flows, but other evidence, such as the thick rhyolite in Greystoke Park, suggests an additional centre active to the east or north-east at that time. Despite the presence of these acidic rocks, the Eycott Group is dominated by the basalt and basaltic–andesite lavas of transitional tholeiitic to calc–alkaline type. These are considered to have erupted in an island arc environment sited closer to the oceanic margin of the plate than the calc–alkaline rocks of the Borrowdale Volcanic Group (Fitton & Hughes 1970).

### Borrowdale Volcanic Group

The Green Slates and Porphyries of Sedgwick (1832), first referred to as the Borrowdale Series by Harkness and Nicholson (Nicholson 1872), are arguably the most widely known rocks of the Lake District. Not only do they form the scenically spectacular fells of the central area (Fig. 4.18), but the green slates, quarried near Coniston, Honister, Kirkstone and Langdale are famed for roofing and as ornamental stones well beyond the confines of the Lake District.

In the absence of fossils age limits have been defined by the underlying Tarn Moor Mudstones (upper Llanvirn) and the overlying Drygill Shales (Longvillian stage of the middle Caradoc) and regarded as Llandeilo–early Caradoc. This view has not been disproved by the discovery of acritarchs suggestive of a Caradoc age in the upper part of the Group (Molyneux *pers. comm.*). However, as pointed out by Wadge (1978), volcanic deposits commonly accumulate very quickly so that the eruptions could have taken place over only a small interval of the broad time span.

**Fig. 4.18.** Typical Lake District scenery in the Borrowdale Volcanic Group. View of Langdale Pikes from Silver Howe, with the peak of Bow Fell at centre right, and Crinkle Crags on the left horizon. (Photo courtesy of F. W. Dunning.)

Differences of opinion over the nature of the junction of the Skiddaw and Borrowdale Volcanic Groups of a few years ago (Ingham & Wright 1972) were clarified by Wadge (1972, 1978) and Jeans (1972) who, after consideration of the relatively few sections where the junction is neither faulted nor covered by drift or scree, concluded that Simpson's (1967) view of an unconformity is the one best borne out by field evidence. As noted above, in the west the Latter-barrow Sandstone lies unconformably on different horizons of the Skiddaw Group (Molyneux & Rushton 1984) and is taken as being closely related to the overlying volcanics which succeed it disconformably in the form of debris-flows or andesites (Allen & Cooper 1986). Eastwards, the differing horizons of the Skiddaw Group are commonly overlain by a variable thickness of conglomerates containing volcanic and mudstone clasts, which culminates in the eastern inliers with over 130 m of the Bampton Conglomerate beneath the tuffs and lavas (Wadge *et al.* 1972).

The Borrowdale Volcanic Group comprises 6,000 m of calc–alkaline volcanics which, although dominated by andesitic lavas and tuffs, show a considerable variation in both the lavas and pyroclastics due to contributions from several volcanic centres. Local successions and their lateral correlatives are, however, reasonably well established (Branney & Soper 1988). Amongst an extensive literature, the works of Mitchell (1929, 1934) and Moseley (1960) for the east, Hartley (1925, 1932, 1942) for the centre and Mitchell (1940, 1956, 1963) and Firman (1957) for the south-west are probably the major contributions to the mapping of the group. In more recent years, detailed mapping has revealed intraformational unconformities, e.g. between the Rainsborrow Tuffs and the overlying Kentmere Dacites (Soper & Numan 1974); but as Moseley and Millward (1982) pointed out, much detailed work is required before the Borrowdale Volcano is fully understood. (See Chapter 16.)

Volcanicity appears to have started in the south-west, where a thick sequence was described by Firman (1957). Although the rocks to the north-east are somewhat younger (Fig. 4.19), these lower parts of the sequence are predominantly basic, consisting of basalts, basaltic andesites and some tuffs. A marked change to acidic magma occurs at the horizon of the Airy's Bridge ignimbrites, the base of which is readily correlated with that of other ignimbrite and dacite units across the Lake District. Higher up, andesitic compositions return with the Wrengill Andesites, where either lava or tuff may be locally dominant; these again give way to acid rocks at the top of the sequence.

The interpretation of these sequences in terms of either two or three variable and complex major cycles in the Borrowdale pile, or three or more major

eruptive centres each with subsidiary vents producing the complexities, was discussed by Millward *et al.* (1978). Positively identified eruptive centres are uncommon, although the Greenscoe vent (Soper 1970) or volcanic neck (Rose & Dunham 1977) is well known. Moseley and Millward (1982) cited the Haweswater complex as being probably the most important centre of the eastern outcrops. The finely preserved sedimentary structures of many tuff horizons clearly indicate subaqueous deposition, while other tuffs which lack such structures and have only a rough bedding are regarded as being subaerial in origin. The record of marine microfossils reported by Numan from mudstones in the Duddon Hall Tuff sequence (Millward *et al.* 1978, Millward 1980a, b) supports the interpretation of the Borrowdale sequence as being formed along a chain of volcanic islands with strato-volcanoes of interbedded lava and pyroclastic deposits, developing into more prominent shield volcanoes characterised by the outpouring of subaerial ignimbrites (see p. 495 and Fig. 16.4).

*Coniston Limestone Group*

In contrast to the earlier Ordovician groups, the mainly sedimentary rocks of the Coniston Limestone Group are better developed, both in terms of thickness and of stratigraphic continuity in the Pennine inliers rather than in the Lake District, despite the length of the main outcrop across the area from Shap to Millom, a distance of some 45 km (Fig. 4.16). Along this outcrop the basal deposits are of Ashgill age and rest with clear unconformity on different horizons of the Borrowdale Volcanics and, in the south-west, on the Skiddaw Group.

The only outcrop in the north of the Lake District is at Carrock Fell, where the Coniston Limestone Group is represented in Dry Gill (Fig. 4.16) by the Drygill Shales (Nicholson & Marr 1887) of much earlier mid-Caradoc age. A presumed unconformable relationship to older rocks is not seen as the small outcrop is faulted against Eycott Volcanics to the north, while in the south it is either faulted against or, in the view of Eastwood *et al.* (1968), has an intrusive contact with part of the younger Carrock Fell Igneous Complex (see Chapter 16). Where not bleached by the Hare-stones Felsite, the Drygill Shales are dark calcareous mudstones and shales, and have yielded tectonically distorted fossils. The fauna, redescribed by Dean (1963), includes the trilobites *Broeggerolithus nichol-soni, Flexicalymene* cf. *caractaci* and *Kloucekia apiculata*, and indicates a Longvillian age for the formation, which is probably of the order of 100 m thick.

The incomplete nature of the Coniston Limestone Group in the main outcrop is partly a result of two major stratigraphical breaks but also from the absence of the earlier part of the succession seen at Cross Fell (p. 119). The accepted oldest part of the sequence, the Stile End Grassing Beds of Harkness and Nicholson (1877), has been regarded as Actonian

(upper Caradoc) following the work of Dean (1963). Later, McNamara (1979) demonstrated that this unit forms part of a laterally varied sequence of rocks of Cautleyan (Ashgill) age extending up to the first major break.

**Fig. 4.19.** Comparative sections in the Borrowdale Volcanic Group (Moseley & Millward 1982). The horizontal line divides the lower from the upper volcanics (see Fig. 4.16).

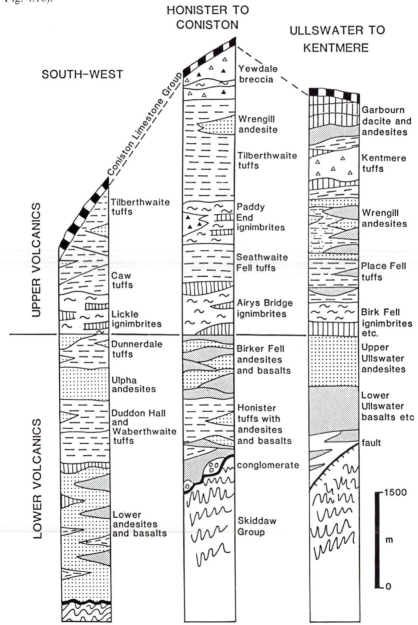

The earliest beds, which mark the submergence of an irregular surface of Borrowdale Volcanics, are of very variable thickness. In the north-east, around Longsleddale, the basal deposits below the Stile End Beds were separated by McNamara as the Longsleddale Formation (35 m thick), with a lower half of poorly sorted conglomerate succeeded by mainly red silts and shales, a coarse green sandstone forming the top 8 m. Fossils are absent in the mainly fluvial sequence, although to the south-west beyond Windermere the basal sandstones are marine and have yielded *Toxochasmops marri*, *Decoroproetus piriceps* and *Encrinurus cornutus*, indicating a Cautleyan Zone 2 age (Ingham 1966). The Stile End Member, a calcareous siltstone up to 28 m thick, has a restricted areal distribution between Longsleddale and Kentmere (Fig. 4.20), with *T. marri*, *D. piriceps* and *Ascetopeltis apoxys* as the most common trilobites (McNamara 1979), grouped as the Proetid Association by McNamara and Fordham (1981).

The overlying member, the Yarlside Rhyolite, is also of restricted distribution, extending from the Longsleddale area towards Shap and attaining a maximum thickness of 185 m on Great Yarlside Crag (Millward & Lawrence 1985). The pink to pale grey flow-banded rock, regarded as a rhyolite *sensu stricto* by Moseley and Millward (1982), was interpreted as a rheomorphic ignimbrite by Millward and Lawrence (1985), a likely interpretation but one not without problems (Branney 1986).

The succeeding Applethwaite Formation is conformable on the underlying sequence except around Longsleddale, where the lowest metre is a rhyolitic-pebble conglomerate, resting on the uneven surface of the Yarlside Rhyolite. Like the Longsleddale Formation, the Applethwaite sediments are traceable throughout the main outcrop although variable in both facies and thickness. On Applethwaite Common, the sequence of about 55 m (first described by Marr 1892), is made up of calcareous mudstones and argil-

**Fig. 4.20.** Lateral variation within the upper Coniston Limestone Group in the main outcrop of the southern Lake District. Lithostratigraphic nomenclature emended from Ingham & McNamara 1978 and McNamara 1979.

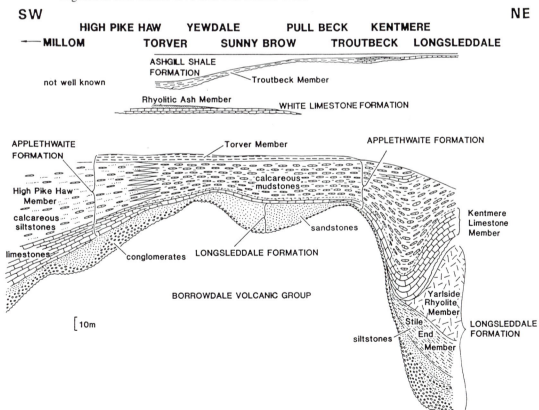

laceous limestones with a diverse shelly fauna, the commonest trilobites being *Calymene subdiademata* and *Gravicalymene susi*. This assemblage, occurs in fine grained carbonate rocks (the *Calymene* Association of McNamara & Fordham (1981)). The fauna is clearly Cautleyan Zone 2, but gives way at the top of the formation to Zone 3 with the replacement of *Tretaspis convergens convergens* by *T.c. deliquus* and the appearance of forms like *Calymene prolata*. The basal 8 m of the succession on Applethwaite Common comprises limestones with shale interbeds (Kentmere Limestone Member of McNamara (1979)) but it thickens to the south-west to form 28 m of sparry limestone near Millom. To the south-west of Coniston, the argillaceous carbonates give way to calcareous siltstones and silty limestones (High Pike Haw Member) with a restricted trilobite fauna comparable with that of the silty Stile End Member. In the Ash Gill section described by Marr (1916), his 'Calymene Beds' are equated with the Applethwaite Formation (Ingham 1966, McNamara 1979).

Overlying the *Calymene* Beds are dark blue mudstones (Marr's 'Phillipsella Beds'). They were termed the Torver Formation by McNamara. They range in thickness up to 7 m, again contain a Zone 3 fauna and are here regarded as a member of the Applethwaite Formation (Fig. 4.20). The member contains the Illaenid Association of McNamara and Fordham (1981), an assemblage which characterises the mudstone facies, with *Stenopareia bowmanni*, *Tretaspis convergens deliqua* and *Primaspis bucculenta* being the dominant trilobites along with *Phillipsinella parabola aquilonia*. The Applethwaite Formation comprises shallow water shelf sediments, with variable carbonate and clastic content being reflected by different faunal associations. The ensuing stratigraphic break marks a period of regression (Ingham & Rickards 1974).

The succeeding units at Ash Gill, the White Limestone and Ash (Marr 1916), are now known to be of Rawtheyan age (Ingham 1966). The former was correlated with Zone 6 of the Cautley succession by Ingham, and confirmed by McNamara (1979) from the sparse fauna, which includes *Tretaspis* aff. *latilimbus distichus* and *Otarion* sp. nov. of Ingham (1970); thus there is a major stratigraphic break below this 4 m micritic limestone unit. The overlying rhyolitic tuff is also thin (5 m) and in the main outcrop is restricted to the Coniston area, although it forms a useful volcanic marker which can be traced further afield in northern England. Despite the restricted distribution, the White Limestone is best regarded as a separate formation to include the overlying Rhyolitic Ash as a member (Figs. 4.20, 4.21). These deposits were laid down under shallow shelf conditions following a temporary Rawtheyan flooding.

The widely distributed 'Phacops mucronatus Beds' and Ashgill Shales of Marr (1916) follow after a further stratigraphic break corresponding to Zone 7 deposition at Cautley (Ingham 1966). The former unit (Troutbeck Formation of McNamara 1979), here regarded as the basal member of the Ashgill Shale Formation, consists of about 5 m of green-grey mudstones. Although laterally persistent, there is considerable variation in carbonate content; in Skelghyll Beck a limestone surface with large orthoconic nautiloids and stromatolites is particularly striking. McNamara (1979) recorded ten trilobite species from the Troutbeck Member including *Mucronaspis olini* and *M. mucronata*. Of these only *M. mucronata* persists into the Ashgill Shales, with a marked change also from the usual Rawtheyan brachiopod fauna (Ingham & Wright 1970) to that of the *Hirnantia* fauna (Temple 1965); the base of the Ashgill Shale *sensu stricto* forms the base of the Hirnantian Stage (Ingham & Wright 1972).

The Ashgill Shales leaden grey mudstones with a maximum thickness of about 20 m in the main outcrop are thicker across the Duddon Estuary and at Cautley. The formation is patchily fossiliferous with a restricted *Hirnantia* fauna (Wright 1968) which, in addition to *Mucronaspis mucronata*, includes *Hirnantia sagittifera*, *Dalmanella testudinaria*, *Eostropheodonta hirnantensis*, *Kinnella kielanae* and *Plectothyrella crassicosta*. After the shallow water calcareous mud at the transgressive base of the formation, the Ashgill Shales are considered to have formed as an offshore belt of shallow water muds (Lawrence *et al.* 1986).

The Ordovician–Silurian boundary lies within the Skelgill Member of the Stockdale Shales Formation of Moseley (1984). It is only in the Coniston area of the Lake District that the *G. persculptus* Biozone is seen to follow the Ashgill Shales, being present in a 30 cm unit of pale grey mudstones (Hutt 1974) in Yewdale Beck. Elsewhere the *G. persculptus* Biozone is unproven, and has been interpreted either as a non-sequence as at Browgill (Rickards 1978) or by the graptolitic facies having changed to a calcareous one with a shelly fauna, that is overlain by black shales of *P. acuminatus*, lower or upper *A. atavus* Biozones. The carbonates are the 'Basal Beds' of Rickards (1970) and include the *Atrypa flexuosa* Limestone of Marr and Nicholson (1888). Comparison with Keisley (Wright 1985), where comparable thin carbonates with a shelly fauna are overlain by *G. persculptus*, suggests that the Basal Beds are of Hirnantian age and that a nonsequence is widely developed.

The Coniston Limestone Group to the east of the Duddon Estuary around Ireleth differs somewhat from the main outcrop. The Applethwaite Formation consists of calcareous mudstones, locally silty, and limestones, referred to by Rose and Dunham (1977) as the High Haume Mudstone Formation with the High Haume Limestone Member. The Formation, with a maximum thickness of 150 m, again has a Cautleyan fauna. Above a stratigraphic break, which encompasses the *Phillipsinella* Beds and White Limestone, a 25 m thick rhyolite flow (the High Haume Rhyolite) is the last recorded Ordovician lava flow, equivalent to the ash horizon of the Ash Gill section. Although the contact is not seen, this lava is succeeded by 9 m of a grey-green mudstone with *Mucronaspis* cf. *olini* (representing the *Phacops mucronatus* beds of Marr) and then the Ashgill Shale Formation, with a maximum estimated thickness of 400 m and the usual *Hirnantia* fauna. These blue-grey mudstones are interrupted within the highest 100 m by a coarse sandy phase of up to 50 m, termed the Rebecca Grit by Rose and Dunham (1977).

Although it is agreed that the sub-Coniston Limestone Group unconformity was followed by a transgression across a mainly positive area (formed from the earlier volcanic pile), there are differing interpretations of the palaeogeography of the Lake District during this transgression. Ingham and McNamara (1978) suggested southerly overstep on to a horst situated to the *south* of the main Coniston Limestone outcrop, while Firman and Lee (1986) regarded the land area as lying to the *north*, interpreting the horst as being a result uplift associated with a suggested Ordovician age for the intrusion of the Lake District batholith (see Chapter 16).

## Pennine Inliers

### Cross Fell

The Cross Fell Inlier, situated along the Pennine Fault line at the western margin of the Alston Block, contains faulted segments of the Skiddaw, Eycott, Borrowdale and Coniston Limestone Groups. The Skiddaw Group, evidently the oldest unit at outcrop across northern England (p. 110), can be traced eastwards (Fig. 4.16) to the small Teesdale Inlier (Johnson 1961, Lister & Holliday 1970) where it occurs with rhyolitic ash of the Borrowdale Volcanics, and has been further proved in boreholes near Crook (Woolacott 1923) and at Allenheads (Burgess 1971; Fig. 4.1).

The Skiddaw Group at Cross Fell is commonly highly cleaved and lacks fossils. The argillaceous Ellergill Beds of Nicholson and Marr (1891) with *D. artus* Biozone graptolites were regarded as older than the Milburn Group of Goodchild (1889) in which shales are interbedded with ashes and andesitic lavas. The latter were regarded by Shotton (1935) as being of high *artus* (i.e. *bifidus*) age and marking the incoming of volcanic facies at the end of Llanvirn times.

Microfossil and graptolite discoveries of the late sixties suggested that although both units were of *artus* age, the Ellergill beds were the younger (Skevington 1970); whilst the recognition by Downie and Soper (1972) that the Eycott Volcanics were of earliest Llanvirn age indicated that the Milburn Beds were the lateral equivalent of these, and not the Borrowdale Volcanics as had been supposed.

The stratigraphy of the Skiddaw Group was revised accordingly by Burgess and Wadge (1974). The new lower division (Murton Formation) consists of argillaceous sediments with subordinate turbidite sandstones and sparse fossils indicating an Arenig age. The shales of the upper division (Kirkland Formation), interbedded with submarine ashes and rare lavas in the lower part (previously the Milburn Beds), are divided into lower and upper *artus* Biozone faunas, the latter (previously the Ellergill Beds) being characterised particularly by abundant *Nicholsonograptus fasciculatus*; the microfossil assemblages support these divisions. As in the Lake District, the base of the Skiddaw Group is not seen, but at least several thousand metres of beds are present (Burgess & Wadge 1974). To the east of Melmerby, at the north end of the inlier, a faulted block of basic andesites with some fine-grained ash includes 55 m of Eycott-type porphyritic andesite. This outcrop is regarded as the easterly extension of the Eycott Volcanic Group of the northern Lake District (Arthurton & Wadge 1981). The current revision of the Skiddaw Group by the British Geological Survey has shown that the turbiditic siltstones and sandstones at the northern end of the Inlier, that is to the north of the extension of the Causey Pike Fault, contain acritarchs of latest Tremadoc or earliest Arenig age (Cooper & Molyneux 1990). This unit, now named the Catterpallot Formation, thus forms the oldest part of the Cross Fell sequence. In the southern half of the inlier, over 1,000 m of acid tuffs and rhyolites crop out sporadically and are correlated with the Upper Rhyolites of the Borrowdale Volcanic Group in the Lake District. The sequence is subdivided into three formations, with the thickest middle unit (Knock Pike Tuff Formation) of welded-tuffs and rhyolites forming the features of Knock Pike and Dufton Pike.

The principal unit of the Coniston Limestone Group is the Dufton Shale Formation (Fig. 4.21), with an estimated thickness of about 400 m (Burgess and Wadge 1974) in a faulted sequence of incompetent rocks. The formation comprises mudstones and siltstones with limestone nodules and locally occurring sandy units near the base and at the top, ranging in age from the mid-Caradoc Longvillian Stage through to the Cautleyan Stage of the Ashgill Series.

Locally, as on Roman Fell, the basal beds consist of ashy sandstones and siltstones termed the *Corona*

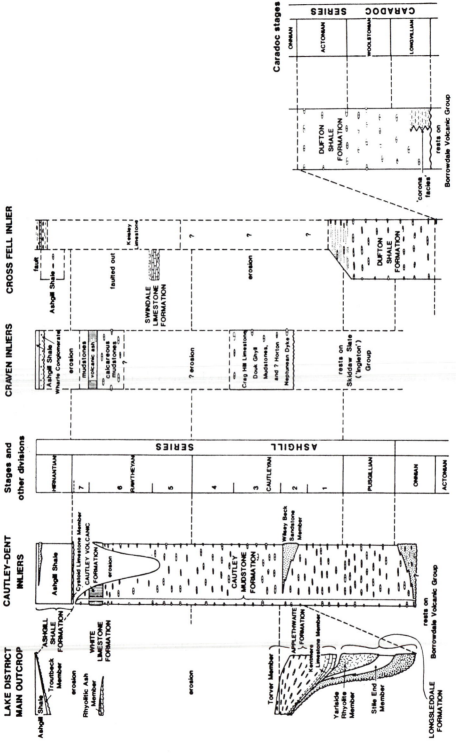

**Fig. 4.21.** Correlation of the Coniston Limestone Group across northern England; small faulted outcrop of Drygill Shales (Longvillian) omitted from figure. In part from Ingham & McNamara (1978) and Ingham & Wright (1972).

Beds by Nicholson and Marr (1891) after the commonly occurring brachiopod *Trematis corona*. Burgess and Holliday (1979) preferred the term '*corona* facies' as they did not regard these sediments as forming a mappable unit. Associated with the *Trematis* is *Lingulasma tenuigranulatum* and various molluscs, particularly bivalves (*Lyrodesma*) and gastropods (*Lophospira, Sinuites*). In places the fauna includes *Dalmanella indica* of Longvillian (*sensu* Hurst 1979c) age. The *corona* facies fauna differs so markedly from that of the purple-stained grey silty mudstones at Melmerby that Dean (1959) distinguished the latter as the 'Melmerby Beds'. The fauna, dominated by brachiopods and trilobites includes *Cremnorthis parva, Dalmanella horderleyensis, Leptestiina oepiki, Broeggerolithus nicholsoni, Brongniartella ascripta* and *Kloucekia apiculata*.

The succeeding Woolstonian (i.e. upper Longvillian of earlier accounts) and later stages are of more uniform dark grey mudstones with a variable carbonate content and, as in the similar but mainly younger sequence at Cautley (fol.), the stages are determined from the contained shelly faunas. These have been well documented and effectively correlated with the stages of the type Caradoc Series in Shropshire by Bancroft (1933) and Dean (1959, 1962). Typical elements of the Woolstonian Stage are *Broeggerolithus nicholsoni longiceps, Bancroftina robusta* and *Sowerbyella sericea*; of the Marshbrookian, *Broeggerolithus transiens* and *Brongniartella bisulcata*; of the Actonian, *Flexicalymene* spp. and *Onniella aspasia*; and of the Onnian, *Flexicalymene onniensis, Onnia* spp., *Onniella* cf. *broeggeri* and abundant *Sericoidea*.

In the overlying Ashgill Series, the Pusgillian sees the arrival of *Tretaspis moeldenensis* and *Gravicalymene jugifera*, with the lithology of the upper part being modified by the incoming of quartz sand that is present also in the thin sequence of Cautleyan (Zone 1) found only in Billy's Beck. The fauna includes *Calymene marginata* and a typically rich Cautleyan brachiopod assemblage including *Glyptorthis, Nicolella, Orthambonites, Plaesiomys, Ptychopleurella* and *Sampo*.

Whilst Cautleyan beds are present in Billy's Beck, the overlying nodular and argillaceous Swindale Limestone in Swindale Beck rests unconformably on Dufton Shales of Pusgillian age. The Swindale Limestone has an extensive fauna including *Phillipsinella parabola, Staurocephalus clavifrons* and *Christiania 'tenuicincta'* and is of Rawtheyan age, at about the Zone 5–Zone 6 boundary according to Ingham and McNamara (1978). Thus there is a marked stratigraphical break as in the Lake District and in contrast to the Cautley District. The argillaceous nature of much of the Swindale Limestone sequence induced Burgess and Wadge (1974) to use the term 'Swindale

Shales' for the formation. This unit, not here accepted, included beds of Hirnantian age identical lithologically and faunally with the Ashgill Shales, even to the extent of containing a more calcareous lithology with *Mucronaspis olini, M. mucronatus* and cystoids as in the basal member of the Ashgill Shale Formation elsewhere.

The Ashgill outcrops at Keisley Bank are noteworthy for a carbonate–mud mound development, the precise age of which has long been in dispute, and for the occurrence of the Ordovician–Silurian boundary in a continuous sequence. Nodular limestones and siltstones at the base give way to massive limestones, some 50 m thick, in places packed with illaenid trilobites and other shelly fossils. The extensive faunal lists (Reed 1896, 1897; Burgess and Holliday 1979) include forms noted above from the Swindale Limestone which may represent a flank facies; certainly the mound is of Rawtheyan and at the top Hirnantian age (Wright 1985); whether the lowest beds are as old as Cautleyan is an open question (Ingham and Wright 1972). Within a metre of the top, the massive limestones change first to a thinly bedded bioclastic limestone–calcareous silt alternation which yields a diverse *Hirnantia* fauna with brachiopods such as craniaceans, *Dolerorthis, Paracraniops, Reuschella, Skenidioides* and *Toxorthis* in addition to more usual elements like *Hirnantia, Hindella* and *Kinnella*; then to green siltstones with graptolites of the *Glyptograptus persculptus* Biozone; and then to dark mottled silts with *Parakidograptus acuminatus* Biozone graptolites of the basal Silurian (Wright 1985). The shallowing and deepening at this boundary, which is evidenced by both sediments and faunas, is attributed to the waxing and waning of the Hirnantian ice-sheet across the southern continents.

## Cautley and Dent

Six small inliers of Ordovician rocks crop out to the east of Sedbergh along the Dent Fault line (Figs. 4.1, 4.16). Apart from one small area of purple andesite ascribed to the Borrowdale Volcanic Group by Ingham (1966), the 640 m of Ordovician rocks all belong to the Coniston Limestone Group.

Marr (1913) designated the district as the type area for the Ashgill Series and it now includes the type successions for the two middle stages of the Series, the Cautleyan and Rawtheyan (Ingham & Wright 1970), based on the biostratigraphical revision of the almost complete Ashgill sequence by Ingham (1966).

The succession is composed largely (580 m) of the Cautley Mudstone Formation (Fig. 4.21), a monotonous grey calcareous mudstone with variably developed limestone nodules, deposited in a relatively quiet muddy shelf sea with a succession of rich faunas of trilobites, brachiopods, bryozoans and corals (Ingham & Rickards 1974). The oldest beds are of Onnian (i.e. latest Caradoc) age and contain abundant specimens of the diagnostic *Onnia superba pusgillensis* and *Flexicalymene onniensis lata* above the basal sandy beds. The succeeding more varied Pusgillian fauna includes *Atractopyge scabra, Gravicalymene jugifera, Paracybeloides girvanensis* and *Tretaspis convergens*, but it is in the Cautleyan that the most diverse faunas occur. Among the trilobites *Calymene prolata, Toxochasmops marri* and *Tretaspis hadelandica brachystichus* appear while the brachiopods include *Catazyga, Dicoelosia, Nicolella, Oxoplecia, Platystrophia, Rhactomena* and *Triplesia* in a varied fauna mainly of orthides and strophomenides. In the southernmost (Gawthrop) inlier an 8 m calcareous sandstone member (Wilsey Beck Sandstone) occurs locally in Zone 2 of this stage.

Faunas in the Cautley Mudstones of Rawtheyan age are much more restricted, with the virtual disappearance of the orthacean brachiopods to leave a dominantly dalmanellid–plectambonitacean assemblage (Ingham & Wright 1970). Among the trilobites *Dindymene hughesiae, Kloucekia robertsi, Tretaspis hadelandica brachystichus* and *Trinodus tardus* become important forms. The mudstones become increasingly ashy in the Rawtheyan, and are interrupted at the top of Zone 6 by the Cautley Volcanic Formation, a predominantly rhyolitic ash unit which thickens up to 24 m in the west (Ingham 1966). This phase of volcanic activity may be correlated with that of the Lake District and of the Craven inliers at this horizon.

The only stratigraphical break recorded in the Cautley succession occurs beneath the Cystoid Limestone, a 3 m thick argillaceous limestone which rests on Cautley Mudstones ranging from Zone 7 down to Zone 5 in age. The limestone is correlated with the mucronatus beds of Marr (1916) at Ash Gill and likewise is characterised by *Mucronaspis olini. M. mucronata* also occurs along with *Calymene drummuckensis*, whilst diploporitan and rhombiferan cystoids are the most commonly found fossils. The brachiopod fauna of dalmanellids and plectambonitaceans is of Rawtheyan aspect, and the overall assemblage indicates the limestone to be of latest Rawtheyan age (Ingham and Wright 1972). Lithostratigraphically, the unit is regarded here as the basal member of the Ashgill Shale Formation.

The Ashgill Shale itself here comprises over 90 m (Ingham & McNamara 1978) of leaden-grey mudstones with only the single trilobite *M. mucronata*, some bryozoans and the typically restricted *Hirnantia* brachiopod fauna. Some 10 m below the top the mudstone sequence is broken by a coarser clastic horizon which varies locally from sandy mudstones to up to 10 m of coarse calcareous sandstones and conglomerates. This horizon was equated by Ingham (1966) with the Wharfe Conglomerate of the Craven Inlier; the presence at this horizon also of the Rebecca Grit at Ireleth (Rose & Dunham 1977) and the conglomerates of the Whicham Valley near Millom noted by Mitchell (1956), suggests a brief phase of relative uplift and erosion resulting in rapid sedimentation over a wide area of northern England at this time. A shallowing at the end of the Ordovician is marked at the top of the Ashgill Shales by a metre of carbonate (the 'Basal Beds' of Rickards 1970); the overlying thin (7 cm) graptolite mudstones of *P. acuminatus* Biozone (Rickards 1970) indicate a deepening at the beginning of the Silurian.

## Craven

A series of small Lower Palaeozoic inliers occur along the southern margin of the Askrigg Block, where they are bounded to the south by the North Craven Fault (Fig. 4.16). The Ordovician is represented at Ingleton, Austwick and Horton-in-Ribblesdale by the enigmatic Ingleton Group, overlain unconformably by an incomplete sequence of the Coniston Limestone Group.

The Ingleton Group (the Ingletonian Series of Rastall 1907) is a mildly metamorphosed (greenschist facies) group of sediments which, apart from a problematic trace fossil recorded by Rayner (1957), lacks both macro and microfossils. These 'Greenish grits and slates' were presumed to be the equivalent of the Ordovician Borrowdale Group by Dakyns *et al.* (1890); subsequently most workers ascribed the rocks to the Precambrian (Rastall 1907, King 1932, Leedal & Walker 1950, Dunham *et al.* 1953). Radiometric evidence provided by O'Nions *et al.* (1973) suggested a Cambrian or early Ordovician age, the latter age being supported by Ingham and Wright (1972).

The folded, cleaved and predominantly green mudstones, siltstones and turbidite sandstones, some 800 m thick, include a very coarse feldspathic turbidite unit popularly known as the 'Ingleton Granite'. Very similar sediments, apart from a lack of the 'granite', have been recorded about 10 km north-east of Horton in the Beckermonds Scar borehole (Fig. 4.1) (Wilson & Cornwell 1982). These are of particular interest as a single fossiliferous horizon has yielded an acritarch assemblage of Arenig age. The rocks at Ingleton may therefore be presumed to be of broadly similar age and essentially the equivalent of the lower part of the Skiddaw Slate Group as suggested by Ingham and Rickards (1974).

The oldest rocks of the Coniston Limestone Group in the inliers are the Douk Ghyll Mudstones and the Crag Hill Limestone (King & Wilcockson 1934), with faunas indicating a mid-Cautleyan age (Fig. 4.21)

(Ingham 1966, Ingham & Rickards 1974). Unnamed calcareous mudstones with limestone nodules recorded from several localities in Crummackdale contain a mid-Rawtheyan (Zone 6) fauna (Ingham 1966), so that Zones 4 and 5 are unrepresented. Immediately overlying these are about 10 m of volcanic ashes, equated with the Cautley Volcanic Formation and succeeded by mudstones of presumably Zone 7 age. A stratigraphic break is followed by up to 2 m of the Wharfe Conglomerate and 12 m of Ashgill Shales with the *Hirnantia* fauna, again corresponding closely to the upper part of the Hirnantian at Cautley (p. 121).

An interesting development described by King (1932) is the Horton 'Neptunean dyke', a fissure filling in the Ingleton Group containing limestone with a diverse facies fauna of Keisley type, which includes the triplesiid brachiopods *Brachymimulus*, *Cliftonia*, *Streptis* and *Triplesia*. No older than mid-Cautleyan, the precise Ashgill age has yet to be determined (Ingham & Wright 1972, Ingham & Rickards 1974).

## Isle of Man

The Manx Slate Group is a complex group of turbiditic sandstones, siltstones and mudstones, with slump breccias and two localities which show minor volcanic activity in the form of andesitic lava, tuff and agglomerate (Simpson 1963). The first complete account of the geology of the island was that of Lamplugh (1903) who, on the basis of *Dictyonema* specimens recorded by Bolton (1899), suggested a Cambrian age for the Manx Slates. No further work was carried out until that of Gillott (1956a, b) and then Simpson (1963), who divided the group into eleven formations with a total mean thickness of 7,600 m. In the absence of additional biostratigraphic evidence, and by interpreting the *Dictyonema*-bearing Cronk Sumark (Cronkshamerk) Slates as being near the top of the sequence, Simpson also believed the Manx Slates to be of Cambrian age. Subsequently, Downie and Ford (1966) described chitinozoans, acritarchs and scolecodonts from the Lonan Flags (in the lower part of the group) which they considered to indicate either an uppermost Tremadoc or lower Arenig age. This study was followed by a comprehensive survey of the acritarchs in the Manx Slates by Molyneux (1980), who recovered diagnostic assemblages from five formations, and from sediments interbedded with volcanics of his Peel Volcanic Formation. Acritarchs were also recovered from three other formations, but were too scarce or too poorly preserved to be of biostratigraphic value. Although Molyneux is cautious in interpretation, two, possibly three, formations (including the Lonan Flags) would appear to be

of late Tremadoc age; two late Arenig; and one late Arenig or early Llanvirn. This last is the Lady Port Banded Formation, which Simpson (1963) interpreted as his oldest formation. Thus, as noted by Molyneux (1980), the stratigraphy of the Manx Slate Group is in need of radical revision. Nevertheless the palaeontological evidence available, with an age range from *A. sedgwicki* to *D. artus* biozones (Fig. 4.3), indicates a close comparison with the Skiddaw Group of the Lake District.

The graded turbidites and sheets of slump breccia throughout the Manx succession, with examples of the latter reaching up to 150 m and 450 m in thickness (e.g. at Ballanayre and Sulby), also invite environmental comparison with the Skiddaw sediments. Deposition took place in an unstable regime on a relatively deep-water slope along the southern margin of the Iapetus Ocean.

## Discussion

The Geological Society's Special Report (No. 3) on the Ordovician (Williams *et al.* 1972) highlighted many gaps in our knowledge of the stratigraphy and correlation of Ordovician areas in Britain. Most of these in England and Wales are now filled due to the efforts of the Geological Survey and many academics during the past two decades. Significant advances have also been made in our understanding of volcanic phenomena and processes and their effect, combined with contemporaneous tectonic activity, on sedimentation. (See also Chapters 16 and 17.)

The advent of plate-tectonic theory and the concept of terranes (see e.g. Soper *et al.* 1989) has stimulated research even more, resulting in a voluminous literature – and problems with palaeogeographic reconstructions.

The Cambro–Ordovician rocks of Britain provided early support for the destruction of the Iapetus Ocean (Wilson 1966). The later closing of this ocean resulted in the Caledonian orogeny (see Ch. 17) originally thought to be caused by the moving together of the Laurentian Plate (which included what are now parts of Newfoundland, Ireland and Scotland) and the Gondwana Plate (which included what are now other parts of Newfoundland and Ireland, and England and Wales) across the ocean until they collided. The collision zone is known as the Iapetus Suture and is thought to cross between Scotland and England in the vicinity of the Solway Firth and to transect Ireland in a NE–SW direction. Recent comprehensive summaries and bibliographies of the development of ideas relevant to England and Wales include for instance, Kokelaar *et al.* (1984), Hutton (1987), Kokelaar (1988), Pickering *et al.* (1988), McKerrow and Soper

(1989). Among early proponents of plate-tectonic theory to England and Wales were Dewey (1969), Fitton and Hughes (1970) and Phillips *et al*. (1976).

Fitton and Hughes (1970) from petrochemical studies of the volcanic rocks of the Lake District and North Wales envisaged a subduction zone dipping downwards to the south-east under the Lake District and Wales 'from a trench lying somewhere north of Girvan' (Phillips *et al. op. cit*.). Mitchell & McKerrow (1975), however, using the geology of western Burma as an analogy, considered that the geology of southern Scotland was consistent with a north-westerly directed subduction zone. Phillips *et al*. (*op. cit*.) proposed a model attempting to reconcile these views.

Since then more palaeontological evidence has confirmed significant differences between the faunas north and south of the Iapetus Suture (see e.g., Cocks & Fortey 1982 and Pickering *et al*. 1988). Allen (1987), however, disputed that the Solway Line was the Iapetus Suture on petrological and structural evidence, emphasising in particular that the terrane concept allows the Lake District terrane to have moved into position 'during the second half of the Silurian Period, becoming the fifth or sixth thin, accreted slice that went into the construction of northern Britain' (Allen *op. cit*., p. 486). He favoured a southern plate margin well south of the Lake District but Fortey *et al*. (1989) supported McKerrow and Soper (1989) who had pointed out (p. 1) that the 'Iapetus Ocean Suture can be traced between Ordovician fossil localities in the Southern Uplands and the Lake District which contain respectively, distinct North American and European faunas. The Southern Uplands contains North American Caradoc faunas in the Northern Belt, and is considered to have been accreted onto the Laurentian (North American) margin between the Llandeilo and the end of the Wenlock. Deep seismic reflection profiling shows that a surface, probably parallel to the suture zone, dips down to the north-west from the Solway Line.' Freeman *et al*. (1988) emphasised the difficulties inherent in the interpretation of much of the geophysical evidence but one of their conclusions is (p. 738). 'During collision the southern continent was partially subducted beneath the northern continent.'

Where then does this leave the Lake District and Wales? Cocks and Fortey (1982) had recognised that the early Ordovician faunas of southern Britain were comparable with those of Gondwana. They and those of Gondwana, however, differed sufficiently from those of the Baltic area, to indicate the existence of another ocean, Tornquist's Sea, between Gondwana and Baltica (a plate which included most of Scandinavia and the Russian Platform). The faunal differences, however, 'became progressively fewer during the Middle Ordovician and by late Ordovician times had largely disappeared, indicating the closure

of Tornquist's Sea' (Cocks & Fortey *op. cit*., p. 476). This implied that in the late Ordovician, Britain suffered a three-plate collision.

Hutton (1987) considered the Caledonides of the British Isles to consist of a collage of terranes which were assembled by sinistral strike–slip in the Devonian along many of the major fault-lines seen today. The terrane relationships therefore bore little resemblance to the palaeogeography that existed during the three-plate collision of Gondwana, Laurentia and Baltica. He did, however, interpret the Lake District–Wexford terrane (see his fig. 1) as part of an arc to south-west subduction under Gondwana and whose activity climaxed in mid-Ordovician (p. 405). Soper *et al*. (1987a) showed that deformation of the slate belts in the Lake District was of early Devonian age (not 'end-Silurian' as previously thought) and interpreted it as due to transpressive strains produced when the southern British terrane accreted northwards on to Laurentia. Branney and Soper (1989) also rejected the notion that the unconformities at the base and top of the Caradoc Borrowdale Volcanic Group were the result of compression, partly on the basis of evidence from the Lake District Batholith (see Soper 1987, Webb *et al*. 1987, Soper *et al*. 1987b, Firman & Lee 1987). They considered tentatively that the basal unconformity was due to buoyancy effects associated with the production of andesitic melt by subduction and that the post-Borrowdale Volcanic Group deformation (pre-Ashgill) is the result of volcano–tectonic faulting and block tilting due to caldera collapse. Ordovician north–south compression is therefore not proved. 'As currently understood the geology of the Lake District provides no evidence of subduction polarity beneath southern Britain during the closure of Iapetus' (Branney & Soper 1988, p. 375).

Kokelaar *et al*. (1982) and Kokelaar (1988), provided wide-ranging syntheses of the marine Welsh Basin in the context of ensialic destructive plate-margin volcanism. Volcanism in South Wales occurred from Arenig to Llanvirn times. In southern Snowdonia volcanism was mainly pre-Caradoc in age, while in central and northern Snowdonia it was mainly intra-Caradoc. Kokelaar (*op. cit*.) produced a well argued case for E–W crustal extension and the existence of predominantly N–S fracture zones (his figs. 3 & 4). These were thought to have developed from a system of deep-seated strike–slip faults. Relatively narrow grabens were formed and changing conditions of stress resulted in relative movements between blocks which channelled magma, hence changing the sites of surface volcanism and affecting sedimentation. Kokelaar emphasised the uncertainty concerning the Ordovician positions of the terranes now making up northern Britain but accepted that a southerly subduction direction under Wales was as good a working hypothesis as was available on present information. He, however, viewed with suspicion 'the present spatial relationships between Wales and the contrasting ensialic Lake District and Leinster (S. Ireland) terrane(s?) to the north . . .' (Kokelaar *op. cit.*, p. 775) and suggested several possible scenarios. Pickering *et al.* (1988) provided a stimulating synthesis of the evidence for the destruction of the Iapetus Ocean from a summary of seismic, stratigraphical, structural, igneous, faunal, palaeomagnetic, sedimentary and palaeolatitude information from Newfoundland, the British Isles and Scandinavia. Avalonia (parts of Newfoundland and southern Britain) was substituted for Gondwana (from which it was separated by the Rheic Ocean) in their three-plate model (cf. Hutton 1987). Their model and Hutton's assumed a Laurentia–Baltica separation after a preliminary closure (Silurian?) before final closure in the early Devonian.

The Ordovician System was defined last century amid much controversy and scenes of personalised acrimony. Through the years many aspects of the system have proved controversial. So they remain today, in particular in the field of plate-tectonics as will be seen from the foregoing, though fortunately, disagreements no longer produce such acrimony; rather they stimulate further lines of research to be followed. As Dewey said in 1982 (p. 410), 'The truth ultimately lies in the rocks and we need a new generation of field work to test hypotheses built from inceptive modelling'. Kokelaar (1988, p. 773) summed up the situation succinctly, 'Clearly we have a great deal to learn'.

*Acknowledgments*

Drs D. A. Bassett, R. E. Bevins, R. M. Owens and Prof. W. T. Deans kindly commented on drafts of parts of this chapter. (M. A. B.)

## REFERENCES

| | | |
|---|---|---|
| ALLEN, J. R. L. | 1981 | An occurrence of dish structures in Caradoc turbidites, Poppit Sands, near Cardigan, Dyfed, Wales. *Proc. Geol. Assoc., London*, **92**, 75–77. |
| ALLEN, P. M. | 1982 | Lower Palaeozoic volcanism in Wales, the Welsh Borderland, Avon, and Somerset. *In* Sutherland, D. S. (Ed.), *Igneous rocks of the British Isles*, 65–91. John Wiley & Sons Ltd, Chichester. |
| | 1987 | The Solway Line is not the Iapetus suture. *Geol. Mag.*, **124**, 485–486. |
| ALLEN, P. M. & COOPER, D. C. | 1986 | The stratigraphy and composition of the Latterbarrow and Redmain sandstones, Lake District, England. *Geol. J.*, **21**, 59–76. |
| ALLEN, P. M. & JACKSON, A. A. | 1985 | Geology of the country around Harlech. *Mem. Br. Geol. Surv.* 112 pp. HMSO, London. |
| ALLEN, P. M., JACKSON, A. A. & RUSHTON, A. W. A. | 1981 | The stratigraphy of the Mawddach Group in the Cambrian succession of North Wales. *Proc. Yorks. geol. Soc.*, **43**(3), 295–329, pls. 16, 17. |
| ANDREW, G. | 1925 | The Llandovery rocks of Garth (Breconshire). *Q. Jl. geol. Soc. Lond.*, **81**, 389–406. |
| ANKETELL, J. M. | 1987 | On the geological succession and structure of South-Central Wales. *Geol. J.*, **22**, 155–165. |
| ARTHURTON, R. S. & WADGE, A. J. | 1981 | Geology of the country around Penrith. *Mem. geol. Surv. GB*, 177 pp. |
| BANCROFT, B. B. | 1933 | *Correlation tables of the stages Costonian–Onnian in England and Wales*, 4 pp., 3 tables. Blakeney, Glos. (privately printed). |
| | 1945 | The brachiopod zonal indices of the stages Costonian to Onnian in Britain. *J. Paleont.*, **19**(3), 181–252. |
| | 1949 | Upper Ordovician trilobites of zonal value in south-east Shropshire. *Proc. R. Soc. London* B, **136**, 291–315. |

BASSETT, D. A.                  1969    Some of the major structures of early Palaeozoic age in Wales and the Welsh Borderland: an historical essay. pp. 76–116. *In* Wood, A. (Ed.), (*q.v.*).

BASSETT, D. A.,                 1966    The stratigraphy of the Bala district, Merionethshire. *Q. Jl. geol. Soc. London*, **122**, 219–271.
WHITTINGTON, H. B. &
WILLIAMS, A.

BASSETT, D. A.,                 1974    Field Excursion Guide, Ordovician System Symposium, Birmingham. *Palaeontol. Assoc. London*, 5–19.
INGHAM, J. K. &
WRIGHT, A. D. (eds)

BASSETT, M. G.                  1981    The Ordovician brachiopods of Cornwall. *Geol. Mag.*, **118**(6), 647–664, pls. 1–4.

                                1982    Ordovician and Silurian sections in the Llangadog–Llandilo area. *In* Bassett, M. G. (ed), *Geological excursions in Dyfed, south-west Wales*. National Museum of Wales, Cardiff, 271–287.

                                1985    Towards a 'common language' in stratigraphy. *Episodes*, **8**(2), 87–92.

BASSETT, M. G. &                1982    *The Cambrian–Ordovician Boundary: sections, fossil distributions, and correlations*. National Museum of Wales, Geological Series No. 3, 1–227.
DEAN, W. T.

BASSETT, M. G. *et al.*         1986    A Geotraverse through the Caledonides of Wales. *In* Fettes, D. J. & Harris, A. L. (Eds.), *Synthesis of the Caledonian Rocks of Britain*. Reidel Publishing Company, Dondrecht, 29–75.

BATES, D. E. B.                 1968    The Lower Palaeozoic brachiopod and trilobite faunas of Anglesey. *Bull. Brit. Mus. nat. Hist. (Geol.)*, **16**(4), 125–199, pls. 1–14.

                                1969    Some early Arenig brachiopods and trilobites from Wales. *Bull. Brit. Mus. nat. Hist. (Geol.)*, **18**, 1–28.

                                1972    The stratigraphy of the Ordovician rocks of Anglesey. *Geol. J.*, **8**, 29–58.

                                1974    The structure of the Lower Palaeozoic rocks of Anglesey, with special reference to faulting. *Geol. J.*, **9**, 39–60.

BATES, D. E. B.,                1969    The geology of Bishops and Clerks Islands, Pembrokeshire. *In* Wood, A. (Ed.), (*q.v.*), 447–449.
BROMLEY, A. V. &
JONES, A. S. G.

BEAVON, R. V.                   1963    The succession and structure east of the Glaslyn River, North Wales. *Q. Jl. geol. Soc. London*, **119**, 479–512.

                                1980    A resurgent cauldron in the early Palaeozoic of Wales, U.K. *J. volcanol. Geotherm. Res.*, **7**, 157–174.

BECKLY, A. J.                   1987    Basin development in North Wales during the Arenig. *Geol. J.*, **22**, thematic issue, 19–30.

BELDERSON, R. H. &              1969    Tidal currents and sand wave profiles in the north-eastern Irish Sea. *Nature, London*, **222**, 74–75.
STRIDE, A. H.

BERGSTRÖM, S. M.                1964    Remarks on some Ordovician conodont faunas from Wales. *Acta Universitatis Lundensis*, Sectio **II**, (3), 1–67.

                                1981    Biostratigraphical and biogeographical significance of conodonts in two British Middle Ordovician olistostromes. *Abstr. Progm. geol. Soc. Am., North-Central Section*, **13**, 271.

BERGSTRÖM, S. M.,               1987    Conodont biostratigraphy of the Llanvirn–Llandeilo and Llandeilo–Caradoc Series boundaries in the Ordovician System of Wales and the Welsh Borderland. (Reprints from Austin.) *Conodonts: Investigative Techniques and Applications*. Ellis Horwood Limited, Chichester.
RHODES, F. H. T. &
LINDSTRÖM, M.

BEVINS, R. E.  1982  Petrology and geochemistry of the Fishguard Volcanic Complex, Wales. *Geol. J.*, **17**, 1–21.

1985  Pumpellyite-dominated metadomain alteration at Builth Wells, Wales – evidence for a fossil submarine hydrothermal system? *Mineralog. Mag.*, **49**, 451–456.

BEVINS, R. E. &  1980  Early Ordovician volcanism in Dyfed, S. Wales. *In* Harris, ROACH, R. A.  A. L. *et al.*, (*q.v.*), 603–609.

1982  Ordovician igneous activity in south-west Dyfed. *In* Bassett, M. G. (ed), *Geological excursions in Dyfed, South Wales*. National Museum of Wales, 65–80.

BEVINS, R. E.,  1984  Petrology and geochemistry of lower to middle KOKELAAR, B. P. &  Ordovician igneous rocks in Wales: a volcanic arc to DUNKLEY, P. N.  marginal basin transition. *In* Bassett, M. G. (ed), Focus on Wales. *Proc. Geol. Ass. London*, **95**(4), 337–347.

BEVINS, R. E.,  1989  Ordovician intrusions of the Strumble Head–Mynydd LEES, G. J. &  Preseli region, Wales: lateral extension of the Fishguard ROACH, R. A.  Volcanic Complex. *Jl. geol. Soc. London*, **146**, 113–123.

BLACK, W. W.,  1971  Ordovician stratigraphy of Abereiddy Bay, Pembroke-BULMAN, O. M. B.,  shire. *Geol. Mag.*, **108**, 546–548. HEY, R. W. & HUGHES, C. P.

BLYTH, F. G. H.  1935  The basic intrusive rocks associated with the Cambrian inlier near Malvern. *Q. Jl. geol. Soc. Lond.*, **91**, 463–478, pl. 29.

1938  Pyroclastic rocks from the Stapeley Volcanic Group at Knotmoor, near Minsterley, Shropshire. *Proc. Geol. Assoc. London*, **49**(4), 392–404. pl. 27.

1944  Intrusive rocks of the Shelve area, south Shropshire. *Q. Jl. geol. Soc. Lond.*, **99** [for 1943], 169–204, pl. 20.

BOLTON, H.  1899  The palaeontology of the Manx Slates of the Isle of Man. *Mem. Proc. Manchr lit. phil. Soc.*, **43**, 1–15.

BRANNEY, M. J.  1986  The Stockdale (Yarlside) Rhyolite – a rheomorphic ignimbrite? Discussion. *Proc. Yorks. geol. Soc.*, **46**, 80–82.

1988  The subaerial setting of the Ordovician Borrowdale Volcanic Group, English Lake District. *Jl. geol. Soc. London*, **145**, 367–376.

BRANNEY, M. J. &  1988  Ordovician volcano-tectonics in the English Lake SOPER, N. J.  District. *Jl. geol. Soc. London*, **145**, 367–376.

BRENCHLEY, P. J.  1964  Ordovician ignimbrites in the Berwyn Hills, North Wales. *Geol. J.*, **4**, 43–54.

1969a  Origin of matrix in Ordovician greywackes, Berwyn Hills, North Wales. *J. sediment. Petrol.*, **39**(4), 1297–1301.

1969b  The relationship between Caradocian volcanicity and sedimentation in North Wales. *In* Wood, A. (Ed.), (*q.v.*), 181–202.

1972  The Cwm Clwyd Tuff, North Wales: a palaeogeographical interpretation of some Ordovician ash-shower deposits. *Proc. Yorks. geol. Soc.*, **39**(2), 199–224.

1978  The Caradocian rocks of the north and west Berwyn Hills, North Wales. *Geol. J.*, **13**(2), 137–164.

BRENCHLEY, P. J. &  1984  The environmental distribution of associations belonging CULLEN, B.  to the *Hirnantia* fauna – evidence from North Wales and Norway. *In* Bruton, D. L. (Ed.), Aspects of the Ordovician System. *Palaeont. Contr. Univ. Oslo*, **295**, 113–125.

BRENCHLEY, P. J. &
NEWALL, G.

1982    Storm-influenced inner-shelf sand lobes in the Caradoc (Ordovician) of Shropshire, England. *J. sedim. Petrol.*, **52**(4), 1257–1269.

1984    Late Ordovician environmental changes and their effect on faunas. *In* Bruton, D. (ed), Aspects of the Ordovician System. *Palaeont. Contr. Univ. Oslo*, **295**, 65–79.

BRENCHLEY, P. J. &
PICKERILL, R. K.

1980    Shallow subtidal sediments of Soudleyan (Caradoc) age in the Berwyn Hills, North Wales, and their palaeogeographic context. *Proc. Geol. Assoc. London*, **91**(3), 177–194.

BROOKS, M. &
JAMES, D. G.

1975    The geological results of seismic refraction surveys in the Bristol Channel, 1970–73. *J. geol. Soc. Lond.*, **131**, 163–182.

BULMAN, O. M. B.

1941    Some Dichograptids of the Tremadocian and Lower Ordovician. *Ann. Mag. nat. Hist.*, (11th Ser.) **7**, 100–121.

1958    The sequence of graptolite faunas. *Palaeontology*, **1**, 159–173.

1971    Some species of *Bryograptus* and *Pseudobryograptus* from Northwest Europe. *Geol. Mag.*, **108**, 361–371.

BULMAN, O. M. B. &
RUSHTON, A. W. A.

1973    Tremadoc faunas from boreholes in Central England. *Bull. geol. Surv. GB*, **43**, 1–39.

BURGESS, I. C.

1971    pp. 33. *In Ann. Rep. Inst. geol. Sci.*: for 1970; 199 pp.

BURGESS, I. C. &
HOLLIDAY, D. W.

1979    Geology of the country around Brough-under-Stainmore. *Mem. geol. Surv. GB*, 131 pp.

BURGESS, I. C. &
WADGE, A. J.

1974    *The geology of the Cross Fell area.* Classical areas of British Geology Series, Institute of Geological Sciences, HMSO, London, 91 pp.

CAMPBELL, S. D. G.

1984    Aspects of dynamic stratigraphy (Caradoc–Ashgill) in the northern part of the Welsh marginal basin. *In* Bassett, M. G. (Ed.), Focus on Wales. *Proc. Geol. Assoc. London*, **95**, 390–391.

CANTRILL, T. C.,
DIXON, E. E. L.,
THOMAS, H. H. &
JONES, O. T.

1916    The Geology of the South Wales Coalfield. Part XII. The country around Milford. *Mem. geol. Surv. GB*, 185 pp.

CAVE, R.

1957    *Salterolithus caractaci* (Murchison) from Caradoc strata near Welshpool, Montgomeryshire. *Geol. Mag.*, **94**, 281–290, pl. 10.

1965    The Nod Glas sediments of Caradoc age in North Wales. *Geol. J.*, **4**(2), 279–298.

1980    Sedimentary environments of the basinal Llandovery of mid-Wales. *In* Harris, A. L. *et al.*, (*q.v.*), 517–526.

1984    *Description 1:50,000, Sheet 163, Aberystwyth.* British Geological Survey. HMSO, London.

CAVE, R. & HAINS, B. A.

1986    Geology of the country between Aberystwyth and Machynlleth. *Mem. geol. Surv. GB*, 148 pp.

CAVE, R. & PRICE, D.

1978    The Ashgill Series near Welshpool, North Wales. *Geol. Mag.*, **115**(3), 183–194, pls. 1, 2.

COCKS, L. R. M.

1985    The Ordovician–Silurian boundary. *Episodes*, **8**(2), 98–100.

COCKS, L. R. M. &
PRICE, D.

1975    The biostratigraphy of the upper Ordovician and lower Silurian of south-west Dyfed, with comments on the *Hirnantia* fauna. *Palaeontology*, **18**, 703–724.

COCKS, L. R. M. &
FORTEY, R. A.

1982    Faunal evidence for oceanic separations in the Palaeozoic of Britain. *J. geol. Soc. London*, **139**, 465–478.

COCKS, L. R. M.,
TOGHILL, P. &
ZIEGLER, A. M.

1970    Stage names within the Llandovery Series. *Geol. Mag.*, **107**, 79–87.

COCKS, L. R. M.,    1984    The Llandovery Series of the Type Area. *Bull. Brit. Mus.*
WOODCOCK, N. H.,            *nat. Hist.* (Geol.), **38**, 131–182.
RICKARDS, R. B.,
TEMPLE, J. T. &
LANE, P. D.

COOPER, A. H. &    1990    The age and correlation of the Skiddaw Group (early
MOLYNEUX, S. G.           Ordovician) sediments in the Cross Fell inlier (northern
          England). *Geol. Mag.*, **127**, 147–157.

COPE, J. C. W.    1980    Early history of the southern margin of the Tjwi anticline in
          the Carmarthen area, South Wales. *In* Harris *et al.* (*q.v.*),
          527–532.

COPE, J. C. W. &    1987    Sediment sources and Palaeozoic history of the Bristol
BASSETT, M. G.           Channel area. *Proc. Geol. Assoc. London*, **98**, 315–330.

COPE, J. C. W.,    1978    Newly discovered Tremadoc rocks in the Carmarthen
FORTEY, R. A. &           district, South Wales. *Geol. Mag.*, **115**(3), 195–198, pl. 1.
OWENS, R. M.

COWARD, M. P. &    1980    The tectonic evolution of the Welsh Caledonides. *In* Harris
SIDDANS, A. W. B.           *et al.* (*q.v.*), 187–198.

COWIE, J. W.,    1972    A correlation of Cambrian rocks in the British Isles. *Geol.*
RUSHTON, A. W. A. &         *Soc. London Special Report*. No. 2, 1–42.
STUBBLEFIELD, C. J.

COX, A. H.    1912    The Pedwardine inlier. *Q. Jl. geol. Soc. Lond.*, **68**, 364–373.
        1916    The geology of the district between Abereiddy and
           Abercastle (Pembrokeshire). *Q. Jl. geol. Soc. London*,
           **71**, 275–342.
        1930    Preliminary note on the geological structure of Pen Caer
           and Strumble Head, Pembrokeshire. *Proc. Geol. Assoc.*
           *London*, **41**, 241-273.

CRIMES, T. P.    1969    Trace fossils from the Cambro–Ordovician rocks of North
           Wales and their stratigraphic significance. *Geol. J.*, **6**(2),
           333–338.
        1970a    The significance of trace fossils in sedimentology, strati-
           graphy and palaeoecology with examples from Lower
           Palaeozoic strata. *In* Crimes, T. P. & Harper, J. C. (eds),
           Trace fossils. *Geol. J. Spec. Issue*, **3**, 101–126.
        1970b    A facies analysis of the Arenig of western Lleyn, North
           Wales. *Proc. Geol. Assoc. London*, **81**(2), 221–240.

CURTIS, M. L. K.    1968    The Tremadoc rocks of the Tortworth inlier, Gloucester-
           shire. *Proc. Geol. Assoc. London*, **79**, 349–362, pls. 8, 9.

DAKYNS, J. R.,    1890    The geology of the country around Ingleborough with
TIDDEMAN, R. H.,           parts of Wensleydale and Wharfedale. *Mem. geol. Surv.*
GUNN, W. &           *GB*, 103 pp.
STRAHAN, A.

DAVIES, A. M. &    1913    On two deep borings at Calvert Station (north Bucking-
PRINGLE, J.           hamshire) and on the Palaeozoic floor north of the Thames.
           *Q. Jl. geol. Soc. Lond.*, **69**, 308–342, pls. 33, 34.

DAVIES, J. H.    1980    A suggested re-interpretation of the Lower Palaeozoic
           stratigraphy around Llanafan-fawr, Central Wales. *Geol.*
           *J.*, **15**, 131–133.

DAVIES, K. A.    1926    The geology of the country between Drygarn and Aberg-
           wesyn (Breconshire). *Q. Jl. geol. Soc. Lond.*, **82**, 436–464.
        1933    The geology of the country between Abergwesyn (Brecon-
           shire) and Pumpsaint (Carmarthenshire). *Q. Jl. geol. Soc.*
           *Lond.*, **89**, 172–201.

DAVIES, K. A. &      1933      The conglomerates and grits of the Bala and Valentian
PLATT, J. I.                   rocks of the district between Rhayader (Radnorshire) and
                               Llansawel (Carmarthenshire). *Q. Jl. geol. Soc. Lond.*, **89**,
                               202–220.

DAVIES, R. G.        1959      The Cader Idris granophyre and its associated rocks. *Q.
                               Jl. geol. Soc. Lond.*, **115**, 189–216.

DEAN, W. T.          1958      The faunal succession in the Caradoc Series of South
                               Shropshire. *Bull. Br. Mus. nat. Hist.* (Geol.), **3**, 191–231,
                               pls. 24–26.

                     1959      The stratigraphy of the Caradoc Series in the Cross Fell
                               Inlier. *Proc. Yorks. geol. Soc.*, **32**, 185–228.

                     1960      The Ordovician rocks of the Chatwell district, Shropshire.
                               *Geol. Mag.*, **97**(2), 163–171.

                     1962      The trilobites of the Caradoc Series in the Cross Fell Inlier
                               of northern England. *Bull. Brit. Mus. nat. Hist.* (Geol.), **7**,
                               67–134.

                     1963      The Stile End Beds and Drygill Shales (Ordovician) in the
                               East and North of the English Lake District. *Bull. Br. Mus.
                               nat. Hist.* (Geol.), **9**, 49–65.

                     1964      The geology of the Ordovician and adjacent strata in the
                               southern Caradoc district of Shropshire. *Bull. Br. Mus. nat.
                               Hist.* (Geol.), **9**, 257–296.

DEAN, W. T. &        1961      The Ordovician and associated Precambrian rocks of the
DINELEY, D. L.                 Pontesford district, Shropshire. *Geol. Mag.*, **98**, 367–376,
                               pl. 20.

DEWEY, J. F.         1969      Evolution of the Appalachian/Caledonian Orogen. *Nature,
                               London*, **222**, 124–129.

                     1982      Plate tectonics and the evolution of the British Isles. *Jl.
                               geol. Soc. London*, **139**, 371–412.

DIXON, E. E. L.      1921      The Geology of the South Wales Coalfield. Part XIII.
                               The country around Pembroke and Tenby. *Mem. geol.
                               Surv. G.B.*, 220 pp.

DOWNIE, C. &         1966      Microfossils from the Manx Slate Series. *Proc. Yorks. geol.
FORD, T. D.                    Soc.*, **35**, 307–322.

DOWNIE, C. &         1972      Age of the Eycott Volcanic Group and its conformable
SOPER, N. J.                   relationship to the Skiddaw Slates in the English Lake
                               District. *Geol. Mag.*, **109**, 259–268.

DUNHAM, K. C.,       1953      A guide to the geology of the district around Ingleborough.
HEMINGWAY, J. E.,              *Proc. Yorks. geol. Soc.*, **29**, 77–115.
VERSEY, H. C. &
WILCOCKSON, W. H.

DUNKLEY, P. N.       1980      Ordovician volcanicity of the SE Harlech Dome. In Harris
                               *et al.* (*q.v.*), 597–601.

EARP, J. R.          1977      Notes on the geology. Llandrindod Wells Ordovician
                               Inlier. *Classical areas of British Geology. Inst. geol. Sci.*
                               HMSO, London.

EASTWOOD, T.,        1931      The geology of the Whitehaven and Workington district.
DIXON, E. E. L.,               *Mem. geol. Surv. GB*, 304 pp.
HOLLINGWORTH, S. E. &
SMITH, B.

EASTWOOD, T.,        1968      Geology of the country around Cockermouth and Cald-
HOLLINGWORTH, S. E.,           beck. *Mem. geol. Surv. GB*, 298 pp.
ROSE, W. C. C. &
TROTTER, F. M.

ELLES, G. L.         1898      The graptolite-fauna of the Skiddaw Slates. *Q. Jl. geol. Soc.
                               Lond.*, **54**, 463–539.

ELLES, G. L.                    1904    Some graptolite zones in the Arenig rocks of Wales. *Geol. Mag.* Decade V, **1**, 199–211.

                                1940    The stratigraphy and faunal succession in the Ordovician rocks of the Builth–Llandrindod inlier, Radnorshire. *Q. Jl. geol. Soc. London*, **95**, 382–445.

EVANS, D. C.                    1906    The Ordovician rocks of western Caermarthenshire. *Q. Jl. geol. Soc. London*, **62**, 597–642.

EVANS, W. D.                    1945    The geology of the Prescelly Hills, north Pembrokeshire. *Q. Jl. geol. Soc. London*, **101**, 89–110.

FIRMAN, R. J.                   1957    The Borrowdale Volcanic Series between Wastwater and Duddon Valley, Cumberland. *Proc. Yorks. geol. Soc.*, **31**, 39–64.

FIRMAN, R. J. &                 1986    Age and structure of the concealed English Lake district
LEE, M. K.                              batholith and its probable influence on subsequent sedimentation, tectonics and mineralization. *In* Nesbitt, R. W. & Nichol, I. (eds), *Geology in the real world – the Kingsley Dunham volume*, 117–127. Institution of Mining and Metallurgy, London.

                                1987    The English Lake District batholith – Ordovician, Silurian, Devonian or . . .? *Geol. Mag.*, **124**, 583–586.

FITCH, F. J.                    1967    Ignimbrite volcanism in North Wales. *Bull. Volcanol.*, **30**, 199–219.

FITTON, J. G. &                 1970    Volcanism and plate tectonics in the British Ordovician.
HUGHES, D. J.                           *Earth planet. Sci. Lett.*, **8**, 223–228.

FORTEY, R. A.                   1984    Global earlier Ordovician transgression and regression and their biological implications. *In* Bruton, D. L. (ed.), *Aspects of the Ordovician System*, Paleontological Contributions, University of Oslo, **295**, 37–50.

FORTEY, R. A. &                 1978    Early Ordovician (Arenig) stratigraphy and faunas of the
OWENS, R. M.                            Carmarthen district, south-west Wales. *Bull. Br. Mus. nat. Hist.* (Geol.), **30**, 225–294, pls. 1–11.

                                1987    The Arenig Series in South Wales: Stratigraphy and Palaeontology. 1. The Arenig Series in South Wales. *Bull. Br. Mus. nat. Hist.* (Geol.), **41**, 69–364.

FORTEY, R. A.,                  1989    The palaeographic position of the Lake District in the
OWENS, R. M. &                          early Ordovician. *Geol. Mag.*, **126**, 9–17.
RUSHTON, W. A.

FREEMAN, B.,                    1988    The deep structure of northern England and the Iapetus
KLEMPERER, S. L. &                      suture zone from BIRPS deep seismic reflection profiles.
HOBBS, R. W.                            *Jl. geol. Soc. London*, **145**, 727–740.

GALE, N. H. &                   1983    Comments on the paper 'Fission-track dating of British
BECKINSALE, R. D.                       Ordovician and Silurian stratotypes' by R. J. Ross and others. *Geol. Mag.*, **120**, 295–302.

GEORGE, T. N.                   1963    Tectonics and palaeogeography in northern England. *Sci. Prog. London*, **51**, 32–59.

GILLOTT, J. E.                  1956a   Breccias in the Manx Slates; their origin and stratigraphic relations. *Lpool Manchr geol. J.*, **1**, 370–380.

                                1956b   Structural geology of the Manx Slates. *Geol. Mag.*, **93**, 301–313.

GOLDRING, R.                    1966    Sandstones of sublittoral (neritic) facies. *Nature, London*, **210**, 1248–1249.

GOODCHILD, J. G.                1889    An outline of the geological history of the Eden valley. *Proc. Geol. Assoc. London*, **11**, 258–284.

GREENLY, E.       1919      The geology of Anglesey. *Mem. geol. Surv. GB*, **1**, 1–138; **2**, 389–980.

      1944      The Ordovician rocks of Arvon. *Q. Jl. geol. Soc. Lond.*, **100**(1/2), 75–83.

GREIG, D. C., WRIGHT, J. E., HAINS, B. A. & MITCHELL, G. H.       1968      Geology of the country around Church Stretton, Craven Arms, Wenlock Edge and Brown Clee. *Mem. geol. Surv. GB*, 379 pp.

GROOM, T. T.       1901      On the igneous rocks associated with the Cambrian beds of the Malvern Hills. *Q. Jl. geol. soc. Lond.*, **57**, 156–184, pl. 7.

      1902      The sequence of the Cambrian and associated beds of the Malvern Hills. With an appendix on the Brachiopoda by Charles Alfred Matley. *Q. Jl. geol. Soc. Lond.*, **58**, 89–149.

GROOM, T. T. & LAKE, P.       1908      The Bala and Llandovery rocks of Glyn Ceiriog (North Wales). *Q. Jl. geol. Soc. Lond.*, **64**, 546–595.

HALLAM, A. & BRADSHAW, M. J.       1979      Bituminous shales and oolitic ironstones as indicators of transgressions and regressions. *Jl. geol. Soc. Lond.*, **136**, 157–164.

HARKNESS, R. & NICHOLSON, H. A.       1877      On the strata and their fossil contents between the Borrowdale Series of the north of England and the Coniston Flags. *Q. Jl. geol. Soc. Lond.*, **33**, 461–484.

HARLAND, W. B. & GAYER, R. A.       1972      The Arctic Caledonides and earlier oceans. *Geol. Mag.*, **109**, 289–314.

HARPER, J. C.       1956      The Ordovician succession near Llanystwmdwy, Caernarvonshire. *L'pool. Manchr. geol. J.*, **1**(4), 385–393.

HARRIS, A. L., HOLLAND, C. H. & LEAKE, B. E. (eds)       1980 [for 1979]      The Caledonides of the British Isles – reviewed. *Spec. Publ. geol. Soc. London*, **8**, 768 pp.

HARTLEY, J. J.       1925      The succession and structure of the Borrowdale Volcanic Series as developed in the area lying between the lakes of Grasmere, Windermere and Coniston. *Proc. Geol. Assoc. London*, **36**, 203–226.

      1932      The volcanic and other igneous rocks of Great and Little Langdale. *Proc. Geol. Assoc. London*, **43**, 32–69.

      1942      The geology of Helvellyn and the southern part of Thirlmere. *Q. Jl. geol. Soc. Lond.*, **97**, 129–162.

HAWKINS, T. R. W.       1966      Boreholes at Parys Mountain, near Amlwch, Anglesey. *Bull. geol. Surv. GB*, no. 24, 7–18.

      1983      Structure of the Ordovician succession around Aberdaron, southwest Llyn, North Wales. *Geol. J.*, **18**(2), 169–181.

HELM, D. G.       1970      Stratigraphy and structure in the Black Combe Inlier, English Lake District. *Proc. Yorks. geol. Soc.*, **38**, 105–148.

HENDRICKS, E. M. I.       1926      The Bala–Silurian succession in the Llangranog district (south Cardiganshire). *Geol. Mag.* **63**, 121–139.

HENNINGSMOEN, G.       1973      The Cambro–Ordovician boundary. *Lethaia*, **6**, 423–439.

HENRY, J.-L.       1980      Trilobites ordoviciens du Massif Armoricain. *Mém. soc. géol. minéral. Bretagne*, **22**, 1–250, pls. 1–48.

HICKS, H.       1981      The classification of the Eozoic and Lower Palaeozoic rocks of the British Isles. *Pop. Sci. Rev.*, **5**, 289–308.

HILLER, N.       1980      Ashgill Brachiopoda from the Glyn Ceiriog District, north Wales. *Bull. Br. Mus. nat. Hist.* (Geol.), **34**(3), 109–216.

      1981      The Ashgill rocks of the Glyn Ceiriog District, north Wales. *Geol. J.*, vol. 16, 181–200.

HOFMANN, H. J.       1975      *Bolopora* not a bryozoan, but an Ordovician phosphatic, oncolitic accretion. *Geol. Mag.*, **112**(5), 523–526.

HOLLAND, C. H., ROSS, R. J. & COCKS, L. R. M. — 1984 — Ordovician–Silurian boundary. *Lethaia*, **17**, 184.

HOWELLS, M. F. & LEVERIDGE, B. E. — 1980 — The Capel Curig Volcanic Formation. *Rep. Inst. Geol. Sci.*, **80/6**, 23 pp. HMSO, London.

HOWELLS, M. F., FRANCIS, E. H., LEVERIDGE, B. E. & EVANS, C. D. R. — 1978 — Capel Curig and Betws-y-Coed. *Classical areas of British geology, Institute of Geological Sciences*, HMSO, London, 73 pp.

HOWELLS, M. F., LEVERIDGE, B. E., ADDISON, R., EVANS, C. D. R. & NUTT, M. J. C. — 1980 — The Capel Curig Volcanic Formation, Snowdonia, North Wales; variations in ash-flow tuffs related to emplacement environment. *In* Harris *et al.* (*q.v.*), 611–618.

HOWELLS, M. F., LEVERIDGE, B. E. & REEDMAN, A. J. — 1981a — *Snowdonia*. Unwins, London, 119 pp.

HOWELLS, M. F., LEVERIDGE, B. E., EVANS, C. D. R. & NUTT, M. J. C. — 1981b — Dolgarrog. *Classical areas of British geology, Institute of Geological Sciences*, HMSO, London, 89 pp.

HOWELLS, M. F., LEVERIDGE, B. E., ADDISON, R. & REEDMAN, A. J. — 1983 — The lithostratigraphic subdivision of the Ordovician underlying the Snowdon and Crafnant volcanic groups, North Wales. *Rep. Inst. Geol. Sci.*, **83/1**, 11–15. HMSO, London.

HOWELLS, M. F., REEDMAN, A. J. & LEVERIDGE, B. E. — 1985 — Geology of the country around Bangor. *Explan. 1:50,000 sheet, Br. Geol. Surv. Sheet 100, England and Wales.* HMSO, London.

HOWELLS, M. F., REEDMAN, A. J. & CAMPBELL, S. D. G. — 1986 — The submarine eruption and emplacement of the Lower Rhyolitic Tuff Formation (Ordovician), N Wales. *Jl. geol. Soc. London*, **143**, 411–423.

HUGHES, C. P., JENKINS, C. J. & RICKARDS, R. B. — 1982 — Abereiddi Bay and adjacent coast. *In* Bassett, M. G. (ed), *Geological excursions in Dyfed, South-west Wales*. National Museum of Wales, 51–63.

HURST, J. M. — 1979a — The environment of deposition of the Caradoc Alternata Limestone and contiguous deposits of Salop. *Geol. J.*, **14**(1), 15–40.

— 1979b — Evolution, succession and replacement in the type upper Caradoc (Ordovician) benthic faunas of England. *Palaeogeogr. Palaeoclimat. Palaeoecol.*, **27**, 189–246.

— 1979c — The stratigraphy and brachiopods of the upper part of the type Caradoc of south Wales. *Bull. Br. Mus. nat. Hist.* (Geol.), **32**, 183–304.

HUTT, J. E. — 1974 — The Llandovery graptolites of the English Lake District. *Palaeontogr. Soc.* [*Monogr.*], **128**, 1–56.

HUTTON, D. H. W. — 1987 — Strike-slip terranes and a model for the evolution of the British and Irish Caledonides. *Geol. Mag.*, **124**, 405–425.

INGHAM, J. K. — 1966 — The Ordovician rocks in the Cautley and Dent districts of Westmorland and Yorkshire. *Proc. Yorks. geol. Soc.*, **35**, 455–505.

— 1970 — The Upper Ordovician trilobites from the Cautley and Dent districts of Westmorland and Yorkshire. *Palaeontogr. Soc.* [*Monogr.*], **124**, 1–58.

INGHAM, J. K. & MCNAMARA, K. J. — 1978 — The Coniston Limestone Group. *In* Moseley, R. (ed), *The geology of the Lake District*, 121–129. Yorkshire Geological Society, Leeds.

INGHAM, J. K. &
RICKARDS, R. B.
1974    Lower Palaeozoic rocks. *In* Rayner, D. H. & Hemingway, J. E. (eds), *The geology and mineral resources of Yorkshire*, 29–44. Yorkshire Geological Society, Leeds.

INGHAM, J. K. &
WRIGHT, A. D.
1970    A revised classification of the Ashgill Series. *Lethaia*, **3**, 233–242.

1972    The North of England. *In* Williams, A. *et al.*, A correlation of Ordovician rocks in the British Isles. *Spec. Rep. geol. Soc. London*, **3**, 1–74.

JAANUSSON, V.
1960    On the Series of the Ordovician System. *Rept. Int. geol. Congr. 21st Session, Norden, pt. vii, Proc. Section 7, Ordovician stratigraphy and correlations*, 70–81.

1979    Ordovician. *In* Robison, R. A. & Teichert, C. (eds), *Treatise on Invertebrate Paleontology, Part A, Introduction*, 136–166. Geological Society of America and University of Kansas Press.

1984a   What is so special about the Ordovician? *In* Bruton, D. L. (ed), Aspects of the Ordovician System. *Palaeont. Contr. Univ. Oslo*, **295**, 1–3.

1984b   Ordovician benthic macrofaunal associations. *In* Bruton, D. L. (ed), Aspects of the Ordovician System. *Palaeont. Contr. Univ. Oslo*, **295**, 127–139.

JACKSON, D. E.
1961    Stratigraphy of the Skiddaw Group between Buttermere and Mungrisdale, Cumberland. *Geol. Mag.*, **98**, 515–528.

1962    Graptolite zones in the Skiddaw Group in Cumberland, England. *J. Paleont.*, **36**, 300–313.

1978    The Skiddaw Group. *In* Moseley, F. (ed), *The geology of the Lake District*, 79–98. Yorkshire Geological Society, Leeds.

1979    A new assessment of the stratigraphy of the Skiddaw Group along the northern edge of the main Skiddaw Inlier. *Proc. Cumberland geol. Soc.*, **4**, 21–31.

JAMES, D. M. D.
1971a   The Nant-y-moch Formation, Plynlimon inlier, west central Wales. *J. geol. Soc. London*, **127**, 177–181.

1971b   Petrography of the Plynlimon Group, west central Wales. *Sediment. Geol.*, **6**, 255–270.

1972    Sedimentation across an intra-basinal slope: the Garnedd-Wen Formation (Ashgillian), west central Wales. *Sediment. Geol.*, **7**, 291–307.

1975    Caradoc turbidites at Poppit Sands (Pembrokeshire), Wales. *Geol. Mag.*, **112**, 295–304.

1983a   Observations and speculations on the northeast Towy "axis", mid-Wales. *Geol. J.*, **18**, 283–296.

1983b   Sedimentation of deep-water slope-base and inner-fan deposits – the Drosgol Formation (Ashgill), west central Wales. *Sediment. Geol.*, **34**, 21–40.

1986    The Rhiwnant Inlier, Powys, Mid-Wales. *Geol. Mag.*, **123**, 585–587.

1987    Ashgill sediments between Plynlimon and Machynlleth, West Central Wales, *Mercian Geol., Nottingham*, **10**, 265–279.

JAMES, D. M. D. &
JAMES, J.
1970
[for 1969]   The influence of deep fractures on some areas of Ashgillian–Llandoverian sedimentation in Wales. *Geol. Mag.*, **106**, 562–582.

JEANS, P. J. F.
1972    The junction between the Skiddaw Slates and Borrowdale Volcanics in Newlands Beck, Cumberland. *Geol. Mag.*, **109**(1), 25–28.

JENKINS, C. J.    1982    *Isograptus gibberulus* (Nicholson) and the isograptids of the Arenig Series (Ordovician) of England and Wales. *Proc. Yorks. geol. Soc.*, **44**(2), 219–248, pls. 16, 17.

1983    Ordovician graptolites from the Great Paxton borehole, Cambridgeshire. *Palaeontology*, **26**(3), 639–651.

JOHNSON, G. A. L.    1961    Skiddaw Slates proved in the Teesdale Inlier. *Nature*, **190**, 996–997.

JONES, O. T.    1909    The Hartfell–Valentian succession in the district around Plynlimon and Pont Erwyd (North Cardiganshire). *Q. Jl. geol. Soc. Lond.* no. 260, **65**, 463–537.

1922    Lead and zinc. The mining district of north Cardiganshire and west Montgomeryshire. *Mem. Geol. Surv. Spec. Min. Res. GB*, **20**, 207 pp.

1925    The geology of the Llandovery District: Part I. The Southern Area. *Q. Jl. geol. Soc. Lond.*, **81**, 344–388.

1949    The geology of the Llandovery District: Part II. The Northern Area. *Q. Jl. geol. Soc. Lond.*, **105**, 43–64.

JONES, O. T. &    1916    The geology of the district around Machynlleth and the
PUGH, W. J.    [for    Llyfnant Valley. *Q. Jl. geol. Soc. Lond.*, no. 282, **71**, 343–
                1915]    385.

1935    The geology of the districts around Machynlleth and Aberystwyth. *Proc. Geol. Assoc. Lond.*, **46**, pt. 3, 247–300.

JONES, O. T. &    1941    The Ordovician rocks of the Builth District. A Preliminary
PUGH, W. J.            Account. *Geol. Mag.*, **78**, 185–191.

1949    An early Ordovician shore-line in Radnorshire, near Builth Wells. *Q. Jl. geol. Soc. Lond.*, **105**, 65–99.

JONES, W. D. V.    1945    The Valentian succession around Llanidloes, Mont-
                   [for    gomeryshire. *Q. Jl. geol. Soc. Lond.*, **100**, 309–332.
                   1944]

KEEPING, W.    1881    The Geology of Central Wales. *Q. Jl. geol. Soc. Lond.*, **37**, 141–177.

1882    On the geology of Cardigan town. *Geol. Mag.*, **19**, 519–522.

KING, W. B. R.    1923    The Upper Ordovician rocks of the south-western Berwyn Hills. *Q. Jl. geol. Soc. Lond.*, **79**, 487–502.

1928    The geology of the district around Meifod (Montgomery-shire). *Q. Jl. geol. Soc. Lond.*, **84**(4), 671–702.

1932    A fossiliferous limestone associated with the Ingletonian beds at Horton-in-Ribblesdale. *Q. Jl. geol. Soc. Lond.*, **88**, 100–111.

KING, W. B. R. &    1934    The Lower Palaeozoic rocks of Austwick and Horton-in-
WILCOCKSON, W. H.            Ribblesdale. *Q. Jl. geol. Soc. Lond.*, **90**, 7–31.

KOKELAAR, B. P.    1980    Tremadoc to Llanvirn volcanism on the southeast side of the Harlech Dome (Rhobell Fawr), N. Wales. *In* Harris *et al. (q.v.)*, 591–596.

1986    Petrology and geochemisry of Rhobell Volcanic Complex: Amphibole-dominated fractionation at an Ordovician arc volcano in Wales. *J. Petrol.*, **27**, 887–914.

1989    Tectonic controls of Ordovician arc and marginal basin volcanism in Wales. *Jl. geol. Soc. London*, **145**, 759–775.

KOKELAAR, B. P.,    1982    A new K–Ar age from uppermost Tremadoc rocks of north
FITCH, F. J. &            Wales. *Geol. Mag.*, **119**(2), 207–211.
HOOKER, P. J.

KOKELAAR, B. P., HOWELLS, M. F., BEVINS, R. E., ROACH, R. A. & DUNKLEY, P. N. 1984a, b The Ordovician marginal basin in Wales. *In* Kokelaar & Howells (eds), Marginal Basin Geology. *Spec. Publ. geol. Soc. London*, **16** (a), 245–269, (b), 291–322.

KOKELAAR, B. P., HOWELLS, M. F., BEVINS, R. E. & ROACH, R. A. 1985 Submarine silicic volcanism and associated sedimentary and tectonic processes, Ramsey Island, SW Wales. *Jl. geol. Soc. London*, **142**, 591–614.

LAMPLUGH, G. W. 1903 The geology of the Isle of Man. *Mem. geol. Surv. UK*.

LAPWORTH, C. 1879 On the tripartite classification of the Lower Palaeozoic rocks. *Geol. Mag.*, New Series, Decade 2, **6**, 1–15.

LAWRENCE, D. J. D., WEBB, B. C., YOUNG, B. & WHITE, D. E. 1986 The geology of the late Ordovician and Silurian rocks (Windermere Group) in the area around Kentmere and Crook. *Rep. Br. geol. Surv.*, **18**(5), ii + 32 pp.

LAWSON, J. D. 1971 Some problems and principles in the classification of the Silurian System. *Mém. Bur. Rech. géol. miniér.*, No. 73, 301–308.

LEEDAL, G. P. & WALKER, G. P. L. 1950 A restudy of the Ingleton Series of Yorkshire. *Geol. Mag.*, **87**, 57–66.

LEGGETT, J. K. 1980 British Lower Palaeozoic black shales and their palaeo-oceanographic significance. *Jl. geol. Soc. London*, **137**, 139–156.

LINDSTRÖM, M. 1959 Conodonts from the Crûg Limestone (Ordovician), Wales. *Micropalaeontology*, **5**, 427–452.

LISTER, T. R., COCKS, L. R. M. & RUSHTON, A. W. A. 1970 [for 1969] The basement beds in the Bobbing borehole, Kent. *Geol. Mag.*, **106**(6), 601–603.

LISTER, T. R. & HOLLIDAY, D. W. 1970 Phytoplankton (Acritarchs) from a small Ordovician inlier in Teesdale (County Durham), England. *Proc. Yorks. geol. Soc.*, **37**, 449–460.

LOCKLEY, M. G. 1978 New evidence on the age of some Lower Palaeozoic rocks near Llanafan-Fawr. *Geol. J.*, **13**, 15–24.

1980 The geology of the Llanuwchllyn to Llanymawddy area, south Gwynedd, North Wales. *Geol. J.*, **15**, 21–41.

1983 Brachiopods from a Lower Palaeozoic mass flow deposit near Llanafan-Fawr, Central Wales. *Geol. J.*, **18**, 93–99.

1988 A review of brachiopod dominated palaeocommunities from the type Ordovician. *Palaeontology*, **26**(1), 111–145.

LOCKLEY, M. G. & WILLIAMS, A. 1981 Lower Ordovician Brachiopoda from mid and southwest Wales. *Bull. Br. Mus. nat. Hist.* (Geol.), **35**, 1–78.

LOWMAN, R. D. & BLOXAM, T. D. 1981 The petrology of the Lower Palaeozoic Fishguard Volcanic Group and associated rocks E. of Fishguard. N. Pembrokeshire (Dyfed), South Wales. *J. geol. Soc. London*, **138**, 47–68.

LYNAS, B. D. T. 1973 The Cambrian and Ordovician rocks of the Migneint area, North Wales. *J. geol. Soc. London*, **129**, 481–503.

1983 Two new Ordovician volcanic centres in the Shelve inlier, Powys, Wales. *Geol. Mag.*, **120**(6), 535–542.

LYNAS, B. D. T., RUNDLE, C. C. & SANDERSON, R. W. 1985 A note on the age and pyroxene chemistry of the igneous rocks of the Shelve Inlier, Welsh Borderland. *Geol. Mag.*, **122**(6), 641–647.

McKERROW, W. S. & SOPER, N. J. 1989 The Iapetus suture in the British Isles. *Geol. Mag.*, **126**, 1–8.

McKERROW, W. S., LAMBERT, R. ST. J. & COCKS, L. R. M. 1983 The Ordovician, Silurian and Devonian periods. *J. geol. Soc. London*, **140**.

McKERROW, W. S.,     1985     The Ordovician, Silurian and Devonian periods. *In* Snell-
    LAMBERT, R. St. J. &           ing, N. J. (ed), The chronology of the geological record.
    COCKS, L. R. M.                 *Mem. geol. Soc. London*, **10**, 73–80.

McNAMARA, K. J.     1979     The age, stratigraphy and genesis of the Coniston Lime-
                               stone Group in the southern Lake District. *Geol. J.*, **14**,
                               41–68.

McNAMARA, K. J. &     1981     Mid-Cautleyan (Ashgill Series) trilobites and facies in
    FORDHAM, B. G.           the English Lake District. *Palaeogeogr. Palaeoclimat.
                               Palaeoecol.*, **34**, 137–161.

MACGREGOR, A. R.     1961     Upper Llandeilo brachiopods from the Berwyn Hills,
                               North Wales. *Palaeontology*, **4**(2), 177–209, pls. 19–23.

                  1962     Upper Llandeilo trilobites from the Berwyn Hills, North
                               Wales. *Palaeontology*, **5**(4), 790–816, pls. 116–118.

MARR, J. E.     1892     The Coniston Limestone Series. *Geol. Mag.*, **29**, 97–110.
              1894     Notes on the Skiddaw Slates. *Geol. Mag.*, **31**, 122–130.
              1913     The Lower Palaeozoic rocks of the Cautley district (York-
                               shire). *Q. Jl. geol. Soc. London*, **69**, 1–18.
              1916     The Ashgillian succession in the tract to the west of
                               Coniston Lake. *Q. Jl. geol. Soc. London*, **71**, 189–204.

MARR, J. E. &     1885     The Lower Palaeozoic rocks of the neighbourhood of
    ROBERTS, T.             Haverfordwest. *Q. Jl. geol. Soc. London.*, **41**, 476–491.

MARR, J. E. &     1888     The Stockdale Shales. *Q. Jl. geol. Soc. London*, **44**,
    NICHOLSON, H. A.          654–732.

MATLEY, C. A.     1928     The Pre-Cambrian Complex and associated rocks of south-
                               western Lleyn (Carnarvonshire), *Q. Jl. geol. Soc. Lond.*,
                               **84**(3), 440–504.

              1932     The geology of the country around Mynydd Rhiw and
                               Sarn, south-western Lleyn, Carnarvonshire. *Q. Jl. geol.
                               Soc. Lond.*, **88**, 238–273.

              1938     The geology of the country around Pwllheli, Llanbedrog
                               and Madryn, south-west Carnarvonshire. *Q. Jl. geol. Soc.
                               Lond.*, **94**(4), 555–606.

MATLEY, C. A.,     1939     Summer field meeting to western part of the Lleyn
    NICHOLAS, T. C. &          Peninsula. *Proc. Geol. Assoc. London*, **50**, 83–100.
    HEARD, A.

MERRIMAN, R. J.,     1986     Petrological and geochemical variations within the Tal-y-
    BEVINS, R. E. &           Fan Intrusion: a study of element mobility during low-
    BALL, T. K.              grade metamorphism with implications for petrotectonic
                               modelling. *J. Petrol.*, **27**, 1409–1436.

MILLWARD, D.     1980a     Ignimbrite volcanism in the Ordovician Borrowdale vol-
                               canics of the English Lake District. *In* Harris *et al.* (*q.v.*),
                               629–634.

             1980b     Three ignimbrites from the Borrowdale Volcanic Group.
                               *Proc. Yorks. geol. Soc.*, **42**, 595–616.

MILLWARD, D. &     1985     The Stockdale (Yarlside) Rhyolite – a rheomorphic ingim-
    LAWRENCE, D. J. D.         brite? *Proc. Yorks. geol. Soc.*, **45**, 299–306.

MILLWARD, D.,     1978     The Eycott and Borrowdale Volcanic Rocks. *In* Moseley,
    MOSELEY, F. &            F. (ed), *The geology of the Lake District*, 99–120. Yorkshire
    SOPER, N. J.             Geological Society, Leeds.

MITCHELL, A. H. G. &     1975     Analogous evolution of the Burma orogen and the
    McKERROW, W. S.          Scottish Caledonides. *Bull. geol. Soc. Amer.*, **86**,
                               305–315.

MITCHELL, G. H.     1929     The succession and structure of the Borrowdale Volcanic
                               Series in Troutbeck, Kentmere and the western part of
                               Long Sleddale (Westmorland). *Q. Jl. geol. Soc. London*,
                               **85**, 9–44.

MITCHELL, G. H. 1934 The Borrowdale Volcanic Series and associated rocks in the country between Long Sleddale and Shap. *Q. Jl. geol. Soc. London*, **90**, 418–444.

1940 The Borrowdale Volcanic Series of Coniston, Lancashire. *Q. Jl. geol. Soc. London*, **96**, 301–319.

1956 The Borrowdale Volcanic Series of the Dunnerdale Fells, Lancashire. *L'pool. Manchr. geol. J.*, **1**, 428–449.

1963 The Borrowdale volcanic rocks of the Seathwaite Fells, Lancashire. *L'pool. Manchr. geol. J.*, **3**, 289–300.

MOLYNEUX, S. G. 1980 [for 1979] New evidence for the age of the Manx Group, Isle of Man. *In* Harris, A. L., Holland, C. H. & Leake, B. E. (eds), The Caledonides of the British Isles – reviewed. *Spec. Publ. geol. Soc. London*, **8**, 415–421.

MOLYNEUX, S. G. & RUSHTON, A. W. A. 1984 Discovery of Tremadoc rocks in the Lake District. *Proc. Yorks. geol. Soc.*, **45**, 123–127.

1988 The age of the Watch Hill Grits (Ordovician), English Lake District: structural and palaeogeographical implications. *Trans. Roy. Soc. Edin.: Earth Sciences*, **79**, 43–69.

MOSELEY, F. 1960 The succession and structure of the Borrowdale Volcanic Series South-East of Ullswater. *Q. Jl. geol. Soc. London*, **116**, 55–84.

1984 Lower Palaeozoic lithostratigraphical classification in the English Lake District. *Geol. J.*, **19**, 239–247.

MOSELEY, F. & MILLWARD, D. 1982 Ordovician volcanicity in the English Lake District. *In* Sutherland, D. S. (ed), *Igneous rocks of the British Isles*, 93–111. John Wiley & Sons Ltd., Chichester.

MURCHISON, R. I. 1834 On the structure and classification of the Transition rocks of Shropshire, Herefordshire, and part of Wales, and on the lines of disturbance which have affected that series of deposits, including the Valley of Elevation of Woolhope. *Proc. geol. Soc. London*, **2**, 13–18.

1839 *The Silurian System, founded on geological researches in the counties of Salop, Hereford, Radnor, Montgomery, Carmarthen, Brecon, Pembroke, Monmouth, Worcester, Gloucester and Stafford; with descriptions of the coalfields and overlying formations*, 768 pp. John Murray, London.

MYERS, J. 1950 Note on the age of the Rocks on the East Side of Newport Bay, Monmouthshire. *Geol. Mag.*, **87**, 263–264.

NEUMAN, R. B. 1984 Geology and paleobiology of islands in the Ordovician Iapetus Ocean: review and implications. *Bull. geol. Soc. Amer.*, **95**, 1188–1201.

NEUMAN, R. B. & BATES, D. E. B. 1978 Reassessment of Arenig and Llanvirn age (early Ordovician) brachiopods from Anglesey, north-west Wales. *Palaeontology*, **21**(3), 571–613.

NICHOLAS, T. C. 1915 The geology of the St Tudwal's Peninsula. *Q. Jl. geol. Soc. London*, **71**(1), 83–143.

NICHOLSON, H. A. 1872 On the Silurian rocks of the English Lake District. *Proc. Geol. Assoc. London*, **3**, 105–114.

NICHOLSON, H. A. & MARR, J. E. 1887 On the occurrence of a new fossiliferous horizon in the Ordovician Series of the Lake District. *Geol. Mag.*, Decade 111, **4**, 339–344.

1891 The Cross Fell Inlier. *Q. Jl. geol. Soc. London*, **47**, 500–512.

O'BRIEN, C., PLANT, J. A., SIMPSON, P. R. & TARNEY, J. 1985 The geochemistry, metasomatism and petrogenesis of the granites of the English Lake District. *J. geol. Soc. London*, **142**, 1139–1157.

O'NIONS, R. K.,
OXBURGH, E. R.,
HAWKESWORTH, C. J. &
MACINTYRE, R. M.
1973 New isotopic and stratigraphical evidence on the age of the Ingletonian: probable Cambrian of northern England. *J. geol. Soc. London*, **129**, 445–452.

OWENS, R. M.
1973 British Ordovician and Silurian Proetidae (Trilobita). *Palaeontogr. Soc. (Monogr.)*, London, 98 pp.

OWENS, R. M.,
FORTEY, R. A.,
COPE, J. C. W.,
RUSHTON, A. W. A. &
BASSETT, M. G.
1982 Tremadoc faunas from the Carmarthen district, South Wales. *Geol. Mag.*, **119**(1), 1–38.

PHILLIPS, W. E. A.,
STILLMAN, C. J. &
MURPHY, T.
1976 A Caledonian plate tectonic model. *J. geol. Soc. London*, **132**, 579–609.

PICKERILL, R. K.
1973 *Lingulasma tenuigranulata* – Palaeoecology of a large Ordovician linguloid that lived within a strophomenid–trilobite community. *Palaeogeogr., Palaeoclimatol., Palaeoecol.*, **13**, 143–156.

1977 Trace fossil from the Upper Ordovician (Caradoc) of the Berwyn Hills, Central Wales. *Geol. J.*, **12**, 1–16.

PICKERILL, R. K. &
BRENCHLEY, P. J.
1979 Caradoc marine benthic communities of the south Berwyn Hills, North Wales. *Palaeontology*, **22**(1), 229–264.

PICKERING, K. T.,
BASSETT, M. G. &
SIVETER, D. J.
1988 Late Ordovician–Early Silurian destruction of the Iapetus Ocean: Newfoundland, British Isles and Scandinavia – a discussion. *Trans. Roy. Soc. Edin.: Earth Sciences*, **79**, 361–382.

POCOCK, R. W.,
WHITEHEAD, T. H.,
WEDD, C. B. &
ROBERTSON, T.
1938 Shrewsbury District, including the Hanwood Coalfield (One-inch Geological Sheet 152 New Series). *Mem. geol. Surv. G.B.*, 297 pp.

PRICE, D.
1973 The *Phillipsinella Parabola–Staurocephalus Clavifrons* fauna and Upper Ordovician correlation. *Geol. Mag.*, **110**(6), 535–541.

1981 Ashgill trilobite faunas from the Llŷn Peninsula, North Wales, UK. *Geol. J.*, **16**, 201–216.

PULFREY, W.
1933 The iron-ore oolites and pisolites of North Wales. *Q. Jl. geol. Soc. London*, **89**, 401–430, pl. 40.

RASTALL, R. H.
1907 The Ingletonian Series of West Yorkshire. *Proc. Yorks. geol. Soc.*, **16**, 87–100.

RASUL, S. M.
1979 Acritarch zonation of the Tremadoc Series of the Shineton Shales, Wrekin, Shropshire, England. *Palynology*, **3**, 53–72.

RASUL, S. M. &
DOWNIE, C.
1974 The stratigraphic distribution of Tremadoc acritarchs in the Shineton Shales succession, Shropshire, England. *Rev. Palaeobot. Palynol.*, **18**, 1–9.

RAYNER, D. H.
1957 A problematical structure from the Ingletonian rocks, Yorkshire. *Trans. Leeds geol. Assoc.*, **7**, 34–42.

REED, F. R. C.
1895 The geology of the country around Fishguard. *Q. Jl. geol. Soc. London*, **51**, 149–195.

1896 The fauna of the Keisley Limestone. Part I. *Q. Jl. geol. Soc. London*, **52**, 407–437.

1897 The fauna of the Keisley Limestone. Part II. Conclusion. *Q. Jl. geol. Soc. London*, **53**, 67–106.

REEDMAN, A. J.,
LEVERIDGE, B. E. &
EVANS, R. B.
1984 The Arfon Group ('Arvonian') of North Wales. *In* Bassett, M. G. (Ed.), Focus on Wales. *Proc. Geol. Assoc. London*, **95**(4), 313–321.

REEDMAN, A. J.
AND FIVE OTHERS
1983   The Cambrian–Ordovician boundary between Aber and Betws Garmon, Gwynedd, North Wales. *Rep. Inst. Geol. Sci.*, No. **83/1**, 7–10.

RENOUF, J. T.
1975   The Proterozoic and Palaeozoic development of the Armorican and Cornubian provinces. *Proc. Ussher Soc. Camborne*, **3**, 6–43.

RICKARDS, R. B.
1970   The Llandovey (Silurian) graptolites of the Howgill Fells, northern England. *Palaeontogr. Soc.* [*Monogr.*], **123**, 1–108.

1976   The sequence of Silurian graptolite zones in the British Isles. *Geol. J.*, **11**, 153–188.

1978   Silurian. *In* Moseley, F. (ed), *The geology of the Lake District*, 130–145. Yorkshire Geological Society, Leeds.

RIDGWAY, J.
1971   The stratigraphy and petrology of Ordovician volcanic rocks adjacent to the Bala Fault in Merionethshire. *Unpubl. Ph.D. Thesis*, Univ. Liverpool.

1975   The stratigraphy of Ordovician volcanic rocks on the southern and eastern flanks of the Harlech Dome in Merionethshire. *Geol. J.*, **10**, 87–106.

1976   Ordovician palaeogeography of the southern and eastern flanks of the Harlech Dome, Merionethshire, North Wales. *Geol. J.*, **11**, 121–136.

ROACH, R. A.
1969   The composite nature of the St. David's Head and Carn Llidi intrusions of North Pembrokeshire. *In* Wood, A. (Ed.) (*q.v.*), 409–433.

ROBERTS, B.
1967   Succession and structure in the Llwyd Mawr syncline, Caernarvonshire, North Wales. *Geol. J.*, **5**, 369–390.

1969   The Llwyd Mawr ignimbrite and its associated volcanic rocks. *In* Wood, A. (Ed.), (*q.v.*), 337–356.

1979   *The Geology of Snowdonia and Llyn: an outline and field guide.* Adam Hilger Ltd, Bristol. 183 pp.

ROBERTS, R. O.
1929   The geology of the district around Abbey-Cwmhir (Radnorshire). *Q. Jl. geol. Soc. Lond.*, **85**, 651–676.

ROSE, W. C. C.
1954   The sequence and structure of the Skiddaw Slates in the Keswick–Buttermere area. *Proc. Geol. Assoc.*, **65**, 403–406.

ROSE, W. C. C. &
DUNHAM, K. C.
1977   Geology and hematite deposits of South Cumbria. *Econ. Mem. Geol. Surv. G.B.*, 170 pp.

ROSS JR, R. J. &
14 OTHERS
1982   Fission-track dating of British Ordovician and Silurian stratotypes. *Geol. Mag.*, **119**(2), 135–153.

RUNDLE, C. C.
1979   Ordovician intrusions in the English Lake District. *J. geol. Soc. London*, **136**, 29–38.

1981   The significance of isotopic dates from the English Lake District for the Ordovician–Silurian time-scale. *J. geol. Soc. London*, **138**, 569–572.

RUSHTON, A. W. A.
1982   The biostratigraphy and correlation of the Merioneth–Tremadoc Series boundary in North Wales. *In* Bassett, M. G. & Dean, W. T. (eds), *The Cambrian–Ordovician boundary: sections, fossil distributions, and correlations*, 41–59. National Museum of Wales, Geol. Ser. **3**, Cardiff.

1985   A Lancefieldian graptolite from the Lake District. *Geol. Mag.*, **122**, 329–333.

RUSHTON, A. W. A. &
HUGHES, C. P.
1981   The Ordovician trilobite fauna of the Great Paxton Borehole, Cambridgeshire. *Geol. Mag.*, **118**(6), 623–646.

RUSHTON, A. W. A. &
MOLYNEUX, S. G.
1989   The biostratigraphic age of the Ordovician Skiddaw Group in the Black Combe Inlier, English Lake District. *Proc. Yorkshire geol. Soc.*, **47**, 267–276.

SADLER, P. M.               1974    Trilobites from the Gorran Quartzites, Ordovician of south
                                    Cornwall. *Palaeontology*, **17**, 71–93, pls. 9, 10.

SAVAGE, N. M. &             1985    Caradoc–Ashgill conodont faunas from Wales and the
    BASSETT, M. G.                  Welsh Borderland. *Palaeontology*, **28**(4), 679–713.

SCHEINER, E. J.            1970    Sedimentology and petrography of three tuff horizons in
                                    the Caradocean sequence of the Bala area (N. Wales).
                                    *Geol. J.*, **7**, 25–46.

SEDGWICK, A.               1832    On the geological relations of the stratified and unstratified
                                    groups of rocks composing the Cumbrian Mountains.
                                    *Proc. geol. Soc. London*, **1**, 399.

                            1845    On the older Palaeozoic (Protozoic) rocks of North Wales.
                                    *Q. Jl. geol. Soc. London*, **1**, 5–22.

SEDGWICK, A. &             1852    On the Classification and Nomenclature of the Lower
    McCoy, F.                       Palaeozoic rocks of England and Wales. *Q. Jl. geol. Soc.
                                    London*, **8**, 136.

SHACKLETON, R. M.          1954    The structural evolution of North Wales. *L'pool. Manchr.
                                    Geol. J.*, **1**, 261–297.

                            1959    The stratigraphy of the Moel Hebog district between
                                    Snowdon and Tremadoc. *L'pool. Manchr. Geol. J.*, **2**,
                                    216–252.

SHARPE, D.                 1842    Sketch of the geology of the South of Westmorland. *Proc.
                                    geol. Soc. London*, **3**, 602–608.

SHEEHAN, P. M.             1973    Brachiopods from the Jerrestad Mudstone (Early Ash-
                                    gillian, Ordovician) from a boring in Southern Sweden.
                                    *Geologica et Palaeontologica*, **7**, 59–76.

SHOTTON, F. W.             1935    The stratigraphy and tectonics of the Cross Fell Inlier. *Q.
                                    Jl. geol. Soc. London*, **91**, 639–704.

SIMPSON, A.                1963    The structure and tectonics of the Manx Slate Series, Isle of
                                    Man. *Q. Jl. geol. Soc. London*, **119**, 367–400.

                            1967    The stratigraphy and tectonics of the Skiddaw Slates and
                                    the relationship of the overlying Borrowdale Volcanic
                                    Series in part of the Lake District. *Geol. J.*, **5**, 391–418.

SKEVINGTON, D.             1966    The lower boundary of the Ordovician System. *Norsk geol.
                                    Tidsskr.*, **46**(1), 111–119.

                            1969    The classification of the Ordovician System in Wales. pp.
                                    161–179. *In* Wood, A. (Ed.), (*q.v.*).

                            1970    A Lower Llanvirn graptolite fauna from the Skiddaw
                                    Slates, Westmorland. *Proc. Yorks. geol. Soc.*, **37**, 395–444.

                            1973    Graptolite fauna of the Great Paxton borehole, Hunting-
                                    donshire. *Bull. geol. Surv. GB*, **43**, 41–57, pls. 8, 9.

SMITH, B.                  1912    The glaciation of the Black Combe district (Cumberland).
                                    *Q. Jl. geol. Soc. London*, **68**, 402–448.

SMITH, S. &                1933    On the occurrence of Tremadoc shales in the Tortworth
    STUBBLEFIELD, C. J.             inlier (Gloucestershire), with notes on the fossils. *Q. Jl.
                                    geol. Soc. London*, **89**, 357–378, pl. 34.

SOPER, N. J.               1970    Three critical localities on the junction of the Borrowdale
                                    Volcanic Rocks and the Skiddaw Slates in the Lake
                                    District. *Proc. Yorks. geol. Soc.*, **37**, 461–493.

                            1987    The Ordovician batholith of the English Lake District.
                                    *Geol. Mag.*, **124**, 481–482.

SOPER, N. J. &             1974    Structure and stratigraphy of the Borrowdale Volcanic
    NUMAN, N. M. S.                 rocks of the Kentmere area, English Lake District. *Geol. J.*,
                                    **9**, 147–166.

SOPER, N. J.,              1987a   Late Caledonian (Acadian) transpression in north-west
    WEBB, B. C. &                   England: timing, geometry and geotectonic significance.
    WOODCOCK, N. H.                 *Proc. Yorks. geol. Soc.*, **46**, 175–192.

SOPER, N. J.,          1987b     (Correspondence), *Geol. Mag.*, **124**, 483–484.
BRANNEY, M. J.,
MATHIESON, N. A. &
DAVIS, N. C.

SOPER, N. J.,          1989      Displaced terranes in Britain and Ireland. *Jl. geol. Soc.*
GIBBONS, W. &                    *London*, **146**, 365–367.
McKERROW, W. S.

STAMP, L. D. &         1923      The igneous and associated rocks of Llanwrtyd (Brecon).
WOOLDRIDGE, S. W.                *Q. Jl. geol. Soc. London*, **79**, 16–46.

STRACHAN, I.           1981      The sequence of graptolite faunas in the Ordovician of the
                                 Shelve Inlier, Welsh Borderland. *Acta paleont. Pol.*,
                                 **26**(1), 19–26.

                       1986      The Ordovician graptolites of the Shelve District, Shrop-
                                 shire. *Bull. Br. Mus. nat. Hist.* (Geol.), **40**(1), 1–58.

STRAHAN, A.,           1909      The geology of the South Wales coalfield. Part X. The
CANTRILL, T. C.,                 country around Carmarthen (Sheet 229). *Mem. geol. Surv.*
DIXON, E. E. L. &                *G.B.*, 177 pp.
THOMAS, H. H.

STRAHAN, A.,           1914      The geology of the South Wales coalfield. Part XI. The
CANTRILL, T. C.,                 country around Haverfordwest (Sheet 228). *Mem. geol.*
DIXON, E. E. L.,                 *Surv. G.B.*, 262 pp.
THOMAS, H. H. &
JONES, O. T.

STUBBLEFIELD, C. J.    1930      A new Upper Cambrian section in south Shropshire (One-
                       [for      inch Sheet 152). *Summ. Prog. Geol. Surv. G.B.*, Part 2,
                       1929]     55–62.
                       1967      Some results of a recent Geological Survey boring in
                                 Huntingdonshire. *Proc. geol. Soc. London*, **1637**, 35–40.

STUBBLEFIELD, C. J. &  1927      The Shineton Shales of the Wrekin district: with notes on
BULMAN, O. M. B.                 their development in other parts of Shropshire and
                                 Herefordshire. *Q. Jl. geol. Soc. London*, **83**(1), 96–146.

SUTHREN, R.            1977      Sedimentary processes in the Borrowdale Volcanic Group.
                                 *In* Conference report: Palaeozoic volcanism in Great
                                 Britain and Ireland. *J. geol. Soc. London*, **133**(4), 410–411.

TAPPIN, D. R. &        1978      New Tremcoloc strata at outcrop in the Bristol Channel.
DOWNIE, C.                       *J. geol. Soc. London*, **135**, 321.

TAYLOR, K. &           1971      The pre-Westphalian geology of the Warwickshire Coal-
RUSHTON, A. W. A.                field with a description of three boreholes in the Merevale
                                 area. *Bull. geol. Surv. G.B.*, **35**, 1–152.

TEMPLE, J. T.          1965      Upper Ordovician brachiopods from Poland and Britain.
                                 *Acta palaeont. pol.*, **10**, 379–450.

THIRLWALL, M. F. &     1983      Sm–Nd garnet age for the Ordovician Borrowdale Vol-
FITTON, J. G.                    canic Group, English Lake District. *J. geol. Soc. London*,
                                 **140**, 511–518.

THOMAS, H. H. &        1924      The volcanic series of Trefgarn, Roch and Ambleston
COX, A. H.                       (Pembrokeshire). *Q. Jl. geol. Soc. London*, **80**, 520–548.

THOMAS, G. E. &        1956      The volcanic rocks of the area between Fishguard and
THOMAS, T. M.                    Strumble Head, Pembrokeshire. *Q. Jl. geol. Soc. London*,
                                 **112**, 291–311.

THORPE, R. S.,         1984      Crustal growth and late Precambrian–early Palaeozoic
BECKINSALE, R. D.,               plate tectonic evolution of England and Wales. *Jl. geol.*
PATCHETT, P. J.,                 *Soc. London*, **141**, 521–536.
PIPER, J. D. A.,
DAVIES, G. R. &
EVANS, J. A.

TOGHILL, P. 1970 A fauna from the Hendrie Shale (Llandeilo) of the Mydrian area, Carmarthenshire. *Proc. geol. Soc. London*, **1663**, 121–129.

TRAYNOR, J-J. 1988 The Arenig in South Wales: sedimentary and volcanic processes during the initiation of a marginal basin. *Geol. J.*, **23**, 275–292.

TREMLETT, W. E. 1969 Caradocian volcanicity in the Lleyn Peninsula. pp. 357–385. In *Wood, A.* (Ed.), (*q.v.*).

TROTTER, F. M. 1937 Gosforth District. *Mem. geol. Surv. G.B.*, 136 pp.
HOLLINGWORTH, S. E.,
EASTWOOD, T. &
ROSE, W. C. C.

TRYTHALL, R. J. B., 1987 Age and controls of ironstone deposition (Ordovician) North Wales. *Geol. J.*, **22**, thematic issue, 31–43.
ECCLES, C.,
MOLYNEUX, S. G. &
TAYLOR, W. E. G.

TURNER, R. E. 1982 Reworked acritarchs from the type section of the Ordovician Caradoc Series, Shropshire. *Palaeontology*, **25**(1), 119–143, pls. 15–17.

TURNER, R. E. & 1979 Acritarch dating of Arenig volcanism in the Lake District. *Proc. Yorks. geol. Soc.*, **42**, 405–414.
WADGE, A. J.

WADE, A. 1911 The Llandovery and associated rocks of north-eastern Montgomeryshire. *Q. Jl. geol. Soc. London*, **57**, 415–459.

WADGE, A. J. 1972 Sections through the Skiddaw–Borrowdale unconformity in eastern Lakeland. *Proc. Yorks. geol. Soc.*, **39**, 179–198.

1978 Classification and stratigraphical relationships of the Lower Ordovician rocks. *In* Moseley, F. (Ed.), (*q.v.*).

WADGE, A. J., 1972 Geology of the Tarn Moor Tunnel in the English Lake District. *Bull. geol. Surv. G.B.*, **41**, 55–73.
NUTT, M. J. C. &
SKEVINGTON, D.

WARD, J. C. 1876 The geology of the northern part of the English Lake District. *Mem. geol. Surv. GB*, 132 pp.

1879 On the physical history of the English Lake District. With notes on the possible subdivisions of the Skiddaw Slates. *Geol. Mag.*, **16**, 110–125.

WATTS, W. W. 1885 On the igneous and associated rocks of the Breidden Hills in east Montgomeryshire and west Shropshire. *Q. Jl. geol. Soc. London*, **41**, 532–546.

1925 The geology of south Shropshire. *Proc. geol. Assoc. London*, **36**, 321–363.

WEBB, B. C. & 1988 Slump folds and gravity slide structures in a Lower Palaeozoic marginal basin sequence (the Skiddaw Group), NW England. *J. Struct. Geol.*, **10**, 463–472.
COOPER, A. H.

WEBB, B., 1987 The Ordovician (?) batholith of the English Lake District. *Geol. Mag.*, **124**, 482–483.
MILLWARD, D.,
JOHNSON, E. &
COOPER T.

WEDD, C. B. 1932 Notes on the Ordovician rocks of Bansley, Montgomeryshire. *Summ. Progr. geol. Surv. G.B.*, Part 2, 49–55.

WEDD, C. B., 1927 The geology of the country around Wrexham. Part I (Sheet 121). *Mem. geol. Surv. G.B.*, 179 pp.
SMITH, B. &
WILLS, L. J.

WEDD. C. B., 1929 The country around Oswestry (Sheet 137). *Mem. geol. Surv. G.B.*, 234 pp.
SMITH, B.,
KING, W. B. R. &
WRAY, D. A.

WELLS, A. K.       1925    The geology of the Rhobell Fawr district (Merioneth). *Q. Jl. geol. Soc. London*, **81**(4), 463–538.

WHITTARD, W. F.     1931a   The geology of the Ordovician and Valentian rocks of the Shelve country, Shropshire. *Proc. Geol. Assoc. London*, **42**(4), 322–344.

1931b   Easter field meeting (extension) to Minsterly, April 8th to 11th, 1931. *Proc. Geol. Ass. London*, **42**(4), 339–344.

1952    A geology of South Shropshire. *Proc. Geol. Assoc. London*, **63**(2), 143–197.

1955–67   The Ordovician trilobites of the Shelve inlier, west Shropshire. Parts 1–9 (1955, 1956, 1958, 1960, 1961 (5 & 6), 1964, 1966, 1967). *Palaeontogr. Soc. (Monogr.)*, London.

1960    Lexique Stratigraphique International. Vol. 1: Europe. Fascicule 3a: England, Wales & Scotland. Part 3aIV: Ordovician. *International Geological Congress, Commission on Stratigraphy*. Centre National de la Recherche Scientifique, 296 pp.

1979    An account of the Ordovician rocks of the Shelve Inlier in west Salop and part of north Powys [by the late W. F. Whittard, compiled by W. T. Dean]. *Bull. Br. Mus. nat. Hist.* (Geol.), **33**(1), 1–69.

WHITTINGTON, H. B.    1938    The geology of the district around Llansantffraid ym Mechain, Montgomeryshire. *Q. Jl. geol. Soc. London*, **94**, 423–457, pls. 38–39.

1966    Trilobites of the Henlan Ash. Arenig Series. Merioneth. *Bull. Br. Mus. nat. Hist.* (Geol.), **11**, 489–505.

WHITTINGTON, H. B. &    1974    The Migneint, Arenig and Bala Districts. *In* D. A. Bassett
WHITWORTH, P. H.                   *et al.* (*q.v.*), 19–25.

WHITTINGTON, H. B. &    1955    The fauna of the Derfel limestone of the Arenig district,
WILLIAMS, A.                       north Wales. *Phil. Trans. R. Soc. London Series* B. Biological Sciences, No. 658, Vol. 238, 397–430.

1964    The Ordovician period. *In* Harland, W. B., Smith, A. G. & Wilcock, B. (eds), The Phanerozoic Time-scale. A symposium dedicated to Professor A. Holmes. A supplement to *Q. Jl. geol. Soc. London*, **120**, 241–254.

WHITTINGTON, H. B.    1984    Definition of the Tremadoc Series and the series of the
AND FIVE OTHERS              Ordovician System in Britain. *Geol. Mag.*, **121**, 17–33.

WILCOX, C. J. &        1981    A reassessment of facies and faunas in the type Llandeilo
LOCKLEY, M. G.             (Ordovician), Wales. *Palaeogeog. Palaeoclimatol. Palaeoecol.*, **34**, 285–314.

WILLIAMS, A.           1953    The geology of the Llandeilo district, Carmarthenshire. *Q. Jl. geol. Soc. London*, **108**, 177–208, pl. 9.

1963    The Caradocian brachiopod faunas of the Bala District, Merionethshire. *Bull. Br. Mus. nat. Hist.* (Geol.), **8**, 330–471.

1969    Ordovician of British Isles. pp. 236–264. *In* Kay, M. (ed), North Atlantic – geology and continental drift, a symposium. *Mem. Am. Assoc. Petrol. Geol.*, **12**, 1082 pp., Tulsa.

1974    Ordovician Brachiopoda from the Shelve district, Shropshire. *Bull. Br. Mus. nat. Hist.* (Geol.), Supplement, **11**, 1–163, pls. 1–28.

WILLIAMS, A. &      1981    The Ordovician–Silurian Boundary in the Garth area of
WRIGHT, A. D.            southwest Powys, Wales. *Geol. J.*, **16**, 1–39.

WILLIAMS, A., LOCKLEY, M. G. & HURST, J. M. 1981 Benthic palaeocommunities represented in the Ffairfach group and coeval Ordovician successions of Wales. *Palaeontology*, **24**(4), 661–694.

WILLIAMS, A., STRACHAN, I., BASSETT, D. A., DEAN, W. T., INGHAM, J. K., WRIGHT, A. D. & WHITTINGTON, H. B. 1972 A correlation of Ordovician rocks in the British Isles. *Spec. Rep. geol. Soc. London*, **3**, 1–74.

WILLS, L. J. 1978 A palaeogeological map of the Lower Palaeozoic floor below the cover of Upper Devonian, Carboniferous and later formations with inferred and speculative reconstructions of Lower Palaeozoic and Precambrian outcrops in adjacent areas. *Mem. geol. Soc. London*, **8**, 1–36.

WILLS, L. J. & SMITH, B. 1922 The Lower Palaeozoic rocks of the Llangollen district. *Q. Jl. geol. Soc. London*, **78**(2), 176–226.

WILSON, A. A. & CORNWELL, J. D. 1982 The Institute of Geological Sciences Borehole at Beckermonds Scar, North Yorkshire. *Proc. Yorks. geol. Soc.*, **44**, 59–88.

WILSON, J. T. 1966 Did the Atlantic close and then re-open? *Nature, London*, **211**, 676–681.

WOOD, A. (ed) 1969 *The Pre-Cambrian and Lower Palaeozoic rocks of Wales.* Univ. Wales Press, Cardiff, 461 pp.

WOODS, D. S. 1974 Ophiolites, melanges, blueschists, and ignimbrites: early Caledonian subduction in Wales? *In* Dott, R. H. Jr. & Shaver, R. H. (eds), Modern and ancient geosynclinal sedimentation. *Spec. Publs. Soc. econ. Paleont. Miner. Tulsa*, **19**, 334–344.

WOODCOCK, N. H. 1984a The Pontesford Lineament, Welsh Borderland. *J. geol. Soc. London*, **141**, 1001–1014.

1984b Early Palaeozoic sedimentation and tectonics in Wales. *In* Bassett, M. G. (Ed.), Focus on Wales. *Proc. Geol. Assoc. London*, **95**(4), 323–335.

WOODCOCK, N. H. & SMALLWOOD, S. D. 1987 Late Ordovician shallow marine environments due to glacio-eustatic regression: Scrach Formation, Mid-Wales. *J. geol. Soc. London*, **144**, 393–400.

WOOLACOTT, D. 1923 On a Boring at Roddymoor Colliery, near Crook, Co. Durham. *Geol. Mag.*, **60**, 50–62.

WRIGHT, A. D. 1968 A westward extension of the Upper Ashgillian *Hirnantia* fauna. *Lethaia*, **1**, 352–367.

1985 The Ordovician–Silurian boundary at Keisley, northern England. *Geol. Mag.*, **122**, 261–273.

ZALASIEWICZ, J. A. 1984 A re-examination of the type Arenig Series. *Geol. J.*, **19**, 105–124.

# 5

# SILURIAN

# C. H. Holland

## Introduction

Silurian biostratigraphy has already reached a substantial refinement,[1] its widely recognised graptolite biozones, for example, representing an individual duration of only about one million years (Rickards 1976). Unfortunately, few isotopic dates have as yet related to the stratigraphical sequence. Thus estimates of the beginning, end, and duration of the Silurian Period are confusingly varied (Spjeldnaes 1978, Holland 1985). Boucot (1975) arrived at a duration of 32 million years (centring at about 420 Ma), allocating a 'relative palaeontological duration' equivalent to approximately 44 per cent for the Llandovery, 21 per cent for the Wenlock, 23 per cent for the Ludlow, and 12 per cent for the Přídolí. Spjeldnaes, in his assessment of the evidence, allowed only 25 million years for the Silurian Period. McKerrow et al. (1980) used all published meaningful evidence, including the fission track data of Ross et al. (1978), to conclude that the beginning of the Silurian was at 438 Ma and its duration 27 million years. Their contribution was criticised by Gale et al. (1980) whose 'compromise' Silurian runs from 425 to 400 Ma. More recent estimates by Harland et al. (1982) and McKerrow et al. (1985) give respectively 438 to 408 and 435 to 412 Ma. During this relatively short span of Silurian time the long-lasting Caledonian cycle of earth history drew to a close.

Increasing similarity of marine faunas throughout the North Atlantic region during the Early Palaeozoic (McKerrow & Cocks 1976) is consistent with the coming together of Baltica and Laurentia (Scotese et al. 1979) as the Proto-Atlantic Ocean (the Iapetus Ocean of Harland & Gayer 1972) closed by subduction. The Old Red Sandstone magnafacies follows appropriately in terms of palaeogeographical evol-

ution as a clastic wedge between the developing continent to the north and marine conditions in the south. Its initiation occurred at different times in different places (Allen 1979, Holland 1979, Bassett 1985a), though it is certainly a late Silurian as well as Devonian phenomenon. Apart from faunal evidence (Holland 1971, McKerrow & Cocks 1976, Ziegler et al. 1977) and a pattern of later sedimentation broadly consistent with shallowing followed by continental conditions, Silurian, as distinct from earlier Palaeozoic, evidence for a Proto-Atlantic Ocean as such is sparse. A dynamic model which involves sequential accretion of an imbricating sedimentary wedge on to the continental margin of Laurentia as developed for the Southern Uplands of Scotland by Legget et al. (1979), remains unconvincing (Holland 1986). This is particularly the case in Ireland, where the pattern is rather one of slices of Ordovician faulted sporadically into place within the widespread Silurian outcrops (Holland 1981).

Berry & Boucot (1967) have noted the relatively insignificant development of 'geosynclines' in the Silurian world and the contrastingly great extent of platforms. Some bear carbonates but others siltstones and mudstones with a characteristic graptolite–bivalve–orthocone assemblage. On the whole the period seems to have been one of relatively warm climate and comparative stability of palaeogeography. Most of the Silurian shelf seas were positioned in low latitudes, which accounts for the cosmopolitanism of Silurian faunas (Ziegler et al. 1977). There is, however, considerable and convincing evidence of Ashgill and possibly earliest Silurian glaciation centred upon what is now north-central Africa (Spjeldnaes 1981). A consequent regression at this time was followed by the well-known marine transgression later in the Llandovery. Silurian rocks on the

[1] A Global Standard for the Silurian System (C. H. Holland & M. G. Bassett (eds), 1989. National Museum of Wales, Cardiff) includes accounts of Silurian stratigraphical classification and international regional treatments, including areas in England and Wales.

southern flank of the Proto-Atlantic have figured much in studies of fossil communities and there does appear to be a relationship with temperature and, especially in detail, with depth of water. The basinal rocks deposited in deeper water to the north-west are rich in evidence of turbiditic deposition, and the splendidly displayed manifestations of marginal submarine slumping led to some of the first, now classical, descriptions of slump phenomena (Jones 1937, 1940; Straw 1937).

In terms of the geology of England and Wales, Silurian palaeogeography is relatively uncomplicated with a basin to the north-west in Wales and the English Lake District and a shelf to the south-east in the Welsh Borderland and Central England, the latter affected by some fluctuations in depth of water after the Llandovery transgression and before the final shallowing and elevation. However, it would be insufficient entirely to ignore what happened to the north-west, across the Proto-Atlantic, though this was now closing, subduction having ceased at a suture possibly along the Solway. The earlier Lake District and Welsh Ordovician volcanicity belonged to a south-easterly dipping subduction zone. There is also additional evidence from the underground Silurian of SE England, which must take us closer to the difficult problems of Palaeozoic plate configuration in Meso-Europe and Neo-Europe (Ager 1975).

## Distribution

Silurian rocks of Llandovery to at least Ludlow age, largely of graptolitic and greywacke facies form an extensive outcrop in the southern part of the English Lake District (Figs. 5.1, 5.12). The scenery developed here, though famed for its beauty, is relatively more subdued than that of the angular crags developed in the Ordovician volcanics. In places in this northern area of outcrop, for instance in the Howgill Fells to the north of Sedbergh, the Silurian does give rise to a splendidly bleak landscape of mountain and moorland. Additionally, there is a small outcrop of Silurian rocks within the Cross Fell Lower Palaeozoic inlier to the east of Penrith. To the south-east of the main outcrop of the eastern Lake District the equivalent strata are seen also in the inliers of Austwick and Horton-in-Ribblesdale, where, immediately north of the Craven fault belt (Ch. 17) erosion has cut down through the Carboniferous Limestone to reveal Silurian rocks in the valleys.

In North Wales, Silurian greywackes, slumped mudstones, and graptolitic rocks form the Denbighshire Moors and the Clwydian Range. From here they pass southwards along the Central Wales syncline between the Harlech and Berwyn domes (Ch. 3, Ch. 4), from whence the outcrop widens to cover much

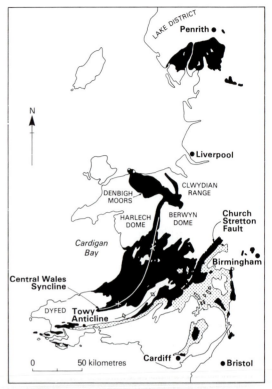

**Fig. 5.1.** Distribution of Silurian rocks in England and Wales. Llandovery, Wenlock, and Ludlow: black; Downton Group (Přídolí) of Wales and the Welsh Borderland: stippled; Přídolí of Northern England not distinguished.

of Central Wales, from the excellent coastal sections along Cardigan Bay to the Long Mountain, Clun Forest, and Radnor Forest in the east (Fig. 5.10).

Farther to the south-west Silurian rocks are seen again in the approximately east-west trending folded and thrust strips of the Hercynian tract of South Pembrokeshire (Dyfed) (Fig. 5.1). They are well exposed, for instance, in Marloes Bay (Fig. 5.6) and the cliffs westwards from here. In Marloes Bay a single Hercynian cleavage is superimposed on rocks previously tilted by Caledonian earth movements (Graham et al. 1977). East of the main mountain and moorland mass of Central Wales, and separated from it by the Towy Anticline, a strip of Silurian strata runs through Llandovery (Fig. 5.6) and bends into a more extensive and complicated outcrop to the south of Builth Wells. Here the thick but shelly sequence was described by Straw (1937) in what was really the pioneering modern treatment of British Ludlow rocks. East of the Church Stretton Fault

**Fig. 5.2.** Aerial view of Wenlock Edge, looking north-eastwards from *c*. SO 442 842[1].
The escarpment is formed mainly by beds of the Farley Member with a capping of
Much Wenlock Limestone Formation. The lower ground to the right is Hope Dale,
occupied by the lower part of the Ludlow sequence. Calcareous beds of the middle part
of the Ludlow form the escarpment in the far right.

(Cambridge University Collection. Photograph by J. K. S. St Joseph, copyright
reserved; reproduced by permission of the Curator of Aerial Photography).

[1] (National Grid Locality).

Complex the course of the main outcrop of the Welsh Borderland is marked by wooded ridges of the Wenlock and Ludlow limestones (Figs. 5.2, 5.3). It bends around the NE-plunging Ludlow anticline to follow the line of Wenlock Edge, disappearing eventually beneath the Carboniferous. To the east yet again there are relatively thin and dominantly shelly Silurian sections in the inliers of Dudley and other small areas in the English Midlands, in the south-eastern inliers of the Malvern and Abberley Hills, of Woolhope, May Hill, Tites Point, and Tortworth, the Mendips, Usk, and Rumney (Cardiff) (Fig. 5.10). If the Silurian portion of the Lower Old Red Sandstone (Přídolí Series) is taken into account (Fig. 5.1), a substantial roughly triangular area of out-

crop must be added between the Towy Anticline and the easterly inliers, extending southwards towards the South Wales coalfield and surrounding the Usk inlier.

Apart from the more obviously extrapolatable concealed Silurian rocks beyond the present outcrops, there is a separate area (Fig. 5.14) of records from deep boreholes in south-eastern and eastern England which will be considered later in this chapter.

## Stratigraphical classification

The history of the establishment of the Silurian System and its principal divisions was reviewed by Whittard

**Fig. 5.3.** Aerial view of Ludlow anticline, looking north-eastwards from *c*. SO 430 680. The central, low lying area of Wenlock Shale is bounded by the parallel wooded ridges of the Much Wenlock Limestone Formation and the middle part of the Ludlow. Ludlow town lies beyond the gap seen at the nose of the anticline. The Wrekin (Precambrian) is visible in the far left. In the distance, to the right of centre, is the synclinal outlier of Brown Clee Hill of Old Red Sandstone and Carboniferous.

(Cambridge University Collection. Photograph by J. K. S. St Joseph, copyright reserved; reproduced by permission of the Curator of Aerial Photography).

(1961), by Cocks *et al.* (1971), and, more recently, by Holland (1987a). Although Murchison had already used the names Wenlock and Ludlow in stratigraphy, the *Silurian System* itself first appeared in publication in 1835, the name taken from that of an ancient Welsh Borderland tribe, the *Silures* (Murchison 1835). Murchison's 'Lower Silurian' then included rocks now regarded as Ordovician. The subsequent controversy with Sedgwick and the eventually satisfying compromise suggested by Charles Lapworth are well known (e.g. see Holland 1974, Secord 1986). The name

*Llandovery* was first used stratigraphically by Murchison in 1859, though it had appeared two years earlier on Geological Survey maps. Thus the Silurian System and three of the four series into which it is now customarily divided relate to a *type* area (as distinct from an internationally agreed standard) in the Welsh Borderland and Wales.

The Subcommission on Silurian Stratigraphy of the International Union of Geological Sciences was constituted from an *ad hoc* body at a meeting held at the University of Birmingham in September 1974. At

| SILURIAN | | | | |
|---|---|---|---|---|
| | PŘÍDOLÍ SERIES | — | | |
| | LUDLOW SERIES | LUDFORDIAN STAGE | *Bohemograptus* proliferation | |
| | | | *leintwardinensis* | |
| | | GORSTIAN STAGE | *tumescens* | *incipiens* |
| | | | *scanicus* | |
| | | | *nilssoni* | |
| | WENLOCK SERIES | HOMERIAN STAGE | *ludensis* | |
| | | | *nassa* | |
| | | | *lundgreni* | |
| | | SHEINWOODIAN STAGE | *ellesae* | |
| | | | *linnarssoni* | |
| | | | *rigidus* | |
| | | | *riccartonensis* | |
| | | | *murchisoni* | |
| | | | *centrifugus* | |
| | LLANDOVERY SERIES | TELYCHIAN STAGE | *crenulata* | |
| | | | *griestoniensis* | |
| | | | *crispus* | |
| | | | *turriculatus* | |
| | | AERONIAN STAGE | *sedgwickii* | |
| | | | *convolutus* | |
| | | | *argenteus* | |
| | | | *magnus* | |
| | | | *triangulatus* | |
| | | RHUDDANIAN STAGE | *cyphus* | |
| | | | *acinaces* | |
| | | | *atavus* | |
| | | | *acuminatus* | |

**Table 5.1.** Silurian chronostratigraphy as now agreed internationally. The terms Lower Silurian and Upper Silurian may be used as subsystems including respectively the Llandovery plus Wenlock, and Ludlow plus Přídolí. The right hand biostratigraphical column is of graptolite biozones.

its meeting in Sydney in August 1976 the Subcommission decided upon a programme of work, the major part of which was to establish international agreement upon Silurian chronostratigraphy. The history of achievement of this is fully reviewed elsewhere (Holland 1985, 1987a). The resulting classification is shown in Table 5.1. Graptolite biozones are also listed here for convenience.

## Llandovery Series

The town of Llandovery in South Wales is situated in the Twyi (Towy) valley and actually upon Ordovician rocks. Immediately to the east is what is justly to be regarded as the *type* area for the Llandovery Series (Fig. 5.6). The outcrop is conveniently divided into two parts, respectively south and north of the River Gwydderig, close to the town itself. The southern area was described in detail by Jones (1925), who used Lower (>730 m), Middle (>240 m), and Upper (>480 m) divisions which he regarded as 'lithological stages'. Within these were thirteen lithological subdivisions, though he did refer to the fossil content of each. The scheme of letters and numbers ($A_1$–$A_4$, $B_1$–$B_3$, and $C_1$–$C_6$) applied to this classification has subsequently been widely used in Llandovery stratigraphy. Later Jones (1949) described the northern area of the Llandovery district, though here he was not able to employ the same detailed lithological subdivision. Williams (1951) added what he modestly described as a 'palaeontological supplement', though it has really been upon the basis of shelly faunal assemblages that the lettered and numbered subdivisions have repeatedly been referred to in Llandovery correlation across the world. His treatment of *Stricklandia* has been much used, both in itself and as a model for later detailed taxonomic treatment of proposed brachiopod lineages.

Cocks *et al.* (1970) established four Llandovery stages (equated respectively with $A_1$–$A_4$, $B_1$–$B_3$, $C_1$–$C_3$, and $C_4$–$C_6$), which were given geographical names based upon localities within the type area. As the lowest ($A_1$) subdivision at Llandovery is unfossiliferous and of uncertain extent (even of uncertain age), these authors used Lapworth's famous locality at Dob's Linn, near Moffat, in the Southern Uplands of Scotland, rather than one in the type area, for the definition of the first, *Rhuddanian Stage*. The basal boundaries of the remaining stages were defined by means of localities in the type area. Unfortunately, detailed faunal changes through continuous boundary stratotypes with chosen boundary marker points were not yet known. Thus the British regional stages for the Llandovery remained subject to further investigation and international comparison before decisions were reached concerning standard chrono-

stratigraphy for the first series of the Silurian System.

The Ordovician–Silurian Boundary Working Group of the IUGS Commission on Stratigraphy was established in 1974 and eventually narrowed its choice for a boundary stratotype to Anticosti Island in Canada and Dob's Linn in Scotland (Cocks 1985). The formal voting procedure resulted in the selection of Dob's Linn. The horizon then chosen was the base of the *Parakidograptus acuminatus* Biozone, rather than the base of the preceding *Glyptograptus persculptus* Biozone. The higher horizon found favour with those involved in research at Dob's Linn and, for instance, with Soviet stratigraphers. It has the undoubted advantage of lifting the base of the Silurian well clear of the widespread *Hirnantia* fauna, whose extent of diachronism remains uncertain. These decisions were ratified by the Commission on Stratigraphy and IUGS in 1985 (Bassett 1985b).

The Subcommission on Silurian Stratigraphy itself eventually decided that three areas in the world: Anticosti, the Oslo region, and the type Llandovery district, offered promise for establishment of boundary stratotypes for the internal division of the Llandovery Series. A working group was established to reinvestigate the Llandovery district, both 'south' and 'north', providing well documented sections and improved graptolitic control. The type area subsequently found formal acceptance, though the Subcommission decided upon three rather than four stages within the Series. The first of these, the Rhuddanian Stage (named originally from Cefn-Rhuddan Farm in the type Llandovery area) was defined by the decision taken in favour of Dob's Linn referred to above. The base of the newly named, second, Aeronian Stage is taken in the Trefawr forestry road section–500 metres north of Cwm-coed-Aeron Farm (GR–SN 8380 3935). The marker point for the base is in a continuous sequence through part of the Trefawr Formation and correlates with the base of the *Monograptus triangulatus* Biozone. The third stage of the Llandovery Series retains the name Telychian of Cocks *et al.* (1970), but its base is now repositioned to correlate with the base of the *Monograptus turriculatus* Biozone. There are also changes in the brachiopod sequence at this level, *Eocoelia intermedia* being replaced by *Eocoelia curtisi*, and *Stricklandia lens progressa* by *Stricklandia laevis*. The name was taken from Pen-lan-Telych Farm. The boundary stratotype is in an old quarry on the west side of the Cefn Cerig road (SN 7743 3232), within part of the Wormwood Formation.

All these decisions of the Subcommission on Silurian Stratigraphy have gone through formal voting procedures culminating in ratification by the Commission on Stratigraphy and IUGS (Bassett 1985b). A full account of the new work on the

Llandovery Series in the Llandovery area is available in Cocks *et al.* (1984). The paper includes maps of both 'southern' and 'northern' areas in which details of the local lithostratigraphy and facies changes are made clear.

## Wenlock Series

The name Wenlock relates especially to the wooded scarp of Wenlock Edge (Fig. 5.2) formed by the limestones of the upper part of the Wenlock Series. The higher beds of the local Llandovery and the lower part of the Wenlock form the lower ground of Ape Dale beneath and to the north-west of the escarpment. The whole narrow strip of country from Coalbrookdale, where the Wenlock rocks appear from beneath the unconformable cover of Carboniferous in the north-east, across the River Severn between Buildwas and Ironbridge, along Wenlock Edge and Ape Dale past Longville-in-the-Dale, Rushbury, and Eaton, south-westwards to the River Onny, forms a compact type and standard area for the Wenlock Series (Fig. 5.4). The higher limestones have long been famous for their shelly fossils but Bassett *et al.* (1975) showed that the somewhat deeper water shales and mudstones below are surprisingly rich in graptolites as well as containing a not inconsiderable shelly fauna.

The standard chronostratigraphy, based upon the area, is shown in Table 5.1. The boundary stratotype for the base of the Wenlock Series and for its lower, Sheinwoodian Stage is taken at a locality in Hughley Brook. (The site is now cared for by the Nature Conservancy.) It shows purple and green mudstones with some siltstone bands of the Llandovery Purple Shales, in which a varied brachiopod fauna is accompanied by trilobites, ostracodes, corals, crinoids,

**Fig. 5.4.** Geological map of the type and standard Wenlock area (after Bassett *et al.* 1975).

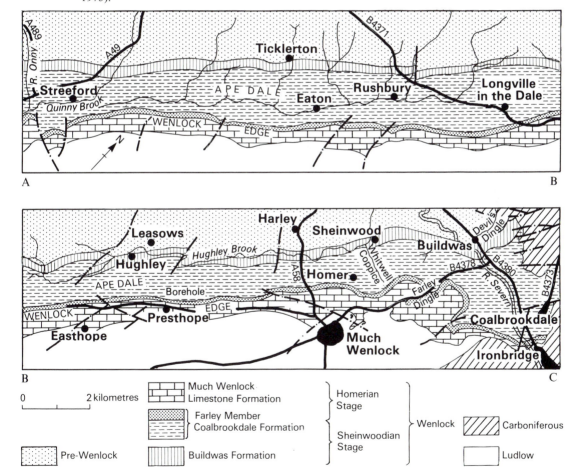

bryozoa, and molluscs. Within the type area these higher beds of the Purple Shales have yielded graptolites indicative of the *Monoclimacis crenulata* Biozone. The succeeding Buildwas Formation (up to 40 m) is of greyish mudstones with bands of calcareous nodules. The fauna is characterised by small brachiopods including *Atrypa reticularis*, *Cyrtia exporrecta*, and *Eospirifer radiatus*. It is distinguished from that of the Llandovery by the presence of such forms as *Dicoelosia biloba*, *Eoplectodonta duvalii*, *Isorthis elegantulina*, and *Resserella sabrinae*. Within the type area the lower beds of the Buildwas Formation yield *Monoclimacis* aff. *vomerina* and *Pristiograptus watneyae*, together indicating correlation with the *Cyrtograptus centrifugus* Biozone.

The standard section for the base of the upper, Homerian Stage of the Wenlock Series is entirely within a continuous sequence of olive to greyish green mudstones of the Coalbrookdale Formation (192–265 m). The boundary stratotype has been precisely defined in a stream section which cuts through Whitwell Coppice half a kilometre north of Homer (Fig. 5.4). The graptolite fauna in the 1.5 m below the boundary comprises *Cyrtograptus ellesae*, *Monograptus flemingii*, and *Pristiograptus dubius*. Above the boundary the fauna includes *Cyrtograptus lundgreni*, *M. flemingii*, and *P. dubius*.

Despite their generally conservative nature, Wenlock brachiopods may be divided into two broad assemblages corresponding approximately to the two stages. The Homerian Stage is characterised by a number of new elements including *Meristina obtusa*, *Resserella canalis*, and *Striispirifer plicatellus*.

Bassett *et al.* (1975) divided the Homerian into two chronozones, the boundary stratotype for the upper being part of the more or less continuous section in the old lane east of Eaton Church, at the point where the boundary between the *Cyrtograptus lundgreni* and *Gothograptus nassa* biozones is marked as crossing the track. These uppermost biozones of the Wenlock are particularly widely identifiable internationally and thus chronostratigraphy has been taken to the very high level of resolution implied by the chronozone.

In terms of lithostratigraphy, the Farley Member (24–27 m) is that part of the succession seen in the higher part of the escarpment where the mudstones of the thick Coalbrookdale Formation have given way to an alternation of grey shaly mudstones and thin bands of limestone nodules. The member comprises most of the *Gothograptus nassa* Biozone. The succeeding *Monograptus ludensis* Biozone extends from the uppermost part of the Farley Member through the whole of the Much Wenlock Limestone Formation (21–29 m). The limestone itself is referred to later in this chapter.

## Ludlow Series

Holland *et al.* (1959) introduced a new classification of the Ludlow succession at Ludlow, employing faunal as well as lithological criteria in recognising nine local subdivisions. Standard sections were described for the basal boundaries of each of these. At this time such rigorously definitive stratigraphy was comparatively unusual and the local 'combined units' were referred to as 'Beds'. It now seems reasonable (Holland *et al.* 1978) to recognise the possible place of fossil content in the recognition of formations, though these divisions must remain strictly mappable units. As such the nine Ludlow subdivisions may reasonably be regarded as formations (Holland *et al.* 1980), their nomenclature modified as shown in Table 5.2.

**Table 5.2.** Formations of the Ludlow Series (Holland *et al.* 1980)

Upper Whitcliffe Formation (approximately 30 m)
Lower Whitcliffe Formation (25 m)
Upper Leintwardine Formation (1·5–5·5 m)
Lower Leintwardine Formation (30 m)
Upper Bringewood Formation (10–45 m)
Lower Bringewood Formation (45–60 m)
Upper Elton Formation (45–75 m)
Middle Elton Formation (45–105 m)
Lower Elton Formation (30–45 m)

Some of the standard sections were also used as basal boundary stratotypes for the four stages into which the Ludlow Series was divided. The graptolite biozones are indicated in Table 5.1, though the recognition of the basal biozone of the Ludlow as that of *Neodiversograptus nilssoni* is a later refinement to be referred to below. The Eltonian/Bringewoodian boundary between what were originally the first and second stages has never been readily recognisable outside the Welsh Borderland and the third Leintwardinian Stage corresponds in range to only one (albeit easily recognisable) graptolite biozone. Accordingly the decision was made (Holland 1980) to recognise only two Ludlow stages in international standard chronostratigraphy as indicated in Table 5.1. Holland *et al.* (1963) gave a detailed account of the faunal succession at Ludlow and hence criteria for correlation elsewhere. Records of these very fossiliferous rocks continue to accumulate (e.g. White & Lawson 1978).

The boundary stratotype for the base of the Gorstian Stage and hence for that of the Ludlow Series is in the old quarry in Pitch Coppice (Fig. 5.5) to the south of the Ludlow–Wigmore road and in the heart of the Ludlow anticline. Here, the uppermost nodular development of the Much Wenlock Limestone is followed by several metres of Lower Elton Formation with its typical lithology of soft, olive, silty shales rich in broken shell fragments. Identifiable fossils are not so common, though there is a characteristic assemblage at this level throughout the Welsh Borderland, its brachiopods such as *Atrypa reticularis, Protochonetes minimus, Leptaena depressa*, and *Resserella canalis* obviously of Wenlock aspect. The outstanding problem of the relationship between this Wenlock/Ludlow boundary and the graptolite sequence was subsequently investigated in a comprehensive study of the Wenlock graptolites of the Ludlow district (Holland *et al.* 1969). It became clear that the *ludensis* (= *vulgaris*) Biozone was present in the top part of the Wenlock Shale and continued into the Much Wenlock Limestone. To have placed the latter within the Ludlow Series would have been intolerable, so the suggestion was made that the succeeding *nilssoni* Biozone should be regarded as the basal biozone of the Ludlow. The discovery of graptolites in the basal beds of the Lower Elton Formation (White 1981) confirms the presence of the *nilssoni* Biozone.

The boundary stratotype for the base of the Ludfordian Stage (and also for the base of the Lower Leintwardine Formation) is at Sunnyhill Quarry on the north-eastern side of the Mary Knoll Valley (Fig. 5.5). The calcareous facies of the middle part of the Ludlow here persists into the Lower Leintwardine Formation and at one time the whole quarry might have been regarded as lying within the 'Aymestry Limestone' of Murchison's classification. The lowest 5 m of the face are of nodular silty limestone with such characteristic fossils of the Upper Bringewood Formation as *Kirkidium knightii, Strophonella euglypha, Poleumita discors*, and certain corals. Their disappearance at this level was taken as critical (Holland *et al.* 1963). The succeeding more evenly bedded and less nodular limestones of the Lower Leintwardine Formation, which mark the base of the Ludfordian Stage, are rich in brachiopods including forms such as *Atrypa reticularis, Isorthis orbicularis, Microsphaeridiorhynchus nucula*, and *Sphaerirhynchia wilsoni* which survive from immediately below, as well as many additional (though not necessarily 'new') forms such as *Dayia navicula, Shagamella ludloviensis, Shaleria ornatella*, and *Lingula lewisii*. Dr L. Cherns has investigated the faunal changes here in great detail and has added the important record of *Saetograptus*

**Fig. 5.5.** Geological map of the type and standard Ludlow area (modified from Holland *et al.* 1963).

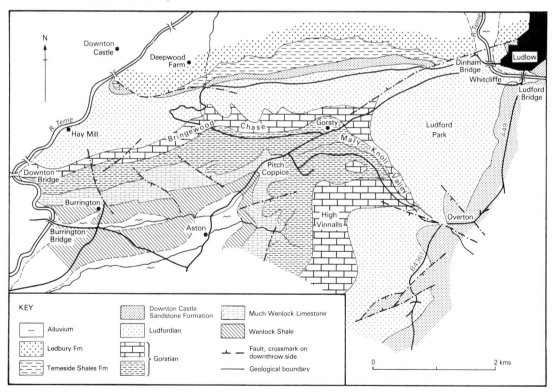

*leintwardinensis* from the Ludfordian. The almost monotypic *leintwardinensis* Biozone is known from shelf to basin in the Welsh Borderland and thence widely about the world. The Ludfordian Stage, particularly in its upper part, has a characteristic shelly fauna in which molluscs are conspicuous. The brachiopods are dominated by the three forms *Microsphaeridiorhynchus nucula*, *Protochonetes ludloviensis*, and *Salopina lunata*. The bivalves include *Fuchsella amygdalina* and *Goniophora cymbaeformis*. A distinctive, if restricted, cephalopod fauna includes *Leurocycloceras whitcliffensis*.

## Přídolí Series

International investigation of, and subsequent agreement upon, the boundary between the Silurian and Devonian systems (Martinsson 1977) can now be looked back upon as something of a case history in stratigraphical procedure – some would certainly regard the work as a model of its kind. One must simply hope that other stratigraphical boundary problems can be settled in a shorter time. Apart from many earlier contributions on the subject, the deliberations of the Silurian–Devonian Boundary Committee, formed under the aegis of the Commission on Stratigraphy of the International Union of Geological Sciences, lasted from 1960 to 1972. But there was a special problem in this particular case, earlier correlations having been found to be in serious error. So much adjustment of view was necessary that an element of compromise (Holland 1965, 1986; Bouček *et al.* 1966) was needed if the final solution was to prove internationally acceptable. There was, as Martinsson so nicely put it, a lost series of the Silurian System – lost, that is to say, in previously faulty correlation tables. The agreed upward extent of this fourth series as defined in the boundary stratotype section at Klonk in Czechoslovakia, is to the base of the *Monograptus uniformis* Biozone.

The Subcommission on Silurian Stratigraphy (Holland 1985, 1987a) decided that the time is not yet ripe for subdivision of the fourth series into stages, but that its basal boundary stratotype should be located in Czechoslovakia, in the Downton district of the Welsh Borderland, or in Podolia in the Soviet Union. It was eventually agreed that the fourth series should be named the Přídolí Series, the name coming from Přídolí in the Barrandian area (Prague Basin) of Bohemia, Czechoslovakia. The boundary stratotype is in the Pôžary section near Prague, with the marker point taken at the base of the *Monograptus parultimus* Biozone. These decisions were ratified by the Commission on Stratigraphy and IUGS (Bassett 1985b), thus completing the programme concerning internal chronostratigraphy which the Subcommission on

Silurian Stratigraphy set itself in 1976 (Holland 1987a).

There is a veritable web of correlation extending across Europe between the marine to brackish 'Downton Series' of the Welsh Borderland and the fully marine Přídolí, which involves the use of ostracodes, conodonts, plants, vertebrates, etc. (Kaljo 1978, Bassett *et al.* 1982). For present purposes the 'Downton Series', better named the *Downton Group*, may be taken as approximately coincident with the Přídolí Series.

The name 'Downtonian' was introduced by Lapworth (1879–80), though Murchison (1839) referred to the Downton-castle building stone. A detailed description of the lower part of the succession (Ludlow Bone Bed, Downton Castle Sandstone, Temeside Shales) formed part of the account by Elles & Slater (1906) of the 'highest Silurian rocks' of the Ludlow to Downton district. King (1934) separated the Downtonian and Dittonian (see Chapter 6), Whitehead & Pocock (1947) introduced a dual nomenclature in which Downtonian was taken to refer to the age of the rocks as dated by vertebrate fossils and the 'Downton Series' was employed for the lithological division. The base of the '*Psammosteus* Limestones' was taken as the boundary between the Downton and Ditton Series. Ball & Dineley (1961) in a study of the Old Red Sandstone of the Brown Clee Hill and adjacent area followed the same arrangement; but Allen & Tarlo (1963) took the boundary at a slightly lower level, where fluviatile sandstones become abundant and the vertebrate fauna changes from the marine-brackish water forms of the Downtonian to the freshwater forms of the Dittonian. Holland & Richardson (1977) in reviewing the problem of recognition of the new Silurian/Devonian boundary in the British Isles, accepted the scheme employed by Allen and Tarlo, and in so doing noted its usage by Jaeger (e.g. 1962) in attempted correlation between the Welsh Borderland and Central Europe. Stratigraphical nomenclature within the 'Downton Series' has continued to vary widely and there has remained a tendency to mix lithostratigraphy and chronostratigraphy. The Ludlow Bone Bed, Downton Castle Sandstone, and Temeside Shales are well understood lithological divisions and the remaining Red Downtonian of various authors is now often referred to as the Ledbury Group or Ledbury Formation – see also Fig. 5.5 and Table 5.4. Ball & Dineley (1961) used the term Grey Downton Formation collectively for the lower units below their Red Downton Formation.

The nose of the Ludlow anticline (Fig. 5.5) is cut through by the River Teme, leaving a small area of the highest Ludlow beds upon which stands Ludlow Castle. Most of the town is built upon the Downton

Group, the thicker Ledbury Formation forming relatively low ground with striking red soils. Such country continues through Corve Dale south-east of, but parallel to, Wenlock Edge. The *'Psammosteus* Limestones' with their associated sandstones form an obvious wooded scarp feature well seen in the splendid panoramic view from above the Whitcliffe to the south-west of Ludlow town.

At Ludlow, Holland *et al.* (1963) provided description of a standard section for the base of the 'Downton Series' (Downton Group) at Ludford Corner near the southern end of Ludford Bridge, and where the position of the Ludlow Bone Bed is marked by an excavation into the steep rock face produced by the activity of generations of geologists. Basset *et al.* (1982) provide a comprehensive account of this important reference section.

# Regional descriptions
## Wales, Welsh Borderland, and Central England

### Llandovery

Building upon earlier work by Williams and St Joseph on *Stricklandia* and by the latter on the pentamerids (St Joseph 1935, 1938; Williams 1951), Ziegler and Cocks studied other brachiopod lineages within the Llandovery, taking respectively *Eocoelia* and *Leptostrophia*. Subsequently Ziegler *et al.* (1968b) following upon the earlier work of Ziegler (1965), used such lineages to correlate various other Llandovery successions with that of the type area. Consequently and additionally, with the recognition of certain brachiopod 'communities' (assemblages or, better, associations) thought to be related to different depth zones in the Llandovery sea, they were able to chart the changing palaeogeography of this region in terms of a marine transgression particularly in late Llandovery times. This transgression and associated shoreline features had earlier been described by Whittard (1932).

The cosmopolitanism of Llandovery shelly shelf faunas is indicated by the recognition of the same 'communities' in Norwegian and North American successions. Ziegler *et al.* (1968a) provided reconstructions of the five communities. The shells are usually found associated as scattered disarticulated forms but much information was collected from the rarer occurrences of communities in position of growth. Only in the shallowest water *Lingula* community (Table 5.3) are both infaunal and epifaunal elements substantially represented, though diversity is low. In the deeper water communities epifaunal elements are dominant. Most of these appear to have

been attached to disarticulated convex-upward shells.

In the original area of investigation in the Welsh Borderland it appears that the shelf region was relatively narrow in Llandovery times and so depth zones and their attendant associations of fossils were perhaps especially clear. Later attempts to recognise such a clear pattern in Wenlock or Ludlow rocks have been much less successful (Lawson 1975, Bassett 1976) and a more complicated arrangement seems to emerge, where associations change frequently from place to place and 'communities', if recognised, would be very numerous indeed. Suggested palaeogeographies for early and late Llandovery times are shown in Figure 5.6A, B.

**Table 5.3.** Communities of Ziegler *et al.* (1968b) listed 1 to 5 in order of increasing depth

(1) *Lingula* Community
(2) *Eocoelia* Community
(3) *Pentamerus* Community [becoming *Pentameroides* in C$_5$ and C$_6$]
(4) *Stricklandia* Community [becoming *Costistricklandia* in C$_5$ and C$_6$]
(5) *Clorinda* Community

A full sequence of communities (Table 5.3) is not always present in any one area. Thus in Shropshire (Ziegler 1965) the Kenley Grit (up to 45 m), *Pentamerus* Beds (up to 120 m), and Purple Shales (or Hughley Shales) (75–105 m), represent (1), (3), and (5) of Table 5.3.

Particularly well known is the outcrop of upper Llandovery rocks bordering the southern margin of the Longmynd (Ch. 2) where the *Pentamerus* Beds are succeeded by the Purple Shales. The typical *Pentamerus* Beds are perhaps the best known of all Llandovery rocks. They are grey shaly siltstones and mudstones with sandy laminae and with shelly limestone bands. Lapworth elsewhere coined the name 'Government Rock' for the variety full of large *Pentamerus oblongus*, their internal septa broken across to give sections shaped like broad arrows.[*]

Whittard (1932) interpreted the basal deposits in this area as a beach with sea stacks, but Ziegler *et al.* (1968b) suggested that most of the sedimentation was sub-tidal. At one locality a *Pentamerus* Community was recognised only 7 m above the unconformity with the Precambrian, while the base of the *Pentamerus* Beds is locally very coarse with variable conglomerates filling hollows in the old topography. When fossiliferous these basal beds contain a *Lingula* Community.

Near the Bog Mine (SO 355 979), where small outliers of the Llandovery rocks lie upon the Llanvirn of the eastern part of the Shelve inlier (Whittard 1979),

[*] The symbol which at one time was used to mark H.M. Government property.

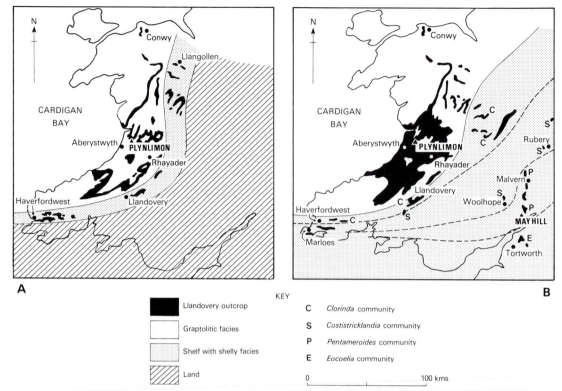

KEY

■ Llandovery outcrop

□ Graptolitic facies

▦ Shelf with shelly facies

▨ Land

C  *Clorinda* community

S  *Costistricklandia* community

P  *Pentameroides* community

E  *Eocoelia* community

0 ————————————— 100 kms

**Fig. 5.6.** Llandovery palaeogeography of Wales and the Welsh Borderland: (A) Rhuddanian and early Aeronian, (B) Telychian (after Ziegler 1965, 1970).

a distinctive local facies, the Bog Quartzite, contains a shelly fauna of unusual abundance. Ziegler *et al.* (1968a) referred to one block of sandstone as having yielded 51 species and 554 individuals. The fauna is a mixture of near-shore *Cryptothyrella* Community (seemingly the equivalent of the *Eocoelia* Community but preceding it in time) and some rocky-bottom elements, such as must be but rarely represented in the stratigraphical record. These quartzites are also rich in the trilobite *Staurocephalus*.

The famous locality in the Onny River near Wistanston (SO 426 853), where the trilobite-rich Onny Shales of the Cardoc are succeeded with slight discordance by the Purple Shales of the upper Llandovery, provides another example of the variable stratigraphical break before the onset of Silurian sedimentation. Farther north-eastwards along this main outcrop of the Llandovery, parallel with Wenlock Edge, lower formations are seen again. The Kenley Grit, a near-shore deposit rests unconformably upon Ordovician or earlier rocks and is distinguished from the succeeding *Pentamerus* Beds by the abundance of coarse clastic material and the absence of carbonates. Fossils from the Kenley Grit all belong

to the *Lingula* Community (Ziegler *et al.* 1968b). At Ticklerton (Fig. 5.4), to the south-east of Church Stretton, the *Pentamerus* Beds belong to the *Pentamerus* Community whilst *Eocoelia hemisphaerica* indicates a $C_1$ to $C_2$ age. At Sheinton (again for example) the deeper water *Stricklandia* Community is recorded, though at this locality *Stricklandia* itself plays a very minor role. The presence here of the form *Stricklandia lens ultima* indicates a $C_3$ to $C_4$ age and such evidence implies a progressive increase in depth during the time of deposition of the *Pentamerus* Beds. Collections along the main outcrop from various localities in the Purple Shales are indicative of the *Clorinda* Community.

In South Wales in Dyfed, there is (unusually) evidence of basic volcanicity in the Llandovery, and Ziegler (1965) used the relationship between thickness of lava and community present to calculate relative depths in the Llandovery sea, the depth ranges of the communities being 'of the order of tens of feet rather than hundreds of feet'.

To the east and south-east of the extensive Welsh outcrop are various inliers with Llandovery rocks. For example those which form the conspicuous landmark of May Hill (Fig. 5.6) are 183 m of coarse greyish sandstones and conglomerates with *Eocoelia hemisphaerica*, passing up into yellowish sandstones and siltstones (152 m) with abundant *Costistricklandia lirata*, *Stricklandia lens*, *Pentamerus oblongus*, *Leptostrophia compressa*, and other shelly fossils. Calcareous beds at the top of the sequence yield the distinctive crinoid *Petalocrinus*. Llandovery rocks also form the core of the Woolhope inlier; 85 m of grey calcareous sandstones and olive mudstones are followed by 6 m of purple and olive muddy siltstones and thin limestones. The succession is capped by 3 m of coarser siltstones and calcareous bands. *Costistricklandia lirata* and *Cyrtia exporrecta* are present and in the top unit there are thin limestone bands packed with *Petalocrinus*.

The Llandovery graptolitic shales of parts of Central and North Wales are good examples of their kind. They comprise black, 'blue', or grey shales with a few interbeds of sandy material. A dark shale with *Monograptus leptotheca* contains a characteristic green band, the so-called 'green streak', about 2 cm thick, widely distributed in Central Wales. The relatively thin succession at Conway (Fig. 5.6), where the Llandovery is only about 100 m thick, originally studied by Elles (1909), is named the Gyffin Shales. Sudbury (1958) undertook a most detailed evolutionary study of the rastritids and triangulate monograptids of the lower Llandovery *Monograptus*

*gregarius* Biozone in the Rheidol Gorge, another well-known Llandovery section, in this case to the east of Aberystwyth (Fig. 5.6).

In graptolitic terms the base of the Silurian approximates to the change from the diplograptid to the monograptid fauna (Bulman 1970). Members of the former, however, linger well into the Llandovery, e.g. *Pseudoclimacograptus*, but in the lowest Silurian these overlap with biserial forms with 'typically Silurian tendencies (e.g. gracilisation of the proximal end: *Akidograptus, Orthograptus, Raphidiograptus*)' (Rickards 1976). The first monograptid *Atavograptus ceryx* (Rickards & Hutt 1970, referring to Northern England) is present even in the *persculptus* Biozone (now uppermost Ordovician) and its successors were to evolve in spectacular fashion through the Silurian and into the Devonian.

There is another important element in the facies patterns of Welsh areas expressed in the presence of greywackes (the Llandovery 'grits' of earlier workers) representing incursion of turbidity currents into the basin and the subject of a pioneering study in the sedimentological work of Wood & Smith (1959) on the Aberystwyth Grits (up to 1,500 m) (Fig. 5.7) along the coastline of Cardigan Bay. The numerous flute moulds and other sedimentary structures indicate current flow north-eastwards. The relative proportion of greywackes to mudstones also decreases markedly

**Fig. 5.7.** Aberystwyth Grits near Aberystwyth. Here greywackes are slightly subordinate to mudstones. Distance across base of cliff about 40 m. (Photograph by A. J. Smith.)

north-eastwards, that is away from the source. The Llandovery greywackes reach a maximum thickness of over 2,500 m in the Plynlimon district of Central Wales (Fig. 5.6).

There is also an important lateral component of sedimentation originating from the south-eastern margin of the Welsh basin and described in terms of deeply eroded submarine channels or fan systems by Kelling & Woollands (1969) for the Rhayader district of Central Wales and from farther south in un-published work by Woollands. Subsequently the notion of contourites, as distinct from turbidites, has been developed, though clearly their recognition is likely to be difficult. Anketell & Lovell (1976)

suggested that the turbidites represented by the Aberystwyth Grits are an upward transition from contourites of the upper Llandovery Grogal Sandstones. The whole complex of sedimentary environments in the Welsh basinal Llandovery has been admirably summarised by Cave (1980). He noted a common rhythm in sedimentation comprising a basal sandy unit with several members of the Bouma (1962) Ta to Td sequence, followed by homogeneous turbiditic mudstone (Te), commonly graded. The top of the tripartite rhythm is of dark grey, pyritous and carbonaceous mudstone with many graptolites and with discontinuous layers of silty mudstone, or of a pale grey to fawn uniform mudstone occasionally

**Fig. 5.8.** Bioherm within bedded Much Wenlock Limestone Formation at Cootes Quarry (SO 459 935). (Photograph by C. H. Holland.)

disturbed by bioturbation. These variable top mudstones are regarded as inter-turbiditic deposits. Turbidity currents became more active in the middle Llandovery of the Welsh basin, and their deposits are especially well developed in the very thick sequences of the upper Llandovery, where they are in striking contrast to the equivalent thin graptolitic succession of Northern England.

## Wenlock

South of Church Stretton there is evidence from boreholes and outcrop of overstep at the base of the Wenlock (Cocks & Rickards 1969), the *centrifugus*, *murchisoni*, and at least part of the *riccartonensis* Biozones being missing as the Coalbrookdale Formation comes to rest directly on the Purple Shales of the upper Llandovery. Near Craven Arms (SO 435 827) it comes to rest directly upon the Ordovician. Near Presteigne (SO 315 645) the algal and bryozoan rich Nash Scar Limestone (*c.* 60 m) rests on the Folly Sandstone of $C_{1-2}$ age. Near Old Radnor (SO 250 590) a similar basal limestone rests unconformably on the Precambrian. In contrast,

in the Long Mountain area (Fig. 5.10) to the west (Palmer 1970) the basal zones of the Wenlock are present.

The widely developed eastern shelf persisted in Wenlock times but with conspicuous development of carbonates. The Woolhope Limestone is present locally at the base of the succession as a lateral equivalent of the more normal calcareous shale facies. At Woolhope (Fig. 5.10) it is some 15–60 m thick and contains a poor fauna of trilobites and brachiopods. It occurs also at May Hill and near Ledbury. Better known, however, is the growth of patch reefs – the so-called 'ballstones' – of the Much Wenlock Limestone Formation with their rich faunas of tabulate corals and stromatoporoids clearly seen in old and still-working quarries along the top of Wenlock Edge (Fig. 5.8). Their setting may be compared with that of the well-studied reefs of Gotland. The famous limestone quarries in Staffordshire as at Wren's Nest near Dudley (Fig. 5.10) are much overgrown and partly obscured by building. They have been over-collected, but a nature trail now displays the interest of the area. The old term 'Dudley locust' for specimens of *Calymene blumenbachi* from this area suggests the

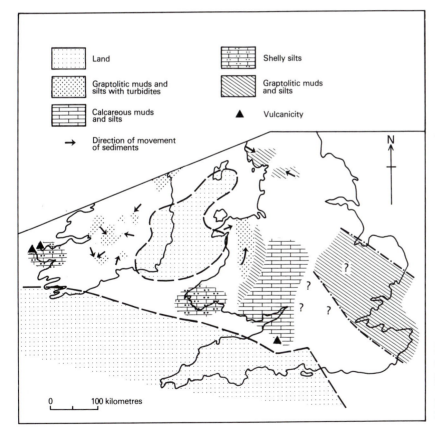

**Fig. 5.9.** Wenlock (Homerian) palaeogeography of England, Wales, and the southern part of Ireland (modified from Holland 1981).

Legend:
Land
Graptolitic muds and silts with turbidites
Calcareous muds and silts
→ Direction of movement of sediments
Shelly silts
Graptolitic muds and silts
▲ Vulcanicity

0    100 kilometres

richness of a particular group such as is found to vary from place to place in the limestone outcrop.

Scoffin (1971), in particular, in a fully illustrated and documented account, has considered the conditions of growth of the Wenlock reefs in Shropshire. As in various parallel situations, the most favourable site for growth (and hence the greatest abundance) was at the seaward fringe of the reef belt, now represented by part of the main outcrop near Easthope. There is also an upward increase of reefal material as the concentration of clay particles becomes less. Shergold & Bassett (1970) described various inter-reef and off-reef facies in argillaceous and crinoidal limestones. Growth of the reefs started on small mounds of crinoidal debris or simply on soft muddy bottoms. The frame builders were especially tabulate corals (*Halysites, Heliolites, Favosites*), together with stromatoporoids, branching rugose corals, and fenestellid bryozoans. Stromatolites, laminar tabulates (*Thecia, Alveolites*), stromatoporoids, algae, and bryozoans acted as reef binders. There were associated crinoids, brachiopods, ostracodes, and gastropods. The patch reefs have dimensions averaging 12 m across with a maximum thickness of some 4·5 m. The maximum depth of the sea was thought to be about 30 m, this assumption being based upon the light requirements of the blue green algae such as *Solenopora*. Seen over the whole outcrop, the Much Wenlock Limestone Formation is now known to be diachronous, ranging from the *lundgreni* Biozone to the *ludensis* Biozone between Dudley and Ludlow. Rider (1981) reviewed the composition, structure and environmental setting of Silurian reefs in the wider north European context.

A suggested Wenlock palaeogeographical map is given in Figure 5.9. Bassett (1974) reviewed the stratigraphy of the Wenlock Series throughout the Welsh Borderland and South Wales. For the early Wenlock he noted the presence of sandy beds in Marloes Bay (Dyfed) and through the small Rumney inlier near Cardiff, past the localised volcanic development of the Mendip area through Tortworth and north-eastwards. The shore line ran to the north of Freshwater West and Freshwater East (SS 020 980) in Dyfed as here the basal part of the Wenlock is missing. A salient extending northwards towards Llandeilo is indicated by shallow water clastic sediments and this persisted as a source of deltaic material later in the Silurian. As the epoch progressed the Much Wenlock Limestone platform increased in area. To the west this resulted in a narrow basinwards slope with associated slumping. Cummins (1957) used the flute moulds seen so well in the Central and North Welsh turbidites (Denbigh Grits), for example at Newtown, to map the configuration of the Welsh basin in Wenlock times, with a Derwent Ridge separating Denbigh and Montgomery troughs.

The probably Wenlock (Ziegler *et al.* 1968b, p. 765) volcanic rocks of the Mendip area consist of tuffs and tuffaceous shales (*c.* 45 m thick) followed by 300–400 m of andesitic lavas, agglomerates, and tuffs. Some 36 m of olive shales, siltstones, and micaceous sandstones follow below the unconformable Old Red Sandstone. Apart from bentonites (or supposed bentonites) and a few tuffs, there had been no records of Wenlock igneous activity in the Welsh Borderland and North Wales until Sanderson & Cave (1980) reported upon small intrusions of microgranite near Bishop's Castle, cutting and altering upper Wenlock sediments. Brecciation and aureole size are indicative of a volatile magma under low confining pressure and these authors concluded that a later Silurian age for the intrusions is more likely than one very much younger.

## Ludlow

The Ludlow Series directly succeeds the Much Wenlock Limestone Formation of Wenlock Edge. The relatively subdued topography above the escarpment is interrupted by a less obvious feature formed by the limestones and calcareous siltstones of the middle part of the Ludlow (Fig. 5.2). Farther south-west, notably at the well-known locality of View Edge, to the south of Craven Arms, famous for its banks of disarticulated *Kirkidium knightii* (Boynton 1979), the Much Wenlock Limestone feature becomes less elevated and the more conspicuous scarp is made by the Ludlow limestones. The continuity of the south-westerly strike of the main Silurian outcrop is broken in the Ludlow district itself by an ENE-plunging asymmetrical anticline. The district is the type area for the Ludlow Series and has been well known as such since Murchison's time.

The plunging anticline is beautifully expressed in the topography (Fig. 5.3), the two limestones referred to above forming conspicuous wooded ridges. In the core of the anticline the Wenlock Shales form low ground partly covered by alluvium. Although their thickness of some 300 m approximates to that of the full sequence at Wenlock Edge, they are confined to the Homerian Stage. The Much Wenlock Limestone Formation thins eastwards through the Ludlow district from about 135 to 60 m. It is less fossiliferous than at Much Wenlock or at Dudley in Central England, corals occurring mainly in the more calcareous nodular beds of the uppermost 15 m of the formation. The more conspicuous, hilly wooded ridge of Bringewood Chase, Mary Knoll, and High Vinnals is formed by the calcareous siltstones and limestones of the middle part of the Ludlow Series.

The Ludlow succession of the type area has already been described. Attempts to relate the richly fossiliferous sequences of the Welsh Borderland to a simple series of communities have met with difficulty. There are many associations at many levels. More broadly, in space there is a pattern of facies from the thinner and sometimes interrupted shelly sequences of the Welsh Borderland and English Midland inliers (e.g. May Hill: Lawson 1955); through the thicker marginal successions at Ludlow and at Leintwardine, the latter with its well-defined submarine channels within the Lower Leintwardine Formation described by Whitaker (1962); to the differing basinal developments of the Builth district and Clun Forest. At Knighton in Central Wales the Ludlow reaches some 2,000 m in thickness. Both this lateral arrangement and the

changing vertical pattern from the graptolitic facies early in the Ludlow to the last phases, when the distinction between shelf and basin becomes largely one of thickness, were summarised by Holland & Lawson (1963) in a series of facies and isopachyte maps. An example of the former is given in Figure 5.10.

Palaeoecological studies are adding to the detailed picture. Thus Watkins (1978) considered bivalve ecology along a gradient of shelf environments from the quietly deposited offshore muds of the earlier part of the sequence to the shallow water, often storm deposited, silts of the uppermost part of the Ludlow.

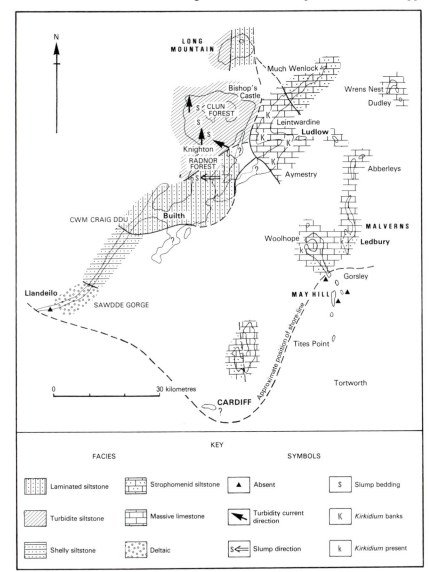

**Fig. 5.10.** Facies pattern in Wales and the Welsh Borderland for the Upper Bringewood Formation (uppermost Gorstian) and its equivalents (modified from Holland and Lawson 1963).

Bivalve abundance is highest at the extremes of this gradient while diversity is greatest in the low-stress, mid-shelf environments. Watkins emphasised the carbonates of the Bringewood Formations as an interruption in his gradient of shallowing environments, and this is obviously so, though there are many other variations in detail.

There is widespread development of shelly conglomeratic limestone beds at the base of the Lower Leintwardine Formation, with evidence of early submarine lithification of both sand-grade and clay-grade carbonates. The beds vary in thickness and character, being thicker and coarser in the south-east where the formation as a whole is thinner. Evidence for hardgrounds includes borings, reworking, encrustation by bryozoans, and the development of laminar stromatolites; the importance of disturbance by storms and tidal currents in the formation of the conglomerates is seen in the presence of lag intraclasts and concentrations of limestone debris (Cherns 1980).

In the lower part of the Ludlow succession in Central and North Wales, Cummins (1959) again used directional indicators to suggest a distinction between the two basinal areas. Slump structures are well known from these rocks and add to the picture of configuration of the basins. Some of the earliest descriptions of these features were by Jones (1937, 1940) from Clwyd (Denbighshire) and by Straw (1937) from the Builth district. Woodcock's (1976) analysis of slump fold asymmetry and facing confirms the configuration of the Montgomery trough, demonstrating a western as well as an eastern margin. Post-*leintwardinesis* Biozone graptoloids in the British Isles are known only from the Long Mountain and Clun Forest (Fig. 5.10) where Holland & Palmer (1974) recognised the *Bohemograptus* proliferation Biozone, more familiar to workers in Central and Eastern Europe. In Central Wales this is followed by a

kind of 'graptolitic siltstone facies' without graptolites, but characteristically with orthoconic nautiloids and bivalves. Elsewhere in the world the graptoloids continued to flourish.

The Ludlow rocks of Llandeilo (Fig. 5.10) have their own distinctive features as a deltaic area of deposition pushed northwards to produce what are now the predominantly red, sandy and conglomeratic Trichrûg Beds with their restricted fauna. They pass laterally north-eastwards into normal greyish beds of top Gorstian age. Farther west in the Marloes Bay area of Dyfed, the Grey Sandstone Group grades upwards into rocks of Old Red Sandstone facies, but of late Wenlock or early Ludlow age (Walmsley & Bassett 1976).

## Přídolí

Three different types of stratigraphical relationship between Downton Group (lower Old Red Sandstone) and older Silurian rocks have been recognised (Bassett *et al.* 1982): I, II, III of Table 5.4 and Figure 5.11.

I. This includes the Ludlow successions of basin facies in Clun Forest, the Knighton district, and other areas in east-central Wales, which are all capped by Downton Group seen as a series of outliers, once obviously part of a continuous sheet. The sequence is very similar in all these areas as summarised by Holland (1962) and revised to a correct formational nomenclature by Bassett *et al.* (1982) as of *Platyschisma helicites* Formation (6–12 m), Yellow Downton Formation (8–11 m), Green Downton Formation (15–90 m), and Red Downton Formation (600 m). (Thicknesses given are for the Knighton district.) The greyish shales and flaggy siltstones of the first of these formations resemble those of the uppermost Ludlow beds but contain a restricted shallow marine fauna of *Turbocheilus* [*Platyschisma*], *Modiolopsis*, *Lingula*,

| I | II | | III |
|---|---|---|---|
| Red Downton Formation | Ledbury Formation | | 'Red Marls' |
| Green Downton Formation | Temeside Shales Formation | | |
| Yellow Downton Formation | *Downton Castle Sandstone Formation* | Sandstone Member | Tilestones Formation |
| *Platyschisma helicites* Formation | | *Platyschisma* Shale Member | |
| | | Ludlow Bone Bed Member | |

**Table 5.4.** Downton successions in Wales and the Welsh Borderland. (For areas of I, II and III see Fig. 5.11.)

and *Londinia*. The succeeding massive yellow siltstones and fine sandstones yield *Lingula minima*. The Green Downton Formation is of distinctive, blocky weathering, irregularly bedded, olive siltstones and calcareous mudstones with a fauna characterised by *Modiolopsis, Londinia*, and leperditiids. The thicker Red Downton Formation is of soft red and purple siltstones, conglomerates, and sandstones, in all of which fossils are very rare.

II. In the Welsh Borderland and Central England the Downton Group begins with thin bone beds or at least a thin phosphatised pebble bed. Apart from this the well-known Downton succession as in the Ludlow District closely resembles lithologically those referred to above. The uppermost Ludlow rocks here are olive grey, often calcareous, siltstones deposited in shallow water and with frequent coquinas (Watkins 1979) representing the effects of storms. The fauna is of low diversity but very rich in individuals, of which the brachiopods *Microsphaeridiorhynchus nucula, Protochonetes ludloviensis*, and *Salopina lunata* and the ostracode *Calcaribeyrichia torosa* are particularly characteristic. Bassett *et al.* (1982) defined the Ludlow Bone Bed Member of the *Downton Castle Sandstone Formation* as of up to 0·30 m of ripple laminated and lenticular bedded siltstones containing five or more thin bone beds each a few mm thick. There are four of these at Kington, Herefordshire (SO 295 567) (Holland & Williams 1985). Antia & Whitaker (1978) found the bone beds to be rippled thelodont sands formed low in the intertidal zone as lag concentrates. In places they can be seen to be an integral part of flaser-bedded sand. There are thelodont scales, acanthodian and brachiopod fragments, quartz grains, and phosphatic nodules, with evidence of diagenetic effects which occurred before and after abrasion by local reworking. The fish debris was concentrated in pockets by tides and currents. The marine invertebrate faunas include the ostracode *Frostiella groenvalliana* together with species of *Londinia*, an association evidently diagnostic of the base of the Downton Group. The succeeding *Platyschisma* Shale Member is of up to 2 m of olive mudstones and siltstones with bands of marine fossils, notably gastropods and bivalves. The Sandstone Member is of about 15 m of yellowish, micaceous, fine-grained sandstone in which *Lingula*, molluscs, and ostracodes are still present along with local concentrations of plant remains.

The succeeding Temeside Shales Formation (12–37 m) resembles the Green Downton Formation of the areas farther west. Rare marine fossils are still present along with plant and fish remains. Turner (1973) showed the usefulness of thelodont assemblages in Downton stratigraphy and correlation. Desiccation cracks and pedogenic carbonates indicate periods of exposure of the mud flats.

The Ledbury Formation varies in thickness from 330 to 460 m. Red mudstones and siltstones ('marls') are accompanied by red and purple sandstones. Marine invertebrates are now rare. Fragmentary fish remains are locally common. Bassett *et al.* (1982) noted that the faunas still indicate a marine influence across tidal flats or the flood plains of rivers. The

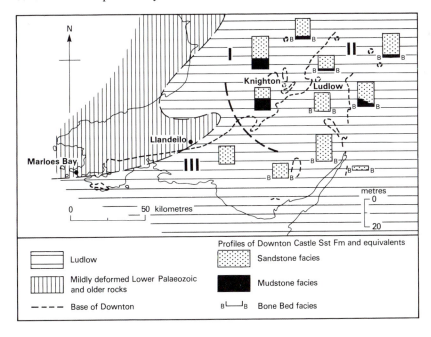

**Fig. 5.11.** Lateral variation of facies in the Downton Castle Sandstone Formation and equivalents (modified from Allen 1974 and Bassett *et al.* 1982). Heavy dashed line separates areas I and II from area III.

sandstones may have been deposited in tidal reaches of the rivers.

Apart from the obvious broad similarity of sequence there are other links between the Downton Group of what was originally the basin and that of the shelf areas of the Ludlow. The uppermost metre or so of the Ludlow succession in the Ludlow district is characterised by an abundance of the small brachiopod *Howellella elegans*. The same fossil is common in the uppermost Ludlow of the basin areas. About 2 m below the Ludlow Bone Bed near Downton Castle Bridge at the western end of the Ludlow district and about 0·6 m below it near Ludford Corner at Ludlow is a thin band crowded with the small inarticulate brachiopod *Craniops implicatus*. An identical band is found in the Knighton district where it

forms a convenient base for the Downton Group. A similar band is known from near Kington in Herefordshire (Holland & Williams 1985, Holland 1987b).

III. In South Central Wales the Downton succession is different from the above in that it oversteps progressively south-westwards across the older Silurian rocks below and finally, near Llandeilo, comes to rest upon the Ordovician. The sequence in these areas begins with the Tilestones Formation (up to 50 m) of very characteristic grey or brown, yellow weathering, micaceous flaggy sandstones. The succeeding red rocks are similar to the Ledbury Formation referred to above.

Although *Cooksonia* is known already from rocks of middle Ludlow age in Central Wales it was in the Přídolí that vascular plants of uniform character

**Fig. 5.12.** Distribution map of Silurian rocks in Northern England (after Rickards 1978).

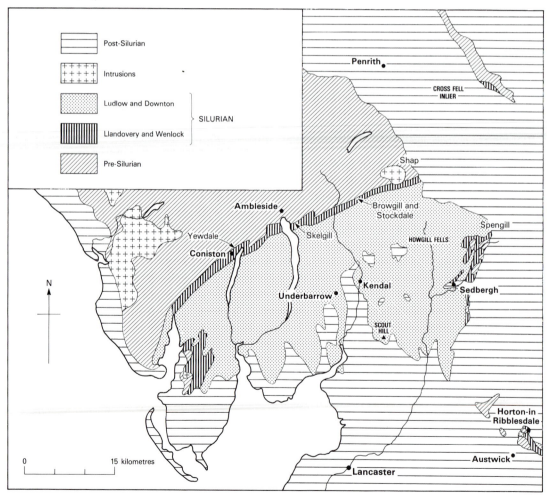

became widespread. They centre upon the genus *Cooksonia* but a review by Edwards (1980) referred also to the more common smooth, dichotomously branching plant axes placed in the form genus *Hostinella*. Richardson & Lister (1969), Richardson & Rasul (1990) and others have made good use of spores in Ludlow–Přídolí stratigraphy. Some of these with triradiate marks along with cuticle-like films, etc., have been used to suggest the presence of terrestrial plant vegetation prior to the appearance of the macroplants. This matter remains controversial (e.g. Gray & Boucot 1980, Smith 1981), but certainly the Downton Group of England and Wales provides vital evidence of early land floras.

Allen & Williams (1981) showed that volcanogenic sediments in the form of graded ash-fall tuffs occur at many horizons in the lower part of the Lower Old Red Sandstone in South Wales. At one particular level there is a distinctive group 2–4 m thick of three almost superposed tuffs widely distributed in South Wales and the Welsh Borderland. This Townsend Tuff Bed lies not far below the '*Psammosteus*' limestone or its equivalent and thus may provide a useful local marker for the base of the Dittonian, though the implication would be that the Ledbury Formation extends upwards into the Dittonian. It occurs in a barren interval between the Silurian marine-brackish vertebrate fauna (*Hemicyclaspis, Sclerodus, Didymaspis,* etc.) and the incoming freshwater *Traquairaspis–Pteraspis* fauna of the upper part of the Ledbury Formation. Allen and Williams saw the Townsend Tuff Bed as originating in distant Plinian eruptions, their dust spreading over extensive coastal mud flats.

## Northern England

The Silurian rocks of the Lake District (Fig. 5.12) as a whole were reviewed by Rickards (1978), who gave many relevant references. The western part has valuable reference sections for the Ordovician–Silurian boundary (Fig. 5.13). The boundary is well seen in Yewdale Beck where the Coniston Limestone of the Ordovician is followed by less than 1 m of Ashgill Shales, 0·3 m of blue-grey mudstones with shelly fossils together with graptolites of the *persculptus* Biozone, and then black shales of the *acuminatus* Biozone. Farther east in Skelgill, Browgill, and Spengill (Fig. 5.12) there is a variously developed basal impure limestone which in the first of these appears to replace the *persculptus, acuminatus,* and part of the *atavus* Biozones. In Browgill, a tributary of Stockdale Beck, at least a metre of black shales with an *acuminatus* Biozone fauna follows non-sequentially upon the Ordovician, though thin calcareous bioturbated beds intervene in another faulted part of the section. In Spengill the *acuminatus* Biozone is thinner.

## Llandovery

The Llandovery rocks of the Lake District, which together with the Wenlock tend to form a distinctive

**Fig. 5.13.** Variations in rock type and thickness about the base of the Silurian in the Lake District. For location of sections see Fig. 5.12 (after Rickards 1978 and unpublished).

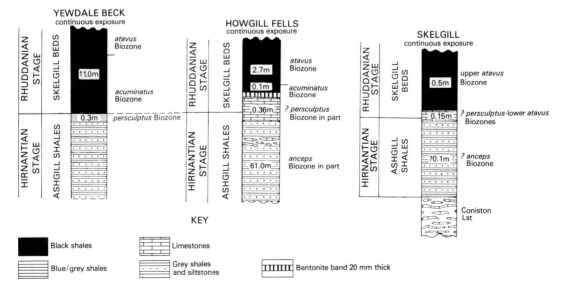

hollow across the topography, are referred to as the Stockdale Shales (60 m), with Skelgill Beds and Browgill Beds, of which the latter represents the Telychian. The Skelgill Beds are black pyritous shales rich in graptolites, regarded by Rickards as a deepening sequence following upon the shallow water conditions of the late Ordovician. There are also grey-blue mudstones formed when the rain of planktonic material lessened, and nodular limestones with sparse trilobite and brachiopod faunas. The beds are thickest locally in the Howgill Fells. The 'green streak' of the *argenteus* Biozone, seen well in Skelgill is a thin band mineralogically the same as the black shales but lacking the pyrite, carbonaceous material, and graptolite fauna of the latter. It provides remarkable evidence of continuity with the Welsh basin where it occurs with *Monograptus leptotheca* at the same horizon. In the Browgill Beds beginning at the base of the *turriculatus* Biozone the black graptolitic shales are very thin in comparison with the pale grey mudstones. Graptolitic bands are especially sparse in the west. Towards the top of the sequence the more widely familiar red beds of the *crenulata* Biozone are locally developed. Grey beds at the top of the sequence are succeeded by the richly fossiliferous *centrifugus* Biozone beds of the basal Wenlock.

## Wenlock

A full sequence of Wenlock graptolite biozones is present in the three formations into which the Wenlock rocks of the Lake District are divided: Brathay Flags (400 m), Lower Coldwell Beds (180 m), and Middle Coldwell Beds (10 m). The two last comprise respectively the *lundgreni* and *ludensis* Biozones. Brathay Quarry, though now flooded, still shows the resistant, blue-grey laminated mudstones with calcareous nodules and bentonites characteristic of the Brathay Flags. They contain orientated orthocones. The Lower Coldwell Beds which thin to the east are proximal turbidites of silt or sand grade with graptolitic horizons of laminated mudstones. The Middle Coldwell Beds form another hollow across the watershed. They are of pale grey mudstones, unlaminated or thoroughly bioturbated. There are limestone lenses rich in trilobites (such as *Delops* and *Struveria*) and brachiopods, and also graptolitic bands of Brathay Flags lithology. It has been suggested that the Lower Coldwell Beds were deposited during a rare episode when turbidites from what is now the Southern Uplands reached the Lake District.

## Ludlow and Přídolí

The Upper Coldwell Beds (450 m) are of Brathay Flags lithology but with more common silty horizons.

They are of *nilssoni* Biozone age, thus taking the succession into the Ludlow. The Coniston Grits (1,700 m) which follow are fine sand-grade turbidites (greywackes) commonly with Bouma (1962) sequences. Thin graptolitic bands allow dating of the succession. Above these beds the Bannisdale Slates (1,500 m), which are mudstones and siltstones with rare laminated graptolitic bands, reach the *leintwardinensis* Biozone. The thick and monotonous Underbarrow Flags (170–900 m) are transitional from these into the Kirkby Moor Flags. They contain in their lower part the association of the ostracode *Neobeyrichia lauensis* and the small brachiopod *Aegiria grayi*, characteristic of the Upper Leintwardine Formation in the Welsh Borderland. The well-sorted siltstones of the Kirkby Moor Flags (430–750 m) crop out widely. They contain calcareous lenses with benthonic shelly faunas of upper Ludfordian age including *Protochonetes ludloviensis*, *Salopina lunata*, and *Acastella prima*. The succession is completed by the Scout Hill Flags. There are now more flaggy horizons and some red beds. Presence of the ostracode *Frostiella groenvallina* has been taken to indicate that the succession here reaches the level of the Downton Group.

## Inliers

Silurian rocks are present also in the Lower Palaeozoic inlier of Cross Fell and in the Craven district of north-west Yorkshire around Austwick and Horton-in-Ribblesdale (Fig. 5.12). In the former, Llandovery Stockdale Shales are followed by Wenlock Brathay Flags. Though structurally complex and poorly exposed the area has yielded more than one hundred graptolite species (Burgess *et al.* 1970).

The Silurian outcrops of Austwick (Crummackdale) and Horton-in-Ribblesdale occupy the valleys below the striking unconformity at the base of the Lower Carboniferous. They are bounded to the south by the North Craven Fault. Poor exposure and faulting result in an incomplete sequence of Skelgill Beds. The succeeding Browgill Beds are lithologically similar to their counterparts in the Lake District. In the Wenlock, the Austwick Formation (140 m, thinning north-eastwards) is of graded greyish sandstone turbidites alternating with mudstones (McCabe & Waugh 1973). The beds are folded and the finer cleaved tops of the greywackes are well displayed. The depositing currents flowed north-westwards. McCabe (1972) recognised the Arcow Formation (18 m, thinning north-eastwards and westwards) of grey calcareous mudstones and siltstones with shelly fossils (orthocones, *Slava*, and *Delops*) between the Austwick Formation and the overlying mudstones with calcareous nodules of the Horton Formation (400 m). He correlated the Arcon Formation with the Middle

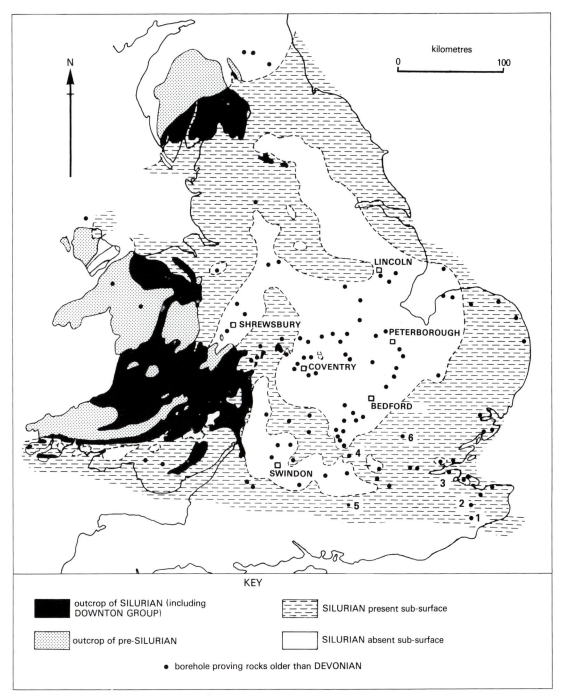

**Fig. 5.14.** Suggested distribution of concealed Silurian rocks of England and Wales (map compiled by Elizabeth Williams with data from Wills 1978). Numbered boreholes are as follows: 1. Brabourne, 2. Chilham, 3. Cliffe, 4. Little Missenden, 5. Shalford, 6. Ware.

Coldwell Beds of the Lake District. The top of the Austwick Formation has graptolites from high in the Wenlock. The Horton Formation yields a Ludlow *nilssoni–scanicus* fauna. The succession is completed by poorly exposed greywacke sandstones of the Studfold Sandstone (100 m), still of Ludlow age, derived from the west. The Horton and Studfold formations resemble lithologically the Upper Coldwell Beds and Coniston Grits of the Lake District.

### Concealed Silurian rocks of eastern and southern England

What amounts to a very substantial knowledge of the pre-Permian floor was charted by Wills (1978) in characteristically stimulating contributions towards the end of a very long career. A number of boreholes for coal or water have penetrated the Silurian rocks Fig. 5.14). A separate marine area over what is now SE England may not have merged with the Welsh basin and its shelf until Přídolí times. There are records (see Fig. 5.14) of both shelly and graptolitic facies of the Llandovery respectively at Shalford and Brabourne, and at Chilham. Cliffe and Ware yielded Wenlock shelly fossils. Of especial interest in Přídolí stratigraphy is the Little Missenden borehole where Straw (1933) recorded a fully marine fauna which he considered to be of this age.

Looked at with continental European eyes the graptolitic sediments relatively poorly exposed in the Ardennes may be seen to be connected via the very many borehole records in the Brabant Massif (see, for example, Walter 1980) to the concealed outcrop referred to above. A narrow Brabant Depression is shown to pass WNW between the 'Ardenne uplift' and a stable shelf to the north.

## Summary

The evolving Silurian palaeogeography of the area which is now England and Wales may be summarised as follows. During the earlier part of the Llandovery Epoch a narrow belt of shelf sea curved from South Wales, through the Welsh Borderland, and thence northwards (Fig. 5.6A). Its shelly sediments suggest a succession of depth-related brachiopod associations in a muddy, silty, or sandy environment. In West and North Wales there is evidence of a basin with graptolitic muds, which can also be traced into the

Lake District. Later in the Llandovery, transgression caused a widening of the south, south-eastern, and eastern shelf area (Fig. 5.6B), with the development of local shelly carbonates and of coarse clastic deposits. Near-shore conditions are evident in places. There was local volcanicity in Dyfed. In the basin the graptolitic mudstones are increasingly accompanied by turbiditic incursions now seen as thick greywackes. The turbidity currents flowed from the south-west, though submarine channels and fan-systems added sediment from the eastern margin of the basin. In the Lake District graptolitic mudstones continued to accumulate, but there are also abundant grey mudstones of distal turbiditic origin.

In the Wenlock the shelf area remained wider (Fig. 5.9) and is characterised by calcareous muds and, in the Much Wenlock Limestone Formation of the Welsh Borderland and Central England, by limestones with their well-known development of bioherms at the edge of the basin. In Central and South Wales the basin, with its introduced turbidites, became split by the Derwent ridge into the Denbigh and Montgomery troughs. In the Lake District laminated muds were followed by proximal turbidites entering the quiet water regime from the north. There was then a return to the earlier muddy conditions.

In the early Ludlow the pattern is similar but variable silts rich in shelly faunas characterise the shelf. In some of the south-eastern inliers there is evidence of local positive effects with thin and broken sedimentary sequences. In the basin areas calcareous siltstones with convolute bedding tend to replace the earlier turbidites (Fig. 5.10) and slumping from the margins of the basin becomes conspicuous. The later part of the Ludlow shows a blurring of the distinction between shelf and basin, which now becomes a matter of thickness of sediment, and hence of degree of subsidence. There are also subtle changes in richness and variety of fauna.

Finally, the Přídolí Series demonstrates the silting up of both shelf and basin with differences again expressed largely as small variations in thickness. The intertidal lag deposit of the Ludlow Bone Bed is seen in the east. Otherwise the succession shows the variable interplay of alluvial plains, mudflats, and the tidal reaches of rivers. Pedogenic carbonates (e.g. Allen 1973, Allen & Williams 1979) signify the temporary stability of the surfaces of the interfluves. However, the Downton Group is best seen as representing the beginning of a long Old Red Sandstone cycle of continental conditions to be detailed in the next chapter.

REFERENCES

AGER, D. V. — 1975 — The geological evolution of Europe. *Proc. Geol. Assoc. London*, **86**, 127–154.

ALLEN, J. R. L. — 1973 — Compressional Structures (Patterned Ground) in Devonian Pedogenic Limestones. *Nature, London*, **243**, 84–86.

1974 — The Devonian rocks of Wales and the Welsh Borderland. *In* Owen, T. R. (Ed.) *The Upper Palaeozoic and post Palaeozoic rocks of Wales.* University of Wales Press, Cardiff, 47–84.

1979 — Old Red Sandstone facies in external basins, with particular reference to southern Britain. *In* House, M. R., Scrutton, C. T. & Bassett, M. G. (Eds.) The Devonian System. *Spec. Pap. Palaeontol. London*, **23**, 65–80.

ALLEN J. R. L. & TARLO, L. B. — 1963 — The Downtonian and Dittonian Facies of the Welsh Borderland. *Geol. Mag.*, **100**, 129–155.

ALLEN, J. R. L. & WILLIAMS, B. P. J. — 1979 — Interfluvial drainage on Siluro–Devonian alluvial plains in Wales and the Welsh Borders. *J. geol. Soc. London*, **136**, 361–366.

1981 — Sedimentology and stratigraphy of the Townsend Tuff Bed (Lower Old Red Sandstone) in South Wales and the Welsh Borders. *J. geol. Soc. London*, **138**, 15–29.

ANKETELL, J. M. & LOVELL, J. P. B. — 1976 — Upper Llandoverian Grogal Sandstones and Aberystwyth Grits in the New Quay area, Central Wales: a possible upwards transition from contourites to turbidites. *Geol. J.*, **11**, 101–108.

ANTIA, D. D. J. & WHITAKER, J. H. McD. — 1978 — A scanning electron microscope study of the Upper Silurian Ludlow bone bed. *In* Whalley, W. B. (Ed.) *Scanning electron microscopy in the study of sediments.* Geo Abstracts, Norwich, England, 119–136.

BALL, H. W. & DINELEY, D. L. — 1961 — The Old Red Sandstone of Brown Clee Hill and the adjacent area. I. Stratigraphy. *Bull. Br. Mus. nat. Hist.* Ser. Geol., **5**, 175–242.

BASSETT, M. G. — 1974 — Review of the stratigraphy of the Wenlock Series in the Welsh borderland and South Wales. *Palaeontology, London*, **17**, 745–777.

1976 — A critique of diachronism, community distribution and correlation at the Wenlock–Ludlow boundary. *Lethaia*, **9**, 207–218.

1985a — Silurian stratigraphy and facies development in Scandinavia. *In* Gee, D. G. and Sturt, B. A. (Eds) *The Caledonide Orogen – Scandinavia and Related Areas*, Part 1. Wiley-Interscience, 283–292.

1985b — Towards a 'Common Language' in Stratigraphy. *Episodes*, **8**, 87–92.

BASSETT, M. G., COCKS, L. R. M., HOLLAND, C. H., RICKARDS, R. B. & WARREN, P. T. — 1975 — The type Wenlock Series. *Rep. Inst. geol. Sci. London.*, **75/13**, 19 pp.

BASSETT, M. G., LAWSON, J. D. & WHITE, D. E. — 1982 — The Downton Series as the Fourth Series of the Silurian System. *Lethaia*, **15**, 1–24.

BERRY, W. B. N. & BOUCOT, A. J. 1967 Continental Stability – A Silurian Point of View. *J. geophys. Res.,* **72**, 2254–2256.

BOUČEK, B., HORNÝ, R. & CHLUPÁČ, I. 1966 Silurian versus Devonian. *Acta Mus. natl. Pragae.,* **22B**, 49–66.

BOUCOT, A. J. 1975 *Evolution and extinction rate controls.* Elsevier, Amsterdam, 427 pp.

BOUMA, A. H. 1962 *Sedimentology of some Flysch Deposits.* Elsevier, Amsterdam, 168 pp.

BOYNTON, H. E. 1979 Studies of *Kirkidium knightii* (J. Sowerby) from the Upper Bringewood Beds near Ludlow, Shropshire. *Mercian Geol.,* **7**, 181–190.

BULMAN, O. M. B. 1970 *In* Teichert, C. (Ed.) *Treatise on Invertebrate Paleontology. Part V, Graptolithina with sections on Enteropneusta and Pterobranchia* (2nd Edit.). Geol. Soc. Amer. and Univ. Kansas Press, 163 pp.

BURGESS, I. C., RICKARDS, R. B. & STRACHAN, I. 1970 The Silurian strata of the Cross Fell area. *Bull. geol. Surv. G.B.,* **32**, 167–182.

CAVE, R. 1980 Sedimentary environments of the basinal Llandovery of mid-Wales. *In* Harris, A. L., Holland, C. H. & Leake, B. E. (Eds.) The Caledonides of the British Isles – reviewed. *Spec. Publ. geol. Soc. London,* **8**, 517–526.

CHERNS, L. 1980 Hardgrounds in the Lower Leintwardine Beds (Silurian) of the Welsh Borderland. *Geol. Mag.,* **117**, 311–326.

COCKS, L. R. M. 1985 The Ordovician–Silurian Boundary. *Episodes,* **8**, 98–100.

COCKS, L. R. M., HOLLAND, C. H., RICKARDS, R. B. & STRACHAN, I. 1971 A correlation of Silurian rocks in the British Isles. *J. geol. Soc. London,* **127**, 103–136. [Also as: *Spec. Rep. geol. Soc. London,* **1**.]

COCKS, L. R. M. & RICKARDS, R. B. 1969 Five boreholes in Shropshire and the relationships of shelly and graptolitic facies in the Lower Silurian. *Q. J. geol. Soc. London,* **124** (for 1968), 213–238.

COCKS, L. R. M. TOGHILL, P. & ZEIGLER, A. M. 1970 Stage names within the Llandovery Series. *Geol. Mag.,* **107**, 79–87.

COCKS, L. R. M., WOODCOCK, N. H., RICKARDS, R. B., TEMPLE, J. T. & LANE, P. D. 1984 The Llandovery Series of the Type Area. *Bull. Br. Mus. nat. hist. Ser. Geol.,* **38**, 131–182.

CUMMINS, W. A. 1957 The Denbigh Grits; Wenlock Greywackes in Wales. *Geol. Mag.,* **94**, 433–451.

1959 The Lower Ludlow Grits in Wales. *Liverpool Manchester geol. J.,* **2**, 168–179.

EDWARDS, D. 1980 Early Land Floras. *In* Panchen, A. L. (Ed.) *The Terrestrial Environment and the Origin of Land Vertebrates.* Systematics Association Special Volume No. 15, Academic Press, London and New York, 55–85.

ELLES, G. L. 1909 The Relation of the Ordovician and Silurian Rocks of Conway (North Wales). *Q. J. geol. Soc. London,* **65**, 169–194.

ELLES, G. L. & SLATER, I. L. 1906 The Highest Silurian Rocks of the Ludlow District. *Q. J. geol. Soc. London,* **62**, 195–222.

GALE, N. H., BECKINSALE, R. D. & WADGE, A. J. 1980 Discussion of a paper by McKerrow, Lambert and Chamberlain on the Ordovician, Silurian and Devonian time scales. *Earth planet. Sci. Lett.,* **51**, 9–17.

GRAHAM, J. R., HANCOCK, P. L. & HOBSON, D. M.
1977 Anomalous bedding-cleavage relationships in Silurian rocks at Marloes Sands, S.W. Dyfed (Pembrokeshire), Wales, *Proc. Geol. Assoc. London*, **88**, 179–181.

GRAY, J. & BOUCOT, A. J.
1980 Microfossils and evidence of land plant evolution. *Lethaia*, **13**, 174.

HARLAND, W. B., COX, A. V., LLEWELLYN, P. S., PICKTON, C. A. G., SMITH, A. G. & WALTERS, R.
1982 *A geologic time scale.* Cambridge University Press, 131 pp.

HARLAND, W. B. & GAYER, B. A.
1972 The Arctic and earlier Oceans. *Geol. Mag.*, **109**, 289–314.

HOLLAND, C. H.
1962 The Ludlovian–Downtonian succession in central Wales and the central Welsh Borderland. *In* Erben, H. K. (Ed.) *Symposiums-Band der 2. Internationalen Arbeitstagung über die Silur/Devon-Grenze und die Stratigraphie von Silur und Devon, Bonn–Bruxelles 1960.* Stuttgart, 87–94.

1965 The Siluro–Devonian boundary. *Geol. Mag.*, **102**, 213–221.

1971 Silurian faunal provinces? *In* Middlemiss, F. A., Rawson, P. F. & Newall, G. (Eds.) *Faunal Provinces in Space and Time.* Seel House Press, Liverpool, 61–76.

1974 The Lower Palaeozoic Systems; an Introduction. *In* Holland, C. H. (Ed.) *Lower Palaeozoic Rocks of the World.* Vol. 2. John Wiley, London, 1–13.

1979 Augmentation and decay of the Old Red Sandstone Continent: Evidence from Ireland. *Palaeogeogr. Palaeoclimatol. Palaeoecol.*, **27**, 59–66.

1980 Silurian series and stages: decisions concerning chronostratigraphy. *Lethaia*, **13**, 238.

1981 Cambrian and Ordovician of the paratectonic Caledonides. *In* Holland, C. H. (Ed.) *A Geology of Ireland.* Scottish Academic Press, Edinburgh, 41–64.

1985 Synchronology. *Phil. Trans. R. Soc., London*, **B309**, 11–27.

1986 Does the golden spike still glitter? *J. geol. Soc. London*, **143**, 3–21.

1989 Principles, History, and Classification. *In* Holland, C. H. and Bassett, M. G. (Eds) *A global standard for the Silurian System.* National Museum of Wales, Cardiff.

*In Press* Concentration of the inarticulate brachiopod *Craniops* near the top of the Ludlow Series in the Central Welsh Borderland. *Mem. New Mexico Bureau of Mines and Mineral Resources.*

HOLLAND, C. H. & LAWSON, J. D.
1963 Facies patterns in the Ludlovian of Wales and the Welsh Borderland. *Liverpool Manchester geol. J.*, 3, 269–288.

HOLLAND, C. H., LAWSON, J. D. & WALMSLEY, V. G.
1959 A revised classification of the Ludlovian succession at Ludlow. *Nature. London*, **184**, 1037–1039.

1963 The Silurian rocks of the Ludlow district, Shropshire. *Bull. Br. Mus. nat. Hist.* Ser. Geol., **8**, 93–171.

HOLLAND, C. H. & PALMER, D. C.
1974 *Bohemograptus*, the youngest graptoloid known from the British Silurian sequence. *In* Rickards, R. B., Jackson, D. E. & Hughes, C. P. (Eds.) Graptolite studies in honour of O. M. B. Bulman. *Spec. Pap. Palaeontol. London*, **13**, 215–236.

| HOLLAND, C. H. & RICHARDSON, J. B. | 1977 | The British Isles. *In* Martinsson, A. (Ed.) *The Silurian–Devonian boundary*. International Union of Geological Sciences, Series A, No. 5, 35–44. |
| HOLLAND, C. H., RICKARDS, R. B. & WARREN, P. T. | 1969 | The Wenlock graptolites of the Ludlow district, Shropshire, and their stratigraphical significance. *Palaeontology, London*, **12**, 663–683. |
| HOLLAND, C. H., LAWSON, J. D., WALMSLEY, V. G. & WHITE, D. E. | 1980 | Ludlow stages. *Lethaia*, **13**, 268. |
| HOLLAND, C. H. & WILLIAMS, E. M. | 1985 | The Ludlow–Downton transition at Kington, Herefordshire. *Geol. J.*, **20**, 31–41. |
| HOLLAND, C. H. *et al.* | 1978 | A guide to stratigraphical procedure. *Spec. Rep. geol. Soc. London*, **11**, 18 pp. |
| INGHAM, J. K. & RICKARDS, R. B. | 1974 | Lower Palaeozoic Rocks. *In* Rayner, D. H. & Hemingway, J. E. (Eds.) *The Geology and Mineral Resources of Yorkshire*. Yorkshire Geological Society, 29–44. |
| JAEGER, H. | 1962 | Das Silur (Gotlandium) in Thüringen und am Ostrand des Rheinischen Schiefergebirges (Kellerwald, Marburg, Giessen). *Symposiums-Band der 2. Internationalen Arbeitstagung über die Silur/Devon-Grenze und die Stratigraphie von Silur und Devon. Bonn–Bruxelles 1960*, Stuttgart, 108–135. |
| JONES, O. T. | 1925 | The geology of the Llandovery district: Part I. The Southern Area. *Q. J. geol. Soc. London*, **81**, 344–388. |
| | 1937 | On the sliding or slumping of submarine sediments in Denbighshire, North Wales, during the Ludlow period. *J. geol. Soc. London*, **93**, 241–283. |
| | 1940 | The geology of the Colwyn Bay district: A study of submarine slumping during the Salopian period. *Q. J. geol. Soc. London*, **95** (for 1939), 335–382. |
| | 1949 | The geology of the Llandovery district: Part II. The Northern Area. *Q. J. geol. Soc. London*, **105**, 43–64. |
| KALJO, D. | 1978 | The Downtonian or Pridolian from the point of view of the Baltic Silurian. *Eesti NSV Tead. Akad. Toim.*, **27**, Geol. Series No. 1, 5–10. |
| KELLING, G. & WOOLLANDS, M. A. | 1969 | The stratigraphy and sedimentation of the Llandoverian rocks of the Rhayader district. *In* Wood, A. (Ed.) *The Pre-Cambrian and Lower Palaeozoic rocks of Wales*. University of Wales Press, Cardiff, 255–282. |
| KING, W. W. | 1934 | The Downtonian and Dittonian strata of Great Britain and North-Western Europe. *Q. J. geol. Soc. London*, **90**, 526–570. |
| LAPWORTH, C. | 1879–80 | On the geological distribution of the Rhabdophora. *Ann. Mag. nat. Hist. London* (5), **3**, 245–257, 449–455 (1879); (5), **5**, 45–62 (1880). |
| LAWSON, J. D. | 1955 | The geology of the May Hill Inlier. *Q. J. geol. Soc. London*, **111**, 85–116. |
| | 1975 | Ludlow benthonic assemblages. *Palaeontology, London*, **18**, 509–525. |
| LEGGETT, J. K., McKERROW, W. S. & EALES, M. H. | 1979 | The Southern Uplands of Scotland; A Lower Palaeozoic accretionary prism. *J. geol. Soc. London*, **136**, 755–770. |
| MARTINSSON, A. (Ed.) | 1977 | *The Silurian–Devonian boundary*. International Union of Geological Sciences, Series A, No. 5, 349 pp. |

| MARTINSSON, A., BASSETT, M. G. & HOLLAND, C. H. | 1981 | Ratification of Standard Chronostratigraphical Divisions and Stratotypes for the Silurian System. *Episodes, Ottawa,* **1981**, 36. |
| MCCABE, P. J. | 1972 | The Wenlock and Lower Ludlow strata of the Austwick and Horton-in-Ribblesdale inlier of north-west Yorkshire. *Proc. Yorkshire geol. Soc.,* **39**, 167–174. |
| MCCABE, P. J. & WAUGH, B. | 1973 | Wenlock and Ludlow sedimentation in the Austwick and Horton-in-Ribblesdale inlier, north-west Yorkshire. *Proc. Yorkshire geol. Soc.,* **39**, 445–470. |
| MCKERROW, W. S. & COCKS, L. R. M. | 1976 | Progressive faunal migration across the Iapetus Ocean. *Nature, London,* **263**, 304–306. |
| MCKERROW, W. S., LAMBERT, R. ST. J. & CHAMBERLAIN, V. E. | 1980 | The Ordovician, Silurian and Devonian time scales. *Earth planet, Sci. Lett.,* **51**, 1–8. |
| MCKERROW, W. S., LAMBERT, R. St. J. & COCKS, L. R. M. | 1985 | The Ordovician, Silurian and Devonian periods. *In* Snelling, N. J. (Ed.) The Chronology of the Geological Record. *Mem. geol. Soc. London,* **10**, 73–80. |
| MURCHISON, R. I. | 1835 | On the Silurian System of rocks. *London Edinburgh phil. Mag.,* **7**, 46–52. |
| | 1839 | *The Silurian System.* London, 768 pp. |
| | 1859 | *Siluria* (2nd Edit. (3rd Edit. on title page)). London. 592 pp. |
| PALMER, D. | 1970 | A stratigraphical synopsis of the Long Mountain, Montgomeryshire and Shropshire. *Proc. geol. Soc. London,* **1660**, 341–346. |
| RICHARDSON, J. B. & LISTER, T. R. | 1969 | Upper Silurian and Lower Devonian spore assemblages from the Welsh Borderland and South Wales. *Palaeontology, London,* **12**, 201–252. |
| RICHARDSON, B. J. & RASUL, S. J. | 1990 | Palynofacies in a Late Silurian regressive sequence in the Welsh Borderland and Wales. *J. geol. Soc. London,* **147**, 675–686. |
| RICKARDS, R. B. | 1976 | The sequence of Silurian graptolite zones in the British Isles. *Geol. J.,* **11**, 153–188. |
| | 1978 | Silurian. *In* Moseley, F. (Ed.) *The Geology of the Lake District.* Yorkshire Geological Society, 130–145. |
| RICKARDS, R. B. & HUTT, J. E. | 1970 | The earliest monograptid. *Proc. geol. Soc. London,* **1663**, 115–119. |
| RIDER, R. | 1981 | Composition, structure and environmental setting of Silurian bioherms and biostromes in Northern Europe. *Spec. Publ. Soc. econ. Paleont. Mineral. Tulsa,* **30**, 41–83. |
| ROSS, R. J. *et al.* | 1978 | Fission-track dating of Lower Paleozoic volcanic ashes in British stratotypes. *In* Zartman, R. E. (Ed.) *Short Papers of the Fourth International Conference, Geochronology, Cosmochronology, Isotope Geology.* U.S. Geol. Surv. Open-file Rep. 78–101, 363–365. |
| ST. JOSEPH, J. K. S. | 1935 | A critical examination of *Stricklandia* (= *Stricklandinia*) *lirata* (J. de C. Sowerby) 1839 *forma typica. Geol. Mag.,* **72**, 401–424. |
| | 1938 | The Pentameracea of the Oslo region. *Nor. geol. Tiddskr.,* **17**, 225–336. |
| SANDERSON, R. W. & CAVE, R. | 1980 | Silurian volcanism in the central Welsh Borderland. *Geol. Mag.,* **117**, 455–462. |
| SCOFFIN, T. P. | 1971 | The conditions of growth of the reefs of Shropshire (England). *Sedimentology,* **17**, 173–219. |
| SCOTESE, C. R., BAMBACH, R. K., BARTON, C., VAN DER VOO, R. & ZEIGLER, A. M. | 1979 | Paleozoic base maps. *J. Geol.,* **87**, 217–277. |

SECORD, JAMES A.                  1986    *Controversy in Victorian Geology: The Cambrian–Silurian Dispute*. Princeton University Press, 364 pp.

SHERGOLD, J. H. &                 1970    Facies and faunas at the Wenlock/Ludlow boundary of
   BASSETT, M. G.                          Wenlock Edge, Shropshire. *Lethaia*, **3**, 113–142.

SMITH, D. G.                      1981    Silurian plant evolution – evidence first, speculation later. *Lethaia*, **14**, 26.

SPJELDNAES, N.                    1978    The Silurian System. In *Contributions to the Geologic Time Scale*. Am. Assoc. Petrol. Geol., Studies in Geology, No. 6, 341–345.

                                  1981    Lower Palaeozoic palaeoclimatology. *In* Holland, C. H. (Ed.) *Lower Palaeozoic Rocks of the World*. Vol. 3. John Wiley, Chichester, 199–256.

STRAW, S. H.                      1933    The fauna of the Palaeozoic rocks of the Little Missenden boring: with a report on the fish remains by Sir Arthur Smith Woodward. *Summ. Prog. geol. Surv. G.B.* for 1932, part 2, 112–142.

                                  1937    The higher Ludlovian rocks of the Builth district. *Q. J. geol. Soc. London*, **93**, 405–456.

SUDBURY, M.                       1958    Triangulate Monograptids from the *Monograptus gregarius* zone (Lower Llandovery) of the Rheidol Gorge (Cardiganshire). *Philos. Trans. R. Soc. London*, Ser. B, **241**, 485–555.

TURNER, S.                        1973    Silurian–Devonian thelodonts from the Welsh Borderland. *J. geol. Soc. London*, **129**, 557–584.

WALMSLEY, V. G. &                 1976    Biostratigraphy and correlation of the Coralliferous Group
   BASSETT, M. G.                          and Gray Sandstone Group (Silurian) of Pembrokeshire, Wales. *Proc. Geol. Assoc. London*, **87**, 191–220.

WALTER, R.                        1980    Lower Paleozoic Paleogeography of the Brabant Massif and its Southern Adjoining Areas. *Meded. Rijks geol. Dienst.*, **32**, 14–25.

WATKINS, R.                       1978    Bivalve ecology in a Silurian shelf environment. *Lethaia*, **11**, 41–56.

                                  1979    Benthic community organization in the Ludlow Series of the Welsh Borderland. *Bull. Br. Mus. nat. Hist.*, Ser. Geol., **31**, 175–280.

WHITAKER, J. H. McD.              1962    The geology of the area around Leintwardine, Herefordshire. *Q. J. geol. Soc. London*, **118**, 319–351.

WHITE, D. E.                      1981    The base of the Ludlow Series in the graptolitic facies. *Geol. Mag.*, **118**, 566.

WHITE, D. E. &                    1978    The stratigraphy of new sections in the Ludlow Series of the
   LAWSON, J. D.                           type area, Ludlow, Salop, England. *Rep. Inst. geol. Sci. London*, No. **78/30**, 10 pp.

WHITEHEAD, T. H. &                1947    Dudley and Bridgenorth. *Mem. geol. Surv. G.B.*, 226 pp.
   POCOCK, R. W.

WHITTARD, W. F.                   1932    The stratigraphy of the Valentian rocks of Shropshire. The Longmynd-Shelve and Breidden Outcrops. *Q. J. geol. Soc. London*, **88**, 859–902.

                                  1961    *Lexique Stratigraphique International*. Vol. I. Europe, Fasc. 3a, Angleterre, Pays de Galles, Ecosse: V Silurien, 273 pp.

WHITTARD, W. F.                   1979    An account of the Ordovician rocks of the Shelve Inlier in
   (Compiled by                            west Salop and part of north Powys. *Bull. Br. Mus. nat.*
   DEAN, W. T.)                             *Hist.*, Ser. Geol., **33**, 1–69.

WILLIAMS, A.                      1951    Llandovery brachiopods from Wales with special reference to the Llandovery district. *Q. J. geol. Soc. London*, **107**, 85–136.

WILLS, L. J.                1978        A palaeogeological map of the Lower Palaeozoic floor
                                        below the cover of Upper Devonian, Carboniferous and
                                        later formations with inferred and speculative recon-
                                        structions of Lower Palaeozoic and Precambrian outcrops
                                        in adjacent areas. *Mem. geol. Soc. London*, **8**, 36 pp., Plates
                                        1 and 2.

WOOD, A. &                  1959        The sedimentation and sedimentary history of the
  SMITH, A. J.                          Aberystwyth Grits (Upper Llandoverian). *Q. J. geol. Soc.
                                        London*, **114** (for 1958), 163–195.

WOODCOCK, N. H.             1976        Ludlow Series slumps and turbidites and the form of the
                                        Montgomery trough, Powys, Wales. *Proc. Geol. Assoc.
                                        London*, **87**, 169–182.

ZIEGLER, A. M.              1965        Silurian marine communities and their environmental
                                        significance. *Nature, London*, **207**, 270–272.

                            1970        Geosynclinal development of the British Isles during the
                                        Silurian Period. *J. Geol.*, **78**, 445–479.

ZIEGLER, A. M.,             1968a       The composition and structure of Lower Silurian marine
  COCKS, L. R. M. &                     communities. *Lethaia*, **1**, 1–27.
  BAMBACH, R. K.

ZIEGLER, A. M.,             1968b       The Llandovery transgression of the Welsh Borderland.
  COCKS, L. R. M. &                     *Palaeontology, London*, **11**, 736–782.
  McKERROW, W. S.

ZIEGLER, A. M.,             1977        Silurian continental distributions, paleogeography, clima-
  HANSEN, K. S.,                        tology and biogeography. *Tectonophysics*, **40**, 13–51.
  JOHNSON, M. E.,
  KELLY, M. A.,
  SCOTESE, C. R. &
  VAN DER VOO, R.

# 6

# DEVONIAN

# D. L. Dineley

## Introduction

The Devonian System presents rocks of two quite different suites of facies in England and Wales. The Old Red Sandstone continental facies is widespread in Scotland and extends into the Scottish Borders while extensive outcrops occur in Wales and the Welsh Borderland. The marine facies of the Devonian occurs in west Somerset, Devon and Cornwall where it includes a variety of clastic and carbonate formations. The rock succession in north and south Cornwall differs conspicuously and only in the south is the lowermost part of the system seen. Volcanic rocks are extensively developed in Cornwall, south Devon and the Cheviot Hills but occur only as rare minor units in Wales. In the subsurface of Central and SE England rocks of both marine and continental facies have been encountered in numerous boreholes. The known extent of the Devonian System in England and Wales, and in southern Scotland is shown in Figure 6.1.

In northern England the system is represented by coarsely clastic and generally unfossiliferous formations in the Cheviot Hills, the Lake District and to the east of the Vale of Eden. Plutonic intrusions of Devonian age form an important part of the Lake District geology. A small outcrop of red beds in Anglesey is the sole representative of the (?Lower) Old Red Sandstone in North Wales and a thin development is present in Central Wales. The Welsh Borderland and English Midlands, however, have a varied succession of Lower Old Red Sandstone formations and thin representatives of the uppermost (Farlovian) stage. The sequences in South Wales are by contrast thick and extensive and may also include rocks of Middle Devonian age.

The continental facies originated from the erosion of the Caledonide uplands and was locally influenced by the accompanying volcanicity. Lacustrine, fluvial and aeolian deposits are recognised, their regimes being interrupted by local or regional earth movements at the end of Early Devonian and Middle Devonian times.

In SW England the Devonian system occurs in two major areas, separated by the central Devon synclinorium of Carboniferous rocks. These major outcrops differ markedly and in each both structural and stratigraphical problems remain. The rock successions in north and south Cornwall are conspicuously unlike one another and only in the south is the base of the system to be seen. The sequence in north Cornwall can be traced eastwards into south Devon where thick carbonate and volcanic formations are present. In north Devon and west Somerset a generally coarser clastic succession occurs, representing the Middle and Upper Devonian series (see Durrance & Laming 1982).

Marine sedimentation in the area ranged from deep water environments in the south and west to local reef and carbonate bank environments in south Devon and littoral environments in north Devon. Although the effects of the contemporaneous earth movements in Wales and SW England are not fully understood, it seems that sedimentation south of what is now the Bristol Channel continued with little or no interruption throughout the period and into the Carboniferous.

## Stratigraphical classification

Although it was founded upon researches in the SW of England, the Devonian System has long been referred to European stratigraphical sections for its component subdivisions. Through the insight of its founders, Sedgwick & Murchison (1839), its equivalence to the Old Red Sandstone formations was recognised at the

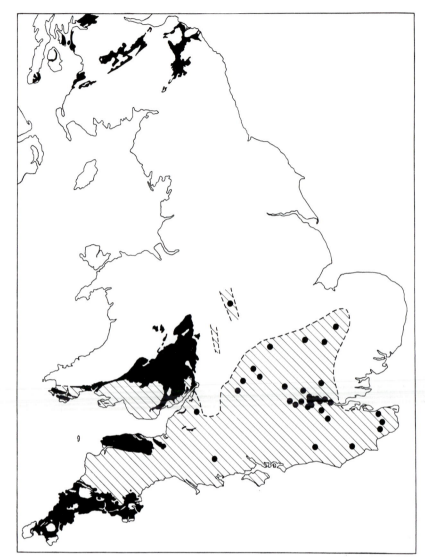

**Fig. 6.1.** Distribution of outcrops of the Devonian system in England and Wales. Devonian or Old Red Sandstone occurs in the sub-surface of central and southern England (diagonal shading) and is encountered in boreholes at the points shown by black dots (in part after L. J. Wills 1973). Old Red Sandstone outcrops in the south of Scotland are also shown.

outset, but the tectonic complications, facies changes, metamorphic imprint and the peculiar distribution of fossils did not then provide unambiguous criteria for the recognition of Devonian series and stages. The presence of locally large thicknesses of limestone in the mid-part of the succession in south Devon with associated volcanic rocks and thick shale formations above and below was recognised and the correlation of these rocks with those of north Devon followed. Rapid strides were made in refining Devonian stratigraphy and palaeontology during the latter half of the nineteenth century and the Geological Survey's attentions to the region were to produce a better understanding

of the Devonian rocks in SW England and their equivalents in the Old Red Sandstone formations of England, Wales and Scotland.

The Devonian System offers such a multitude of facies and has been so affected by diastrophism to the extent that many problems of correlation remain to this day. Through the work in recent years of e.g. George (1970), Earp & Haines (1971), Edmonds *et al.* (1975), House *et al.* (1977) and Allen (1979), the Devonian System of England and Wales is known in detail comparable with that of other Palaeozoic systems and its broad relationships to the Devonian of Europe are apparent. House *et al.* (1977) provide the most complete correlation of Devonian rocks to date while two field guides, one covering the Devonian of Scotland, the Welsh Borderland and South Wales (Friend & Williams 1978) the other SW England (Scrutton 1978) include much new information. The *Proceedings of the Ussher Society* (1964 to date) contain many papers and short notes on Devonian stratigraphy and palaeontology.

Subdivision and correlation on the basis of lithology is fraught with difficulties but the use of this criterion is unavoidable in the Old Red Sandstone continental facies. Palynomorphs, plants and vertebrates are currently being pressed into biostratigraphic service in these rocks and the discovery of widespread ash bands in the Lower Old Red Sandstone of South Wales and the Welsh Borderland provides a further criterion for correlation there.

In the marine facies of Western and Central Europe ammonoids and conodonts have proved to be especially useful in subdivision and correlation (Figs. 6.2, 6.3) and these forms have been recovered locally in SW England in sufficient numbers to be applied successfully for correlation purposes.

The long dispute over the Siluro/Devonian boundary was all but ended when international agreement was reached on the matter at the International Geological Congress in Montreal in 1972 (see also Ch. 5). The boundary horizon was designated as within a bed containing the first and immediately abundant occurrence of *Monograptus uniformis* and *M. uniformis augustidens* in the succession at the village of Klonk in the Barrandian area of Czechoslovakia. This is a far cry from the base of the Ludlow Bone Bed which had been customarily used by British workers

| STAGES | | STUFEN | | AMMONOID ZONES | CONODONT ZONES |
|---|---|---|---|---|---|
| LOWER DEVONIAN | FAMENNIAN | WOCKLUMERIA  | | *Prionoceras* sp. | *Protognathodus* |
| | | | | *Cymaclymenia euryomphala* | |
| | | | | *Wocklumeria sphaeroides* | *Spathognathodus costatus* |
| | | | | *Kalloclymenia subarmata* | |
| | | CLYMENIA  | | *Gonioclymenia speciosa* | *Polygnathus styriacus* |
| | | | | *Gonioclymenia hoevelensis* | |
| | | PLATYCLYMENIA 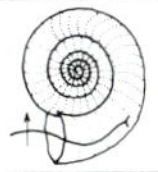 | | *Platyclymenia annulata* | *Scaphignathus velifer* |
| | | | | *Prolobites delphinus* | |
| | | | | *Pseudoclymenia sandbergeri* | *Palmatolepis marginifera* |
| | | CHEILOCERAS 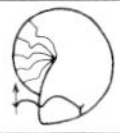 | | *Sporadoceras pompeckji* | *Palmatolepis rhomboidea* |
| | | | | *Cheiloceras curvispina* | *Palmatolepis crepida* |
| | FRASNIAN | MANTICOCERAS  | | *Manticoceras* sp | *Palmatolepis triangularis* |
| | | | | *Crickites holzapfeli* | *Palmatolepis gigas* |
| | | | | *Manticoceras cordatum* | *Ancyrognathus triangularis* |
| | | | | | *Polygnathus asymmetricus* |
| | | | | *Pharciceras lunulicosta* | *S. hermanni-P. cristatus* |
| | GIVETIAN | MAENIOCERAS 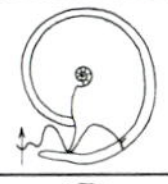 | | *Maenioceras terebratum* | *Polygnathus varcus* |
| | | | | *Maenioceras molarium* | |
| | | | | *Cabrieroceras crispiforme* | *Polygnathus kockelianus* |
| | EIFELIAN | ANARCESTES 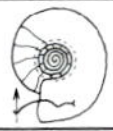 | | *Pinacites jugleri* | *Polygnathus ensensis* |
| | | | | *Anarcestes lateseptatus* | *P. costatus costatus and P. costatus patulus* |
| UPPER DEVONIAN | EMSIAN | MIMOSPHINCTES 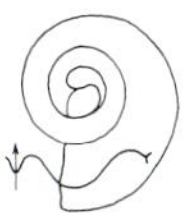 | | *Sellanarcestes wenkenbachi* | *Polygnathus serotinus* |
| | | | | *Mimagoniatites zorgensis* | *Polygnathus inversus* |
| | | | | *Anetoceras hunsrueckianum* | *Polygnathus gronbergi* |
| | SIEGENIAN | GRAPTOLITE ZONES | | *Monograptus hercynicus* | *Polygnathus dehiscens* |
| | | | | | *Ancyrodelloides-Ic. pesavis* |
| | GEDINNIAN | | | | *Icriodus w. postwoschmidti* |
| | | | | *Monograptus uniformis* | *Icriodus woschmidti* s.s. |

**Fig. 6.2.** Stratigraphical subdivision of the Devonian system showing the stages erected in western Europe for the Rhenish facies, the German ammonoid *stufen* and the currently recognised ammonoid zones and conodont zones (after House 1973).

| | | STAGES USED HERE | FRANCE/ BELGIUM | GERMANY | CZECHO-SLOVAKIA |
|---|---|---|---|---|---|
| | | | TOURNAISIEN | | |
| DEVONIAN | UPPER | FAMENNIAN | FAMENNIEN | WOCKLUM / DASBERG / HEMBERG / NEHDEN | |
| DEVONIAN | UPPER | FRASNIAN | FRASNIEN | ADORF | |
| DEVONIAN | MIDDLE | GIVETIAN | GIVETIEN | GIVET | SRBSKO |
| DEVONIAN | MIDDLE | EIFELIAN | COUVINIEN | EIFEL | CHOTEČ / DALEJE |
| DEVONIAN | LOWER | EMSIAN | COBLENCIEN | EMS (KOBLENZ) | ZLICHOVIAN |
| DEVONIAN | LOWER | SIEGENIAN | COBLENCIEN | SIEGEN | PRAGIAN |
| DEVONIAN | LOWER | GEDINNIAN | GEDINNIEN | GEDINNE | LOCHKOVIAN |

**Fig. 6.3.** The stage names employed for the Devonian system in different parts of Europe. The sequence in Czechoslovakia falls within the so-called Bohemian or Hercynian facies province, the faunas of which show many distinctions from those of the same age in the west (after House *et al.* 1977).

(White 1950, House *et al.* 1977) but in the marine facies and the world at large it may prove to be a more useful horizon.

The recommendations of the I.U.G.S. Subcommission on Devonian Stratigraphy seek to define the Series and Stages of the System on the basis of biostratigraphy with designated stratotype sections (Zeigler & Klapper 1985). These have little impact on stratigraphic classification in Britain (House & Dineley 1986). Several new formation names in South Devon and Cornwall have been introduced of late for the tectonically discrete contemporaneous Middle and Upper Devonian units there (see Dineley 1986). The upper boundary of the system is drawn at the base of the *Gattendorfia* Zone (approximating to the base of the conodont *Siphonodella sulcata* Zone), a practice now widely adopted by British geologists (George *et al.* 1969, Paproth & Streel 1985).

## Palaeontology

It is unfortunate that so few species of fossils from the Old Red Sandstone facies are known in the marine Devonian and that there is so little intercalation of the marine and non-marine formations in the British Isles. This naturally makes correlation difficult, and until recently the situation has not been greatly helped by reference to sequences in adjacent parts of Europe. In recent years, however, much has been done to establish the detailed stratigraphic ranges of many fossil groups, none perhaps being more important than the palynomorphs. By means of spores a correlation of Old Red Sandstone formations within Europe and, at least partly, with the marine sequence is now possible.

Apart from a small number of invertebrates the commonest fossils in the continental deposits are plants and vertebrates. The latter are widely distributed in the Downtonian and Dittonian formations and are also present though less common in the Breconian and the Farlovian (Fig. 6.4). No Middle Devonian vertebrates are known south of Scotland. Thelodont denticles are well known from horizons in the Downtonian (Silurian) and Dittonian (Turner 1973). Ostracoderm fragments, principally from the heterostraci, are also locally common. The pteraspids in particular have been pressed into stratigraphic service in the Lower Old Red Sandstone (White 1956, Ball *et al.* 1961). Acanthodian and arthrodire taxa occur throughout but are poorly understood and difficult to identify from very fragmentary material.

The fauna of the Dartmouth Beds (pp. 191–2) includes pteraspids and other ostracoderms, and is possibly as rich if not richer in species than are the faunas of the Anglo-Welsh areas to the north. Although identifications are difficult, forms akin to the *crouchi–rostrata* group of pteraspids have been recognised in addition to the widespread *Rhinopteraspis cornubica* and a *Protaspis*. *Traquairaspis* sp. has recently been found in the Dartmouth Beds near Plymouth.

In the Upper Old Red Sandstone the commonest forms are species of *Bothriolepis* and *Holoptychius*. Large indeterminate arthrodires also occur. By comparison with the Lower Old Red Sandstone

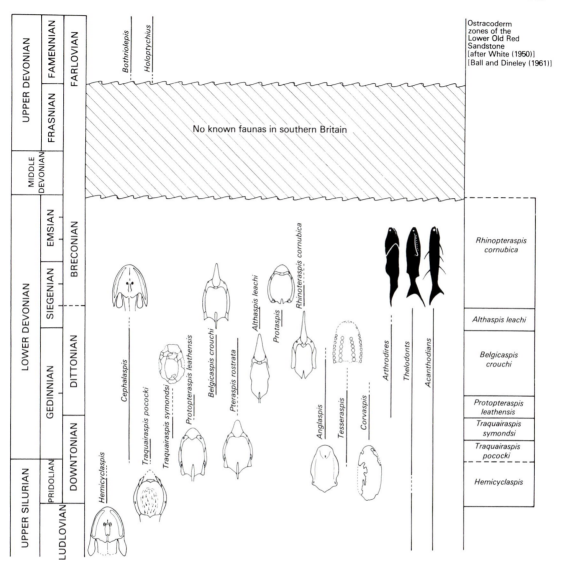

**Fig. 6.4.** The stratigraphic ranges of the more common vertebrates in the Lower Old Red Sandstone of Wales and the Welsh Borderlands.

faunas the Upper are impoverished and they appear to be less varied than their contemporaries in Scotland.

The invertebrate faunas include bivalves, ostracods, eurypterids and miriapods. *Lingula* spp. occur in the lowest Downtonian rocks together with a small number of articulate brachiopods and other marine or brackish water invertebrates. Trace fossils are locally abundant and remarkable. Problematic, large burrow-like structures resembling *Beaconites* occur at several widespread localities in the Dittonian rocks of South

Wales and small ichnofossils occur abundantly immediately below the tuff bands in the Upper Downtonian (Allen & Williams 1981).

The floras of the Lower Old Red Sandstone in the Anglo–Welsh area are not extensive but, locally, material is abundant. Macroscopic plants include *Gosslingia, Cooksonia, Sawdonia, Zosterophyllum, Steganotheca* and *Pachytheca* (Lang 1937; Croft & Lang 1942; Edwards 1979, 1980a, b). The palynology of the Old Red Sandstone is now known well in outline and several detailed investigations have provided local detail (Richardson 1967, Richardson & Lister 1969, Dolby 1970, Thomas 1978).

The marine faunas of the Devonian are extensive and stratigraphically interesting. Those of the Lynton Beds (p. 192) are dominated by bivalves but elsewhere brachiopods are the commonest shelly fossils. The coral faunas of the Middle and Early Upper Devonian of Devonshire are distinctive and include upwards of fifteen genera. Other invertebrates of prime stratigraphic importance include trilobites, ammonoids, conodonts and ostracodes. Trace fossils (Goldring 1962, 1971), especially *Chondrites* sp., and rugosa (Scrutton 1977a, b) are locally abundant.

Detailed stratigraphic data and general discussion of the palaeontology are given in e.g. Simpson (1961), House & Selwood (1966), House *et al.* (1977) and see also Dineley (1986).

# Regional descriptions

The stratigraphic terminology used by House *et al.* (1977) is employed in most of the sections in the following account; where later terminology is used the source is given.

**Fig. 6.5.** Old Red Sandstone successions in northern England (after House *et al.* 1977).

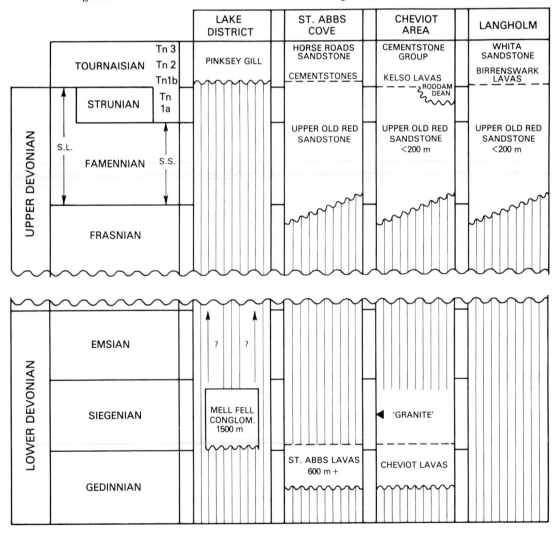

## Northern England

Lower Old Red Sandstone lavas and continental clastic rocks are exposed in the Cheviot Hills, which span the Anglo–Scottish Border and at St Abbs Head (Berwickshire), where they rest with strong unconformity upon highly folded Lower Palaeozoic rocks. They are unconformably overlain by Upper Old Red Sandstone strata which pass upwards into limestone or sandstone formations of Carboniferous age (Fig. 6.5).

The Lower Old Red Sandstones includes thin red siltstones with palaeosol calcretes and silcretes, intraformational conglomerates and sandstones and polymictic conglomerates. Thick, mainly andesitic, pyroclastic and effusive volcanic rocks, the Cheviot Volcanic Group, may be 1,000 m or more thick. These Lower Old Red Sandstone formations are intruded by the Cheviot 'granite' which pre-dates the Upper Old

Red Sandstone. Its age has been given as 380 Ma (Mitchell 1972). No Middle Old Red Sandstone formations have been identified.

West of the Cheviots further variegated marls, siltstones, sandstones and conglomerates occur with palaeosols ('cornstones') and rare scattered Upper Old Red Sandstone vertebrate fossils (Smith 1967, Leeder 1973).

Red beds resting unconformably upon Ordovician strata in the Lake District and east of the Vale of Eden include thick conglomeratic units, up to 1,500 m thick. Of these the Mell Fell Conglomerate is the best known (Capewell 1955). Its exact age is uncertain but it rests between Ordovician and Dinantian rocks. The Polygenetic Conglomerate occurring between Lower Palaeozoic and Carboniferous rocks of the eastern side of the Vale of Eden and a similar isolated conglomeratic unit near Sedburgh are also of possible Old Red Sandstone age (House et al. 1977, p. 68).

**Fig. 6.6.** Stratigraphic successions of Old Red Sandstone in Wales, the Welsh Borderlands and the English Midlands (after House et al. 1977 and Allen & Williams 1978).

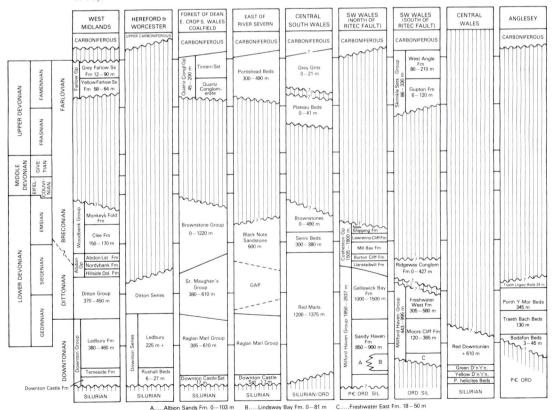

## Anglesey

The Old Red Sandstone of Anglesey is isolated from
the other Welsh outcrops. Its origin may have been
somewhat different from that of the southern outcrops
and its site of deposition a separate basin. Four
formations of probably Early Old Red Sandstone age
occur on the NE coast of the island. At the base of the
sequence the Bodafon Beds (3–45 m) are pebbly
sandstones and conglomerates (Fig. 6.6). They are
overlain by and intertongue with the Traeth Bach Beds
(130 m) and Porth y Mor Beds (345 m). The Traeth
Bach unit comprises red mudstone with some conglomerates and several dolomitic palaeosols. The
Porth y Mor Beds exhibit a sequence of upward-fining
cycles of red sandstones, siltstones and mudstones.
Intraformational conglomerate layers are common
and in the lowest 100 m dolomitised palaeosols are
thick and numerous. The succeeding Traeth Lligwy
Beds (24 m) are a distinctive fine-grained unit and
possibly lacustrine in origin. Allen (1965) regarded the
two lower formations as correlatives respectively of
the Downton Series of the Welsh Borderlands, and the
upper two are Dittonian in age.

## South and Central Wales and Welsh Borderlands

Rocks of Old Red Sandstone facies in this region crop
out over an area of more than 4,000 km² (Fig. 6.7) and
actually range in age from late Silurian to early
Carboniferous. They have been most closely studied in
the Clee Hills (Ball *et al.* 1961, Allen 1974) and in
SW Dyfed (Allen & Williams 1978, Marshall 1978,
Thomas 1978). By far the greater part of the succession
(up to 2,000 m) consists of the Lower Old Red
Sandstone Series. The Upper Series, reaching a
thickness of 330 m in Dyfed, skirts the Carboniferous
outcrop of the South Wales and Forest of Dean
coalfields and occurs as small outcrops in the Clee
Hills and near Bristol. Between the western margins of
the outcrop and the outliers of Central Wales lies the
Church Stretton disturbance. In Dyfed the Ritec Fault
appears to separate different successions of Lower Old
Red Sandstone and no doubt exerted some influence
on facies development during the Devonian period.

Figure 6.6 shows the stratigraphic units currently
recognised in this region (Williams, 1978). Virtually
everywhere the base of the Old Red Sandstone rests
upon an erosion surface or angular unconformity (but

**Fig. 6.7.** Sketch map of Old Red Sandstone outcrops of South and Central Wales and
the Welsh Borderland.

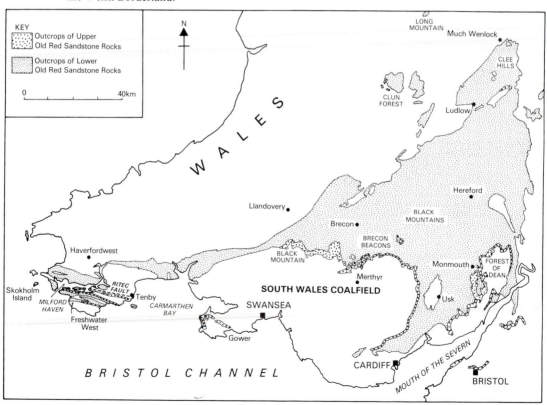

see Walmsley & Bassett 1976). There follows a thin passage sequence in most areas which effects the change into red-bed, partly continental, facies. This succession in the Borderland contains locally abundant invertebrate fossils as well as scattered or segregated vertebrate fragments. The red beds above are singularly unfossiliferous, except for rare vertebrates and spores.

The Downtonian rocks are essentially fine-grained siltstones and mudstones. Sandstone sheets and lenses occur at intervals with intraformational conglomeratic layers and local thin palaeosol limestone nodule bands. The succeeding Dittonian strata contain a higher proportion of sandstone and intraclast beds with conspicuous cross-bedding, intraformational erosion surfaces and other sedimentary features of note. Within a thickness of about 150 m from the base of the series several palaeosol limestones or calcretes, the 'Psammosteus Limestones', are more or less continuous over wide areas and some may extend over virtually the entire basin.

Other similar limestones up to 3 m thick occur higher in the succession (Abdon Limestone, Fynnon Limestone) in Shropshire and SE Wales.

A number of ash bands or tuffs occur in the higher Downtonian rocks throughout the entire region from SW Wales to the West Midlands (Allen & Williams 1978, 1981). The value of these marker beds in the correlation of successions which are generally very variable in lithologies and impoverished in fossils is very high. They also afford the only indication of Siluro–Devonian volcanic activity between Devonshire and the Cheviot area on the borders of Scotland. The possibility exists that they may be associated in origin with part of the volcanic suite in the Lower Devonian of south Devon.

Breconian rocks follow without break upon the Dittonian (Figs. 6.6, 6.8) except in Dyfed south of the Ritec Fault where the Ridgeway Conglomerate Formation may, in part, be of Breconian age. They are, on the whole, rocks of coarser lithologies than those below, with a smaller proportion of siltstones and mudstones. They are remarkably unfossiliferous except for rare instances of ostracoderms, acanth-

**Fig. 6.8.** North face of Pen y Fan, Brecon Beacons. The escarpment shows the cyclical alluvial clastic sequences of the 'Brownstones' (Fig. 6.6) capped unconformably by the fluviatile Plateau Beds (B.G.S. A.11105).

odians, and arthrodires. Plants, however, have been recorded from several localities in the Senni Beds of south Central Wales (Croft & Lang 1942, Edwards & Richardson 1974).

The upper limit of these Lower Old Red Sandstone formations is everywhere an unconformity upon which rest Upper Old Red Sandstone or Carboniferous strata. Upper Old Red Sandstone beds are similarly coarsely clastic in general character. In Dyfed south of the Ritec Fault (pp. 543–5) the Skrinkle Sandstones Group is comprised of the Gupton Formation of tabular and lenticular sandstones and conglomerates and the highly conglomeratic West Angle Formation which passes upwards into the Lower Limestone Shales of the Carboniferous (Marshall 1978).

## Central Wales

Three outliers beyond the main Old Red Sandstone outcrop of the Welsh Borderland (Fig. 6.7) occur at Long Mountain (Palmer 1970), Clun Forest (Earp 1940) and near Knighton (Holland 1959). Each overlies Ludlovian strata in apparent conformity. The successions in each are comparable, thin and entirely of a fine siltstone, shale and sandstone facies with ostracods, eurypterids, species of *Lingula* and *Modiolopis complanata*, *Turbocheilas* [*Platyschisma*] and vertebrate fragments. There is overall a close similarity to the Downtonian succession in the Clee Hills area with a passage from grey-green to red fine clastic units and a late Silurian age is probable.

## Forest of Dean–Severn Valley

In outline and in many of its details the Old Red Sandstone succession in the Forest of Dean area compares closely with that in the Welsh Borderland (Fig. 6.6). The Downton Castle Sandstone is overlain by the predominantly argillaceous Raglan Marl Group which in turn is followed by the St Maughan's Group with its many upward-fining cyclothems, and the strongly arenaceous Brownstones Group. Near Bristol the Black Nore Sandstone appears as a possible correlative of the Brownstones Group inserted between the St Maughan's Group and the (Upper Old Red Sandstone to lowermost Carboniferous) Portishead Beds.

## South-west Wales

Between Skokholm Island and Carmarthen Bay (Fig. 6.7) the Old Red Sandstone of Dyfed is present in sharp E–W folds as a thick succession of well-differentiated local formations. Two rather different sequences may be distinguished, viz. on the northern

and the southern sides of the Ritec Fault (see Fig. 6.6). North of the fault the rocks are apparently confined to the Lower Old Red Sandstone while to the south the Skrinkle Sandstone Group has long been known to belong in part to the Upper Old Red Sandstone. The succession in places rests with marked unconformity and a basal conglomerate upon deformed Lower Palaeozoic rocks. Recently Allen & Williams (1978) recorded about 3,200 m of red beds (Silurian and L. Devonian) between the Ritec and Benton Faults, of alluvial and fluviatile origins and including tuffs as well as clastics and pedogenic carbonates. As with previous studies, theirs are largely based on the extensive cliff and foreshore exposures, and they emphasise the marked difference in successions in successive E–W trending zones (see also Allen *et al.* 1981).

### North of the Ritec Fault

Here a thickness of about 4,000 m is embraced by the Milford Haven Group and the overlying Cosheston Group (Allen & Williams 1978, Thomas 1978). The basal unit, the Red Cliff Formation (Fig. 6.6), is mainly of red mudstones interbedded with sandstones in alternations on scales of decimetres to metres. The sandstones are unusual for the conspicuous bioturbation within them. A few calcretes are also present. The overlying Albion Sands Formation consists of thick pale, yellow or buff, cross-bedded sandstones with conspicuous quantities of igneous debris and, commonly, large intraformational mud clasts. Extraformational conglomerates are also present, but calcretes are very rare. In the Lindsway Bay Formation thick conglomerates rich in igneous debris alternate with granule-rich mudstones, and this distinctive facies appears to replace the Albion Sands Formation laterally eastwards. Above this unit lies the Sandy Haven Formation with characteristic bright red mudstones in which occur calcretes with *gilgai* or patterned-ground anticline-like features, thin intercalated fossiliferous sandstones, pebbly sandstones and conglomerates. Near the middle of this formation the Townsend Tuff Bed and other tuffs form regional marked beds of great significance (Allen & Williams 1978, 1981). The upper 1,000–1,500 m of the Milford Haven Group comprises the Gelliswick Bay formation, a succession of closely alternating red mudstones, fine-grained sandstones and intraformational conglomerates. Calcrete beds are common.

The Cosheston Group is comprised of five formations, all of fluviatile origin. Mudstones, sandstones and conglomerates both intraformational and extraformational occur. Cyclothems a few metres to tens of metres thick are clearly distinguishable and are especially characteristic of the lowest, the Llanstadwell, Formation. In the succeeding formation within the Group abundant sedimentary deformation structures include load-casts, ball-and-pillow structures and convolute laminations.

### South of the Ritec Fault

Two major rock units occur south of the Ritec Fault (Fig. 6.6) – the Ridgeway Conglomerate Formation above, and the Milford Haven Group below. The lowest member of the Milford Haven Group, the highly conglomeratic Freshwater East Formation, rests upon an irregular surface of Wenlockian rocks and is probably local in origin. The Moor Cliffs formation comprises thick red and green mudstones, with abundant calcrete bands, correlatable in part to the 'Psammosteus' Limestones (House et al. 1977, pp. 44–47). Other lithologies present include minor fine-grained sandstones, intraformational conglomerates and air-fall tuffs. On top of the Moor Cliffs Formation the sequence is predominantly one of closely alternating red mudstones, sandstones and intraformational conglomerates in upward-fining cyclotherms. Vertebrate remains are locally not uncommon and bioturbation features, cross-bedding and other features occur in many of the sandier or coarser units.

The Ridgeway Conglomerate Formation (Williams 1971), of local alluvial fan origin, is made up of coarse polymictic conglomerates, coarse to fine sandstones and mudstones. A few calcrete layers occur but the bulk of the formation consists of massive conglomerates with poorly developed cross-bedding. It

rests with a hiatus of unknown significance upon the Milford Haven Group thus its age can only be postulated as post-Dittonian and pre-Farlovian.

Resting disconformably upon the Ridgeway Conglomerate Formation, the Skrinkle Sandstone Group constitutes a passage from Devonian into Carboniferous strata. The lower of its two formations, the Gupton, contains lenticular and tabular sandstones and conglomerates and has yielded scales similar to Holoptychius. Above this the West Angle Formation contains numerous conglomeratic units and calcretes followed by red and grey fining-upwards sequences and shales with rare beds rich in plant remains, calcretes and shelly limestone (Marshall 1978). Spores indicate a lower Carboniferous age for these shelly units.

The striking differences between the sequences to the north and south of the Ritec Fault indicate that different depositional regimes occurred in Lower Devonian times there. Pre-Carboniferous erosion of the Old Red Sandstone to the north of the fault suggests that uplift on that side occurred while deposition continued to the south.

## The concealed Devonian

Over thirty deep boreholes in the Midlands and SE England have penetrated Devonian strata (Fig. 6.9). Overlying rocks in contact with the Devonian range from Carboniferous to Cretaceous in age. A

**Fig. 6.9.** Devonian rocks in the subsurface of southern England (after Allen 1979). Boreholes are numbered on the map and section and the boxes indicate the identified intervals in each borehole; the unbracketed figures give the vertical interval in metres through which the rocks are known.

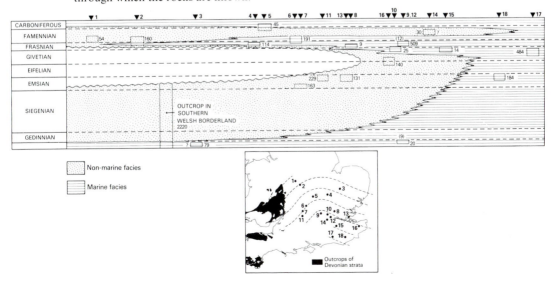

wide variety of lithologies similar to those known in the Old Red Sandstone outcrops occurs and in the boreholes some marine formations are found. Palaeontological evidence of their ages is restricted to spores, plant remains and vertebrate fragments with some information also being provided by acritarchs, chitinozoa and invertebrates. The Lower Old Red Sandstone/marine Devonian is here represented by variously coloured siltstone, marls and sandstones, and an equal number of Middle Upper Old Red Sandstone occurrences is known. A good account of these subsurface Devonian materials is given by Chaloner and Richardson (*in* House *et al.* 1977). While it is clear that Lower, Middle and Upper Devonian rocks are present, the geographical location of the boundaries of the Middle Devonian in the subsurface is uncertain and the distribution of the continental

facies is also problematic. Wills (1973, Pl. 1, fig. 1) postulated the feather edge of the Lower Old Red Sandstone/Devonian subcrop as running NE from north Oxfordshire to near Cambridge and Norwich and thence south to the Thames estuary and east Kent. The Middle Old Red Sandstone/Devonian, he suggests, lies south of a line from the Bristol Channel to Hampshire and thence NE to join the feather edge at Cambridge.

## South-west England

In this region (Fig. 6.10) 5,000 m or more of Devonian strata may be present but the relationships of the different environments of deposition to one another and to the Hercynian (Variscan) geosyncline of western Europe are still not entirely clear. The rock

**Fig. 6.10.** General map of the Devonian formations of the south-west of England. Post-Devonian formations unshaded.

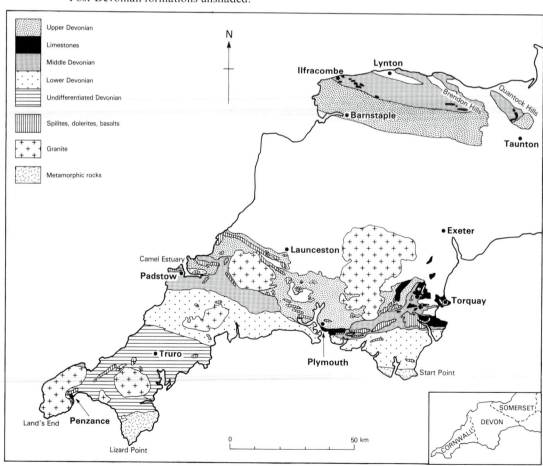

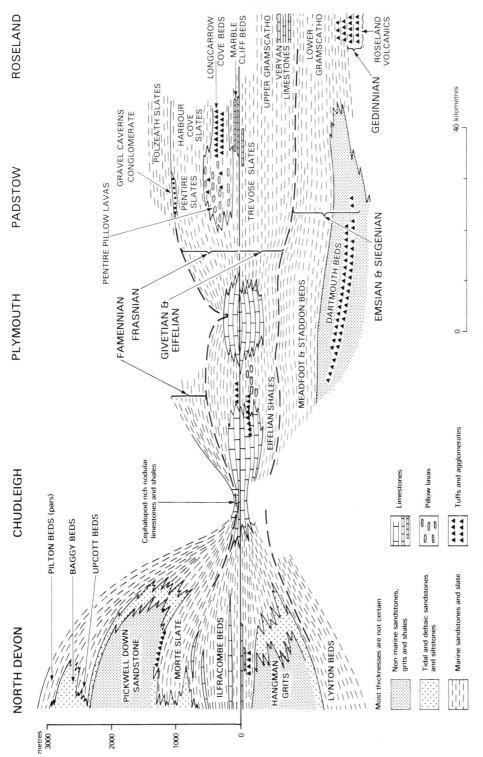

**Fig. 6.11.** Facies relationships of the Devonian formations in south-west England (after House 1975).

successions in north Devon, south Devon, north Cornwall, and south Cornwall differ significantly from one another and appear to have evolved under different tectonic conditions. Those of north Devon and west Somerset expose mainly members of the Middle and Upper Devonian Series, with the Lower Devonian represented only by part of the Lynton Beds. All these are of littoral, neritic and continental facies, to which no base is seen (Fig. 6.11). Throughout the rest of the peninsula marine facies dominate except for the Dartmouth Beds which in south Devon and east Cornwall are also of continental aspect. Volcanic units, virtually absent in north Devon are widespread and thick in the south of the county and in Cornwall.

In south Cornwall rocks of Ordovician and Silurian age are known (Sadler 1973a, b) and it is probable that a shallow shelf sea persisted throughout this region until late Silurian–Early Devonian times when the Old Red Sandstone continent rose to the north. Rocks at the Lizard in south Cornwall may represent part of the ?oceanic sub-Devonian floor (Ch. 2). At Start Point in south Devon the local schists are regarded as most probably altered Devonian rocks while at Dodman Point in Cornwall metamorphosed sub-greywackes and slates closely resemble the nearby Devonian Gramscatho beds.

## North Devon–west Somerset

The Devonian of this region, which includes Exmoor Forest, the Brendon Hills and the Quantock Hills, presents a succession of alternating marine and continental deposits (Fig. 6.12) and lies on the northern limb of the synclinorium of central Devon (Sellwood & Durrance 1982; see also Ch. 17). The overall succession is given in Table 6.1. These divisions are best exposed along the cliffs and shoreline (Fig. 6.13). Inland exposures are poor and difficult to interpret.

**Table 6.1.** Devonian succession in north Devon–west Somerset

| Pilton Shales | c. 400 m | Famennian to L. Dinantian |
|---|---|---|
| Baggy Sandstones | 400 m | Famennian |
| Upcott Beds | 250 m | Famennian |
| Pickwell Down Sandstones | c. 600–1,200 m | Famennian |
| Morte Slates | c. 600–1,500 m | Famennian |
| Ilfracombe Beds | c. 350–630 m | Givetian–Frasnian |
| Hangman Sandstone Group | c. 800 m | ?Eifelian |
| Lynton Beds | c. 200–400 m | Emsian |

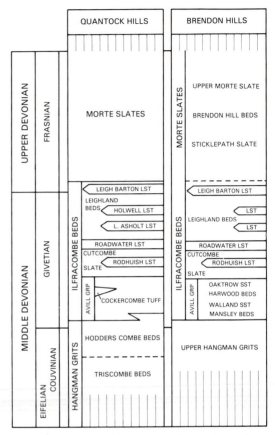

**Fig. 6.12.** Devonian successions in the Quantock and Brendon Hills, West Somerset (after House *et al.* 1977).

The Lynton Beds form the core of a faulted anticline, the axis of which strikes ESE through Lynton. They are mostly fine-grained, laminated and locally bioturbated sandstones and mudstones with thin shell beds, containing mostly brachiopods and bivalves. The Hangman Sandstone Group (Goldring *et al.* 1978; Tunbridge 1980, 1983; B.G.S. Sheets 277, 292, 295) is made up of coarser sediments than the formation below and shows both large and small sedimentary structures indicative of littoral environments. It is barren, except for rare plant remains such as *Psilophyton* sp.

Near the top of this group are cross-bedded sandstones coarsening upwards; mud-drapes and reactivation surfaces are known and a temporary estuarine or fan-delta environment is postulated. Above the Hangman Group the Ilfracombe Slates comprise slates with minor sandstones and limestones, indicating a return to fully marine conditions of

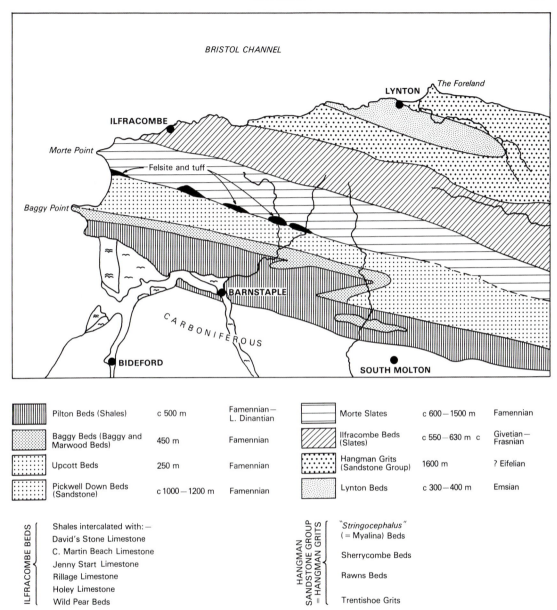

| | | | |
|---|---|---|---|
| Pilton Beds (Shales) | c 500 m | Famennian – L. Dinantian |
| Baggy Beds (Baggy and Marwood Beds) | 450 m | Famennian |
| Upcott Beds | 250 m | Famennian |
| Pickwell Down Beds (Sandstone) | c 1000 – 1200 m | Famennian |

| | | | |
|---|---|---|---|
| Morte Slates | c 600 – 1500 m | Famennian |
| Ilfracombe Beds (Slates) | c 550 – 630 m c | Givetian – Frasnian |
| Hangman Grits (Sandstone Group) | 1600 m | ? Eifelian |
| Lynton Beds | c 300 – 400 m | Emsian |

ILFRACOMBE BEDS
Shales intercalated with: —
David's Stone Limestone
C. Martin Beach Limestone
Jenny Start Limestone
Rillage Limestone
Holey Limestone
Wild Pear Beds

HANGMAN SANDSTONE GROUP = HANGMAN GRITS
"*Stringocephalus*" ( = Myalina) Beds
Sherrycombe Beds
Rawns Beds
Trentishoe Grits

**Fig. 6.13.** Stratigraphical succession and map of North Devon area (after Ussher 1907, Hambling 1910, Webby 1965 and Goldring *et al.* 1978).

deposition. Several authors have offered a stratigraphic subdivision of this unit (Evans 1922, 1929; Holwill 1962; B.G.S. Sheets 277, 292, 295) on the basis of two fossiliferous limestones within this otherwise clastic formation.

The Morte Slates are grey and purple slates with

rare thin sandstones and intraformational conglomerates. A few levels contain brachiopods and pectinacean and pterineid bivalves. The Morte Slates are believed to be the lowest formation within the Upper Devonian, the bulk of which is occupied by coarser sedimentary rocks. The Pickwell Down

Sandstones, though poorly exposed, appear to be made up of largely wavy, laminated and cross-laminated siltstones and sandstones. Rare intra-formational conglomerates occur and vertebrate fragments (?*Bothriolepis* sp.) have been found. Tuff bands have been recorded. Lithologically distinct, the unfossiliferous Upcott Beds are a pale mudstone–siltstone formation, perhaps representing a muddy littoral environment. The dark mudstones and fine sandstones of the Baggy Sandstones mark another phase of shallow-water sedimentation, with repeated shoreline advances producing sequences from sub-marine delta platform to fluvial deposition (Goldring 1971). Many beds contain abundant trace fossils

(Goldring 1962). At the top of the Baggy Sandstone in the coastal section a conspicuous slump of sand(stone) masses has intruded into mudstone. The uppermost Devonian formation, the Pilton Shales, are typically neritic siltstones and shales with coarser siltstone intercalations. Load structures are common in many of the beds, and there are abundant signs of pene-contemporaneous erosion. The formation contains an extensive marine fauna of brachiopods, bryozoans, gastropods, echinoderms and trilobites, but trace fossils are rare (Goldring 1970). The uppermost 250–350 m of the Pilton shales include the Devonian–Dinantian boundary and horizons with a fauna characteristic of the *Gattendorfia* Zone.

**Fig. 6.14.** Stratigraphical successions in South Devon (after House *et al.* 1977). Shaded areas represent volcanic rocks.

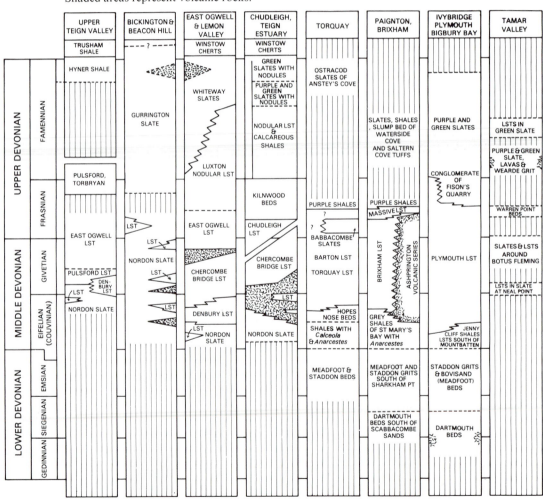

## South Devon

Here the Devonian rocks are not only different from those in the north of the county but also vary laterally along strike. Two areas where limestones form an important and fossiliferous part of the sequence are Torquay and Plymouth. In the intervening country the limestone units are greatly diminished in thickness or are absent, the succession being, however, largely of slates and volcanic rocks. Marginal reef facies have been recognised around Tor Bay and platform carbonate facies crop out in a broad belt inland (B.G.S. Sheet 339).

The region abounds in local stratigraphic units (Fig. 6.14) but a general sequence is shown in Table 6.2.

**Table 6.2.** Devonian succession of south Devon (Torbay)

| | |
|---|---|
| Slates and shales with nodules up to 1,000 m | Famennian–Frasnian |
| Slates with thin bedded or massive limestones up to 1,000 m | Frasnian |
| Massive limestones and subordinate volcanics and up to 800 m | |
|     Dartington Tuffs | Eifelian–Frasnian |
|     Includes at Torbay: | |
|         Barton Limestone | Givetian |
|         Walls Hill Limestone | early Givetian |
|         Daddyhole Limestone | Eifelian |
| Meadfoot Beds and Staddon Grits up to 450 m | Siegenian–Emsian |
| Dartmouth Beds (slates siltstones and volcanics) | Gedinnian–Siegenian |

No regional angular unconformities have been recognised within the succession, but there are signs of crustal unrest locally and the volcanic episodes presumably indicate instability on a wide scale. Local non-sequences and disconformities may be revealed as detailed work proceeds but post-depositional deformation may obscure such features.

The Dartmouth Beds (Dartmouth Slates), the oldest Devonian rocks of the region, comprise at least 700 m of reddish or dark sandstones, siltstones, slates and volcanic rocks and have yielded ostracoderm and other vertebrate remains (Dineley 1966). They also contain sedimentary structures and bedding which similarly indicate an origin akin to that of the Lower Old Red Sandstone of South Wales, though the coarse agglomerates and thick basaltic tuffs of the Dartmouth Beds are unknown to the north.

Overlying the Dartmouth Beds in south Devon and eastern Cornwall, the Meadfoot Beds form a succession of grey to buff, marine, slates, siltstones and subordinate sandstones. Local thin limestones of fossil detritus occur and the coarser clastic beds show many sedimentary features associated with vigorous current action in shallow waters. Trace fossils, too, are abundant in some siltstones and shales. Rhythmic alternations of fine and coarse sediments and graded bedding are also conspicuous (Richter 1967). The Staddon Grits are a coarser version of the same type of succession. Volcanic rocks such as agglomerates and tuffs within the Meadfoot division reach thicknesses of over 100 m in parts of south Devon and east Cornwall (B.G.S. Sheet 346).

The limestones of the middle part of the Devonian include several massive and also thin-bedded varieties (Scrutton 1977a, b). Many of the former contain abundant crinoid debris and coral and stromatoporoids in reefal associations, or masses with pockets of shelly fossils. The thin-bedded limestones are generally thought of as shallow-water carbonate bank deposits. The earliest limestones are dark, well-bedded and bear a rich fauna of mainly solitary and fasciculate corals. Large stromatoporoid colonies follow in more massive limestones. Thick basaltic ashes and lava flows, the Ashprington Volcanic Series and the

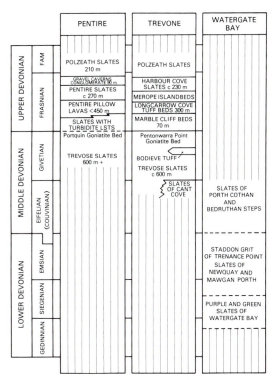

**Fig. 6.15.** Stratigraphic sections in north Cornwall (after House *et al.* 1977).

Dartington Tuffs, occur in this part of the succession. Richly fossiliferous bioclastic limestones and biohermal masses of limited extent complete the sequence.

Carbonate deposition continued into the Late Devonian when it gave way to argillaceous sedimentation, producing purple, green and grey shales or slates with thin nodular limestones. At Chipley, Famennian pillow lavas occur. While the Upper Devonian is undoubtedly several hundred metres thick throughout most of south Devon, in the Chudleigh area it is not only complete in terms of European zones but is extraordinarily thin in consequence of having been deposited over a sharp submarine rise (House & Butcher 1973, House 1975). The apparent thickness of the succession west and south of Dartmoor may be due to fold and nappe repetition (Dineley 1986).

## North Cornwall

The Devonian strata adjacent to the southern margin of the Culm (Ch. 7) outcrop from near Tintagel to Launceston and Dartmoor are highly deformed and difficult to interpret (B.G.S. Sheets 335 & 336) (see Ch. 17). They are almost entirely argillaceous rocks and volcanics. Only on the coast are exposures extensive enough to allow the structure and succession to be determined. Thicknesses are uncertain but broad stratigraphical units can be distinguished (Fig. 6.15), some with nodular and thin-bedded cherts, sandstone or limestone turbidites or limestone nodules.

The lowest rocks, the equivalents of the Dartmouth Beds, dark slates with vertebrate remains, can be traced westwards from Looe Bay in eastern Cornwall to Watergate Bay on the north Cornish coast. Staddon Grits follow and in turn give way to the Middle Devonian, an essentially argillaceous accumulation with very minor arenaceous units. It continues upwards into Upper Devonian and, eventually, Carboniferous formations without apparent stratigraphic break. Only within the Eifelian and higher levels are fossils locally plentiful and well preserved enough to establish the precise age and correlation of these formations. Lateral variation becomes conspicuous in the Upper Devonian for which two contrasting developments, the Trevone succession and the Pentire succession have been recognised. The Pentire succession includes some 450 m of pillow lavas and mud-flow conglomerates not found at Trevone (Gauss & House 1972). Both geological and palaeontological evidence suggests that in north Cornwall the Devonian succession is remarkably thinner than the rocks of equivalent age in north Devon about 60 km to the north. In central Devon a range of volcanic rocks of late Devonian age occurs amidst the slates of the Brent Tor and Tamar Valley area. Schrod and others

noted the late Devonian–early Carboniferous variations in south and west Devon and Cornwall and the possible 'exotic terrane' nature of much of the area (see Dineley 1986).

## South Cornwall

Between Perranporth on the north Cornish coast and the Lizard Complex on the south coast the Devonian rocks comprise a mass of inter-bedded greywackes and slates, with other lithologies in very minor proportions (Fig. 6.16). They are relatively unfossiliferous, lithologically monotonous and highly deformed (Fig. 6.17) (see Wilson & Taylor 1976).

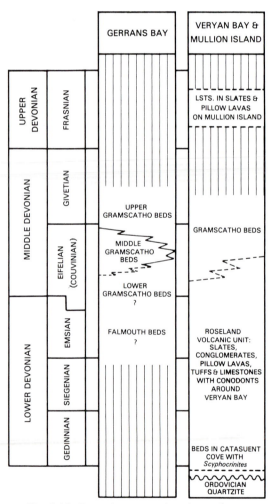

**Fig. 6.16.** Devonian successions in southern Cornwall (after House *et al.* 1977).

**Fig. 6.17.** Recumbent folds in Mylor slates, foreshore east of Marazion, Cornwall (B.G.S. A.13122).

Nevertheless, a general grouping has been adopted for many years, viz.:

Gramscatho Beds (about 900 m–1,200 m)
Mylor Beds (mudstones)
Veryan Series or Gidley Well Beds (about 370 m)

Near Land's End lower Devonian tholeitic pillow lavas and pyroclastics occur within slates of un-identified thickness.

Sadler (1973a, b) in the Roseland area suggested two major divisions, a lower Roseland Volcanics Unit and a higher Gramscatho Unit.

The Roseland Unit, 400 m thick, includes slates, volcanics, thin limestones and some conglomerates. It rests upon slates possibly of Silurian age. Ages determined on the basis of conodonts from within the Roseland Unit range from Gedinnian to Eifelian and underlying the ?Silurian slates are Llandeilo quartzites. The Gramscatho Unit, 1,200 m thick, includes coarse greywackes and slates and embraces the Veryan Series or Gidley Well Beds of previous usage. The limestones of the Veryan Series are of Eifelian age and at Gervans Bay they are 400 m thick and lie between 500 m and 300 m of greywackes below and above respectively.

The origin of these deposits, so different from those of the rest of the south-west Peninsula, seems to be related more to the Hercynian geosyncline than to the presence of the tectonically active landmass to the north.

Several radical re-interpretations of the structure of this and of the Peninsula conform to a model of exotic terrane moved from the south during the Carboniferous (Dineley 1986) (see also Ch. 17).

# Palaeogeography

The Devonian period was a time of major geographical change throughout the world, with attendant changes taking place in both animal and plant kingdoms. In Britain the Caledonide mountains perhaps reached their zenith, despite vigorous contemporaneous erosion. Their major extent and development were largely to the north and west of the areas in which the Devonian rocks discussed here were deposited, but they exerted a prime influence upon the accumulation of the stratigraphic succession.

At the same time the inception and evolution of the Cornubian–Rhenish geosyncline afforded the tectonic setting for the deposition of the marine Devonian formations in SW England. Plate tectonic theory appears to explain the formation of the Caledonian mountain-belt and its forelands but it has yet to provide a widely acceptable mechanism to account for the evolution of the Cornubian–Rhenish geosyncline and its subsequent orogenic deformation (see Dineley 1986).

The Anglo–Welsh area of Old Red Sandstone deposition (Fig. 6.18) was an external basin or cuvette, probably merging with the littoral zone in the vicinity of what is now north Devon but with a strand line that was subject to continuous and sometimes extensive shifts to north or south. There can be little doubt that highlands in the general area of Central–North Wales and the St George's Channel area shed detritus to the south and south-east throughout Lower Old Red Sandstone times. The sediment was transported largely by fluvial agencies to be deposited across a prograding alluvial plain with shifting, braided and meandering river systems. Water flow may have fluctuated vigorously, flooding was occasionally violent and widespread. Between the channels were widespread alluvial flats and local, short-lived lakes or pools periodically invaded by floodwaters or desiccated. During Downtonian time the marine environments that had existed in late Ludlovian time were progressively and swiftly pushed southwards as the sediment influx increased in response to the growing topographic relief to the north.

In this way the major upward-coarsening sequence of the Lower Old Red Sandstone was initiated. Towards the end of the Downtonian the basin of deposition extended from Dyfed to mid and northern Wales and across much of southern and Central England. Phases of relative stability are indicated by the presence of widespread calcrete deposits such as the '*Psammosteus*' Limestones. These were followed by more active fluvial deposition during which an increasing range of detrital minerals and rock fragments contributed to the sediment supply. The provenance of the vast quantities of mica and of garnet in the Lower Old Red Sandstone remain unknown. To the west tectonism increasingly influenced the deposition of these clastic sediments, as witnessed by the succession in SW Wales. The effect upon deposition of the Ritec Fault, a major tectonic element, is obvious and there may have been other such elements. Several authors have put forward evidence to suggest that an E–W uplift in the vicinity of the present Bristol Channel and S Wales provided much of the detritus for the late Lower Devonian–?Middle Devonian formation of South Wales and North Devon. Although not located in the subsurface, it may have been the source of much of the clastic mineral and rock input clearly not derived from the north (Dineley 1986, Tunbridge 1986).

Across North–Central England vigorous erosion of Caledonian structures and the deposition of red continental beds proceeded but the full extent of deposition here is unknown. Middle Devonian uplift may have removed much that had previously been deposited. Volcanic activity in the Cheviot area was on the margins of the volcanic province of the Midland Valley of Scotland.

To the south the passage to marine deposition lay beyond south Devon in Dittonian–Breconian times and marine volcanic activity additionally provided

**Fig. 6.18.** Stratigraphical summary of the Old Red Sandstone and marine Devonian rocks of south-west England, Wales and the Welsh Borderland (after Allen 1979).

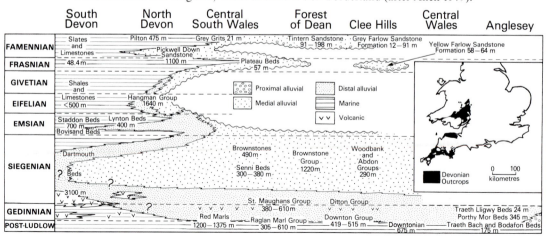

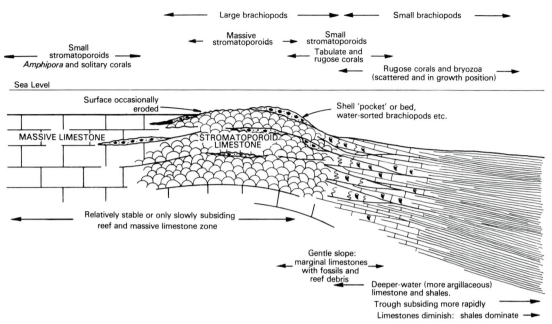

**Fig. 6.19.** Diagrammatic section through part of the mid-Givetian Tor Bay reef complex (after Dineley 1961).

large masses of debris. No sign of this latter, except possibly the tuff bands high in the Downtonian rocks, occurs, north of the Bristol Channel.

In the main area of continental deposition, stands of vegetation flourished along the margins of the streams, channels and lakes, and occasionally over wider areas. In these areas and in the more permanent bodies of water, arthropods and other animals were present;

bivalves and vertebrates were adapted to several such environments.

South of the zone of continental deposition, sandstones and mudstones were deposited in shallow water with estuarine to pro-delta environments extending beyond. Volcanic activity continued into Middle Devonian time in the south.

During the Middle Devonian epoch erosion and

**Fig. 6.20.** Hypothetical arrangement of facies belts across the sea floor of the south-west of England in Middle Devonian times. (Principally after C. T. Scrutton 1977 and Edwards *et al.* 1975.)

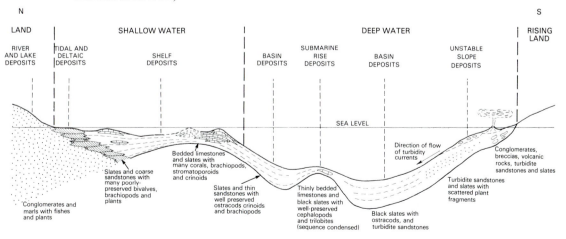

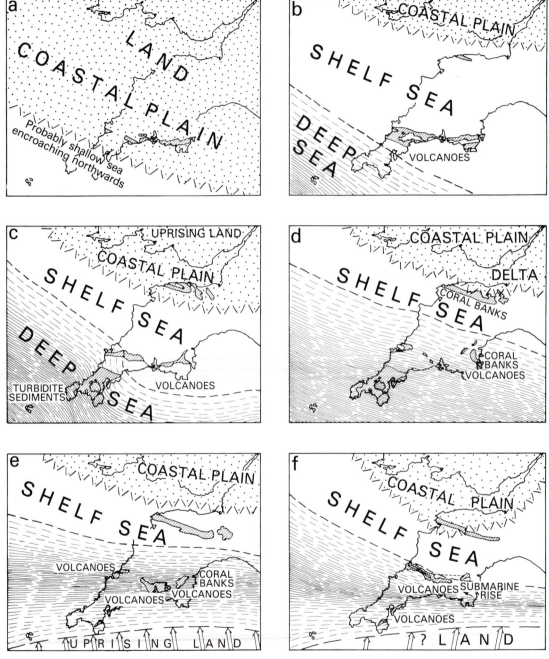

**Fig. 6.21.** Impressions of the Devonian palaeogeography of south-west England: a, Lower Devonian–Siegenian; b, Emsian; c, Middle Devonian–Eifelian; d, Givetian; e, Upper Devonian–Frasnian; f, Early Famennian (after Edmonds *et al.* 1975). Cross-shaded areas indicate present-day outcrops.

deposition north of the present Bristol Channel continued but no sedimentary record of this remains. In the north Devon–west Somerset area thick sandstone masses accumulated prior to the inception of a strong transgression northwards (Webby 1965, 1967 and see Dineley 1986).

In Eifelian times carbonate deposition began to extend across the south Devon area and, in the Givetian reef barriers of the carbonate platforms of the Torbay area formed (Fig. 6.19). These barriers finally subsided in the late Givetian but carbonate deposition persisted in the north, bioclastic limestone sheets being capped ultimately by small bioherms in mid–late Frasnian times. A similar carbonate and reef development took place in the Plymouth area upon a local shoal, but elsewhere shales and basaltic volcanic rocks (Richter 1965) were laid down (Figs. 6.20, 6.21). Farther south the deposition of the Roseland and Gramscatho rocks appears to have taken place in a different environment where greywacke deposition and volcanic episodes occurred. The coarseness of these clastic rocks and the unique character of the Gramscatho beds suggests an origin different from that of the rocks to the north. Several authors (e.g. Wilson & Taylor 1976, and see Anderton et al. 1979) have postulated a rising tectonic element to provide a southern sediment source.

The late Devonian Farlovian is represented by widely separated outcrops of continental sedimentary rocks in northern England, the English Midlands and Welsh Borderlands and South Wales. These units are of Famennian age and indicate a wide range of Old Red Sandstone environments. Deposition generally continued without break into the Carboniferous. In the north the Old Red Sandstone facies persisted into the early Carboniferous but south of the English Midlands there was a rapid passage into Carboniferous marine deposits.

During late Devonian times a second great wedge of sandstone penetrated southwards from the Bristol Channel area (Pickwell Down Sandstone) but by mid-Famennian time the sea once more transgressed northwards. South of this advancing strand line sands and silts were deposited across Devon and Cornwall, covering all the carbonate banks with fine silt or mud. In Cornwall the limestone turbidites of the Marble Cliff Beds appear to have been derived from a shallow water environment to the south-west (B.G.S. Sheets 335 & 336, Tucker 1969). Elsewhere siliceous muds (now cherts and slates) with carbonate nodules accumulated. Limestone deposition seems to have been restricted to a relatively few topographic 'highs' or *schwellen* which provided areas of more slowly subsiding sea floor. The Upper Devonian at Chudleigh demonstrates such a 'high' (Tucker & Van Straaten 1970, House & Butcher 1973). To the south-west volcanic activity once again was in spate.

The Devonian rocks of England and Wales may in summary be viewed as the infillings of sedimentary basins that ranged from the fault-bound continental (or internal) basins of the north to the epicontinental and geosynclinal (external) basins of the south-west. The later phases of the Caledonian earth movements influenced their geometries, infillings and duration while the initiation of Variscan sedimentation and earth movements can be seen in the south-west (*see e.g. Gutteridge et al. 1989*). As study continues, comparisons with Devonian stratigraphy and palaeontology in France, Belgium and Germany emphasise the strong palaeogeographical links with Western Europe (see Dineley 1986).

*Acknowledgements*

Dr E. J. Loeffler assisted in the preparation of this chapter; both she and Dr B. P. Williams kindly read the typescript and offered useful criticisms.

REFERENCES

ALLEN, J. R. L.          1965     Sedimentation and palaeogeography of the Old Red Sandstone of Anglesey, North Wales. *Proc. Yorkshire geol. Soc.*, **35**, 139–185.

                         1974     Sedimentology of the Old Red Sandstone (Siluro–Devonian) in the Clee Hills area, Shropshire, England. *Sediment. Geol.*, **12**, 73–167.

                         1979     Old Red Sandstone facies in external basins, with particular reference to southern Britain. *In* House, M. R., Scrutton, C. T. & Bassett, M. G. (Eds.) The Devonian System. *Special Papers in Palaeontology*, No. **23**, 65–80.

ALLEN, J. R. L.,           1981   Field Meeting: The facies of the Lower Old Red Sandstone
  THOMAS, R. G. &                north of Milford Haven, S.W. Dyfed, Wales. *Proc. Geol.*
  WILLIAMS, B. P. J.             *Assoc. London*, **92**, 251–267.

ALLEN, J. R. L. &          1978   The sequence of the earlier Lower Old Red Sandstone
  WILLIAMS, B. P. J.             (Siluro–Devonian), north of Milford Haven, Southwest
                                  Dyfed (Wales). *Geol. J.*, **13**(2), 113–136.

                           1981   Sedimentology and stratigraphy of the Townsend Tuff Bed
                                  (Lower Old Red Sandstone in South Wales and the Welsh
                                  Borders). *J. geol. Soc. London*, **138**, 15–29.

ANDERTON, R.,              1979   *A Dynamic Stratigraphy of the British Isles.* George Allen &
  BRIDGES, P. H.,                Unwin, London, 301 pp.
  LEEDER, M. R. &
  SELLWOOD, B. W.

BALL, H. W.,               1961   The Old Red Sandstone of Brown Clee Hill and the
  DINELEY, D. L. &               adjacent area. *Bull. Br. Mus. nat. Hist.*, **5**, 177–310.
  WHITE, E. I.

CAPEWELL, J. G.            1955   The Post-Silurian Pre-marine Carboniferous Sedimentary
                                  rocks of the eastern side of the English Lake District. *Q. J.
                                  geol. Soc. London*, **111**, 23–46.

CHALONER, W. G. &          1977   South-east England. *In* House, M. R. *et al.* (Eds.) A
  RICHARDSON, J. B.              correlation of Devonian rocks of the British Isles. *Spec.
                                  Rep. Geol. Soc. London*, **8**, 110 pp.

CROFT, W. N. &             1942   The Lower Devonian flora of the Senni Beds of Mon-
  LANG, W. H.                    mouthshire and Breconshire. *Phil. Trans. R. Soc. London*,
                                  **B231**, 131–163.

DINELEY, D. L.             1961   The Devonian system in south Devon. *Field Studies*, **1**,
                                  121–140.

                           1966   The Dartmouth Beds of Bigbury Bay, south Devon. *Quart.
                                  J. geol. Soc. London*, **122**, 187–217.

                           1986   Cornubian Quarter-century: Advances in the geology of
                                  south-west England. *Proc. Ussher Soc.*, **6**, 275–290.

DOLBY, G.                  1970   Spore assemblages from the Devonian–Carboniferous
                                  transition measures in southwest Britain and southern Eire.
                                  *Congr. Coll. Univ. Liège*, **55**, 269–275.

DURRANCE, E. M. &          1982   *The Geology of Devon.* Univ. of Exeter. 346 pp.
  LAMING, D. J. C. (Eds.)

EARP, J. R.                1940   The Geology of the southwestern part of the Clun Forest.
                                  *Q. J. geol. Soc. London*, **96**, 1–11.

EARP, J. R. &              1971   *The Welsh Borderland.* Brit. Reg. Geol., 3rd Edit.,
  HAINS, B. A.                   H.M.S.O., London.

EDMONDS, E. A.,            1975   *South-West England.* Brit. Reg. Geol., 4th Edit., H.M.S.O.,
  McKEOWN, M. C. &               London.
  WILLIAMS, M.

EDWARDS, D.                1979   A late Silurian flora from the Lower Old Red Sandstone of
                                  south-west Dyfed. *Palaeontology*, **22**, 23–52.

                           1980a  Early Land Floras. *In* Panchen, A. L. (Ed.) *The Terrestrial
                                  Environment and the Origin of Land Vertebrates.* Academic
                                  Press, London, 55–86.

                           1980b  The early history of land plants based on late Silurian and
                                  Lower Devonian floras of the British Isles. *In* Harris, A. L.,
                                  Holland, C. H. & Leake, B. E. (Eds.) The Caledonides of
                                  the British Isles – reviewed. *Sp. Publ. geol. Soc. London*, **8**,
                                  pp. 405–410.

EDWARDS, D. &              1974   Lower Devonian (Dittonian) plants from the Welsh
  RICHARDSON, J. B.              Borderland. *Palaeontology*, **17**, 311–324.

EVANS, J. W.        1922      The Geological Structure of the country around Combe Martin, North Devon. *Proc. Geol. Assoc. London*, **33**, 201–228.

           1929      Devonian. A – Sedimentary Rocks. *In* Evans, J. W. & Stubblefield, C. J. (Eds.), *Handbook of the Geology of Great Britain*, Murby, London, 136.

FLOYD, P. A.        1981      The Hercynian trough; Devonian and Carboniferous volcanism in South West England. *In* Sutherland, D. S. (Ed.) *Igneous Rocks in the British Isles*, 227–242. John Wiley & Sons Ltd.

FRIEND, P. F. &        1978      *A Field Guide to selected outcrop areas of the Devonian of*
   WILLIAMS, B. P. J. (Eds.)            *Scotland, the Welsh Borderland and South Wales* (for the International Symposium on the Devonian System). Palaeontological Association, London.

GAUSS, G. A. &        1972      The Devonian successions in the Padstow area, North
   HOUSE, M. R.            Cornwall. *J. geol. Soc. London*, **128**, 151–172.

GEORGE, T. N.        1970      *South Wales*. Brit. Reg. Geol., 3rd Edit., H.M.S.O., London.

GEORGE, T. N. *et al.*        1969      Recommendations on stratigraphical usage. *Proc. geol. Soc. London*, No. **1656**, 139–166.

GOLDRING, R.        1962      The bathyal lull; Upper Devonian and Lower Carboniferous sedimentation in the Variscan geosyncline. *In* Coe, K. (Ed.) *Some aspects of the Variscan Fold Belt*. University Press, Manchester, 75–91.

           1970      The stratigraphy about the Devonian–Carboniferous boundary in the Barnstable area of North Devon, England. *C.R. 6ᵉᵐᵉ Congr. Internat. Strat. Geol. Carbonif.*, Sheffield 1967, 807–816.

           1971      Shallow water sedimentation as illustrated in the Upper Devonian Baggy Beds. *Mem. geol. Soc. London*, **5**, 1–80.

GOLDRING, R.,        1978      2, North Devon. *In* Scrutton, C. T. (Ed.) *A field guide to*
   TUNBRIDGE, I. P.,            *selected areas of the Devonian of south-west England*. The
   WHITTAKER, A. &            Palaeontological Association, 18–27.
   WILLIAMS, B. J.

GUTTERIDGE, P.,        1989      *The role of tectonics in Devonian and Carboniferous*
   ARTHURTON, R. S. &            *sedimentation in the British Isles*. Yorks. geol. Soc. Occ.
   NOLAN, S. C. (Eds)            Pub., **6**.

HAMLING, J. G. &        1910      Excursion to north Devon, Easter 1910. *Proc. Geol. Assoc.*
   ROGERS, I.            *London*, **21**, 457–472.

HOLLAND, C. H.        1959      The Ludlovian and Downtonian rocks of the Knighton district, Radnorshire. *Q. J. geol. Soc. London*, **114**, 449–478.

HOLWILL, R. W. J.        1962      The succession of limestones within the Ilfracombe Beds (Devonian) of North Devon. *Proc. Geol. Assoc. London*, **73**, 281–293.

HOUSE, M. R.        1975      Facies and Time in Devonian Tropical Areas. *Proc. Yorkshire geol. Soc.*, **40**, 233–288.

HOUSE, M. R. &        1973      Excavations in the Devonian and Carboniferous rocks near
   BUTCHER, N. E.            Chudleigh, South Devon. *Trans. Roy. geol. Soc. Cornwall*, **20**, 199–220.

HOUSE, M. R. &        1985      Devonian Series Boundaries in Britain. *Cour. Forsch.-Inst.*
   DINELEY, D. L.            *Senckenberg*, **75**, 301–310.

HOUSE, M. R.        1977      A correlation of Devonian rocks of the British Isles. *Spec.*
   AND FIVE OTHERS            *Rep. geol. Soc. London*, **8**, 110 pp.

HOUSE, M. R. &          1966    Palaeozoic palaeontology in Devon and Cornwall. *In*
  SELWOOD, E. B.                 Present views of some aspects of the geology of Cornwall
                                 and Devon. Hosking, K. F. G. & Shrimpton, G. J. (Eds.)
                                 *Roy. Geol. Soc. Cornwall* (Anniversary Vol. for 1964),
                                 48–66.

LANG, W. H.             1937    On the Plant-remains from the Downtonian of England
                                 and Wales. *Phil. Trans. Roy. Soc. London*, B, 227,
                                 245–291.

LEEDER, M. G.           1973    Sedimentology and palaeogeography of the Upper Old
                                 Red Sandstone in the Scottish Border basin. *Scott. J. Geol.*,
                                 **9**, 117–144.

LOEFFLER, E. J. &       1980    A new pteraspid ostracoderm from the Devonian Senni
  THOMAS, R. G.                  Beds Formation of South Wales, and its stratigraphic
                                 significance. *Palaeontology*, London, **23**, 287–296.

MARSHALL, J. D.         1978    *Sedimentology of the Skrinkle Sandstone Group (Devonian–
                                 Carboniferous), South-west Dyfed.* Unpublished Ph.D.
                                 thesis, University of Bristol.

MITCHELL, J. G.         1972    Potassium–Argon ages from the Cheviot Hills, Northern
                                 England. *Geol. Mag.*, **109**, 421–426.

PALMER, D. C.           1970    A stratigraphical synopsis of the Long Mountain, Mont-
                                 gomeryshire. *Proc. geol. Soc. London*, No. **1660**, 341–346.

PAPROTH, E. &           1985    In Search of a Devonian–Carboniferous Boundary.
  STREEL, M.                     *Episodes*, **8**, 110–111.

RICHARDSON, J. B.       1967    Some British Lower Devonian spore assemblages and their
                                 stratigraphic significance. *Rev. Palaeobot. Palynol.*, **1**,
                                 11–29.

RICHARDSON, J. B. &     1969    Upper Silurian and Lower Devonian spore assemblages
  LISTER, T. R.                  from the Welsh Borderland and South Wales. *Palaeon-
                                 tology*, London, **12**, 201–252.

RICHARDSON, J. B. &     1979    Palynological evidence for the age and provenance of the
  RASUL, S. M.                   Lower Old Red Sandstone from the Apley Barn Borehole,
                                 Witney, Oxfordshire. *Proc. Geol. Assoc. London*, **90**, 27–42.

RICHTER, D.             1965    Stratigraphy, igneous rocks and structural development of
                                 the Torquay area. *Trans. Devon. Assoc.*, **97**, 57–70.

                        1967    Sedimentology and facies of the Meadfoot Beds (Lower
                                 Devonian) in south-east Devon (England). *Geol. Rundsch.*,
                                 **56**, 543–561.

SADLER, P. M.           1973a   *A proposed stratigraphical succession for the Roseland area
                                 of South Cornwall.* Unpublished Ph.D. thesis, University of
                                 Bristol.

                        1973b   An interpretation of new stratigraphic evidence from south
                                 Cornwall. *Proc. Ussher Soc.*, **2**, 535–550.

SCRUTTON, C. T.         1977a   Reef facies in the Devonian of eastern South Devon,
                                 England. *Men. Bur. Rech. geol. minieres.*, **89**, 125–135.

                        1977b   Facies variations in the Devonian limestones of eastern
                                 South Devon. *Geol. Mag.*, **114**, 165–193.

SCRUTTON, C. T. (Ed.)   1978    *A field guide to selected areas of the Devonian of south-west
                                 England.* The Palaeontological Association, 73 pp.

SEDGWICK, A. &          1839    On the classification of the older rocks of Devon and
  MURCHISON, R. I.              Cornwall. *Proc. geol. Soc. London*, **3**, 121–123.

SELWOOD, E. B. &        1982    Chapter Two – The Devonian Rocks. *In* Durrance, E. E. &
  DURRANCE, E. M.              Laming, D. J. L. (Eds.) *The Geology of Devon.* University of
                                 Exeter, pp. 15–41.

SIMPSON, S.             1951    Some solved and unsolved problems of the stratigraphy of
                                 the marine Devonian in Great Britain. *Abh. Senckenb.
                                 Naturf. Ges.*, **485**, 53–66.

| SMITH, T. E. | 1967 | A preliminary study of sandstone sedimentation in the Carboniferous of Berwickshire. *Scott. J. Geol.*, **3**, 282–305. |

THOMAS, R. G. | 1978 | *The Stratigraphy, Palynology and Sedimentology of the Lower Old Red Sandstone, Cosheston Group, S.W. Dyfed, Wales.* Unpublished Ph.D. thesis, University of Bristol.

TUCKER, M. E. | 1969 | Crinoidal turbidites from the Devonian of Cornwall and their palaeogeographical significance. *Sedimentology*, **13**, 281–290.

TUCKER, M. E. & VAN STRAATEN, P. | 1970 | Conodonts and facies on the Chudleigh Schwelle. *Proc. Ussher Soc.*, **2**, 160–170.

TUNBRIDGE, I. P. | 1977 | Notes on the Hangman Sandstones (Middle Devonian) of north Devon. *Proc. Ussher Soc.*, **3**, 339.

| 1980 | The Yes Tor Member of the Hangman Sandstone Group (North Devon). *Proc. Ussher Soc.*, **5**, 7–12.

| 1983 | The Middle Devonian shoreline of North Devon, England. *J. geol. Soc. London*, **140**, 147–158.

| 1986 | Mid-Devonian tectonics and sedimentation in the Bristol Channel area. *J. geol. Soc. London*, **143**, 107–116.

TUNBRIDGE, I. P. & WHITTAKER, A. | 1978 | 2. North Devon: 2a. Lower and Middle Devonian. *In* Scrutton, C. T. (Ed.) A field guide to selected areas of the Devonian of South-West England. *Palaeont. Assoc. London*, 8–13.

TURNER, S. | 1973 | Siluro–Devonian Thelodonts from the Welsh Borderland. *J. geol. Soc. London*, **129**, 557–584.

USSHER, W. A. E. | 1907 | The geology of the country around Plymouth and Liskeard. *Mem. geol. Surv. Gt. Br.*

WALMSLEY, V. G. & BASSETT, M. G. | 1976 | Biostratigraphic correlation of the Coralliferous Group and Gray Sandstone Group (Silurian) of Pembrokeshire, Wales. *Proc. Geol. Assoc. London*, **87**, 191–220.

WEBBY, B. D. | 1965 | The Middle Devonian marine transgression in north Devon and west Somerset. *Geol. Mag.*, **102**, 478–488.

| 1966 | Middle Devonian Palaeogeography of North Devon and West Somerset, England. *Palaeogeogr. Palaeoclimatol. Palaeoecol.*, **2**, 27–46.

WHITE, E. I. | 1950 | The vertebrate faunas of the Lower Old Red Sandstone of the Welsh Borders. *Bull. Br. Mus. nat. Hist.*, **A1**, 57–67.

| 1956 | Preliminary note on the range of pteraspids in western Europe. *Bull. Inst. Roy. Sci. nat. Belg.*, **32(10)**, 1–10.

WILLIAMS, B. P. J. | 1971 | Sedimentary features of the Old Red Sandstone and Lower Limestone Shales of South Pembrokeshire, South of the Ritec Fault. *In* Bassett, D. M. & M. G. (Eds.) *Geological Excursions in south Wales and the Forest of Dean.* Geol. Assoc. (South Wales Group), Cardiff, 222–239.

| 1978 | The Old Red Sandstone of the Welsh Borderland and South Wales. *In* Williams, B. P. J. & Friend, P. F. (Eds.) *A field guide to selected outcrop areas of the Devonian of Scotland, the Welsh Borderland and South Wales.* The Palaeontological Association, 55–106.

WILLS, L. J. | 1973 | A palaeogeographic map of the Palaeozoic floor below the Permian and Mesozoic formations in England and Wales. *Mem. geol. Soc. London*, **7**, 23 pp.

WILSON, N. A. C. & TAYLOR, R. T. | 1976 | Stratigraphy and Sedimentation in West Cornwall. *Trans. Roy. geol. Soc. Cornwall*, **20**, 246–259.

ZIEGLER, W. & KLAPPER, G. | 1985 | Stages of the Devonian System. *Episodes*, **8**, 104–109. |

# 7

# DINANTIAN

# M. R. Leeder

## Introduction

Dinantian outcrops cover approximately 12 per cent of the 'solid' land area of England and Wales. Within these outcrops and also in the extensive pre-Permian subcrop may be found lime mudbank, 'Cementstone', 'Yoredale Cycle', platform limestone, braided river sand-sheet, limestone turbidite and radiolarian chert facies. This incomplete list is remarkable in its diversity of depositional themes. No less remarkable would be an accompanying list of faunal and floral communities (e.g. see Ramsbottom *in* McKerrow 1978).

The Dinantian shows *par excellence*, the effects of 'tilt-blocks' and 'half-grabens' upon sedimentary facies and their thickness. Early recognition of basin/block relations from stratigraphic mapping (e.g. Trotter & Hollingworth 1932) and from other 1930's wartime and postwar oil company geophysical surveys (Lees & Cox 1937, Lees & Taitt 1946, Falcon & Kent 1960, Kent 1966) was followed by a number of detailed gravity surveys by Bott and co-workers in northern England (e.g. Bott & Masson Smith 1957; Bott 1961, 1967, 1974; Bott *et al.* 1978). These surveys established that some block sequences were underlain by unexposed Caledonian 'Newer Granite' plutons (Fig. 7.1) whose crustal mass deficiency was evidently accompanied by a buoyant tendency in Dinantian times. The surveys also confirmed earlier stratigraphic studies which deduced that blocks were separated from adjacent sedimentary basins by hinge-lines and normal faults active during sedimentation. The basins were not thus simply filled-in hollows inherited from erosional regimes of Devonian age. More recently it has become clear (e.g. Miller & Grayson 1982, Leeder 1982, 1987a, Smith *et al.* 1985, Gawthorpe 1987, Gutteridge 1987, *Besly & Kelling 1988 and Gutteridge et al. (Eds) 1989)* that the predominant subsidence style is that of the tilt-block of half-graben, as seen in many Mesozoic and Tertiary extensional basin systems.

The tensional stress regime of the Dinantian upper crust was accompanied by scattered alkaline basalt eruptions (Ch. 16) though the most important eruptive centres lay in Scotland, in the Midland Valley and Scottish Borders (Francis 1967, 1970, 1978; Leeder 1974b). The origin of Dinantian basins has been attributed to the effects of a phase of lithospheric stretching (Figs. 7.1A, 7.1B) induced by Liguro-Bretonic subduction in Brittany and central France (Leeder 1982, 1987a).

North of the positive massif known as St George's Land the early Dinantian basins were gulf-like (Fig. 7.2) in form and developed distinctive facies and faunal associations indicative of deposition and colonisation in geographically restricted, tideless, hypersaline gulfs with a variable input of river drainage. These basins generally show thick (1·5–3 km), fairly complete Dinantian successions. Tilt-block highs and horsts in northern England, by way of contrast, show thinned successions of late Dinantian age, usually dominated by thick, open marine platform carbonate facies showing minor cyclicity with frequent evidence of exposure (Walkden 1987) and deltaic clastic intercalations.

To the south of St George's Land lay the South Wales/Mendips and 'Culm' basins (see Figs. 7.9, 7.13). The former basin has a fill dominated by carbonate-ramp facies up to 1·5 km thick. The 'Culm' basin contains a dominantly fine grained pelagic succession (but important distal flysch) a few hundred metres thick at most. The junction of the two basins cannot be recognised because of a cover of Permo-Trias in the Bristol Channel and by outcrops of Devonian strata lying above a suspected northerly-directed Exmoor thrust (Brooks & Thomson 1973) in north Devon and Somerset. As we shall discuss below the Bristol Channel Landmass, formerly active in the Middle Devonian, became active once more in the late Dinan-

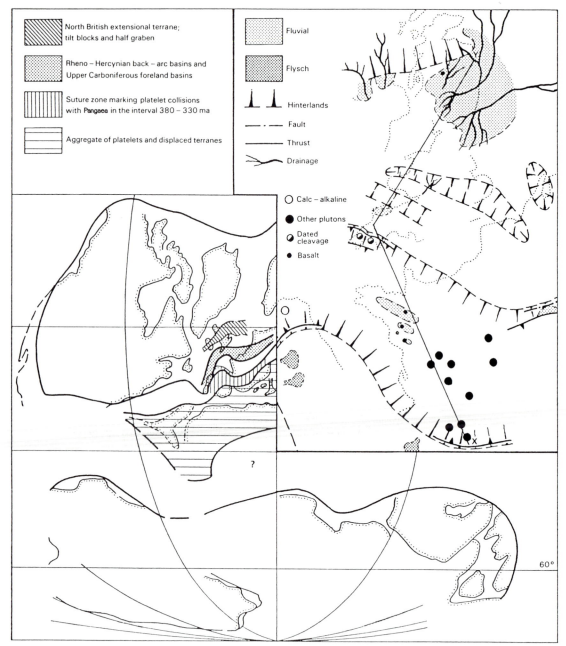

**Fig. 7.1a.** World and regional Dinantian tectonic setting (after Leeder 1987; Leeder and McMahon 1988).

tian before its major palaeogeographic 'comeback' as a thrust-related feature in the Silesian (Ch. 8) when it effectively separated the S. Wales basin from the 'Culm' basin.

## Stratigraphic classification

Since the Dinantian is a subsystem of the Carboniferous then the Tournaisian and Visean must be series of

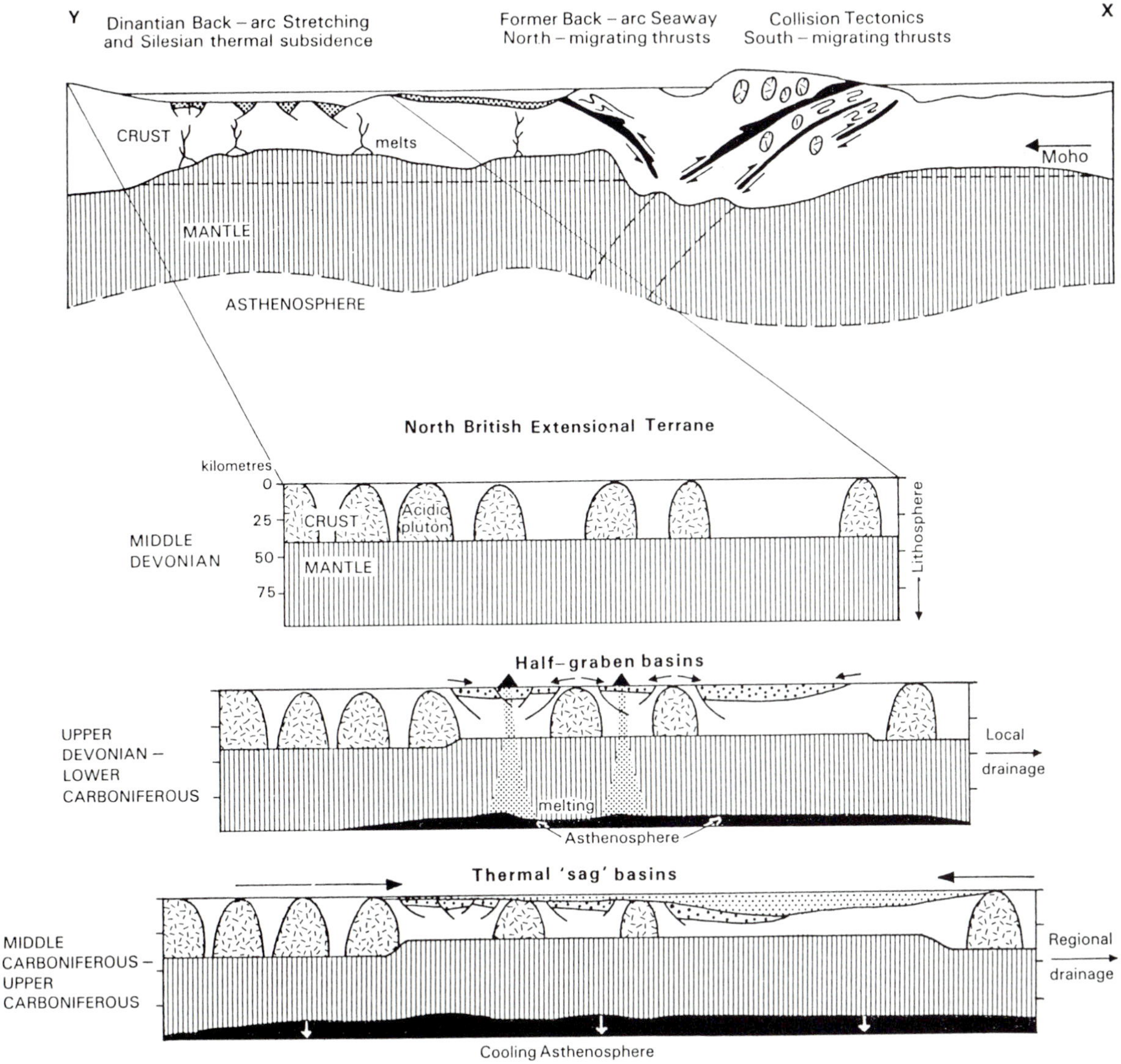

Dinantian Back – arc Stretching
and Silesian thermal subsidence

Former Back – arc Seaway
North – migrating thrusts

Collision Tectonics
South – migrating thrusts

**Fig. 7.1b.** Section along the line X-Y of Figure 7.1a to show the extensional setting of Anglo-Welsh Dinantian half-graben basins (after Leeder 1982; Leeder and McMahon 1988).

the subsystem (George & Wagner 1972). George *et al.* (1976) argued that since the lowest part of the type Tournaisian in Belgium was of Fammenian age, the base of the Tournaisian could not correspond to the base of the Dinantian and so the Tournaisian could not be used as a division of the Dinantian. This problem has recently been overcome following the redefinition of the base of the Tournaisian in Belgium (Conil *et al.* 1977) which now corresponds to the base of the Dinantian in Britain (Ramsbottom & Mitchell

1980). Additional problems arise concerning the definition of the base to the Visean in Britain and the chaotic state of the various 'Vaughnian' zonal schemes adopted in different areas. These difficulties were partly surmounted by the establishment of a set of six regional stages (Table 7.1) for the Dinantian of the British Isles (George *et al.* 1976). These chronostratigraphic subdivisions were based upon stratotype stage boundaries. The base of each stage is defined, the top being then automatically determined by the base of the

| CORAL BRACHIOPOD ZONES 1973 SMITH et al 1967 (P₂–D₁) GREEN + WELCH 1965 (S₂–K) | | CONODONT ZONES RHODES et al 1969 and AUSTIN 1979 | MIOSPORE ZONES NEVES et al 1972 and CLAYTON et al 1974 | GONIATITE ZONES PRENTICE & THOMAS 1965 (Devon & Cornwall) | | MAJOR CYCLES RAMSBOTTOM 1973 | STAGES | BELGIUM | U.S.A. |
|---|---|---|---|---|---|---|---|---|---|
| Upper Posidonia (P₂) Zone | VISEAN | Gnathodus girtyi collinsoni / Gnathodus monodosus | NC (pars) / VF | P₂ c/b/a | IIIY | | | | Hombergian (CHESTER) |
| Upper Dibunophyllum (D₂) Zone | | Mestognathus beckmanni / Gnathodus bilineatus | (Tripartites vetustus — Rotaspora fracta) | P₁ d/c/b/a | IIIβ | 6 group | BRIGANTIAN | V3c | Gasperian / St Genevieve |
| Lower Dibunophyllum (D₂) Zone | | Cavusgnathus—Apatognathus (pars) | NM (Raistrickia nigra Traquitrites marginatus) / TC (Perotrilites tesseliatus Schulzospora campyloptera) | B₂ / B₁ | IIIα / IIδ | 5 group | ASBIAN | V3b | St Louis |
| Seminula (S₂) Zone | | Cavusgnathus—Apatognathus (pars) | PU | | | 4 | HOLKERIAN | V3a / V2b | |
| Upper Caninia (C₂ S₁) Zone (pars) | | Cavusgnathus—Apatognathus (pars) | (Lycospora pusilla) | Pe₃ | IIγ | 3 | ARUNDIAN | V2a / V1b | Salem / Warsaw |
| Upper Caninia (C₂ S₁) Zone (pars) | | no conodonts / Mestognathus beckmanni — Polygnathus bischoffi | | | | 2 | CHADIAN | V1a | Keokuk |
| Lower Caninia (YC₁) Zone — Upper Fauna / Middle Fauna | TOURNAISIAN | Gnathodus antetexanus—Polygnathus lacinatus / Polygnathus lacinatus Pseudopolygnathus longiposticus | CM (Schopfites clavigar Avroraspora macra) | Pe₂ / Pe₁ | IIβ / IIα | | IVORIAN | Tn3 | Burlington / Fern Glen / Meppen |
| Zaphrentis (Z) Zone (= Lower Fauna) | | Spathognathodus costatus costatus / Gnathodus delicatus | PC | Ga | | 1 | | Tn2 | Chouteau |
| Cleistopora (K) Zone | | Spathognathodus cf S.robustus S.tridentatus / Siphonodella—Polygnathus inornatus / Patrognathus variabilis—S.plumulus | VI (Vallatisporites valatus Retusotriletes incohatus) | | Iα | | HASTARIAN | Tn 1b | Hannibal |
| O.R.S. Facies | | | | | | | | | Glen Park |

**Table 7.1.** Comparison of Dinantian zoning schemes advanced in recent years, with the Dinantian stages proposed by George et al. 1976 (after George et al. 1976), as modified by Ramsbottom & Mitchell 1980. (K, Z, C, S, D in column 1 refer to Vaughan's (1905) zones).

overlying stage. The base of most of the stages is taken at the first change in lithology occurring below the entry of a particular faunal group. The Courceyan stage is known to be exactly equivalent to the redefined Tournaisian (Ramsbottom & Mitchell 1980) but is now replaced by the Hastarian and Ivorian stages.

The original Vaughnian Zones (Vaughan 1905) applied only to the Bristol area (Table 7.1). Although Vaughan regarded his zones as firmly based upon evolutionary lineages they are now regarded as assemblage biozones strongly influenced by facies type. Difficulties of correlation of the Avon Gorge zones north of St George's Land led Garwood (1907,

1913) to define a number of biozones based upon faunal marker bands in the Ravenstonedale area. However, these zones were only of use within the limits of occurrence of the marker bands and were not traceable outside the Stainmore Trough.

In recent years microfossil groups (Table 7.1) have proved of very great value in intra-Dinantian correlation, particularly conodonts (Rhodes et al. 1969, Austin 1973, Metcalfe 1976, 1980), miospores (Neves et al. 1972, 1973) and foraminifera (Cummings 1961, George et al. 1976). Miospores are valuable zonal indicators in Scotland and the Northumberland basin where thick non-marine successions dominate and where specialised quasi-marine macrofaunas occur in the marine facies. Future detailed application of the Belgian foraminiferal zonal schemes (Conil & Lys 1968) to the British marine carbonate successions should lead to a considerable increase in the precision of correlation. Although uncommon over much of the Dinantian shelf, goniatites from the Culm basin may be correlated with the German zonal schemes (Prentice & Thomas 1960, Mathews 1971).

In 1973 Ramsbottom proposed that six major cyclothems of transgression could be recognised in England and Wales. These major cyclothems are approximately coincident with the new Dinantian stages noted above since it was Ramsbottom's thesis that each major transgressional episode was eustatic and accompanied by new migratory faunas. George (1979), however, argued that the recognition of such cycles is impossible in areas subject to differential vertical crustal movements such as the Anglo-Welsh basins in Dinantian times. Recent sedimentological studies (Barraclough 1983) have shown there is little support for significant regional facies changes at these supposed cycle boundaries in northern England.

# Regional descriptions

## Northern gulfs and basins

North of the positive massif known as St George's Land a number of half-grabens were initiated in the early Dinantian. Some of these basins were gulf-like (Fig. 7.2) in morphology during Tournaisian to Arundian times and developed distinctive facies and faunal associations indicative of deposition and colonisation in geographically restricted, tideless, hypersaline gulfs with a variable input of river drainage. These hanging-wall portions of the basins generally show thick (1.5–3 km), fairly complete Dinantian successions (Table 7.2).

## Northumberland/Solway basin

The structural limits to this basin (Fig. 7.2) are defined on the south by marked increases of sediment thickness from the Alston horst over the Stublick/90 Fathom Faults (Trotter & Hollingworth 1932, Johnson 1960). To the north, on the Scottish side of

Table 7.2. Stratigraphic successions in northern England and north Wales (after George et al. 1976; Mitchell 1978). Shading denotes hiatus.

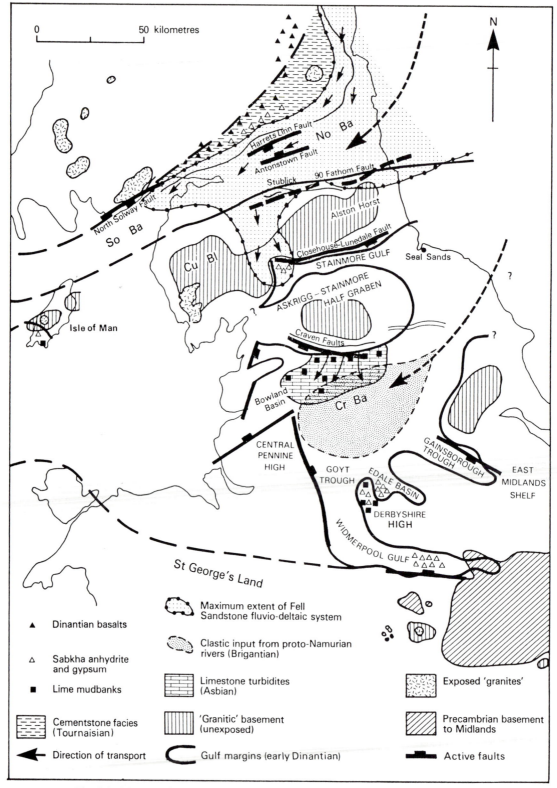

**Fig. 7.2.** Map to show generalised palaeogeographic elements in northern/central England together with the distribution of some important lithofacies groups. (Note that the map does not imply a static distribution of Dinantian lithofacies). Cu Bl – Cumbrian Block; Cr Ba – Craven Basin; No Ba – Northumberland Basin; So Ba – Solway Basin. Data from many sources (see text).

the border, the North Solway Fault (Deegan 1973, Leeder 1974a, Ord *et al.* in press), active during deposition, and the Border Volcanic Line (Leeder 1974b, 1976a) define suitable structural margins, although the latter had little effect upon Dinantian sedimentation. The main basin is thus best viewed as a complex half-graben with a faulted southern margin. It should be noted, however, that these southern bounding-structures seen today are both Permian or younger (Leeder 1987b). They are inferred to overlie presumed syn-depositional Dinantian structures present at depth. BGS mapping (Day 1970, Frost & Holliday 1980) has revealed an axial inner graben (Fig. 7.2) superimposed upon this half-graben in which very great thicknesses of clastic sediments accumulated particularly in Asbian times. This area shows, in fact, the thickest sequences of British Dinantian strata.

Dinantian events began in the mid-Tournaisian with the eruption of the alkaline basaltic Birrenswark/Kelso Lavas (5–150 m) over the defunct and palaeosol-encrusted alluvial plains of the upper Old Red Sandstone (Leeder 1973b, 1974b, 1976a). The lavas are dated at 360 Ma by K-Ar techniques (de Souza 1975). Although the lavas now outcrop as a narrow strip along the Scottish–English Border and show elongate NE–SW isopachs (Leeder 1974b) they probably have a larger subcrop than previously thought. Thus miospore zoning of clastic sediments overlying and underlying the Cockermouth Lavas in Cumbria (Butcher *in* Mitchell 1978) reveals a Tournaisian age and raises the question of whether the two lava groups are continuous under the western part of the basin. These early Dinantian lavas are thought to be due to lithospheric extension via a cycle of mantle partial-melting, crustal upwarp and lava eruption. These events caused half-graben collapse, enabling Tournaisian marine transgression and establishment of new axial drainage systems (Leeder 1974a, 1974b, 1976a, 1982).

Early Dinantian sedimentation was dominated by influxes of mature clastic detritus from Southern Uplands sourcelands, deposited in river courses, alluvial fans and fluvio-deltaic locii along the northern basin margin (Deegan 1973, Leeder 1974a). A widespread Tournaisian development were the poorly fossiliferous 'Cementstone' facies (Fig. 7.2), used in the sense of Bell *et al.* (1967), comprising clastic fining-upwards cycles of river channel origin and clastic coarsening-upwards cycles of lake-fill origin with dolomitised, clay-rich cementstone limestones in their basal parts (Scott 1971, 1986; Leeder 1974a; Anderton 1985). Desiccated lake and floodplain facies include pseudomorphs after halite, gypsum and anhydrite. Rootlet beds and very thin coals also occur. The Cementstone environment was a startling juxtaposition of hypersaline lakes and freshwater river channels

on an arid coastal plain. It represents an almost perfect arid-zone analogue to the Coal Measures coastal plains (*contra* Shearman 1966) and is an example of a truly schizohaline (*sensu* Folk & Siedlecka 1977) environment.

The late Tournaisian and early Chadian Lower Border Group of Bewcastle and Liddesdale, together with correlatives in the Cheviot foothills, and Northumberland, shows numerous Yoredale-type 'marine' limestone/deltaic clastic cycles. The carbonate members contain a variety of peritidal stromatolite growth forms and vermetid gastropod (formerly thought to be serpulid) bioherms and biostromes (Leeder 1973a, 1975a, 1975b; Burchette & Riding 1977). Quasi-marine shelly faunas with abundant *Dictyoclostus teres* (rare outside the basin) occur locally, particularly in western areas. Foraminifera, conodonts and colonial corals are very rare, indicating that the gulf was separated from marine Dinantian areas by some ecological barrier, possibly greater than normal salinity. Palaeocurrent evidence and facies analysis shows that high-constructive lobate deltas periodically advanced along the basin axis fan from NE to SW (Leeder 1974a) to form the cycles. The diachronous spread (Fig. 7.2) of a major braided fluvial system along this axis is recorded in the Middle Border Group Fell Sandstone and its correlatives from Berwickshire to the Solway Firth (Robson 1956, Hodgson 1978). The resulting multistorey sandbody, an important aquifer in Northumberland (Hodgson & Gardiner 1971), splits and intercalates with marine facies to the south-west (Day 1970).

The basal Asbian transgression is represented by the Clattering Band of Bewcastle and its correlatives (Ramsbottom 1973) with their distinctive fauna of *Lithostrotion martini*, *L. portlocki* and the gigantoproductid *Semiplanus*. The Upper Border Group is again composed of Yoredale-type cycles but in the central 'mid-graben' bounded by the Antonstown/Harretts Linn (Fig. 7.2) hinge lines an abnormal thickness (c. 2 km) of dominantly deltaic clastic facies is found (Day 1970, Frost & Holliday 1980), again derived from a NE source (Leeder 1987b). In Berwickshire and Northumberland the Asbian Scremerston Coal Group is largely composed of deltaic facies with numerous thick coals. An homogenous regime of Yoredale-type cyclicity was established over the whole basin by mid-Asbian times, with the transgressive Dun Limestone of Berwickshire finding correlatives with the transgressive Melmerby Scar Limestone of the Alston Block indicating that at this time the gulf-like morphology of the Northumberland/Solway basin was no more. Brigantian sedimentation, although similar in both block and basin, still shows some thickness differences across the Stublick/90 Fathom fault lines (Johnson 1960, 1967; Holliday *et al.* 1979)

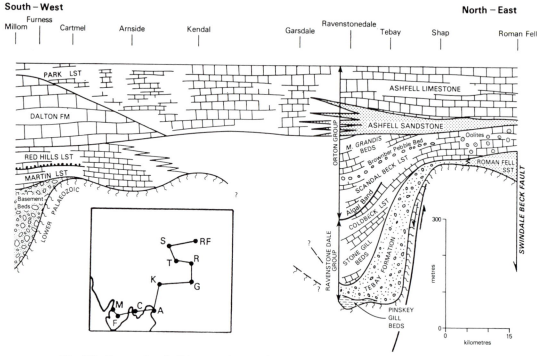

**Fig. 7.3.** Extrapolated ribbon section to show stratigraphy of the Stainmore basin/
Furness area (after Ashton 1970; Mitchell 1978).

though the evidence of a thickened 'basin-type' succession in the Harton Borehole (Ridd *et al.* 1970) *south* of the 90 fathom fault indicates that the bounding hinge-lines were probably gradational, stepped features rather than sharp single lines.

## Askrigg–Stainmore half-graben

The limits to this basin and its inferred connection to the Craven basin via North Yorkshire have been determined largely by detailed geophysical surveys (Bott 1967). The northern and southern margins are drawn along structural lines, the Closehouse–Lunedale–Swindale Beck faults (Fig. 7.2) to the north and the Stockdale Monocline to the south, which coincide with steep gravity gradients. The northern margin was undoubtedly fairly abrupt, as proven by stratigraphic studies (Burgess & Harrison 1967, Burgess & Mitchell 1976) and the very thick sequence (3100 m) revealed by the Seal Sands well on Teesside. The southern margin seems to have been transitional, particularly from Chadian times onwards when the basin form approximated to a half-graben (Burgess & Mitchell 1976). Johnson (1967) postulated that the basin was blind-ended on its western margin but it remains possible that there was connection to

Furness/Grange in Cumbria (Fig. 7.3) where the successions show similarity to those of Ravenstonedale at the head of the main basin (Table 7.2).

The main outcrops around Ravenstonedale (Garwood 1913; Turner 1950, 1959; Ashton 1970; Mitchell 1978; Holliday *et al.* 1979, Barraclough 1983) provide a more-or-less complete Dinantian succession, mainly in carbonate facies (Fig. 7.3). The Basement Beds are locally derived coarse clastic successions which rest unconformably upon frequently red-stained and sometimes calcified cornstones (of soil origin?) Lower Palaeozoic basement (Ashton 1970, Holliday *et al.* 1979). Rapid local thickness changes and facies variations in the Basement Beds may be related to local valley-fill processes and to movements upon contemporaneous growth faults such as the Swindale Beck Fault (Burgess & Harrison 1967, Burgess & Holliday 1979). At Ravenstonedale the nearshore-gulf Pinskey Gill beds rest directly upon Lower Palaeozoic beds. They yield Tournaisian spores (Johnson & Marshall 1971) and conodonts (Varker & Higgins 1979) and are *overlain* by coarse clastic deposits of alluvial fan aspects. Above, the Stone Gill Beds of the Ravenstonedale Group contain limestones and thin mudstones of a variety of 'gulf' facies (Ashton 1970) including the predominant dolomitised, restricted-biota micrites containing calcareous algal nodules and clasts, algal mat stromatolites, ostracods, vermetid gastropods and *Syringopora*. Calcretes (Barraclough 1983), birds-eye desiccation structures, pseudomorphs after halite, gypsum and celestite and quartz-replaced evaporite nodules with traces of original anhydrite all occur (Ashton 1970, Holliday *et al.* 1979). These features indicate deposition in a variety of nearshore to peritidal, quiet water, gulf margin environments which find modern parallels in Bahamian tidal flats (Barraclough 1983).

Towards the top of the Ravenstonedale Group and in the lower Arundian Scandal Beck Limestone of the Orton Group, unrestricted-biota micrites and intra-clast-rich biosparites become common (Ashton 1970, Barraclough 1983). These contain 'normal' shallow water shelly faunas, including abundant *Thysano-phyllum pseudovermiculare* and other corals in the Scandal Beck Limestone. Agitated-water oolitic and pelletal limestones are commoner in the thinned sequences of the Shap area, towards the presumed northern shoreline of the gulf. The Ashfell Sandstone is a complex fluvio-deltaic incursion of Fell Sandstone facies, probably derived from the extension of a series of delta lobes originating in the Northumberland gulf and which travelled south and south-westwards (Fig. 7.2) down the Eden/Shap 'low'. Final delta abandonment was followed by the Holkerian marine transgression which deposited a 'blanket' suite (Fig. 7.3) of well-washed offshore bioclastic carbonates with *Litho-strotion minus* and *Davidsonina carbonaria*. These facies continue up into the Asbian/Brigantian Alston Group where they are the chief 'scar' formers in the area. As in the Northumberland basin the combined effects of transgression and block subsidence destroyed the Stainmore trough as the geomorphic expression of a structural basin.

In the Grange/Furness area (Nicholas 1967, Ashton 1970, Ramsbottom 1973, Rose & Dunham 1978, Mitchell 1978) the Chadian Martin Limestone overlies a variable thickness of Tournaisian Basement Beds and contains facies and faunas very similar to the Stone Gill member of the Ravenstonedale Group. The overlying Arundian Red Hill Oolite, with a fauna including *Michelinia megastroma, Palaeosmilia mur-chisoni* and *Stenocisma isorhyncha*, is actually a pelletal limestone. It was deposited after a hiatus involving lithification and pedogenesis (Adams & Cossey 1981) of the topmost Martin Limestone. There is no straightforward correlation here between this event and the Brownber Pebble Bed 'regression' of Ravenstonedale (Ramsbottom 1973) since ar-chaeodiscid foraminifera occur below the 're-gressional' horizon in Ravenstonedale but above it at Furness/Grange (George *et al.* 1976). Above, there is an upward transition to the dark, well-bedded

skeletal wackestones with clastic mud partings of the Arundian Dalton Beds with the brachiopod *Delepinea carinata* and a rich coral fauna and then a transition to the Holkerian Park Limestone, a monotonous series of pale grey skeletal grainstones equivalent to the 'blanket' Ashfell Limestones of Ravenstonedale (Fig. 7.3).

## Craven, Bowland and E. Midland basins

The Craven basin (Figs. 7.2, 7.4a, 7.5, 7.7) is defined to the north by facies changes from the Askrigg Block over the late-Dinantian 'reef' belt (Hudson 1930, 1932, 1933; Bond 1950a, 1950b, 1950c, Black 1950, 1954, 1958; Parkinson 1957) associated with the Craven fault system. The precise location of this boundary in the early Dinantian is not known and must await detailed geophysical surveys and deep boreholes. This northern basin margin may only be approximately traced westwards across Lancashire into the Irish Sea to the southern Isle of Man where a somewhat attenuated basinal sequence, some 350 m thick outcrops (Dickson *pers. comm.* 1979). Eastwards, gravity surveys (Kent 1967) and a few deep boreholes reveal a complex of apparently blind-ended half-grabens of gulf form bounded by active syndepositional faults (e.g. Normanton, Morley-Campsall). To the south the Derbyshire block (Fig. 7.2) forms a marked promontory of the Midland Block, with the linear E–W Widmerpool Gulf separating the Derbyshire Block from the northern margin of St George's Land (Kent 1966, Llewellyn & Stabbins 1970). The southern basin margin must have lain north of the present outcrop in North Wales in early Dinantian times but late-Dinantian successions indicate rapid subsidence in the area and an extension of the subsidiary basin south-wards over St George's Land. The approximate boundaries to the Craven basin (Fig. 7.4) are likely to be considerably modified as and when data from oil company geophysical surveys and deep boreholes become available.

In the Clitheroe–Skipton area (Figs. 7.4, 7.5) of the Bowland Basin the successions are dominated by clastic mudstones, argillaceous limestones and the spectacular lime mudbank complexes whose tops reached well into the photic zone. Macrofaunas are of only limited use in basinal correlation. Recent con-odont studies (Metcalfe 1976) have established good correlations and have solved several outstanding problems (Fig. 7.5). In particular it is now known that numerous postulated infra-Dinantian unconformities (Hudson & Mitchell 1937, Hudson 1944, Parkinson 1944, Hudson & Dunnington 1945), previously thought due to compressional stresses within the basin, cannot be upheld. The evidence for uncon-formity has been attributed to (1) lateral facies changes

216

(Metcalfe 1976), (2) development of turbiditic limestones with carbonate intraformational debris and (3) the effects of disharmonic folding (Mosley 1962).

By way of contrast to the above deductions, recent studies (Gawthorpe & Clemmey 1985; Gawthorpe 1987) in the Bowland area reveal angular unconformities within the Clitheroe Limestone

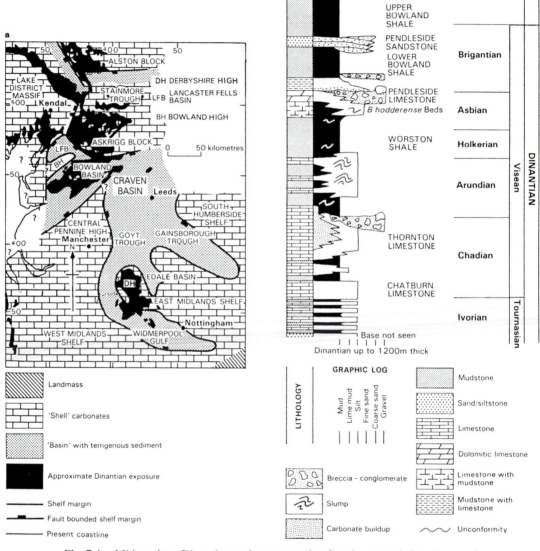

**Fig. 7.4a.** Mid to late Dinantian palaeogeography for the central Pennines and surrounding areas showing the location and orientation of shelf-to-basin transitions. The strong NW–SE and NE–SW trend to these transitions suggests underlying structural control is related to transfer and extensional fault systems within the basement (after Gawthorpe 1987).

**Fig. 7.4b.** Log shows the succession of the Bowland Basin showing the main intervals of carbonate, terrigenous mud and sand deposition. Also shown are the positions of major limestone breccia-conglomerate debris flows. Waulsortian carbonate buildups, unconformities and sedimentary slump and slide structures (all after Gawthorpe 1987).

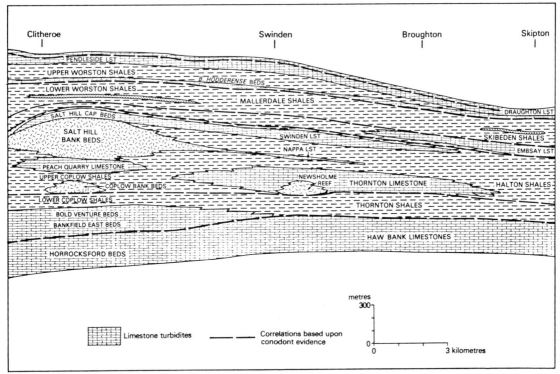

**Fig. 7.5.** Extrapolated ribbon section from Skipton to Clitheroe in Craven basin (after Metcalfe 1976). Dashed lines indicate condont zone boundaries.

Group, major synsedimentary slumped horizons in the Worston Series and spectacular local developments of debris flows, turbidites and foundered slabs within the Thornton and Pendleside Limestone Group (Fig. 7.4). These features all testify to extensive synsedimentary tectonic movements. The Visean sediments were clearly tilted prior to deposition of the Bowland Shales, producing a mappable unconformity between the two. The foundered blocks, slide masses and debris flows of the Pendleside Limestone, the thickness variations within the Bowland Shale Group and the bands of limestone debris in the lower part of the Bowland Shales all record the effects of tilt-block extensional tectonics on sedimentation (Gawthorpe & Clemmey 1985; Gawthorpe 1987).

The Chadian lime-mudbanks of the Clitheroe area (Miller & Grayson 1972) contain micritic core, flank and marginal mound facies rich in crinoid debris. High sedimentary dips (30°) are indicated by geopetal cavities. Superimposed upon original relief are the effects of a prominent emersion and erosive interval which caused intra-mudbank depressions to be cut and filled by boulder beds from adjacent areas. Similar mudbanks, with additional apron forms, occur along the Craven fault belt in the Cracoe area where they mark a transition from the Great Scar carbonate platform of the Askrigg area into the Craven basin. Mundy (*in* McKerrow 1978) has defined a number of 'reef' communities, usually dominated by brachiopods, from these mudbanks. The shallow water algal community, forming in the topmost mound areas, consisted of a calcareous algal framework colonised by an encrusting fauna of bryozoans, sponges, corals and brachiopods. Evidence from both the Clitheroe and Cracoe mudbanks indicates that the bases of the mudbanks may have been as much as 150 m below the photic zone.

Major palaeogeographic episodes in the basin are indicated by the limestone turbidites and spectacularly coarse debris-flows of the Embsay and Draughton/Pendleside Limestones (Fig. 7.5). These extensive deposits cannot simply be reef–talus gravity deposits but must indicate massive sediment slumps from broad uplift and fault dissection of the Cracoe 'reef' belt to the north. Local unconformities and boulder beds are also known in the topmost Dinantian of the Settle area (Garwood & Goodyear 1924) and in the Isle of Man (J. A. D. Dickson pers. comm. 1979).

The abundant clastic mud deposited in the Dinantian of the Craven basin must have bypassed the Askrigg carbonate platform to the north (Fig. 7.2). A clue to the origin of the mud is provided by a single sharp-based sandstone in the Holkerian/Asbian Skibeden Shales of the Skipton area (Metcalfe 1976). This sandstone is of turbiditic origin and represents a lone precursor to the late Dinantian (Pendleside sandstones) and early Namurian clastic turbidite fans that were soon to prograde into the basin from the NE, sourced from the mouths of the major 'Millstone Grit' rivers (see Ch. 8).

Geophysical and deep borehole evidence indicates that the gulfs of the English Midlands all lead off from the Craven basin, but little is known about their stratigraphical evolution. Boreholes in the Edale (Dunham 1973) and Widmerpool (Llewellyn & Stabbins 1970) gulfs both reveal early Dinantian sabkha anhydrite facies, showing a 'restricted gulf' pedigree at this time. Later Dinantian facies in many boreholes comprise mudstone and argillaceous limestone facies similar to the main basin outcrops to the NW.

## Northern and Midland tilt-block 'highs'

### Alston–Cumbrian and Askrigg 'highs'

These positive, granite-based blocks (Fig. 7.2) formed low-lying land areas undergoing weathering during early Dinantian times (Dunham et al. 1965, Johnson 1967). The southern Askrigg area was gradually transgressed over during the early Arundian, followed by the remaining block areas in the early Holkerian (Fig. 7.6). Initially, thick marine platform carbonate sediments accumulated on the blocks. These now make up the Great Scar (Askrigg) and Melmerby Scar (Alston) limestones. These examples are predominantly well-washed, biogenic carbonates often highly bioturbated, with major crinoid banks, shelly/coral biostromes and algal (*Girvanella*) bands in places. Eight major bedding surfaces may be mapped over large areas in the Kingsdale and Knipe Scar Limestones in the upper part of the Great Scar Limestone Group (Schwarzacher 1958, Doughty 1974). These penecontemporaneously potholed and scalloped bedding surfaces, covered by thin mud-

**Fig. 7.6.** Flat-lying Holkerian-age limestone unconformable on Silurian Horton Flags, Combes Quarry, near Helwith Bridge, Horton-in-Ribblesdale, Yorkshire (B.G.S. L.599).

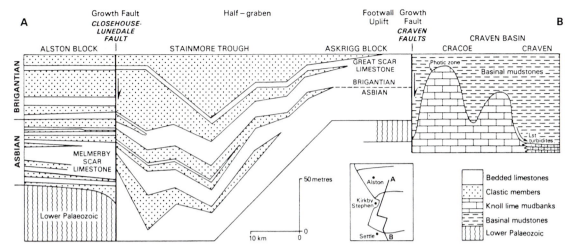

**Fig. 7.7.** Extrapolated ribbon section from Alston Block to Craven Basin to show Asbian/Brigantian facies changes, particularly in the Yoredale cycles (after Burgess & Mitchell 1976; Ramsbottom 1974).

stones in subcrops (Waltham 1971), are probably emersion surfaces subject to karstification, and early diagenetic cementation. They record 'minor', perhaps eustatically controlled, cyclicity on the blocks (Ramsbottom 1973) and find possible correlatives in both Asbian and Brigantian carbonate successions in Derbyshire, North Wales and Bristol/Mendips (see review by Walkden 1987).

The Brigantian stage is marked all over the block areas and in basin areas also, by the periodic progradation, abandonment and retreat of delta lobes (Moore 1958, 1959). The marine carbonate/deltaic clastic cycles thus produced (4–9 in number) are termed 'Yoredale' cyclothems from their type locality in the valley of the River Ure on the Askrigg block. We should, however, note that essentially similar cycles comprise more-or-less the whole Dinantian succession in the central Northumberland basin (Day 1970, Leeder 1974a). Traced from blocks to adjacent basins the Brigantian cycles thicken, usually more in the clastic members (Fig. 7.7). Traced southwards over the northward-tilted Askrigg block the clastic members of the cycles thin rapidly into carbonate/clastic mudstone alterations (Ramsbottom 1974). There is evidence here that the delta distributaries (with low feldspar-abundancies) that deposited the Yoredale clastic members must have been separated from the major proto-Millstone Grit fluvial system (rich in feldspar) that effectively by-passed the northern English blocks to deposit fine clastic facies and turbidites into the Craven Basin (Fig. 7.2) during Brigantian times. Despite many years of study there is still no convincing hypothesis to account for the periodicity of delta lobe

advance in Yoredale-type cycles, although tectonic (Bott & Johnson 1967), eustatic (Ramsbottom 1973, 1977; Walkden 1987) and sedimentary (Moore 1959, Allen 1960, Duff *et al.* 1967) controls have all been proposed in recent years. A recent model invoked a combined tectonic/sedimentary theory (Leeder & Strudwick 1987).

## Derbyshire 'high'

Outcrops of block character in Derbyshire are of Asbian to Brigantian age, but deep boreholes at Woo Dale (Cope 1973) and sections on the block/basin margin in the Manifold Valley reveal lithologies as old as Chadian (George *et al.* 1976). The exposed block-type sequences are entirely of shallow water carbonate types (Wolfenden 1958, Sadler 1966, Butcher & Ford 1973), rich in brachiopod/coral bioceonoses, with spectacular examples of knoll and apron lime mudbanks in the Castleton, Matlock and Dovedale areas (Wolfenden 1958, Smith *et al.* 1967, Stevenson & Gaunt 1971, Broadhurst & Simpson 1973), some of which show block-to-basin transitions. Smith *et al.* (1985) and Gutteridge (1987) have recently interpreted such facies variations as being due to the development of extensional faulting and half-graben structures. The Castleton mudbanks show important algal contributions to bank growth (Wolfenden 1958, Broadhurst & Simpson 1973) and evidence in the form of fissuring and boulder-beds for post-mudbank karstification, and erosion (Simpson & Broadhurst 1969).

The shelf platform carbonates show numerous emersion surfaces, with hummocky relief and potholes, discovered by Walkden (1972, 1974, 1977; see also Bridges 1982). The surfaces are usually covered by thin clay 'wayboards' of volcanically derived dusts. Walkden (1977) demonstrated that major erosive features at the Asbian/Brigantian boundary, hitherto regarded as channels, are attributable to karstic solution during a lengthy period of subaerial exposure of the northern part of the block platform. The emergent surface was covered in some areas by a subaerial lava flow. Other thin basaltic lava flows and tuffs, not all necessarily subaerial, occur in the Matlock and Castleton areas (Smith *et al.* 1967, Stevenson & Gaunt 1971).

## Northern fringes of St George's Land

Geophysical evidence (Kent 1966), together with the attenuated and condensed sequences seen at Breedon Cloud, Leicestershire (Mitchell & Stubble-field 1941), Little Wenlock, Shropshire (Pocock *et al.* 1938) and in boreholes serve to define roughly the northern boundary of St George's Land (George 1958, Llewellyn & Stabbins 1970). To the west, outcrops in North Wales are of Asbian/Brigantian age and suggest a gradational block/basin boundary running north-westwards between Oswestry and Llangollen to Anglesey in late-Dinantian times. These outcrops indicate an important accelerated subsidence phase in the late-Dinantian, as witnessed by the accumulation of almost 1 km of shallow water bioclastic limestones at Prestatyn (George 1958). Well-developed 'minor' cycles of Schwarzacher (1958)/Walkden (1984) type are seen in the Asbian limestones of Llangollen and their correlatives at Colwyn Bay and Anglesey (Power & Somerville 1975, Somerville 1979). In his detailed study around Llangollen, Somerville recognised 9–10 shoaling-upwards cycles with emergent tops showing palaeokarstic relief, calcite laminated calcite crusts

**Fig. 7.8.** Avon Gorge looking northwards from Clifton Suspension Bridge. Massive limestone (Arundian-Holkerian) forms cliff on right-hand side and lies above southwards-dipping Avon Thrust. Wooded area beyond cliff consists of younger (Asbian-Brigantian) contorted mudstones, sandstones and limestones. The older massive limestones appear again below the Thrust (centre) while vertical cliff in distance is composed of crinoidal limestones (also Arundian-Holkerian in age). (B.G.S. A.9763).

and associated K-bentonites similar to Walkden's (1972, 1974) clay 'wayboards' of Derbyshire. Perhaps the most spectacular palaeokarstic surface in the British Dinantian are seen at Lligwy Bay, Anglesey where deep (2–3 m) abraded potholes are infilled by continental clastic facies, the fill now exhumed and eroded into surrealistic pedestals on the modern foreshore (Walkden 1983).

## East Midlands Shelf

Over 140 wells have penetrated the Dinantian subcrop of the East Midlands. Strank (1987), in an important study, established the biostratigraphy of these wells. Tournaisian strata have not been penetrated in many wells but seem to have a wide distribution, though absent over local tilt-block highs. Widespread volcanics have been recorded on geophysical logs. Thick Chadian successions with build-ups have been penetrated in many areas, particularly around Lincoln (> 500 m of shallow water limestones). This constrasts with only 46 m recorded in the area of the Stixwold/Nocton highs. Arundian strata are also markedly thinned along the Askern–Grantham axis. Holkerian sediments are the first to transgress the Nocton, Foston and Stixwold Highs. The South Humberside 'shelf' was transgressed in Asbian times when extensive carbonate sedimentation also occurred in most areas. Thick 'basinal' Asbian sequences are recorded in the Edale Gulf. Brigantian times were marked by uplift and erosion of the area east of Lincoln along the Askern, Spittal, Stixwold and Nocton Highs but slow subsidence occurred over the South Humberside Shelf.

## Southern basins

### Bristol–Mendips area

The classic Avon Gorge section (Fig. 7.8) (Vaughan 1905, Kellaway & Welch 1955) is now known to be incomplete (Table 7.3), with at least four non-sequences (Mitchell 1972, Ramsbottom 1973). A full Dinantian sequence is seen in the Mendips to the south. There is a general trend in these basins (Figs. 7.9, 7.10, 7.11) for N–S facies changes from shallow northern carbonate facies with intercalactions of continental clastics to offshore southern carbonate facies (Kellaway & Welch 1955, Ramsbottom 1977). These may be properly deduced to have been deposited on a southwards-prograding carbonate ramp (Fig. 7.12), following the major early Dinantian ?eustatic sea level rise (Wright 1987). Plotting time-rock units on stratigraphic sections (Fig. 7.10) shows clearly the increasing duration and extent of hiatuses in northern areas close to the emergent flanks of St George's Land (Ramsbottom 1977, fig. 2).

The Famennian Old Sandstone facies (Ch. 6) pass transitionally upwards into the Tournaisian Lower Limestone Shales, the Devono-Carboniferous boundary being defined by the first appearance of Dinantian spores within ORS facies (Dolby & Neves 1970, Utting & Neves 1970). The Lower Limestone Shales comprise alternating detrital limestone members and thick fine-grained clastics interpreted as shelf-basin facies (Burchette 1987). The limestones are lenticular, poorly sorted, cross-stratified and show sharp, sometimes scoured, bases (Butler 1972, 1973; Butler et al. 1972). They predominantly contain crinoidal debris, often haematised, phosphate nodules and fish remains. Some units are stromatolitic with pseudomorphs after gypsum, desiccation cracks and 'birdseyes'. Bioherms and biostromes of vermetid gastropods are also common. These features indicate carbonate deposition in a spectrum of peritidal and shallow subtidal environments, including carbonate barriers. Three depositional cycles may be recognised within the Lower Limestone Shale Group (Burchette 1987). Each represents the evolution and termination of a barrier-lagoon complex. The earliest barrier was reworked by shoreface retreat, the second represents an oolite tidal-delta, the third developed on the second as an unstable oolite barrier spit. Barriers 1 & 2 were preserved during successive transgressions. Thick time-equivalent units to these transgressive/regressive cycles comprise shelf basin/ramp clastic mudrocks with storm beds.

There is an upward passage from the Lower Limestone Shales to the Black Rock Limestone. The dark, large-scale cross stratified limestones at the base still show muddy flasers (Mathews et al. 1973). These muddy limestones pass upwards into well-washed grainstones and fine-grained carbonates. The fine-grained carbonate debris becomes dominant upwards causing dark wackestones to develop (Butler 1972). This upward trend was probably caused by progressive water deepening. Extensive cherts in the middle of the unit are the lateral equivalents to the tuffs and lavas that comprise the Middlehope Volcanics seen in the western Mendips (Butler 1973). The topmost Black Rock Limestone reverts to a shallow, well-washed facies once more, with abundant corals and benthonic foraminifera.

The top of the Black Rock Limestone is extensively dolomitised. The dolomitic development (the laminosa dolomite) thickens northwards until in the Chepstow/Forest of Dean (Fig. 7.10) area the entire thickness of carbonates between the Lower Limestone Shales and the Crease Limestone (≡ Gully Oolite, see below) is dolomitised. In addition to this northward thickening of the dolomite member there is increasing disconformity at the top of the member. Thus the upper coral fauna of the topmost 120 m of the Limestone in the Mendips is absent in the Avon Gorge

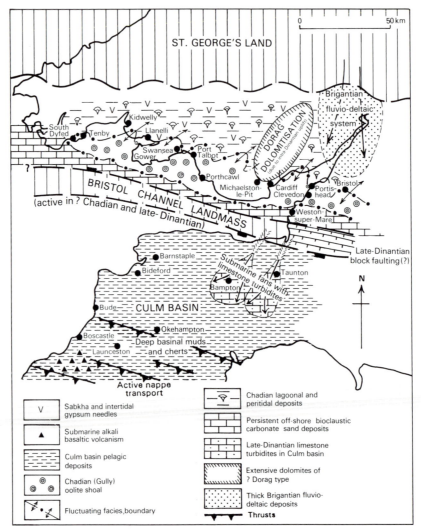

**Fig. 7.9.** Map to show a generalised palaeogeography and facies distribution in SW England and S Wales. Note that the map does not imply a static distribution of lithofacies throughout Dinantian times (see legend). Data from very many sources (see text).

(Mitchell 1972), a conclusion supported by conodont studies (Butler 1973). This major discontinuity defines the top of the Ivorian stage in the area and may be interpreted as due to the diachronous southward spread of a major regressional episode (Ramsbottom 1973, 1977). There is no associated evidence of supratidal evaporite facies to indicate that the dolomitisation was due to sabkha brines. The 'mixing' or Dorag dolomitisation model (Hanshaw *et al.* 1971, Badiozamani 1973) better fits the observed sediment-

ological and stratigraphic data (see Hird *et al.* (1987) whose additional data support this idea).

The succeeding Chadian, Arundian and Holkerian stages all show a similar broad motif of (1) basal transgressive bioclastic limestones thinning northwards (2) poorly fossiliferous cross-stratified oolitic limestones and (3) micritic limestones often with abundant stromatolites (e.g. Murray & Wright 1971). Cycles involving these facies seem indicative of repeated transgression followed by southward pro-

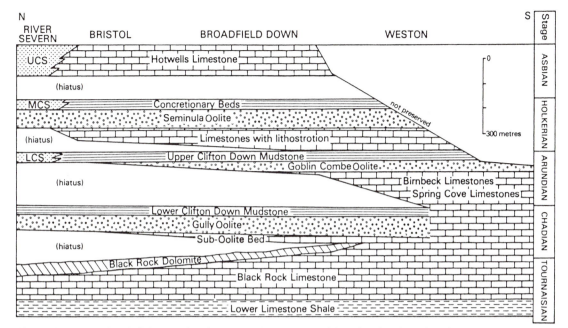

**Fig. 7.10.** Extrapolated ribbon section from Bristol to Weston with rock units plotted against time-stratigraphic intervals. Undashed intervals represent postulated hiatuses (after Ramsbottom 1977).

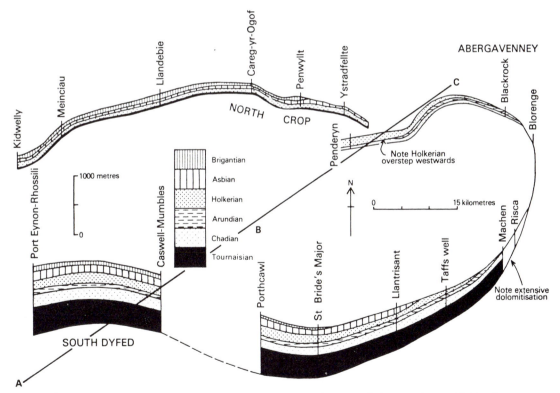

**Fig. 7.11.** Extrapolated ribbon section for the north and south crops of the S Wales coalfield (after George 1970). Note that the details of North Crop stratigraphy must be reconsidered in the light of current studies (Wright *et al.* 1981). (A–B–C is line of section on Fig. 7.12).

224 M. R. LEEDER

gradation of progressively shallower water facies. A very extensive and persistent tidal oolite shoal, with contemporary water depth and subsidence delicately adjusted, is indicated for the area south of Broadfield Down. Here the Lower Clifton Down Mudstone lenses out (Kellaway & Welch 1955) and the Gully and Goblin Combe oolites merge to eventually form the Burrington Oolite, some 250 m thick. These oolites indicate development of a broad shallow platform morphology during Holkerian times. South of the outcrop area, extremely thick successions of carbonate ramp-toe facies are known from the Cannington Borehole (up to 1 km – Whittaker & Scrivener and Lees & Hennebert 1982).

A further feature of the Arundian to Holkerian successions is the northward increase in fluvio-deltaic and shallow marine clastic facies (e.g. Cromhall, Drybrook Sandstones) at the expense of carbonates. The Holkerian/Asbian Hotwells Group always rests with non-sequence upon the underlying Clifton Down Group in the area and comprises at least 14 well-defined cyclothems. At least five upward shoaling carbonate sequences, ending in subaerial 'fireclays' with plant remains, occur in the Hotwells Limestone while at least nine Yoredale-type clastic cycles make up the Upper Cromhall Sandstone, the carbonate members of the cycles dying out northwards (Dearnley in Kellaway 1967). Asbian lithofacies are of shelf aspects with frequent emersion surfaces. They testify to the development of a shelf slope to the south (Wright 1987).

## South Wales

The Dinantian succession in South Wales (Figs. 7.11, 7.12) is thickest and most complete in South Dyfed (Pembrokeshire). Northwards the succession thins and shows increasing evidence of stratigraphic breaks and nearshore, non-clastic facies changes (George 1958, Sullivan 1966, George et al. 1976). These northward changes reflect (1) Intra-Dinantian major cyclicity due to tectonic (George 1979) and/or

**Fig. 7.12.** Carboniferous Limestone outcrop in southern Wales and stratigraphic cross-section (see Fig. 7.11) showing major lithofacies types. A – south Dyfed; B – Gower and Vale of Glamorgan (west of Cardiff); C – north limb, north east of Merthyr Tydfil. The Lower Limestone Shale Group (Stage I) represents 'start-up' phase. The Ivorian to Arundian period (Stage II) represents a ramp with fringing high energy oolite shoals. The Holkerian (Stage III) represents a barrier-type ramp with a thick oolitic development. These two phases constitute the 'catch-up' phase of the evolution of the carbonate province. The Asbian (Stage IV) represents deposition on a broad 'shelf' (after Wright 1987).

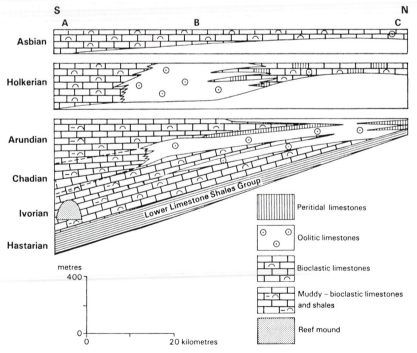

eustatic (Ramsbottom 1973, 1977, 1979, 1981) causes, (2) generalised northward shoaling on to the flanks of the positive St George's Land, (3) Intra-Dinantian uplift, tilting and erosion and (4) major late-Dinantian uplift and erosion.

The episodic 'Tournaisian' marine transgression, with lagoonal, oolite shoal and shelf facies (Burchette, *in* Wright *et al.* 1981) is recorded in the Lower Limestone Shales which may be mapped over the whole basin (Burchette 1987; see previous section). Phosphatic lag deposits underlie the oolite facies and

common vermetid gastropod biostromes (Burchette & Riding 1977) occur in the lagoonal facies. Overlying the Lower Limestone Shales are the lateral equivalents to the Black Rock Limestone with correlatives of the *laminosa* dolomite at the top (Dixon 1921; George 1933, 1940). As in the Avon Gorge there is probably a non-sequence. The dolomites are absent in South Dyfed.

The Chadian transgressive facies are of tidal shoal oolite type ($\equiv$ Caninia Oolite of Bristol/Mendips) over much of the basin other than in South Dyfed where the

**Table 7.3.** Stratigraphic succession in SW England and S Wales (after George *et al.* 1976). Shading denotes hiatuses.

| STAGES | 'CULM' BASIN | | | SOUTH WALES | | BRISTOL/MENDIPS | | |
|---|---|---|---|---|---|---|---|---|
| | CORNWALL Boscastle Launceston | N. DEVON Barnstaple Bideford | N.E. DEVON Westleigh Bampton | S. WALES Gower | S. WALES Clydach Valley | MENDIPS Burrington | BRISTOL Avon Gorge | |
| | Namurian | Namurian | Namurian | Namurian | Namurian | Triassic | Namurian | |
| BRIGANTIAN | Fire Beacon Chert Formation; Upper limestone with P, goniatites & III β conodonts (Buckator Formation) | Black Shale Formation; Chert | Upper Westleigh Limestone | Oystermouth Beds (20 metres); Oxwich Head Limestone (90 metres) | | (not seen); Hotwells Limestone (30 metres) | Upper Cromhall Sandstone; Hotwells Limestone | Hotwells Group (210 m) |
| ASBIAN | | Formation | Lower Westleigh Limestone | | | | | |
| HOLKERIAN | Lower limestone with II δ conodonts | | | Hunts Bay Oolite (260 metres) | Limestones with oolites (27 metres) | Clifton Down Limestone (220 metres) | Upper Clifton Down Limestone; Semynula Oolite | Clifton Down Group (300 metres) |
| ARUNDIAN | Trambley Cove Formation; Tintagel Volcanic Formation (Tintagel Group) | Basement Limestone | | ? ; High Tor Limestone (110 metres) | | Quarry 2 Lst; Rib Mdst; Aveline's Hole Limestone | Lower Clifton Down Lst; Upper C D Mdst; Goblin Combe Oolite | Burrington Oolite (230 m) |
| CHADIAN | Barras Nose Formation with II β-δ conodonts | | Formation | Caswell Bay Mdst; Caswell Bay Oolite (35 metres) ? | Llanelly Fm (18 metres); Gilwern Oolite (12 metres); Clydach Beds | ? ; Ham Mdst; Gully Oolite; Quarry 3 Lst | Lower C D Mdst; Gully Oolite; Suboolite Bed | |
| IVORIAN | Yeolmbridge Formation | Pilton Beds | | Penmaen Burrows Limestone (300 metres) | Blaen Onnen Oolite (30 metres); Pwll-y-Cwm Oolite (7 metres) | Black Rock Limestone (270 metres) | Black Rock Limestone (130 metres) | |
| HASTARIAN | | | | Cefn Bryn Shales (110 metres) | Lower Limestone Shale (12 metres) | Lower Limestone Shale (150 metres) | Lower Limestone Shale (70 m); Palate Bed; Bryozoa Bed; Shirehampton Beds (35 metres) | |
| | Devonian | Devonian | Devonian | Upper Old Red Sandstone | Upper Old Red Sandstone | Upper Old Red Sandstone | Upper Old Red Sandstone | |

Berry Slade Formation is dominantly bioclastic with dolomitised algal/coral reefs at the top. Detailed study of the Caswell Bay Oolite of Gower (George 1977; Ramsay 1987) show that the well-sorted, cross-stratified oosparitic facies contains channel-like features with steep walls indicative of early diagenetic cementation and erosion. Layers of bioclastic debris were probably storm-produced, although pockets of *in situ* coralliferous facies also occur. The unit is capped by a laterally persistent stromatolite horizon with pseudomorphs after gypsum, representing tidal flat progradation over the defunct tidal shoal channels. Similar facies to the Caswell Bay Oolite may be widely traced eastwards, though thoroughgoing dolomitisation obscures the facies present around Risca. In Gower George (1977) has evidence that late Chadian uplift caused minor tilting and erosion of part of the lagoonal/peritidal Caswell Bay Mudstone which overlies the Caswell Bay Oolite.

The Chadian–Arundian shallow subtidal to supratidal successions in the North Crop around Abergavenny (Figs. 7.11, 7.12) show evidence of repeated non-sequences with abundant development of calcretes and karstic surfaces (Raven & Wright *in* Wright *et al.* 1981).

The Arundian transgression deposited transgressive shoreline facies and offshore dark bioclastic limestones over much of South Wales, with overstep

recorded in Gower (George 1977). Post-Chadian/pre-Holkerian tectonic uplift (George 1979) is also prominently recorded along the north crop, where Arundian rocks are absent and there is marked progressive westward overstep of the succeeding Holkerian, first on to the truncated Chadian and then on to the Lower Limestone Shales (Robertson & George 1929, George 1970). Holkerian facies in South Dyfed continue as the richly fossiliferous bioclastic Stackpole Limestone. Over much of the rest of the basin tidal shoal oolitic facies (*Seminula* phase) occur with lagoonal facies in the Dowlais Limestone of the North Crop (Wright *in* Wright *et al.* 1981).

Offshore bioclastic carbonates continued their dominance in South Dyfed during Asbian and Brigantian times. In Gower and over the North and South Crops (where not cut out by the basal Namurian unconformity) clastic mudstones form the topmost Dinantian facies, together with important spreads of littoral and sublittoral quartz conglomerates and sandstones sometimes containing marine faunas (Owen & Jones 1961) along the North Crop. These developments of clastic facies suggested to Kelling (1974) that a major phase of uplift occurred in St George's Land in late-Dinantian times, leading to a shrinkage of the South Wales basin.

The Dinantian history of South Wales presents a fascinating interplay between demonstrable differen-

**Fig. 7.13a.** Locality map for South Devon and North Cornwall showing distribution of nappes. [337–339 are numbers of B.G.S. 1:50,000 Geological Sheets].

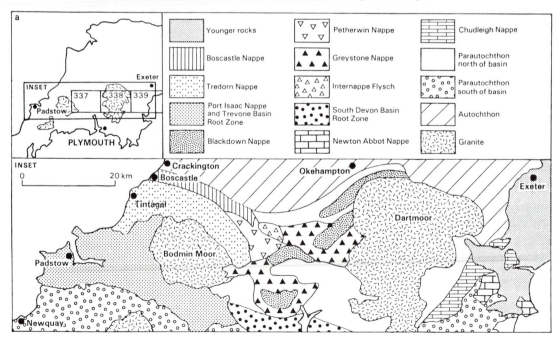

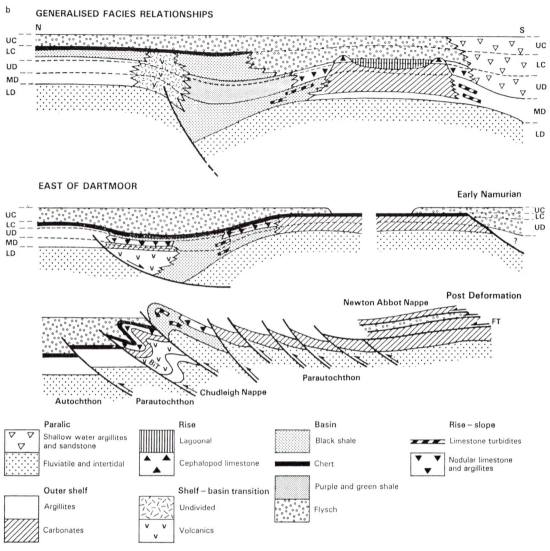

**Fig. 7.13b.** Sections show generalised facies relationships and tectono-stratigraphic development in a N–S section East of Dartmoor. UC – Upper Carboniferous; LC – Lower Carboniferous; UD – Upper Devonian; MD – Middle Devonian; LD – Lower Devonian; FT – Forder Green Thrust (all after Sellwood & Thomas 1986).

tial subsidence and uplift and possible eustatic sea-level migrations (Ramsbottom 1973, George 1979). A major problem concerns the widespread development of dolomitisation in northern and eastern areas throughout the Dinantian (Fig. 7.9). Bhatt's (1976) hypothesis of dolomitisation due to subsurface reflux of high Mg/Ca brines from a northern sabkha through a central reef belt cannot be substantiated on field evidence. The writer considers, as in the *laminosa* case above, that the Dorag dolomitisation model

is more appropriate. Periodic hinterland uplift causing continental and marine waters to mix in the subsurface plays a key role in this model and is amply demonstrated in the Dinantian successions of eastern areas.

## Culm basin of Devon and Cornwall

According to the most simplistic view the southward deepening trends deduced above for South

Wales and the Mendips carry on until the northern margin of the Culm basin is reached in north Devon. Here (Table 7.3, Fig. 7.9) the Devono-Carboniferous boundary is located by trilobite faunas and lies in the deep basinal mudstone facies of the Pilton Beds (Goldring 1955). An upward deepening trend is deduced for the area, from the coastal Famennian Baggy Beds (Goldring 1971) to the Pilton Beds. Problems remain concerning the source of the basinal muds by this hypothesis, a similar dilemma to that of muddy 'Culm' facies in southern Ireland. The critical question concerns the palaeogeographic nature of the Bristol Channel area in the Dinantian. It is known from sedimentary evidence and palaeocurrents that a sizeable landmass, termed the Bristol Channel Landmass (Fig. 7.9), must have separated South Wales/Mendips from Devon during both the Middle Devonian (Tunbridge 1986) and the Namurian/West-phalian (Williams in Owen et al. 1971, Kelling 1974, Anderton et al. 1979). There is no other evidence for the existence of this landmass during Upper Devonian/Lower Carboniferous times other than indications of southerly uplift in Gower (George 1977) during the Chadian, the above possibly of mud derivation, and the existence of platform carbonate clasts in limestone turbidites from the eastern margins of the Culm basin. It should be noted that any estimate of the palinspastic extent of the postulated landmass should take into account possibly 25 km of northward thrusting under the Bristol Channel as deduced from gravity surveys (Brooks & Thompson 1973).

Although the Culm basin may be said to contain everywhere 'deep' basinal facies there are marked local facies contrasts. Polyphase deformation makes thickness difficult to estimate in the area but it is likely that nowhere does the Dinantian exceed a few hundred metres. As noted above the northern outcrop of the Culm synclinorium, from Barnstaple to near Taunton, begins with the Tournaisian basinal mudstones of the Pilton Beds. These pass upwards into chert–limestone–mudstone lithologies comprising radiolarian cherts, micritic laminated limestones and laminated silty mudstones (Prentice 1958, 1960). These contain the pelagic Posidonia, sphaerocone goniatites and pelagic trilobites. In the eastern outcrops around Bampton and Westleigh, ?Chadian to ?Brigantian fine-grained turbiditic limestones occur above the Pilton Beds and are interbedded with non-calcareous shales containing pelagic faunas (Mathews

& Thomas 1974). Similar distal turbidites occur in the Westleigh area but in the topmost Lower Westleigh Limestone and in the Upper Westleigh Limestone, coarse carbonates contain large limestone clasts of obviously shallow water provenance. Most of these clasts are carbonate–mud supported, ungraded, structureless and with no sole structures (Mathews & Thomas 1974). The beds closely resemble debris-flow deposits. Rarer, graded beds of sand-size calcareous debris show sole structures and internal laminations indicative of true turbidite origin. Careful study has shown that reworked conodonts occur in the clasts and matrix of the mid Asbian/Brigantian Upper Westleigh Limestones. These conodonts indicate contributions from mid-Tournaisian to upper Holkerian limestones of shelf facies (Mathews & Thomas 1974). These reworked clasts give evidence for a phase of platform margin extensional tectonics in the Mendips carbonate platform to the north.

Recent detailed mapping in South Devon and North Cornwall had radically changed views on Dinantian environments (Isaac et al. 1982, 1983; Selwood et al. 1984, 1985; Selwood and Thomas 1986a, b, 1987). The area (Fig. 7.13) is now recognised as part of a southerly-derived nappe-and-thrust terrane which developed from former extensional Middle to Upper Devonian basins. During Dinantian and Namurian times these basins were progressively inverted by shortening deformation, a process accompanied by volcanicity, olistostromes and northwards-prograding flysch.

The Boscastle Nappe, transported from the south, includes nearshore, shallow-water Dinantian sediments. The Tredorn Nappe includes the Tintagel Group (Chadian–Arundian), comprising pyritic slates with laminated siltstone bands and the tuffs. agglomerates and lavas of the Tintagel Volcanic Formation (Dearman 1959, Selwood 1961, Edmonds et al. 1968, Freshney et al. 1972). East of Dartmoor a condensed Dinantian succession includes extensive black shales and cherts, now dismembered by the Chudleigh Nappe. Between Dartmoor and Bodmin Moor the Greystone Nappe contains Dinantian black shales within which are intercalated flysch turbidites in southern areas. These latter are southerly derived. In all areas, northward nappe transport was accompanied by wedges of southerly-derived flysch of comparable types during the Namurian (see also Chapter 17).

## REFERENCES

| ADAMS, A. E. & COSSEY, P. J. | 1981 | Calcrete development at the junction between the Martin Limestone and the Red Hill Oolite (Lower Carboniferous), South Cumbria. *Proc. Yorkshire geol. Soc.*, **43**, 411–431. |
| ALLEN, P. | 1960 | 'Deltaic' cyclothems. *Geol. Mag.*, **97**, 524–525. |

ANDERTON, R.                    1985    Sedimentology of the Dinantian of Foulden, Berwickshire, Scotland. *Trans. Roy. Soc. Edinb., Earth Sci.*, **76**, 7–12.

ANDERTON, R.,                   1979    *A Dynamic stratigraphy of the British Isles.* Allen & Unwin,
  BRIDGES, P. H.,                       London. 301 pp.
  LEEDER, M. R. &
  SELLWOOD, B. W.

ASHTON, P.                      1970    Sedimentology of the Ravenstonedale Limestone and its correlatives in Westmorland and the Furness District of Lancashire. *Unpubl. PhD. thesis, Univ. of Southampton.*

AUSTIN, R. L.                   1973    Modification of the British Avonian conodont zonation and reappraisal of European Dinantian conodont zonation and correlation. *Annls. soc. geol. Belge.*, **96**, 523–532.

BADIOZAMANI, L.                 1973    The Dorag dolomitisation model – application to the middle Ordovician of Wisconsin. *J. sediment. Petrol.*, **43**, 965–984.

BARRACLOUGH, R.                 1983    Tectonic versus eustatic sea level changes in the Lower Carboniferous of northern England: a sedimentary study. *Unpubl. Ph.D. thesis, University of Leeds.*

BELT, E. S., FRESHNEY,          1967    Sedimentology of Carboniferous Cementstone Facies,
  E. C. & READ, W. A.                   British Isles and eastern Canada. *J. Geol.*, **75**, 7551–721.

BESLY, B. M. &                  1988    *Sedimentation in a synorogenic basin complex: The*
  KELLING, G. (Eds)                     *Carboniferous of northwest Europe.* Blackie & Sons, Glasgow, 276 pp.

BHATT, J. J.                    1976    Geochemistry and petrology of the Main Limestone Series (Lower Carboniferous) South Wales, U.K. *Sediment. Geol.*, **15**, 55–86.

BLACK, W. W.                    1950    The Carboniferous geology of the Grassington area, Yorkshire. *Proc. Yorkshire geol. Soc.*, **28**, 29–42.

                                1954    Diagnostic characters of the Lower Carboniferous knoll-reefs in the north of England. *Trans. Leeds Geol. Ass.*, **6**, 262–297.

                                1958    The structure of the Burnsall-Cracoe district and its bearing on the origin of the Cracoe knoll-reefs. *Proc. Yorkshire geol. Soc.*, **31**, 391–414.

BOND, G.                        1950a   The Lower Carboniferous reef limestone of Cracoe, Yorkshire. *Q. J. geol. Soc. London*, **105**, 157–188.

                                1950b   The Lower Carboniferous reef limestones of northern England. *J. Geol.*, **58**, 313–329.

                                1950c   The nomenclature of the Lower Carboniferous 'reef' limestones in the north of England. *Geol. Mag.*, **87**, 267–278.

BOTT, M. H. P.                  1961    A gravity survey off the coast of NE England. *Proc. Yorkshire geol. Soc.*, **33**, 1–20.

                                1964    Formation of sedimentary basins by ductile flow of isostatic origin in the upper mantle. *Nature, London*, **201**, 1082–1084.

                                1967    Geophysical investigations of the northern Pennine basement rocks. *Proc. Yorkshire geol. Soc.*, **36**, 139–168.

                                1974    The geological interpretation of a gravity survey of the English Lake District and the Vale of Eden. *J. geol. Soc. London*, **130**, 309–331.

BOTT, M. H. P. &                1957    The geological interpretation of a gravity survey of the
  MASSON-SMITH, D.                      Alston Block and the Durham coalfield. *Q. J. geol. Soc. London*, **113**, 93–117.

BOTT, M. H. P. &                1967    The controlling mechanism of Carboniferous cyclic
  JOHNSON, G. A. L.                     sedimentation. *Q. J. geol. Soc. London*, **122**, 421–441.

BOTT, M. H. P.,              1978    Granite beneath Market Weighton, east Yorkshire. *Q. J.*
  ROBINSON, J. &                    *geol. Soc. London*, **135**, 535–544.
  KOHNSTAM, M. A.

BRIDGES, P. H.               1982    The origin of cyclothems in the late Dinantian platform
                                     carbonates at Crich, Derbyshire. *Proc. Yorkshire geol.
                                     Soc.*, **44**, 159–180.

BROOKS, M. &                 1973    The geological interpretation of a gravity survey of the
  THOMPSON, M. S.                   Bristol Channel. *J. geol. Soc. London*, **129**, 245–274.

BROADHURST, F. M. &          1973    Bathymetry on a Carboniferous reef. *Lethaia*, **6**, 367–381.
  SIMPSON, J. M.

BURCHETTE, T. P.             1987    Carbonate-barrier shorelines during the basal Carbon-
                                     iferous transgression: the Lower Limestone Shale Group,
                                     South Wales and Western England. *In* Miller, J., Adams,
                                     A. E. & Wright, V. P. (Eds.) *European Dinantian Environ-
                                     ments*, Wiley, Chichester, 239–264.

BURCHETTE, T. P. &           1977    Attached vermiform gastropods in Carboniferous mar-
  RIDING, R.                        ginal marine stromatolites and biostromes. *Lethaia*, **10**,
                                     17–28.

BURGESS, I. C. &             1967    Carboniferous Basement Beds in the Roman Fell District,
  HARRISON, R. K.                   Westmorland. *Proc. Yorkshire geol. Soc.*, **36**, 203–225.

BURGESS, I. C. &             1976    Visean lower Yoredale limestones on the Alston and
  MITCHELL, M.                      Askrigg blocks, and the base of the $D_2$ zone in northern
                                     England. *Proc. Yorkshire geol. Soc.*, **40**, 613–630.

BURGESS, I. C. &             1979    Geology of the country around Brough-under-Stainmore.
  HOLLIDAY, D. W.                   *Mem. geol. Surv. G.B.*

BUTCHER, N. J. &             1973    The Carboniferous Limestone of Monsal Dale,
  FORD, T. D.                       Derbyshire. *Mercian Geol.*, **4**, 179–195.

BUTLER, M.                   1972    Conodont faunas and stratigraphy of certain Tournasian
                                     sections in the Bristol–Mendip area. *Unpubl. Ph.D. thesis,
                                     Univ. of Bristol.*

BUTLER, M.                   1973    Lower Carboniferous conodont faunas from the eastern
                                     Mendips, England. *Palaeontology*, **16**, 477–517.

BUTLER, M.,                  1972    A new exposure of the Old Red Sandstone–Lower
  WILLIAMS, B. P. J. &              Limestone Shale transition at Portishead, Somerset. *Proc.
  BRADSHAW, R.                      Bristol Nat. Soc.*, **32**, 151–155.

CLAYTON, G., HIGGS, K.,      1974    Palynological correlations in the Cork Beds (Upper
  GUEINN, K. J. &                   Devonian–?Upper Carboniferous) of southern Ireland.
  VAN GELDER, A.                    *Proc. Roy. Irish Acad.*, **74**, 145–155.

CONIL, R., GROESSENS,        1977    Nouvelle charte stratigraphique du Dinantien type de la
  E. & PIRLET, H.                   Belgique. *Ann. Soc. geol. Nord*, **96**, 363–371.

CONIL, R. & LYS, M.          1968    Utilisation stratigraphique des foraminiferes due
                                     Dinantien. *Ann. Soc. geol. Belg.*, **91**, 491–558.

CONIL, R., AUSTIN, R.,       1977    International correlation of Dinantian strata (table).
  BLESS, M., DIL, N.,               *Meded. Rijks. Geol. Dienst.*, NS **27–3** (Appendix).
  GROESSENS, E., LEES, A.,
  LONGERSTACY, P.,
  LYS, M., PAPROTH, E.,
  PIRLET, H., POTY, E.,
  RAMSBOTTOM, W. H. C. &
  SEVASTOPULO, G.

COPE, F. W.                  1973    Woo Dale borehole near Buxton, Derbyshire. *Nature,
                                     London*, **243**, 29–30.

CUMMINGS, R. H.              1961    The foraminiferal zones of the Carboniferous sequence of
                                     the Archerbeck borehole, Canonbie, Dumfriesshire. *Bull.
                                     geol. Surv. Gt. Br.*, **18**, 107–128.

DAY, J. B. W. — 1970 — Geology of the country around Bewcastle. *Mem. geol. Surv. G.B.*

DEARMAN, W. R. — 1959 — The structure of the Culm Measures at Meldon near Okehampton, north Devon. *Q. J. geol. Soc. London*, **115**, 65–106.

DEEGAN, C. E. — 1973 — Tectonic control of sedimentation at the margin of a Carboniferous depositional basin in Scotland. *Scott. J. Geol.*, **9**, 1–28.

DIXEY, F. & SIBLY, T. F. — 1918 — The Carboniferous Limestone Series in the south-eastern part of the South Wales Coalfield. *Q. J. geol. Soc. London*, **73**, 111–164.

DIXON, E. E. L. — 1921 — The country around Pembroke and Tenby. *Mem. geol. Surv. G.B.*

DOLBY, G. & NEVES, R. — 1970 — Palynological evidence concerning the Devonian/Carboniferous boundary in the Mendips, England. *C.R. 6me Cong. Int. Strat. Geol. Carb.*, Sheffield 1967, **2**, 631–642.

DOUGHTY, P. S. — 1974 — *Davidsonina (Cyrtinia) septosa* (Phillips) and the structure of the Visean Great Scar Limestone north of Settle. Yorkshire. *Proc. Yorkshire geol. Soc.*, **40**, 41–48.

DUFF, P. McL. D., HALLAM, A., & WALTON, E. K. — 1967 — *Cyclic Sedimentation.* Elsevier, Amsterdam, London, New York, 280 pp.

DUNHAM, K. C. — 1973 — A recent deep borehole near Eyam in Derbyshire. *Nature, London*, **241**, 84–85.

DUNHAM, K. C., DUNHAM, A. C., HODGE, B. L. & JOHNSON, G. A. L. — 1965 — Granite beneath Visean sediments with mineralisation at Rookhope, northern Pennines. *Q. J. geol. Soc. London*, **121**, 383–414.

EDMONDS, E. A., WRIGHT, J. E., BEER, K. E., HAWKES, J. R., WILLIAMS, M., FRESHNEY, E. C., & FENNING, P. J. — 1968 — Geology of the country around Okehampton. *Mem. geol. Surv. G.B.*

FALCON, N. F. & KENT, P. E. — 1960 — Geological results of petroleum exploration in Britain 1945–1957. *Mem. geol. Soc. London*, **2**, 1–56.

FOLK, R. L. & SIEDLECKA, A. — 1977 — Sedimentary and diagnetic fabrics of the schizohaline environment as exemplified by late Palaeozoic rocks of Bear Island, Svalbard. *Sediment. Geol.*, **11**, 1–15.

FRANCIS, E. H. — 1967 — Review of Carboniferous–Permian Volcanicity in Scotland. *Geol. Rundsch.*, **57**, 219–245.

— 1970 — Review of Carboniferous volcanism in England and Wales. *J. Earth Sci.*, Leeds, **8**, 41–56.

— 1978 — Igneous activity in a fractured craton – Carboniferous volcanism in N. Britain. *In* Bowes, D. R. & Leake, B. E. (Eds.) Crustal evolution in northwestern Britain and adjacent regions. *Geol. J. Spec. Issue*, **10**, 279–296.

FRESHNEY, E. C., McKEOWN, M. C. & WILLIAMS, M. — 1972 — Geology of the coast between Tintagel and Bude. *Mem. geol. Surv. G.B.*

FROST, D. V. & HOLLIDAY, D. W. — 1980 — Geology of the Country around Bellingham. *Mem. geol. Surv. UK.*

GARWOOD, E. J. — 1907 — Notes on the faunal succession in the Carboniferous Limestone of Westmorland and neighbouring portions of Lancashire and Yorkshire. *Geol. Mag.*, **44**, 70–74.

— 1913 — The Lower Carboniferous succession in the north west of England. *Q. J. geol. Soc. London*, **68**, 449–596.

GARWOOD, E. J. & GOODYEAR, E. 1924 The Lower Carboniferous succession in the Settle district and along the line of the Craven faults. *Q. J. geol. Soc. London*, **80**, 184–273.

GAWTHORPE, R. L. 1987 Tectono-sedimentary evolution of the Bowland Basin, N. England, during the Dinantian. *J. geol. Soc. London*, **144**, 59–71.

GAWTHORPE, R. L. & CLEMMEY, H. 1985 Geometry of submarine slides in the Bowland Basin (Dinantian) and their relation to debris flows. *J. geol. Soc. London*, **142**, 555–565.

GEORGE, T. N. 1933 The Carboniferous Limestone Series in the West of the Vale of Glamorgan. *Q. J. geol. Soc. London*, **89**, 221–272.

1940 The structure of Gower. *Q. J. geol. Soc. London*, **96**, 131–198.

1958 Lower Carboniferous palaeogeography of the British Isles. *Proc. Yorkshire geol. Soc.*, **31**, 227–318.

1970 South Wales. *Br. reg. Geol.* (3rd Ed.). H.M.S.O.

1977 Mid-Dinantian (Chadian) Limestones in Gower. *Phil. Trans. Roy. Soc.*, **B.280**, 411–463.

1979 Eustasy and tectonics: sedimentary rhythms and stratigraphical units in British Dinantian correlation. *Proc. Yorkshire geol. Soc.*, **42**, 229–262.

GEORGE, T. N. & WAGNER, R. H. 1972 I.U.G.S. Subcommission on Carboniferous Stratigraphy. Proceedings and report on the general assembly at Krefeld, August 21–22. *C.R. 7me Cong. Int. Strat. Geol. Carb.* Krefeld 1971, **1**, 139–147.

GEORGE, T. N. AND SIX OTHERS 1976 A correlation of Dinantian rocks in the British Isles. *Geol. Soc. Lond. Spec. Rept.*, **7**, 87 pp.

GOLDRING, R. 1955 The upper Devonian and lower Carboniferous trilobites of the Pilton Beds in north Devon. *Sen. Leth.*, **36**, 27–48.

1971 Shallow water sedimentation as illustrated in the upper Devonian Baggy Beds. *Mem. geol. Soc. London*, No. 5.

GREEN, G. W. & WELCH, F. B. A. 1965 Geology of the country around Wells and Cheddar. *Mem. geol. Surv. G.B.*

GUTTERIDGE, P. 1987 Dinantian sedimentation and the basement structure of the Derbyshire Dome. *Geol. J.*, **22**, 25–41.

GUTTERIDGE, P., ARTHURTON, R. S. & NOLAN, S. C. (Eds) 1989 *The role of tectonics in Devonian and Carboniferous sedimentation in the British Isles.* Yorkshire geol. Soc. Occ. Pub., **6**.

HANSHAW, B. B., BACK, W. & DEIKE, R. G. 1971 A geochemical hypothesis for dolomitisation by ground water. *Econ. Geology*, **66**, 710–724.

HIRD, K., TUCKER, M. E. & WATERS, R. A. 1987 Petrography, geochemistry and origin of Dinantian dolomites from SE Wales. *In* Miller, J., Adams, A. E. & Wright, V. P. (Eds.) *European Dinantian Environments*, Wiley, Chichester, 369–378.

HODGSON, A. V. 1978 Braided river bedforms and related sedimentary structures in the Fell Sandstone Group (Lower Carboniferous) of North Northumberland. *Proc. Yorkshire geol. Soc.*, **41**, 509–532.

HODGSON, A. V. & GARDINER, M. D. 1971 An investigation of the aquifer potential of the Fell Sandstone of Northumberland. *Q. J. Engng. Geol.*, **4**, 91–109.

HOLLIDAY, D. W., NEVES, R. & OWENS, B. 1979 Stratigraphy and palynology of early Dinantian strata in shallow boreholes near Ravenstonedale, Cumbria. *Proc. Yorkshire geol. Soc.*, **42**, 343–356.

HUDSON, R. G. S. 1930 The Carboniferous of the Craven reef belt; the Namurian unconformity at Scaleber, near Settle. *Proc. Geol. Assoc., London*, **41**, 290–322.

HUDSON, R. G. S. 1932 The pre-Namurian knoll topography of Derbyshire and Yorkshire. *Trans. Leeds Geol. Assoc.*, **5**, 49–64.

1933 The scenery and geology of north-west Yorkshire. pp. 228–255 in 'The geology of the Yorkshire Dales'. *Proc. geol. Assoc. London*, **44**, 227–269.

1944 The Carboniferous of the Broughton Anticline. *Proc. Yorkshire geol. Soc.*, **25**, 190–214.

HUDSON, R. G. S. & DUNNINGTON, H. V. 1945 The Carboniferous rocks of the Swinden Anticline, Yorkshire. *Proc. Geol. Assoc. London*, **45**, 195–215.

HUDSON, R. G. S. & MITCHELL, G. H. 1937 The Carboniferous geology of the Skipton Anticline. *Summ. Prog. geol. Surv. Gt. Br.* for 1935, pt. **2**, 1–35.

ISAAC, K. P., TURNER, P. J. & STEWART, I. J. 1982 The evolution of the Hercynides of central SW England. *J. geol. Soc. London*, **139**, 521–531.

ISAAC, K. P., CHANDLER, P., WHITELEY, M. J. & TURNER, P. J. 1983 An excursion guide to the geology of central SW England: report of the field meeting to West Devon and E.Cornwall. *Proc. Geol. Assoc. London*, **94**, 357–376.

JOHNSON, G. A. L. 1960 Palaeogeography of the northern Pennines and part of NE England during deposition of Carboniferous cyclothemic deposits. *Rep. 21st Int. Geol. Congr. Copenhagen*, Part **XII**, 118–128.

1967 Basement control of Carboniferous sedimentation in northern England. *Proc. Yorkshire geol. Soc.*, **36**, 175–194.

JOHNSON, G. A. L. & MARSHALL, A. E. 1971 Tournaisian beds in Ravenstonedale, Westmorland. *Proc. Yorkshire geol. Soc.*, **38**, 261–280.

KELLAWAY, G. A. 1967 The Geological Survey Ashton Park borehole, and its bearing on the geology of the Bristol district. *Bull. geol. Surv. Gt. Br.*, **27**, 49–153.

KELLAWAY, G. A. & WELCH, F. B. A. 1955 The Upper old Red Sandstone and Lower Carboniferous rocks of Bristol and the Mendips compared with those of Chepstow and the Forest of Dean. *Bull. geol. Surv. Gt. Br.*, **9**, 1–21.

KELLING, G. 1974 Upper Carboniferous sedimentation in South Wales. *In* Owen, T. R. (Ed.) *The Upper Palaeozoic and post-Palaeozoic rocks of Wales*. Cardiff (Univ. of Wales Press), 185–224.

KENT, P. E. 1966 The structure of the concealed Carboniferous rocks of NE England. *Proc. Yorkshire geol. Soc.*, **35**, 323–352.

1967 A contour map of the sub-Carboniferous floor in the northeast Midlands. *Proc. Yorkshire geol. Soc.*, **36**, 175–194.

LE BAS, M. J. 1972 Caledonian igneous rocks beneath central and eastern England. *Proc. Yorkshire geol. Soc.*, **39**, 71–86.

LEEDER, M. R. 1973a Lower Carboniferous serpulid patch reefs, bioherms and biostromes. *Nature, London*, **242**, 41–42.

1973b Sedimentology and palaeogeography of the Upper Old Red Sandstone in the Scottish Border Basin. *Scott. J. Geol.*, **9**, 117–144.

1974a Lower Border Group (Tournaisian) fluvio-deltaic sedimentation and palaeogeography of the Northumberland basin. *Proc. Yorkshire geol. Soc.*, **40**, 129–180.

1974b Origin of the Northumberland basin. *Scott. J. Geol.*, **10**, 288–296.

1975a Lower Border Group (Tournaisian) limestones from the Northumberland basin. *Scott. J. Geol.*, **11**, 151–167.

LEEDER, M. R.                    1975b    Lower Border Group (Tournaisian) stromatolites from the
                                          Northumberland basin. *Scott. J. Geol.*, **11**, 207–226.

                                 1976a    Palaeogeographic significance of pedogenic carbonates in
                                          the topmost Upper Old Red Sandstone of the Scottish
                                          Border Basin. *Geol. J.*, **11**, 21–28.

                                 1976b    Sedimentary facies and the origins of basin subsidence
                                          along the northern margin of the supposed Hercynian
                                          ocean. *Tectonophysics*, **36**, 167–179.

                                 1982     Upper Palaeozoic basins of the British Isles – Caledonide
                                          inheritance versus Hercynian plate margin processes. *J.
                                          geol. Soc. London*, **139**, 479–491.

                                 1987a    Tectonic and palaeogeographic models for Lower Carbon-
                                          iferous Europe. *In* Miller, J., Adams, A. E. & Wright,
                                          V. P. (Eds.) *European Dinantian Environments*, Wiley,
                                          Chichester, 1–20.

                                 1987b    Sediment deformation structures and the palaeotectonic
                                          analysis of extensional sedimentary basins. *In* Jones, N.
                                          & Preston, R. M. F. (Eds.) Deformation of Sediments
                                          & Sedimentary Rocks. *Spec. Publ. geol. Soc. London*,
                                          137–146.

LEEDER, M. R. &                  1987     Delta-Marine interactions: a discussion of sedimentary
    STRUDWICK, A. E.                      models for Yoredale-type cyclicity in the Dinantian of
                                          northern England. *In* Miller, J., Adams, A. E. & Wright,
                                          V. P. (Eds.) *European Dinantian Environments*, Wiley,
                                          Chichester, 115–130.

LEEDER, M. R. &                  1988     Upper Carboniferous (Silesian) basin subsidence in northern
    McMAHON, A. H.                        Britain. *In* Besly, B. M. & Kelling, G. (Eds.) *Sedimenta-
                                          tion in a synorogenic basin complex – The Upper Carboni-
                                          ferous of Northwest Europe.* Blackie, Glasgow, 43–52.

LEES, A. &                       1982     Carbonate rocks of the Knap Farm Borehole at Canning-
    HENNEBERT, N.                         ton Park, Somerset. *Rept. Inst. Geol. Sci.*, **82/5**, 18–36.

LEES, G. M. &                    1937     The geological basis of the present search for oil in Great
    COX, P. T.                            Britain by the D'Arcy Exploration Company Ltd. *Q. J.
                                          geol. Soc. London*, **93**, 156–194.

LEES, G. M. &                    1946     The geological results of the search for oilfields in Great
    TAITT, A. H.                          Britain. *Q. J. geol. Soc. London*, **101**, 253–313.

LLEWELLYN, P. G. &               1970     The Hathern Anhydrite Series, Lower Carboniferous,
    STABBINS, R.                          Leicestershire, England. *Trans. Instn. Min. Metall.*, **79B**,
                                          B1–15.

MATHEWS, S. C.                   1971     Comments on palaeontological standards for the
                                          Dinantian. *C.R. 6me Cong. Int. Strat. Géol. Carb.* Sheffield
                                          1967, **3**, 1159–1163.

MATHEWS, S. C., BUTLER,          1973     Field meeting: Lower Carboniferous successions in N.
    M. & SADLER, P. M.                    Somerset. *Proc. Geol. Assoc. London*, **84**, 175–179.

MATHEWS, S. C. &                 1974     Lower Carboniferous conodont faunas from NE
    THOMAS, J. M.                         Devonshire. *Palaeontology*, **17**, 371–385.

McKERROW, W. S. (Ed.).           1978     *The Ecology of Fossils.* Duckworth, London. 384 pp.

METCALFE, I.                     1976     The conodont biostratigraphy of the Lower Carboniferous
                                          sediments of the Skipton Anticline and Craven Lowlands.
                                          *Unpubl. Ph.D. thesis. Univ. of Leeds.*

                                 1980     Conodont zonation and correlation of the Dinantian and
                                          early Namurian strata of the Craven Lowlands, N.
                                          England. *Rept. Inst. Geol. Sci.*, **80/10**.

MILLER, J. &                     1972     Origin and structure of the Lower Visean 'reef' limestones
    GRAYSON, R. F.                        near Clitheroe. Lancashire. *Proc. Yorkshire geol. Soc.*, **38**,
                                          607–638.

MILLER, J. &
GRAYSON, R. F.
1982
The regional context of Waulsortian facies in N. England. *In* Bolton, K., Lane, H. R. & Le Mone, D. V. (Eds.) *Sympos. on Paleoenvironmental Setting & Distribution of Waulsortian Facies.* El Paso Geol. Soc. & Univ. of Texas, 17–33.

MITCHELL, G. H. &
STUBBLEFIELD, C. J.
1941
The Carboniferous Limestone of Breedon Cloud, Leicestershire, and the associated inliers. *Geol. Mag.*, **78**, 207–219.

MITCHELL, M.
1972
The base of the Visean in south-west and north-west England. *Proc. Yorkshire geol. Soc.*, **39**, 151–160.

1978
Dinantian. *In* Moseley, F. (Ed.) *The geology of the Lake District.* Yorkshire Geological Society, Leeds.

MOORE, D.
1958
The Yoredale Series of upper Wensleydale and adjacent parts of north-west Yorkshire. *Proc. Yorkshire Geol. Soc.*, **31**, 91–148.

1959
Role of deltas in the formation of some British Lower Carboniferous cyclothems. *J. Geol.*, **67**, 522–539.

MOSELEY, F.
1962
The structure of the SW part of the Sykes anticline, Bowland, W. Yorkshire, *Proc. Yorkshire geol. Soc.*, **33**, 287–314.

MURRAY, J. W. &
WRIGHT, C. A.
1971
The Carboniferous limestone of Chipping Sodbury and Wick, Gloucestershire. *Geol. J.*, **7**, 255–270.

NEVES, R., GUEINN, K. J.,
CLAYTON, G.,
IOANNIDES, N. &
NEVILLE, R. S. W.
1972
A scheme of miospore zones for the Dinantian. *C.R. 7me Cong. int. Strat. Géol. Carb.*, Krefeld 1971, **1**, 347–353.

NEVES, R., GUEINN, K. J.,
CLAYTON, G.,
IOANNIDES, N.,
NEVILLE, R. S. W. &
KRUSZEWSKA, K.
1973
Palynological correlations within the Lower Carboniferous of Scotland and northern England. *Trans. Roy. Soc. Edinb.*, **69**, 23–70.

NICHOLAS, C.
1967
The stratigraphy and sedimentary petrography of the Lower Carboniferous rocks SW of the Lake District. *Unpubl. Ph.D. thesis. Univ. of London.*

ORD, D., CLEMMEY, H. &
LEEDER, M. R.
1988
Interaction between faulting and sedimentation during Dinantian extension of the Solway Basin, SW Scotland. *J. geol. Soc. London.* **145**, 249–259.

OWEN, T. R. &
JONES, D. G.
1961
The nature of the Millstone Grit–Carboniferous Limestone junction of a part of the North Crop of the South Wales Coalfield. *Proc. Geol. Assoc. London*, **72**, 239–249.

OWEN, T. R., BLOXAM,
T. W., JONES, D. G.,
WALMSLEY, V. G. &
WILLIAMS, B. P. J.
1971
Summer (1968) Field Meeting in Pembrokeshire, South Wales. *Proc. Geol. Assoc. London*, **82**, 17–60.

PARKINSON, D.
1944
The origin and structure of the lower Visean reef-knolls of the Clitheroe district, Lancashire. *Q. J. geol. Soc. London*, **99**, 155–168.

1957
Lower Carboniferous reefs of northern England. *Bull. Amer. Assoc. Petrol. Geol.*, **41**, 511–537.

POCOCK, R. W.,
WHITEHEAD, T. H.,
WEDD, C. B. & ROBERTON, T.
1938
Geology of the Shrewsbury district. *Mem. geol. Surv. G.B.*

POWER, G. &
SOMERVILLE, I. D.
1975
A preliminary report on the occurrence of minor sedimentary cycles in the 'Middle White Limestone' ($D_1$ Lower Carboniferous) of North Wales. *Proc. Yorkshire geol. Soc.*, **40**, 491–497.

PRENTICE, J. E.              1958    The radiolarian cherts of N Devonshire, England. *Eclog. Geol. Helvet*, **51**, 706–711.

                             1960    The Dinantian, Namurian and Lower Westphalian rocks of the region south-west of Barnstaple, north Devon. *Q. J. geol. Soc. London*, **115**, 261–289.

PRENTICE, J. E. &           1960    The Carboniferous goniatites of north Devon. *Abs. Proc. 3rd Conf. Geol. Geomorph. S.W. England, Roy. geol. Soc. Cornwall*, 6–8.
THOMAS, J. M.

RAMSAY, A. T. S.             1987    Depositional environments of Dinantian limestones in Gower, S. Wales. *In* Miller, J., Adams, A. E. & Wright, V. P. (Eds.) *European Dinantian Environments*, Wiley, Chichester, 265–308.

RAMSBOTTOM, W. H. C.        1973    Transgressions and regressions in the Dinantian: a new synthesis of British Dinantian stratigraphy. *Proc. Yorkshire geol. Soc.*, **39**, 567–607.

                             1974    Dinantian. *In* Rayner, D. H. & Hemingway, J. E. (Eds.) *The geology and mineral resources of Yorkshire.* Yorkshire Geological Society, Leeds.

                             1977    Major cycles of transgression and regression (mesothems) in the Namurian. *Proc. Yorkshire geol. Soc.*, **24**, 261–291.

                             1979    Rates of transgressions and regressions in the Carboniferous of NW Europe. *J. geol. Soc. London*, **136**, 147–154.

                             1981    Eustacy, Sea Level and Local Tectonism, with examples from the British Carboniferous. *Proc. Yorkshire geol. Soc.*, **43**, 473–482.

RAMSBOTTOM, W. H. C. &      1980    The recognition and division of the Tournaisian Series in Britain. *J. geol. Soc. London*, **137**, 61–63.
MITCHELL, M.

RHODES, F. H. T.,            1969    British Avonian (Carboniferous) conodont faunas, and their value in local and international correlation. *Bull. Br. Mus. (Nat. Hist.) Geol.* Suppl. **5**.
AUSTIN, R. L. &
DRUCE, E. C.

RIDD, M. F.,                 1970    A deep borehole at Horton on the margin of the Northumberland Trough. *Proc. Yorkshire geol. Soc.*, **38**, 75–103.
WALKER, D. B. &
JONES, J. M.

ROBERTSON, T. &             1929    The Carboniferous Limestone of the North Crop of the South Wales Coalfield. *Proc. Geol. Assoc. London*, **40**, 18–40.
GEORGE, T. N.

ROBSON, D. A.                1956    A sedimentary study of the Fell Sandstones of the Coquet valley, Northumberland. *Q. J. geol. Soc. London*, **112**, 241–262.

ROSE, W. C. C. &            1978    Geology and hematite deposits of south Cumbria. *Mem. geol. Surv. G.B.*
DUNHAM, K. C.

SADLER, H. E.                1966    A detailed study of microfacies in the Mid-Visean ($S_2$–$D_1$) limestones near Hartington, Derbyshire, England. *J. sediment. Petrol.*, **36**, 864–879.

SCHWARZACHER, W.            1958    The stratification of the Great Scar Limestone in the Settle district of Yorkshire. *L'pool. Manchr. geol. J.*, **2**, 124–142.

SCOTT, W. B.                 1971    The sedimentology of the Cementstone Group in the Tweed Basin: Burnmouth and Merse of Berwick. *Unpubl. Ph.D. thesis, Sunderland Polytechnic.*

                             1986    Nodular carbonates in the Lower Carboniferous, Cementstone Group of the Tweed Embayment, Berwickshire: evidence for a former sulphate evaporite facies. *Scott. J. Geol.*, **22**, 325–345.

SELWOOD, E. B.               1961    The Upper Devonian and Lower Carboniferous stratigraphy of Boscastle and Tintagel, Cornwall. *Geol. Mag.*, **98**, 161–167.

| | | |
|---|---|---|
| SELWOOD, E. B. &<br>THOMAS, J. M. | 1984 | Structural models of the geology of the north Cornwall coast: a reinterpretation. *Proc. Ussher. Soc.*, **6**, 134–136. |
| | 1986a | Variscan facies and structure in central SW England. *J. geol. Soc. London*, **143**, 199–208. |
| | 1986b | Upper Palaeozoic successions and nappe structures in north Cornwall. *J. geol. Soc. London*, **143**, 75–82. |
| | 1987 | Sedimentation in SW England. *In* Miller, J., Adams, A. E. & Wright, V. P. (Eds.) *European Dinantian Environments*, Wiley, Chichester, 189–198. |
| SELWOOD, E. B.,<br>STEWART, I. J. &<br>THOMAS, J. M. | 1985 | Upper Palaeozoic sediments and structure in north Cornwall – a reinterpretation. *Proc. Geol. Assoc. London*, **96**, 129–141. |
| SELWOOD, E. B. & others | 1984 | Geology of the country around Newton Abbot. *Mem. Geol. Surv. U.K.* |
| SHEARMAN, D. J. | 1966 | Origin of marine evaporites by diagenesis. *Trans. Instn. Min. Metall.*, **75B**, 208–215. |
| SIMPSON, I. M. &<br>BROADHURST, F. M. | 1969 | A Boulder Bed at Treak Cliff, North Derbyshire. *Proc. Yorkshire geol. Soc.*, **37**, 141–151. |
| SMITH, K.,<br>SMITH, N. J. P. &<br>HOLLIDAY, D. W. | 1985 | The deep structure of Derbyshire. *Geol. J.*, **20**, 215–225. |
| SMITH, E. G., RHYS, G. H.<br>& EDEN, R. A. | 1967 | Geology of the country around Chesterfield, Matlock and Mansfield. *Mem. geol. Surv. U.K.* |
| SOMERVILLE, I. D. | 1979 | Minor sedimentary cyclicity in late Asbian (Upper $D_1$) limestones in the Llangollen district of N Wales. *Proc. Yorkshire geol. Soc.*, **42**, 317–341. |
| DE SOUZA, H. A. F. | 1975 | K-Ar ages of Carboniferous igneous rocks from E Lothian and the south of Scotland. *Unpubl. M.Sc. thesis. Univ. of Leeds.* |
| STEVENSON, I. P. &<br>GAUNT, G. D. | 1971 | Geology of the country around Chapel en le Frith. *Mem. geol. Surv. G.B.* |
| STRANK, A. R. E. | 1987 | The stratigraphy and structure of Dinantian strata in the East Midlands, UK. *In* Miller, J., Adams, A. E. & Wright, V. P. (Eds.) *European Dinantian Environments*, Wiley, Chichester, 157–175. |
| SULLIVAN, R. | 1966 | The stratigraphical effects of the mid-Dinantian movements in SW Wales. *Palaeogeogr., Palaeoclimatol., Palaeoecol.*, **2**, 213–244. |
| TROTTER, F. M. &<br>HOLLINGWORTH, S. E. | 1932 | The geology of the Brampton District. *Mem. geol. Surv. G.B.* |
| TUNBRIDGE, I. P. | 1986 | Mid-Devonian tectonics and sedimentation in the Bristol Channel area. *J. geol. Soc. London*, **143**, 107–116. |
| TURNER, J. S. | 1950 | Notes on the Carboniferous Limestone of Ravenstonedale, Westmorland. *Trans. Leeds. Geol. Assoc.*, **6**, 124–134. |
| | 1959 | Pinskey Gill Beds in the Lune Valley, Westmorland, and Ashfell Sandstone in Garsdale, Yorkshire. *Trans. Leeds Geol. Assoc.*, **7**, 151–174. |
| UTTING, J. & NEVES, R. | 1970 | Miospores from the Devonian/Carboniferous transition beds of the Avon Gorge, Bristol, England. *Cong. Coll. Univ. Liege*, **55**, 411–422. |
| VARKER, W. J. &<br>HIGGINS, A. C. | 1979 | Conodont evidence for the age of the Pinskey Gill Beds of Ravenstonedale, NW England. *Proc. Yorkshire geol. Soc.*, **42**, 357–369. |
| VAUGHAN, A. | 1905 | The palaeontological sequence in the Carboniferous Limestone of the Bristol area. *Q. J. geol. Soc. London*, **61**, 181–307. |

WALKDEN, G. M.            1972    The mineralogy and origin of interbedded clay wayboards in the Lower Carboniferous of the Derbyshire Dome. *Geol. J.*, **8**, 143–160.

1974    Palaeokarstic surfaces in upper Visean (Carboniferous) Limestones of the Derbyshire Block, England. *J. sediment. Petrol.*, **44**, 1232–1247.

1977    Volcanic and erosive events on an Upper Visean carbonate platform, North Derbyshire. *Proc. Yorkshire geol. Soc.*, **41**, 347–366.

1983    Polyphase erosion of subaerial omission surfaces in the late Dinantian of Anglesey, North Wales. *Sedimentology*, **30**, 861–878.

1984    Cyclicity in late-Dinantian marine carbonates of Britain. *C.R. 9me Int. Cong. Strat. Geol. Carbonif.*, Illinois, 1979, 561–569.

1987    Sedimentary and diagenetic styles in late Dinantian Carbonates of Britain. *In* Miller, J., Adams, A. E. & Wright, V. P. (Eds.) *European Dinantian Environmnts*, Wiley, Chichester, 131–156.

WALTHAM, A. C.            1971    Shale units in the Great Scar Limestone of the southern Askrigg Block. *Proc. Yorkshire geol. Soc.*, **38**, 285–292.

WHITTAKER, A. & SCRIVENER, R. C.            1982    The Knap Farm Borehole at Cannington Park, Somerset. *Rept. Inst. Geol. Sci.*, **82/5**, 1–7.

WOLFENDEN, E. B.            1958    Palaeoecology of the Carboniferous reef complex and shelf limestones in NW Derbyshire, England. *Bull. geol. Soc. Amer.*, **69**, 871–898.

WRIGHT, V. P.            1987    The evolution of the early Carboniferous Limestone province in SW Britain. *Geol. Mag.*, **124**, 477–480.

WRIGHT, V. P., RAVEN, M. & BURCHETTE, T. P.            1981    A field guide to the Carboniferous Limestone around Abergavenny. *University of Cardiff.*

# 8

# SILESIAN

# G. Kelling and J. D. Collinson

## Introduction

The Silesian is one of the most extensive and arguably the most economically important geological unit in mainland Britain (Fig. 8.1). These attributes have stimulated the interest of naturalists from the earliest days of scientific enquiry, particularly where coals are exposed (e.g. Whitehurst 1778) and many of the basic principles of stratigraphic division and geological structure were established in these rocks.

Silesian sediments record a wide variety of depositional environments, from deep marine basins to alluvial fans, and being predominantly clastic in character, they contrast with the cabonate-rich sequences of the Dinantian. Moreover, the Silesian succession of England and Wales records a gradual progression from the environmental and morphological diversity of the late Dinantian to the more uniform conditions of late Silesian times when widespread fluviatile and deltaic sediments accumulated at or near the prevailing sea level. A conspicuous feature of this gross trend is a *cyclicity* that is the long-recognised hallmark of the British Upper Carboniferous. This cyclicity can be discerned at a variety of scales and in varying degrees of complexity (Duff *et al.* 1967). The largest 'cyclothems' occur at the scale of scores to hundreds of metres in thickness and are commonly delimited by 'marine band' horizons that play a critical role, particularly in the Namurian, in stratigraphic division and correlation (Ramsbottom 1977, 1978a, 1979; Ramsbottom *et al.* 1978). These cyclothems have a broadly coarsening-upward character and most are attributed to major deltaic advance. In addition thinner cyclic sequences (scale of metres to tens-of-metres) are commonly present and have been traditionally regarded as typical of the Coal Measures in NW Europe (Robertson 1948, Trueman 1954). They commonly also coarsen upwards, with a capping of seatearth and coal, representing a phase of emergence, soil formation and plant growth on the floodplain or delta tops. However, many fining-upward sequences are also present and these commonly attributed to sedimentation in channels. The vertical ordering of the lithologies in both cyclothems and smaller sequences is highly variable, both vertically and laterally.

While the repetition of most of the intermediate-scale sequences can be ascribed to sedimentary causes (delta-switching, fluvial channel filling and abandonment, etc.), the mechanisms controlling the major cyclothems remain a matter of energetic debate. Both eustatic and tectonic factors have been cited as the ultimate cause of the marine transgressions which initiated and terminated successive regressive cyclothems (Bott & Johnson 1967, Duff *et al.* 1967, Ramsbottom 1979).

Whether or not diastrophism played a part in controlling Silesian cyclicity, the rapid lateral changes of thickness and lithofacies in these rocks demonstrate that differential subsidence and uplift were at least locally as important as they had been in Dinantian times. Contemporaneous tectonism is attested by localised warping and vertical fault-movements, especially in the early Namurian of northern England (Wilson 1960) and in the late Westphalian of the Midlands. (*See Besly & Kelling 1988 and Gutteridge* et al. *1989 for detailed reviews.*)

The crustal architecture on which the first Silesian deposits were formed was inherited from the Dinantian (Ch. 7), with continuity of sedimentation across the Dinantian–Silesian boundary in basinal areas. On Dinantian 'shallows' or 'highs' such as the eastern part of the South Wales Basin and on the northern and southern flanks of the rejuvenated St George's Land or Wales–Brabant Barrier, Silesian sands and gravels overstep on to older rocks.

To the north the wider effects of renewed uplift in

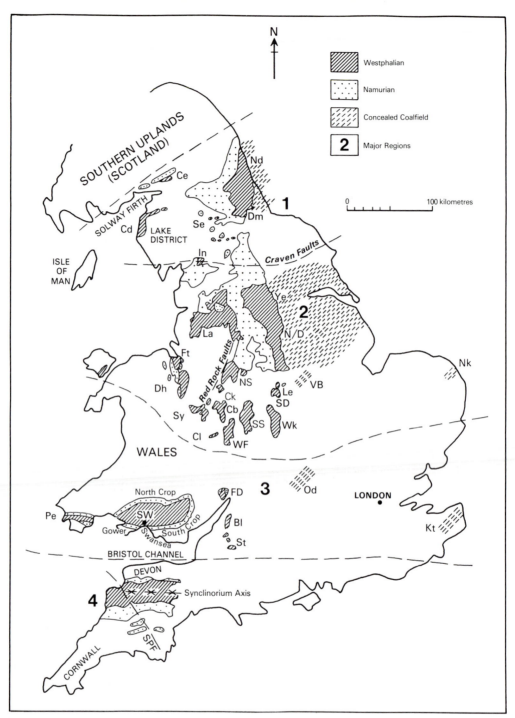

**Fig. 8.1.** General location map of Silesian outcrops and subcrops in England and Wales with coalfields lettered as follows: Bl – Bristol; Cb – Coalbrookdale; Cd – Cumberland; Ce – Canonbie; Ck – Cannock; Cl – Clee; Dh – Denbigh; Dm – Durham; FD – Forest of Dean; Ft – Flintshire; In – Ingleton; Kt – Kent; La – Lancashire; Le – Leicestershire; Nd – Northumberland; NS – North Staffordshire; N/D – Nottingham-shire – Derbyshire*; Od – Oxfordshire; Pe – Pembrokeshire; SD – South Derbyshire; Se – Stainmore; SS – South Staffordshire; St – Somerset; SW – South Wales; Sy – Shrewsbury; VB – Vale of Belvoir; WF – Wyre Forest; Wk – Warwickshire; Y – Yorkshire.*

Other abbreviations: Nk – Somerton Borehole; SPF – Sticklepath Fault; 1 – Northern England; 2 – Central Province; 3 – South Wales and Southern England; 4 – South-west England.

* Collectively known as East Pennine Coalfield.

source-areas can be detected in the widespread development of the typical 'Millstone Grit' coarse sand facies in the Namurian of northern England and North Wales. However, the 'block-and-basin' bathymetry of the Dinantian and early Silesian was replaced gradually by more uniform patterns of regional subsidence which permitted the development of rather constant 'Coal Measures' facies throughout central and northern England and North Wales. Moreover in late Silesian times an important new tectonic element appeared in southern Britain as the increasing tempo of Hercynian tectonism led to folding, uplift and northward thrusting of the Culm troughs of the south-west province. This had consequent effects on the structure

and facies development in adjacent regions promoting, for example, the development of the coal basins of South Wales, Somerset and Kent through crustal flexure resulting from thrust loading (Dewey 1982). Northward expansion of this compressive stress-field into the areas of central and northern England, previously characterised by a tensional regime (see p. 253), is reflected in the Midlands by rejuvenation of boundary faults, with local development of coarse alluvial sediments.

North of St George's Land (Fig. 8.2) the interplay between these N–S compressive stresses and the pre-existing block and basin features gave rise to a complex pattern of Hercynian structures. In this

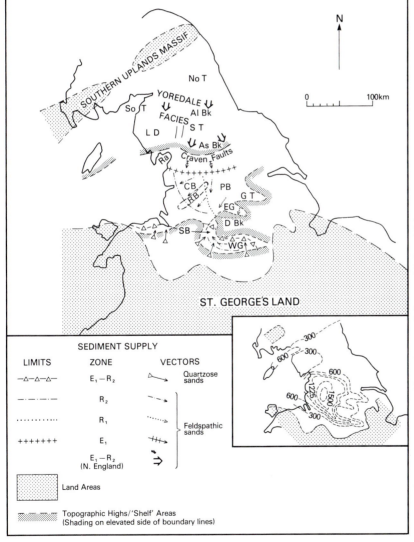

**Fig. 8.2.** Generalised Namurian palaeogeography of Northern and Central England and North Wales.

Al Bk – Alston Block; As Bk Askrigg Block; CB – Craven Basin; D Bk – Derbyshire Block; EG – Edale Gulf; G T – Gainsborough Trough; L D – Lake District massif; No T – Northumberland Trough; PB – Pennine Basin; Ra T – Ravestonedale Trough; R Bk – Rossendale Block; SB – Staffordshire Basin; So T – Solway Trough; S T – Stainmore Trough; WG – Widmerpool Gulf.

*Inset* shows isopachs (in metres) for Namurian (after Ramsbottom 1969).

region few of the Hercynian fold-axes (or the Upper Carboniferous facies belts) conform to the broadly E–W orientation of more frontal regions in southern Britain. Thus the Late Silesian rocks of England and Wales record a phase of broad uplift and continental sedimentation, with inherited structural elements important in the localised resolution of crustal stresses. This overall pattern of sedimentation and structure is plausibly explained by the Mackenzie (1978) model of lithospheric stretching (U. Devonian–Dinantian) and later thermal recovery, inducing upper crustal 'sagging' during the Silesian. Whether these conditions are best attributed to a Hercynian orogenic scenario in NW Europe involving transform-megashear tectonics (Badham 1982, Dewey 1982), to a pattern of northwards subduction and Rheno-Hercynian back-arc basin development (Leeder 1982), or to early rifting along N–S lines (Haszeldine 1984a) is a question that still awaits resolution. (See also Ch. 17.)

## Stratigraphic divisions and correlation

The venerable terms Millstone Grit (Fig. 8.3) and Coal Measures were first formally assigned (with the underlying Carboniferous or Mountain Limestone) to the System now known as the Carboniferous by Conybeare & Phillips (1822). These lithostratigraphic divisions were later allotted to the Upper Carboniferous by Green et al. (1878) and became broadly equated respectively with the Namurian and Westphalian 'Stages' of continental Europe. The Silesian, defined on essentially biostratigraphic grounds (van Leckwijk 1960, 1984) as a term corresponding to the upper Carboniferous of western Europe, has now been designated as a Subsystem and the constituent biostratigraphic divisions (Namurian, Westphalian and Stephanian) are now regarded as Series (George & Wagner 1972, Ramsbottom et al. 1978). With respect to the North American succession the lower Stages of the British Namurian fall within the late Mississippian (Saunders 1973) while the later Stages correspond to part of the Morrowan, of Pennsylvanian age (see Table 8.1). The base of the European Westphalian Series probably lies near the top of the Morrowan, but as yet only a general correlation of the higher Westphalian and Pennsylvanian stratigraphic units as possible.

The base of the Silesian is defined at the lower limit of the Namurian, namely at the base of the Pendleian Stage. This is determined on biostratigraphic grounds at the base of the *Cravenoceras leion* goniatite zone, a horizon recognised in many parts of Britain.

**Fig. 8.3.** The course cross-bedded Chatsworth Grit forms a scarp characteristic of the outcrops of the "Millstone Grit" (Namurian) sandstones in the central Pennines. Stanage Edge, west of Sheffield (R. H. Roberts).

**Table 8.1.** Stratigraphical divisions of the Silesian in England and Wales, and generalised correlation with eastern United States and Russian successions. (After Ramsbottom *et al.* 1978.)

| SUB SYSTEM | SERIES | STAGES | INDEX | GONIATITES ZONES (chronozones) | Marker goniatite bands | NON-MARINE BIVALVE ZONES | GEOLOGICAL SURVEY DIVISIONS | U.S.A. SUB SYSTEM | U.S.A. SERIES | U.S.S.R. SUB SYSTEM | U.S.S.R. SERIES |
|---|---|---|---|---|---|---|---|---|---|---|---|
| SILESIAN | STEPH-ANIAN | STEPHANIAN C B A — CANTABRIAN | | | | | | DES MOINESIAN ATOKAN MORROWAN | MIDDLE CARBONIFEROUS | MOSCOVIAN | MOSCOVIAN |
| | WESTPHALIAN | WESTPHALIAN D | | | | A prolifera | Upper Coal Measures | PENNSYLVANIAN | MIDDLE CARBONIFEROUS | MIDDLE CARBONIFEROUS | MOSCOVIAN |
| | | | | | | A. tenuis | Upper Coal Measures | | | | |
| | | WESTPHALIAN C | | 'Anthracoceras' | 'A' cambriense | A. phillipsi | | | | | |
| | | | 'A' | | {'A' hindi / 'A' aegiranum} | Upper similis-pulchra | Middle Coal Measures | | | | BASHKIRIAN |
| | | WESTPHALIAN B | | | | Lower similis-pulchra | | | | | |
| | | | | | 'A' vanderbeckei | A. modiolaris | | | | | |
| | | WESTPHALIAN A | | | | C. communis | Lower Coal Measures | | | | |
| | | | $G_2$ | G. listeri / G. subcrenatum | G. subcrenatum | C. lenisulcata | | | | | |
| | NAMURIAN | YEADONIAN | $G_1$ | Gastrioceras cumbriense / Gastrioceras cancellatum | G. cancellatum | | Millstone Grit 'Series' | PENNSYLVANIAN | MORROWAN | LOWER CARBONIFEROUS | BASHKIRIAN |
| | | MARSDENIAN | $R_2$ | Reticuloceras superbilingue / R. bilingue / R. gracile | R. gracile | | | | | | SERPUKHOVIAN |
| | | KINDER-SCOUTIAN | $R_1$ | R. reticulatum / R. nodosum / R. circumplicatile | H. magistrorum | | | | | | |
| | | ALPORTIAN | $H_2$ | Homoceratoides prereticulatus / Homoceras undulatum / Hudsonoceras proteus | Hd. proteus | | | MISSISSIPPIAN (pars) | CHESTERIAN | LOWER CARBONIFEROUS | SERPUKHOVIAN |
| | | CHOKIERIAN | $H_1$ | Homoceras beyrichianum / Homoceras subglobosum | H. subglobosum | | | | | | |
| | | ARNSBERGIAN | $E_2$ | Nuculoceras nuculum / Cravenoceratoides nitidus / Eumorphoceras bisulcatum | C. cowlingense | | | | | | |
| | | PENDLEIAN | $E_1$ | Cravenoceras malhamense / Eumorphoceras pseudobilingue / Cravenoceras leion | C. leion | | | | | | |

The upper boundary of the Silesian is internationally agreed to be the base of the Permian, though defining this presents problems in England and Wales (Chapter 9).

Of the three Series which comprise the Silesian Subsystem, only the Namurian and Westphalian are unequivocally widespread in England and Wales. Rocks of undoubted Stephanian age, once thought to be relatively widespread (Trueman 1946), are now believed to be rare (see discussion in Ramsbottom *et al.* 1978, p. 8, and Wagner 1983). The Namurian is divided into seven internationally recognised stages and the Westphalian into a further four, all designated on biostratigraphic grounds. Marine faunas (especially goniatites) furnish the basis of a series of biozones that (at least in the Namurian) have been redefined to include the entire sequence in which diagnostic goniatites occur (Ramsbottom *et al.* 1962) and have thus acquired a chronostratigraphic status. This status extends to the stages that have been created by combining two or more zones. The relative paucity of marine fossils in the Westphalian precludes the zonal precision possible in the Namurian although certain of the remarkably widespread marine bands demarcate the bases of the Westphalian A, B and C stages (Figs. 19.1a and b). The base of the Westphalian D stage is less readily identified, being defined on floral grounds (Ramsbottom et al. 1978, Cleal 1984).

For correlative purposes, therefore, goniatite assemblages reign supreme in the Silesian, particularly in the Namurian. Conodont (Higgins 1975) and miospore (Owens et al. 1977) zonations have proved useful supplementary techniques while plants are a long-standing aid in the broad subdivision of these Series (Dix 1934). Within the Westphalian, the 'mussel bands' (non-marine bivalves) were recognised as useful in inter-regional correlation and gave rise to a bizonal scheme (Davies & Trueman 1927) which was extended into an essentially chronostratigraphic zonation with the aid of prominent coals or marine bands as zonal boundaries (see Calver 1969). Locally, tuffs and bentonites (Trewin & Holdsworth 1973) are useful in Namurian correlation while in the Silesian several tonsteins, considered to be of volcanic origin (see e.g. Price & Duff 1969, Williamson 1970, Spears 1971, Spears & Rice 1973) are probably corrrelatable with horizons in continental Europe (Spears & Kanaris-Sotiriou 1979).

The following account deals with the Silesian sequence, in broadly regional fashion which takes account of major palaeogeographical divisions and of the present-day distribution of outcrops (Fig. 8.2). In the four regions thus delineated the Namurian and Westphalian successions are outlined and interpreted in terms of the evolving palaeogeography.

# Regional descriptions

## Northern England

For the present purpose the southern boundary of this region is the Craven Faults, while the northern limit is arbitrarily defined, though less precisely, by the Solway Firth and the Scottish border, as there is geological continuity through the Solway–Northumberland trough and into the basins of southern Scotland.

In palaeogeographical terms the southern boundary effectively marks the abrupt transition (during the greater part of the Namurian) from a northern zone of platform or shelf aspect, dominated by more positive palaeorelief, to the subsiding basins of the Central Province (Fig. 8.2). On the north, the elevated massif of the Southern Uplands persisted from the Dinantian into the early Namurian but diminished in importance as a source-land and was extensively breached and inundated by mid-Silesian time. Thus later Westphalian successions throughout most of northern England are broadly homogeneous.

## Namurian

The early Namurian platforms of northern England inherited their basic structural pattern from the Dinantian. The 'blocks', such as those of Askrigg and Alston (Figs. 7.2, 7.4, 8.2 (note that the traditional 'block' and 'basin' terminology is retained in this chapter (cf. Ch.7) (Eds.)), apparently rest on Caledonian granites (Bott & Masson-Smith 1957, Bott 1967, Bott et al. 1978). They are separated by zones of greater subsidence, such as the Solway–Northumberland and Stainmore 'Troughs' (Fig. 8.4) in which Namurian sequences are significantly thicker than on the blocks, but comparable in facies. Although the margins of the major blocks are now defined by faults, most of them probably acted as hinge-lines during sedimentation (Kent 1966), partly in response to differential tilting of the blocks (Ch. 7). Evidence for contemporaneous fault-movement is provided by an uncomformity developed below the late Pendleian Grassington Grits in the area just north of the Craven Faults but which disappears north of Swaledale (Wilson 1960, Ramsbottom 1974a).

In the lower Namurian sequences the southern margin of the Askrigg Block is marked by a conspicuous change in lithological character. The thicker successions of the Central Province (Craven Basin) are of the typically coarse 'Millstone Grit' facies described later whereas the thinner platform sequences are similar to the 'Yoredale' facies typically developed in the Dinantian. The Yoredale sequences (p. 000) involve repetition (upwards) of marine limestones, marine shales and relatively thin sandstones, on which lie seatearths or ganisters often topped with thin coals. The shales are frequently bioturbated and most of the cycles coarsen upwards, suggesting deltaic progradation. The extensive limestones are generaly thin and biomicritic, with a relatively restricted benthonic fauna suggestive of shallow, possibly hypersaline conditions (Ramsbottom 1974a). Goniatites are rare in these limestones, in contrast to their abundance in the mudstones of the basinal facies of the Central Province. Although the limestone horizons of the Namurian Yoredale sequences *may* represent the acme of marine transgression in a particular region, studies by T. Elliott (1974, 1975, 1976) demonstrate that a substantial part of the intervening clastic sequence may be attributed to post-abandonment reworking by shallow marine processes, rather than conforming to the simple model of deltaic progradation previously favoured (Moore 1959, Johnson et al. 1962).

Throughout much of the platform area the base of the Namurian (the *Cravenoceras leion* goniatite zone) occurs at the base of a thick bioclastic (locally biostromal) limestone – the Great or Main Limestone (Fig. 8.4). The frequency and thickness of the

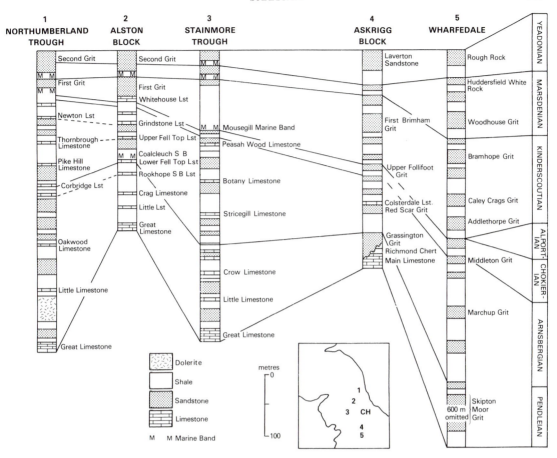

**Fig. 8.4.** Representative Namurian sequences in Northern England, with suggestive correlations. *Inset* indicates locations of measured sections: CH – Cleveland Hills (after fig. 10, Ramsbottom *et al.* 1978).

Yoredale limestones decline upwards, but they persist longer on the block areas (into the Kinderscoutian on the Alston Block). Around Richmond, thick cherts occur in the lower part of the Namurian, suggesting either enhanced marine solution of the locally abundant sponges or unusually silica-rich fluvial waters associated with the Pendleian deltas (Wells 1955). Similar Pendleian cherts in Yoredale facies, recorded from the subsurface in the Cleveland Hills area, have been interpreted as indicating the eastern margin of the Askrigg Block (Kent 1966, Ramsbottom 1974a).

The intra-Pendleian unconformity (Fig. 8.4), on the southern margin of the Askrigg Block, equates with a northward increase in thickness of early Namurian sequences (Rowell & Scanlon 1957), suggesting contemporaneous tilting of the block in this general direction. Moreover, the Lake District massif persisted as a positive area, as indicated by the probable absence of Namurian rocks from the Gosforth area and the attenuated Namurian succession of the Whitehaven and Caldbeck districts (Ramsbottom 1977). Here only the basal ($E_1$ and $E_2$) and topmost ($G_1$) stages are known and the intervening Hensingham Grit Group is a typical Yoredale sequence. It is unclear whether the palaeontological discontinuity here results from a genuine break in sedimentation or from highly condensed deposition, for which other evidence is lacking. North of Maryport more than 500 m of Hensingham Group sediments have been detected in boreholes, presumably thickening north-ward into the Solway trough.

On the eastern fringe of the Alston Block, the lower Namurian is divisible into two facies, separable by a boundary trending approximately NNE–SSW through Barnard Castle. West of this line 'channel' sandstones are common whereas they are lacking in the generally more muddy sequences to the south-east.

The lower Namurian cyclothems of northern England were probably formed by progradation and abandonment of high-constructive, lobate deltas (Elliott 1974, 1975, 1976; cf. Farmer & Jones 1969) which appear to have advanced broadly towards the south-east to south-west quadrant. Thus the 'channel' sandstones of the blocks probably represent distributaries of these early Namurian deltas.

The higher Namurian stages (post-Arnsbergian) are thin and poorly represented throughout the platform area (Fig. 8.4) and it has been suggested that the entire Alportian stage is missing north of the Craven Faults (Ramsbottom 1977). However, it appears that the coarser 'Millstone Grit' facies, confined to the Central Province in the early Namurian, migrated progressively northwards. Thus several of the highest sandstone units in the basinal sequence of the Central Province may be traced across most of the northern platform. Accordingly, by the end of the Namurian the facies distinctions between basins and blocks had been effectively eliminated and the only evidence for the positive character of the platform is the thinned later Namurian on the Alston Block and the recurrence of shelly, shallow marine units in the Millstone Grit sequences along the line of the still active Craven Faults (Ramsbottom 1974a).

Although the broad outlines of the Namurian evolution of northern England are now well established, several problems remain unresolved. One of the most important is the true palaeogeographical relationships between tectonic platform and basin. Apart from their finer grain-size, the sandstones of the Yoredale facies of the platform are petrographically similar to the coarse feldspathic sandstones of the Central Province whose provenance has been assigned to a metamorphic terrain (Gilligan 1920). Sedimentary sequences and directional structures indicate successive advances of the Millstone Grit deltas from the north and north-east while the sparse evidence from the finer platform deltaics tentatively suggests overall fluvial transport in a southerly direction (Fig. 8.2). It thus remains uncertain whether the delta-lobes of the platform are simply the lateral extensions of major systems prograding from the north-east and subjected to more intense marine reworking or are the products of direct input from the Scottish Highlands, as the sedimentological continuity of later Namurian sequences into the Scottish Midland Valley might

suggest. If the latter interpretation is correct there remains a major problem in explaining how the coarse Millstone Grit detritus 'by-passed' the northern region to accumulate in the Central Province.

## Westphalian

By the early Westphalian, a broadly uniform style of sedimentation had become established from the south Midlands to the Scottish Highlands.

Although all the principal Westphalian marine bands (Table 8.1) have been recognised in the basins of northern England they do not always yield the characteristic goniatites. Instead they commonly contain a benthonic fauna of calcareous and/or horny brachiopods, sometimes accompanied by cephalopods and foraminifera (almost invariably contained in a mudstone or shale). Combinations of these forms into faunal facies has enabled Calver (1968, 1969) to interpret the history of the successive marine incursions in terms of relative 'marine-ness' and thus, secondarily, of distance from the contemporary shoreline. Broadly, the progression from non-marine to most fully marine conditions is marked successively by the *Planolites*, *Ammodiscus*, *Lingula*, *Myalina*, productoid, pectinoid and finally, goniatite facies. The pectinoid and goniatite facies are typically developed in the Westphalian A marine bands whereas the *Myalina* and productoid facies are more characteristic of the Westphalian B and C. In northern England the index goniatites of the Subcrenatum Marine Band on the platform are replaced by the mixed benthonic productoid fauna of the equivalent Quarterburn Marine Band and its correlatives. This persistence of shallow marine faunal facies on the blocks provides virtually the only evidence for tectonic differentiation of blocks and basins during the early Westphalian. At this time a land area or zone of non-deposition may have lain in the vicinity of the Cleveland Hills or north Yorkshire (Calver 1968) while general thinning of the Coal Measures towards the Lake District suggests that this massif still influenced late Silesian sedimentation (Fig. 8.5). Westphalian rocks are most extensively exposed in the Northumberland–Durham coalfield and in a narrow zone in west Cumberland, fringing the northern Lake District. Small outliers occur in the Ingleton, Stainmore and Midgeholm areas and extend into the Canonbie basin of southern Scotland (Fig. 8.1).

The principal difference between the Namurian and the early Westphalian in northern England is a gradual decrease in the number and thickness of marine bands and a corresponding increase in the importance of coals and seatearths. Cyclic sequences of Coal Measures type are thus thinner and more numerous than in the Namurian.

The salient features of Westphalian sedimentation

and palaeoenvironments are discussed more fully later in the account of the Central Province. Only the local variations in this general pattern are outlined here.

In the thick succession (900 m) of the Northumberland–Durham coalfield the lowest part of Westphalian A displays a facies that is virtually identical with the underlying late Namurian. Coarsening-upward sequences, 15–30 m thick, capped by relatively thick, sheet-like coarse sandstones are reminiscent of the shallow marine delta progradation sequences of the late Namurian. Several marine bands occur in the basal 60 m but the coals are generally thin. Near the top of Westphalian A the sequence, while still dominantly sandy, has few marine horizons and coals and seatearths are more abundant, capping thinner cycles with a better-developed fluvial component. Subsurface mapping of the channel sandstones in the uppermost cycles (below the Harvey or Vanderbeckei Marine Band (Table 8.1 and Fig. 19.1b) reveals a southerly directed fluvial trend (Smith & Francis 1967) and a similar pattern obtains in broadly coeval sandstones around Tynemouth (Farmer & Jones 1969, Land 1974). Fielding (1984a, 1985) has suggested that the Westphalian succession was formed in a lower delta-plain setting, which evolved, though southward deltaic progradation, into an upper delta-plain by mid-Westphalian A times.

Above the Harvey marine band the Westphalian B or Middle Coal Measures cyclic sequences are thin but numerous. This part of the succession contains the largest proportion of thick coals and also the most prolific non-marine mussel bands. The cycles display great lateral variability with chanelling and seam-splitting more common than in the lower cycles. Haszeldine & Anderton (1980) suggested that the prominent sand-bodies within this mid-Westphalian sequence were formed as braided river sheet-sands, deposited on an alluvial coastal plain of low relief, dominated by tectonically generated palaeoslopes and local sediment sources, although details of this interpretation were questioned by Heward & Fielding (1980). The upper part of Westphalian B is marked by more muddy, coarsening-upward sequences, a decrease in the frequency and thickness of coals and a concomitant increase in the number of marine horizons – essentially a reversion to the lower delta plain conditions of the basal Westphalian. Fielding (1984b, 1985) and Haszeldine (1983, 1984b) later emphasised the lacustrine, inter-distributary character of much of the Westphalian A and B sequence in northeast England.

Westphalian C sediments, consisting of 'grey' facies mudstones, siltstones, thin coals and subordinate fine sandstones, are preserved near Durham and Sunderland, where their thickness is now considered to be about 350 m. These sediments are considered to mark a further phase of southwards deltaic progradation, leading to re-establishment of upper delta-plain conditions (Fielding 1984a).

To the west, the lower Westphalian successions of the Midgeholme and Stainmore coalfields are similar to those of NE England, but no Silesian higher than the basal part of the Westphalian B is preserved. However, the Canonbie Coalfield, just north of the Scottish border, contains, in addition, a thick succession of Westphalian C and possibly Westphalian D, sandstone-dominated and in 'primary' red-bed facies, with several thin freshwater limestone bands carrying coiled shells of the annelid, *Spirorbis* (Lumsden *et al.* 1967).

The West Cumberland Basin to the north and west of the Lake District displays a condensed Westphalian succession (450 m for Westphalian A and B stages) but is broadly similar in character to the Northumberland sequence. The Upper Coal Measures (Westphalian C), however, are represented by a thick group (300 m) of 'red-beds', including at least two *Spirorbis* limestones.

The small Ingleton coalfield includes an early Westphalian sequence (A and B stages) some 450 m thick, unconformably succeeded by 550 m of red sediments assigned to the upper part of Westphalian C on the basis of a non-marine bivalve fauna indicating the Phillipsii Chronozone (Ford 1954). The lower sequence is comparable in general character with the Westphalian A and B of the northern coalfields but the overlying red beds commence wth a thick intra-formational rudite followed by red and green sandstones, shales and fireclays with Spirorbis limestones and no coals. It is tempting to equate this red-bed development with the late Westphalian red measures encountered in Central Province successions but it is not yet clear how much of the reddening should be ascribed to leaching and oxidation below the arid Permo-Triassic land surface. Distinctive marginal facies, such as occur in the Etruria Marl and Keele Formations of more southerly areas (pp. 253–4), are rare in the north and the only evidence for possible marginal instability is the pre-Westphalian C discordance at Ingleton.

## Central Province

The Central Province lay to the south of the Askrigg Block and to the north of St George's Land (Fig. 8.2). The northern margin is defined by the line of the Craven Faults, but the southern margin is obscured by younger sediments. However it was probably a diffuse and rather variable hinge line. Within the Central Province, several topographic features were subject to variable rates of subsidence during the Namurian but were inherited from the Dinantian. The main positive element was the Derbyshire Block which played a major role in controlling Namurian deposition (Fig.

8.3). This area of Dinantian shallow water carbonate deposition stood as a submerged ridge in early Namurian times, flanked to the north and west by steep slopes (Broadhurst & Simpson 1973). To the north lay the main Pennine and Craven Basin separated by the poorly defined Rossendale Block (Miller and Grayson 1982), and subsidiary relief features such as the Edale Gulf and the Gainsborough Trough (Fig. 8.2). to the west lay the Staffordshire Basin, bounded to the west by a margin in the region of the present-day Red Rock Fault (Trewin & Holdsworth 1973). A submerged ridge probably separated the Staffordshire Basin from the Craven Basin. To the south of the Derbyshire Block, the Widmerpool Gulf was a further basinal area which may have been separated from the Staffordshire Basin by a ridge near the south-western corner of the Block (Trewin & Holdsworth 1973). The surface of the Derbyshire

Block probably sloped gently to the east.

To the west, the Craven Basin was probably connected with basins in Ireland but some shallowing and thinning in this direction is suggested by isopach data (Fig. 8.2 inset) and by the faunal facies of the marine bands (Ramsbottom 1969).

The theme of Namurian sedimentation is the infilling of this complex topography so that by the early Westphalian, more or less uniform coal measure environments were established. Silesian sedimentation was further controlled by a pattern of regional sibsidence which had its main centre in Lancashire (Fig. 8.5 inset).

## Namurian

Refined zonation based upon goniatites permits the detailed correlation of Namurian sequences required

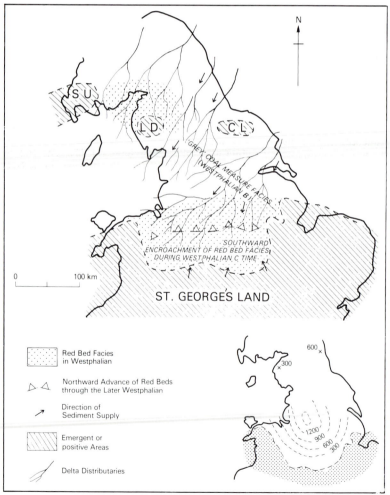

**Fig. 8.5.** Generalised Westphalian palaeo-geography of Northern and Central England and North Wales. Depiction of delta-distributary system is diagrammatic.

CL – Cleveland massif; L D – Lake District massif; S U – Southern Uplands massif.

*Inset* shows isopachs (in metres) of total Westphalian (after Calver 1969).

for precise reconstruction of the evolving palaeo-geography (e.g. Ramsbottom 1969, Bisat 1924). The goniatites and their associated marine fossils occur in discrete bands, separated by unfossiliferous sediment even in uniform mudstone successions (e.g. Rams-bottom *et al.* 1967). This suggests that water salinity fluctuated episodically and in some cases may have been associated with eustatic fluctuations in sea level (Ramsbottom 1977, 1978a, b, 1979, 1981). Deposition of unfossiliferous sediment which forms the greatest part of virtually all sequences took place in water which may have been almost fresh at times.

The Namurian sediments of the Central Province are almost exclusively detrital and range from mudstones to coarse pebbly sandstones. Two principal sources of sand were active. One, lying to the north or north-east supplied abundant feldspathic debris throughout Namurian times (Sorby 1859, Gilligan 1920) whilst the other, to the south and presumably within St George's Land, provided a less prolific supply of quartzitic material. The quartzose sands were confined to the southern part of the Province and were effectively swamped by the feldspathic supply in late Namurian times.

*The Craven/Pennine Basin*

The Namurian sediments of the Craven/Pennine Basin (Fig. 8.2) record the advance of a sequence of deltas from the north and east. Periods of deltaic progradation are recorded by sequences from tens to hundreds of metres thick which show a broadly up-wards coarsening trend. Periods of delta abandonment and transgression are recorded by thin bands of marine

mudstone which abruptly overlie the coarsening up-wards units. However not all the marine bands are associated with major transgressive events (McCabe 1975). The general succession has commonly been des-cribed as 'cyclic', the coarsening-upwards sequences between marine bands initially recognised by Wright *et al.* (1927), being examples of larger-scale 'cyclothems'. To label sequences in this way, however, is to imply similarity and at least two distinct types of sequence are recognisable, each representing the advance of a different type of delta (Fig. 8.6). The two types of sequence have been termed:

1. Deep-water, turbidite-fronted delta sequences,
2. Shallow-water, sheet delta sequences (Collinson 1976).

In addition, the Yeadonian Haslingden Flags of Lancashire appear to be an example of a third, minor type of delta comparable to the modern birdfoot deltas of the Mississippi (Collinson & Banks 1975). The characteristics of the two major types are outlined below:

1. Deep-water deltaic sequences (Fig. 8.6A) are usually several hundreds of metres thick and comprise a lower turbidite sequence, a middle, dominantly silty, upwards-coarsening sequence and an upper part dominated by mutually erosive channels filled with coarse pebbly sandstone (Walker 1966a, 1966b; Collinson 1969, 1970; McCabe 1975, 1978; Baines 1977). The full se-quence records the advance of deltas fed by large distributary channels. The most characteristic and

**Fig. 8.6.** Comparative sections through (A) Turbidite-fronted deep-water delta sequence (Kinderscoutian), and (B) Shallow-water, elongate delta (U. Haslingden Flags), which is similar to the sequence of a shallow-water sheet delta. Note the difference in the vertical scales.

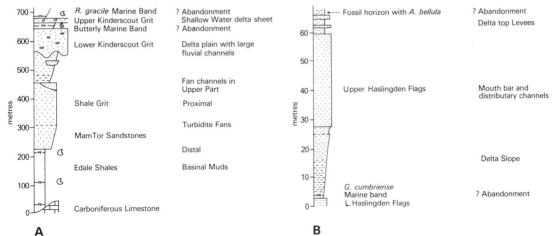

unusual feature is the delta-front apron of coalescing turbidite lobes composed of coarse sediment much of which by-passed the delta slope in feeder channels.

2. Shallow-water sheet delta sequences (Fig. 8.6B) are commonly tens of metres thick and consist of a lower unit coarsening upwards from mudstone to sandstone and a laterally extensive upper sandstone unit comprising channel fills and unchannelised deposits (Benfield 1969, Mayhew 1967, Okolo 1983). The sandstones may be capped by seatearths and thin coal seams. These sequences probably result from the advance of deltas lacking significant turbidite precursors, the sandstones representing a complex of mouth-bar and distributary channel sands such as that of the Lafourche delta of the Holocene Mississippi (Kolb & van Lopik 1966).

The oldest Namurian sandstones of the Pennine basin are the Pendleian Skipton Moor Grits and their correlatives (Fig. 8.7). These occur in the north of the basin and constitute a deep-water turbidite-fronted delta sequence. The turbidites rapidly thicken southwards away from the foot of the pre-Namurian topographic slope north of Skipton and extend into the basin at least as far as the Preston–Blackburn area. In the Roosecote Boreholes in Furness, thin turbidites

are present at this level (Aitkenhead & Jones 1971). The upper part of the Skipton Moor Grit sequence comprises distributary channel sandstones which link upstream (northwards) with the Grassington Grit which fills channels incised into the Dinantian limestones of the Askrigg Block (Wilson 1960).

In the Arnsbergian, the main site of deposition seems to have shifted to the west, where 400 m of delta front turbidites and slope deposits were laid down in the Bowland Fells area. These show evidence of slope instability in the form of slump scars, slump folds and sandstone dykes (Johnson 1981). After the early phases of vigorous supply, the northern source area seems to have become less active. Throughout the Chokierian and Alportian only thin sequences of probable sheet delta type were developed along the northern edge of the basin mainly north of Wharfedale (Ramsbottom 1974a). The causes of this quiescence are unknown but may include diversion of the main sediment supply to areas in the east or possibly to the Gainsborough Trough (Steel 1986), or changes in source area tectonics. In the south of the basin, none of these minor phases of deltaic activity can be detected prior to the late Kinderscoutian. In North Derbyshire, the Edale Shales accumulated in quiet deep water conditions, the only variation being in salinity as attested by the many marine bands (Hudson & Cotton

**Fig. 8.7.** Ribbon diagram showing distribution of principal sand-bodies in the Craven/ Pennine Basin Namurian and their postulated depositional environments. (After Ramsbottom 1969.)

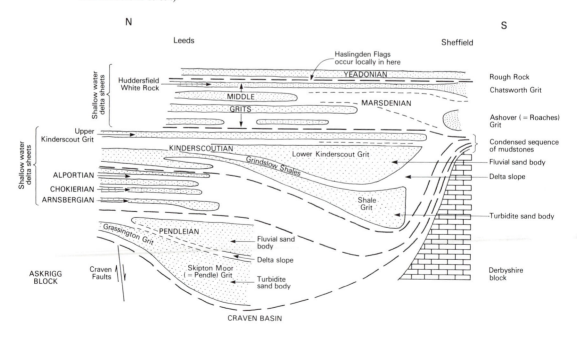

1945, Ramsbottom *et al.* 1967). Thin bentonites also punctuate the lower part of the mudstones which are banked up against the relief formed by the Derbyshire Block limestones (Trewin 1968).

A new phase of major deltaic advance began in the Kinderscoutian, again involving a turbidite-fronted delta (Figs. 8.6A, 8.7). In Wharfedale, the sequence is relatively thin with only 15–40 m of turbidites underlying the slope deposits and the distributary channel sandstones of the Caley Crags Grit (McCabe 1975). To the south the sequence becomes thicker, and especialy the basal turbidite unit, so that in the Edale area the total delta sequence is some 600 m thick. The Edale Shales are overlain abruptly by the turbidites of the Mam Tor Sandstones (Allen 1960) which pass up into the thicker, more erosional turbidites of the Shale Grit (Walker 1966a). The overlying Grindslow Shales and the lower Kinderscout Grit represent the prograding delta slope and the fluvial distributary channels respectively (Collinson 1969, McCabe 1978). The southwards thickening of the sequence can be related in part to the growth of delta beyond the limits of the Pendleian delta system. The Kinderscoutian delta advanced southwards to the Derbyshire Block but its westwards advance ended near Blackburn. Here delta top conditions were not established until Marsdenian times (Collinson *et al.* 1977) and the Marsdenian and Yeadonian sediments record the advance of a series of shallow water sheet deltas, each advance separated by deposition of a marine band. The Rough Rock, which ends the Namurian sequence, is a sheet sandstone which extends over not only the Pennine Basin but also most of the Central Province (Shackleton 1962).

### Staffordshire Basin and Widmerpool Gulf

These two basins have much in common. Until late Kinderscoutian times, clastic input was entirely from the south. Early Namurian sediments consist of mudstones and siltstones with thin quartzitic sandstones in the more central parts of the Staffordshire Basin, and in the Widmerpool Gulf (where the evidence is confined to boreholes (Aitkenhead 1977)). Palaeocurrents from south-derived sandstones in the Staffordshire Basin fan out to the north and the sandstones thin down-current. The turbiditic sandstones are interbedded in the Pendleian and early Arnsbergian with calcareous siltstones, thought to have been derived by density underflows from a shelf area which probably lay to the west. The calcareous siltstones form part of a simple cyclic pattern in both basins over this restricted time interval. The cycle: marine mudstone–calcareous siltstone–quartzitic turbidites, may represent a response to eustatic basin level change but the marginal deposits of the basins are not well enough known to test fully this suggestion

(Trewin & Holdsworth 1973). Towards the Derbyshire Block in the Staffordshire Basin, turbidite sandstones die out and mudstones and siltstones drape the Dinantian topography. The thin, shallow water sequence seen at Astbury, south of Congleton (Evans *et al.* 1968, Trewin & Holdsworth 1973) and the absence of pre-Yeadonian Namurian sediments in the Bowsey Wood Borehole near Madeley (Earp 1961) suggest that a tectonically active western margin of the basin may have lain close to the line of the present Red Rock Fault.

Only along the southern edge of the Namurian outcrop is there any evidence of deltaic advance. Distributary channel sandstones and other shallow water sediments occur sporadically in the Arnsbergian to early Marsdenian interval (Ashton 1974, Bolton 1978). Simultaneously the elevated area separating the Pennine Basin from the Widmerpool gulf accumulated a condensed sequence of mudstones within which the marine faunal horizons are well developed where developed in boreholes at Ashover, Derbyshire (Ramsbottom *et al.* 1962).

Not until the uppermost Kinderscoutian times did feldspathic sandstones of northern provenance arrive in the Staffordshire Basin (Fig. 8.2). The Longnor Sandstone (equivalent to the Upper Kinderscout Grit of the Pennine Basin) is a feldspathic turbidite unit with southerly palaeocurrents (Ashton 1974). It probably records overtopping of the barrier postulated for the NW corner of the Derbyshire Block and is the precursor of the early Marsdenian main basin-filling sequence – the Roaches Grit of Staffordshire and the Ashover Grit of Derbyshire. This turbidite-fronted deep-water deltaic sequence is anomalous in that palaeocurrents in both the Staffordshire Basin and in the Ashover area indicate flow to the NW despite the northern provenance suggested by the petrography. The paradox is perhaps resolved if the main river system which delivered sediment from the north flowed to the east of the present outcrop before swinging westwards along the line of the Widmerpool Gulf, filling it in the process and eventually feeding sediment into the Staffordshire Basin from its southeastern corner (Jones 1977, 1980).

Later Namurian sediments, such as the Chatsworth Grit and the Rough Rock sequences follow the pattern established in the Pennine Basin and are shallow-water sheet delta sequences. Palaeocurrents suggest that by this time the Derbyshire Block was no longer a significant influence on deposition (Mayhew 1967, Shackleton 1962, Kerey 1978).

To the east, in Nottinghamshire, the upper Namurian sandstones are reservoirs in several small oil fields and high Namurian sediments probably lapped southwards on to the northern margins of St George's Land (Ramsbottom *et al.* 1978).

*North Wales*

The Namurian sediments are exposed in a narrow N–S outcrop between Point of Ayr and Oswestry (west of the Flint and Denbigh coalfields, Fig. 8.1). They were deposited near the SW margin of the Central Province though they are now separated from the main outcrop by the Permo-Triassic of the Cheshire Plain. The lower sandstones of the sequence are quartzitic but feldspathic sands occur near the top. The Namurian section thickens northwards and becomes more complete and also less sandy, suggesting more basinal conditions (Ramsbottom 1974b). The lowest Namurian sandstone, the partly Arnsbergian Middle Cefn-y-Fedw sandstone, dies out northwards from its maximum development around Ruabon Mountain. The higher Dee Bridge Sandstone is feldspathic and of Kinderscoutian or possibly Marsdenian age. The more extensive Aqueduct Grit of the Ruabon area and its probable correlative the Lower Gwespyt Sandstone of northern Flintshire are Yeadonian and probably represent the furthest known extension of the Rough Rock of the Central Province.

## Westphalian

Westphalian sediments of the Central Province include some of the main coal deposits of England and Wales (Chapter 19). They have also provided economically important quantities of ironstone, ceramic clays and refractories. This intense economic exploitation has provided an unrivalled wealth of local stratigraphical information from which it is possible here to identify only a few broad trends of palaeogeographical evolution.

The Westphalian Coal Measures of the Central Province were laid down in one large depositional area (Fig. 8.5). The topographical variation seen in the early Namurian had been eliminated, though the same broad pattern of subsidence persisted with its maximum in south Lancashire (Calver 1969). Present outcrops of the Coal Measures (Fig. 8.1) thus are fragments of this area preserved in discrete coalfields whose position is determined by later folding and faulting. The main coalfields of Yorkshire–north Derbyshire–Nottinghamshire (East Pennine) and Lancashire–north Staffordshire lie on either side of the Pennine Anticline whilst other areas of preserved and accessible Coal Measures occur on horsts, as in the Leicestershire, Warwickshire, south Staffordshire, Coalbrookdale and the Wyre Forest Coalfields and in concealed extensions beneath Mesozoic cover such as the Vale of Belvoir and Oxfordshire (Poole 1971). The Flint–Denbigh coalfield, detached from the others by a thick Triassic cover, has a similar stratigraphical sequence to other Midlands areas.

Westphalian sediments extend well beyond the area of Namurian deposition, particularly in the south where Coal Measures overstep on to older (i.e. pre-Carboniferous) rocks in areas which had previously been part of St George's Land. This depositional encroachment to the south took place gradually throughout the Westphalian. The earliest sediments in these southern areas are normally of Coal Measure facies but local red beds of early Westphalian age are known from areas close to St George's Land, such as South Staffordshire and the Wyre Forest (Fig. 19.1a). However, in the main basin to the north the base of the Westphalian is not marked by a sudden change in depositional conditions as compared with the Namurian. In the Lancashire, north Staffordshire and east Pennine Coalfields the base of the Lower Coal Measures (Westphalian A) is marked by the Subcrenatum Marine Band (so named because of the occurrence of *Gastrioceras subcrenatum*). The measures occur as widespread upwards-coarsening sequences sometimes several tens of metres thick (but in the lowermost measures much thinner) and with well-developed marine bands at their bases. Bivalve escape shafts in interbedded sandstone–siltstone sequences suggest locally rapid deposition, possibly under the control of a monsoonal climatic regime (Broadhurst *et al.* 1980). The cycles can be capped by thick sheet sandstones such as, near the base of the measures, the Crawshaw Sandstones or Woodhead Hill Sandstone of Derbyshire and Staffordshire and, higher in the succession the Elland Flags or Old Lawrence Rock of Yorkshire and Lancashire. The thicker sequences compare with the shallow-water sheet deltas of the underlying upper Namurian and appear to be fed from the same northern source area.

The Lower Coal Measures succession, however, does mark a gradual transition towards the more typical pattern of coal measure sedimentation best developed in the Middle Coal Measures (Westphalian B). Here the sequences between coal seams or seat earth horizons are on a smaller scale, seldom more than a few tens of metres thick and averaging 9–10 m in the east Pennine Coalfield (Duff & Walton 1962). The coals themselves are up to several metres thick. Marine bands here are much less important than in the Lower Coal Measures and the inverse relationship between the importance of marine bands and occurrence of thick coal seams reflects progressive removal of marine influence from the area. The sequences between the coal seams and seat earths are highly variable both between sequences and laterally within the same sequence. They seldom match the 'ideal' sequences or 'cycles' used to typify Coal Measure sedimentation and the most common sequence is made up entirely of mudstone (Duff & Walton 1962). The individual sequences are commonly less extensive laterally than are the thicker Millstone Grit 'cycles'.

Faunal mudstones often occur at the bases of sequences and form the roof rock of the underlying coal seam. They most commonly contain non-marine bivalves ('mussels') and, more rarely, marine faunas (Calver 1969) (see Table 8.1). The latter fauna occur in distinct bands which are all-important in correlation. In the East Pennine Coalfield there are 19 but fewer have been recognised in the other coalfields (see Figs. 19.1a and b).

Cutting through the succession of coal seams and finer sediments are elongate channel sandstone bodies which may be more than 10 m thick and commonly are of limited width (up to a few hundreds of metres). Where these have been eroded down into coal seams they may constitute 'washouts' which form mining hazards.

This lithofacies assemblage reflects the development of extensive swamps through which flowed large distributary streams. Between the channels, shifting and changing freshwater lakes and lagoons were filled by the advance of small deltas and crevasse splays as river flow and sediment load were diverted (Elliott, R. E. 1969; Fielding 1984b, 1986a; Guion 1984). Such episodes of lake-filling gave rise to clastic sediment sequences and to the development of near-emergent surfaces suitable for plant colonisation, the plant communities being closely related to sedimentary setting (Scott 1978). Once established, plant growth and productivity was commonly able to keep pace with subsidence and compaction, allowing a thick peat to accumulate and to be preserved below a high water table. Eventual death of the swamp vegetation led rapidly to re-establishment of a lake which non-marine bivalves were able to colonise.

In areas of lower subsidence, thick coals were able to develop whilst with more rapid subsidence two or more equivalent seams were separated by clastic sediment sequences (e.g. Fielding 1984a, 1986b). The 'splitting' so 'produced is particularly well seen along the southern margin of the main basin where, for example, four thick coals in south Staffordshire between the 'Bottom' and the 'Thick' coals equate with nine thinner seams in the thicker equivalent succession near Cannock (Mitchell 1954). Areas of low subsidence, on the southern fringes of the basin are also characterised by abundant seatearth horizons representing near-emergent surfaces which were not sinking rapidly enough for peat to accumulate.

The base of Westphalian C is marked by the extensive Aegiranum Marine Band (*Anthracoceras aegiranum*) and the period as a whole is characterised by the gradual disappearance of important coal seams and the incoming of red beds. The distribution of red beds within the Coal Measures is complex (Fig. 19.1b) and made more confusing by the difficulty of distinguishing between pene-contemporaneous Coal

Measure reddening and later deep reddening below the Permo-Triassic unconformity (Trotter 1953, Hoare 1959). Pene-contemporaneous red beds were formed as early as Westphalian A and B in small areas on the southern fringes of the basin (e.g. Wyre Forest, Worcestershire) (Poole 1965). Throughout Westphalian C times this red bed facies gradually extended northwards until it occupied most of the Central Province (Fig. 8.5). 'Normal' Coal Measure conditions persisted in north Staffordshire long enough for economically important coals (e.g. Great Row) and ironstones (Black Band Formation) (Fig. 19.1b) to be laid down during Late Westphalian C whilst to the south, the red mudstones of the Etruria Formation and to the west (in present-day North Wales) the Ruabon 'Marls' were forming. The Etruria Formation red beds are dominantly non-calcareous mudstones and siltstones with seatearths and complex soil horizons but lacking in coal. They indicate more oxidising conditions within the sediment soon after deposition possibly due to a lowered water table (Besly 1983, Besly and Turner 1983). Those of north Staffordshire are thicker than elsewhere even though their base is younger (Gibson 1905, Cope 1954), reflecting more rapid subsidence in the basin. Occurring unpredictably within the fine-grained sequence are coarse and highly immature sandstones characterised by volcanic and sedimentary rock fragments of southerly derivation (Williamson 1942). These so-called 'Espley' sandstones are generally of limited lateral extent and many appear to be products of sinuous river channels. The northwards extension of red beds, accompanied by the increasing importance of a southern source area, is probably related to increased tectonic activity during latest Westphalian C. This activity led, within the depositional area of the Midlands, to local block-faulting and folding of earlier and pre-Westphalian rocks so that over much of Shropshire, Staffordshire and Warwickshire an unconformity separates sediments of Westphalian D age from older rocks (Ramsbottom *et al.* 1978). This unconformity is at its most spectacular in the Coalbrookdale (Fig. 8.1) area where, as the 'Symon' Unconformity or 'fault' it separates folded Middle Coal Measures from the Coalport Formation. This phase of tectonic activity was also marked by the intrusion of dolerites in south Staffordshire at Rowley Regis and Wednesfield but these intrusions are too late to account for the volcanic rock fragments of the 'Espley' sandstones. Evidence of earlier Westphalian igneous activity (apart from 'Tonsteins') comes from the Vale of Belvoir and elsewhere in the East Midlands where basalt lava flows are associated with thick coals of Westphalian A age (Ramsbottom *et al.* 1978, Plate 2; Kirton 1984).

The base of Westphalian D is arbitrarily equated (Fig. 19.1b) with the base of the Newcastle ( = Halesowen, = Coed yr Allt, = Coalport) Formation. In the south this Formation rests on the intra-Westphalian unconformity, and records a return of more normal, coal-measure conditions throughout the Central Province and an extension of the area of deposition beyond what had previously prevailed. In Shropshire and Oxfordshire, the Westphalian D sediments rest unconformably on older Palaeozoic rocks and a link may have been established with the then highly active 'Pennant' basins of South Wales (Kellaway 1970). Subsequently deposition of red beds was re-established, resulting in accumulation of the Keele (= Erbistock) Formation, a poorly exposed but thick sequence of alluvial sediments, possibly of Stephanian or Autunian age (Wagner 1983).

## South Wales and southern England

This region comprises the main coal basin of South Wales, with Pembrokeshire (Dyfed), the Forest of Dean and Bristol areas and the concealed basins of Oxfordshire and Kent (Fig. 8.2). Older Palaeozoic rocks of the St George's Land massif define the northern limit of this region and are known both at outcrop and (east of the Severn) from borehole and geophysical evidence (Dunham & Poole 1974, Bisson et al. 1967). The southern edge of this massif acted as a hinge-line throughout much of the Silesian, mirroring the behaviour of the northern flank of St George's Land (Fig. 8.8). The southern limit of the basinal region

is concealed below the Bristol Channel to the south or younger sediments to the east. Relationships with the Culm successions of the south-west region are further obscured by suspected tectonic dislocation (see later).

Important transverse structural elements (the N–S aligned Usk and Severn axes) influenced Silesian sedimentation in the vicinity of the Severn Estuary and similar 'Malvernoid' structures probably existed farther to the east (Kellaway & Hancock 1983).

## Namurian

### South Wales

In this area goniatite zonal chronostratigraphy has demonstrated the diachronous character of the traditional lithostratigraphic divisions of Basal Grits, Middle Shales and Farewell Rock (in ascending order), which constituted the Millstone Grit of earlier writers (see Jones 1974). Part of the Plastic Clay Beds and possibly part of the Upper Limestone Shales of areas west of the Swansea Valley may belong to the Pendleian stage (Woodland & Evans 1964, Kelling & George 1971) while quartzose conglomerates of Basal Grits facies occur within the *Dibunophyllum* subzones (Brigantian) of the Neath Valley (Owen & Jones 1961). Much of the Farewell Rock sandstone on the North Crop of the main coal-basin overlies the Subcrenatum band and is thus Westphalian (Leitch et al. 1958). The tripartite rock-stratigraphic division outlined above is thus misleading and should now be discarded.

**Fig. 8.8.** Generalised Namurian palaeography for South Wales, Southern and South-west England. Note that Early Westphalian data from Devon and north Cornwall are included.

B C L – Bristol Channel landmass. RF – Ritec Fault.

*Inset* shows isopachs (in metres) of total Namurian in South Wales (after Ramsbottom 1969).

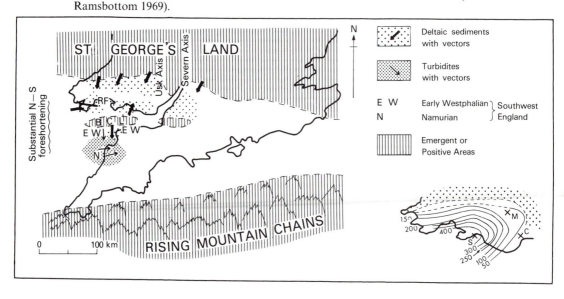

The most complete Namurian successions (up to 650 m) occur in Gower (see e.g. Ramsbottom 1978a) and the area north of Swansea (Fig. 8.8). Progressive thinning and overstep of the basal conglomerates and sandstones on to successively lower stratigraphic horizons occurs to the east and has been ascribed to contemporaneous uplift of the Usk Axis (Fig. 8.8; see George, T. N. 1956, 1970). Quartzose sands and gravels within the late Dinantian of these eastern and northern areas (p. 226) also reflect premonitory uplift of St George's Land, followed by an early Namurian transgression. This resulted in diachronous deposition of the coarse basal facies on the bevelled surface of Dinantian and older units (Jones 1974).

Within the northern part of this more restricted marine basin a basal sandy and pebbly facies is assigned to the Pendleian–Alportian stages in the upper Swansea Valley–Ammanford area but only the Kinderscoutian and Marsdenian stages are represented in the thin basal sandstones farther to the west, near Haverfordwest (Jones 1974). These texturally mature quartzose sandstones and conglomerates were formed by littoral and sublittoral sand-bodies advancing northwards on to the flanks of St George's Land. On the southern limb of the South Wales syncline the early Namurian is represented by thick mudstone successions which locally contain thin radiolarian cherts. Here, also, the marine bands belong to the goniatite–pectenoid biofacies (Ramsbottom 1969), consistent with an offshore muddy shelf environment (Kelling 1974).

In SW Wales the early Namurian sediments were deposited in a NW-trending embayment, flanked by deltaic lobes to the west and north. Muds and silts of lagoonal to prodelta character dominated the embayment and accumulated to considerable thickness, for example near Tenby, where the mudstones coarsen upwards into distributary mouth-bar sand-bodies (George and Kelling 1982).

Major 'cyclothems' (the *mesothems* of Ramsbottom 1978a) can be identified in the later Namurian (Kinderscoutian–Yeadonian) successions of the main basin on the basis of widespread and thick marine bands. They generally coarsen upwards but are irregular in detail and vary in thickness from about 25 to more than 200 m. They are dominated by south-flowing fluvial and localised delta-plain sediments in the northeastern areas (around Brynmawr), while littoral and sublittoral sequences formed farther to the west and south (Kelling 1974). Following the *Reticuloceras superbilingue* marine transgression a major delta lobe advanced southwards from the Llandebie area of the North Crop (Fig. 8.1). Simultaneously, the offshore marine muds of the South Crop (Fig. 8.1) were interrupted in the area east of Port Talbot by the northward progradation of a small sandy delta,

presumably receiving detritus from a resuscitated Bristol Channel land-mass (Fig. 8.8) (see p. 256).

The development of littoral sands and lagoonal muds in the middle Marsdenian marks a major regressive phase. However, the succeeding *Gastrioceras cancellatum* marine incursion heralded basin-wide deposition of muds of pene-marine to brackish character. In the north-eastern and eastern areas of the main basin, sand-filled fluvial channels of Late Namurian age, indicating flow to the south or south-west, are incised into older littoral sediments. Co-evally, the Aberkenfig delta-lobe of the South Crop advanced north-westwards, ultimately merging with the southward prograding Llandebie system in the period following the short-lived *G. cumbriense* transgression (Kelling 1974).

In Pembrokeshire, deposition during the later Namurian occurred within the NW-trending embayment described above. Following the *R. superbilingue* marine transgression prolonged deposition of silty or muddy sediments took place within lagoonal, mudflat and possible mouth-bar environments and a small, southwards-advancing delta lobe (near Marros). The *G. cancellatum* marine incursion gave rise to widespread offshore muds and, locally, prodelta silt-turbidites, succeeded abruptly in the Tenby area by wave-agitated and channelled sands marking the eastwards progradation of another delta lobe. During the later Yeadonian regional uplift in Pembrokeshire resulted in the establishment of fluvial channel facies throughout most of this sub-basin, forming the local Farewell Rock. Internal features and lateral relationships of these channels suggest that they represent elements within a delta-distributary network (George 1982). A zone of tectonic instability, along the line of the Ritec Fault (Fig. 8.8), separated a northern region, marked by south-flowing fluvial complexes, from a southern region with a single river-system flowing eastwards (George, G. T. 1970, Kelling 1974).

### Forest of Dean and Bristol–Somerset

Namurian rocks appear to be absent from the Forest of Dean, presumably because of continued uplift along the line of the Usk Axis. The poorly exposed Namurian sequence of the Bristol area formed in a discrete basin and is represented by a dominantly sandy sequence, the Quartzitic Sandstone Group (Kellaway & Welch 1955). Only one goniatite specimen (*Eumorphoceras*) is known from near the base of this formation but floral evidence from the Ashton Park borehole indicates the Pendleian to Marsdenian stages (Kellaway 1967). The facies appears to be littoral to sublittoral. The Quartzitic Sandstone Group becomes thinner to the south of Bristol and thickens northwards to about 300 m at Yate, Gloucestershire.

Recently the Rodway Beds, near Cannington, Somerset, formerly considered Devonian, have yielded goniatites indicative of the *G. cancellatum* horizon. In the Withiel Farm borehole some 138 m of these beds are in faulted contact with Dinantian rocks (Whittaker 1975). The abrupt southerly attenuation of the Quartzitic Sandstone Group of the Bristol district together with the evidence from Somerset suggests that in the intervening region there may be an easterly extension of that Bristol Channel landmass which was intermittently uplifted during the later Namurian (Owen 1964).

## Westphalian

### South Wales

The structural and morphological framework of early Westphalian (Lower and Middle Coal Measures) deposition was broadly similar to that of the later Namurian (cf. George, T. N. 1962). Thus there is progressive thinning of the Lower and Middle Coal Measures as they are traced eastwards (towards the still active Usk Axis), from a maximum of about 800 m in the Ammanford–Swansea area to less than 200 m on the East Crop near Pontypool (see Fig. 8.9 and Thomas 1974). However, this attenuated eastern sequence still retains its full stratigraphic integrity.

The Subcrenatum Marine Band attains a thickness of nearly 25 m in the western part of the main basin. This marine incursion, together with those represented by several thinner marine bands in the basal Westphalian of South Wales, interrupted the establishment of paralic conditions throughout most of the areas on the southern flank of St George's Land, a process which had commenced in the late Namurian.

Compared to the Namurian the lower Westphalian sequences reveal an increase in the number and thickness of coals and seatearths and a corresponding decrease in the frequency of marine bands (Fig. 19.1a). The Lower and Middle Coal Measures successions are generally muddy and thick, widespread sandstones are rare. The few thicker sand-bodies represent localised activity of deltaic distributary channels (Fig. 8.10a).

In the north-eastern part of the main basin the channelled Farewell Rock (post-Subcrenatum) records southward progradation of a comparatively persistent delta-lobe, while several lenticular coarse and pebbly sandstones on the East and South Crops formed as littoral and sublittoral sand-bodies. In Pembrokeshire the early Westphalian is represented by fluvio-deltaic facies, with well-developed delta-plain sequences fed by meandering distributaries flowing mainly to the west and south-west (Williams 1966).

The Westphalian B stage was marked throughout South Wales by deposition of mudstones, on a low-lying coastal plain subject to frequent, often local, invasion by a shallow sea, together with thick, extensive coals, seatearths and numerous ironstone bands and non-marine 'mussel'-bearing units deposited in swamps and fresh or brackish lakes and lagoons. Minor deltaic input from a degraded St George's Land to the north has been identified from the north-eastern part of the main basin (Thomas, L. P. 1967), but evidence from western areas, including Pembrokeshire, indicates the northwards progradation of a fluvio-deltaic complex derived from an uplifted Bristol Channel source-land (Williams 1968, Kelling 1974).

This southerly source was also responsible for the

**Fig. 8.9.** Generalised cross section of Westphalian deposits in the main coalfield of South Wales, illustrating the marked easterly thinning of the succession and the development of the "Pennant-type" sandstones above the *A. cambriense* marine band horizon. The progressive stages of change in coal rank (from bituminous coals in the east to anthracites in the west) are also illustrated. (After George 1970.)

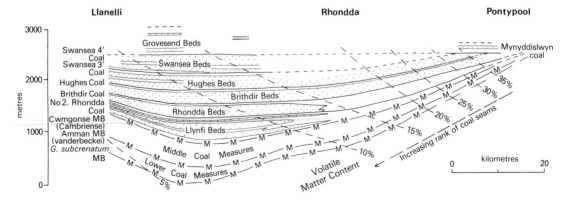

(a)

(b)

**Fig. 8.10.** Typical Westphalian sedimentary features in South Wales (G. Kelling):
  (a) Channel margin with associated soft-sediment deformation features, delta-front
      sequence; Middle Coal Measures (Westphalian B), coast N. of Wiseman's Bridge,
      Sandersfoot, Dyfed, South Wales.
  (b) Medium-scale cross-stratification grading up into large-scale cross-strata, from
      south-derived fluvial sandstones (aggrading lateral bar sequence); middle Pennant
      Measures (Westphalian C), Tonypandy, Mid-Glamorgan, South Wales (height of
      crag is 5 m).

localised supply of fluvial sand to the developing Margam–Maesteg 'sag' towards the end of Westphalian B time (just prior to the Cefn Coed (Aegiranum) marine transgression).

The early part of the Westphalian C (prior to the Upper Cwmgorse (Cambriense) Marine Band) is again marked by muddy sediments associated with backswamp and lacustrine delta-plain environments. South-transported channel sands are confined to eastern and central regions of the main basin. However, the earliest fluvial sandstones of lithic character ('Pennant type') appear in Mid Glamorgan at this time and indicate a renewal of supply from the south and south-west (Fig. 8.10b).

Elliott & Ladipo (1981) recognised syn-sedimentary gravity slides (growth faults) within these mid-Westphalian sediments on the South Crop, NW of Cardiff. These features are listric normal faults of ephemeral character which affect a discrete sedimentary interval during deposition, and are common in major present-day deltas, such as the Mississippi. Although the examples described are modest in scale, some larger strike-faults in the southern part of the coalfield (such as the Jubilee Slide and perhaps the Moel Gilau) may have originated in a similar syn-sedimentary fashion.

Following the short-lived G. cambriense marine transgression (the last such event recorded in the Silesian rocks of South Wales), a profound change occurred in the character and environment of sedimentation. This change is attested by the appearance of thick bodies of highly lithic sandstone: the so-called 'Pennant' sandstones, giving rise to the term 'Pennant Measures' as a synonym for the Upper Coal Measures in this region. These distinctive sandstones appear earliest in the west and south-west and spread progressively north-eastwards.

The Pennant Measures achieve their maximum thickness of 1,600 m in the Swansea area and thin towards the north-east, but not as conspicuously as the earlier Silesian formations (see Fig. 8.9 and Owen 1964), suggesting that the influence of the Usk Axis to the east was waning.

The Pennant Measures are dominated by thick sandstones with subordinate mudstones and a few thick coals. They were formed in the channels and floodplains of a broad alluvial tract as north-flowing streams of low to moderate sinuosity transported huge volumes of lithic detritus into the actively subsiding South Wales basin. This material evidently was derived from the rapidly rising frontal chains of the Hercynian foldbelt to the south (Fig. 8.11). Red and green lutites and quartz-arenites (some pebbly) that occur in the lower part of the Pennant Measures of the eastern area have been ascribed to penecontemporaneous derivation from red upland soils

(Downing & Squirrell 1965). It has been argued (Kelling 1974) that these lower sequences represent littoral and lagoonal environments, possibly formed on the shores of a shrinking body of marine (or more probably brackish) water flooding the southern margins of a now-peneplaned St George's Land.

Westphalian D sequences, especially those in Pembrokeshire, record the further withdrawal of the sea and the establishment of a major NW-flowing alluvial system, while development of the localised sub-Mynyddislwyn unconformity towards the East Crop suggests some recrudescence of the Usk Axis. Some of the uppermost Pennant Measures are of red-bed facies and, moreover, have yielded a flora considered to belong to the Cantabrian stage of the Stephanian (Ramsbottom et al. 1978, p. 8). Such beds presumably reflect the onset of the conditions which were to characterise the succeeding Permo-Triassic period.

An interesting and economically important aspect of the South Wales Westphalian is the marked increase in coal rank (decrease in volatiles) which occurs when individual seams are, or the general succession is, traced from the south-east to the north-west part of the main basin, culminating in the anthracite belt of the Ammanford area (Fig. 8.9). Since this is the most intensely deformed part of the coalfield tectonic processes are a plausible explanation for the rapid increase in rank (Trotter 1948) although geochemical studies suggest that anthracitisation may be linked to the Hercynian phase of mineralisation (Davies & Bloxam 1974) but see Chapter 19).

### Forest of Dean, Oxfordshire and Bristol–Somerset

Separated from the Bristol basin by the Severn estuary, the Forest of Dean (Fig. 8.1) is essentially a Silesian outlier resting unconformably on the Dinantian or Old Red Sandstone. The oldest Silesian in this area formerly was considered to be the upper part of the Drybrook Sandstone, miospores from which were thought to suggest a Westphalian A age (Sullivan 1964). However, recent work indicates a Visean age for this sequence (Cleal 1986). The overlying Trenchard Formation (Fig. 19.1b) is regarded as Westphalian C on rather inadequate evidence, pointing to the possibility of a mid-Silesian unconformity. Two distinctive Trenchard lithofacies have been identified: a conglomeratic and red-bed facies in the north, and a grey fluvial sandstone facies in the south, the latter yielding palaeocurrents from the south (Stead 1974). Coal-seams in this interval split towards the north.

The succeeding Pennant and Supra-Pennant Groups attain a maximum thickness of about 580 m (in the south) and belong to the Westphalian D stage. These formations are lithologically and compositionally identical with the Pennant Measures of South

Wales and, like the latter, they record the rapid northward advance of a fluvial complex conveying detritus from the newly uplifted orogenic belts to the south (Fig. 8.11).

To the north-east, boreholes into the concealed Oxfordshire coalfield have revealed a succession of up to 1,000 m of Westphalian D sandstones, with subordinate shales and up to eight coal seams of potential economic interest (Dunham & Poole 1974). These sediments rest discordantly on Devonian and Silurian rocks and include (at Steeple Aston) a series of basic intrusives and basalt-breccias, thought to represent a volcanic plug injected in late Carboniferous or early Permian times. Through the Withycombe Park borehole (Poole 1978) this sequence, spectacularly reddened in its upper part, can be traced north into the exposed Warwickshire coalfield. The Forest of Dean and Oxfordshire successions clearly indicate that by late Westphalian times St George's Land was no longer an effective barrier to sedimentation and had been overwhelmed by the flood of detritus from the south.

The Westphalian A and B stages of the poorly exposed Bristol–Somerset area are represented principally by mudstones with several economic coals, and attain a thickness of about 420 m. The Subcrenatum band is locally thick and contains a varied fauna (Calver *in* Kellaway 1967) and the general aspect of the sequence is consistent with formation on a low-lying coastal plain subject to periodic marine invasion. The Upper Coal Measures, by contrast, are a thick (nearly 2,000 m) typical 'Pennant-type' alluvial sandstone sequence, closely comparable with the Pennant Measures of South Wales and the Forest of Dean (Stead 1974), and also derived from the south. Red and green clays and lenticular pebbly sandstones similar to the northern facies of the Trenchard Group, occur below the main Pennant succession in the north Bristol basin.

*Kent*

First discovered in the last decade of the 19th century, this structural basin of Carboniferous rocks (Fig. 8.1) is concealed beneath a minimum of 320 m of Mesozoic and Cainozoic sediments. A maximum of about 900 m of Westphalian rocks in present, resting unconformably on the Dinantian or Devonian (Fig. 19.1a). Namurian rocks are absent.

The Westphalian A and B successions (225 m approx.) are predominantly muddy, with several marine bands, while coals and sandstones are more frequent to the west (Stubblefield 1954). Above the local equivalent of the Aegiranum marine band the Westphalian C succession is represented by little more than 60 m of mudstones and thin sandstones capped by an important coal (Kent No. 6). Bivalve faunas associated with this seam indicate a position within the *tenuis* chronozone of the Westphalian D stage and indicate a local non-sequence involving the missing *phillipsii* chronozone (Ramsbottom *et al.* 1978).

The 600 m succession assigned to the Westphalian D stage is dominated by lithic sandstones of Pennant-type but includes some thick shales and several economic coals. Rocks of red-bed facies are absent from this sequence.

Thus, in the Kent coalfield the early Westphalian was dominated by deltaic deposition of relatively fine-grained sediments. Following the Aegiranum marine transgression, however, alluvial processes introduced great volumes of coarser sediment into this basin, giving rise to thick, multistorey channel-fill sandstones in the Westphalian D succession (Read 1979). This palaeoenvironmental evolution is comparable to that described from the Bristol and South Wales coalfields and suggests that broadly similar conditions obtained along the southern flank of St George's Land, at least as far east as Kent (Fig. 8.11).

*Norfolk*

Boreholes at Gimmingham, East Ruston and Somerton in Norfolk, along with offshore drilling and geophysical evidence have revealed the presence of a concealed Carboniferous basin, extending into the North Sea, which contains Coal Measures (Allsop 1984). Details are as yet not readily available.

## South-west England

The Silesian of Devon and Cornwall is a clastic sequence which occurs in a variety of facies and which represents a development of the palaeogeographic pattern which had been evolving since Devonian times (see Chapters 6 and 7). The intense Hercynian deformation, the scarcity of fossils and the poor inland exposure have complicated the unravelling of the stratigraphy and palaeogeography of this region. However, the poor inland exposure is to some extent compensated by the fine coastal exposures of north Cornwall and north Devon between Boscastle and the Bideford estuary (Freshney *et al.* 1979, Edmonds *et al.* 1979).

The Silesian sediments lie in a broad synclinorium (Fig. 8.1) trending E–W and flanked to north and south by Dinantian rocks. Most of the Silesian outcrop is occupied by Westphalian rocks but the Namurian is sparingly exposed to both north and south. Major strike-slip faulting has displaced blocks along roughly NW–SE lines (Dearman 1963, 1971). The Silesian is thus best described in terms of two areas, corresponding roughly to the two limbs of the synclinorium (Table 8.2).

## Northern sequence

The northern sequence crops out on the coast between Barnstaple estuary and Greencliff, 3 km south of Westward Ho! At Fremington, west of Barnstaple, the Limekiln Beds, a thin sequence of black mudstones, some 30 m above known Visean rocks have yielded goniatites of Kinderscoutian and Marsdenian age (Moore 1929; Prentice 1960a, 1960b).

Inland, to the east at Bampton, goniatites of Arnsbergian age are known (Thomas 1962, 1963; House & Selwood 1966) but exposure is poor. The Limekiln Beds are overlain by the Instow Beds, a sequence of thin turbidite sandstones deposited by currents flowing to the east (Prentice 1962). These lower deposits reflect a continuation of the basinal conditions established in the Dinantian.

Turbidite deposition continued into the overlying

**Fig. 8.11.** Generalised late Westphalian palaeogeography for South Wales, Southern and South-west England (but including the southern part of the Central Province). Period illustrated is mid-Westphalian C, succeeding the final (*A. cambriense*) marine invasion. B C L – Bristol Channel Landmass; FD – Forest of Dean.

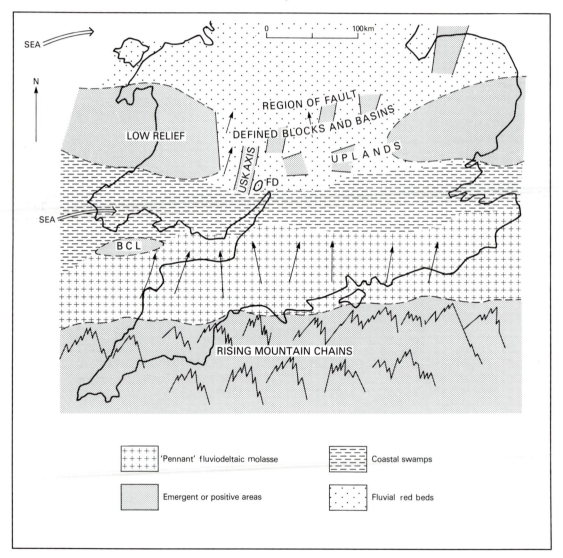

550 m thick Westward Ho! Formation, which is mainly thin-bedded but complicated by the occurrence of large channels, some filled with thick sandstones and some with muddy sediment. These channels may have been cut and filled by turbidity currents though the mud-filled features may be slump scars (Walker 1966b).

The background, non-turbidite, sediment shows an upwards increase in bottom-current activity, suggesting a general shallowing. This trend is confirmed by the overlying Northam and Abbotsham Formations (Bideford Group) which have yielded non-marine bivalves, of Westphalian A age, and which occur as nine 'cycles' or sequences which vary in thickness from 20 to 200 m (de Raaf et al. 1965, Walker 1969). Whilst all the sequences coarsen upwards, they vary not only in thickness but also in the detailed sequence of facies. Some involve basal turbidites whilst others do not. The variability of sequence in the six Abbotsham Formation 'cycles' more recently has been interpreted in terms of the successive progradations of a river-dominated elongate delta (Elliott 1976). Sequences capped by major channel sandstones may represent axial parts of the delta whilst those with non-channelised sandstones reflect more lateral positions. The tops of cycles are marked by intense bioturbation resulting from low sedimentation rates following abandonment. The deltas built out from north to south, providing further evidence for a Bristol Channel Landmass (Fig. 8.9 and p. 256).

The highest sandstone of the Abbotsham Formation is overlain by a thin coal seam (the Culm) which is succeeded by deltaic sediments of the Greencliff Beds. Near Bideford, these have produced a fauna which suggests an age around late *lenisulcata* or early *communis* (i.e. Westphalian A) (Prentice 1960b).

The northern sequence therefore bridges the Namurian–Westphalian boundary and records a transition from basinal conditions with axial flow to deltaic conditions fed by lateral supply from the north.

The northern sequence is separated from the southern sequence at the coast by a major dextral strike-slip fault, the Sticklepath Fault, which trends NW–SE and intersects the coast at Greencliff (Fig. 8.2).

## Southern sequence

This sequence is entirely basinal and is dominated by turbidite sandstones. Correlations within it have been the subject of considerable debate over the last twenty years but recent syntheses (Burne & Moore 1971, Freshney & Taylor 1972, Freshney et al. 1979) have provided a basis for agreement.

The southern sequence occupies the coast between Greencliff and Boscastle. The southern limit, near Boscastle, is a complex tectonic zone. To the south of the zone, a sequence of northerly-transported nappes involve mainly Devonian strata whilst north of the zone, south-facing folds in the Silesian sediments record backfolding against the thrusts (Sellwood and Thomas 1986). The Silesian was laid down during a period of north–south shortening involving nappe development further south, although the precise relationship between the tectonic activity and the sedimentation awaits a more detailed analysis.

The oldest Silesian sediments in the south are the Crackington Formation, a 300 m turbidite sequence which has yielded goniatites ranging in age between the Arnsbergian and the Marsdenian (Mackintosh 1964, Freshney et al. 1979). The turbidity currents flowed axially from west to east (Fig. 8.11). The beds are now folded in spectacular zig-zag folds particularly well seen at Millook Haven and are separated by faults from Dinantian sediments to the south and younger Westphalian rocks to the north. The Wanson Mouth Fault which separates the Crackington Formation from the younger (partly Westphalian) Bude Formation, may have caused the southward disappearance of some beds (Black Mudstone Beds, Welcombe Beds and Hartland Beds) seen below the Bude Formation in the Hartland area (Burne & Moore 1971). This sequence equates to some extent with the Cockington Formation seen east of Clovelly (Prentice 1960b, Moore 1968). Shales low in the sequence have yielded goniatites of Yeadonian and possibly Marsdenian age and Freshney & Taylor (1972) consider that the Welcombe Beds are equivalent to part of the Crackington Formation. Recent work suggests that the base of the Bude Formation should be taken at the top of the Hartland Quay Shales (Freshney et al. 1979).

Whichever scheme is preferred, the Bude Formation is the youngest unit in this southern area and the highest exposed beds have yielded goniatites indicating a level just above the listeri marine band (i.e. Westphalian A).

The Bude Formation consists of interbedded shales and sandstones of turbidite or other mass flow origin (Burne 1969, 1970), with some evidence of slumping. There are several thick shales in the sequence, which have been used along with the 'slump' beds to correlate across this structurally complicated zone (e.g. Lovell 1965; King 1966, 1971; Freshney & Taylor 1972). Palaeocurrents from the Bude Formation are broadly from north to south, contrasting with those of the Crackington Formation. Axial flow along the basin in the Namurian is therefore succeeded by lateral supply in Westphalian times (Burne 1969, Ashwin 1957). Studies (by e.g. Goldring & Seilacher 1971 and Higgs 1984) concluded that the Bude basin was relatively shallow (? outer shelf or perhaps lacustrine (Higgs

**Table 8.2.** Correlation diagram for the Namurian and Westphalian sequences of South-west England as exposed in the north and south limbs of the main synclinorium (see Fig. 8.7).

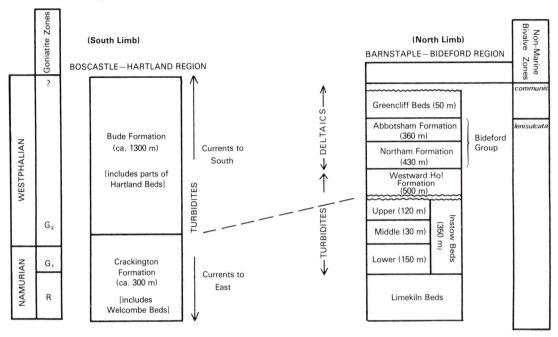

1986, 1987, Melvin 1986, 1987) and was filled by southwards progradation of the Bideford Delta, which fed turbidite channel-levee complexes downslope. These features probably reflect the increasing influence of the Bristol Channel Landmass and the slumps may record increasing tectonic activity associated with oversteepening of the basin margin. Such effects probably relate to migration of nappe fronts from the south.

### Correlation and comparison of the areas

Current ideas and stratigraphic data allow broad correlation between northern and southern areas but not in the detail possible for the Silesian elsewhere (Table 8.2). Both areas show a transition from axial to lateral sediment supply which coincides roughly with the Namurian–Westphalian boundary. In the south, this is entirely in the context of basinal deposition but in the north it coincides with a phase of basin filling and the establishment of deltaic conditions.

It has been suggested that the Bude Formation turbidites are distal equivalents of the Bideford Group deltaics (Freshney & Taylor 1972). Whilst turbidites occur in the Bideford Group cyclic sequences, they do not resemble the Bude Formation turbidites and this

model therefore seems unlikely. It seems more probable that the differences in the Westphalian of the two areas reflect differences in topography and drainage along the southern margin of the Bristol Channel Landmass, but a final interpretation must await a more confident palinspastic reconstruction of the area.

## Summary

During all but latest Silesian times, NW Europe appears to have been located in the contemporary tropical belt and most of the British Isles area lay within a vast, humid fluvio-deltaic plain that intermittently was flooded by the sea. More persistent deeper marine conditions characterised southern parts of Britain but this marine trough was eliminated in the early Westphalian, presumably as a consequence of the uplift of the juvenile Hercynian fold belt.

In the north the appearance of penecontemporaneous red beds in the mid-Westphalian of basins on both flanks of St George's Land (Forest of Dean, Oxford, Warwickshire, Wyre Forest) is most plausibly attributed to fluctuations in the local water-table. However, this could in turn denote a more pronounced seasonal variation in relative precipitation.

This change may have ensued from the continued northward drift of north-west Europe, from a roughly equatorial Carboniferous latitude into a position within the more arid trade-wind latitudes of the Northern hemisphere, as attested by the reddened desert dune sands of the following Permian period.

The changes engendered by climatic and eustatic factors were, however, less significant than those attributed to the tectonic and magmatic events that are normally subsumed under the general heading of the Hercynian orogeny. The folded and thrust terrains of SW England provide the most spectacular manifestation of the Late Silesian orogenic movements, which imposed a pan-European E–W structural grain upon this area. The northern limit of this highly deformed region is traditionally drawn at the so-called Variscan (Hercynian) Front, which runs through southernmost Wales and eastwards through the southern margin of the Bristol and Somerset Coalfield (and is mainly a belt of marked contrasts in thickness and depth to basement, according to Matthews (1974).

North of the Front the structural fragmentation and separation of the late Carboniferous basins was accomplished by tectonic movements that were manifested more commonly by fracturing than by intense folding. These movements were by no means strictly synchronous and their trends were strongly influenced by pre-existing structural trends. Such inherited features induced the development of the localised sediment sources and of the coarse marginal fault-generated facies recognised in certain of the Midland basins. They also promoted the diversity of trend that is a conspicuous feature of Hercynian structures in central and northern England. Many of the faults formed in the Silesian (and older) sedimentary cover at the close of the Carboniferous period were to remain active for a considerable time, influencing the extensional tectonic style of New Red Sandstone sedimentation and enhancing the irregular relief created by the end-Carboniferous tectonic events.

## REFERENCES

| | | |
|---|---|---|
| AITKENHEAD, N. | 1977 | The Institute of Geological Sciences borehole at Duffield, Derbyshire. *Bull. geol. Surv. G.B.*, **59**, 1–38. |
| AITKENHEAD, N. & JONES, C. M. | 1971 | The Roosecote borehole. *Ann. Rep. Inst. geol. Sci. London*, 32–34. |
| ALLEN, J. R. L. | 1960 | The Mam Tor Sandstones, a 'turbidite' facies of the Namurian deltas of Derbyshire, England. *J. sediment. Petrol.*, **30**, 193–208. |
| ALLSOP, J. M. | 1984 | Geophysical appraisal of a Carboniferous basin in north-east Norfolk, England. *Proc. geol. Assoc. London*, **95**, 175–180. |
| ASHTON, C. A. | 1974 | Palaeontology, stratigraphy and sedimentology of the Kinderscoutian and Lower Marsdenian of North Staffordshire and adjacent areas. *Unpublished Ph.D. thesis, University of Keele*, 377 pp. |
| ASHWIN, D. P. | 1957 | The structure and sedimentation of the Culm sediments between Boscastle and Bideford, north Devon. *Unpublished Ph.D. thesis, University of London*. |
| BADHAM, J. P. N. | 1982 | Strike-slip orogens – an explanation for the Hercynides. *J. geol. Soc. London*, **139**, 495–506. |
| BAINES, J. G. | 1977 | The stratigraphy and sedimentology of the Skipton Moor Grits (Namurian, $E_{1C}$) and their lateral equivalents. *Unpublished Ph.D. thesis, University of Keele*, 226 pp. |
| BENFIELD, A. C. | 1969 | The Huddersfield White Rock cyclothem in the Central Pennines. Report of field-meeting, 20–23 September, 1968, *Proc. Yorkshire geol. Soc.*, **37**, 181–187. |
| BESLY, B. M. | 1983 | Sedimentology and stratigraphy of red beds in the Westphalian A and C of Central England. *Unpub. Ph.D. thesis, University of Keele*, 831 pp. |
| BESLY, B. M. & KELLING, G. (Eds) | 1988 | *Sedimentation in a synorogenic basin complex: The Carboniferous of northwest Europe.* Blackie & Sons, Glasgow, 276 pp. |

BESLY, B. M. & TURNER, P. 1983 Origin of redbeds in a moist tropical climate (Etruria Formation, Upper Carboniferous, UK). *In* Wilson, R. C. L. (Ed.) *Residual Deposits, Spec. Pub. geol. Soc. London*, **11**, 131–147.

BISAT, W. S. 1924 The Carboniferous goniatites of the north of England and their zones. *Proc. Yorkshire geol. Soc.*, **20**, 40–124.

BISSON, G., LAMB, R. K. & CALVER, M. A. 1967 Boreholes in the concealed Kent coalfield between 1948 and 1959. *Bull. geol. Surv. Gt. Britain*, **26**, 99–166.

BLESS, M. J., BOUCKAERT, J., CALVER, M. A., GRAULICH, J. M. & PAPROTH, E. 1977 Palaeogeography of upper Westphalian deposits in NW Europe with reference to Westphalian C north of the mobile Variscan belt. *Meded. Rijks geol. Dienst.*, N.S., **28**, 101–147.

BOLTON, T. 1978 The palaeontology, sedimentology and stratigraphy of the upper Arnsbergian, Chokierian and Alportian of the North Staffordshire Basin. *Unpublished Ph.D. thesis, University of Keele*, 382 pp.

BOTT, M. H. P. 1967 Geophysical investigations of the northern Pennine basement rocks. *Proc. Yorkshire geol. Soc.*, **36**, 139–168.

BOTT, M. H. P. & JOHNSON, G. A. L. 1967 The controlling mechanism of cyclic sedimentation. *Q. J. geol. Soc. London*, **122**, 421–441.

BOTT, M. H. P. & MASSON-SMITH, D. J. 1957 The geological interpretation of a gravity survey of the Alston Block and the Durham coalfield. *Q. J. geol. Soc. London*, **113**, 93–117.

BOTT, M. H. P., ROBINSON, J. & KOHNSTAMM, M. A. 1978 Granite beneath Market Weighton, east Yorkshire. *J. geol. Soc. London*, **135**, 535–544.

BROADHURST, F. M. & SIMPSON, I. M. 1973 Bathymetry of a Carboniferous reef. *Lethaia*, **6**, 367–381.

BROADHURST, F. M., SIMPSON, I. M. & HARDY, P. G. 1980 Seasonal Sedimentation in the Upper Carboniferous of England. *J. Geol.*, **88**, 639–651.

BURNE, R. V. 1969 Sedimentological studies of the Bude Formation. *Unpublished D.Phil. thesis, University of Oxford.*

1970 The origin and significance of sand volcanoes in the Bude Formation (Cornwall). *Sedimentology*, **15**, 211–228.

BURNE, R. V. & MOORE, L. J. 1971 The Upper Carboniferous rocks of Devon and north Cornwall. *Proc. Ussher Soc.*, **2**, 288–298.

CALVER, M. A. 1968 Distribution of Westphalian marine faunas in northern England and adjoining areas. *Proc. Yorkshire geol. Soc.*, **37**, 1–72.

1969 Westphalian of Britain. *C.r. 6eme Cong. Int. Strat. Geol. Carb. (Sheffield, 1967)*, **1**, 233–254.

CLEAL, C. J. 1984 The recognition of the base of the Westphalian D stage in Britain. *Geol. Mag.*, **121**, 125–129.

1986 Plant macrofossils from the Edgehills Sandstone, Forest of Dean. *Bull. Brit. Mus. Nat. Hist. (Geol.)*, **40**, 235–246.

COLLINSON, J. D. 1969 The Sedimentology of the Grindslow Shales and the Kinderscout Grit: a deltaic complex in the Namurian of northern England. *J. sediment. Petrol.*, **39**, 194–221.

1970 Deep channels, massive beds and turbidity current genesis in the Central Pennine Basic. *Proc. Yorkshire geol. Soc.*, **37**, 495–520.

1976 Deltaic evolution during basin fill – Namurian of the Central Pennine Basin, England. *Bull. Am. Assoc. Petrol. Geol.*, **60**, 52 (Abstract).

| | | |
|---|---|---|
| COLLINSON, J. D. & BANKS, N. L. | 1975 | The Haslingden Flags (Namurian $G_1$) of south-east Lancashire; bar-finger sands in the Pennine Basin. *Proc. Yorkshire geol. Soc.*, **40**, 431–458. |
| COLLINSON, J. D., JONES, C. K. & WILSON, A. A. | 1977 | The Marsdenian (Namurian $R_2$) succession west of Blackburn; implication for the evolution of Pennine delta systems. *Geol. J.*, **12**, 59–76. |
| CONYBEARE, W. D. & PHILLIPS, W. | 1822 | *Outlines of the Geology of England and Wales.* William Phillips, London, 471 pp. |
| COPE, F. W. | 1954 | The North Staffordshire coalfield. *In* Trueman, A. E. (Ed.) *The Coalfields of Great Britain.* Arnold, London, 219–243. |
| DAVIES, H. G. | 1967 | The lithofacies of a Lower Coal Measures sandstone unit between Sheffield and Brighouse. *In* Neves, R. & Downie, C. H. (Eds.) *Geological Excursions in the Sheffield Region.* Sheffield, 129–139. |
| DAVIES, J. H. & TRUEMAN, A. E. | 1927 | A revision of the non-marine lamellibranchs of the Coal Measures and a discussion of their zonal sequence. *Q. J. geol. Soc. London*, **83**, 210–259. |
| DAVIES, M. & BLOXAM, T. W. | 1974 | The geochemistry of some South Wales coals. *In* Owen, T. R. (Ed.) *The Upper Palaeozoic and Post-Palaeozoic Rocks of Wales.* Univ. of Wales Press, Cardiff, 225–261. |
| DEARMAN, W. R. | 1963 | Wrench faulting in Devon and south Cornwall. *Proc. Geol. Assoc. London*, **74**, 265–287. |
| | 1971 | A general view of the structure of Cornubia. *Proc. Ussher Soc.*, **2**, 220–236. |
| DEWEY, J. F. | 1982 | Plate tectonics and the evolution of the British Isles. *J. geol. Soc. London*, **139**, 371–414. |
| DIX, E. | 1934 | The sequence of floras in the Upper Carboniferous with special reference to South Wales. *Trans. R. Soc. Edinburgh*, **57**, 789–821. |
| DUFF, P. McL. D. & WALTON, E. K. | 1962 | Statistical basis for cyclothems: a quantitative study of the sedimentary succession in the East Pennine coalfield. *Sedimentology*, **1**, 235–255. |
| DUFF, P. McL. D., HALLAM, A. & WALTON, E. K. | 1967 | *Cyclic Sedimentation.* Elsevier, Amsterdam, 280 pp. |
| DOWNING, R. A. & SQUIRRELL, H. C. | 1965 | On the red and green beds in the Upper Coal Measures of the eastern part of the South Wales coalfield. *Bull. geol. Surv. G.B.*, **23**, 45–56. |
| DUNHAM, K. C. & POOLE, E. G. | 1974 | The Oxfordshire Coalfield. *J. geol. Soc. London*, **130**, 387–391. |
| EARP, J. R. | 1961 | Exploratory boreholes in the North Staffordshire coalfield. *Bull. geol. Surv. G.B.*, **17**, 153–191. |
| EDMONDS, E. A., WILLIAMS, B. J. & TAYLOR, R. T. | 1979 | Geology of Bideford and Lundy Island. *Mem. geol. Surv. G.B.*, 143 pp. |
| ELLIOTT, R. E. | 1969 | Deltaic processes and episodes; the interpretation of productive Coal Measures occurring in the East Midlands, Great Britain. *Mercian Geol.*, **3**, 111–135. |
| ELLIOTT, T. | 1974 | Abandonment facies of high-constructive lobate deltas with an example from the Yoredale Series. *Proc. Geol. Assoc. London*, **85**, 359–365. |
| | 1975 | The sedimentary history of a delta lobe from a Yoredale (Carboniferous) cyclothem. *Proc. Yorkshire geol. Soc.*, **40**, 505–536. |
| | 1976 | Upper Carboniferous sedimentary cycles produced by river-dominated, elongated deltas. *J. geol. Soc. London*, **132**, 199–208. |

ELLIOTT, T. &          1981    Syn-sedimentary gravity slides (growth faults) in the Coal
  LADIPO, K. O.                 Measures of South Wales. *Nature, London*, **291**, 220–222.
EVANS, W. B.,          1968    Geology of the country around Macclesfield, Congleton,
  WILSON, A. A.,                Crewe and Middlewich. *Mem. geol. Surv. G.B.*, 328 pp.
  TAYLOR, B. J. & PRICE, D.
FALCON, N. F. &        1960    Geological results of petroleum exploration in Britain
  KENT, P. E.                   1945–1957. *Mem. geol. Soc. London*, **2**, 1–56.
FARMER, N. &           1969    The Carboniferous, Namurian rocks of the coast section
  JONES, J. M.                  from Howick Bay to Foxton Hall, Northumberland.
                                *Trans. nat. Hist. Soc. Northumberland*, **17**, 1–27.
FIELDING, C. R.        1984a   A coal depositional model for the Durham Coal Measures
                                of NE England. *J. geol. Soc. London*, **141**, 919–921.
                       1984b   Upper delta-plain lacustrine and fluviolacustrine facies
                                from the Westphalian of the Durham coalfield, NE
                                England. *Sedimentology*, **31**, 547–567.
                       1985    Coal depositional models and the distinction between
                                alluvial and delta plain environments. *Sediment Geol.*, **42**,
                                41–48.
                       1986a   Fluvial channel and overbank deposits from the West-
                                phalian of the Durham coalfield, NE England. *Sedimen-
                                tology*, **33**, 119–140.
                       1986b   The anatomy of a coal-seam split, Durham coalfield,
                                northwest England. *Geol. J.*, **21**, 45–57.
FORD, T. D.            1954    The Upper Carboniferous rocks of the Ingleton coalfield.
                                *Q. J. geol. Soc. London*, **110**, 231–265.
FRESHNEY, E. C.        1970    Cyclical sedimentation in Petrockstow Basin. *Proc. Ussher
                                Soc.*, **2**, 179–189.
FRESHNEY, E. C. &      1967    The Petrockstow Basin. *Proc. Ussher Soc.*, **1**, 278–280.
  FENNING, P. J.
FRESHNEY, E. C. &      1971    The structures of Mid-Devon and north Cornwall. *Proc.
  TAYLOR, R. T.                 Ussher Soc.*, **2**, 241–248.
                       1972    The Upper Carboniferous stratigraphy of north Cornwall
                                and Devon. *Proc. Ussher Soc.*, **2**, 464–471.
FRESHNEY, E. C.,       1979    Geology of the country around Bude and Bradworthy.
  EDMONDS, E. A.,               *Mem. geol. Surv. G.B.*, 62 pp.
  TAYLOR, R. T. & WILLIAMS, B. J.
FRESHNEY, E. C.,       1972    Geology of the coast between Tintagel and Bude. *Mem.
  McKEOWN, M. C. &              geol. Surv. G.B.*, 87 pp.
  WILLIAMS, M.
GEORGE, G. T.          1970    The sedimentology of Namurian sequences in South
                                Pembrokeshire. *Unpublished Ph.D. thesis, University of
                                Wales (Swansea)*, 202 pp.
                       1982    Sedimentology of the Uppper Sandstone Group &
                                (Namurian, G1) in south-west Dyfed: a case study. *In*
                                Bassett, M. G. (Ed.) *Geological Excursions in Dyfed, South-
                                West Wales*, National Museum of Wales, Cardiff, 203–214.
GEORGE, G. T. &        1982    Stratigraphy and sedimentology of Upper Carboniferous
  KELLING, G.                   sequences in the coalfield of south-west Dyfed. *In* Bassett,
                                M. G. (Ed.) *Geological Excursions in Dyfed, South-West
                                Wales*), National Museum of Wales, Cardiff, 175–201.
GEORGE, T. N.          1956    The Namurian Usk anticline. *Proc. geol. Assoc. London*,
                                **66**, 297–316.
                       1962    Devonian and Carboniferous foundations of the Variscides
                                in northwest Europe. *In* Coe, K. (Ed.) *Some aspects of the
                                Variscan fold belt*. Manchester Univ. Press, 19–47.
                       1970    *British Regional Geology: South Wales*. 3rd Ed. H.M.S.O.,
                                London.

GEORGE, T. N. & WAGNER, R. H. 1972 I.U.G.S. Subcommission on Carboniferous stratigraphy. Proceedings and report on the general assembly at Krefeld, August 21–22. *C.R. 7eme Cong. Int. Strat. geol. Carb. (Krefeld, 1971)*, **1**, 139–147.

GIBSON, W. 1905 Geology of the North Staffordshire coalfield. *Mem. geol. Surv. G.B.*, 523 pp.

GILLIGAN, A. 1920 The petrography of Millstone Grit of Yorkshire. *Q. J. geol. Soc. London*, **75**, 251–294.

GOLDRING, R. & SEILACHER, A. 1971 Limulid undertracks and their sedimentological implications. *Neues Jahrb. Geol. Palaeontol. Abhandlungen*, **137**, 422–442.

GREEN, A. H. AND SIX OTHERS 1878 The Geology of the Yorkshire coalfield. *Mem. geol. Surv. G.B.*, 823 pp.

GUION, P. D. 1971 A sedimentological study of the Crawshaw Sandstone (Westphalian A) in the East Midlands coalfield area. *Unpublished M.Sc. thesis, University of Keele*.

1984 Crevasse splay deposits and roof-rock quality in the Threequarters Seam (Carboniferous) in the East Midlands coalfield, UK. *In* Rahmani, R. A. and Flores, R. M. (Eds.) Sedimentology of Coal and Coal-bearing sequences, *Spec. Pub. Inter. Assoc. Sedimentologists*, 291–308.

GUTTERIDGE, P., ARTHURTON, R. S. & NOLAN, S. C. (eds) 1989 *The role of tectonics in Devonian and Carboniferous sedimentation in the British Isles.* Yorkshire geol. Soc. Occ. Pub., **6**.

HASZELDINE, R. S. 1983 Fluvial bars reconstructed from a deep, straight channel, Upper Carboniferous coalfield of northeast England. *J. sediment. Petrol.*, **53**, 1233–1247.

1984a Carboniferous North Atlantic palaeogeography: stratigraphic evidence of rifting, not megashear or subduction. *Geol. Mag.*, **121**, 443–463.

1984b Muddy deltas in freshwater lakes, and tectonism in the Upper Carboniferous coalfield of NE England. *Sedimentology*, **31**, 811–822.

HASZELDINE, R. S. & ANDERTON, R. 1980 A ? braidplain facies model for the Westphalian B Coal Measures of north-east England. *Nature, London*, **284**, 51–53.

HAUBOLD, H. & SARJEANT, W. A. S. 1973 Tetrapodenfahrten aus den Keele und Enville Groups von Shropshire und South Staffordshire. *Z. geol. Wiss. DDR.*, **1**, 895–933.

HEWARD, A. P. & FIELDING, C. R. 1980 A ? braidplain facies model for the Westphalian B Coal Measures of north-east England. *Nature, London*, **287**, 87–88.

HIGGINS, A. C. 1975 Conodont zonation of the late Visean – early Westphalian strata of the south and central Pennines of northern England. *Bull. geol. Surv. G.B.*, **53**, 1–90.

HIGGS, R. 1984 Possible wave-induced sedimentary structures in the Bude Formation (Lower Westphalian, south-west England) and their environmental implications. *Proc. Ussher Soc.*, **6**, 88–94.

1986 'Lake Beds' (early Westphalian, S.W. England), storm-dominated siliclastic shelf sedimentation in an equatorial lake. *Proc. Ussher Soc.*, **6**, 417–418.

1987 The fan that never was? Discussion of 'Upper Carboniferous fine-grained turbiditic sandstones from southwest England. A model for growth in an ancient, delta-fed subsea fan'. *J. sediment. Petrol.*, **57**, 378–379.

HOARE, R. H.                     1959    Red beds in the Coal Measures of the west Midlands. *Trans. Instn. Ming. Eng.*, **119**, 185–198.

HOLDSWORTH, B. K.               1963    Pre-fluvial, autogeosynclinal sedimentation in the Namurian of the southern Central Province. *Nature*, **199**, 133–135.

HOUSE, M. R. &                   1966    Palaeozoic palaeontology in Devon and Cornwall. *In* Hosking, K. G. & Shrimpton, J. G. (Eds.) *Present views of some aspects of the geology of Cornwall and Devon.* Roy. geol. Soc. Cornwall Commem. Vol., 45–86.
SELWOOD, E. B.

HUDSON, R. G. S. &              1945    The Carboniferous rocks of the Edale Anticline, Derbyshire. *Q. J. geol. Soc. London*, **101**, 1–37.
COTTON, G.

JOHNSON, E. W.                   1981    A tunnel section through a prograding Namurian (Arnsbergian, $E_2a$) delta, in the western Bowland Fells, north Lancashire. *Geol. J.*, **16**, 93–110.

JOHNSON, G. A. L.,               1962    The base of the Namurian and of the Millstone Grit in north-eastern England. *Proc. Yorkshire geol. Soc.*, **33**, 341–362.
HODGE, B. L. &
FAIRBAIRN, R. A.

JONES, C. M.                     1977    The sedimentology of Carboniferous fluvial and deltaic sequences, the Roaches Grit Group of the south-west Pennines and the Pennant Sandstone of the Rhondda Valleys. *Unpublished Ph.D. thesis, University of Keele*, 206 pp.

                                 1980    Deltaic sedimentation in the Roaches Grit and associated sediments (Namurian $R_2b$) in the south-west Pennines. *Proc. Yorks. geol. Soc.*, **43**, 39–67.

JONES, D. G.                     1974    The Namurian Series in the South Wales. *In* Owen, T. R. (Ed.) *The Upper Palaeozoic and post-Palaeozoic rocks of Wales.* Univ. of Wales Press, Cardiff, 117–132.

KELLAWAY, G. A.                  1967    The Geological Survey Ashton Park borehole and its bearing on the geology of the Bristol district. *Bull. geol. Surv. G.B.*, **27**, 49–153.

                                 1970    The Upper Coal Measures of south-west England compared with those of South Wales and the southern Midlands. *C.R. 6eme Cong. Inter. Strat. Geol. Carb. (Sheffield, 1967)*, **3**, 1039–1055.

KELLAWAY, G. A. &               1983    Structure of the Bristol District, the Forest of Dean and the Malvern Fault Zone. *In* Hancock, P. D. (Ed.) *The Variscan Fold Belt in the British Isles.* Adam Hilger Ltd, Bristol, 88–107.
HANCOCK, P. L.

KELLAWAY, G. A. &               1955    The Upper Old Red Sandstone and Lower Carboniferous rocks of Bristol and the Mendips compared with those of the Forest of Dean. *Bull. geol. Surv. G.B.*, **9**, 1–21.
WELCH, F. B. A.

KELLING, G.                      1974    Upper Carboniferous sedimentation in South Wales. *In* Owen, T. R. (Ed.) *The Upper Palaeozoic and the post-Palaeozoic rocks of Wales.* Univ. of Wales Press, Cardiff, 185–224.

KELLING, G. &                    1971    Upper Carboniferous sedimentation in the Pembrokeshire coalfield. *In* Bassett, D. A. & Bassett M. G. (Eds.) *Geological excursions in South Wales and the Forest of Dean.* Cardiff, 240–259.
GEORGE, G. T.

KENT, P. E.                      1966    The structure of the concealed Carboniferous rocks of north-eastern England. *Proc. Yorkshire geol. Soc.*, **35**, 323–352.

KEREY, I. E.                     1978    Sedimentology of the Chatsworth Grit Sandstone in the Goyt–Chapel en le Frith area. *Unpublished M.Sc. thesis, University of Keele*, 132 pp.

KING, A. F.           1965    Xiphasurid trails from the Upper Carboniferous of Bude, North Cornwall. *Proc. geol. Soc. London*, No. 1626, 162–165.

1966    Structure and stratigraphy of the Upper Carboniferous Bude Sandstones, north Cornwall. *Proc. Ussher Soc.*, **1**, 229–232.

1971    Correlation in the Upper Carboniferous Bude Formation, north Cornwall. *Proc. Ussher Soc.*, **2**, 285–288.

KIRTON, S. R.      1984    Carboniferous volcanicity in England with special reference to the Westphalian of the E and W Midlands. *J. geol. Soc. London*, **141**, 161–170.

KOLB, C. R. & LOPIK, J. R. VAN.      1966    Depositional environments of the Mississippi River deltaic plain, south-eastern Louisiana. *In* Shirley, M. L. (Ed.) *Deltas in their geologic framework*. Houston geol. Soc., 17–61.

LAND, D. H.        1974    Geology of the Tynemouth district. *Mem. geol. Surv. G.B.*, 236 pp.

LEEDER, M. R.      1982    Upper Palaeozoic basins of the British Isles – Caledonide inheritance versus Hercynian plate margin processes. *J. geol. Soc. London*, **139**, 481–494.

LEITCH, D., OWEN, T. R. & JONES, D. G.      1958    The basal Coal Measures of the South Wales coalfield from Llandybie to Brynmawr. *Q. J. geol. Soc. London*, **113**, 461–483.

LOVELL, J. P. B.      1965    The Bude Sandstones from Bude to Widemouth, north Cornwall. *Proc. Ussher Soc.*, **1**, 172–174.

LUMSDEN, G. I., TULLOCH, W., HOWELLS, M. F. & DAVIES, A.      1967    The geology of the neighbourhood of Langholm. *Mem. geol. Surv. G.B.*, 255 pp.

MCCABE, P. J.      1975    The sedimentology and stratigraphy of the Kinderscout Grit Group (Namurian $R_1$) between Wharfedale and Longdendale. *Unpublished Ph.D. thesis, University of Keele*, 172 pp.

1976    Deep distributary channels and giant bedforms in the Upper Carboniferous of the Central Pennines, northern England. *Sedimentology*, **24**, 271–290.

1978    The Kinderscoutian delta (Carboniferous) of northern England: A slope influenced by density currents. *In* Stanley, D. J. & Kelling, G. (Eds.) *Sedimentation in Submarine Canyons, Fans and Trenches*. Dowden, Hutchinson and Ross, Stroudsburg, 116–126.

MCKENZIE, D. P.      1978    Some remarks on the development of sedimentary basins. *Earth planet. Sci. Lett.*, **40**, 25–32.

MACKINTOSH, D. M.      1964    The sedimentation of the Crackington Measures. *Proc. Ussher Soc.*, **1**, 88–89.

MATHEWS, S. C.      1974    Exmoor Thrust? Variscan Front? *Proc. Ussher Soc.*, **3**, 82–94.

MAYHEW, R. W.      1967    The Ashover and Chatsworth Grits in north-east Derbyshire. *In* Neves, R. & Downie, C. H. (Eds.) *Geological Excursions in the Sheffield Region*. Univ. of Sheffield, Sheffield, 94–103.

MELVIN, J.      1986    Upper Carboniferous fine-grained turbiditic sandstones from southwest England: A model for growth in an ancient, delta-fed subsea fan. *J. sediment. Petrol.*, **56**, 19–34.

1987    Reply to Higgs, 1987, *q.v. J. Sediment. Petrol.*, **57**, 380–382.

MILLER, J. &
GRAYSON, R. F.
1982 The regional context of Waulsortian facies in Northern England. *In* Bolton *et al.* (Eds.) *Symposium on the Palaeoenvironmental setting and distribution of the Waulsortian facies.* El Paso Geol. Soc. and Univ. Texas at El Paso, 17–33.

MITCHELL, G. H.
1954 The South Staffordshire coalfield. *In* Trueman, A. E. (Ed.) *The Coalfields of Great Britain.* Edward Arnold, London, 273–282.

MOORE, D.
1959 The Yoredale Series of Upper Wensleydale and adjacent parts of north-west Yorkshire. *Proc. Yorkshire geol. Soc.,* **31**, 91–148.

MOORE, E. W. J.
1929 The occurrence of *Reticuloceras reticulatum* in the Culm of north Devon. *Geol. Mag.,* **66**, 356–358.

MOORE, L. J.
1968 The stratigraphy of Carboniferous rocks of the Hartland area of north Devon. *Proc. Ussher Soc.,* **2**, 18–21.

OKOLO, S. A.
1983 Fluvian distributary channels in the Fletcher Bank Grit (Namurian R) at Ramsbottom, Lancashire, England. *In* Collinson, J. D. and Lewin, J. (Eds.) *Modern and Ancient Fluvial Systems.* Inter. Assoc. Sedimentol. Spec. Pub. **6**, 421–433.

OWEN, T. R.
1964 The tectonic framework of Carboniferous sedimentation in South Wales. *In* van Straaten, L. M. J. U. (Ed.) *Deltaic and Shallow Marine Deposits.* Elsevier, Amsterdam, 301–307.

OWEN, T. R. &
JONES, D. G.
1961 The nature of the Millstone Grit/Carboniferous Limestone junction of a part of the North Crop of the South Wales coalfield. *Proc. Geol. Assoc. London,* **72**, 239–249.

OWENS, B., NEVES, R.,
GUEINN, K. J.,
MISHELL, D. R. F.,
SABRY, H. S. M. Z. &
WILLIAMS, J. E.
1977 Palynological division of the Namurian of northern England and Scotland. *Proc. Yorkshire geol. Soc.,* **41**, 381–398.

POOLE, E. G.
1965 Trial boreholes on the site of a reservoir at Eymore Farm, Bewdley, Worcestershire. *Bull. geol. Surv. G.B.,* **24**, 151–156.

1971 The Oxfordshire Coalfield. *Nature, London,* **232**, 394–395.

1978 Stratigraphy of the Withycombe Farm borehole, near Banbury, Oxfordshire. *Bull. geol. Surv. G.B.,* **68**, 1–63.

PRENTICE, J. E.
1960a Dinantian, Namurian and Westphalian rocks of the district south-west of Barnstaple, north Devon. *Q. J. geol. Soc. London,* **115**, 261–289.

1960b The stratigraphy of the Upper Carboniferous rocks of the Bideford region, north Devon. *Q. J. geol. Soc. London,* **116**, 397–408.

1962 The sedimentation history of the Carboniferous in Devon. *In* Coe, K. (Ed.) *Some aspects of the Variscan fold belt.* Univ. of Manchester Press, Manchester, 93–108.

PRICE, N. B. &
DUFF, P. McL. D.
1969 Mineralogy and Chemistry of tonsteins from Carboniferous sequences in Great Britain. *Sedimentology,* **13**, 45–69.

RAFF, J. F. M. DE,
READING, H. G. &
WALKER, R. G.
1965 Cyclic sedimentation in the Lower Westphalian of North Devon, England. *Sedimentology,* **4**, 1–52.

RAMSBOTTOM, W. H. C.
1969 Interim report of the Namurian Working Group. *C.R. 6eme Inter. Strat. Geol. Carb.,* **1**, 71–77.

1974a Namurian. *In* Rayner, D. H. & Hemingway, J. E. (Eds.) *The Geology and Mineral Resources of Yorkshire.* Yorkshire geol. Soc., Leeds, 73–87.

| | | |
|---|---|---|
| RAMSBOTTOM, W. H. C. | 1974b | The Namurian in North Wales. *In* Owen, T. R. (Ed.) *The Upper Palaeozoic and post-Palaeozoic rocks of Wales.* Univ. of Wales Press, Cardiff, 151–167. |
| | 1977 | Major cycles of transgression and regression (mesothems) in the Namurian. *Proc. Yorkshire geol. Soc.*, **41**, 261–291. |
| | 1978a | Namurian mesothems in South Wales and Northern France. *J. geol. Soc. Lond.*, **135**, 307–312. |
| | 1978b | Namurian. *In* Moseley, F. (Ed.) *The Geology of the Lake District.* Yorkshire geol. Soc., Leeds, 178–180. |
| | 1979 | Rates of transgression and regression in the Carboniferous of NW Europe. *J. geol. Soc. London*, **136**, 147–154. |
| | 1981 | Eustacy, sea level and local tectonism with examples from the British Carboniferous. *Proc. Yorkshire geol. Soc.*, **43**, 473–482. |
| RAMSBOTTOM, W. H. C., CALVER, M. A., EAGAR, R. M. C., HODSON, F., HOLLIDAY, D. W., STUBBLEFIELD, C. J. & WILSON, R. B. | 1978 | Silesian (Upper Carboniferous). *Geol. Soc. London Spec. Rept. No. 10*, 81 pp. |
| RAMSBOTTOM, W. H. C., RHYS, G. H. & SMITH, E. G. | 1962 | Boreholes in the Carboniferous rocks of the Ashover district, Derbyshire. *Bull. geol. Surv. G.B.*, **19**, 75–168. |
| RAMSBOTTOM, W. H. C., STEVENSON, I. P. & GAUNT, G. D. | 1967 | Fossiliferous localities in the Namurian rocks of Edale. *In* Neves, R. & Downie, C. (Eds.) *Geological excursions in the Sheffield Region*, Univ. of Sheffield, Sheffield, 75–90. |
| READ, W. A. | 1979 | A quantitative analysis of an Upper Westphalian fluviodeltaic succession and a comparison with earlier Westphalian deposits in the Kent Coalfield. *Geol. Mag.*, **116**, 431–443. |
| ROBERTSON, T. M. | 1948 | Rhythm in sedimentation and its interpretation with particular reference to the Carboniferous sequence. *Trans. geol. Soc. Edinburgh*, **14**, 141–175. |
| ROWELL, A. J. & SCANLON, J. E. | 1957 | The Namurian of the north-west quarter of the Askrigg Block. *Proc. Yorkshire geol. Soc.*, **31**, 1–38. |
| SAUNDERS, W. B. | 1973 | Upper Mississippian ammonoids from Arkansas and Oklahoma. *Spec. Pap. geol. Soc. America*, **145**, 1–110. |
| SCOTT, A. C. | 1978 | Sedimentological and ecological control of Westphalian B plant assemblages from West Yorkshire. *Proc. Yorkshire geol. Soc.*, **41**, 461–508. |
| SELLWOOD, E. B. & THOMAS, J. M. | 1986 | Variscan facies and structure in central SW England. *J. geol. Soc. London*, **1432**, 199–208. |
| SHACKLETON, J. S. | 1962 | Cross-strata of the Rough Rock (Millstone Grit Series) in the Pennines. *Liverpool Manchester geol. J.*, **3**, 109–118. |
| SHACKLETON, R. M., RIES, A. C. & COWARD, M. P. | 1982 | An interpretation of the Variscan structures in S.W. England. *J. geol. Soc. London*, **139**, 535–544. |
| SIMPSON, S. | 1971 | The Variscan structure of north Devon. *Proc. Ussher Soc.*, **2**, 249–252. |
| SMITH, D. B. & FRANCIS, E. A. | 1967 | Geology of the country between Durham and West Hartlepool. *Mem. geol. Surv. G.B.* 208 pp. |
| SMITH, D. B., BRUNSTROM, R. G. W., MANNING, P. I., SIMPSON, S. & SHOTTON, F. W. | 1974 | A Correlation of Permian rocks in the British Isles. *Geol. Soc. London Spec. Rept. No. 5*, 45 pp. |

SORBY, H. C.                 1859    On the structure and origin of the Millstone Grit in South
                                     Yorkshire. *Proc. Yorkshire geol. polytech. Soc.*, **3**, 669–675.

SPEARS, D. A.                1971    The mineralogy of the Stafford tonstein. *Proc. Yorkshire
                                     geol. Soc.*, **38**, 497–516.

SPEARS, D. A. &              1973    An Upper Carboniferous tonstein of volcanic origin.
  RICE, C. M.                        *Sedimentology*, **20**, 281–294.

SPEARS, D. A. &              1979    A geochemical and mineralogical investigation of some
  KANARIS-SOTIRIOU, R.               British and other European tonsteins. *Sedimentology*, **26**,
                                     407–425.

STEAD, J. T. G.              1974    The sedimentology of the Upper Coal Measures of the
                                     Forest of Dean and adjacent areas. *Unpublished Ph.D.
                                     thesis, University of Wales (Swansea)*, 228 pp.

STEEL, R. P.                 1986    The Namurian sedimentary history of the Gainsborough
                                     Trough. *In* Besly, B. M. and Kelling, G. (Eds.) *Abstracts
                                     Volume, Conference on Controls of Upper Carboniferous
                                     Sedimentation, Northwest Europe*, Univ. of Keele, 25–26.

STUBBLEFIELD, C. J.          1954    The Kent Coalfield and other possible concealed coalfields
                                     of the English Midlands. *In* Trueman, A. E. (Ed.) *q.v.* 154–
                                     166.

SULLIVAN, H. J.              1964    Miospores from the Drybrook Sandstone and associated
                                     measures in the Forest of Dean Basin. *Palaeontology
                                     London*, **7**, 351–392.

THOMAS, J. M.                1962    The Culm Measures in Devon and north-west Somerset,
                                     east of Bampton. *Unpublished Ph.D. thesis, University of
                                     London*. 315 pp.

                             1963    The Culm Measures succession in north-east Devon and
                                     north-west Somerset. *Proc. Ussher Soc.*, **1**, 63–64.

THOMAS, L. P.                1967    A Sedimentary study of the sandstones between the
                                     horizons of the Four-Feet Coal and the Gorllwyn Coal of
                                     the Middle Coal Measures of the South Wales coalfield.
                                     *Unpublished Ph.D. thesis, University of Wales (Swansea)*,
                                     176 pp.

                             1974    The Westphalian (Coal Measures) in South Wales. *In*
                                     Owen, T. R. (Ed.) *The Upper Palaeozoic and post-
                                     Palaeozoic rocks of Wales*. Univ. of Wales, Cardiff,
                                     131–160.

TREWIN, N. H.                1968    Potassium bentonites in the Namurian of Staffordshire and
                                     Derbyshire. *Proc. Yorkshire geol. Soc.*, **37**, 73–91.

TREWIN, N. H. &              1973    Sedimentation in the Lower Namurian rocks of the North
  HOLDSWORTH, B. K.                  Staffordshire Basin. *Proc. Yorkshire geol. Soc.*, **39**, 371–
                                     408.

TROTTER, F. M.               1948    The devolatilization of coal seams in South Wales. *Q. J.
                                     geol. Soc. London*, **104**, 387–437.

                             1953    Reddened beds of Carboniferous age in northwest England
                                     and their origin. *Proc. Yorkshire geol. Soc.*, **29**, 1–20.

TRUEMAN, A. E.               1946    Stratigraphical problems in the Coal Measures of Europe
                                     and North America. *Q. Jl. geol. Soc. London*, **102**, 49–86.

TRUEMAN, A. E. (Ed.)         1954    *The Coalfields of Great Britain*. Edward Arnold, London,
                                     396 pp.

VAN LECKWIJK, W. P.          1960    Report of the Subcommission on Carboniferous Strat-
                                     igraphy. *C.R. 4eme Cong. Inter. Strat. Geol. Carb.
                                     (Heerlen, 1958)*, **1**, xxiv–xxv.

                             1964    Rapport d'ensemble sur les travaux du 5me congrès du
                                     Carbonifière. *C.R. 5me Cong. Int. Strat. Géol. Carb.* (Paris,
                                     1963), **1**, xxvii–xl.

WAGNER, R. H.                1983    A lower Rotliegend flora from Ayrshire. *Scott. J. Geol*, **19**,
                                     135–155.

WALKER, R. G.　　　　1966a　Shale Grit and Grindslow Shales: transition from turbidite to shallow-water sediments in the upper Carboniferous of northern England. *J. sediment. Petrol.*, **36**, 90–114.

1966b　Deep channels in turbidite-bearing formations. *Bull. Am. Assoc. Petrol. Geol.*, **50**, 1899–1917.

1969　The juxtaposition of turbidite and shallow-water sediments: Study of a regressive sequence in the Pennsylvanian of North Devon, England. *J. Geol.*, **27**, 125–143.

WELLS, A. J.　　　　　1955　The development of chert between the Main and Crow limestones in north Yorkshire. *Proc. Yorkshire geol. Soc.*, **30**, 177–196.

WHITEHEAD, T. H., ROBERTSON, T. M., POCOCK, R. W. & DIXON, E. E. L.　　1928　Geology of the Country between Wolverhampton and Oakengates. *Mem. geol. Surv. G.B.*, 236 pp.

WHITEHURST, J.　　　　1778　*An enquiry into the original state and formation of the earth.* W. Bent, London, 283 pp.

WHITTAKER, A.　　　　1975　Namurian strata near Cannington Park, Somerset. *Geol. Mag.*, **112**, 325–326.

WILLIAMS, P. F.　　　　1966　The sedimentation of the Pembrokeshire Coal Measures. *Unpublished Ph.D. thesis, University of Wales (Swansea)*, 259 pp.

1968　The sedimentation of Westphalian (Ammanian) Measures in the Little Haven–Amroth coalfield, Pembrokeshire. *J. sediment. Petrol.*, **38**, 332–362.

WILLIAMSON, I. A.　　　1970　Tonsteins – Their nature, origins and uses. *Ming. Mag.*, **122**, 119–126, 203–211.

WILLIAMSON, W. O.　　1942　Some grits and associated rocks in the Etruria Marl of North Staffordshire. *Geol. Mag.*, **80**, 20–32.

WILSON, A. A.　　　　　1960　The Carboniferous rocks of Coverdale and adjacent valleys in the Yorkshire Pennines. *Proc. Yorkshire geol. Soc.*, **32**, 285–316.

WOODLAND, A. W. & EVANS, W. B.　　　　1964　The geology of the South Wales coalfield, Part IV: Pontypridd and Maesteg. 3rd Ed. *Mem. geol. Surv. G.B.*, 391 pp.

WRIGHT, W. B., SHERLOCK, R. L., WRAY, D. A., LLOYD, W. & TONKS, L. H.　　　1927　The Geology of the Rossendale anticline. *Mem. geol. Surv. G.B.*, 282 pp.

# 9

# PERMIAN

# D. B. Smith

## Introduction

The Permian scene in England and Wales was set by the late Carboniferous to early Permian Hercynian earth movements which, by gentle folding, block faulting, and uplift, caused profound changes in the contemporary geography. Gone were the enormous almost featureless sedimentary plains of the Westphalian, and in their place was established, in tropical latitudes deep within the vast Laurasian continent, a diverse and inhospitable landscape sculptured by the harsh processes of desert erosion. Great thicknesses of Carboniferous rocks were removed from large areas during the Stephanian and early Permian, and the almost ubiquitous unconformity between Carboniferous and Permian strata generally spans at least one stage. Reddening of the desert surface thus created was widespread at this time, and little of the earliest sediment formed upon it has survived.

Desert erosion and sedimentation continued intermittently for the whole of the Permian in much of England and Wales and surrounding areas and in time almost barren desert sediments accumulated to considerable thickness in a number of low-lying and subsiding areas. In the north-west and north-east, however, desert conditions were abruptly ended by the formation of the extensive epicontinental Bakevellia and Zechstein seas wherein up to 1,200 m of carbonate and evaporite rocks were formed in the closing phases of the period.

## Distribution, classification and correlation

The distribution of Permian rocks in and around England and Wales is shown in Figure 9.1. They comprise a lower and upper series, conventionally divided at the incoming of marine strata in the north (Smith *et al.* 1974) and at supposedly equivalent levels elsewhere. No subdivision of the Lower Permian into

the Autunian and Saxonian stages of continental Europe has been recognised, although a general equivalence has been demonstrated. More detailed reviews of the Lower Permian in Britain than are possible here have been given by Smith (1972a, 1976), and Permian strata in NW England and adjoining undersea areas have been reviewed by Arthurton *et al.* (1978), Colter (1978), and Ebbern (1981). Lower Permian strata in NE England and the southern North Sea have been summarised by Glennie (1972, 1984) and Marie (1975), and Upper Permian strata in these areas have been detailed by Smith (1970a, 1974, 1979), Taylor & Colter (1975) and Taylor (1984).

The great lateral variability and absence of fossils in most of the continental Permian rocks of England and Wales make correlation between strata of the various basins particularly difficult, although broadly similar lithological sequences in several basins probably evolved in parallel as landscapes and climate changed. Enough fossils have been found in early Permian rocks of the English Midlands, however, to allow those strata to be assigned to the Autunian stage with some confidence and thus tentatively to stages recognised in the marine facies of USA and USSR. Correlation between the marine sequences in the north-west and north-east is firmer than that between the various continental rocks, there being a measure of palaeontological control, and is reasonably firm between English Zechstein strata and those of Holland, Germany, and Poland. Correlation with Permian rocks in the type area of USSR remains uncertain in the absence of fusulinids and useful ammonoids in Zechstein strata.

Because of the uncertainty of correlation between British Permian strata and international stratotypes, the identification of Permian system and stage boundaries in England and Wales is generally unsatisfactory. Thus, for example, the base of the Permian rocks is placed in the English Midlands on the evidence of scanty plant and reptilian remains, and in

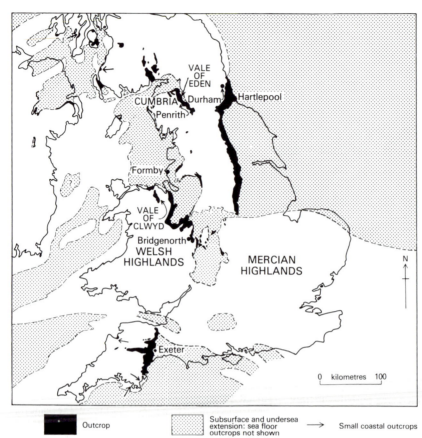

**Fig. 9.1.** Distribution of Permian rocks in England, Wales and adjoining areas showing exposed (black) and concealed or undersea (stippled) strata. Based on many sources including Whittaker (Ed.) (1985).

SW England by reference to the radiometric ages of about 285 million years of volcanic rocks near the base of the local continental sequence; elsewhere, only the evidence of unconformity and lithology is available. The useful convention whereby the base of the Upper Permian in England is taken at the marine Zechstein incursion is unsatisfactory in that it has not been established that this incursion coincided with the basal Upper Permian Kazanian transgression of the Russian Platform; the view based on scanty palynological evidence (Visscher 1971) that the Zechstein rocks are late Kazanian to Tatarian in age now seems to be gaining acceptance over that based on invertebrate evidence that they might be Kungurian to Kazanian in age. Controversy also surrounds the placing of the top of the Permian in rocks in England and Wales (see Chapter 10); in an almost barren sequence some authors (e.g. Smith *et al.* 1974) have arbitrarily taken the junction at the base of the Pebble Beds of the Sherwood Sandstone Group and their equivalents, but an alternative view is that it should be taken at the top of the youngest Zechstein evaporites (Rhys 1974). Both views are difficult to

apply in areas where neither lithological division is present.

## Lower Permian
### The evolving scene

For perhaps 30–40 million years, from late Carboniferous time until the late Permian marine incursions, England and Wales formed part of one of the great deserts of world history. Comparable in size and variety with the Sahara, and occupying a similar position relative to the contemporary equator and climatic belts, this desert was the scene for much of Lower Permian time of mass downwasting of an initially immature landscape of which little survived and, in later Lower Permian time, of the patchy accumulation of thick sequences of continental sediments. There are indications that differential uplift and movement along some major faults may have continued from Carboniferous into early Lower Permian time, and locally affected the course of erosion and sedimentation, but undoubtedly the most significant early Permian earth movement was tension and attenuation associated with the formation and persistent subsidence of a number of troughs (several of the trap-door type, faulted on the east) and inland drainage basins in the areas of the Irish Sea, Western England, and the North Sea; these became the main repositories for later Lower Permian sediments, including the principal reservoir rocks of the southern North Sea gas fields, and their formation may have been connected with the onset of rifting and the break-up of Laurasia (Russell 1976). Volcanicity, which contributed so prominently to the late Carboniferous and early Permian scene in SW Scotland, was restricted in England and Wales to the south-west, especially in an area around Exeter (Fig. 9.1).

The early Lower Permian, judging from the scanty surviving evidence, appears to have been a time of stark contrasts, in which areas of Lower Palaeozoic rocks that had not been covered by Carboniferous strata (or from which such strata had subsequently been removed) remained as wild uplands whilst desert weathering of Carboniferous rocks in remaining areas resulted in landscapes dominated by immature tabular landforms. Rainfall, attracted by relief that may locally have reached 3,000 m, was sufficient to maintain a sparse population of plants and animals, but, like that of many modern deserts, was probably episodic and with a high rate of runoff. Dewfall was probably appreciable and the diurnal temperature range extreme.

In this harsh setting, erosion was probably mainly physico–chemical, with detritus riven from soil-free and almost unvegetated rocks accumulating as screes before being swept into adjacent valleys by runoff, sheet-floods and debris-flows; streams at this time would be almost all short and ephemeral, and first cycle sediments accumulating on innumerable alluvial fans and valley bottoms were all locally derived. Most of this sediment was coarse and angular, but with a higher proportion of finer material in more distal fan and valley-bottom environments and with varying amounts of infiltrated fines in the more proximal parts; reddening of the sediments, as in comparable modern environments (e.g. Walker et al. 1978) probably began soon after deposition. Wind-blown sand, though probably quantitatively unimportant, doubtless accumulated locally, but wind at this stage was probably mainly an erosive and transporting agent by which, for example, coarse material was abraded (ventifacts are locally common) and fines selectively winnowed and perhaps largely removed from the basins. Pediplains were probably not extensive at this time.

This partial reconstruction of the early Permian scene in England and Wales in the light of modern desert environments is based on the limited evidence of buried topography and overlying sediments in the Vale of Eden, West Cumbria, parts of the English

Midlands, and in SW England, and accords with similar evidence from SW Scotland and much of Western Europe. It is suggested that the pattern revealed by these four areas was typical of much of England and Wales, but that vistas gradually changed as landscapes lowered and matured below the level of earlier valley fill; much of the evidence having thus been removed, only the sketchiest outlines of early Lower Permian palaeogeography can be inferred.

Later Lower Permian palaeogeography may be reconstructed with rather more confidence (Fig. 9.2), for, despite contemporaneous erosion and reworking of sediments as landscapes continued to evolve, more deposits of this age and larger areas of the buried land surface have survived. This evidence shows that patterns of sedimentation at this time differed considerably between west and east.

In Wales (where the only surviving direct evidence is from the Vale of Clwyd in the north (Fig. 9.1)) and much of western England, continental sedimentation continued within the framework established earlier and thick sequences of Lower Permian sediments accumulated in valleys and in a series of mainly Charnoid troughs and subsiding inland drainage basins. Initially these valleys, troughs, and basins were separate, and the course of sedimentation in each was probably governed by a unique combination of local factors; deposits of this age in the various western basins were therefore extremely variable and are difficult to correlate in the almost total absence of useful fossils. Later, however, some of the valleys, troughs, and basins merged as intervening barriers wasted or were buried and pediplains extended and matured; because of this many late Lower Permian sediments are continuous from basin to basin, include a proportion of far-travelled debris, and are more readily correlated with each other.

The evolution of the Lower Permian landscape of Wales and western England and perhaps also of the climate as mountainous areas were reduced and lowered, is reflected in a common sequence of sediment types in each of the larger depositional areas. Earliest sediments were almost everywhere of coarse angular local detritus deposited on alluvial fans and plains, and in the wider basins these pass laterally into sands, silts, and clays. Evaporitic muds accumulated in saline playas on the floor of the large Irish Sea Basin and perhaps elsewhere. Water-laid sediment completely filled the smaller valleys and troughs, but was subordinate to, and commonly buried by, onlapping wind-blown sand in higher parts of the sequence in all the larger basins. These sands are up to 715 m thick at Formby (Fig. 9.1), near Liverpool, and exceed 400 m in parts of the Vale of Eden, the Midlands and SW England; investigations by Shotton (1937, 1956) suggested that they accumulated on transverse dunes

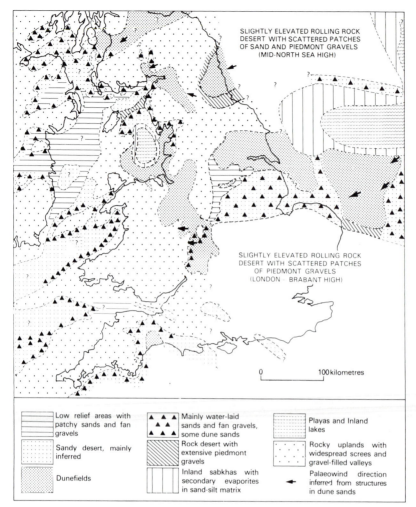

SLIGHTLY ELEVATED ROLLING ROCK
DESERT WITH SCATTERED PATCHES
OF SAND AND PIEDMONT GRAVELS
(MID-NORTH SEA HIGH)

SLIGHTLY ELEVATED ROLLING ROCK
DESERT WITH SCATTERED PATCHES
OF PIEDMONT GRAVELS
(LONDON - BRABANT HIGH)

0                    100 kilometres

| | | |
|---|---|---|
| Low relief areas with patchy sands and fan gravels | Mainly water-laid sands and fan gravels, some dune sands | Playas and Inland lakes |
| Sandy desert, mainly inferred | Rock desert with extensive piedmont gravels | Rocky uplands with widespread screes and gravel-filled valleys |
| Dunefields | Inland sabkhas with secondary evaporites in sand-silt matrix | Palaeowind direction inferred from structures in dune sands |

**Fig. 9.2.** Palaeogeography of Britain and adjoining areas during the late Lower Permian. Based on Smith (1976) and Smith & Taylor (1991). (N.B. The wind direction evidence has been challenged by Sneh (1988).)

under the influence of prevailing easterly and north-easterly winds but work by Sneh (1988) has suggested that many of the Early Permian dunes were of the oblique type and were formed under the influence of mainly northerly winds. Whether or not the whole of the western part of England became one great sand sea, as suggested by Shotton (1956), cannot now be determined, but there can be little doubt that the desert dunes of the late Lower Permian were much more extensive than the present outcrops of dune sandstone. The abundance of breccia wedges in marginal parts of several of the dune fields, however, shows that sub-stantial elevated sediment source areas persisted in Western Britain throughout the Lower Permian despite the long continued levelling effects of down-wasting, pediplanation and infilling of depressions.

Landscapes in the eastern part of England and the southern North Sea, as suggested by the morphology of the Permian land surface beneath Permian and Triassic deposits, evolved to far greater maturity than in the west before being overwhelmed in places by sediment. The evidence suggests that virtually the entire area was reduced to a vast rolling pediplain by the coalescence of a complex of earlier pediments, and that this surface was subsequently gently warped by subsidence in the southern North Sea Basin so as to form a major area of internal drainage. Here, probably somewhat later than in the west of Britain, continental sediments typical of a wide range of desert environments (Glennie 1972) ac-cumulated to thicknesses locally exceeding 300 m. Evaporitic playa deposits formed extensive spreads on the floor of the basin, which may have lain well below contemporary sea level (Smith 1970b), whilst bare rock desert or thin piedmont gravels occupied large parts of the marginal area.

## Palaeontology

In common with most modern deserts, it seems likely that the Lower Permian desert in England and Wales supported a relatively diverse specialised flora and fauna. However, the inferred slow rate of deposition, the abrasion and repeated reworking of sediment, and the prevailing strongly oxidising environment all conspired against the preservation of plant and animal remains and only trace fossils and the most durable organic debris were preserved. With this unpromising background it is unlikely that the few fossils found in the Lower Permian rocks of England and Wales are representative of the full contemporary biota.

Fossils have been recorded in the Vale of Eden, the Warwickshire area of the English Midlands, and in SW England. Only the Midlands have yielded an assemblage of species, some represented by a single specimen, and the full stratigraphic range of these is poorly known. Body fossils from here comprise the pelycosaur reptiles *Haptodus grandis* Paton, *Sphenacodon britannicus* von Huene, and *Ophiacodon* sp. (Paton 1974) together with the labyrinthodont *Dasycleps bucklandi*, all from the early Lower Permian Kenilworth Breccia Group, and the conifers *Lebachia* or *Walchia piniformis* (Schlotheim) pars Florin, and *L. frondosa* (Renault) var *zeilleri* Florin from the overlying Ashow Group. Footprints from the Kenilworth Breccias have been assigned to the Autunian by Haubold & Katzung (1975) who also ascribe an Autunian age to rocks of the underlying Gibbet Hill Group which are more commonly regarded as late Carboniferous (e.g. Ramsbottom *et al.* 1978). Footprints are the only fossils found in the Lower Permian rocks of the Vale of Eden, whilst reptilian or amphibian burrows (Ridgway 1974, Pollard 1976) constitute the only reported evidence of animal life in the later Carboniferous or early Lower Permian rocks of SW England.

The unrepresentativeness of the exceptionally short faunal list in the Midlands was noted by Paton (1974), who pointed out that the few forms found suggest a fairly varied fauna; 'there must', she wrote, 'have been plenty of food to support two species of carnivorous pelycosaurs'. Presumably some of the lower members of this food chain were herbivorous, but nevertheless direct evidence of root structures, for example, has yet to be reported from Lower Permian rocks of England and Wales.

# Regional descriptions

## The East Irish Sea Basin and the Vale of Clwyd

The broad outlines of the group of basins collectively known as the East Irish Sea Basin are known from marine geophysical and sampling surveys (e.g. Bott 1964, 1968; Wright *et al.* 1971), but the few available details of the stratigraphy come mainly from hydrocarbon exploration boreholes (Kent 1948, Falcon & Kent 1960, Colter & Barr 1975, Colter 1978, Ebbern 1981, Jackson *et al.* 1987). These show that the red aeolian Collyhurst Sandstone is up to 715 m thick at Formby near the down-faulted eastern margin of the basin, and that central areas are occupied by a thick sequence of red sandstones, siltstones, and subordinate mudstones of which some and perhaps most were probably deposited in and around an extensive playa. It seems likely that aeolian sandstones similar to those at Formby extend in a belt around the southern and south-western margin of the basin, and there inter-tongue with an outer belt of water-laid sandstones and breccias such as those exposed on the Cumbrian coast near Whitehaven. The presence of abnormally thick aeolian sandstones so near the eastern margin of the basin may point to differential synsedimentary subsidence, perhaps associated with movement on the major boundary fault or faults. Thin piedmont breccias are extensive at the base of the Lower Permian sequence. The narrow Vale of Clwyd in North Wales is probably best regarded as a marginal cuvette of the East Irish Sea Basin. Geophysical surveys of the southern part of the Vale (Collar 1974) and field survey (Warren *et al.* 1973) show that the cuvette is an asymmetric 'trap-door' basin, bounded on the east by the Vale of Clwyd Fault; from his geophysical data, Collar estimated that the vertical displacement on this fault locally exceeded 900 m and that the sedimentary fill (including some strata of probable Triassic age) is more than 525 m thick. Shotton (1956) described the continental sediments here assigned to the Lower Permian as red aeolian sandstones deposited as barchans by a prevailing east-north-easterly wind; he suggested a thickness of 210–300 m.

## Vale of Eden and Carlisle Basin

Thick sequences of Lower Permian continental red sedimentary rocks exposed in the Vale of Eden comprise early and marginal breccias (locally known as 'Brockrams') and the rather thicker and more continuous aeolian Penrith Sandstone (Fig. 9.11). Breccias and aeolian sandstones of this age also occur in parts of the somewhat larger Carlisle Basin to the north, but little is known of their detailed distribution and mutual relationships. The various deposits in the Vale of Eden unconformably overlie Carboniferous rocks ranging from Dinantian to Westphalian C in age, and are overlain by early Upper Permian marine strata, so that a Lower Permian age is strongly indicated. Supporting fossil evidence is confined to reptilian footprints found high in the Penrith Sandstone near Penrith.

In the Vale of Eden the breccias are thickest in the south where they are banked against a markedly uneven buried landscape. Lower parts are composed mainly of fragments of local Dinantian limestones in a red matrix of wind-abraded sand, but higher parts, as in contemporaneous breccias elsewhere in Britain, contain a more varied suite and include basal Carboniferous conglomerate, rhyolite, quartzite, and rare fragments of dolerite thought to be related to the Whin Sill, which has been radiometrically dated at about $301 \pm 5$ Ma; dolomite and some gypsum has replaced much of the limestone in these younger breccias.

The Penrith Sandstone (0–400 m) crops out mainly in central and northern parts of the Vale of Eden and interdigitates with fan breccias in the south. It features large-scale cross-lamination consistent with accumulation as barchans under a prevailing easterly wind (Shotton 1956, Waugh 1974), but uppermost Penrith Sandstone strata near Hilton have been described as sheetflood deposits (Burgess & Holliday 1974). Petrological examination by Waugh (1970b) of Penrith Sandstone from the type area showed that the rock there is generally a texturally and mineralogically mature coarse well-sorted orthoquartzite composed of well-rounded quartz grains (90–95 per cent) with subordinate feldspar and rock fragments; a limited suite of heavy minerals is also present. Where breccias are absent, in the north of the Vale, most of the sandstone has been silicified and some authigenic potassium-feldspar is also present (Waugh 1970b, 1978); Waugh attributed the formation of the silica to contemporaneous dissolution and reprecipitation of quartz dust by alkaline groundwaters. The mineralogical content of the Penrith Sandstone is consistent with derivation from upwind supply areas, especially outcrops of Millstone Grit in NE England. Derivation by reworking of early Permian fan-breccias composed of these materials, perhaps in adjoining parts of the North Sea Basin, also appears likely.

## English Midlands

Continental mainly red sedimentary rocks believed to be of Lower Permian age occur in a chain of interconnected basins from Lancashire to Warwickshire and beyond. At least some of these basins came into being in late Westphalian time, and intermittent movement along boundary faults and synsedimentary subsidence and tilting are suggested by numerous non-sequences and unconformities.

The Lower Permian sequence appears to be most complete in the central parts of the Warwickshire Basin, where the base of the Permian is placed on

scanty fossil evidence at the base of the Kenilworth Breccia Group (Smith *et al.* 1974). The Kenilworth Breccias and associated sandstones have been shown to pass into the Clent Breccias (Shotton 1929) and, by inference, are probably also generally equivalent to similar breccias which, under a variety of local names (Alberbury, Clent, Enville, Nechells, etc.) overlie a major unconformity at the margins of several of the other basins; they are accumulations of detritus derived from the Welsh and Mercian highlands (Fig. 9.1), the Clent Breccias, for example, being composed mainly of angular fragments of Precambrian volcanic rocks. The breccias were deposited on coalescing fans and in wadis at the edge of the various depositional basins and pass into sandstones, siltstones, and perhaps mudstones in basin centres.

In the north of the Cheshire–Shropshire Basin (Fig. 9.2), continental breccias near the base of the assumed Lower Permian sequence are interbedded with, but subordinate to, the red aeolian Collyhurst Sandstone (0–180 m), which also locally includes red mudstones; despite the proximity to the northernmost limit of the outcrop, it seems probable that this mixed sequence was originally formed some kilometres from the margin of the depositional basin. No such mixing is found along the western and southern boundaries of this basin, where the thick (275 m+) Bridgnorth Sandstone rests unconformably on underlying strata including the Enville and Alberbury breccias; the aeolian origin of this spectacular sandstone formation (Fig. 9.3) was proved convincingly by the pioneer work of Shotton (1937). Similar sandstones are present beneath Triassic rocks in the Worcestershire Basin to the south.

The youngest Permian deposits of the Midlands are a number of patches of thin breccia and conglomerate such as the Stockport Conglomerate in the north and the Barr Beacon Beds and the Hopwas, Moira, and Quartzite breccias farther south. All contain a mixture of local and far-travelled rock fragments and may have formed as residual lag gravels on a mature desert pavement. Only the Stockport Conglomerate can be dated as Lower Permian, although a generally similar age appears likely for the others also.

## South-west England

The outcrop of Late Carboniferous and Permian continental red beds in SW England stretches across Somerset and Devonshire and beneath the English Channel. Their eastward extension is concealed by younger strata, but they probably extend east to Purbeck and beyond and underlie much of the adjoining English Channel. At outcrop they are seen, perhaps more clearly than anywhere else in England and Wales, to have accumulated in a series of valleys

**Fig. 9.3.** Bridgnorth Sandstone at Kinver Edge, west Midlands, showing large-scale barchanoid aeolian dune-bedding. The face is about 10 m high.

and troughs (some fault-bounded) and to have been banked against and progressively buried a reddened hilly land surface. As in other areas, the earliest sediments are locally derived fan and wadi breccias (Fig. 9.4), including sheetflood deposits, but merging of basins as fill built up and interfluves declined permitted influxes of more distant materials in later parts of the breccia sequences. Aeolian sandstones were initially subordinate but came to dominate higher parts of the sequence in the large basins as gradients lowered and hilly source areas dwindled.

Landscape rejuvenation caused by synsedimentary earth movements and sporadic outbreaks of volcanism is recorded by numerous non-sequences and disconformities.

The large quantity of volcanic debris in the continental breccias near the base of the sequence in the Exeter area suggests that volcanism was a major feature of the contemporary local scene and its products were once much more widespread than now; Permian volcanic rocks in a number of other areas in SW England are probably slightly younger than those

**Fig. 9.4.** Coarse imbricated red sandy fan-breccia composed of locally derived fragments of Carboniferous rocks, mainly limestone. Scale 15 cm. Torbay Breccia, Tor Bay, Devon.

around Exeter. It was formerly thought (Smith *et al.* 1974) that the radiometric age of (291±6 Ma) lavas low in the sequence near Exeter (Miller *et al.* 1962; Miller & Mohr 1964) helped to establish the position of the base of the Permian here, but this evidence is now thought to be unreliable (Warrington & Scrivener 1988).

As in the Midlands, the early Permian breccias and sandstones have many local names (see Smith *et al.* 1974, for summary) and in each area contain a suite of Devonian or Carboniferous rocks, or both, with or without contemporary volcanic debris; in places the mixture is so distinctive that individual fans may be recognised (Laming 1966). Clasts are smaller and more rounded in distal parts of fans than in proximal parts, though poor grading is prevalent, and sandstones are relatively more abundant there. Despite active downwasting, hilly areas to the west continued to supply coarse detritus that accumulated as fan breccias at the margins of all the basins: breccia was, however, quantitatively overwhelmed in time by huge amounts of mainly aeolian sand, which accumulated to thicknesses of up to 500 m in the western part of the main basin and there forms the youngest part of the presumed Lower Permian sequence. Breccias and

sands interdigitate at basin margins, and isolated dunes on breccia fans are preserved locally (Fig. 9.5).

## Southern North Sea Basin and adjoining areas

Although similar in its general evolution to the smaller basins of western Britain, that of the southern North Sea differs in that almost all traces of the immature phase of desert erosion were removed and a mature pediplain was established before the earliest Lower Permian sediments now preserved were deposited. Judging from the relief displayed at outcrop in NE England (Fig. 9.6), this pediplain was gently rolling, with elevated tracts generally coinciding with resistant Lower Carboniferous rocks and with scattered hills up to perhaps 200 m high; boreholes farther east show that the basin extended far across the present North Sea (Fig. 9.2) and was bounded to the north and south by topographic and structural highs which continued to supply sediment throughout the late Lower Permian. No indigenous fossils have been reported in any of the Lower Permian rocks of this region.

The earliest widespread Lower Permian deposit in NE England is a multicoloured breccia or breccio-conglomerate (Fig. 9.7), locally 10 m thick but generally less than 1 m, which has also been reported in many North Sea boreholes. It contains abundant far-travelled (probably recycled) pebbles, some of them wind-abraded and coated with desert varnish, in a sandy calcareous matrix locally rich in frosted quartz grains. Like the Stockport Conglomerate and similar breccias in the Midlands, this thin breccia may once have been a stabilised lag gravel similar to those commonly found covering modern desert pediplains.

Later Lower Permian sediments, proved by the intensive search for hydrocarbons, comprise a full suite of desert sediments but, because of the generally lower relief, have less coarse detritus than in the western basins. Accounts by Glennie (1972, 1982), Marie (1975), and Glennie et al. (1978) show that the sediments that accumulated on the south-west flank of

the basin comprise a complex of interdigitating sheet flood and fluviatile (wadi and wadi-fan) sandstones and fine breccias with some aeolian sand, and that these mainly water-laid rocks are peripheral to a belt of thick (locally 200 + m) and generally younger wind-blown sandstones that are the main gas reservoir of the southern North Sea Basin and extend on land as the Basal Permian Sands of Yorkshire (0–30 m); the equivalent Yellow Sands of County Durham (0–60 m) may have been deposited in a marginal embayment of the northern North Sea Basin (Fig. 9.2) but details of their distribution offshore are scanty and Marie (1975), favours a connection with the southern North Sea Basin. In the latter, these water-laid and aeolian sandy deposits together comprise the Leman Sandstone Formation (Rhys 1974). Glennie (1972)

**Fig. 9.5.** Aeolian red sandstone in distal fan breccias, probably representing a single small barchan. Roundham Head, Tor Bay, Devonshire. The face is about 6 m high.

**Fig. 9.6.** Carboniferous–Permian unconformity near Pontefract Station, Yorkshire. Here Cadeby Formation (formerly the Lower Magnesian Limestone) is separated from Westphalian sandstone by a thin discontinuous layer of Lower Permian breccia.

also revealed that the most rapidly subsiding interior of the basin was occupied by a great inland sabkha plain composed of evaporite-bearing clays, silts, and fine sands (the Silverpit Formation of Rhys 1974), with an extensive but perhaps ephemeral central playa. From published descriptions it seems probable that some of the halite and gypsum was precipitated directly in the playa but that much of it crystallised and grew displacively from saline groundwater within the sabkha and playa sediments. Smaller sabkhas may, according to Glennie, also have been present in deflation hollows between barchans of the dune belt, as in many modern deserts.

The petrology of the aeolian Yellow Sands in Durham (equivalent to grey and white strata at the top of the red beds of the southern North Sea Basin) has been studied by Versey (1925), Hodge (1932) and Pryor (1971) who showed them to be composed mainly of subrounded monocrystalline and polycrystalline quartz grains with less than 10 per cent of feldspars and a varied suite of heavy minerals;

ferromagnesian minerals are uncommon, possibly through post-depositional leaching, and detrital micas occur in trace quantities only. Laminae and isolated grains of coarse well-rounded quartz are conspicuous minor components in Durham, and up to 20 per cent of small rock fragments are locally present in Yorkshire where the sands were redistributed during the Zechstein transgression (Versey 1925) and therefore are Upper Permian in age. Only the uppermost part of the Durham Yellow Sands are generally thought to have been thus redistributed, although in some places in the North Sea a considerable thickness of Permian continental rocks are believed to have been so affected (Pryor 1971, Rhys 1974). The characteristic colour of the Yellow Sands at outcrop results from the surface oxidation of pyrite and ferrous oxides that pervade the formation in the sub-surface where it is grey. It has been suggested that much or all of the formation may at one time have been red like its counterparts in the other Lower Permian basins and the recent discovery of red Yellow Sands off the coast at Sunderland

(Smith 1984) tends to support this suggestion. Research into the complex diagenetic history of the desert rocks of the southern North Sea (Glennie *et al.* 1978) has shown that rocks of the various facies respond differently to burial, probably because of their different primary and secondary mineralogy and initial porosity; porosity in all the rocks diminished progressively with increasing depth of burial, and further mineralogical changes and reduction in porosity continued during subsequent Cenozoic uplift.

# Upper Permian

## The evolving scene

In much of England and Wales the continental conditions that characterised the Lower Permian persisted throughout the Upper Permian, but they ended abruptly in the NW and NE of England and much of the North Sea area when the Bakevellia and Zechstein seas (Fig. 9.8) were formed. These two seas are thought to have originated when the Boreal Ocean broke in from the north and flooded the inferred sub sea-level inland drainage basins there (Smith 1970b), but for most of their existence they were separated by a persistent Pennine land barrier.

The deposits of both the Bakevellia and Zechstein seas are strongly cyclic in response to alternating phases of marine expansion and contraction; those of the Bakevellia Sea comprise three incomplete cycles and those of the much larger and deeper Zechstein Sea comprise four relatively complete cycles (EZ1 to EZ4; see Table 9.1) and a number of minor incomplete cycles. Where complete, each cycle comprises an initial carbonate member succeeded gradationally (commonly with diachronism) by a sulphate member

**Fig. 9.7.** Imbricated sandy basal Permian piedmont breccia of fragments of Carboniferous rock (mainly Namurian and Dinantian limestone) resting unconformably on Namurian mudstone. Borehole near Sedgefield, County Durham. Scale bar 3 cm.

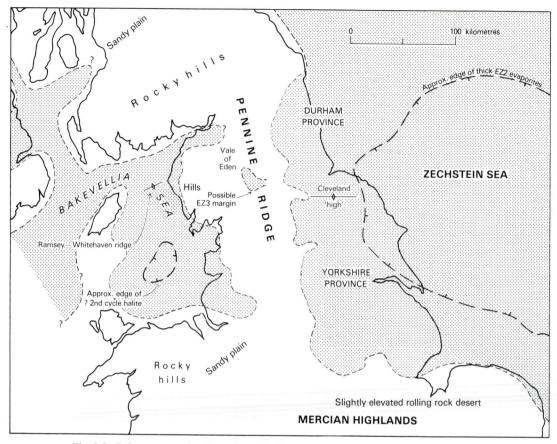

**Fig. 9.8.** Palaeogeography of northern England and adjoining areas in the early Upper Permian. Based on Smith & Taylor (in press).

(gypsum or anhydrite), a chloride member (generally halite), and finally a member composed of the highly soluble salts of potassium and magnesium; where incomplete, it is generally the later members that are absent. The three cycles of the Bakevellia Sea sequence (Table 9.2) probably all pre-date the third cycle of the Zechstein Sea sequence, and the time of deposition of the latter may not be represented by marine strata in most of the Bakevellia Sea Basin.

The early marine deposits of the Bakevellia and Zechstein seas are generally similar to those being formed today in shallow epicontinental tropical and suptropical seas, and contemporary conditions in northern England must have been very like those in and around the modern Persian Gulf. Evaporites make up a large proportion of the rocks of some cycles in each basin, however, and it is apparent that both seas on several occasions became highly saline and probably partly or wholly isolated from world oceans. Isolation of the Zechstein Sea may have resulted from

re-emergence of the original rim of the inferred chain of Lower Permian North Sea basins, and it has been argued that Zechstein sea levels periodically fluctuated by more than 100 m (Smith 1970b, 1979, 1980b). Evidence of lesser changes of level in the Bakevellia Sea has been advanced by Arthurton & Hemingway (1972) and Arthurton et al. (1978). In both basins the main cause of such changes was probably slight oscillations of world ocean level relative to the original basin rims. At times of low ocean level, complete or almost complete emergence of the rims would have led to a reduction in the rate of sea water influx below the rate of evaporation and a decline in basinal sea level ('evaporative drawdown') would have ensued; a subsequent rise in world ocean level relative to that of the basin rims would have resulted in reflooding and the beginning of the next major cycle. During phases of drawdown the two seas probably dwindled to a series of shallow hypersaline playas and salinas surrounded by enormous salt flats; only the most

| VALE OF EDEN | SERIES | GROUPS | | YORKSHIRE PROVINCE (OUTCROP AREA) | DURHAM PROVINCE (County Durham, East Tyne and Wear, County Cleveland) | YORKSHIRE PROVINCE (East and North Yorks and Humberside and adjoining North Sea) | NORTH GERMANY AND HOLLAND | |
|---|---|---|---|---|---|---|---|---|
| EDEN SHALES (WITH BROCKRAM) | UPPER PERMIAN | ESKDALE GROUP | EZ5 | UPPER MARLS | ROXBY FORMATION | LITTLEBECK ANHYDRITE<br>SLEIGHTS SILTSTONE | ? GRENZANHYDRIT | Z5 |
| BLOCKY FACIES | | STAINTON-DALE GROUP | EZ4 | SHERBURN ANHYDRITE<br>UPGANG FORMATION<br>ROTTEN MARL | SHERBURN ANHYDRITE<br>ROTTEN MARL | SNEATON HALITE<br>SHERBURN ANHYDRITE<br>UPGANG FORMATION<br>CARNALLITIC MARL | ALLER SALZE<br>PEGMATITANYHYDRIT<br>Thin unnamed carbonate<br>ROTER SALZTON | Z4 |
| D-BED<br>BELAH DOLOMITE | | TEESSIDE GROUP | EZ3 | BILLINGHAM ANHYDRITE<br>BROTHERTON FORMATION | BOULBY HALITE<br>BILLINGHAM ANHYDRITE<br>SEAHAM FORMATION | BOULBY HALITE<br>BILLINGHAM ANHYDRITE<br>BROTHERTON FORMATION<br>GRAUER SALZTON | LEINE SALZE<br>HAUPTANHYDRIT<br>PLATTENDOLOMIT<br>GRAUER SALZTON | Z3 |
| C-BED | | AISLABY GROUP | EZ2 | EDLINGTON FORMATION | SEAHAM RESIDUE<br>ROKER DOLOMITE AND CONCRETIONARY LIMESTONE | FORDON EVAPORITES<br>KIRKHAM ABBEY FORMATION | STASSFURT SALZE AND BASALANHYDRIT<br>HAUPTDOLOMIT AND EQUIVALENTS | Z2 |
| B-BED<br>A-BED AND HILTON PLANT BEDS | | DON GROUP | EZ1 b<br>EZ1 a | b Sprotbrough Member, a Wetherby Member } CADEBY FORMATION<br>MARL SLATE | HARTLEPOOL ANHYDRITE<br>FORD FORMATION<br>RAISBY FORMATION<br>MARL SLATE | HAYTON ANHYDRITE<br>CADEBY FORMATION<br>MARL SLATE | WERRAANHYDRIT<br>ZECHSTEINKALK<br>KUPFERSCHIEFER | Z1 |
| PENRITH SANDSTONE AND BROCKRAM | LOWER PERMIAN | | | BASAL PERMIAN (YELLOW) SANDS AND BRECCIAS | YELLOW (BASAL PERMIAN) SANDS AND BRECCIAS | LEMAN SANDSTONE AND SILVERPIT FORMATIONS; BASAL BRECCIAS | ROTLIEGENDES | |

**Table 9.1.** Classification and correlations of Upper Permian strata in north-east England, the Vale of Eden and north Germany and Holland. Little beck marls above the Top Anhydrite are now arbitrarily assigned to the Triassic. The abbreviations EZ1 to EZ5 refer to the main evaporite cycles of the English Zechstein sequence and Z1 to Z5 refer to those of the Zechstein sequence in northern Germany and Holland. In terms of the Russian type sequence, these Upper Permian strata are probably mainly Kazanian and Tatarian in age. (Based on Smith 1989, Table 1.)

| MANX-FURNESS BASIN (CENTRAL AREA) | SOUTH CUMBRIA | WEST CUMBRIA | VALE OF EDEN |
|---|---|---|---|
| ST BEES EVAPORITES AND SHALES | ST BEES SHALES (WITH BROCKRAM) | ST BEES SHALES (WITH BROCKRAM) | EDEN SHALES (WITH BROCKRAM) |
| ANHYDRITE | | BLOCKY FACIES | BLOCKY FACIES |
| ANHYDRITE | | | D-BED |
| DOLOMITE | | | BELAH DOLOMITE |
| HALITE | ROOSECOTE ANHYDRITE | ST BEES EVAPORITES: FLESWICK ANHYDRITE | C-BED |
| | ROOSECOTE DOLOMITE | FLESWICK DOLOMITE | |
| ANHYDRITE | HAVERIGG HAWS ANHYDRITE | SANDWITH ANHYDRITE | B-BED |
| DOLOMITE | MAGNESIAN LIMESTONE | SANDWITH DOLOMITE | |
| | | SALTOM DOLOMITE | |
| DOLOMITE, ANHYDRITE, MUDSTONE, ETC. | 'GREY BEDS' | SALTOM SILTSTONE | A-BED AND HILTON PLANT BEDS |

**Table 9.2.** Classification and correlation of representative Upper Permian sequences in north-west England and the Irish Sea (after Smith *et al.* 1974 and Arthurton *et al.* 1978).

adaptable and specialised organisms could have survived in these unremittingly harsh environments. Phases of freshening of the Zechstein sea water have also been inferred (e.g. Turner & Magaritz 1986).

The formation of the Upper Permian seas led to a marked rise in the regional ground water level and to a climatic amelioration that profoundly changed the pattern of sedimentation and consequently the landscape in surrounding areas; these changes were augmented by the cutting off of the supply of wind-blown sand from the newly inundated areas and by the establishment of a sediment-stabilising plant cover. In the rock sequences the changes are indicated by the incoming of extensive deposits of water-laid continental sediments, but oxidising conditions apparently continued to prevail for these rocks, like those of the Lower Permian, are almost devoid of fossils and are ubiquitously red.

Reconstruction of the early Upper Permian scene from the evidence of the distribution and character of the continental sediments suggests that much of England at this time had a gently rolling landscape, with a progressively diminishing relief as sedimentary plains extended, merged, and gradually buried many of the surviving eminences. Upland areas such as most of Wales and the Welsh Borders, the Lake District, the Pennine Ridge, Mercian Highlands (Fig. 9.8), and much of the old St George's land (Fig. 8.3) nevertheless remained uncovered except locally, and doubtless continued to supply sediment to the surrounding plains throughout the Upper Permian. Few, if any, of the Lower Permian sand seas survived into the Upper Permian but there are indications that sandy desert conditions were temporarily re-established in several areas when, through periodic shrinkage and desiccation of the northern seas, marine influences waned; extensive deposits of wind-blown silt and fine sand were formed in NW England after the sea and lakes there finally dried up (Burgess & Holliday 1974), and similar deposits interdigitate with younger parts of the Zechstein sequence in NE England. With the ending of marine influences as the Zechstein Sea too dwindled and dried up, and with a continuing decline in the diversifying influence of uplands, the whole of England and Wales once again became dominated by desert conditions in the late Upper Permian. Under the influence of a generally high water table resulting from the low relief, the desolate and virtually lifeless sedimentary plains of the earlier Upper Permian extended into the former marine basins and vast spreads of uniform mainly water-laid clays, silts, and fine-grained sands were formed; ephemeral saline lakes in which gypsum and perhaps halite were precipitated lay only in the most rapidly subsiding depocentres, and these too eventually disappeared as influxes of sand (destined to become the Sherwood

Sandstone Group of probable Triassic age) were swept diachronously into the basins.

## Palaeontology

Marine invertebrate fossils and the remains of land plants are present in two of the three cyclic sequences of the Bakevellia Sea and in three of the four main Zechstein cycles, and fish and reptiles are found in the rocks of the first and second Zechstein cycles. Most of these fossils were described by King (1850) in a monograph that remains the standard work for several fossil groups; discoveries by later workers have slightly extended the list of species identified. The Upper Permian faunas of northern England and adjoining areas have been summarised by Pattison (1970) and in Pattison et al. (1973) and Smith et al. (1974) whilst Stoneley (1958) and Schweitzer (1986) have reviewed the plant assemblages; the microfloras were described by Clarke (1965) and have been summarised by Warrington (in Pattison et al. 1973). Lack of space precludes full discussion of these biotas but it is important to note the close relationship between the facies and faunas of the northern seas and that a full and varied marine macrofauna is found only in marginal carbonate rocks of the first cycle (EZ1) of the English Zechstein sequence; here brachiopods (especially productids such as *Horridonia*), gastropods, and bivalves are the most abundant macrofossils, but bryozoans (especially pinnate forms such as *Fenestella* and ramose forms such as *Acanthocladia*) are abundant in shelf and shelf-edge reefs where crinoids are also present. Equivalent inshore Zechstein strata, and rocks formed in the more restricted parts of the Bakevellia Sea, have a limited molluscan fauna of which the bivalve *Bakevellia binneyi* (Brown) is the most widespread form and in places occurs alone (Pattison 1970, p. 160). Rocks of the second cycle of the English Zechstein (EZ2) contain only a limited range of indigenous gastropods and bivalves and one species of brachiopod, whilst those of EZ3 contain only a few species of gastropods and bivalves.

# Regional descriptions
## Bakevellia Sea Basin

Geophysical evidence (Bott 1964, 1968) has shown that the outlines of the Bakevellia Sea depositional basin coincided roughly with those of the present eastern Irish Sea (Fig. 9.8) and that it extended on to present land areas only in Ulster, Cumbria, W Lancashire, and Cheshire; its former extent in Ireland and to the west of the Isle of Man is unknown. The main basin was divided by the Ramsey–Whitehaven Ridge (Fig. 9.8), over which subsidence is thought to have been slower than in the Solway Firth–Carlisle Basin to the north and in the Manx–Furness Basin to the south. Apart from limited offshore data in the Manx–Furness Basin where the thickest (c. 580 m) sequence is present (Ebbern 1981, Jackson *et al.* 1987), almost all knowledge of the Upper Permian rocks of the Bakevellia Sea Basin comes from exposures of thin and generally incomplete sequences scattered around the basin margins; in most of these sequences epicontinental sediments are predominant. The evidence of cyclicity and sea-level oscillation comes mainly from these sensitive marginal rocks but falls far short of that necessary to establish the full depositional history of the basin. Different names and classifications are in use for each of the main outcrops, and are shown in Table 9.2 together with their probable correlatives.

## West and south Cumbria

Upper Permian strata in west Cumbria (Figs. 9.9, 9.10) comprise the St Bees Evaporites and the overlying St Bees Shales (Arthurton & Hemingway 1972); in south Cumbria the name St Bees Shales is applied to the whole sequence. The formations are known mainly from boreholes but additional information comes from limited surface exposures and from a former anhydrite mine.

In west Cumbria, Arthurton & Hemingway (1972, p. 570) recognised three onlapping cyclic sequences in the St Bees Evaporites and defined these in terms of the succession proved in a cored borehole near St Bees Head (Fig. 9.9) on the edge of the Ramsey–Whitehaven Ridge. Each cyclic sequence has, at its base, siltstone and/or dolomite overlying a

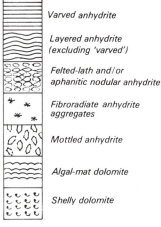

**Fig. 9.9.** St Bees Evaporites, type section in borehole near St Bees Head, west Cumbria. (Reproduced from Arthurton & Hemingway (1972, Fig. 1), by permission of the Yorkshire Geological Society.)

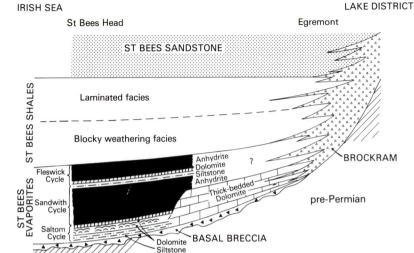

IRISH SEA

St Bees Head

LAKE DISTRICT

Egremont

**Fig. 9.10.** Permian strata at the north-eastern margin of the Manx–Furness Basin. About 150 m of strata are represented in a horizontal distance of about 10 km. (After Moseley, F. (Ed.) 1978.)

discontinuity and the first and second of these basal members contain a marine molluscan fauna in their lower parts. The presence of algal lamination, fenestral fabric and displacive gypsum or anhydrite in upper parts of these members indicates shallowing of the water and periodic emergence, and peritidal conditions are also indicated by the varied fabrics of the evaporites in the younger parts of each cycle. Fully marine environments were thus short-lived on this eastern coastline of the Bakevellia Sea, being replaced by progradational marginal sabkhas as the sea receded and perhaps becoming lacustrine in the later stages of some cycles. Although the sequence in the type borehole is generally representative, the lithology and relative thickness of the component beds is variable and the whole sequence thickens westwards (Fig. 9.10).

The St Bees Evaporites of south Cumbria include the lateral equivalents of the St Bees Shales of west Cumbria, and contain dolomite and anhydrite members of at least two evaporite cycles; lateral variation is extreme, and beds of continental breccia are present at many levels (Dunham & Rose 1949, Rose & Dunham 1977). Early Upper Permian plant and molluscan remains have been found in marginal red siliciclastic rocks and dolomite near the base of the sequence, and the anhydrites include both primary (layered) and secondary (displacive and replacive) types.

The St Bees Shales of west Cumbria are up to 200 m thick and comprise a lower unit of dark brownish red blocky siltstone and silty mudstone and an overlapping upper unit of red micaceous siltstone, mudstone, and fine-grained sandstone (Arthurton et al. 1978); both types interdigitate with thin fan breccias at the

basin margin. The blocky facies contains calcareous concretions and laminae of coarse millet-seed sand and also, in its lower part, nodules and beds of gypsum and anhydrite and veins of fibrous gypsum; Arthurton et al. (1978) regard it as mainly aeolian in origin. The upper unit features upward-fining minor cycles, ripple lamination, and desiccation cracks and breccias, and is probably a distal sedimentary-plain deposit. It passes by alternation into the overlying fluviatile St Bees Sandstone, conventionally taken as Triassic.

## Eastern Irish Sea and west Lancashire

Information from boreholes offshore (Smith et al. 1974, Colter & Barr 1975, Colter 1978, Jackson et al. 1987) shows that the general sequences in more central parts of the East Irish Sea Basin are generally similar to those onshore in Cumbria; the thicknesses of individual dolomite and anhydrite members are comparable in the several areas, and only the presence of thick halite in the presumed second cycle in several of the offshore boreholes is distinctive. Without sedimentological detail or information on strata between the boreholes and Cumbria, few conclusions on depositional trends may be drawn. The similarity between the onshore and offshore sequences is nevertheless striking and may suggest that, even in the middle, the East Irish Sea Basin may have been relatively shallow and could at times have dried up completely.

The Upper Permian sequence in the east of the basin was proved by boreholes at Formby (Kent 1948, Falcon & Kent 1960); it comprises more than 100 m of red mudstone with subordinate red and grey fine-grained sandstone and with thin beds of limestone and dolomite concentrated in the lowest and highest parts of the formation. From the lowest 9 m of these strata Pattison (1970, p. 131) identified a limited early Upper Permian foraminifer–bivalve–ostracod fauna.

## Solway Firth–Carlisle and Vale of Eden Basins

The existence of a thick sedimentary sequence under the Carlisle plain and the Solway Firth is known from geophysical surveys (Bott 1964) but details of the strata present are scanty; however an Upper Permian marine sequence of carbonate rocks and perhaps evaporites is almost certainly present. The only exposures of Upper Permian strata are in the north-east, where red St Bees Shales are exposed, and in the south-east where St Bees Shales include a persistent sulphate bed and merge with the Eden Shales (Fig. 9.11) of the Vale of Eden.

The Upper Permian sequence and sedimentary history of the Vale of Eden has been investigated in considerable detail, partly because of its workable gypsum deposits, and was summarised by Arthurton et al. (1978) who quote full references. The generalised sequence shown in Table 9.1 conceals marked lateral variation in the predominantly continental deposits of this remote intermontane desert basin, and this variability is more fully depicted in Figure 9.11. Four main depositional cycles are recognised in the Eden Shales, into which all the Upper Permian rocks are

grouped, and the first three of these are probably generally equivalent to the three cycles recognised in west Cumbria; the remains of transported Upper Permian land plants are present in each of these cycles and are locally abundant. Basal rocks of the fourth cycle are thought from their marine fossils to be more closely related to the third cycle of the English Zechstein sequence (Burgess 1965, Pattison 1970).

Arthurton et al. (1978) draw a vivid picture of early Upper Permian sedimentation in the Vale of Eden Basin, with successively younger strata, including fan breccias around the margins, overlapping earlier deposits as the basin gradually filled. From the cyclic repetition of grey and purplish red strata and from the vertical distribution of plant remains, evaporites and carbonate-rich rocks, these authors infer alternately high and low ground water levels, and it is clear from the displacive character of much of the gypsum and anhydrite that at times the ground water was highly saline. Environmental reconstructions based on the lithology and sedimentary structures of the Eden Shales suggest that the Vale of Eden then was occupied by an almost flat elongate sedimentary plain flanked by slightly elevated rocky desert, and that from time to time the sub-facies in the plain included alluvial fans, continental sabkhas, deflation surfaces, playas and minor dunefields. It would have been a hot, dry, and desolate place for most of Upper Permian time, and

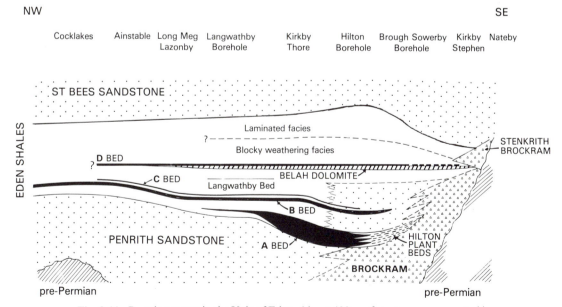

**Fig. 9.11.** Permian strata in the Vale of Eden. About 400 m of strata are represented in a horizontal distance of about 55 km. (After Moseley, F. (Ed.) 1978.)

the lack of evidence of abundant former halite is surprising.

The supposed general equivalence of the three depositional cycles in west Cumbria with the first three cycles of the Eden Shales is based on some similarities of sequence and on the probability that the major oscillations in ground water level coincided with transgressions and regressions in the Bakevellia and Zechstein seas. No marine fossils have been found in the Eden Shales below the Belah Dolomite (Fig. 9.11), however, and marine incursions from the Bakevellia Sea into the Vale of Eden have not been proved. In contrast, the coincidence there of the scanty fauna of the thin Belah Dolomite at the base of the fourth depositional cycle with that of the Seaham Formation of the third cycle of the English Zechstein sequence is taken to indicate a brief incursion of the Zechstein Sea into the southern part of the Vale (Burgess 1965, Burgess & Holliday 1974). This inundation may have resulted from a combination of gentle local synsedimentary subsidence and the long-continued downwasting of the Pennine barrier, aided by the tendency for the transgression at the beginning of the third

Zechstein cycle in NE England to extend beyond the limits of its two predecessors.

Upper Permian strata above the D-Bed Evaporite and Belah Dolomite are generally similar to those of west Cumbria and comprise two main lithological units. The lower of these units comprises more than 30 m of brick-red massive sandstone and blocky argillaceous siltstone, the lithology and scarce sedimentary structures of which are consistent with aeolian deposition on the damp surface of an inland sabkha (Burgess & Holliday 1974); this unit is probably generally synchronous with the Carnallitic Marl and fourth cycle evaporites of the English Zechstein sequence. The upper unit succeeds the lower by alternation and comprises brick-red finely cyclic alluvial plain mudstone, siltstone, and sandstone (Burgess & Holliday 1974); these are typical of youngest Upper Permian strata throughout northern England, suggesting that by this time local influences on sedimentation had become relatively minor except at the margins of the basin where fan-breccias continued to be formed. The dominance of water-laid sediment may indicate a slight increase in rainfall, but

**Fig. 9.12.** The Zechstein, showing the major sub-basins. Reproduced from Smith (1980b), by kind permission of the publishers (Schweizerbart'sche Verlagsbuchhandlung, Stuttgart).

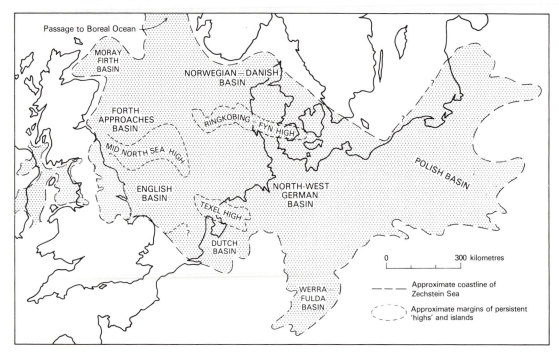

the red colour nevertheless shows that the rainfall was periodic and that oxidising conditions generally prevailed.

## South Lancashire and North Cheshire

Upper Permian sedimentary rocks in S Lancashire and adjoining areas comprise the Manchester Marls; they are up to 100 m thick and although known in detail mainly from boreholes, they are exposed in a number of places around the northern rim of the Cheshire sub-basin. Lower parts of the Manchester Marls are of interbedded mudstones and limestones, and have yielded an Upper Permian specialised fauna of foraminifers, bivalves, gastropods, and ostracods (Pattison 1970, p. 132); plant remains are also locally abundant. The sedimentology of the Manchester Marls has not been studied in detail and depositional cycles have not been distinguished. From their fossils, Pattison (1970) suggested that the lower parts of the marls are probably rough correlatives of the lowest carbonates at Formby and in west Cumbria. Upper parts of the Manchester Marls are of interbedded red mudstone, siltstone, and fine-grained sandstone, and pass diachronously into the overlying red fluviatile sandstones assigned to the Triassic. According to Colter & Barr (1975), the Manchester Marls are increasingly sandy to the south and may pass laterally into red continental sandstone in central or south Cheshire.

## Zechstein Sea Basin

The Zechstein Sea Basin occupied most of the present North Sea and extended eastwards into parts of Holland, Germany, Denmark, and Poland (Fig. 9.12).

It was divided into several sub-basins by ridges ('highs', 'swells') where subsidence was relatively slower than in the centres of the sub-basins, and the Upper Permian rocks of eastern England and the southern North Sea were formed in the English Basin which initially measured some 250 km by 350 km and was generally coincident with the existing Lower Permian Basin. The early Zechstein Sea here is thought to have been 200–250 m deep, increasing to perhaps 300 m as subsidence outstripped sedimentation, and it appears rapidly to have become starved and at times euxinic in its deeper parts during the formation of the carbonates of the first and second of the four main cycles (Smith 1980b); subsequently the basin filled when thick evaporites were formed in the interior at a rate exceeding that of subsidence and, with the original topographic hollow thus virtually eliminated, space for sediment accumulation in the third and later cycles was created mainly by continuing subsidence.

The broad ridges between the various sub-basins were probably also high points on the Lower Permian deserts, and became shoal areas where the thin Zechstein rocks are mainly shallow-water and peritidal carbonates and evaporites. As such they probably emerged first when water levels fell and sequences over them are incomplete. Because the sub-basins were freely intercommunicating during the early phases of each cycle, the deposition of sediments at these times was influenced by basin-wide as well as by more local factors, and faunas and lithological sequences are similar in places as far apart as Greenland (in its Permian position) and Poland. Increasing isolation and perhaps eventual separation of the basins in later phases of the evaporite cycles resulted in reduced continuity of the more saline rocks, which were thus influenced to a greater degree by local conditions and

| YORKSHIRE PROVINCE | | DURHAM PROVINCE | |
|---|---|---|---|
| EXISTING NAMES | PROPOSED NEW NAMES | EXISTING NAMES | PROPOSED NEW NAMES |
| Top Anhydrite | Littlebeck Anhydrite | | |
| Sleights Siltstone | | | |
| Upper Halite and Potash | Sneaton Halite and Potash | | |
| Upper Anhydrite | Sherburn Anhydrite | Upper Anhydrite | Sherburn Anhydrite |
| Upgang Formation | | Upgang Formation | |
| Carnallitic Marl | | Rotten Marl | |
| Boulby Halite and Potash | | Boulby Halite | |
| Billingham Main Anhydrite | Billingham Anhydrite | Billingham Main Anhydrite | Billingham Anhydrite |
| Upper Magnesian Limestone | Brotherton Formation | Seaham Formation | |
| Middle Marls | Edlington Formation | Middle Marls | Edlington Formation |
| Lower Magnesian Limestone (Upper subdivision / Lower subdivision) | Cadeby Formation (Sprotbrough Member / Wetherby Member) | Middle Magnesian Limestone | Ford Formation |
| | | Lower Magnesian Limestone | Raisby Formation |
| Marl Slate | | Marl Slate | |

**Table 9.3.** Revised nomenclature for units of English Zechstein sequences.

are less readily correlated between sub-basins.

Zechstein strata crop out almost continuously in a narrow belt near the western edge of the English Zechstein Basin (Fig. 9.1) and extend across the North Sea beneath later rocks. At outcrop they comprise a series of progradational belts, mainly of limestone and dolomite, which are separated by emersion and erosion surfaces or by red beds containing evaporites or evaporite solution residues; in these rocks is indelibly recorded the evidence of repeated expansion and contraction of the Zechstein Sea, and its eventual demise. A minor barrier (the Cleveland High, Fig. 9.8) lying across North Yorkshire partly separated a Durham Province to the north from a Yorkshire Province, Zechstein rocks exposed in Durham having been formed, in general, farther offshore than those at outcrop in Yorkshire. The classification of Upper Permian strata in the two provinces is shown in Table 9.1, which also shows their supposed equivalents in the Vale of Eden and in Holland and northern Germany. The nomenclature used here for units of the English Zechstein sequence is one in which some formerly confusing names have been replaced by others based

geographically; the new and old names are shown in Table 9.3. An interpretation of the mutual relationships of the various Zechstein formations is shown in Figure 9.13; their broad distribution and thickness trends are given by Taylor & Colter (1975).

*Cycle EZ1* The lithology and biota of the Marl Slate clearly indicate that euxinic conditions were rapidly established in deeper parts of the newly formed Zechstein Sea. This initial marine deposit is a thin sapropelic silty laminated limestone or dolomite with a fauna of *Lingula*, reptiles, and palaeoniscid fish; its content of metallic elements is much lower than that of the equivalent Kupferschiefer of the Zechstein sequence in Germany (Deans 1950, Hirst & Dunham 1963) and the formation is absent from the more freely aerated shallow-water margins and some of the shoal areas such as the Cleveland High and parts of the Mid-North Sea High. Anoxic conditions and slow deposition apparently persisted on the basin floor throughout deposition of the remaining carbonates of the first cycle, but the warm aerated shallow waters of the basin margin and over the ridges were ideal for the formation of a wide range of mainly oolitic carbonate

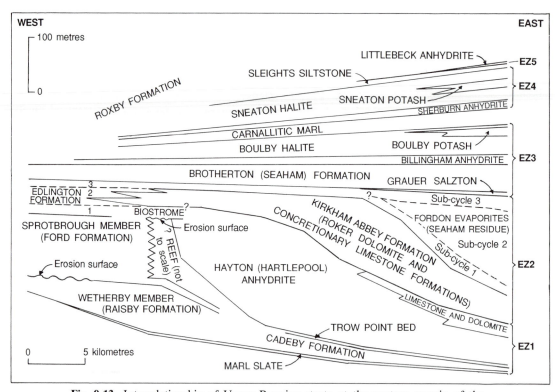

**Fig. 9.13.** Interrelationship of Upper Permian strata at the western margin of the English Zechstein Basin (Yorkshire Province nomenclature, Durham names in brackets), showing the approximate thicknesses of the main stratigraphic units. (Reproduced from Smith 1989 by permission of the Yorkshire Geological Society.)

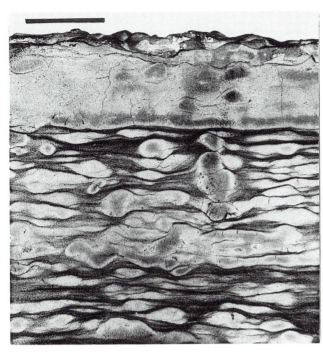

**Fig. 9.14.** Mottling in calcitic dolomite of the Raisby Formation (Lower Magnesian Limestone) from a borehole at Hartlepool, Cleveland. Scale bar 2 cm. (B.G.S. MLD 803.)

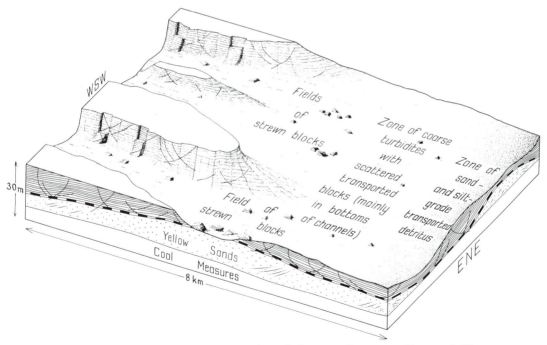

**Fig. 9.15.** Diagrammatic reconstruction of the scene in eastern Tyne and Wear immediately after late Raisby Formation (Lower Magnesian Limestone) submarine sliding. Reproduced from Smith (1970c, fig. 8) by permission of the Yorkshire Geological Society.

sediments which rapidly built a flat-topped shelf up to 50 km across. At outcrop in Yorkshire these shelf rocks are widely divided into two sub-cycles (a & b, Table 9.1, Fig. 9.13) by an erosion surface that is thought to have been cut when sea level fell by at least 5 m. Below this surface the shelf oolites of the first sub-cycle have a predominantly bivalve fauna and contain large numbers of bryozoan-algal patch reefs (Smith 1981a) but reefs are absent where the oolites pass landward into the argillaceous dolomites of the Lower Marls. Equivalent strata formed in deeper water on outer parts of the shelf and on

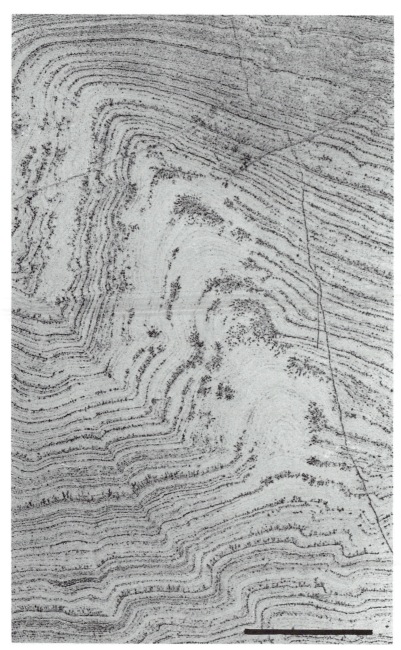

**Fig. 9.16.** Algal-stromatolitic lamination picked out by dendritic manganese dioxide in fine-grained dolomite. Slabbed core of borehole through biostrome, Hawthorn County, Durham. Scale bar 2 cm. (Reproduced from Smith 1981a, by permission of the Society of Economic Paleontologists & Mineralogists).

the basin slope are more widely exposed in Durham than in Yorkshire and contain a relatively varied brachiopod–bryozoan–bivalve fauna; they are commonly coarsely mottled (Fig. 9.14) and display evidence of extensive and locally massive downslope sliding (Fig. 9.15) (Smith 1970c, Kaldi 1980). Rocks of the second sub-cycle at outcrop in Yorkshire are mainly of sparingly shelly, bioturbated, fine oolitic dolomite and, from their sedimentary structures, were deposited mainly as oolite sand waves on a broad shelf under at least 12 m of water (Smith 1968, Kaldi 1986); they pass eastwards, on outer parts of the shelf, into partly pelleted dolomite mudstones. Equivalent strata in the Durham Province comprise a complex of mainly oolitic shelf dolomites protected on its seaward flank by a massive bryozoan-algal shelf-edge reef up to 100 m thick and more than 30 km long (Smith 1981b); biostromal algal-stromatolitic dolomite (Fig. 9.16) covers the flat top of this reef. The youngest carbonate unit of this sub-cycle is a thin distinctive bed of oncoids and columnar algal-stromatolites that may have been continuous across the Zechstein Sea into Germany

and Poland (Smith 1970a, 1986). The first cycle is completed by the massive Hartlepool (= Hayton) Anhydrite, which rings the Zechstein basin and has a displacive fabric that was at first thought to be consistent with formation in a sabkha environment (Taylor & Fong 1969; Taylor 1980, 1984); the writer (Smith 1989) suspects that it may have been a primary gypsum sea-floor precipitate. The seaward face of the anhydrite plunges steeply to the basin floor where the formation passes into about 20 m of anhydrite–carbonate laminates in which Taylor (1980) recognised four widespread minor cycles. Halite is generally absent from the rocks of the EZ1 cycle but it forms lenses within the Werra (= Hartlepool) Anhydrite in the southern North Sea (Rhys 1974).

*Cycle EZ2* Marine rocks of this cycle are slightly less extensive than those of the first cycle and comprise a relatively narrow peripheral belt of shelf and basin-slope carbonate rocks and exceptionally thick evaporites that filled the basin. The shelf carbonates are mainly of shallow-water oolitic dolomite with a scanty bivalve–gastropod–ostracod fauna, and were

**Fig. 9.17.** Calcitic concretionary banding on bedding plane of laminated dolomitic limestone. Slope-facies of Cycle EZ2 shelf carbonates, Fulwell Quarry, Sunderland. The coin is 26 mm across.

probably formed near sea level in a shelf-edge shoals or barrier bars; they pass landwards into an almost barren sequence of red and grey coastal plain and lagoonal mudstones and siltstones (the Edlington Formation or Middle Marls, Table 9.1, Fig. 9.13) that contain both primary and secondary anhydrite and some (mainly secondary) halite. The equivalent basin-slope rocks are mainly of fine-grained partly laminated dolomite and limestone and feature both extraordinary calcite concretions (Fig. 9.17) and evidence of repeated submarine slumping (Smith 1971a); they pass on the basin floor into a thin sequence of brown fetid bituminous calcite laminates that may be a hydrocarbon source rock. Widespread dissolution of the underlying Hartlepool Anhydrite (Smith 1972b) has led to brecciation, dedolomitisation, and foundering of the second cycle carbonate rocks and to the spectacular collapse-breccias of the Durham coast (Fig. 9.18). The thick progradational second cycle evaporites of the basin comprise several sub-cycles and consist mainly of halite with marginal belts rich in anhydrite and polyhalite (Stewart 1963, Taylor &

Colter 1975, Colter & Reed 1980, Taylor 1981). Some of the salts in this cycle may be of deep-water origin, but others have fabrics consistent with formation in shallow water or on salt flats and thus may indicate major falls in sea level and phases of desiccation. The thick salts of this cycle are the main component and driving force of the many salt domes of the southern North Sea.

*Cycle EZ3*    Renewed marine transgression, generally beyond the shorelines of the second cycle, led to the deposition of the fine-grained limestone, now partly dolomitic, of the Seaham Formation and its Yorkshire Province equivalent, on a mainly shallow-water shelf; this formation has a distinctive biota of two species of bivalves and the supposed alga *Calcinema*. The extraordinary width (up to 100 km) of this carbonate shelf is probably a result of the low depositional slope following infilling of the original basin by EZ2 salts, and indicates unusually uniform conditions around the margins of the basin; the shelf edge lies well seaward of that of the two preceding cycles, but the shelf rocks there similarly thin sharply

**Fig. 9.18.** Cycle EZ2 shelf and basin-slope carbonate rocks, brecciated and dedolomitised by dissolution of former underlying Hartlepool Anhydrite. Coastal cliffs, Ryhope, Sunderland.

eastwards and pass into a few metres of argillaceous limestone and calcareous mudstone on the basin floor. Subsequent emergence of much of the shelf resulted in the establishment there of a widespread sabkha on which the succeeding nodular Billingham Anhydrite was formed and this in turn was followed by the extensive Boulby Halite and (near the end of the cycle), by the highly variable Boulby Potash. An origin close to sea level, in and on vast salt flats and in shallow brine pools, is envisaged for both these chloride deposits in basin-margin areas (Smith 1973), but their present mineralogy owes much to a long and complex series of diagenetic changes (Stewart 1951a); little is known of the mineralogy or depositional environment of these two deposits in the broad interior of the English Zechstein basin but their combined thickness there (more than 150 m) implies access to a large source of salts and at least limited connection with the main part of the Zechstein Sea and world oceans.

Rocks of the third and fourth cycles of the English Zechstein sequence are separated by the barren red blocky siltstone of the *Carnallitic Marl*, a salt-rich unit comparable with and perhaps equivalent to the blocky aeolian siltstone and sandstone unit of the Cumbrian

**Fig. 9.19.** Displacive halite cubes in argillaceous siltstone. Scale bar 1 cm. Unit 'D' of Sneaton Halite (EZ4); slabbed core from borehole near Whitby, North Yorkshire.

sequences. It records a phase when marginal flats extended far into the basin and the Zechstein Sea dwindled to a series of shallow playas; the evaporites of these playas form a large part of the Carnallitic Marl in the basin centre (Taylor & Colter 1975, Smith & Crosby 1979).

*Cycle EZ4* At its base a thin carbonate bed – the Upgang Formation – has yielded no fossils and appears to be the product of a highly saline but extensive shallow inland sea. It is succeeded across much of the basin by the striking halite-rich Sherburn Anhydrite (the equivalent of the Pegmatitanhydrit of the Zechstein in sequence in Germany) and is overlain by the thick Sneaton (Upper) Halite with which the Sneaton (Upper) Potash is interbedded. The Sneaton Halite and the Sneaton Potash are more uniform in thickness and lithology than the Boulby Halite and Boulby Potash of the third cycle, but both bear features consistent with deposition on extensive marginal salt flats (Smith 1973) and evidence of a complex diagenetic history (Stewart 1951b). It may be that the Sneaton Potash is the final product of an evaporite sub-cycle, for the succeeding halite has a high content of red siltstone (Fig. 9.19) suggesting that it may have been formed mainly displacively on a subaerially exposed brine-soaked plain (Smith 1971b). The silty halite contains abundant potash salts in the interior of the basin.

At the end of the fourth cycle, Zechstein salt sedimentation was again interrupted by a phase when silts and muds (now red) spread across much of the basin to form the *Sleights Siltstone*; conditions generally similar to those prevailing during deposition of the Carnallitic Marl seem likely, with a similarly high content of wind-blown dust and secondary evaporite minerals.

*Cycle EZ5* The Sleights Siltstone underlies a minor fifth cycle which comprises a thin anhydrite on land and a few metres of anhydrite and halite under the southern North Sea (Taylor & Colter 1975).

The completion of the fifth cycle is generally taken to mark the end of the depositional history of the Zechstein Sea, although extensive thin lenses of anhydrite in succeeding red beds beneath the North Sea clearly point to continuing episodic subsidence and flooding of former Zechstein depocentres. Its completion with the ensuing covering of the entire basin with thick, almost barren, clastic deposits is also taken by many (see Rhys 1974, for example) as marking the end of the Permian period. It must be emphasised, however, that this view is a convention and that here, as elsewhere in Britain, the placing of the junction of the Permian and Triassic systems in rock sequences is not based on close faunal control and there is no reason to suppose that it would be coincident with widespread sedimentary changes.

## Acknowledgement

I wish to record my thanks to Dr D. H. Raynor for her many helpful suggestions after her reading of the draft of this chapter.

## REFERENCES

| | | |
|---|---|---|
| ARTHURTON, R. S. & HEMINGWAY, J. E. | 1972 | The St Bees evaporites – a carbonate–evaporite formation of Upper Permian age in West Cumberland, England. *Proc. Yorkshire geol. Soc.*, **38**, 565–591. |
| ARTHURTON, R. S., BURGESS, I. C. & HOLLIDAY, D. W. | 1978 | Permian and Triassic. *In* Moseley, F. (Ed.) *The Geology of the Lake District* (*q.v.*), 189–266. |
| BOTT, M. H. P. | 1964 | Gravity measurements in the north-eastern part of the Irish Sea *Q. J. geol. Soc. Lond.*, **120**, 369–396. |
| | 1968 | The geological structure of the Irish Sea Basin. *In* Donovan, D. T. (Ed.) *Geology of Shelf Seas*, Oliver & Boyd, Edinburgh, 93–113. |
| BURGESS, I. C. | 1965 | The Permo–Triassic rocks around Kirkby Stephen, Westmorland. *Proc. Yorkshire geol. Soc*, **35**, 91–101. |
| BURGESS, I. C. & HOLLIDAY, D. W. | 1974 | The Permo–Triassic rocks of the Hilton Borehole, Westmorland. *Bull. geol. Surv. Gt. Br.*, **46**, 1–34. |
| CLARKE, R. F. A. | 1965 | British Permian saccate and monosulcate miospores. *Palaeont.*, **8**, 322–354. |
| COLLAR, F. A. | 1974 | A geophysical interpretation of the structure of the Vale of Clwyd, North Wales. *Geol. J.*, **9**, 65–76. |
| COLTER, V. S. | 1978 | Exploration for gas in the Irish Sea. *Geologie en Mijnbouw*, **57**, 503–516. |

COLTER, V. S. &
BARR, K. W.
1975
Recent developments in the geology of the Irish Sea and Cheshire Basins. *In* Woodland, A. W. (Ed.) *Petroleum and the Continental Shelf of North-West Europe*, 1, *Geology*. Applied Science Publishers, Barking, 61–75.

COLTER, V. S. &
REED, G. E.
1980
Zechstein 2 Fordon Evaporites of the Atwick No. 1 borehole, surrounding areas of N.E. England and the adjacent southern North Sea. *In* Fuchtbauer, H. & Peryt, T. M. (Eds) The Zechstein Basin with Emphasis on Carbonate Sequences. *Contr. Sedimentology*, 9, 115–129.

DEANS, T.
1950
The Kupferschiefer and associated lead–zinc mineralization in the Permian of Silesia, Germany and England. *Rep XVIII Int. geol. Congr.*, 7, 340–352.

DUNHAM, K. C. &
ROSE, W. C.
1949
Permo–Triassic geology of south Cumberland and Furness. *Proc. Geol. Ass. London*, 60, 11–40.

EBBERN, J.
1981
The geology of the Morecambe Gas Field. *In* Illing, L. V. & Hobson, G. D. (Eds.) *Petroleum Geology of the Continental Shelf of North-West Europe*. Heyden & Son Ltd., London, 485–493.

FALCON, N. L. &
KENT, P. E.
1960
Geological results of petroleum exploration in Britain 1945–1957. *Mem. geol. Soc. Lond.*, 2, 56 pp.

GLENNIE, K. W.
1972
Permian Rotliegend of northwest Europe interpreted in the light of modern desert sedimentation studies. *Bull. Amer. Ass. Petrol. Geol.*, 56, 1047–1071.

1984
Early Permian–Rotliegend. *In* Glennie, K. W. (Ed.) *Introduction to the Petroleum Geology of the North Sea*. Blackwell, Oxford, 41–60.

GLENNIE, K. W. &
BULLER, A. T.
1983
The Permian Weisslegend of NW Europe: the partial deformation of aeolian dune sands caused by the Zechstein transgression. *Sedimentary Geology*, 35, 43–81.

GLENNIE, K. W.,
MUDD, G. C. &
NAGTEGAL, P. J. C.
1978
Depositional environment and diagenesis of Permian Rotliegendes sandstones in Leman Bank and Sole Pit areas of the UK, southern North Sea. *J. geol. Soc. Lond.*, 135, 25–34.

HAUBOLD, H. &
KATZUNG, G.
1975
Die position der Autun/Saxon–Grenze (Unteres Perm) in Europa Und Nordamerika. *Schriftenr. geol. Wiss. Berlin*, 3, S.87–138.

HIRST, D. M. &
DUNHAM, K. C.
1963
Chemistry and petrography of the Marl Slate of S.E. Durham, England. *Econ. Geol.*, 58, 912–940.

HODGE, M. R.
1932
The Permian Yellow Sands of north-east England. *Proc. Univ. Durham phil. Soc.*, 8, 410–458.

JACKSON, D. I.,
MULHOLLAND, P.,
JONES, S. M. &
WARRINGTON, G.
1987
The geological framework of the East Irish Sea Basin. *In* Brooks, J. & Glennie, K. W. (Eds.) *Petroleum Geology of North West Europe*, Graham & Trotman, London, 191–203.

KALDI, J.
1980
The origin of nodular structures in the Lower Magnesian Limestone (Permian) of Yorkshire. *In* Fuchtbauer, H. & Peryt, T. M. (Eds.) The Zechstein Basin with emphasis on Carbonate Sequences. *Contr. Sedimentology*, 9, 45–60.

1986
Sedimentology of sandwaves in an oolite shoal complex in the Cadeby Magnesian Limestone Formation (Upper Permian) of Eastern England. *In* Harwood, G. M. & Smith, D. B. (Eds.) The English Zechstein and related topics. *Spec. Publ. Geol. Soc. London*, 22, 63–74.

KENT, P. E.
1948
A deep borehole at Formby, Lancashire. *Geol. Mag.*, 5, 253–264.

KING, W.                         1850   A monograph of the Permian fossils of England. *Palaeontogr. Soc. (Monogr.)*, 258 pp.

LAMING, D. J. C.                 1966   Imbrications, palaeocurrents and other sedimentary features in the Lower New Red Sandstone, Devonshire, England. *J. sedim. Petrol.*, **36**, 940–959.

MARIE, J. P. P.                  1975   Rotliegendes stratigraphy and diagenesis. *In* Woodland, A. W. (Ed.) *Petroleum and the Continental Shelf of North-West Europe*, **1**, *Geology*. Applied Science Publishers, Barking, 205–210.

MILLER, J. A.,                   1962   The potassium–argon age of the lava at Killerton Park,
   SHIBATA, K. &                        near Exeter. *Geophys. J.*, **6**, 394–396.
   MONRO, M.

MILLER, J. A. &                  1964   Potassium–argon measurements on the granites and some
   MOHR, P. A.                          associated rocks from south-west England. *Geol. J.*, **4**, 105–126.

MOSELEY, F. (Ed.)                1978   *The Geology of the Lake District.* Yorkshire Geological Society, Leeds, 284 pp.

PATON, ROBERTA L.                1974   Lower Permian pelycosaurs from the English Midlands. *Palaeontology*, **17**, 541–552.

PATTISON, J.                     1970   A review of the marine fossils from the Upper Permian rocks of Northern Ireland and north-west England. *Bull. geol. Surv. Gt. Br.*, **32**, 123–165.

PATTISON, J.,                    1973   A review of late Permian and early Triassic biostratigraphy
   SMITH, D. B. &                       in the British Isles. *In* Logan, A. & Mills, L. V. (Eds.) The
   WARRINGTON, G.                       Permian and Triassic Systems and Their Mutual Boundary. *Mem. Can. Soc. Petrol. Geol.*, **2**, 220–260.

POLLARD, J. E.                   1976   A problematical trace fossil from the Tor Bay Breccias of South Devon (Comment on paper by J. M. Ridgway). *Proc. Geol. Assoc. London*, **87**, 105–108.

PRYOR, W. A.                     1971   Petrology of the Permian Yellow Sands of northeastern England and their North Sea Basin equivalents. *Sedimentary Geol.*, **6**, 221–254.

RAMSBOTTOM, W. H. C.,            1978   A correlation of Silesian Rocks in the British Isles. *Spec.*
   CALVER, M. A.,                       *Rep. Geol. Soc. London*, **10**, 81 pp.
   EAGAR, R. M. C.,
   HODSON, F.,
   HOLLIDAY, D. W.,
   STUBBLEFIELD, C. J. &
   WILSON, R. B.

RIDGWAY, J. M.                   1974   A problematical trace fossil from the New Red Sandstone of South Devon. *Proc. Geol. Assoc. London*, **85**, 511–517.

RHYS, G. H. (Compiler)           1974   A proposed standard lithostratigraphic nomenclature for the southern North Sea and an outline structural nomenclature for the whole of the (UK) North Sea. *Rep. Inst. Geol. Sci.*, **74/8**, 14 pp.

ROSE, W. C. &                    1977   Geology and hematite deposits of South Cumbria.
   DUNHAM, K. C.                        *Econ. Mem. Geol. Surv. Gt. Br.*, Sheets 58, part 48, 170 pp.

RUSSELL, M. J.                   1976   A possible Lower Permian age for the onset of ocean floor spreading in the northern North Atlantic. *Scott. J. Geol.*, **12**, 315–323.

SCHWEITZER, H-J.                 1986   The land flora of the German and English Zechstein. *In* Harwood, G. M. & Smith, D. B. (Eds.) The English Zechstein and related topics. *Spec. Publ. Geol. Soc. London*, **22**, 31–54.

SHOTTON, F. W.      1929   The geology of the country around Kenilworth (Warwickshire). *Q. J. geol. Soc. London*, **85**, 167–220.

SHOTTON, F. W.      1937   The Lower Bunter Sandstone of north Worcestershire and east Shropshire. *Geol. Mag.*, **74**, 534–553.

SHOTTON, F. W.      1956   Some aspects of the New Red Desert in Britain. *L'pool. Manch. geol. Jl.*, **1**, 450–465

SMITH, D. B.       1968   The Hampole Beds – a significant marker in the Lower Magnesian Limestone of Yorkshire, Derbyshire and Nottinghamshire. *Proc. Yorkshire geol. Soc.*, **36**, 463–477.

                   1970a  Permian and Trias. *In* Hickling, G. (Ed.) Geology of Durham County. *Trans. nat. Hist. Soc. Northumb.*, **41**, 66–91.

                   1970b  The palaeography of the British Zechstein. *In* Rau, J. L. & Dellwig, L. F. (Eds.) *Third Symposium on Salt*, 1. N. Ohio geol. Soc., Cleveland, 20–23.

                   1970c  Submarine slumping and sliding in the Lower Magnesian Limestone of Northumberland and Durham. *Proc. Yorkshire geol. Soc.*, **38**, 1–36.

                   1971a  The stratigraphy of the Upper Magnesian Limestone in Durham – a revision based on the Institute's Seaham Borehole. *Rep. Inst. geol. Sci.*, **71/3**, 12 pp.

                   1971b  Possible displacive halite in the Permian Upper Evaporite Group of northeast Yorkshire. *Sedimentology*, **17**, 221–232.

                   1972a  The British Isles. *In* Falke, H. (Ed.) *Rotliegend, essays on European Lower Permian*. Brill, Leiden, 1–33.

                   1972b  Foundered strata, collapse-breccias and subsidence features of the English Zechstein. *In* Richter-Bernburg, G. (Ed.) Geology of saline deposits. *Unesco, Earth Sci.*, **7**, Paris, 255–269.

                   1973   The origin of the Permian Middle and Upper Potash deposits of Yorkshire: an alternative hypothesis. *Proc. Yorkshire geol. Soc.*, **39**, 327–346.

                   1974   Permian. *In* Rayner, D. H. & Hemingway, J. E. (Eds.) *The Geology and Mineral Resources of Yorkshire*. Yorkshire Geological Society, Leeds, 115–144.

                   1976   A review of the Lower Permian in and around the British Isles. *In* Falke, H. (Ed.) *The Continental Permian in Central, West and South Europe*. Reidel Publishing Co., Dordrecht, Holland, 13–22.

                   1979   Rapid marine transgressions of the Upper Permian Zechstein Sea. *J. geol. Soc. Lond.*, **136**, 155–156.

                   1980a  Permian and Triassic rocks. *In* Robson, D. A. (Ed.) *The Geology of north east England*. Spec. Publ. Nat. Hist. Soc. Northumbria, Newcastle upon Tyne, 36–48.

                   1980b  The evolution of the English Zechstein basin. *In* Füchtbauer, H. & Peryt, T. M. (Eds.) The Zechstein Basin with Emphasis on Carbonate Sequences. *Contr. Sedimentology*, **9**, 7–33.

                   1981a  Bryozoan–algal patch reefs in the Upper Permian Lower Magnesian Limestone of north-east England. *In* Toomey, D. H. (Ed.) European Fossil Reef Models. *S.E.P.M. Spec. Publ.*, **30**, 187–202.

                   1981b  The Magnesian Limestone (Upper Permian) reef complex of north-east England. *In* Toomey, D. H. (Ed.) European Fossil Reef Models (*q.v.*).

SMITH, D. B.                    1984   Red Basal Permian Sands; a new discovery in National
                                       Coal Board Boreholes off Sunderland, North-east
                                       England. *Proc. Yorkshire geol. Soc.*, **44**, 497–500.

                                1986   The Trow Point Bed: a deposit of Upper Permian Marine
                                       oncoids, peloids and columnar stromatolites in the Zech-
                                       stein of north-east England. *In* Harwood, G. M. & Smith,
                                       D. B. (Eds.) The English Zechstein annd related topics.
                                       *Spec. Publ. geol. Soc. London*, **22**, 113–125.

                                1989   The late Permian palaeogeography of north-east England.
                                       *Proc. Yorkshire geol. Soc.*, **47**, 285–312.

SMITH, D. B. &                  1979   The regional and stratigraphical context of Zechstein 3
  CROSBY, A.                            and 4 potash deposits in the British sector of the southern
                                       North Sea and adjoining land areas. *Econ. Geol.*, **74**,
                                       397–408.

SMITH, D. B. &                  1991   Permian. *In* Cope, J. C. W. *et al.* (Eds.) *Atlas of
  TAYLOR, J. C. M.                      Palaeogeography and Lithofacies*. Geological Society
                                       Publishing House, Bath.

SMITH, D. B. &                  1974   A correlation of Permian rocks in the British Isles. *Spec.
  FOUR OTHERS                          Rep. Geol. Soc. London*, **5**, 45 pp.

SNEH, A.                        1988   Permian dune patterns in northwestern Europe challenged.
                                       *J. Sediment. Petrol.*, **58**, 44–51.

STEWART, F. H.                  1951a  The petrology of the evaporites of Eskdale No. 2 boring,
                                       East Yorkshire. Part 2. The Middle Evaporite Bed. *Miner.
                                       Mag.*, **29**, 445–475.

                                1951b  Part 3. The Upper Evaporite Bed. *Miner. Mag.*, **29**,
                                       557–572.

                                1963   The Permian Lower Evaporites of Fordon in Yorkshire.
                                       *Proc. Yorkshire geol. Soc.*, **34**, 1–34.

STONELEY, H. M.                 1958   The Upper Permian flora of England. *Bull. Br. Mus. nat.
                                       Hist., Geol.*, **3**, 295–337.

TAYLOR, J. C. M.                1980   Origin of the Werraanhydrit in the U.K. Southern North
                                       Sea – a reappraisal. *In* Füchtbauer, H. & Peryt, T. M.
                                       (Eds.) *The Zechstein Basin with Emphasis on Carbonate
                                       Sequences*. Contr. Sedimentology, **9**, 91–113.

                                1981   Zechstein facies and petroleum prospects in the central and
                                       northern North Sea. *In* Illing, L. V. & Hobson, G. D. (Eds.)
                                       *Petroleum Geology of the Continental Shelf of North-West
                                       Europe*. Heyden & Son Ltd., London, 176–185.

                                1984   Late Permian–Zechstein. *In* Glennie, K. W. (Ed.) *Introduc-
                                       tion to the Petroleum Geology of the North Sea*. Blackwell,
                                       Oxford, 61–83.

TAYLOR, J. C. M. &              1969   Correlation of Upper Permian strata in East Yorkshire and
  FONG, G.                             Durham. *Nature, London*, **224**, 173–175.

TAYLOR, J. C. M. &              1975   Zechstein of the English sector of the Southern North Sea
  COLTER, V. S.                        Basin. *In* Woodland, A. W. (Ed.) *Petroleum and the
                                       Continental Shelf of North-West Europe*, **1**, *Geology*.
                                       Applied Science Publishers, Barking, 249–263.

TURNER, P. &                    1986   Chemical and isotopic studies of a core of Marl Slate from
  MAGARITZ, M.                         NE England: influence of freshwater influx into the
                                       Zechstein Sea. *In* Harwood, G. M. & Smith, D. B. (Eds.)
                                       The English Zechstein and related topics. *Spec. Publ.
                                       geol. Soc. London*, **22**, 19–29.

VERSEY, H. C.                   1925   The beds underlying the Magnesian Limestone in York-
                                       shire. *Proc. Yorkshire geol. Soc.*, **20**, 200–214.

VISSCHER, H.                    1971   The Permian and Triassic of the Kingscourt Outlier,
                                       Ireland. *Spec. Pap. geol. Surv. Irl.*, **1**, 114 pp.

WALKER. T. R.,
 WAUGH, B. &
 CRONE, A. J.
1978 Diagenesis in first-cycle desert alluvium of Cenozoic age, southwestern United States and northwestern Mexico. *Geol. Soc. Amer. Bull.*, **89**, 19–32.

WARREN, P. T.,
 WILSON, H. W.,
 HAWKINS, T. R. W.,
 SMITH, E. G. &
 MILLS, D. A. C.
1973 Geological Survey 1 : 50,000 Geological Sheet 107 (Denbigh).

WARRINGTON, G. &
 SCRIVENER, R. C.
1988 Late Permian fossils from Devon: regional geological implications. *Proc. Ussher Soc.*, **7**, 95.

WAUGH, B.
1970a Formation of quartz overgrowths in the Penrith Sandstone (Lower Permian) of northwest England as revealed by scanning electron microscopy. *Sedimentology*, **14**, 309–320.

1970b Petrology, provenance and silica diagenesis of the Penrith Sandstone (Lower Permian) of northwest England. *J. sedim. Petrol.*, **40**, 1226–1240.

1978 Authigenic feldspar in British Permo–Triassic sandstones. *J. geol. Soc. Lond.*, **135**, 51–56.

WHITTAKER, A. (Ed.)
1985 *Atlas of onshore sedimentary basins in England and Wales, post-Carboniferous tectonics and stratigraphy.* Blackie, Glasgow.

WRIGHT, J. E.,
 HULL, J. H.,
 MCQUILLIN, R. &
 ARNOLD, S. E.
1971 Irish Sea Investigations 1969–70. *Rep. Inst. geol. Sci.*, **71/19**, 45 pp.

# 10

# TRIASSIC

# M. G. Audley-Charles

## Introduction

The close of the Palaeozoic Era was marked by an extinction of many fossil groups and the beginning of the Mesozoic Era was remarkable for the great changes in marine invertebrate faunas. The evidence for these dramatic changes has been found in the rocks which accumulated in the shallow seas marginal to the great circum-Pacific and Tethys oceans (Kummel 1970). The British Isles lay beyond reach of the Tethyan ocean waters for most of the Triassic period although occasional brief incursions reached England and Wales after the middle Triassic. Like the Permian System the Triassic rocks of England and Wales are mainly red continental siliciclastic facies with some locally important evaporite sequences. These Permo–Triassic rocks were formerly called the New Red Sandstone.

In many parts of Britain the beginning of the Triassic system seems to be associated either with an increase in grain size of the red siliciclastic facies or with Permian marls giving way to Triassic sandstones. The Triassic rocks are readily divisible into a lower sandstone facies and an upper mudstone facies with intercalations of sandstone and siltstone. They pass up into Jurassic mudstones with limestones and marls at the top of the Triassic sequence reflecting the incoming of marine waters.

A notable feature often dominating Triassic accumulations was the influence of horst and graben fault systems (Audley-Charles 1970a), which controlled the subsidence of many basins both onshore and offshore (Zeigler 1975, Selley 1976). Sedimentation was more widespread than during the Permian reflecting the continuing subsidence of the basins and grabens as well as the gradual wearing down of the Hercynian (Variscan) uplands and the remnant uplands of the Caledonide mountain chain. This gradual process of extending sedimentary basins continued into the Jurassic. Eustatic rises in sea level and basin subsidence resulted in Triassic rocks being overlapped by Jurassic deposits.

The main stratigraphical features of the Triassic of England and Wales may be summed up as: (1) a general poverty of stratigraphically diagnostic fossils, especially in the red sandstone facies; (2) facies variations and diachronism within the successions; (3) the general overstepping nature of the base of the Triassic (except where it rests on Permian deposits) and, in general, an overlapping relationship by successively younger members, hence the Triassic rocks of England and Wales rest on a basement that ranges down to the Precambrian. The distribution of Permo-Triassic rocks is shown in Figure 10.1.

## Classification and correlation

The Triassic rocks of England and Wales comprise a lower sandy series and an upper mudstone series formerly called respectively the Bunter and Keuper Series on the basis of general comparison with the mainly siliciclastic sequences of Germany and Holland which are traditionally divided into Bunter, Muschelkalk and Keuper Series. It has long been recognised that the typical German Muschelkalk facies of marine limestone and marls is not developed in England and Wales. During the last decade evidence from palynology has revealed that there are great differences in age between German and British Bunter and Keuper lithofacies. The stratigraphical nomenclature applied to the Triassic rocks of Britain has been recently revised (Warrington *et al.* 1980), bringing it into line with modern practice. This revision takes account of palynology and the results of deep drilling both

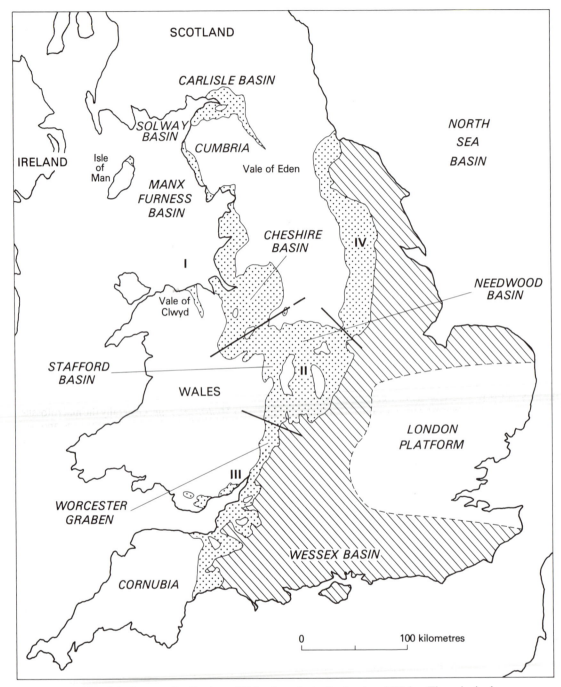

**Fig. 10.1.** Present distribution of Triassic rocks in England and Wales. The principal outcrops of the Sherwood Sandstone, Mercia Mudstone and Penarth Groups are indicated by dotted area, and their subsurface distribution is indicated by striped ornament. (I-IV are the regions described in the text.)

onshore and offshore (Fig. 10.2). The revised litho-stratigraphic divisions are shown below:

| Traditional Nomenclature (to be abandoned) | Warrington et al. (1980) Revised Nomenclature |
|---|---|
| | approximate equivalent |
| Rhaetic | Penarth Group |
| Keuper Marl | Mercia Mudstone Group |
| Keuper Sandstone ⎫ Bunter Sandstone ⎭ | Sherwood Sandstone Group |

The two main impediments to Triassic stratigraphic correlation are the scarcity of stratigraphically diagnostic fossils and the strong diachronism of the lithofacies. As Warrington et al. (1980) have stated 'The relationships of British palynomorph assemblages to the ammonite-based stages of the Tethyan Triassic are at present imprecisely known and though, as yet, it is not possible to define the exact positions of stage boundaries within the British rock-sequences, it seems likely that all the Tethyan stages are represented in Britain'.

The Permo–Triassic boundary cannot be defined with fossils in any part of Britain (see Fig. 10.2) or in the Germanic facies of Europe. Even in the fully marine sequences there are very few places where this important boundary between the Palaeozoic and Mesozoic can be precisely defined (Tozer 1967, Silberling & Tozer 1968). Warrington et al. (1980) proposed an arbitrary lithostratigraphic boundary at the base of the Bröckelschiefer in Germany which can be traced in seismic reflection profiles and in boreholes across the North Sea Basin into north-east England.

In marked contrast to the lithostratigraphic criteria used to define the base of the Trias in England, the top was defined by Warrington et al. (1980) following Cope et al. (1980) on palaeontological evidence, viz. in rocks bearing the first appearance of ammonites of the genus Psiloceras, marking the base of the Hettangian stage in Britain.

## Palaeontology

The palaeontology of the British Triassic rocks has been reviewed recently by Warrington (1976). The marine invertebrate fauna characterising the Triassic of the Tethyan realm are almost entirely absent from England and Wales and ammonites are completely unknown in the Triassic rocks of Britain. Instead the fossils of the British Trias reflect the mainly continental desert environments that dominated this period. Four kinds of fossils have been found: (1) invertebrates such as branchiopod crustaceans, inarticulate brachiopods, bivalves and ostracodes; (2) fish; (3) amphibians, reptiles and early mammals; (4) plant remains and palynomorphs (including miospores, acritarchs and dinoflagellates).

The lower part of the Sherwood Sandstone Group is particularly poor in fossils and much of the Mercia Mudstone Group is also poorly fossiliferous. The most fossiliferous part of the Triassic sequence in England and Wales is the Penarth Group that comprises marine limestones bearing bivalves, brachiopods, crustaceans, coelenterates, gastropods, echinoids, fish, amphibians, reptiles, plants, palynomorphs, and foraminifers.

*Sherwood Sandstone Group.* Branchiopod crustaceans (*Euestheria*) are known from Staffordshire (Cantrill 1913), inarticulate brachiopods (*Lingula*) from Nottinghamshire (Rose & Kent 1955), bivalves like *Modiolus* and *Mytilus* and ostracodes occur in the central Midlands (Warrington et al. 1980). Arachnids (belonging to Scorpionida) have been found in Warwickshire and Worcestershire (Wills 1910, 1947). Fish remains are represented by *Ceratodus* teeth in Warwickshire and Worcestershire (Wills 1910), *Gyrolepis* (Walker 1969), *Woodthropes wilsonii* and *Senionotus metcalfei* (Swinnerton 1925) from the central Midlands. Amphibia *Cyclotosaurus leptognathas* and *Mastodonsaurus lavaii* have also been found in the Midlands. Reptiles *Rynchosaurus articeps* and *archosaurus* have been found in the Midlands and in Devon (Walker 1969).

Plants from Worcestershire (Wills 1910) include *Equisetites, Schizoneura, Voltzia, Yuccites* as well as lycopsids, pteropsids and cycadopsids (Warrington 1976).

*Mercia Mudstone Group.* Generally the macrofossils have been found in the sandy and silty beds and not in the red mudstones that dominate this group. *Euestheria* and an insect wing have been found in Cheshire (Thompson 1966). Bivalvia such as *Nuncula, Pholadomya* and *Thracia* have been reported from Warwickshire (Newton 1887) and *Modiola* from Somerset (Green & Welch 1965). Fish spines and teeth have been found in the Midlands and Somerset (Warrington 1976). Amphibian remains have been found in Warwickshire.

According to Warrington (1970a) the plant fossils from the Arden Sandstone Formation include representatives of the same sphenopsid and coniferopsid genera as occur in the underlying Sherwood Sandstone Group. The great changes in marine invertebrates that are such an evolutionary feature of the passage from the Palaeozoic to Mesozoic are not reflected in the land plants. These showed little change until the late Triassic when Mesozoic features began to appear in the flora.

The Tea Green Marls facies, now regarded as part of the Blue Anchor Formation (Warrington et al. 1980), developed towards the top of the Mercia Mudstone Group, have yielded miospores, acritarchs and dinoflagellate cysts from Nottinghamshire (Morbey 1975).

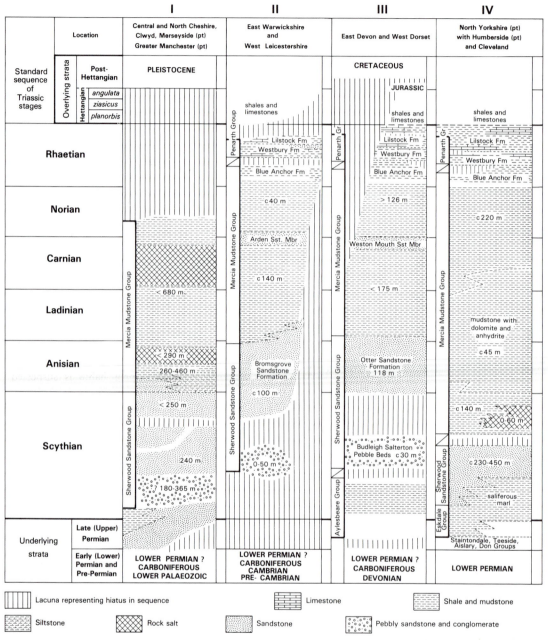

**Fig. 10.2.** Summary of correlation of the principal Triassic lithostratigraphic divisions in England and Wales (after Warrington *et al.* 1980).

The Grey Marls facies (now part of the Blue Anchor Formation) near the top of the Mercia Mudstone Group of SW England contains bivalves, gastropods, fish, amphibians, reptiles, early mammals and plant remains (Warrington 1976).

*Penarth Group.* This group deposited under the influence of marine incursions has yielded more marine elements in the fauna: foraminifers, coelenterates, branchiopods, gastropods, bivalves, brachiopods, ostracodes, insects, ophiuroids, echinoids, fish, amphibia, reptiles and plants (Warrington 1976).

# Regional descriptions

The Triassic deposits of England and Wales are described under four geographical areas: (I) North-west England and the north-east Irish Sea basin, (II) Central Midlands, (III) Southern England and South Wales, (IV) Eastern England (Fig. 10.1).

## North-west England and the north-east Irish Sea (I)

This area includes the Solway, Manx–Furness Basins (Wright *et al.* 1971, Colter & Barr 1975) and extends onshore in the Vale of Clwyd, Isle of Man, west Cumbria, west Lancashire and Merseyside where they connect with the inland basin of Cheshire and the English Midlands and to the north-east with the Carlisle basin and Vale of Eden. The Triassic is not well exposed onshore in this region owing to the cover of glacial and other superficial deposits. The rock-sequences are best known from boreholes onshore and offshore (Colter & Barr 1975, Colter & Ebbern 1978, Ebbern 1981).

The Sherwood Sandstone Group and Mercia Mudstone Group are widespread but the Mercia Mudstone Group and Penarth Group have only been found well-developed in the Cheshire Basin (Figs. 10.1, 10.2). The revised regional nomenclature (Warrington *et al.* 1980) is as follows:

| Traditional Trias Nomenclature (used to date in Geological Survey publications) | Revised Nomenclature |
| --- | --- |
| Keuper Sandstone | Helsby Sandstone Formation |
| Upper Mottled Sandstone | Wilmslow Sandstone |
| Bunter Pebble Beds | Chester Pebble Beds |
| Lower Mottled Sandstone* | Kinnerton Sandstone Formation |

* Permian in part.

Offshore differences in facies have led to a different scheme being proposed (Rhys 1974).

In Cumbria and the Isle of Man, the Sherwood Sandstone Group of largely unfossiliferous sand-stones has gradational boundaries with both the underlying Permian marls and shales and the overlying Mercia Mudstone Group. In the Vale of Clwyd, the poorly exposed unfossiliferous sandy deposits are very difficult to correlate stratigraphically and could belong to the Permian or the base of the Triassic succession (Smith *et al.* 1974).

Stratigraphically useful fossils are absent from much of the Sherwood Sandstone Group of north-west England and the north-east Irish Sea Basin (Sarjeant 1974), although an early Triassic miospore assemblage was reported from north-east Cheshire (Warrington 1970b).

The Sherwood Sandstone–Mercia Mudstone Group boundary has been regarded as Middle Triassic (Anisian) age in the southern part of the Cheshire Basin on the basis of the reptile *Rhynchosaurus articeps* (Warrington *et al.* 1980).

The Wilkesley borehole (Pugh 1960) sited on the Lias outlier SW of Audlem in the southern part of the Cheshire Basin penetrated a thick and complete succession of the Mercia Mudstone Group containing two rock-salt-bearing deposits (Poole & Whiteman 1966) which separate the parts of the typical red mudstones formerly called the Keuper Marl.

The stratigraphically useful fossils in the Mercia Mudstone Group are miospore assemblages found in borehole cores. The ages indicate ranges from the highest Röt (late Scythian or early Anisian) of the early Triassic to Ladinian or Middle Triassic although a Carnian age for the upper halite is suspected (Warrington *et al.* 1980).

The Penarth Group overlying the Tea Green Marl facies (of the Blue Anchor Formation) of the Mercia Mudstone Group is present in the southern part of the Cheshire Basin below the Lias.

The Penarth Group contains a variety of macro- and microfossil assemblages thought to be Rhaetian (late Triassic) in age (Poole & Whiteman 1966).

## Central Midlands (II)

Although the Triassic rocks are widely exposed in this area, much of the succession is concealed beneath the onlapping younger Mesozoic rocks to the east and south. A number of deep basins and grabens are present in the Midlands, e.g. the Stafford and Needwood basins and the Worcester graben (Fig. 10.1). The Triassic rocks are unconformable on a topographically irregular surface of various Palaeozoic formations and the influence of faulting on the development of the Triassic basins of this region is strongly indicated (Audley-Charles 1970a, b; Wills 1976).

The lowest strata regarded as Triassic are the various pebble conglomerates of the Sherwood

Sandstone Group (formerly called the Bunter Pebble Beds) which overlie a variety of Permian and pre-Permian deposits. There is no palaeontological evidence for this Permo–Triassic boundary. On the flanks of the London Platform in Warwickshire, Leicestershire and Northamptonshire the Triassic deposits are thin but to the north and west Triassic graben subsidence allowed thicker accumulations, particularly during the early Triassic. However, late Triassic deposits spread over a wider area (Audley-Charles 1970b) so that the Mercia Mudstone Group was more extensive than the Sherwood Sandstone Group. One consequence of this is that the marginal coarse-grained facies of breccias, conglomerates, sandstones and siltstones can occur at any stratigraphical level reflecting the spreading of the basin margins over a landscape being gradually buried by the overlapping erosional products of its own denudation. These breccias tend to have clasts of fairly local origin and are generally interpreted as gravel fans disgorged by wadis into the depositional basins. In contrast conglomerates with large, rounded polymict pebbles (Fig. 10.3) indicate various sources transported by a major river system having a large catch-

**Fig. 10.3.** Typical polymict conglomerate of Sherwood Sandstone Group showing rounded cobbles and pebbles with occasional sandstone lenses. Kidderminster Formation, Marlbrook Gravel Pit, Worcestershire (B.G.S. A13504).

ment area; it has even been argued that some pebbles in the Bunter Pebble Bed were derived from the Armorican massif of northern France (Wills 1948, 1956). A series of sedimentary cycles (approximating to fining-upward cycles) have been identified in various parts of the English Midlands (Wills 1970, 1976). These features suggest fluvial deposition under meandering river-channel and flood-plain as well as braided-channel conditions.

In parts of the Midlands the transition from the sandstone to mainly mudstone facies (Sherwood Sandstone Group to Mercia Mudstone Group) takes place in the so-called 'Waterstone facies' of mudstone–siltstone–sandstone intercalated sequences.

The Bromsgrove Sandstone Formation of the Sherwood Sandstone Group is relatively rich in fossils (Warrington *et al.* 1980), indicating a late Scythian to Ladinian age.

The Mercia Mudstone Group shows typical red mudstones and silty mudstones with local sandstone–siltstone developments. Some of these such as the Arden Sandstone are widespread.

The top of the Mercia Mudstone Group is of Tea Green Marl facies composed of grey and green, rhythmically interbedded laminated mudstones and unlaminated fine-grained dolomites which are overlain by the Penarth Group. The fossils of the Mercia Mudstone Group in the Midlands indicate a Carnian (late Triassic) age although it seems likely that the Group extends from the middle Triassic (Ladinian) into the late Triassic. The overlying Penarth Group which is probably of Rhaetian age, comprised of mudstones, with sandstones, calcareous mudstones, bioclastic limestones and pure calcilutites, all relatively rich in macrofossil and microfossil assemblages.

Some of the important changes in stratigraphic nomenclature recommended by Warrington *et al.* (1980) are:

| Former Trias Nomenclature | Revised Nomenclature |
|---|---|
| Tea Green Marls ⎫<br>Grey Marls ⎬ | Blue Anchor Formation of Mercia Mudstone Group |
| Arden Sandstone Group | Arden Sandstone Member of Sherwood Sandstone Group |
| Keuper Sandstone of Central Midlands including Basement Beds, Building Stones and most of Waterstones ⎫⎬⎭ | Bromsgrove Sandstone Formation of the Sherwood Sandstone Group |

## Southern England and South Wales (III)

The Sherwood Sandstone Group is well exposed in Devon with strong subsurface development in Somerset, Dorset (where the lower sandstone is known as the Aylesbeare Group), Wiltshire, Hampshire,

Gloucestershire and Oxfordshire where it is covered by the Mercia Mudstone Group, Penarth Group, Jurassic and younger strata but not known in South Wales. It is thought to extend offshore into the English Channel and Western Approaches.

The Budleigh Salterton Pebble Beds and the overlying Otter Sandstone Formation have been generally and arbitrarily regarded as the lower part of the Triassic succession in south Devon. Henson (1970) argued that the underlying red marls and the red sandstones below them exposed west of Budleigh Salterton, traditionally regarded as Permian, could belong to the Triassic system. In the absence of any palaeontological evidence the Permo–Triassic boundary cannot be defined in this region.

The Mercia Mudstone Group in Devon and Somerset is composed mainly of red mudstones (Fig. 10.4) with a few local sandstones and one or more evaporite developments. The sandstones and locally breccias and conglomerates are particularly well developed at the basin margins where the Mercia Mudstones Group overlaps the Sherwood Sandstone Group as, for example, on to the Mendip Hills where the well-known Dolomitic Conglomerate (Fig. 10.5) occurs filling in the buried landscape and is now being re-excavated by erosion. Of particular palaeontological interest are the reptilian remains in the Dolomitic Conglomerate facies of the Mendips. Associated with these deposits are the infillings in karst solution depressions and fissures containing reptiles and early mammals and mammal-like reptiles. These fissure fillings seem to range in age from Triassic (belonging to both Mercia Mudstone and Penarth Groups) to post-Triassic (Robinson 1957, 1971; Warrington 1976).

The Penarth Group of mudstones, sandstones, siltstones, calcareous mudstones, calcilutites and fissile shales is well developed in Devon and Somerset. As in other parts of England the relatively rich microfauna and macrofauna indicate a late Triassic, probably Rhaetian age but are not sufficiently diagnostic stratigraphically to allow close correlation with the Tethyan marine facies. Lithostratigraphically the Penarth Group of southern England and South Wales is composed of two formations. The lower is the Westbury Formation (equivalent to the Westbury Beds and the Black Shale and include the famous Bone Beds). Its contact on the Blue Anchor Formation is generally non-sequential. Above the Westbury Formation is the Lilstock Formation which includes the Cotham and Langport Members. According to Warrington *et al.* (1980) the top of the Penarth Group is separated from the base of the Jurassic by a few metres of alternating fissile shales and limestones of the 'Ostrea Beds' or pre-*planorbis* Beds of Richardson (1906). The changes in nomenclature and revision of

**Fig. 10.4.** Mercia Mudstone Group. 'Red Marls' overlain by 'Tea Green Marls' facies (pale band) and black shales (Blue Anchor/Westbury Formations). Aust Cliff, immediately south of Severn Bridge, Avon. (B.S.G. A10669.)

the stratigraphy in these type sections of SW England can be summarised as:

| Traditional Trias Nomenclature (used to date in Geological Survey publications) | Revised Nomenclature |
|---|---|
| RHAETIC SERIES | PENARTH GROUP |
| Cotham Beds, White Lias, Langport Beds and Watchet Beds | Lilstock Formation |
| Westbury Beds, Black Shales, Bone Beds | Westbury Formation |

The Triassic deposits of the Wessex Basin appear to thicken considerably below younger cover between the Mendip Hills and the London Platform. What little borehole evidence there is suggests the influence of Triassic faulting giving rise to N–S trending graben connecting the Worcester graben with the Armorican massif of northern France (Audley-Charles 1970a, b). As in the other Triassic grabens of England, this Wessex extension of the Worcester graben contains halite deposits within the Mercia Mudstone Group (Audley-Charles 1970a, b; Warrington *et al*. 1980).

Farther eastwards on the southern and western flanks of the London Platform the Mercia Mudstone Group and the Penarth Group wedge out, the younger Penarth Group probably overlapping the Mercia Mudstone Group.

### Eastern England (IV)

The Sherwood Sandstone, Mercia Mudstone and Penarth Groups are widely exposed in eastern England, and extend, covered by younger Mesozoic strata, to the east and south-east of their crop into the North Sea Basin and mainland Europe. They are underlain by marls and saliferous marls of Permian age. The Permo–Triassic boundary cannot be distinguished palaeontologically and the generally accepted boundary is arbitrary (Warrington *et al*. 1980). Offshore it is taken at the base of the Bröckelschiefer, regarded as equivalent to the base of the Saliferous Marl division of the Eskdale Group (Smith *et al.* 1974).

The Triassic deposits of eastern England can be regarded as having accumulated at the western margin of the North Sea Basin (Kent 1968, Taylor 1968).

**Fig. 10.5.** Dolomitic Conglomerate (Sherwood Sandstone Group) resting unconformably on cross-bedded Black Nore Sandstone (Devonian). Woodhill Bay, Portishead, Severn Estuary, Avon. (R. H. Roberts.)

Facies changes are present, particularly sandy facies related to the basin margin which generally become muddier and thicken eastwards (Audley-Charles 1970a, b). The overlapping trends of successively higher parts of the Mercia Mudstone sequence which oversteps the underlying Palaeozoic formations of the London Platform are associated with sandy and pebbly margin deposits accumulated on the buried landscape.

The Sherwood Sandstone Group comprises sandy formations which locally are pebbly (formerly called the Bunter Pebble Beds). The base of the overlapping Mercia Mudstone Group is marked by a sharp erosional contact the 'Hardegsen disconformity' correlated across the North Sea Basin into northern Germany (Geiger & Hopping 1968).

In part of Nottinghamshire, Elliott (1961) recognised eight formations on the basis of detailed lithological characters among which were several skerries (dolomitic sandy beds). One of the characteristic features of the Mercia Mudstone Group is the presence of beds and nodules of gypsum and anhydrite. In the northern part of this embayment of the North Sea Basin into eastern England halite deposits are preserved in the lowest part of the Mercia Mudstone Group. These evaporites are clearly related to the Röt halite of the southern halites of other English onshore and offshore basins and grabens where the rock-salt deposits occur at lithostratigraphically higher levels in the Mercia Mudstone Group (Warrington et al. 1980).

The presence of a dolomitic facies within the Mercia Mudstone Group of eastern England may be the onshore extension of the Muschelkalk facies of the southern North Sea Basin (Balchin & Ridd 1970). Onshore it may crop out in some of the skerries (Warrington 1974a, b).

The Penarth Group of eastern England is represented by the Westbury Formation (generally 5–11 m thick) and the overlying Lilstock Formation (5–11 m thick). The Westbury Formation comprises dark shales with bone-bearing sandstones and a thin limestone which passes upward into grey-green calcareous mudstones of the Cotham Member of the Lilstock Formation with its overlying chocolate brown mudstones.

Thinly bedded limestones comparable with the 'White Lias Limestone' of the Langport Member of the Lilstock Formation are present locally in the south-western part of this area. Generally the Cotham Member is the highest element of the Penarth Group overlain non-sequentially by basal beds of the Lias.

Ostracodes from the Penarth Group of Nottinghamshire were correlated with the German 'Middle Röt' by Anderson (1964), while palynomorphs from the Penarth Group of eastern England have been correlated with the late Triassic of Austria (Morbey 1975).

## Regional tectonic setting

The Triassic deposits of England and Wales accumulated in grabens, half-grabens and basins (Fig. 10.6) whose depocentres were located entirely on a cratonic basement. Generally the sediments are found above various Palaeozoic platform deposits. Burke (1976) summarised the occurrence of grabens that formed at the margins of the developing Atlantic Ocean during the Mesozoic thus placing the complex array of grabens and horsts that characterise much of the Triassic history of England and Wales (Audley-

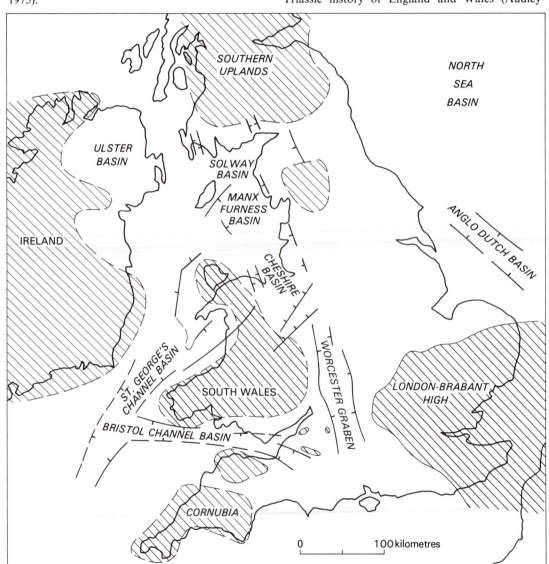

**Fig. 10.6.** Sketch map to show principal graben structures of Triassic age in England and Wales. Striped ornament indicates areas of Triassic erosion and non-accumulation.

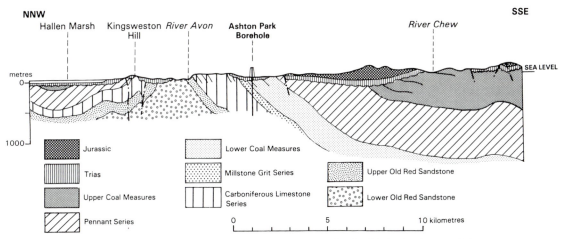

**Fig. 10.7.** Section across Geological Survey Map of Bristol District (parts of sheets Nos. 250, 251, 264, 265, 280 and 281) to show the amount of pre-Triassic erosion of the underlying Carboniferous and Devonian strata.

Charles 1970a, b) into a context of extensional lithospheric plate margins. The orientation and precise location of these fault-bounded basins are probably influenced not only by the orientation of the tensional forces, but also by inhomogeneities, such as old lines of weakness, in the continental basement.

The development of horst and graben structures in the British Isles and North Sea Basin region during the Triassic period provide a good example of the first stage of the 'rift and drift' behaviour of continental margins associated with the formation of new oceans, in this case with the opening of the north Atlantic ocean in the Late Jurassic and Early Cretaceous.

Examination of the Geological Survey maps of onshore England and Wales where Triassic deposits have been preserved in contact with their basement and where the younger Triassic rocks are overlain by

Jurassic strata may tell us something of the tectonic processes that were associated with this Triassic phase of rifting. The dominant stratigraphic patterns discernible from the geological maps are that the base of the Triassic non-marine sequence oversteps the underlying Palaeozoic and Precambrian rocks with strong angular unconformity and the younger members of the Triassic sequence are overlapped conformably by the Jurassic marine strata. The basal Triassic rocks in several regions rest unconformably on older members of the folded Carboniferous succession. For example, in the Mendip hills as displayed on the Geological Survey sheet Nos. 250, 251, 264, 265, 280 and 281 (Fig. 10.7) and also in other areas of Hercynian folds displayed on Geological Survey sheet No. 125 (Fig. 10.8). Where the older Triassic rocks overlie various members of the Carboniferous Limestone these

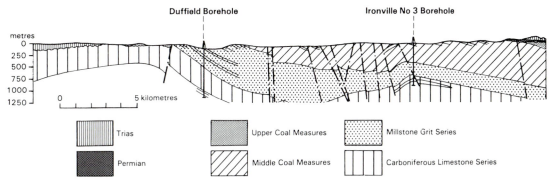

**Fig. 10.8.** Section across Geological Survey Map of Derby Sheet No. 125 to show the amount of pre-Triassic erosion of the underlying Carboniferous strata.

maps imply that 1–3 km of the Carboniferous succession had been eroded before the Triassic sandstones were deposited. The only alternative explanation is that the Namurian (Millstone Grit) and Westphalian (Coal Measure) deposits were originally very thin in these areas from which they are now absent. This seems highly unlikely because only a few kilometres away these deposits are approximately 2–3 km thick (Figs. 10.7 and 10.8).

The evidence of pre-Triassic and syn-Triassic erosion of the Carboniferous and older rocks, amounting locally to probably 1–3 km, suggests regional uplift (Fig. 10.9). The pattern (Fig. 10.6) of Triassic basins and grabens accords with Falvey's (1974) general concept of the faulted collapse of crustal domes that are generated by a thermal expansion of the lithosphere that precedes the splitting of a continent with the formation of a new ocean. Bott (1981) proposed a mechanism for doming of continental crust generally preceding rifting by normal faulting. He noted that the uplift appears to be related to the isostatic response to the development of a low density, high temperature region in the upper mantle. The uplift and erosion phase preceding the subsidence of the Triassic on-shore grabens and basins seems to indicate significant differences from the Jurassic rifting and subsidence in the North Sea basins (Wood & Barton 1983) that appears to support the uniform extension of the lithosphere model (McKenzie 1978) for basin formation.

The presence of marine Jurassic deposits overlying the grabens and basins filled with Triassic mainly nonmarine sediments suggests either that crustal subsidence continued through the Jurassic and/or, that eustatic rise of sea level persisted through the Jurassic. Vail et al. (1977) inferred global sea-level movements that do not include any notable eustatic rises during the Mesozoic until the late Jurassic rise that preceded the great Cretaceous transgressions. If the Vail et al. (1977) sea-level curves are a reliable guide for the Jurassic it suggests that continuing crustal subsidence over 90 Ma is mainly responsible for the accumulations of Triassic–Jurassic successions in England and Wales. Such subsidence accords with the concept of a late Carboniferous/early Permian thermal uplift accompanied by a strongly erosional phase of the uplifted region followed by a prolonged period of subsidence associated with cooling and contracting of the lithosphere. This mechanism was discussed by Fischer (1975) and Kinsman (1975) in their isostatic models of basin subsidence. Bott (1976) proposed a mechanism of continental graben subsidence in excess of 5 km involving tensional forces that result in grabens between 24 and 48 km wide. This corresponds well (Fig. 10.10) with the width of Triassic grabens in England and Wales (Audley-Charles 1970a, b). The

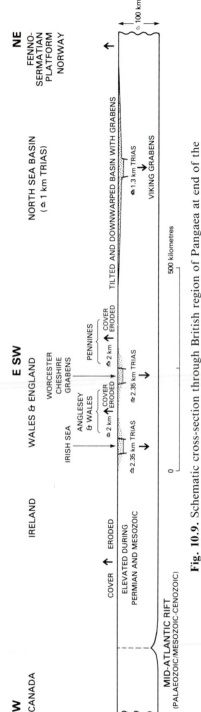

**Fig. 10.9.** Schematic cross-section through British region of Pangaea at end of the Triassic period. Note the grabens filled with Triassic deposits contain between 2·3 km–1·3 km of compacted sediments and that the elevation of the intervening larger horst-like regions amounts to about 2 km during the Triassic.

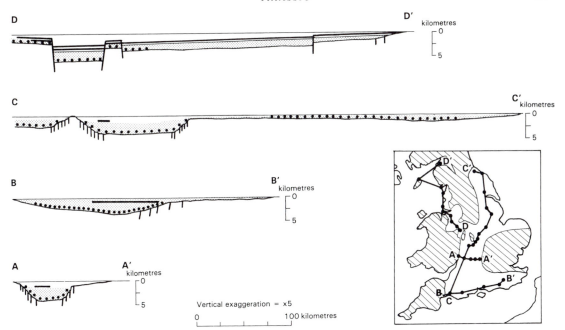

**Fig. 10.10.** Schematic cross-section through some Triassic grabens and basins of England and Wales based mainly on boreholes (after Audley-Charles 1970a). On the inset map present-day coastline is shown for reference. Areas with few or no Triassic deposits shown by striped ornament. Principal salt deposits shown in solid black line. Basal pebbly facies indicated by coarse dots. Four lines of section are indicated.

Bott and Falvey models can accommodate the spasmodic subsidence of the graben indicated by the sedimentological and stratigraphical evidence of the British Triassic deposits with their periodic coarse gravels and notable thickening along faulted basin margins.

The thermal doming hypothesis would imply increased heat flow to the British region during Permo–Triassic times. This heat could have contributed to the maturation of organic matter in the Carboniferous. The locally overlying porous/ permeable arenites provide potential hydrocarbon traps of the type that have been exploited in the offshore region of England.

## Palaeogeography

That the Triassic climate of England and Wales was mainly hot and arid is indicated by continental red bed clastic facies with only sporadic plant remains. The sedimentological evidence in the form of evaporites at most stratigraphical levels and the occurrence of desiccation cracks, calcretes, exfoliated pebbles, sheeted bedrock and diagenetic clay mineralogy was well summarised by Anderton et al. (1979). Spasmodic rainfall must have persisted throughout the Triassic because the great majority of Triassic rocks were transported and deposited from rivers. Alluvial fans, braided streams, mudflows have been interpreted from the sedimentological features of the Sherwood Sandstone Group (Fig. 10.11).

The transitional facies between the Sherwood Sandstone Group and the overlying Mercia Mudstone Group was formerly called the Waterstones and was regarded by some (Audley-Charles 1970a, b) as a relatively thin but very widespread deposit in the British Isles related to a temporary marine transgression. Rose & Kent (1955) found *Lingula* in these deposits in Nottinghamshire. Warrington (1970a) reported marine microplankton in the Midlands. Ireland et al. (1978) reported intertidal sediments with trace fossils in the Waterstones of Cheshire which they interpreted as a consequence of the Röt marine

320     M. G. AUDLEY-CHARLES

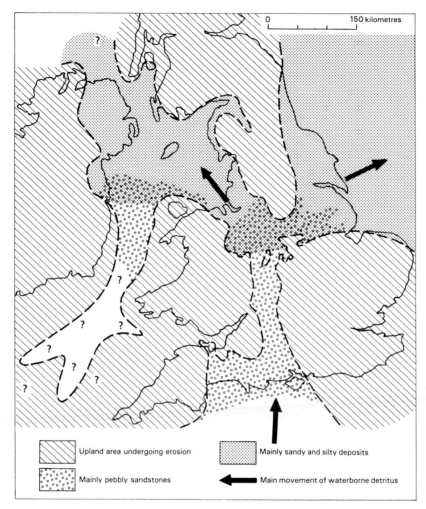

**Fig. 10.11.** Palaeogeographic sketch map of England and Wales for the early Triassic (after Audley-Charles, 1970b and Warrington 1974b).

Legend:
- Upland area undergoing erosion
- Mainly pebbly sandstones
- Mainly sandy and silty deposits
- Main movement of waterborne detritus

0    150 kilometres

transgression. Evidence seems to be growing for a very flat plain occupying Central England and perhaps extending to distant parts of the British Isles allowing the Röt marine transgression to penetrate far into Britain depositing a diachronous 'Waterstones' facies.

The Mercia Mudstone Group is composed of red dolomitic mudstones and siltstones with greenish bands, minor beds of mainly fine-grained dolomitic sandstones, minor beds of gypsum, anhydrite and halite and locally celestite deposits. The fossils and sedimentary structures in these rocks do not permit a unique interpretation of the environment of deposition. The two main interpretations are (i) desert plains with playa lakes, or (ii) an inland sea or hypersaline lake. Evidence seems to be weighted in favour of the first interpretation, largely on the negative argument of the absence of distinctive lacustrine or marine features such as laminated muds. The origin of the halite deposits remains uncertain and this has lent weight to the inland sea or hypersaline lake model (Fig. 10.12). The red sediments of the Mercia Mudstone Group gave way to grey and green dolomitic mudstones due perhaps to increasing organic content with anaerobic conditions following the long history of oxidising conditions of the red facies. These so-called 'Tea Green Marls' and 'Grey Marls' are composed of interbedded mudstones and fine-grained dolomites locally with nodular gypsum. Anderton et al. (1979) compared these deposits, which contain rare U-shaped burrows and acritarchs, with modern marine sabkhas.

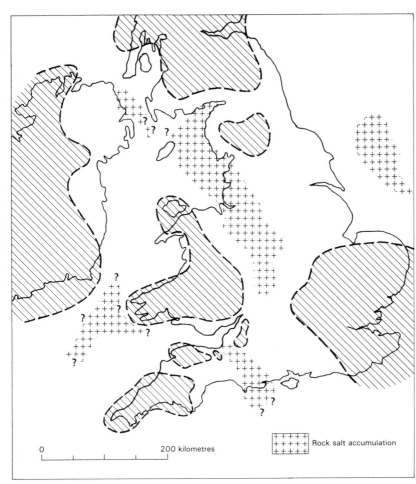

**Fig. 10.12.** Palaeogeographic sketch map of England and Wales for the late Triassic showing location of main halite accumulations (crosses). Areas of few or no late Triassic deposits shown by striped ornament.

The Penarth Group contains marine shelly fossils such as echinoids, pectinids and oysters, indicating that marine waters were spreading across large parts of England and Wales. Locally near the base is a strong burrowed horizon with bones, teeth and scales and spines of marine and freshwater animals – the famous 'Rhaetic bone bed'. Kent (1970) suggested the English Midlands were rapidly covered by marine waters. After this transgression the dark shales of the Westbury Formation with their rather restricted shelly fauna including *Rhaetavicula contorta* accumulated widely over England and locally in South Wales. These deposits are considered to indicate oxygen-depleted conditions in the bottom waters related perhaps to poor circulation and restricted connections with the open sea water of what is now the Alpine region. The Cotham Member of the Lilstock Formation, composed of non-marine marls, testifies to the regression of the marine waters over eastern England and Wales while in south-western England the oysters in the Cotham suggest marine connections persisted in the south-west. The Cotham stromatolites (Hamilton 1961) may indicate intertidal conditions. The youngest beds of the Penarth Group, the Langport Member limestones and shales with their marine shelly fauna of oysters, bivalves, echinoids and solitary corals, represented the re-advance of marine waters – a marine invasion which was to persist and dominate much of England and Wales for the remainder of the Mesozoic.

## REFERENCES

ANDERSON, F. W.                1964    Rhaetic ostracoda. *Bull. geol. Surv. G.B.*, **21**, 133–174.

ANDERTON, R.,                  1979    *A Dynamic Stratigraphy of the British Isles.* G. Allen &
    BRIDGES, P. H.,                    Unwin, London. 301 pp.
    LEEDER, M. R. &
    SELLWOOD, B. W.

AUDLEY-CHARLES, M. G.          1970a   Stratigraphical correlation of the Triassic rocks of the
                                       British Isles. *Q. J. geol. Soc. London*, **126** 19–47.

                               1970b   Triassic palaeogeography of the British Isles. *Q. J. geol.
                                       Soc. London*, **126**, 48–89.

BALCHIN, D. A. &               1970    Correlation of the younger Triassic rocks across eastern
    RIDD, M. F.                        England. *Q. J. geol. Soc. London*, **126**, 91–101.

BOTT, M. H. P.                 1981    Crusted doming and the mechanism of continental rifting.
                                       *Tectonophysics*, **73**, 1–8.

                               1976    Formation of sedimentary basins of graben type by
                                       extension of the continental crust. *In* Bott, M. H. P. (Ed.)
                                       Sedimentary Basins of Continental Margins and Cratons,
                                       *Tectonophysics*, **36**, 77–86.

BURKE, K.                      1976    Development of graben associated with the initial ruptures
                                       of the Atlantic Ocean. *In* Bott, M. H. P. (Ed.) Sedimentary
                                       Basins of Continental Margins and Cratons, *Tectono-
                                       physics*, **36**, 93–112.

CANTRILL, T. C.                1913    *Estheria* in the Bunter of South Staffordshire. *Geol. Mag.*,
                                       **50**, 518–519.

COLTER, V. S. &                1975    Recent developments in the geology of the Irish Sea and
    BARR, K. W.                        Cheshire Basins. *In* Woodland, A. (Ed.) *Petroleum and the
                                       Continental Shelf of Northwest Europe* 1, Geology. Applied
                                       Science Publishers Ltd., London, 61–73.

COLTER, V. S. &                1978    The petrography and reservoir properties of some Triassic
    EBBERN, J.                         sandstones of the Northern Irish Sea basin. *J. geol. Soc.
                                       London*, **135**, 57–62.

COPE, J. C. W.,                1980    A correlation of Jurassic rocks in the British Isles. Part I:
    GETTY, T. A.,                      Introduction and lower Jurassic. *Spec. Rep. Geol. Soc.
    HOWARTH, M. K.,                    London*, **14**, 73 pp.
    MORTON, N. &
    TORRENS, H. S.

EBBERN, J.                     1981    The geology of the Morecambe gas field. *In* Illing, L. V. &
                                       Hobson, G. D. (Eds.) *Petroleum Geology of the Continental
                                       Shelf & North-West Europe.* Heydon & Son Ltd., London,
                                       485–493.

ELLIOTT, R.                    1961    The stratigraphy of the Keuper Series in southern Nott-
                                       inghamshire. *Proc. Yorkshire Geol. Soc.*, **33**, 197–234.

FALVEY, D.                     1974    The development of continental margins in plate tectonic
                                       theory. *Aust. Pet. Explor. Assoc. J.*, **14**, 95–106.

FISCHER, A. G.                 1975    Origin and growth of basins. *In* Fischer A. G. & Judson, S.
                                       (Eds.) *Petroleum and Global Tectonics.* Princeton Univer-
                                       sity Press, Princeton, 47–79.

GEIGER, M. E. &                1968    Triassic stratigraphy of the southern North Sea Basin. *Phil.
    HOPPING, C. A.                     Trans. R. Soc. London*, B.254, 1–36.

GREEN, G. W. &                 1965    Geology of the country around Wells and Cheddar. *Mem.
    WELCH, F. B. A.                    geol. Surv. G.B.*, 225 pp.

HAMILTON, D.                   1961    Algae growths in the Rhaetic Cotham Marble of Southern
                                       England. *Palaeontology, London*, **4**, 324–333.

HENSON, M. R. 1970 The Triassic rocks of south Devon. *Proc. Ussher Soc.*, **2**, 172–177.

IRELAND, R. J., POLLARD, J. E., STEEL, R. J. & THOMPSON, D. B. 1978 Intertidal sediments and trace fossils from the Waterstones (Scythian–Anisian?) at Dansbury, Cheshire. *Proc. Yorkshire geol. Soc.*, **41**, 399–436.

KENT, P. E. 1968 The Rhaetic Beds. *In* Sylvester-Bradley, P. C. & Ford, T. D. (Eds.) *The Geology of the East Midlands.* Leicester University Press, 174–187.

1970 Problems of the Rhaetic Beds in the East Midlands. *Mercian Geol.*, **3**, 361–373.

KINSMAN, D. J. J. 1975 Rift valley basins and sedimentary history of trailing continental margins. *In* Fischer, A. G. & Judson, S. *Petroleum of Global Tectonics.* Princeton University Press, 83–126.

KUMMEL, B. 1970 *History of the Earth* (2nd Edit.). W. H. Freeman & Co., San Francisco.

MCKENZIE, D. P. 1978 Some remarks on the development of sedimentary basins. *Earth Planet. Sci. Lett.*, **40**, 25–32.

MORBEY, S. J. 1975 The palynostratigraphy of the Rhaetian Stage, Upper Triassic in Kendelbuchgraben, Austria. *Palaeontogigraphica*, B.152, 1–75.

NEWTON, E. T. 1887 On the remains of fishes from the Keuper of Warwick and Nottingham. *Q. J. geol. Soc. London*, **43**, 537–540.

POOLE, E. G. & WHITEMAN, A. J. 1966 Geology of the country around Nantwich and Whitchurch. *Mem. geol. Surv. G.B.*, 154 pp.

PUGH, W. 1968 Triassic salt: discoveries in the Cheshire–Shropshire basin. *Nature, London*, **187**, 278–279.

RHYS, G. H. 1974 A proposed standard lithostratigraphic nomenclature for the southern North Sea and an outline structural nomenclature for the whole of the (U.K.) North Sea. A report of the joint oil industry – Institute of Geological Sciences Committee on North Sea Nomenclature. *Rep. Inst. geol. Sci.*, No. 74/8, 14 pp.

RICHARDSON, L. 1906 On the Rhaetic and contiguous beds of Devon and Dorset. *Proc. Geol. Assoc. London*, **19**, 401–409.

ROBINSON, P. L. 1957 The Mesozoic fissures of the Bristol Channel area and their vertebrate faunas. *J. Linn. Soc. (Zool.)*, **43**, 260–282.

1971 A problem of faunal replacement on Permo–Triassic continents. *Palaeontology, London*, **14**, 131–153.

ROSE, G. N. & KENT, P. E. 1955 A *Lingula*-Bed in the Keuper of Nottinghamshire. *Geol. Mag.*, **92**, 476–480.

SARJEANT, W. A. S. 1974 A history and bibliography of the study of fossil vertebrate footprints in the British Isles. *Palaeogeogr. Palaeoclimatol. Palaeoecol.*, **16**, 265–378.

SELLEY, R. 1976 The habitat of North Sea Oil. *Proc. Geol. Assoc. London*, **87** (4), 359–387.

SILBERLING, N. J. & TOZER, E. T. 1968 Biostratigraphic classification of the marine Triassic in North America. *Spec. Pap. Geol. Soc. Am.*, **110**, 63 pp.

SMITH, D. B., BRUNSTROM, R. G. W., MANNING, P. I., SIMPSON, S. & SHOTTEN, F. W. 1974 A correlation of the Permian rocks of the British Isles. *Spec. Rpt. geol. Soc. London*, **5**, 45 pp.

SWINNERTON, H. H. 1925 A new catopterid fish from the Keuper of Nottingham. *Q. J. geol. Soc. London*, **81**, 87–99.

TAYLOR, F. M.     1968     Permian and Triassic formations. *In* Sylvester-Bradley, P. C. & Ford, T. (Eds.) *The Geology of the East Midlands.* Leicester University Press, 149–173.

THOMPSON, D. B.     1966     The occurrence of an insect wing and brachiopods (*Euestheria*) in the Lower Keuper Marl at Styal, Cheshire. *Mercian Geol.*, **1**, 237–245.

TOZER, E. T.     1967     A standard for Triassic time. *Bull. geol. Surv. Can.*, **156**, 103 pp.

VAIL, P. R., MITCHUM, R. M. & THOMPSON, S.     1977     Seismic stratigraphy and global changes of sea level, Part 4: Global cycles of relative changes of sea level. *Mem. Am. Assoc. Petrol. Geol.*, **26**, 83–97.

WALKER, A. D.     1969     The reptile fauna of the 'Lower Keuper' Sandstone. *Geol. Mag.*, **106**, 470–476.

WARRINGTON, G.     1970a     The stratigraphy and palaeontology of the 'Keuper' Series of the central Midlands of England, *Q. J. geol. Soc. London*, **126**, 183–223.

    1970b     The 'Keuper' Series of the British Trias in the Northern Irish Sea and neighbouring areas. *Nature, London*, **226**, 254–256.

    1974a     Triassic. *In* Rayner, D. H. & Hemingway, J. E. (Eds.) *The Geology and Mineral Resources of Yorkshire.* Yorkshire Geological Society, 145–160.

    1974b     Les evaporites du Trias Britannique. *Bull. Soc. geol. Fr.* (7th Ser.), XLV, 708–723.

    1976     British Triassic palaeontology. *Proc. Ussher Soc.*, **3**, 341–353.

WARRINGTON, G., AUDLEY-CHARLES, M. G., ELLIOTT, R. E., IVIMEY-COOK, H. C., KENT, P. R., ROBINSON, P. L., SHOTTEN, F. W. & TAYLOR, F. M.     1980     A correlation of Triassic rocks in the British Isles. *Spec. Rep. geol. Soc. London*, **13**, 78 pp.

WILLS, L. J.     1910     On the fossiliferous Lower Keuper rocks of Worcestershire. *Proc. Geol. Assoc., London*, **21**, 249–331.

    1947     A monograph of British Triassic Scorpions. *Monogr., palaeontogr. Soc.*, Pt. 1 & Pt. 2, 137 pp.

    1948     *The Palaeogeography of the Midlands.* University Press of Liverpool, 144 pp.

    1956     *Concealed Coalfields.* Blackie, London & Glasgow, 208 pp.

    1970     Triassic succession in the Central Midlands in its regional setting, *Q. J. geol. Soc. London*, **126**, 225–285.

    1976     The Trias of Worcestershire and Warwickshire. *Rep. Inst. geol. Sci. London*, No. 76/2, 211 pp.

WOOD, R. & BARTON, P.     1983     Crustal thinning and subsidence in the North Sea. *Nature, London*, **302**, 134–136.

WRIGHT, J. E., HULL, J. H., McQUILLIN, R. & ARNOLD, S. E.     1971     Irish Sea investigations 1969–1970. *Rep. Inst. geol. Sci. London*, No. 71/19, 55 pp.

ZIEGLER, W.     1975     Outline of the geological history of the North Sea. *In* Woodland, A. W. (Ed.) *Petroleum and the Continental Shelf of N.W. Europe.* Applied Science Publishers Ltd., London, **1**, 165–167.

# 11

# JURASSIC
# A. Hallam

## Introduction

From the time of William Smith the English Jurassic rocks have been of considerable interest to stratigraphers because of their general accessibility, frequently rich fossil content and lack of severe tectonic disturbance. Long before the time that W. J. Arkell

(1933) wrote his famous treatise England had become classic ground for students of the Jurassic, especially the two magnificent coastal sections of Dorset and Yorkshire. Although Arkell's book remains an invaluable reference work and includes an absorbing discussion on the evolution of stratigraphic principles, it is a measure of the amount of research progress in

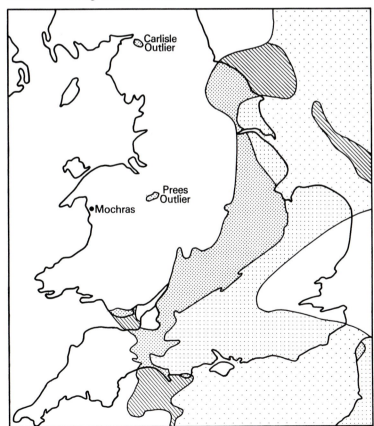

Fig. 11.1. Outcrop and subcrop distribution of Jurassic. Fine stippling – onshore outcrop; intermediate stippling – onshore subcrop; oblique lines – offshore outcrop; coarse stippling – offshore subcrop. Western Approaches and Cardigan Bay excluded because of lack of published data.

the last few decades that much of it is now out of date. Surprising though it may seem after more than a century of research, a great deal of stratigraphic revision and refinement has taken place and the increased attention paid in recent years to facies analysis, involving both sedimentological and palaeoecological investigations in a stratigraphic context, has given us a far more comprehensive and detailed understanding of depositional environments than obtained hitherto. Finally, the new stratigraphic and facies information from a large number of onshore and offshore boreholes has greatly extended and improved our knowledge of the underlying tectonic framework of southern Britain.

Jurassic outcrops dipping gently in a generally eastward direction extend across England in a wide belt from Cleveland and East Yorkshire through the East and South Midlands to Somerset and Dorset. Maps of subsea outcrops extending offshore from the Yorkshire and Dorset coasts are contained respectively in Dingle (1971) and Donovan & Stride

(1961). West of the erosional limit of the main outcrop, Liassic outliers (Fig. 11.1) occur on the Shropshire and Cheshire borders (Prees), Cumbria (Carlisle), Glamorgan and the margin of Cardigan Bay (Mochras). Middle Jurassic has been recognised offshore in Cardigan Bay and a more extensive sequence ranging almost to the top of the Jurassic in the Bristol Channel. Subsurface Jurassic rocks extend eastward beneath the Cretaceous cover over the whole of the eastern part of England except for an area extending from the London region into north Kent and East Anglia (Fig. 11.1). In the southern North Sea a substantially complete Jurassic sequence is preserved only in the Sole Pit Trough (p. 343); elsewhere Lower Cretaceous (locally with topmost Jurassic) oversteps progressively on to Liassic or even pre-Jurassic strata.

Establishing the structural framework of deposition (Fig. 11.2) requires taking into account evidence from the underlying Triassic rocks; where data are available the thickness of Triassic and Liassic strata appears to vary in sympathy. To take just one example, the

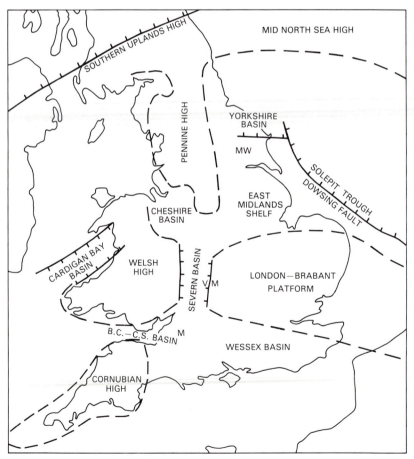

**Fig. 11.2.** Principal structural features controlling Jurassic sedimentation. (BC–CS = Bristol Channel–Central Somerset. MW = Market Weighton, VM = Vale of Moreton, M = Mendips)

Wilkesley borehole, in which the youngest Jurassic is only Lower Sinemurian, reveals the thickest Hettangian sequence in England (Poole & Whiteman 1966), confirming that the Cheshire Basin, first recognised from the full Triassic sequence, persisted at least locally into the early part of Jurassic time.

Much attention has been paid in the past to zones of incomplete and relatively thin successions such as the Market Weighton, Vale of Moreton and Mendips Swells (Fig. 11.2). These are, however, very limited in extent. In a broader picture, also embracing offshore regions, it is more appropriate to refer to *platforms* such as the London–Brabant Platform, where extensive Mesozoic cover rocks allow the palaeogeography to be reconstructed with some precision, or more loosely to *highs* where tectonically positive areas occur but where most or all of the Mesozoic cover has subsequently been stripped away by erosion. Speculation will probably never cease as to the extent to which the Pennine, Welsh and Cornubian 'highs' were

inundated by sea at different times during the Jurassic, but it is almost certain that deposition rates over them were substantially lower than in the intervening basins or troughs. For an area such as the East Midlands, which was a shallow marine zone intermediate between a basin and a high or platform, the term *shelf* is deemed the most appropriate.

References to older literature are contained in Arkell (1933, 1956) and are in consequence mostly excluded here. Such is the flood of literature in the last two decades, however, that citations in this chapter cannot hope to be exhaustive; emphasis in selection is placed upon wide-ranging reviews and more recent works or those which illustrate a particular point I wish to make. The regional surveys of Arkell (1947a, b), Sylvester Bradley & Ford (1968) and Rayner & Hemingway (1974) contain substantial sections devoted to the Jurassic. For a detailed account of regional lithostratigraphy and ammonite correlation the two Geological Society reports by Cope *et al.*

**Fig. 11.3.** Bedding plane in core of Lower Kimmeridge Clay covered with partially crushed specimens of the ammonite *Rasenia s.p.* (B.G.S. borehole at North Wooton, Norfolk). (B.G.S. F 6485.)

(1980a, b) are invaluable. Another valuable data source is the British Geological Survey Deep Geology Unit's atlas of onshore sedimentary basins (Whitaker 1985).

# Zonal stratigraphy

It was recognised well back into the last century, notably by Oppel, that ammonites (Fig. 11.3) provide by far the best means of correlation of Jurassic strata. Because of their exceptionally high rate of evolutionary turnover, which is decidedly not a mere artifact of the admittedly high degree of taxonomic splitting to which they have been subjected, they will undoubtedly continue to be the standard by which (relative) Jurassic time is measured, and against which zonations based on other types of fossil must be calibrated.

Table 11.1 gives a list of the currently established ammonite zones and their relationship to the eleven stages of the system. Where there are differences from the zonal scheme of Arkell (1956) reference is made to the relevant articles outlining the reasons for such changes. These works should be consulted for precise definitions of zones and subzones. Some English stratigraphers still follow Arkell in expanding the Bajocian downwards to include the Aalenian, which is widely used on the continent. I prefer to follow this continental practice, with the Aalenian being exactly equivalent to Arkell's Lower Bajocian; his Middle Bajocian then becomes the Lower Bajocian. For the topmost Jurassic stage, the Portlandian is presented as an alternative to the Volgian, with which it is substantially equivalent, now that the ammonite sequence of the Portland Beds is better known (Wimbledon & Cope 1978).

Of course, ammonites are rare or absent in a number of very shallow marine and marginal to non-marine facies and, as megafossils, unsuitable for the increasingly important subsurface exploration unless good cores are available. Hence a variety of microfossils have been utilised for stratigraphic subdivision. Ostracods have long proved useful in subdividing marginal marine or non-marine deposits like the Purbeck Beds and the utility of pollen and spores has more recently been investigated for such rocks. Foraminiferal, ostracod and coccolith zonations have also been erected, but the species utilised are almost invariably much longer ranging than ammonite species, so that obtaining precision even to the nearest stage is not always possible. The most promising microfossils for future investigations appear to be dinoflagellates, although they are similarly confined to marine facies. Most of the micropalaeontological biostratigraphic research has been done for oil companies and consequently much of the work has not been published.

# Regional stratigraphy

Table 11.2 is a correlation chart for the most important regions of England and Wales, namely Yorkshire, the East and South Midlands and Dorset. Each of these regions, and others of lesser importance, will be dealt with successively in the ensuing account. Figures 11.4, 11.5 and 11.6 illustrate some of the variations in thickness and dominant lithology across the country.

## Yorkshire

A certain amount of information is available from scattered inland outcrops and boreholes but by far the best exposed sequence is seen in the superb coastal cliffs centred on Whitby and Scarborough and extending between Redcar and Filey, a distance of some 40 miles. The deposits were laid down in what is generally known as the Yorkshire Basin, sharply delimited to the south by the Market Weighton Swell.

*Lower Jurassic.* The base of the Lias is nowhere exposed on the coast, where the oldest beds, belonging to the Angulata and Bucklandi zones, occur at Redcar. The only other outcrop of Lower Lias is in the Robin Hoods Bay structural dome, where a complete sequence of zones down to the Semicostatum can be studied. Middle and Upper Lias occur at a number of places between the Staithes region and Blea Wyke. As elsewhere in England the Lower Lias is predominantly argillaceous, with an invertebrate megafauna dominated by ammonites and bivalves. The principal lithological changes up the succession involve the proportion of quartz silt and fine sand and the presence of calcite or siderite (as the major early diagenetic carbonate occurring in nodule horizons).

The Hettangian and early Sinemurian contain thin bands of calcite microsparite and resemble the Blue Lias of Southern and Central England. A more shaly sequence follows, becoming progressively sandier up the Oxynotum and Raricostatum zones (the 'Siliceous Shales'). An abrupt drop in sand content immediately precedes the start of the Pliensbachian and the succeeding 'Pyritous Shales' in turn become progressively sandier, with conspicuous reddish weathering siderite nodules, culminating in a distinctive sandstone unit spanning the Carixian–Domerian boundary. A new sedimentary cycle commences with shales of the Margaritatus Zone, which pass up into progressively more ferruginous deposits of the Cleveland Ironstone Formation which caps the Domerian. In this formation partly coalescing layers of nodular siderite mudstone and intervening shales give way up the succession to massive bioturbated ironstone (the so-called Main Seam), well exposed at Old Nab and Kettleness. Eleven km to the SSE, at Hawsker

**Table 11.1.** Ammonite zones recognised in Britain

There are a number of changes from the scheme presented by Arkell (1956). for an explanation of these changes see the following publications. Hettangian to Toarcian: Dean *et al.* (1961); Aalenian–Bajocian: Mouterde *et al.* (1971), Parsons (1974); Bathonian: Torrens (1965, 1969); Oxfordian: Sykes & Callomon (1979); Kimmeridgian: Ziegler (1962), Cope (1967, 1978); Portlandian/Volgian: Casey (1974), Wimbledon & Cope (1978).

### PORTLANDIAN STAGE
( ≡ Volgian (pars))

Subcraspedites lamplughi
Subcraspedites preplicomphalus
Subcraspedites primitivus
Paracraspedites oppressus
Titanites anguiformis
Galbanites kerberus
Galbanites okusensis
Glaucolithites glaucolithus
Progalbanites albani

### KIMMERIDGIAN STAGE

UPPER
Virgatopavlovia fittoni
Pavlovia rotunda
Pavlovia pallasioides
Pectinatites pectinatus
Pectinatites hudlestoni
Pectinatites wheatleyensis
Pectinatites scitulus
Pectinatites elegans

LOWER
Aulacostephanus autissiodorensis
Aulacostephanus eudoxus
Aulascostephanus mutabilis
Rasenia cymodoce
Pictonia baylei

### OXFORDIAN STAGE

UPPER
*Tethyan ammonite zones*
Ringsteadia pseudocardata
Decipia decipiens
Perisphinctes cautisnigrae
Gregoryceras transversarium

*Boreal ammonite zones*
Amoeboceras rosenkrantzi
Amoeboceras regulare
Amoeboceras serratum
Amoeboceras glosense

MIDDLE
*Tethyan ammonite zones*
Perisphinctes plicatilis

*Boreal ammonite zones*
Cardioceras tenuiserratum
Cardioceras densiplicatum

LOWER
Cardioceras cordatum
Quenstedtoceras mariae

### CALLOVIAN STAGE

UPPER
Quenstedoceras lamberti
Peltoceras athleta

MIDDLE
Erymnoceras coronatum
Kosmoceras jason

LOWER
Sigaloceras calloviense
Macrocephalites macrocephalus

### BATHONIAN STAGE

UPPER
Clydoniceras discus
Oxycerites aspidoides
Prohecticoceras retrocostatum

MIDDLE
Morrisiceras morrisi
Tulites subcontractus

LOWER
Procerites progracilis
Zigzagiceras zigzag

### BAJOCIAN STAGE

UPPER
Parkinsonia parkinsoni
Garantiana garantiana
Strenoceras subfurcatum

LOWER
Stephanoceras humphriesianum
Emileia (Otoites) sauzei
Witchellia laeviuscula
Hyperlioceras discites

### AALENIAN STAGE

Graphoceras concavum
Ludwigia murchisonae
Leioceras opalinum

### TOARCIAN STAGE

UPPER (YEOVILIAN)
Dumortieria levesquei
Grammoceras thouarsense
Haugia variabilis

LOWER (WHITBIAN)
Hildoceras bifrons
Harpoceras falciferum
Dactylioceras tenuicostatum

### PLIENSBACHIAN STAGE

UPPER (DOMERIAN)
Pleuroceras spinatum
Amaltheus margaritatus

LOWER (CARIXIAN)
Prodactylioceras davoei
Tragophylloceras ibex
Uptonia jamesoni

### SINEMURIAN STAGE

UPPER
Echioceras raricostatum
Oxynoticeras oxynotum
Asteroceras obtusum

### SINEMURIAN STAGE

LOWER
Caenisites turneri
Arnioceras semicostatum
Arietites bucklandi

### HETTANGIAN STAGE

Schlotheimia angulata
Alsatites liasicus
Psiloceras planorbis

**Table 11.2.** Correlation chart of Jurassic rock formations in the principal English sections

| STAGES | DORSET | SOUTH MIDLANDS (Avon to Oxon.) | EAST MIDLANDS (Northants. to Lincs.) | YORKSHIRE |
|---|---|---|---|---|
| PORTLANDIAN | Lulworth Beds<br>Portland Limestone<br>Portland Sand | 'Purbeck' Beds<br>Portland Beds | Lower Spilsby Sandstone | |
| KIMMERIDGIAN | Kimmeridge Clay | Kimmeridge Clay | Kimmeridge Clay | Kimmeridge Clay |
| OXFORDIAN | Upper Calcareous Grit<br>Osmington Oolite<br>Berkshire Oolite<br>Lower Calcareous Grit<br>Upper Oxford Clay | Coral Rag, Wheatley Limestones, etc.<br>Lower Calcareous Grit<br>Upper Oxford Clay | Ampthill Clay<br>West Walton Beds<br>Upper Oxford Clay | Ampthill Clay<br>Upper Calcareous Grit<br>Coralline Oolite<br>Lower Calcareous Grit<br>Oxford Clay |
| CALLOVIAN | Middle Oxford Clay<br>Lower Oxford Clay<br>Kellaways Beds<br>Upper Cornbrash | Middle Oxford Clay<br>Lower Oxford Clay<br>Kellaways Beds<br>Upper Cornbrash | Middle Oxford Clay<br>Lower Oxford Clay<br>Kellaways Beds<br>Upper Cornbrash | Hackness Rock<br>Langdale Beds<br>Kellaways Rock<br>Cornbrash |
| BATHONIAN | Lower Cornbrash<br>Forest Marble<br>Upper Fuller's Earth Clay<br>Fuller's Earth Rock<br>Lower Fuller's Earth Clay<br>Crackment Limestones | Lower Cornbrash<br>Forest Marble<br>Great Oolite<br>Upper F. E. Clay<br>Fuller's Earth Rock<br>Lower F. E. Clay<br>Anabacia Limestone<br>White Limestone<br>Hampen Marly Fm.<br>Taynton Limestone<br>Sharp's Hill Fm.<br>Chipping Norton Fm. | Lower Cornbrash<br>Blisworth Clay<br>Blisworth Limestone<br>Rutland<br>Formation | Scalby Formation<br>? |
| BAJOCIAN | Upper Inferior Oolite<br>Middle Inferior Oolite | Upper Inferior Oolite<br>Middle Inferior Oolite | Lincolnshire<br>Limestone | Scarborough Formation<br>Cloughton Formation<br>Eller Beck Formation |
| AALENIAN | Lower Inferior Oolite<br>Bridport Sand | Lower Inferior Oolite | Grantham Formation<br>Northampton Sand<br>Formation | Saltwick Formation<br>Dogger Formation |
| TOARCIAN | Downcliff Clay<br>Junction Bed<br>Thorncombe Sand<br>Eype Clay, etc. | Upper Lias Clay<br>with Cotswold Sand, etc. | Upper Lias<br>Clay | Blea Wyke Sand<br>Peak Shales, etc.<br>Alum Shale<br>Jet Rock, etc.<br>Grey Shales |
| DOMERIAN | Marlstone<br>Middle Lias Clays | Marlstone<br>Middle Lias Clays | Marlstone<br>Middle Lias Clays | Cleveland Ironstone<br>Staithes<br>Formation |
| CARIXIAN | Green Ammonite Beds<br>Belemnite Marls | Lower<br>Lias<br>Clays | Lower Lias<br>Clays | Ironstone Shales, etc. |
| SINEMURIAN | Black Ven Marls<br>Shales with Beef | | Frodingham Ironstone, etc.<br>Bucklandi Clays<br>Granby Limestones | Siliceous Shales<br>Calcareous Shales |
| HETTANGIAN | Blue<br>Lias | Blue<br>Lias | Angulata Clays<br>Hydraulic Limestones | Shales at<br>Redcar |

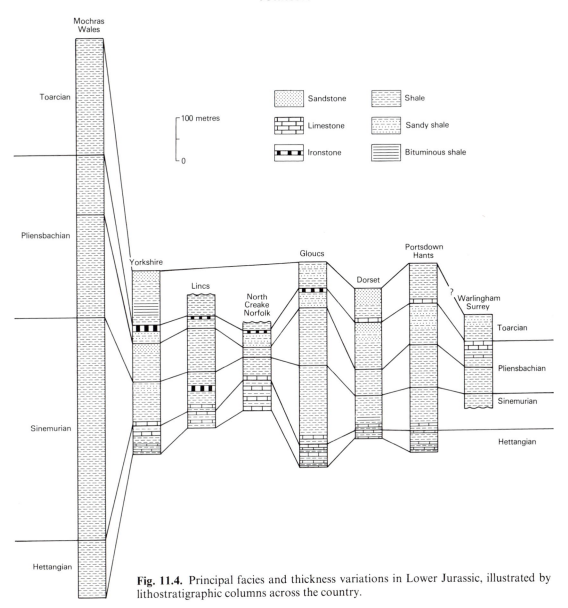

**Fig. 11.4.** Principal facies and thickness variations in Lower Jurassic, illustrated by lithostratigraphic columns across the country.

Bottoms, the ironstone has passed into ferruginous sandy shale and sandstone.

The Upper Lias marks a sharp reversal to more argillaceous facies. The fairly normal, slightly silty 'Grey Shales' of the Tenuicostatum Zone pass up into one of the most interesting units of the whole Yorkshire sequence, the Jet Rock. The jet consists of scattered fragments of lignitised driftwood in a matrix of finely laminated bituminous shales, a facies which continues to the top of the Falciferum Zone. A series of horizons of striking concretionary nodules is composed of calcite, in contrast to the siderite of the Middle Lias. Some of these exhibit multiple growth, and isotopic evidence indicates a transition from the sulphate-reduction to the fermentation zone of diagenesis (Campos & Hallam 1979). Scattered calcite nodules also occur in the overlying Alum shales of the Bifrons Zone; many contain well-preserved ammonites, as also in the Grey Shales. The Upper Toarcian exhibits a further coarsening upward sequence between Ravenscar and Blea Wyke, so that bioturbated sands cap the Lias and pass into the

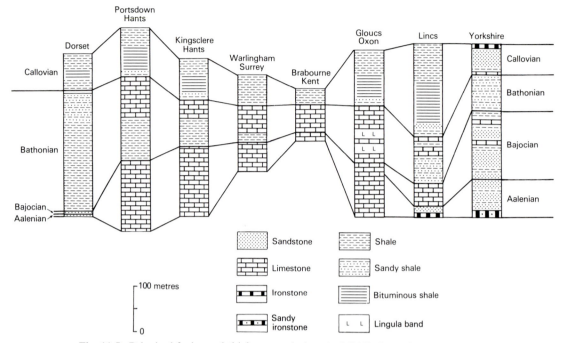

**Fig. 11.5.** Principal facies and thickness variations in Middle Jurassic.

Middle Jurassic without a marked break. Inland in Rosedale the Upper Toarcian sequence culminates in a thin oolitic ironstone but in most areas NW of the Peak Fault (north of Scarborough) sandy Middle Jurassic rests on an eroded surface of Alum Shales.

The detailed researches of Howarth (1955, 1962, 1973) on ammonite stratigraphy make the Upper Pliensbachian (Domerian) – Lower Toarcian (Whitbyan) sequence the best known and probably the most complete anywhere in the world.

*Middle Jurassic.* In contrast to the Lias the Middle Jurassic is dominantly non-marine and contains an abundance of medium- to coarse-grained sandstones that are responsible for the attractive scenery of the Yorkshire Moors (or Cleveland Hills). Some of the shaly beds have yielded world-famous plant fossils. The rocks are best seen on the coast at Cloughton Wyke, around Hundale Point, Long Nab and the margins of Cayton Bay.

In place of the old 'Estuarine' or 'Deltaic Series' nomenclature Hemingway & Knox (1973) have proposed a modern-style stratigraphic nomenclature for the pre-Callovian sequence. Above the Dogger, which is a bioturbated marine chamositic sandstone, comes a succession of non-marine sandstones, siltstones and shales with several subordinate marine units in between, all of which are of Bajocian age. The Eller Beck Formation of marine shale, sandstone and ironstone and the thicker Scarborough Formation,

with an important limestone component, separate the Saltwick, Cloughton and Scalby non-marine formations, but the Cloughton Formation contains thin marine units in addition. On the coast the unit traditionally known as the Millepore Bed, rich in bryozoans and represented inland by the Whitwell Oolite, is overlain by the Yons Nab Beds. There are no ammonites known between the top Lower Bajocian forms of the Scarborough Formation (Parsons 1977) and the Lower Callovian of the Upper Cornbrash, and it is unlikely that the intervening 60 m of the Scalby Formation, which bear all of the signs of rapid deposition, span the large time gap indicated. It has indeed recently been suggested that the formation may represent only the topmost Bathonian (Leeder & Nami 1979). Recent palynological work has produced different results. Fisher and Hancock (1985) claimed a late Bajocian age for the Scalby Formation. On the other hand Riding and Wright (1989) have argued for a more conventional age of ?late Bajocian to Bathonian, with possibly a fairly complete succession without a major stratigraphic gap, in contrast to the previous two interpretations. Detailed examination of the junction with the overlying Cornbrash shows a sharp change to marine beds, with evidence that the Scalby Formation sediments were compacted or lithified, slightly uplifted, eroded and burrowed, before the Cornbrash was deposited.

Above the thin sandy limestone of the Upper

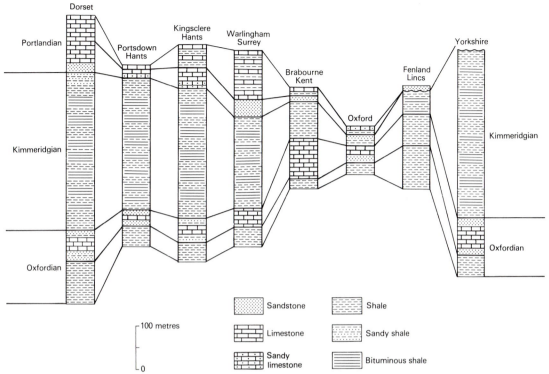

**Fig. 11.6.** Principal facies and thickness variations in Upper Jurassic.

Cornbrash, and directly overlying 'Shales of the Cornbrash' (Wright 1977), the Callovian consists predominantly of bioclastic friable marine sandstones capped by a thin sandy limestone and chamosite oolite, the Hackness Rock. Wright's (1968) detailed research revealed a minor unconformity at the base of the Hackness Rock and another within the underlying sands, which led him to create the new term Langdale Beds for Coronatum Zone sandstones resting on Calloviense Zone Kellaways Rock. Wright's (1978) research on the Callovian recognised over most of the Yorkshire Basin three sedimentary cycles, with marine influence increasing upwards.

*Upper Jurassic.* The Yorkshire 'Corallian' compares quite closely with that in Southern England in consisting of an alternation of marine calcareous sandstones and dominantly oolitic limestones; it forms a prominent north-facing escarpment running west from Scarborough (Fig. 11.7). Wright (1972) improved our knowledge of the ammonite sequence and proposed a refined stratigraphic subdivision. Above the widespread Lower Calcareous Grit Formation of the Cordatum Zone comes the Coralline Oolite Formation comprising successively the Hambleton Oolite and Malton Oolite members, separated in the east by a thin sandstone unit; also in the east a 'coral rag' facies

occurs at the top. The Corallian sequence concludes with the Upper Calcareous Grit Formation which passes westwards in the Malton region into cementstones. Most of the Upper Oxfordian in Yorkshire is represented by clays that can be correlated with the Ampthill Clay of the Midlands.

The highest Jurassic in Yorkshire, the Kimmeridge Clay, underlies the Vale of Pickering and is only poorly exposed on the coast. Cope (1974) established that all ammonite zones up to the Pectinatus are present except the Hudlestoni. The younger Pavlovia zones were presumably deposited but removed by subsequent erosion, because the Kimmeridge Clay is abruptly overlain by the Lower Cretaceous Speeton Clay. The facies of the Kimmeridge Clay, which measures 385 m in the Fordon borehole, bears a general resemblance to that of Dorset and lacks any sandy horizons such as are found in the South Midlands.

## Wales

Jurassic outcrops are confined to the Hettangian and Lower Sinemurian and are well exposed on the Glamorgan coast. The most interesting feature is the lateral passage westwards from normal offshore Blue Lias facies into nearshore facies which overlaps the

**Fig. 11.7** Corallian rocks (capped by glacial till), near Scarborough, Yorkshire. (B.G.S. L 1348.)

Rhaetic on to Carboniferous Limestone, providing one of the best documented and most readily observed examples of the basal Liassic transgression (Hallam 1960, Wobber 1965). In the Lavernock region in the east thin alternating beds of argillaceous microsparite and shale of the Planorbis Zone pass up into a more shaly unit of the Liasicus Zone (the so-called Lavernock Shales). In the succeeding Angulata and Bucklandi zones the proportion of limestones increases significantly. In the Southerndown region 27 km to the west, 12 m of the coral-bearing conglomeratic biosparite known as the Sutton Stone (equivalent to at least part of the Planorbis Zone) rests unconform-

ably on the Carboniferous Limestone. Ager (1986) interpreted it as a debris-flow deposit laid down in a 'geological instant' but this is disputed by Fletcher (1988), who argues for the more conventional view invoking deposition over a long period of time. The Sutton Stone is overlain sharply by the Southerndown Beds of the younger Hettangian and basal Sinemurian, these being limestones and marls of Blue Lias facies containing numerous thin beds of cherty conglomerate derived locally from the eroded Carboniferous Limestone.

One of the most interesting boreholes drilled in recent years was that at Llanbedr, on the coast of

Cardigan Bay and better known as the Mochras borehole (Woodland 1971). This penetrated, beneath Tertiary sediments, a virtually complete sequence of some 1,300 m of Liassic rocks, by far the greatest thickness known in the British area, comparable indeed to the thickness of the whole Jurassic elsewhere. The facies consists almost entirely of monotonous calcareous and silty mudstones even including the Middle Lias which elsewhere in the country is much sandier; there are very subordinate thin layers and nodules of microsparite. These rocks are close to the fault-bounded edge of the Cardigan Bay Basin, which geophysical surveys had indicated earlier to contain a thick, probably Mesozoic, sedimentary sequence. Combined offshore borehole and geophysical data indicate some 1,400 m of Middle and Upper Jurassic underlying the bay (Penn & Evans 1976, Millson 1987). At the base, fluviodeltaic sediments comparable with the Bajocian of Yorkshire rest unconformably on Lias. These are overlain by Bathonian of lagoonal and marginal marine facies like that of the South and East Midlands but ten times greater in thickness, implying as for the Lias by far the highest rate of subsidence and sedimentation in the British area.

## Midlands

The English Midlands are here treated rather broadly to embrace the whole region between Humberside and Avon. Bradshaw & Penney (1982) give a valuable review of stratigraphic and facies variations in the northern part of the region.

*Lower Jurassic.* Being predominantly argillaceous, the Lias is only poorly exposed apart from a few brick- and cementworks and consequently much of our knowledge comes from old records and boreholes. As in Glamorgan a shaly Liasicus Zone intervenes between thin bedded argillaceous limestones and shales, some bituminous, of the Hettangian and Lower Sinemurian. The higher part of the Lower Lias consists almost entirely of clays, which pass up into sandy clays with sideritic nodules in the Margaritatus Zone of the Middle Lias. The Upper Lias shows a reversal to more normal clays like the Lower Lias but with a widespread thin horizon of laminated bituminous shales and marly limestone in the Falciferum Zone, locally known as the 'Fish and Insect Beds' because of their well preserved, unusual faunal content. Towards the Bath region some sands (the Cotswolds Sands) enter the Upper Lias section, which thus compares more closely with SW England.

Probably the most interesting deposits are a number of thin horizons of sideritic chamosite or limonite oolite which have long been exploited as ironstones (Chapter 19). The most widespread of these is the

Marlstone Rock Bed spanning the Pliensbachian–Toarcian boundary (Howarth 1978, 1980). This locally passes into a sandy bed low in iron, but is always recognisable from its resistant lithology and distinctive brachiopod and bivalve fauna. In north Lincolnshire/south Humberside economic interest has centred on the Frodingham Ironstone (Turneri and Obtusum zones) (Hallam 1963), and the Northampton Sand Ironstone of the Corby region. The Frodingham Ironstone disappears southwards and in NE Leicestershire is represented by thin ferruginous limestones separated by thicker clays (Hallam 1968). The Ibex Zone of north Lincolnshire contains another, thinner, ironstone, the Pecten Bed (Sellwood 1972).

The thickness of the Lias diminishes sharply eastwards from the Severn and Cheshire Basins towards the London Platform, where borehole records demonstrate a progressive thinning to zero and overlap of successive Lower Lias zones (Donovan *et al.* 1979). Thus the (incomplete) Lower Lias of the Prees Outlier in the Cheshire Basin is over 600 m and that of the Stowell Park borehole in the Severn Basin 360 m thick (Green & Melville 1956). Thicknesses over the East Midlands Shelf are normally less than 250 m. Only in the Severn Basin is a complete Toarcian sequence present. From Oxfordshire to Lincolnshire Middle Jurassic lies unconformably on different horizons within the Bifrons Zone (Hallam 1968, Howarth 1978).

*Middle Jurassic.* Aalenian, Bajocian and Bathonian rocks are developed in much more varied lithologies than the Lias and are generally better exposed. The rocks are poor in ammonites and consequently there has been a good deal of confusion and disagreement about correlation, which is only just beginning to be cleared up. For these various reasons the deposits of these stages must be dealt with in rather more detail than the Lias and Oxford Clays respectively below and above. For the terminology of the East Midlands succession and the correlation of the Bathonian with the South Midlands Bradshaw's (1978) work is used.

The Aalenian of the East Midlands is represented by the Northampton Sand and overlying Grantham Formations. The petrology of the Northampton Sand Ironstone is the most fully described of all the British Jurassic ores (Taylor 1949). It consists of several metres of sideritic chamosite oolite and siderite mudstone with a marine fauna dominated by bivalves. Locally it passes up into a few metres of the so-called Variable Beds, consisting of cross-bedded calcite–quartz sandstones, limestones and mudstones. These lithologies dominate the entire formation SW of Towcester. The ironstone belongs to the Opalinum Zone and passes south-westwards into the sandy limestone of the Scissum Beds, which are well developed in the Cotswolds and occur patchily in

Oxfordshire. The Northampton Sand Formation is best developed in the region extending between Northamptonshire and south Lincolnshire and is impersistent in north Lincolnshire.

The Grantham Formation (formerly called the Lower Estuarine Series) consists of several metres of poorly fossiliferous sands, silts and clays with plant rootlets and marks the regressive upper part of the Aalenian. It is succeeded by a major Lower Bajocian limestone unit, the Lincolnshire Limestone Formation, which reaches a maximum of 35 m in south Lincolnshire and passes northwards into the Cave Oolite between the Humber and Market Weighton; it disappears south of Kettering (Sylvester Bradley 1968). The stratigraphy is outlined by Ashton (1980).

Although a creamy white-weathering oolite is the dominant lithology there is considerable variation up the succession. The Lower Lincolnshire Limestone is composed at the base of micritic beds including the Collyweston 'Slate' (a fissile limestone formerly used for roofing) and the Nerinea Beds, which pass up into the typical cross-bedded oolite of the Ketton Beds. The Upper Lincolnshire Limestone which succeeds unconformably, with local channelling, has a more uniform lithology of coarse, shelly cross-bedded oolite and includes some of the best building stones of Britain, notably the freestones of Ancaster, Clipsham and Weldon. As regards the inferred environment of deposition, the transgressive Lincolnshire Limestone succession is thought to reflect the landward migration of an offshore barrier complex across lagoonal and tidal flat deposits (Ashton 1980, Marshall & Ashton 1980).

In north Lincolnshire the Lincolnshire Limestone is overlain by sandstones and coals and then by a marine unit identified by its ostracod fauna as an equivalent of the Scarborough Formation. Therefore the Lincolnshire Limestone cannot range above the earliest Sauzei Zone (Bradshaw & Bate 1982).

Bradshaw (1978) has given the name Rutland Formation to a series of sands, silts and clays overlying the Lincolnshire Limestone and traditionally known as the Upper Estuarine Series. These deposits, which show an overall fining upwards tendency, contain several rhythmic units with sandy beds containing marine fossils truncating rooted clays below. A medial Wellingborough Limestone Member becomes increasingly sandy north-eastwards. The invertebrate fauna is a low diversity marginal marine one dominated by bivalves and becomes progressively more restricted north-eastwards.

South of a line passing through Peterborough and Kettering the Lincolnshire Limestone disappears and the age of the so-called White Sands of north Oxfordshire, which rests on Scissum Beds or equivalents, has been disputed. Horton (1977) has recently

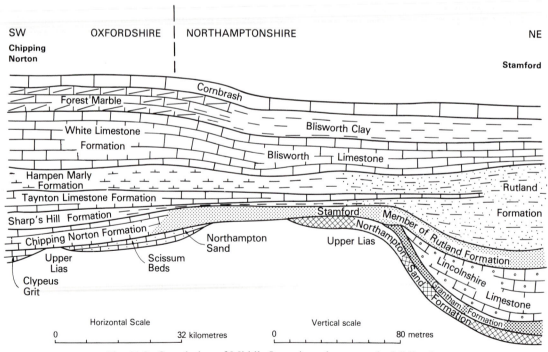

**Fig. 11.8.** Correlation of Middle Jurassic rocks across the Midlands.

argued for an equivalence to the 'Lower Estuarine Series' or Grantham Formation, but the detailed facies analysis of Bradshaw (1978), backed up by an Upper Bajocian to Bathonian palynological age determination, shows conclusively that they must in fact belong to the basal Stamford Member of the Rutland Formation. This is a very important key to correlation of the rather different Bathonian successions of the East and South Midlands (Fig. 11.8).

The Rutland Formation is succeeded by the thin, predominantly fine-grained Blisworth Limestone Formation which, according to Bradshaw but not to Torrens (1968) is an exact equivalent of the White Limestone Formation of Oxfordshire. Bradshaw's correlation, for which he offers good arguments, implies the existence of a continuous lithostratigraphic unit extending all the way from east Gloucestershire to south Humberside. The overlying Blisworth Clay Formation passes south-westwards into the Forest Marble Formation, which contains shelly cross-bedded oolitic limestone, with the boundary being located in south Northamptonshire and north Buckinghamshire.

At the summit of the Aalenian–Bathonian sequence is the well-known Cornbrash Formation, a thin biomicrite which embraces the top zone of the Bathonian and bottom zone of the Callovian.

The age equivalent deposits of the South Midlands are developed in a more calcareous and marine facies than that of the East Midlands. The Aalenian and Bajocian are only fully developed in the south Cotswolds, where some 80 m of limestone form the celebrated west-facing escarpment, and wedge out rapidly towards Oxfordshire. Table 11.3 gives the succession of long-established lithostratigraphic units with the age revisions of Parsons (1976). Mudge (1978) has recently proposed formation status for the Lower Inferior Oolite and created a series of component members. Thus in his scheme the well-known Pea Grit becomes the Crickley Oncolite. Above the sandy micrite of the Scissum Beds (or Leckhampton Limestone in Mudge's scheme) the bulk of the Aalenian consists of creamy-weathering oolitic 'freestone' of classical type, with the regressive facies of the Harford Sands and Tilestone at the top. The Middle and Upper Inferior Oolite contain an alternation of oolitic freestones and shelly micrites such as the Gryphite and Trigonia 'Grits'.

Baker (1981) undertook a detailed facies analysis of the Upper Aalenian Oolite Marl and Upper Free-

**Table 11.3.** Litho- and biostratigraphy of the Aalenian and Bajocian of the south Cotswolds, based on the recent revision of Parsons (1976). Mudge (1978) has revised the lithostratigraphy of the Lower Inferior Oolite and proposed a series of members; nearly all the familiar names below are abandoned.

| MAJOR LITHOSTRAT. SUBDIVISIONS (Formation status ?) | MINOR LITHOSTRAT. SUBDIVISIONS (Member status ?) | ZONES |
|---|---|---|
| Upper Inferior Oolite | Clypeus Grit<br>Upper Caul Bed | } Parkinsoni |
| | Upper Trigonia Grit | Garantiana |
| × × × × × × × MAJOR NON-SEQUENCE × × | | |
| | Bourguetia Beds | Sauzei |
| Middle Inferior Oolite | Witchellia Grit<br>Notgrove Freestone<br>Gryphite Grit, etc. | } Laeviuscula |
| | Lower Trigonia Grit | Discites |
| × × × × × × × MINOR NON-SEQUENCE × × | | |
| | Tilestone<br>Snowshill Clay<br>Harford Sands | } ? Concavum |
| Lower Inferior Oolite | Upper Freestone and<br>Oolite Marl<br>Lower Freestone<br>Pea Grit<br>Lower Limestone | } ? Murchisonae |
| | Scissum Beds | Opalinum |

stone, which are grouped together as the Oolite Marl Member. This is interpreted as a former lagoonal carbonate mud, rich in fauna, which was eventually overwhelmed by an oolite shoal spreading south-eastwards. Small coral patch reefs occur in the micrite, and an extensive reef flat is inferred to have existed to the north-west of the present Cotswold escarpment.

A major erosion surface separates the Upper Bajocian (or Upper Inferior Oolite) from the underlying strata and both north-eastwards and eastwards into Oxfordshire and south-westwards towards the Mendips it oversteps on to Lias or even Carboniferous (Fig. 11.9).

It is only in recent years that the well-known Bathonian sequence of north Gloucestershire and Oxfordshire has been subjected to detailed facies analysis (Sellwood & McKerrow 1974, Palmer 1979). Borehole records show the Lower Bathonian to overlap the Upper Bajocian Clypeus Grit on to Aalenian and Liassic deposits towards the London Platform and eventually come to rest directly on Palaeozoic (Fig. 11.10). Directly east of the Moreton Swell on the Gloucestershire–Oxfordshire border, marine limestones of the Chipping Norton Formation pass eastwards into marginal marine sands (the Swerford Member) and then into the basal, Stamford Member of the Rutland Formation (Fig. 11.8). Above the succeeding marine clays of the Sharp's Hill Formation follows the oolitic Taynton Limestone Formation and then the finer-grained sediments of the Hampen Marly and White Limestone Formations. The Forest Marble and Cornbrash Formations complete the sequence.

In contrast to the underlying Middle Jurassic the Callovian sequence is very uniform in facies and thickness over the whole of the Midlands (Callomon 1968). Above the basal limestone of the Upper Cornbrash (Macrocephalus Zone) the Kellaways Clay and Rock (a shelly, calcareous sandstone) range from 2 to 7 m in thickness. The bituminous, ammonite-rich Lower Oxford Clay of the Middle Callovian is the principal source of bricks in the United Kingdom and virtually the only exposures are in the huge brick-works principally located around Bedford and Peterborough. It varies in thickness from 16 to 25 m and the overlying *Gryphaea*-bearing, more calcareous, non-bituminous Middle Oxford Clay (Upper Callovian) averages about 16 m. Subsurface information indicates an overall thinning towards London. Perhaps the most interesting bed is the Lamberti Limestone which caps the Callovian. This is a condensed microsparite crammed with a very diverse fauna of ammonites including the earliest cardio-ceratids, the last cosmoceratids, both boreal families, together with Tethyan elements such as perisphinctids, oppeliids and phylloceratids. It can still be observed in a disused quarry in Buckinghamshire (Hudson & Palframan 1969).

*Upper Jurassic.* This is developed in most of the Midlands almost entirely in marine argillaceous facies and was in consequence only poorly known until recent borehole information became available.

The basal two zones of the Oxfordian comprise marly, non-bituminous Upper Oxford Clay, which has a uniform thickness of 30–40 m except where it is overstepped by Middle Oxfordian in Buckinghamshire and Cambridgeshire. The succeeding Corallian Group of the Oxford district was made famous by the researches of Arkell (1947a) and need not be dealt with here more than briefly. Compared with elsewhere, its varied sequence of sandstones, clays and bioclastic limestones with local coralline developments is condensed and incomplete, the Middle Oxfordian Coral Rag being overlain directly by Kimmeridge Clay. A few miles east of Oxford the coarser-grained limestones and sandstones disappear and an uninterrupted argillaceous sequence extends from the Oxford to the Kimmeridge Clay until the Corallian facies reappears

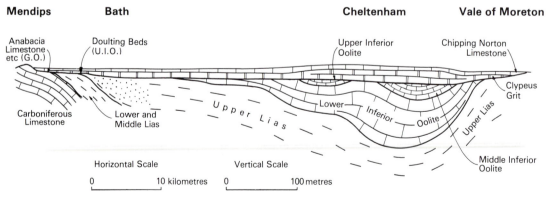

**Fig. 11.9.** Diagram to illustrate the Upper Bajocian transgression from Somerset to Oxfordshire. (G.O. = Great Oolite; U.I.O. = Upper Inferior Oolite.)

in Yorkshire. The Ampthill Clay is the lithostratigraphic term traditionally given to the Upper Oxfordian clay facies of the Midlands. Torrens & Callomon (1968) proposed the term Elsworth Rock Series for clays and limestones of Cordatum to Plicatilis Zone age in the south (the Elsworth Rock is a condensed limestone with scattered limonite ooliths); the term formation would seem more appropriate however. In Cambridgeshire the isolated Upware oolite and 'coral rag' is probably Plicatilis to Transversarium Zone in age.

New borehole information from the Fenland shows the Ampthill Clay to range between 36 and 55 m (Gallois & Cox 1977). A new term, the West Walton Beds, is proposed for 10–15 m of Middle Oxfordian silty and calcareous mudstones and correlated with the Elsworth Rock farther south. Gallois and Cox were able to apply to this region the boreal ammonite zonation of Sykes & Callomon (1979), earlier established in Scotland and Greenland.

The Fenland boreholes also indicate that the overlying Kimmeridge Clay is, at 43 m, appreciably thinner than in southern England but very similar in lithological and faunal characteristics, for example in the presence of coccolith-rich limestones at the horizon of the White Stone Band of Dorset (Gallois & Cox 1974). The youngest Kimmeridgian of Norfolk and Lincolnshire belongs to the Pectinatus Zone. Northwards and southwards from the Wash the Kimmeridge Clay is progressively overstepped by younger deposits and cut out entirely at Market

Weighton and the northern margin of the London Platform. In contrast to this facial development, the sequence around Oxford contains an arenaceous unit in the Pectinatus Zone, the Shotover Sands, and the more localised Wheatley Sands in the Pallasioides Zone (Arkell 1947a, Cope 1978). The intervening Hartwell Clay has long been celebrated for its superbly preserved aragonitic ammonite fauna, much better than the crushed specimens found in Dorset. The Pallasioides Zone is overlain non-sequentially by a condensed, glauconitic version of the Portland and Purbeck Beds, with a basal 'lydite' (black chert) pebble bed.

Perhaps the most important stratigraphic advance in recent years has been the conclusive establishment by Casey (1974) of the Volgian age of the lower part of the Spilsby Sandstone of Lincolnshire, which rests unconformably on the Kimmeridge Clay, thereby reviving an old, discredited idea of the Russian palaeontologist Pavlov. Some 10 m of glauconitic sandstone has a lateral equivalent in north Norfolk in the lower part of the Sandringham Sands. Casey recognised no fewer than five new ammonite zones in the Lower Spilsby Sandstone ranging in age from the upper part of the Middle to the Upper Volgian, though these might have only subzonal status compared with other Jurassic stages. A non-sequence separates it from the Upper Ryazanian of the Upper Spilsby Sandstone (Table 11.2). The more general importance of Casey's work is that it establishes eastern England as the region with the most complete

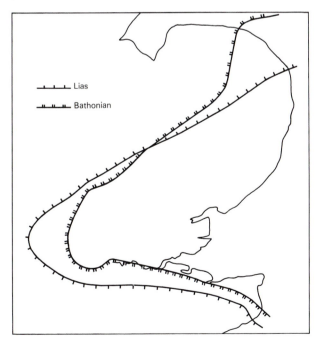

**Fig. 11.10.** Limits of Liassic and Bathonian around the western margins of the London–Brabant Platform.

Jurassic–Cretaceous marine boundary beds in NW Europe, and a key area for correlating the Russian Platform succession with those of Greenland and Siberia.

## Southern England

This region extends eastwards from the Bristol Channel through Avon, Somerset, Dorset, Wiltshire and Hampshire into Sussex and Kent. In the Bristol Channel there is known to be a 2,170 m predominantly argillaceous sequence (Fig. 11.11) ranging from the Hettangian to the Kimmeridgian (Lloyd *et al.* 1973, Evans & Thompson 1979). This pronounced basinal development extends onshore into Somerset, where the bottom four zones on the coast between Watchet and Lilstock (Palmer 1972) are appreciably thicker (179 m) and more shaly than the otherwise similar sequences of Glamorgan (119 m) and Dorset (47 m). By comparison with the Central Somerset–Bristol Channel Basin the so-called Wessex Basin farther east is hardly a basin at all; the sequence is thick only in comparison with parts of the Dorset sequence and the southern margin of the London Platform (Hallam & Sellwood 1976). A 'Portsdown Swell' in Hampshire separating

western and eastern parts of the Wessex Basin was recognised by Wilson (1968a) in his analysis of late Oxfordian palaeogeography but considering the Jurassic as a whole this was not a particularly significant or persistent topographic feature (Hallam & Sellwood 1976). East of Dorset and Wiltshire the Jurassic is only known from borehole records.

Two works by Arkell (1933, 1947b) remain by far the best of the detailed guides to the Jurassic geology of the most celebrated area, 80 km. of the Dorset coast between Lyme Regis and Swanage.

*Lower Jurassic.* The basal formation of the Blue Lias, belonging to the Hettangian and Lower Sinemurian, is excellently exposed in the coastal cliffs west of the Lyme Regis. It comprises a thin-bedded alternation of shelly, bioturbated argillaceous microsparites and marls with thin bituminous shales. Long celebrated for its rich invertebrate and vertebrate fauna, it has also provoked a controversy about the primary or secondary nature of the lithological rhythms (Hallam 1960, 1964; Campos & Hallam 1979). The overlying 'Shales with Beef' (Table 11.2), dominantly composed of bituminous shales with scattered calcareous concretions, pass up into the more normal clays of the Black Ven Marls, which

**Fig. 11.11.** Blue Lias, Somerset coast, near inland village of Kilve (B.G.S. A 11687).

range to the top of the Sinemurian. The Lower Pliensbachian consists of the more calcareous and paler-coloured Belemnite Marls, with fine banding of trace-fossil mottled marls and marly limestones bearing some resemblance to the Blue Lias (Sellwood 1972), and the Green Ammonite Beds, lithologically closer to the Black Ven Marls.

The earliest sandstone unit in the sequence, the Three Tiers, almost exactly coincides with the Lower–Middle Lias boundary. The Middle Lias (Domerian or Upper Pliensbachian) resembles that elsewhere in England in being sandier than the Lower Lias. Some 60 m of Eype Clay are succeeded by 27 m of Downcliff Sands which are capped by a condensed ammonite-rich limestone bed, the Margaritatus Stone (Howarth 1957). Above 23 m of Thorncombe Sands comes the so-called Junction Bed, a condensed micrite limestone with conglomeratic base, rarely more than 1 m thick, which ranges in age from the Spinatum to the Bifrons Zone, thus spanning the Pliensbachian–Toarcian boundary. The Upper Toarcian is represented by silty clay (Downcliff Clay) passing up into regularly alternating beds of friable and calcite-cemented fine sand, the Bridport Sands, which span the Toarcian–Aalenian boundary without apparent break and form vertical cliffs in the neighbourhood of Bridport. Isotopic evidence indicates that the calcite layers have segregated late in diagenesis (Campos & Hallam 1979).

Northwards towards Avon and the Cotswolds the most important facies change takes place in the Toarcian. The Bridport Sand facies continues as the Yeovil Sands and then almost disappears over the Mendip Hills. From Bath northwards the Midford and Cotswold Sands descend to progressively older horizons down to the Bifrons Zone and are overlain by a condensed ammonite-rich limestone of Upper Toarcian age, the Cephalopod Bed, which is likewise diachronous (Davies 1969).

Throughout the whole region the sandstone and condensed ammonitiferous-limestone facies of the Upper Lias disappears rapidly eastwards and in Hampshire the strata are more or less entirely argillaceous as in most of England (Hallam & Sellwood 1976). Likewise no Toarcian sands occur in the Bristol Channel (Evans & Thompson 1979). Among other features of note in the region from north Somerset to south Gloucestershire, the Marlstone Rock bed reappears as a discrete bed at the top of the Middle Lias, in sandy iron-shot limestone facies, and the lower Lias is highly condensed though facially similar to Dorset (Donovan 1956, Sellwood 1972).

*Middle Jurassic.* In Dorset, above the Aalenian in Bridport Sand facies, comes a highly condensed 'Inferior Oolite' Bajocian sequence only 4·5 m thick at maximum, consisting of ammonite-rich micritic lime-stones with scattered limonite ooids and one horizon of limonitic oncoids (the so-called 'snuff-boxes'); an erosional 'hardground' separates the Lower from the Upper Bajocian. This distinctive development compares closely with the type Bajocian in Normandy but is different from anything else in England. A borehole put down a short distance offshore, in Lyme Bay, has revealed a thicker Bajocian sequence, almost 20 m. This has been very comprehensively analysed both petrographically and palaeontologically (Penn *et al.* 1980). Eastwards into Hampshire and Sussex the Aalenian–Bajocian expands into a much thicker series of dominantly oolitic limestones, a Cotswolds-type development (Hallam & Sellwood 1976).

The Bathonian of Dorset is also very different from that in the Midlands. It consists predominantly of a fully marine clay of substantial thickness, reaching according to Martin (1967) an astonishing maximum of 400 m in the Lulworth Banks borehole. This indicates a basinal development quite unlike the underlying Inferior Oolite. The most important formation is the Fuller's Earth Clay, which splits up towards north Dorset into lower and upper units separated by an expanding micritic limestone formation, the Fuller's Earth Rock. The overlying Forest Marble is also a predominantly clay formation, with thin sandstones and a thin central bioclastic lime-stone. Northwards towards Bath the Upper Fuller's Earth Clay and Forest Marble are split by an expanding Great Oolite limestone unit with the overlying Bradford Clay, long celebrated for its rich and well-preserved fauna (Palmer & Fürsich 1974). The Lower Fuller's Earth Clay can be traced north of Avon into Gloucestershire, where it eventually passes into the Chipping Norton Formation. Stratigraphic details for Dorset and the ammonite zonation are given by Torrens (1969) and an account of the oolitic and coraliferous formations of the Great Oolite of the Bath region given by Green & Donovan (1969), while Penn *et al.* (1979) described the type Bathonian.

The Hampshire and Sussex Bathonian sequence is thinner than in Dorset with a marine argillaceous unit ('Fuller's Earth Clay') overlain by limestones ('Great Oolite'). Approaching the London Platform in Surrey and Kent the sequence thins, the clays disappear and limestones eventually overstep on to Palaeozoic (Hallam & Sellwood (1976) and Fig. 11.10). A recent stratigraphic and sedimentological analysis of the Great Oolite of the Humbly Grove Oilfield in Hampshire has demonstrated the establishment of a south-westward-descending carbonate ramp following the early Bathonian transgression marked by the Fuller's Earth Clay; the ramp subsided at the start of the Callovian (Sellwood *et al.* 1985).

The facies of the Callovian is virtually the same as that in the Midlands and hence does not warrant extra comment. The stage's maximum thickness in England of some 200 m is recorded from borehole in Wiltshire.

*Upper Jurassic.* Following the rather poorly exposed Upper Oxford Clay the Corallian Group is splendidly exhibited in coastal cliffs between Weymouth and Ringstead Bay, Dorset and has received much detailed attention since Arkell's investigations (Wilson 1968b, Talbot 1973, Fürsich 1973, 1977, Brookfield 1978). The sequence tends towards a cyclic development, shallow marine clays passing up into marginal marine sands which are characteristically overlain sharply by shell-packed limestones signifying slow deposition in offshore shelf conditions. Thus the Oxford Clay passes up into the Nothe Grit overlain by the 'Trigonia' hudlestoni Bed. Above this the Nothe Clay passes up into Bencliff Grit overlain by further shell beds, while in the Upper Oxfordian the Sandsfoot Clay passes up into Sandsfoot Grit. The most important limestone unit is a Middle Oxfordian oolite within the Osmington Oolite Formation; unlike in Wiltshire and Oxfordshire 'coral rag' facies is absent. The deposits are rich in a variety of trace fossils and body fossils notably bivalves, the most prominent of which belong to the trigoniid genus *Myophorella*.

A mixed limestone–sand–clay facies persists in the Middle and Upper Oxfordian across southern England but the sandy facies in the lower part of the sequence, for example the Bencliff Grit in Dorset and Highworth Grit in Wiltshire, is conspicuous only in the west. Wilson (1968a) attributed a major role to the Portsdown Swell of Hampshire in controlling the rate and type of sedimentation. Over the swell only high energy carbonates and quartz sands accumulated.

Although there is no ambiguity in defining the Oxfordian–Kimmeridgian boundary in Dorset (on the basis of replacement of *Ringsteadia* by *Pictonia*) Brookfield (1978) pointed out that the major facies change takes place somewhat higher up the sequence with the widespread onset of bituminous shales in the Mutabilis Zone. He therefore proposed that the more varied shallow marine sequence ranging from the Ringstead Waxy Clay to the Grey Clays, passing laterally into the Abbotsbury Ironstone (Cymodoce Zone) should be excluded from the Kimmeridge Clay proper, which is accorded lithostratigraphic group status, and put into the underlying Upper Calcareous Grit Group.

The Dorset coast in the Isle of Purbeck has by far the best exposures of Kimmeridge Clay in the whole of England (Fig. 11.12). Although the ammonites are invariably crushed they are extremely abundant, and important advances in stratigraphic zonation have been made in recent years by Ziegler (1962) and Cope (1967, 1978). Close attention is now being paid to the sedimentology and diagenesis, especially because of the interest provoked by the oil discoveries in the

northern North Sea, where Kimmeridge Clay is thought to be the likely source rock (Tyson *et al.* 1979, Irwin 1980).

The great bulk of the sequence from the Mutabilis to the Rotunda Zone consists of decimetre to metre-thick alternations of normal and laminated bituminous shales. Both types of shale contain, besides the ubiquitous ammonites, a low diversity, high density fauna of small bivalves including *Protocardia*, lucinoids and protobranchs. A number of secondary dolomite beds form hard ledges on the foreshore, and the Pectinatus Zone contains several bands of distinctive whitish coccolith limestone (Downie 1957), lateral equivalents of which can be traced as far north as the Fenland (Gallois & Cox 1974). The Fittoni Zone at the top of the sequence consists of silty clays without bituminous horizons (Hounstout Clay and Marl) and form a facies transition to the overlying Portland Beds, into which indeed they have been transferred by Townson (1975).

A maximum thickness of 530 m of Kimmeridge Clay is attained in the Weald, as revealed by the Ashdown borehole, although the thickness in Dorset boreholes are not much less. Northwards towards London the sequence thins considerably as a result of condensation rather than erosion, and the Warlingham borehole reveals a facies generally similar to Dorset (Worssam & Ivimey-Cook 1971). The Kimmeridge Clay of the Swindon region of Wiltshire compares with the Oxford region in being thin, with significant non-sequences, and containing sand within the Pectinatus Zone. An excellent and richly fossiliferous section in the Lower Kimmeridge Clay (Cymodoce to Eudoxus Zones) of Westbury, Wiltshire has recently been described by Birkelund *et al.* (1983). The succession can easily be correlated with both eastern England (Wash boreholes) and Dorset, and is remarkably constant over the whole area, individual beds being traceable for distances of 100–200 km. This lateral constancy, implying a very flat sea floor, matches that recognised in the Blue Lias (Hallam 1960).

The Portland Beds are a shallow water, dominantly carbonate, marine sequence excellently exposed in the Isles of Portland and Purbeck and long quarried for building stone. The lithostratigraphy has recently been revised, and the sedimentary environments interpreted, by Townson (1975) and a new ammonite zonation put forward by Wimbledon & Cope (1968). In Townson's scheme the Portland Group, averaging 75 m thick, comprises successively the Portland Sand and Portland Limestone Formations, each divided into several members.

The Portland Sand Formation passes up from silty dolomitic clay to slightly sandy bioturbated dolomite rock. The Portland Limestone Formation in its lower

**Fig. 11.12.** Kimmeridge Clay with calcareous bands (the prominent bed half-way up the cliff is the Yellow Ledge Stone Band), east of Kimmeridge Bay, Isle of Purbeck, Dorset (R. H. Roberts).

part consists of fine-grained whitish limestones rich in calcitised spicules of the sponge *Rhaxella*, the silica of which has migrated in diagenesis to produce a series of discontinuous layers of black cherts. The upper part (the old 'Freestone Series') is coarser grained, with oolitic beds and a higher proportion of shells, especially *Laevitrigonia* and *Aptyxiella*, though sponge spicules persist. A feature of especial ecological interest is the occurrence of small *Solenopora*–oyster–bryozoan patch reefs.

The term Purbeck Beds has traditionally been applied to the marginal- and non-marine deposits of thin-bedded limestone and shale in east Dorset between the Portland Group and siliciclastic Wealden; though ammonite evidence is lacking they have been placed entirely in the Jurassic. Following Casey's (1974) work, only the lower part of the sequence, up to the oyster lumachelle known as the Cinder Bed, is now accepted as equivalent to the topmost Jurassic in the boreal chronostratigraphic scheme. (In the Tethyan scheme even the lower part of the Purbecks probably belongs to the 'basal Cretaceous' Berriasian stage, as indicated by both ostracod and palynological evidence.)

The lower, 'Jurassic', part of the Purbecks are now known as the Lulworth Beds and range up to a maximum of 60 m in the type area. They comprise an extremely interesting series of micritic and pellet limestones with stromatolitic and soil horizons, evaporite pseudomorphs and collapse breccias, with a high-density, low-diversity fauna of ostracods and small-sized bivalves.

Both the Portland and Purbeck Beds form a distinctive carbonate group throughout southern England, wedging out northwards against the London Platform and disappearing completely farther west, north of the Swindon and Oxford districts.

## Southern North Sea

Bradshaw (1978) synthesised the Jurassic geology of the UK sector of the southern North Sea from a study of oil company borehole logs. The only area with a full Jurassic sequence is the Sole Pit Trough (Fig. 11.2), 250 × 50 km in area, with a well-defined SW margin formed by the Dowsing Fault Zone, which separates it from the East Midlands Shelf and London–Brabant Platform. Its NE boundary is less sharply defined by numerous small NW–SE faults. Unlike Yorkshire, strata thicknesses have been much affected by halo-kinesis. Away from the Sole Pit Trough, Lower Cretaceous progressively oversteps Jurassic down to Permo–Trias.

Bradshaw's most important conclusion was that the Sole Pit Trough Jurassic compares closely with that of Yorkshire in general thickness and facies. The Yorkshire Basin, in fact, can be considered as a westerly offshoot or prolongation of the trough. Thus the Lias, often 300 m thick, is predominantly argillaceous (and less silty southwards) but with Domerian silts and sands, some containing chamo-site ooids. Similarly there are local Upper Toarcian sands and ferruginous Aalenian siliciclastics. The Bajocian–Bathonian sequence contains deltaics but the Bajocian overall becomes more marine south-wards. The Callovian deposits are similar to those in Yorkshire. There is a 'Corallian' facies in the mid to late Oxfordian, unlike over the East Midlands Shelf.

## Tectonic and igneous activity

Despite the occasional claims made for Jurassic folding episodes there is nothing in the sedimentary facies, stratal thickness or structures to suggest an orogenic zone in England. Tectonic activity was instead of the taphrogenic kind implying regional tension rather than compression and any slight angular discordances within Jurassic strata, as rec-ognised in Yorkshire and elsewhere, can easily be attributed to renewed sedimentation following eros-ional planing of tilted fault blocks.

It has become increasingly apparent since the Second World War that many of the major geographic features of the British area were established as early as Triassic times, when local fault-controlled collapse of parts of the Palaeozoic basement created a series of horsts and grabens (Audley-Charles 1970, Kent 1975). That at least some of these grabens persisted into the early Jurassic is apparent from the rapid lateral changes in stratal thickness and facies at the margins (Whittaker 1975). Thus the Cheshire, Severn and Central Somerset Basins contain unusually thick Lias as well as Trias. In the circumstances it is not surprising that some of the most basinal develop-ments, as inferred from high subsidence rates, occur in marginal areas or even offshore, such as the Cardigan Bay and Bristol Channel Basins and Sole Pit Trough. Whittaker (1975) argued that the Central Somerset Basin (or Graben) was active at least until the Middle Jurassic and Mudge (1978) has cited evidence for persistence of the Severn Basin, bounded on the east and west by faults, into the Aalenian–Bajocian.

Further evidence of taphrogenic activity comes from the recognition of neptunean sills and dykes in the Junction Bed of Dorset (Jenkyns & Senior 1977) and from the pronounced changes in sedimentary rate and facies up the Dorset sequence, most notably from the Bajocian to the Bathonian, when a 'swell' was evidently converted rapidly into a 'basin'. Sellwood & Jenkyns (1975) interpreted such facts in terms of adjustment of the Mesozoic cover rocks to episodes of normal faulting in the Hercynian basement. Hallam & Sellwood (1976) followed up this interpretation by undertaking a graphical analysis of Jurassic stage thickness changes throughout southern England. By this means they were able to show that the margins of the Cornubian–Armorican Platform High were more affected by basement faulting than the corresponding borders of the London–Brabant Platform; also that such faulting progressively died out during the course of the period so that lateral changes of thickness became less pronounced from Callovian times onward.

The Market Weighton structure has long attracted special interest, as has the nearby Peak Fault north of Scarborough, which is now interpreted as a transcurrent fault (Hemingway 1974). Sellwood & Jenkyns (1975) argued for control of the structure by halokinesis. There is, however, no evidence for the necessary thick Zechstein salts and the geophysical arguments of Bott et al. (1978) for a fault-bounded block underlain by a basement granite appear to be more plausible. The structure can also be considered as having originated as a positive rebound response to the faulted downwarp which marks is northern margin. The evidence for stratal attenuation over the structure is limited, and overall thinning of the pre-Chalk sequence seems to be primarily the result of a series of phases of uplift and erosion (Kent 1980). Structures similar to the Market Weighton Block occur along the margin of the Sole Pit Trough to the south-east (Bradshaw 1978).

It is now apparent that the pre-war enthusiasm for 'axes of uplift', eloquently espoused by Arkell (1933), was wide of the mark insofar as posthumous move-ment was invoked along old Hercynian fold axes rather than faults. Within the last few years there has been a considerable increase in the amount of sub-surface information in southern England, principally because of oil company exploration, such that a comprehensive picture of extensional tectonics is emerging. Thus Chadwick (1986) recognised two periods of Wessex Basin crustal extension in the early and late Jurassic marked by rapid subsidence of fault-bounded basins and commonly by erosion of adjacent upfaulted blocks; these phenomena are superimposed on thermally induced regional subsidence. Sellwood et al. (1986) utilised data from numerous boreholes to produce a series of isopach maps and sections to illustrate Mesozoic basin evolution in southern England. Further oil company exploration in the Irish Sea has revealed coarse non-marine siliciclastic deposits, of late Oxfordian–Kimmeridgian age, resting with marked unconformity on marine Lower and Middle Jurassic strata (Millson 1987). This is strikingly similar to Asturias in northwest Spain, where late Jurassic coarse and fine red siliciclastics rest with marked disconformity on marine Lias. In both cases, early late Jurassic uplift of tilted fault blocks, probably bound up with the early opening of the Atlantic Ocean, is the likely cause.

Until recently none thought of the British area as a site of igneous activity in Jurassic times. This situation has been drastically changed by the discovery of several hundred metres of porphyritic olivine basalt in the Bathonian of the North Sea off eastern Scotland (Gibb & Kanaris-Sotiriou 1976), which supports the overall tectonic interpretation of regional taphrogeny. Hallam & Sellwood (1968) interpreted the Upper Bathonian fuller's earth seam near Bath as a bentonite. This is because it is composed of pure or almost pure smectite in sharp contrast to the illitic–kaolinitic clays that characterise the normal Jurassic argillaceous deposits (including most of the Fuller's Earth Clay Formation) and because of the presence of certain distinctive accessory minerals. This interpretation has now been fully confirmed by Jeans et al. (1977), who have discovered devitrified glass shards in calcareous concretions together with high temperature feldspar, sphene and fragments of igneous rock.

Bradshaw (1975) has also found abundant smectite in more or less contemporaneous deposits in the East Midlands and argues for an ash-fall volcanic origin on the basis of its episodic occurrence up the sequence. Smectite-rich clays with associated zeolites, apatite and biotite are also widespread in late Oxfordian deposits and signify a later volcanic episode (Brown et al. 1969, Ali 1977, Chowdhury 1982).

The location of the source volcano or volcanoes remains an unsolved problem. The North Sea Forties Oilfield seems too far away and the petrology is wrong, at least for the Bath deposits, for which a trachytic composition is inferred. A location somewhere within the Western Approaches is perhaps the most reasonable. It is noteworthy, therefore, that a dolerite sill has been located in the Fastnet Basin and dated at 166 million years, indicating a Bajocian age of emplacement (Harrison et al. 1979). If this or a neighbouring area operated as a minor volcanic centre in the mid and perhaps also early Jurassic, an explanation might be provided for the marked increase in smectite content of Toarcian clays westwards across southern England and up the Pliensbachian succession of Dorset (Corbin 1980).

# Depositional environments

This subject is dealt with in more detail in Hallam (1975) and hence warrants only brief treatment here. Sellwood (1978) gives a good general and well-illustrated account of sediment–fauna relationships and describes a number of faunal associations which are termed communities, but without citing examples from given rock formations. If the term community is to be used at all it should surely be based on detailed quantitative work, as undertaken for the Lower Oxford Clay and Corallian respectively by Duff (1975) and Fürsich (1977).

Most of the English and Welsh Jurassic rocks are marine but there are none for which water depths significantly in excess of 100 m need be invoked; many were probably deposited in much shallower depths. Moderate to very low rates of sedimentation and hence subsidence can be inferred, with a mean rate for the whole of Great Britain of about $1 \, \text{m}/5 \times 10^4$ years and a maximum (for the Mochras Lias) of $1 \, \text{m}/2 \times 10^4$ years (Hallam & Sellwood 1976). A high proportion of limestones signify correspondingly low rates of siliciclastic influx from the neighbouring land areas. The absence of debris-flow deposits, turbidites and slumping phenomena confirms the general inference of extremely low depositional slopes (an apparent exception may be the basal conglomerate of the Sutton Stone in Glamorgan, reinterpreted (Ager 1986) as a storm-induced debris-flow deposit (but see Fletcher (1988)). Such general considerations allied with detailed facies analysis render it unlikely that wave activity, apart from that produced by occasional storms, and tidal currents were of any great strength in inshore waters; tidal range was probably in general very low and sands were never carried far from land (e.g. Bradshaw 1978). On the other hand, Brasier & Brasier (1978) argued for tidal influence on sand bodies of the Kellaways Beds on the Market Weighton Swell; see also Wilson (1968b).

The Jurassic rocks of England and Wales may for the most part readily be subdivided into three major environmental categories.

1. *Deeper shelf*. This category includes most of the Lias clay formations together with the Fuller's Earth, Oxford, Ampthill and Kimmeridge Clays. The clays, composed almost entirely of illite with subordinate kaolinite and mixed-layer illite-smectite, may contain a moderate proportion of silt and fine sand in the proximity of land or in regressive intervals such as the Middle Lias. Low sedimentation rates correlate with an increase in the proportion of calcite both in the form of marls and microsparite layers and concretions. Periodic variations in the oxygen content of bottom waters are signified by alternations of normal shales or clays with horizons of finely laminated

bituminous shales. The most notable examples of such 'anoxic' deposits are the Lower Toarcian Jet Rock of Yorkshire and its thinner lateral equivalents in the Midlands, the Middle Callovian Lower Oxford Clay and many horizons in the Kimmeridge Clay, together with beds in the Hettangian and Lower Sinemurian, especially the Shales with Beef in Dorset.

The invertebrate megafauna is dominated by ammonites and bivalves, the most characteristic of which are small, thin-shelled forms including proto-branchs, certain pterioids and astartids (Hallam 1976). Larger forms may occur at limited intervals (e.g. Sellwood 1972). Apart from the ubiquitous *Chondrites* there are few distinctive types of trace fossil. The bituminous shales are not entirely devoid of benthos as is sometimes alleged but in the more shelly beds the fauna is typically of a low-diversity high-density type with a high proportion of juveniles, signifying high stress conditions. A burrowing fauna is totally absent from these shales in most cases presumably because of insufficient oxygen, but abundant nuculids in the Lower Oxford Clay have churned up the sediment to destroy the otherwise ubiquitous fine lamination. The best preserved marine vertebrates come from these bituminous deposits in which the normal processes of organic decay have been arrested until complete burial has been accomplished.

Morris (1980), in his comparison of the principal English clay formations, argues that the Kimmeridge Clay probably accumulated in an environment that periodically fluctuated between mildly oxygenated and totally anoxic. In contrast, the Lower Oxford Clay signifies deposition in mildly oxygenated bottom waters, while the Toarcian Jet Rock was probably laid down in very poorly oxygenated bottom waters, with reducing conditions extending up to the sediment surface. The small-scale cyclic alternations of normal and bituminous shale in the Kimmeridge Clay have been interpreted in terms of varying degrees of oxygenation of bottom waters related to cyclic ascent and descent of the $O_2/H_2S$ interface. Thermoclines are thought to have probably been the principal agent in controlling stratification and stagnation of the water column (Tyson *et al.* 1979).

Storms are now thought to have been an important influence on Kimmeridge Clay deposition. Thus Wignall (1989) has described a number of storm-produced beds in the Dorset section, including such features as graded rip-up clasts, silt laminae and shell pavements. Palaeoecological research indicates that the pavements represent brief benthic colonisation events in a predominantly anaerobic–dysaerobic environment. A feedback mechanism of storm-induced benthic oxygenation and temperature-stratified inhibition of storm mixing may account for the abrupt nature of the small-scale cyclicity referred to above.

2. *Shallower, open shelf.* The rocks of this category signify deposition in more agitated, wave or current-disturbed waters than the first category and consequently are on the whole coarser grained, though fine-grained deposits are not rare. They are predominantly carbonates and include the basal Jurassic Sutton Stone of South Wales, most formations in the Inferior and Great Oolite of the South Cotswolds, most of the Lincolnshire Limestone, various Corallian limestones and the Portland Limestone. Comparisons with the Bahamas have been attempted and a variety of environments have been inferred, ranging from oolite shoals and open shelf bioclastic and ooid sand-flats to more protected quieter water areas with coral patch reefs (Wilson 1968b, Green & Donovan 1969, Townson 1975, Mudge 1978, Baker 1981, Sellwood *et al.* 1985).

In the same major environmental category can be included a number of fine- to medium-grained quartz sandstones such as the Liassic Thorncombe and Bridport Sands of Dorset, the Kellaways Rock, the various 'Calcareous Grits' of the Oxfordian and the upper part of the Portland Sand. These are normally strongly bioturbated so that inorganically produced sedimentary structures are rather exceptional. Where they do occur there is uncertainty about the extent to which they were induced by wave or tidal current activity. The presence of hummocky cross stratification in such deposits as the Bencliff Grit of Dorset and Staithes Formation of Dorset indicates the activity of periodic storm waves in a shoreface regime.

Oolitic ironstones also fall into this category. As discussed at some length elsewhere (Hallam 1975, Hallam & Bradshaw 1979) they probably formed on low shoals in the proximity of river mouths. The most characteristic elements of the megafauna are large thick-shelled bivalves such as *Gryphaea*, *Lopha*, *Liostrea*, *Cardinia*, *Ctenostreon*, *Plagiostoma*, *Chlamys*, *Isognomon* and trigoniids. (Such shells may also be common in some fine-grained deposits such as the Blue Lias.) Large gastropods including *Bourguetia* and nerineids are also characteristic but less common and diverse, while ammonites are rare. Locally compound corals, echinoids, bryozoans, brachiopods, sponges and the red alga *Solenopora* may be common in the calcareous facies. The abundant trace fossils include *Diplocraterion*, *Rhizocorallium* and *Thalassinoides* (or *Spongeliomorpha*).

3. *Marginal marine to non-marine.* The environments in this category range from 'lagoons' to supratidal flats, coastal plain swamps and marshes to river deltas and flood plains. The lagoonal category includes those finer-grained calcareous and argillaceous deposits with a reduced marine fauna characterised by a combination of high-density (individual abundance) with low-diversity, in which stenohaline

elements such as cephalopods, echinoderms, brachiopods, corals and bryozoans are almost totally absent. Evidence of blue-green algal activity in the form of stromatolites and discrete fossils may be widespread, as in the Lulworth Beds of Dorset (Pugh 1968). The general association in these beds of pseudomorphs of evaporite minerals and brackish water molluscs and ostracods suggests strongly fluctuating salinities, provoking comparison with Florida Bay (Brown 1963, 1964). Palmer (1979) also made such a comparison in his interpretation of the Bathonian Hampen Marly and White Limestone Formations of Oxfordshire.

One should bear in mind that because of the extremely low oceanward gradients and extensive areas of very shallow sea in the British area, 'lagoonal' conditions probably extended for considerable distances offshore and that there was no physical barrier separating these extensive stretches of water from the open sea.

The silts and clays of the Bathonian Rutland Formation of the East Midlands, containing a number of truncated rootlet horizons, were interpreted by Bradshaw (1978) as signifying a series of prograding coastal marsh and swamp deposits interrupted by periodic marine transgression, with fully marine conditions never being achieved.

It has been customary to interpret the non-marine Bajocian–Bathonian sequence of Yorkshire as signifying a river delta, with the sandstone units as distributary channel and the intervening siltstones and carbonaceous shales as delta-top interdistributary swamp deposits (Hemingway 1974). Leeder & Nami (1979) challenged this interpretation for the Scalby Beds and made a plausible case for deposition in a braided river channel system. Thus the basal, giant cross-stratified Moor Grit Member of the Scalby Formation, cutting down into the underlying marginal marine Scarborough Formation, is seen not as the product of delta progradation but as a fluvial channel sand deposited after a long phase of uplift and erosion.

## Palaeogeography

It is generally agreed that the Jurassic climate of England and Wales was appreciably warmer than today, with temperatures comparable to the tropical zone. This is clearly signified in the marine deposits by the presence of reef corals and giant molluscs, such as some of the ammonites in the Blue Lias, 'Corallian' and Portland Beds. Support is provided by the presence in terrestrial floras such as the celebrated Middle Jurassic plant beds of Yorkshire, of abundant ferns with modern relatives that cannot tolerate frost (Barnard 1973). Such ferns also signify humid conditions, as does the abundance of kaolinite throughout the Jurassic succession (Sellwood and Sladen 1981), this being a mineral that only forms on land in conditions of intense chemical weathering. The same conclusion can be drawn from the occurrence of ironstones, which occur in almost every Jurassic stage (Hallam 1975, 1984). The only exception to this general pattern of humidity concerns the end of the Period. The co-existence in the Lower Purbeck Beds of Dorset of $CaSO_4$-evaporites with in situ coniferous tree stumps with highly variable growth-rings, fresh-water insects, ostracodes and molluscs, points to semi-arid Mediterranean-type climates with strong seasonal variations in precipitation. A modern analogue is a region of ephemeral lakes and coastal lagoons in South Australia (Francis 1984).

It is extremely difficult to infer the position of coastlines even for restricted time intervals because too many strata have been removed by subsequent erosion and, even when deposits are still preserved, there may be no unambiguous interpretation. Furthermore if, as seems likely, all or most of the land areas had very subdued topographic relief, even modest rises and falls of sea level or variations in rates of regional uplift and subsidence could have caused extensive advances and retreats of coastlines in geologically short time intervals.

Where deposits of the deeper shelf facies abut against the inferred positive areas indicated in Figure 11.2, there is generally little change of facies. Thus the Lias, Oxford and Kimmeridge Clays persist more or less unchanged except for condensation right up to the erosional termination around the margin of the London Platform, suggesting that the feather edge of subcrop falls some distance short of the former coastline, or even that most of the platform was temporarily inundated. The location of land masses is easier to determine, at least approximately, from the facies distributions of shallower water, more regressive deposits. Thus the quartz sandy character of the lower part of the Great Oolite Group around the margin of the London Platform suggests that in early Bathonian times this area was acting as a sediment source (Bradshaw 1978). Similarly the Southern Uplands Massif appears to have been a significant source of siliciclastic sediments (probably reworked late Palaeozoic rocks) for the Yorkshire Basin during the Upper Pliensbachian and Upper Toarcian to Bajocian time intervals. The same can be inferred for its seaward extension, the Mid-North Sea High, from the regressive Middle Jurassic deposits of the Sole Pit Trough. It can also be confidently inferred that the Cornubian Platform was an emergent source of siliciclastic sediments during the late Pliensbachian and late Toarcian to Aalenian.

There is less certainty about other likely positive

areas. Kent (1975) considered that the Pennines once bore an attenuated Mesozoic succession, but the abundance of fragments of black chert (lydite) in the mid-Volgian pebble beds of Central England points to a Carboniferous Limestone source to the north, as Neaverson (1925) inferred, because the Palaeozoic basement of the London Platform, the only likely alternative, consists mainly of pre-Carboniferous rocks (Bradshaw 1978). An emergent siliciclastic source area opposite the Market Weighton Block is suggested by the Aalenian, Great Oolite, Kellaways and Lower Oxfordian facies distributions, while westward and northward attenuation of Upper Lias and Oxford Clay suggests the existence of a Pennine High (Bradshaw & Penney 1982). Wilson (1986a) likewise inferred a westerly source for the sands in the lower part of the Corallian sequences of the South Midlands and southern England. In support of this, the 'Corallian' facies of the Bristol Channel consists almost entirely of sands (Evans & Thompson 1979). The Kimmeridgian sands of Oxfordshire and Wiltshire also probably came from the west.

There is also doubt concerning the Welsh High. Apart from the basal Liassic overstep on to Carboniferous Limestone in Glamorgan, there is no evidence whatsoever of its nature throughout the rest of the Jurassic; the Liassic sequences of the Cardigan Bay and Cheshire Basins respectively to the west and east are notably deficient in quartz sand. However, sands in the Westbury Beds (Rhaetic) of the Severn Valley were probably derived locally from the west (Arkell 1933).

Transgressions and regressions, and corresponding changes in depth of water in the marine sector, provide a *leitmotif* for Jurassic history. The first important transgression was in the early Lias. The basal Hettangian rests almost everywhere with minor unconformity on the Rhaetic and oversteps it on to Palaeozoic in South Wales and the Mendips. A more or less progressive overlap of succeeding Lower Lias zones up to the early Pliensbachian on to the Palaeozoic massif of the London Platform is recorded from borehole data by Donovan *et al.* (1979). A second transgressive event around this same massif is recorded from early Bathonian deposits which overstep Lias on to Palaeozoic (Fig. 11.10). The event in question appears to mark a continuation of the Upper Bajocian transgression which is clearly recorded over the Mendip and Moreton Swells (Fig. 11.9). The celebrated 'Callovian transgression', which really began in the late Bathonian, had the effect of replacing a complex geography of shallow lagoons and coastal plains by a rapidly deepening sea, until a uniform Oxford Clay facies extended over much of England.

Important transgressions in other regions are recorded in England by deposits signifying sea deepening or retreat of shorelines. The most notable of these are the replacement of the varied shallow water Domerian sediments by Lower Toarcian shales and the extensive spread of limestones at the expense of siliciclastic and ferruginous deposits in the Lower Bajocian and Middle Oxfordian.

The earliest significant regression was in the early Middle Jurassic. This is best expressed by the replacement of Liassic marine shales by 'deltaic' facies in Yorkshire but is also apparent in Central England. Although the sea did not retreat from southern England a widespread shallowing from the Lower to the Middle Jurassic can be inferred from the facies. Finally, there was at the end of the period a major regressive phase, expressed in southern England as a shallowing upward sequence from Kimmeridge Clay through the Portland Beds to the marginal marine or non-marine Lulworth Beds, and in the rest of the country by a widespread erosional unconformity between the Kimmeridge Clay and late Volgian to early Cretaceous deposits.

Viewed in the world context it appears that the Hettangian, early Sinemurian, Carixian, early Toarcian, early Bajocian, early Callovian and mid-Oxfordian transgressions or deepening events reflect eustatic rises of sea level, and the end Jurassic event a fall, at least in part (Hallam 1978, 1981, 1988). On the other hand, the early Middle Jurassic regressive event runs counter to the world trend of continuing transgressions and appears to be connected with an epeirogenic uplift centred in the North Sea region and presumably connected with collapse of the central graben (Hallam & Sellwood 1976). The fact that the Kimmeridge Clay is widely transgressive in the southern North Sea region, overstepping on to Permo–Triassic along the Mid-North Sea High and on to Lias or Triassic over anticlines and salt domes in the Sole Pit Trough (Bradshaw 1978) is probably related both to sea-level rise and to local tectonism (Hallam & Sellwood 1976).

Careful examination of any Jurassic sedimentary sequence usually indicates alternations of (1) high- and low-energy deposits, (2) laminated and non-laminated deposits or (3) depositional and erosional events, on a scale ranging from tens of metres down to centimetres. There is general agreement that the underlying control in at least the majority of cases is varying depth of sea, but stratigraphic resolution is unfortunately not sufficiently precise, especially for the smaller-scale instances, to decide readily between eustatic and local epeirogenic alternatives. Further study of such rhythmic sequences, especially their relationship with faunal change and rate and amount of transgression, can confidently be expected to yield interesting results in the future.

# REFERENCES

AGER, D. V.     1986     A reinterpretation of the basal 'Littoral Lias' of the Vale of Glamorgan. *Proc. Geol. Ass. Lond.*, **97**, 29–35.

ALI, O. E.     1977     Jurassic hazards to coral growth. *Geol. Mag.*, **114**, 63–64.

ARKELL, W. J.     1933     *The Jurassic System in Great Britain*. Clarendon Press, Oxford, 681 pp.

    1947a     *The Geology of Oxford*. Clarendon Press, Oxford, 267 pp.

    1947b     *The geology of the country around Weymouth, Swanage, Corfe and Lulworth*. Mem. geol. Surv. G.B.

    1956     *Jurassic Geology of the World*. Oliver & Boyd, Edinburgh, 806 pp.

ASHTON, M.     1980     The stratigraphy of the Lincolnshire Limestone Formation (Bajocian) in Lincolnshire and Rutland (Leicestershire). *Proc. Geol. Ass. Lond.*, **91**, 203–233.

AUDLEY-CHARLES, M. S.     1970     Triassic palaeogeography of the British Isles. *Quart. J. geol. Soc. Lond.*, **126**, 49–73.

BAKER, P. G.     1981     Interpretation of the Oolite Marl (Upper Aalenian, Lower Inferior Oolite) of the Cotswolds, England. *Proc. Geol. Ass. Lond.*, **92**, 169–187.

BIRKELUND, T., CALLOMON, J. H., CLAUSEN, C. K., NOHR HANSEN, H. & SALINAS, I.     1983     The Lower Kimmeridge Clay at Westbury, Wiltshire, England. *Proc. Geol. Ass. Lond.*, **94**, 289–309.

BOTT, M. H. P., ROBINSON, J. & KOHNSTAMM, M. A.     1978     Granite beneath Market Weighton, East Yorkshire. *J. geol. Soc. London*, **135**, 535–564.

BRADSHAW, M. J.     1975     Origin of montmorillonite bands in the Middle Jurassic of eastern England. *Earth Planet, Sci. Letters*, **26**, 345–352.

    1978     *A facies analysis of the Bathonian of eastern England*. Unpubl. D.Phil. thesis, University of Oxford.

BRADSHAW, M. J. & BATE, R. H.     1982     Lincolnshire borehole proves greater extent of the Scarborough Formation (Jurassic: Bajocian). *J. Micropalaeontol.*, **1**, 141–147.

BRADSHAW, M. J. & PENNEY, S. R.     1982     A cored Jurassic sequence from north Lincolnshire, England: stratigraphy, facies analysis and regional content. *Geol. Mag.*, **119**, 113–134.

BRASIER, M. D. & BRASIER, C. J.     1978     Littoral and fluviatile facies in the 'Kellaways Beds' on the Market Weighton Swell. *Proc. Yorkshire geol. Soc.*, **42**, 1–19.

BROOKFIELD, M. E.     1978     The lithostratigraphy of the upper Oxfordian and lower Kimmeridgian Beds of south Dorset. *Proc. Geol. Ass. Lond.*, **89**, 1–32.

BROWN, G., CATT, J. A., & WEIR, A. H.     1969     Zeolites of the clinoptilolite–heulandite type in sediments of south-east England. *Min. Mag.*, **37**, 480–488.

BROWN, P. R.     1963     Algal limestones and associated sediments in the basal Purbeck of Dorset. *Geol. Mag.*, **100**, 565–573.

    1964     Petrography and origin of some Upper Jurassic beds from Dorset. *J. sediment. Petrol.*, **34**, 254–269.

CALLOMON, J. H.     1968     The Kellaways Beds and Oxford Clay. *In* Sylvester Bradley, P. C. & Ford, T. D. (Eds.) *The Geology of the East Midlands*. Leicester Univ. Press, 400 pp.

CAMPOS, H. S. & HALLAM, A.     1979     Diagenesis of English Lower Jurassic limestones as inferred from oxygen and carbon isotope analysis. *Earth Plant. Sci. Lett.*, **45**, 2331.

CASEY, R.                      1974    The ammonite succession at the Jurassic–Cretaceous boundary in eastern England. *In* Casey, R. & Rawson, P. F. (Eds.) *The Boreal Lower Creataceous.* Seed House Press, Liverpool, 448 pp.

CHADWICK, R. A.                1986    Extension tectonics in the Wessex Basin, southern England. *J. geol. Soc. Lond.*, **143**, 465–488.

CHOWDHURY, A. N.               1982    Smectite, zeolite, biotite and apatite in the Corallian (Oxfordian) sediments of the Baulking area in Berkshire, England. *Geol. Mag.*, **119**, 487–496.

COPE, J. C. W.                 1967    The palaeontology and stratigraphy of the lower part of the Upper Kimmeridge Clay of Dorset. *Bull. Brit. Mus. (Nat. Hist.), Geol.*, **15**, 1–79.

                               1974    New information on the Kimmeridge Clay of Yorkshire. *Proc. Geol. Ass. Lond.*, **85**, 211–221.

                               1978    The ammonite faunas and stratigraphy of the upper part of the Upper Kimmeridge Clay of Dorset. *Palaeontology*, **21**, 469–533.

COPE, J. C. W.,                1980a   A correlation of Jurassic rocks in the British Isles. Part One:
   GETTY, T. A.,                       Introduction and Lower Jurassic. *Geol. Soc. Lond. Spec.*
   HOWARTH, M. K.                      *Rep.*, No. **14**, 1–73.
   MORTON, N. &
   TORRENS, H. S.

COPE, J. C. W., DUFF, K. L.,   1980b   A correlation of Jurassic rocks in the British Isles. Part
   PARSONS, C. F.,                     Two: Middle and Upper Jurassic. *Geol. Soc. Lond. Spec.*
   TORRENS, H. S.,                     *Rep.* No. **15**, 1–109.
   WIMBLEDON, W. A. &
   WRIGHT, J. K.

CORBIN, S. G.                  1980    *A facies analysis of the Lower–Middle Jurassic boundary beds of north-west Europe.* Unpubl. Ph.D. thesis, University of Birmingham.

DAVIES, D. K.                  1969    Shelf sedimentation: an example from the Jurassic of Britain. *J. Sediment. Petrol.*, **39**, 1344–1370.

DEAN, W. T.,                   1961    The Liassic ammonite zones and subzones of the north-
   DONOVAN, D. T. &                    west European Province. *Bull. Brit. Mus. (nat. Hist.),*
   HOWARTH, M. K.                      *Geol.*, **4**, 438–505.

DINGLE, R. V.                  1971    A marine geological survey of the north east coast of England (Western North Sea). *J. geol. Soc. Lond.*, **127**, 303–338.

DONOVAN, D. T.                 1956    The zonal stratigraphy of the Blue Lias around Keynsham, Somerset. *Proc. Geol. Ass. Lond.*, **66**, 182–212.

DONOVAN, D. T. &               1961    An acoustic survey of the sea floor south of Dorset and its
   STRIDE, A. H.                       geological interpretation. *Phil. Trans. roy. Soc. Lond.*, **244B**, 299–330.

DONOVAN, D. T.,                1979    The transgression of the Lower Lias over the northern
   HORTON, A. &                        flank of the London Platform. *J. geol. Soc. Lond.*, **136**,
   IVIMEY-COOK, H. C.                  165–173.

DOWNIE, C.                     1957    Microplankton from the Kimmeridge Clay. *Quart. J. geol. Soc. Lond.*, **112**, 413–433.

DUFF, K. L.                    1975    Palaeoecology of a bituminous shale – The Lower Oxford Clay of Central England. *Palaeontology*, **18**, 443–482.

EVANS, D. J. &                 1979    The geology of the central Bristol Channel and the Lundy
   THOMPSON, M. S.                     area, South Western Approaches, British Isles. *Proc. Geol. Ass. Lond.*, **90**, 1–14.

FISHER, M. J. &                1985    The Scalby Formation (Middle Jurassic, Ravenscar Group)
   HANCOCK, N. J.                      of Yorkshire: reassessment of age and depositional environment. *Proc. Yorkshire geol. Soc.*, **45**, 293–298.

FLETCHER, C. J. N.  1988  Tidal erosion, solution cavities and exhalative mineralisation associated with the Jurassic unconformity at Ogmore, South Glamorgan. *Proc. Geol. Ass. Lond.*, **99**, 1–14.

FRANCIS, J. E.  1984  The seasonal environment of the Purbeck (Upper Jurassic) fossil forests. *Palaeogeog., Palaeoclimatol., Palaeoecol.*, **48**, 285–307.

FÜRSICH, F. T.  1973  *Thalassinoides* and the origin of the nodular limestones in the Corallian Beds (Upper Jurassic) of Southern England. *N. Jb. Geol. Paläont, Mh*, 136–156.

1977  Corallian (upper Jurassic) marine benthic associations from England and Normandy. *Palaeontology*, **20**, 337–385.

GALLOIS, R. W. &  1974  Stratigraphy of the Upper Kimmeridge Clay of the Wash
COX, B. M.  area. *Bull geol. Surv. G.B.*, No. **47**, 1–28.

1977  The stratigraphy of the Middle and Upper Oxfordian sediments of Fenland. *Proc. Geol. Ass. Lond.*, **88**, 207–228.

GIBB, F. G. F. &  1976  Jurassic igneous rocks of the Forties Field. *Nature, London*,
KANARIS-SOTIRIOU, R.  **260**, 23–25.

GREEN, G. W. &  1956  The stratigraph of the Stowell Park borehole. *Bull. geol.*
MELVILLE, R. V.  *Surv. G.B.*, No. **11**, 1–66.

GREEN, G. W. &  1969  The Great Oolite of the Bath area. *Bull. geol. Surv. G.B.*,
DONOVAN, D. T.  No. **30**, 1–63.

HALLAM, A.  1960  A sedimentary and faunal study of the Blue Lias of Dorset and Glamorgan. *Phil. Trans. roy. Soc. Lond.*, **243B**, 1–44.

1963  Observations on the palaeoecology and ammonite sequence of the Frodingham Ironstone. *Palaeontology*, **6**, 554–574.

1964  Origin of the limestone–shale rhythm in the Blue Lias of England – a composite theory. *J. Geol.*, **72**, 157–169.

1968  The Lias. *In* Sylvester Bradley, P. C. & Ford, T. D. (Eds.) *The Geology of the East Midlands*. Leicester Univ. Press, 188–210.

1975  *Jurassic Environments*, Cambridge Univ. Press, 269 pp.

1976  Stratigraphic distribution and ecology of European Jurassic bivalves. *Lethaia*, **9**, 245–259.

1978  Eustatic cycles in the Jurassic. *Palaeogeog., Palaeoclimatol., Palaeoecol.*, **23**, 1–32.

1981  A revised sea-level curve for the early Jurassic. *J. geol. Soc. Lond.*, **138**, 735–743.

1984  Continental humid and arid zones in the Jurassic and Cretaceous. *Palaeogeog., Palaeoclimatol., Palaeoecol.*, **46**, 195–223.

1988  A re-evaluation of Jurassic eustasy in the light of new data and the revised Exxon curve. *In* Wilgus, C. K. (Ed.) *Sea Level Changes – An Integrated Approach*. Spec. Publ. Soc. econ. Palaeontol. Mineral. Tulsa, **42**, 261–273.

HALLAM, A. &  1968  Origin of Fuller's Earth in the Mesozoic of Southern
SELLWOOD, B. W.  England. *Nature, Lond.*, **220**, 1193–1195.

1976  Middle Mesozoic sedimentation in relation to tectonics in the British area. *J. Geol.*, **84**, 301–321.

HALLAM, A. &  1979  Bituminous shales and oolitic ironstones as indicators of
BRADSHAW, M. J.  transgressions and regressions. *J. geol. Soc. Lond.*, **136**, 157–164.

HARRISON, R. K.,  1979  Mesozoic igneous rocks, hydrothermal mineralisation and
JEANS, C. V. &  volcanogenic sediments in Britain and adjacent areas. *Bull.*
MERRIMAN, R. J.  *geol. Surv. G.B.*, No. **70**, 57–69.

HEMINGWAY, J. E.            1974    Jurassic. *In* Rayner, D. H. & Hemingway, J. E. (Eds.) *The Geology and Mineral Resources of Yorkshire.* Yorkshire geol. Soc., 161–223.

HEMINGWAY, J. E. &          1973    Lithostratigraphic nomenclature of the Middle Jurassic
  KNOX, R. W. O'B.                  strata of the Yorkshire basin of north-east England. *Proc. Yorkshire geol. Soc.,* **39**, 527–535.

HORTON, A.                  1977    The age of the Middle Jurassic 'white sands' of north Oxfordshire. *Proc. Geol. Ass. Lond.,* **88**, 147–162.

HOWARTH, M. K.              1955    Domerian of the Yorkshire coast. *Proc. Yorkshire geol. Soc.,* **30**, 147–175.

                            1957    The Middle Lias of the Dorset coast. *Quart. J. geol. Soc. Lond.,* **113**, 185–203.

                            1962    The Jet Rock Series and Alum Shale Series of the Yorkshire coast. *Proc. Yorkshire geol. Soc.,* **33**, 381–422.

                            1973    The stratigraphy and ammonite fauna of the Upper Liassic Grey Shales of the Yorkshire coast. *Bull. Brit. Mus. (nat. Hist.), Geol.,* **24**, 235–277.

                            1978    The stratigraphy and ammonite fauna of the Upper Lias of Northamptonshire. *Bull. Brit. Mus. (nat. Hist.), Geol.,* **29**, 235–288.

                            1980    The Toarcian age of the upper part of the Marlstone Rock Bed of England. *Palaeontology,* **23**, 637–656.

IRWIN, H.                   1980    Early diagenetic carbonate precipitation and pore fluid migration in the Kimmeridge Clay of Dorset, England. *Sedimentology,* **27**, 577–591.

JEANS, C. V.,               1977    Origin of Middle Jurassic and Lower Cretaceous Fuller's
  MERRIMAN, R. J. &                 earths in England. *Clay Minerals,* **12**, 11–44.
  MITCHELL, J. G.

JENKYNS, H. C. &            1977    A Liassic palaeofault from Dorset. *Geol. Mag.,* **114**, 47–52.
  SENIOR, J. R.

KENT, P. E.                 1975    The tectonic development of Britain and the surrounding seas. *In* Woodland, A. W. (Ed.) *Petroleum and the Continental Shelf of North-West Europe,* **1**, *Geology.* Applied Science Publishers Ltd., Barking, Essex, 3–28.

                            1980    Subsidence and uplift in east Yorkshire and Lincolnshire: a double inversion. *Proc. Yorkshire geol. Soc.,* **42**, 505–524.

LEEDER, M. R. &             1979    Sedimentary models for the Scalby Formation (U. Deltaic
  NAMI, M.                          Series; M. Jurassic) and evidence for major Bathonian uplift of the Yorkshire Basin. *Proc. Yorkshire geol. Soc.,* **42**, 461–482.

LLOYD, A. J.,               1973    The geology of the Bristol Channel floor. *Phil. Trans. roy.*
  SAVAGE, R. J. G.,                 *Soc. Lond.,* **274A**, 595–626.
  STRIDE, A. H. &
  DONOVAN, D. T.

MARSHALL, J. D. &           1980    Isotopic and trace element evidence for submarine
  ASHTON, M.                        lithification of hardgrounds in the Jurassic of eastern England. *Sedimentology,* **27**, 271–289.

MARTIN, A. J.               1967    Bathonian sedimentation in southern England. *Proc. Geol. Ass. Lond.,* **78**, 473–488.

MILLSON, J. A.              1987    The Jurassic evolution of the Celtic Sea Basins. *In* Brooks, J. & Glennie, K. W. (Eds.) *Petroleum Geology of North West Europe,* Graham & Trotman, London, 599–610.

MORRIS, K. A.               1980    A comparison of major sequences of organic-rich mud deposition in the British Jurassic. *J. geol. Soc.,* **137**, 157–170.

MOUTERDE, R. *et al.*       1971    Les zones du Jurassique en France. *C. R. Somm. Séances Soc. géol. France,* fasc. **6**, 1–27.

| | | |
|---|---|---|
| MUDGE, D. C. | 1978 | Stratigraphy and sedimentation of the Lower Inferior Oolite of the Cotswolds. *J. geol. Soc. Lond.*, **135**, 611–627. |
| NEAVERSON, E. | 1925 | The petrography of the Upper Kimmeridge Clay and Portland Sand in Dorset, Wiltshire, Oxfordshire and Buckinghamshire. *Proc. Geol. Ass. Lond.*, **36**, 240–256. |
| PALMER, C. P. | 1972 | The Lower Lias (Lower Jurassic) between Watchet and Lilstock in North Somerset (United Kingdom). *Newsl. Stratigr.*, **2**, 1–30. |
| PALMER, T. J. | 1979 | The Hampen Marly and White Limestone formations: Florida-type carbonate lagoons in the Jurassic of central England. *Palaeontology*, **22**, 189–228. |
| PALMER, T. J. & FÜRSICH, F. T. | 1974 | The ecology of a Middle Jurassic hardground and crevice fauna. *Palaeontology*, **17**, 507–524. |
| PARSONS, C. F. | 1974 | On the *Sauzei* and so-called *Sowerbyi* zones of the Middle Bajocian. *Newsl. Stratigr.*, **3**, 153–180. |
| | 1976 | Ammonite evidence for dating some Inferior Oolite sections in the north Cotswolds. *Proc. Geol. Ass. Lond.*, **87**, 45–66. |
| | 1977 | A stratigraphic revision of the Scarborough Formation. *Proc. Yorkshire geol. Soc.*, **41**, 203–222. |
| PENN, I. E., DINGWALL, R. G. & KNOX, R. W. O'B. | 1981 | The Inferior Oolite (Bajocian) sequence from a borehole in Lyme Bay, Dorset. *Rep. Inst. geol. Sci.*, **79/3**, 1–27. |
| PENN, I. E. & EVANS, C. D. R. | 1976 | The Middle Jurassic (mainly Bathonian) of Cardigan Bay and its palaeogeographic significance. *Rep. Inst. geol. Sci.*, **76/6**, 6 pp. |
| PENN, I. E., MERRIMAN, R. J. & WYATT, R. J. | 1979 | The Bathonian strata of the Bath–Frome area. *Rep. Inst. geol. Sci.*, **78/22**, 1–87. |
| POOLE, E. G. & WHITEMAN, A. J. | 1966 | Geology of the country around Nantwich and Whitchurch. *Mem. geol. Surv. G.B.* |
| PUGH, M. E. | 1968 | Algae from the Lower Purbeck limestones of Dorset. *Proc. Geol. Ass. Lond.*, **79**, 513–523. |
| RAYNER, D. H. & HEMINGWAY, J. E. (Eds.) | 1974 | *The Geology and Mineral Resources of Yorkshire.* Yorkshire geol. Soc., 405 pp. |
| RIDING, J. B. & WRIGHT, J. K. | 1989 | Palynostratigraphy of the Scalby Formation (Middle Jurassic) of the Cleveland Basin, north-east Yorkshire. *Proc. Yorkshire geol. Soc.*, **47**, 349–354. |
| SELLWOOD, B. W. | 1972 | Regional environmental changes across a Lower Jurassic stage boundary in Britain. *Palaeontology*, **15**, 125–157. |
| | 1978 | Jurassic. *In* McKerrow, W. S. (Ed.) *Ecology of Fossils.* Duckworth, London, 383 pp. |
| SELLWOOD, B. W. & MCKERROW, W. S. | 1974 | Depositional environments in the lower part of the Great Oolite Group of Oxfordshire and north Gloucestershire. *Proc. Geol. Ass. Lond.*, **85**, 189–210. |
| SELLWOOD, B. W. & JENKYNS, H. C. | 1975 | Basins and swells and the evolution of an epeiric sea (Pliensbachian–Bajocian of Great Britain). *J. geol. Soc. Lond.*, **131**, 373–388. |
| SELLWOOD, B. W., SCOTT, J., MIKKELSEN, P. & AKROYD, P. | 1985 | Stratigraphy and sedimentology of the Great Oolite Group in the Humbly Grove Oilfield. *Marine Petrol. Geol.*, **2**, 44–55. |
| SELLWOOD, B. W., SCOTT, J. & LUNN, G. | 1986 | Mesozoic basin evolution in southern England. *Proc. Geol. Ass. Lond.*, **97**, 259–290. |
| SYKES, R. M. & CALLOMON, J. H. | 1979 | The *Amoeboceras* zonation of the Boreal Upper Oxfordian. *Palaeontology*, **22**, 839–903. |
| SYLVESTER BRADLEY, P. C. | 1968 | The Inferior Oolite Series. *In* Sylvester Bradley, P. C. & Ford, T. D. (Eds.) *The Geology of the East Midlands.* Leicester Univ. Press, 211–226. |

SYLVESTER BRADLEY, P. C. & FORD, T. D. (Eds.) 1968 *The Geology of the East Midlands.* Leicester Univ. Press, 400 pp.

TALBOT, M. R. 1973 Major sedimentary cycles in the Corallian beds (Oxfordian) of southern England. *Palaeogeog., Palaeoclimatol. Palaeoecol.,* **14**, 293–317.

TAYLOR, J. H. 1949 Petrology of the Northampton Sand Ironstone Formation. *Mem. geol. Surv. G.B.*

TORRENS, H. S. 1965 Revised zonal scheme for the Bathonian stage of Europe. *Rep. Carpatho–Balkan geol. Ass. 7th Congr. Sofia,* Part 2, vol. **1**, 47–55.

1968 The Great Oolite Series. *In* Sylvester Bradley, P. C. & Ford, T. D. (Eds.) *The Geology of the East Midlands.* Leicester Univ. Press, 227–263.

1969 Field meeting in the Sherborne–Yeovil district. *Proc. Geol. Ass. Lond.,* **80**, 301–330.

TORRENS, H. S. & CALLOMON, J. H. 1968 The Corallian Beds, the Ampthill Clay and the Kimmeridge Clay. *In* Sylvester Bradley, P. C. & Ford, T. D. (Eds.) *The Geology of the East Midlands.* Leicester Univ. Press, 291–299.

TOWNSON, W. G. 1975 Lithostratigraphy and deposition of the type Portlandian. *J. geol. Soc. Lond.,* **131**, 619–638.

TYSON, R. V., WILSON, R. C. L. & DOWNIE, C. 1979 A stratified water column environmental model for the type Kimmeridge Clay. *Nature, London,* **277**, 377–380.

WHITAKER, A. 1975 A postulated post-Hercynian rift valley system in southern Britain. *Geol. Mag.,* **112**, 137–149.

WHITAKER, A. (Ed.) 1985 *Atlas of Onshore Sedimentary Basins in England and Wales.* Blackie, Glasgow and London, 71 pp.

WIGNALL, P. B. 1989 Sedimentary dynamics of the Kimmeridge Clay: tempests and earthquakes. *J. geol. Soc. Lond.,* **146**, 273–284.

WILSON, R. C. L. 1968a Upper Oxfordian palaeogeography of southern England. *Palaeogeog., Palaeoclimatol., Palaeoecol.,* **4**, 5–28.

1968b Carbonate facies variation within the Osmington Oolite Series in southern England. *Palaeogeog., Palaeoclimatol., Palaeoecol.,* **4**, 89–123.

WIMBLEDON, W. A. & COPE, J. C. W. 1978 The ammonite faunas of the English Portland Beds and the zones of the Portlandian Stage. *J. geol. Soc. Lond.,* **135**, 183–190.

WOBBER, F. J. 1965 Sedimentology of the Lias (Lower Jurassic) of South Wales. *J. Sediment. Petrol.,* **35**, 683–703.

WOODLAND, A. W. (Ed.) 1971 The Llanbedr (Mochras Farm) borehole. *Rep. Inst. geol. Sci.,* **71/18**, 115 pp.

WORSSAM, B. C. & IVIMEY-COOK, H. C. 1971 The stratigraphy of the Geological Survey borehole at Warlingham, Surrey. *Bull. geol. Surv. G.B.,* No. **30**, 1–111.

WRIGHT, J. K. 1968 The stratigraphy of the Callovian rocks between Newtondale and the Scarborough coast, Yorkshire. *Proc. Geol. Ass. Lond.,* **79**, 363–399.

1972 The stratigraphy of the Yorkshire Corallian. *Proc. Yorkshire geol. Soc.,* **39**, 225–264.

1977 The Cornbrash Formation (Callovian) in North Yorkshire and Cleveland. *Proc. Yorkshire geol. Soc.,* **41**, 325–346.

1978 The Callovian succession (excluding Cornbrash) in the western and northern parts of the Yorkshire Basin. *Proc. Geol. Ass. Lond.,* **89**, 239–261.

ZIEGLER, B. 1962 Die Ammoniten–Gattung *Aulacostephanus* im Oberjura. *Palaeontographica,* **119A**, 1–72.

# 12

# THE CRETACEOUS
## P. F. Rawson

### Introduction

Outcropping in England's white cliffs and rolling downs, the Chalk symbolises, and gives its name to, the Cretaceous system (Fig. 12.1). In Cretaceous times England and Wales lay about ten degrees south of their present position and enjoyed a noticeably warmer climate. They also experienced significant tectonic activity as a result of plate movements to the west and south. The tectonic evolution of the area was controlled primarily by the progressive opening of the northern Atlantic (Ziegler 1982). During the early Cretaceous the resultant tensional stresses led to the continuation of rifting and block faulting, which

**Fig. 12.1.** The Chalk escarpment at Bratton Hill, Westbury, Wiltshire, marks the northwestern edge of the Salisbury Plain. The White Horse is one of many carvings on the Chalk escarpments of southern England; such carvings are said to have great antiquity testifying to the importance of the downs as ancient routeways (B.G.S. A.97111).

355

affected sedimentation significantly. Then in the mid-Cretaceous the rifting effectively ceased and the region began to behave as a passive continental margin. At about the same time plate convergence began in western Tethys and eventually resulted in a marked compressional pulse across north-west Europe during the Senonian (late Cretaceous: see Table 12.1 for stages).

**Table 12.1.** The Cretaceous stages

| | |
|---|---|
| UPPER CRETACEOUS | Maastrichtian ⎫<br>Campanian ⎪<br>Santonian ⎬ *<br>Coniacian ⎪<br>Turonian ⎭<br>Cenomanian |
| LOWER CRETACEOUS | Albian<br>Aptian<br>Barremian<br>Hauterivian<br>Valanginian<br>Berriasian/Ryazanian |

Note: "Berriasian" is the standard lowest Cretaceous stage term. Because of correlation problems at the Jurassic/Cretaceous boundary "Ryazanian" is often used for the basal Cretaceous stage in the Boreal Realm (including England). (* Senonian.)

These major tectonic events were accompanied by eustatic changes in sea level and overall the Cretaceous was an interval of significant sea-level rise. The combination of these two factors had a profound influence on the palaeogeographic evolution of England and Wales.

## Structural controls on Cretaceous sedimentation

During the early Cretaceous rifting phase a number of partially fault-bounded structural features, formed during the Triassic or Jurassic, continued to affect sedimentation (Fig. 12.2). The most conspicuous of these is the *London–Brabant Massif*, which formed a landmass during the early Cretaceous until it was submerged by the late Albian transgression. The western (English) part is generally referred to as the *East Anglian Massif* or *London Platform*; this was fringed to the west by an intermittent seaway, the '*Bedfordshire Straits*'. To the north of the massif lay the *East Midlands Shelf* and *Yorkshire Basin* while to the south the *Wessex Basin* (*sensu* Kent 1975 and Rawson *et al.* 1978) included two closely linked subsiding regions, the *Wealden* and *Vectian* ( = Wessex *sensu* Allen 1976 and others) *Basins*.

The late Albian transgression established a continuous marine regime from the North Sea across eastern England to the Channel (Fig. 12.3d). The transgression coincided closely with the cessation of faulting in many areas, so as the former highs were buried their influence became more passive, limited primarily to a continuing influence on sediment thickness. In general, the late Cretaceous was a tectonically quiet phase during which there was a tendency towards simple basinal downwarp in the North Sea and adjacent onshore areas foreshadowing the Tertiary basinal pattern (Kent 1975). Basin inversion (the change of an area from one of subsidence to one of relative uplift) commenced during very late Cretaceous (Senonian) times as a response to a compressional pulse from western Tethys. The offshore Sole Pit Basin was inverted at this time. The role of the East Anglian Massif reversed a little earlier, during the mid-Cretaceous: its late Albian covering of Gault Clay is a deeper water sediment than the pebbly chalk (Hunstanton Formation) facies over the East Midlands Shelf to the north, and the Lower Chalk facies is of deeper water type compared with that over the former shelf (Jeans 1980).

## Palaeogeography

A very late Jurassic (mid-Volgian) eustatic fall in sea level was accompanied by a sudden increase in tectonic activity (late Cimmerian movements). Hence by the beginning of the Cretaceous much of England and Wales had become land. The palaeogeographical pattern thus established (Fig. 12.3a) continued through the early Cretaceous until the end of the Barremian. During this 30 million year interval sedimentation was restricted to two main areas separated by the East Anglian Massif. To the north, marine deposition extended from north Norfolk to Yorkshire and over the adjacent southern North Sea Basin, while to the south a thick pile of non-marine sediments accumulated in the Wessex Basin and contiguous areas. Landmasses adjacent to the northern area were probably low-lying, and there is little evidence of a major river system. In contrast, to the south rivers flowed into the non-marine basin off highland areas (the 'London uplands') intermittently developed along the southern part of the East Anglian Massif, and from the Cornubian and Armorican highlands.

Sediment and faunal distributions suggest that in this early phase of Cretaceous history the westerly limits of deposition may have lain not far from the western margins of the present-day outcrop (Figs. 12.3a, 12.3b). However, non-marine and probably some marine sediments accumulated in restricted areas in the Celtic Sea and Western Approaches Basins. The pattern began to change at the beginning

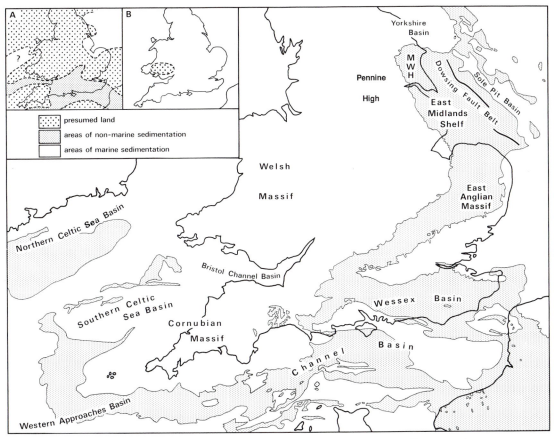

**Fig. 12.2.** Outcrop and pre-Pleistocene subcrop of Cretaceous sediments and their structural setting. Inset A: latest Jurassic – earliest Cretaceous palaeogeography (minimum transgression). Inset B: late Campanian palaeogeography (maximum transgression). MWH = Market Weighton High.

of the Aptian, when a marine transgression flooded southern England. Subsequent sea-level rises from then until late in the Cretaceous resulted initially in the re-establishment of links between the areas north and south of the East Anglian Massif by late Aptian times (Fig. 12.3c). This was followed by submergence of the Market Weighton High in mid-Albian and the East Anglian Massif in late Albian times, by which time the western shoreline was also retreating (Fig. 12.3d). Westward encroachment of the sea continued through much of the late Cretaceous until by the time of maximum transgression in the late Campanian probably only the highest parts of the Welsh Massif remained above sea level (Fig. 12.2B: Cope 1984).

## The stratigraphic framework

In this chapter the geology of the onshore areas is described first, to provide a framework for the succeeding discussion of Cretaceous events. An extensive spate of studies in the last 20 years has established a detailed and reliable lithostratigraphy and biostratigraphy for the Lower Cretaceous, but this has still been only partially achieved for the Upper Cretaceous. Much primary stratigraphical information was published in early Geological Survey sheet memoirs, culminating in Jukes-Browne and Hill's (1900–1904) collation on the Gault, Upper Greensand and Chalk. For some of the key areas from Cambridge south-

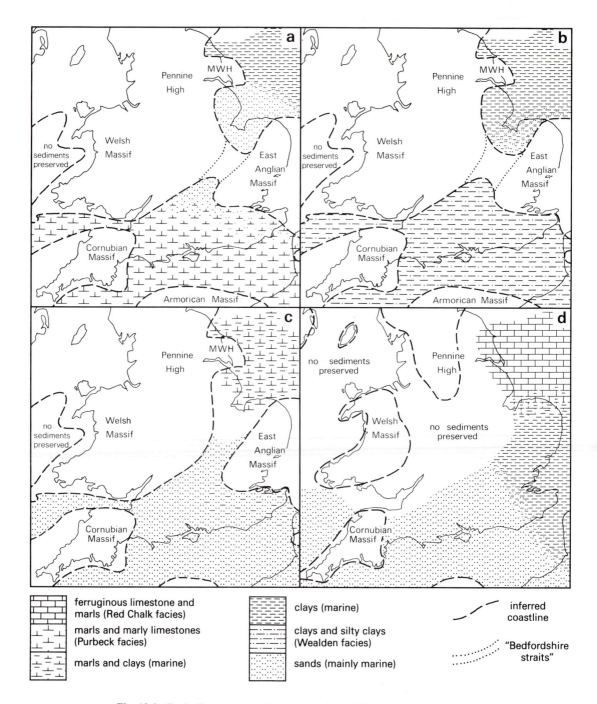

**Fig. 12.3.** Early Cretaceous palaeogeography and facies.

    **a.** mid-Ryazanian time
    **b.** mid-Hauterivian time
    **c.** late Aptian time
    **d.** late Albian time

Note: the geographical reconstructions are 'conservative'. 'Radical' reconstructions would leave only the core of highs as land. Facies distributions generally support the 'conservative' reconstructions.

wards there are recent British Geological Survey memoirs. Rawson *et al.* (1978) summarised the correlation of the whole Cretaceous but did not attempt to rename the lithological units to accord with a strictly lithostratigraphical nomenclature. The spirit of the Holland *et al.* (1978) 'permissive' guide to stratigraphical procedure is followed here: most of the familiar, long-established and generally well-understood names are retained but used in a strictly lithostratigraphical sense, with the appropriate hierarchical term added. This follows a procedure already begun, for example, by Allen (1976) for the Wealden, and appears infinitely preferable to the invention of new names based on newly defined type localities. The major units into which the English Cretaceous is divided are summarised in Tables 12.2 and 12.3.

All localities mentioned in the text are shown on one or other of the text-figures.

# THE FOSSIL RECORD

## Fossil assemblages

Molluscs, brachiopods and echinoids are probably the most conspicuous macrofossils in the English Cretaceous, and there are many important microfossil groups. On a global scale, the most significant biological events were the diversification of angiosperms during the early Cretaceous, the rapid increase in numbers and diversity of planktonic foraminifera and calcareous nannofossils during the middle of the Cretaceous and the decline and eventual extinction of many fossil groups (especially the ammonites and giant reptiles) during the late Cretaceous. Most of these events are reflected in the English sequences and one was of profound importance, namely the 'blooming' of calcareous nannofossils to form a thick ooze – the Chalk.

In general the most thoroughly studied fossil groups are those which have been used in correlation (see Tables 12.2 and 12.3). In contrast with the Jurassic, the study of Cretaceous assemblages and their relationships with lithofacies is in its infancy, though Kennedy (1978) has made a pilot qualitative review of selected assemblages from southern England.

## Biogeography

The geographical relationships of many faunas and floras have attracted more interest. On a broad scale the distribution of taxa in the English Cretaceous reflects global biogeography. In the northern hemisphere two faunal realms are recognised, the Tethyan and Boreal. They were distinguished on the basis of ammonite distributions but many other fossil groups show a comparable distributional pattern. Primary controls on distribution were probably climate and palaeogeography.

In pre-Aptian times north-west Europe, including eastern England, formed a peripheral area of the Boreal Realm, the West European Province (Rawson 1981). Although faunas were dominantly boreal there were a few endemic forms and there is clear Tethyan influence among many fossil groups. The occurrence of Tethyan ammonites facilities correlation with standard sections in southern France (Kemper *et al.* 1981).

Southern England at that time was non-marine, but transgressions in the Aptian flooded the area and introduced faunal influences from Spain and the Paris Basin. As transgressive pulses continued over Europe during the Aptian and Albian to flood increasingly larger areas, so the biogeographical pattern changed. Thus by the middle Albian the whole of western Europe belonged to the European (or hoplitinid) ammonite province of the Boreal Realm (Owen 1973). Tethyan-derived ammonites co-existed with boreal ones, especially during the late Albian. A similar pattern occurred during the late Cretaceous, when western Europe occupied part of a single province within the Boreal Realm. Two sub-provinces can be recognised, a northern one embracing Yorkshire to north Norfolk (and extending through north Germany to eastern Europe) and a southern one including the rest of England and the Paris Basin. The sub-provinces are distinguished by differences in ratios of various invertebrate groups and by differing echinoid faunas. Again, some Tethyan influence is apparent at times, though not as well documented as for the early Cretaceous.

## Stages

The standard Cretaceous stages are summarised in Tables 12.1, 12.2 and 12.3; there are no stratotype sections in Britain and because most are in the Tethyan Realm there are problems in defining stage boundaries accurately within the English sequences. Current usage of stage terms was discussed fully by Rawson and co-workers (1978) and those usages are followed here with only minor modification (indicated on Tables 12.2, 12.3).

## Zones

Ammonites provide a standard zonal scheme for the English Lower Cretaceous (Table 12.2) and in practice are common enough in most marine strata for correlation to be attempted with tolerable accuracy. Belemnites are beginning to be used too (e.g. Rawson & Mutterlose 1983). Among microfossils, foraminiferal assemblages of Ryazanian to Barremian age are fairly long-ranging, while some younger forms are more short-lived; detailed range charts are given by Hart *et*

**Table 12.2.** Subdivision and correlation of English Lower Cretaceous rocks (partly after Rawson *et al.* 1978, fig. 3).

| STAGE/ SUBSTAGE | AMMONITE ZONE | YORKSHIRE BASIN — Speeton | EAST MIDLANDS SHELF — Lincolnshire | EAST MIDLANDS SHELF — North Norfolk | WEALDEN BASIN — The Weald | VECTIAN BASIN — Isle of Wight | AMMONITE ZONE |
|---|---|---|---|---|---|---|---|
| UPPER ALBIAN | dispar | Hunstanton Fm 13m | Hunstanton Formation 7m | Gault Clay Fm 18m | Gault Clay Formation 90m (nodule beds at base) | Upper Greensand Formation 36m | dispar |
| UPPER ALBIAN | inflatum | | | | | | inflatum |
| MIDDLE ALBIAN | lautus | A beds 9m — minimus marls 6m; Greensand Streak | | | | Gault Clay Formation 30m | lautus |
| MIDDLE ALBIAN | loricatus | | | | | | loricatus |
| MIDDLE ALBIAN | dentatus | | | | | | dentatus |
| LOWER ALBIAN | mammillatum | | Carstone Formation 18m | Carstone Formation 18m | | Carstone Formation 22m | mammillatum |
| LOWER ALBIAN | tardefurcata | | | | Folkestone Fm 78m | Sandrock Formation 56m | tardefurcata |
| UPPER APTIAN | jacobi | ewaldi beds 3m (sequence incomplete) | sands, clays 5m | | Sandgate Formation 45m | Ferruginous Sands Formation 80m | jacobi |
| UPPER APTIAN | nutfieldiensis | | Sutterby Marl 3.5m | | | | nutfieldiensis |
| UPPER APTIAN | martinioides | | | | Hythe Formation 90m | | martinioides |
| LOWER APTIAN | bowerbanki | | | | | | bowerbanki |
| LOWER APTIAN | deshayesi | | | | | | deshayesi |
| LOWER APTIAN | forbesi | | | | Atherfield Clay Formation 50m | | forbesi |
| BARREMIAN | fissicostatus | upper B beds 9m | Skegness Clay 2m | | | | fissicostatus |
| BARREMIAN | bidentatum | | Roach Formation 15m | | | Vectis Formation 58m | Cypridea Assemblage — tenuis |
| BARREMIAN | stolleyi | | | | upper division | | tenuis |
| BARREMIAN | innexum | | Tealby Clay Formation (upper member) | | | | |
| BARREMIAN | denckmanni | cement beds 10m | | Dersingham Formation (with Snettisham Clay facies) 25m | | | spinigera |
| BARREMIAN | elegans | | | | | | spinigera |
| BARREMIAN | fissicostatum | lower B beds 21m | Tealby Lst Mbr 5m | | | | |
| BARREMIAN | rarocinctum | | | | | | |
| BARREMIAN | variabilis | | | | | | |
| UPPER HAUTERIVIAN | marginatus | beds C1 – C11 39m | Tealby Clay Formation (lower member) | | Weald Clay Formation 450m | Wessex Formation 400+m | clavata |
| UPPER HAUTERIVIAN | gottschei | | | | (with large 'Paludina' limestones) | | clavata |
| UPPER HAUTERIVIAN | speetonensis | | | | | | |
| LOWER HAUTERIVIAN | inversum | | | | | | |
| LOWER HAUTERIVIAN | regale | beds D1 – D2D 1m | Claxby Formation 6m | | lower division (with small 'Paludina' limestones) | | dorsispinata |
| LOWER HAUTERIVIAN | noricum | | | | | | dorsispinata |
| LOWER HAUTERIVIAN | amblygonium | | | | | | |
| UPPER VALANGINIAN | (faunal gap) | beds D2E – D8 13m | | Leziate Member 35m | Tunbridge Wells Formation 120m | | aculeata |
| UPPER VALANGINIAN | Dichotomites | | | | | | aculeata |
| LOWER VALANGINIAN | Polyptychites | | Hundleby Member 5m | | Wadhurst Fm 70m | | paulsgrovensis |
| LOWER VALANGINIAN | Paratollia | | | | Ashdown Formation 210m | | setina |
| UPPER RYAZANIAN | albidum | | Spilsby Fm (upper mbr) 11m | Mintlyn Member 15m | | 104m of undifferentiated Purbeck Group in the Arreton Borehole | granulosa fasciculata |
| UPPER RYAZANIAN | stenomphalus | | | | Durlston Formation 70m | | granulosa fasciculata |
| UPPER RYAZANIAN | icenii | | | | | | |
| LOWER RYAZANIAN | kochi | | | | | | |
| LOWER RYAZANIAN | runctoni | | | | | | |

Yorkshire Basin vertical labels: Speeton Clay Formation 102m; A beds 9m; B beds 40m; C beds 39m; D beds 14m.

Weald vertical labels: Weald Clay Formation 450m; Hastings Group 400m.

North Norfolk vertical label: Sandringham Formation (pars).

Notes: a. Correlation between the non-marine *Cypridea* (ostracod) zones and the standard (ammonite) zonal sequence is extremely tentative.

b. Solid lines between rock units indicate reasonably firm correlation with the appropriate zonal scale; dashed lines indicate a provisional correlation.

c. The Barremian ammonite zones and the position of the Hauterivian/ Barremian boundary follows Rawson & Mutterlose (1983).

*al.* (1981). Ostracods are also of value: eight marine Lower Cretaceous zones were recognised by Neale (1978) and the non-marine Purbeck/Wealden facies are zoned by cypridean ostracods (Anderson 1973; Kilenyi & Neale 1978). Correlation with non-marine facies in other parts of Europe has been attempted using the cyprideans but the only evidence for correlating between these non-marine beds and marine strata comes from plant microfossils (e.g. Worssam 1978). Two groups of marine microplankton that are being used increasingly for correlation are the calcareous nannofossils (e.g. Taylor 1982) and dinocysts (e.g. Rawson & Riley 1982).

The Upper Cretaceous macrofossil zones in current use (Table 12.3) are essentially those first worked out in France and introduced to England over 100 years ago by Barrois (1876). Index taxa include ammonites, belemnites, a bivalve, echinoids, crinoids and a bra-chiopod: this strange mixture primarily represents the vagaries of preservation and occurrence in the Chalk. In the past 20 years, work in Germany and North America in particular has demonstrated the value of inoceramid bivalves in late Cretaceous biostratigraphy, and there is little doubt that these will be used increasingly in England (Mortimore 1983, table 2). Belemnites and echinoids are also useful for more refined correlation of parts of the European Chalk. Among microfossils, planktonic foraminifers have provided an international scheme; its use in England is limited by the scarcity of the phylogenetically more advanced species characteristic of Tethys. Benthonic smaller foraminifers are being utilised more successfully (e.g. Carter & Hart 1977), but their ranges cannot be used south of the Paris Basin. Crux (1982) has produced a broad Cenomanian to Campanian calcareous nannofossil zonation.

**Table 12.3.** Subdivisions and correlation of the Upper Cretaceous Chalk Group.

| NORTHERN PROVINCE | | STAGE | SOUTHERN PROVINCE | | |
|---|---|---|---|---|---|
| LITHOLOGICAL UNIT | ZONE | | ZONE | LITHOLOGICAL UNIT | |
| | | MAASTRICHTIAN | | | |
| Flamborough Chalk Formation | higher zones beneath drift | CAMPANIAN | *mucronata* [a] | Portsdown Member 30m | Sussex White Chalk Formation |
| | | | *quadrata* [a] | Culver Member 115+ | |
| | *lingua* [b] | | *pilula* [d] | Newhaven Member 75m | |
| 300+m | *testudinarius* [e] | SANTONIAN | *testudinarius* [e] | | |
| | *socialis* [e] | | *socialis* [e] | | |
| Burnham Chalk Formation | *rostrata* [d] | CONIACIAN | *coranguinum* [d] | Seaford Member 90m | |
| | *cortestudinarium* [d] | | *cortestudinarium* [d] | Lewes Member 90m | |
| 150m | *planus* [d] | TURONIAN | *planus* [d] | | |
| Welton Chalk Formation | *lata* [f] | | *lata* [f] | Ranscombe Member 85m | |
| | *labiatus* [b] | | *labiatus* [b] | | |
| 53m | *geslinianum* [c] | UPPER CENOMANIAN | *geslinianum* [c] | Plenus Marls Formation | |
| Ferriby Chalk Formation | *trecensis* [d] | | *naviculare* [c] | Abbotts Cliff Chalk Formation 22m | |
| | *subglobosus* [d] | MIDDLE CENOMANIAN | *rhotomagense* [c] | East Wear Bay Formation 58m | |
| 28m | | LOWER CENOMANIAN | *mantelli* [c] | | |

Notes: a. All correlations are provisional because of uncertainties in the biostratigraphic scales.

b. The *geslinianum* Zone and the position of the Cenomanian/Turonian boundary are after Wright & Kennedy (1981).

c. The letters a–f against zonal indices indicate: a = belemnite; b = bivalve; c = ammonite; d = echinoid; e = crinoid; f = brachiopod.

# THE LOWER CRETACEOUS SERIES

Almost till the end of early Cretaceous times the East Anglian Massif separated two areas with completely different depositional histories: the 'Bedfordshire Straits' provided an intermittent link round the western margin of the massif.

# The northern region: East Midlands Shelf and Yorkshire Basin

North of the East Anglian Massif lay the East Midlands Shelf (Fig. 12.4), bounded to the north and east by active fault systems – the Howardian–Flamborough and contiguous Dowsing fault belts. Over the southern part of the shelf shallow marine sediments accumulated: this depositional area is generally referred to as the Spilsby Basin. The northern part of the shelf was emergent and formed the Market Weighton High, a hinge with a fault-bounded northern margin

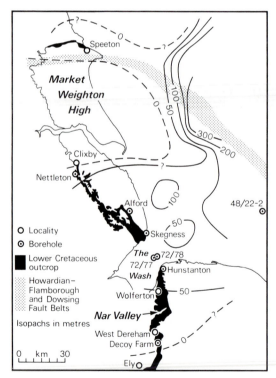

**Fig. 12.4.** Lower Cretaceous rocks of the East Midlands Shelf: outcrop; isopachs of pre-Albian sediments (modified from Kent 1980, fig. 5); localities.
For location of Fig. 12.4 see Fig. 12.6.

and a gentle southern slope merging into the Spilsby Basin.

## The Speeton Clay Formation

The downfaulted northern margin of the Market Weighton High formed a southern boundary to the Yorkshire Basin. Here an argillaceous sequence was deposited through most of early Cretaceous time, the Speeton Clay Formation. The formation crops out in the southern part of Filey Bay where it rests non-sequentially on the Kimmeridge Clay Formation, the boundary being marked by a 10 cm thick phosphatic nodule band. All six Lower Cretaceous stages are represented within the Speeton Formation, here only about 100 m thick, but there are several breaks in the succession and the higher beds (mid-Barremian to mid-Albian) are mainly hidden beneath landslips and are poorly known. Fossils are common at many levels and the Speeton section is the type locality for many of the English Valanginian to early Barremian ammonite zones.

Lamplugh (1889) divided the sequence, and his lithological divisions are still used (Table 12.2) with more detailed refinements summarised by Rawson et al. (1978) and Rawson and Mutterlose (1983). Phosphatic nodule beds and glauconite concentrates indicate pauses in sedimentation, and these together with mottled beds (Chondrites), silty layers and rapid vertical colour changes facilitate the lithological subdivision. Small-scale coarsening-upward cycles occur in the lower B beds and correlate with the north German Blatterton facies (Rawson & Mutterlose 1983).

The Speeton Formation is in slipper-clay facies (sensu Hancock 1976, Table 1) but water depths may have been less than 100 m, the clay accumulating close to a low-relief landmass (the Market Weighton High) supplying only clay-grade detritus to the adjacent basin. In the latest Ryazanian there was a brackish interlude, and in early Barremian times the bottom waters may have been oxygen-deficient during deposition of the 'Blatterton'.

Inland, the Speeton Formation thickens dramatically as a result of contemporaneous listric faulting, then thins further west to disappear near Knapton, where the overlying Hunstanton Member comes to rest on the Kimmeridge Formation.

## The pre-Albian rocks of the Spilsby Basin

There are no pre-Albian sediments over the Market Weighton High; they reappear in a different facies at Clixby, some 75 km south of Knapton at the northern margin of the Spilsby Basin. Within this basin accumulated a thin (maximum about 50 m) but varied series of shallow marine sediments whose interrelationships are summarised in Table 12.2. Facies and faunal distributions indicate nearshore environments in north Norfolk and south Humberside with a broad intervening area of offshore, inner to mid-shelf sediments. The western margin of the basin may have been only a few kilometres west of the present outcrop (Fig. 12.3) which forms a narrow band through Lincolnshire before passing beneath the Fen deposits to re-emerge as a short north-south strip in north Norfolk (Fig. 12.4). A limited number of boreholes penetrate the Chalk cover east of the outcrop so that a generalised cross-section of the whole embayment can be constructed (Fig. 12.5).

The succession begins with the widely distributed sands of the Spilsby Formation and its Norfolk correlatives. At the base a phosphatic nodule bed rests on the Kimmeridge Clay and similar beds higher in the sands mark widespread pauses in deposition (Fig. 12.5). The sands are generally medium-grained, glauconitic, incoherent sediments showing few sedi-

mentary structures. At some levels they are more argillaceous while calcareous concretions also occur and are the main source of the diverse bivalve faunas (Kelly 1983). The bivalve assemblages generally suggest an unstable substrate. Ammonites occur through most of the sequence and indicate a Volgian (latest Jurassic) to Ryazanian age (Casey 1973).

In the Skegness to Wash area the overlying strata are predominantly sandy clays with scattered iron ooliths and some sandy or calcareous horizons. To the north-west there is a gradual thinning with accompanying facies changes. Iron ooliths become more abundant and even form ironstones (e.g. the Claxby Ironstone Formation), impure limestones appear (the Tealby Limestone Member) and non-sequences become more significant as individual units lap out against the Market Weighton High. Erosion horizons associated with the non-sequences may extend over much of the basin (Fig. 12.5); those of mid-

**Fig. 12.5.** Section across the uppermost Jurassic and Lower Cretaceous rocks (Albian excluded) of the East Midlands Shelf. The position of the control points is shown in Fig. 12.4.

Age of the numbered unconformities:

9 early Albian ('mid-*tardefurcata* break') – base of Carstone Fm.
8 base late Aptian (base of *martinioides* Zone)
7 base Aptian (base of *fissicostatus* Zone)
6 mid-Hauterivian (base of *inversum* Zone)
5 latest Ryazanian (high *albidum* Zone)
4 late Ryazanian (base of *stenomphalus* Zone)
3 early Ryazanian (base of *kochi* Zone)
2 late Volgian (base of *preplicomphalus* Zone)
1 mid-Volgian (base of *oppressus* Zone)

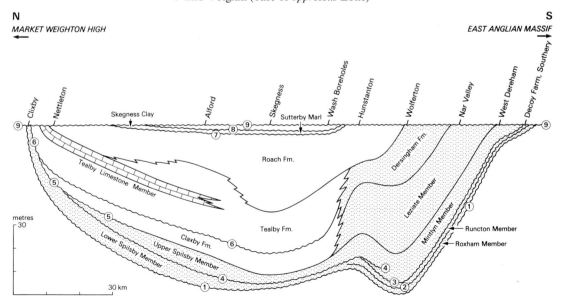

Hauterivian and early and late Aptian age are particularly prominent. Above the mid-Hauterivian erosion surface the Lower Tealby Clay Member represents the maximum spread of clays over the shelf (Fig. 12.3b) following a major sea-level rise. It is the nearest analogue to the Speeton Formation.

Southwards into Norfolk most horizons become more sandy until the (incomplete) southernmost sequence is almost wholly arenaceous. The main facies changes take place over the short distance between the Wash and Hunstanton boreholes (Fig. 12.5) and this together with the local thinning of the sequence at Hunstanton suggests that the Wash Line ( = North Norfolk 'Swell'), which influenced Cenomanian facies and thicknesses (Carter & Hart 1977; Jeans 1980), was already active in early Cretaceous times. The line may be related to a pre-Mesozoic rise beneath north Norfolk which brings Precambrian rocks close to the surface in the North Creake borehole (Kent 1947; Dunning *in* Harris *et al.*, p. 95, 1975).

Offshore from north Norfolk and Lincolnshire, the Spilsby Formation sand facies continues towards the Dowsing Fault Line at the edge of the shelf (Fig. 12.3a) but the overlying sediments are argillaceous (Figs. 12.3b, 12.3c) and are assigned to the Speeton Clay Formation (Rhys 1974). Thicknesses vary considerably but in well section 48/22-2 off the north-east Norfolk coast and almost on the Dowsing Fault Line the Spilsby Formation is unusually thick (24 m) for the offshore area, and is overlain by 172 m of clay assigned to the Speeton Formation.

The source of sediment supply to the East Midlands Shelf remains problematic. Cross-bedding in the Sandringham Formation (mainly the Leziate Member) shows a predominantly easterly dip (Schwarzacher 1953). Heavy mineral suites in the Spilsby Formation (Versey & Carter 1926; Ingham 1929) include abundant fresh kyanite and staurolite which diminish northwards as if they were derived from the London Brabant Massif. In contrast, Allen (1967) suggested a Scottish-Scandinavian origin for these metamorphic minerals while regarding the sands as 'a hotch-potch from far and near, mixed together by longshore, inshore and offshore processes'. The mixing may have been in part under tidal influence for herringbone cross-bedding is common in the well-sorted, quartzose sands of the Leziate Member in north Norfolk.

## The Carstone and Hunstanton Formations

An important early Albian transgression marked the beginning of a new and more widespread phase of sedimentation. A basal sand, the Carstone Formation (maximum thickness 12 m), oversteps earlier Cretaceous beds to rest on the Jurassic at both ends of the outcrop (Fig. 12.5): it extends from the northern margin of the East Anglian Massif to the southern part of the Market Weighton High, and scattered patches occur over the crest of the high. The same event is represented in the Speeton Clay Formation by the 9 cm thick 'greensand streak'.

The Carstone Formation is generally a medium to coarse, pebbly, glauconitic, quartz sand stained brown by a limonitic cement. Distinct pebble bands occur and towards the margins of the outcrop the base is conglomeratic. Cross-bedding is just visible at Hunstanton, where dips are apparently to the south-west (Versey & Carter 1926), and Nettleton, where there is a southerly dip. Heavy mineral suites are similar to those of the earlier Cretaceous sands and some of the sediment may have been reworked from them as derived Lower Spilsby ammonites occur in the basal Carstone conglomerate in south Humberside. Indigenous fossils are known only from Humberside, where early Albian brachiopods occur in a slightly argillaceous, finer-grained facies.

Over most of the shelf the Carstone Formation grades up into the Hunstanton Formation (formerly 'Red Chalk'), but at its southern extent it is overlain by the Gault Formation. The Hunstanton Formation was regarded as the basal member of the Ferriby Chalk Formation by Wood and Smith (1978) but as a separate formation by Jeans, Kent and Rawson (in discussion of Wood & Smith), and the latter view is followed here. The formation extends from north Norfolk to Speeton; its continuity over the Market Weighton High marks the final submergence of that feature by the mid-Albian transgression (Fig. 12.3d).

The Hunstanton Formation occurs in two distinct facies. Over the East Midlands Shelf it is a thin (generally 1–2 m) sequence of impure limestones and calcareous mudstones, usually of a bright red colour. Penecontemporaneous reworking is indicated by included Carstone sand grains and pebbles, which diminish in quantity upward, and by epizoan-encrusted chalk pebbles. Algal mats indicate deposition in the photic zone (Jeans 1973; Gallois & Morter 1982). The occurrence of both Middle and Upper Albian ammonites and belemnites shows that the sequence is highly condensed.

The facies changes in character as the Hunstanton Formation expands into the Yorkshire Basin, where at Speeton it is about 13 m thick. Here a gradual passage from Speeton Formation clays through red marls to more calcareous beds marks an important change from terrigenous to pelagic sedimentation which coincides closely with the late Albian transgression. 'Griotte' chalks (nodular chalks in which calcareous 'nodules' are surrounded by anastamosing marl seams) are common, and overall the facies invites comparison with the *Ammonitico Rosso* of the Mediterranean Mesozoic (Eller 1981). Of particular interest are some thin 'breccia nodule bands' composed of allochthonous nodules, apparently transported downslope as a result of earthquakes along the Howardian–Flamborough fault system (Jeans 1973, Eller 1981).

Correlative red marls with subordinate chalks are widely distributed over the North Sea region (Red Chalk and Rødby Formations).

In north Norfolk the Hunstanton Formation passes laterally into the Gault Clay Formation. Much of the East Anglian Gault is composed of small-scale (generally 1–2 m) fining-upward cycles, each with a basal erosion surface (Gallois & Morter 1982). The Lower Gault is only a few metres thick, but it oversteps the Carstone Member before being overstepped in turn towards the East Anglian Massif by transgressive Upper Gault. The Upper Gault (late Albian) transgression flooded the massif to link with the seas over southern England (Fig. 12.3d).

## The 'Bedfordshire Straits'

The name 'Bedfordshire Straits' is applied to the low-lying area fringing the western margin of the East Anglian Massif (Figs. 12.3, 12.6). Periodic spillovers from the north across this region are indicated by marine or near-marine horizons showing an apparent northward increase in salinity in the Purbeck and Wealden beds of the Wessex Basin. The earliest pulse, the Cinder Beds horizon, may be represented within the straits by *remanié* Spilsby Formation fossils at the base of the Aptian deposits of Upware and Potton.

The latter locality is only 40 km north-east of Stewkley, where the Whitchurch Sands Formation may be of Cinder Beds age.

During the Aptian a complex pattern of deposition and erosion set in within the straits. Early Aptian deposition took place as far south as Potton; the sediments were then reworked and their remnant faunas incorporated in the base of late Aptian sands. These, the Woburn (or Potton) Sands Formation, occur from near Upware to Leighton Buzzard (Fig. 12.6) and are up to 60 m thick. They are mainly coarse, yellow or silver sands, often with large-scale cross-bedding. They contain local fuller's earth (see page 380) seams and beds of iron-cemented 'carstone'. Nodule beds at the base have yielded rich derived and indigenous faunas at Little Brickhill and Upware (Keeping 1883; Casey 1961).

Sedimentation also commenced south of the straits at this time, notably in the vicinity of Faringdon (Fig. 12.7), so clearly they had opened again by the beginning of the late Aptian. They probably remained open during the early Albian, though deposits of that age are known only from the vicinity of Leighton Buzzard. Here a thin but complex series of sediments (Owen 1972) sandwiched between the Woburn Formation and the Gault Formation include typical Carstone facies, with lenticles of fossiliferous Shenley Limestone (*regularis* Subzone). Elsewhere, any Aptian/Albian Lower Greensand patches that may occur have been overstepped by the Gault. With the Gault transgressions the separate identity of the straits was no longer maintained as they merged with the newly flooded region to the east.

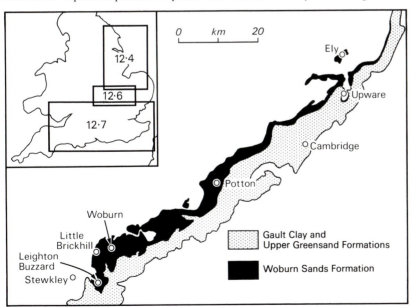

**Fig. 12.6.** Lower Cretaceous of the 'Bedfordshire Straits'; outcrop and localities. Inset shows boundaries of Figs. 12.4; 12.6; 12.7.

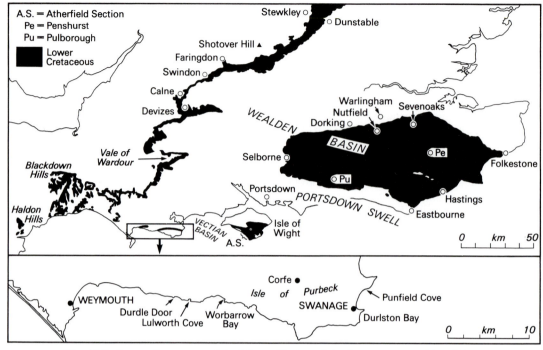

**Fig. 12.7.** Lower Cretaceous of the Wessex Basin: sub-basins, outcrop and localities.

# The southern region: Wessex Basin

Throughout the Wessex Basin (Fig. 12.7) there is the same general sequence, non-marine beds (Purbeck and Wealden facies) being followed by the marine Lower Greensand Group and Gault and Upper Greensand Formations. However, the development of more rapidly subsiding regions within the main basin resulted in significant differences in thickness and facies, and sedimentation may have been influenced by contemporaneous faulting, especially along the southern margin of the East Anglian massif (Stoneley 1982, Chadwick 1986).

Within the Wealden Basin maximum sedimentation occurred along what is now the axis of the Wealden anticline, especially in the more westerly part (Figs. **12.7a, 12.8b)**, so that the greatest thicknesses of preserved sediments are in the western Weald. In the Vectian Basin, maximum sedimentation probably occurred in the area from east of Swanage to south of the Isle of Wight. Between the Wealden and Vectian Basins lay the Portsdown Swell, which influenced sedimentation as far east as Eastbourne (Young & Monkhouse 1980); it was normally submerged but may have formed a land area during part of Lower

Greensand times and on other occasions hindered the movement of sediment from one basin to the other.

Wessex Basin exposures are limited to parts of the Weald, the western outliers and a narrow strip from the southern part of the Isle of Wight to Dorset and south Devon, but a few deep boreholes through the Chalk give glimpses of the covered area.

## The non-marine sequence: Purbeck and Wealden facies

The non-marine sediments of the Wessex Basin fall into two divisions, the lower calcareous and argillaceous and the upper argillaceous and arenaceous. To these the names 'Purbeck' and 'Wealden' have long been applied, either as lithostratigraphic names or as facies terms. The latter usage remains valuable though 'Purbeck' is also used as a group name. The boundary between the Purbeck and Wealden facies is transitional and therefore hard to define, while interpretations are further complicated by the significant lateral variation in facies between the Wealden and Vectian Basins. Lithostratigraphic subdivisions currently used are summarised in Table 12.2 and discussed below.

Biostratigraphic correlation within the Purbeck and Wealden facies is based primarily on ostracods. In addition to broad zones based on *Cypridea* (Table 12.2) a much finer division is founded on the occurrence of faunicycles. Each (named) faunicycle consists of a less saline C-phase (*Cypridea* phase) and a more saline S-phase fauna (see Anderson 1971). Most of the faunicycles occur in both the Wealden and Vectian Basins, and give a clear indication of the fluctuating salinities at this time. Palynofacies also suggest salinity fluctuation, especially in the Wealden Formation (Batten 1982). Both ostracods and palynofacies indicate increasing salinity towards the 'Bedfordshire Straits' to the north-west.

The ostracod assemblages have also been used to correlate with non-marine facies in other parts of Europe and even farther afield (Anderson 1973). Correlation between the non-marine and marine sequences is much more difficult. Palynomorphs, together with near-marine ostracods in the S-phases,

provide the firmest clues. In addition some of the quasi-marine bands may be local reflections of broader-scale transgressions: these become more marked north-westwards and are seen most clearly in the Warlingham borehole (Worrsam & Ivimey-Cook 1971). Their use has to be treated with great caution, as evidenced by the varying recent interpretations on some of the Purbeck quasi-marine beds.

Combining the various lines of evidence suggests that the Purbeck Group spans the Jurassic-Cretaceous boundary (mid-Volgian to earliest late Ryazanian), the Hastings Group is approximately late Ryazanian and Valanginian and the Weald Clay Formation Hauterivian to Barremian. An important brackish-marine band at the base of Worssam's (1978) upper division of the Weald Clay Formation may represent a major mid-Hauterivian sea-level rise (Table 12.2).

*The Purbeck Group.* The Purbeck Group rests comfortably but with sharp contact on the Portland Stone

**Fig. 12.8.** The Wealden Basin: isopachs and current directions.

**a.** Reconstructed Purbeck Group isopachs (simplified from Howitt 1964, fig. 7a).

**b.** Reconstructed Weald Clay Formation isopachs (after Worssam 1978, fig. 2).

**c.** Hastings Group/Weald Clay Formation current directions (after Allen 1976, fig. 3).

**d.** Folkestone Formation current directions (after Narayan 1971, fig. 9). (Fa = Farnham;  Fo = Folkestone;  Ha = Hastings;  W.B. = Warlingham borehole.)

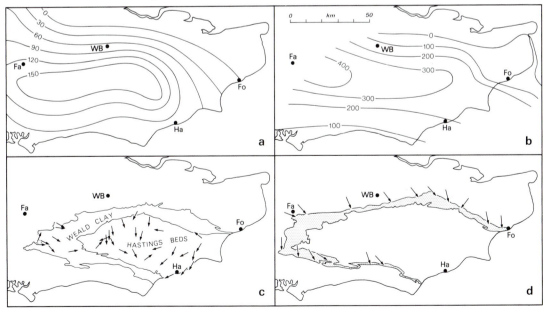

Formation. It consists of a varied series of non-marine limestones, marls and shales, with evaporites near the base. Two main divisions are recognised, the Lulworth and Durlston Formations (Casey 1963; Townson 1975). For a long time these rocks were placed in the Jurassic and the Jurassic/Cretaceous boundary drawn at the base of the Hastings Goup (itself an arbitrary line). However, Casey (1963) suggested that the Cinder Bed at the base of the Durlston Formation represents a brief marine incursion from the north which probably marks the basal Cretaceous transgression, and that the Durlston Formation is therefore wholly Cretaceous in age. That position is followed here though the exact age of the Cinder Beds event is debatable (Casey 1973; Rawson & Riley 1982; Wimbledon & Hunt 1983).

Purbeck strata occur throughout the Wessex Basin but the influence of the Portsdown Swell was already apparent, forming a region of reduced sedimentation separating two subsiding basins. Exposure in the Wealden Basin is very limited but there are good coastal exposures in the Vectian Basin, especially in the type area of the Isle of Purbeck. Here the thickest development is at Swanage (Fig. 12.9) where numerous lithological subdivisions are distinguished (see Melville & Freshney 1982); despite lateral facies changes and thinning, some of these can be traced westward at least as far as Lulworth Cove.

The Cretaceous Durlston Formation consists mainly of limestones and shales with several distinctive fossil horizons. The basal Cinder Bed is an oyster (*Liostrea distorta* J. de C. Sowerby) lumachelle which, in Durlston Bay, has yielded the regular echinoid *Hemicidaris purbeckensis* Forbes. Equivalent beds in the Vale of Wardour are sandy and have a richer molluscan fauna, while farther north still, in outliers of the supposedly contemporaneous Whitchurch Formation extending as far north as Stewkley in Buckinghamshire, the faunas are more-or-less marine (Casey & Bristow 1964).

Some higher horizons, especially the Scallop Bed, also yield marine or quasi-marine bivalves, but the bulk of the Durlston Formation is characterised by non-marine ostracods and gastropods.

In the Wealden Basin the Purbeck Group is visible only in three small inliers in Sussex (Howitt 1964), though it has been penetrated by several boreholes. The maximum thickness (171 m) was in the Penshurst borehole. Strata above the Cinder Bed correlative are placed in the Durlston Formation but lithologies differ noticeably from those in the Vectian Basin, especially in the occurrence of cross-bedded sandstones capping small-scale cycles in the 'Arenaceous Beds' and 'Beef Beds' (Anderson & Bazley 1971, p. 13). In many aspects they more closely resemble Wealden lithologies, which led Allen (1976) to place the Durlston

Formation in the Hastings Group. The current correlation between the two basins (Table 12.2) by means of facies and ostracods is questioned by Wimbledon and Hunt (1983), who point out discrepancies between ostracod and palynological correlations and suggest that the Purbeck beds of the Wealden Basin are slightly younger than previously supposed.

*The Wealden facies.* Because of the facies differences between the Wealden and Vectian Basins, different lithostratigraphic names are used in the two basins. In the Wealden Basin two units are distinguished, a lower of ferruginous sandstones, siltstones and shales called the Hastings Group, and an upper, more argillaceous unit, the Weald Clay Formation. In the Hastings Group the thick, arenaceous Ashdown and Tunbridge Wells Formations are separated by a transgressive argillaceous unit, the Wadhurst Clay Formation. This clay extends over the whole Weald. A later transgression in the western Weald is marked by the Grinstead Clay Formation, which splits the Tunbridge Wells sands there before petering out eastwards. Hence in the eastern Weald the upper and lower members of the Tunbridge Wells Formation cannot be separated. Although these major units can be placed in vertical succession their boundaries are diachronous: the vertical lithological changes represent three major

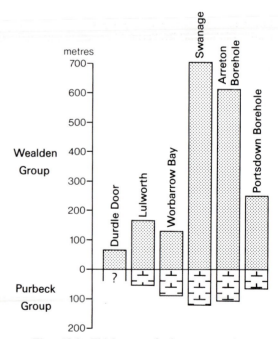

**Fig. 12.9.** Thickness of the non-marine strata. (Purbeck–Wealden) in the Wessex Basin.

coarsening-upward cycles. Superimposed on this broad cyclicity is a smaller-scale cyclicity which, coupled with lateral facies changes, results in the recognition of numerous minor, local subdivisions within the Hastings Group.

The Weald Clay Formation is less varied but two major cycles occur, the lower characterised by small-'*Paludina*' limestones and the higher by large '*Paludina*' limestones. Small-scale rhythms are superimposed on these (Worssam 1978). Red clay bands are developed in more marginal areas and are useful for local correlation. Sand bodies still occur but are much less prominent than in the Hastings Group.

The petrography and sedimentology of the Wealden facies of the Wealden Basin have attracted much interest. Many individual sands are of limited geographical extent, but some pebble beds are of remarkably wide occurrence and provide regional markers at the base of major cycles (e.g. top Ashdown pebble bed). Provenance studies on clasts and heavy minerals provide information on the origin of the sediments and the nature of the sourcelands (e.g. Allen 1967). This, together with evidence of bedload transport directions (Fig. 12.8c) and other sedimentological evidence, has resulted in a model for the depositional environment discussed below.

In the western outliers the Wealden may be represented locally, most probably by the Shotover Ironsands Formation of the Shotover region, Oxfordshire. This area probably lay near the western margin of the Wealden Basin.

The Wealden of the Vectian Basin is well exposed in coastal sections. The succession is again divided into two units, a lower of marls and sands and an upper of shales. Daley and Stewart (1979) named the former the Wessex Formation (formerly 'Wealden Marls') and the latter the Vectis Formation (formerly 'Wealden Shales'). In the Isle of Wight the Wessex Formation may be as much as 550 m thick (only the top 170 m are exposed), and is composed mainly of marls and marly shales with subordinate sands. The latter include the Sudmoor Point Member, which consists predominantly of point bar sands deposited in a meander belt at least 1·8 km wide (Stewart 1981). Traced westward into Dorset the formation becomes more varied – the 'variegated marls and sands' (see Arkell 1947). Sandstones, grits and conglomerates appear and these coarsen and increase in proportion westward as the overall thickness decreases. There is rapid and irregular variation in lithologies so that most subdivisions cannot be traced far laterally, but a band of coarse grit with iron cement extends from Swanage to Durdle Door, a distance of 23 km (Arkell 1947). The formation disappears west of Durdle Door, through erosion and Gault overstep.

Throughout the Vectian Basin the sediments are rich in kaolinite, felspars and tourmaline from the Cornubian highlands, and there is evidence of derived Kimmeridgian material in the Isle of Wight. Fossils are not common, but ostracods, *Unio* and gastropods occur. The dinosaurs *Iguanodon* and *Hypsilophodon* and footprints of *Megalosaurus* have been recorded.

The Vectis Formation consists of dark shales with occasional thin shelly limestones and minor sands. It is fairly uniform through most of the basin but in the vicinity of Corfe the shales pass laterally into sandy clays and shales difficult to separate from the variegated beds below. There is a gradual westerly thinning from a maximum of 58·5 m at Atherfield to only 3·5 m at Corfe: the beds are not known farther west.

In the Isle of Wight much of the formation consists of thin (70–90 cm) fining-upward cycles (sandy mud to mud). The Barnes High Sandstone Member (6 m), towards the base of the formation, is a coarsening-upward sequence: bedding structures change upwards from lenticular bedding through wavy bedding to large-scale cross-bedding.

Ostracods are abundant in the Vectis Formation, while the thin shelly limestones are composed mainly of *Viviparis*. Brackish beds with *Liostrea* and other mollusca occur in the top of the Shepherds Chine Member.

While the Purbeck facies was deposited in a semi-arid environment, Wealden sediments must have accumulated under a much wetter regime. Increased rainfall apparently resulted from periodic rejuvenation (by block faulting) of the highs surrounding the Wessex Basin, so that rivers discharged sediment over the basin at very fluctuating rates.

The nature of the Wealden environment is still debatable: deltas prograding into a landlocked lake (e.g. Allen 1959; Taylor 1963) became an inadequate model with the recognition of exposure and shallow water conditions even in the 'pro-delta' clays. Hence Allen (1976, 1981) has postulated that the Wessex Basin was an embayment opening out towards the 'Boreal sea' (Fig. 12.10). In this model, arenaceous sediments resulted from phases of uplift of surrounding massifs, when major streams deposited a broad zone of braided flood-sands traversed by channels. Towards the centre of the basin this 'braidplain' zone passed into a meander plain region of silts and clays deposited in sinuous streams and numerous lakes. During particularly wet (flood) phases the braidplain sands from the north (London Uplands) and south (Armorica) transgressed over the meander plain sediments to meet and even overlap along the axis of the basin. In the north-western part of the basin (i.e. nearer to the sea) lay lagoons and bays, where argillaceous sedimentation predominated. Occasionally this clayey facies transgressed rapidly over the sandy facies to cover much of the Wealden Basin

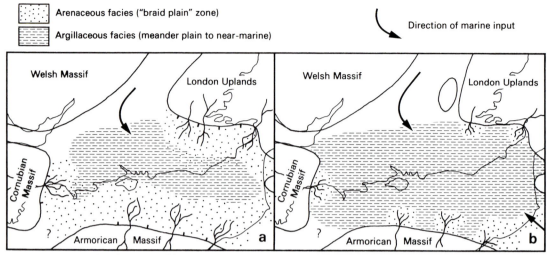

**Fig. 12.10.** Wealden environments during formation of (a) arenaceous facies and (b) argillaceous facies.

(Wadhurst, Grinstead and Wealden Formations). The transgressions may have resulted from a reduced input of sand due to down-faulting of bordering highs rather than from rises in sea level, though it is tempting to link the beginning of Wealden clay sedimentation with the basal Hauterivian transgression.

The fluctuating importance of the three main sources of sediment, the London Uplands, Cornubia and Armorica, is documented from pebble and heavy mineral distributions. Through most of Hastings Group times, Cornubian detritus poured into the Vectian Basin while the Wealden Basin received sediment from the London Uplands and Armorica. Towards the end of this interval Cornubian detritus began to spill into the western Weald too and it became the dominant component there in Wealden Formation times. There was also a 'boreal' input which diminishes southward from the north-west Weald, complementing the facies evidence of transgressions from the same direction and the faunal evidence of diminishing salinity southwards.

## The Lower Greensand Group

When the sea returned to the Wessex Basin at the beginning of the Aptian after a 30 million year absence it gently planed the top of the Wealden. Hence the Lower Greensand is disconformable and derived Wealden (and older) fossils occur at the base. Initially, offshore muds and silts (Atherfield Formation) were deposited, but sands soon prograded over the area to form a thick (maximum 200 m) pile of shallow marine sediments. Despite the name of the group, the glauconite which imparts the green colour has usually been

oxidised so that the bulk of the sands are yellow or brown.

The Portsdown Swell continued to affect facies and thicknesses (Fig. 12.11) and may have been emergent at times for it is a likely source of land-fall ash redeposited as pockets of fuller's earth in the South Downs (Young & Morgan 1981).

Casey's (1961) synthesis of Lower Greensand stratigraphy and palaeontology forms the starting point for all subsequent work. Of the many successions he described, the 5·5 km long cliff exposure from Atherfield Point to Rocken End in the Isle of Wight is probably the most important. This, the Atherfield section, not only serves as a standard of reference for the Vectian Basin but is the type locality for many of the standard ammonite zones of the English Aptian. It is also the stratotype of the lowest subdivision of the Lower Greensand Group, the Atherfield (Clay) Formation.

The Atherfield Formation (Simpson 1985) consists predominantly of silty clays, with local seams of fuller's earth in west Kent. The base generally belongs to the *obsoletus* Subzone (*fissicostatum* Zone), but in east Kent sedimentation started a little later. In the Atherfield section the formation is divided into several units, the thin, richly fossiliferous Perna beds marking the base. In Dorset there is an interesting horizon at the top of the formation, the 'Punfield Marine Band', here renamed the Punfield Bed. At Punfield Cove this consists of 0·3 m of argillaceous and sandy limestones with lignite streaks; at Corfe it is a thin clay ironstone; at Worbarrow Bay a fossiliferous ironstone only 0·07 m thick; and possibly at Lulworth Cove a ferruginous limestone 0·15 m thick. At Punfield the bed contains a diverse shallow marine fauna (Simpson 1983) including decapod crustaceans, gastropods, bivalves and the ammonites *Deshayesites* and *Roloboceras*. Farther west the ammonites and gastropods eventually disappear, until at Lulworth Cove there is a purely bivalve fauna (*Ceratostreon*, *Eomiodon* and *Filosina*) of brackish-water aspect. The faunal change may represent a passage from Wealden to Lower Greensand conditions along the estuary of an eastward-flowing river which briefly discharged sand (the Crackers Member) as far as the Isle of Wight.

The sandy beds above the Atherfield Formation (Table 12.2) show significant vertical and lateral variation. In the Wealden Basin, the Hythe Formation occurs in two distinct facies. In east Kent and part of Sussex the 'rag and hassock' consists of alternating layers (15–60 cm thick) of hard, sandy blue-hearted limestones (the Kentish Rag) and loamy sand speckled with glauconite (the Hassock). West of a line from Pulborough the beds become much less calcareous, and massive lenticular cherts appear in the western Weald.

Early in the late Aptian, Hythe Formation sediments in marginal areas of the basin were reworked and their faunas reincorporated in the basal nodule

beds (*nutfieldiensis* Zone) of the transgressive Sandgate Formation. This consists predominantly of loams, silts and silty clays, often glauconitic. Around Nutfield fuller's earth and cherty and glauconitic sandstones and limestones occur, overlain by glauconitic loams. West of Dorking the Sandgate facies change again. Here, bands of doggers and calcareous stone (the Bargate Member) are overlain by ferruginous loams (Puttenham Member). The Bargate facies continues south-eastward into Sussex until it thins east of Pulborough and apparently passes into glauconitic loams again, though in much of Sussex the Sandgate and Hythe Formations cannot be separated with confidence (Young *et al.* 1978).

At its type locality on the Kent coast, the Folkestone Formation consists of 20 m of coarse, yellowish greensands with bands of calcareous and glauconitic sandstone, sometimes strongly bioturbated. These probably represent nearshore sediments. Inland, within 16 km of the coast, appears the typical, more distal, facies of generally unfossiliferous, unconsolidated, ironstained, cross-bedded sands with occasional pebbly and silty seams. Irregular veins of secondary iron form boxstones. Within the formation

**Fig. 12.11.** Sections through the marine Lower Cretaceous rocks of the Wessex Basin.

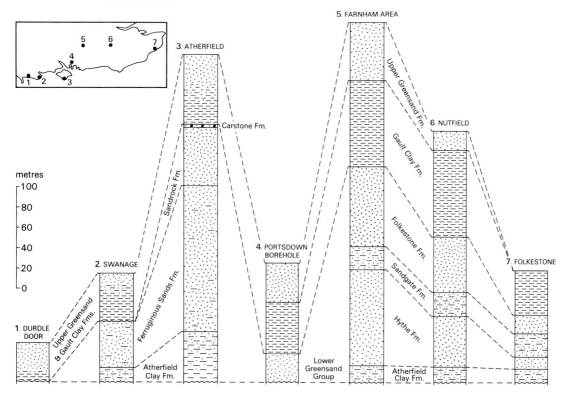

a significant erosion surface marks the 'mid-*tardefurcata* break' (Casey 1961, p. 501). This repres-ents a phase of widespread minor folding along east-west axes, followed by transgression which can be traced from the Wessex Basin to the Yorkshire Basin (Table 12.2).

In the western outliers of the Wealden Basin (Fig. 12.7) there are local deposits representing the feather edge of the late Aptian (*nutfieldiensis* Zone) trans-gression. The best known is the Faringdon (Sponge Gravels) Formation, preserved in a shallow, NNW–SSE trending trough cut into the late Jurassic (Poole *et al.* 1971). This consists of cross-bedded sands and gravels composed largely of sponge fragments. In Wiltshire, the Seend (Ironsands) Formation around Devizes and Calne has yielded a rich molluscan and brachiopod fauna (Casey 1961, p. 563). This *nut-fieldiensis* Zone transgression also reopened connec-tions through the 'Bedfordshire Straits' to the East Midlands Shelf (Fig. 12.3c). A second, early Albian (*mammillatum* Zone) transgression is represented in Wiltshire by loams at the base of the Gault.

In the Vectian Basin, the Ferruginous Sands Form-ation is the lowest of three sandy units (Table 12.2) and the only one which occurs towards the basin margin, in Dorset. It shows considerable lithological variation and numerous subdivisions have been recognised in the Isle of Wight. Here the lithology varies from grits and sands to interlaminated sands and clays. At Atherfield the formation is about 145 m thick. In Dorset it thins rapidly and at Worbarrow and farther west the development of brilliant colours and thin ironstone seams gives a distinct Wealden aspect (Arkell 1947, p. 176). One thick (maximum 6 m) pebbly sandstone near the top is apparently traceable from Punfield (bed 13) to Worbarrow (bed 14) and may be the correlative of the Walpen and Blackgang Members ('Groups' XIII and XIV) of the Isle of Wight (Arkell 1947, p. 174) which in turn represent the transgressive Sandgate Formation of the Wealden Basin.

The Sandrock Formation (maximum 57 m) is the approximate equivalent of the Folkestone Formation. It consists of white and yellow quartz sands often compacted into 'sandrock'. The overlying Carstone Formation (maximum 22 m) is a gritty, red-brown, ferruginous sand with pebbles and phosphatic nodules. It has a sharp, erosive base: the non-sequence between the two formations is another reflection of the mid-*tardefurcata* break. The formation is not con-tinuous with the Carstone Formation of eastern England though it is of comparable facies and age. Relics of the same facies occur also in the interven-ing area, around Shenley in Bedfordshire.

Fossils have generally been leached out of the sandier facies of the Lower Greensand Group, though

they are preserved in calcareous nodules and in more argillaceous sediments. Careful collecting has yielded a rich and diverse fauna, dominantly molluscan, with a local abundance of brachiopods and sponges. The group has also yielded some of the earliest flowering plants. Casey (1961, pp. 601–611) has listed several hundred fossil species, but a detailed biofacies analysis has yet to be published.

The Lower Greensand sands were probably derived from a variety of sources, including the slightly reju-venated London Uplands (Middlemiss 1976). Some of the rivers that had discharged Wealden sediment from the London Uplands and Cornubia were still active during Lower Greensand times. In addition, the re-establishment of a marine connection to the north from late Aptian times is marked by the appearance of abundant fresh kyanite of 'boreal' provenance in the late Aptian–early Albian sediments of southern Eng-land and Normandy (Juignet *et al.* 1973, p. 320). The sands were deposited predominantly in shallow marine waters and show evidence of strong tidal influence. This is first apparent in the Hythe Form-ation (Narayan 1963) but is particularly well shown in the Folkestone Formation (Fig. 12.8d) and correlative Woburn Formation farther north. Here large-scale cross-bedding, often with bioturbated clay drapes over individual foresets, indicates the migration of sand-wave complexes over the sea floor in a south-easterly direction. The initiation of strong tides in the Aptian may be linked with graben collapse in the Western Approaches which would have allowed the penetra-tion of currents from the opening Atlantic Ocean (Anderton *et al.* 1979, p. 237).

## The Gault Clay and Upper Greensand Formations

There is a rapid transition from the sands of the Lower Greensand to the clays of the Gault Formation (Fig. 12.12), often marked by a metre or two of rather condensed silty clays and fine sands with bands of phosphatic nodules. Throughout the Wessex Basin the typical Gault Formation facies is a pale to dark clay with phosphatic, pyritic and calcareous nodules which often occur in discrete bands. The lowest beds (to at least the *spathi* Subzone) thin over the Ports-down Swell, so that there are lithological differences between the Wealden and Vectian Basins. Within the Weald, Lower Gault sedimentation was also influenced by local axes (Owen 1971).

Lithological subdivision of the Gault is based on minor colour differences (in part representing small-scale cycles) and distinctive nodule bands: thus the type section at Copt Point, Folkestone, is divided into 13 beds (see Owen 1971), some of which can be grouped into larger units recognisable over a broad area of the Weald. Phosphatic nodule beds indicate

**Fig. 12.12.** Gault overlying the white sands of the Folkestone Beds (Lower Greensand) at Covers Farm, Westerham, 6 km west of Sevenoaks, Kent (B.G.S. SE.1382a).

pauses in deposition and the most significant one forms the Lower/Upper Gault boundary: this marks an important trasngression in early *cristatum* Subzone times throughout the Anglo-Paris Basin. The same transgression finally submerged the East Anglian Massif and hence there is a lateral passage from the Upper Gault to the Hunstanton Formation in north Norfolk.

The Gault Formation is mainly an outer shelf mud (slipper clay facies) with faunal assemblages similar to those of the Speeton Clay, and probably accumulated in 125–200 metres of water. However, towards the western margins of the Wessex Basin parts of the Lower Gault become more silty, and this foreshadows the westerly-derived influx of sand in late Albian times to form the Upper Greensand Formation. The sand prograded progressively eastward so that the diachronous base of the formation youngs in that direction. Eventually the sand extended approximately to a line drawn through Dunstable, Sevenoaks and Eastbourne (Fig. 12.3d): east of this the late Albian sediments are wholly in Gault Clay facies. In the Weald the Upper Greensand reaches a maximum thickness in the vicinity of Selborne; it thins over the

Portsdown Swell and thickens again in the Vectian Basin. Eventually it oversteps the Lower Gault clays and in the Haldon Hills outlier rests on the Permian Teignmouth Breccias (Hamblin & Wood 1976). Within the Vectian Basin, the mid-Dorset Swell (Fig. 12.13) first made its influence apparent late in the Albian; the highest beds of the Upper Greensand Formation are missing over it though facies are similar on either side (Drummond 1970). The swell had more profound effects on Cenomanian sedimentation.

Upper Greensand facies vary, but the sands are generally fine-grained, becoming coarser and even pebbly in the Haldon and Blackdown Hills (the 'Blackdown facies'). The 'greensand' facies is a glauconitic calcareous sandstone while the 'malmstone' is a fine sandstone (often calcareous) with abundant sponge spicules and a high proportion of colloidal silica.

In both the Gault and Upper Greensand Formations molluscs generally dominate the macrofauna. Ammonites are common in the former formation but sparse in the latter, while bivalves occur everywhere. However, Gault bivalves are predominantly infaunal forms while epifaunal, often strongly ornamented,

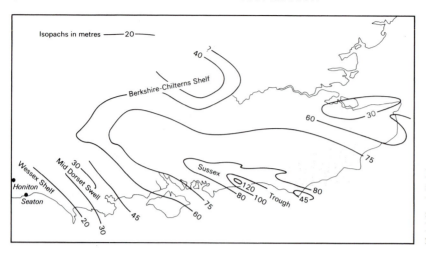

**Fig. 12.13.** Reconstructed isopachs for the 'Lower Chalk' of southern England (after Mortimore 1983, fig. 3a).

species are common in the Upper Greensand. The Blackdown Sands are famed for their beautifully preserved silicified fossils, among which bivalves and gastropods are predominant: the fauna is typical of an offshore sandy bottom and shows some of the earliest known evidence of predatory gastropod drill-holes (Taylor *et al.* 1983). In the Haldon Hills, littoral faunas (including corals) indicate closer proximity to the shore.

# THE UPPER CRETACEOUS SERIES

The Cenomanian transgression proved an integrating force: for the first time a broadly uniform sediment, the chalk, was deposited from Yorkshire to Devon (Fig. 12.14) and in all the offshore basins. With continuing sea-level rise through most of late Cretaceous times almost the whole of England and Wales was submerged (Fig. 12.2B). Clastic input was minimal, being limited to a little clay over most of the area, apparently derived from the east or south-east (Jeans 1968). In Dorset and Devon, Cenomanian sand deposition reflected both local tectonics and proximity to the shoreline. By the Turonian the shoreline had retreated enough for chalk sedimentation to start in this region too, and the only parts of Britain where younger nearshore sediments are known are in Northern Ireland and Argyll, Scotland where greensands were deposited until early Santonian times.

Chalk sedimentation continued into the Maastrichtian in Norfolk, the North Sea and the Channel, but elsewhere the higher levels have been removed by erosion so that the former extent of the Maastrichtian remains speculative. On a global scale there was a Maastrichtian regression, so that within the British

Isles the depositional area may have shrunk after the Campanian.

## Chalk petrology

Despite the search for modern analogues (e.g. Black 1980; Hancock 1980) the Chalk facies remains unique to the Cretaceous and Danian. Its petrology has been reviewed by Hancock (1976, 1980). The typical white-chalk facies is a pure limestone ($CaCO_3$ *c.* 98 per cent), with thin interbeds of marl and bands or scattered nodules of flint. The limestone is a bioclastic micrite

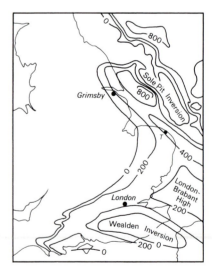

**Fig. 12.14.** Isopachs (in metres) of preserved Upper Cretaceous sediments (after Hancock 1984, fig. 7.8).

composed essentially of detritus from calcareous algae, mainly in the form of simple, plate-like crystals but sometimes as coccoliths (a ring of plates) or even complete coccospheres (composed of 7–20 overlapping coccoliths). The coarser fraction (10–100 $\mu$m) include foraminifers, ostracods, calcispheres, bryozoans, echinoid plates and bivalve fragments (especially inoceramid prisms).

The marl interbeds may be several centimetres thick and some appear to be laterally continuous for several hundred kilometres. Such bands often lie on an erosion surface and are clearly primary. Some contain Mg-rich smectite and may represent altered volcanic ash (Pacey 1984). Lesser, generally discontinuous, marls also occur, often flasers separating lenticles of chalk (the griotte facies: Hancock 1976, plate 2a; Jeans 1980, plates 6, 9); their origin is more problematic.

Flint occurs predominantly in layers parallel to the bedding, where it may form continuous sheets at one extreme and small, scattered, discrete nodules at the other. It sometimes forms cross-cutting veins or vertical cylinders with a burrow in the core (paramoudras). Its origin is still debatable but is probably biogenic, formed during several diagenetic stages commencing soon after deposition of the chalk. That flint formation continued for some time and was not usually penecontemporaneous with deposition is suggested by its general absence in intraformational conglomerates and the lack of epizoan-encrusted nodules. Concentration of biogenic silica into discrete flint bands often occurred at burrowed horizons, and many flint nodules are casts of burrows (Bromley 1967; Bromley & Ekdale 1986); the silica both infilled pore spaces and replaced calcium carbonate.

### Chalk environments

Chalk was formerly regarded as a very homogeneous rock deposited more or less continuously in a stable environment (hence the significance of its faunas in early palaeoevolutionary studies). Now a variety of chalk facies and sedimentary structures have been recognised (e.g. Kennedy & Garrison, 1975) and it has become apparent that not only did environments vary considerably but there were frequent breaks in sedimentation. Much of the Chalk is composed of small-scale rhythms, either marl/limestone couplets or fining-upward cycles. The boundaries between cycles are sharp and even erosive. Sedimentation was also affected locally by syndepositional faulting and uplift. In Yorkshire, 'shatter belts' in the Chalk running east–west along the Flamborough fault belt probably reflect both contemporaneous movement and later reactivation, while in southern England slumped horizons and facies/thickness changes are well documented on individual folds along the line of the Portsdown Swell (Gale 1980; Mortimore 1986b).

The typical white chalk is neither a deep oceanic ooze nor a deposit of very shallow waters. In general, it was probably deposited under normal marine salinity at a depth of 100–600 m. The sea floor was usually so soft that some bivalves were specially adapted to 'float' on the sediment surface or to lie half buried within it. However, during major pauses in deposition (marked by 'hard-ground' surfaces, Hancock 1976, p. 511) it was often cemented, especially in the shallower-water areas over submerged highs. Phosphatisation and glauconitisation are associated with the cementation phase. Faunas adapted to it: crustaceans modified their life habits (Bromley 1967) and encrusting organisms such as oysters plastered the surface of the cemented layer.

## The Chalk Group

The Chalk Group was deposited in two lithological/faunal provinces, northern and southern, with an intermediate region across East Anglia. The northern province Chalks are generally harder than their southern correlatives (due to two phases of calcite cementation, the earlier possibly under bacterial control (Jeans 1980)) and show strong faunal links with north Germany. The southern province chalks are faunally and lithologically linked with the Paris Basin.

Because of the difficulty of making a purely lithological subdivision of a rock so apparently homogeneous as the Chalk, the long established division into Lower, Middle and Upper Chalk was based effectively on a mixture of biostratigraphical and lithological criteria. In practice, ever since the recognition of a zonal sequence for the English Chalk over 100 years ago (Barrois 1876) lithological units have played a subordinate role in Chalk stratigraphy, though proving useful in field mapping. Unfortunately, the fossil zones are often poorly defined, index species may range far outside the zone and the range of a species often has been calibrated against that of another rather than against a lithological log, so that there is a great danger of circular arguments being applied in correlation. Furthermore, some of the zones have always proved unusable in the northern province Chalk. Even the boundaries between Lower, Middle and Upper Chalk may have been drawn at different levels in the two provinces.

Recognition of the lateral continuity of some lithological markers (e.g. marl and flint bands, bioclastic horizons and small-scale sedimentary cycles) coupled with the increasing recognition of several distinctive chalk facies is providing a welcome stimulus to lithological studies, and recent work in the northern and southern provinces represents a major step forward in British Chalk stratigraphy. For the first time, wholly lithostratigraphical subdivisions of much of the Chalk have been proposed (Table 12.3), and these will provide a sound framework for the determination of individual fossil ranges and hence for the elucidation of zones.

## The northern province: Yorkshire to Norfolk

Wood and Smith (1978) divided the Chalk Group of Yorkshire and Humberside into four formations, the Ferriby Chalk, Welton Chalk, Burnham Chalk and Flamborough Chalk Formations (Table 12.3) and these can be traced southward through Lincolnshire into north Norfolk. The basic criteria for distinguishing them are that the Ferriby and Flamborough Formations are flintless, the Welton Chalk contains flint nodules and the Burnham Chalk has tabular and related flints. Numerous marl and flint bands form distinctive, widespread markers and these have been given formal names (Fig. 12.15) (Wood & Smith 1978). The maximum thickness of the group exceeds 500 m.

The Ferriby Chalk Formation as defined here is a slightly marly, whitish chalk with finely disseminated iron, and hence is a little darker than overlying chalks. It is about 25 m thick over the East Midlands Shelf, thickening a little at Speeton. Wood (1980) has described in detail the main lithological subdivisions; some, such as the Paradoxica Bed at the base, have long been recognised and traced over a wide area. There are well-marked shell bands, bioclastic horizons and erosion surfaces. Jeans (1980) recorded five fining-upward cycles (maximum thickness 7 m) for which he used member names though these have yet to be formally described. Fossils are common at some horizons and inoceramids and some brachiopods occur through most of the formation.

The Welton Chalk Formation consists principally of thick-bedded to massive chalks, but rare laminate chalks (sensu Hancock 1976) appear. The flint nodules which characterise much of the formation are typically of thalassinidean burrow-fill type: they do not occur in the lowest 4–5 m. The base of the formation is defined by a pebble-strewn erosion surface just beneath the Black Band. The latter forms a distinctive marker, consisting of variegated, occasionally laminated marls, which may equate with part of the Plenus Marl Formation of the southern province (Hart & Bigg 1982). Several other named marls occur higher in the succession. There is a low-diversity fauna of inoceramids, small brachiopods and thin-shelled echinoids.

The Burnham Chalk Formation is predominantly thin-bedded with numerous layers of laminate chalk. A distinctive series of tabular and semi-tabular flint bands forms a well-marked mappable feature in the lower part of the formation and represents a late Turonian flint maximum. There are also several well-defined named marls. The fauna is generally more abundant and diverse than in adjacent formations but unevenly distributed.

The Flamborough Chalk Formation consists of well-bedded flintless chalk, with stylolitic surfaces and bands and partings of marl. It is more than 300 m thick and has not been studied in the same detail as have the underlying formations. Faunas are varied; the Flamborough sponge beds (about 10 m thick) appear about 120 m above the base and higher still there are horizons rich in scaphitids. The upper part of the formation is concealed beneath the drift, but log data from uncored boreholes suggests that it may be overlain by flinty chalks (Mortimore & Wood 1986).

## East Anglia and the southern province

The East Anglian chalks were deposited over the buried London Brabant Massif and are thinner than chalks to the north and south. Both faunally and lithologically the sequence is transitional between the two provinces (Mortimore & Wood 1986). Further south, the chalks of the southern province (Sussex Trough and marginal shelves) represent the north-western part of the Anglo–Paris Basin.

The southern province successions differ from those of the north; the chalk is generally softer, hardgrounds are much commoner and continuous tabular flints are uncommon. However, both provinces share some general trends; discrete marl seams are best developed in the Turonian and lower Campanian chalks and there is a high Turonian level of maximum flint development (Mortimore & Wood 1986). In addition, some distinctive lithological markers extend from the Yorkshire to Dorset or Sussex coasts and provide a valuable check on correlation between the two provinces.

Although chalk is the dominant lithofacies, marginal sands and sandy limestones occur close to the Cornubian massif. The most rapidly subsiding area lay over south-east England (Fig. 12.13) where over 400 m of predominantly soft chalk accumulated in the Sussex Trough. Hardgrounds and hard nodular chalks are typical principally of the condensed, shallower-water facies exposed along the western outcrops from the Berkshire–Chilterns shelf to Dorset (Bromley & Gale 1982; Mortimore 1983).

The Lower, Middle and Upper Chalk divisions were first proposed for southern province chalks. The bases of the Middle and Upper Chalk were defined primarily by rock bands, but fossils were utilised also, especially where the rock bands are absent or difficult to recognise. The resultant confusion was discussed by Wood

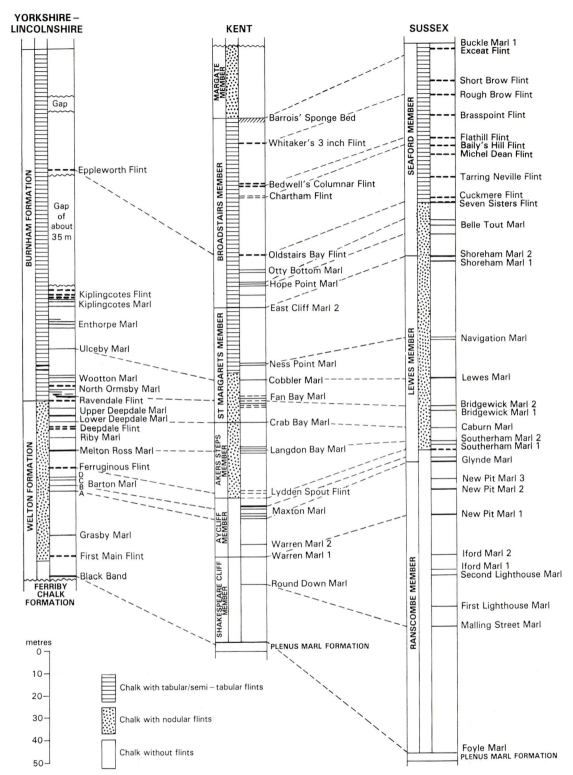

**Fig. 12.15.** Correlation of marker bands in the Turonian and lower Coniacian chalks of the northern and southern provinces. (Based on data in Wood & Smith 1978, Mortimore 1986a; Mortimore & Wood 1986; Robinson 1986; Wood, Ernst & Rasemann 1984.)

and Smith (1980) and Mortimore (1983): the Chalk Rock (forming the base of the Upper Chalk) in particular has often been misidentified (Bromley & Gale 1982). Hence, formal lithostratigraphic subdivisions have now been proposed for both the southeast and south-west of England (Table 12.3). As in the northern province, distinctive marl and flint bands are named (Fig. 12.15).

The 'Lower Chalk' subdivisions are virtually flintless. Almost everywhere the sequence commences with 1–5 m of marly, glauconitic chalk with indigenous phosphatic nodules, the Glauconitic Marl. This generally has an erosive base but locally there is a transition from the underlying Gault or Upper Greensand Formations. From Hertfordshire to East Anglia the Cambridge Greensand occupies a similar stratigraphical position: it is a silty, glauconitic marl well known for its extraordinary variety of boulders and pebbles (Jukes-Browne & Hill 1903) which apparently drifted into the sea entangled in tree roots. Many of the derived phosphatic nodules in the Cambridge Greensand are remanié late Albian ammonites, some not known elsewhere in England. Indigenous fossils suggest that the matrix sediment may range in age from latest Albian to earliest Cenomanian (Morter & Wood 1983).

Above these basal beds the bulk of the 'Lower Chalk' is a marly chalk, the marl content decreasing upward. Hence two lithological units have long been recognised, the Chalk Marl and the Grey Chalk. The former shows rhythmic alternations of marl and chalk while the latter is more massive. From East Anglia to the Berkshire–Chilterns shelf they are separated by a well-marked, hard chalk which forms a mappable feature. This, the Totternhoe Stone, is a brownish, gritty (comminuted inoceramid shells) chalk with scattered phosphatic pebbles. It reaches its maximum thickness of 6·5 m in Bedfordshire.

Towards the deeper-water areas of south-east England the Totternhoe Stone disappears. It then becomes difficult to draw a boundary between the Chalk Marl and Grey Chalk so that their definition has varied from author to author. Thus for the North Downs Robinson (1986) has proposed a new division (Table 12.3). The Glauconitic Marl forms a basal member to the East Wear Bay Formation, the remainder of which is composed of rhythmic units up to 3 m thick. The formation becomes less marly upward. The overlying Abbots Cliff Chalk Formation is a white chalk, the lower member of which is characterised by lenticular laminated structures. These have been traced over to Sussex and the Isle of Wight (Kennedy 1969, fig. 16) so that Robinson's divisions may prove more widely applicable.

The highest unit in the 'Lower Chalk' is a thin sequence of alternating marls and chalks collectively known as the Plenus Marls after the occurrence in the upper part of the belemnite *Actinocamax plenus*. The marls (0·75–8·5 m) have an erosive base and form a very distinctive lithological and geophysical marker. They were ranked as a formation by Robinson (1986) to accord with formal North Sea terminology.

Above the Plenus Marl Formation the 'Middle-Upper' Chalk shows a broadly similar character over most of the province, consisting of flintless chalks overlain by nodular chalks with flints which in turn pass up into smoother chalks with numerous flint bands. Early workers used these lithological changes to divide the Kentish chalk, but this promising approach was soon overshadowed by the increasing emphasis on the use of fossils in Chalk stratigraphy and it was over a hundred years before it was revived (Mortimore 1983, 1986a, 1986b; Robinson 1986).

While the Kent coast sections have long played a prominent role in Chalk studies, those of the Sussex coast are thicker and the succession is more complete beneath the Tertiary unconformity (Table 12.3). Hence when Mortimore (1983) grouped the former Middle and Upper Chalk of south-east England into the Sussex White Chalk Formation, with six members, he relied primarily on stratotypes in the Sussex Trough. He later expanded and modified his scheme (Mortimore 1986a, 1986b), naming several 'hypostratotypes' from the Kent coast.

Concurrently with the latter work, Robinson (1986) proposed an alternative nomenclature for the Kentish chalk which incorporated a revival of Dowker's (1870) lithological divisions. Robinson claimed that the boundaries of Mortimore's (1983) members and beds do not correspond with the main lithological boundaries there. However, many individual marker beds can be traced from Sussex and the broad lithological characteristics of correlative members are very similar. Minor discrepancies in the placing of boundaries (Fig. 12.15) reflect both lateral change and differing approaches: Mortimore (1986a) placed emphasis on the value of marl bands as boundary markers because some are so distinctive on geophysical logs, while Robinson sometimes utilised other criteria. Mortimore (1987) has synthesised the two schemes, and further research should reduce the plethora of alternative names currently available for individual marker beds: the degree of correlation suggested in Fig. 12.15 indicates the feasibility of this.

The lowest division (the Ranscombe, formerly Caburn, Member) is a flintless chalk corresponding with all but the highest part of the former 'Middle Chalk'. At the base is the Melbourn Rock, a complex of nodular marls and incipient hardgrounds. Above are nodular chalks with griotte marls, passing up into smoother, sometimes softer chalks. Sedimentation of smoother chalks apparently started earlier in Sussex (New Pit Beds) than in Kent ('Aycliff Member').

Flints first appear at about the base of the second main division, the Lewes Member. This unit is characteristically a nodular chalk, consisting mainly of rhythmic units composed of soft chalk passing up into nodular chalk and terminating in an omission surface or hardground, sometimes with an overlying marl band. In Kent the high Turonian flint maximum is well marked by several closely-spaced flint courses occurring around the level of the Bridgwick (= Fan Bay) Marls and contrasting markedly with the less flinty chalk below; in Sussex the contrast is much less (Mortimore & Wood 1986).

The remaining members are predominantly of soft white chalks with numerous flint bands. The flints vary in character through the sequence, but the members are separated primarily on the presence or absence of well-marked marl bands. Thus the Seaford Member has only a few, weak marl seams while the overlying Newhaven Member characteristically has numerous seams, with some nodular chalks. However, Mortimore (1986a) stressed the marked local absence of both marl seams and flints above Whitaker's 3″ Flint in the Thanet sections (see Robinson 1986, fig. 11), including Buckle Marl 1 which normally marks the base of the Newhaven Member.

In Kent the remainder of the sequence is truncated by the Tertiary erosion surface, but in the Culver (formerly Whitecliff) Member of Sussex marls occur only in the lowest few metres. Conspicuous marl bands reappear in the Portsdown Member, which is the youngest (late Campanian) chalk known *in situ* in the southern province. However, derived Maastrichtian faunas occur in post-Cretaceous flint gravels in Hampshire and the stage is represented offshore.

In the more marginal, sometimes condensed, facies extending south-westwards from East Anglia, rock bands are generally more strongly developed and may form mappable features. The Melbourn Rock occurs immediately above the Plenus Marl Formation but extends into the deeper-water areas to the south-east. It is a rubbly-weathering, conglomeratic chalk (intraclastic chalk, sensu Robinson 1986, p. 146). Higher in the sequence, the Chalk Rock (accorded formation status by Bromley & Gale 1982) is a condensed nodular chalk embracing several hardgrounds – some of which are traceable for up to 300 km, from Hertfordshire to Dorset. South-eastward, the Chalk Rock facies passes into typical Lewes Member nodular chalks. Some hardgrounds still persist: in particular the Ogbourne hardground at the base of the Chalk Rock is probably represented in the Isle of Wight by the 'Spurious Chalk Rock'. The Top Rock hardground complex is well developed in the Chilterns

and is represented in the upper part of the Lewes Member of the Sussex Trough.

The Cenomanian succession in south Dorset and Devon differs considerably from that elsewhere, reflecting in part the proximity of land but also the influence of a north-west–south-east trending high, the mid-Dorset Swell (Drummond 1970). The swell delimits a shelf area (the Wessex Shelf) to the south-west from the main Sussex Trough to the north-east (Fig. 12.13). In the trough sediments comprise the standard Chalk Marl/Grey Chalk succession. In contrast, the sequence on the shelf (preserved now in a series of scattered outliers west of the main outcrop) is complex and highly condensed, comprising sandy limestones (the Beer Head Limestone Formation of Jarvis & Woodruff 1984) and correlative sands. The latter include the Wilmington Sands near Honiton and the upper part of the Haldon Sands Formation of the Haldon Hills (Hamblin & Wood 1976). In Dorset a yellow quartzose grit, the Eggardon Grit, is at least in part of Cenomanian age though the lower beds may be Albian.

While the Chalk Marl is virtually restricted to the Sussex Trough, younger chalks lap on to the flanks of the mid-Dorset Swell and eventually onto the adjoining shelf to cover the more clastic facies beneath. The base of the chalk facies is diachronous (Kennedy 1970), becoming progressively younger westwards until in the Seaton district it is basal Turonian. Here the Turonian chalks, together with remnants of Coniacian chalk preserved from erosion in local downwarps, form the Seaton Chalk Formation (Jarvis & Woodruff 1984); this rests on a hardground surface at the top of the Beer Head Formation.

# CRETACEOUS EVENTS

The distribution of Cretaceous facies reflects not only the relationship between the individual basins and surrounding highs but also the superimposed effects of sea-level changes, vulcanicity and 'anoxic' events. Some of these effects can be of considerable value in correlation if carefully interpreted.

## Sea-level changes

Sea level rose through most of the Cretaceous but there was a significant fall in Maastrichtian times (Fig. 12.16). There is some dispute over whether the rises were synchronous in different areas and whether they were step-like, with intervals of still-stand or fall between rises, or progressive. One result of sea-level changes was the development at basin margins of unconformities and transgressive/regressive sequences. However, the same events are often detectable away from the margins by sedimentary breaks or sharp facies changes (Fig. 12.5), rapid faunal turn-overs etc. Using such criteria a large number of sea-level changes have been recognised in north-west Europe, most representing short-lived, isochronous events.

From the North Sea region Rawson and Riley (1982) documented twelve early Cretaceous events. Most can be recognised farther afield and the late Ryazanian, mid-Valanginian, mid-Hauterivian and some of the Aptian and Albian ones appear to be of global significance. For the late Cretaceous, Hancock and Kauffman (1979) produced a north European sea-level curve with several significant peaks. In a more detailed study in north-west Germany, Ernst, Schmid and Seibertz (1983) distinguished at least 11 'eustato-events' in the Cenomanian and Turonian alone. Some of these are apparently present in England and may be of value in broader correlation.

From the evidence so far it is likely that some 30–40 short-lived, isochronous sea-level events may be represented in north-west Europe. Many may mark regional events and probably form the boundaries of sedimentary cycles on the mesothem scale (sensu Ramsbottom 1977), but at least six early Cretaceous and four late Cretaceous ones are of much greater lateral extent and appear to mark 'global' sea-level changes. Others may also do so. Any mechanism invoked to explain such eustatic changes must also take into account a mean interval of not more than 8 million years and conceivably as little as 5 m.y., and a possible difference in minimum and maximum sea-level height of over 300 m. Rate of change in sea-floor spreading may not be an adequate explanation, and there is no evidence of Cretaceous ice sheets – though the possibility has been suggested (Kemper 1983).

## Vulcanicity

Until recently it was assumed that north-west Europe was free of volcanic activity until late Cretaceous times. Now several Lower Cretaceous bentonitic horizons have been recognised, probably derived from sources in offshore areas of Norway, the Netherlands and south-west England. In late Ryazanian times four thin volcanogenic mudstones (bentonites) were deposited at Speeton (Knox & Fletcher 1978). A possible source is the Zuidwal vent in the Netherlands though the radiometric age of 144±1 Ma (Dixon et al. 1981) would date the agglomerate there as early Ryazanian. Two further bentonites occur in the Hauterivian, and all six (plus a Valanginian one) have been discovered in an offshore borehole 80 km to the east-north-east (BGS 81/43; Lott et al. 1986).

A suspected tuff band in the upper part of the Weald Clay Formation of the Warlingham borehole is prob-

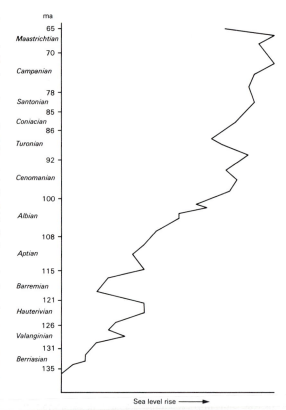

**Fig. 12.16.** Cretaceous sea-level changes in north-west Europe (late Cretaceous after Hancock & Kauffman 1979).

ably of Barremian age, but the next important volcanic episode occurred from late Aptian to mid-Albian times. Evidence for this is widespread in south-eastern England (Jeans et al. 1982): volcanogenic glauconite grains occur in the greensand facies, while secondary bentonites represent local accumulations of ash swept from adjacent land areas (East Anglian Massif and Portsdown Swell). The source of the ash probably lay in the southern North Sea, where wells in the Dutch sector indicate that a phase of undersaturated basic magmatism occurred in mid-Cretaceous times (Dixon et al. 1981, p. 132). The heaviest ashfalls are represented by the well-known fuller's earths in the Sandgate Formation, which contain well-preserved glass shards (see Jeans et al. 1977). The horizon (nutfieldiensis Zone, Upper Aptian) correlates with bentonites in BGS borehole 81/40 in the central North Sea (Lott et al. 1985), and with tuffs at Sarstedt, near Hannover in north Germany (Gaida et al. 1978). A secondary bentonite from Aptian/Albian boundary beds at Goodstone, Surrey (Morgan et al. 1979) correlates with a well-dated occurrence at

Vohrum, near Hannover (Kemper & Zimmerle 1978).

Smectite-rich clays occur also in the lower part of the Gault Clay Formation, and other smectite-rich horizons in the Chalk Group suggest that vulcanism may have continued through late Cretaceous times (Harrison et al. 1979; Pacey 1984). Jeans and co-workers (1982) recorded volcanogenic glauconite from the Lower Chalk and Harrison and co-workers (1979, Fig. 1) indicate derived volcanic material from a sample of Upper Chalk in Norfolk. Further study of such horizons is needed, especially in the light of well-established records of tuffs in the Turonian, Campanian and Maastrichtian in north Germany (Gaida et al. 1978) and of late Cretaceous magmatic activity in the Rockall region (Harrison et al. 1979). The widespread occurrence and mineralogical characteristics of the latest Cenomanian 'Black Band' horizon across eastern England and the southern and central North Sea marks it as the most obvious volcanogenic level.

### 'Anoxic' events

On three occasions during the Cretaceous extensive areas of ocean floor became stagnant, and during these 'oceanic anoxic events' organic-rich laminated shales were deposited (Jenkyns 1980). The first and most important phase lasted from the late Barremian to the end of the Albian and is marked by alternations of anoxic and normal shales. A second spanned the Cenomanian-Turonian boundary and a third (less well developed) occurred in Coniacian and Santonian times. They may have developed as a result of nutrient-rich, oxygen-depleted water accumulating on shelves during transgression before being swept into adjacent ocean basins.

On the north-west European shelf bottom waters were generally well oxygenated, but phases of low oxygen ('kenoxic') or oxygen deficient ('anoxic') conditions are indicated by thin developments of dark, laminated, bituminous shales. Two such levels are well documented in the marine Lower Cretaceous rocks of north Germany (Kemper & Zimmerle 1978; Gaida et al. 1981) where the sediments were probably deposited in a barred basin during a regional fall in sea level. The lower ('Blatterton') horizon is of late early Barremian age while the upper ('Fischschiefer') is early Aptian. In England, both events may be represented in the Yorkshire Basin though apparently not farther south. In the lower B beds at Speeton a few metres of kerogen-rich shales are the time equivalent of the 'Blatterton' and the 'Fischschiefer' may be represented by poorly exposed laminated black shales in the highest B beds. Both of these horizons probably represent part of the broader 'late' Barremian to late Albian oceanic event and their firm stratigraphic dating may help to sharpen the dating of individual oceanic anoxic levels.

A brief interval of reduced oxygenation at the end of the Cenomanian is the local representative of the second oceanic anoxic event. Its effects are apparent across much of north-west Europe (Hart & Bigg 1981) but they were accentuated in Humberside and adjacent offshore areas by volcanic activity. This produced the 'Black Band' discussed above, a thin, laminated organic mudstone with a sparse, agglutinated benthonic foraminiferal fauna. Its areal distribution was apparently controlled by the position of submerged highs. This event is important as a seismic marker over parts of the North Sea.

# A CATASTROPHIC CONCLUSION

The end of the Cretaceous marks the end of the Mesozoic era. The supposedly abrupt disappearances at that time of major fossil groups such as dinosaurs and ammonites has long caught the eye and attracted speculation from diverse authorities whose terrestrial or extra-terrestrial explanations become more bizarre the further they stand away from the rocks.

A realistic appraisal of many different fossil groups (see papers in Birkelund & Bromley 1979) has shown wide variation in rates of evolution and patterns of extinction through the late Cretaceous. The oft-quoted extinction of the Ammonoidea at the end of the Maastrichtian was the final event in a 30 million year decline, while many 'Tertiary' bivalve groups are now known to have evolved during the Cretaceous. On the other hand some planktonic calcareous foraminifera and nannofossils show a rapid turnover at the end of the Cretaceous, more abrupt in oceanic than in shelf forms. Thus there is evidence of both long-term decline and a catastrophic event. For such a varied pattern a 'multicausal scenario is required in order to explain the documented biological changes' (Birkelund & Hakansson 1982, p. 383). Such a scenario must allow for major environmental changes which had varying effects on terrestrial, shelf and oceanic communities. The changes were a result of normal planetary processes (such as changing sea levels) coupled with a catastrophic event at the end of the Cretaceous – though whether the latter reflects spectacular vulcanicity or a bolide impact is still open to debate (Hallam 1988).

## REFERENCES

| | | |
|---|---|---|
| ALLEN, P. | 1959 | The Wealden environment: Anglo-Paris Basin. *Philos. Trans. R. Soc. London*, **B242**, 283–346. |
| | 1967 | Origin of the Hastings facies in north-western Europe. *Proc. Geol. Assoc. London*, **78**, 27–105. |
| | 1976 | Wealden of the Weald: a new model. *Proc. Geol. Assoc. London*, **86** (for 1975), 389–437. |
| | 1981 | Pursuit of Wealden models. *J. geol. Soc. London*, **138**, 375–405. |
| ANDERSON, F. W. | 1971 | Appendix B. The sequence of ostracod faunas in the Wealden and Purbeck of the Warlingham borehole. *Bull. geol. Surv. G.B.*, **36**, 122–138. |
| | 1973 | The Jurassic-Cretaceous transition: the non-marine ostracod faunas. *In* Casey, R. & Rawson, P. F. (Eds.) *The Boreal Lower Cretaceous*. Geol. J. Spec. Issue **5**, 101–110. |
| ANDERSON, F. W. & BAZLEY, R. A. B. | 1971 | The Purbeck Beds of the Weald (England). *Bull. geol. Surv. G.B.*, **34**, 1–173. |
| ANDERTON, R., BRIDGES, P. H., LEEDER, M. R. & SELLWOOD, B. W. | 1979 | *A dynamic stratigraphy of the British Isles.* Allen & Unwin, London. 301 pp. |
| ARKELL, W. J. | 1947 | The geology of the country around Weymouth, Swanage, Corfe and Lulworth. *Mem. geol. Surv. G.B.*, 386 + xii pp. |
| BARROIS, C. | 1876 | *Recherches sur le Terrain Crétace Supérieur de l'Angleterre et de l'Irlande.* Lille, 232 pp. |
| BATTEN, D. J. | 1982 | Palynofacies and salinity in the Purbeck and Wealden of southern England. *In* Banner, F. T. & Lord, A. R. (Eds.) *Aspects of micropalaeontology*. George Allen & Unwin, 278–295. |
| BIRKELUND, T. & BROMLEY, R. G. (Eds.) | 1979 | *Cretaceous-Tertiary boundary events: vol. 1, The Maastrichtian and Danian of Denmark.* University of Copenhagen, 210 pp. |
| BIRKELUND, T. & HAKANSSON, E. | 1982 | The terminal Cretaceous extinction in Boreal shelf seas – a multicausal event. *Spec. Pap. geol. Soc. Am.*, **190**, 373–385. |
| BLACK, M. | 1980 | On Chalk, Globigerina Ooze and Aragonite Mud. *In* Jeans, C. V. & Rawson, P. F. (Eds.) *Andros Island, Chalk and Oceanic Oozes*. Yorkshire geol. Soc. Occ. Publ. **5**, 54–85. |
| BROMLEY, R. G. | 1967 | Some observations on burrows of thalassinidean Crustacea in chalk hardgrounds. *Q. J. geol. Soc. London*, **123**, 157–182. |
| BROMLEY, R. G. & EKDĀLE, A. A. | 1986 | Flint and fabric in the European Chalk. *In* Sieveking, G. de G. & Hart, M. B. (Eds.) *The scientific study of flint and chert*. Cambridge University Press, Cambridge, 71–82. |
| BROMLEY, R. G. & GALE, A. S. | 1982 | The Lithostratigraphy of the English Chalk Rock. *Cretaceous Research*, **3**, 273–306. |
| CARTER, D. J. & HART, M. B. | 1977 | Aspects of mid-Cretaceous stratigraphical micropalaeontology. *Bull. Br. Mus. nat. Hist. (Geology)*, **29**, 1–135. |
| CASEY, R. | 1961 | The stratigraphical palaeontology of the Lower Greensand. *Palaeontology*, London, **3**, 487–621. |
| | 1963 | The dawn of the Cretaceous period in Britain. *Bull. S-E Union Sci. Socs.*, **117**, 1–15. |
| | 1973 | The ammonite succession at the Jurassic-Cretaceous boundary in eastern England. *In* Casey, R. & Rawson, P. F. (Eds.) *The Boreal Lower Cretaceous*. Geol. J. Spec. Issue **5**, 193–266. |

CASEY, R. & | 1964 | Notes on some ferruginous strata in Buckinghamshire and
BRISTOW, C. R. | | Wiltshire. *Geol. Mag.*, **101**, 116–128.

CHADWICK, R. A. | 1986 | End Jurassic–early Cretaceous sedimentation and subsidence (late Portlandian to Barremian), and the late-Cimmerian unconformity. *In* Whittaker, A. (Ed.) *Atlas of Onshore Sedimentary Basins in England and Wales: Post-Carboniferous Tectonics and Stratigraphy.* Blackie, Glasgow, 52–56.

COPE, J. C. W. | 1984 | The Mesozoic history of Wales. *Proc. Geol. Assoc. London*, **95**, 373–385.

CRUX, J. A. | 1982 | Upper Cretaceous (Cenomanian to Campanian) calcareous nannofossils. *In* Lord, A. R. (Ed.) *A stratigraphical index of calcareous nannofossils.* Ellis Horwood Ltd., London, 81–135.

DALEY, B. & | 1979 | Week-end field meeting: the Wealden Group in the Isle of
STEWART, D. J. | | Wight, 17–19 June 1977. *Proc. Geol. Assoc. London*, **90**, 51–54.

DIXON, J. E., | 1981 | The tectonic significance of post-Carboniferous igneous
FITTON, J. G. & | | activity in the North Sea Basin. *In* Illing, L. V. & Hobson,
FROST, R. T. C. | | G. D. (Eds.) *Petroleum Geology of the Continental Shelf of North West Europe.* Institute of Petroleum, London, 121–137.

DOWKER, G. | 1870 | On the Chalk of Thanet, Kent, and its connection with the Chalk of East Kent. *Geol. Mag.*, **7**, 466–472.

DRUMMOND, P. V. O. | 1970 | The Mid-Dorset Swell. Evidence of Albian–Cenomanian movements in Wessex. *Proc. Geol. Assoc. London*, **81**, 679–714.

ELLER, M. | 1981 | The Red Chalk of eastern England: a Cretaceous analogue of Rosso Ammonitico. *In* Farinacci, A. & Elmi, S. (Eds.) *Rosso Ammonitico Symposium Proceedings*, Milan, 207–231.

ERNST, G., | 1983 | Event-Stratigraphie im Cenoman und Turon von NW-
SCHMID, F. & | | Deutschland. *Zitteliana*, **10**, 531–554.
SEIBERTZ, E. | |

GAIDA, K.-H., | 1978 | Das Oberapt von Sarstedt und Seine Tuffe. *Geol. Jahrb.*
KEMPER, E. & | | **A45**, 43–123.
ZIMMERLE, W. | |

GALE, A. S. | 1980 | Penecontemporaneous folding, sedimentation and erosion in Campanian Chalk near Portsmouth, England. *Sedimentology*, **27**, 137–151.

GALLOIS, R. W. & | 1982 | The stratigraphy of the Gault of East Anglia. *Proc. Geol.*
MORTER, A. A. | | *Assoc. London*, **93**, 351–368.

**HALLAM, A.** | 1988 | A compound scenario for the end-Cretaceous mass extinctions. *Revista Española de Paleontologia*, n'Extraordinairo, 7–20.

HAMBLIN, R. J. O. & | 1976 | The Cretaceous (Albian–Cenomanian) stratigraphy of the
WOOD, C. J. | | Haldon Hills, South Devon, England. *Newsl. Stratigr.*, **4** (3), 135–149.

HANCOCK, J. M. | 1976 | The petrology of the Chalk. *Proc. Geol. Assoc. London*, **86**, 499–535.

| 1980 | The significance of Maurice Black's work on the Chalk. *In* Jeans, C. V. & Rawson, P. F. (Eds.) *Andros Island, Chalk and Oceanic Oozes.* Yorkshire geol. Soc. Occ. Publ. **5**, 86–97.

| 1984 | Cretaceous. *In* Glennie, K. W. (Ed.) *Introduction to the Petroleum Geology of the North Sea.* Blackwell Scientific Publications, Oxford, 133–150.

HANCOCK, J. M. &          1979   The great transgressions of the Late Cretaceous. *J. geol.*
  KAUFFMAN, E. G.                 *Soc. London*, **136**, 175–186.

HARRIS, A. L.,            1975   A correlation of the Precambrian rocks in the British Isles.
  SHACKLETON, R. M.,             *Spec. Rep. geol. Soc. London*, **6**, 135 pp.
  WATSON, J.,
  DOWNIE, C.,
  HARLAND, W. B. &
  MOORBATH, S.

HARRISON, R. K.,          1979   Mesozoic igneous rocks, hydrothermal mineralisation and
  JEANS, C. V. &                 volcanogenic sediments in Britain and adjacent regions.
  MERRIMAN, R. J.                *Bull. geol. Surv. G. B.*, **70**, 57–69.

HART, M. B. &             1981   Anoxic events in the late Cretaceous Chalk seas of North-
  BIGG, P. J.                    West Europe. *In* Neale, J. W. & Brasier, M. D. (Eds.)
                                 *Microfossils from Recent and Fossil Shelf Seas*. Ellis
                                 Horwood Ltd., 177–185.

HART, M. B.,              1981   Cretaceous. *In* Jenkins, D. G. & Murray, J. W. (Eds.)
  BAILEY, H. W.,                 *Stratigraphical Atlas of Fossil Foraminifera*. Ellis Horwood
  FLETCHER, B. N.,               Ltd., 149–227.
  PRICE, R. & SWEICICKI, A.

HOLLAND, C. H.            1978   A guide to stratigraphical procedure. *Spec. Rep. geol. Soc.*
  AND 17 OTHERS                  *London*, **10**, 18 pp.

HOWITT, F.               1964   Stratigraphy and structure of the Purbeck inliers of Sussex,
                                 England. *Q. J. geol. Soc. London*, **120**, 77–113.

INGHAM, F. T.            1929   The Petrography of the Spilsby Sandstone. *Proc. Geol.
                                 Assoc. London*, **40**, 1–17.

JARVIS, I. &             1984   Stratigraphy of the Cenomanian and basal Turonian
  WOODROOF, P. B.                (Upper Cretaceous) between Branscombe and Seaton, SE
                                 Devon, England. *Proc. Geol. Assoc. London*, **95**, 193–215.

JEANS, C. V.             1968   The origin of the montmorillonite of the European chalk
                                 with special reference to the Lower Chalk of England. *Clay
                                 Miner*, **7**, 311–329.

                         1973   The Market Weighton Structure: tectonics, sedimentation
                                 and diagenesis during the Cretaceous. *Proc. Yorkshire geol.
                                 Soc.*, **39**, 409–444.

                         1980   Early submarine lithification in the Red Chalk and Lower
                                 Chalk of Eastern England: a bacterial control model and its
                                 implications. *Proc. Yorkshire geol. Soc.*, **43**, 81–157.

JEANS, C. V.,            1977   Origin of Middle Jurassic and Lower Cretaceous Fuller's
  MERRIMAN, R. J. &              Earths in England. *Clay Miner*, **12**, 11–44.
  MITCHELL, J. G.

JEANS, C. V.,            1982   Volcanic clays in the Cretaceous of southern England and
  MERRIMAN, R. J.,               northern Ireland. *Clay Miner*, **17**, 105–156.
  MITCHELL, J. G. & BLAND, D. J.

JENKYNS, H. C.           1980   Cretaceous anoxic events: from continents to oceans. *J.
                                 geol. Soc. London*, **137**, 171–188.

JUIGNET, P.,             1973   Boreal influences in the Upper Aptian–Lower Albian beds
  RIOULT, M. &                   of Normandy, northwest France. *In* Casey, R. & Rawson,
  DESTOMBES, P.                  P. F. (Eds.) *The Boreal Lower Cretaceous*. Geol. J. Spec.
                                 Issue **5**, 303–326.

JUKES-BROWNE, A. J. &    1900–4  The Cretaceous rocks of Britain. *Mem. geol. Surv. G.B.*, 3
  HILL, W.                       vols.

KEEPING, W.             1883   *The fossils and palaeontological affinities of the Neocomian
                                 deposits of Upware and Brickhill*. Sedgwick Prize essay for
                                 1879. Cambridge, 167 pp.

KELLY, S. R. A.         1983   Boreal influences on English Ryazanian bivalves. *Zit-
                                 teliana*, **10**, 285–292.

| | | |
|---|---|---|
| KEMPER, E. | 1983 | Uber Kalt- und Warmzeiten der Unterkreide. *Zitteliana*, **10**, 359–369. |
| KEMPER, E., RAWSON, P. F. & THIEULOY, J.-P. | 1981 | Ammonites of Tethyan ancestry in the early Lower Cretaceous of north-west Europe. *Palaeontology*, London, **24**, 251–311. |
| KEMPER, E. & ZIMMERLE, W. | 1978 | Die anoxischen sedimente der praoberaptischen Unterkreide NW-Deutschlands und ihr palaeogeographischer Rahmen, *Geol. Jahrb.*, **45A**, 3–41. |
| KENNEDY, W. J. | 1970 | A correlation of the uppermost Albian and the Cenomanian of south-west England. *Proc. Geol. Assoc. London*, **81**, 613–677. |
| | 1978 | Cretaceous. *In* McKerrow, W. S. (Ed.) *The Ecology of Fossils*. Duckworth, 280–322. |
| KENNEDY, W. J. & GARRISON, R. E. | 1975 | Morphology and genesis of nodular chalks and hardgrounds in the Upper Cretaceous of southern England. *Sedimentology*, **22**, 311–386. |
| KENT, P. E. | 1947 | A deep boring at North Creake, Norfolk. *Geol. Mag.*, **84**, 2–18. |
| | 1975 | The tectonic development of Great Britain and the surrounding seas. *In* Woodland, A. W. (Ed.) *Petroleum and the Continental Shelf of North West Europe*. **1**, *Geology*. Applied Science Publishers Ltd., Barking, 501 pp., 3–28. |
| | 1980 | *Eastern England from the Tees to the Wash*. British Regional Geology, H.M.S.O., London, 155 pp. |
| KILENYI, T. & NEALE, J. W. | 1978 | The Purbeck/Wealden. *In* Bate, R. & Robinson, E. (Eds.) *A stratigraphical index of British Ostracoda*. Geol. J. Spec. Issue **8**, 299–324. |
| KNOX, R. W. O'B. & FLETCHER, B. N. | 1978 | Bentonites in the Lower D Beds (Ryazanian) of the Speeton Clay of Yorkshire. *Proc. Yorkshire geol. Soc.*, **42**, 21–27. |
| LAMPLUGH, G. W. | 1889 | On the subdivisions of the Speeton Clay. *Q. J. geol. Soc. London*, **45**, 575–618. |
| LOTT, G. K., BALL, K. C. & WILKINSON, I. P. | 1985 | Mid-Cretaceous stratigraphy of a cored borehole in the western part of the Central North Sea. *Proc. Yorkshire geol. Soc.*, **45**, 235–248. |
| LOTT, G. K., FLETCHER, B. N. & WILKINSON, I. P. | 1986 | The stratigraphy of the Lower Cretaceous Speeton Clay Formation in a cored borehole off the coast of north-east England. *Proc. Yorkshire geol. Soc.*, **46**, 39–56. |
| MELVILLE, R. V. & FRESHNEY, E. C. | 1982 | *The Hampshire Basin*. British Regional Geology, HMSO, London, 146 pp. |
| MIDDLEMISS, F. A. | 1976 | Studies in the sedimentation of the Lower Greensand of the Weald, 1875–1975: a review and commentary. *Proc. Geol. Assoc. London*, **86**, 457–473. |
| MORGAN, D. J., HIGHLEY, D. E. & BLAND, D. J. | 1979 | A montmorillonite, kaolinite association in the Lower Greensand of south-east England. *In* Mortland, M. M. & Farmer, V. C. (Eds.) *International clay conference 1978*, Elsevier, London, 301–310. |
| MORTER, A. A. & WOOD, C. J. | 1983 | The biostratigraphy of Upper Albian–Lower Cenomanian *Aucellina* in Europe. *Zitteliana*, **10**, 515–529. |
| MORTIMORE, R. N. | 1983 | The stratigraphy and sedimentation of the Turonian-Campanian in the Southern Province of England. *Zitteliana*, **10**, 27–41. |
| | 1986a | Stratigraphy of the Upper Cretaceous White Chalk of Sussex. *Proc. Geol. Assoc. London*, **97**, 97–139. |

MORTIMORE, R. N.            1986b    Controls on Upper Cretaceous sedimentation in the South
                                    Downs, with particular reference to flint distribution. *In*
                                    Sieveking, G. de G. & Hart, M. B. (Eds.) *The scientific study
                                    of flint and chert*. Cambridge University Press, Cambridge,
                                    21–42.

                           1987    Upper Cretaceous Chalk in the North and South Downs,
                                    England: a correlation. *Proc. Geol. Assoc. London*, **98**,
                                    77–86.

MORTIMORE, R. N. &         1986    The distribution of flint in the English Chalk, with
    WOOD, C. J.                     particular reference to the 'Brandon Flint Series' and the
                                    high Turonian flint maximum. *In* Sieveking, G. de G. &
                                    Hart, M. B. (Eds.) *The scientific study of flint and chert*.
                                    Cambridge University Press, Cambridge, 7–20.

NARAYAN, J.                1963    Cross-stratification and palaeogeography of the Lower
                                    Greensand of south-east England and Bas Boulonnais,
                                    France. *Nature, London*, **199**, 1246–1247.

                           1971    Sedimentary structures in the Lower Greensand of the
                                    Weald, England, and Bas Boulonnais, France. *Sedim.
                                    Geol.*, **6**, 73–109.

NEALE, J. W.               1978    The Cretaceous. *In* Bate, R. & Robinson, E. (Eds.) *A
                                    stratigraphical index of British Ostracoda*. Ellis Horwood,
                                    325–384.

OWEN, H. G.                1971    Middle Albian stratigraphy in the Anglo-Paris Basin. *Bull.
                                    Br. Mus. nat. Hist.* (Geology), Suppl. **8**, 164 pp.

                           1972    The Gault and its junction with the Woburn Sands in the
                                    Leighton Buzzard area, Bedfordshire and Buckingham-
                                    shire. *Proc. Geol. Assoc. London*, **83**, 287–312.

                           1973    Ammonite faunal provinces in the Middle and Upper
                                    Albian and their palaeogeographical significance. *In* Casey,
                                    R. & Rawson, P. F. (Eds.) *The Boreal Lower Cretaceous*.
                                    Geol. J. Spec. Issue **5**, 131–144.

PACEY, N. R.               1984    Bentonites in the Chalk of central eastern England and
                                    their relation to the opening of the northeast Atlantic.
                                    *Earth planet. Sci. Lett.*, **67**, 48–60.

POOLE, E. G.,              1971    Calcium montmorillonite (fullers' earth) in the Lower
    KELK, B.,                      Greensand of the Farnham Area, Berkshire. *Rep. Inst. geol.
    BAIN, J. A. &                   Sci. London*, **71/12**.
    MORGAN, D. J.

RAMSBOTTOM, W. H. C.        1977    Major transgressions and regressions (Mesothems) in the
                                    Namurian. *Proc. Yorkshire geol. Soc.*, **41**, 261–291.

RAWSON, P. F.              1981    Early Cretaceous ammonite biostratigraphy and biogeo-
                                    graphy. *In* House, M. R. & Senior, J. R. (Eds.) *The
                                    Ammonoidea. System Assoc. Spec. Volume* **18**.

RAWSON, P. F.,             1978    A correlation of Cretaceous rocks in the British Isles. *Spec.
    CURRY, D.,                      Rep. geol. Soc. London*, **9**, 70 pp.
    DILLEY, F. C.,
    HANCOCK, J. M.,
    KENNEDY, W. J.,
    NEALE, J. W.,
    WOOD, C. J. &
    WORSSAM, B. C.

RAWSON, P. F. &            1983    Stratigraphy of the Lower B and basal Cement Beds
    MUTTERLOSE, J.                  (Barremian) of the Speeton Clay, Yorkshire, England.
                                    *Proc. Geol. Assoc. London*, **94**, 133–146.

RAWSON, P. F. &            1982    Latest Jurassic–Early Cretaceous events and the 'Late
    RILEY, L. A.                    Cimmerian Unconformity' in North Sea area. *Bull. Am.
                                    Assoc. Petrol. Geol.*, **66**, 2628–2648.

RHYS, G. H. (Ed.) 1974 A proposed standard lithostratigraphic nomenclature for the southern North Sea and an outline structural nomenclature for the whole of the (UK) North Sea. *Rep. Inst. geol. Sci., London*, **74/8**, 14 pp.

ROBINSON, N. D. 1986 Lithostratigraphy of the Chalk Group of the North Downs, south-east England. *Proc. Geol. Assoc. London*, **97**, 141–170.

SCHWARZACHER, W. 1953 Cross-bedding and grain size in the Lower Cretaceous sands of East Anglia. *Geol. Mag.*, **90**, 322–330.

SIMPSON, M. 1983 Decapod Crustacea and associated fauna of the Punfield Marine Band (Lower Cretaceous; Lower Aptian), Punfield, Dorset. *Proc. Dorset nat. Hist. archaeol. Soc.*, **104**, 143–146.

SIMPSON, M. I. 1985 The stratigraphy of the Atherfield Clay Formation (Lower Aptian, Lower Cretaceous) at the type and other localities in southern England. *Proc. Geol. Assoc. London*, **96**, 23–45.

STEWART, D. J. 1981 A meander-belt sandstone of the Lower Cretaceous of Southern England. *Sedimentology*, **28**, 1–20.

STONELEY, R. 1982 The structural development of the Wessex Basin. *J. geol. Soc. London*, **139**, 545–554.

TAYLOR, J. D., CLEEVELY, R. J. & MORRIS, N. J. 1983 Predatory gastropods and their activities in the Blackdown Greensand (Albian) of England. *Palaeontology*, London, **26**, 521–553.

TAYLOR, J. H. 1963 Sedimentary features of an ancient deltaic complex: the Wealden rocks of south-eastern England. *Sedimentology*, **2**, 2–28.

TAYLOR, R. 1982 Lower Cretaceous (Ryazanian to Albian) calcareous nannofossils. *In* Lord, A. R. (Ed.) *A stratigraphical index of calcareous nannofossils*, Ellis Horwood Ltd., 40–81.

TOWNSON, W. G. 1975 Lithostratigraphy and deposition of the type Portlandian. *J. geol. Soc. London*, **131**, 619–638.

VERSEY, H. C. & CARTER, C. 1926 The Petrography of the Carstone and associated beds in Yorkshire and Lincolnshire. *Proc. Yorkshire geol. Soc.*, **20**, 349–365.

WIMBLEDON, W. A. & HUNT, C. O. 1983 The Portland-Purbeck junction (Portlandian-Berriasian) in the Weald, and correlation of latest Jurassic–early Cretaceous rocks in southern England. *Geol. Mag.*, **120**, 267–280.

WOOD, C. J. 1980 Upper Cretaceous. *In* Kent, P. E., *Eastern England from the Tees to the Wash*. British Regional Geology, HMSO, London, 92–105.

WOOD, C. J. & SMITH, D. 1978 Lithostratigraphical nomenclature of the Chalk in North Yorkshire, Humberside and Lincolnshire. *Proc. Yorkshire geol. Soc.*, **42**, 263–287.

WOOD, C. J., ERNST, G. & RASEMANN, G. 1984 The Turonian–Coniacian stage boundary in Lower Saxony (Germany) and adjacent areas: the Salzgitter–Salder Quarry as a proposed international standard section. *Bull. geol. Soc. Denmark*, **33**, 225–238.

WORSSAM, B. 1978 The stratigraphy of the Weald Clay. *Rep. Inst. geol. Sci. London*, **78/11**, 23 pp.

WORSSAM, B. C. & IVIMEY-COOK, C. H. 1971 The stratigraphy of the Geological Survey borehole at Warlingham, Surrey. *Bull. geol. Surv. G.B.*, **36**, 1–111.

WRIGHT, C. W. & KENNEDY, W. J. 1981 The Ammonoidea of the Plenus Marls and the Middle Chalk. *Monogr. palaeontogr. Soc. London*, 148 pp.

YOUNG, B. & MONKHOUSE, R. A. 1980 The geology and hydrogeology of the Lower Greensand of the Sompting borehole, West Sussex. *Proc. Geol. Assoc. London*, **91**, 307–313.

YOUNG, B. &            1981    The Aptian Lower Greensand fuller's earths of Bognor
  MORGAN, D. J.                Common, West Sussex. *Proc. Geol. Assoc. London*, **92,**
                              33–37.

YOUNG, B.,             1978    New fuller's earth occurrences in the Lower Greensand of
  MORGAN, D. J. &              south-eastern England. *Trans. Instn. Ming Metall.*, **B87,**
  HIGHLEY, D. E.               B93–B96.

ZIEGLER, P.            1982    *Geological Atlas of Western and Central Europe.* Shell
                              International Petroleum Maatschappij B.V., 130 pp., 40
                              pls.

# 13

# TERTIARY

# D. Curry

## Introduction

The Tertiary rocks of south-east England have excited the interest of naturalists from an early date because of their easy accessibility and the beauty and variety of their fossils. The first description of British fossils (Solander in Brander (1766)) to use the Linnaean system deals with the molluscs of the Barton Beds. It is beautifully illustrated and is still a standard reference work. In addition, some of the earliest British stratigraphical work relates to the beds of the Tertiary. In 1814, for example, Webster described the Isle of Wight sequences and compared their succession of faunas with that established in the Paris Basin by Cuvier and Brongniart (1810); thus providing possibly the first ever example of international stratigraphical correlation.

The simple structures and easy accessibility of the British Tertiary beds enabled their sequences and interrelations to be worked out in broad terms at an early date. Amongst important early workers were Charlesworth (1835), Prestwich (1847), Lyell (1852), Fisher (1862) and Gardner et al. (1888). In addition, many early monographs of the Palaeontographical Society were devoted to Tertiary fossils. Although interest in the Tertiaries waned somewhat around the start of the present century, research has been given a new impetus recently by the study of microfossils and radiometric dating, both of which have given a new precision to correlation. Further, the analysis of off-shore data has provided a much more assured picture of the palaeogeography of Tertiary times.

The main areas of outcrop of Palaeogene rocks in England and Wales occur in the synclines of the London and Hampshire Basins. Both of these out-crops have a continuation seawards; the Hampshire Basin extending south-eastwards into the Dieppe Basin, which occupies the greater part of the eastern English Channel, and the London Basin forming part of the North Sea Basin. Beds of latest Palaeocene to mid-Eocene age are present throughout this area. Older Palaeocene beds (Thanet Formation) occur in addition in Kent, and late Eocene and early Oligocene beds are present in the Hampshire Basin. Earliest Palaeocene (Danian) beds have been recorded in place only in the western part of the English Channel, but derived fossils of that age have been found in the Thanet, Reading, and Wittering Formations.

Another series of outcrops, this time exclusively continental and of mid-Tertiary age, is present locally in the western half of southern Britain. These include the Bovey and Petrockstow beds of Devon, and sequences NE of Lundy Island in the Bristol Channel, ESE of the Lizard in the English Channel and on the coast and off-shore from mid-Wales (Mochras). Late Oligocene to mid-Miocene marine sequences are unknown in mainland Britain, the nearest occurrences on land in that time-span being in Belgium and western France.

The main area of occurrence of Neogene marine rocks in England and Wales is in East Anglia and includes the post-mid-Pliocene shelly sands known as 'Crags'. There are in addition isolated occurrences in Cornwall (St Erth) and Kent (e.g. Lenham). Continental Neogene deposits occur locally in fissures, particularly in Derbyshire and N Wales. The occurrences mentioned above are summarised in Table 13.1; localities mentioned in the text are indicated on Figure 13.1.

## Nomenclature, Zonal Stratigraphy, Correlation, Palaeoecology

The formal nomenclature of the time-divisions of the Tertiary is in a state of confusion. There is disagreement both as to the levels at which major time-

**Table 13.1.** Occurrences of Tertiary beds in England and Wales.

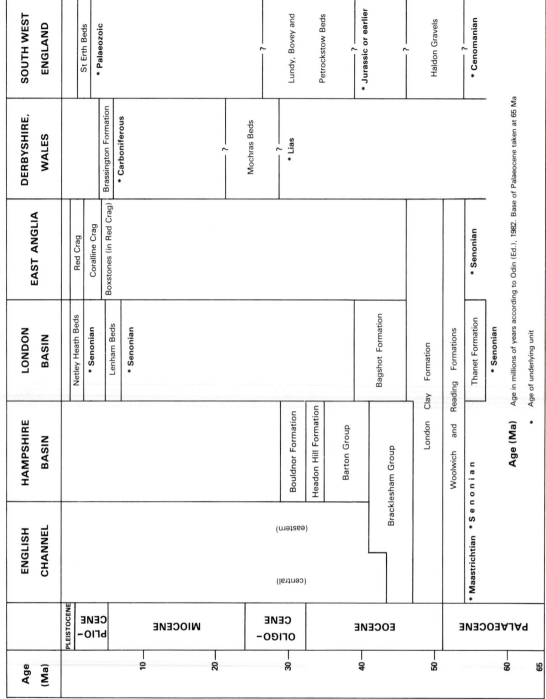

boundaries should be taken and also as to a suitable selection of stage-names to relate to these. Details of the problem are given in Curry *et al.* (1978), and the nomenclatorial framework recommended in that publication is adopted here. Stratigraphical analysis based on the occurrence of biozones has, until recently, been very imprecise because no group of marine fossils was known to earlier workers to have the proved value for that purpose displayed, for instance, by the graptolites and ammonites of earlier eras. Much early work, based mostly on gastropods and bivalves, has proved unreliable because of the restrictions which facies and provinciality impose in these groups. In the past twenty years, however, planktonic foraminiferids, coccolithophores and palynomorphs (especially dino-flagellates) have been employed to construct biozonations of relatively wide geographical application. Numbered zonal sequences based alternatively on planktonic foraminiferids (P(Palaeogene) and N(Neogene) Zones) or coccolithophores (NP and NN Zones: prefix N = nannoplankton) have come into wide use. These are adopted, as appropriate, in the

present chapter (for literature on these zonal schemes see references in Curry *et al.* 1978). Additional information on correlation is provided by the large benthonic foraminiferid *Nummulites* (late Palaeocene to mid-Oligocene), which underwent rapid evolution in the Tethyan area. It migrated thence from time to time, mostly via the proto-Atlantic, into NW Europe, and its local appearances and extinctions are valuable as time-markers. Assemblages of mammals (Hooker & Insole, 1980) have been of great value in correlating continental sequences, whilst pollen and spores, which may be present in both marine and continental deposits, have provided links between marine and continental zonations and have given information about contemporary climates.

Benthonic foraminiferids (Murray & Wright 1974) and ostracods (Keen 1977) were used, in association with molluscs, to provide detailed palaeoecological interpretations, and leaves and seeds of land-plants (e.g. Reid & Chandler 1933) have supplemented the information from pollen and spores to give glimpses of the vegetable cover.

**Fig. 13.1.** Outcrops of Tertiary Beds in Wales and Southern England, with names of places of geological interest.

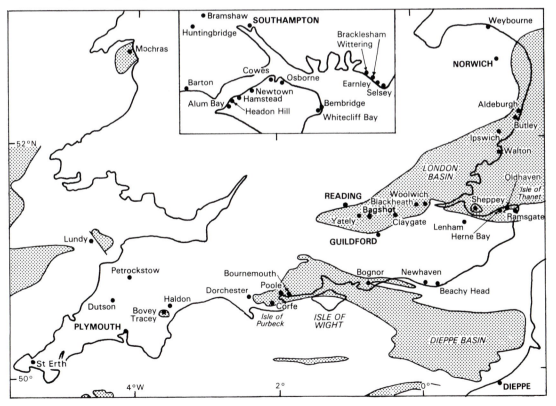

| Magneto-stratigraphy. Normal episodes in black | NP | THAMES VALLEY AND ESTUARY | ISLE OF WIGHT (EAST) | ISLE OF WIGHT (WEST) | BOURNEMOUTH AND STUDLAND | Dinoflagellate zones (Bujak et al 1980, and others) | | | Age (Ma) |
|---|---|---|---|---|---|---|---|---|---|
| (unnamed) | ?22 | | | *Hamstead Beds* 78 m (Bouldnor Formation) | | *gochti* | EARLY | OLIGO-CENE | 30 |
| | 20 | | *Bembridge Beds* 34 m | 25 m | Bembridge Limestone 5m | *perforata* | LATE | EOCENE | 35 |
| | 17 | | *Osborne Beds* 21 m | 23 m (Headon Hill Formation) | Hengistbury Beds 15m / Boscombe Sand 10m | *variab. – cong.* / *R. porosum* | | | |
| Huntingbridge | 16 | | *Headon Beds* 65 m | 43 m | | *fenestratum* / *H. porosa* | | | |
| | 16 | Bagshot Formation *(Beds)* 115 m | *Barton Beds* 137 m | 91 m (Barton Group) | *Bournemouth Marine Beds* 20 m | *intricatum* | MIDDLE | | 40 |
| Earnley | 15 | | (*Huntingbridge Beds*) | | *Bournemouth Freshwater Beds* 70 m (Bournemouth Group 237 m) | *arcuatum* | | | |
| Wittering | 14 | | Selsey Sand 29 m / Marsh Farm 15 m / Earnley Sand 28 m | | 100 m ? | *comatum* | | | 45 |
| | 13 | | Wittering Formation 77 m | | Poole Formation / Agglestone Grit | *laticinctum* | EARLY | | |
| London Clay 3 / 2 / 1 | 12 | | (Bagshot Sands 42 m) | (Bagshot Sands 10 m) | Corfe Clay | *abbreviatum* | | | |
| | 11 | London Clay Formations 115-155 m | London Clay Formation 140 m | 81 m (Reading and Woolwich) | Redend Sandstone / London Clay Formation 30 m / 85 m ? | *reticulata* / *ursulae* | | | 50 |
| Oldhaven | 9 | | Woolwich Formations 47 m | 26 m (Reading and Woolwich) | 30 m | *phosphoritica* | LATE | PALAEOCENE | |
| | 8 | 20 m | | | | *hyperacanthum* | | | 55 |
| Thanet | | Thanet Formation *(Beds)* 0-30 m | | | | *Deflandrea* spp. | | | |
| | | | | | | | | | 60 |

Brackleshamd Group

**Age (Ma)**   Age in millions of years according to Odin (Ed.), 1982. Base of Palaeocene taken at 65 Ma

**NP**   Nannoplankton Zone identified in the British succession, with Zone number (mostly from Aubry, 1983).

**Table 13.2.**  Palaeogene sequences of the London and Hampshire Basins.

## Sedimentology, Radiometric Dating, Palaeomagnetism

The sedimentary style of the English Palaeogene conforms with that of NW Europe in general. Earliest Palaeocene (Danian) marine beds are calcareous and contain little or no glauconite. From mid-Palaeocene to late Eocene times marine limestones are subordinate and the sequences, mostly silts and fine sands, are typically glauconitic. Throughout the Tertiary, continental sequences are dominantly paralic, limestones being of infrequent occurrence. Marine Neogene units are mostly sandy, and typically give evidence of tidal action.

Studies of the distribution of heavy minerals (Blondeau & Pomerol 1969, Morton 1982a, b) have led to the identification of possible areas of source rocks and of periods of transgression and regression.

The presence of glauconite at many levels has provided a means of absolute dating, using both the K-Ar and Rb-Sr systems. About 20 sets of determinations have been made on British samples, with a standard error of the order of $\pm 2$ per cent (Odin (ed.) 1982). Preliminary studies of the Palaeogene sequences of the Hampshire and London Basins have revealed the presence of at least 20 magnetic reversals. Table 13.2 summarises the data and is based on Townsend & Hailwood (1985) and E. A. Hailwood (pers. comm.). Aubry et al. (1986) have proposed correlations with the marine magnetic anomaly sequence as follows: Thanet, 26; London Clay (1, 2, 3), 24; Wittering, 23; Earnley, 21; Huntingbridge, 20. The Oldhaven magnetozone remains uncorrelated.

## Regional Stratigraphy

Most of the stratigraphical nomenclature of the English Tertiary was established in the early part of the 19th century and is now obsolescent. It has only recently (Stinton 1975, Insole & Daley 1985, Edwards & Freshney 1987) been revised to produce a formal hierarchy based on members, formations and groups. To provide continuity with earlier publications, a comparison of old and new nomenclatures is included in Table 13.2 (old names are in italics).

# Palaeogene successions of the London and Hampshire Basins

The sequences are only slightly consolidated and in many areas have been weathered by oxidation and percolation of ground-water to a depth of several metres. As a result, fossiliferous exposures are infrequent, being mostly confined to sea-shore sections and the rare deep man-made excavations. Table 13.2 lists the successions.

## Late Palaeocene

A large time-gap (some 20Ma) locally separates the latest Mesozoic and earliest Tertiary strata in the London and Hampshire Basins. Although the contact everywhere appears to be conformable, the lowest Tertiary beds rest on Campanian Chalk in Norfolk and Dorset and overlap progressively from both directions on to the Coniacian in the area of the London Platform (Fig. 13.2a). The lowest Tertiary unit (whether at the base of the Thanet or of the later Reading Formation) is almost without exception a glauconitic sand containing a mixture of unrolled and battered, well-rounded flints. This was developed on a marine erosion surface.

The Thanet Formation (type of the Thanetian (Renevier 1873–1874)) occurs only in the London Basin, although close equivalents are known from N France and Belgium. It consists of marine glauconitic clayey silts and fine sands, which vary in thickness from a maximum of 30 m in Kent to zero near Guildford and Ipswich (Hester 1965). The wedge of sediments so formed is so regular that it has been speculated that it is not an original feature but is the result of tilting and planation prior to the deposition of the Woolwich/Reading Formations. The lower and middle parts of the Thanet Formation are well exposed near Ramsgate and the middle and upper parts near Herne Bay. Its basal conglomerate (Thanet Base Bed) has yielded only derived fossils, so its precise age is uncertain. At higher levels a molluscan fauna is present which includes several genera now characteristic of cool seas (Wrigley 1949), including *Arctica*, *Astarte*, *Aporrhais*. Small foraminiferids are abundant and likewise indicate a cool sea, with a depth of less than 50 m (Haynes 1958). This depth estimate is supported by the presence, in the highest beds, of calcareous algae which, on the other hand, suggest relatively warm water. The upper part of the Thanet Formation yields *Heliolithus riedeli*, indicating the presence of nannoplankton Zone NP8. At lower levels *H. riedeli* appears to be absent, suggesting an older age (NP6–7?) for these (Siesser, Ward & Lord 1987). Contemporaneous volcanic activity is indicated by the existence of volcanic ash locally in the Base Bed (Knox 1979), and this also may be responsible (through the release of silica into the sea) for the unusual abundance of sponge spicules higher in the formation. The heavy minerals of the Thanet Formation point to derivation from Scottish and Scandinavian sources by longshore drift, a situation repeated in the London Clay Formation (Morton 1982a).

The Woolwich and Reading Formations, which are,

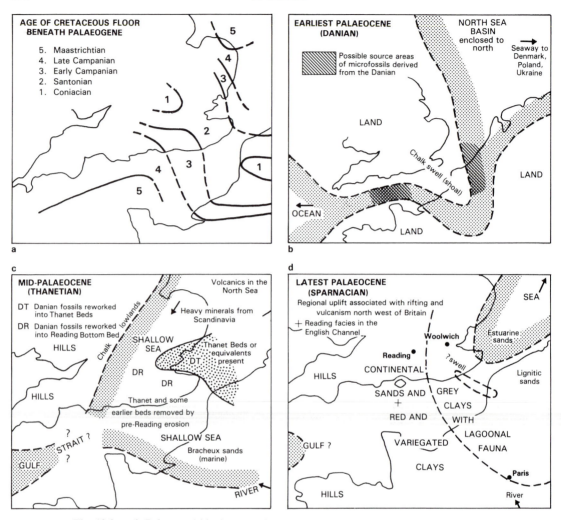

**Fig. 13.2.** a–h Palaeographical maps of Southern Britain and nearby regions during the Tertiary.

approximately at least, lateral equivalents, are present throughout the area, and can be matched in the eastern and central English Channel, and N France, including the Paris Basin. In contrast to the preceding and succeeding units, their heavy minerals now include components from the south (Ardennes, and possibly Armorica). In addition, they show a strong regional pattern, a continental facies (Reading Formation, *argile plastique*) being developed W of London and S of Paris, with an estuarine or lagoonal facies (Woolwich Formation, *lignites* = type Sparnacian) to the east and north of these areas respectively. In England, the lowest bed associated with both facies is however

marine, as mentioned earlier. Towards E Kent the estuarine sequence dies out and the whole sequence becomes marine. The Woolwich/Reading Bottom Bed has yielded sharks' teeth (Gurr 1963) and, locally, banks of *Ostrea bellovacina* associated with a microfauna which resembles that of the Thanet Formation rather than that of the succeeding London Clay. *Discoaster multiradiatus* (NP 9/10, Table 13.2) is present locally.

The main mass of the Woolwich Formation is typically developed along the N Kent coast and in S London, with an outlier at Newhaven. It comprises dark lagoonal or estuarine clays and sands with a

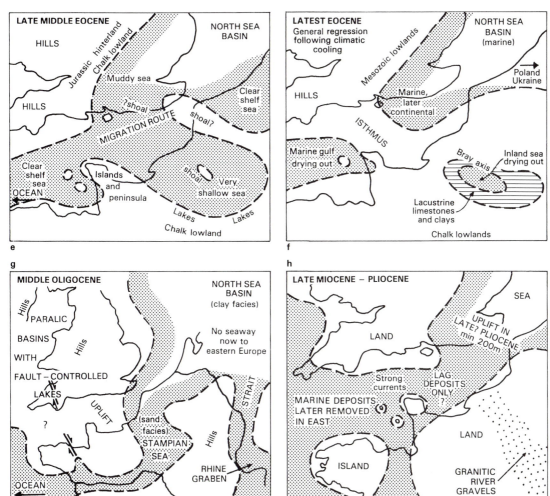

restricted molluscan fauna including *Brotia mel-anioides*, *Tympanotonos*, *Corbicula* and *Ostrea tenera*. Freshwater limestone with *Viviparus* is present in S London. Overlying the beds of Woolwich facies in the south of the London Basin, and deeply ravining them locally, is a series of sands and flint-pebble beds, the Oldhaven and Blackheath Beds. In E Kent their fauna is wholly marine, but restricted in content (Oldhaven facies), with *Protocardia* and *Glycymeris*, but in the area SE of London its contained mollusca are mostly of estuarine types, including many species present in the Woolwich clays. Some of the estuarine species are certainly reworked, but others (e.g. *Tellinocyclas*)

appear to be indigenous. Vertebrates, including mammals, sharks, and the freshwater ganoid *Lepisos-teus*, are present (White 1931), suggesting, with the molluscs, an open estuarine environment in which the Oldhaven and Blackheath sands may represent channel fills in nearly contemporaneous Woolwich clay flats. In an alternative explanation (King 1981) the Oldhaven Beds are regarded as wholly younger than the Blackheath Beds, and as the equivalent of beds of London Clay facies to the north and west.

In the western half of the London Basin and in the Hampshire Basin the Woolwich Formation is replaced laterally by the reddish and variegated clays, with associated pale dune-sands, of the Reading Formation, though some interdigitation of the Woolwich and Reading facies occurs locally. The main mass of the Reading Formation is typically unfossiliferous, but locally has yielded continental fossils (leaves, charophytes, *Bithynia* (Curry 1959)). It is thought to have been laid down in an area of freshwater marshes, with dunes and temporary pools (Hawkins 1946, Hester 1965). An alternative hypothesis (Buurman 1980) suggested that the Reading Formation is a local modification of the Woolwich Formation due to penecontemporaneous uplift and pedogenesis.

## Earliest Eocene

The London Clay Formation is the most uniform and at the same time the most widespread of the English Palaeogene deposits. It occurs throughout the London and Hampshire Basins, except near Dorchester, where it is cut out by the overlap of the Poole Formation. Offshore, closely similar units are present in the central English Channel and the Dieppe Basin, extending thence across N France and Belgium (Argile d'Ypres) into N Germany (Eozän 1–4). It comprises brown and blue-grey marine clays with a variable admixture of silt and fine sand. Glauconite is present only sporadically. In the east of the London area the succession is almost wholly of clays, but to the west and in the Hampshire region several alternations of clays and fine sands can be made out, and thin beds of rounded flint pebbles are present locally and are of value in correlation (King 1981). Layers of septarian concretions occur in the more clayey units. Ash bands are present at the base in Essex (Elliott 1971) and are widespread in the North Sea (Jacqué & Thouvenin 1975) and in Denmark and Germany at about the same level.

Macrofossils occur sparingly, as a rule, but are common locally. About 350 species of molluscs, mostly gastropods and bivalves, are known, but include also the cephalopods *Nautilus*, *Aturia* and *Belosaepia*. The cool-water molluscan genera noted as occurring in the Thanet Formation persist, but are accompanied by warm-water forms, such as *Athleta*, *Tibia* and *Ficus*. Worms (*Rotularia*, *Ditrupa*) occur frequently. Microfossils, especially foraminiferids and dinoflagellates, are widespread. Several dinoflagellate zones have been recognised (Bujak *et al.* 1980) and provide correlations with other parts of NW Europe and with NE Spain. Study of the dinoflagellates indicates that the lowest beds are slightly diachronous from east to west, the ash levels being probably the earliest of all. The top of the London Clay Formation is substantially diachronous in the reverse sense, the highest clays in the London area being possibly the lateral equivalents of the lowest part of the Bracklesham Group of the Hampshire Basin (Eaton 1976, but cf. Bujak *et al.* 1980), and, probably, of some part of the Poole Formation of Dorset.

The London Clay Basement Bed rests with a sharp contact on the Reading or Woolwich Formation beneath, is usually sandy and glauconitic, and may contain abundant marine molluscs of relatively shallow-water aspect. Similar sandy shell-beds with a somewhat similar fauna occur locally at higher levels in the Hampshire Basin (Bognor, Portsmouth). In the eastern part of the London area, the middle part of the London Clay Formation provides evidence of deposition in relatively deep water (200 m perhaps). It contains crinoids, *Terebratulina* and the mollusc *Thyasira*, together with a foraminiferal fauna rich in nodosariids and, at some levels, a high proportion of planktonics. A great variety of seeds of land-plants has been found at Sheppey (Reid & Chandler 1933). Fruits of the swamp-palm *Nipa* are common, the whole flora suggesting a tropical rain-forest with a mean annual temperature of about 25° C. There are few other indications of proximity to land in the London Clay Formation, however, and it seems that this flora may have floated for some tens of kilometres before entrapment in the sediment.

Overlying typical London Clay almost everywhere is a series of mainly unfossiliferous sands and clays which have so far defied accurate correlation. They include the Claygate Beds, marine alternations of sand and clay to the SW of London, and certain units in the Hampshire Basin which have been referred to the (Lower) Bagshot Beds. The latter include 30–40 m of pale sands in the Isle of Wight, and the Redend, Corfe, and Agglestone units of Purbeck (Arkell 1947). These Purbeck units (now grouped in the Poole Formation) include ferruginous sandstones, overlain by a series of pipeclays of economic importance, and these by grits, the whole being apparently of continental origin. Almost their only fossils are plant remains, including seeds, which however resemble the flora of succeeding beds (Chandler 1962), rather than that of the London Clay. The grit series contains pebbles of radiolarian chert and other ancient rocks, presumably derived from Cornubia. In addition, in the area between Dorchester and Wareham, silicified Purbeck limestone and Upper Greensand chert are present, probably of southern origin (Plint 1982).

## Later Early and Middle Eocene

At Whitecliff Bay, Isle of Wight, and on the opposite coast of West Sussex the London Clay Formation (with its attendant sands) is seen to be overlain by the thick marine sequence of the Bracklesham Group, which can be traced north-westwards into the New Forest. The Bracklesham Group (Curry *et al.* 1977,

Edwards & Freshney 1987) is a series of clayey sands which is typically more coarse-grained than the London Clay and contains notable amounts of glauconite at most levels. It contains four major sedimentary cycles, named successively as the Wittering Formation, Earnley Sand, Marsh Farm Formation and Selsey Sand, which yield a series of characteristic molluscan faunas. The cool-water molluscs noted in earlier units are now absent, and warm-water forms, including *Athleta*, *Ancilla*, *Architectonica*, *Mesalia*, and *Chama*, are present. The large bivalve *Venericor planicosta* is abundant at some levels and the large foraminiferid *Nummulites* occurs in England for the first time.

The base of the Wittering Formation is marked by a series of laminated clays, probably of intertidal origin, which have yielded no macrofossils except indistinct moulds of marine molluscs. Dinoflagellates are present however, and hint at correlation with the highest levels of the London Clay in the London area, as already stated. Overlying these laminated clays are bioturbated glauconitic sandy clays with many molluscs including small *Venericor* and *Turritella*, together with rare nummulites (*N. planulatus, N. aquitanicus*). The return of intertidal conditions is marked by more laminated clays, with lignite. The lower part of the overlying Earnley Sand comprises bioturbated sandy marine clays which may be very rich in glauconite. Large molluscs (*Venericor, Ispharina* etc.) are abundant, but nummulites appear to be absent. At higher levels, however, *N. laevigatus* occurs in rock-forming abundance, with many molluscs (especially bivalves) and a NP14 nannoflora (Aubry 1983). In the overlying Marsh Farm Formation the presence of laminated clays suggests a return to intertidal conditions once more, the fauna becomes restricted and includes rare estuarine molluscs.

A pebble bed occurs at the base of the Selsey Sand which, as a unit, is somewhat more fine-grained and less glauconitic, and yields, at Selsey, four distinct molluscan faunules, the oldest of which includes the very large gastropod *Campanile*. The youngest of these faunules is unique in the English Palaeogene, although of a type well known in the Cotentin of France and the western English Channel. In this, small marine gastropods are abundant, together with microfossils suggesting a seaweed environment, and, hence, clear shallow water (Murray & Wright 1974). *Nummulites variolarius*, which first appeared in the upper part of the Earnley Sand, where it is rare and very small, becomes very abundant in the upper part of the Selsey Sand. At this level a NP15 nannoflora has been found, with NP16 suspected at the summit.

In the western half of the Hampshire Basin the Bracklesham Group passes progressively westwards into deposits of a continental type. At Alum Bay (Fig. 13.3), glauconitic levels are infrequent and the series includes brightly coloured sands and grey or black clays. Leaves occur at several levels and almost the only other fossils recorded are dinoflagellates, of which five faunules have been recognised and provide a means of correlation with Whitecliff Bay (Eaton 1976). In Dorset, lateral equivalents of the Bracklesham Group are to be sought in the Branksome Sand and (possibly) the underlying Poole Formation, which together comprise the Bournemouth Group. The Branksome Sand comprises pale sands, pipeclay and lignitic clays which have yielded marine fossils only in its upper part (Bournemouth Marine Beds). The dinoflagellates of that unit correlate with those of the Selsey Sand and Huntingbridge Beds (Costa *et al.* 1976). In its lower part (Bournemouth Freshwater Beds) the Branksome Sand includes several leaf-beds which give no clue as to age. This unit must presumably equate with the lower part of the Bracklesham Group. Creechbarrow Hill, near Corfe, Dorset is capped by a freshwater limestone resting on sands and gravels of the Poole Formation. On the basis of its molluscs this was formerly thought to be a local equivalent of the Bembridge Limestone. Its mammal fauna (Hooker 1977) is of Middle/Upper Eocene age, however, and suggests correlation with the lower part of the Barton Beds.

The lateral equivalent of the Bracklesham Group in the London Basin is the Bagshot Formation, which occurs in an area of heathlands to the west of London. This is lithologically similar to the Bracklesham Group, but is more sandy and is almost completely lime-free. As a result, fossils are infrequent, usually occurring as moulds which are difficult to identify. Its lower part (Lower Bagshot Beds) comprises fine sands, mostly current-bedded, which presumably correlate with some part of the Wittering Formation, although there is as yet no fossil evidence for this. Overlying these are laminated clays, followed by very glauconitic sandy clays (Middle Bagshot Beds). Palynomorphs from the laminated clays suggest a correlation with the Wittering Formation or, more probably, the Earnley Sand (Potter 1977). At Yateley the glauconitic clays have yielded solid fossils in a marine fauna which clearly equates with that of the *laevigatus* Bed of the Earnley Sand. The highest unit of the Bagshot Formation (Upper Bagshot Beds) is of fine marine sands, with some clay levels. Current bedding is rare. At its base is a thin, but persistent, pebble bed, in which Lower Greensand chert occurs (Dines & Edmunds 1929). Molluscs are present locally as moulds, but their identification is difficult and it is not certain whether they indicate correlation with the uppermost part of the Bracklesham Group or with the lower part of the Barton Group (Curry 1965).

## Late Eocene

The Palaeogene sequences of the London Basin terminate with the Bagshot Formation. In the Hampshire Basin the Bracklesham Group is overlain by the new sedimentary cycle of the Barton Group ($\approx$ *Beds*). The base of this unit was formerly taken at a thin nummulite bed (*N. prestwichianus*), but is now (Edwards & Freshney 1987) redefined on a lithological basis. As a result, the Huntingbridge (New Forest) level, formerly included on faunal grounds in the Bracklesham Group (*Beds*), now falls, in part at least, into the Barton Group. The Barton Beds are the type of the Bartonian (Mayer-Eymar 1858). In its lower part the Barton Formation consists of more or less glauconitic sandy clays with a large mollusc fauna (over 600 species) which is almost entirely marine, but includes rare estuarine forms, especially in the most westerly outcrops. In the upper part there is a transition through estuarine or lagoonal beds (highest Barton Beds) to the fluviatile and lacustrine deposits which are dominant in the succeeding Solent Group.

The fauna of the Barton Group is notably endemic, there being few species in common with beds of the same age on the European continent. Amongst the molluscs, *Corbula* and *Turritella* are abundant, and many species of turrids and volutes are present. The solitary coral *Turbinolia* is common. A rich nannoflora has been recognised, indicating the presence of Zone NP16 (also occurring at the Huntingbridge level) in the Lower Barton Beds; and the presence of Zone NP17 in the Middle and Upper Barton Beds (Aubry 1983). The overlying non-marine unit (the Totland Bay member of the Headon Hill Formation) has yielded, near Milford-on-Sea, a large series of mammals, including primates, rodents, creodonts, and the herbivores *Palaeotherium* and *Dichodon* (Cray 1973). This fauna is not known in the Paris Basin, but can be matched at Euzet (Gard), France, Carbonised seeds of fresh water plants (*Limnocarpus, Stratiotes*) are common, together with the calcareous fruits of stoneworts (*Chara*), whose remains are common at many levels in the succeeding deposits also and have been used for correlation with French sequences (Feist-Castel 1977).

**Fig. 13.3.** Alum Bay, Isle of Wight, provides a spectacular exposure of Upper Cretaceous Chalk through the Palaeogene to the Bembridge Marls of the Oligocene exposed on the top of Headon Hill (right centre). Flanked by the white Chalk, the vertically-dipping variegated Tertiary sands and clays present the most colourful cliff scenery in Britain (R. H. Roberts).

The depositional sequence of the Barton Group, plus the Totland Bay member, is consistent with the gradual infilling of an originally marine sedimentary basin which started at a depth of about 100 m and finished only a few metres above sea level. It may be noted in passing that all the continental units above the Barton Group contain calcareous fossils in abundance. In this respect they contrast strongly with continental units in the Bournemouth Group and with the mid-Tertiary sequences in western Britain, in which calcareous fossils are almost unknown. The explanation for this difference may lie in the source-areas in each case. Material for the Dorset and western units came mainly from the non-calcareous uplands of Cornubia and Wales, whereas the Headon Hill and later formations in Hampshire may have received lime-rich water draining from chalk lowlands to the north-west and (possibly) the south.

The Colwell Bay member of the Headon Hill Formation marks the beginning of a new sedimentary cycle. In an area NE of a line from Brockenhurst in the New Forest to Newport in the Isle of Wight the Totland Bay member is succeeded by dark clays (Brockenhurst Beds) with scattered flint pebbles and a purely marine fauna. Elsewhere it is followed by the Venus Beds, greenish or grey sandy clays with a mixed fauna, mostly of estuarine molluscs, which also overlie the Brockenhurst Beds where these are present. It is not clear whether the Venus Beds are wholly younger than the Brockenhurst Beds or whether they may in part represent a lateral variation in a south-westerly direction. The fauna of the Brockenhurst Beds is noteworthy for the presence of several genera of corals, although *Turbinolia* is absent. Its molluscan fauna, too, is distinctive. As with the Barton fauna, this cannot be matched in the Paris region. It has, however, many species in common with those of the Latdorf sands (Lattorfian, basal division of the type Oligocene) of N Germany and the Grimmertingen sands of Belgium. Characteristic species include *Ostrea ventilabrum*, *Venericardia deltoidea* and *Athleta dunkeri*. The nannoflora is that of the NP20 Zone (Martini 1971). The fauna of the Venus Beds includes *Sinodia suborbicularis* ('*Venus*'), with *Corbicula obovata*, *Ostrea velata* and several species of cerithiids in a typical brackish-water assemblage. Continental conditions recur in higher units of the Headon Hill Formation, with a series of pale marls, fluviatile sands, and thin lacustrine limestones. The sands typically yield *Viviparus* and the bivalve *Potamomya*, whereas the marls, deposited no doubt in still water, are dominated by the pulmonates *Lymnaea* and *Planorbis*. *Chara* occurs in both facies and several species are present. Ostracods are abundant at some levels, but of few species (Keen 1977). Near Ryde, these higher units contain grits with scattered flint-pebbles, and a clay

with abundant complete fishes (*Diplomystus*). Conformably overlying the Headon Hill Formation is the Bembridge Limestone, once important as a building stone. It contains abundant moulds of freshwater molluscs and (locally) land molluscs. A new group of mammals occurs in the Bembridge Limestone, and has been recognised as equivalent to the well-known fauna of the *gypse* of Montmartre, Paris (Savage *in* Davies 1975). This includes *Anoplotherium*, *Dichobune* and more evolved species of *Palaeotherium*. This correlation between the Bembridge Limestone and the *gypse* is confirmed by the species of *Chara* of the two formations.

## Early Oligocene

The Oligocene Series was created by Beyrich (1854) with the Latdorf sands as its lowest marine unit. Because of the resemblances between the faunas of the Latdorf sands and the Brockenhurst Beds, the base of the Oligocene in England was customarily taken at the base of the Totland Bay member (later, of the Colwell Bay member). Martini (1971), however, stated that the nannoplankton of the Brockenhurst Beds (NP20) indicate that this unit is distinctly older than the Latdorf sands (NP21) and so should be referred to the Eocene. Châteauneuf (1980), on the other hand, saw no important difference in their palynofloras. The facies of the English and German successions are so different that it has not proved possible to identify an English level above the Brockenhurst Beds which could be said with assurance to be of the same age as the Latdorf sands. Two marine units occur above the Brockenhurst Beds in England; they are the Bembridge Oyster Beds and the Cranmore member, the lowest and highest units respectively of the Bouldnor Formation. The fauna of the Cranmore member is found in NW Europe in beds which have always been regarded as Middle Oligocene. The base of the Oligocene in England is therefore taken here at the base of the Bouldnor Formation.

The Bouldnor Formation commences with a restricted marine incursion, which rapidly gives way once more to continental deposits. The highest levels mark a return to restricted marine conditions. The Bembridge Oyster Beds, which are best developed at the eastern end of the Isle of Wight, are thin sands, with scattered flint pebbles. Fossils include the molluscs *Sinodia*, *Mytilus* and *Ostrea vectensis*, together with the cirripedes *Balanus* and *Aporolepas*. The Bembridge Oyster Beds become less marine when traced westwards and are absent west of Newtown River, indicating that the associated marine incursion, like that of the Brockenhurst Beds, probably came from the east. The Bembridge Oyster Beds form the lowest unit of the Bembridge Marls member, which comprises grey clays with subordinate sands, formed in brackish lagoons or flood-plain and coastal lakes (Daley 1973). The mollusc fauna of these clays includes the brackish-water genera *Corbicula, Potamides,* and *Tarebia,* with freshwater *Viviparus* and *Potamaclis.* Near their base, between Cowes and the Newtown River, the Bembridge Marls contain a lithographic limestone, which has yielded a variety of insects and freshwater arthropods (Reid & Strahan 1889).

The succeeding Hamstead member sees a continuation of the facies and sedimentary style established in the beds beneath. Its lowest unit ('Black Band') is a clay with freshwater molluscs and seeds of aquatic plants. It overlies a weathered surface of rootlet-penetrated Bembridge Marls, and contains scattered angular flints. Mammals have been found at a slightly higher level and include *Anthracotherium, Brachyodus,* and *Entelodon,* members of a wave of immigrants from Asia (Gabunia 1964). The remainder of the Hamstead member is a monotonous series of grey and greenish marls with layers of freshwater molluscs and seeds, though a few levels yield brackish-water molluscs and dinoflagellates.

The Cranmore member comprises thin clays in which the conditions of deposition are seen to become increasingly saline upwards. At their base appears an estuarine fauna dominated by cerithiids (*Pirenella monilifera, Tympanotonos labyrinthus*), with *Nystia* and *Corbicula.* At higher levels marine genera (*Corbula, Athleta, Sinodia, Ostrea*) occur in addition. The environment must still have been hyposaline, however, as no animals (such as echinoderms or corals) restricted to sea water of normal salinity are present. The succession ends at this point and a time-gap of some 25 M.y. separates these highest beds from the next younger marine unit known on the British mainland, the Lenham Beds.

# Continental Mid-Tertiary Beds of Wales and South West England

The units dealt with in this section range in age from late Palaeocene to early Miocene and occur in a number of scattered outliers, most of which are more or less fault-bounded. All of the units are non-calcareous, and the only fossils recorded are of plants, including palynomorphs. For this reason they have proved difficult to date accurately.

## Haldon gravels

These occupy the highest parts of the Haldon Hills (Devon) and include a lower unit with unabraded flints, overlain by gravels with somewhat battered flints, together with a suite of exotic rocks. All of the latter could have been derived originally from Dartmoor, perhaps by recycling through basement beds of the Chalk. The Haldon gravels are thought to be flood-plain deposits laid down in a savannah environment, and on the pattern of occurrence of their heavy minerals have been dated tentatively in the range late Palaeocene to early Eocene (Hamblin 1974).

## Bovey Formation and Petrockstow and Dutson Beds

The Bovey and Petrockstow sequences occupy depressions situated on the line of the Sticklepath–Marland dextral wrench-fault (Dearman 1964). They include pottery-clays of economic importance, associated with thin lignites (Fig. 13.4), overlain and underlain by sands and gravels derived variously from nearby Palaeozoic rocks or from the Dartmoor Granite. The sequences are exceptionally thick (660 m proved at Petrockstow, over 1,100 m deduced from geophysical evidence in the Bovey Basin), and include units laid down in flood-plain, lacustrine, and, possibly, alluvial fan environments, but not in a deep lake (Edwards 1976). No doubt the basins were formed and deepened in association with movement along the wrench-fault, and the absence of evidence for a deep lake implies that sedimentation kept pace with the deepening, and that the episode (or episodes) was not of a short duration. The upper part of the Bovey Formation has yielded seeds of species known from the Oligocene of the Isle of Wight, together with pollen thought to indicate an early to middle Oligocene age. The lower (unexplored) part is probably Eocene. The Petrockstow sequence was tentatively dated as late Eocene to early Oligocene (Boulter in Curry *et al.* 1978). The Dutson Beds (Freshney *et al.* 1982) occupy a tiny basin on an associated fault-line near Launceston and have been dated as late Eocene.

## Lundy Tertiary basin

A basin similar to those of Bovey, Petrockstow, and Dutson is known in the Bristol Channel NE of Lundy (Fletcher 1975) and has been dated provisionally as mid-Oligocene (Boulter in Curry *et al.* 1978), see also Chapter 15.

## Mochras Farm, Llanbedr, Wales

An exploratory borehole at this site penetrated over 500 m of clastic Tertiary deposits, ranging from clays to conglomerates, which included thin lignites. They rest on an even thicker sequence of Liassic beds (Woodland 1971). The whole sequence is less than 3 km from Lower Palaeozoic rocks at the surface and it is clear that a major fault separates the two. The nature of the Liassic and Tertiary rocks indicates that movement took place along the fault contemporaneously in each case, with sedimentation keeping in step with the movement. The Tertiary sequences clearly point to deposition above sea level. They have yielded a palynoflora, on the basis of which various ages in the range mid-Oligocene to early Miocene have been proposed (Boulter in Curry *et al.* 1978).

## The Neogene Period in southern Britain

The only British rocks which are post-mid-Oligocene and pre-late Miocene in age were deposited on land, as discussed in the previous section. To encounter unequivocally marine rocks between these ages it is necessary to go to Belgium, NW Germany, or Denmark, on the fringes of the subsiding North Sea, or to the western English Channel and the Loire valley. It seems probable therefore that the whole of Britain was above sea level during this period. It should be noted,

however, that the remains of the oldest marine Neogene deposits, the Lenham Beds, occur at the considerable height of 180 m, and that the somewhat younger Netley Heath Beds are present only above 150 m, so that it is not excluded that there may have been substantial mid-Tertiary marine deposition in SE England, whose sequences have subsequently been removed by erosion (see also the sections on palaeogeography and tectonic structure).

### Late Neogene marine deposits in England

Table 13.1 lists the occurrences and indicates their relation to the geological time-scale.

### Lenham Beds

These are ferruginous marine fine sands with some glauconite and rounded flint-pebbles which rest on and fill pipes in the Chalk on the crest of the North Downs near Lenham, Kent, and are also known at

**Fig. 13.4.** Clays and lignites of the Southacre Member of the Bovey Formation (Oligocene) exposed in the working face (1970) of East Golds clay pit, Newton Abbot, Devon. The contortions, including low-angle normal faults, are considered to be of periglacial origin. The ball clays of the Bovey basin have long been used in the pottery industry. (B.G.S. A 11564.)

Beachy Head, Eastbourne (Worssam *et al.* 1963). Fossils occur only as moulds, but over 100 species, mostly of molluscs, have been recognised. The age of the fauna is controversial. Most of the species occur also in the Pliocene Coralline Crag, but a few suggest a greater age. Recent opinion (Funnell in Curry *et al.* 1978) favours a correlation with the Deurnian (late Miocene) of Belgium.

## Netley Heath Beds

These occur between Guildford and Dorking and, like the Lenham Beds, rest at high levels on the Chalk. They consist of structureless masses of gravel and sand, suggesting flow and mixing under glacial conditions. Amongst the gravel is material derived from the Lower Greensand, together with ferruginous sandrock with moulds of marine molluscs, including abundant *Corbulomya complanata*, a species typical of the Red Crag and of the late Pliocene of Belgium. The heavy minerals of the bulk rock resemble those of the Lenham Beds and are quite distinct from those of the ferruginous sand-rock, which themselves are unlike those of the Red Crag of East Anglia (Dines & Edmunds 1929). The implication is that the Netley Heath Beds are a glacial (no doubt, Pleistocene) reworking of a series of late Miocene and Pliocene marine sands, which have otherwise been completely destroyed.

## The East Anglian Crags

These mark the westerly edge of the North Sea sedimentary basin and comprise a series of shelly sands which are unequivocally marine at the base, but contain a somewhat restricted fauna in their highest levels. The sea temperature indicated by the successive faunas changes concurrently from warm temperate to subarctic.

The oldest of these deposits, the Coralline Crag, occurs in a narrow strip to the W of Aldeburgh and comprises white or buff calcareous marine sands, rich in bryozoans ('corallines') and foraminiferids, and locally containing rich concentrations of molluscs. Current-bedding structures are common and the resultant rock is very similar to materials now accumulating at moderate depths in the English Channel. At some localities the substance of the aragonite fossils has been dissolved and remobilised as calcite which, on precipitation, has cemented the rock into a building stone. About 300 species of molluscs are known, of which about 60 per cent are extinct (Boswell 1952). Comparison of the fauna with those of Belgium shows the closest correlation to be with the Luchtbal sands (Upper Scaldisian, late Pliocene) (Cambridge in Curry *et al.* 1978).

The Red Crag rests on an eroded surface of the Coralline Crag and oversteps it on to the London Clay. At its base is a conglomerate which includes pebbles of flint and of Coralline Crag, with rolled bones of mammals, both contemporaneous and derived. Amongst the pebbles are brown sandstones with moulds of molluscs ('Boxstones'), thought to be older than the fauna of the Lenham Beds (Newton 1916–17), and so of Miocene age. The main body of the Red Crag is of clean sands, grey and pyritic at depth, but brick-red at surface, due to oxidation. It is commonly current-bedded, especially in its lower part. Fossils, mostly molluscs, are abundant, and much of the molluscan material has been comminuted and polished before final burial. Some 250 species of molluscs are known, of which about half are extinct. The proportion of cool-water species is somewhat higher than in the Coralline Crag, and rises from the south (Walton) to the north (Butley). Microfossils are not common, presumably due to contemporary sorting. Some derived microfossils are present, including material from the Jurassic and Chalk, which, in some cases, must have travelled some tens of kilometres before incorporation. The molluscan fauna indicates that the Red Crag correlates with the Kruisschans and Merksem sands of Belgium.

The Norwich Crag rests on the Chalk to the east of Norwich. It comprises pale sands with thin clay seams, and with pebbles near the base. Molluscs occur sporadically throughout. The molluscan fauna contains notably fewer species than those of the Coralline and Red Crags and a much smaller proportion of southern forms. Land- and freshwater molluscs are not uncommon and suggest that the coastline was not far away.

The Weybourne Crag is a pebbly sand which occurs locally immediately above the Chalk on the north Norfolk coast. It contains a sparse cool-water fauna of marine molluscs and foraminiferids, which is notable for the first occurrence in Britain of *Tellina balthica*.

The relative stratigraphical positions of the Red, Norwich, and Weybourne Crags are not clear from surface exposures, and were originally deduced from the composition of their respective molluscan faunas. Confirmatory evidence as to their relative position is provided by a study of the foraminiferids from a borehole at Ludham, ENE of Norwich (Funnell 1961), which passed through 40 m of assorted Crag deposits before penetrating London Clay. The lowest 14 m compared closely with the Red Crag, and was overlain by 8 m referred to the *Scrobicularia* Crag, which overlies the Red Crag locally in Suffolk. Most of the upper part of the bore was recognised as Norwich Crag, whilst the fauna of the top 2 m resembled that of Weybourne.

Overlying the Norwich and Weybourne Crags are

tills which are clearly attributable to the Pleistocene. It is not clear, however, where within the series of Crags the Pliocene–Pleistocene boundary should be placed. This boundary is defined at the base of the Calabrian in Italy, and correlation with British sequences is difficult to establish. Recent opinion (Funnell in Curry *et al.* 1978) favoured the base of the Norwich Crag or, possibly, some level within that formation. (This subject is discussed further at the beginning of Chapter 14.)

### St Erth Beds, Cornwall

This unit rests in a notch cut laterally into Palaeozoic slates at about 30 m above sea level. It comprises up to 6 m of sands which contain lenticular masses of fossiliferous marine clay. Its molluscan fauna contains about 60 per cent of extant species, which now either live in the area or have southern affinities. There is a rich foraminiferal fauna, and both molluscs and foraminiferids show many similarities with faunas known in northern France near Rennes and in the south of the Cotentin, but rather few links with the East Anglian Crags. The St Erth Beds appear to be late Pliocene, and of about the same age as the Coralline Crag (Mitchell 1973).

### Neogene continental deposits

As already stated, the Palaeogene sequence at Mochras may extend up into the Miocene. A number of solution-hollows, mainly in Carboniferous Limestone, contain probable Neogene deposits. However, the only ones which have been dated occur at Brassington, Derbyshire (SK2354). These occur at a height of 300 m and are sands and clays with a palynoflora indication deposition in or near a slow-moving river (presumably near to sea level), and a late Miocene–early Pliocene age (Walsh *et al.* 1972).

# Palaeogeography

In the Late Cretaceous, the proto-Atlantic, which was opening around a hinge in the Canadian Arctic, was already over 1,000 km wide at our latitudes, and Britain was situated near the western edge of a continent to its east. Greenland was still attached to Scandinavia and, via the future Rockall Bank, to Britain. Iceland did not exist. India and Africa (including Arabia) were separated from the rest of Eurasia by a continuous seaway, the Tethys, and western Europe was about 10° nearer to the equator than it is now. There was no connection between the Atlantic and Arctic oceans, and so no southward flow of polar seawater. Britain probably still lay in the zone of

persistent south-westerly winds and would thus have possessed an equable climate with a mean annual temperature perhaps 10° C higher than now. The North Sea had been an area of substantially continuous subsidence and marine deposition since early Jurassic times. It appears that there was no ice-sheet at the North Pole, although an ice-cap of limited area may have been present on land near the South Pole. Most of the world's present mountain ranges had not yet been formed, although the circum-Pacific chains were growing actively, and the stumps of many ancient chains, like those of Scandinavia and the north-western parts of the British Isles, formed areas of high land.

Much less water was locked up in ice than at the present day and as a result sea levels were relatively high and large areas on the margins of the continental blocks were inundated by seas in which calcilutites were the most common deposits, to be replaced in earliest Palaeocene times by calcarenites. The presence of a land-link between North America and Europe accounts for the close similarities between the faunas of mammals and marine invertebrates of these two continents up to about 50 Ma ago. At the end of the Cretaceous, however (65 Ma), came the first step in a chain of events which was to break this link. A SW–NE rift system developed across the Greenland/Scandinavian plateau, with associated igneous activity, and, at about 55 Ma, ocean spreading began along this new line. The newly-created Denmark Strait progressively isolated America from Europe and, as a result, the land faunas of the two areas rapidly became dissimilar, as did much of the benthic marine fauna.

Ash from the volcanic sequences found its way into the North Sea area, where it occurs in deposits of mid-Palaeocene to early Eocene age, including the Thanet Beds in Kent and the basal London Clay in Essex. The rifting and the development of the new seaway exposed the edges of the plateau to rapid denudation, which may account for the sudden change in sedimentary style throughout NW Europe from limestones to glauconitic sandy clays at the end of the Dano-Montian, and the appearance in the Thanet Beds of heavy minerals (hornblende in particular) from a Scottish or Scandinavian source (Morton 1982a).

The English Channel and Celtic Sea regions remained basinal areas in which shallow marine sequences developed, and the former was at times an important migration route from the Tethys and Atlantic oceans to the North Sea and its extension eastwards south of Scandinavia to Poland (Pożaryska 1971).

The late Cretaceous had been a period of world-wide transgression and marine deposition in which, for the most part, only the ancient massifs stood above sea level. In and near Britain these included parts of

Brittany, the Cotentin and Cornubia, Wales, Scotland, and most of Ireland and, possibly, the Pennines. The Palaeocene, by contrast, was a period of regression. Deposition became concentrated in the slowly-subsiding areas of the North Sea and its dependencies, the Low Countries and SE England, and in the Paris Basin.

Late Cretaceous marine deposits were eroded away over large areas as the pattern of Chalk zones beneath the sub-Tertiary unconformity indicates (Fig. 13.2a). The area of greatest erosion coincides approximately with the London–Brabant massif and the result may be a composite controlled by tilting to the NE, compaction with sagging of the pre-Tertiary sequences on either side of the Palaeozoic platform and possible uplift of the latter. The post-Cretaceous surface is probably a submarine erosion floor as the succeeding beds in England appear all to be marine.

Owing to the general absence of deposition, Danian palaeogeography is difficult to elucidate, but presumably most of Britain was land. However, the marine limestones of the western English Channel resemble both in lithology and fauna those of the type Danian of Denmark, implying a marine connection between the two (Fig. 13.2b). This could have followed the line of the scattered deposits in the Paris Basin. Such deposits no doubt covered a much larger area at one time, as derived Danian microfossils occur at several higher Palaeogene levels (Fig. 13.2c).

Whether the English Channel was closed in Thanetian times is uncertain. The Thanet Formation thins westwards to a feather-edge near Guildford, but displays no sign of an approach to a shoreline as it does so. This wedging can reasonably be explained by regional tilt, associated with bevelling, in which case no clue is provided to the position of the shore. Beds of Thanetian age are, of course, absent from the Hampshire Basin; in the English Channel they have only been recorded west of 4°. It does not follow from this that a Thanetian seaway was not present, though; the English Channel might then, as it is now, have been one essentially of non-deposition. The higher beds of the Thanet Formation contain, in addition to Danian planktonics, microfossils derived from the Chalk. Both flints and Chalk microfossils occur sporadically throughout the Palaeogene succession both in England and in the Paris Basin, indicating that the Chalk underwent at least intermittent denudation during this period, probably from lowlands in the region of the present earlier Mesozoic outcrops. Jurassic dinoflagellates, too, also occur in the London Clay (Davey *et al.* 1966), indicating that Jurassic beds also came under erosional attack from time to time.

Sparnacian beds show a strong geographical pattern. Continental (mostly red) clays occur to the west of London, in the Hampshire Basin, the central English Channel and in the Paris Basin west and south of Paris. In a band to the east (West Kent, Newhaven, Dieppe, remainder of the Paris Basin, Flanders) occur brackish lagoonal clays, and in East Kent a restricted marine succession is present (Fig. 13.2d). It seems clear from the above that the English Channel seaway was then closed. There is no evidence that this was due to a eustatic retreat of the sea and a possible mechanism for this closure is regional uplift associated with the episode of rifting and vulcanicity to the NW of Britain.

The base of the London Clay is marked by a substantial rise in sea level. This was probably of eustatic origin and associated with a rise in world temperature, as a similar phenomenon is present in SW France and NE Spain. The accompanying transgression was probably the most widespread of the Palaeogene in England and carried the London Clay across on to the Chalk in Norfolk and Dorset. Davis and Elliott (1958) concluded on faunal grounds that the London Clay sea was closed to the west and only opened through the Channel in late London Clay times. From their analysis it follows that SE England and the Paris Basin formed the westernmost arm of a tideless Baltic-style sea which was an appendage of the North Sea which connected via the North German plain and Poland to the Tethys. An alternative oceanic connection, almost as long, might have been via the northern North Sea and the developing Denmark Strait to the Atlantic. However, Curry (1965) placed the new connection with the Atlantic somewhat later. The English Channel region was certainly a seaway during most of Bracklesham Group times, when it provided a migration route for several genera of larger foraminiferids (*Nummulites, Cuvillierina, Alveolina, Orbitolites*) into SE England and the Paris and Belgian areas (Fig. 13.2e). Evidence of local uplift appears in the west of the Hampshire Basin, however, probably in association with the Purbeck tectonic line (Plint 1982). The sequences around Poole are mostly continental, but at Selsey, in the east of the Hampshire Basin, almost entirely marine. Finely-laminated intertidal units are present at some levels and Murray and Wright (1974) used foraminiferids to elucidate the local palaeogeography. The equivalent in the London area of the Bracklesham Group is the Bagshot Formation. There are striking similarities between the two units, especially at the level of the *Nummulites laevigatus* bed and one may conclude from this that the coastline ran SW–NE and that the Weald axis was then not active locally. However, uplift and erosion of some part of the Weald area are established by the presence of Lower Greensand cherts (of a type not known NW of the Chilterns) in the Upper Bagshot Beds.

There is evidence of uplift in relation to tectonic

lines in other areas. The presence of reworked Danian planktonics in the Wittering Formation hints at uplift along the Start-Alderney line, the probable source area, and shoaling along the Portsdown axis may be responsible for the presence of reworked nummulites in the Selsey Sand Formation at Selsey. The Bray and Artois axes were active in mid-Eocene times and the latter, which separates the Belgian and Paris Basins, was thought by Pomerol (1973) to have isolated the North Sea area from the south-western seas during the Palaeogene from mid-Eocene times onward. Complete isolation seems unlikely, however, in view of the resemblances between the sequences in the eastern English Channel and Belgium and between the mollusc faunas of Brockenhurst, Grimmertingen (Belgium), and Latdorf (N Germany).

The marine faunas of the Bracklesham Group have many species in common with those of the same age in the Paris Basin and Belgium. This is much less true of the Barton Group. The fauna of the Brockenhurst Beds, by contrast, resembles those of the sands of Grimmertingen and Latdorf. Beds of that facies and age are absent from the Paris Basin. They occur in the Cotentin (argile à corbules de Rauville, dated by Châteauneuf (1980) on their palynoflora), but the mollusc and ostracod faunas of that unit are quite unlike those of Brockenhurst. There is thus evidence of progressive isolation of the Hampshire Basin and of the complete closure of the English Channel seaway east of the Cotentin, possibly by eustatic control linked to a contemporary fall in world temperatures. Such a barrier would be consistent with the restricted occurrence of the Brockenhurst Beds and Bembridge Oyster Beds (Curry 1965), and the close similarities between the land biota (charophytes, molluscs, mammals) of the Hampshire and Paris Basins in latest Eocene and early Oligocene times (Fig. 13.2f).

The youngest beds in the Isle of Wight show only the beginnings of the next marine episode (Fig. 13.2g), well represented in the Middle Oligocene clays of Belgium and NW Germany, and the Stampian sands south of Paris. No marine beds of this age are known from the Cotentin or the English Channel region, and the latter seaway was probably closed. The Stampian sea appears to have retreated to the south-west via the Loire valley, and late Oligocene sequences of the Paris Basin are all continental, like the scattered deposits of that age in England.

Marine beds of early Miocene age are known only from the western English Channel, but those of late Miocene age occur locally in NW France and also at several points along the crest of the Downs (Lenham Beds) and as derived boulders ('boxstones') in East Anglia (Fig. 13.2h). All of these late Miocene units are relatively coarse-grained and suggest the presence of currents, and it seems certain that a seaway connected

the North Sea to the Bay of Biscay via the eastern English Channel. Whether the central and western parts of the Channel region formed a seaway is less certain, however. The high proportion of molluscs of southern affinities in the Coralline Crag suggests that this southward connection remained open well into the Pliocene, and possibly up the earliest Pleistocene, as the Red Crag (but not the succeeding Norwich Crag) shows evidence of strong tidal currents. The Straits of Dover were finally closed by uplift along the Weald–Artois line, no doubt aided by eustatically-controlled regression as world temperatures continued to fall.

## Tertiary tectonic activity

Some aspects of this have been dealt with in the sections on stratigraphy and palaeogeography. The main manifestations on land are the Purbeck–Isle of Wight monoclinal line and the elongated dome of the Weald, both of which have traditionally been dated as

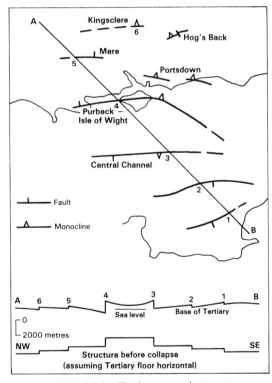

**Fig. 13.5.** Major Tertiary tectonic structures in the Central English Channel and neighbouring regions (precise age of structure No. 1 is not confirmed). Postulated doming and collapse.

Miocene and linked with the Alpine mountain-building episode. Other features, all north-facing, are the Mere fault, the Pewsey–Kingsclere line, the Portsdown axis, and the Hog's Back monocline.

Mapping in the English Channel has revealed the presence of three other E–W lines, two of which are seen to involve Palaeogene rocks. These are south-facing and form a symmetrical pattern with others to the north (Fig. 13.5). The whole may tentatively be interpreted as the result of compression resulting in the uplift of basement blocks into a dome, and the subsequent collapse of this dome. The dating of this group of six E–W structures is debatable. The presence of Purbeckian cherts in the Poole Formation of Dorset implies faulting and uplift to the south in mid-Eocene times or earlier (Plint 1982), and further uplift in late Eocene times is suggested by the occurrence of the continental Creechbarrow Beds. Pebbles of late Maastrichtian flint (Curry 1986) in the Boscombe Sand point to prior or contemporaneous activity along the line of the Central Channel Structure (Fig. 13.5), and from late Eocene times there are suggestions of land to the south-west in the pattern of the marine incursions of the Brockenhurst and Bembridge Oyster Beds.

Whilst Eocene tectonic activity thus appears to be established in the Isle of Purbeck and Central Channel regions, there is on the other hand no clear indication of similar activity in the Isle of Wight, where the thickness and lithology of the late Eocene and early Oligocene sequences show no significant variation from north to south. The main tectonic activity in the Isle of Wight thus appears to be post-early Oligocene. If indeed the faulting and doming in and to the south of the Hampshire Basin is the result of forces related to Alpine mountain-building, it would be reasonable to expect the first signs of movement to appear sometime in the Eocene, with the maximum activity in the late Oligocene and the early Miocene, as in the Alps themselves.

The doming in the Weald area appears to have a longer and more complicated history (Fig. 13.6). Firstly it should be noted that the anticlinal structure of the Weald does not persist at depth, where early Jurassic beds are synclinal. The pattern of thicknesses of succeeding units indicates that it was a subsiding basin at least through Jurassic and early Cretaceous times, which underwent inversion at some time in the late Cretaceous or early Palaeogene. The timing of the

**Fig. 13.6.** Postulated tectonic evolution of the Weald during the Cenozoic.

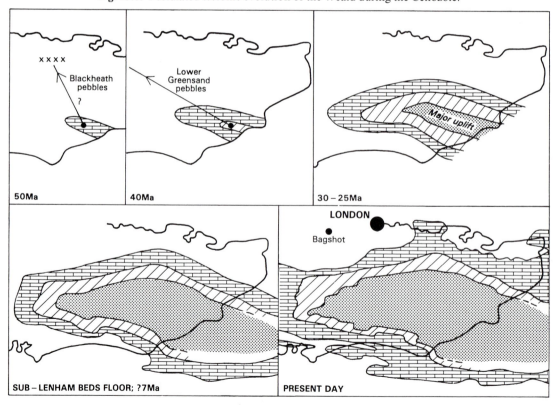

associated doming is uncertain, but probably comprised several episodes. The local concentration of flint pebbles in the Blackheath Beds has been thought by some to indicate the presence of some doming in the Weald by earliest Eocene times. However, there are alternative explanations for their origin, and other authors (e.g. King 1981) have pointed out that the preceding and succeeding units (Woolwich Beds, London Clay) give no indication of a land-mass or shoal in the Weald area. If the supposed Lower Greensand cherts in the Upper Bagshot Beds are correctly identified these must have come from the Weald, as no similar cherts have been reported from the Lower Greensand of the Chilterns. Doming of at least 250 m would thus have occurred by the late Eocene. The greater part of the uplift of the Weald was probably of later date, however, and had at least two components. The occurrences of Lenham Beds at about 180 m O.D. at several points on the North Downs and near Eastbourne appear to form part of a marine late Miocene peneplain which can be traced via Calais and Cassel (near the Belgian–French border) to dip gently to sea level near Antwerp. Northwards this peneplain appears to link with the sub-Crag surface in Suffolk, also near sea level. Late Miocene marine beds are known at low altitudes in the Cotentin and near Rennes, and all of the occurrences mentioned are of shallow-water deposits. The implication is that a considerable part of the area between the Cotentin and Suffolk, and eastwards to Antwerp has been domed upwards to a maximum of at least 200 m since late Miocene times. Note that this doming covers an area considerably greater than that of the Weald itself. However, the amount of uplift is small in relation to the total seen in the Weald (1,200 m), and indeed the combined Eocene and post-Miocene episodes may represent appreciably less than half this total. There is no direct evidence for the dating of the remainder of the Weald uplift and it may well be of late Oligocene/ early Miocene age, as postulated for much of the movement on the Isle of Wight–Purbeck structure.

## Palaeoclimate

Many analyses, both general and detailed, have been carried out into the climatic situation during the Cenozoic. The methods used have been geochemical (oxygen isotope determination of seawater temperatures), sedimentological (presence of evaporites, red beds, indication of wind direction), and palaeontological (comparative analysis of floras and land and marine faunas of temperature-sensitive groups). Indications world-wide suggest a sharp fall in temperature at the end of the Mesozoic, followed by a rise throughout the Palaeocene. Temperatures remained high dur-

ing the Eocene, after which there was another sharp fall in the earliest Oligocene, to be followed by further, but less substantial peaks in the Middle Miocene and Pliocene (Fig. 13.7). Global results are modified locally by the effects of latitudinal change or the opening or closing of major seaways, due to continental drift.

In Britain the molluscs and foraminiferids suggest temperate conditions in the late Palaeocene, as do the rare plant remains. The London Clay flora (early Eocene) was thought by Reid and Chandler (1933) to indicate tropical (rain-forest) conditions, although Daley (1972) expressed reservations because of the presence of a minority of plant genera thought to indicate no more than warm temperate conditions. Indications from Britain in the middle Eocene are uniformly in favour of a subtropical environment

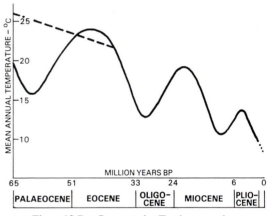

**Fig. 13.7.** Suggested Tertiary palaeo-temperatures in Britain (based mainly on Reid & Chandler (1933) and Buchardt (1978).

(swamp-palms and sea-snakes, known from the London Clay, occur also in the Bracklesham Group). However, some cool-water indicators amongst the molluscs reappear in latest Eocene and early Oligocene times (return of *Arctica* and *Aporrhais* (absent since London Clay times) in the Brockenhurst Beds). Crocodiles are still present, however, and persist right up into the Bouldnor Formation. Mean temperatures were probably still some 10°C higher than at present, therefore.

No evidence is yet available as to climate from the scattered later Oligocene and Miocene deposits of the English mainland, but the molluscs of the Coralline Crag and St Erth Beds have a definitely 'southern' cachet, indicating a mean water temperature of perhaps 3–5°C above the present. Successively later marine faunas suggest lower and lower temperatures as the first glacial episode approached.

# REFERENCES

ARKELL, W. J. — 1947 — The geology of the country around Weymouth, Swanage, Corfe, and Lulworth. *Mem. geol. Surv. G.B.* 386 pp.

AUBRY, M. P. — 1983 — Biostratigraphie du Paléogène épicontinental de l'Europe du nord-ouest. Étude fondée sur les nannofossiles calcaires. *Docum. Lab. Géol. Lyon*, No. 89, 317 pp., 8 pl.

AUBRY, M. P., HAILWOOD, E. A. & TOWNSEND, H. A. — 1986 — Magnetic and calcareous–nannofossil stratigraphy of the lower Palaeogene formations of the Hampshire and London basins. *J. geol. Soc. London*, **143**, 729–735.

BEYRICH, E. — 1854 — Ueber die Stellung des hessischen Tertiärbildungen. *Mber. K. preuss. Akad. Wiss.* (1854), 640–666.

BLONDEAU, A. & POMEROL, Ch. — 1969 — A contribution to the sedimentological study of the Palaeogene of England. *Proc. Geol. Assoc. London*, **79**, 441–455.

BOSWELL, P. G. H. — 1952 — The Pliocene–Pleistocene boundary in the East of England. *Proc. Geol. Assoc. London*, **63**, 301–312.

BRANDER, G. — 1766 — *Fossilia Hantoniensia collecta, et in Musaeo Britannico deposita.* London. 43 pp., 9 pl.

BUCHARDT, B. — 1978 — Oxygen isotope palaeotemperatures from the Tertiary period in the North Sea area. *Nature, London*, **275**, 121–123.

BUJAK, J. P., DOWNIE, C., EATON, G. L. & WILLIAMS, G. L. — 1980 — Dinoflagellate cysts and acritarchs from the Eocene of southern England. *Spec. Pap. Palaeontol. London*, No. 24. 100 pp., 22 pl.

BUURMAN, P. — 1980 — Palaeosols in the Reading Beds (Paleocene) of Alum Bay, Isle of Wight, U.K. *Sedimentology*, **27**, 593–606.

CHANDLER, M. E. J. — 1962 — *The Lower Tertiary floras of southern England. II. Flora of the Pipe-clay Series of Dorset (Lower Bagshot).* British Museum (Natural History), London. 176 pp., 29 pl.

CHARLESWORTH, E. — 1835 — Observations on the Crag formation and its organic remains, with a view to establish a division of the Tertiary strata overlying the London Clay in Suffolk. *Philos. Mag.* (3), **7**, 81.

CHÂTEAUNEUF, J.-J. — 1980 — Palynostratigraphie et paléoclimatologie de l'Éocène supérieur et de l'Oligocène du Bassin de Paris. *Mém. Bull. Rech. géol. minièr.* No. 116, 357 pp., 31 pl.

COSTA, L., DOWNIE, C., & EATON, G. L. — 1976 — Palynostratigraphy of some Middle Eocene sections from the Hampshire Basin (England). *Proc. Geol. Assoc. London*, **87**, 273–284.

CRAY, P. E. — 1973 — Marsupialia, Insectivora, primates, Creodonta and Carnivora from the Headon Beds (Upper Eocene) of Southern England. *Bull. Brit. Mus. nat. Hist.* A. Geol. **23**, No. 1, 102 pp., 6 pl.

CURRY, D. — 1959 — Opercula of calcite in the Bithyniinae. *J. Conch. London*, **24**, 349–350.

— 1965 — The Palaeogene beds of south-east England. *Proc. Geol. Assoc. London*, **76**, 151–173.

— 1986 — Foraminiferids from decayed Chalk flints and some examples of their use in geological interpretation. *In* Sieveking, G. de G. & Hart, M. B. (Eds.) *The scientific study of flint and chert.* Cambridge University Press, Cambridge, 99–103.

CURRY, D., ADAMS, C. G.,    1978    A correlation of Tertiary rocks in the British Isles. *Geol.*
   BOULTER, M. C.,      *Soc. London Spec. Rep.* **12**. 72 pp.
   DILLEY, F. C.,
   EAMES, F. E.,
   FUNNELL, B. M., &
   WELLS, M.K.

CURRY, D., KING, A. D.,    1977    The Bracklesham Beds (Eocene) of Bracklesham Bay
   KING, C. & STINTON, F. C.      and Selsey, Sussex. *Proc. Geol. Assoc. London*, **88**, 243–
     254.

CUVIER, G. &    1810    Essai sur la géographie minéralogique des environs de
   BRONGNIART, A.      Paris. *Mém. Cl. Sci. Math. Phys. Inst. imp. France* (1810).
     278 pp.

DALEY, B.    1972    Some problems concerning the early Tertiary climate
     of southern Britain. *Palaeogeogr. Palaeoclimatol.*
     *Palaeoecol.*, **11**, 177–190.

   1973    The palaeoenvironment of the Bembridge Marls
     (Oligocene) of the Isle of Wight, Hampshire. *Proc. Geol.*
     *Assoc. London*, **84**, 83–93.

DAVEY, R. J., DOWNIE, C.,    1966    Studies on Mesozoic and Cainozoic dinoflagellate cysts.
   SARJEANT, W. A. S., &      *Bull. Brit. Mus. nat. Hist.* A. Geol. Supplement 3. 248 pp.,
   WILLIAMS, G. L.      26 pl.

DAVIES, A. M.    1975    *Tertiary faunas*, Volume II: the sequence of Tertiary
     faunas. (Revised by F. E. Eames & R. J. G. Savage.)
     George Allen & Unwin, London. 447 pp.

DAVIS, A. G. &    1958    The palaeogeography of the London Clay sea. *Proc. Geol.*
   ELLIOTT, G. F.      *Assoc. London*, **68**, 255–277.

DEARMAN, W. R.    1964    Wrench-faulting in Cornwall and South Devon. *Proc.*
     *Geol. Assoc. London*, **74**, 265–287.

DINES, H. G. &    1929    The geology of the country around Aldershot and Guild-
   EDMUNDS, F. H.      ford. *Mem. geol. Surv. U.K.* 182 pp.

EATON, G. L.    1976    Dinoflagellate cysts from the Bracklesham Beds of the Isle
     of Wight, southern England. *Bull. Brit. Mus. nat. Hist.* A.
     Geol. **26**, 227–332.

EDWARDS, R. A.    1976    Tertiary sediments and structure of the Bovey Basin, south
     Devon. *Proc. Geol. Assoc. London*, **87**, 1–26.

EDWARDS, R. A. &    1987    Lithostratigraphic classification of the Hampshire Basin
   FRESHNEY, E. C.      Palaeogene Deposits (Reading Formation to Headon
     Formation). *Tertiary Res.*, **8**, 43–73.

ELLIOTT, G. F.    1971    Eocene volcanics in south-east England. *Nature, London*,
     **230**, 9.

FEIST-CASTEL, M.    1977    Evolution of the charophyte floras in the Upper Eocene
     and Lower Oligocene of the Isle of Wight. *Palaeontology,*
     *London*, **20**, 143–157.

FISHER, O.    1862    On the Bracklesham Beds of the Isle of Wight Basin. *Q. J.*
     *geol. Soc. London*, **18**, 65–94.

FLETCHER, B. N.    1975    A new Tertiary basin east of Lundy Island. *J. geol. Soc.*
     *London*, **131**, 223–225.

FRESHNEY, E. C.,    1982    A Tertiary basin at Dutson, near Launceston, Cornwall,
   EDWARDS, R. A.,      England. *Proc. Geol. Assoc. London*, **93**, 395–402.
   ISAAC, K. P., WITTE, G.,
   WILKINSON, G. C.,
   BOULTER, M. C.,
   & BAIN, J. A.

FUNNELL, B. M.    1961    The Palaeogene and Early Pleistocene of Norfolk. *Trans.*
     *Norfolk & Norwich Nat. Soc.*, **19**, 340–364.

GABUNIA, L. C.    1964    Sur la corrélation des faunes de Mammifères de l'Oligocène d'Europe et d'Asie. *Mém. Bur. Rech. géol. minièr.*, **28**, 979–984.

GARDNER, J. S., KEEPING, H., & MONCKTON, H. W.    1888    The Upper Eocene, comprising the Barton and Upper Bagshot Formations. *Q. J. geol. Soc. London*, **44**, 578–635.

GURR, P. R.    1963    A new fish fauna from the Woolwich Bottom Bed (Sparnacian) of Herne Bay, Kent. *Proc. Geol. Assoc. London*, **73**, 419–447.

HAMBLIN, R. J. O.    1974    The Haldon Gravels of south Devon. *Proc. Geol. Assoc. London*, **84**, 459–476.

HAWKINS, H. L.    1946    Field meeting at Reading. *Proc. Geol. Assoc. London*, **57**, 164–171.

HAYNES, J.    1958    Certain smaller British Foraminifera. Part V. Distribution. *Contr. Cushman Found. foram. Res.*, **9**, 83–92.

HESTER, S. W.    1965    Stratigraphy and Palaeogeography of the Woolwich and Reading Beds. *Bull. geol. Surv. G.B.* **23**, 117–137.

HOOKER, J. J.    1977    The Creechbarrow Limestone – its biota and correlation. *Tertiary Res.*, **1**, 139–145.

HOOKER, J. J. & INSOLE, A. N.    1980    The distribution of mammals in the English Palaeogene. *Tertiary Res.*, **3**, 31–45.

INSOLE, A. N. & DALEY, B.    1985    A revision of the Lithostratigraphic Nomenclature of the Late Eocene and Early Oligocene Strata of the Hampshire Basin, Southern England. *Tertiary Res.*, **7**, 67–100.

JACQUÉ, M. & THOUVENIN, J.    1975    Lower Tertiary tuffs and volcanic activity in the North Sea. *In* Woodland, A. W. (Ed.) *Petroleum and the continental shelf of north-west Europe* **1**, *Geology*. Applied Science Publishers, London, 455–465.

KEEN, M. C.    1977    Ostracod assemblages and the depositional environments of the Headon, Osborne and Bembridge Beds (Upper Eocene) of the Hampshire Basin. *Palaeontology, London*, **20**, 405–445.

KING, C.    1981    The stratigraphy of the London Clay and associated deposits. *Tertiary Research spec. Pap.* **6**. Backhuys, Rotterdam. 158 pp.

KNOX, R. W.    1979    Igneous grains associated with zeolites in the Thanet Beds of Pegwell Bay, north-east Kent. *Proc. Geol. Assoc. London*, **90**, 55–59.

LYELL, C.    1852    On the Tertiary strata of Belgium and French Flanders. *Q. J. geol. Soc. London*, **8**, 277–370.

MARTINI, E.    1971    Standard Tertiary and Quaternary calcareous nannoplankton zonation. *In* Farinacci, A. (Ed.) *Proceedings of the II Planktonic Conference, Roma 1970*. Edizioni Tecnoscienza, Rome. 739–785.

MAYER-EYMAR, K.    1858    Versuch einer neuen Klassifikation des Tertiär-Gebilde Europa's. *Verh. schweiz. naturf. Ges.*, **42**, 165–199.

MITCHELL, G. F.    1973    The late Pliocene marine formation at St Erth, Cornwall. *Philos. Trans. R. Soc. London, Ser. B*, **266**, 1–37.

MORTON, A. C.    1982a    The provenance and diagenesis of Palaeogene sandstones of southeast England as indicated by heavy mineral analysis. *Proc. Geol. Assoc. London*, **93**, 263–274.

   1982b    Heavy minerals of Hampshire Basin Palaeogene strata. *Geol. Mag.*, **119**, 463–476.

MURRAY, J. W. & WRIGHT, C. A.    1974    Palaeogene Foraminiferida and palaeoecology, Hampshire and Paris Basins and the English Channel. *Spec. Pap. Palaeontol. London*, **14**. 171 pp. 20 pl.

| | | |
|---|---|---|
| NEWTON, R. B. | 1916–17 | On the conchological features of the Lenham sandstones of Kent, and their stratigraphical importance. *J. Conch. London*, **15**, 56–84, 97–118, 137–149. |
| ODIN, G. S. (Ed.) | 1982 | *Numerical dating in stratigraphy.* John Wiley & Sons, Chichester, 2 vols., 1040 pp. |
| PLINT, A. G. | 1982 | Eocene sedimentation and tectonics in the Hampshire Basin. *J. geol. Soc. London*, **139**, 249–254. |
| POMEROL, Ch. | 1973 | *Ère Cénozoïque.* Doin, Paris. |
| POTTER, J. F. | 1977 | Eocene, lower Bracklesham Beds, iron workings in Surrey. *Proc. Geol. Assoc. London*, **88**, 229–241. |
| POŻARYSKA, K. | 1971 | La limite Crétacé-Tertiaire en Pologne. *Conférences*, **88**, 15–30. |
| PRESTWICH, J. | 1847 | On the probable age of the London Clay and its relations to the Hampshire and Paris Tertiary systems. *Q. J. geol. Soc. London*, **3**, 354–377. |
| REID, C. & STRAHAN, A. | 1889 | The geology of the Isle of Wight, 2nd Edition. *Mem. geol. Surv. U.K.* 349 pp., 5 pl. |
| REID, E. M. & CHANDLER, M. E. J. | 1933 | *The London Clay Flora.* British Museum (Natural History), London. |
| RENEVIER, E. | 1873–4 | *Tableau des terrains sédimentaires (in 4°) plus un texte explicatif.* Rouge & Dubois, Lausanne. 34 pp., 9 pl. |
| SIESSER, W. G., WARD, D. J. & LORD, A. R. | 1987 | Calcareous nannoplankton biozonation of the Thanetian Stage (Palaeocene) in the type area. *J. Micropalaeontol.*, **6**, 85–102. |
| STINTON, F. C. | 1975 | Fish otoliths from the English Eocene, Part 1. *Monogr. palaeontogr. Soc. London.* 56 pp., 3 pl. |
| TOWNSEND, H. A. & HAILWOOD, E. A. | 1985 | Magnetostratigraphic correlation of Palaeogene sediments in the Hampshire and London Basins, Southern UK. *J. geol. Soc. London*, **142**, 957–982. |
| WALSH, P. T., BOULTER, M. C., IJTABA, M., & URBANI, D. M. | 1972 | The preservation of the Neogene Brassington Formation of the southern Pennines and its bearing on the evolution of Upland Britain. *J. geol. Soc. London*, **128**, 519–559. |
| WHITE, E. I. | 1931 | *The vertebrate faunas of the English Eocene.* Vol. **1**, Brit. Mus. Nat. Hist., London, 123 pp. |
| WEBSTER, T. | 1814 | On the freshwater formations in the Isle of Wight, with some observations on the strata over the Chalk in the south-east part of England. *Trans. geol. Soc. London*, **2**, 161–254. |
| WOODLAND, A. W. (Ed.) | 1971 | The Llanbedr (Mochras Farm) Borehole. *Rep. Inst. geol. Sci.*, **71/8**, 115 pp. |
| WORSSAM, B. C. (with contributions by J. INESON and others) | 1963 | Geology of the country around Maidstone, (288). *Mem. geol. Surv. G.B.* 152 pp. |
| WRIGLEY, A. G. | 1949 | The Thanet Sands. *South-east. Nat. Canterbury*, **54**, 41–46. |

# 14

# QUATERNARY

# G. S. Boulton

## Introduction

The Quaternary era represents the latest stage in a trend of general global cooling which had begun by Eocene times (Kurten 1968). Since the early Tertiary mean annual temperatures in north-west Europe have fallen by about 10°C (Buchardt 1978), a change accompanied by an increasing degree of global glacierisation. Alaska may have been glaciated by Eocene times; the Antarctic ice sheet, a fundamental influence on the earth's thermal regime, had built up by mid-Miocene times; and the Greenland ice sheet by Pliocene times.

The record of oxygen isotope variation through time in the skeletons of deep ocean benthonic foraminifers, which we believe to be an index of global ice volume change and oceanic temperatures, shows a cyclical fluctuation throughout the Neogene (Shackleton 1986). After 2·4 Ma the amplitude of this fluctuation increases, suggesting a greater fluctuation in ice volume, and possibly temperature, between warm peaks and cold troughs. The major climatic cycle reflected in this oxygen isotope record for the Lower Quaternary shows a dominant period of approximately 40,000 years with a subsidiary 20,000 year cycle (Fig. 14.1). However, between 0·9 and 0·6 Ma, the length of the dominant cycle appears to increase to about 100,000 years (with subsidiary 40,000 and 20,000 year cycles), whilst the amplitude of fluctuation also increases. This persists at least up to the last climatic cycle (Fig. 14.1).

These natural periodicities coincide with major periods of fluctuation of incident solar radiation predicted by Milankovic (1941) and reflecting changes in the earth's orbit around the sun due to orbital eccentricity (100,000 period), the obliquity of the ecliptic (40,000 year period) and the precession of the equinoxes (20,000 year period). They are now regarded as the external driving force for climatic change.

How then do we account for the change in dominant periodicity between 0·9 Ma and 0·6 Ma and the increased amplitude of change both then and at 2·4 Ma? It is probable that this reflects a change in the earth's topography, which is an important boundary condition influencing fluid circulation (atmosphere, oceans) over the earth's surface. There has been strong Late Tertiary and Quaternary uplift in the Himalayan and Tibetan areas, and other mountain areas such as the Western Cordilleras of America (Ruddiman & Raymo 1988). Experiments with atmospheric general circulation models (Manabe & Terpstra 1974) show that an earth without mountain ranges is one in which Rosby waves, the standing atmospheric waves which strongly influence patterns of meridional heat transfer, are poorly developed. It is possible that strong Late Tertiary and Quaternary uplift in these mountain areas produced enhanced atmospheric waves which in turn increased the magnitude of the climate system's response to solar forcing, leading to development and intensification of late Cenozoic glaciation.

In Britain, we have direct evidence of three major glacial periods, the first of which was in the Anglian stage, and indirect evidence of three earlier stages during which Britain *may* have been glaciated.

The basis of the European terrestrial subdivisions was laid by the work of Penck and Brückner (1909) in the northern Alps, where four great *glacial* episodes (Gunz, Mindel, Riss and Würm) were identified separated by temperate *interglacials*, a concept introduced by Geikie (1895). Jessen and Milthers (1928) demonstrated by pollen analysis that interglacial beds showed a characteristic vegetational sequence, and that within glacial periods, *interstadial* vegetational sequences occurred, reflecting climatic

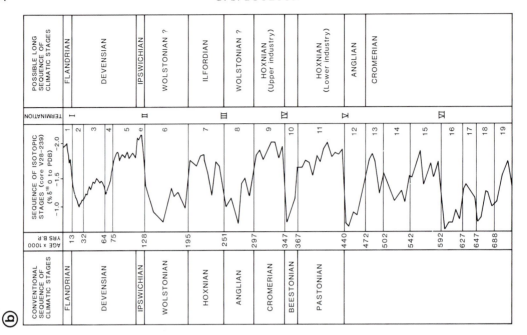

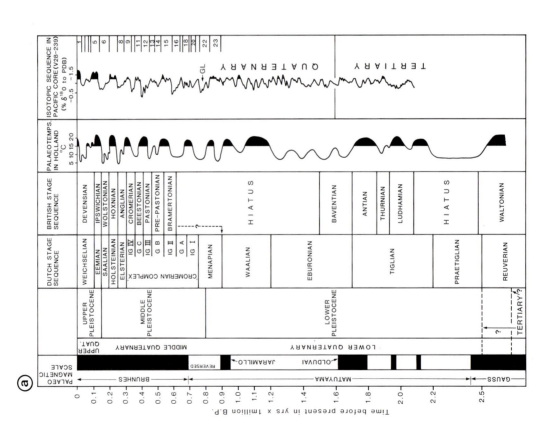

ameliorations which did not allow development of full interglacial floras. It is primarily on the basis of pollen analysis of organic sequences and their relationships to beds reflecting glacial or periglacial conditions that the sequence of alternating temperate and cold stages shown in Fig. 14.1a has been built up in Britain.

Dating is achieved primarily by [14]C dating for the last 40,000 years. For earlier periods some dating is possible using uranium-series, thermo-luminescence and amino-acid racemisation. Most correlations are based on unique floral assemblages thought to be characteristic of particular periods; on vertebrate faunas which have evolved relatively rapidly; and sedimentary correlation, involving distinctive till units and river terrace sequences.

Fourteen to seventeen alternately cold and temperate stages are recognised in Britain in the last 2·4 million years (Fig. 14.1a). However, the oxygen isotope record from the deep oceans, which is believed to reflect major changes in global glacier ice volume, shows at least twice that number of major fluctuations

---

**Fig. 14.1.** Stages and correlations of the British Quaternary.

**(a)** The British chronology is correlated with the Dutch sequence on the basis of palaeo-botany by Zagwijn and Staalduinen (1975). The palaeomagnetic calibration of the Dutch sequence allows this in turn to be correlated with the deep ocean record. Palaeotemperatures in Holland, inferred from palaeobotanical evidence (Zagwijn & Staalduinen 1975) are compared with Shackleton and Opdyke (1973) curve showing global ice volume changes reflected in the oxygen isotope composition of deep ocean foraminifers

$$\left( \delta = \frac{{}^{18}O/{}^{16}O \text{ sample}}{{}^{18}O/{}^{16}O \text{ standard}} - 1 \right)$$

expressed in parts per thousand.

The standard is PDG, a belemnite from the Pee Dee formation of south Carolina.

Workers on deep sea cores place the Tertiary/Quaternary boundary at the top of the Olduvai event, substantially later than those working on land.

The conventional correlation between the northern European sequence and the classical Alpine is: Weichselian = Würm; Saalian = Riss; Elsterian = Mindel. The equivalent of the alpine Günz is uncertain.

**(b)** Possible correlations with the deep ocean record of alternative long and short stage sequences of the Middle and Upper Quaternary in Britain.

during the same period (Fig. 14.1a). This contrast serves to demonstrate the incomplete and fragmentary nature of the terrestrial record which consists of numerous temporally short and spatially impersistent phases of sedimentation. Solving this jig-saw puzzle and relating it to a continuous oceanic standard is one of the major unfinished tasks of Quaternary geology, and has led to serious questioning of the conventionally accepted sequence in Britain.

# Palaeontology

## (a) Palaeobotany

There is no evidence of the production of new plant species in the Quaternary. However, the progressive restriction in distribution during successive temperate stages of such genera as *Carpinus* (Hornbeam), *Tsuga* (Hemlock) and *Picea* (Spruce) may be a reflection of change in the gene pool. The great expansion in the area of arctic and alpine environments during the Quaternary could be expected to lead to speciation of herbaceous plants as they colonised new areas and niches, though this cannot be demonstrated from the fossil record.

There is evidence of some extinction. An extinct species of *Pterocarya* has been identified in the Lower Quaternary and of *Stratiotes* (Water Soldier) in the Upper Tertiary. At the beginning of the Quaternary there is large scale local extinction of late Tertiary genera, such as *Sequoia*, *Taxodium*, *Carya* and *Nyssa* which are found in Eastern Asia and North America. Some genera, for instance *Tsuga* which occurs in Ludhamian, Antian and Pastonian Stages survive until the beginning of the Middle Quaternary. These local extinctions may reflect progressively more severe glacial conditions which extinguished these genera in Europe and Western Asia whose fringing barriers of seas, mountains and deserts prevented re-colonisation (Sparks & West 1972).

One of the most striking aspects of British Quaternary history is the recurrent vegetation cycle during interglacial stages, schematically divided by Turner and West (1968) into the four sub-stages shown in Table 14.1

This repetitive cycle is illustrated in Fig. 14.2 for the sequence of Quaternary interglacials. Although the natural floral sequence in the latter part of the present, Flandrian, interglacial has been substantially disturbed by man, it is clear that we are currently in zone III.

Whereas the interglacial vegetational cycle clearly reflects major climatic changes, much of its apparent simplicity is a response to the slow tempo of soil development, as a raw mineral soil produced by glacial action and periglacial soil churning during glacial

**Table 14.1**

| ZONES | CLIMATE | SOILS | VEGETATION |
|-------|---------|-------|------------|
| IV | Cooling | Leached soils Bog formation | Coniferous woodland |
| III | Warm optimum | Brown earths | Closed deciduous forest and thermophilous herbs |
| II | Warm | Stable | Thermophilous park or open woodland |
| I | Cool but ameliorating | Unstable Immature Base rich | Herbaceous |

stages is matured by enhanced biological and chemical activity during the succeeding interglacial.

Particular interglacials may be recognised from the details of the plant succession and from the presence or absence of individual taxa, which is strongly influenced by the isolation of Britain from the European mainland by rising early interglacial sea levels. For instance, Ipswichian sea levels rose above present-day levels early in the interglacial, which probably accounts for the lack of Ipswichian *Picea* (spruce) and *Abies* (Silver Fir) in England though they were common on the continent, whereas a relatively late sea level rise during the Hoxnian allowed them in.

The floras of the glacial stages are generally unlike any single modern phytogeographical group, but comprise a mixture of groups, from arctic and alpine species, to those of currently southerly distribution, to maritime groups (Godwin 1977). They are generally treeless, though *Salix* (Willow), *Betula* (Birch) (often the dwarf forms) *Juniperus* (Juniper) and *Pinus* (Pine) may occur. Indicators of continentality such as *Hippophae* (Sea Buckthorn) and *Artemisia* (Mugwort) are common, and give a steppe-like character to the flora. The mixture of species groups probably reflects the occurrence at unusually low latitudes of the variety of microclimates typical of a periglacial environment, whilst the absence of trees may reflect windiness, or permafrost and associated waterlogging and solifluction, or grazing by large herbivores (West 1977).

Interstadial forest floras are rare. An example is in the Chelford Birch–Pine–Spruce forest flora of early Devensian age. It probably represents a climate intermediate between interglacial and glacial conditions, and accounts for the frequent difficulty in defining the end of the Ipswichian interglacial.

## (b) Fauna

Although compared with earlier systems the Quaternary gives us an almost unique insight into relatively short-term faunal responses to environmental change, the record is still relatively poor compared with that of land plants.

## (i) Vertebrates

Quaternary fishes, reptiles, amphibians and birds so far described in Britain all belong to living species. The mammals however have attracted most attention and exhibit strong evolutionary changes through the Quaternary.

Fig. 14.3a shows the stratigraphic distribution of mammalian species. It is biased in several ways. Cave faunas, which are primarily of upper Quaternary age because of the low preservation potential of cave sites, largely consist of the prey remains of carnivora such as man, owl, wolf, bear, fox and hyaena. Fluviatile deposits give the least distorted picture of contemporary communites, provided that a variety of fluviatile environments are represented, whilst lacustrine beds yield large mammal remains, amphibia and fish. Lower or early Middle Quaternary vertebrates are almost exclusively known from East Anglia. Unfortunately the source of many fine collections cannot be sufficiently precisely located to allow them to be placed accurately within the modern stratigraphic scheme. Thus, because of this uncertainty several species in Fig. 14.3a are given a temporal range within the pre-Anglian Quaternary which is probably larger than their true range. These problems reduce the significance of the histogram in Fig. 14.3a although one obvious feature is the relative paucity of Flandrian species, which may largely reflect human hunting activity.

There is a major change between the earliest Quaternary faunas of Tertiary aspect and those of the Ludhamian stage, which lends support to the concept that they are separated by a hiatus of 0·5 million years, and which may reflect climatic deterioration. There is comparatively little change through the rest of the Lower Quaternary, but in Middle and Upper Quaternary time, when climatic fluctuations were intensified, faunal change accelerated, probably as a result of intensified selection pressures due to rapid restrictions and expansions of ecological niches (Stuart 1974). Thus, glacial and interglacial stage faunas can be distinguished, as can the faunas of successive environmentally similar stages.

The only British mammalian species known to have survived from earliest Quaternary times into the Flandrian, and which has only recently become extinct in Britain, is the beaver (*Castor fiber*).

Using present distributions as a guide one might conclude that interglacial stages with lion, rhinoceros, hippopotamus, hyaena and elephant indicate a climate verging on the tropical, a conclusion not supported by floral evidence (Sparks & West 1972). Zeuner (1959) has suggested however, that the primary adaption of many of these species is to vegetation and that climate plays a secondary role.

## (ii) Molluscs

Marine mollusc faunas are abundant in the Lower Quaternary marine beds of East Anglia, in beds laid down by marine transgressions in post-Anglian

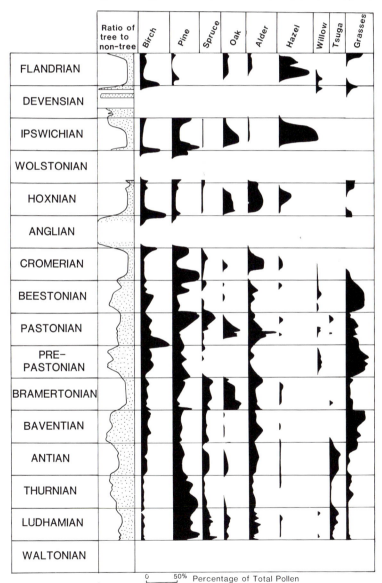

Fig. 14.2. A schematic diagram showing the percentage representation of selected pollen taxa through the stages of the British Quaternary from sites in south-east England. No attempt is made to show the relative duration of stages. Warm stages are most readily distinguished by the higher tree:non-tree ratios and relatively high percentage of Oak and Hazel. Note the post-Pastonian absence of Tsuga.

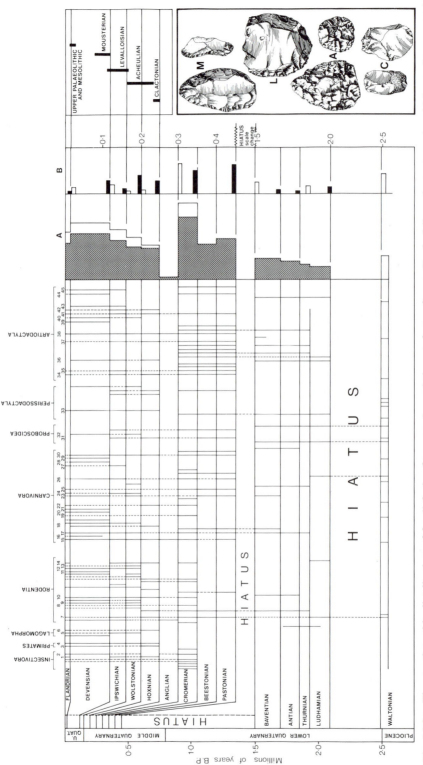

**Fig. 14.3a.** The stratigraphic distribution of fossil finds of mammals, and the artefacts of man. Dashed lines show stages in which species present in overlying and underlying stages have not been found. The relative abundances identified in each stage are shown in column A. In column B, the black histogram columns show the relative frequency of new species introductions in Britain, and the white columns the relative frequency of species extinctions in Britain. Some common mammals are identified individually: 1, Shrew; 2, Mole; 3, Man; 4, Man's artefacts; 5, Hare; 6, Rabbit; 7, Beaver; 8, Hamster; 9, Lemming; 10, Bark Vole; 11, Water Vole; 12, Tundra Vole; 13, Field Vole; 14, Steppe Lemming; 15, Wolf; 16, Arctic Fox; 17, Red Fox; 18, Bear; 19, Weasel; 20, Stoat; 21, Polecat; 23, Glutton; 25, Otter; 26, Spotted Hyaena; 27, Wild Cat; 28, Lynx; 29, Leopard; 30, Lion; 31, Elephant; 32, Woolly Mammoth; 33, Horse; 34, Wild Boar; 35, Hippopotamus; 36, Giant Elk; 37, Fallow Deer; 38, Red Deer; 39, Elk; 40, Reindeer; 41, Roe Deer; 42, Aurochs; 43, Bison; 44, Saiga Antelope; 45, Musk Ox.

Many species have been left unidentified because of space problems. Typical artefacts belonging to the principal stone tool assemblages are shown with their stratigraphic distribution.

temperate stages; and as secondary, derived fossils in glacial sequences, where they have played an important role in dating Late Devensian glacial fluctuations.

Temperate faunas of similar facies show clear evolutionary changes involving replacement of Tertiary faunas by ones of modern aspect. Earliest Quaternary changes, in particular the contrast between the Coralline Crag and the Red Crag show a replacement of southern faunas by northern and boreal forms, a widespread change in Europe which has been seen as marking the beginning of the Quaternary. A major period of extinction occurred between Pastonian and Hoxnian stages, but subsequent changes are relatively small.

Amongst non-marine molluscs evolution has been relatively slow. Of indigenous Flandrian species, 72 per cent are known from pre-Anglian stages, whilst only nine species known from pre-Devensian stages are now extinct (Kerney 1977). Faunal changes are primarily explained by changing palaeoenvironments, and thus non-marine molluscs can be valuable palaeoenvironmental indicators (e.g. Evans 1972, Kerney & Turner 1977, Sparks & West 1972).

As is found amongst other groups, many fossil communities of non-marine molluscs appear to be made up of elements whose modern counterparts appear climatically incompatible, as in Late Devensian and Early Flandrian time, when mobile thermophilous species took advantage of newly created niches to co-exist with surviving cold forms (Kerney 1977).

## (iii) Insects

The exoskeletons of insects, in particular the Coleoptera, are highly resistent to decay and fossilise well. Study of British coleopteran fossils suggests that there has been no significant morphological evolution during the Quaternary (Coope 1977), but that Quaternary populations in temperate latitudes are largely the result of late Tertiary speciation (Mathews 1976). They occupy a wide range of ecological niches. Many species have fastidious requirements, and are relatively mobile. Evolutionary stability, environmental sensitivity, and rapidity of response, make them ideal palaeoenvironmental and palaeoclimatic indicators. Statistical comparisons between modern insect distributions and climate have made it possible to reconstruct summer and winter temperatures for series of sites in Britain during late Devensian and early Flandrian times. They give an almost unique picture of continental temperature changes during this time (Fig. 14.2).

The relatively little palaeoentomological work which has been done on the interglacial stages tends to bear out environmental conclusions drawn from study of the flora. For instance the Ipswichian (interglacial) Stage site of Bobbitshole (GR TM 165432) near Ipswich has produced assemblages of pollen and Coleoptera both of which suggest a eutrophic lake partially overgrown by reeds and surrounded by lightly wooded country (West 1957, Coope 1975) and the abundance of dung beetles is explained by numerous large mammal bones from a correlative horizon (Coope 1977). Coope (1977) also suggested on the basis of coleopteran assemblages that whereas the Cromerian and Ipswichian interglacials were substantially warmer than the present day, Hoxnian temperatures were very similar.

The coleopteran faunas of the Devensian stage have been studied in detail, and give us an insight into processes of faunal and floral migration during glacial stages when palaeoenvironments appear to have changed much more rapidly than during interglacial stages. Components of the biota appear at times to be strikingly out of step. For instance during the early part of the Late Devensian Windermermere Interstadial at about 13,000 B.P. (Coope & Pennington 1977), when insect faunas indicate a climate similar to that of the present day, the flora remained that of a cold relatively treeless environment, but by the time birch woodland had taken hold, the insects suggest that climate had deteriorated substantially (Fig. 14.10b).

These imbalances between different faunal and floral groups and the peculiar mixing of assemblages are typical of times of rapid environmental changes in glacial stages and reflect different rates of migration and the varying rates of equilibration of faunal and floral elements to new habitats.

## (iv) Man and his cultural remains

The rarity of the pre-Flandrian skeletal remains of man probably reflects his numerical insignificance as well as a mode of life which rendered fossilisation unlikely. Most evidence of his activity comes from his artefacts.

The earliest evidence of man comes from south-west England. Associated together in a cave at Westbury-sub-Mendip (Bishop 1974, Wymer 1988) there are crudely worked flints, and charcoal, which may be evidence for fire, and a mammalian fauna of probable Cromerian age ($\approx$300–500,000 B.P.). In Kents Cavern near Torquay a series of crude hand axes have been found which may be contemporary with a Cromerian fauna.

Until Flandrian times man's artefacts are largely restricted to flint tools. From Late Anglian times widespread and more or less culturally distinctive species of flint implements can be identified. Four principal pre-Late Devensian groups are identified. Their range is shown in Fig. 14.3a.

*Clactonian*   Industries dominated by chopper cores and flakes with unspecialised flake tools.

*Acheulian*   Industries dominated by hand axes and associated with specialised and unspecialised flake tools.

*Levalloisian*   Industries dominated by flakes struck from prepared cores, sometimes associated with hand axes.

*Mousterian*   Flake tools with well made flake points and scrapers predominating.

The earliest skeletal remains of man in Britain, comprising the occipital and left and right parietal bones of a cranium estimated to have had a capacity of about 1,325 cm³, comes from late Hoxnian fluviatile deposits of the Thames at Swanscombe (GR 559174) where it is associated with Acheulian hand axes and bones of large grassland mammals. The skull is that of a female who probably belonged to early Neanderthal populations (Stringer 1974) which were adapted to life in open grassland rather than wooded country (Oakley 1952). Other evidence of Hoxnian occupation by man comes from Hoxne itself where Acheulian implements are associated with a period of deforestation during late zone II.

A cave in Jersey, La Cotte de St. Brelade, contains teeth and an occipital bone of a juvenile attributed to *Homo sapiens neanderthalensis*, associated with a Mousterian industry, probably of late Ipswichian age (McBurney 1972). The Mousterian artefact assemblage, generally associated with neanderthal man, persists in a series of cave sites in Wales, Somerset, Devon, Derbyshire and Kent until about 35,000 B.P.

Remains of man of entirely modern aspect have

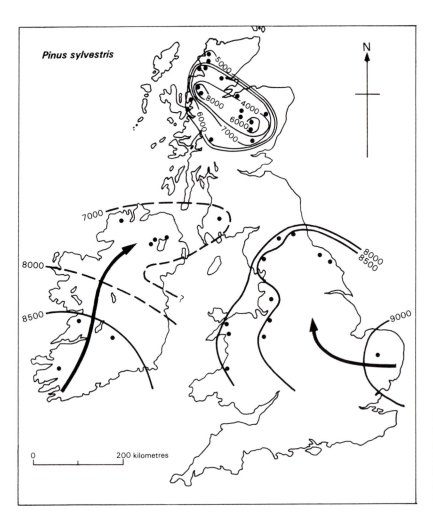

**Fig. 14.3b.** Isochrons on the spread of *Pinus sylvestris* (Scots Pine) in Britain during the Flandrian period (redrawn from Birks 1982).

been found at Kents Cavern and dated at about 28,000 B.P., an interstadial period immediately prior to the most extensive phase of Devensian ice sheet growth. They are associated with a fauna of woolly rhinoceros and rheindeer, and an Upper Palaeolithic industry including bone implements. Upper Palaeolithic industries are also found both during and after the Late Devensian glacial maximum. Of particular interest is the site at Paviland, south Wales, where part of an *in situ* human skeleton of modern aspect has been dated at 18,460 ± 340 B.P., suggesting that men dwelt in caves here when glacier ice may only have been a few kilometres to the north.

Evidence of cave dwelling in England persists until the early Flandrian forest expansion.

# Distribution of Quaternary beds

Quaternary beds in England and Wales can usefully be divided into four major groups on the basis of their stratigraphical position and geographical distribution.

## (a) Pre-Anglian beds in East Anglia

In East Anglia there are exposed on land a wedge of sediments which result from deposition in the Upper Tertiary and Lower and Middle Quaternary in the western marginal zone of the North Sea Basin. They provide almost the only sedimentary record of these times in Britain, and include the well-known 'Crags' (Fig. 14.4) which form part of a series of marine,

**Fig. 14.4.** The stages of the Pre-Anglian Quaternary and important environmental changes. The biostratigraphic zones are those established in the Ludham borehole (Funnell & West 1962). (Although pollen, foraminifers and molluscs have been studied in other stages, their interrelations are not yet so fully worked out.) The Quaternary/Tertiary boundary is normally placed at the top or bottom of the Red Crag in Britain.

| Chronostratigraphic stages | Common Lithostratigraphic terms | Biostratigraphic zones | | | Environment | Movement of sea level | Important events in East Anglia |
|---|---|---|---|---|---|---|---|
| | | pollen | foraminifera | molluscs | | regression / transgression | |
| ANGLIAN | LOWESTOFT TILL, CORTON BEDS, NORTH SEA DRIFT, NORWICH BRICKEARTH, MUNDESLEY SANDS, CROMER TILLS | | | | GLACIAL | CORTON BEDS | FIRST DIRECT EVIDENCE OF GLACIATION |
| CROMERIAN | CROMER FOREST BED SERIES (for details see West 1980) | | | | TEMPERATE | | |
| BEESTONIAN | | ? KESGRAVE sands & gravels | | | PERIGLACIAL | | FIRST DIRECT EVIDENCE OF PERMAFROST. WELSH ERRATICS INTRODUCED INTO THAMES BASIN |
| PASTONIAN | | ? | | | TEMPERATE | | |
| PRE-PASTONIAN a – d | | | | | COLD | | MIDLANDS ERRATICS INTRODUCED VIA THAMES SYSTEM (POSSIBLY BY GLACIERS?) |
| BRAMERTONIAN | WESTLETON BEDS, CHILLESFORD CLAY, CHILLESFORD CRAG, NORWICH CRAG | Lp5 | | | TEMPERATE | HIATUS | |
| BAVENTIAN | EASTON BAVENTS CLAY | Lp4 c b a | Lf 7 6 | Lm6 | PERIGLACIAL | | FIRST TRULY GLACIAL CLIMATE. POSSIBLE GLACIATION SOMEWHERE IN THE NORTH SEA BASIN |
| ANTIAN | | Lp3 | Lf5 | Lm6 Lm5 | TEMPERATE | | |
| THURNIAN | | Lp2 | Lf4 | Lm5 Lm4 | COLD | | |
| LUDHAMIAN | LUDHAM CRAG | Lp1 b a | Lf3 Lf 2 1 | Lm3 Lm2 Lm1 | TEMPERATE | | |
| WALTONIAN | RED CRAG | | | | COOL TEMPERATE | HIATUS | |
| TERTIARY ? | CORALLINE CRAG | | | | | | |

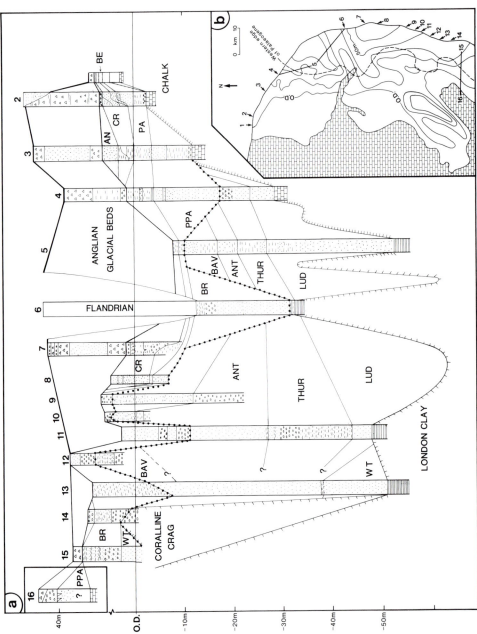

**Fig. 14.5. a.** A schematic section through the pre-Anglian Quaternary basin in East Anglia showing the distribution of stages. Terrestrial horizons are shaded. Notice how the distribution of stages strongly suggests an important post-Baventian hiatus. 1, Beeston; 2, West Runton; 3, Mundesley; 4, Happisburgh; 5, Ludham; 6, Great Yarmouth; 7, Corton; 8, Pakefield; 9, Covehithe; 10, East Bavents; 11, Southwold; 12, Dunwich; 13, Sizewell; 14, Aldburgh; 15, Chillesford; 16, Barham. **b.** A map showing contours on the base of the Quaternary and the line of section and the location of sites in (a).

estuarine and freshwater deposits. Floral correlations with the more continuous Dutch sequence which accummulated in the southern end of the subsiding North Sea Graben suggests that major erosional gaps occur within the sequence. Sedimentation in this part of the basin was brought to an end by Britain's most extensive glacial event, which occurred in the Anglian stage.

### (b) Anglian to Ipswichian beds in central and southern England – the 'Older Drifts'

South of the relatively well-defined limit of the Late Devensian ice sheet (Fig. 14.6a) there is a broad zone of relatively deeply eroded and weathered suites of older glacigenic sediments. They are often restricted to interfluves; lacustrine beds containing Hoxnian pollen assemblages occur on them; and terrace sediments in the river valleys often contain organic beds dating from the Ipswichian interglacial or early- and mid-Devensian interstadials. The stratigraphy of this zone is a major current problem, in particular the number of interglacials and the number and extent of glacial episodes.

### (c) The unglaciated zone – southern England

There has been little net sedimentation in this area. The Quaternary record is preserved primarily in river terraces, above all the Thames, and in records of high sea levels at many points along the coastline. In this zone, as in the previous one, evidence of intense periglacial activity is to be found, in the form of solifucted sediments, loess, and patterned ground.

### (d) The area of Devensian glaciation – the 'Newer Drifts'

A Devensian ice sheet reached its maximum extent a little after 20,000 B.P. in England and Wales approximately along the line shown in Fig. 14.6a. North of this line, glacial erosion has ensured that few earlier Quaternary sediments have survived. Most of the Quaternary record reflects Late Devensian deglaciation. Many lowland areas are draped with thick sequences of glacigenic sediments and show primary glacial depositional landforms. In the uplands of Wales and the Lake District small glaciers occupied corries and valleys during the dying phases of the Devensian stage.

The succeeding Flandrian stage, the present inter-glacial (for there is no reason to suppose that another glacial will not follow), left its record in all the above areas, a record best read through fluviatile, lacustrine, and shallow marine sediments.

A fifth area of great potential interest is that of the continental shelf, where in the central graben of the North Sea, Quaternary sediments exceed 1,000 m in thickness.

# Tectonic framework and the form of Britain

Britain primarily owes its form to two geologically recent events, the tectonic changes which follow regression of the Chalk sea and intense climatic changes of the Quaternary.

The Early Tertiary Laramide orogeny associated with the onset of sea floor spreading in the North Atlantic caused profound palaeogeographical changes within and around Britain. Most important were strong downward tilting to the east and enhanced subsidence in the North Sea graben (Clarke 1973). Relative uplift in western England and Wales is probably related to isostatic uplift of basement ridges intruded by low density granitic masses. This uplift has been active throughout Late Cenozoic times and has produced a characteristic pattern in the west of fault-bounded uplifted blocks and intervening basins.

The eastward tilting appears to be associated more with net uplift in the west than subsidence in the east, for, contrary to a frequently stated view (e.g. Mitchell 1977), there is no evidence of subsidence in the Quaternary of eastern England, but rather the reverse, as Lower Quaternary marine beds are succeeded by Middle and Upper Quaternary terrestrial sequences rather than by evidence of net subsidence. There is as yet no evidence that subsidence reflected by the 1,000 m of Quaternary sediment in the Central Graben of the North Sea Basin (McCave et al. 1977) extended as far as the coast of England.

There are many highland landscape facets in England and Wales which appear to be eroded remnants of a surface which carried a dominant east and south-easterly trending drainage pattern developed on the easterly-tilted sub-Eocene surface after the post-Cretaceous emergence. In many places, as in the Wealden area, the antecedent pattern of drainage continued to run across rising Tertiary folds. Linton (1951) identified a Proto-Trent rising in the high ground of north-west Wales and flowing to the east into the North Sea in the vicinity of the Wash. Many landscape facets have now been recognised in central and eastern England which may be related to the late Tertiary and early Quaternary representatives of this major drainage basin (see Straw & Clayton 1979). One of the few fragments of sedimentary evidence of this system is the Brassington Formation (Walsh et al. 1972) – Pliocene fluviatile sediments interpreted as deposited by a southward-flowing river and subsequently let down into a sink hole in the Carboniferous Limestone of Derbyshire, and interpreted as a relic of a surface originally developed near sea level and now raised by about 450 m (King 1977).

Major erosion surfaces at 250–300 m on the Chalk plateau of south-eastern England were regarded by Wooldridge and Linton (1955) as equivalents of the surfaces in the Midlands and Pennines, although they have recently been suggested to be of pre-Oligocene

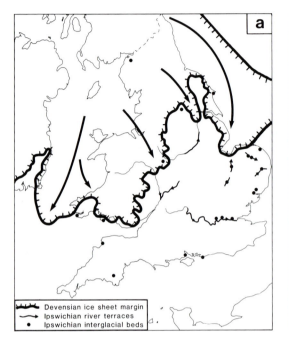

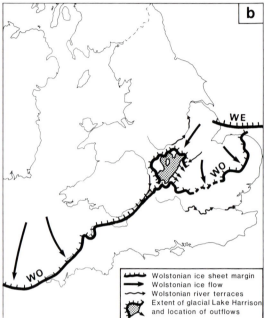

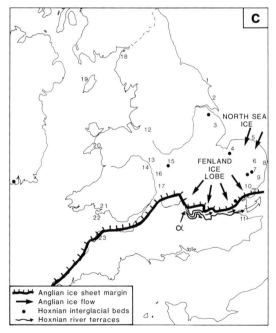

**Fig. 14.6.** Inferred maximum extensions of the ice sheets in Britain:

(a) Devensian.

(b) Wolstonian/'Weltonian'. WO shows the Wolstonian margin based on Straw (1978) in East Anglia, Shotton (1953) in the Midlands and a concensus from Kidson (1977) in western England. Bowen *et al.* (1986) deny the reality of the Wolstonian and identify a 'Weltonian' margin WE.

(c) Anglian. There is a concensus about the margin east of (a). Bowen *et al.* believe that the Wolstonian margin in western England shown in (b) is really the Anglian margin.

Numbers show locations of sections in Figure 14.6.

age and the terrestrial time equivalents of marine Tertiary beds of the southern North sea (Jones 1980).

## The pre-Anglian Quaternary

In late Tertiary and early Quaternary times, the drainage of England and Wales was dominated by rivers rising on the up-tilted high ground of Wales and western England and draining towards the North Sea basin and the English Channel. The most complete record of the pre-Anglian Quaternary in Britain is represented by marine and marginal-marine sediments in East Anglia which accumulated near the western margin of the southern North Sea basin. However, terrestrial sediments in southern and central England, although isolated in time and space, give invaluable evidence of contemporary terrestrial environments.

Sediments accumulating in East Anglia in the western margin of the North Sea basin comprise a wedge, thickening in a seaward direction, up to 70 m in thickness. Fig. 14.5(a) shows a section through these Lower or Middle Quaternary beds. The lower part of the sequence is marine, but terrestrial sediments become more frequent higher in the sequence. This may reflect infilling of the basin, progressive uplift, or eustatic fall in sea level as the intensity of global glacierisation increases through the Quaternary.

Study of pollen, foraminifers and molluscs (West 1961, 1980; Funnell 1961; Norton & Beck 1972; Funnell et al. 1979) has enabled a series of temperature and cold episodes of different intensity to be recognised, which form the basis of the stage names shown in Fig. 14.4. Palynological correlation with the more complete sequence of the Netherlands (Zagwijn & Staalduinen 1975; West 1980, fig. 2) suggests the existence of at least two major gaps in the sequence. One is of about 0·4 million years between the Waltonian and Ludhamian stages, and the other a possible gap of up to a million years between the Baventian and the Pastonian. The problem lies in the fact that though the Ludhamian–Baventian sequence is correlated with the Dutch Tiglian and Eburonian stages on the basis of specific pollen taxa, many of the overlying stages are currently correlatable only on the basis of magnetic polarity and the presence or absence of thermophyllous pollen. Thus the pre-Pastonian a cold episode may be correlated with the basal (normally polarised) Menapian, or with Glacials A or B of the Dutch Cromerian complex, or even with the Eburonian stage. Funnell (1987) prefers this latter correlation, suggesting that the pre-Pastonian a is a facies variant of the Bavention and that the Bramertonian is a facies variant of the Antian. The temperate Pastonian stage may correlate with Waalian C or Interglacial III of the Dutch Cromerian complex; whilst the cold Beestonian stage may correlate with Glacials A, B or C of the Dutch Cromerian complex.

Between the Baventian and English Cromerian stages we therefore appear to have four units, Bramertonian (temperate), pre-Pastonian (cold), Pastonian (temperate) and Beestonian (cold), which cannot yet be firmly correlated with the Dutch sequence, though they clearly fit somewhere between Glacial C of the Dutch Cromerian Complex and the Eburonian/late Tiglian stage.

A major break in the sequence appears to occur above the Baventian, which, together with the underlying sequence appears to be cut out by the Westleton Beds, the Norwich Crag of the Bramertonian stage and the presumed overlying pre-Pastonian beds (see Fig. 14.5). This major break in the sequence would therefore represent a long erosional interval, and is succeeded by littoral and sub-littoral beds.

Boreal molluscan faunas from the base of the Red Crag at Walton-on-the Naze indicate a striking phase of cooling compared with the faunas of the underlying Coralline Crag. This interface is therefore recommended by some as a base for the Quaternary (Mitchell et al. 1973).

The Red Crag is a shallow-water facies of the Waltonian, a deeper water facies of which is represented by silty clays revealed by boring at Stradbrooke, Suffolk, in a deep north-easterly trending trough in the basin (Beck et al. 1972, fig. 5b). After a regression indicated by sub-aerially eroded Coralline Crag below sea level in Suffolk (Dixon 1978) a major marine transgression occurred at the base of the Red Crag, enlarging the basin westwards as indicated by the outliers of Red Crag at Rothamstead at 131 m O.D. and Netley Heath at 183 m O.D. On the basis of evidence from the late Tertiary St Erth beds in the Scilly Islands (GR 356352) Mitchell (1972) suggested that local sea level at the Tertiary/Quaternary boundary was at least 45 m above O.D., whilst the planations in the lower Thames basin, in south-west Lincolnshire, and north of the Fens at around 200 m O.D., may have been produced by high early Quaternary sea levels (Wooldridge & Linton 1955, Straw & Clayton 1979). However the exposed Red Crag in East Anglia suggests that the local maximum height of the transgression in lower Red Crag times was 25 m O.D. (Dixon 1978) followed by a regression in the upper Red Crag. Apart from Waltonian clays at −20 m in the Stradbrooke borehole there is no evidence of deep water conditions in East Anglia, which suggests post-Waltonian tilting of eastern England towards the east.

Sediments belonging to the succeeding Ludhamian and Thurnian stages are known only from boreholes. The Ludhamian comprises a series of shelly marine sands with a temperate fauna and flora which is correlated with the Tiglian of the Netherlands (Zag-wijn & Staalduinen 1975). The correlation with the Netherlands shown in Fig. 14.1 suggests a major non-sequence between the Waltonian Red Crag and Ludhamian sediments, although at no place is a contact between the two known. As much of the Praetiglian cool period is missing, it is possible that the non-sequence reflects a eustatic fall in sea level due to a global increase in glacier ice volume. Weathering and oxidation of the Red Crag to give its present colour may have taken place during this regression.

The base of the Ludhamian reflects a major transgression. Funnell's (1961) work on the Ludham Crag foraminifera suggests that this transgression represents at least 15 m of relative sea level rise, and that subsequently sea level fell throughout the Ludhamian stage and into the succeeding Thurnian where there is fossil and sedimentary evidence of a diconformity at the top of foraminiferal stage LIV. This coincides with faunal and floral evidence of cold climate, suggesting that part of the Ludhamian/Thurnian regression may be due to glacial eustasy.

The succeeding temperate Antian and cold Baventian stages are known from sub-aerial exposures at Easton Bavents in Suffolk. Work on molluscs by (Norton & Beck 1972) suggests an increasing water depth with a change from littoral to sublittoral facies at Easton Bavents during deposition of the temperate Antian and cooler early Baventian. Funnell and West (1962) and West et al. (1980) showed that climate deteriorated markedly during Baventian times to produce the first English cold stage of truly glacial intensity and also suggested that the high proportions of alkaline amphibole found in the Easton Bavents clays indicated the presence of a contemporaneous glaciation somewhere in the North Sea area, the first British evidence for quaternary glaciation. A relative fall of sea level is suggested at this time (Funnell & West 1962), which may represent a eustatic decline of sea level due to glacier build-up (Fig. 14.4).

The major post-Baventian hiatus suggested by the correlation with the Dutch sequence and by the distribution of beds in Fig. 14.5 was presumably a period of relatively low sea level. During an ice age such as the Quaternary, a dominant control on shallow water sedimentation is glacial eustasy. Thus the post-Baventian hiatus may reflect low sea level produced by glacier growth during the later part of the Eburonian cold stage, ushered in in East Anglia, by Baventian sediments. The subsequent temperate transgression is represented by the Norwich Crag deposits in the vicinity of Norwich (Funnell 1961) which suggest that it reached a height of at least 27 m O.D., whilst the correlative Westleton Beds of Suffolk, which are regarded as beach plain deposits by Hey (1967), occur up to 15 m O.D.

In the upper part of the Norwich Crag at Norwich, we find the first introduction of erratic quartz and quartzite pebbles, which become widespread in East Anglia in pre-Pastonian times (Fig. 14.4). At Happisburgh (Fig. 14.5) there is evidence that the pre-Pastonian sea dropped to at least $-4$ m O.D. This regression may well be a major eustatic sea level fall in response to increase of global ice volume and the quartz and quartzite erratics may reflect a change in the Thames river system in response to this regression, or to glacial nourishment of that system. A transgression occurs at the beginning of the Pastonian, which is a temperate stage characterised by broad-leaved woodland. There is a possible temporary regression in the middle of the stage, and strong regression at the end. The so-called Bridlington Crag, enclosed as a raft in the Basement till of Holderness, East Yorkshire has been correlated with the Pastonian stage of the basis of its pollen content (Reid & Downie, 1973).

The succeeding Beestonian cold stage contains marine and fluviatile sediments with ice-wedge casts indicating a periglacial climate, the first definite evidence of Quaternary permafrost in Britain (Fig. 14.4).

Although evidence of terrestrial environments contemporary with the East Anglian marine basin is sparse, recent evidence suggests that there may have been extensive glaciation at least in Wales and the Welsh borders during pre-Pastonian and Beestonian times. Rose et al. (1976) and Rose and Allen (1977) drew attention to a major body of sand and gravel (Kesgrave sands and gravels) in southern East Anglia, up to 30 m thick and persistent over an area of 3,000 km², representing a series of bars and dunes deposited by a north-easterly flowing river of variable regime, in which numerous intraformational ground ice structures demonstrate a contemporary periglacial climate. They suggested that these sediments were deposited by a periglacially swollen ancestor of the Thames, and that their fall in altitude towards the south-east reflects the formation of terraces as the river migrated in that direction. They are the earliest deposits which can be related to a river system flowing in a course similar to its modern one, and mark an important phase in the geomorphological and hydrological evolution of Britain. At Barham (Fig. 14.5) these sediments are overlain by a soil presumed to be of Cromerian age, and which elsewhere overlies Westleton beds. The Kesgrave Formation may thus correlate with the Pre-Pastonian or Beestonian stages (Fig. 14.4).

It is possible that this pre-Pastonian or Beestonian ancestor of the River Thames was partially fed by an ice sheet as the Kesgrave Formation contains a suite of volcanic erratics which Hey and Brenchley (1977) suggested was derived from North Wales, and which could have been introduced into the upper Thames catchment by an ice sheet. At Sugworth south of Oxford a large fluviatile channel filling contains Cromerian interglacial fossils (Briggs *et al.* 1975) (Fig. 14.7). Pebbles within the Sugworth channel are similar to those in the Plateau Drifts of the Oxford region (Sandford 1945), the Northern Drifts of the upper Thames region (Hey 1986) and the Middle Thames Gravel Formation of the middle Thames area (Gibbard 1985). These were deposited by a river whose catchment extended well to the north of the Thames catchment and whose sediments show a dramatic increase in quartzites derived from the Trias of the west Midlands.

The Cromerian interglacial stage is represented at many points along the East Anglian coast. An interglacial woodland flora develops during the stage (Fig. 14.2). It contains a rich fauna (see Fig. 14.3) and evidence of an important marine transgression. Other beds belonging to the Cromerian stage occur at Sugworth and in cave deposits at Westbury – sub-Mendip (Bishop 1974, 1975). The Cromerian interglacial is an important datum as it can be correlated with reasonable confidence with other European sequences, and has been equated with Interglacial IV of the Dutch Cromerian Complex, the Rhume Interglacial at Bilshausen near Göttingen (Lüttig 1965)

and an interglacial site at Voigstedt near Dresden (Erd 1973, West 1980).

## Anglian to Ipswichian Stages

In 1973 a synthesis by Mitchell, Penny, Shotton and West of several previous decades of research in the British Quaternary suggested the following post-Anglian sequence of glacial and interglacial stages (Fig. 14.1b):

Flandrian (interglacial)
Devensian (glacial)
Ipswichian (interglacial)
Wolstonian (glacial)
Hoxnian (interglacial)
Anglian (glacial)

Since then, considerable doubts have been thrown on the reality of a distinctive Wolstonian glacial stage, whilst evidence has been produced of additional post-Hoxnian, pre-Ipswichian warm 'interglacial' stages (Fig. 14.1b).

### (a) Anglian Stage

The integrity and stratigraphic position of the Anglian stage is well-defined in East Anglia.

**Fig. 14.7** Schematic correlations between important Quaternary sequences in England and Wales. The shaded areas show correlations between inferred Devensian (DEV), Wolstonian (WO)/Weltonian (WE) and Anglian (ANG) glacial beds. The top row shows east coast sections, the middle row central England sections, and the bottom row west coast sections. The site locations are shown in Fig. 14.6c.

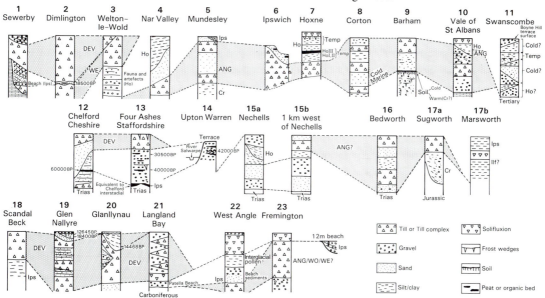

428        G. S. BOULTON

Cromerian interglacial beds exposed along the East Anglian coastline are overlain directly by a suite of glacigenic deposits (Figs. 14.4, 5, 7) which in turn are overlain by distinctive Hoxnian interglacial beds. The glacigenic beds define an Anglian stage, and are the first unequivocal evidence of Quaternary glaciation in Britain, probably equivalent to the Elsterian stage of northern Europe and the Riss glaciation of the Alps.

The most southerly extent of glacier ice was achieved during Anglian times (Fig. 14.6). It largely brought to an end active late Cenozoic sedimentation in the southern and central North Sea (Cameron, Stoker & Long 1987).

At its maximum extent in the Vale of St. Albans there is evidence for two ice advances separated by a retreat phase (Gibbard 1977). At Corton near Lowestoft there are two Anglian tills separated by sands containing cold climate pollen and an indigenous marine fauna (West & Wilson 1968), reflecting high glacial sea level and probable isostatic depression of the crust due to ice loading. Further north at Happisburgh there are three till units separated by waterlain beds. Thus it is clear that the Anglian stage contained at least two, and possibly three major glacier fluctuations.

The Anglian glacial margin can be traced with diminishing confidence to the north of Oxford (Straw & Clayton 1979) and may correlate with the most extensive phase of glaciation in the Severn Valley identified by Wills (1938) as lying near to Gloucester (Fig. 14.6) and marked by predominantly Welsh erratics. In south-west England glacial deposits occur at several points along the coast of Somerset, the north coast of Devon (Hawkins & Kellaway 1971, Kidson 1977) and as far south as the northern Scilly Isles (Mitchell 1972), and although their attribution is contentious (e.g. Kidson 1977, Bowen et al. 1986) their extent makes an Anglian age most likely.

The pattern of grain size distribution, mineralogical variation (Perrin et al. 1979) and pebble orientations in Anglian tills in East Anglia (West & Donner 1956) suggest that the Anglian ice sheet fanned out from the Wash region. There is also evidence that it produced substantial erosion, lowering the chalk scarp in the East Anglia by 100 m or more (Boulton et al. 1977)

The Thames river system and possibly the Severn were major courses along which Anglian outwash flowed, and the Upper Gravel Trains and Harefield and Winter Hill terraces gravels of the middle and upper Thames (Figs. 14.6, 8) are thought to have been laid down by this glacial meltwater (Hare 1947, Clayton 1977). The Anglian ice sheet in its first advance into the area flowed up the course of the Thames in the Vale of St Albans, and dammed it to form a proglacial lake in which varve like sediments accumulated, with a minimum of 485 couplets (Gibbard 1977). The old course was re-established during glacier retreat, but the subsequent, more extensive Anglian readvance permanently deflected the Thames further south, near to its modern course (Fig. 14.6c), with a southward-flowing tributary in the Vale of St. Albans.

The relationship of the Anglian glacial stage to the succeeding Hoxnian interglacial stage is clearly demonstrated near Hatfield by organic sequences in kettle holes produced during glacier decay which indicate a transition from an open late-glacial flora with dwarf birch and tree birch to boreal forest and then to temperate mixed oak forest with a sequence characteristic of the Hoxnian. Similar sequences resting on Anglian till at Hoxne in Suffolk (Fig. 14.7), and Marks Tey in Essex, show a transition in

**Fig. 14.8** The elevation and ages of the major terraces of the Thames, Severn and Trent. (Amended after Clayton 1977.)

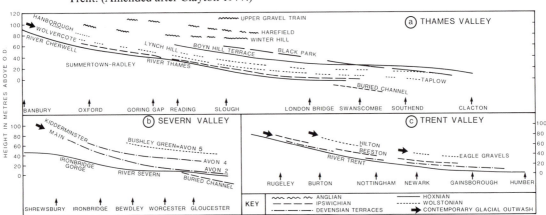

lacustrine beds from Anglian late-glacial to Hoxnian interglacial pollen sequences.

## (b) Hoxnian Stage

The floral characteristics which allow the Hoxnian to be differentiated from other temperate phases include a high frequency of *Hippophaë* (buckthorn) in the immediately preceding part of the Anglian, the abundance of *Tilia* (lime), the late rise of *Ulmus* (elm) and *Corylus* (hazel) in zone HoII, and the importance of *Abies* (silver fir) in zone HoIII (Figs. 14.2, 14.9). The flora indicates a more oceanic climate than any other Quaternary interglacial stage.

At Marks Tey in Essex (Turner 1970) a record of

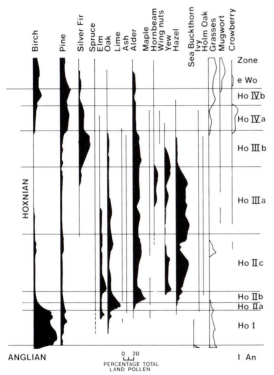

**Fig. 14.9.** Pollen diagram through the Hoxnian interglacial lake deposit at Marks Tey, Essex, redrawn from Turner (1970). All four zones of the Hoxnian (H.I–H.IV) are represented, as are the preceding and succeeding transitions from and to cold stages which are presumed to be the Anglian and Wolstonian respectively. A duration of about 20,000 years has been estimated for the interglacial beds at this site.

the whole Hoxnian stage is preserved in a series of lake muds (Fig. 14.9). The muds formed during the early and middle parts of the interglacial (zone HoI to HoIIIa) and show well-developed lamination caused by annual flushes of the diatom *Stephanodiscus astraea* var. *minutula* into the lake. Counts of these laminations, and extrapolation, suggest a duration of about 20,000 years for the Hoxnian stage.

The vertebrate fauna of the Hoxnian stage is summarised in Fig. 14.3a. The stage is also important in containing the first evidence of hominid colonisation of Britain. Acheulian artefacts (Fig. 14.3) and evidence of occupation occur in zone HoIII at Hoxne during a temporary phase of deforestation (Wymer 1974). There is also evidence of cold conditions separating upper and lower temperate beds. At Clacton-on-Sea an earlier, Clactonian industry has been found in Hoxnian beds, whilst at Swanscombe, in Hoxnian gravels of the River Thames, a Clactonian industry has been found succeeded by an Acheulean industry associated with skull fragments ascribed to an early *Homo sapiens* (Wymer 1977, Day 1977). Bridgeland (1980) has suggested that the Clactonian industry lies in gravels belonging to an early Hoxnian Black Park Terrace aggradation which includes the Clacton beds whilst the Acheulian industry lies in the Boyn Hill Terrace gravels deposited by a Thames of much lower gradient (Fig. 14.8). Between these temperate sedimentary episodes at Swanscombe there is evidence of a mid-Hoxnian cool phase (Waechter *et al.* 1971), suggesting that the Hoxnian may be a two-phase interglacial (Fig. 14.1b).

As in the earlier Quaternary, the Hoxnian fluviatile record is limited to the Thames basin. The major Hoxnian aggradation of the Middle Thames is the Boyn Hill Terrace, which contains the Hoxnian sediments at Swanscombe within it, and which can be correlated with fluviatile gravels near sea level at Clacton. The location of the Boyn Hill terrace system (Wooldridge 1938) and its downstream correlative demonstrates that the Thames had developed its present course by Hoxnian times.

There is good evidence of high Hoxnian sea levels around the coasts of Britain. In the Nar Walley just south of the Wash, estuarine beds give evidence of sea level up to 20–23 m between zones HoII and HoIII (Stevens 1960); the Upper Speeton Shell Bed north of Flamborough Head contains estuarine molluscs up to 32 m O.D. and is probably of Hoxnian age (Penny 1974); at Clacton, Pike and Godwin (1953) suggested a marine transgression up to 9 m O.D. in zone HoIII; a raised beach at Slindon in Sussex at 35 m O.D. contains an Acheulean artefact assemblage and is presumed to be of Hoxnian age; whilst there are many shore platforms around the southern margin of the Celtic Sea at about 20–25 m O.D. which may be of

Hoxnian age (Mitchell 1977, Kidson 1977). On the other hand, if we accept the interpretation of the deep ocean isotopic record as a record of global ice volume change it suggests that only during the last interglacial were global eustatic sea levels substantially higher than modern ones (Shackleton & Opdyke 1973). This implies that high Hoxnian sea levels must be a local phenomenon: a product of residual isostatic uplift, a local tectonic effect, or the result of a net increase in continental freeboard in north-west Europe. Szabo and Collins (1975) dated bones from Hoxnian deposits at Clacton as 245,000 ± 25,000 years B.P.

## (c) Post-Hoxnian, pre-Devensian interglacials

Recent research has demonstrated that the conventional stage sequence of Hoxnian–Wolstonian–Ipswichian is now untenable, and that at least one further warm, interglacial stage existed between the Hoxnian and Ipswichian stages.

Hoxnian interglacial deposits tend to lie on interfluves, suggesting substantial post-Hoxnian erosion. By contrast the older parts of terrace sequences lying within the valleys of several modern rivers contain clear evidence of pre-Flandrian, post-Hoxnian temperate conditions (Fig. 14.7). Unfortunately none of these beds contains a pollen sequence which reflects a full interglacial cycle, but careful study of the mosaic of pollen records from interglacial sediments in this geomorphological position has led to the conclusion that all these sites could belong to a single interglacial cycle (Sparks & West 1972, West 1977) (Fig. 14.7) correlated with the type site of the Ipswichian stage at Bobbitshole (West 1957) near Ipswich in East Anglia.

Studies of vertebrate faunas however show two very different assemblages at sites which have similar palynofacies and which are conventionally ascribed to the Ipswichian (Table 2). One assemblage is characterised by Hippopotamus, straight-toothed Elephant and Fallow-deer and has been dated at Stutton in Suffolk by U/Th as 125,000 ± 20,000 years B.P. (Szabo & Collins 1975), whilst the other is characterised by Horse and forest Mammoth (Sutcliffe & Kowalski 1976, Sutcliffe 1976) and has been dated at 174,000 ± 30,000.

The Ipswichian type-site at Bobbitshole shows greatest floral affinities with the Hippopotamus-bearing interglacial beds, suggesting that the Ipswichian is the last interglacial, correlating with stage 5e of the deep ocean isotopic record (Fig. 14.1).

The relationships between two florally similar pre-Devensian interglacials are best demonstrated in fluviatile sequences in southern England. The local equivalent of the Taplow terrace of the Thames (Fig. 14.8) at Ilford contains a Horse/forest Mammoth fauna, whereas the Upper Flood Plain Terrace, which lies below it and is therefore later, contains, at Trafalgar Square, a phylogenetically younger fauna with Hippopotamus and straight-toothed Elephant. At Marsworth in Buckinghamshire a Horse/forest

**Table 14.2** Characteristic Ipswichian and "Ilfordian" faunas, and sites at which they are found

| Ipswichian (Trafalgar Square Fauna) | "Ilfordian" (Ilford Fauna) |
|---|---|
| Hippopotamus | Mammoth (*Mammuthus trogontherii*?) |
| Straight-tusked elephant | Straight-tusked elephant |
| Rhinoceros (*Dicerorhinus hemitoechus*?) | Rhinoceros |
| Fallow deer | (*D. hemitoechus*) |
| Red deer | (*D. kirchbergenis*) |
| Giant deer | Horse |
| Bison | |
| Lion | |
| *Similar faunas are found at:* | *Similar faunas are found at:* |
| Terrace no. 3 at the Avon, Warwickshire | Aveley, Essex |
| Beeston terrace of the Trent, Nottinghamshire | Stoke Tunnel beds, Ipswich, Suffolk |
| Allenton Terrace of Derwent, Nottinghamshire | Brandon, Suffolk |
| Joint Mitnor Cave, Devon | Stutton, Suffolk |
| Barrington, Cambridgeshire | Lexden, Essex |
| Tornewton Cave, Devon | Marsworth (lower deposit) |
| Kirkdale Cave, Yorkshire | Stanton Harcourt (lower deposit) |
| Bacon Hole, S. Wales | |
| Marsworth, Buckinghamshire (upper deposit) | |
| Victoria Cave, Yorkshire | |
| Stanton Harcourt, Oxford (upper deposit) | |

Mammoth fauna is overlain by a Hippopotamus/ straight-toothed Elephant fauna with an intervening bed showing strong cryoturbation of the type produced in periglacial climates (Green *et al.* 1984), indicating that the two interglacials are separated by a cold period. Table 14.2 shows the sites at which faunas of the two post-Hoxnian pre-Ipswichian temperate periods are found. The earlier of the two is informally referred to here as 'Ilfordian'.

Along the coast of Wales and south-west England there occur platforms of marine erosion up to 15 m above modern sea level. Amino acid determinations on limpets and winkles from these platforms show that they fall into two series of different age. These, together with U/Th determinations suggest ages of 132,000 and 210,000 years B.P. (Bowen *et al.* 1986). The younger is clearly equivalent to the Ipswichian (isotope stage 5e) and the older may be equivalent to the 'Ilfordian', and isotope stage 7 of the deep ocean record.

## (d) Status of the Wolstonian stage and post-Hoxnian, pre-Ipswichian cold periods

In the area around Coventry and Leicester Shotton (1953, 1968) and Rice (1968) mapped a sequence of beds which reflect a major glacial episode punctuated by a retreat phase (Fig. 14.7), and ascribed to a Wolstonian (glacial) stage. At Quinton near Birmingham, interglacial deposits, thought to be of Hoxnian age, are overlain by presumed Wolstonian tills and underlain by tills which may be of Anglian age (Horton 1974), whilst at Nechells, Hoxnian interglacial beds were interpreted as stratigraphically underlying a till sequence (Kelly 1974). The evidence for a pre-Ipswichian age of the Wolstonian glacial sequence comes from the River Avon (a tributary of the Severn) which cuts through the Wolstonian glacial sequence and which contains an Ipswichian fauna at the base of No. 3 terrace gravels (Shotton 1968; Fig. 14.8).

Wolstonian ice is thought to have penetrated the area around Birmingham and Coventry from the north-west and north-east, thus damming up a major lake, Lake Harrison (Fig. 14.6), against the Jurassic escarpment of the Cotswolds (Bishop 1957). Water from this proglacial lake overflowed into the upper Thames catchment and introduced northern erratics into the sediments of the Wolvercote terrace of the rivers Evenlode and Cherwell (Fig. 14.8). The lake expanded in Mid-Wolstonian times, diminished during a subsequent ice readvance, and drained finally during the Wolstonian retreat.

Bowen *et al.* (1986) have recently suggested that the Wolstonian drifts of the Midlands are lithologically continuous with the Anglian drifts of East Anglia and thus that the term Wolstonian should be abandoned, and that the events inferred from so-called Wolstonian deposits should be ascribed to the Anglian. This interpretation discounts the interpretations of the Nechells and Quinton sites.

There is however evidence of a post-Hoxnian, pre-Ipswichian glaciation in north-eastern England. At Welton le Wold a Hoxnian fauna and Acheulean artefacts lie beneath the Welton Till (Alabaster & Straw 1976) which has been correlated with the Basement Till of Holderness (Catt & Penny 1966) in east Yorkshire, which in turn is overlain by Ipswichian deposits at Sewerby (Boylan 1967) (Fig. 14.7).

It seems likely that this 'Weltonian' glacial episode reached no further south than southern Lincolnshire, as Perrin *et al.* (1979) have demonstrated that the mineralogy and grain-size distribution of surface tills further south all have a distinctively Anglian character.

The precise stratigraphic position of the 'Weltonian' glacial episode in relation to the 'Ilfordian' and Ipswichian temperate stages is not clear. The major erosional phase which intervenes between the Hoxnian and 'Ilfordian' stages seems the most likely candidate for a substantial episode of glaciation. However, the Saalian glaciation of the western European mainland, which immediately pre-dates the continental equivalent of the Ipswichian, was the most extensive of west European Quaternary glaciations, and it would be very surprising if its equivalent in Britain were not extensive, suggesting a pre-Ipswichian post-'Ilfordian' age for the 'Weltonian'.

## Devensian Stage

The fluctuations of ice volume on earth estimated from oxygen isotope studies on deep sea cores (Shackleton 1977; Fig. 14.10) suggest that the last interglacial was characterised by a sea level higher than present and oceanographic conditions in the eastern Atlantic (Gardner & Hays 1977) similar to today, and that it ended about 125,000 years ago (stage 5d/5e boundary). It was succeeded by a phase of increased ice volume and cooling in the Atlantic (stage 5a–d), and at 75,000–80,000 B.P. (stage 4/5 boundary) a further phase of dramatic ice volume increase coupled with oceanic cooling. A reduction in ice volume in stage 3 was followed by the ice volume maximum of stage 2 at about 20,000 B.P. and in turn by the ice volume minimum of the Flandrian (stage 1).

A major late-Devensian ice sheet reached a maximum extent over England and Wales between about 18,000–20,000 B.P., equivalent to the stage 2 peak in

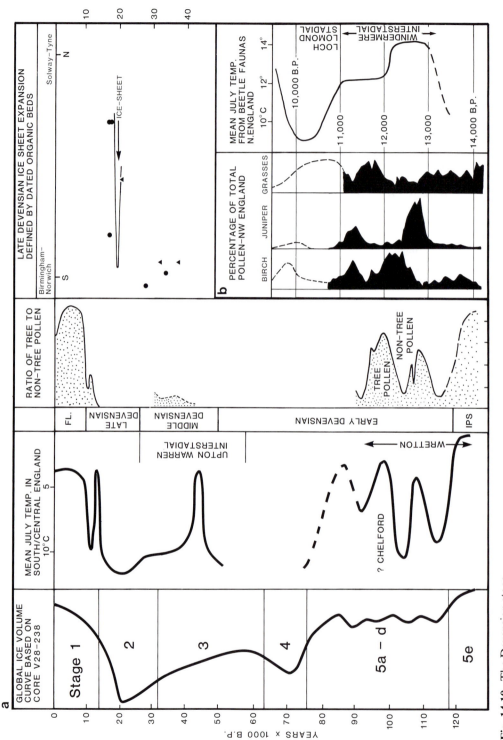

**Fig. 14.10.** The Devensian stage.

**a.** Comparison between the deep sea oxygen isotopic record and known ice sheet fluctuations. The latter part of the palaeotemperature record is based on beetle evidence (Coope 1977) and the early part on pollen (West 1977). This record has been adjusted prior to 40,000 B.P. to suggest a fit with the deep ocean record. The knowledge of ice sheet fluctuations is limited to a very rapid advance to and retreat from a maximum at about 18,000 B.P.

**b.** The presumed imbalance of a response to rapidly changing climate between flora and beetle fauna during the Late Devensian. Beetles respond

the ice-volume curve (Figs. 14.10, 11). The effectiveness of glacial erosion has ensured that almost all pre-Late Devensian Quaternary sediments have been removed from within the area occupied by the ice sheet. The few exceptions are near to the ice sheet margin, locations subjected to the shortest glacial occupation.

The area beyond the Devensian ice sheet as it waxed and waned was subject to an intense periglacial climate in which strong ground freezing and the development

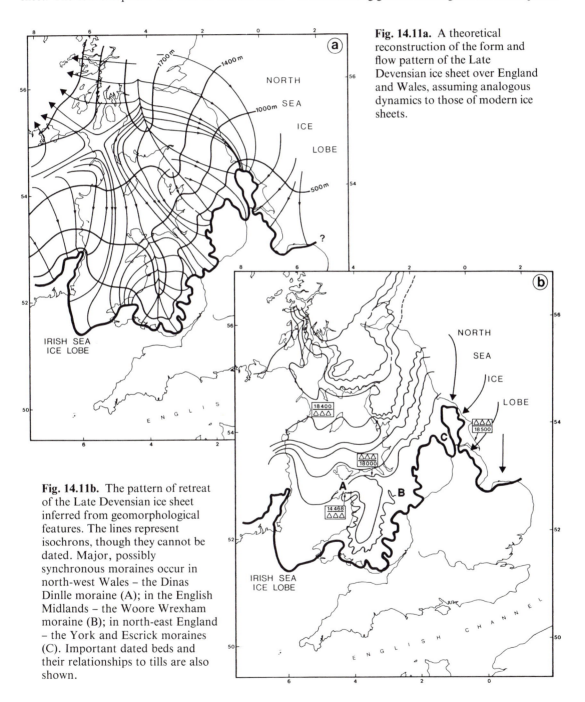

**Fig. 14.11a.** A theoretical reconstruction of the form and flow pattern of the Late Devensian ice sheet over England and Wales, assuming analogous dynamics to those of modern ice sheets.

**Fig. 14.11b.** The pattern of retreat of the Late Devensian ice sheet inferred from geomorphological features. The lines represent isochrons, though they cannot be dated. Major, possibly synchronous moraines occur in north-west Wales – the Dinas Dinlle moraine (A); in the English Midlands – the Woore Wrexham moraine (B); in north-east England – the York and Escrick moraines (C). Important dated beds and their relationships to tills are also shown.

of permafrost led to typical ground-ice structures, whilst strong winds, including katabatic winds from the ice sheet transported much sediment over the poorly vegetated landscape (Catt 1977, 1978) (Fig. 14.13a). Evidence suggests two major Late Devensian periods in which such structures formed; the first roughly coeval with the maximum ice sheet extent, the second with the glacier readvance at about 10–11,000 B.P. (Fig. 14.10).

At Scandal Beck in Cumbria, silts with Ipswichian interglacial pollen lie beneath till (Huddart *et al.* 1977). At Chelford in Cheshire, organic beds lying beneath a glacial complex (Worsley 1980) reveal evidence for a pine–birch–spruce forest (similar to the boreal forests of southern Finland (Simpson and West 1958), which has been correlated with the Brørup interstadial in Denmark. A range of $^{14}$C dates from 24,000 to 60,800 B.P. has been obtained for these beds, the consensus being that the last gives a minimum age, At Four Ashes in Staffordshire, a few miles north of the Devensian ice sheet margin, a series of fluviatile gravels lie on bedrock and are overlain by till. They contain organic beds from the Ipswichian interglacial at the base, succeeded by correlatives of the Chelford interstadial beds, and then by representatives of a treeless Middle Devensian interstadial (now known as the Upton Warren Interstadial Complex; Morgan 1973) (Fig. 14.10). Beds from this interstadial have been found at many sites in river terraces lying south of the Devensian ice sheet margin and study of their bettle faunas allowed Coope (1975) to reconstruct the pattern of temperature change (Fig. 14.10).

The youngest organic beds which pre-date the Late-Devensian ice sheet maximum occur at Dimlington on the Yorkshire coast where silts overlying the Basement Till (Wolstonian/Weltonian) contain a fauna and flora indicating an extreme periglacial climate and have yielded $^{14}$C dates of 18,240 B.P. and 18,500 B.P. (Penny *et al.* 1969). Thus the succession of post-Ipswichian floras suggests a phase of post-interglacial boreal forest represented by sites at Chelford. Four Ashes, and Wretton in Norfolk (West *et al.* 1974) (and probably equivalent to parts of stage 5a–d of the deep ocean record – Fig. 14.10); a mid-Devensian treeless interstadial represented by many terrace sites in the Midlands of England, with frequent remains of mammoth, bison, horse reindeer and Arctic fox (probably equivalent to parts of stage 3); and a final phase of extreme floral depletion immediately prior to the ice sheet maximum in stage 2. There may have been a major expansion of glaciers in stage 4, between the early forested and later treeless interstadials, but as yet there is no evidence for it in England and Wales.

The pattern of glacierisation of Britain is clearly indicated by the form of the ice sheet margin, although some parts, such as the lobe which thrust down the east coast of England, are difficult to understand in terms of our knowledge of modern ice sheet dynamics. At the maximum, there was an ice cap over Wales, a major dome over Scotland and major ice streams flowing down the Irish Sea both into St Georges Channel and the Cheshire Plain, and another down the east coast of England.

The Late Devensian ice sheet left few marginal moraines during retreat, but the pattern of retreat can be reconstructed from other evidence. Observations on modern glaciers indicate clearly that exposed rock surfaces acquire their most recent striae immediately beneath the retreating margin of the glacier, and that drumlins and surficial fabrics in tills also acquire their final orientation in that position. As the ice front will generally be at right angles to these features, its pattern of retreat can be reconstructed from striae, drumlins and fabrics, in addition to any frontal moraines which may occur. The resultant pattern (Fig. 14.11b) indicates an extremely rapid rate of retreat in St Georges Channel and on the east coast of England compared with adjacent land areas, hastened perhaps by rising eustatic sea level or because of relatively low surface profiles of glaciers on unstable soft sediment beds (Boulton & Jones 1979). Major moraines occur both in Cheshire (Woore–Wrexham moraine) and Yorkshire (York and Escrick moraines) some 30 km from the Late Devensian limit, and may represent synchronous stand-still or readvance phases, and may equate with a push moraine at Dinas Dinlle in North Wales (Saunders 1968). This phase may also have seen a readvance of the Welsh ice cap, as Welsh till overlies Irish Sea till near Shrewsbury, and push structures produced by Welsh ice are seen at Ellesmere in Cheshire.

Fig. 14.10b shows the general pattern of Devensian ice sheet retreat. Dating such a retreat pattern would normally rely upon dates of the first floral and faunal immigration into the areas vacated by the retreating ice sheet. Unfortunately most of the earliest post-glacial organic accumulations are in kettle holes from which stagnant ice may have melted long after deglaciation (Boulton 1977), which may be the reason why the pattern of dates on basal organic accumulations clearly does not reflect patterns of deglaciation. The earliest dates are from the base of a kettle hole at Glen Ballyre in the Isle of Man (Thomas 1977) which gave four $^{14}$C dates in the range 18,400 to 18,900, although there is some doubt about their validity (Shotton 1977) and new determinations (Tooley 1977) give an oldest date of 15,150 B.P.; and one of 14,468 B.P. from near the base of a kettle hole in North Wales (Coope & Brophy 1972, Boulton 1977).

After 14,500 B.P. and until about 12,000 B.P. pollen

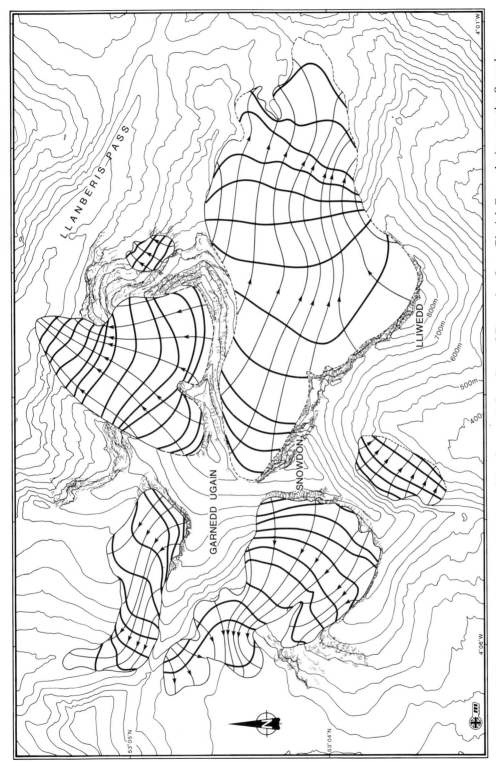

**Fig. 14.12.** A glaciological reconstruction of the form and flow lines of Devensian Late-Glacial Cwm glaciers on the Snowdon massif (north Wales) at their maximum extent. The positions of the margins were identified from moraines, striae, drumlins and *roches moutonnées*.

evidence from western England and Wales (Pennington 1977) indicates a continuous invasion of plant communities and development of richer soils. Snow bed communities and tolerant mountain plants are followed by pioneer species on immature soils and varied communities demanding base-rich soils. A widespread increase in juniper at 13,000 B.P. is taken to reflect a general climatic amelioration. Apart from a temporary deterioration of climate between about 12,000 and 11,800 B.P. interstadial conditions were

**Fig. 14.13.** The distribution of Devensian periglacial sediments and structures in England and Wales. There were probably two major periods of formation. The first roughly contemporary with the Late Devensian ice sheet maximum phase, and the second with the Loch Lomond stadial period. Most of those north of the Late Devensian ice sheet maximum position probably formed during the latter period.

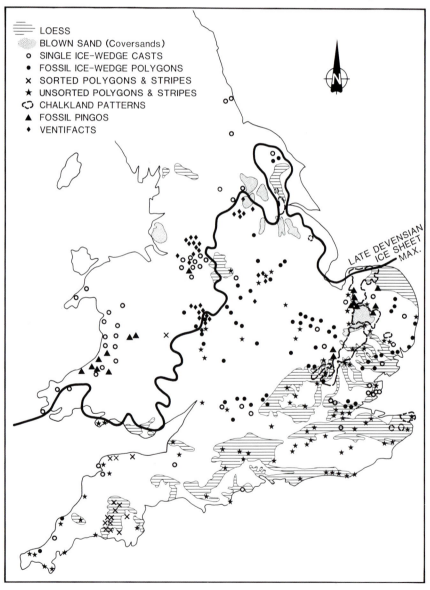

sustained until a little before 11,000 B.P. at which time a major regression of vegetation indicates a startling reversion to glacial conditions until about 10,000 B.P. (the Loch Lomond stadial). Fossil coleoptera show an essentially similar picture (Coope 1977) although their response to climatic change appears to lead that of the flora, which may reflect a more rapid response of mobile insects to environmental change. A Late-Glacial interstadial, the Windermere Interstadial, has been defined by Coope and Pennington (in discussion of Coope 1977) beginning at about 13,000 B.P. and ending at 11,000 B.P. (Fig. 14.10). Sissons (1974) believed that glaciers disappeared entirely from Scotland during the Windermere Interstadial.

During the Loch Lomond stadial between 11,000 and 10,000 B.P. there is evidence of strong regeneration of glaciers in Britain. A major ice cap formed in western Scotland, and mountain areas such as those of Wales (Fig. 14.12) and the Lake District developed valley or corrie glaciers or small ice caps (Fig. 14.11).

This stadial was terminated by a rapid climatic amelioration which ushered in the present interglacial.

## Flandrian Stage

After the end of the Loch Lomond stadial, the climate of Britain ameliorated rapidly, ushering in the Flandrian interglacial stage, taken to begin at 10,000 B.P. Rapid colonisation of the hitherto open landscape by trees followed the normal interglacial pattern of birch–pine–hazel–elm–oak–alder. A large number of [14]C dates on the first appearance of specific tree pollen permit us to date the progressive invasion of the landscape by individual species (Birks 1982). Fig. 14.3b shows the migration of pine into Britain.

The simple pattern of evolution of soil types and introduction of plant species occurs as in previous interglacials until about 5,500 to 5,000 B.P. when a

**Fig. 14.14a, b.** The form of deformed sea-level planes of several Flandrian ages, reconstructed from the altitudes of regression or transgression surfaces at sites down the west coast (a) and east coast (b) of England and Wales.

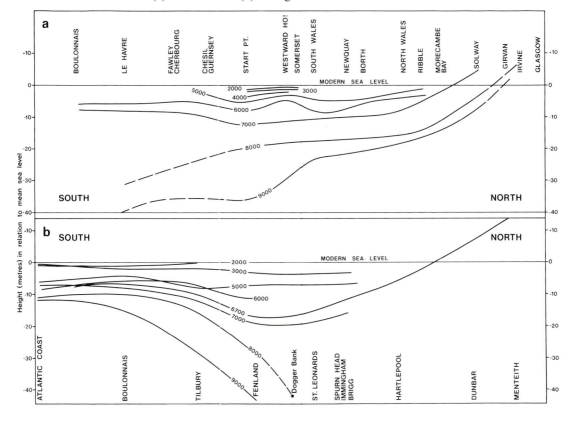

sudden and dramatic decline in elm pollen frequencies appears to coincide with a phase of active forest clearance and use of leaf fodder by Neolithic farmers. From this time on the history of vegetation in England and Wales is dominated by the pattern of forest clearance by man, and planting of crop species. Some climatically controlled processes are however dis-

cernable which indicate cooling and increased precipitation after a so-called climatic optimum at 5,000 B.P.

It is presumed that global sea levels were 120–150 m below present levels during the maximum extent of Devensian ice sheets, suggesting that much of the English Channel and southern North Sea was land.

**Fig. 14.14c.** Net isostatic changes in north-west Europe since 9,500 B.P. assuming a uniform eustatic sea-level rise. Note the uplift in the areas once occupied by ice-sheet centres. The subsidence in southern England and the southern Baltic region may reflect collapse of a 'marginal bulge' formed by extrusion of mantle from beneath the ice sheets.

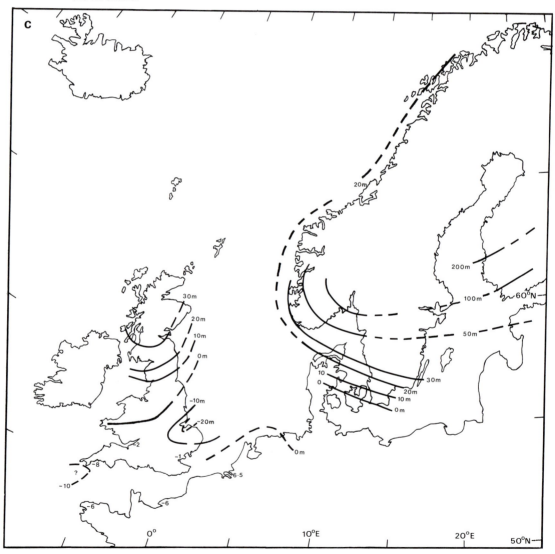

Major channels on the bed of the Channel reflect Late Devensian and early Flandrian river systems (see Chapter 15) and buried cliff lines at −42 m and −28 m off south Devon (Hails 1975) probably reflect halt stages during the marine transgression which followed ice sheet decay. In north Britain the great mass of glacier ice isostatically depressed the earth's crust by an amount greater than the eustatic sea-level fall (Fig. 14.14c). On deglaciation, eustatic sea levels rose, and isostatic rebound of the crust took place. In the south the eustatic component was dominant, producing a major transgression; in the north there was an initial regression due to dominance of the isostatic uplift component, followed by a transgression as the rate of eustatic sea-level rise overtook the crustal uplift rate. These processes continued into the Flandrian (Fig. 14.14a–b) and the two areas are separated by a well-defined hinge-line near to the Scottish border.

Warped surfaces reflecting mid- to early-Flandrian sea level planes on the east coast of England and possibly the west coast of Wales are lower than contemporary surfaces to the north and south. One possible explanation is that they reflect the subsidence of a 'marginal bulge' produced by flow of mantle material from the isostatically depressed area beneath the centre of the ice sheet. Alternatively it may reflect retardation of uplift in the southern North Sea area because of crustal loading by the transgression which inundated the southern North Sea area between 8,700 and 7,500 B.P.

The pattern of isobases, whether produced by errors of measurements or real regional geological trends suggests that no area of England or Wales can be used as an indicator of global eustatic change.

# Correlation with the deep-ocean isotopic record

The deep-ocean oxygen–isotope record suggests an increasing intensity of global glacierisation during lower to middle Quaternary times which continues a trend apparent during the Tertiary. There are several features of the middle and lower Quaternary in England which may reflect these trends. They are the possibility of glaciation somewhere in the North Sea basin in Baventian times, the evidence of permafrost by at least Beestonian times if not pre-Pastonian times, and, as a culmination, the greatest extent of glacier ice in Britain during Anglian stage.

Unfortunately there are as yet few direct means of correlation between terrestrial and oceanic sequences, which leaves little alternative but to count back from the present interglacial. If we assume a short, conventional stratigraphy of a succession of simple interglacials comprising Ipswichian, Hoxnian, and Cromerian, then the Ipswichian would be equivalent to isotope stage 5, or 5e, Hoxnian to stage 7, and Cromerian to stage 9. However, a longer stratigraphy can be constructed in which we split conventional Ipswichian sites into an Ipswichian *sensu stricto* and an earlier, Ilfordian; and recognise a possible intra-Hoxnian cold stage at Swanscombe and between the upper and lower Acheulian industries at Hoxne (Wymer 1974), which would give a correlation of Flandrian ≃ stage 1, Devensian = stages 2–4 (or 5d), Ipswichian = stage 5 (or 5e), 'Weltonian' = stage 6 (or 8), 'Ilfordian' = stage 7, later Hoxnian = stage 9, earlier Hoxnian = stage 11, Anglian = stage 12, Cromerian = stage 13 (Fig. 1b).

## REFERENCES

| ALABASTER & STRAW | 1976 | The Pleistocene context of faunal remains and artefacts discovered at Welton-le-Wold, Lincolnshire. *Proc. Yorkshire geol. Soc.*, **41**, 75–94. |
| BECK, R. B., FUNNELL, B. M. & LORD, A. R. | 1972 | Correlations of Lower Pleistocene at depth in Suffolk. *Geol. Mag.*, **109**, 137–139. |
| BIRKS, H. J. B. | 1982 | *Flandrian isochron maps and tree migrations in the British Isles.* Cambridge Univ. Press. |
| BISHOP, M. J. | 1974 | A preliminary report on the Middle Pleistocene mammal bearing deposits of Westbury-sub-Mendip, Somerset. *Proc. Univ. Bristol Speleolog. Soc.*, **13**, 301–318. |
| | 1975 | Earliest record of man's presence in Britain. *Nature, London*, **253**, 95–97. |
| BISHOP, W. W. | 1957 | The Pleistocene geology and geomorphology of three gaps in the Midlands Jurassic scarp. *Phil. Trans. Roy. Soc. Lond. B.*, **241**, 255–305. |

BOULTON, G. S.                    1977   A multiple till sequence formed by a late-Devensian Welsh
                                         ice-cap: Glanllynnau, Gwynedd. *In* Bowen, D. Q. (Ed.)
                                         *Studies in the Welsh Quaternary* (*Cambria*, special issue).

BOULTON, G. S.,                   1977   A British ice sheet model and patterns of glacial erosion
    JONES, A. S.,                        and deposition in Britain. *In* Shotton, F. W. (Ed.) *British
    CLAYTON, K. M. &                     Quaternary Studies*. Oxford University Press, Oxford, 231–
    KENNING, M. J.                       246.

BOULTON, G. S. &                  1979   Stability of temperate ice caps and ice sheets resting on beds
    JONES, A. G.                         of deformable sediment. *J. Glaciol, London*, **24**, 29–43.

BOWEN, D. Q.,                     1986   Correlation of Quaternary Glaciations in England,
    ROSE, J.,                            Ireland, Scotland and Wales. *Quat. Sci. Revs*, **5**, 299–340.
    McCABE, A. M. &
    SUTHERLAND, D. G.

BOYLAN, P. J.                     1967   The Pleistocene mammalia of the Sewerby–Hessle buried
                                         cliff, East Yorkshire. *Proc. Yorkshire geol. Soc.*, **36**,
                                         115–125.

BRIDGLAND, D.                     1980   A reappraisal of Pleistocene stratigraphy in north Kent and
                                         eastern Essex, and new evidence concerning former courses
                                         of the Thames and Medway. *Quaternary Newsletter, No.
                                         32*, 15–24.

BRIGGS, D. J.                     1975   New Interglacial site at Sugworth. *Nature, London*, **257**,
    AND SEVEN OTHERS                     477–479.

BUCHARDT, B.                      1978   Oxygen isotope palaeotemperatures from the Tertiary
                                         period in the North Sea area. *Nature*, **275**, 121–123.

CAMERON, T. D. J.,                1987   The history of Quaternary Sedimentation in the North Sea
    STOKER, M. S. &                      basin. *J. geol. Soc. London*, **144**, 43–58.
    LONG, D.,

CATT, J. A.                       1977   Loess and coversands. *In;* Shotton, F. W. (Ed.) *British
                                         Quaternary Studies*. Oxford University Press, Oxford, 221–
                                         229.

                                  1978   The contribution of loess to soils in lowland Britain. *In*
                                         Limbrey, S. & Evans, J. G. (Eds.), The effects of Man on
                                         the Landscape: the lowland zone. *Counc. Br. archaeol.
                                         Res. Rept.*, **21**, 12–20.

CATT, J. A. &                     1966   The Pleistocene deposits of Holderness, East Yorkshire.
    PENNY, L. F.                         *Proc. Yorkshire geol. Soc.*, **35**, 375–400.

CLARKE, R. H.                     1973   Cainozoic subsidence in the North Sea. *Earth Planet. Sci.
                                         Lett.*, 329–332.

CLAYTON, K. M.                    1977   River terraces. *In* Shotton, F. W. (Ed.) *British Quaternary
                                         Studies*. Oxford University Press, Oxford, 153–167.

COOPE, G. R.                      1975   Climatic fluctuations in north-west Europe since the last
                                         interglacial indicated by fossil assemblages of Coleoptera.
                                         *In* Wright, A. D. & Moseley, F. (Eds.) *Ice Ages Ancient
                                         and Modern*, Geol. Jour. Special Issue 6, Liverpool,
                                         153–168.

                                  1977   Fossil coleopteran assemblages as sensitive indicators of
                                         climatic changes during the Devensian (last) cold stage.
                                         *Phil. Trans. R. Soc. Lond., B.*, **280**, 313–340.

COOPE, G. R. &                    1972   Late glacial environmental changes indicated by coleop-
    BROPHY, J. A.                        teran succession from North Wales. *Boreas*, 97–142.

COOPE, G. R. &                    1977   The Windermere Interstatial of the Late Devensian. *Phil.
    PENNINGTON, W.                       Trans. Roy. Soc. Lond.*, **280**, 337–339.

DAY, M. H.                        1977   *Guide to Fossil Man*. Cassell, London.

DIXON, R. G.                      1978   Deposits marginal to the Red Crag basin. *Bull. geol. Soc.
                                         Norfolk*, **30**, 92–104.

ERD, K.                           1973   Pollenanalytische gliederung des Pleistozans der Deuts-
                                         chen Demonkratischen Republik. *Z. grol. Wiss. Berlin*, **1**,
                                         1087–1103.

| | | |
|---|---|---|
| EVANS, J. G. | 1972 | *Land snails in archaeology*. Seminar Press of Norfolk. *Trans. Norfolk Norwich Nat. Soc.*, **19**, 340–356. |
| FUNNELL, B. M. | 1961 | The Palaeogene and early Pleistocene of Norfolk. *Trans. Norfolk Norwich Nat. Soc.*, **19**, 340–356. |
| | 1987 | Late Pliocene and early Pleistocene stages of East Anglia and the adjacent North Sea. *Quaternary Newsletter*, Univ. East Anglia, Norwich, UK, 1-11. |
| FUNNELL, B. M., NORTON, P. E. P. & WEST, R. G. | 1979 | The Crag at Bramerton, near Norwich, Norfolk. *Phil. Trans. R. Soc. Lond. B.*, **287**, 489–534. |
| FUNNELL, B. M. & WEST, R. G. | 1962 | The Early Pleistocene of Easton Bavents, Suffolk. *Q. J. geol. Soc. Lond.*, **118**, 125–141. |
| GARDNER, J. V. & HAYS, J. D. | 1977 | Responses of sea-surface temperature and circulation to global climatic change during the past 200,000 years in the eastern equatorial Atlantic. *In* Cline *et al.* (Eds.). *Geol. Soc. Amer. Mem.*, **145**. |
| GEIKIE, J. | 1895 | The classification of European glacial deposits. *J. Geol.*, **3**, 241–270. |
| GIBBARD, P. L. | 1977 | Pleistocene history of the Vale of St. Albans. *Phil. Trans. Roy. Soc. Lond., B.*, **280** 445–483. |
| | 1985 | *The Pleistocene History of the Middle Thames Valley.* Cambridge University Press, Cambridge, 155 pp. |
| GODWIN, H. | 1977 | Quaternary history of the British flora. *In* Shotton F. W. (Ed.) *British Quaternary Studies.* Oxford University Press. |
| GREEN, C. P. AND EIGHT OTHERS | 1984 | Evidence of two temperate episodes in Late Pleistocene deposits at Marsworth, UK. *Nature, London*, **309**, 778–781. |
| HAILS, J. R. | 1975 | Sediment distribution and Quaternary history of Start Bay. *J. geol. Soc. London*, **131**, 19–36. |
| HAWKINS, A. M. & KELLAWAY, G. A. | 1971 | Field meeting at Bristol and Bath with special reference to new evidence of glaciation. *Proc. Geol. Ass. London.*, 267–292. |
| HEY, R. W. | 1967 | The Westleton Beds reconsidered. *Proc. geol. Ass. London*, **78**, 427–445. |
| | 1986 | A re-examination of the Northern Drift of Oxfordshire. *Proc. geol. Ass. London*, **97**, 291–301. |
| HEY, R. W. & BRENCHLEY, P. J. | 1977 | Volcanic pebbles from Pleistocene gravels in Norfolk and Essex. *Geol. Mag.*, **114**, 219–225. |
| HORTON, A. | 1974 | The sequence of Pleistocene deposits proved during the construction of the Birmingham motorways. *Inst. geol. Sci. Rept.*, **74/11**. |
| HUDDART, D., TOOLEY, M. J. & CARTER, P. S. | 1977 | The coasts of north-west England. *In* Kidson, C. and Tooley, M. J. (Eds.) *The Quaternary History of the Irish Sea.* Seal House Press, Liverpool, 119–154. |
| JESSEN, K. & MILTHERS, V. | 1928 | Stratigraphical and palaeontological studies of interglacial freshwater deposits in Jutland and north-west Germany. *Danm. Geol. Unders. II Raeke*, **48**. |
| KELLY, M. R. | 1974 | The Middle Pleistocene of north Birmingham. *Phil. Trans. R. Soc. London, B.*, **247**, 533–592. |
| KERNEY, M. P. | 1977 | British Quaternary non-marine Mollusca: a brief review. *In* Shotton, F. W. (Ed.) *British Quaternary Studies.* Oxford University Press, Oxford, 31–42. |
| KERNEY, M. P. & TURNER, C. | 1977 | The biostratigraphy of Late Devensian and Flandrian deposits at Folkestone, Kent. *Proc. geol. Ass. London.*, **88**. |
| KIDSON, C. | 1977 | The coast of south-west England. *In* Kidson, C. and Tooley, M. J. (Eds.) *The Quaternary History of the Irish Sea.* Seal House Press, Liverpool, 257–298. |

KING, C. A.                   1977    The early Quaternary Landscape with consideration of
                                      Neotectonic matters. *In* Shotton, F. W. (Ed.) *British
                                      Quaternary Studies*. Oxford University Press, Oxford, 137,
                                      152.

KURTEN, B.                    1968    *Pleistocene Mammals of Europe*. Weidenfeld and Nichol-
                                      son, London.

LINTON, D. L.                 1951    Midland drainage: some consideration bearing on its
                                      origin. *Adv. Sci. London*, **7**, 449–456.

LUTTIG, G.                    1965    The Bilshausen type section, West Germany. *Geol. Soc.
                                      America Special Paper*, **84**, 159–178.

McBURNEY, C. B. M.            1972    The Cambridge excavation at La Cotte de St. Brelade
                                      Jersey – preliminary report. *Proc. prehis. Soc. East
                                      Anglia*, **437**, 167–207.

McCAVE, I. N.,               1977    The Quaternary of the North Sea. *In* Shotton, F. W. (Ed.)
    CASTON, V. N. D. &                *British Quaternary Studies*. Oxford University Press,
    FANNIN, N. G. T.                  Oxford, 187–204.

MANABE, S. &                 1974    The effects of mountains on the general circulation of the
    TERPSTRA, T. B.                   atmosphere as identified by numerical experiments. *J. atm.
                                      Sci.*, **31**, 3–42.

MATHEWS, J. V.               1976    Evolution of the sub-genus *Cyphelophorous*. Description
                                      of two new fossil species and discussion of *Helphorus
                                      tuberculatus*. *Can. J. Zool.*, **54**, 652–676.

MILANKOVIC, M.               1941    Kanon der Erdbestrahlung. *T. Serb. Acad. Spec. Publ.*,
                                      **132**.

MITCHELL, G. F.              1972    The Pleistocene history of the Irish Sea; second
                                      approximation. *Sci. Proc. Roy. Dublin Soc. A.*, **4**,
                                      181–199.

                             1977    Raised beaches and sea levels. *In* Shotton, F. W. (Ed.)
                                      *British Quaternary Studies*. Oxford University Press,
                                      Oxford, 169–186.

MITCHELL, G. F.,             1973    A correlation of Quaternary deposits in the British Isles.
    PENNY, L. F.,                     *Special Report Geol. Soc. Lond., No.* **4**, 99 pp.
    SHOTTON, F. W. &
    WEST, R. G.

MORGAN, A. V.                1973    The Pleistocene geology of the area north and west of
                                      Wolverhampton, Staffordshire, England. *Phil. Trans.
                                      Roy. Soc. Lond. B.*, **265**, 233–297.

NORTON, P. E. P. &           1972    Lower Pleistocene molluscan assemblages and pollen from
    BECK, R. B.                       the Crag of Aldeby, and Easton Bavents, Suffolk. *Bull.
                                      geol. Soc. Norfolk*, **22**, 11–31.

OAKLEY, K. P.                1952    Swanscombe Man. *Proc. Geol. Assoc. London*, **63**,
                                      271–300.

PENCK, A. &                  1909    *Die Alpen im Eiszeitalter*. Tauchnitz, Leipzig.
    BRUCKNER, E.

PENNINGTON, W.               1977    The Late Devensian flora and vegetation of Britain. *Phil.
                                      Trans. Roy. Soc. Lond. B.*, **280**, 247–271.

PENNY, L. F.,                1969    Age and insect fauna of the Dimlinton Silts, East
    COOPE, G. R. &                    Yorkshire. *Nature, London*, **24**, 65–67.
    CATT, J. A.

PENNY, L. F.                 1974    Quaternary. *In* D. H. Rayner and J. E. Hemingway (Eds.),
                                      *The Geology and Mineral Resources of Yorkshire*.
                                      Yorkshire Geological Society, 245–264.

PERRIN, R. M. S.,            1979    The distribution, variation and origins of pre-Devensian
    ROSE, J. &                        tills in eastern England. *Phil. Trans. Roy. Soc. Lond. B.*,
    DAVIES, H.                        **287**, 535–570.

PIKE, K. &                   1953    The interglacial at Clacton-on-Sea, Essex. *Q. J. geol. Soc.
    GODWIN, H.                        London*, **108**, 261–272.

| | | |
|---|---|---|
| REID, P. C. & DOWNIE, C. | 1973 | The age of the Bridlington Crag. *Proc. Yorkshire geol. Soc.*, **39**, 315–318. |
| RICE, R. J. | 1968 | The Quaternary deposits of central Leicestershire. *Phil. Trans. Roy. Soc. London A.*, **62**, 459–509. |
| ROSE, J., ALLEN, P. & HEY, R. W. | 1976 | Middle Pleistocene stratigraphy in southern East Anglia. *Nature, Lond.*, **263**, 49–94. |
| ROSE, J. & ALLEN, P. | 1977 | Middle Pleistocene stratigraphy in south-east Suffolk. *J. geol. Soc. Lond.*, **133**, 83–107. |
| RUDDIMAN, W. F. & RAYMO, M. E. | 1988 | Climate regimes during the past 3 Ma. *Phil. Trans. roy. Soc. Lond. B.*, **318**, 411–430. |
| SANDFORD, K. S. | 1945 | River development and superficial deposits. *In* Martin, A. R. and Steel, R. W. (Eds.) *The Oxford Region.* Oxford University Press. |
| SAUNDERS, G. E. | 1968 | A fabric analysis of the ground moraine deposits of the Lleyn Peninsula. *Proc. Geol. Assoc. London*, **79**, 305–324. |
| SHACKLETON, N. J. | 1977 | The oxygen isotope stratigraphic record of the Late Pleistocene. *Phil. Trans. Roy. Soc. Lond., B.*, **280**, 169–182. |
| | 1986 | The Plio-Pleistocene ocean: stable isotope history. *Mesozoic and Cenozoic oceans.* Geodynamics Series, **15**, American Geophysical Union, 141–153. |
| SHACKLETON, N. J. & OPDYKE, N. D. | 1973 | Oxygen isotope and palaeomagnetic stratigraphy of Equatorial Pacific core B28–238. *Quat. Res.*, **3**, 39–55. |
| SHOTTON, F. W. | 1953 | The Pleistocene deposits of the area between Coventry, Rugby and Leamington, and their bearing upon the topographic development of the Midlands. *Phil. Trans. Roy. Soc. London B.*, **237**, 209–260. |
| | 1968 | The Pleistocene succession around Brandon, Warwickshire. *Phil. Trans. Roy. Soc. Lond. B.*, **254**, 387–400. |
| | 1977 | The English Midlands. *X INQUA Congress Excursions Guide A2.* Geol Abstracts, Norwich. |
| SIMPSON, I. M. & WEST, R. G. | 1958 | On the stratigraphy and palaeobotany of the late Pleistocene organic deposit at Chelford, Cheshire. *New Phytologist*, **57**, 239–250. |
| SISSONS, J. B. | 1974 | The Quaternary in Scotland; a review. *Scott. J. Geol.*, **10**, 311–337. |
| SPARKS, B. W. & WEST, R. G. | 1972 | *The Ice Age in Britain.* Methuen, London. |
| STEVENS, L. A. | 1960 | The interglacial of the Nar Valley. *Q. J. geol. Soc. Lond.*, **115**, 291–319. |
| STRAW, A. & CLAYTON, K. M. | 1979 | *Eastern and Central England* in *Geomorphology of the British Isles.* Methuen, London, 247 pp. |
| STRINGER, C. | 1974 | Population relationships of Later Pleistocene hominids: a multivariate study of available crania. *Jl. Archaeol. Sci.*, **1**, 317–342. |
| SUTCLIFFE, A. J. | 1976 | The British Glacial–Interglacial sequence – a reply. *Quaternary Newsletter*, **18**, 1–7. |
| SUTCLIFFE, A. J. & KOWALSKI | 1976 | Pleistocene rodents of the British Isles. *Bull. Br. Mus. Nat. Hist. (Geol.)*, **27**, 33–147. |
| SZABO, B. J. & COLLINS, D. | 1975 | Ages of fossil bones from British interglacial sites. *Nature, Lond.*, **254**, 680–682. |
| THOMAS, G. S. P. | 1977 | The Quaternary of the Isle of Man. *In* Kidson, C. (Ed.) *The Quaternary History of the Irish Sea.* Seal House Press, Liverpool. |
| TOOLEY, M. J. (Ed.) | 1977 | *The Isle of Man, Lancashire East and Lake District.* X INQUA Congress Excursion Guide Geo. Abstracts, Norwich. |
| TURNER, C. | 1970 | The Middle Pleistocene deposits at Marks Tey, Essex. *Phil. Trans. R. Soc. London, B.*, **257**, 373–440. |

| TURNER, C. & WEST, R. G. | 1968 | The subdivision and zonation of interglacial periods. *Eiszeit. u. Gegenw.*, **19**, 93–101. |

WAECHTER, J. D'A., NEWCOMER, M. H. & CONWAY, B. W. — 1971 — Swanscombe 1970. *Proc. R. anthropol. Inst. 1970*, 43–64.

WALSH, P. T., BOULTER, M. S., IJTABA, M. & URBANI, D. M. — 1972 — The preservation of the Neogene Brassington formation of the southern Pennines and its bearings on the evolution of upland Britain. *J. Geol. Soc. Lond.*, **128**, 5-9-59.

WEST, R. G. — 1957 — Interglacial deposits of Bobbitshole, Ipswich. *Phil. Trans. R. Soc. London B.*, **241**, 1–31.

1961 — Vegetational history of the early Pleistocene of the Royal Society borehole at Ludhom, Norfolk. *Proc. Roy. Soc. Lond. B.*, **155**, 437–453.

1977 — *Pleistocene Geology and Biology*. Longman, London and New York.

1980 — *The pre-glacial Pleistocene of the Norfolk and Suffolk Coasts*. Cambridge University Press.

WEST, R. G. & DONNER, J. J. — 1956 — The glaciations of East Anglia and the East Midlands: a differentiation based on stone orientation measurements of the tills. *Q. J. geol. Soc. London*, **112**, 69–91.

WEST, R. G. & WILSON, D. G. — 1968 — Cromer Forest Bed Series. *Nature, London*, **209**, 497–498.

WEST, R. G., DICKSON, C. A., CATT, J. A., WEIR, A. J. & SPARKS, B. W. — 1974 — Late Pleistocene deposits at Wretton, Norfolk II. Devensian deposits. *Phil. Trans. Roy. Soc. Lond. B.*, **247**, 185–212.

WEST, R. G., FUNNELL, B. M. & NORTON, P. E. P. — 1980 — An early Pleistocene cold marine episode in the North Sea; pollen and faunal assemblages at Covehithe, Suffolk, England. *Boreas.*, **9**, 1-10.

WILLS, L. J. — 1938 — The Pleistocene development of the Severn from Bridgenorth to the sea. *Q. J. geol. Soc. Lond.*, **94**, 161–242.

WOOLDRIDGE, S. W. — 1938 — The glaciation of the London basin and the evolution of the lower Thames drainage system. *Q. J. geol. Soc. Lond.*, **94**, 627–667.

WOOLDRIDGE, S. W. & LINTON, D. L. — 1955 — *Structure, surface and drainage in south-east England*. Philip, London.

WORSLEY, P. W. — 1980 — Problems in radiocarbon dating the Chelford interglacial of England. *In* Cullingford, R. A., Davidson, D. A. & Lewin, J. (Eds.) *Timescales in Geomorphology*. Wiley, Chichester, 287–304.

WYMER, J. J. — 1974 — Clactonian and Achenlian industries in Britain; their chronology and significance. *Proc. Geol. Assoc. London.*, **85**, 391–421.

1977 — The archaeology of man in the British Quaternary. *In* Shotton, F. W. *British Quaternary Studies*. Oxford University Press, Oxford.

1988 — Palaeolithic Archaeology and the British Quaternary sequence. *Quat. Sci. Revs.*, **7(1)**, 79–97.

ZAGWIJN, W. H. & STAALDUINEN, C. J. VAN (Eds.) — 1975 — *Toelichting bij Geologische Overzichtzkaarten van Nederland*. Rijks Geologische Dienst. Haarlem.

ZEUNER, F. E. — 1959 — *The Pleistocene period: its climate, chronology and faunal successions* (2nd ed.). Hutchinson, London.

# 15

# THE OFFSHORE GEOLOGY OF ENGLAND AND WALES

## A. J. Smith

## Introduction

Beyond the shorelines of England and Wales lies the continental shelf, a portion of which is now politically a part of Britain. Our detailed knowledge of the offshore is the result of intensive studies by academic, industrial, and government scientists over the last twenty-five to thirty years though the earliest studies reach back into the early 19th century (for a review see Donovan 1968). In this chapter no precise attention is paid to political boundaries, rather the geology is described as an entity and as an extension of the geology of England and Wales across that most transient of geological features, the shoreline. The sea areas described are the North Sea, especially the southern North Sea, the English Channel and its Western Approaches, the Celtic Sea (here, as in common usage, limited to the sea north of the WSW extension of the Cornubian peninsula in spite of a clearly defined larger extent in an original description), the Bristol Channel, the St George's Channel and Cardigan Bay (both part of the Irish Sea of some writers), and the Irish Sea, particularly east of the Isle of Man and north to the Solway Firth (see Fig. 15.1 for localities mentioned in text). The ensuing account of the offshore area is based on the major stratigraphic divisions and, where appropriate, each division is dealt with by sea areas in a clockwise manner, i.e. North Sea, through the English Channel to the Irish Sea.

Each sea area has its own distinctive if not unique geological history, indeed even parts of individual seas have separate histories. That this is so is well illustrated by the sharp contrast between the English Channel where rocks of many systems are exposed on the sea floor or covered with only a thin veneer of recent sediments a major unconformity in the making (Curry 1989) and the North Sea which has been the site of continued subsidence through the Tertiary and Quaternary and where evidence of its long history is hidden by late-Cenozoic sediments. For this reason whilst the former was first explored by using surface and near-surface investigative methods, the North Sea, except for a relatively small area off the NE coast of England (Dingle 1971), yielded its geological history only through the interpretation of geophysical data and deep drilling. Such methods have removed the physical barrier once presented to investigation by the move seawards – indeed for some methods the sea provides a better medium through which to investigate the subsurface than the land.

It is customary when dealing with the offshore, to describe not only the known surface geology (Figs. 15.2, 3) but also the known and surmised subsurface geology. The geological development of the offshore areas, whilst obviously being linked to events found on the present dry land areas of Britain and continental Europe, does have aspects which at the least may be described as separate. The suturing of plates during the Caledonian orogeny and subsequent fracturing, the extensional and compressional events related to the Hercynian orogeny, the responses to the opening of the Atlantic, and Tethyan events all played their part in the tectonic evolution of north-western Europe (see, for example, Ziegler 1975, 1978, 1981, 1982, 1986, 1990; McQuillin 1986; Karner *et al*. 1987) and especially of the offshore area.

The academic and economic interest in the offshore area is well illustrated by the enormous number of publications which have appeared in the last two decades. Several major works have appeared and these

445

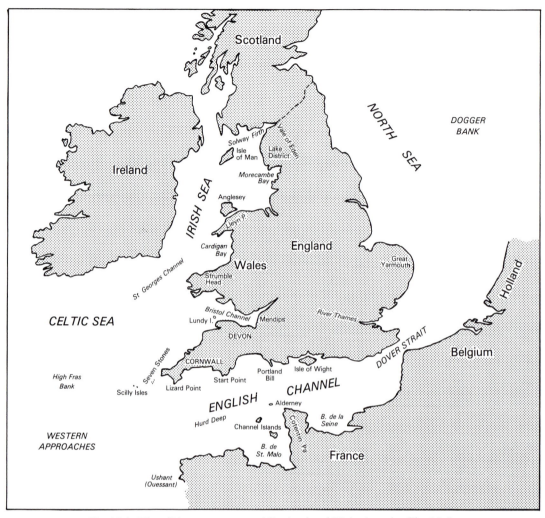

**Fig. 15.1.** The offshore areas around England and Wales (see also Fig. 15.4).

include Pomerol (1972), Dunham & Smith (1975), Woodland (1975), Banner *et al.* (1979), Illing & Hobson (1981), Naylor & Shannon (1982), Parkin & Crosby (1982), Ziegler (1982, 1990), Glennie (1984), Brooks *et al.* (1986), Brooks & Glennie (1987). A shorter review by Jenkins and Twombley (1980) was concerned particularly with petroleum geology. More significantly has been the monumental task undertaken by the British Geological Survey since the early 1960s (see Walmsley 1987). (A map showing the coverage and type of B.G.S. maps at a scale of 1:250,000 available in January 1991 together with other maps concerned with the shelf around England and Wales is given in an appendix to this chapter.)

## Precambrian and Lower Palaeozoic

Our detailed knowledge of the Precambrian and Lower Palaeozoic of the offshore areas of England and Wales is mainly limited to the observed seaward extension of known land exposures of rocks of these ages. Such offshore information as exists links these submarine outcrops directly to the land. So far only a few deep boreholes have yielded new information on rocks of these ages from farther offshore, since the Lower Palaeozoic and older rocks are below *economic basement*, i.e. the depth below which hydrocarbons are not expected to be discovered in economically viable quantities.

Nowhere in the offshore regions have any inliers of

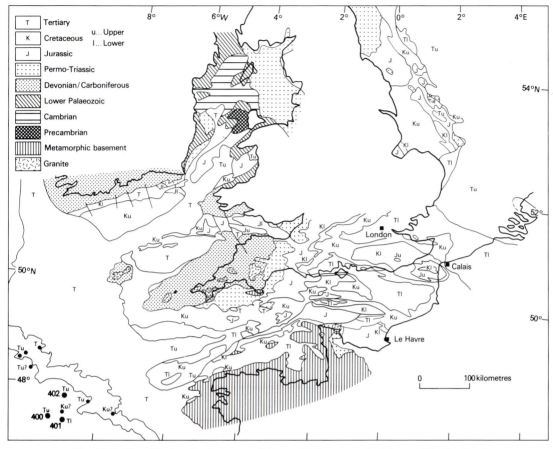

**Fig. 15.2.** Simplified geological map of the pre-Pleistocene of offshore of England and Wales based on the author's own work and published maps of the Institute of Geological sciences.

Precambrian or Lower Palaeozoic rocks been discovered except immediately adjacent to the shore off southern Cornwall and Devon and the Channel Islands. In the North Sea metamorphic rocks of 410 Ma to 450 Ma have been recovered at depth from a number of places whilst the Mid-North Sea–Ringkøbing–Fyn High (see Fig. 15.4 for structural features mentioned in text) and the sub-North Sea extension of the London–Brabant massif have yielded rocks with Precambrian ages, 690–870 Ma and 700 Ma respectively (Larsen 1971, Le Grand 1968, Frost *et al.* 1981). The former feature was, throughout the Mesozoic, a persistently buoyant mass which reached from Northern England eastwards across the North Sea into North Central Europe. The buoyancy may, at least in part, be due to buried Palaeozoic granites which have been postulated on the basis of gravity anomalies (Donato *et al* 1983). The London–Brabant massif, for its part, appears to have

been a micro-continent surrounded on all sides by Caledonian fold belts, even though the boundary between the massif and the fold belts is poorly defined, and, later, to have been the northern limit of Hercynian fold belts.

It seems clear from the few samples and the mass of gravity (Donato & Tully 1981) and magnetic data that the southern North Sea is underlain by a NW–SE trending arm of the Caledonides and Barton & Matthews (1984) believed that reactivation of the Caledonian thrust-zone was to affect deposition of Permian sediments. Whilst in places these rocks have been metamorphosed to amphibolite facies, much of the rock is of greenschist facies derived from the predominantly shaly facies – Lower Palaeozoics known beneath the eastern Midlands of England, Eastern England into the Ardennes and the North German–Polish Caledonides. Similar rocks can be predicted at depth beneath the eastern English

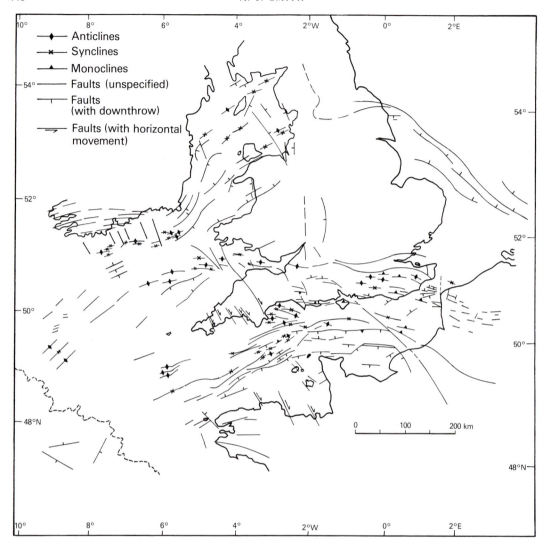

**Fig. 15.3.** The principal structural features seen in the surface geology of the offshore of England and Wales.

Channel but the character of these older basement rocks changes in the vicinity and west of the Cotentin Peninsula (Fig. 15.1) to a highly metamorphosed and intruded complex. Rocks of this type are extensively exposed on the sea floor around the Channel Islands and off the north coast of Brittany; they include the highly metamorphosed Pentevrian and the Brioverian, the latter folded in two episodes of the Cadomian orogeny, the whole crossed by acid, intermediate and basic intrusions as well as being affected by Hercynian events (Lefort 1975, 1977). On the north side of the Channel the Lizard–Start complex is exposed in a

series of tectonic inliers of small extent, though how extensive and at what depths these rocks continue beneath younger strata is a matter of conjecture. A number of writers (including Shackleton *et al.* 1982) have expressed the view that this complex is related to a high-level thrust zone (see Chapter 17) which is the product of the Hercynian orogeny. Davies (1984) and Hailwood *et al.* (1984) were among a number of workers who gave Silurian to Devonian ages for events in the formation of the Lizard complex, and Rattey & Sanderson (1984) gave a Carboniferous age for its tectonic emplacement. In mid-Channel and

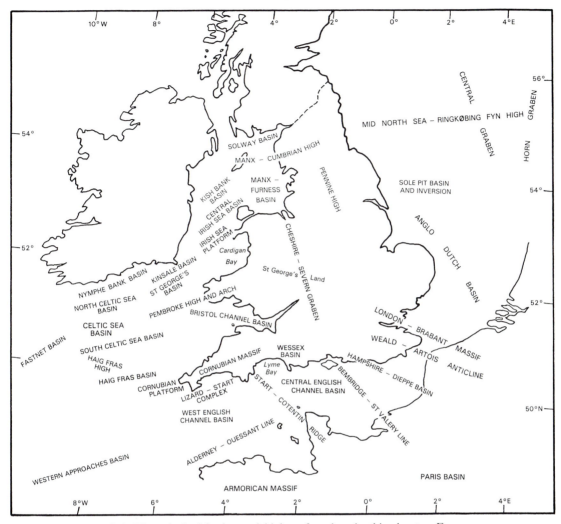

**Fig. 15.4.** The principal basins and highs referred to in this chapter. For a more detailed structural map see Fig. 19.9.

westwards, geophysical interpretations by Hill & King (1953), Day *et al.* (1956) and Avedik (1975a, b) placed the depth to metamorphic basement, here also of late as well as early Palaeozoic age, at as little as a kilometre between Start and Cotentin and at four kilometres north of Ushant (Ouessant), shallowing somewhat westwards towards the shelf margin. The metamorphic 'floor' is uneven providing a series of irregular depressions (see Avedik 1975b, Fig. 3). More recent, but unpublished, work must have redefined the shape of the depressions, perhaps into a more angular form. The Alderney–Ouessant Line (Smith & Curry 1975) separates the shallow or exposed basement to the south from the relatively deep basement to the

north. In the vicinity of this line Lefort (1977) postulated a Caledonian ophiolite complex related to a southward subducting plate.

West-southwest of Land's End a ridge of basement rocks exists at no great depth though where it outcrops only rocks of Upper Palaeozoic age are exposed (see Figs. 15.5, 6). In the Celtic Sea the Precambrian and Lower Palaeozoic basement is deeper, perhaps at depths similar to those suggested by Avedik (1975a, b) for the Western Approaches. Off south-west Wales, Lower Palaeozoic rocks outcrop for a short way offshore and may be at no great depth south-westwards on the Pembroke Arch of Dobson (1979), or the Welsh Spur, which separates the South and

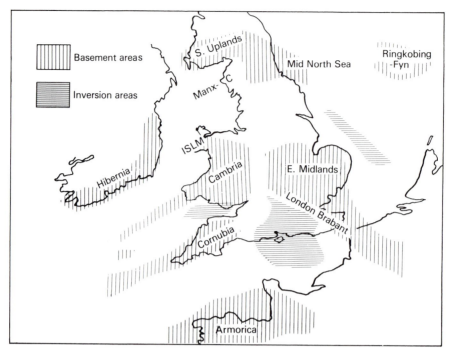

**Fig. 15.5.** The principal massifs and inversion areas which substantially affected the geological evolution of the offshore of England and Wales. (ISLM = Irish Sea Landmass.)

North Celtic Sea Basins (Fig. 15.4). Lower Palaeozoic strata are known in Cardigan Bay; these strata are succeeded by Upper Palaeozoic and Mesozoic sediments which, in the north of the Bay, are overstepped by Tertiary strata. The last abut the Cambrian at a faulted contact near Harlech (Woodland 1971). From Lleyn and Anglesey a ridge, the Irish Sea platform, extends south-westwards at relatively shallow depth. A similar but broader ridge, again covered by younger strata, may extend northwards from Anglesey to the Isle of Man. East of this ridge, the Lower Palaeozoic and Precambrian basement gets deeper eastwards towards Lancashire and Cumbria (Colter 1978).

All the Lower Palaeozoic rocks off Wales and north-western England were part of the succession affected by the Caledonian orogeny; the rocks off eastern England were, presumably, also affected by Caledonian events. To what extent any Lower Palaeozoics off southern England were affected by Caledonian as opposed to Hercynian events remains a matter of conjecture (Zwart & Dornsiepen 1978, Ziegler 1981).

Deep seismic reflection profiling and other geophysical techniques confirm this general setting and reveal something of the deep crustal structure (BIRPS and ECORS 1986; Matthews 1986; McGeary *et al.*

1987) while in the vicinity of the Solway Firth and the Northumberland Basins features relating to the Iapetus suture can be identified penetrating the Moho (Beamish & Smythe 1986, Klemperer & Matthews 1987).

## Upper Palaeozoic

The offshore rocks of the Upper Palaeozoic, particularly of the later Upper Palaeozoic, are far more widely known than those of the Lower Palaeozoic – even so, they have only a limited exposure on the sea floor. The Caledonian orogeny modified the depositional pattern of NW Europe and subsidence of the North Sea Basin as a definable entity probably began at this time. Post-orogenic sediments derived from the newly uplifted Caledonides of northern and western Britain were deposited in inter-montane (often fault-controlled) basins which were the product of interference between the Arctic–North Atlantic wrench system and the central European rift system (Ziegler 1981, McQuillin 1986). Red sandstones of the Devonian are extensive in the North Sea, though it should be noted that some red sandstones found in boreholes in the southern North Sea may be of Lower Carboniferous age. In the North Sea off Scotland, the Old Red

Sandstone (Caithness) facies of the Devonian was deposited over a wide area and has been preserved. Farther south, Old Red Sandstone facies can be predicted as far south as the London–Brabant massif; however, some marine carbonate-rich Middle Devonian strata found in the vicinity of the Argyll oilfield (see Chapter 19) north of the Mid-North Sea–Ringkøbing–Fyn High preceded and succeeded by red beds indicate a marine incursion, though by what route is not clear. Beneath London, boreholes reveal a change hereabouts to a marine facies. Some details may be construed from our knowledge of the Boulonnais (Ager & Wallace 1966) where a basal transgression with non-marine facies is succeeded by Givetian marine fossiliferous limestones, shales and hypersaline lagoonal deposits followed by more fossiliferous limestones, by turbidites and, finally, by shallow water sandstones.

The English Channel may be postulated to have tectonised and metamorphosed marine Devonian and Carboniferous rocks of the Rheno-Hercynian and Saxo–Thuringian facies at depth. Rocks of the Old Red Sandstone facies are found off Tregor in Brittany, presumably on the south side of the Normanian high. Marine Devonian rocks can be expected in the western part of the Western Approaches as the basin extended into the Devonian basins of what is now eastern North America. During late Devonian times it seems feasible that parts of the Western Approaches were a high-part of the Bretonic uplift (Ziegler 1978). Off SW England and westwards beyond Haig Fras, marine Devono–Carboniferous slaty beds are exposed on the sea floor (Smith et al. 1965). This ridge-like feature, the Cornubian Platform, extends to the continental margin. North of it, in the Fastnet Basin, red Upper Devonian sands and tuffs have been drilled (Robinson et al. 1981, Ainsworth & Horton 1986) and Old Red Sandstone facies may be expected at depth below the younger sediments of the Celtic Sea, the Bristol Channel, and the approaches to the St George's Channel. These were probably the sites of interplay between marine and continental conditions though during the Middle Devonian a Bristol Channel landmass must have existed: fault bounded, it gave rise to coarse sediments to the north, the Ridgeway and Llanishen Conglomerates, and to the south, the Hangman Grits (Chapter 6 and Allen 1974) or the Rawns Formation (Tunbridge 1986). Tunbridge showed the landmass consisted of lower Palaeozoic and '?Pre-Cambrian' sediments, volcanic rocks and meta-sediments. Cope & Bassett (1987) favoured a persistent Precambrian source. A similar evolution probably also occurred in the vicinity of the Pembroke Arch. Much of the Cardigan Bay area must have remained an upland source area throughout most of the Devonian (Owen 1978), it being an

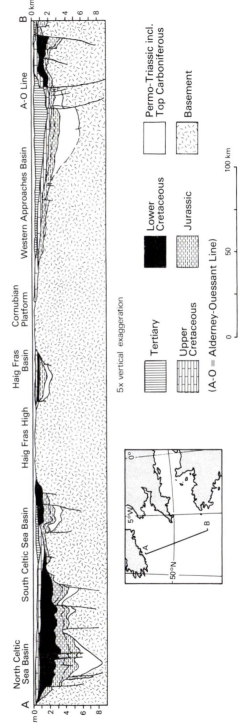

**Fig. 15.6.** Section to Basement across the Celtic Sea and Western Approaches. Basement includes all pre-Upper Carboniferous strata, intrusions, and metamorphic complexes.

extension of the Welsh Massif. Old Red Sandstone facies rocks of the Devonian may exist beneath the known Carboniferous of the Irish Sea, though how extensive they are remains a matter of conjecture, and whilst Devonian strata may still exist in the Solway Basin it is possible that none now exist on the Manx–Cumbrian High.

During the Devonian period volcanic activity seems to have been widespread in the offshore regions: submarine spilitic keratophyre volcanism occurred in the central North Sea, perhaps related to rifting, and similar volcanism occurred in and off south-west England. Mid or late Devonian intrusives were also emplaced in the North Sea (Frost *et al.* 1981) and similar intrusives may occur in the North Celtic and Irish Seas too.

By early Carboniferous times, the Caledonian mountains would have been much reduced and marine incursion reached into Ireland, Scotland, Northern England and the adjacent seas. Lower Carboniferous deposits in the southern North Sea range from shallow marine carbonates to shales with sandier paralic facies, including coal seams farther north. South of the London–Brabant massif, Lower Carboniferous limestones must underlie the eastern Channel presumably of facies similar to those of the Boulonnais limestones which include black argillaceous dolomite and thick algal-rich limestones. Westwards, such deposits have given way to marine shales and then turbidites with local source areas. These sediments form part of the Cornish–Rhenish facies already mentioned. In the Fastnet Basin, tectonically fractured limestones of shallow water character were reached in boreholes and similar sediments may extend to the Bristol Channel and the St George's Channel. In the inner Bristol Channel, Carboniferous Limestone is exposed (Evans 1981). North of St George's Land, a ridge which extended from Cardigan Bay to the London–Brabant massif, more limestones and shales lie beneath the Irish Sea, their landward equivalents being clearly displayed in Anglesey, the Great Orme, the Isle of Man, and north and south Cumbria. To what extent the sediment types exposed on land change under what is now the sea is not clear, but some change of facies, i.e. to a deeper water, shaly facies, can be predicted as the basins which were to become the seas around the British Isles evolved.

In late Carboniferous times, the North Sea was an area of submergence and the facies resemble those of the surrounding land areas (Leeder *et al.* 1990) with successions thickening towards the offshore of today reaching a thickness of more than a kilometre. Sediments fed the North Sea from the Fenno–Scandian continent to the north and from smaller source areas such as the persistent London–Brabant Massif. Since the Carboniferous is generally regarded as below economic basement, most boreholes do not penetrate far into this system. Namurian sandstones, silts, and grey mudstones with some coals, the whole being 737 m thick, were drilled east of Scarborough while farther south, off Yarmouth, Namurian rocks rest on Dinantian rocks. Westphalian coal-bearing rocks have been drilled farther east from Yarmouth and in one place (BH 52/12–3) these rocks rest on Dinantian strata (Ramsbottom *et al.* 1979). Westphalian rocks near the sub-crop are frequently reddened, implying deep pre-Permian weathering. Eames (1975) presented a pre-Permian subcrop map showing the distribution of Carboniferous strata and he noted that the gas of the southern North Sea fields is related to higher rank Carboniferous coals devolatilased by igneous activity in late Cretaceous–Tertiary times, an opinion restated by Kirton & Donato (1985). Intra-Carboniferous unconformities in the successions off the north slope of the London–Brabant Massif (Tubb *et al.* 1986) suggest continuous tectonic activity in the region.

Concealed Coal Measures exist south of the London–Brabant Massif in France and beneath south-eastern England thus more Coal Measures, though perhaps much thrusted, can be expected to lie beneath the eastern English Channel. To what extent late Carboniferous sediments may exist under the central and western parts of the English Channel remains in doubt, though the possibility of coal seams must be remote. Hercynian movements and rapid sedimentation in early and mid-Carboniferous times must have completed much of the sedimentation hereabouts by late Carboniferous times.

The Celtic Sea basins may not have been involved in Hercynian events to the degree which is commonly accepted (Fig. 15.7), instead the area may have been subject to block faulting and the possibility of paralic Upper Carboniferous sediments existing in this area cannot be discounted (Gardiner & Sheridan 1981).

Some Upper Carboniferous sediments may exist beneath parts of the Bristol Channel at depth, though during Namurian and 'Pennant' (mid-Westphalian) times the Bristol Channel landmass of Devonian times was rejuvenated and acted as a source area (Kelling 1974, Tunbridge 1986). The Cardigan Bay area continued as a source area in late Carboniferous times though off Pembrokeshire and north of the Lleyn peninsula Upper Carboniferous sediments have been found in I.G.S. and commercial boreholes (Barr *et al.* 1981). To the north of Wales, the Irish Sea Basin contains Upper Carboniferous sediments with Coal Measures (Colter & Barr 1975, Colter 1978). As in the case in the North Sea, coal seams thicken seawards from known coal workings thus thick seams may be expected beneath the Irish Sea, though in the Morecambe gasfield (see Chapter 19) only reddened

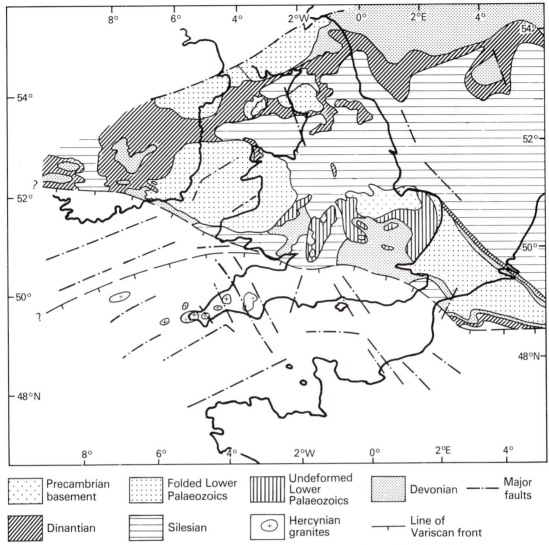

**Fig. 15.7.** The pre-Permian geology north of the Variscan Front. The Front has two possible locations west of South Wales; both are shown. The area south of the Front is assumed to comprise deformed Palaeozoic and older rocks.

beds 'of uncertain age, but of Namurian and West-phalian aspect' (Ebbern 1981) have been drilled. In the Kish Bank Basin, between 10 and 30 km east of Dublin a thick sequence of Upper Carboniferous strata exists with many, some up to 4 m thick, coal seams (McArdle & Keary 1986).

### The pre-Permian surface

By late Carboniferous times, rift tectonics were affecting the whole of the north-west European area and

major climatic changes were taking place. The pre-Permian situation presented here (Fig. 15.7) is as postulated by Wills (1973), W. H. Ziegler (1975), P. A. Ziegler (1978, 1981) and Sellwood & Scott (1986). The Mid-North Sea–Ringkøbing–Fyn High, consisting of cratonic basement and Lower Palaeozoic and Old Red Sandstone strata and intrusives, became by the end of the Carboniferous a major positive feature. Though intermittently breached along faults, it was to remain a barrier until early Tertiary times (Jenyon *et al.* 1984).

The floor of the southern North Sea was composed of thick Carboniferous deposits the topmost layers of which were reddened by deep weathering. Some Hercynian folding is seen (Tubb *et al.* 1986). Subsidence centred on the north-east side of this region and a series of faults developed parallel to the NE margin of the London–Brabant Massif. This had on it a relatively thin and undisturbed Lower Palaeozoic and Old Red Sandstone succession.

The Variscan (Hercynian) Front, with overthrust successions from the south, crosses the eastern end of the English Channel. Taylor (1986) proposed that 'thin-skinned' tectonics characterises the Variscan deformation of southern England with flat-lying thrust slices overriding relatively unmetamorphosed and undeformed rocks. The depth to the basement is not large in SE England where the pre-Mesozoic basement shallows from about 1,500 m off the Sussex coast to about 300 m in the Dover Strait (Wallace 1982). The geology of the Western English Channel and the Western Approaches at this time consisted of highly deformed and altered pre-mid-Carboniferous rocks with the intensity of deformation decreasing northwards and there is an obvious relationship between the tectonic style, the basement control and the Variscan Front (Shackleton 1984). Isaac *et al.* (1982) suggested that the entire region, including the Armorican Massif, may be allochthonous on a mid-crustal plane which dips at a low angle to the south, the whole being faulted as a graben or half-graben by early Permian times by extensional tectonics with the relaxation of pressure (Leeder 1982, Beach 1987). Seismic evidence (Day & Edwards 1984, Leveridge *et al.* 1984) indicates a number of south-dipping thrusts off SW England and some of these continue to mid-crustal levels (Hillis & Day 1987, Hobbs *et al.* 1987, Gibbs 1987). In late Carboniferous times intrusions had cut through the Devonian strata off the south-west of England – the Scillies, Seven Stones, and Haig Fras granites (Jones *et al.* 1988). These, like the granites of SW England, may also in part be allochthonous (Isaac *et al.*, loc. cit.). In the Bristol Channel the pre-Permian surface probably consists of eroded paralic Carboniferous and Old Red Sandstone strata. Mechie & Brooks (1984) described a pre-Upper Carboniferous supra-basement sequence thickening southward from a series of pre-Variscan Precambrian basement highs which, presumably, formed part of the southern margin of the St George's Land Massif. The latter had, by early Permian times, been separated from the London–Brabant Massif by the Severn or Worcester–Cheshire Graben. The pre-Permian floor of Cardigan Bay consisted of undifferentiated Lower Palaeozoic rocks near the present coast with the possibility of Carboniferous strata further west (Dobson & Whittington 1987). By Permian times the area was affected

by graben or 'trap door' faulting permitting Permian sedimentation to occur (Barr *et al.* 1981). Such faulting is usually related to pre-existing structures and some continue to be active to the present (Blenkinsop *et al.* 1986). The Irish Sea, floored by paralic Carboniferous sediments, was similarly affected by faulting (Bushell 1986); in part this was a northerly extension of the Worcester–Cheshire Graben. In the north, in the Solway Firth, another graben-like feature existed in the vicinity of the Iapetus Suture and here, too, extensive Carboniferous Coal Measures as well as Lower Carboniferous strata can be postulated as comprising the pre-Permian surface.

## Permian

The late Carboniferous events marked a major change in the tectonic development of NW Europe and particularly so in the offshore areas. From this time onwards vertical movements, mainly the product of extensional tectonics probably controlled by re-activated deep strike-slip movements, were to control the ensuing evolution (Gibbs 1986, Glennie 1986, McQuillin 1986, Tubb *et al.* 1986) (Fig. 15.8). Basins which began at this time were to have long histories. The Permian Southern North Sea basin, south of the recently developed Manx (Isle of Man)–Mid-North Sea–Ringkøbing–Fyn High, extended, with some interruptions, from the Irish Sea to Poland. To the south lay the St George's Land–London–Brabant Massif and south of this there were smaller basins undergoing graben-like development in the area of the central and western English Channel and its Western Approaches. North of the Cornish Platform two other elongate basins extended from the present continental margin, one into Cardigan Bay, the other south of the Pembroke High towards the Bristol Channel (Fig. 15.9). Permian sediments in the Irish Sea are found south-west of Anglesey and on the east side of the Isle of Man–Anglesey ridge in the Manx–Furness Basin and in the Solway Basin.

Nearly all these Permian basins around England and Wales are fault bounded and the earliest sediments in the succession are usually coarse, being derived from the newly uplifted mountains following Hercynian events. Desert conditions prevailed. In the Southern North Sea Basin, the Lower Permian Rotliegendes sediments consist of conglomerates and sandstones, deposited in wadis, fans and dunes, and silts associated with sabkhas. Borehole evidence has enabled a clear picture of the palaeogeography to be established (Glennie 1972, Marie 1975). These successions, which rest unconformably on Carboniferous strata, are succeeded by the Kupferschiefer; the lateral equivalent of the Marl Slate of NE England, and the bituminous, basal, shaly member of the Upper Per-

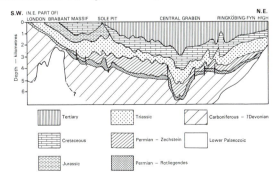

reviews by Oele and co-workers (eds.) in 1979, Stoker *et al.* (1986) and Cameron *et al.* (1987). Offshore data on Early and Middle Pleistocene deposits are scarce from the North Sea; indeed the same is true of the other seas around England and Wales, but it would appear that the major factor was the rapid growth of the Rhine delta (Zagwijn 1979), which gives a complete succession as the distributaries spread sediment westwards and northwards. The deltaic and associated sediments exceed 1 km in thickness in places. Cameron *et al.* (1987) described northward prograding Lower and early Middle Pleistocene sediments up to 500 m thick across the southern North Sea. They describe a late Middle Pleistocene unconformity after which glacial erosion and sedimentation processes predominate. Deep valleys were cut, some as plunge pools (see Wingfield 1990), in three recognisable Middle and Upper Pleistocene glaciations. These valleys are filled by glacio-lacustrine, glacio-marine and interglacial sediments. Saalian (Wolstonian) deposits are said to be widespread on the sea floor beneath a thin cover of Holocene sediments.

The Weichselian (Devensian) glaciation was not as extensive as the Saalian and reached only to the vicinity of the Dogger Bank (Fig. 15.18). Oele *et al.* (1979) suggested that much of the area may have been dry land, Stoker and Long (1984), however, considered that sea-ice may have been involved. The reasons for sea-level change may be many, but principally they must be glacio-isostatic readjustments and eustatic sea-level changes.

In the main part of the English Channel, Pleistocene deposits are missing entirely. Kellaway *et al.* (1975) favoured a major glaciation during the Saalian with ice moving eastwards to breach the Dover Strait. Though numerous features exist in the Channel which remain

**Fig. 15.18.** Limits of the Quaternary glaciation.

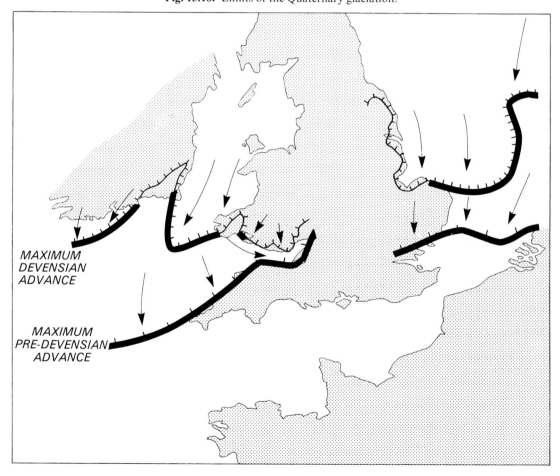

MAXIMUM
DEVENSIAN
ADVANCE

MAXIMUM
PRE-DEVENSIAN
ADVANCE

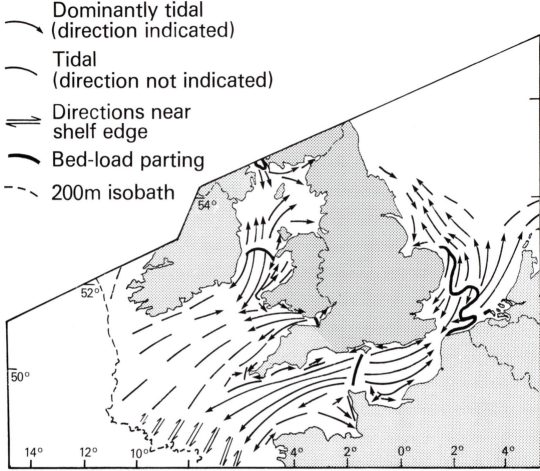

**Dominantly tidal**
(direction indicated)

**Tidal**
(direction not indicated)

**Directions near**
shelf edge

**Bed-load parting**

**200m isobath**

**Fig. 15.19.** Tidally moved sediments around England and Wales after Lee & Ramster (1981).

Cardigan Bay and the Irish Sea similar conditions to those which have existed in the North Sea have led to an assemblage of sediments which reflect the response of late Quaternary deposits to the present energy regime (Pantin 1977).

Stride (1982) reviewed the origin and distribution of bed forms assumed by tidally propelled sediments and concluded that forms are a product of sediment supply and energy regime. In addition to the sandwaves, dunes, and ribbons, older, late Quaternary and early Holocene, banks abound. Large banks exist in the Western Approaches and Celtic Sea, e.g. Labadie and Cockburn Banks, others are known in Cardigan Bay and the Bristol Channel. Other banks exist near the mouth of the Seine and the greatest number occur in the eastern English Channel, the Dover Strait, off eastern and south-eastern England and in the southern North Sea off NE France, Belgium, and Holland. All

of the foregoing are strongly linear. The largest feature, the Dogger Bank, in the middle North Sea, has an irregular shape (Fig. 15.21). The origins of these are still a matter of debate – some bars are related to earlier sea-level stands while others have a glacial or glacially-related origin, being composed of morainic or out-wash material.

Elongate depressions in the floor of the continental shelf (deeps, pits, *fosses* of the French) occur in the English Channel, the North Sea, and the western Irish Sea and though partially filled with Recent sediments they make striking features. Their origin has been the subject of much debate. The largest, the Hurd Deep (the Fosse Central of the French) (Figs. 15.1, 21), is 100 km long, 10 km wide at its widest and, though now partially filled, is cut nearly 300 m into the strata of the shelf. Hamilton & Smith (1972) gave a detailed de-scription and reviewed hypotheses presented for its

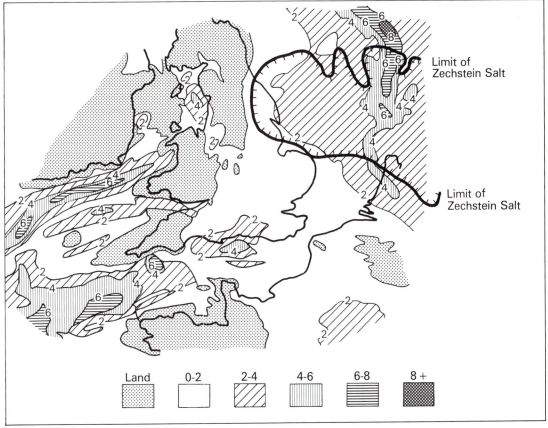

**Fig. 15.8.** Isopachytes of total thickness of Late Permian to Recent strata, based on Ziegler (1982). (Thicknesses in kilometres.)

mian Zechstein sequence. The Zechstein sequence of the southern North Sea is characterised by four major cycles (Fig. 15.10) (see also Chapter 9); the base of each cycle sometimes being marked by widespread bituminous shales, the Kupferschiefer representing the base of the first cycle. In the first cycle the Zechstein sea dried out giving carbonates and evaporites. Subsequent cycles required replenishment of the evaporated sea water, possibly across the intermittently breached barrier to the north, and it has been suggested that water levels were below contemporary global sea level. The evaporites of the later cycles consist of progradational wedges (Taylor & Colter 1975, and Taylor 1981) comprising thick evaporites and laminated anhydrites in basinal areas and thick shallow water carbonates and sulphate banks on the basin margins and on highs, the whole preserving a record of continued subsidence and continued overstepping of the margins. The evaporite sequences, though being areally widespread and thick

(1,500 m) in the present offshore, only just extend into the onshore in eastern Yorkshire. Diapirism has altered thicknesses considerably.

Dixon *et al.* (1981) described basic and ultrabasic rocks found in the Lower Permian and Zechstein of the eastern sectors of the North Sea: there are no records of similar rocks on the English side of the southern North Sea.

The Southern North Sea Basin was separated from the English Channel area by the London–Brabant Massif. No Permian deposits have been found in the eastern English Channel but they may exist at depth; Permian deposits are now found in the western Baie de la Seine and off the south coast of Dorset, Devon, and Cornwall. The earliest deposits are usually coarse, linking shoreward to breccias and additionally in southern England to volcanics, the Exeter Volcanic Series (Chapter 16), which follows the Culm successions. The Volcanic Series is represented by potassium-rich vesicular lavas with subordinate olivine basalts

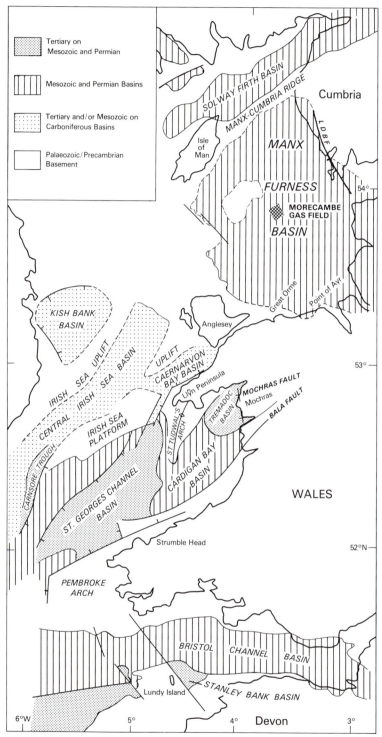

**Fig. 15.9.** The principal offshore features to the south, west, and north of Wales (after Naylor & Mounteney 1975). (LDBF = Lake District Boundary Fault.)

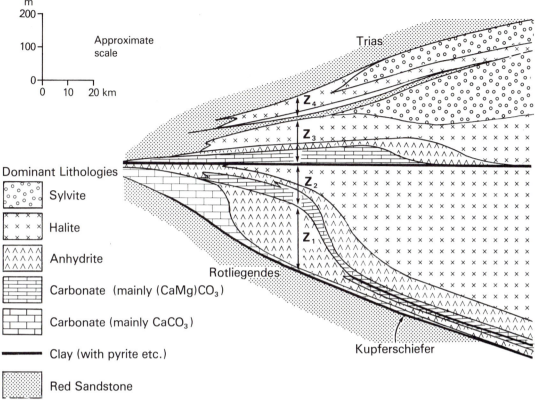

**Fig. 15.10.** The four major cycles of the Zechstein of the southern North Sea between England (left hand) and towards the central part of the North Sea Basin (after Taylor 1981).

and thus may be related to extensional tectonics with crustal thinning. It is not known if the entire Permian period is represented in these more southern successions. As with the North Sea, progradational wedges of sediments thickening towards the present offshore may be conjectured. No evidence, however, of Permian evaporites has yet been revealed in the western English Channel or in the Western Approaches and Celtic Sea Basins. At 48°47′N 9°06′W, off the southern flank of the Cornubian Platform, a 3.7 km deep commercial borehole (BNOC 72/10-1A) penetrated more than 900 m of 'undifferentiated' coarse sediments. Composed of acid igneous clasts and containing tourmaline and zircon they were thought to be comparable with the Permian deposits of SW England (Bennet *et al.* 1985) (Fig. 15.11).

Lloyd *et al.* (1973) did not record Permian deposits in the Bristol Channel, but noted that the distinction between Permian and Triassic deposits in cores is well nigh impossible. A thin Permian succession, deposited to the accompaniment of contemporaneous faulting and erosion, may be only conjectural for Cardigan Bay

(Barr *et al.* 1981). Here again, no Permian evaporites are known though B.G.S. samples 71/47 and 71/55 from south of St Tudwals Peninsula are said to be gypsiferous marls and mudstones of 'Permo-Triassic age' (B.G.S. 1972). It is in the Manx–Furness Basin of the Irish Sea, a trapdoor-type basin downfaulted on its eastern margin, that thicker Permian deposits are again known. The Lower Permian is represented by the Penrith Sandstone with brockram, the latter derived from the stripping of adjacent Carboniferous Limestone, and the peripheral Collyhurst Sandstones (up to 700 m) which pass laterally into shales in the centre of the basin. The Upper Permian comprises sandstones with shales, dolomites, and anhydrites as well as halites which give way to shales towards the centre of the basin (Colter & Barr 1975; Colter 1978). In the Morecambe gas-field the Collyhurst Sandstone is represented by 280 m of silts and mudstones and the St Bees Evaporites, 300 m of halite, dolomite, and shale. The succeeding St Bees Sandstone is included in the Sherwood Sandstone Group (Ebbern 1981).

Bott (1968, and *in* Bott & Young 1971) estimated

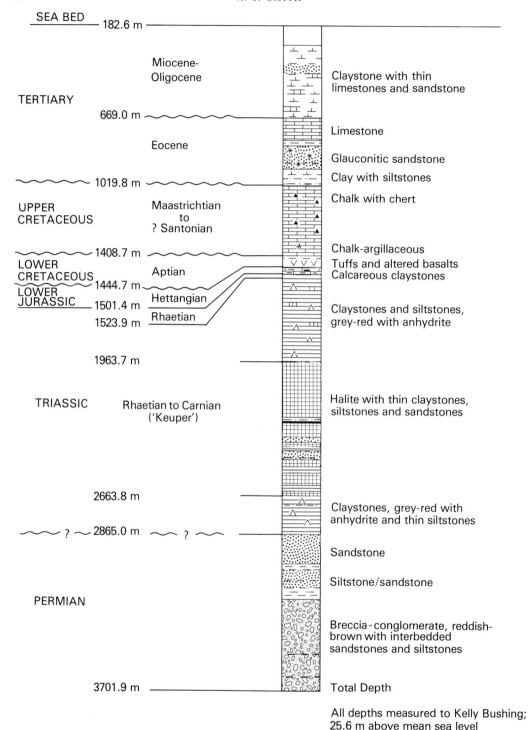

SEA BED — 182.6 m

Miocene-
Oligocene

TERTIARY

669.0 m

Eocene

1019.8 m

UPPER
CRETACEOUS

Maastrichtian
to
? Santonian

1408.7 m

LOWER
CRETACEOUS          Aptian

1444.7 m

LOWER
JURASSIC          Hettangian

1501.4 m

1523.9 m          Rhaetian

1963.7 m

TRIASSIC          Rhaetian to Carnian
('Keuper')

2663.8 m

? — 2865.0 m — ?

PERMIAN

3701.9 m

Claystone with thin
limestones and sandstone

Limestone

Glauconitic sandstone

Clay with siltstones

Chalk with chert

Chalk-argillaceous

Tuffs and altered basalts
Calcareous claystones

Claystones and siltstones,
grey-red with anhydrite

Halite with thin claystones,
siltstones and sandstones

Claystones, grey-red with
anhydrite and thin siltstones

Sandstone

Siltstone/sandstone

Breccia-conglomerate, reddish-
brown with interbedded
sandstones and siltstones

Total Depth

All depths measured to Kelly Bushing;
25.6 m above mean sea level

**Fig. 15.11.** Lithological log at Britoil well 72/10–1A at 48°47′23·4″N : 09°06′01·9″W.
Reproduced, with permission, from Bennet *et al.* (1985).

that the offshore Solway Basin has about 4 km of fill and that Permo–Triassic sediments account for much of this total. The Permo–Triassic succession may resemble those found onshore north of the Lake District, in the Vale of Eden, and in southern Scotland, though evaporites could be present.

## Mesozoic

Where Triassic rocks overlie the Permian successions there is little sign of an unconformity, though when they rest on pre-Permian rocks there is usually a marked unconformity. In Triassic times the Permian basins continued to subside but their framework was modified by a new set of fault patterns. The Mid-North Sea–Ringkøbing–Fyn Ridge continued to exercise an important control but it was cut by what some authors term the Southern Graben (a southward extension of the Central Graben) and, nearer Denmark, the Horn Graben. A set of fault-controlled basins developed parallel to the north-eastern margin of the London–Brabant Massif. In addition, Permian salt bodies began to move diapirically, adding further complexities to the situation (Fig. 15.12). Fault movements also occurred in the Western Approaches Basin and in the Celtic Sea basins and their northern and eastern extensions. North of St George's Land, now separated from the London–Brabant Massif by the Worcester Graben, Triassic sediments accumulated in the Manx–Furness Basin and in the Solway Basin.

The early Triassic sediments of the Anglo-Dutch Basin of the southern North Sea exceed 1,500 m in thickness. The lower sediments, the Bacton Group, comprise shales followed by sandstones (equivalent to the Bunter Sandstones or the Sherwood Sandstone Group of the English Midlands) and these give way laterally to evaporites. The depositional conditions thus copy those of the Permian and, indeed, the Rot Halite Member of the succeeding Hailsborough Group closely resembles the Zechstein deposits. The lower part of the Haisborough Group including the Rot Salt is in part coincident with the Muschelkalk division of the Trias. The succeeding shales and evaporites of the Southern North Sea are the offshore equivalents of the Mercia Mudstone Group (Lott & Warrington 1988) (Chapter 10). Thicknesses of the offshore sedimentary formations vary laterally reflecting contemporary and subsequent diastrophic and non-diastrophic movements. Brennand (1975) discussed the variations in thickness and lithology in the Trias of the North Sea and noted that, in the north-western part of the southern North Sea, the Rhaetic represented by the Winterton Group oversteps eroded earlier Trias sediments.

In the eastern English Channel, Triassic sediments may exist below the Jurassic deposits, whilst further

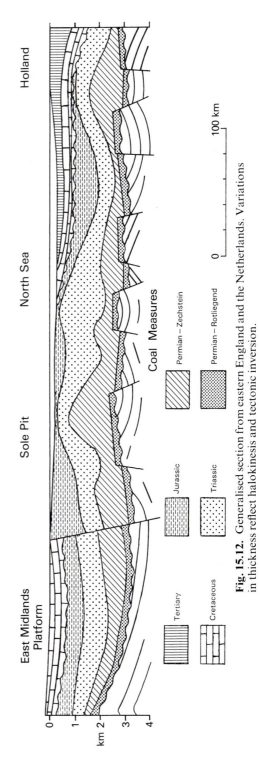

**Fig. 15.12.** Generalised section from eastern England and the Netherlands. Variations in thickness reflect halokinesis and tectonic inversion.

west Triassic sediments, or sediments believed to be Triassic (Curry *et al.* 1970), outcrop in the eastern part of the Western Approaches Basin and in the Celtic Basin. Pebble affinities suggest that sediments were transported across the English Channel from France northwards to the Midlands of England in the Triassic. Triassic sediments more than 600 m thick were recorded on the south side of the North Celtic Sea Basin and in the Fastnet Basin (Robinson *et al.* 1981) with silts and evaporites dominating the upper part of the succession. Ainsworth and Horton (1986) described this succession as of late Triassic age and noted that it rested directly on the Devonian basement. On the south flank of the Cornubian Platform at 9°W, about 700 m of Upper Triassic (Carnian) halite and poly-halite were drilled in a commercial borehole (Fisher & Jeans 1982). These evaporites rested on about 200 m of silty anhydritic claystones again of Carnian age. The evaporites were succeeded by more than 400 m of Keuper red siltstones with anhydrite. Upwards the red colouring gave way to olive green and there are 22 m of dark grey claystones with thin limestones which were given a Rhaetian age (Bennet *et al.* 1985) (Fig. 15.11).

Lloyd *et al.* (1973) referred to Triassic deposits (Keuper marl) in the Bristol Channel and these pass into the Triassic succession of S Wales and the Mendip region (see Chapter 10). Whittington (1980) estimated thick deposits ($\pm$ 2,000 m) together with evaporites, in the St George's Channel and Cardigan Bay Basins, a view supported by Dimitropoulos and Donato (1983). Rocks of inferred Triassic age are reported from the Mochras borehole (Woodland 1971) and from off-shore wells in the southern part of the basin (Barr *et al.* 1981). The full extent of Triassic evaporites west of Britain has yet to be clearly defined but a salt wall parallel to the Bala Fault in Cardigan Bay confirms the existence of halites hereabouts. In the Manx–Furness Basin, north of St George's Land, the Triassic is represented by the St Bees and Mottled Sandstones and the Lower Keuper Sandstones, Keuper Water-stones, Saliferous Beds and Upper Keuper Marls; an unconformity, the Hardegsen Unconformity, pre-cedes the Lower Keuper Sandstones. Ebbern (1981) gave details of this succession in the Morecambe gas field: the St Bees Sandstone Formation consists of nearly 1,300 m of mainly fluviatile argillaceous sand-stones, the Bunter Pebble Beds ( = Chester Pebble Beds of Warrington *et al.* 1980) are missing, the Keuper Sandstone Formation ( = Helsby Sandstone Form-ation (ibid)), some 200 m thick, consists of alluvial deposits and the Keuper Waterstones Formation, between 40 and 70 m thick, is composed of alternating clean sands and muddy sands. The succeeding Keuper Saliferous Beds ( = part Mercia Mudstone Group) outcrop on the seafloor. Bushell (1986) noted the

importance of the fluviatile, braided stream, facies as reservoir rocks in the Morecambe field.

As has been described in the Trias Chapter of this book (Chapter 10), the continental conditions of the Triassic gave way late in the period to Tea Green Marls and eventually marine, admittedly of a re-stricted facies, sediments – the Penarth Beds and the White Lias. Late Triassic times might have been coincident with a global rise in sea level (Vail 1977); thus by the end of the Triassic much of the conti-nental hinterland was flooded and marine conditions were re-established. Frequently, however, the top-most Triassic has been eroded and with it evidence of Rhaetic formations.

Well to the south of Britain, movements were occurring in the Tethyan region, whilst west of south-ern England the Biscay rift had developed in Triassic times as the Atlantic Ocean developed. Stemming from and associated with these, a series of tectonic movements were to ensue, many given the distinction of a regional correlatable event. Just how precisely these events can be correlated remains to be seen, but throughout the late Triassic, the Jurassic, and into the early Cretaceous, movements occurred which have been grouped into the Cimmerian 'orogeny'.

Barton and Wood (1984) suggested that the North Sea Basin was subjected to 70 km extension, out of a total of 100 km extension concentrated across the Central Graben, during mid-Jurassic to early Cretaceous times. In the northern North Sea major upwarping occurred in mid-Jurassic times with sediments being transported in several directions. However, in the area which is our main concern a few major rifts continued as sites of marked subsidence and this condition lasted until mid-Cretaceous times. Particularly important were the Southern Graben across the Mid-North Sea–Ringkøbing–Fyn High, the Sole Pit Basin, the Western English Channel Graben, the Celtic Sea Basins, the Bristol Channel Basin, and the St George's Channel Basin, which reached into Cardigan Bay. During the Jurassic the Irish Sea Basin, including the Manx–Furness Basin, may have been sinking only in early Jurassic times and the Solway Basin may also contain Lower Jurassic sediments. Offshore, only the Mid-North Sea High, the London–Brabant Massif, Armorica, and the Cornubian platform were positive features for much of the Jurassic.

During the Sinemurian, fully marine conditions were established and in the southern North Sea, cyclical Liassic series of open-marine shaly facies were developed. A nearly complete, but thin, Jurassic suc-cession exists in the westernmost of the two Sole Pit Basins. In the eastern basin a thick Liassic succession exists, though there is evidence of erosion in Toarcian times (Glennie & Boegner 1981). Limestones occur in the lower Lias, and though some ironstones with ferruginous ooids, preceded by some silty mudstones, occur in middle Lias, shale facies dominate. The exposed and cored Lias of the English Channel is similar to that of the Dorset coast (Chapter 11) but the full extent of the Lias is not yet known in the western part of the Channel. Liassic rocks (nearly 500 m) have been cored south of Start Point but are missing in cores from beneath the Chalk in mid-Channel south of Cornwall and on the extension of the covered Cornubian Platform west of 8° West (Evans *et al.* 1981). Nevertheless, Liassic sand, shale and limestone sequences exist near the Atlantic margin in the Fastnet Basin (Robinson *et al.* 1981, Ainsworth & Horton 1986) where they are intruded by olivine-dolerite sills possibly related to updoming prior to rifting (Caston *et al.* 1981) and south of the Cornubian Platform at 9° W where 60 m of Hettangian dark grey claystones and limestones and calcareous claystones rest conformably on the Permo–Triassic succession (Bennet *et al.* 1985). In the Bristol Channel, the Lias, similar to that found on land, is well developed and in the St George's Channel and Cardigan Bay Basins it reaches a maximum development. At Mochras, the borehole showed that the Lias is particularly thick, exceeding 1,300 m, due to contemporaneous movements on the Mochras Fault (Woodland 1971). This lower Jurassic thins rapidly to the Irish Sea Platform. In the Manx–Furness Basin some inversion of movement must have occurred for no Lias has been found there; however, some Lower Liassic sediments resting on Rhaetic shales are known in the Vale of Eden–Solway Firth Basin and might thus be expected in the offshore Solway Basin. Throughout the Jurassic and Cretaceous, this region may have received sediments; however, inversion movements and subsequent erosion of sediments means that no record of post-Triassic sedimentation remains.

The commencement of the Mid-Jurassic was accompanied by a regressive phase; only in SW England and in the English Channel was open-marine deposition maintained. In Bajocian–Bathonian times the Mid-North Sea–Ringkøbing–Fyn High was part of a major positive dome-like feature centred on the central North Sea. Fluvio-deltaic sedimentation spread southwards as far as the northern part of the southern North Sea with deltaic episodes being interrupted by marine transgressions. Locally, non-marine and lagoonal mudstones occur. The Mid-North Sea High and the London–Brabant Massifs were major sources of detrital sediments. These uplifts and associated fault movements comprise P. A. Ziegler's (1981) Mid-Cimmerian tectonic phase. In post-Jurassic times the movements of the Mid-Jurassic were to be reversed – a feature which characterises a number of offshore localities of England and Wales. It can be assumed that before the Mid-Cimmerian movements the Lias

of the North Sea had been more extensive and perhaps thicker. In the eastern Sole Pit Basin, the eroded Lower Jurassic sediments are succeeded by transgressive marine Callovian carbonates and clays and the deltaic facies is missing.

In the English Channel, particularly in the central part, the Middle Jurassic successions are extensions of those seen on the Dorset coast (Dingwall & Lott 1979) (Fig. 15.13) whilst in the east and in the southern part of the central Channel, the facies resemble those of the Boulonnais and Normandy respectively. There are local variations in thickness. Farther west the extent, thickness, and facies of the Middle Jurassic, indeed of the entire Jurassic, are again as yet unpublished. It seems likely, however, that the Jurassic occurs in fault-bounded basins with, consequently, local variations in thickness and facies (Millson 1987). The western part of the Western Approaches, the Cornubian Platform and Armorica underwent uplift associated with major tectonic developments at what was to be the margin of the European continent and all became sediment sources. In addition there were other smaller sources linked to more local faulting and possibly to halokinesis. Robinson *et al.* (1981) reported that there are Bajocian and Bathonian sediments – shales and clean sands – in the Fastnet Basin and that there may have been local erosion.

Farther east and north of the Cornubian Platform, the Celtic Sea, Bristol Channel, and St George's Channel Basins, the last with its extension into Cardigan Bay, continued to receive sediments, though in the Cardigan Bay Basin geophysical records suggest an unconformity between the Lower and Middle Jurassic strata. Penn and Evans (1976) inferred that most of Cardigan Bay had lagoonal and marginal marine conditions with a delta in one place in Mid-Jurassic times. Though Ziegler (1982) showed in his palaeogeographic maps an emergent St George's Land throughout the Jurassic, the evidence from around Wales discredits this for Lower Jurassic times and implies it to be unlikely for the Mid-Jurassic (Hallam & Sellwood 1976). Lloyd *et al.* (1973) and Kammerling (1979) wrote that there is a complete, thick, predominantly argillaceous Jurassic sequence in the Bristol Channel though the Bajocian–Bathonian is extremely thin. Kammerling also stated that south-west of the Bristol Channel the Upper Jurassic is missing. The Irish Sea Platform appears from geophysical evidence to have only a thinned Mid-Jurassic succession on it whilst, as already stated, no Jurassic sediments have yet been recovered from the Irish Sea though it is possible that some local, perhaps fault-bounded, Jurassic depocentres existed there.

Volcanic activity occurred in middle-Jurassic times (Hallam & Sellwood 1968) and in the northern North Sea undersaturated alkali-basalts have been cored in

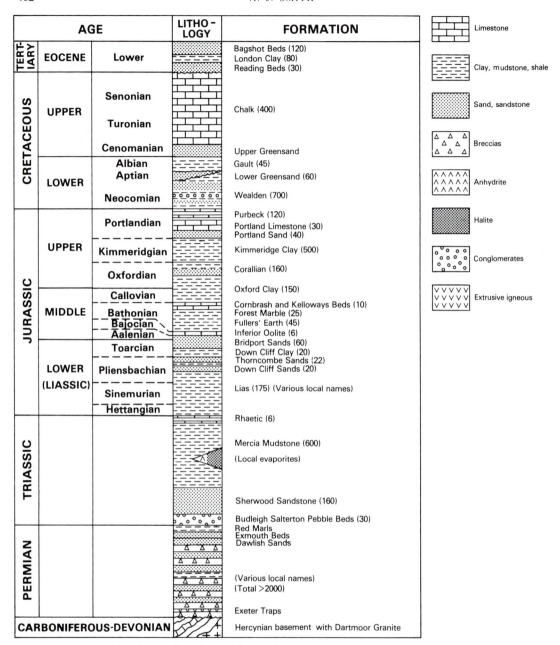

| AGE | | | LITHO-LOGY | FORMATION |
|---|---|---|---|---|
| TERT-IARY | EOCENE | Lower | | Bagshot Beds (120) / London Clay (80) / Reading Beds (30) |
| CRETACEOUS | UPPER | Senonian / Turonian / Cenomanian | | Chalk (400) |
| | | | | Upper Greensand |
| | LOWER | Albian / Aptian | | Gault (45) / Lower Greensand (60) |
| | | Neocomian | | Wealden (700) |
| JURASSIC | UPPER | Portlandian | | Purbeck (120) / Portland Limestone (30) / Portland Sand (40) |
| | | Kimmeridgian | | Kimmeridge Clay (500) |
| | | Oxfordian | | Corallian (160) |
| | MIDDLE | Callovian | | Oxford Clay (150) |
| | | Bathonian | | Cornbrash and Kelloways Beds (10) / Forest Marble (25) / Fullers' Earth (45) |
| | | Bajocian | | Inferior Oolite (6) |
| | | Aalenian | | |
| | | Toarcian | | Bridport Sands (60) / Down Cliff Clay (20) / Thorncombe Sands (22) / Down Cliff Sands (20) |
| | LOWER (LIASSIC) | Pliensbachian | | |
| | | Sinemurian | | Lias (175) (Various local names) |
| | | Hettangian | | |
| | | | | Rhaetic (6) |
| TRIASSIC | | | | Mercia Mudstone (600) |
| | | | | (Local evaporites) |
| | | | | Sherwood Sandstone (160) |
| | | | | Budleigh Salterton Pebble Beds (30) / Red Marls / Exmouth Beds / Dawlish Sands |
| PERMIAN | | | | (Various local names) / (Total >2000) |
| | | | | Exeter Traps |
| CARBONIFEROUS-DEVONIAN | | | | Hercynian basement with Dartmoor Granite |

Legend:
- Limestone
- Clay, mudstone, shale
- Sand, sandstone
- Breccias
- Anhydrite
- Halite
- Conglomerates
- Extrusive igneous

**Fig. 15.13.** Succession of strata in the Central English Channel–Wessex Basin area, thicknesses in metres. Based on unpublished material prepared by R. Stoneley and reproduced here with his permission.

the Forties and Piper fields (Howitt *et al.* 1975). Other volcanic centres may have existed in the outer Western Approaches and even on the London–Brabant Massif.

After regression at the close of the Callovian, the Upper Jurassic of the North Sea commenced with paralic sediments followed by a marine facies – glauconite-rich Oxfordian sediments marking the transgression. These are succeeded by calcareous Corallian strata and Kimmeridge Clay (Fig. 15.14). Together these form the upper part of the Humber Group of the

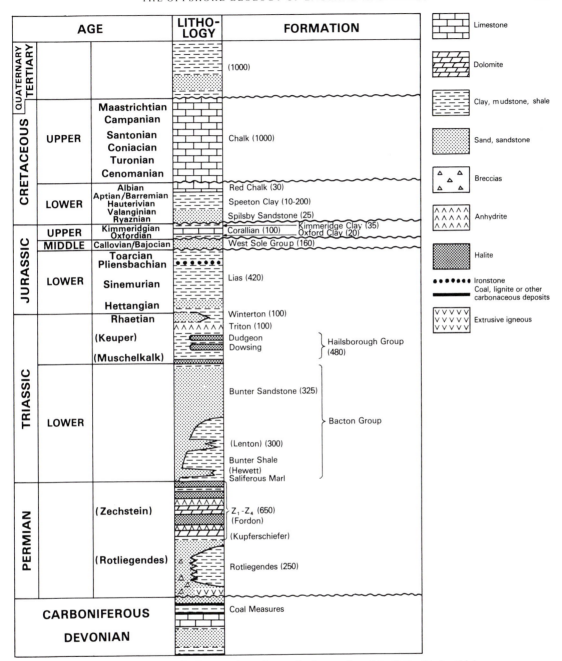

| AGE | | | LITHO-LOGY | FORMATION |
|---|---|---|---|---|
| QUATERNARY | | | | (1000) |
| TERTIARY | | | | |
| CRETACEOUS | UPPER | Maastrichtian | | Chalk (1000) |
| | | Campanian | | |
| | | Santonian | | |
| | | Coniacian | | |
| | | Turonian | | |
| | | Cenomanian | | |
| | LOWER | Albian | | Red Chalk (30) |
| | | Aptian/Barremian | | Speeton Clay (10-200) |
| | | Hauterivian | | |
| | | Valanginian | | Spilsby Sandstone (25) |
| | | Ryaznian | | |
| JURASSIC | UPPER | Kimmeridgian | | Kimmeridge Clay (35) |
| | | Oxfordian | | Corallian (100) Oxford Clay (20) |
| | MIDDLE | Callovian/Bajocian | | West Sole Group (160) |
| | LOWER | Toarcian | | |
| | | Pliensbachian | | Lias (420) |
| | | Sinemurian | | |
| | | Hettangian | | |
| TRIASSIC | | Rhaetian | | Winterton (100) |
| | | (Keuper) | | Triton (100) |
| | | | | Dudgeon |
| | | | | Dowsing |
| | | (Muschelkalk) | | |
| | LOWER | | | Bunter Sandstone (325) |
| | | | | (Lenton) (300) |
| | | | | Bunter Shale (Hewett) Saliferous Marl |
| PERMIAN | | (Zechstein) | | Z₁-Z₄ (650) (Fordon) (Kupferschiefer) |
| | | (Rotliegendes) | | Rotliegendes (250) |
| CARBONIFEROUS | | | | Coal Measures |
| DEVONIAN | | | | |

Hailsborough Group (480)

Bacton Group

Legend:
- Limestone
- Dolomite
- Clay, mudstone, shale
- Sand, sandstone
- Breccias
- Anhydrite
- Halite
- Ironstone
- Coal, lignite or other carbonaceous deposits
- Extrusive igneous

**Fig. 15.14.** Generalised succession of strata in the southern North Sea Basin; thickness in metres.

Anglo-Dutch Basin, and are about 300 m thick in places. Over much of the southern part of the North Sea the topmost Jurassic is missing, producing the late-Cimmerian unconformity which separates the Jurassic from the Cretaceous hereabouts. Near the northern margin of the London–Brabant Massif, local sandy deposits, e.g. the Lower Spilsby Sandstone, occur before the close of Jurassic times.

South of the London–Brabant Massif, the Upper Jurassic found offshore resembles the successions of similar age seen on land. In the eastern part of the Channel, inliers in the Weald-Artois anticline reveal Kimmeridge Clay followed by sandier lithologies than those found farther west, the coarser sediments presumably derived from the London–Brabant Massif. In the northern sector of the central part of the Channel, the Kimmeridgian clays pass up into the sandy and limy Portland facies of the Volgian and the Portland–Purbeck limestones are recognisable in seismic records over a large area. The Upper and, possibly, Middle Jurassic are absent southwards, cut out by the mid-Cretaceous unconformity and overlap.

The full extent of the Upper Jurassic farther westwards is not known, though Masson and Roberts (1980) pointed to reef developments at what is now the continental margin in Late Jurassic times indicating open marine conditions before the development of the Bay of Biscay. From the Fastnet Basin, Robinson *et al.* (1981) described a 400 + m sequence of Kimmeridgian to Purbeckian shallow water, possibly even brackish water, carbonates and in the Kinsale field Upper Jurassic sediments have been penetrated. South of the Cornubian Platform at 9° W there are no Middle or Upper Jurassic deposits, the Lower Jurassic sediments being unconformably overlain by Lower Cretaceous deposits (Bennet *et al.* 1985).

In the Bristol Channel, Lloyd *et al.* (1973) described an Upper Jurassic sequence which appears to thin towards Wales. Whittington (1980) on the basis of geophysical investigations favoured an Upper Jurassic sequence in the St George's Channel but stated that there are no Upper Jurassic sediments in Cardigan Bay or on the Irish Sea Platform. Blundell *et al.* (1971) and Dobson *et al.* (1973) postulated rocks of this age in the

western part of Cardigan Bay as well as St George's Channel and this has been proved by deep drilling (Barr *et al.*, 1981); nearly 1,500 m of lagoonal to marginal marine (including anhydritic) sediments being drilled some 20 km NW of Strumble Head. No sediments of this age have been reported from the Manx–Furness Basin of the Irish Sea or the Solway Basin.

Globally, sea level was falling during the Late Jurassic and in the British area vertical movements continued. Kent (1975) stated that they were either halokinetic in origin or related to differential movement of tensional fault blocks. Gibbs (1983) believed that listric structures are important in the northern North Sea and Smith (1984) envisaged listric structures reaching mid-crustal depths in the English Channel (Fig. 15.15). For the southern North Sea, Gibbs (1986) and McQuillin (1986) indicated that strike-slip faults influenced the movements throughout the region. Fyfe *et al.* (1981) presented a Cretaceous subcrop map for British waters which illustrated the degree of movement and erosion represented by the mid-Mesozoic (= late Cimmerian of some authors) unconformity. Sedimentation continued through from Jurassic to Cretaceous times in the basinal areas while erosion is most marked on the structurally highest parts (see, for example, Milsom 1986).

During the Cretaceous the previously distinct Northern and Southern basins of the North Sea became a single unit but whilst the Upper Cretaceous deposits are generally both uniform and widespread, the Lower Cretaceous deposits of the North Sea, and indeed the remainder of offshore England and Wales, are varied. The Lower Cretaceous stratigraphy of the southern North Sea is complicated by tectonic inversions and this is particularly true for the Sole Pit area

**Fig. 15.15.** Cross-section of the Central English Channel. This interpretation makes use of the hypothesis of high-level low-angle movements.

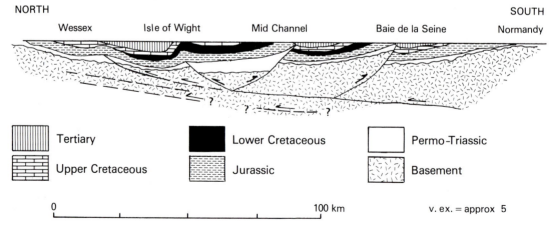

NORTH                                                                                                              SOUTH

Wessex          Isle of Wight          Mid Channel          Baie de la Seine          Normandy

Tertiary    Lower Cretaceous    Permo-Triassic

Upper Cretaceous    Jurassic    Basement

0                                                   100 km                    v. ex. = approx 5

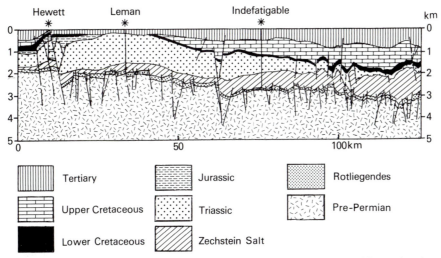

**Fig. 15.16.** Section across the Sole Pit Basin of the southern North Sea illustrating the effects of the tectonic inversion which characterises this Basin.

(Glennie & Boegner 1981, Jenyon 1987) (Fig. 15.16): see Chapters 12 and 19. Though the Lower Cretaceous deposits are generally marine – mainly shallow water and of no great thickness – more coastal aspects of sedimentation occur lower in the succession and near the basin margin – particularly against the London–Brabant Massif. The upper part of the Spilsby Sandstone is taken to be lowest Cretaceous and is succeeded by the Speeton Clay which attains a thickness of 175 m (Rhys 1974) in the southern North Sea. These Lower Cretaceous successions capped by the Albian Red Chalk thicken eastwards, but thin northwards. When cored 80 km NE of Speeton, the succession closely resembled the onshore Speeton sequence and contained seven bentonite horizons, six of which were recognised subsequently onshore (Lott et al. 1986). Throughout these times the London–Brabant Massif was a buoyant source area (Allen 1981) and though links around the western margin of this feature existed, the character of the sediments is markedly different to the south of it. The continental nature of the Wealden successions is well demonstrated (Allen 1981) and they are thicker than anything of similar age in the southern North Sea.

In the English Channel the Wealden Series, here characterised by sandy and clayey detritic formations, are widespread, being as much as 1,000 m thick south of the Isle of Wight (Curry & Smith 1975, Smith & Curry 1975). Lower Cretaceous beds of similar character have been found in the western English Channel in the vicinity of the Hurd Deep (about 150 km thick) and in the eastern part of the Channel between the Weald and the Boulonnais. In the central English Channel,

the Aptian has only been cored in the northern part. The Albian is possibly present in the southern part and the Aptian is presumed to be cut out by the 'sub-Cenomanian' unconformity. Where cored the lithologies are identical to those of the south coast of England and northern France. The full extent of representatives of these stages in the western part of the Channel is not yet well known.

Geophysical evidence and sampling from the continental slope off the Western Approaches and the Celtic Sea suggest that sea-floor spreading in the Bay of Biscay started during the Barremian to early Aptian (Montadert et al. 1977, 1979) and these events led to a marked change in the depositional picture of the western basins. Near the developing continental margin reefal limestones developed (Masson & Roberts 1981) early in the Cretaceous while farther east parts of the Celtic Sea and Western Approaches troughs subsided differentially (Colin et al. 1981). The Lower Cretaceous of the Fastnet Basin, some 50 km east of the margin resembles the Wealden facies of England (Ainsworth et al. 1985) and the non-marine and brackish Wealden Series reaches an estimated thickness of 300 m in the central part of the North Celtic Basin while more than 1,000 m thickness has been recorded in the Nymphe Bank–Fastnet Basins. A regional unconformity marks the base of the Aptian–Albian Greensand Series.

The movements were accompanied by volcanism: the Wolf Rock phonolite has been given a Lower Cretaceous date and Hallam & Sellwood (1968) suggested that the bentonite-rich fuller's earth of the onshore Lower Cretaceous successions was derived from volcanoes somewhere in the Western Approaches. Later work (Harrison *et al.* 1979), however, gave the Wolf rock phonolites a very early early-Cretaceous age (120–130 Ma); even so, evidence for igneous activity throughout the Lower Cretaceous remains strong. The recent borehole on the south flank of the Cornubian Platform at 9°W (Bennet *et al.* 1985) revealed tuffaceous sediments and an intrusive olivine basalt, the latter K/Ar dated at 110 Ma. A Late Aptian age is suggested for the extrusives.

In the St George's Channel Basin, syntectonic deposition of Lower Cretaceous sediments took place and the Wealden Series may be well represented in the main axis of the basin. However, farther landwards in Cardigan Bay, the Lower Cretaceous seems to have been a time of erosion with the possibility that Middle and Lower Jurassic and Triassic rocks were eroded off Wales and Ireland. During the Aptian–Albian, a transgression spread across the Celtic Sea Basin, but no evidence of a transgression into the St George's Channel Basin has been found.

Lloyd and co-workers (1973) and Evans & Thompson (1979) stated that no Cretaceous strata occur in the Bristol Channel Basin; however, this seems to be a product of post-Cretaceous erosion rather than non-deposition for it would seem that the Bristol Channel had a long history of tectonic inversion (Kammerling 1979, Brooks *et al.* 1988). In contrast, the absence of Cretaceous strata in the Irish Sea is probably a consequence of non-deposition rather than post-depositional erosion.

The Upper Cretaceous of the southern North Sea follows the events of Albian and Aptian tectonic activity and, in common with patterns elsewhere, the sediments change from shales and clastics to carbonates. These carbonate sediments, commencing in the late Albian, are the most extensive Mesozoic deposits of the offshore of England and Wales. In the southern North Sea three areas of inversion of subsidence developed in late Cretaceous times (the Central Graben Inversion, the Sole Pit Inversion, and the Central Netherlands Inversion), thus thick (± 1,000 m) Upper Cretaceous successions rest on thinner Jurassic successions, whilst thick Jurassic sequences previously deposited in subsiding troughs have only thin Upper Cretaceous successions (Glennie & Boegner 1981, Ziegler 1982). In the Sole Pit area Jenyon (1987) recognised a buried topography at a Trias/Chalk unconformity.

Hancock and Scholle (1975) summarised the sedimentary history of the Chalk successions of the southern North Sea. The calcite of the coccolith plates is of the low magnesium variety which is stable at ordinary pressures and temperatures, protecting it from early lithification. It is believed that the Chalk of the North Sea was deposited in deep water, possibly as deep as 1,000 m in places. All the stages of the Upper Cretaceous are represented in the southern North Sea with nearly half the total thickness represented by the chalks of the Maastrichtian stage alone. Total thicknesses of the Chalk reach 1,400 m east of the Sole Pit Inversion axis. Sequences thin towards the London–Brabant Massif though this feature, which had played a major role since Devonian times, was submerged by the Upper Cretaceous seas and nearly 250 m of Chalk remain on it even after post-Cretaceous erosion.

In the Western Approaches and the English Channel, as indeed in the Celtic Sea Basin, all stages of the Upper Cretaceous are again represented. Throughout, the facies of the Chalk closely resemble those known in the land areas. South of the London–Brabant Massif the Chalk thickens towards the Channel. Detailed accounts of the Cretaceous of SE England and France abound: for a recent comparison see Robaszynski and Amedro (1986). In this area, a number of significant inversions of tectonic movements occurred at this time – in the Weald region and in the central English Channel – all part of the Permian through Cretaceous episodic crustal extension which affected southern England and the Channel region (Stoneley 1982, Chadwick 1986, Beach 1987). The Chalk is widespread in the English Channel, some 250 to 300 m thick in the central and eastern Channel, and 500 and even 800 m in the Western Approaches though there is some thinning over the 'swell' between Start and Cotentin peninsula (Lefort 1975), Curry *et al.* (1970, 1971) and Smith & Curry (1975) suggested that the higher zones of the Upper Cretaceous overstep each other towards the Start–Cotentin line, but more recent evidence suggests a thinning of all zones rather than a series of oversteps. Details of the Chalk in the western Channel were given by Evans *et al.* (1981) based on cores south of Cornwall and west of 8°W: all stages are represented and the lower chalks are usually harder than higher post-Coniacian chalks.

Clearly some form of transgression from the margin of the new Atlantic Ocean seems possible for the western part of the Channel, but there are no reasons to separate the sediments of this Cenomanian transgression from those of other parts of Europe. Weighell *et al.* (1981) showed that the Upper Cretaceous of the Celtic Sea is marine, dominated by chalk. Starting with paralic sands at the base of the Cenomanian, deeper water sediments follow. Up to end-Coniacian the chalks are hard; later chalks are soft. Maastrichtian chalks are the youngest represented. In the borehole at 9°W on the south flank of the Cornubian Platform, however, only Campanian (or just possibly Santonian) to Late Maastrichtian Chalk exists and there may be a hiatus with Lower Maastrichtian sediments missing (Bennet *et al.* 1985). The total thickness of Chalk drilled here is about 400 m.

An unconformity can be identified at the base of the Chalk and traced over much of the Celtic Sea (Bleakley 1985). Below the unconformity records show a folded succession in a series of synclinal basins while between the basins the post-unconformity sediments rest on what appears to be the basement. Strata of the Upper Cretaceous are not known in the St George's Channel, Bristol Channel, and Irish Sea (Manx–Furness) Basins, though the possibility of Upper Cretaceous rocks being found in these areas, even as thin and interrupted successions, cannot be dismissed entirely.

# Tertiary

Off England and Wales, Danian strata occur in both the southern North Sea and in the western part of the English Channel and its Western Approaches. Always of more restricted areal extent than the preceding Maastrichtian stage strata, the Danian chalk of the North Sea is limited to the northern part of the region; it is well documented from the vicinity of the Ekofisk and Dan oil fields and from the subsided Ringkøbing–Fyn High. In the Western Approaches the distinctive Danian calcarenites ( > 100 metres) lie near the centre of the Western Approaches/western English Channel Basin where they are nearly completely overlapped by the succeeding Eocene strata. Farther west there is up to 25 m of post-Danian Palaeocene silty calcareous clays (Evans et al. 1981). In all areas uplift and erosion occurred after the deposition of the Danian.

During Tertiary times the North Sea underwent widespread subsidence permitting more than three kilometres of sediments to be deposited (Nielson et al. 1986) (Fig. 15.17). The maximum subsidence was north of that part of the North Sea which is the subject

of this chapter. The main areas of sediment supply were Scotland and Norway, though at times local sources were significant. At first subsidence was related to reactivated faults causing Upper Cretaceous Chalk to be eroded from the scarps: Jenyon (1987) described buried karstic features in the top of the chalk in the Sole Pit area. Subsequently, broader but still rapid downwarping ensued. The influence of the Mid-North Sea–Ringkøbing–Fyn High was nowhere in evidence but the London–Brabant High continued as a feature in Palaeogene times. Published descriptions of the Tertiary sediments of the North Sea concentrate on conditions and events in the northern North Sea but Ziegler (1978) stated that the Palaeocene and Eocene series of the southern North Sea are represented mainly by clays and silts. From the character of these sediments it follows that subsidence may have exceeded rates of sediment supply in this part of the North Sea. A detailed succession from one site, a 151·8 m deep borehole about 100 km off NE Yorkshire, presented details of the local Palaeogene succession (Lott et al. 1983). Here away from the main area of subsidence there is evidence of transgressions and regressions but the locality was probably exceptional in being relatively positive with little subsidence.

Ash-bearing rocks occur widely (Knox & Ellison 1979, Knox & Morton 1983, Knox 1984) and Jacqué & Thouvenin (1975) correlated these on the basis of dinoflagellate floras. The earliest ash is of Thanetian age but a later, more extensive episode occurred. Dated, by inference, at about 53·5 Ma (Fitch et al. 1978), this is a reliable correlation layer, and as Francis (Chapter 16) states, this must represent an event of magnitude comparable with the explosion of Santorini. Evidence of this event may exist in the English Channel but so far nothing has been recorded.

Regional breaks probably caused by eustatic sealevel changes mark both the base and the top of the Oligocene, though again subsidence was at a faster

**Fig. 15.17.** Section across the southern North Sea Basin showing principal structures and the effect of subsidence during the Tertiary.

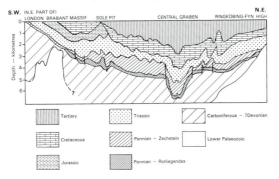

rate in the centre of the region. As in the Eocene, the general character of these sediments is monotonous silts and clays.

From the eastern English Channel, Curry and Smith (1975) described the large extent of Tertiary sediments in the Dieppe Basin, a south-easterly extension of the Hampshire Basin. The succession, which strongly resembles those of the London, Hampshire, and Belgian Basins, but contrasts with the more marine succession of the Paris Basin, commences with a Thanetian limestone which is succeeded by the equivalents of the Woolwich and London Clay Formations, the Bracklesham Group and, possibly, the Barton Formation. There is no evidence of breaks in the succession. Farther west, south of the Isle of Wight, the Palaeogene succession in an outlier starts with the Reading Formation and includes the London Clay and the lowest part of the Bracklesham Group (Curry 1962).

West, beyond the Start–Cotentin swell, there is an extensive area covered by Palaeogene strata. These have been locally described by Curry and co-workers (1965, 1970) and Andreieff and co-workers (1975) though in many places the Tertiary strata, which thicken towards the shelfbreak, have not been differentiated. There is only one recorded Thanetian site and, so far as is known, Early Eocene and early Middle Eocene strata only exist west of 5°W. Late Middle Eocene strata are widespread; all the Eocene sediments are marine and shallow-water limestones characterise the Middle Eocene succession which outcrops extensively in the Bay of St Malo. The remainder of the Palaeogene succession is poorly represented in the Western Channel: Upper Eocene calcareous sandstones occur only locally and the occurrence of true marine Oligocene sediments is in doubt. Two outliers of continental beds have been identified about 60 km WSW of Plymouth but the only known outcrop of Oligocene age, a freshwater limestone, occurs some 100 km west of Guernsey. The possibility of Oligocene strata exists west of 8° (Evans et al. 1981), but the evidence was equivocal until the results of a commercial borehole at approximately 48°47'N 9°06'W were published (Bennet et al. 1985). There, no Palaeocene sediments were found but early Eocene claystones rich in planktonic foraminifers occur, succeeded at a sharp break by a 350 m Middle Eocene sequence of soft glauconitic shelly sandy microfossiliferous limestones which became purer chalky limestones upwards. No Lower Oligocene deposits were recorded but the Late Oligocene is represented by 150 m of calcareous clays with occasional sandstone horizons which were probably deposited in a relatively restricted environment.

In the Celtic Sea, continuous seismic profiling suggested the extensive spread of Tertiary strata in two synclinal areas though more recent work (see below) has questioned this, preferring some deposits to be of Quaternary age. In both cases the Tertiary, possibly Neogene, strata have been said to rest conformably on the Upper Cretaceous successions (Hamilton & Blundell 1971, and Doré 1976) and there is no evidence of any Danian strata in this area. Naylor and Mounteney (1975) suggested that the chalk sea environment of deposition persisted through the Palaeocene and into the Eocene. Ziegler (1978) stated that there is no break, the Laramide deformation (an early Tertiary set of movements, generally quite widespread in NW Europe) not being identified. However, Weighell and co-workers (1981) wrote that the end of the Cretaceous regression caused erosion particularly over the more positive features. Tertiary sediments are said to mask the Upper Cretaceous rocks: west of 8°W the Tertiary strata thicken significantly, but east of 8°W they tend to be patchy. Some doubt, then, remains concerning the exact nature of the Mesozoic–Cenozoic boundary and this may, in part, be due to Tertiary faulting, some of it on NW–SE lines. Perhaps towards the centres of the basins, conformable successions exist, whilst towards the margins unconformities with overstepping occur. On this basis Danian sediments, of limited areal extent, may exist towards the centres of the present synclines.

The Eocene strata are well bedded and show up well on continuous seismic profiles. The few samples recovered show clays of early Eocene age and lignitic sands and silts of late Eocene or Oligocene age. Several authors suggest a break between the Eocene and Oligocene successions with a short period of earth movements related to the movement of Iberia (Masson & Parson 1983) and erosion. Delanty et al. (1980) postulated Oligocene sediments in the Nymphe Bank area and Dobson and Whittington (1979) and Whittington et al. (1981) favoured them in the south and central Irish Sea and the Kish Bank Basin. Non-marine deposits appear to characterise the Oligocene sediments hereabouts.

Though the Eocene successions thin eastwards and northwards in the Celtic Sea, St George's Channel, and Cardigan Bay regions and exhibit features of shallower water marine deposition, the Oligocene is thick in Cardigan Bay and always non-marine. A similar facies is postulated in the St George's Channel Basin and for some way south-westwards.

Fletcher (1975) described an outlier of mid-Oligocene sediments found in the Stanley Bank Basin, in the Bristol Channel just east of Lundy. The non-marine deposits are preserved in a sag basin on the eastern side of a dextral wrench fault aligned northwest–south-east, the Sticklepath/Lustleigh fault zone (Holloway & Chadwick 1986). The deposits are of limited extent and the original area of deposition may not have been much greater than that of the present basin. Tertiary deposits are thin on the Irish Sea Platform and into the Irish Sea. No Tertiary deposits have been described from the Manx–Furness Basin.

In the southern North Sea the Neogene strata are composed of shallow-marine and paralic sands and clays (the clays predominating), with some lignites, especially in the Pliocene. The total thickness may be of the order of about 200 m though thicknesses vary and the evidence suggests that major salt structures were still moving in the Neogene. Details are missing, for as Curry and co-workers (1978) stated, 'the Neogene deposits of the North Sea are too deep to be of interest in connection with platform and pipeline foundation problems and too shallow to be of interest for oil and gas production'.

No Miocene or Pliocene strata have been recorded in the central and eastern parts of the English Channel. After some late Oligocene earth movements, the area seems to have been subjected to mild uplift. There is disconformity between the Palaeogene and the Neogene in the western part of the English Channel and the Western Approaches. Aquitanian or Burdigalian Miocene rests on late Lutetian or earlier Eocene. The Miocene, which reaches a thickness of more than 100 m, has been described by Curry et al. (1962, 1965). It is represented by *Globigerina* silts; the contained nannoplankton and foraminifera have enabled Martini (1974), Jenkins (1977) and Jenkins and Martini (1986) to give the samples Burdigalian and Langhian ages. There is evidence of transgression from the west. At 9°W in the borehole on the south flank of the Cornubian Platform, the Miocene is represented by a succession of soft glauconitic calcareous clays with occasional limestone and sandstone beds. Evidence of Early, Middle and Late Miocene dates exists in the rich assemblage of planktonic globigerinid foraminifers. The thickness of the Miocene strata drilled is about 200 m though the precise nature and depth of the Miocene/Oligocene boundary has not been determined. Curry et al. (1962) reported Miocene chalks dredged from several localities beyond the shelf break off the Western Approaches but, though Eocene sediments were found at one site, no Oligocene strata were reported. Warrington and Owens (1977) reported similar sediments in the Celtic Sea Basin though the extensive Neogene deposits based on the interpretation of seismic records (Blundell et al. 1971, Doré 1976, Blundell 1979) may in fact be of Quaternary age. In the St George's Channel Basin and into Cardigan Bay, early Miocene sediments are postulated by Whittington (1980) and he believed these to be a non-marine continuation of the Oligocene sediments. His evidence was based on seismic records and information from the Mochras borehole. On other evidence, mainly from oil drilling, he stated that marine

Miocene and indeed Pliocene sediments may have been deposited in Cardigan Bay and the St George's Channel but lost during glacial erosional episodes.

Curry and co-workers (1970) described a sample of Plio-Pleistocene age collected at a position about 100 km due south of Land's End. This marine clay contained *Globigerina bulloides* and *G. pachyderma*. More extensive Pliocene deposits may exist towards the margin of the continental shelf, but these lie beneath a cover of recent sediments. There is no information to suggest Miocene or Pliocene deposition in the Manx–Furness Basin of the Irish Sea or in the Solway Basin and it would appear that the region was one of slight uplift and erosion in Neogene times.

Tertiary volcanic activity in the NW British Igneous Province and Ireland can be linked to dykes inferred beneath the sea floor around England and Wales (Kirton & Donato 1985). In the North Sea a belt of devolatised coals may be associated with dykes radiating from the Mull centre. These coals reach southeastwards across the North Sea from Northumberland. Dykes related to an Irish centre occur in the eastern Irish Sea off SW Cumbria and west and northwest of Anglesey. The Lundy Granite and the dykes to the NW are part of this episode. The dyke swarms are thought to relate to NW–SE compressive stress. The same stress system is the cause of the fracturing affecting the Upper Cretaceous and Palaeogene rocks of S England and N France and, presumably, the rocks on the floor of the E English Channel (Bevan & Hancock 1986).

## Quaternary

The Pleistocene geology of NW Europe is dominated by glacial events even though these occupied only the last 600,000 years. As described by Boulton (Chapter 14), the pre-glacial Pleistocene history goes unrepresented in much of Britain, there being several major hiatuses. However, on the Dutch side of the North Sea the pre-glacial Pleistocene history is almost complete and borehole information suggests that the same may be true for much of the central portion of the North Sea and the Southern Bight (Stoker et al. 1983). This deposition was close to the depo-centres which had been active throughout the Tertiary, and Caston (1979) has shown that in several places more than 1 km thickness of Pleistocene sediments exist. Elsewhere around Britain, Pleistocene deposits rarely exceed 100 m thickness except when they occupy over-deepened hollows and nearly everywhere the sediments are those of glacial or interglacial stages. Only the English Channel shows an apparent absence of Quaternary deposits.

The Quaternary of the North Sea was described by McCave and co-workers (1977) with more complete reviews by Oele and co-workers (eds.) in 1979, Stoker *et al.* (1986) and Cameron *et al.* (1987). Offshore data on Early and Middle Pleistocene deposits are scarce from the North Sea; indeed the same is true of the other seas around England and Wales, but it would appear that the major factor was the rapid growth of the Rhine delta (Zagwijn 1979), which gives a complete succession as the distributaries spread sediment westwards and northwards. The deltaic and associated sediments exceed 1 km in thickness in places. Cameron *et al.* (1987) described northward prograding Lower and early Middle Pleistocene sediments up to 500 m thick across the southern North Sea. They describe a late Middle Pleistocene unconformity after which glacial erosion and sedimentation processes predominate. Deep valleys were cut, some as plunge pools (see Wingfield 1990), in three recognisable Middle and Upper Pleistocene glaciations. These valleys are filled

by glacio-lacustrine, glacio-marine and interglacial sediments. Saalian (Wolstonian) deposits are said to be widespread on the sea floor beneath a thin cover of Holocene sediments.

The Weichselian (Devensian) glaciation was not as extensive as the Saalian and reached only to the vicinity of the Dogger Bank (Fig. 15.18). Oele *et al.* (1979) suggested that much of the area may have been dry land, Stoker and Long (1984), however, considered that sea-ice may have been involved. The reasons for sea-level change may be many, but principally they must be glacio-isostatic readjustments and eustatic sea-level changes.

In the main part of the English Channel, Pleistocene deposits are missing entirely. Kellaway *et al.* (1975) favoured a major glaciation during the Saalian with ice moving eastwards to breach the Dover Strait. Though numerous features exist in the Channel which remain

**Fig. 15.18.** Limits of the Quaternary glaciation.

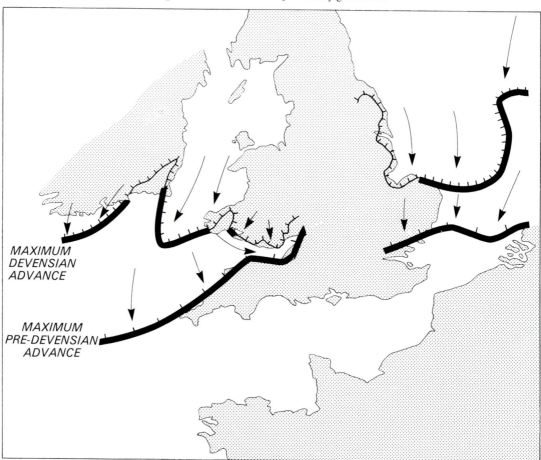

enigmas, including the occurrence at Selsey Bill on the south coast of England of erratics believed to be derived from Brittany, the glacial theory is generally dismissed (Kidson & Bowen 1976). Some of the extensive palaeovalley fill (see later) may be of youngest Pleistocene age as may the infill of the palaeovalley of the Seine.

In the far south-west of the Western Approaches there is a thin wedge of unconsolidated sediment landward of the shelf break – this, too, may be late Pleistocene.

Farther north in the vicinity of the Labadie, Cockburn and other banks, cores have revealed two Late Pliocene–Quaternary formations composed of sands with gravels and muds (Pantin & Evans 1984), and the banks themselves may have a glacial or fluvioglacial origin. Indeed, if ice reached into the area from Irish glacial centres then the banks themselves may have a glacial origin and may even have been ice-moulded.

Closer to the south coast of Ireland there is clear evidence of two stages of glacial deposits. Delanty and Whittington (1977) reinterpreted what had previously been thought to be Neogene deposits (Blundell et al. 1971) as Quaternary. At least two boulder clay horizons separated by marine clays of an interglacial (Ipswichian) are recognised here and through the St George's Channel into Cardigan Bay. Off north Cornwall and west of Devon, Doré (1976) recognised Pleistocene deposits resting on Neogene strata. Delanty and Whittington would regard these as late glacial deposits resting on interglacial sediments. There seems little doubt that Devensian (Weichselian) deposits are widespread, reaching from Wales across to the south side of the Bristol Channel along the north Devon and Cornish coasts to beyond the Scilly Isles (Kidson 1977a, b).

The Irish Sea and Welsh ice masses scoured the relatively soft Tertiary and Mesozoic strata of Cardigan Bay. At the seafloor, only the evidence of the Devensian ice remains though boreholes have indicated pre-Devensian, interglacial, marine sediments below glacial deposits (Delanty & Whittington 1977).

The late Quaternary history of the Irish Sea basin is dominated by the Devensian ice sheets (Pantin 1977) which gave a general cover of boulder clay on the earlier sediments. Irish, Scottish, and Lake District sources all contributed to the sediments. The possibility of late Quaternary freshwater lake conditions before the establishment of the present marine regime has been commented upon by several authors (see Pantin 1977).

Numerous authors have commented on the post-Devensian rise in sea level around the British Isles and the picture presented is one of an eustatic rise in sea level influenced by regional isostatic readjustments – for a discussion see Tooley (1982) and associated papers in the International Geological Correlation Project 61 report.

## Holocene

In broad terms, and in common with continental shelves everywhere, the floor of the shelf sea around England and Wales deepens only gradually to the shelf break. In detail, however, there exists a wide range of features which render the sea floor far from smooth. In addition to islands and submerged islands, such as Haig Fras (Fig. 15.1) (Smith et al. 1965), there are features which are the products of erosion and deposition. Many of the former are wholly or partially filled in; some of the latter were created before the present regime was established. Today the surface sediment is in motion in response to strong diurnal tidal streams, storm-induced ground swell and, less effectively, temperature-difference-induced water movements (Figs. 15.19, 15.20).

The most detailed knowledge concerns the English Channel. There, the sediment cover is often thin or non-existent in the western province while in the central and eastern provinces there are many local variations in thickness. Modification of Quaternary bar deposits occurred as sea levels changed (Le Fournier 1980). Overall the composition of the sediments is not related to the sea-floor geology, rather it is a mixture of two end members: the robust clastic component, mainly flint, and the organic carbonate component derived mainly from Bryozoa, Echinoidea, Foraminifera, and Mollusca (Larsonneur et al. 1980). The size distribution correlates well with the tidal energy input, the coarser sediment occurring near the areas of higher tidal forces. The sediment distribution pattern of the North Sea (Jarke 1956, Houbolt 1968) does not have the same obvious explanation. There the composition presumably reflects the input from the Quaternary ice sheets and related melt waters and the Rhine system, the whole modified by tidal forces and by near shore processes as sea level rose and fell. Holocene deposition rates are low (Caston 1979). Thicknesses may only be of a few metres except near coasts; Greensmith and Tucker (1973) estimated 36 m of Holocene sediments in the Thames estuary, though special factors such as local subsidence may apply there.

In the Celtic Sea and Bristol Channel the same controls are at work. In the west, carbonate sediments, often fine grained, predominate towards the shelf margin but gravels and coarse sands occur to the north and east. The coarser sediment was derived from ice sheets and meltwaters (Hamilton 1979). In the Bristol Channel, reworked Late Pleistocene deposits comprise much of the sea floor while fine (clay and silt-grade) material, usually marine-derived, predominates in the bays and estuaries (Evans 1981). In

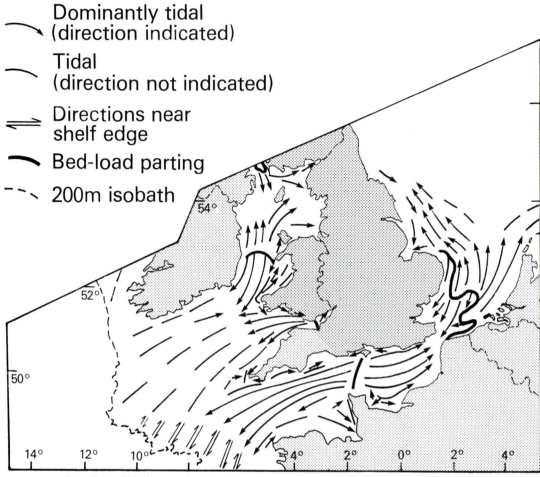

**Fig. 15.19.** Tidally moved sediments around England and Wales after Lee & Ramster (1981).

Cardigan Bay and the Irish Sea similar conditions to those which have existed in the North Sea have led to an assemblage of sediments which reflect the response of late Quaternary deposits to the present energy regime (Pantin 1977).

Stride (1982) reviewed the origin and distribution of bed forms assumed by tidally propelled sediments and concluded that forms are a product of sediment supply and energy regime. In addition to the sandwaves, dunes, and ribbons, older, late Quaternary and early Holocene, banks abound. Large banks exist in the Western Approaches and Celtic Sea, e.g. Labadie and Cockburn Banks, others are known in Cardigan Bay and the Bristol Channel. Other banks exist near the mouth of the Seine and the greatest number occur in the eastern English Channel, the Dover Strait, off eastern and south-eastern England and in the southern North Sea off NE France, Belgium, and Holland. All

of the foregoing are strongly linear. The largest feature, the Dogger Bank, in the middle North Sea, has an irregular shape (Fig. 15.21). The origins of these are still a matter of debate – some bars are related to earlier sea-level stands while others have a glacial or glacially-related origin, being composed of morainic or outwash material.

Elongate depressions in the floor of the continental shelf (deeps, pits, *fosses* of the French) occur in the English Channel, the North Sea, and the western Irish Sea and though partially filled with Recent sediments they make striking features. Their origin has been the subject of much debate. The largest, the Hurd Deep (the Fosse Central of the French) (Figs. 15.1, 21), is 100 km long, 10 km wide at its widest and, though now partially filled, is cut nearly 300 m into the strata of the shelf. Hamilton & Smith (1972) gave a detailed description and reviewed hypotheses presented for its

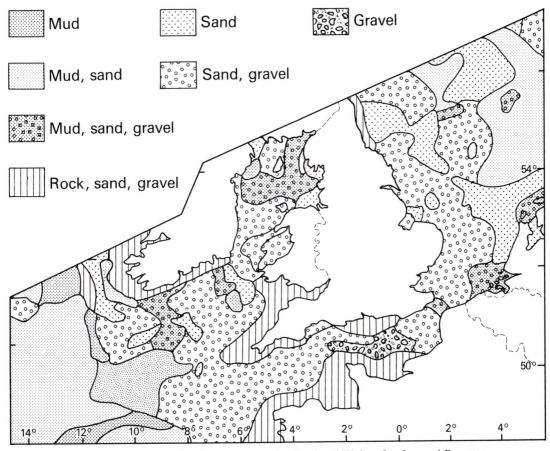

Mud

Mud, sand

Mud, sand, gravel

Rock, sand, gravel

Sand

Sand, gravel

Gravel

**Fig. 15.20.** Sediment distribution around England and Wales after Lee and Ramster (1981).

origin. These include fluviatile erosion, active tectonism, cavern collapse, and tidal scour: none is entirely satisfactory. Kellaway and co-workers (1975) favoured subglacial erosion when the Channel was filled by a glacier. The filled depression in the Dover Strait, the Fosse Dangeard, was explained by the same hypothesis (Destombes *et al.* 1975).

Interpreting continuous seismic profiles has revealed a system of valley-like forms cut in the floor of the central and eastern provinces of the English Channel. Dingwall (1975) linked these into a drainage pattern of what he thought was a late-Tertiary origin. Auffret *et al.* (1980) provided a more detailed map. The system has three main parts, one reaches from the Lobourg Channel, in the Dover Strait, and parallels the south coast of England before swinging SW, a second is in mid-Channel and is more complex: both show links to present-day rivers. The third part reaches from the present mouth of the Seine and all three coalesce at the eastern end of the Hurd Deep.

Parts of the valleys are cut deep, up to 100 m, into the Channel floor, often showing local overdeepening at confluences, and in most instances they are filled with apparently coarse and rudimentarily layered sediments. Smith (1985, 1989) explained most of the features by one mechanism, though he implied that the same mechanism may have worked more than once. He argued that when the northern part of the North Sea was dammed by glacial ice the Dover Strait did not exist, instead a barrier composed of the permafrost-sealed Chalk of the northern limb of the Weald–Artois anticline caused a lake to develop in the Southern Bight of the North Sea. Some water escaped northwards beneath the ice front, creating the Silver Pit and other eroded features in the central North Sea by subglacial erosion, but eventually the Dover Strait ridge was overtopped. Enormous quantities of water were involved, eroding the Fosse Dangeard, which parallels the lower boundary of the Chalk, as a series of plunge pools. As the breach increased the flood waters

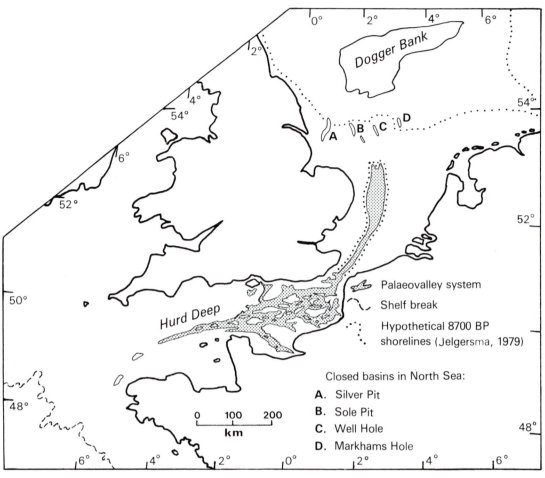

**Fig. 15.21.** The extent of the palaeovalley system around England and Wales based on Smith (1985).

scoured the eastern and central Channel, which at the time of lower sea level was a rolling plain cut by the lower reaches of the rivers established on what is now the land area of southern England and northern France. The flood waters had great erosional power, cutting deeply into the floor of the Channel uninfluenced by the outcrop pattern and modifying existing valleys to create an anastomosing pattern of channels deeply incised at points of confluence. At the major position of confluence, north of the Cotentin Peninsula, the combined erosional power created a markedly overdeepened and at the same time partially infilled feature – the present Hurd Deep. The entire episode, or possibly succession of episodes, Smith argued, was of extremely short, possibly only of weeks, duration. A more 'conventional' origin for the Dover Strait is, however, argued by Gibbard (1988).

A broad trough (see Fig. 15.21) extends northwards into the North Sea from the Dover Strait which is, presumably, related to events in the eastern English Channel. Cornwell and Carruthers (1986) have described a buried valley system found on-shore near the Suffolk coast and D'Olier (1975) similar deep and infilled channels off Suffolk and Essex. He suggested a subglacial 'tunnel valley' origin. A more integrated account of the filled and unfilled valleys of the region has yet to be made, but datum-level changes of a rapid, and perhaps local, nature can be implied.

Farther north a variety of depressions exists in the North Sea between the eastern English coast and the Dogger Bank. Donovan (1972) explained the Silver Pit as due to tidal scour but, as already mentioned, Smith (1985) preferred a sub-glacial origin. Long & Stoker (1986) explained the asymmetrical form of unfilled valleys cut into the Pleistocene deposits of the central North Sea as due to periglacial conditions.

In the western Irish Sea in the Kish Bank area an elongate and partially filled valley system some 60 km long has been described by Whittington (1977) and given a sub-aerial, fluviatile origin. However, while overdeepened valleys can be found close to the shore lines of Wales, south-west England, and southern Ireland, the whole does not appear to be part of an Irish Sea River Valley as envisaged by Hull (1912). Indeed, even the Hurd Deep and the relatively small fosses of the Western Approaches do not form part of an English Channel River Valley in the style proposed by Hull. All these erosional features were presumably cut at times of a lower sea level which formed the limit of effective down-cutting. Tidal scour is not thought to have created these depressions but it can be, locally, an erosive force and will, when conditions are suitable, inhibit the infilling of features created by other agencies.

Now, as described above, the tide redistributes the sediments which are the products of erosion and biogenic productivity.

## REFERENCES

AGER, D. V. & WALLACE, P. — 1966 — The environmental history of the Boulonnais, France. *Proc. Geol. Assoc. London,* **77,** 385–418.

AINSWORTH, N. R. & HORTON, N. F. — 1986 — Mesozoic micropalaeontology of exploration well Elf 55/30–1 from the Fastnet Basin, offshore southwest Ireland. *Journ. of Micropalaeontology,* **5,** 19–29.

AINSWORTH, N. R., HORTON, N. F. & PENNEY, R. A. — 1985 — Lower Cretaceous micropalaeontology of the Fastnet Basin, offshore SW Ireland. *Marine and Petroleum Geology,* **2,** 327–340.

ALLEN, J. R. L. — 1974 — The Devonian rocks of Wales and the Welsh Borderlands. *In* Owen, T. R. (Ed.) *The Upper Palaeozoic and post-Palaeozoic rocks of Wales.* University of Wales Press, Cardiff, 47–84.

ALLEN, P. — 1981 — Pursuit of Wealden models. *J. geol. Soc. London,* **133,** 375–405.

ANDREIEFF, P. AND SIX OTHERS — 1975 — The stratigraphy of the post-Palaeozoic sequences in part of the Western Channel. *Philos. Trans. R. Soc. London, Ser. A,* **279,** 79–97.

ANON — 1985 — Offshore operating companies. *Irish Offshore Review Newsletter.* 16/05/85, p. 2.

AUFFRET, J. P., ANDRE, D., LARSONNEUR, C. & SMITH, A. J. — 1980 — Cartographie du réseau de palaeovallées at de l'épaisseur des formations superficiels meubles de la Manche orientale. *Ann. Inst. océanogr. Paris,* **56,** 21–35.

AVEDIK, F. — 1975a — Seismic refraction survey in the Western Approaches. Preliminary results. *Philos. Trans. R. Soc. London, Ser. A.,* **279,** 29–40.

— 1975b — The seismic structure of the Western Approaches and the South Armorican continental shelf and its geological interpretation. *In* Woodland, A. W. (Ed.) *Petroleum and the continental shelf of North-west Europe.* Applied Science Publishers, London, 29–44.

BAIN, J. A. — 1986 — British geological maps and their availability. *J. geol. Soc. London,* **143,** 569–576.

BANNER, F. T., COLLINS, M. B. & MASSIE, K. S. (Eds.) — 1979 — *The North-west European shelf seas,* Volumes I & II. Elsevier, Amsterdam, 300 pp.

BARR, K. W., COLTER, V. S. & YOUNG, R. — 1981 — The geology of the Cardigan Bay–St George's Channel Basin. *In* Illing, L. V. & Hobson, G. D. (Eds.) *Petroleum and the Continental shelf of North-west Europe.* Heyden & Son Ltd., London, 432–443.

BARTON, P. J. &
MATTHEWS, D. H. — 1984 — Deep structure and geology of the North Sea region interpreted from a seismic refraction profile. *Ann. geophys.*, **2**, 663–668.

BARTON, P. J. &
WOOD, R. — 1984 — Tectonic evolution of the North Sea Basin, crustal stretching and subsidence. *Geophys. J. R. astron. Soc. London*, **79**, 987–1022.

BEACH, A. — 1987 — A regional model for linked tectonics in North West Europe. *In* Brooks, J. & Glennie, K. W. (Eds.). *Petroleum Geology of North West Europe*. Graham & Trotman, London, 43–48.

BEAMISH, D. &
SMYTHE, D. K. — 1986 — Geophysical images of the deep crust: the Iapetus Suture. *J. geol. Soc. London*, **143**, 489–497.

BENNET, G.,
COPESTAKE, P. &
HOOKER, N. P. — 1985 — Stratigraphy of the Britoil 72/10–1A well, Western Approaches. *Proc. Geol. Assoc. London*, **96**, 255–261.

BEVAN, T. G. &
HANCOCK, P. L. — 1986 — A late Cenozoic regional mesofracture system in southern England and northern France. *J. geol. Soc. London*, **143**, 355–362.

BIRPS & ECORS — 1986 — Deep seismic reflection profiling between England, France and Ireland. *J. geol. Soc. London*, **143**, 45–52.

BLENKINSOP, T. G.,
LONG, R. E.,
KUSZNIR, N. J. &
SMITH, M. J. — 1986 — Seismicity and tectonics in Wales. *J. geol. Soc. London*, **143**, 327–334.

BLUNDELL, D. J. — 1979 — The geology and structure of the Celtic Sea. *In* Banner, F. T., Collins, M. B. & Massie, K. S. (Eds.) *The North-west European shelf seas: I. Geology and sedimentology*. Elsevier, Oxford, 43–60.

BLUNDELL, D. J., DAVEY,
F. J. & GRAVES, L. J. — 1971 — Geophysical surveys over the south Irish Sea and Nymphe Bank. *J. geol. Soc. London*, **127**, 339–375.

BOTT, M. H. P. — 1968 — The geological structure of the Irish Sea Basin. *In* Donovan, D. T. (Ed.) *The geology of shelf seas*. Oliver & Boyd, Edinburgh, 93–113.

BOTT, M. H. P. &
YOUNG, D. G. G. — 1971 — Gravity measurements in the north Irish Sea. *J. geol. Soc. London*, **126**, 413–434.

BRENNAND, T. P. — 1975 — The Triassic of the North Sea. *In* Woodland. A. W. (Ed.) *Petroleum and the continental shelf of North-west Europe*. Applied Science Publishers, London, 295–313.

BROOKS, J. &
GLENNIE, K. W. (Eds.) — 1987 — *Petroleum Geology of North West Europe*. Graham & Trotman, London. 1 – 598 pp; 2 – 599–1219.

BROOKS, J.,
GOFF, J. C. &
van HOORN, B. (Eds.) — 1986 — Habitat of Palaeozoic gas in NW Europe. *Geol. Soc. London Special Publication*, **23**. Scottish Academic Press, Edinburgh, 276 pp.

BROOKS, M.,
TRAYNER, P. M. &
TRIMBLE, T. J. — 1988 — Mesozoic reactivation of Variscan thrusting in the Bristol Channel area. *J. geol. Soc. London*, **145**, 439–444.

BUSHELL, T. P. — 1986 — Reservoir geology of the Morecambe field. *In* Brooks, J., Goff, J. C. & van Hoorn, B. (Eds.) *Habitat of Palaeozoic gas in NW Europe. Geol. Soc. London Spec. Publication*, **23**. Scottish Academic Press, Edinburgh, 189–208.

CAMERON, T. D. J.,
STOKER, M. S. & LONG, D. — 1987 — The history of Quaternary sedimentation in the UK sector of the North Sea Basin. *J. geol. Soc. Lond.*, **144**, 43–58.

CASTON, V. N. D. — 1979 — The Quaternary sediments of the North Sea. *In* Banner, F. T., Collins, M. B. & Massie, K. S. (Eds.) *The North-west European shelf seas: I. Geology and sedimentology*. Elsevier, Oxford, 195–270.

CASTON, V. N. D., DEARNLEY, R., HARRISON, R. K., RUNDLE, C. C. & STYLES, M. T. 1981 Olivine-dolerite intrusions in the Fastnet Basin. *J. geol. Soc. London*, **138**, 31–46.

CHADWICK, R. A. 1986 Extension tectonics in the Wessex Basin, southern England. *J. geol. Soc. London*, **143**, 465–488.

COLIN, J. P., LEHMANN, R. A. & MORGAN, P. E. 1981 Cretaceous and Late Jurassic biostratigraphy of the North Celtic Sea basin, offshore southern Ireland. *In* Neale, J. W. & Brasier, M. D. (Eds.) *Microfossils from Recent and Fossil shelf seas.* Ellis Horwood Ltd., 122–155.

COLTER, V. S. 1978 Exploration for gas in the Irish Sea. *Geol. Mijnbouw*, **57**, 503–516.

COLTER, V. S. & BARR, K. W. 1975 Recent developments in the geology of the Irish Sea and Cheshire basin. *In* Woodland, A. W. (Ed.) *Petroleum and the continental shelf of North-west Europe.* Applied Science Publishers, London, 61–76.

COPE, J. C. W. & BASSETT, M. G. 1987 Sediment sources and Palaeozoic history of the Bristol Channel area. *Proc. Geol. Assoc. London*, **98**, 315–330.

CORNWELL, J. D. & CARRUTHERS, R. M. 1986 Geophysical studies of a buried valley system near Ixworth, Suffolk. *Proc. Geol. Assoc. London*, **97**, 357–364.

CURRY, D. 1962 A Lower Tertiary outlier in the central English Channel with notes on the beds surrounding it. *Q. J. geol. Soc. London*, **118**, 177–205.

1989 The rock floor of the English Channel and its significance for the interpretation of marine unconformities. *Proc. Geol. Assoc. London*, **100**, 339–352.

CURRY, D. & SMITH, A. J. 1975 New discoveries concerning the geology of the central and eastern parts of the English Channel. *Philos. Trans. R. Soc. London, Ser. A.*, **279**, 155–167.

CURRY, D., HAMILTON, D. & SMITH, A. J. 1970 Geological and shallow subsurface geophysical investigations in the Western Approaches to the English Channel, *Rep. Inst. geol. Sci. London*, **70/3**, 12 pp.

1971 Geological evolution of the western English Channel and its relation to the nearby continental margin. *In* Delany, F. M. (Ed.) The Geology of the East Atlantic continental margin. *Rep. Inst. geol. Sci. London*, **70/14**, 129–142.

CURRY, D., MURRAY, J. W. & WHITTARD, W. F. 1965 The geology of the Western Approaches to the English Channel, III, The Globigerina Silts and associated rocks. *Colston Pap.*, **17**, 239–264.

CURRY, D., MARTINI, E., SMITH, A. J. & WHITTARD, W. F. 1962 The geology of the Western Approaches of the English Channel, I. Chalky rocks from the upper reaches of the continental slope. *Philos. Trans. R. Soc. London, Ser. B*, **245**, 267–290.

CURRY, D. AND SIX OTHERS 1978 A correlation of Tertiary rocks in the British Isles. *Spec. Rep. geol. Soc. London*, **12**, 72 pp.

DAVIES, G. R. 1984 Isotopic evolution of the Lizard Complex. *J. geol. Soc. London*, **141**, 3–14.

DAY, G. A. & EDWARDS, J. W. F. 1984 Variscan thrusting in the basement of the English Channel and SW Approaches. *Proc. Ussher Soc.*, **5**, 432–436.

DAY, A. A., HILL, M. N., LAUGHTON, A. S. & SWALLOW, J. C. 1956 Seismic prospecting in the Western Approaches of the English Channel. *Q. J. geol. Soc. London*, **112**, 15–42.

DELANTY, L. J. & WHITTINGTON, R. J. 1977 A re-assessment of the 'Neogene' deposits of the south Irish Sea and Nymphe Bank. *Mar. Geol.*, **24**, M23–M30.

DELANTY, L. J., WHITTINGTON, R. J. & DOBSON, M. R.    1980    The geology of the North Celtic Sea west of 7° longitude. *Proc. R. Irish Acad.*, **81B**, 37–54.

DESTOMBES, J. P., SHEPHARD-THORN, E. R. & REDDING, J. H.    1975    A buried valley system in the Strait of Dover with appendix on pollen analysis by M. T. Morzadec-Kerfourn. *Philos. Trans. R. Soc. London, Ser. A*, **279**, 243–256.

DIMITROPOULOS, K. & DONATO, J. A.    1983    The gravity anomaly of the St George's Channel Basin, southern Irish Sea – a possible explanation in terms of salt migration. *J. geol. Soc. London*, **140**, 239–244.

DINGLE, R. V.    1971    A marine geological survey off the north-east coast of England (western North Sea). *J. geol. Soc. London*, **127**, 303–338.

DINGWALL, R. G.    1975    Sub-bottom infill channels in an area of the eastern English Channel. *Philos. Trans. R. Soc. London, Ser. A*, **279**, 233–242.

DINGWALL, R. G. & LOTT, G. K.    1979    I.G.S. boreholes drilled from M.V. *Whitethorn* in the English Channel 1973–75. *Rep. Inst. geol. Sci. London*, **79/8**, 45 pp.

DIXON, J. E., FITTON, J. G. & FROST, R. T. C.    1981    The tectonic significance of post-Carboniferous igneous activity in the North Sea Basin. *In* Illing, L. V. & Hobson, G. D. (Eds.) *Petroleum geology of the continental shelf of North-west Europe*. Heyden & Son Ltd., London, 121–137.

DOBSON, M. R.    1979    Aspects of the post Permian history of the aseismic continental shelf to the west of the British Isles. *In* Banner, F. T., Collins, M. B. & Massie, K. S. (Eds.) *The North-west European shelf seas: I. Geology and sedimentology*. Elsevier, Oxford, 25–41.

DOBSON, M. R. & WHITTINGTON, R. J.    1979    The geology of the Kish Bank Basin. *J. geol. Soc. London*, **136**, 243–249.

DOBSON, M. R., EVANS, W. E. & WHITTINGTON, R. J.    1973    The geology of the South Irish Sea. *Rep. Inst. geol. Sci. London*, **77/25**, 36 pp.

D'OLIER, B.    1975    Some aspects of Late Pleistocene–Holocene drainage of the River Thames in the eastern part of the London Basin. *Philos. Trans. R. Soc. London, Ser. A*, **279**, 269–277.

DONATO, J. A. & TULLY, M. C.    1981    A regional interpretation of North Sea gravity data. *In* Illing, L. V. & Hobson, G. D. (Eds.) *Petroleum geology of the Continental Shelf of North-west Europe*. Heyden & Son Ltd., London, 65–75.

DONATO, J. A., MARTINDALE, W. & TULLY, M. C.    1983    Buried granites within the Mid North Sea High. *J. geol. Soc. London*, **140**, 825–837.

DONOVAN, D. T.    1968    The geology of the continental shelf around Britain. *In* Donovan, D. T. (Ed.) *Geology of Shelf Seas*. Oliver & Boyd, Edinburgh, 1–14.

   1972    The geology and origins of the Silver Pit and other basins in the North Sea. *Proc. Yorkshire geol. Soc.*, **39**, 267–293.

DORE, A. G.    1976    Preliminary geological interpretation of the Bristol Channel approaches. *J. geol. Soc. London*, **132**, 453–459.

DUNHAM, K. C. & SMITH, A. J.    1975    A discussion on the geology of the English Channel. *Philos. Trans. R. Soc. London, Ser. A*, **279**, 295 pp.

EAMES, T. D.    1975    Coal rank and gas source relationships – Rotliegendes reservoirs. *In* Woodland, A. W. (Ed.) *Petroleum and the continental shelf of North-west Europe*. Applied Science Publishers, London, 191–201.

EBBERN, J. 1981 The geology of the Morecambe gas field. *In* Illing, L. V. & Hobson, G. D. (Eds.), *Petroleum geology of the continental shelf of North-west Europe*. Heyden & Son Ltd., London, 485–493.

EVANS, C. D. R. 1981 The geology and superficial sediments of the Inner Bristol Channel and Severn Estuary in *The Severn Barrage*, Inst. Civ. Engineers, London, 27–34.

EVANS, C. D. R., LOTT, G. K. & WARRINGTON, G. 1981 The Zephyr (1977) wells, South-Western approaches and Western English Channel. *Rep. Inst. geol. Sci. London*, **81/8**, 44 pp.

EVANS, D. J. & THOMPSON, M. S. 1979 The geology of the central Bristol Channel and the Lundy area, South Western Approaches, British Isles. *Proc. Geol. Assoc. London*, **90**, 1–14.

FISHER, M. J. & JEANS, C. V. 1982 Clay mineral stratigraphy in the Permo-Triassic Red Bed sequences of BNOC 72/10–1A, Western Approaches and the South Devon coast. *Clay Miner.*, **17**, 79–89.

FITCH, F. J., HOOKER, P. J., MILLER, J. A. & BRERETON, N. R. 1978 Glauconite dating of Palaeocene–Eocene rocks from East Kent and the time-scale of Palaeogene volcanism in the North Atlantic region. *J. geol. Soc. London*, **135**, 499–512.

FLETCHER, B. N. 1975 A new Tertiary basin east of Lundy Island. *J. geol. Soc. London*, **131**, 223–225.

FROST, R. T. C., FITCH, F. J. & MILLER, J. A. 1981 The age and nature of the crystalline basement of the North Sea basin. *In* Illing, L. V. & Hobson, G. D. (Eds.) *Petroleum geology of the continental shelf of North-west Europe*. Heyden & Son Ltd., 43–57.

FYFE, J. A., ABBOTTS, I. & CROSBY, A. 1981 The subcrop of the mid-Mesozoic unconformity in the U.K. area. *In* Illing, L. V. & Hobson, G. D. (Eds.) *Petroleum geology of the continental shelf of North-west Europe*. Heyden & Son Ltd., London, 236–244.

GARDINER, P. R. R. & SHERIDAN, D. J. R. 1982 Tectonic framework of the Celtic Sea with special reference to the location of the Variscan Front. *J. Struct. Geol.*, **3**, 317–332.

GIBBARD, P. L. 1988 The history of the great northwest European rivers during the last three million years. *Philos. Trans. R. Soc. London, Ser. B.*, **318**, 559–602.

GIBBS, A. D. 1983 Constraints on seismo-tectonic models for the North Sea. *In* Ritsema, A. R. & Gurpinar, A. (Eds.) *Seismicity and Seismic risk in the offshore North Sea Area*. D. Reidel, 31–34.

1986 Strike-slip basins and inversions: a possible model for southern North Sea gas areas. *In* Brooks, J., Goff, J. C. & van Hoorn, B. (Eds.) Habitat of Palaeozoic gas in NW Europe. *Geol. Soc. London Spec. Publication*, **23**. Scottish Academic Press, Edinburgh, 23–25.

1987 Basin development, examples from the United Kingdom and comments on hydrocarbon prospectivity. *Tectonophysics*, **133**, 189–198.

GLENNIE, K. W. 1972 Permian Rotliegendes of North-West Europe interpreted in the light of modern desert sedimentation studies. *Bull. Amer. Assoc. Petrol. Geol.*, **56**, 1046–1071.

1984 *Introduction to the Petroleum Geology of the North Sea*. Blackwells, Oxford, 236 pp.

GLENNIE, K. W.                    1986   Development of NW Europe's Southern Permian gas basin. *In* Brooks, J. *et al.* (Eds.) Habit of Palaeozoic gas in NW Europe. *Geol. Soc. London Spec. Publication*, **23**. Scottish Academic Press, Edinburgh, 3–22.

GLENNIE, K. W. &                  1981   Sole-Pit inversion tectonics. *In* Illing, L. V. & Hobson,
  BOEGNER, P. L. K.                       G. D. (Eds.) *Petroleum geology of the continental shelf of North-west Europe.* Heyden & Son Ltd., London, 110–120.

GREENSMITH, J. T. &               1973   Holocene transgressions and regressions on the Essex coast
  TUCKER, E. N.                           outer Thames estuary. *Geol. Mijnbouw*, **52**, 193–202.

HAILWOOD, E. A.,                  1984   Palaeomagnetism of the Lizard Complex, S.W. England.
  GASH, P. T. R.,                         *J. geol. Soc. London*, **141**, 27–35.
  ANDERSON, P. C. &
  BADHAM, J. P. N.

HALLAM, A. &                      1968   Origin of Fuller's Earth in the Mesozoic of southern
  SELLWOOD, B. W.                         England. *Nature, London*, **220**, 1193–1195.

                                  1976   Middle Mesozoic sedimentation in relation to tectonics in the British area. *J. Geol.*, **84**, 301–321.

HAMILTON, D.                      1979   The geology of the English Channel, South Celtic Sea and Continental margin, south-western approaches. *In* Banner, F. T., Collins, M. B. & Massie, K. S. (Eds.) *The North-west European Shelf Seas: I. Geology and sedimentology.* Elsevier, Oxford, 61–87.

HAMILTON, D. &                    1971   Submarine geology of the approaches to the Bristol Chan-
  BLUNDELL, D. J.                         nel. *Proc. geol. Soc. London*, **1664**, 297–300.

HAMILTON, D. &                    1972   The origin and sedimentary history of the Hurd Deep,
  SMITH, A. J.                            English Channel. Colloque sur la géologie de la Manche. *Mém. Bur. Rech. géol. minières*, **79**, 59–78.

HANCOCK, J. M. &                  1975   Chalk of the North Sea. *In* Woodland, A. W. (Ed.)
  SCHOLLE, P. A.                          *Petroleum and the continental shelf of North-west Europe.* Applied Science Publishers, London, 413–428.

HARRISON, R. K.,                  1979   Mesozoic igneous rocks, hydrothermal mineralisation and
  JEANS, C. V. &                          volcanogenic sediments in Britain and adjacent regions.
  MERRIMAN, R. J.                         *Bull. geol. Surv. G.B.*, **70**, 57–69.

HILL, M. N. &                     1953   Seismic prospecting in the English Channel and its geo-
  KING, W. B. R.                          logical interpretation. *Q. J. geol. Soc. London*, **109**, 1–19.

HILLIS, R. R. &                   1987   Deep events in UK Southwestern Approaches. *Geophys J.*
  DAY, G. A.                              *R. astron. Soc. London*, **89**, 243–250.

HOBBS, R. W.,                     1987   Is lower crustal layering related to extension? *Geophys. J.*
  PEDDY, C. &                             *R. astron. Soc. London*, **89**, 239–242.
  BIRPS GROUP

HOLLOWAY, S. &                    1986   The Sticklepath–Lustleigh fault zone: Tertiary sinistral
  CHADWICK, R. A.                         reactivation of a Variscan dextral strike-slip fault. *J. geol. Soc. London*, **143**, 447–452.

HOUBOLT, J. J. H. C.              1968   Recent sediments in the Southern Bight of the North Sea. *Geol. Mijnbouw*, **47**, 245–273.

HOWITT, F.,                       1975   The occurrence of Jurassic volcanics in the North Sea. *In*
  ASTON, E. R. &                          Woodland, A. W. (Ed.) *Petroleum and the continental shelf
  JACQUE, M.                              of North-west Europe.* Applied Science Publishers, London, 379–388.

HULL, E.                          1912   *Monograph on the Sub-oceanic physiography of the North Atlantic Ocean.* Stanford, London. 41 pp & 11 pls.

I.G.S. (Institute of              1972   Boreholes. *Annual Report for 1971.* Inst. geol. Sci. London,
  Geological Sciences)                    128–129.

ILLING, L. V. &                   1981   *Petroleum geology of the continental shelf of North-west*
  HOBSON, G. D. (Eds.)                    *Europe.* Heyden & Sons Ltd., 521 pp.

| | | |
|---|---|---|
| ISAAC, K. P., TURNER, P. J. & STEWART, I. J. | 1982 | The evolution of the Hercynides of central S.W. England. *J. geol. Soc. London*, **139**, 521–531. |
| JACQUE, M. & THOUVENIN, J. | 1975 | Lower Tertiary tuffs and volcanic activity in the North Sea. *In* Woodland, A. W. (Ed.) *Petroleum and the continental shelf of North-west Europe.* Applied Science Publishers, London, 455–466. |
| JARKE, J. | 1956 | Der Boden der sudlichen Nordsee. 1. Beitrag: Eine neue Bodenkarte der sudlichen Nordsee. *Dtsch. Hydrogr. Z.*, **9**, 1–9. |
| JELGERSMA, S. | 1979 | Sea-level changes in the North Sea Basin. *In* Oele, E. *et al.* (Eds.) The Quaternary History of the North Sea. *Acta Univ. Uppsala. Symp. Univ. Ups. Annum Quingentesimum Celebrantis* **2**, Uppsala 233–248. |
| JENKINS, D. A. L. & TWOMBLEY, B. N. | 1980 | Review of the petroleum geology of offshore northwest Europe. *Trans. Instn. Min. Metall.* (Spec. Issue – Petroleum), 6–23. |
| JENKINS, D. G. | 1977 | Lower Miocene planktonic foraminifera from a borehole in the English Channel. *Micropalaeontology*, **23**, 297–318. |
| JENKINS, D. G. & MARTINI, E. | 1986 | Age determinations of Neogene sediments from the English Channel and Western Approaches. *J. micropalaeontol.*, **5**, 5–6. |
| JENYON, M. K. | 1987 | Seismic expression of real and apparent buried topography. *J. Petroleum Geol.*, **10**, 41–58. |
| JENYON, M. K., CRESWELL, P. M. & TAYLOR, J. C. M. | 1984 | Nature of the connections between the Northern and Southern Zechstein Basins across the Mid-North Sea High. *Marine Petrol. Geol.*, **1**, 355–363. |
| JONES, D. G., MILLER, J. M. & ROBERTS, P. D. | 1988 | A seabed radiometric survey of Haig Fras, S. Celtic Sea, UK. *Proc. Geol. Assoc. London*, **99**, 193–203. |
| KAMMERLING, P. | 1979 | The geology and hydrocarbon habitat of the Bristol Channel Basin. *J. Petrol. Geol.*, **2**, 75–93. |
| KARNER, G. D., WEISSEL, J. K., DEWEY, J. F. & MUNDAY, T. J. | 1987 | Geotectonic imaging of the north-western European continent and shelf. *Marine Petrol. Geol.*, **4**, 94–102. |
| KELLAWAY, G. A., REDDING, J. H., SHEPHARD-THORN, E. R. & DESTOMBES, J. P. | 1975 | The Quaternary history of the English Channel. *Philos. Trans. R. Soc. London, Ser. A*, **279**, 189–218. |
| KELLING, G. | 1974 | Upper Carboniferous sedimentation in South Wales. *In* Owen, T. R. (Ed.) *The Upper Palaeozoic and post-Palaeozoic rocks of Wales.* University of Wales Press, Cardiff, 185–224. |
| KENT, P. E. | 1975 | Review of North Sea Basin development. *J. geol. Soc. London*, **131**, 435–468. |
| KIDSON, C. | 1977a | Some problems of the Quaternary of the Irish Sea. *In* Kidson, C. & Tooley, M. J. (Eds.) *The Quaternary History of the Irish Sea.* Seel House Press, Liverpool, 1–12. |
| | 1977b | The coast of South-West England. *In* Kidson, C. & Tooley, M. J. (Eds.) *The Quaternary History of the Irish Sea.* Seel House Press, Liverpool, 257–298. |
| KIDSON, C. & BOWEN, D. Q. | 1976 | Some comments on the history of the English Channel. *Quaternary Newsletter*, **18**, 8–10. |

KIRTON, S. R. & DONATO, J. A.    1985    Some buried Tertiary dykes of Britain and surrounding waters deduced by magnetic modelling and seismic reflection methods. *J. geol. Soc. London*, **142**, 1047–1058.

KLEMPERER, S. L. & MATTHEWS, D. H.    1987    Iapetus suture beneath the North Sea by BIRPS deep seismic reflection profiling. *Geology*, **15**, 195–198.

KNOX, R. W. O'B.    1984    Nannoplankton zonation and the Palaeocene/Eocene boundary beds of NW Europe – an indirect correlation by means of volcanic ash layers. *J. geol. Soc. London*, **141**, 993–999.

KNOX, R. W. O'B. & ELLISON, R. A.    1979    A lower Eocene ash sequence in S.E. England. *J. geol. Soc. London*, **136**, 251–254.

KNOX, R. W. O'B. & MORTON, A. C.    1983    Stratigraphical distribution of early Palaeogene pyroclastic deposits in the North Sea Basin. *Proc. Yorkshire geol. Soc.*, **44**, 355–363.

LARSEN, O.    1971    K/Ar age determinations from the Precambrian of Denmark. *Geol. Surv. Denmark, 2nd Series*, 1–34.

LARSONNEUR, C. VASLET, D., AUFFRET, J. P. & BOUYSSE, P.    1980    Les sédiments superficiels de la Manche. *Carte Géologique Bur. Rech. géol. minières*, scale 1 : 500,000.

LEE, A. J. & RAMSTER, J. W.    1981    *Atlas of the seas around the British Isles*. Ministry of Agriculture, Fisheries and Food. 7 pp. + 67 sheets.

LEEDER, M. R. J.    1982    Upper Palaeozoic basins of the British Isles – Caledonide inheritance versus Hercynian plate margin processes. *J. geol. Soc. London*, **139**, 479–491.

LEEDER, M., RAISWELL, R., AL-BIATTY, H., McMAHON, A. & HARDMAN, M.    1990    Carboniferous stratigraphy, sedimentation and correlation of well 48/3-3 in the southern North Sea Basin. *J. geol. Soc. London*, **147**, 287–300.

LEFORT, J. P.    1975    Le contrôle du socle dans l'évolution de la sédimentation en Manche occidentale après le Paléozoïque. *Philos. Trans. R. Soc. London, Ser. A*, **279**, 137–144.

   1977    Possible 'Caledonian' subduction under the Domnonean domain, North Armorica area. *Geology*, **5**, 523–526.

LE FOURNIER, J.    1980    Modern analogue of transgressive sand bodies off eastern English Channel. *Bull. Centres Rech. Exploration-Production Elf-Aquitaine*, **4**, 99–118.

LE GRAND, R.    1968    Le Massif du Brabant. *Serv. géol. Belg. Mem.*, **9**, 148 pp.

LEVERIDGE, B. E., HOLDER, M. T. & DAY, G. A.    1984    Thrust nappe tectonics in the Devonian of S Cornwall and the western English Channel. *In* Hutton, D. H. W. & Sanderson, D. J. (Eds.) *Variscan tectonics of the North Atlantic region. Spec. Pub. geol. Soc. London*, **14**, 103–112.

LLOYD, A. J., SAVAGE, R. J. G., STRIDE, A. H. & DONOVAN, D. T.    1973    The geology of the Bristol Channel floor. *Philos. Trans. R. Soc. London, Ser. A.*, **274**, 595–626.

LONG, D. & STOKER, M. S.    1986    Valley asymmetry: evidence for periglacial activity in central North Sea. *Earth Surface Processes and Landforms*, **11** (5), 525–532.

LOTT, G. K. & WARRINGTON, G.    1988    A review of the latest Triassic succession in the UK sector of the Southern North Sea Basin. *Proc. Yorkshire geol. Soc.*, **47**, 39–147.

LOTT, G. K., FLETCHER, B. N. & WILKINSON, J. P.    1986    Stratigraphy of the Lower K Speeton Clay formation in a cored borehole off coast of NE England. *Proc. Yorkshire geol. Soc.*, **46**, 39–56.

LOTT, G. K., KNOX, R. W. O'B., HARLAND, R. & HUGHES, M. J. 1983 The stratigraphy of Palaeogene sediments in a cored borehole off the coast of north-east Yorkshire. *Rep. Inst. Geol. Sci. London*, **83/9**, 9.

MCARDLE, P. & KEARY, R. 1986 Offshore coal in the Kish Bank Basin: its potential for commercial exploitation. *Geol. Survey of Ireland Report Series*, **86/3**, Mineral Resources 45.

MCCAVE, I. N., CASTON, V. N. D. & FANNIN, N. G. T. 1977 The Quaternary of the North Sea. *In* Shotton, F. W. (Ed.) *British Quaternary Studies – recent advances*. Clarendon Press, Oxford, 187–204.

MCGEARY, S., WARNER, M. R., CHEADLE M. J. & BLUNDELL, D. J. 1987 Crustal structure of the continental shelf around Britain derived from BIRPS deep seismic profiling. *In* Brooks, J. & Glennie, K. W. (Eds.) *Petroleum Geology of North West Europe*. Graham & Trotman, London, 33–42.

MCQUILLIN, R. 1986 Extensional basins and ancient transcurrent fault zones in UK continental shelf. *In* Nesbitt, R. W. & Nichol, I. (Eds.) *Geology in the real world – the Kingsley Dunham volume*. Inst. Min. Metallurgy, London, 285–293.

MARIE, J. P. P. 1975 Rotliegendes stratigraphy and diagenesis. *In* Woodland, A. W. (Ed.) *Petroleum and the continental shelf of North-west Europe*. Applied Science Publishers, London, 205–211.

MARTINI, E. 1974 Calcareous nannoplankton and the age of the *Globigerina* Silts of the western approaches of the English Channel. *Geol. Mag.*, **111**, 303–306.

MASSON, D. G. & PARSON, L. M. 1983 Eocene deformation on the continental margin SW of the British Isles. *J. geol. Soc. London*, **140**, 913–920.

MASSON, D. G. & ROBERTS, D. G. 1980 Later Jurassic–early Cretaceous reef trends on the continental margin SW of the British Isles. *J. geol. Soc. London*, **138**, 437–443.

MATTHEWS, D. H. 1986 Seismic reflections from the lower crust around Britain. *In* Dawson, J. B., Carswell, D. A., Hall, J. K. & Wedepol, K. H. (Eds.) The nature of the lower continental crust. *Spec. Pub. geol. Soc. London*, **24**. Blackwells, Oxford, 11–22.

MECHIE, K. & BROOKS, M. 1984 A seismic study of deep geological structure in the Bristol Channel area, SW Britain. *Geophys. J. R. astron. Soc. London*, **78**, 661–689.

MILLSON, J. 1987 The Jurassic evolution of the Celtic Sea basins. *In* Brooks, J. & Glennie, K. W. (Eds.). *Petroleum Geology of North West Europe*, Graham & Trotman, London, 599–610.

MILSOM, J. 1986 Small inversion structures on seismic sections from the southern North Sea. *Proc. Geol. Assoc. Lond.*, **97**, 379–382.

MONTADERT, L., ROBERTS, D. G., CHARPAL, O. de & GUENNOC, P. 1979 Rifting and subsidence of the northern continental margin of the Bay of Biscay. *In* Montadert, L., Roberts, D. G. *et al.* (Eds.) *Initial Rep. Deep Sea drill. Proj.*, **48**, US Govt. Printing Office, Washington, 1025–1060.

MONTADERT, L. AND 13 OTHERS 1977 Rifting and subsidence on passive continental margins in the North East Atlantic. *Nature, London*, **268**, 305–309.

NAYLOR, D. & MOUNTENEY, S. N. 1975 *Geology of the North-west European continental shelf*. Graham & Trotman, London, 162 pp.

NAYLOR, D. & SHANNON, P. M. 1982 *The geology of offshore Ireland and West Britain*. Graham & Trotman, London, 161 pp.

NIELSEN, O. B., SORENSEN, S., THIEDE, J. & SKARBO, O. 1986 Cenozoic differential subsidence of the North Sea. *Bull. Amer. Assoc. Petrol. Geol.*, **70**, 276–298.

OELE, E.,                        1979    The Quaternary History of the North Sea. *Acta Univ.*
    SCHUTTENHELM, R. T. E.               *Uppsala, Symp. Univ. Ups. Annum Quingentesimum Cel-*
    & WIGGERS, A. J.                     *brantes*, **2**, Uppsala, 248.

OWEN, T. R.                      1978    *The geological evolution of the British Isles.* Pergamon
                                         Press, Oxford, 160 pp.

PANTIN, H. M.                    1977    Quaternary sediments of the northern Irish Sea. *In* Kidson,
                                         C. & Tooley, M. J. (Eds.) *The Quaternary history of the Irish
                                         Sea.* Seel House Press, Liverpool, 27–54.

PANTIN, H. M. &                  1984    The Quaternary history of the central and southwestern
    EVANS, C. D. R.                      Celtic Sea. *Marine Geology*, **57**, 259–293.

PARKIN, M. &                     1982    Geological results of boreholes drilled on the southern
    CROSBY, A.                           United Kingdom continental shelf by the I.G.S.
                                         1969–1982. *Inst. geol. Sci. Unit Internal Rept.*, **82/2**,
                                         3 vols, 358 pp.

PENN, I. E. &                    1976    The Middle Jurassic (mainly Bathonian) of Cardigan Bay
    EVANS, C. D. R.                      and its palaeogeographical significance. *Rep. Inst. geol. Sci.
                                         London*, **76/6**, 8.

POMEROL, C.                      1972    Colloque sur la géologie de la Manche. *Mém. Bur. Rech.
                                         géol. minières*, **79**, 326 pp.

RAMSBOTTOM, W. H. C.,            1978    A correlation of Silesian rocks in the British Isles. *Spec.
    AND SIX OTHERS                       Rep. geol. Soc. London*, **10**, 81 pp.

RATTEY, P. R. &                  1984    The structure of S.W. Cornwall and its bearing on the
    SANDERSON, D. J.                     emplacement of the Lizard Complex. *J. geol. Soc. London*,
                                         **141**, 87–95.

RHYS, G. H.                      1974    A proposed standard lithostratigraphic nomenclature for
                                         the southern North Sea and an outline structural nomen-
                                         clature for the whole of the (U.K.) North Sea. *Rep. Inst.
                                         geol. Sci. London*, **74/8**, 36 pp.

ROBASZYNSKI, F. &                1986    The Cretaceous of the Boulonnais (France) and a com-
    AMEDRO, F.                           parison with the Cretaceous of Kent (United Kingdom).
                                         *Proc. Geol. Assoc. London*, **97**, 171–208.

ROBINSON, K. W.,                 1981    The Fastnet Basin: An integrated analysis. *In* Illing, L. V. &
    SHANNON, P. M. &                     Hobson, G. D. (Eds.) *Petroleum geology of the continental
    YOUNG, D. G. G.                      shelf of North-west Europe.* Heyden & Son Ltd., London,
                                         444–454.

SELLWOOD, B. W. &                1986    A geological map of the sub-Mesozoic floor beneath
    SCOTT, J.                            Southern England. *Proc. Geol. Soc. London*, **97**, 81–85.

SHACKLETON, R. M.                1984    Thin-skinned tectonics, basement control and the Variscan
                                         Front. *In* Hutton, D. H. W. & Sanderson, D. J. (Eds.)
                                         *Variscan tectonics of the North Atlantic region. Spec. Pub.
                                         geol. Soc. London*, **14**, 125–129.

SHACKLETON, R. M.,               1982    An interpretation of the Variscan structures in SW
    REIS, A. C. &                        England. *J. geol. Soc. London*, **139**, 535–544.
    COWARD, M. P.

SMITH, A. J.                     1984    Structural evolution of the English Channel region. *Bull.
                                         Soc. geol. Nord.*, **103**, 253–264.

                                 1985    A catastrophic origin for the palaeovalley system of the
                                         eastern English Channel. *Marine Geology*, **64**, 65–75.

                                 1989    The English Channel – by geological design or catastrophic
                                         accident? *Proc. Geol. Assoc. London*, **100**, 325–337.

SMITH, A. J. &                   1975    The structure and geological evolution of the English
    CURRY, D.                            Channel. *Philos. Trans. R. Soc. London, Ser. A*, **279**, 3–20.

SMITH, A. J.,                    1965    The geology of the Western Approaches of the English
    STRIDE, A. H. &                      Channel. IV. A recently discovered Variscan granite west
    WHITTARD, W. F.                      northwest of the Scilly Isles. *In* Whittard, W. F. & Brad-
                                         shaw, R. (Eds.) *Submarine Geology and Geophysics. Col-
                                         ston Pap.*, **17**, Butterworths, London, 287–302.

STOKER, M. S. &  1984  A relict ice-scoured erosion surface in the central North
LONG, D.  Sea. *Marine Geology*, **61**, 85–93.

STOKER, M. S.,  1986  *A revised Quaternary Stratigraphy for the central North*
LONG, D. & FYFE, J. A.  *Sea.* British Geological Survey Report **17/2**, 35.

STOKER, M. S.,  1983  Palaeomagnetic evidence for early Pleistocene in the central
SKINNER, A. C.,  and northern North Sea. *Nature, London*, **304**, 332–334.
FYFE, J. A. & LONG, D.

STONELEY, R.  1982  The structural development of the Wessex Basin. *J. geol.
Soc. London*, **139**, 543–554.

STRIDE, A. H. (Ed.)  1982  *Offshore tidal sands.* Chapman & Hall, London, 222 pp.

TAYLOR, J. C. M.  1981  Zechstein facies and petroleum prospects in the central and
northern North Sea. *In* Illing, L. V. & Hobson, G. D. (Eds.)
*Petroleum geology of the continental shelf of North-west
Europe.* Heyden & Son Ltd., London, 176–185.

1986  Prospects in the Variscan thrust province of southern
England. *In* Brooks, J. *et al.* (Eds.) Habitat of Palaeozoic
gas in NW Europe. *Spec. Pub. Geol. Soc. London*, **23**.
Scottish Academic Press, Edinburgh, 37–53.

TAYLOR, J. C. M. &  1975  Zechstein of the English sector of the southern North Sea
COLTER, V. S.  Basin. *In* Woodland, A. W. (Ed.) *Petroleum and the
continental shelf of North-west Europe.* Applied Science
Publishers, London, 249–263.

TOOLEY, M. J.  1982  I.G.C.P. Project 61: Sea-level movements during the last
deglacial hemicycle (about 15000 years). Final report of the
UK Working Group. *Proc. Geol. Assoc. London*, **93**, 3–125.

TUBB, S. R.,  1986  Palaeozoic prospects on the northern flanks of the London–
SOULSBY, A. &  Brabant massif. *In* Brooks, J. *et al.* (Eds.) Habitat of
LAWRENCE, S. R.  Palaeozoic gas in NW Europe. *Spec. Pub. Geol. Soc.
London*, **23**. Scottish Academic Press, Edinburgh, 55–72.

TUNBRIDGE, I. P.  1986  Mid-Devonian tectonics and sedimentation in the Bristol
Channel area. *J. geol. Soc. London*, **143**, 107–116.

VAIL, P. R.  1977  Seismic stratigraphy and global changes in sea level part 6:
stratigraphical interpretation of seismic reflection patterns
in depositional systems. Seismic stratigraphy: applications
to hydrocarbon exploration. *Mem. Amer. Assoc. Petrol.
Geol.*, **26**, 117–133.

WALLACE, P.  1982  The subsurface Variscides of Southern England and their
continuation into Continental Europe. *In* Hancock, P. L.
(Ed.) *The Variscan fold belt in the British Isles.* Adam
Hilger, Bristol, 198–208.

WALMSLEY, P. J.  1987  The British Geological Survey contribution to the exploration of the continental shelf. *J. geol. Soc. London*, **144**,
207–212.

WARRINGTON, G. &  1977  Micropalaeontological biostratigraphy of offshore samples
OWENS, B. (Compilers)  from southwest Britain. *Rep. Inst. geol. Sci. London*, **69/1**,
49.

WARRINGTON, G.  1980  A correlation of Triassic rocks in the British Isles. *Spec.
AND EIGHT OTHERS  Rep. geol. Soc. London*, **13**, 78.

WEIGHELL, A. J.,  1981  Upper Cretaceous geology of the Celtic Sea. The Geo-
DOBSON, M. R.,  logical history of the North Atlantic borderlands. *Mem.
WHITTINGTON, R. J. &  Can. Soc. Petrol. Geol.*, **7**, 727–741.
DELANTY, L. J.

WHITTINGTON, R. J.  1977  A late glacial drainage pattern in the Kish Bank area and
post-glacial sediments in the Central Irish Sea. *In* Kidson,
C. & Tooley, M. J. (Eds.) *The Quaternary History of the
Irish Sea.* Seel House Press, Liverpool, 55–68.

WHITTINGTON, R. J.                 1980      *Geophysical studies in the southern Irish Sea.* Unpublished
                                                    Ph.D. Thesis, University of Wales, 364.

WHITTINGTON, R. J.,               1981      Aspects of the geology of the South Irish Sea. *Geol. J.*, **16,**
    CROCKER, P. F. &                          85–88.
    DOBSON, M. R.

WINGFIELD, R.                      1990      The origin of major incisions within the Pleistocene
                                                    deposits of the North Sea. *Marine Geology*, **91,** 31–52.

WILLIS, L. J.                      1973      A palaeogeological map of the Palaeozoic floor below the
                                                    Permian and Mesozoic formations of England and Wales.
                                                    *Mem. geol. Soc. London*, **17,** 16.

WOODLAND, A. W. (Ed.)             1971      The Llanbedr (Mochras Farm) borehole. *Rep. Inst. geol.
                                                    Sci. London*, **71/18,** 115.

                                   1975      *Petroleum and the continental shelf of North-west Europe
                                                    Vol. 1 Geology.* Applied Science Publishers, London, 501 pp.

ZAGWIJN, W. H.                     1979      Early and Middle Pleistocene coastlines. *Univ. Uppsala,
                                                    Symp. Univ. Ups. Annum Quingentesimum Celbrantes*, **2,**
                                                    31–42.

ZIEGLER, P. A.                     1975a     North Sea Basin history in the tectonic framework of
                                                    North-Western Europe. *In* Woodland, A. W. (Ed.) *Pet-
                                                    roleum and the continental shelf of North-west Europe.*
                                                    Applied Science Publishers, London, 131–149.

                                   1975b     Outline of the geological history of the North Sea. *In*
                                                    Woodland, A. W. (Ed.) *Petroleum and the continental shelf
                                                    of North-west Europe.* Applied Science Publishers, London,
                                                    165–190.

                                   1978      North-Western Europe: Tectonics and basin development.
                                                    *Geol. Mijnbouw*, **57,** 589–626.

                                   1981      Evolution of sedimentary basins in North-West Europe. *In*
                                                    Illing, L. V. & Hobson, G. D. (Eds.) *Petroleum geology of
                                                    the continental shelf of North-west Europe.* Heyden & Son
                                                    Ltd., London, 3–39.

                                   1982      *Geological Atlas of western and central Europe.* Shell Inter-
                                                    national Petroleum Maatschappij B.V., The Hague,
                                                    130 pp. & 40 pls.

                                   1986      Geodynamic model for the Palaeozoic crustal consolida-
                                                    tion of western and central Europe. *Tectonophysics*, **126,**
                                                    303–328.

                                   1990      *Geological Atlas of western and central Europe, 2nd
                                                    Edition.* Shell International Petroleum Maatschappij B.V.,
                                                    The Hague, 244 pp. & 56 pls.

ZWART, H. J. &                     1978      The tectonic framework of Central and Western Europe.
    DORNSIEPEN, U. F.                          *Geol. Mijnbouw*, **57,** 627–654.

# Appendix

## Maps concerned with offshore geology of England and Wales

DUNNING, F. W. (Editor), 1985. Geological structure of Great Britain, Ireland and surrounding seas. *Geol. Soc. London.* 1 sheet, scale 1:3,000,000.

LARSONNEUR, C., VASLET, D., AUFFRET, J. P. & BOUYSSE, P. 1980. Les Sédiments superficiels de la Manche. *Carte Géologique, Bur. Rech. géol. minières.* Scale 1:500,000.

LEE, A. J. & RAMSTER, J. W. 1981. *Atlas of the seas around the British Isles.* Ministry of Agriculture, Fisheries and Food. 7 pp + 67 sheets.

BRITISH GEOLOGICAL SURVEY MAPS at scale of 1:250,000. (Available January 1991.) Information from British Geological Survey, Keyworth. (See also Bain, 1986 and Walmsley, 1987).

BRITISH GEOLOGICAL SURVEY, 1979. *Sub-Pleistocene Geology of the British Isles and the adjacent continental shelf:* 1 sheet; scale 1:2,500,000.

BRITISH GEOLOGICAL SURVEY, 1986. *Pre-Permian Geology of the United Kingdom (South):* 2 sheets; scale 1:1,000,000.

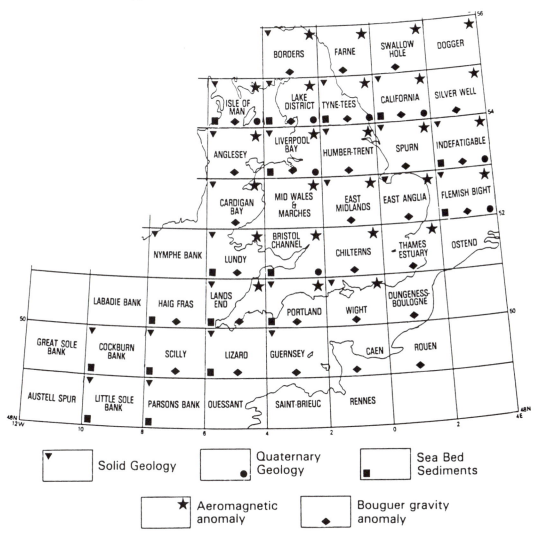

**Appendix Fig.** British Geological Survey maps of England and Wales and the adjacent continental shelf at scale of 1:250,000. (Map reproduced by kind permission of the British Geological Survey.) As available January 1991.

# 16

# IGNEOUS ROCKS

# E. H. Francis

## Introduction

When Geikie reviewed the igneous rocks of Britain in 1897, their distribution in time and space seemed more limited than we now know them to be. Current evidence of previously unsuspected activity in Silurian, Silesian, Jurassic and Cretaceous times has accumulated partly as a result of refinement in palaeontological and radiometric age-dating techniques, but more from exploration for hydrocarbons on shore and on the continental shelf. Volcanic rocks have been recognised in boreholes drilled by various techniques, and cored holes have been particularly valuable for the sampling and recognition of thin ash-fall layers in sedimentary sequences. Improvements in laboratory techniques and current understanding of diagenetic processes have facilitated reinterpretation of their more decomposed, and hence more cryptic correlatives at outcrop.

The revised stratigraphy of volcanic activity, together with petrochemistry and analogies drawn with modern counterparts, offers a continuous reference against which the location of the region relative to ancient lithospheric plates and plate movements can be assessed. The pattern so determined is relatively simple. The earliest, volumetrically largest accumulations, extending from Precambrian to Lower Palaeozoic times, indicate the intermittent proximity of a destructive plate margin or margins; subduction beneath island arcs and possibly Andean type continental edges generated magmas which were typically, and mainly, calc–alkaline in composition. With the closure of the Proto–Atlantic (Iapetus) Ocean during the end-Caledonian orogeny in Siluro–Devonian times, however, the whole of the region on and off shore, with the possible exception of SW England, became cratonised, that is underlain by continental crust; thus the bulk of the volcanic products thereafter

were alkali–basalts with subordinate tholeiitic types – most of them showing an affinity to modern varieties associated with continental rifting.

The Upper Palaeozoic rocks of SW England do not conform to that pattern. In particular the late Carboniferous granites were emplaced as part of a magmatism which was generated by a Hercynian orogeny that had a much diminished impact farther to the north. Whether the orogeny resulted from strike-slip or from closure by subduction of either a mid-European sea or even a marginal basin floored by oceanic crust remains a matter of such uncertainty as to merit separate treatment in the account which follows.

## Precambrian and Eocambrian magmatism

Precambrian and Eocambrian rocks, including both extrusive and intrusive igneous complexes, form scattered outcrops in Wales, the Borders and the Midlands (Fig. 16.1); they have been proved by boreholes in Central England. Most of the exposed rocks fall within the 700–450 Ma age range (Thorpe et al. 1984) and moreover, Rb–Sr and U–Pb data indicate that these igneous rocks mainly represent new additions to the crust, containing little or no recycled material. Indeed, although as noted in Chapter 18, the seismic evidence suggests the presence of older continental crust at depth, that basement may be only about 800 Ma old with an upper limit of 1,300 Ma (Hampton & Taylor 1983, Davies et al. 1985).

The late Precambrian and Eocambrian can be regarded as having been a period of major crustal growth generated by widespread calc–alkaline activity related to a complex system of island arcs, most of them with a thin and slightly older sialic basement.

This implies subduction of ancient oceanic lithosphere with micro-continental collision, and postulated models of the positions of old Benioff zones and sutures are various enough to have exhausted most possibilities. What can be inferred, however, is that the plate movements and sutures of the time (they should, as Harris observes in Chapter 2, be regarded as forerunners of the Caledonian orogeny) left a legacy of north-easterly crustal fracturing which had a major and continuing influence on subsequent Palaeozoic sedimentation and volcanism (Woodcock 1984).

## North Wales

Precambrian igneous activity can be viewed as a simple sequence in spite of the continuing controversy over structure and stratigraphy reviewed elsewhere in this book (Ch. 2). It can be stated as comprising earlier (late Proterozoic, Monian) and later (Eocambrian, Arvonian) periods of calc–alkaline activity with three pulses of related granitic intrusion; island-arc or continental marginal setting seems to have been generally assumed though the location and direction of subduction remains arguable.

The bulk of the volcanic rocks within the Monian are spilitic pillow lavas up to 800 m thick and occurring at two "stratigraphic" levels corresponding to Greenly's (1919) New Harbour and Gwna groups (see Chapter 2). The trace element composition of the former falls within the field of volcanic arc lavas (Thorpe *et al.* 1984), but that of the latter is comparable to that of modern oceanic crust (Thorpe 1972) and so is the composition of associated serpentinite masses derived from harzburgites, lherzolites, pyroxenites and gabbros. As an alternative to interpreting the basic suite as a whole as allochthonous ophiolite, Maltman (1975) argued that the serpentinites at least are magmatic intrusions. All the other volcanic rocks within the Monian sequence are clastic. They include the Church Bay Tuff Formation of the Skerries Group – massive tuffs with angular feldspars and a little quartz in an epidote–sericite base – and the Fydlyn and Gwyddel formations of slightly altered acid and intermediate tuffs at the top of the supergroup. These clastics, as well as the allochthonous ophiolite are consistent with many earlier interpretations involving subduction, though as Gibbons (1983) pointed out in rejecting such models as oversimplistic, evidences of Monian accretionary prism, forearc basin or volcanic arc have yet to be found.

The Arvonian (Arfon Group) rocks, unconformably overly the Monian in eastern Anglesey, but crop

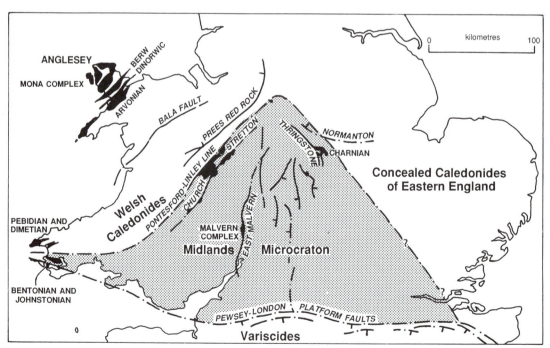

**Fig. 16.1.** Location of Precambrian volcanics within the framework of the Caledonides of Southern England and Wales (after Pharaoh *et al.* 1987a).

out most extensively on the mainland. The Padarn Tuff Formation at the base forms the Bangor–Padarn ridge and consists mainly of acid ash-flow tuffs, many of them welded (Fig. 16.2). The formation is of calc–alkaline affinity (Thorpe 1982) and is more than 2 km thick between the Dinorwic Fault and Llanberis, where the flows are inferred to have been ponded in a tectono–volcanic depression (Reedman *et al.* 1984). Overlying formations contain substantial proportions of tuffite and other volcanogenic sediments. It is uncertain how far these Arfon Group rocks can be correlated with the 159 m of calc–alkaline rocks, including andesites and dacitic lavas and basaltic tuffs drilled beneath Cambrian sediments of the Harlech Dome (Allen & Jackson 1978).

The oldest intrusion in the region is the unmetamorphosed Coedana Granite which cuts the Monian on Anglesey and has a radiometric age of $603 \pm 34$ Ma (Beckinsale & Thorpe 1979). By contrast the Sarn adamellite cropping out over an area of 3 km$^2$ in Llŷn is metamorphosed in spite of having the younger age of $549 \pm 19$ Ma (Beckinsale *et al.* 1984). By way of further apparent anomaly the Twt Hill granite near Caernarvon, thought by Thorpe (1982) to have provided debris to the lowest certainly Cambrian conglomerate appears to have a Cambro–Ordovician Rb–Sr age of $498 \pm 7$ Ma (Thorpe *et al.* 1984).

## South Wales

In two areas of south-west Wales, Precambrian igneous rocks comprise an earlier volcanic and a later intrusive suite. The Pebidian volcanics around St David's and Haycastle consist of at least 1,500 m of andesitic and rhyolitic lavas, tuffs and volcaniclastics. In the Talbenny–Johnston area, the Benton volcanic suite, less well-exposed and represented only by two

**Fig. 16.2.** Welded ignimbrite with large fiamme (flattened obsidian). Padarn Volcanic Formation south of Llŷn Padarn (B.G.S. L.2501).

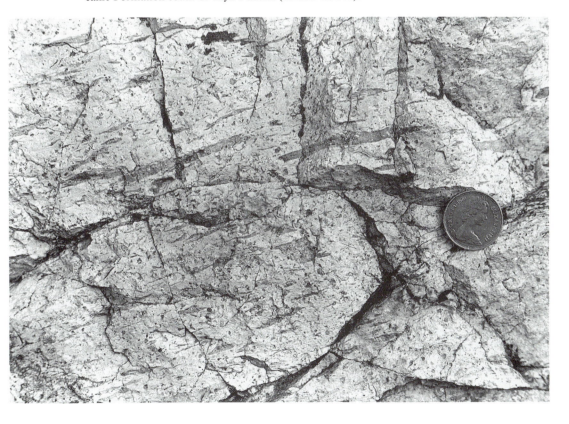

groups, consists mainly of K-rich rhyolitic lavas and pyroclastics (Thorpe 1982).

The Dimetian complex of St David's–Haycastle includes intermediate and acid intrusives, the largest being the St David's granophyre, which may be mantle-derived judging by its initial $^{87}Sr/^{86}Sr$ ratios (Patchett & Jocelyn 1979). All are characteristically Na-rich and calc–alkaline in character (Thorpe 1982). The intrusions of the Johnston complex range in composition from quartz-diorite through tonalite to albite granite with late pegmatites and quartz-normative basic dykes. They resemble the Dimetian plutonics in being calc–alkaline and Na-rich, K-poor, but have slightly higher normative plagio-clase/quartz ratios suggesting a greater depth of crystallisation (Thorpe op. cit.). U–Pb ages of between 650 and 570 Ma for the St David's granophyre and $643^{+5}_{-28}$ Ma for the Johnston rocks (Patchett & Jocelyn 1979) provide a link between the intrusive complexes of the two areas.

## Malverns

As in South Wales, the igneous rocks of the Malvern area comprises a volcanic suite – the Warren House Group – and a highly deformed intrusive complex – the Malvernian. The contact between them is faulted, but the complex was probably emplaced as the sub-volcanic phase of the extrusives.

The Warren House Group forms a 1 km² outcrop and consists mainly of spilites, keratophyres and Na-rich rhyolites cut by dolerite dykes. The basalts are mainly olivine tholeiites and the group as a whole has some trace-element affinity with marginal basin and island arc suites (Pharaoh et al. 1987b).

The Malvernian is formed mainly of diorites and tonalites with subordinate granites and ultramafics – the whole intruded by basic dykes and a single K-rich trachyte. The chemistry of the complex shows it to be relatively Mg-rich overall. The intermediate and acid members have calc-alkaline affinities inviting comparison with the Johnston complex and with younger orogenic belts (Lambert & Holland 1971; Thorpe 1972). Their age of $681 \pm 53$ Ma (Beckinsale et al. 1981) is also comparable with the Johnston complex.

## Shropshire

'Fine dolerites', gabbros, acid intrusions and 'late dolerites' emplaced in that order, form an intrusive complex in the Stanner–Hanter area, 50 km southwest of Church Stretton (Holgate & Hallows 1941). Dated at $702 \pm 8$ Ma (Patchett et al. 1980) they are, together with Johnston and Malvern complexes, the earliest expressions of volcanism in England and Wales.

Ages of Uriconian rocks shows them to be younger ($558 \pm 16$ and $533 \pm 13$) possibly as late as early

Cambrian (Patchett et al. 1980). Outcrops of both the Western Uriconian in the Pontesford and Linley Hills and of the Eastern Uriconian at the Wrekin, Lawley, Caer Caradoc and Wart Hill are fault-bounded. Correlation is thus difficult and the total thickness of the Uriconian uncertain, though the 1,300 m measured on Caer Caradoc defines the lower limit for these lavas and tuffs of basic, intermediate and acid composition (Greig et al. 1968). Allowing for alteration the group has a trace element distribution of both within-plate and subduction-related affinities (Pharaoh et al. 1987b). It differs from the earlier Warren House Group in its abundance of calc–alkaline andesites.

## Central England

The Charnian Supergroup, forming a scatter of small outcrops in Leicestershire and also proved in boreholes (Allsop et al. 1987), has the 900 m tuffaceous Blackbrook Formation at the base. It is overlain by the 1,400 m Maplewell Group – a sequence of intermediate to acid agglomerates, fine lithic and coarse crystal tuffs, volcaniclastics and slump breccias with a few porphyroids which are partly lavas and partly intrusions (Evans 1968). The porphyroids have between 52 per cent and 78 per cent $SiO_2$ and are rich in $Na_2O$, poor in $K_2O$ and $TiO_2$, plotting on a calc–alkaline trend which differs from the Uriconian rocks; their trace element distribution is typically island arc (Pharaoh et al. 1987b).

Diorites in the area fall into two groups. Those undisputedly related to, and intruded into the Charnian are markfieldites which have given an Rb–Sr age of $552 \pm 8$ Ma (Cribb 1975). They are relatively K-rich and calc–alkaline, typical of subduction zones (Le Bas 1981). The South Leicester diorites have given apparently comparable Eocambrian ages but are so similar geochemically to the local Caledonian rocks (Le Bas 1981, Pharaoh et al. 1987a) as to raise suspicion that they may be early Ordovician intrusions.

Farther south, in the small inlier near Nuneaton, the Caldecote Volcanic Formation, mainly of acidic composition, comprises a 120 m sequence of tuffs and agglomerates overlain by a coarse welded ash-flow.

# Lower Palaeozoic volcanism

Early plate-tectonic reconstructions relating to England and Wales imply that the consumption of the crust of the Iapetus Ocean began with south-easterly subduction in late Cambrian times and continued, accompanied by extensive volcanic activity, throughout the Ordovician and early Silurian. An analogy drawn with modern plate margins by Fitton et al.

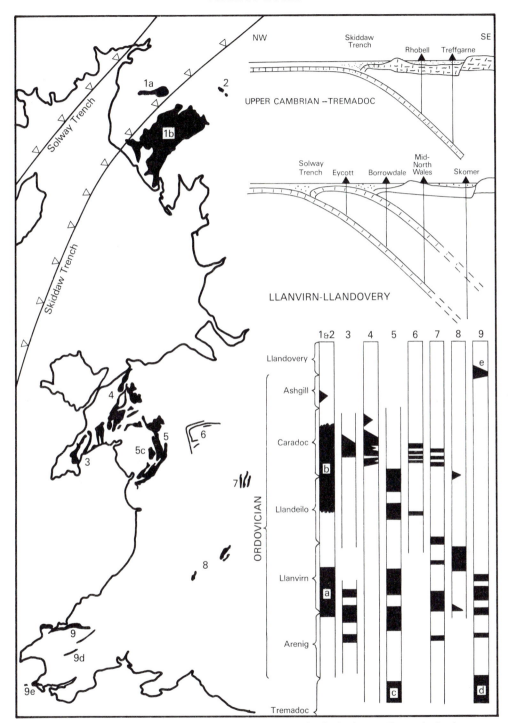

**Fig. 16.3.** Map and sections showing distribution of principal extrusive rocks in the Lower Palaeozoic. Plate tectonic models (top right) after Fitton *et al.* (1982).

1. Lake District (*a.* Eycott, *b.* Borrowdale); 2. Cross Fell; 3. Llŷn; 4. Snowdonia; 5. Arans and Cader Idris (*c.* Rhobell Fawr); 6. Berwyns; 7. Shelve; 8. Builth–Llanwrtyd; 9. Strumble–Fishguard (*d.* Treffgarne, *e.* Skomer).

(1982) assumed a simple progression in magma type from tholeiitic through calc–alkaline relative to time and distance from a destructive margin which, together with related island arcs, migrated with time (Fig. 16.3). Viewed critically, these arcs seem to be much smaller in scale than modern circum-Pacific arcs and, as noted by Stillman & Francis (1979) it is frequently difficult to recognise magmatic affinities in regionally metamorphosed rocks. Such reason for suggesting that this interpretation of tectono–magmatic activity along the south-eastern margin of Iapetus may be over-simplified is even more persuasively underlined by the introduction of the 'exotic' (suspect) terrane concept to the British Caledonides (Barber 1985). The configuration most favoured at the time of writing (Soper et al. 1987; Pharaoh et al. 1987a) is of a wedge-shaped Midlands microcraton indented into the Caledonide Orogen so as to divide the NE–SW Welsh Basin from the mirror-image NW–SE trending concealed Eastern England sector which links the Lake District with the Brabant and Ardennes massifs (Figs. 16.1, 16.8) (see also Chapter 4).

Thus although the simple model has been retained here (Fig. 16.3) because it serves to distinguish the calc–alkaline island arc setting of the Lake District from the mainly back-arc extensional activity of the Welsh Basin it should be noted that some of the faults in Wales and the Midlands may well represent terrane boundaries as well as reactivated Precambrian lineaments which provided access for ascending magma.

The presence of hyaloclastites and spilites, coupled with the frequent sediment-volcanic alternation indicates that much of the activity in Wales was submarine. Hyaloclastites and sediments with marine fossils also occur within volcanic piles of the Lake District, but there they form a much smaller proportion of the sequence, perhaps reflecting a higher rate of effusion relative to subsidence.

## Lake District

Volcanism began in the Lake District at or near the end of the Arenig, following the deposition of the thick sequence of Skiddaw Group slates and greywackes. It is represented mainly by an earlier Eycott Group, of early Llanvirn age in the north and the younger mid-

**Fig. 16.4.** The lower part of the Borrowdale volcanic succession between Buttermere and Haweswater (Millward *et al.* 1978, fig. 37, p. 108).

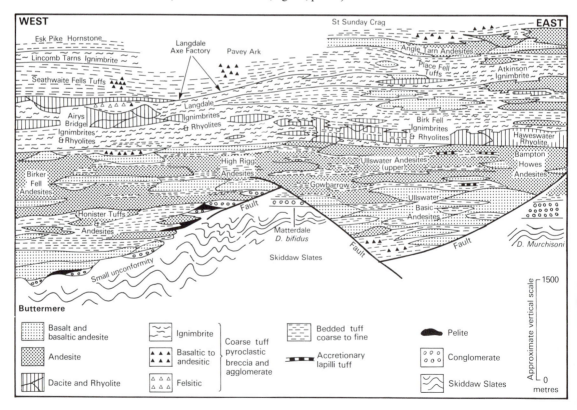

Llandeilo to mid-Caradoc (diachronous at base and top) Borrowdale Group to the south (Moseley 1984). A final episode in Ashgill times gave the Cautley Volcanic Formation tuffs (see also Chapter 4).

The Eycott Group, about 2·6 km thick, consists mainly of basalts and basaltic andesites with subordinate rhyolites and acid ignimbrites; pyroclastic rocks form about 25 per cent of the total, occurring mainly in the lower part of the sequence, where some of them appear to be subaqueous, though other unbedded layers may represent unwelded ignimbrites (Millward et al. 1978). They are correlated with the spilites, andesites and tuffs of the Kirkland Formation of Cross Fell (Burgess & Wadge 1974). The Carrock Fell gabbro complex is penecontemporaneous ($468 \pm 10$ Ma) with the group and together with the smaller Castle Head, Keswick and Embleton basic intrusions, is almost certainly comagmatic with it; indeed, its original intrusive form was a sub-horizontal body which may have fed the lavas (Harris & Dagger 1987). On the basis of low values of Fe, Ti and V as well as $K_2O/Na_2O$ and La/Y ratios, Fitton et al. (1982) suggested an affinity between the Eycott Group and island-arc tholeiites, regarding the group, as transitional between these and calc–alkaline volcanics.

The Borrowdale Volcanic Group comprises about 6 km thickness of lavas, ash-fall tuffs and ignimbrites in the Lake District (Fig. 16.4), but less in Cross Fell. Millward et al. (1978) describe the lavas as ranging mainly from high-alumina basalts to high-potash andesites, but with subordinate dacites and rhyolites so distinctive as to form a bimodal grouping. They further recognised three basic-to-acid magmatic cycles erupted from three major and several minor centres. There is agreement between Millward et al. (1978) and Fitton et al. (1982) that the basic and intermediate rocks are quartz- to hypersthene-normative, that the acid rocks are corundum-normative, and that values of Fe, Ti and V as well as ratios of La/Y and $K_2O/Na_2O$ are consistent with modern continental margin (Andean type) calc–alkali provinces. Both agree, too, on models invoking south-eastwards subduction, but whilst Millward et al. (1978) postulated volcanic island arcs supported by continental crust, Fitton et al. (1982) believed the volcanic islands to have been supported by a Skiddaw Slate-filled trench (Fig. 16.3) resting on oceanic lithosphere. Evidence of subaerial eruption provided by many of the lavas, welded tuffs and unconformities within the sequence has to be reconciled with evidence of subaqueous accumulation in the form of hyaloclastites, and intercalated fossiliferous sediments. Palaeogeographic reconstructions have invoked a Mexican-type andesitic plateau built up by multiple small eruptions and with sporadic lakes (Suthren 1977);

three or more overlapping shield or strato-volcanoes in an island arc (Millward 1979); and a late Ordovician caldera collapse which might account for the post-Borrowdale unconformity (Branney & Soper 1988). Further uncertainty is introduced to such reconstructions by the reinterpretation of many andesite sheets formerly described as blocky lavas as high-level peperitic sills intruded into wet sediments; Branney and Suthren (1988) have gone so far as to suggest that a substantial part of the Borrowdale Volcanic Group is composed of such sills rather than of lavas, thereby casting doubt on existing lithostratigraphic schemes.

The last phase of volcanic activity began with extrusion of the 70-m thick Yarlside (Stockdale) Rhyolite (Gale et al. 1979) reinterpreted (Millward & Lawrence 1985) as a rheoignimbrite, in the early Ashgill sediments of the main Lake District outcrop; it ended with the late Ashgill rhyolitic tuffs, 20–30 m thick, comprising the Cautley Volcanic Formation of the Dent–Cautley inliers (Ingham & McNamara 1978).

## Wales and the Welsh borderland

The volcanic stratigraphy (Allen 1982) shows that in Wales activity continued throughout the period, though the shift of locality with time was random (see also Chapter 4). Fitton et al. (1982) assumed a compositional trend with time, from early Ordovician tholeiites through dominantly calc–alkaline to final (early Silurian) alkaline types, but this now seems oversimplified. Bevins et al. (1984) and Kokelaar et al. (1984) emphasised that tholeiitic rocks are much more abundant than calc–alkaline among compositions which are essentially bimodal basalt-and-rhyolite. They infer that whilst the earliest (Tremadoc) activity was comparable with that of island arcs, the subsequent more voluminous mid-Ordovician outpourings are more closely akin geochemically to the volcanics of marginal basins underplated by continental crust. Partial melting of that crust is assumed by most workers to have been the cause of the abundant acid rocks.

Earliest activity, during the Tremadoc, was centred at Rhobell Fawr (Fig. 16.3) in the north and Treffgarne in the south. The Rhobell sequence comprises nearly 4 km of mainly basaltic lavas, with subordinate late-stage intrusive andesites and a rhyolite. Kokelaar (1986) described the basalts as mainly calc–alkaline similar to modern island-arc types, but with some low-K tholeiites. The 150 m Treffgarne sequence similarly reflects a basalt–andesite–rhyolite fractionation (Bevins et al. 1984).

During Arenig and Llanvirn times, activity spread eastwards and northwards from Rhobell Fawr. The Aran Group (Ridgway 1975, 1976) of Cadair Idris (Fig.

16.5) and the Arans comprises 0·5 km of basal rhyolites and andesites separated by sediments from 1·2 km of spilites and hyaloclastites which are in turn overlain first by sediments, then by late acid, mainly ignimbritic extrusives up to 1 km thick. According to Bloxam & Lewis (1972), the basic rocks are of island-arc tholeiitic affinity. Dunkley (1979) believed the rhyolitic rocks to be derived in part by crustal remelting and he equated their emplacement with periods of uplift and unconformity; he described the basic rocks, however, as having more tholeiitic trends than those of Rhobell.

In Llŷn, Fitch (1967) described some of the pillow lavas occurring within the Llanvirn volcanic succession as being andesitic. In the Shelve district of the borders, too, andesitic types are recorded among the minor intrusions, lavas and volcaniclastics of the Stapely Volcanic Formation (Lynas 1983); magmatic affinities are described as transitional between tholeiitic and alkaline (Lynas et al. 1985) or subduction-related tholeiitic (Leat & Thorpe 1986a).

The earliest (Arenig) activity farther south, at Ramsay Island in Dyfed, gave rise to an unusual sequence of distal turbiditic silicic tuffs and submarine debris flows; they were followed during the Llanvirn by welded and unwelded ash-flows and both intrusive and extrusive proximal rhyolites – all indications of

major tectonic sea-floor disturbance (Kokelaar et al. 1985). Penecontemporaneous Llanvirn activity nearby, around Fishguard and Strumble Head (Lowman & Bloxam 1981) is represented by a tripartite sequence comparable with the Arans, comprising a lower 0·5 m of rhyolitic lavas and tuffs, a median suite of pillow lavas, hyaloclastites and dolerites and an upper 0·5 km of rhyolitic rocks. Bevins (1982) further distinguished basaltic andesites and dacites in the suite, agreeing with Lowman & Bloxam (1981) as to overall tholeiitic affinities – a conclusion elaborated by Fitton et al. (1982) who pointed to the presence of both quartz-normative and olivine-normative varieties suggesting, perhaps, two discrete magma pulses.

Llanvirn volcanic rocks at Builth, 0·6 km thick, comprise a range of basic, intermediate and acid intrusions and extrusions including hyaloclastites and ash-flows. They are highly aluminous and calc–alkaline in their affinities – possibly transitional between ocean-floor and island-arc environments (Bevins et al. 1984). The thinner (100 m + ) rhyolites and tuffs of the same age at Llandeilo (Allen 1982) may be the lateral equivalent of the Builth rocks. Late Ordovician volcanism was centred entirely in the north, apart from early Caradoc pillow lavas near Llanwrtyd. Even on the south-east flank of the Harlech Dome, activity ceased early in Caradoc times

**Fig. 16.5.** Cadair Idris (Ordovician volcanic rocks intercalated with sediments) viewed from the north-west (B.G.S. L.1260).

with the eruptions of three acid ash-flow tuffs, up to 460 m thick (Ridgway 1975) – a phase probably coeval with the emplacement of the calc–alkaline adamellitic granophyre of Cadair Idris (Bloxam & Lewis 1972).

Welsh activity reached its acme during the Caradoc, extending from Llŷn through Snowdonia to the Border country. In Llŷn outpourings of basalts trachybasalts and rhyolites with subordinate ignimbrites are estimated by Leat & Thorpe (1986b) to have covered an original area of 244 km$^2$ with a volume of up to 560 km$^3$. Their subvolcanic expression takes the form of more than 20 stocks of tonalite and microgranite of subalkaline and peralkaline composition indicative of derivation from transitional tholeiitic magmas evolved under ensialic crustal tension (Croudace 1982). Picrite and diorite differentiates are recorded from a basic igneous complex at Rhiw (Cattermole 1976).

In Snowdonia two volcanic sequences, each more than 1 km thick and separated by a sedimentary interval, accumulated within the limits of the *Diplograptus multidens* graptolite zone. In the south the earliest of these is represented by the rhyolitic lavas and tuffs of the Moelwyn and Y Glog formations. In the north the approximately equivalent, but much thicker Llewellyn Volcanic Group accumulated at a number of centres, including nearly 1 km of rhyolitic lavas and breccias near Conway and variously intermediate and basic lava-tuff complexes in the Carneddau (Howells *et al.* 1985). The Capel Curig Volcanic Formation at the top of the group is the most extensive division; it is 175 m thick, consisting of acid ash flows, partly welded and showing evidence of having been emplaced partly sub-aerially, and partly sub-aqueously (Francis & Howells 1973, Howells & Leveridge 1980).

The upper of the two major sequences is the Snowdon Volcanic Group, traditionally subdivided into Lower and Upper Acid formations separated by the dominantly basaltic Bedded Pyroclastic Formation containing pillow lavas as well as tuffs and hyaloclastites (Williams, H. 1927; Williams, D. 1930;

**Fig. 16.6 a.** Diagrammatic section through the Lower Rhyolitic Tuff formation showing distribution of lithologies.

**b.** Section to scale of the primary ash-flow tuffs palinspastically restored to indicate the form of the Lower Rhyolitic Tuff Formation caldera (after Howells *et al.* 1986).

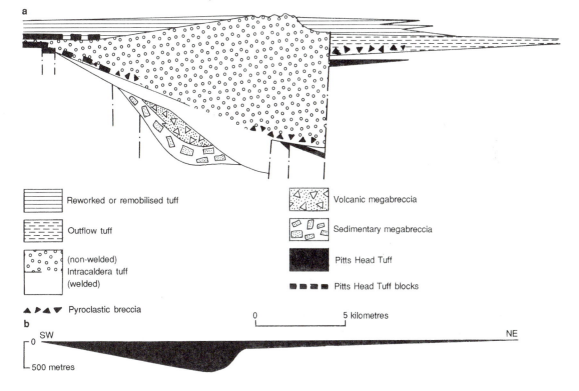

Shackleton 1959; Beavon 1963). The lowest – the Lower Rhyolitic Tuff Formation – is a submarine eruptive cycle comprising pyro-breccias, welded and non-welded ash flows, tuffites, intrusive and extrusive rhyolites as well as subordinate subaqueous basalts. They reach a maximum of 600 m in what was interpreted by Howells *et al.* (1986) and Reedman *et al.* (1987) as the filling of an asymmetric downsag caldera (Fig. 16.6) – a modification of Beavon's (1980) vision of a resurgent caldera which encompassed the whole of the Snowdon Volcanic Group. Traced north-eastwards, the equivalent Crafnant Volcanic Group is reduced in thickness to 500 m of mainly acid tuffs and tuffites

(Howells *et al.* 1978, 1981, 1985) with small lenses of basic tuffs, hyaloclastites and pillow lavas scattered throughout. Thicker accumulations of such subaqueous basic rocks represent the last products of volcanism in NE Snowdonia (Fig. 16.7), where they give rise to the Dolgarrog and Tal-y-Fan formations (*op. cit.*).

Dolerites, mainly as sills, are intruded throughout (though not above) the Snowdon volcanic sequences. Their distribution has suggested to Campbell *et al.* (1988) incipient filling of a NE–SW rift which failed to form an oceanic basin in late Ordovician times. The dolerites include little-differentiated high-alumina

**Fig. 16.7.** Basic volcanics from Dolgarrog area, northern Snowdonia.

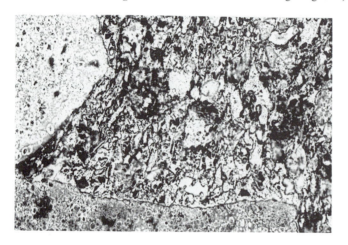

**A.** Hyaloclastite from the Bedded Pyroclastic Formation, Cwm Idwal, showing shards and fragments of chloritised, vesiculated basalt. (Plane polarised light ×8.)

**B.** Pillow breccia from the Dolgarrog Volcanic Formation, Pont Newydd, 1 km west of Dolgarrog. (E. H. Francis)

varieties as well as types allied to ocean-floor basalts (Floyd *et al.* 1976). Using elements such as Ti, Zr, Nb and Y, unlikely to have been mobilised by regional low-grade metamorphism, Leat *et al.* (1986) inferred that the basalts of the region are transitional between volcanic arc and within-plate. They identified the Snowdon rhyolites as high-Zr peralkaline rocks like those above subduction zones, thereby implying a complex tectonic setting transitional between active margin or arc and extensional environments. However, according to Campbell *et al.* (1987) these peralkaline rocks are merely the fractionated last samples of several rhyolitic batches tapped successively from an evolving sub-volcanic Snowdon magma system.

In the Berwyn Hills to the east, the Caradoc sequences show considerably reduced activity in terms of volume and variety of volcanic rocks; three acid ash-flow units, the thickest 180 m thick, are intercalated with sediments (Brenchley 1972). More distal representatives farther south in Powys are volcaniclastic sediments interbedded with shales (Dixon 1988). The late Caradoc volcanism of the Shelve district is represented by andesitic lavas and tuffs forming the Hagely and Whittery volcanic groups, both about 100 m thick and of calc–alkaline affinity (Leat & Thorpe 1986a).

When volcanic activity recommenced in the Llandovery, following a long interval represented by uninterrupted Ashgill sedimentation, it differed in location and composition from that of the Caradoc. The Skomer Volcanic Group of south-west Wales is about 1·0 km thick, including sediments intercalated with 0·76 km of alkaline lavas and tuffs. The volcanics are referred to the alkali–hawaiite – mugearite – trachyte – soda–rhyolite series by Ziegler *et al.* (1969) who assumed the sodic composition to be primary and thus explained K-rich ignimbrites in the sequence as being anomalous. The anomaly may be due to crustal melting associated with the ascent of the Na-rich alkaline magma according to Fitton *et al.* (1982).

Llandovery volcanics otherwise crop out only in two small outliers. In the Mendips, near Shepton Mallet, up to 500 m of dacites with subordinate andesites and pyroclasts are, in contrast to the Skomer rocks, calc–alkaline in composition; volcanics of the same age in the Tortworth inlier of Gloucestershire contain higher Ni and Zn, but are otherwise similar (Van de Kamp 1969).

Thin beds of bentonite interbedded with Wenlock and Ludlow sediments in Shropshire (Ross *et al.* 1982) testify to continuing activity – possibly on a large scale – throughout the Silurian. The last (Siluro–Devonian) and best documented ash-fall is the Townsend Tuff, extending over $10^4$ km$^2$ in South Wales (Allen & Williams 1981). By comparison with isopachs of

modern plinean activity, the recorded thicknesses of up to 3 m suggest that there may have been one or more large stratovolcanoes within, or no great distance from, the Welsh Basin. None has yet been located, though it may be significant that small intrusions of microgranite cut late Wenlock sediments in the Borderland (Sanderson & Cave 1980).

## Central and Eastern England

Beyond the eastern margin of the Midlands microcraton (as depicted in Fig. 16.1) Ordovician activity is represented by intrusions, partly at outcrop, though largely concealed, and by extrusives known only from drilling. The earliest intrusions are the South Leicester (Enderby and Croft) diorites which have given possibly aberrant Eocambrian radiometric ages (p. 25), but are geochemically similar to the Mountsorrel granodiorite farther north. These are partially exposed, partly met in boreholes and have given a recalculated age of $442 \pm 17$ Ma (Allsop 1987); they appear from their Nb–Y ratios to have been generated in a volcanic island arc (Pharaoh *et al.* 1987a).

The Warboys granodiorite, entirely concealed, is probably contemporaneous with the early Caledonian South Leicester diorites though it has given radiometric ages which suggest Hercynian overprinting (Allsop 1987). The extrusive rocks proved in boreholes are mainly acid to intermediate in composition, comprising lavas, ignimbrites and other volcaniclastics. Samples from the Glinton borehole have given ages of $448 \pm 30$ Ma and plot in the within-plate field as do rocks from the Woo Dale borehole of Derbyshire, though specimens from the Great Osgrove Wood and Cox's Walk boreholes overlap on the volcanic arc field adjacent to the Mountsorrel granodiorites.

# Late Caledonian magmatism

Most models for the closure of the Iapetus Ocean (e.g. Phillips *et al.* 1976) assume suturing to have been completed by mid-Silurian times along a lineament coincident with the present-day Northumberland Trough (the 'Solway Trench' of Fig. 16.3); the oceanic lithosphere is believed to have been consumed by subduction both to north-west and south-east. The volcanic rocks of the Cheviot Hills thus lie to the north-west of, but very close to, the postulated suture and represent the southernmost of the otherwise Scottish suite of Siluro–Devonian calc–alkaline extrusives; the other northern England granites lie to the south-east of the postulated suture.

The Cheviot volcanics were deposited on an eroded

surface of folded Ordovician and Silurian sediments. Of uncertain thickness, they must originally have extended well beyond the 600 km² which they cover now. Above a 60-m basal agglomerate, they consist mainly of acid andesites and dacites of limited compositional range (61 per cent to 67 per cent $SiO_2$) comparable with many rocks in modern calc–alkaline suites. If they are post-suturing they cannot be strictly Andean in type. Moreover, they do not conform to the geochemical pattern now apparent in the correlative calc–alkaline volcanics of the Scottish Midland Valley and Highlands. The Scottish rocks show progressive north-westward increases in Sr, K/Th and La/Y ratios and on this evidence Fitton *et al.* (1982) discussed the possibility of an underlying detached slab of oceanic lithosphere (former Iapetus) dipping to north-west – an association comparable with that of the modern Cascade Range of north-west USA where subduction has similarly ceased. They nevertheless suggest that as Ni and Cr concentrations and lithophile elements are too high to have been derived from the slab alone, there must have been mixture with materials from the mantle. Because the Sr, K/Th and La/Y values of the Cheviot lavas, however, do not fall on the same linear trend, and because, moreover, the Cheviot rocks are younger (389 to 393 Ma) than their Scottish 'counter-parts' ($407\pm$ to $411\pm6$ Ma) Thirlwall (1981, 1982, 1983) postulated a different subduction regime or no subduction at all, relating the generation of the magma there perhaps to the close proximity to the suture.

There are no known extrusive equivalents of the other plutonic rocks of Northern England. These include the Eskdale, Skiddaw and Shap granites in the Lake District (Firman 1978) where they form separate outcrops, but are postulated to merge at depth (Chapter 18) so as to form a single batholith extending eastward beneath the Vale of Eden to join the Weardale Granite (Dunham *et al.* 1965) concealed beneath the Carboniferous sediments of the Alston Block (Fig. 16.8), though Lee (1986) has expressed doubts about this. Farther south, the Wensleydale Granite, first located by gravity survey (Bott 1967) has since been proved by drilling and has given a similar Devonian (*c.* 400 Ma) age to that of Weardale.

On radiometric grounds (Rundle 1979, Wadge *et al.* 1979) as well as from geophysical measurements (Lee 1986) it is clear that the Lake District Batholith was emplaced in at least three stages represented by the Threlkeld Microgranite (Ordovician), the Eskdale Granite and Ennerdale and Carrock Fell granophyres (Silurian), and the Shap and Skiddaw granites (early Devonian). The distribution of trace elements least mobilised by hydrothermal alteration indicate a trend with time from early tholeiitic to increasingly evolved calc–alkaline granites; this might imply magmagenesis beneath an evolving arc or continental margin, involving little or no melting or sedimentary protolith according to O'Brien *et al.* (1985), though Brown & Locke (1979) envisage some contamination during ascent. The geochemistry of minor intrusions related

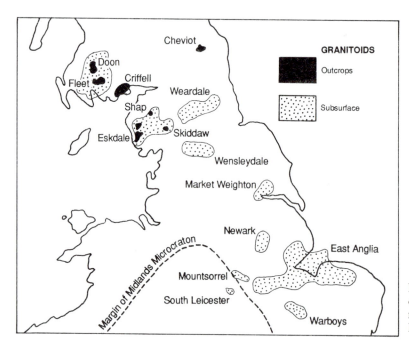

**Fig. 16.8.** Caledonian granitoids of southern Scotland, northern and eastern England.

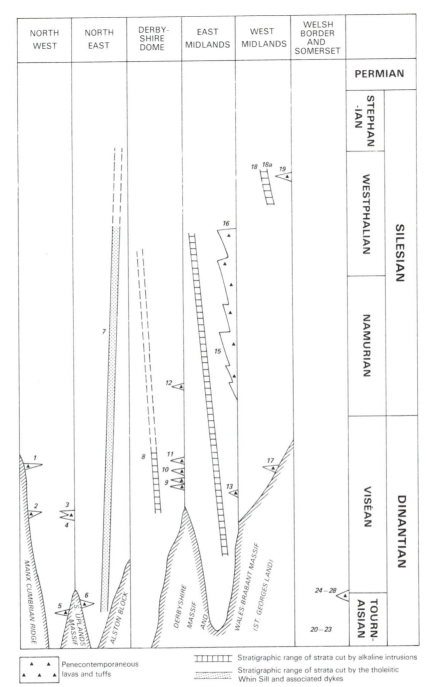

**Fig. 16.9.** Stratigraphical distribution of Carboniferous volcanic rocks in the cratonic parts of England and Wales: 1. Scarlet volcanics; 2. Cockermouth lavas; 3. Oakshaw Tuff; 4. Kershopefoot Basalt; 5. Rawney Tuff; 6. Cottonshope (Kelso Traps); 7. Whin Sill; 8. Olivine-dolerite sills of Derbyshire; 9. Lower Miller's Dale Lava, Cave Dale Lava, Nunlow Tuffs, Ashover basalts and breccias; 10. Upper Miller's Dale Lava, Lower Matlock Lava, Ashover Tuff (part); 11. Upper Matlock Lava, Litton Tuff, Ashover Tuff (part); 12. K-bentonites in Namurian tuffs, Bothamsall etc.; 16. Westphalian lavas and tuffs, Belvoir, Screverston and Sproxton; 17. Little Wenlock Basalt; 18. Olivine-dolerite and teschenites, West Midlands; 18a. Whitwick Dolerite; 18b. Oxfordshire Coalfield; 19. Tuff at Barrow Hill; 20. Brockhill Dyke; 21. Bartestree Dyke; 22. Monochiquitic dyke at Llanllywel; 23. Monochiquitic neck at Golden Hill; 24. Twickenham Lava; 25. Goblin Coombe lava and tuff; 26. Uphill lava; 27. Woodspring lava and tuff; 28. Spring Cove lava (after Francis 1970).

to the batholith similarly reflect its multicomponent origin (Macdonald *et al.* 1985).

In Eastern England three further granites centred on Market Weighton (Bott 1978), Newark (Rollin 1982) and East Anglia (Chroston *et al.* 1987) have been identified by geophysical anomalies (Fig. 16.8) though they have still to be proved by drilling: Evans and Allsop (1987) have suggested that they may be a late-Caledonian chain linked with the northern England batholiths.

# Upper Palaeozoic Volcanism of the Craton

After the early Devonian volcanism represented by the Cheviot lavas and the Townsend Tuff, mentioned previously, there was no further activity until Carboniferous time. There then began a period of almost continuous thick accumulation of shallow-water sediments (Chap. 7), the subsidence in part controlled

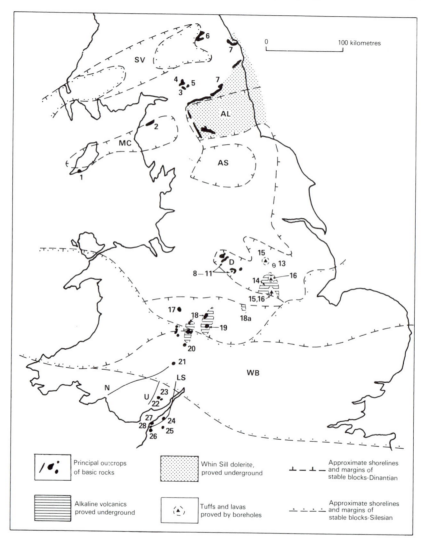

**Fig. 16.10.** Distribution of Carboniferous rocks relative to palaeogeography: AL, Alston Block; AS, Askrigg Block; D, Derbyshire Dome; LS, Lower Severn axis; MC, Manx–Cumbrian massif; N, Vale of Neath Disturbance; SV, Southern Uplands massif; U, Usk anticline; WB, Wales Brabant massif (after Francis 1970). For key to numbers see caption to Figure 16.9.

by continuing movement along deep-seated fractures, several of which were probably reactivated from time to time, partly because of the buoyancy of the low-density granites emplaced during the Devonian (Bott *et al.* 1984, Leeder 1982) and partly in response to lithospheric stretching. The fractures of this block-and-basin tectonism seems equally to have controlled the location of igneous activity, for volcanic rocks are virtually absent from the centres of the synde-positional basins. There is thus no need to postulate subduction; indeed the geochemical evidence contra-dicts it (Francis 1978).

## Dinantian to Westphalian basaltic lavas and sills

Figures 16.9 and 16.10 show a distribution pattern of generally small-scale localised effusions of lavas (mainly basalt) from centres which were at first scattered throughout England but were later restricted almost entirely to the Midlands. The rocks erupted sporadically during this long period (*c.* 65 Ma) are mainly alkaline to sub-alkaline in composition.

The lower density of the young Carboniferous sediments imposed hydrostatic control on the ascent of basaltic columns, so that apart from the minor plugs and dykes in feeder pipes, the intrusions consan-guineous with the lavas consist almost entirely of sills or sill-complexes. They are closely related both geographically and chemically to the extrusives and, like them, are restricted to marginal areas of Carbon-iferous sedimentation, being particularly extensive in the Dinantian and early Westphalian of the East Midlands.

The largest single volcanic field is of early West-phalian age lying concealed beneath Mesozoic sediments in the East Midlands. The imperfect record earlier obtained by drilling for oil has now been ex-panded by the NCB's[1] exploration programme in the Vale of Belvoir so as to show a 150-m pile (Fig. 16.11) consisting mainly of basalt lavas (flows up to 30 m thick) with subordinate volcaniclastics and dolerite sills (Burgess 1982). These rocks are mainly alkaline (basanite-hawaiite) in composition, but also include rocks of sub-alkaline (tholeiitic) affinity (Kirton 1984). This compositional range is comparable with that of the less voluminous Dinantian volcanics of Derbyshire where basalt lavas range from tholeiitic to mildly undersaturated (Macdonald *et al.* 1984) and where Ineson *et al.* (1983) recorded a sill of tholeiitic affinity contrasting with one which had provided the first English example of a fully differentiated teschenite-picrite (Harrison 1977). Late Westphalian volcanics appear to be entirely alkaline, both in the west Midlands where they consist mainly of sills

**Fig. 16.11.** Diagrammatic section to show relationships between volcanic rocks and sediments in the Vale of Belvoir Coalfield (Burgess 1982, fig. 6). (Initials (other than CX) are abbreviations for coal seam names.)

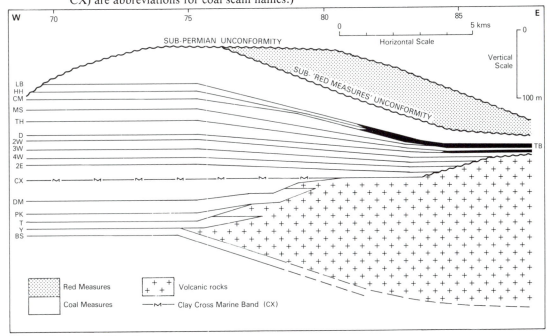

[1] NCB (National Coal Board).

(Kirton 1984), in the minor dyke occurrences along the Welsh Borders (Lawrence *et al.* 1981), and also in the Oxfordshire Coalfield where up to 165 m of such basalts and basalt breccias (age $298 \pm 6$ Ma) cut late Westphalian sediments (Poole 1977).

In the very similar alkali-basaltic suite of Scotland, Macdonald *et al.* (1977) discerned several geographically distinct lineages with chemical variation related to fractionation in high-level magma chambers. They also inferred progressively more undersaturation with time, implying derivation by partial melting of mantle periodotite at correspondingly greater depths ranging from 50 to 100 km. The absence of acid differentiates in England apart from minor tholeiitic andesites among the Cockermouth lavas of Cumbria (Macdonald & Walker 1985), can be attributed to the generally smaller scale and shorter duration of activity at most centres. Otherwise the available evidence appears to be consistent with the pattern in Scotland and with mid-plate continental volcanism in general.

## Dinantian to Westphalian pyroclastics

The influence of sedimentary environments is apparent from the contrast between the partly hyaloclastite eruptions in the marine Dinantian of Derbyshire (Cheshire & Bell 1977) and subaerially laterised tops of some of the lavas in the Westphalian of the east Midlands (Burgess 1982). Water at the surface and within the growing pile of Carboniferous sediments was important, for it determined that much of the activity was phreatomagmatic (Surtseyan) as evidenced by the textures of the pyroclastic rocks in and around the feeder pipes. A high degree of fragmentation and dispersion is characteristic of this kind of activity. Hence the distribution of thin layers of ash over wide areas throughout the period. As the ash fell in various depositional environments there were differences in diagenesis so that the present composition of the layers is related to the lithology of the adjacent sediments. Thus the layers found in the massive Dinantian limestones of Derbyshire (Walkden 1972, Harrison & Aitkenhead 1982) and North Wales (Somerville 1979) are blue, green, yellow and orange coloured clays ('wayboards') composed of mixed-layer illite and smectite with a little disorganised kaolinite; like similar clays in the Namurian marine shale sequences of Staffordshire they are allied to K-bentonites (Trewin 1968). They contrast with the hard layers of ordered kaolinite, seldom more than 5

**Fig. 16.12.** Whin Sill: quartz-dolerite dipping left to right (North to South), Roman wall district of Northumberland (B.G.S. L.1512).

cm thick, occurring in or adjacent to Westphalian coal seams; these are the *tonsteins* of the Ruhr and adjacent coalfields, each layer having a sufficiently distinctive texture to be used in correlation. The basaltic origin of early Westphalian *tonsteins* is most readily discernible near volcanic centres like those of the east Midlands where relict vitroclastic textures and lateral passage into coarse tuffs can be made out (Francis 1969); farther away, where textures are more cryptic, the $TiO_2$ content is diagnostic of basaltic origin (Price & Duff 1969). Trace elements of some late Westphalian *tonsteins* indicate derivation from acid magmas (Spears & Kanaris-Sotiriou 1979), but it is clear from the distribution of those layers that the sources were centres of known Plinean rhyolitic activity located far to the south, possibly in the Bohemian massif (Francis 1988).

## Late Carboniferous tholeiite intrusions

The long term pattern of magmatism and associated stress systems was briefly interrupted towards the end of the Carboniferous by a single phase of tholeiitic intrusion in Northern England. The main part of the suite comprises the Whin Sill complex (Fig. 16.12), compositely up to 100 m thick and providing evidence that its emplacement was controlled by syn-sedimentary structures in what were then young low-density Carboniferous deposits (Francis 1982). The complex was fed by dykes emplaced along penecontemporaneously formed E–W extensional fractures which cut all earlier structures. Trace element data (Thorpe & Macdonald 1985) indicate that some of these were the channels of distinct magmatic pulsing and that the complex overall was derived by polybaric fractionation of olivine-tholeiite from partially melted heterogenous mantle peridotite and low crustal contamination of mantle-derived magmas.

The sill has no extrusive equivalents and the related dykes, extending outside the areas of Carboniferous sedimentation, indicate a stress regime which was entirely different from that of both the earlier and later volcanics. Not least, the radiometric dating of both sill-complex (Fitch *et al.* 1970) and dykes (Wadge *et al.* 1972) consistently gave an age of about 296 Ma, indicating a single, apparently late Westphalian episode in contrast to the extended period of alkaline activity. That late episode affected a belt of country 200 km wide, which extended northwards into Scotland and might also be extrapolated eastwards for 1,000 km to Sweden, where a swarm of similar tholeiitic dykes associated with sills gives similar K-Ar ages. The suggestion by Russell & Smythe (1983) and Haszeldine (1984) that the episode may reflect rifting related to incipient early Atlantic opening is controversial, though it is notable that all other comparable

large-scale tholeiitic intrusion complexes of the world have accompanied continental break-up.

## Latest Carboniferous–Permian activity

The tholeiitic phase, like the emplacement of granites in SW England is assumed to be related to the final (Asturian) phase of the Hercynian orogeny. There followed a resumption of alkali–basaltic extrusion which had structural associations different from any which had obtained before. Rifting along entirely new NNW to N–S lines exerted control over the growth of several new syn-sedimentary basins and also provided access for renewed small-scale sporadic alkali-basaltic volcanism. In the North Sea, the initiation of the Central Graben (Fig. 15.17) was accompanied by the extrusion of alkali-basalt lavas in sediments of Rotliegendes age (Pennington 1975). The west Midlands sills, described above, may also be partly of this phase judging by their largely Permian K-Ar ages of 265–285 Ma (Fitch *et al.* 1970) and their occurrence on the alignment of the Stranraer–Cheshire rift. This, however, must be balanced against the stratigraphical evidence of late Westphalian phreatic eruption of ash in the area (Francis 1970).

# Upper Palaeozoic (and earlier) magmatism in the SW England Hercynian orogen

## Lizard Complex

The age and mode of emplacement of the Lizard Complex (Fig. 16.13) remain uncertain. The geological history proposed by Edmonds *et al.* (1975) based on the work of Flett (1946) and Green (1964, 1966) is as follows:

1. Precambrian sedimentation and basaltic intrusion, now represented by the Old Lizard Head Series and by the Landewednack and Traboe Schists.
2. Precambrian intrusion of acid sills, now the Man O'War Gneiss.
3. Eocambrian orogeny and intrusion of peridotite, now altered to serpentine.
4. Emplacement of gabbro.
5. Mid-Devonian orogeny and emplacement of basic dykes as well as of the intrusive microgranite which is now the Kennack Gneiss.
6. Permo-Carboniferous thrusting and later faulting.

On this reading the periodotite is seen as a diapiric plug-like mass with a few sill-like apophyses; it is assumed to have generated a high-temperature metamorphic aureole during an episode of amphibolite-facies regional metamorphism.

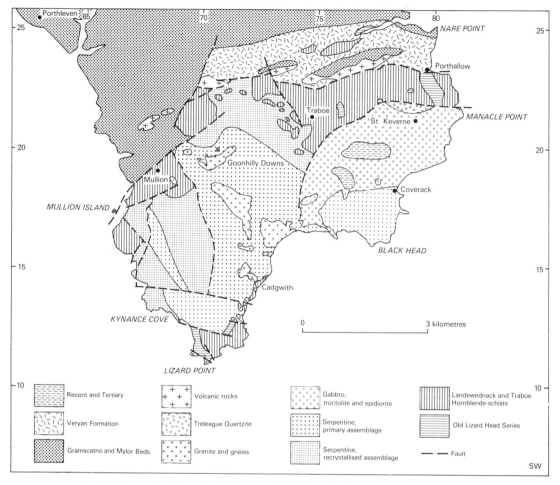

**Fig. 16.13.** Simplified geological map of the Lizard after Flett (1946) and Green (1966).

The complex was re-interpreted by Strong *et al.* (1975) as a fragment of ancient oceanic lithosphere obducted on to continental crust. Exposures in the east coast are compared with ophiolite sequences, comprising from north to south (supposedly downwards) sheeted dykes, layered gabbros and ultramafic cumulates. The absence of pillow lavas which might have been expected at the top is explained in terms of thrusting; they are identified as the pillow-lavas now seen in the Meneage breccias. The re-interpretation has been followed by conflicting reviews of the available geochemical evidence. On the one hand Floyd *et al.* (1976) determined compositions within the complex to be diverse, with only the Landewednack Schists akin to ocean-floor basalts. By contrast Kirby (1979) found the basic members of the complex to be consistently of oceanic tholeiitic affinity so as to suggest that the complex represents Silurian oceanic crust obducted during early Devonian times.

Davies (1984) obtained an Sm–Nd age of $375 \pm 34$ Ma for the gabbro, suggesting that oceanic crust as wide as the present-day Red Sea was first formed and then promptly obducted during the Devonian. Similar ages from acid veins in the Kennack Gneiss were taken by Styles & Rundle (1984) to relate to metamorphism accompanying the emplacement of the complex. Tectonic emplacement is certainly supported by geophysical survey (Rollin 1986) indicating separate allochthonous sheets of schist and ophiolite. Devonian timing of emplacement is consistent with the palaeomagnetic evidence (Hailwood *et al.* 1984) though structural analysis would allow for it being as late as Carboniferous (Rattey & Sanderson 1984).

## Devonian and Carboniferous basalts

Basaltic piles of limited thickness and extent accumulated in Devon and Cornwall throughout the Devonian and early Carboniferous, though activity ended before the Namurian. To the distribution of the principal Devonian rocks shown in Fig. 16.14, must be added the many thin layers of volcaniclastic debris which are interbedded almost throughout the Devonian sequence and which indicate that activity may have been continuous, albeit sporadic, almost throughout the period.

Most of the eruptions were submarine, giving rise to pillow-lavas and hyaloclastite sequences. Among the thickest of these is the 450-m Pentire Pillow Lava Group in the Devonian (Gauss & House 1972) and the 91-m Tintagel Volcanic Formation in the Carboniferous (McKeown *et al.* 1973). All these basaltic rocks are sheared and have undergone hydrous alteration (spilitization) to 'greenstone'. In spite of this regional metamorphism Floyd (1976) inferred from the distribution of rare earth elements (REEs) that during the Devonian magmatism was generally tholeiitic in south Cornwall as compared with both tholeiitic and alkaline elsewhere, whereas in the early Carboniferous activity was entirely alkaline (hawaiites and mugearites); this implies a progressive change of fractionation with time, possibly reflecting an underlying slab subducted at a low angle from far to the south, beyond the Moldanubian Zone (Floyd 1982). Floyd

further believed the basalts to be mainly of continental affinity because of the relatively low K/Rb, Sr/Rb, Na/K and K/Ba ratios as well as that the Ti and Zr values are too high for oceanic basalts. Chandler and Isaac (1982), however, saw ocean-floor affinities in some of the Dinantian basalts and inferred a setting within a small oceanic basin. Floyd (1984) reconciled the conflicting evidence by interpreting the basalts with oceanic affinities as abortive attempts to produce oceanic crust in a rifted continental margin lying to the north of the short-lived ocean floor now represented by the Lizard ophiolite. Isaac's (1985) reconciliation requires an oceanic marginal basin persisting in SW England as late as early Viséan times, subsequently emplaced as tectonic units. Both models accommodate the presence of a mid-European Sea on the site of what was about to become the Rheno–Hercynian Zone of the orogen (Barnes & Andrews 1986).

## Cornubian granites

Five large outcropping plutons and several small bosses are linked at depth to form a single granite intrusion – the Cornubian Batholith – extending E–W for at least 250 km (Fig. 16.15) and downwards for 10 to 12 km where it merges with the Lower Crust (Holder & Bott 1971). The granites contain abundant xenoliths, not only of local metasediments, but of diorite and diabase too, and they are extremely 'greisened' by

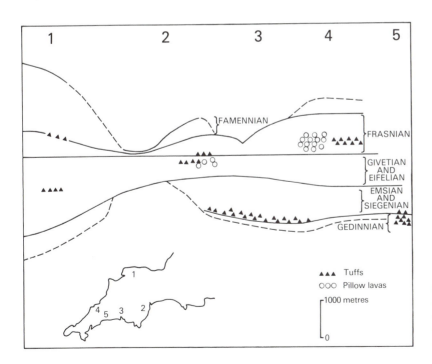

**Fig. 16.14.** Distribution of Devonian eruptive rocks in SW England (after House 1975, House *et al.* 1977).

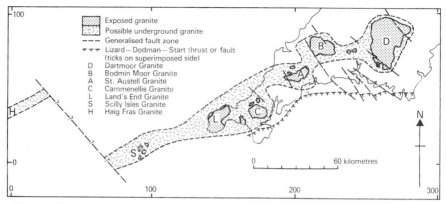

**Fig. 16.15.** Outcrops and possible underground continuation of the Cornubian batholith (after Edmonds *et al.* 1975). (See also Fig. 19.13.)

late-stage hydrothermal alteration. Associated aplite dykes and veins are Na-rich and mostly thin, though the Meldon Aplite is 20 m wide. There is also a surrounding suite of quartz-porphyry dykes (elvans) up to 40 m wide and an outer suite of lamprophyre dykes. Edmonds *et al.* (1975) suggested that the elvans were feeders to subaerial dacites and rhyolites now preserved as pebbles in the overlying New Red Sandstone molasse and as possible flows cropping out near Plymouth (Cosgrove & Elliott 1976).

The batholith is not uniform in texture. More than 90 per cent of the rocks are coarse-grained with megacrysts (feldspars up to 170 mm) or mesocrysts (15–43 mm); most of the remaining granites are fine-grained, with or without megacrysts, whilst a minor proportion are medium grained granites and aplites distinguished by their content of pale brown lithium-mica (Hawkes & Dangerfield 1978, Stone 1979). The major plutons show an unexplained spatial alternation in texture, such that the Dartmoor, St Austell and Lands End bodies are mainly megacrystic, whereas the Bodmin Moor, Isles of Scilly and Carnmellis granites are mainly mesocrystic. There is also some variation within plutons.

The differences in texture are only partly reflected in the chemistry. The REE profiles of Bodmin and Carnmenellis differ from those of Lands End and Dartmoor (Darbyshire & Shepherd 1985) and some of the lithium-mica rocks are enriched in Na and Li relative to K (Stone 1984). However, the pervasive effects of metsomatism make the chemistry of the original magmas uncertain (Manning & Exley 1984, Jefferies 1985).

There is uncertainty too as to the age and mechanism of emplacement. Radiometric ages extending over late Carboniferous to early Permian times have been narrowed by Darbyshire & Shepherd (1985) to 290–280 Ma while Hawkes & Dangerfield (1978) suggested that

the best estimate for time of emplacement might be the 295 Ma age of the K–Ar contact metamorphism of the doleritic dykes at Meldon. Work by Sheppard (1977) on D/H and $^{18}O/^{16}O$ in kaolinites and other altered minerals indicated that the granites are probably crustal in origin rather than mantle-derived as suggested by Badham (1976). As the Cornubian rocks are analogous more to the European Variscan granites than to those of the Andes, Exley (1979) similarly discounted both oceanic mantle and plate collision as originating factors, though he accepts the possibility that the lateral movements advocated by Badham (op. cit.) might have generated differential heat flow which in turn gave rise to density layering and to diapiric ascent accompanied by stoping, assimilation and recrystallisation. The controversy continues up to the time of writing, with Charoy (1986) and Stone & Exley (1986) advocating crystal melting while Thorpe (1987), Thorpe *et al.* (1986) and Leat *et al.* (1987) infer a source in subducted oceanic lithosphere. (Metalliferous mineralisation is discussed in Chapter 19.)

## Permian lavas

On most interpretations of Hercynian tectonics, SW England can be assumed to have been added to the cratonised part of the country to the north by Permian times. The new N–S lineaments which affected northern and midlands Britain as well as the North Sea do not, however, seem to have extended to SW England; there the earlier Hercynian structures were reactivated as controls over the accumulation of early Permian sediments (Edmonds *et al.* 1975). As those sediments are largely unfossiliferous, correlation depends largely on radiometric ages obtained from lavas of the 'Exeter Traps' (Thorpe *et al.* 1986). These lavas, erupted from agglomerate-filled pipes, may once have been extensive, but are now represented only by small scattered outcrops. They consist mainly of highly potassic (up to 13 per cent $K_2O$) trachybasalts, mafic syenites, minettes, syenitic lamprophyres and leucitites with subordinate olivine-basalts (Knill 1969, Cosgrove 1972). They are variously explained as the result of sub-crustal melting generated by the rise of the Cornubian batholith, by differentiation of olivine-basalts at depth followed by potash-granite contamination (Knill 1969), or most recently as granite-related magmas generated by oblique subduction of oceanic lithosphere (Thorpe *et al.* 1986, Thorpe 1987, Leat *et al.* 1987).

## Mesozoic volcanism

### Triassic

The NNW and N–S aligned rifting continued to operate in the Triassic, but is associated with volcanism only in the northern North Sea, where basaltic lavas are intercalated with red beds (Woodall & Knox 1979). The eruptive phase has no expression on land, nor are there intrusions comparable with the low-alkali (? tholeiite) dykes of NW Brittany and alkaline dykes of Norway, with the possible exception of the highly altered Butterton–Swynnerton group of post-'Bunter' dykes of Staffordshire (Scott 1925).

### Jurassic

The volcanic origin of the montmorillonite-rich Bathonian Fuller's Earth was confirmed by Jeans *et al.* (1977), who identified pyroclasts of flow-banded, devitrified trachytic lava, glass shards and crystals in calcareous nodules. The angularity of the pyroclasts indicates deposition from direct air-fall into a shallow water environment; and as the bed is 2 to 3 m thick at Combe Hay, near Bath, it must represent an eruption of considerable magnitude particularly if the source was as far away as the Northern North Sea graben, where heavily weathered alkali-basalts and tuffs are interbedded with Bathonian–Bajocian sediments (Howitt *et al.* 1975, and Fall *et al.* 1982). Bathonian sediments nearer to that source in East England contain montmorillonite and a high proportion of smectite which is almost entirely derived by alteration of volcanic material (Bradshaw 1975). A south-westerly source cannot, however, be ruled out, for a biotite–olivine–gabbro giving a comparable (Bathonian) K–Ar age of $166 \pm 4$ Ma (Rundle *in* Harrison *et al.* 1979) was dredged in the off-shore Western Approaches Basin.

### Cretaceous

At least two major volcanic episodes are now recognised during the Lower Cretaceous. In the Barremian Speeton Clay of Yorkshire, Knox and Fletcher (1978) reported four thin bentonitic mudstones up to 20 cm thick and totalling 50 cm. They contain smectite, kaolinite and residual igneous debris, and represent ash-falls believed to emanate from a source now buried beneath younger sediments in the southern North Sea rather than from the Wolf Rock – now dated (*c.* 120–130 Ma) as Barremian (Harrison *et al.* 1977). Both the Wolf Rock and the neighbouring Epson Shoal, off Cornwall, are micro-porphyritic fluxioned phonolites with nosean locally replaced by analcime and cancrinite (Harrison *et al.* 1979): circular outcrop suggests emplacement as a feeder neck.

Subsequent Lower Cretaceous (Aptian) volcanism – possibly related to rifting and sea-floor spreading in the Bay of Biscay – is represented by four beds of Fuller's Earth clay (Ca–bentonite) cropping out

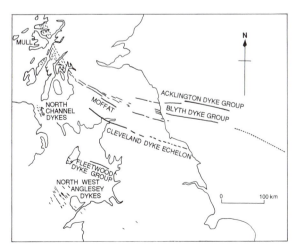

**Fig. 16.16.** Tertiary dykes.

sporadically, individually up to 3·5 m thick, in south-eastern England (Jeans *et al.* 1977, Young *et al.* 1978). The clays contain zeolitised shards, pumice, high-temperature sanidine and lithic pyroclasts; these and the abundances of immobile elements are indicative of alkaline to peralkaline trachytic magma sources. One is possibly the Zuidwall volcanic plug of Holland. An alternative source area is the Fastnet Basin – the SW extension to the N Celtic Sea Basin – where six major olivine-dolerite sill complexes are intruded into Lower Lias sediments and where three igneous centres (? plugs) were located in seismic traverses (Caston *et al.* 1981). They appear to represent a mid-Jurassic to early Tertiary activity perhaps indicative of early rifting between the American and European plates (see also Chapter 12).

# Tertiary volcanism

The main Hebridean volcanism, which reached its acme *c.* 58–60 Ma, is represented only peripherally on the mainland of England and Wales by distal members of dyke swarms. The most persistent is the tholeiitic Armathwaite–Cleveland dyke, traceable for 400 km from the Mull centre through Cumberland to Durham. Its K–Ar age of 58·4 ± 1·1 Ma (Evans *et al.* 1973) matches ages obtained from the main Mull activity. Several other persistent olivine-free to olivine-poor members of the Mull swarm follow similar WNW trends in the Northumberland Coalfield, the Acklington Dyke being the northernmost; they can be traced thence into the North Sea (Kirton & Dunato 1985) – altogether 800 km from Mull (Fig. 16.16).

In Anglesey and northern Snowdonia, a group of north-westerly alkali-dolerite dykes up to 12 m wide may have been emplaced in two phases. Most belong to an earlier phase which gives K–Ar ages earlier than 63·5 Ma (Evans *et al.* 1973) thus identifying the swarm with the Antrim Plateau Basalts rather than with the Mourne centre (Fig. 16.16).

Geographically separated from the other Tertiary centres, the Lundy Granite in the Bristol Channel was shown by Meighan (1979) to resemble the Hebridean granites in being highly fractionated, with a trace element content suggesting differentiation in the upper continental crust from a mantle-derived tholeiitic magma of variable composition. This thesis requires

that there should be a complementary mass of basic rocks and whilst none is known at the surface, an adjacent belt lying at shallow depth is indicated by gravity and magnetic surveys (Brooks & Thompson 1973). The *c.* 52 Ma age of the granite (Dodson & Long 1962) corresponds with the onset of sea-floor spreading between Greenland and northwest Europe according to correlations of Anomaly 24 (Hailwood *et al.* 1979).

In and around the North Sea Basin, there is growing evidence of quite different Tertiary volcanism expressed by beds of tuff and volcanogenic sediments. The earliest are tuffs of Thanetian age found in a restricted area of the North Sea (Jacqué & Thouvenin 1975) and represented on shore by igneous grains and montmorillonite in the Thanet Beds of Kent (Knox 1979).

A later, more important episode is represented by early Palaeogene (*c.* 53·5 Ma, Fitch *et al.* 1978) pyroclastic rocks which extend over an area of 500,000 km$^2$ (Fig. 16.16) and include the 'ash-marker' – one of the most reliable indices of correlation used in North Sea exploration. Jacqué & Thouvenin (1975) described it as consisting of a number of graded layers, up to 8 cm thick, containing shards and plagioclase phenocrysts indicative of an alkali–basalt magma. They pointed out that over such an area even a 1-cm layer represents 4 km$^3$ of ash, so that the episode as a whole may be comparable in magnitude to historic eruptions of Santorini, Katmai and Hekla particularly as no less than 179 successive grain-sorted cycles have been identified in the equivalent Mo-Clay of the Danish mainland (Pederson *et al.* 1975). The tuff extends to East Anglia where it was recorded first by Elliott (1971) in the 'Hard Stone Band' of the London Clay and is now known (Knox & Ellison 1979) to extend through a thickness of strata including at least 40 layers of ash up to 8 cm thick.

Measurements in Denmark show a thinning from north to south and also west to east, indicating a source in the Skagerrak. The wide distribution throughout the North Sea, as well as the thickness of some of the individual layers in East Anglia, suggest at least one additional source, perhaps in the Hebridean Province. Knox and Morton (1983) believed that this phase of volcanic activity, like the emplacement of the Lundy Granite, may be associated with rifting that preceded the Anomaly 24 spreading between Greenland and Rockall.

## REFERENCES

ALLEN, J. R. L. &          1981      Sedimentology and stratigraphy of the Townsend Tuff Bed
WILLIAMS, B. P. J.                   (Lower Old Red Sandstone) in South Wales and the Welsh
                                     Borders. *J. geol. Soc. London.*, **138**, 15–29.

| ALLEN, P. M. | 1982 | Lower Palaeozoic volcanic rocks of Wales. *In* Sutherland, Diane (Ed.) *Igneous rocks of the British Isles.* Wiley, London, 65–91. |

ALLEN, P. M. & JACKSON, A. A. — 1978 — The Bryn-Teg Borehole, North Wales. *Bull. geol. Surv. G.B.,* **61**, 1–27.

ALLSOP, J. M. — 1987 — Patterns of late Caledonian intrusive activity in Eastern and Northern England from geophysics, radiometric dating and basement geology. *Proc. Yorkshire geol. Soc.,* **46**, 335–353.

ALLSOP, J. M., AMBROSE, K. & ELSON, R. J. — 1987 — New data on the stratigraphy and geophysics in the area around Hollowell, Northamptonshire, provided by a coal exploration borehole. *Proc. Geol. Assoc. London,* **98**, 157–170.

BADHAM, J. P. N. — 1976 — Cornubian geotectonics: lateral thinking. *Proc. Ussher Soc.,* **3**, 448–454.

BARBER, A. J. — 1985 — A new concept of mountain building. *Geology Today,* **1**, 116–121.

BARNES, R. P. & ANDREWS, J. R. — 1986 — Upper Palaeozoic ophiolite generation and subduction in south Cornwall. *J. geol. Soc. London,* **143**, 117–124.

BEAVON, R. V. — 1963 — The succession and structure east of the Glaslyn River, North Wales. *Q. J. geol. Soc. London,* **119**, 479–512.

— 1980 — A resurgent cauldron in the early Palaeozoic of Wales, UK. *J. volcanol. Geotherm. Res.,* **7**, 157–174.

BECKINSALE, R. D., EVANS, J. A., THORPE, R. S., GIBBONS, W. & HARMON, R. S. — 1984 — Rb–Sr whole-rock isochron ages, $^{18}O$ values and geochemical data for the Sarn Igneous complex and the Parwyd gneisses of the Mona Complex of Llŷn, N. Wales. *J. geol. Soc. London,* **141**, 701–709.

BECKINSALE, R. D., THORPE, R. S., PANKHURST, R. J. & EVANS, J. A. — 1981 — Rb–Sr whole-rock isochron evidence for the age of the Malvern Hills complex. *J. geol. Soc. London,* **138**, 69–73.

BECKINSALE, R. D. & THORPE, R. S. — 1979 — Rubidium-strontium whole-rock isochron evidence for the age of metamorphism and magmatism in the Mona Complex of Anglesey. *J. geol. Soc. London,* **136**, 433–440.

BEVINS, R. E. — 1982 — Petrology and geochemistry of the Fishguard Volcanic Complex, Wales. *Geol. J.,* **17**, 1–21.

BEVINS, R. E., KOKELAAR, B. P. & DUNKLEY, P. N. — 1984 — Petrology and geochemistry of lower to middle Ordovician igneous rocks in Wales: a volcanic arc to marginal basin transition. *Proc. Geol. Assoc. London,* **95**, 337–347.

BLOXAM, T. W. & LEWIS, A. D. — 1972 — Ti, Zr and Cr in some British pillow lavas and their petrogenetic affinities. *Nature, Phys. Sci. London,* **237**, 134.

BOTT, M. H. P. — 1967 — Geophysical investigations of the northern Pennine basement rocks. *Proc. Yorkshire geol. Soc.,* **36**, 139–168.

BOTT, M. H. P., ROBINSON, J. & KOHNSTAMM, M. D. — 1978 — Granite beneath Market Weighton, east Yorkshire. *J. geol. Soc. London,* **135**, 535–543.

BOTT, M. H. P., SWINBURNE, P. M. & LONG, R. E. — 1984 — Deep structure and origin of the Northumberland and Stainmore troughs. *Proc. Yorkshire geol. Soc.,* **44**, 479–495.

BRADSHAW, M. J. — 1975 — Origin of montmorillonite bands in the Middle Jurassic of eastern England. *Earth Planet. Sci. Letts.,* **26**, 245–252.

BRANNEY, M. J. & SOPER, N. J. — 1988 — Ordovician volcano-tectonics in the English Lake District. *J. geol. Soc. Lond.,* **145**, 367–376.

BRANNEY, M. J. & 1988 High-level Peperitic Sills in the English Lake District;
SUTHREN, R. J.                  Distinction from Block Lavas, and Implications for Bor-
                                rowdale Volcanic Group Stratigraphy. *Geol. J.*, **23**, 171–
                                187.

BRENCHLEY, P. J. 1972 The Cwm Clwyd Tuff, North Wales: a palaeographical
                                interpretation of some Ordovician ash-shower deposits.
                                *Proc. Yorkshire geol. Soc.*, **39**, 199–330.

BROOKS, M. & 1973 The geological interpretation of a gravity survey of the
THOMPSON, M. S.                 Bristol Channel. *J. geol. Soc. London*, **129**, 245–274.

BROWN, G. C. & 1979 Space–time variations in British Caledonian granites: some
LOCKE, C. A.                    geophysical correlations. *Earth Planet. Sci. Lett.*, **45**, 69–79.

BURGESS, I. C. 1982 The stratigraphical distribution of Westphalian volcanic
                                rocks to the east and south of Nottingham. *Proc. Yorkshire
                                geol. Soc.*, **44**, 29–43.

BURGESS, I. C. & 1974 The geology of the Cross Fell area. Classical areas of
WADGE, A. J.                    British Geology. *Inst. geol. Sci. London*, 92 pp.

CAMPBELL, S. D. G., 1987 The emplacement of geochemically distinct groups of
REEDMAN, A. J.,                 rhyolites during the evolution of the Snowdon caldera
HOWELLS, M. F. &                (Ordovician), N. Wales, UK. *Geol. Mag.*, **124**, 501–511.
MANN, A. C.

CAMPBELL, S. D. G., 1988 A Caradoc failed-rift within the Ordovician marginal basin
HOWELLS, M. F.,                 of Wales. *Geol. Mag.*, **125**, 257–266.
SMITH, M. &
REEDMAN, A. J.

CASTON, V. N. D., 1981 Olivine–dolerite intrusions in the Fastnet Basin. *J. geol.
DEARNLEY, R.,                   Soc. London*, **138**, 31–46.
HARRISON, R. K.,
RUNDLE, C. C. &
STYLES, M. T.

CATTERMOLE, P. J. 1976 The crystallisation and differentiation of a layered intru-
                                sion of hydrated olivine basalt parentage at Rhiw, North
                                Wales. *Geol. J.*, **11**, 45–70.

CHANDLER, P. & 1982 The geological setting, geochemistry and significance of
ISAAC, K. P.                    Lower Carboniferous basic volcanic rocks in central south-
                                west England. *Proc. Ussher. Soc.*, **5**, 279–88.

CHAROY, B. 1986 The genesis of the Cornubian batholith (South-west
                                England): the example of the Carnmenellis pluton. *J.
                                petrol.*, **27**, 571–604.

CHESHIRE, S. G. & 1977 The Speedwell Vent, Castleton, Derbyshire: a Carbon-
BELL, J. D.                     iferous littoral cone. *Proc. Yorkshire geol. Soc.*, **41**, 173–
                                184.

CHROSTON, P. N., 1987 New seismic refraction evidence on the origin of the Bouger
ALLSOP, J. M. &                 anomaly low near Hunstanton, Norfolk. *Proc. Yorkshire
CORNWELL, J. D.                 geol. Soc.*, **46**, 311–319.

COSGROVE, M. E. 1972 The geochemistry of potassium-rich Permian volcanic
                                rocks of Devonshire, England. *Contrib. Mineral. Petrol.*,
                                **36**, 155–170.

COSGROVE, M. E. & 1976 Supra-batholithic volcanism of the south-west England
ELLIOTT, M. H.                  granites. *Proc. Ussher Soc.*, **3**, 381–401.

CRIBB, S. J. 1975 Rubidium–strontium ages and strontium isotope ratios
                                from the igneous rocks of Leicestershire. *J. geol. Soc.
                                London*, **131**, 203–212.

CROUDACE, I. W. 1982 The geochemistry and petrogenesis of the Lower
                                Palaeozoic granitoids of the Lleyn Peninsula, North
                                Wales. *Geochim. Cosmochim. Acta.*, **46**, 609–622.

DARBYSHIRE, D. P. F. & SHEPHERD, T. J.    1985    Chronology of granite magmatism and associated mineralization, S.W. England. *J. geol. Soc. London*, **142**, 1159–1177.

DAVIES, G. R.    1984    Isotopic evolution of the Lizard Complex. *J. geol. Soc. London*, **141**, 3–14.

DAVIES, G. R., GLEDHILL, A. & HAWKESWORTH, C.    1985    Upper crustal recycling in southern Britain: evidence from Nd and Sr isotopes. *Earth Planet. Sci. Lett.*, **75**, 1–12.

DIXON, R. J.    1988    The Ordovician (Caradoc) Volcanic Rocks of Montgomery, Powys, N. Wales. *Geol. J.*, **23**, 149–155.

DIXON, J. E., FITTON, J. B. & FROST, R. T. C.    1981    The tectonic significance of post-Carboniferous igneous activity in the North Sea basin. *In* Illing, L. V. & Hobson, D. V. (Eds.) *Petroleum Geology of the Continental Shelf of north-west Europe*. Institute of Petroleum, London, 121–137.

DODSON, M. H. & LONG, L. E.    1962    Age of the Lundy Granite, Bristol Channel. *Nature, London*, **195**, 975–976.

DUNHAM, K. C.    1974    Granite beneath the Pennines in north Yorkshire. *Proc. Yorkshire geol. Soc.*, **40**, 191–194.

DUNHAM, K. C., DUNHAM, A. C., HODGE, B. L. & JOHNSTON, G. A. L.    1965    Granite beneath Viséan sediments with mineralisation at Rookhope, northern Pennines. *Q. J. geol. Soc. London*, **121**, 383–417.

DUNKLEY, P. N.    1979    Ordovician volcanicity of the SE Harlech Dome. *In* Harris, A. L., Holland, C. H. & Leake, B. E. (Eds.) The Caledonides of the British Isles – reviewed. *Geol. Soc. London Spec. Publ.*, **8**, 597–601.

EDMONDS, E. A., McKEOWN, M. C. & WILLIAMS, M.    1975    South-west England (4th Edit.) *Brit. reg. Geol.*, 136 pp.

ELLIOT, G. F.    1971    Eocene volcanics in south-east England. *Nature, Phys. Sci. London*, **230**, 9.

EVANS, A. M.    1968    Precambrian rocks. A, Charnwood Forest. *In* Sylvester-Bradley, P. C. & Ford, T. D. (Eds.) *The geology of the East Midlands*. Leicester University Press, 1–12.

EVANS, C. J. & ALLSOP, J. M.    1987    Some geophysical aspects of the deep geology of eastern England. *Proc. Yorkshire geol. Soc.*, **46**, 321–333.

EVANS, A. L., FITCH, F. J. & MILLER, J. A.    1973    Potassium–argon age determinations on some British Tertiary igneous rocks. *J. geol. Soc. London*, **129**, 419–444.

EXLEY, C. S.    1979    Speculations on the nature of the SW England Batholith at depth. *Proc. Ussher Soc.*, **4**, 362–369.

FALL, H. G., GIBB, F. G. F. & KANARIS-SOTIRIOU, R.    1982    Jurassic volcanic rocks of the northern North Sea. *J. geol. Soc. London*, **139**, 277–292.

FIRMAN, R. J.    1978    Intrusions. *In* Moseley, F. (Ed.) The geology of the Lake District. *Yorkshire geol. Soc. Occ. Publ. No. 3*, 146–163.

FITCH, F. J.    1967    Ignimbrite volcanism in North Wales. *Bull. Volcanol.* **30**, 199–219.

FITCH, F. J., HOOKER, P. J., MILLER, J. A. & BRERTON, N. R.    1978    Glauconite dating of Palaeocene-Eocene rocks from east Kent and the time-scale of Palaeogene volcanism in the North Atlantic region. *J. geol. Soc. London*, **135**, 499–512.

FITCH, F. J., MILLER, J. A. & WILLIAMS, S. C.    1970    Isotopic ages of British Carboniferous rocks. *C. R. 6me Congr. Int. Stratigr. géol. Carbonif.*, Sheffield, 1967, **2**, 771–789.

FITTON, J. G.,
THIRLWALL, M. F. &
HUGHES, D. J.
1982
Volcanism in the Caledonian orogenic belt of Britain. *In* Thorpe, R. S. (Ed.) *Orogenic andesites and related rocks.* Wiley, 611–636.

FLETT, J. S.
1946
Geology of Lizard and Meneage (2nd Edit.) *Mem. geol. Surv. G.B.*

FLOYD, P. A.
1976
Geochemical variations in the greenstones of SW England. *J. Petrol.,* **17,** 522–545.

1982
The Hercynian Trough: Devonian and Carboniferous volcanism in south-western Britain. *In* Sutherland, D. S. (Ed.) *Igneous Rocks of the British Isles.* Wiley, 227–242.

1984
Geochemical characteristics and comparison of the basic rocks of the Lizard complex and the basaltic lavas within the Hercynian troughs of S.W. England. *J. geol. Soc. London,* **141,** 61–70.

FLOYD, P. A.,
LEES, G. J. & PARKER, A.
1976
A preliminary geochemical twist to the Lizard's new tale. *Proc. Ussher Soc.,* **3,** 414–425.

FLOYD, P. A.,
LEES, G. J. &
ROACH, R. A.
1976
Basic intrusions in the Ordovician of North Wales – geochemical data and tectonic setting. *Proc. Geol. Assoc. London,* **87,** 389–400.

FRANCIS, E. H.
1969
Les tonstein du Royaume-Uni. *Ann. Soc. géol. Nord.,* **89,** 209–214.

1970
Review of Carboniferous volcanism in England and Wales. *J. Earth Sci.* (Leeds) **8,** 41–56.

1978
Igneous activity in a fractured craton: Carboniferous volcanism in northern Britain. *In* Bowes, D. R. & Leake, B. E. (Eds.) Crustal evolution in northwestern Britain and adjacent areas. *Geol. J. Spec. Issue,* **10,** 279–296.

1982
Magma and Sediment – I: Emplacement mechanism of late Carboniferous tholeiite sills in northern Britain. *J. geol. Soc. London,* **139,** 1–20.

1988
Mid-Devonian to early Permian volcanism – Old World. *In* Harris, A. L. & Fettes, D. J. (Eds.) The Caledonian–Appalachian Orogen. *Geol. Soc. London Spec. Publ.,* **38,** 573–584.

FRANCIS, E. H. &
HOWELLS, M. F.
1973
Transgressive welded tuffs among the Ordovician sediments of NE Snowdonia, N Wales. *J. geol. Soc. London,* **129,** 621–641.

GALE, N. H.,
BECKINSALE, R. D. &
WADGE, A. J.
1979
A Rb–Sr whole rock isochron for the Stockdale Rhyolite of the English Lake District and a revised mid-Palaezoic time-scale. *J. geol. Soc. London,* **136,** 235–242.

GAUSS, G. A. &
HOUSE, M. R.
1972
The Devonian succession in the Padstow area, north Cornwall. *J. geol. Soc. London,* **128,** 151–171.

GEIKIE, A.
1897
*The ancient volcanoes of Great Britain.* London. 2 vols.

GIBBONS, W.
1983
Stratigraphy, subduction and strike-slip faulting in the Mona Complex of North Wales – a review. *Proc. Geol. Assoc. London,* **94,** 147–163.

GREEN, D. H.
1964
The petrogenesis of the high temperature peridotite in the Lizard area, Cornwall. *J. Petrol.,* **5,** 134–188.

1966
A re-study and re-interpretation of the geology of the Lizard peninsula, Cornwall. *In* Present views of some aspect of the geology of Cornwall and Devon, *Roy. geol. Soc. Cornwall. Commem. Vol. for 1964.* 87–114.

GREENLY, E.
1919
The geology of Anglesey. *Mem. geol. Surv. G.B.* 2 vols.

GREIG, D. C.,
WRIGHT, J. E.,
HAINS, B. A. &
MITCHELL, G. H.
1968
Geology of the country around Church Stretton, Craven Arms, Wenlock Edge and Brown Clee. *Mem. geol. Surv. G.B.* 379 pp.

| | | |
|---|---|---|
| HAILWOOD, E. A., GASH, P. J. R., ANDERSON, P. C. & BADHAM, J. P. N. | 1984 | Palaeomagnetism of the Lizard Complex, S.W. England. *J. geol. Soc. London*, **141**, 27–35. |
| HAILWOOD, E. A., BOCK, W., COSTA, LUCY, DUPUEBLE, P. A., MÜLLER, CARLA & SCHNITKER, D. | 1979 | Chronology and biostratigraphy of north-east Atlantic sediments. *Initial Rep. Deep Sea drill. Proj.*, **48**, 1119–1141. |
| HAMPTON, C. M. & TAYLOR, P. N. | 1983 | The age and nature of the basement of Southern Britain: evidence from Sr and Pb isotopes in granites. *J. geol. Soc. London*, **140**, 499–509. |
| HARRIS, P. & DAGGER, G. W. | 1987 | The intrusion of the Carrock Fell Gabbro Series (Cumbria) as a sub-horizontal tabular body. *Proc. Yorkshire geol. Soc.*, **46**, 371–380. |
| HARRISON, R. K. | 1977 | Petrology of the intrusive rocks in the Duffield Borehole, Derbyshire. *Bull. geol. Surv. G.B.*, **59**, 41–59. |
| HARRISON, R. K. & AITKENHEAD, N. | 1982 | Tuffaceous clays of the Widmerpool Formation, Lower Carboniferous, north western Underwood and in the IGS Duffield borehole, Derbyshire. *Rep. Inst. geol. Sci.*, **82/1**. |
| HARRISON, R. K., JEANS, C. V. & MERRIMAN, R. J. | 1979 | Mesozoic igneous rocks, hydrothermal mineralisation and volcanogenic sediments in Britain and adjacent regions. *Bull. geol. Surv. G.B.*, **70**, 57–69. |
| HARRISON, R. K., SNELLING, N. J., MERRIMAN, R. J., MORGAN, G. E. & GOODE, A. J. | 1977 | The Wolf Rock, Cornwall: new chemical, age determination and palaeomagnetic data. *Geol. Mag.*, **114**, 249–264. |
| HASZELDINE, R. S. | 1984 | Carboniferous North Atlantic palaeogeography: stratigraphic evidence for rifting not megashear or subduction. *Geol. Mag.*, **121**, 443–463. |
| HAWKES, J. R. & DANGERFIELD, J. | 1978 | The Variscan granites of SW England: a progress report. *Proc. Ussher Soc.*, **4**, 158–171. |
| HOLDER, A. P. & BOTT, M. H. P. | 1971 | Crustal structure in the vicinity of south-west England. *Geophys. J. R. astr. Soc.*, **23**, 465–489. |
| HOLGATE, N. & HALLOWES, K. A. K. | 1941 | The igneous rocks of the Stanner–Hanter district, Radnorshire. *Proc. Geol. Assoc. London*, **78**, 241–267. |
| HOUSE, M. R. | 1975 | Facies and time in Devonian tropical areas. *Proc. Yorkshire geol. Assoc.*, **40**, 233–288. |
| HOUSE, M. R., RICHARDSON, J. B., CHALONER, W. G., ALLEN, J. R. L., HOLLAND, C. H. & WESTOLL, T. S. | 1977 | A correlation of Devonian rocks of the British Isles. *Geol. Soc. London Spec. Rept.*, **7**, 110 pp. |
| HOWELLS, M. F., FRANCIS, E. H., LEVERIDGE, B. E. & EVANS, C. D. R. | 1978 | Capel Curig and Betws-y-Coed: Description of 1:25,000 Sheet SH75. Classical areas of British geology. *Inst. geol. Sci. London*. 73 pp. |
| HOWELLS, M. F. & LEVERIDGE, B. E. | 1980 | The Capel Curig Volcanic Formation. *Rep. Inst. geol. Sci.*, **80/6**, 23 pp. |
| HOWELLS, M. F., LEVERIDGE, B. E., EVANS, C. D. R. & NUTT, M. J. C. | 1981 | Dolgarrog: Description of 1:25,000 Geological Sheet SH76. Classical areas of British geology. *Inst. geol. Sci. London*, 87 pp. |

HOWELLS, M. F., REEDMAN, A. J. & CAMPBELL, S. D. G. 1986 The submarine eruption and emplacement of the Lower Rhyolitic Tuff Formation (Ordovician), N. Wales. *J. geol. Soc. London*, **143**, 411–423.

HOWELLS, M. F., REEDMAN, A. J. & LEVERIDGE, B. E. 1985 Geology of the country around Bangor. Explan. 1:50,000 Sheet. *Br. geol. Surv. Sheet 106, England and Wales.*

HOWITT, F., ASTON, E. R. & JACQUÉ, M. 1975 The occurrence of Jurassic volcanics in the North Sea. *In* Woodland, A. W. (Ed.) *Petroleum and the Continental Shelf of north-west Europe*, **1**, *Geology*. Applied Science Publishers, London, 379–386.

INESON, P. R., WALTERS, S. G. & SIMON, R. M. 1983 The petrology and geochemistry of the Waterswallows Sill, Buxton, Derbyshire. *Proc. Yorkshire geol. Soc.*, **44**, 341–354.

INGHAM, J. K. & MCNAMARA, K. J. 1978 The Coniston Limestone Group. *In* Moseley, F. (Ed.) The geology of the Lake District. *Yorkshire geol. Soc. Occ. Publ. No. 3*, 121–129.

ISAAC, K. P. 1985 Discussion of papers on the Hercynian back-arc marginal basin of S.W. England. *J. geol. Soc. London*, **142**, 927–929.

JACQUÉ, M. & THOUVENIN, J. 1975 Lower Tertiary tuffs and volcanic activity in the North Sea. *In* Woodland, A. W. (Ed.) *Petroleum and the Continental Shelf of North-west Europe*, **1**, *Geology*. Applied Science Publishers, London, 455–465.

JEANS, C. V., MERRIMAN, R. J. & MITCHELL, J. G. 1977 Origin of Middle Jurassic and Lower Cretaceous fuller's earths in England. *Clay Minerals*, **12**, 11–44.

JEFFERIES, N. L. 1985 The distribution of rare earth elements within the Carnmellis Pluton, Cornwall. *Mineral. Mag.*, **49**, 495–504.

KIRBY, G. A. 1979 The Lizard complex as an ophiolite. *Nature, London*, **282**, 58–61.

1984 The petrology and geochemistry of dykes of the Lizard Ophiolite Complex, Cornwall. *J. geol. Soc. London*, **141**, 53–59.

KIRTON, S. R. 1984 Carboniferous volcanicity in England, with special reference to the Westphalian of the east and west Midlands. *J. geol. Soc. London*, **141**, 161–170.

KIRTON, S. R. & DONATO, J. A. 1985 Some buried Tertiary dykes of Britain and surrounding waters deduced by magnetic modelling and seismic reflection methods. *J. geol. Soc. London*, **142**, 1047–1057.

KNILL, DIANE C. 1969 The Permian igneous rocks of Devon. *Bull. geol. Surv. G.B.*, **29**, 115–138.

KNOX, R. W. O'B. 1979 Igneous grains associated with zeolites in the Thanet Beds of Pegwell Bay, north-east Kent. *Proc. geol. Assoc. London*, **90**, 55–59.

KNOX, R. W. O'B. & ELLISON, R. A. 1979 A Lower Eocene ash sequence in SE England. *J. geol. Soc. London*, **136**, 251–253.

KNOX, R. W. O'B. & FLETCHER, B. N. 1978 Bentonites in the Lower D Beds (Razanian) of the Speeton Clay of Yorkshire. *Proc. Yorkshire geol. Soc.*, **42**, 21–27.

KNOX, R. W. & MORTON, A. C. 1983 Stratigraphical distribution of Early Palaeogene pyroclastic deposits in the North Sea Basin. *Proc. Yorkshire geol. Soc.*, **44**, 355–363.

KOKELAAR, B. P. 1986 Petrology and geochemistry of the Rhobell Volcanic Complex: amphibole-dominated fractionation at an early Ordovician arc volcano in North Wales. *J. Petrol.*, **27**, 887–914.

KOKELAAR, B. P.,
BEVINS, R. E. &
ROACH, R. A.

1985

Submarine silicic volcanism and associated sedimentary and tectonic process, Ramsay Island, S.W. Wales. *J. geol. Soc. London*, **142**, 591–613.

KOKELAAR, B. P.,
HOWELLS, M. F.,
BEVINS, R. E.,
ROACH, R. A. &
DUNKLEY, P. N.

1984

The Ordovician marginal basin of Wales. *In* Kokelaar, B. P. & Howells, M. F. (Eds.) Marginal Basin Geology: Volcanic and associated sedimentary and tectonic processes in modern and ancient marginal basins. *Geol. Soc. Lond. Spec. Publ.* **16**, 245–269.

LAMBERT, R. St. J. &
HOLLAND, J. G.

1971

The petrography and chemistry of the igneous complex of the Malvern Hills, England. *Proc. Geol. Assoc. London*, **82**, 323–352.

LAWRENCE, D. J. D.,
SANDERSON, R. W. &
WATERS, R. A.

1981

A ?Lower Carboniferous dyke from Castleton, Gwent, South Wales. *Proc. Geol. Assoc. London*, **92**, 125–127.

LE BAS, M. J.

1981

The igneous basement of southern Britain with particular reference to the geochemistry of the pre-Devonian rocks of Leicestershire. *Trans. Leics. Lit. Phil. Soc.*, **75**, 41–57.

LEAT, P. T.,
JACKSON, S. E.,
THORPE, R. S. &
STILLMAN, C. J.

1986

Geochemistry of bimodal basalt-subalkaline/peralkaline rhyolite provinces within the southern British Caledonides. *J. geol. Soc. London*, **143**, 259–273.

LEAT, P. T.,
THOMPSON, R. N.,
MORRISON, M. A.,
HENDRY, G. L. &
TRAYHORN, T. C.

1987

Geodynamic significance of post-Variscan intrusive and extrusive potassic magmatism in S.W. England. *Trans. roy. Soc. Edinb.: Earth Sci. (for 1986)*, **77**, 349–360.

LEAT, P. T. &
THORPE, R. S.

1986a

Ordovician volcanism in the Welsh Borderland. *Geol. Mag.*, **123**, 629–640.

1986b

Geochemistry of an Ordovician basalt–trachybasalt–subalkaline/peralkaline rhyolite association from the Lleyn Peninsula, North Wales, UK. *Geol. J.*, **21**, 29–43.

LEE, M. K.

1986

A new gravity survey of the Lake District and three-dimensional model of the granite batholith. *J. geol. Soc. London*, **143**, 425–435.

LEEDER, M. R.

1982

Upper Palaeozoic basins of the British Isles – Caledonide inheritance versus Hercynian plate margin process. *J. geol. Soc. London*, **139**, 481–491.

LOWMAN, R. D. W. &
BLOXAM, T. W.

1981

The petrology of the Lower Palaeozoic Fishguard Volcanic Group and associated rocks East of Fishguard, N Pembrokeshire (Dyfed), South Wales. *J. geol. Soc. London*, **138**, 47–68.

LYNAS, B. D. T.

1983

Two new Ordovician volcanic centres in the Shelve Inlier, Powys, Wales. *Geol. Mag.*, **120**, 535–542.

LYNAS, B. D. T.,
RUNDLE, C. C. &
SANDERSON, R. W.

1985

A note on the age and pyroxene chemistry of the igneous rocks of the Shelve Inlier, Welsh Borderland. *Geol. Mag.*, **122**, 641–647.

MACDONALD, R.,
GASS, K. N.,
THORPE, R. S. &
GASS, I. G.

1984

Geochemistry and petrogenesis of the Derbyshire Carboniferous basalts. *J. geol. Soc. London*, **141**, 147–159.

MACDONALD, R.,
THOMAS, J. E. &
RIZZELLO, S. A.

1977

Variations in basalt chemistry with time in the Midland Valley province during the Carboniferous and Permian. *Scott. J. Geol.*, **13**, 11–22.

MACDONALD, R.,
THORPE, R. S.,
GASKGARTH, J. W. &
GRINDROD, A. R.

1985

Multicomponent origin of Caledonian lamprophyres of northern England. *Mineral. Mag.*, **49**, 485–494.

MACDONALD, R. &          1985    Geochemistry and tectonic significance of the Lower
  WALKER, B. H.                  Carboniferous Cockermouth Lavas, Cumbria. *Proc.*
                                 *Yorkshire geol. Soc.*, **45**, 141–146.

McKEOWN, M. C.,          1973    Geology of the country around Boscastle & Holsworthy.
  EDMONDS, E. A.,                *Mem. geol. Surv. G.B.,* 148 pp.
  WILLIAMS, M.,
  FRESHNEY, E. C. &
  MASSON SMITH, B. J.

MALTMAN, A. J.           1975    Ultramafic rocks in Anglesey – their non-tectonic emplace-
                                 ment. *J. geol. Soc. London,* **131**, 593–606.

MANNING, D. A. C. &      1984    The origins of late-stage rocks in the St Austell granite – a
  EXLEY, C. S.                   reinterpretation. *J. geol. Soc. London*, **141**, 581–591.

MEIGHAN, I. G.           1979    The acid igneous rocks of the British Tertiary Province.
                                 *Bull. geol. Surv. G.B.,* **70**, 10–22.

MILLWARD, D.             1979    Ignimbrite volcanism in the Ordovician Borrowdale
                                 Volcanics of the English Lake District. *In* Harris, A. L.,
                                 Holland, C. H. & Leake, B. E. The Caledonides of the
                                 British Isles – reviewed. *Geol. Soc. London Spec. Publ.*, **8**,
                                 629–634.

MILLWARD, D. &           1985    The Stockdale (Yarlside) Rhyolite – a rheomorphic ignim-
  LAWRENCE, D. J. D.             brite. *Proc. Yorkshire geol. Soc.*, **45**, 299–306.

MILLWARD, D.,            1978    The Eycott and Borrowdale volcanic rocks. *In* Moseley, F.
  MOSELEY, F. &                  (Ed.) The geology of the Lake District. *Yorkshire geol. Soc.*
  SOPER, N. J.                   *Occ. Publ. No. 3*, 99–120.

MOSELEY, F.              1984    Lower Palaeozoic lithostratigraphical classification in the
                                 English Lake District. *Geol. J.*, **19**, 239–247.

O'BRIEN, C., PLANT, J.,  1985    The geochemistry, metasomatism and petrogenesis of the
  SIMPSON, P. R. &               English Lake District. *J. geol. Soc. London*, **142**, 1139–
  TARNEY, J.                     1157.

PATCHETT, P. J. &        1979    U–Pb zircon ages for late Precambrian igneous rocks in
  JOCELYN, J.                    South Wales. *J. geol. Soc. London*, **136**, 13–19.

PATCHETT, P. J.,         1980    Rb–Sr whole-rock isochron ages of late Precambrian to
  GALE, N. H.,                   Cambrian igneous rocks from southern Britain. *J. geol.*
  GOODWIN, R. &                  *Soc. London,* **137**, 649–656.
  HUMM, M. I.

PEDERSON, A. K.,         1975    Early Tertiary volcanism in the Skagerrak: new chemical
  ENGELL, J. &                   evidence from ash-layers in the Mo–Clay of northern
  RØNSBO, J. G.                  Denmark. *Lithos*, **8**, 255–268.

PENNINGTON, J. J.        1975    The geology of the Argyll Field. *In* Woodland, A. W. (Ed.)
                                 *Petroleum and the Continental Shelf of North-west Europe.*
                                 **1**, *Geology* Applied Science Publishers, London, 285–291.

PHARAOH, T. C.,          1987a   The concealed Caledonides of Eastern England: prelimin-
  MERRIMAN, R. J.,               ary results of a multidisciplinary study. *Proc. Yorkshire*
  WEBB, P. C. &                  *geol. Soc.*, **46**, 355–369.
  BECKINSALE, R. D.

PHARAOH, T. C.,          1987b   Geochemical evidence for the tectonic setting of Late
  WEBB, P. C.,                   Proterozoic volcanic suites in Central England. *In*
  THORPE, R. S. &                Pharaoh, T. C., Beckinsale, R. D. & Rickard, D. (Eds.)
  BECKINSALE, R. D.             Geochemistry and mineralisation of Proterozoic volcanic
                                 suites. *Geol. Soc. Lond. Spec. Publ.*, **33**, 541–552.

PHILLIPS, W. E. A.,      1976    A Caledonian plate tectonic model. *J. geol. Soc. London,*
  STILLMAN, C. J. &             **132**, 579–609.
  MURPHY, T.

PIDGEON, R. T. &         1978    Cogenetic and inherited zircon U-Pb systems in granites:
  AFTALION, M.                   Palaeozoic granites of Scotland and England. *In* Bowes,
                                 D. R. & Leake, B. E. (Eds.) Crustal evolution in north-
                                 western Britain and adjacent regions. *Geol. J. Spec. Issue*
                                 *No. 10*, 183–220.

POOLE, E. C.                    1977    Stratigraphy of the Steeple Aston Borehole, Oxfordshire. *Bull. geol. Surv. G.B.*, **57**, 89 pp.

PRICE, N. R. &                  1969    Mineralogy and chemistry of tonsteins from Carboniferous
    DUFF, P. McL. D.                    sequences in Great Britain. *Sedimentology*, **13**, 45–69.

RATTEY, P. R. &                 1984    The structure of S.W. Cornwall and its bearing on the
    SANDERSON, D. J.                    emplacement of the Lizard Complex. *J. geol. Soc. London*, **141**, 87–95.

REEDMAN, A. J.,                 1984    The early Cambrian Arfon Basin of N.W. Wales. *Proc.
    LEVERIDGE, B. E. &                  geol. Assoc. London*, **95**, 313–324.
    EVANS, R. B.

REEDMAN, A. J.,                 1987    The Pitts Head Tuff Formation: a subaerial to submarine
    HOWELLS, M. F.,                     welded ash-flow tuff of Ordovician age, North Wales. *Geol.
    ORTON, G. &                         Mag.*, **124**, 427–439.
    CAMPBELL, S. D. G.

RIDGWAY, J.                     1975    The stratigraphy of Ordovician volcanic rocks on the southern and eastern flanks of the Harlech Dome in Merionethshire. *Geol. J.*, **10**, 87–106.

                                1976    Ordovician palaeogeography of the southern and eastern flanks of the Harlech Dome, Merionethshire, North Wales. *Geol. J.*, **11**, 121–136.

ROLLIN, K. E.                   1986    Geophysical surveys on the Lizard Complex, Cornwall. *J. geol. Soc. London*, **143**, 437–446.

                                1982    *Investigation of the geothermal potential of the U.K.: a review of data relating to hot dry rock and a selection of targets for detailed study.* Institute of Geological Sciences Geothermal Report Series.

ROSS, R. J., NAESER, C. W.,     1982    Fission track dating of British Ordovician and Silurian
    IZETT, G. A.,                       stratotypes. *Geol. Mag.*, **119**, 135–153.
    OBRADOVICH, J. D.,
    BASSET, M. G., HUGHES,
    C. P., COCKS, C. R. M.,
    DEAN, W. T., INGHAM,
    J. K., JENKINS, C. J.,
    RICKARDS, R. B.,
    SHELDON, P. R., TOGHILL,
    P., WHITTINGTON, H. B. &
    ZALASIEWICZ, J.

RUNDLE, C. C.                   1979    Ordovician intrusions in the English Lake District. *J. geol. Soc. London*, **136**, 29–38.

RUSSELL, M. J. &                1983    Origin of the Oslo Graben in relation to the Hercynian–
    SMYTHE, D. K.                       Alleghenian orogeny and rifting of the North Atlantic. *Tectonophysics*, **94**, 457–472.

SANDERSON, R. W. &              1980    Silurian volcanism in the Central Welsh Borderland. *Geol.
    CAVE, R.                            Mag.*, **117**, 455–462.

SCOTT, A.                       1925    The intrusive igneous rocks of north Staffordshire. *In* Gibson, W., The geology of the country around Stoke-upon-Trent. *Mem. geol. Surv. G.B.*, 86–99.

SHACKLETON, R. M.               1959    The stratigraphy of the Moel Hebog district between Snowdon and Tremadoc. *Liverpool Manchester Geol. J.*, **2**, 216–251.

SHEPPARD, S. M. F.              1977    The Cornubian batholith, SW England: D/H and $^{18}O/^{16}O$ studies of kaolinite and other alteration minerals. *J. geol. Soc. London*, **133**, 573–591.

SOMERVILLE, I. D.               1979    A cyclicity in the early Brigantian (D2) limestones east of the Clwydian Range, North Wales, and its use in correlation. *Geol. J.*, **14**, 69–86.

SOPER, N. J.,
WEBB, B. C. &
WOODCOCK, N. J.

1987

Late Caledonian (Acadian) transpression in North West England: timings, geometry and geotectonic significance. *Proc. Yorkshire geol. Soc.*, **44**, 175–192.

SPEARS, D. A. &
KANARIS-SOTIRIOU, R.

1979

A geochemical and mineralogical investigation of some British and other European tonsteins. *Sedimentology*, **26**, 407–425.

STILLMAN, C. J. &
FRANCIS, E. H.

1979

Caledonide volcanism in Britain and Ireland. *In* Harris, A. L., Holland, C. H. & Leake, B. E. (Eds.) The Caledonides of the British Isles – reviewed. *Geol. Soc. London Spec. Publ.* **8**, 557–577.

STONE, M.

1979

Textures of some Cornish granites. *Proc. Ussher Soc.*, **4**, 370–379.

1984

Textural evolution of lithium mica granites in the Cornubian batholith. *Proc. geol. Assoc. London*, **95**, 29–41.

STONE, M. &
EXLEY, C. S.

1986

High heat production granites of southwest England and their associated mineralisation: a review. *Trans. Inst. Min. Metall.*, **95**, 25–36.

STRONG, D. F.,
STEVENS, R. K.,
MALPAS, J. &
BADHAM, J. P. N.

1975

A new tale for the Lizard. *Proc. Ussher Soc.*, **3**, 252.

STYLES, M. T. &
RUNDLE, C. C.

1984

The Rb–Sr isochron age of the Kennock Gneiss and its bearing on the age of the Lizard Complex, Cornwall. *J. geol. Soc. London*, **141**, 15–19.

SUTHREN, R.

1977

Sedimentary processes in the Borrowdale Volcanic Group. *In* Stillman, C. J. (Ed.) Palaeozoic volcanism in Great Britain and Ireland. *J. geol. Soc. London*, **133**, 401–411.

THIRLWALL, M. F.

1981

Plate tectonic implications of chemical data from volcanic rocks of the British Old Red Sandstone. *J. geol. Soc. London*, **138**, 123–138.

1982

Systematic variation in chemistry and Nd–Sr isotopes across a Caledonian calk–alkaline volcanic arc: implications for source materials. *Earth Plan. Sci. Lett.*, **58**, 27–50.

1983

Discussion on implications for Caledonian plate tectonic models of chemical data from volcanic rocks of the British Old Red Sandstone. *J. geol. Soc. London*, **140**, 315–318.

THORPE, R. S.

1972

The geochemistry and correlation of the Warren House, the Uriconian and the Charnian volcanic rocks from the English Pre-Cambrian. *Proc. geol. Assoc. London*, **83**, 269–286.

1974

Aspects of magmatism and plate tectonics in the Precambrian of England and Wales. *Geol. J.*, **9**, 115–136.

1982

Precambrian igneous rocks of England, Wales and southeast Ireland. *In* Sutherland, Diane, (Ed.) *Igneous rocks of the British Isles*, Wiley, London, 19–35.

1987

Permian K-rich volcanic rocks of Devon: petrogenesis, tectonic setting and geological significance. *Trans. R. Soc. Edinb: Earth Sci.*, **77** (for 1986), 361–366.

THORPE, R. S.,
BECKINSALE, R. D.,
PATCHETT, P. J.,
PIPER, J. D. A.,
DAVIES, G. R. &
EVANS, J. A.

1984

Crustal growth and late Precambrian–early Palaeozoic plate tectonic evolution of England and Wales. *J. geol. Soc. London*, **141**, 521–536.

| | | |
|---|---|---|
| THORPE, R. S., COSGROVE, M. E. & CALSTEREN, P. W. C. | 1986 | Rare-earth element, Sr- and Nd-isotope evidence for petrogenesis of Permian basaltic and K-rich volcanic rocks from southwest England. *Mineral. Mag.*, **50**, 481–490. |
| THORPE, R. S. & MACDONALD, R. | 1985 | Geochemical evidence for the emplacement of the Whin Sill Complex of Northern England. *Geol. Mag.*, **122**, 389–396. |
| TREWIN, N. H. | 1968 | Potassium bentonites in the Namurian of Staffordshire and Derbyshire. *Proc. Yorkshire geol. Soc.*, **37**, 73–91. |
| VAN DE KAMP, P. C. | 1969 | The Silurian volcanic rocks of the Mendip Hills, Somerset; and the Tortworth area, Gloucestershire, England. *Geol. Mag.*, **106**, 542–553. |
| WADGE, A. J., HARRISON, R. K. & SNELLING, N. J. | 1972 | Olivine-dolerite intrusions near Melmerby, Cumberland and their age determination by the potassium–argon method. *Proc. Yorks. geol. Soc.*, **39**, 59–68. |
| WADGE, A. J., GALE, N. H., BECKINSALE, R. D. & RUNDLE, C. C. | 1979 | A Rb–Sr isochron age for the Shap granite. *Proc. Yorkshire geol. Soc.*, **42**, 297–306. |
| WALKDEN, G. M. | 1972 | The mineralogy and origin of interbedded clay wayboards in the Carboniferous Limestone of the Derbyshire Dome. *Geol. J.*, **8**, 143–149. |
| WILLIAMS, D. | 1930 | The geology of the country between Nant Peris and Nant Ffrancon, Snowdonia. *Q. J. geol. Soc. London*, **86**, 191–203. |
| WILLIAMS, H. | 1927 | The geology of Snowdon (North Wales). *Q. J. geol. Soc. London*, **83**, 346–431. |
| WOODALL, D. & KNOX, R. W. O'B. | 1979 | Mesozoic volcanism in the North Sea and adjacent areas. *Bull. geol. Surv. G.B.*, **70**, 34–56. |
| WOODCOCK, N. H. | 1984 | The Pontesford Lineament, Welsh Borderland. *J. geol. Soc. London*, **141**, 1001–1014. |
| YOUNG, B., MORGAN, D. J. & HIGHLEY, D. E. | 1978 | New fuller's earth occurrences in the Lower Greensand of southeastern England. *Trans. Instn. Min. Metal.*, **87**, B93–96. |
| ZIEGLER, A. M., McKERROW, W. S., BURNE, W. V. & BAKER, P. | 1969 | Correlation and environmental setting of the Skomer Volcanic Group, Pembrokeshire. *Proc. geol. Assoc., London*, **80**, 407–439. |

# 17

# STRUCTURE

# F. W. Dunning

## Introduction

The geotectonic subdivision of the upper continental crust in Britain is mainly based, as elsewhere, on contrasts of deformation and lithofacies between major rock assemblages (for geophysical evidence see Chapter 18). These contrasting assemblages may either be lateral equivalents in age, such as foreland formations and their correlatives in an orogenic belt, or superposed formations in the classical 'basement-cover' relationship. The basement may form part of an orogenic belt or it may be an older folded cover series, as for example the Carboniferous beneath Permo-Trias in northern England.

There have been ten, possibly eleven, major tectonic episodes in the 3000 million-year history of the earth's crust in Britain as a whole; of these, the following are known with reasonable certainty to have affected the crust beneath England and Wales:

1. One or more unidentified Proterozoic events possibly involved in the formation of older metamorphic rocks now found in Anglesey, Llŷn and Shropshire.
2. The Cadomian orogeny (c. 600 Ma), which probably affected all parts of the Precambrian crust south of the Solway Firth.
3. The Caledonian orogeny (c. 400 Ma) in which Lower Palaeozoic and some older Devonian sequences were deformed.
4. The Hercynian or Variscan orogeny (c. 290 Ma), which deformed the Upper Palaeozoic of south-west England and the mainly Carboniferous cover in the foreland.
5. The Alpine orogeny (c. 25 Ma), which deformed the Permian to Palaeogene cover in southern England, though earlier, late Jurassic to mid-Cretaceous movements (Cimmerian *sensu lato*) were respon-

sible for some of the deformation ascribed until recently to the Alpine orogeny.

These events have shaped and delimited the main geotectonic units of England and Wales (Figs. 17.1, 17.2). These units are:

1. The *Precambrian basement* consisting mainly of supracrustal assemblages: the mainly low-grade metamorphic Monian Supergroup and the mainly non-metamorphic Uriconian–Longmyndian rocks and their equivalents. Gneissic rocks within the Mona Complex in Anglesey and Llŷn (Gibbons 1983a, b) and the Rushton Schists in the Church Stretton area are not substantially older than the supracrustal rocks (Table 17.1). It is doubtful whether any ancient basement of Pentevrian–Icartian type occurs under England and Wales, but any that does would probably comprise small, circumscribed microcontinental terranes accreted to the Cadomian orogenic complex (Thorpe *et al.* 1984).
2. The *Caledonides* – strictly speaking the non-metamorphic late Caledonides – comprising the folded Lower Palaeozoic sequences of the Lake District and Wales. Very little is known about the concealed Caledonides between Wales and the Lake District and under north-east England. In south-east England north and south of the Thames Estuary and in Norfolk, boreholes have revealed what may be an extension of the Caledonian-folded Brabant Massif which might possibly connect with the main North Atlantic Caledonides.
3. The *Hercynides* (*Variscides*) are conventionally regarded as the continuation into England of the Rhenohercynian Zone of the European Variscides. They comprise strongly folded Devonian and Carboniferous formations closely resembling their Rhenish equivalents in many respects but not in structural detail, though both may be large-scale

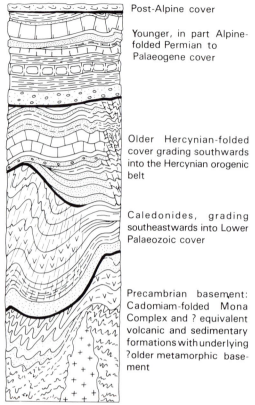

Post-Alpine cover

Younger, in part Alpine-folded Permian to Palaeogene cover

Older Hercynian-folded cover grading southwards into the Hercynian orogenic belt

Caledonides, grading southeastwards into Lower Palaeozoic cover

Precambrian basement: Cadomiam-folded Mona Complex and ? equivalent volcanic and sedimentary formations with underlying ?older metamorphic basement

**Fig. 17.1.** Main structural storeys composing the crust beneath England and Wales.

décollement structures. The Hercynides of southwest England probably continue eastwards under the Isle of Wight–Purbeck region. The concealed extension of the Foreland Thrust Belt to the north which runs eastwards to the Boulonnais involves mainly formations of the following unit.

4. The *Older Cover* (Hercynian-folded Upper Palaeozoic) whose southern border-zone consists of Old Red Sandstone and Carboniferous formations involved in the Hercynian Foreland Thrust Belt bounded by the so-called 'Variscan Front', the remainder consisting mainly of Carboniferous formations whose folding was controlled by older basement structures of north–west, north–east, north–south trends, i.e. 'Charnoid', 'Caledonoid' and 'Malvernoid' respectively.

5. The *Younger Cover* (Permian to Palaeogene) affected to varying extent by late Jurassic and early- to mid-Cretaceous ('Cimmerian', 'late Cimmerian') and Alpine (Oligocene–Miocene) movements.

6. The *post-Alpine Cover* (Neogene–Quaternary), with virtually no deformation.

# Precambrian basement

The Precambrian basement of England and Wales (see Ch. 2) is poorly exposed in many small scattered outcrops, only three of which (Anglesey, Long Mynd–Caer Caradoc, Charnwood Forest) are big enough to give even a hint of regional structure. Seismic velocities correspond reasonably well with upper crustal velocities in Brittany (Bamford *et al.* 1977), while heat-flow studies (Richardson & Oxburgh 1978) suggest low-grade metamorphic basement down to 15 km. From this and the evidence from outcrops and boreholes, it is possible to build up a geotectonic picture in which low-grade metamorphic rocks of Brioverian affinities, affected by older Cadomian events, with possibly in places some enclaves of older, high-grade Pentevrian–Icartian-type orthogneisses, are covered by an extensive dislocated sheet of non-metamorphic late Precambrian (Vendian) ignimbritic volcanics and subordinate molassic sediments with associated calc–alkaline igneous complexes, the whole strongly deformed by later Cadomian events straddling the Cambrian–Precambrian boundary.

## Metamorphic Precambrian

The structure of the largest and best-known area of metamorphic Precambrian in England and Wales – the Mona Complex in Anglesey – is still highly controversial (Chapter 2). Shackleton's (1969) re-reading of the structural evidence on Holy Island has been generally accepted: the main first-phase folds, such as the major Rhoscolyn Anticline which dominates the north-west of the island, face upwards to the SE and have a NE 'Caledonoid' trend. These folds are superimposed on a generally SE-dipping succession and are modified by second-phase minor folds with horizontal axial–planar crenulation cleavage and by later conjugate kink-bands. Greenly's (1919) hypothesis of major recumbent folding is contradicted by the abundant way-up evidence, and his succession must be reversed. However, his estimate of the thickness of the 'Bedded Succession' (c. 6000 m) still stands if it is accepted that the succession (now termed the Monian Supergroup, Gibbons 1983) is continuous from the South Stack Formation to the Fydlyn Felsitic Formation. Shackleton's interpretation of Greenly's 'Autoclastic Mélange', the extensive chaotic material with exotic clasts within the Gwna Group, as an olistostrome or submarine slide-breccia which he compared to the celebrated *argile scagliose* of the Apennines is generally accepted though it may not apply to all outcrops of the mélange. Shackleton also suggested that the 'Basal Gneisses', which Greenly thought to be an older basement to his 'Bedded Succession', were simply a prograde migmatitic facies of the Gwna

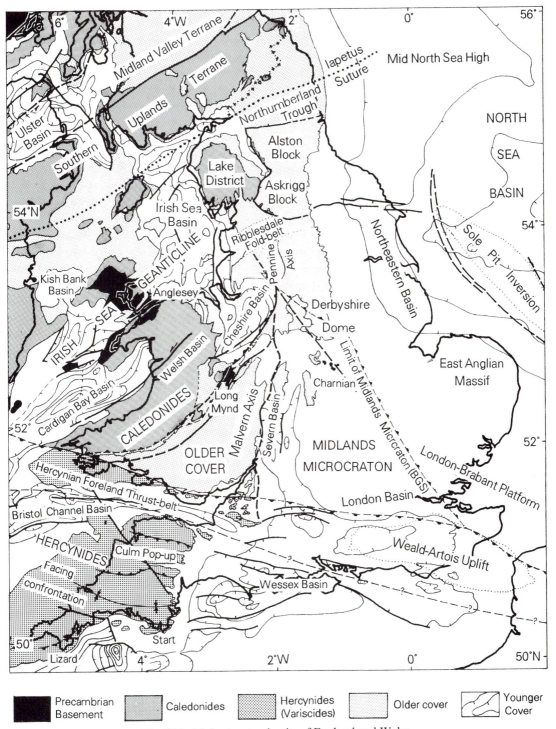

**Fig. 17.2.** Main structural units of England and Wales.

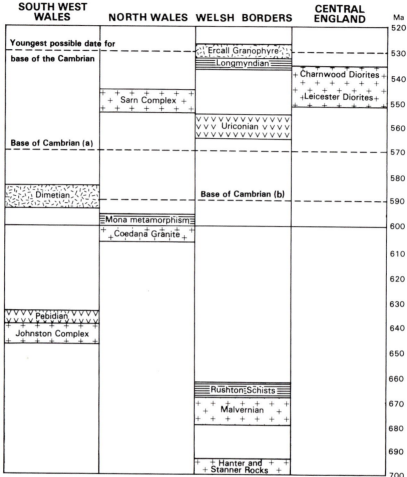

**Table 17.1.** Ages of English and Welsh Precambrian formations from isotope and palaeomagnetic data, after Thorpe *et al.* (1984). Base of Cambrian (a) after Snelling (1985); (b) after Harland *et al.* (1982).

Group, and Thorpe and co-workers (1984) considered this to be supported by the isotope evidence. This suggestion was challenged by Baker (*in* Shackleton 1969, p. 20), who claimed that the 'Gneisses' were retrograded and affected by more deformations than the 'Bedded Succession'. None of the junctions between gneiss and Supergroup can be identified as an unconformity. The gneisses are also taken to be older basement in a radically different synthesis of Mona geology put forward by Barber & Max (1979) on the basis of structural and palynological researches in Anglesey and Llŷn (Fig. 17.3). The main observations and inferences of this synthesis are:

1. The Bedded Succession in northern Anglesey is apportioned to three superposed tectonic units: the lowermost *South Stack Unit* which was overthrust while still undeformed by the middle *New Harbour*

*Unit*, which in turn is probably in tectonic contact with the uppermost *Cemlyn Unit*. The last comprises a variety of formations including some of Greenly's New Harbour Group.

2. The above units are separated from the gneisses in central Anglesey by a pre-Ordovician mylonite zone extending from SW to NE across northern Anglesey.

3. In central Anglesey south of the main gneiss outcrop and Coedana Granite, the metamorphic rocks of Greenly's Penmynydd Zone were in the condition of mylonitised amphibolite-facies gneiss *before* the deposition of the Gwna Group, which, as in Llŷn, is in fault contact with mylonitised gneisses.

4. The Gwna Group, which has yielded allegedly Lower Cambrian acritarchs (Muir *et al.* 1979), is

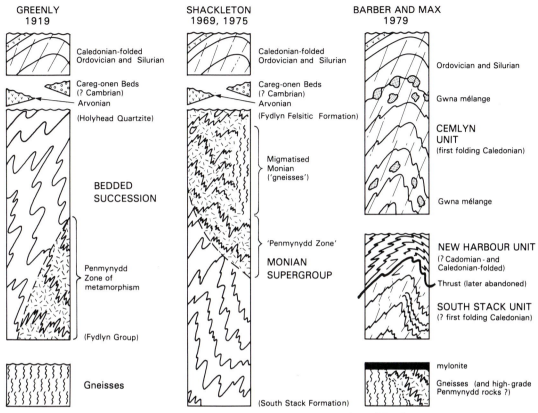

**Fig. 17.3.** Interpretations of the tectonic succession of Anglesey.

conformably overlain by the Ordovician, the base of which rests on the irregular upper surface of the Gwna olistostrome.

5. The main-phase folding of all Groups of the Bedded Succession *other than the New Harbour Group* is the same as that in the Ordovician, i.e. *Caledonian.*

6. The main-phase Caledonian deformation folds an older, intensely folded and refolded planar and mylonitic pressure-solution fabric in the New Harbour Group. This fabric, along with the Penmynydd metamorphism and the Coedana Granite, is of Precambrian age.

Kohnstamm (1980) in a written discussion extended Barber & Max's (1979) recognition of widespread mylonitisation to the point of regarding the whole of the New Harbour Group as a mylonitic–phyllonitic facies of other Mona Groups in the so-called Cemlyn Unit. In the same discussion, Barber & Max preferred no longer to regard the contact between the South Stack Unit and the New Harbour Unit as a thrust, accepting that the New Harbour Unit is in normal stratigraphical contact with the underlying South Stack Unit, and though highly tectonised, has the

same deformational history as the South Stack Unit. Barber and co-workers (1981) later interpreted the Rhoscolyn fold as coeval with congruent second-phase structures which fold an earlier cleavage, considering the latter to be related to a large-scale recumbent anticline overturned to the NW in the South Stack Unit. A major problem had always been posed for Barber & Max by the apparently Precambrian age of the Gwna mélange south of the main gneiss outcrop. Outliers of Arvonian-type volcanics and Cambrian conglomerate rest unconformably on Gwna rocks, which are also overstepped by the Ordovician (Wood *in* Barber & Max 1979, discussion). There also appears to be a metamorphic transition from Gwna greenschists into Penmynydd gneisses in the southern Aethwy block (Chapter 2). Barber & Max (1979) admitted the pre-Ordovician age of the deformation of the Gwna mélange in the Aethwy block and suggested that the mélange there may be of different age to that in northern Anglesey. Gibbons (1983a) regards the Penmynydd schists in Llŷn as mylonitic schists formed in a zone of ductile shear between low-grade Gwna mélange and the granites and gneisses of the Sarn Complex; the Penmynydd rocks are thus post-Gwna as

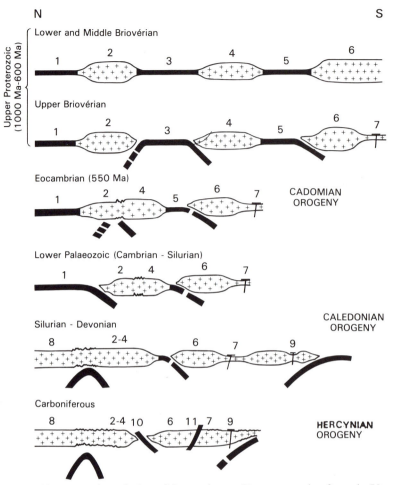

**Fig. 17.4.** Plate-tectonic evolution of the north-west European region from the Upper Proterozoic to the Carboniferous: (1) Iapetus, (2) Mona microcontinent, (3) Longmyndian ocean, (4) Midland block, (5) Channel ocean, (6) Domnonian-Mancellian block, (7) Mid-Armorican domain (thinned crust and volcanics of Belle-Isle-en-Terre), (8) American plate, (9) Trough of St Georges-sur-Loire, (10) Channel suture, (11) Imbricate volcanics of Belle-Isle-en-Terre wedged between the Domnonian-Mancellian and Mid-Armorican domains (after Auvray 1979).

opposed to pre-Gwna in the scheme of Barber & Max.

The occurrence in Anglesey of rocks (serpentinites, gabbros, pillow lavas and cherts) suggestive of oceanic derivation, the presence of blueschists, and the location in relation to Anglesey of calc–alkaline suites such as the Johnston complex and the Arvonian, have inspired a number of plate-tectonic syntheses for the late Precambrian orogenesis of the region (Baker 1973, Thorpe 1974, Wood 1974, Wright 1976, Anderton et al. 1979, Auvray 1979, Barber & Max 1979, Roberts 1979, Cogné & Wright 1980, Barber et al. 1981, Thorpe et al. 1984). All involve a Mona

(Anglesey) microplate separated by a marginal sea from a 'Longmyndian' continent, with subduction to the SE or NW or both under Anglesey, with or without SE subduction under the Longmyndian continent. In Auvray's (1979) synthesis (Fig. 17.4), three oceans are separated by three continents which respectively contract and collide at various times: *Iapetus ocean* / Mona microcontinent / *Longmyndian ocean* / Midland block microcontinent / *Channel ocean* / Domnonian–Mancellian continent (North Brittany–Normandy). The Channel ocean is marked by a NE–SW ophiolite belt based mainly on geophysical evidence (Lefort 1977); this belt could, according to Cogné & Wright (1980), represent the NE–SW 'Celtic' suture north-west of Anglesey, offset hundreds of kilometres to the east by an E–W transform skirting the south coast of Britain.

In the synthesis of Barber *et al.* (1981), a Precambrian continental fragment of older basal gneiss and younger Penmynydd metasediments was swept into a S-dipping Cambrian subduction zone in central Anglesey. In northern Anglesey, on the other hand, they envisage a *northward* directed subduction zone beneath a northern continent, concomitant with the formation of the Gwna mélange in the Cambrian.

Gibbons (1983a) carefully evaluated the evidence for Monian subduction in a most valuable review of Monian geology, underlining the lack of cohesion in the otherwise 'typically plate-tectonic' lines of evidence. He also considers that the mélanges are more suggestive of continental collision than of subduction complexes. The 'ophiolites' are stratigraphically and geochemically disparate rock-units which cannot realistically be equated with the components of Monian oceanic crust. Evidence for the presence of other components of a subduction regime (a forearc basin, an accretionary prism or a Monian volcanic arc) is non-existent. Gibbons underlines the disruptive nature of all major contacts in the Mona Complex and suggests an analogy with Cordilleran 'suspect terranes', the Coedana granite/hornfels/Sarn complex/gneisses representing slivers of continental crust sheared into the Monian Supergroup by strike–slip faulting.

Several Precambrian complexes may arguably be included as 'metamorphic' types. The **Rushton Schists**, brought up in the Church Stretton fault-zone, are greenschist-facies metasediments of pre-Uriconian age. Their foliation strikes NW to NNE and some specimens are microscopically crenulated. The nearby **Primrose Hill Gneisses** comprise cataclastically deformed granite, hybrid gneiss and schists which resemble the Malvernian.

The **Malvernian** is an extraordinarily varied assemblage of highly altered calc–alkaline plutonic rocks, mainly diorite and tonalite, cut by many pink granite and pegmatite veins and two generations of basic dykes. The N–S outcrop is a string of fault-blocks in which almost every mappable boundary is a shear-zone (Penn & French 1971). Internally, the fault-blocks are foliated, sheared and fractured, the shearing being most intense in the south. Shear-belts of quartzose chlorite- and biotite-schists, mylonite zones and also the rough foliation in the diorites all trend more or less E–W. The schists are folded in a few places into small-scale angular or isoclinal folds. Baker (1971) has made the important observation that the N–S Malvernoid trend is not reflected in the

internal fabric of the complex, though some N–S shearing and fracturing may be seen in the northern blocks.

The petrogenesis and metamorphic status of the **Johnston Complex** in south-west Dyfed are uncertain; the rocks, mainly quartz-diorites of possible hybrid origin (Wright 1969, Owen *et al.* 1971, p. 42), are commonly gneissose. The foliation, which is cut by undeformed basic dykes, trends WSW and NNW, oblique to the Hercynian thrusting (Jones *in* Cantrill *et al.* 1916), and is almost certainly of Precambrian age.

## Non-metamorphic Precambrian

Wherever they are seen, the non-metamorphic Precambrian volcanic and sedimentary formations (Uriconian, Charnian, Pebidian, Longmyndian) are strongly folded. The folding in the outcrops may be local and related to the fault-lines along which several of the occurrences are brought up. A large area of the Phanerozoic subcrop in central–eastern England is also composed of the same formations, possibly in an eroded Basin-and-Range-type fault-block structure.

The **Uriconian** volcanics in the Caer Caradoc and Cardington outcrops near Church Stretton are disposed in an anticline and two synclines trending NW. In detail the formations are affected by much minor folding, crushing and shearing, some of it associated with thrusts which are themselves folded (Mitchell *in* Greig *et al.* 1968). The thrusting may be pre-Longmyndian and is to be distinguished from other large-scale thrusts (Sharpstones, etc.) of post-Caradoc/pre-Silurian age affecting the same outcrops. In other Uriconian outcrops in the Church Stretton area, the strike is parallel to the Church Stretton fault-zone, whilst in the Wrekin and Wrockwardine ranges the Uriconian formations strike E–W.

The well-known reconstruction of the sub-Triassic surface in Charnwood Forest by Watts (1947) showed the **Charnian** disposed in a complex strike- and cross-faulted anticline with parasitic minor folds plunging SE. According to Evans (1968), the folding is concentric with much bedding slip. The most conspicuous tectonic effect is the pervasive strong cleavage which affects all types of rock including the intrusive 'porphyroids'. Quartz and feldspar grains are recrystallised and conglomerate pebbles flattened and elongated in the plane of cleavage. The latter does not coincide exactly with the trend of the main anticline – it is more westerly – and according to Evans (1979) may be a superimposed Caledonian cleavage.

The **Caldecote Volcanic Formation** at Nuneaton strikes NW and dips SW at about 10° less than the overlying Cambrian (Allen 1968). Though structurally disturbed and affected by Precambrian faulting, the

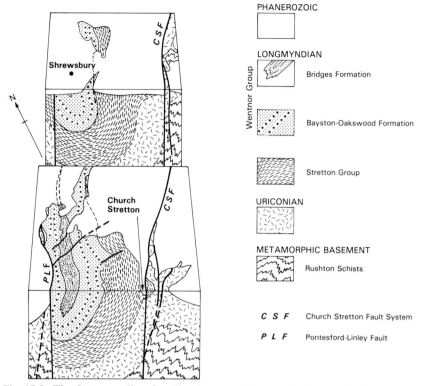

**Fig. 17.5.** The Longmyndian Syncline between Church Stretton and Haughmond Hill.

rocks, as Baker (1971) has pointed out, are quite uncleaved, in striking contrast to the Charnian.

The **Pebidian** volcanic rocks exposed in the Caledonian St David's Anticline (Shackleton 1975) are apparently folded into open NE-trending Precambrian folds on the coast, but inland they trend NW. 'Dimetian' granophyre intrusions are heavily fractured. In the Hayscastle Anticline, the Pebidian pyroclastics under the basal Cambrian conglomerate are highly cleaved. Dips recorded in the **Benton Volcanic Group** (Jones *in* Strahan *et al.* 1914) suggest a northerly or northwesterly grain. The work of James (1956), supported by Greig and co-workers 1966, strongly suggested that the very thick **Longmyndian** sequence was disposed in a large isoclinal syncline with a gentle southerly plunge (Fig. 17.5). The syncline is overturned to the east in the Long Mynd and to the west in the Shrewsbury area to the north. The core, in the Bridges Formation, may comprise small, tightly packed isoclines. The more shaly beds are cleaved, though nowhere as strongly as in the Charnian. The older Stretton Group is missing on the overturned western limb of the syncline, presumably cut out by westward overstep of the unconformable Wentnor Group. James

described an inverted unconformity of Wentnor grits on 'Western' Uriconian in the Chittol area on the western margin of the Long Mynd. Greig *et al.* (1968) considered the Longmyndian sedimentation and folding to be related to the Church Stretton and Pontesford–Linley faults, but Baker (1971) suggested that the major fold is a gravity structure possibly off the flank of the Monian tectogene.

## Caledonian Fold-belt

The correspondence of the seismic velocity of the pre-Caledonian basement south of the Southern Uplands with upper crustal velocities in Brittany suggests that the Cadomian orogeny may have affected the entire basement of England and Wales. This would imply that the Caledonian sequence (Table 17.2) starts everywhere with Lower Palaeozoic formations resting unconformably on Precambrian basement. The only exception may be the small Silurian inliers on the Scottish border which are perhaps part of a quite distinct geotectonic unit lying to the north of the Iapetus Suture. Only in Wales is the complete Lower

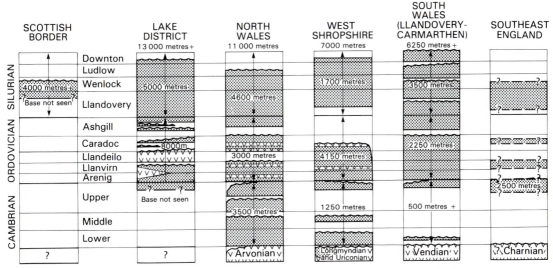

**Table 17.2.** Sequences that build the Caledonides. Figures give aggregate thicknesses of strata assigned to the Cambrian, Ordovician and Silurian systems (v v ore mainly volcanic sequences).

Palaeozoic sequence exposed; elsewhere, as for example in the Lake District and in the concealed 'Brabant Massif', rocks no older than Ordovician and Silurian are exposed at the surface or reached by drilling. But by analogy with their extensions in south-eastern Ireland, south–central England and in Belgium, the presence of Cambrian at depth may reasonably be inferred.

The exposed and concealed Caledonides in England and Wales fall into five major geotectonic units. The Scottish border outcrops form part of the *Southern Uplands–Longford Down–Central Irish Inliers Belt*. The Lake District and the Cross Fell and Austwick inliers belong to the *Lake District–Isle of Man–Leinster Belt*. The *Irish Sea 'Horst'* (or Uplift) separates the latter from the *Welsh Basin*; both are probably constricted or faulted out north-eastwards against a sub-Pennine Precambrian block. Under the Thames Estuary and East Anglia is the north-western portion of the *Brabant–Ardennes Fold Zone* (Brabant Massif), which according to Wills (1978) may be connected around the Midland 'Craton' to the main Caledonian belt of north-western Britain.

The two small Silurian outcrops at the Scottish Border SW of the Cheviot massif and at Berwick-upon-Tweed are clearly part of the Southern Uplands structure. The steeply dipping and inverted Wenlock greywackes and mudstones in the Cheviot Hills are on strike with rocks in the Riccarton inlier (Warren 1964) that regionally young to the south-east whilst locally displaying NW-younging sequences in N-facing strike-faulted asymmetric folds. This is, of course, the characteristic structural style of the Southern Uplands, explained in terms of imbricate thrusting related either to a deep décollement (Fyfe & Weir 1976) or to an accretionary prism (McKerrow *et al.* 1977) or perhaps both. The Berwick-upon-Tweed outcrop is adjacent to the Eyemouth section described by Shiells & Dearman (1963, 1966) and is likely to have the same structure of early N-facing monoclines with corrugated flat limbs.

In the 30 to 40 km gap between the Southern Uplands and the Lake District across the Solway Firth, a major structure or structures must intervene: the near vertical dip of the Eycott Group volcanics in the northern Lake District would, if repeated northwards in younger formations, bring in high Silurian rocks within at most 10 km, whilst steeply dipping Wenlock greywackes crop out in the southernmost Southern Uplands. The sub-Solway structures could be either a large complementary syncline and anticline or, more likely, one or more N-dipping imbricate thrusts on the Southern Uplands pattern. If the *Iapetus Suture* is present beneath the Solway Firth, as suggested by the BIRPS WINCH and NECLINE deep seismic profiles (Klemperer & Matthews 1987, Freeman *et al.* 1988), it would be represented by the most southerly imbricate thrust or would perhaps be a steep Ordovician structure terminating upwards against overstepping Silurian.

The structure of the **Lake District** has been admirably summarised by Soper & Moseley (1978) and the following is taken largely from their account. The three main stratigraphic units – the *Skiddaw Group*, the *Borrowdale Volcanic Group* and the *Windermere Group* are also structurally distinct (Fig. 17.6). Though all are profoundly affected by the end-Silurian orogeny, the Tremadoc–Llanvirn Skiddaw Group displays N-trending pre-Borrowdale, pre-cleavage folds, whilst the open ENE folds of the Borrowdales themselves have a significant pre-Windermere component of mid-Ordovician deformation, which Branney and Soper (1988) now interpret in terms of volcano-tectonic faulting and block tilting in association with caldera-collapse. The main Lake District anticline may partly date from the same mid-Ordovician episode. In the ductile, strongly layered pelitic and semipelitic lithologies of the Skiddaw Group, complicated outcrop patterns result from the interference of main-phase end-Silurian, ENE upright folds with the W-facing, N–S pre-Borrowdale folds.

Early $F_1$ folds in the Caldew Valley (Fig. 17.7) clearly result from soft-sediment deformation which is increasingly recognised as the principal mechanism of folding on all scales in the Skiddaw Group. Banham and co-workers (1981) described a major N-facing, pre-Eycott (pre-Llanvirnian) recumbent fold, the Gillbrea 'nappe', formed by gravity-gliding down a N-facing palaeoslope in the Cockermouth area. The main Caledonian cleavage, which is not quite congruous with the ENE folds in the Borrowdales, appears to have formed immediately before the emplacement of the Lake District batholith. It is stronger in the Arenig slates in the west than in the younger Llanvirn slates in the east. The cleavage is locally deformed, where steep and well-developed, by a sub-horizontal crenulation or fracture cleavage associated with recumbent folds ($F_2$ of Simpson 1967)

**Fig. 17.6.** Block diagram of the central Lake District.

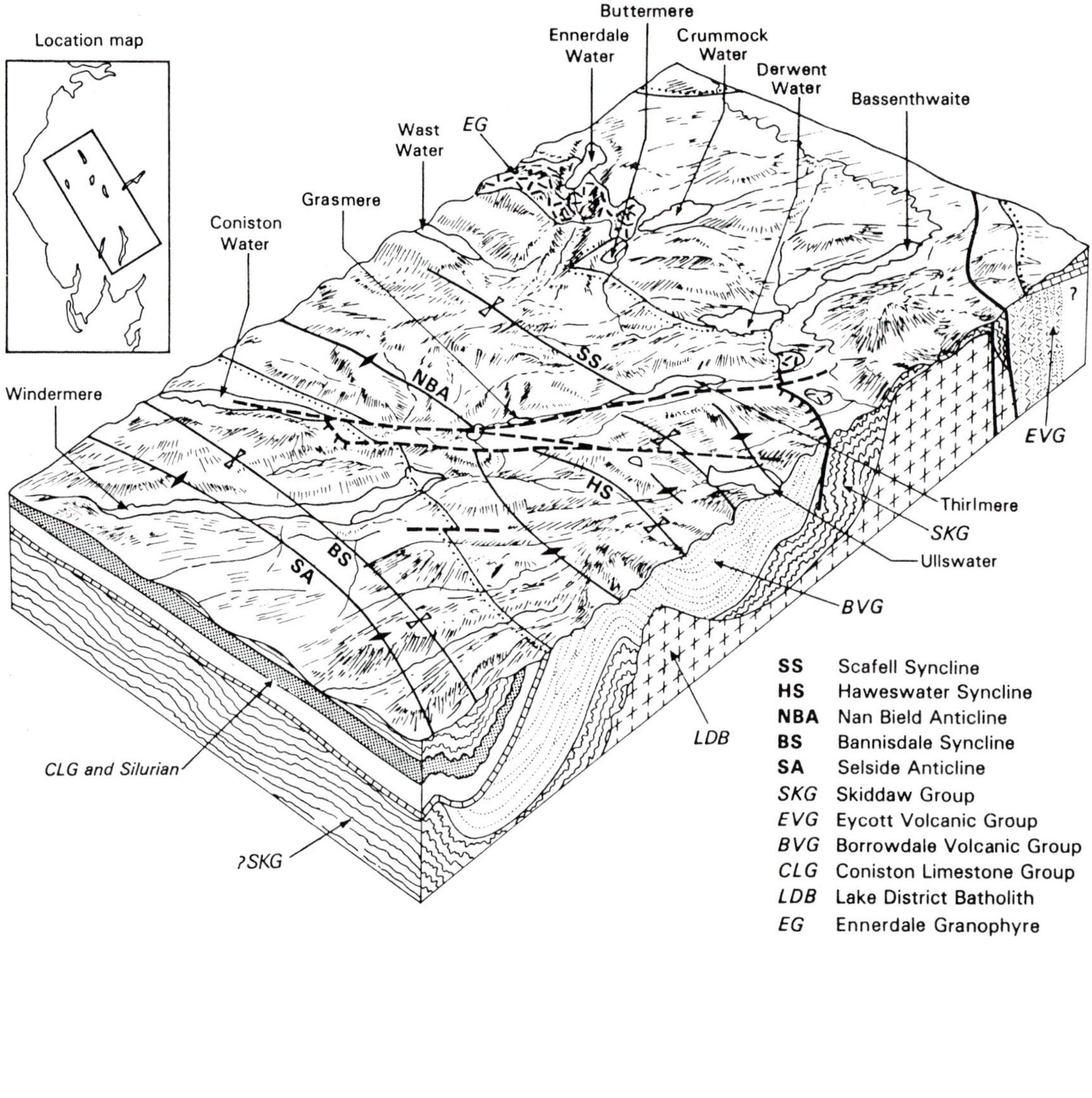

SS  Scafell Syncline
HS  Haweswater Syncline
NBA Nan Bield Anticline
BS  Bannisdale Syncline
SA  Selside Anticline
SKG Skiddaw Group
EVG Eycott Volcanic Group
BVG Borrowdale Volcanic Group
CLG Coniston Limestone Group
LDB Lake District Batholith
EG  Ennerdale Granophyre

**Fig. 17.7.** Early cleavage-free pre-Borrowdale folds of probable synsedimentary type preserved in hornfelsed Skiddaw slates, River Caldew (B.G.S. A.6672).

which post-date andalusite in the Skiddaw granite aureole. Younger brittle-style kink-bands and monoclines are widespread.

The Borrowdale Volcanic Group, with its massive, competent lava and ignimbrite units interbedded with thick ashfall tuffs, is folded into major open buckles affecting the whole thickness of the volcanic pile. These buckles must embody a sizeable end-Silurian component of deformation, but the latter is most evident in the non-congruous cleavage intensely developed in the tuffs and in other zones of high strain such as the axial region of the Nan Bield Anticline. The base of the Borrowdale Volcanic Group is in some places an unconformity (Simpson 1967) where the Group rests on eroded synsedimentary folds in the Skiddaw Group; in other places there is a concordant contact and in still others it is disrupted by faulting and décollement/disharmonic thrusting.

The end-Silurian deformation in the Silurian rocks is, as elsewhere, lithology-dependent: open concentric folds characterise massive arenites such as the Ludlovian Coniston Grits, whereas in well-layered turbidite and mudstone/shale sequences, as in the A6–M6 roadway sections, the folds are tighter and rather straight-limbed. The cleavage, which diminishes towards the south, is again not quite congruous with the folds. On the regional scale, belts of minor folding alternate with homoclinal steep zones which are parallel to the two main Caledonian folds, the Bannisdale Syncline and Selside Anticline. The incompetent Llandoverian Stockdale Shales are surprisingly unfolded but appear to be a zone of décollement thrusting between the Silurian greywackes and the underlying Borrowdale Volcanic Group.

Lake District rocks form the basement of the Alston and Askrigg Blocks and crop out in the Cross Fell,

Austwick and Teesdale inliers. In the Cross Fell inlier (Burgess & Wadge 1974), the Arenigian greywacke-siltstones of the Murton Formation are folded in the south into asymmetric E–W isoclines with strong slaty cleavage which is deformed by a subhorizontal crenulation cleavage associated with folding parallel to the NNW Pennine Line. In the north, where cleavage is weak or absent, the rocks are folded into NE Caledonide folds. A comparable style of folding is seen in the Llanvirnian Kirkland Formation. The Borrowdale and Upper Ordovician–Silurian rocks display open E–W folding. In the Austwick inliers, the 'Ingletonian' (equivalent to older Skiddaw Group rocks) comprises greywackes and slates that are strongly indurated and isoclinally folded in marked contrast to the open folding of the Upper Ordovician–Silurian succession (Fig. 17.8), though the trend of folding in both is parallel to the North Craven Fault (Leedal & Walker 1950). Soper *et al.* (1987) explained the arcuate trend of slaty cleavage in northern England, both in outcrop and as suggested by trends in the overlying cover, as the result of Early Devonian (Acadian) northward drift of the Midland Microcraton acting as a rigid indenter. This northward drift into an embayment produced by suturing of Laurentia and Baltica created transpressive strain that resulted in the clockwise transection of cleavage relative to fold traces.

The **Welsh Basin** consists of a Lower Palaeozoic sequence up to 10 km thick which occupies a steep-sided trough with a floor of Precambrian continental crust around 18 km thick. To the north-west, the Lower Palaeozoic cover on the exposed Precambrian horst of Anglesey is strongly folded and overthrust with SE vergence; a similar vergence of Caledonian folds characterises most of Wales with the exception of parts of the Nantlle slate belt and the coastal Aberystwyth Grits, some folds in which face NW. The age of the main folding in central, NW and SW Wales in post-Ludlow and possibly post-Downtonian; in eastern Wales, Shropshire and extreme SW Wales, it is post-Ordovician but pre-Llandovery. The rocks have undergone low-grade zeolite-facies metamorphism but are unaffected by batholithic plutonism.

The main structures in Wales (Fig. 17.9) are, south

**Fig. 17.8.** Structure of the Austwick inliers, after Leedal & Walker (1950). Section after King & Wilcockson (1934).

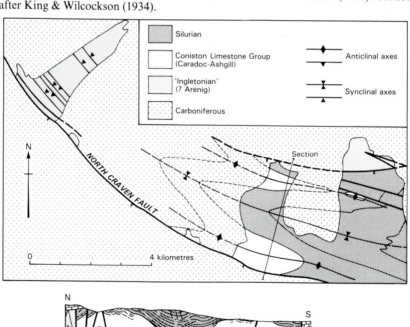

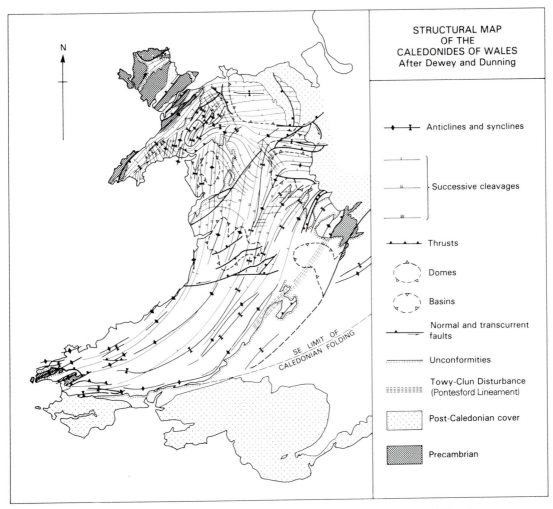

**Fig. 17.9.** Structural map of the Caledonides of Wales (after Dunning 1966 and Dewey 1969).

of a line from Aberdovey to Church Stretton, a simple alternation of anticlines and synclines roughly parallel to the basin contours. North of this line, the structure is complicated by domes, depressions, major faults, sharp inflexions of strike and by polyphase deformation. Central and SW Wales are dominated by the Towy and Teifi Anticlines separated by the Central Wales Syncline. The Towy Anticline (Fig. 17.10) has a history of pre-Llandovery and pre-Wenlock uplift (George 1963). Although the Towy Anticline and other folds west of Carmarthen swing into a latitudinal trend, as if diverted by Hercynian pressure, the St David's and Hayscastle Anticlines further west show no such tendency. The main structures of North Wales are the Padarn Ridge, the Snowdon and Lleyn

synclinoria, the Harlech Dome, the Tarannon, Llanderfel and Llangollen Synclines, the Berwyn Dome and the Denbigh–Clwyd depression (useful descriptions are given in Bassett 1969). Shackleton (*in* Beavon 1963, discussion) and later Rast (1969) and Bromley (1969) suggested the existence in Snowdonia of a Caradocian subvolcanic magma dome with surrounding rim-syncline later folded within the Snowdon synclinorium (Fig. 17.11). Rast (1969) further suggested that the Harlech and Berwyn Domes might also be due to volcanic doming and that the quaquaversal cleavage around the two domes may be related to their growth. However, Fitch (*in* Bromley 1969, discussion) and Roberts (1979) cited key stratigraphical, tectonic and geophysical evidence that militates against the

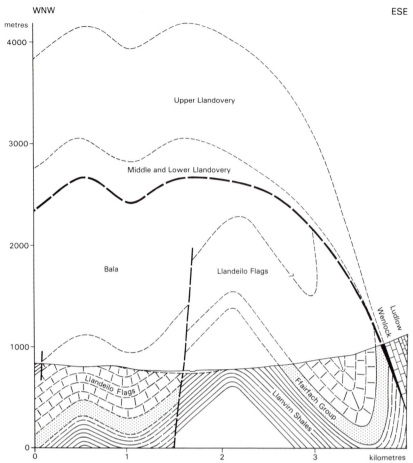

**Fig. 17.10.** Diagrammatic reconstruction of the Towy Anticline in the neighbourhood of Llandeilo showing a pre-Llandovery stage of development (after George 1970).

concept of the magma dome and rim-syncline. Lynas (1970) proposed that the quaquaversal cleavage is a dewatering-induced, diagenetic fabric developed in wet sediments during the early volcano-tectonic doming. The gently dipping quaquaversal cleavage at Ffestiniog ($S_1$ of Lynas 1979; $S_0$ of Roberts 1979) is subparallel to bedding and unrelated to folds. In the view of Hawkins & Jones (1981), it may be due mainly to flexural slip in an early concentric stage of folding prior to tighter folding accompanied by steep axial planar cleavage which cuts the quaquaversal cleavage and can be equated with the main cleavage of Snowdonia ($S_1$ of Helm *et al.* 1963 and Roberts 1979; $S_2$ of Lynas 1970 and Hawkins & Jones 1981). Bromley (1971) on the other hand believed that the low-angle quaquaversal cleavage steepens to become the Snowdonia cleavage whilst the steep cleavage that intersects

the quaquaversal cleavage – giving rise to pencil slates – is of purely local extent. The steep cleavage is axial planar to N-plunging folds on the northern flank of the Harlech Dome which are co-linear with the periclinal folds of the Harlech Dome and with the main folds and cleavage of central and south-west Wales (see Dewey 1969, fig. 9 and discussion). In NE Snowdonia, the main NE-trending $S_1$ cleavage and folds are flexed by NW-trending folds with a related crenulation cleavage; the same deformation was claimed by Helm *et al.* (1963) to be responsible for the great deflection of folds from NNE to ESE around the Berwyn Dome, but other authors, for example Dewey (1969) and Roberts (1979), regarded the deflection as a primary structure. Coward & Siddans (1979) ascribe the overall structural pattern of North Wales to impingement on a rigid block underlying the Berwyn Dome.

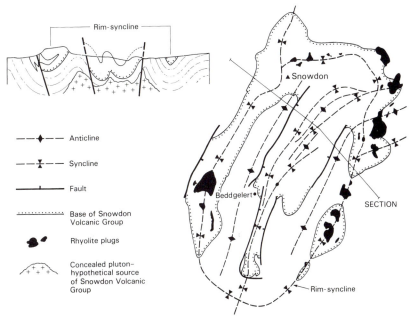

Anticline

Syncline

Fault

Base of Snowdon Volcanic Group

Rhyolite plugs

Concealed pluton-hypothetical source of Snowdon Volcanic Group

Rim-syncline

Snowdon

Beddgelert

SECTION

Rim-syncline

**Fig. 17.11.** The Snowdon rim-syncline (after Dewey 1969).

Campbell and co-workers (1985), in a useful discussion of North Wales cleavage and fold development, invoke ramping over the inclined roof and sidewall of the Tan y grisiau Microgranite to explain the local geometry of the main cleavage, as well as moulding around basement blocks under the Harlech and Berwyn Domes for the larger arcuate structures. Kokelaar (1988) related the Welsh Ordovician vulcanism to N–S grabens produced by E–W extension associated with N–S plate convergence. The latter resulted in strike-slip movements that brought the Lake District–Leinster terrane into its present juxtaposition with Wales, in the manner suggested by Nutt & Smith (1981).

Bulk strain studies by Coward & Siddans (1979) indicated vertical extensions up to as much as 157 per cent (in the Nantlle Slate Belt) which, to be compatible with the present-day crustal thickness beneath North Wales, require a deep décollement under all of North Wales, above which there has been an overall shortening of 43 km. It is of interest to note that Shackleton (1954) rejected the possibility of décollement on the grounds that Palaeozoic/Precambrian contacts in Anglesey and Llŷn are undisturbed. Woodcock (1984a) has proposed several ways of achieving shortening, amongst them the appealing one of reverse movement on old listric faults associated with the formation of the Welsh basin, though he himself favours a strike-slip regime for the latter.

East of a line from Llandeilo to Welshpool, which also marks the major facies change in Silurian rocks, slaty cleavage dies out and the Silurian, along with the Downtonian and Dittonian are only very gently folded, whereas the Ordovician beneath transgressive Upper Llandoverian and Wenlockian sediments is more strongly folded and faulted against the Long Mynd horst (George 1963). The latter, defined by the Pontesford–Linley and Church Stretton fault-systems, probably passes under the Llandyfaelog (Careg Cennen) Disturbance, possibly to underlie the Benton block, across which the characteristic Welsh Borderland Ordovician–Silurian relations assert themselves. Woodcock (1984b) identifies a *Pontesford Lineament* embracing the Clun Forest Disturbance (Holland 1959), the Pontesford–Linley Fault and the belt of folding and faulting which extends down the west side of the Llandrindod inlier into the east flank of the Towy Anticline. The Pontesford Lineament was intermittently active from the mid-Ordovician to the Triassic, and Woodcock believes it to have a large strike-slip component.

In the Old Red Sandstone and older cover of the Midland Craton in south-east Wales and Herefordshire, a curving line joining the Llandyfaelog Disturbance to the Abberley Hills separates a zone of weak Caledonian platform folding on the north-west from a zone of weak to strong (along the Malvern Line) Hercynian folding on the south-east. The Ordovician and Silurian formations diminish across the Long Mynd horst from basinal thicknesses (4 km and 2·7 km respectively) to platform thicknesses of around 1 km each, thus defining the north-west margin of the Midland Craton (Watson 1975). The southern margin of the Craton is the buried Hercynian 'Front', whilst on the east, the Craton subsides beneath folded Silurian formations which may form part of the Brabant–Ardennes fold-zone or perhaps the more extensive 'Eo-North Sea Geosyncline' (Wills 1978). Wills saw the structure of the Craton and its unconformable cover as one of NW-trending horst-like ridges and graben-like troughs in which augmented thicknesses of Cambrian accumulated (2,750 m+ in the Minety Borehole near Swindon) before the Upper Llandovery transgression and planation. In many areas Ordovician rocks were largely eroded or not deposited. Evans (1979) interpreted the Precambrian and Lower Palaeozoic rocks under the East Midlands and southern Pennines as the deformed and intruded infill of a Caledonian NW-trending aulacogen (a graben of quasi-geosynclinal character). Results of a BGS multidisciplinary study (Pharaoh *et al.* 1987) involving isotope, geochemical and dipmeter studies on boreholes in eastern England supported the idea of a NW–SE Caledonian belt meeting the main NE–SW Caledonides at a syntaxis in central England, as originally proposed in a classic paper by Turner (1949).

Although the palaeomagnetic evidence for Caledonian plate movement is equivocal and some Caledonian sedimentation clearly took place on a continental basement, other evidence, particularly the provinciality of Cambrian and early Ordovician shallow-water faunas and the occurrence of oceanic radiolarian cherts within the fold-belt, favours a plate-tectonic mechanism for the Caledonian orogenic cycle. A multiplicity of plate tectonic evolutionary theories has been produced for the British Caledonides; the difficulties have been stressed in an excellent review essay by Dewey (1982). It is likely, however, that the Lower Palaeozoic in England and Wales was deposited on a sialic shelf underlain from the Arenig onwards by a subduction zone of moderate dip, which was responsible for extensive vulcanism in the Ordovician but apparently became inactive when the Iapetus Ocean closed at the end of the Ordovician. The granites intruded into the Caledonian crust of northern England – the Lake District/Weardale batholith – are seemingly too young to be ascribed to Caledonian subduction and are more likely to have resulted from deep melting in the collision zone, coupled with lateral flow.

## Hercynian Fold-belt

Enormous strides have been made in the past decade or so in understanding the structure of the Hercynian fold-zone of south-west England. Starting with the seminal synthesis of Sanderson & Dearman (1973), there has come the recognition of extensive nappe tectonics in north Cornwall and south Devon (Isaac *et al.* 1982, Coward & McClay 1983, Selwood *et al.* 1984), and indeed the interpretation of the whole fold-belt in terms of 'thin-skinned' décollement tectonics (Shackleton *et al.* 1982, Coward & Smallwood 1984).

The broad outcrop pattern of SW England – the Culm synclinorium flanked by Devonian rocks dominated in the south by the great Dartmouth–Watergate anticline, with upfaulted metamorphic massifs forming the two southern promontories of the Lizard and Start – is deceptively simple. The true complexity first emerged in the regional structural scheme of Sanderson & Dearman (1973), whose structural zones correlate approximately with the isotope age provinces of Dodson & Rex (1971) (Fig. 17.12). N-facing folds in the Devonian and Dinantian rocks of Zone 1 pass southwards into upright folds in the Namurian and Westphalian of Zone 2. The Devonian beds of North Devon mainly dip off a major asymmetric anticline with a thrust northern limb which runs NNW through Lynton. The Silesian Bude and Crackington Formations of Zone 2 are disposed in an anticlinorium and two synclinoria. In Zone 3 the folds fan over to verge south and assume in Zone 4 the well-known recumbent chevron style, the area of outcrop of which is much expanded by N-dipping lag-faults which Freshney (1977) related to the emplacement of the Bodmin granite. Lloyd & Whalley (1986) relate these faults to narrow zones of simple shear which modify originally upright chevron folds, leading to the apparent S-facing overfolding. Zones 1–4 form a superstructure resting on the major N-dipping Rusey Fault Zone above a zone of intense deformation (Zone 5) characterised by large-scale $F_1$ recumbent isoclinal folds (Hobson & Sanderson 1975), variable high strain and anomalous N–S minor fold directions associated with a stretching lineation in shallow-water rocks shown to be of Lower Carboniferous and Upper Devonian age (Selwood *et al.* 1985). Below this zone, recumbent folds in Devonian slates reputedly face south (Zone 6), but along a line from Polzeath on the coast eastwards to the southern margin of the Dartmoor granite, a remarkable confrontation was thought to take place: south of the 'confrontation line', folds in Zone 7 now face *north* and maintain this facing direction to the Lizard Boundary Thrust. Roberts & Sanderson (1971) suggested that the north-facing folds were older than the south-facing folds; the horizontal slaty cleavage ($S_1$) in the south is folded into a secondary crenulation cleavage ($S_2$) equivalent to the primary $S_1$ slaty cleavage in the north. Gauss (1973) on the other hand considered that the north- and south-facing folds were coeval and were brought together by

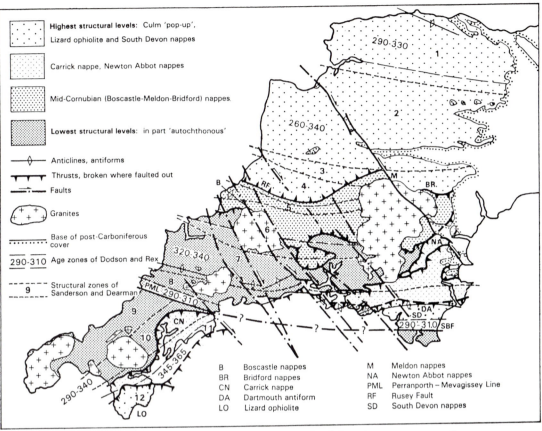

**Fig. 17.12.** Hercynian structure of south-west England, based on numerous authors.

| | |
|---|---|
| B | Boscastle nappes |
| BR | Bridford nappes |
| CN | Carrick nappe |
| DA | Dartmouth antiform |
| LO | Lizard ophiolite |
| M | Meldon nappes |
| NA | Newton Abbot nappes |
| PML | Perranporth–Mevagissey Line |
| RF | Rusey Fault |
| SD | South Devon nappes |

a large N-dipping thrust at Polzeath marking the confrontation line. Selwood & Thomas (1985; 1986a, b) claim that the direction of tectonic transport is everywhere northwards and that the confrontation line does not exist. However, Andrews *et al.* (1988) have re-affirmed the southward-facing character of $D_1$ structures north of Polzeath, thus supporting the facing confrontation, but consider the high-strain zone to be a NW-directed $D_2$ structure. Seago & Chapman (1988), working in the Tamar Valley north of Plymouth, have also discovered S-facing $D_1$ structures which confirm Sanderson & Dearman's original extension of the facing confrontation into South Devon. They interpret the facing confrontation as backthrust Carboniferous flysch under which a north-ward-directed thrust-pile 'chiselled' its way. The Dart-mouth antiform in Zone 8 according to Hobson (1976) is a N-facing $F_1$ fold modified by later coaxial folding and normal faulting between Dartmouth and Looe; Chapman *et al.* (1984) offer an alternative model in which the normal fault becomes a reverse fault in a nappe-pile containing the Dartmouth anticline. In

Watergate Bay, the fold seems to be an antiform in the lower limb of a recumbent $F_1$ syncline. The axis of the Dartmouth–Watergate antiform is repeatedly offset by major wrench-faults which in places abruptly cut off structure. Turner (1986) believes that, together with E–W fault-bounded highs, these faults controlled basin evolution and subsequent deformation. Zones 9–11 have a Cadomian (NE) trend and terminate northwards at the Perranporth–Mevagissey line, which may be a major structural break correlatable with the Start Boundary Fault. Zone 11, bounded on the south-east by the Lizard Boundary Thrust, is a belt of intense deformation with anomalous fold-trends resembling Zone 5. Rattey & Sanderson (1984) recog-nised 6 phases of deformation in the Middle Devonian Gramscatho and Mylor Flysch sequence north of the Lizard. $D_1$ and $D_2$ are coeval with the emplacement of the Lizard ophiolite (see below) which might thus be of post-Famennian date. $D_1$ folds in Zone 9 are recum-bent and face NNW; those in Zone 10 also face NNW but are gently to moderately inclined. In Zones 9 and 10 and in part of Zone 11, $F_2$ folds are coaxial with $F_1$;

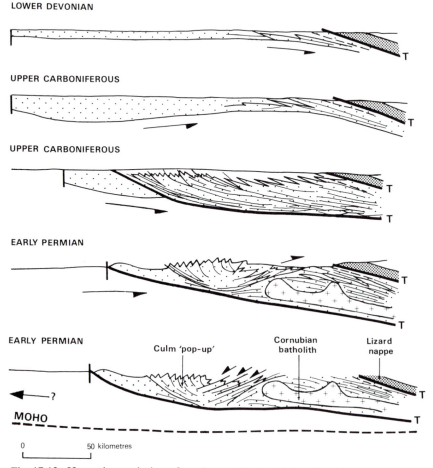

**Fig. 17.13.** Hercynian evolution of south-west England (after Shackleton *et al.* 1982).

in other parts of Zone 11 immediately NW and NE of the Lizard massif, $F_1$ folds verge to the W and form interference structures with the $F_2$ folds. These anomalous zones may be wrench-type shear belts related to the emplacement of the ophiolite slab. Flat-lying $F_3$ folds appear to be related to the margins of the granite cupolas which do, however, cut the folds. $F_{4-6}$ are minor, post-granite deformations. Holder & Leveridge (1986) identify a major flysch nappe, the Carrick nappe, in Zones 10 and 11, and offshore from seismic data. The Middle Devonian flysch was derived from an older, higher crystalline nappe which had picked up a slice of oceanic crust – the Lizard ophiolite. The latter was carried NNW 'piggy-back' successively on the Dodman and Carrick nappes.

In the complex allochthonous terrane between Bodmin Moor and Dartmoor, the earliest event ($D_1$) is the emplacement in the late Viséan/early Namurian of a gravity nappe – the Greystone Nappe – composed of

Lower Carboniferous basinal sediments and volcanics (Isaac *et al.* 1982, Turner 1985). Subsequent $D_2$ thrusting in late Lower Namurian time emplaced 3 further nappes which include Upper Devonian basin, rise margin and rise sediments. The entire nappe sequence was then sliced by $D_3$ north-dipping faults comparable to those in Zones 5 and 6 on the coast. It is now becoming clear that much of the terrane to the north of the Bodmin and Dartmoor granites and south of the Rusey Fault (Zones 5 and 6) is also allochthonous and has been re-folded by younger events affecting the Culm Synclinorium.

A zone of N-verging nappes at the coast (in descending order, the Boscastle, Tredorn and Port Isaac nappes – Selwood & Thomas 1984a, b; 1985) can be traced as far as Okehampton and the Sticklepath fault where it is sharply cut off. The nappes plunge beneath the Culm Synclinorium along the Rusey Fault, which in this interpretation becomes the new 'confrontation line'. East and southeast of Dartmoor, N-facing nappes reappear; they are thrust over high-level upright structures comparable to those north of the Rusey Fault Zone, but the latter cannot be recognised with any certainty (Selwood *et al.* 1984).

In the thin-skinned model of Shackleton *et al.* (1982), the entire fold-belt is to a greater or lesser degree allochthonous (Fig. 17.13). By totalling the estimated shortening in the various tectonic zones of Devon and Cornwall, Shackleton and co-workers suggested that the overall shortening cannot be less than 150 km. They considered the whole Hercynian zone of SW England to be underlain by a décollement dipping gently southwards, related to a S-dipping subduction zone south of the Lizard. The displacement on the décollement ranges from zero in the north to 150 km in the south. In this scheme, the Culm Synclinorium is envisaged as a giant 'pop-up' structure resting on a curved décollement of its own. Shackleton *et al.* believed that the crustal thickness of the region (c. 27 km) was insufficient to allow the generation of granites by crustal melting and suggested that the granite magma was generated to the south and injected northwards above the décollement. However, Sanderson (1984) considers the crust to have been 6 km thicker at the end of the Carboniferous and therefore sufficient for the normal generation of granite magma. Despite Sanderson's view that Hercynian tectonic shortening was due mainly to transpression related to strike-slip faulting and Turner's (1986) view that the thrust-faults steepen with depth, the general consensus seems to favour the thin-skinned model. However, the importance of strike-slip faulting in compartmentalising structure is generally accepted (Burton & Tanner 1986, Turner 1986). The BIRPS 'SWAT' lines 6-7, 8 and 9 clearly demonstrate a thrust-stack structure extending down to the lower crust under the SW Approaches and Plymouth Bay basins (BIRPS & ECORS 1986).

A spate of recent research including geochronological and geophysical work seems to confirm the Upper Palaeozoic and therefore Hercynian age and emplacement of the Lizard Complex. The Start Schists, though possibly pre-Devonian, appear to possess a predominantly Hercynian tectonic imprint and so are also dealt with here.

The **Lizard Complex** in south Cornwall (see also Ch. 16) consists of variably deformed and partially serpentinised peridotite associated with gabbro and overlain and underlain by amphibolite-facies pelitic schists (the 'Old Lizard Head Series') and hornblendic schists (the Landewednack Hornblende Schists and the Traboe schists). Sm–Nd studies by Davies (1984) on the gabbro strongly indicate a primary igneous cooling age of 375 Ma, i.e. Middle Devonian (Givetian). Many workers now consider

the complex to be an obducted ophiolite comprising three essential constituents in correct sequence: peridotite/gabbro/sheeted-dyke complex, with only the pillow lava element missing, though the Landewednack Hornblende Schists may be their metamorphosed equivalent. The Kennack Gneiss, an anatectic migmatised basic gneiss syntectonically emplaced along the base of the peridotite sheet, has given a Rb–Sr whole-rock isochron age of 369 Ma, i.e. late Devonian (Frasnian), thus dating the obduction of the complex (Styles & Rundle 1984) if this interpretation of the Kennack Gneiss is correct.

Though complicated in its internal structure, the Lizard Complex as a whole forms a simple sheet-like mass, a proposal originally made by Sanders (1955). This has been confirmed by seismic refraction lines by Doody & Brooks (1986) suggesting that the ophiolite is a very thin sheet no more than a few hundred metres thick and overlying 3 km of Devonian sediments resting on metamorphic basement which rises slowly eastwards to crop out in the Eddystone and Start. The reverse fault forming the Lizard boundary may be a purely superficial structure. Bromley (1979) believed the Complex to be made up of three juxtaposed thrust units, but doubt has now been cast on this interpretation as a result of inland geochemical mapping by Smith & Leake (1984).

Much of the Lizard complex is strongly deformed. The peridotite has a subvertical foliation trending NW or N–S and the gabbro an E–W subvertical foliation. The Old Lizard Head Series is incompetently folded with varying foliation trends whilst the Landewednack Hornblende Schists have a very regular subhorizontal foliation and strong lineation parallel to rare minor NNW-trending recumbent isoclines; the Traboe schists possess a coarse subvertical foliation locally parallel to that in the ultrabasic rocks. Leake & Styles (1984) showed from borehole evidence (from the badly exposed central Lizard area) that the Traboe schists, thought by Green (1966) to be Landewednack Hornblende Schists metamorphosed by the Lizard Peridotite, and more realistically by Bromley (1979) and Kirby (1979) to be deformed and retrograded Lizard (Crousa) gabbro, are a complex disrupted succession of recrystallised ultramafic and mafic cumulates overlying the main Lizard peridotite. Though geochemically close, the precise relationship between the cumulates and the Crousa Gabbro is not clear.

The Meneage succession immediately north of the Lizard Boundary and in Roseland, variously interpreted as a tectonic crush-breccia and an ophiolitic mélange, was thought by Barnes (1983) to be a well-stratified sedimentary mélange (olistostrome) possibly related to the opening of a small strike-slip basin, the Gramscatho basin. Le Gall & Darboux (1986) claim to have discovered a tectonic lens of fossiliferous Middle Carboniferous limestone in the Meneage zone directly beneath Lizard hornblende schists.

The **Start Schists** are low-grade grey quartz–muscovite-schists containing a prominent zone of greenschist which picks out the E–W antiformal structure of the outcrop. Hobson (1977) believed that the entire deformation sequence may be a deeper-level version of the Hercynian events of south Cornwall and suggested that the Start Boundary may

**Fig. 17.14.** Plate-tectonic model for the evolution of the European Hercynides (after Bromley 1976).

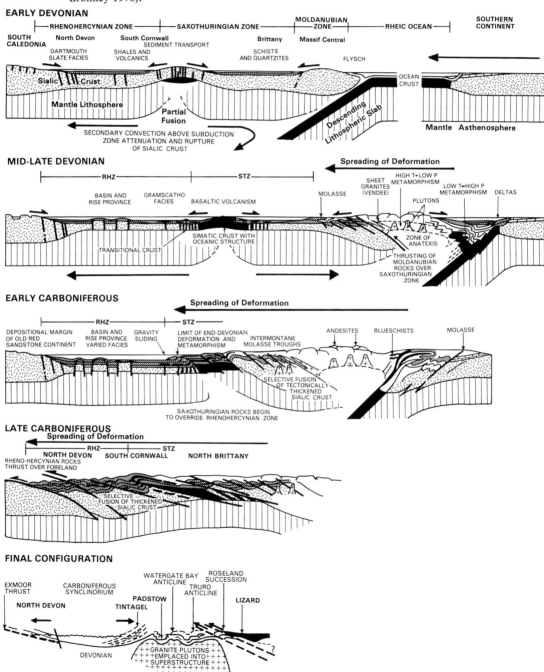

be an extension of the Perranporth–Mevagissey line. According to Hobson, the main antiform and the numerous small folds which fold quartz veins following the schistosity of the grey schists are second-phase $F_2$ folds, while the schistosity itself is related to $F_1$ isoclines. Later small-scale $F_3$ recumbent folds deform the schistosity, as do kink-bands. The attitude of small $F_1$ folds in the greenschists suggests that the greenschist–greyschist boundaries are slides and that the whole sequence is inverted. Deformation of the Start Schists and of the Devonian formations on the north side of the Start Boundary Fault began in the Late Lower Namurian according to Turner (1986).

The bathyal facies, recumbent folding, thrusting and granitic magmatism of SW England seem to many to argue for a plate-tectonic orogenic mechanism in which the crust of a 'Mid-European' or 'Rheic' ocean was subducted beneath the region, resulting in eventual continental collision. However, Floyd's (1984) geochemical study of Cornubian greenstones showed that they are in fact tholeiitic and alkali-basalts, which would seem to preclude the existence of a subduction zone in or near SW England, but may be consistent with Reading's (1973) proposal that the region was an extending marginal basin of Japan Sea type, possibly 200–300 km across. Alkaline vulcanism may be expected in such a basin (Floyd 1973). The same solution is embodied in Bromley's (1976) plate-tectonic model (Fig. 17.14) which assumed that the Lizard Complex is an upthrust section of the oceanic crust of the marginal basin and that the 'Rheic Suture' lies far to the south within the Moldanubian Zone of the European Hercynides. A quite different model was advocated by Badham (1982) who saw the dominant mechanism as the emplacement of microplates by strike–slip motions along the margin of a northern Euramerican continent closely analogous to the 'docking' mechanisms of exotic terranes in the Western Cordillera of North America. According to Badham, the Lizard ophiolite was formed at one such strike-slip margin and obducted during a compressional phase. Barnes & Andrews (1986) interpret the ophiolite as formed at a NNW spreading axis within the Gramscatho pull-apart basin; the latter originated in an intra-continental E–W dextral transform system which generated the Rhenohercynian Zone. Holder & Leveridge's (1986) solution differs from all others in requiring *southward* subduction of the Lizard oceanic crust under the Mid-German Crystalline Rise microplate.

## Older Hercynian-folded cover: Upper Palaeozoic in the Hercynian Foreland

The Hercynian-folded Old Red Sandstone and Carboniferous rocks between the SW England Hercynides and 'St. George's Land' occupy a triangle, the south side of which is the Mendips–Gower–Pembroke fold-belt of WNW ('Armorican') trend and the east side the Malvern (N–S) Axis of folding. Between the two, folds accommodate in various directions, curving from the 'Armorican' into the 'Malvernoid' trend, or bisecting the angle between the two. The junction with the SW England Hercynides may be a major thrust-fault near Cannington Park, where Namurian rocks and a thick Dinantian limestone sequence closely adjoin Middle Devonian outcrops (Whittaker 1975a). However, Brooks and co-workers (1977) suggested on the basis of seismic refraction lines in the Bristol Channel that the Exmoor gravity gradient can be explained in terms of a basement high and a density decrease with depth, without recourse to a thrust. The thrust periclines of the southern Mendips, with their overturned northern limbs and associated *klippen*, are Jura-type structures which have been attributed to décollement on Dinantian Lower Limestone Shales (Green & Welch 1965) but may possibly also be related to a deeper décollement in Cambrian shales, which might crop out as the Farmborough Fault Belt (Green pers. comm.). Williams & Chapman (1986) have constructed balanced sections across the Bristol–Mendip region. They offer a thin-skinned foreland thrust belt interpretation in which the klippen ahead of the periclines are infolds of piggy-back thrust sheets emplaced by low-angle ramp climb. In this scheme, no appeal is made to incompetent lubricating formations: the basal décollement dips at a constant low angle (Fig. 17.15) through ORS, Silurian and older rocks. Total shortening amounts to around 20 km.

In Pembrokeshire south of the Pembrokeshire Coal-field, the Old Red Sandstone and Carboniferous are disposed in large-amplitude WNW folds with Lower Palaeozoic cores on which a Hercynian cleavage has been impressed (Hancock *et al.* 1983). In the western half of the Coalfield, the Precambrian Johnston Complex and Benton Volcanic Group are carried northwards on the Johnston Thrust as a detached slice of the basement for a distance of 4 km over the Westphalian of the Coalfield. The southern half of the Coalfield is heavily thrust and overfolded, whilst the less deformed northern strip displays along its northern flanks small 'disturbances' of Caledonoid trend. Coward & Smallwood (1984) postulate a floor-thrust beneath piggy-back imbricate thrusting in the coalfield N of the Ritec Thrust (Fig. 17.16).

In addition to this high-level thin-skinned tectonics, Powell (1987) sees at deeper levels an inversion mechanism in which major growth faults (Ritec, Benton) established in a pre-Hercynian extensional regime develop contractional geometries in their upper levels during Hercynian compression (Fig. 17.17).

The Variscan Front, still geophysically identifiable

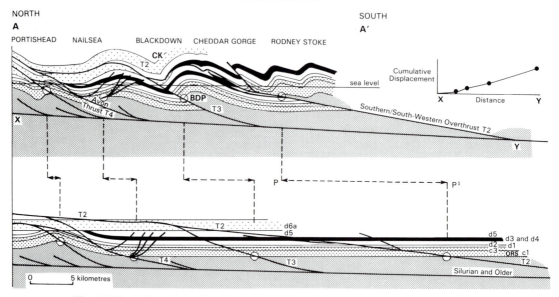

**Fig. 17.15.** Balanced section through the Western Mendips (top) with restored stratigraphy (below). Note emplacement of the Blackdown klippe as a downfolded piggy-back overthrust in the balanced section. BDP = Blackdown Pericline; CK = along-strike projection of Churchill klippe; $T_1$–$T_4$ = major thrusts – subscripts increase in order of piggy-back thrust development from hinterland to foreland; ORS = Old Red Sandstone ($c_1$ Lower, $c_3$ Upper); $d_1$–$d_6$ = Carboniferous (Dinantian–Westphalian). Vertical scale = horizontal. (After Williams & Chapman, Fig. 6, 1986).

as the *Faille du Midi* under Romney Marsh, disappears beneath the thick Mesozoic cover of the Wessex Basin. The Kent Coalfield immediately adjoining on the north the *Faille du Midi* is a gentle basin with its long axis in the NW 'Charnoid' direction. This is almost perpendicular to the ENE trend of the London–Reading syncline of Upper Old Red Sandstone and also to the older Devonian subcrop between London and Norwich. The structures in the concealed Devonian cover are presumably older and produced in response to a different stress-field.

It is a reasonable assumption that south of the Variscan Front in Kent a series of thrust sheets will be found similar to those in the Pas de Calais (Wallace

1983). Farther west, deep seismic reflection surveys by IGS and commercial operators on roughly N–S lines between Cirencester and Devizes and continuing as far south as Dorchester (Chadwick *et al.* 1983) revealed gently dipping reflectors interpreted as S-dipping thrusts overridden by N-verging asymmetric folds extending as far north as Calne. When extrapolated SW, the most northerly thrust – the Variscan Front in this locality – lines up with the Farmborough Fault Belt in the Radstock Basin. A major fold at the southern end of the profile might carry Culm facies in its upper limb (Fig. 17.18) and may be analogous to the main anticline of North Devon (p. 193). A most significant result of this survey is the identification of

**Fig. 17.16.** Balanced section along the eastern side of the Pembrokeshire Coalfield, showing a 'thin-skinned' interpretation. (After Coward & Smallwood 1984, by permission of the Geological Society of London.)

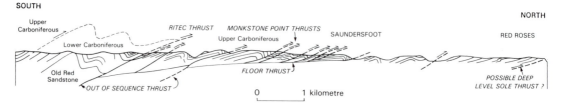

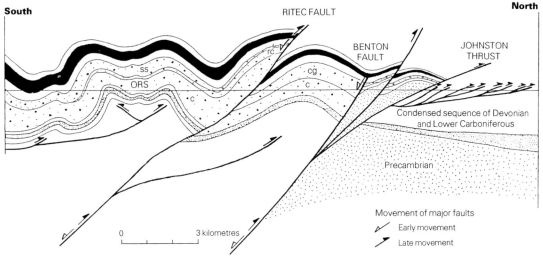

South        RITEC FAULT        North

BENTON FAULT

JOHNSTON THRUST

Condensed sequence of Devonian and Lower Carboniferous

Precambrian

Movement of major faults

⤢ Early movement

⤢ Late movement

0      3 kilometres

**Fig. 17.17.** Section through central Pembrokeshire showing Hercynian reactivation of pre-Hercynian extensional fractures. (See pp. 188–9 for O.R.S. lithological units, Ss = Skrinkle Sandstone which passes upwards into Carboniferous Strata) (after Powell 1987).

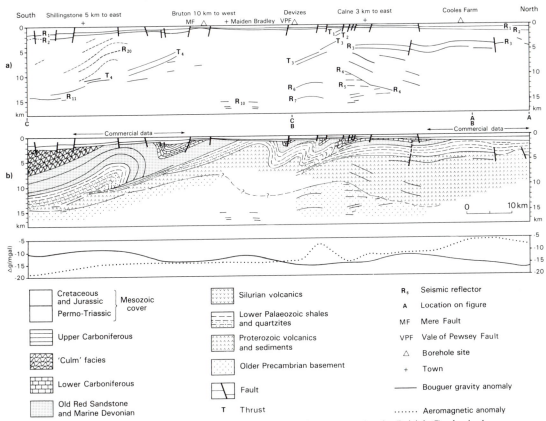

| | | |
|---|---|---|
| ▢ Cretaceous and Jurassic ⎱ Mesozoic ▢ Permo-Triassic ⎰ cover | Silurian volcanics | R₆   Seismic reflector |
| Upper Carboniferous | Lower Palaeozoic shales and quartzites | A   Location on figure |
| 'Culm' facies | Proterozoic volcanics and sediments | MF   Mere Fault |
| Lower Carboniferous | Older Precambrian basement | VPF   Vale of Pewsey Fault |
| Old Red Sandstone and Marine Devonian | ▢ Fault | △   Borehole site |
| | T   Thrust | +   Town |
| | | ——   Bouguer gravity anomaly |
| | | ······   Aeromagnetic anomaly |

**Fig. 17.18.** Deep section based on seismic reflection surveys by the British Geological Survey from Cirencester to Dorchester, showing Hercynian overfolds and thrust-sheets and a deeper Proterozoic sequence resting on possible crystalline basement. (From Chadwick *et al.* 1983.)

deep reflectors north of Calne suggesting the existence of a 5 km-thick late Proterozoic sequence resting on possible crystalline basement at a depth of 9 km deepening northwards to 14 km. This is the first real glimpse we have had of the deep structure of the Midland Microcraton.

From the Mendips to Pembroke the Variscan Front is no longer definite: a zone of folding dominated by 'Hercynian' trends merges imperceptibly into a region to the north dominated by pre-Devonian basement trends (Dunning 1966). Shackleton (1984) has redefined the Variscan Front as the northern exposed limit of the décollement underlying the Variscan thin-skinned tectonic zone. Support for this interpretation comes from the BIRPS 'SWAT' lines 2-3 and 4 which clearly show inclined reflectors extending down to the lower crust where the Variscan Front décollement should be located (BIRPS & ECORS 1986).

The **Malvern Axis** is the most conspicuous feature in a zone of N–S folding which extends from Bath to Burnley. The Axis has a long history of movement in the Phanerozoic, starting with post-Cambrian pre-Llandoverian fault-controlled uplift and erosion (Brooks 1970), but the main deformation is Hercynian, some of it intra-Westphalian but mostly post-Westphalian. It is essentially an anticlinal flexure complicated by axial faulting and thrusting (Fig. 17.19). In the south, it forms the eastern rim of the

**Fig. 17.19.** Block diagram of the Malvern structure.

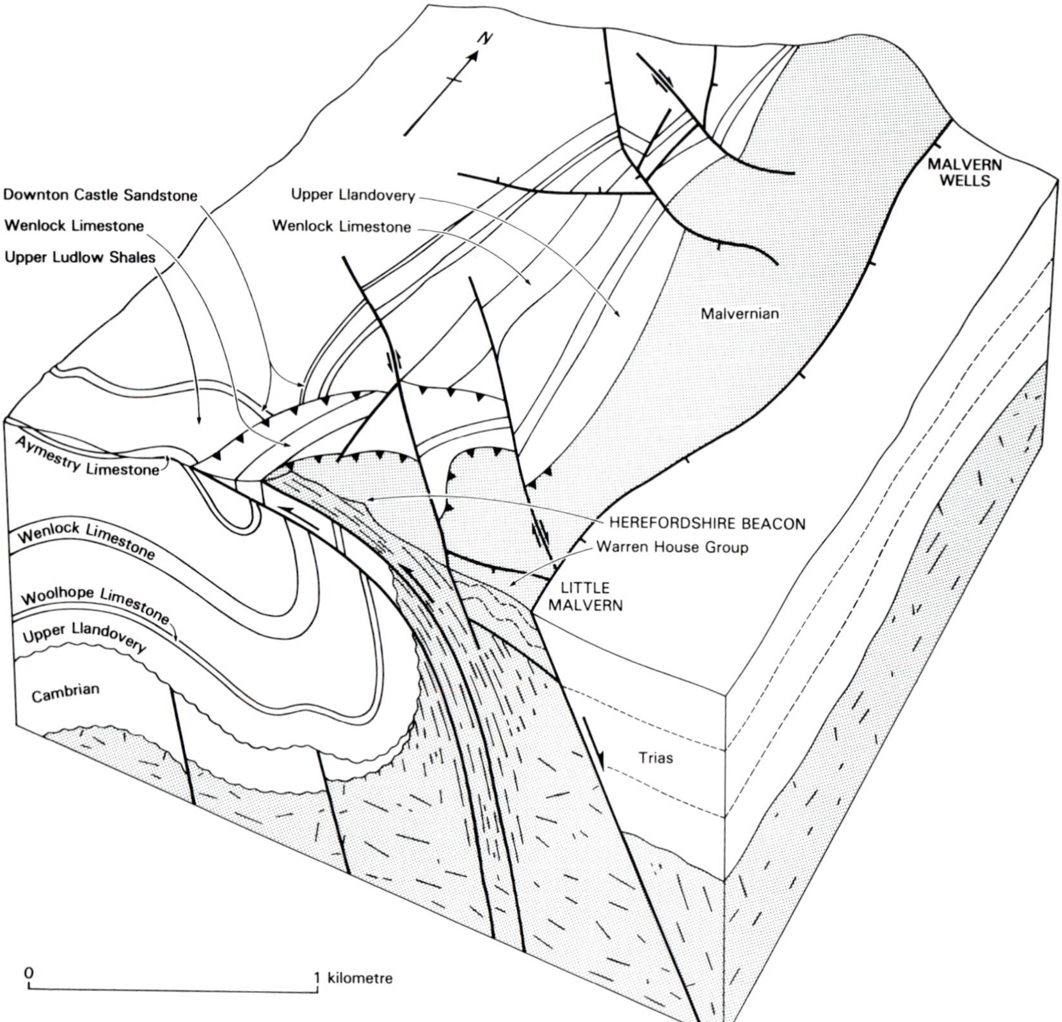

Radstock Basin. In the Malvern Hills, the Precambrian (Malvernian) basement breaks through along the crest; the west limb is overturned and the Malvernian in places carried over Silurian on W-directed overthrusts, whilst the structurally higher east limb is largely faulted out by the Eastern Boundary Fault defining the Mesozoic Severn Basin. The Axis continues northwards to the Abberley Hills where W-facing isoclinal folds involve Westphalian D of the Forest of Wyre (Mamble) Coalfield. It may be conjectured that it thence strikes NNE under the faults at the NW boundary of the South Staffordshire Coalfield to

**Fig. 17.20.** Evidence for the Derby–Bolton deep fracture.

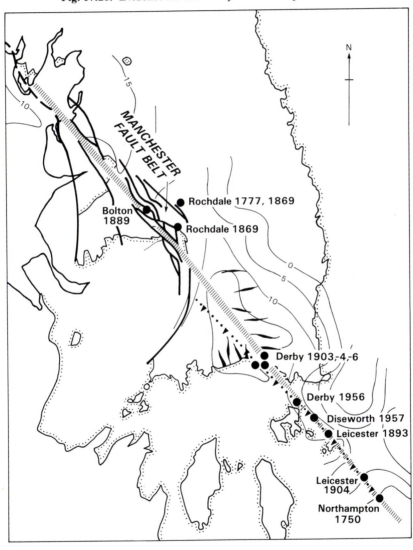

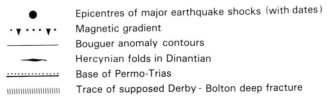

Epicentres of major earthquake shocks (with dates)

Magnetic gradient

Bouguer anomaly contours

Hercynian folds in Dinantian

Base of Permo-Trias

Trace of supposed Derby - Bolton deep fracture

the North Staffordshire Coalfield where the Axis and the extrapolated Pontesford–Linley Line converge. The Malvernoid trend is thereafter taken up by the Goyt Trough and the Pennine Anticline which swings into parallelism at its northern end with the Caledonoid Ribblesdale fold-belt, possibly marking the position of the concealed south-eastern Caledonian 'Front'. The Malvernoid trend is also evident in the Oxfordshire and Warwickshire Coalfields but gives place northwards to 'Charnoid' (NW) trends in the Leicestershire and Nottinghamshire Coalfields. N–S and WNW fold-trends meet in the Derbyshire Dome along a major concealed NW-trending basement fracture marked by a linear magnetic gradient and seismic epicentres (Dunning 1977). This fracture, the 'Derby–Bolton Deep Fracture' (Fig. 17.20) was originally identified by Turner (1936) as the western limit of the Precambrian basement beneath the southern Pennines.

The Ribblesdale fold-belt is sited on the Craven/Bowland Basins (Fig. 7.2 and 17.21) and may be an inversion structure related to its augmented basinal sequence, but is perhaps more likely to be a rotational transpression zone between the rigid Precambrian foundation of the southern Pennines and the granite-stiffened Askrigg Block. Arthurton (1984) has shown that the *en echelon* fold pattern and growth are closely associated with wrench movements of Dinantian–early Namurian date within a zone of regional dextral shear. However, Gawthorpe (1987) regards the present-day structure as due to late Carboniferous (Hercynian)

compression/transpression whereas the Dinantian regime was one of extension/transtension leading to the formation of a fault-controlled half-graben, the Bowland Basin. The Ribblesdale folds are often intensely disharmonic as between crumpled and thrust shale cores and concentric grit envelopes. Northerly fold-trends appear in the broadly synclinal Namurian trough of the Lancaster Fells (which nevertheless has a Caledonoid trend) and pass into E-facing N–S or NNE monoclines in the southern Lake District, the largest being the steep-limbed, complexly faulted, east-facing monocline constituting the Dent Line. Though strikes in the Silurian locally parallel these monoclines, suggesting Hercynian reactivation of Caledonian structures (Moseley 1972), the Dent Fault, to which the Dent Line gives place southwards, cuts across the regional Caledonian strike in the Howgill Fells. The thin Carboniferous cover of the Askrigg Block is very slightly domed about the Craven inliers (Dunham & Wilson 1985), whilst the cover of the Alston Block is similarly domed about the Teesdale and Cross Fell inliers, dipping away eastwards towards the N–S-trending Durham Coalfield which contrasts strongly with the E–W-trending structures in the concealed Carboniferous in east Yorkshire (Kent 1966). In the far more conspicuous Lake District Dome, the Carboniferous dips outwards from the Caledonian core to the NW, N and NE, whilst dips in the south are easterly away from the N–S monoclines. The Pennine fault-system limiting the Alston Block on the west was initiated as a late Caledonian normal

**Fig. 17.21.** Section from the Scottish Border to the English Channel showing the principal divisions of the platform cover in England. Vertical scale about 12 times the horizontal.

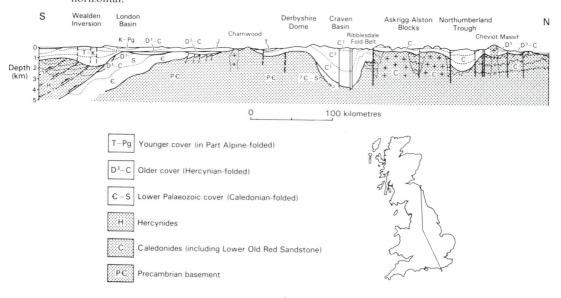

fault with westerly downthrow prior to the E–W Hercynian pressures which thrust the Lower Palaeozoic of the Cross Fell inlier eastwards along the inner compressional faults of the system, and at the same time formed the shallow syncline between the Lake District and the Alston Block (Burgess & Wadge 1974). To the accompaniment of late Hercynian normal faulting along the Pennine line (Outer Pennine Fault), the Carboniferous syncline evolved into the early Permian cuvette of the Vale of Eden. Folding in the Northumberland Trough between the Stublick Faults and the Southern Uplands massif follows Caledonoid trends in the south-west (Bewcastle Anticline and Canonbie Coalfield) whilst to the north-east, folds are tangential to the Cheviot massif (Shiells 1963).

Much of the Carboniferous cover in England and Wales is intensely faulted. A great many faults, especially the big NW-faults in the Lancashire Coalfield, also affect the Permo–Trias; their post-Triassic component of movement may, according to Kent (1975), be pre-Tertiary. The exceedingly regular conjugate fault systems in the Yorkshire Coalfields and the Alston Block, and the great wrench-faults in northeast Wales (Bala–Bryneglwys) and South Wales (Neath Disturbance) with their histories of prolonged movement, have no analogues in the Younger Cover and are presumably Hercynian.

## Younger cover

The post-Hercynian sedimentary cover of England and Wales comprises a pre-Alpine (pre-Neogene) sequence ranging in age from Permian to Palaeogene, with important discordances in the mid-Cretaceous and early Palaeogene, and a thin post-Alpine sequence of Pliocene–Pleistocene age in East Anglia.

East of a line curving SSW from Sunderland to Torquay, the younger cover consists essentially of the onshore extensions of the southern North Sea and Anglo–Paris Basins separated by the East Anglia massif (the 'London Platform'). West of this line, a number of small, steep-sided basins are accommodated between the upland positive areas of south-west England, Ireland, Wales, the Pennines, the Lake District and the Isle of Man (Fig. 17.2). South of a line from the Mendips to the Thames Estuary, the cover formations are thrown into a shoal of monoclines and periclinal folds by post-Oligocene/pre-Miocene, Alpine compression. North of this line, the cover is broadly flexed, domed, tilted and fractured by movements which are perhaps more likely to be of mid- or end-Cretaceous (Austric or Laramide respectively) than Alpine (Savic) date.

Perhaps the most conspicuous geotectonic feature of the younger cover south of the East Anglia massif is the discrepancy between the structure of the cover at the surface and the morphology of the underlying basement. The basement (Hercynian fold-belt and adjacent foreland) is depressed to a depth of more than 4 km west of the Isle of Wight to form the Wessex Basin (Kent 1949), which is connected across sills to the Vale of Pewsey Basin, the Severn or Worcester Basin and the Central Somerset–Bristol Channel Basin (Smith 1985). A trough extends ENE from the Wessex Basin under the Weald uplift. It is largely filled with a thick (>1·5 km) Jurassic sequence, though Whittaker (1975b) speculated that it may also conceal a Permo–Triassic graben. The onshore structures in the cover are simply extensions of those in the Channel, which Smith & Curry (1975) apportioned to three provinces: a western province of ENE-trending structures not in evidence on land; a central province of E–W structures, of which the Isle of Wight and Purbeck monoclines are conspicuous examples; and an eastern province dominated by the broad, flat-bottomed Hampshire–Dieppe Basin and the Weald–Artois uplift, which have NW-curving-to-EW, Hercynian trends.

The **Weald–Artois** uplift is an inversion structure in which originally downwarped strata are now upwarped. Within the broad uplift of the Weald, the central area of more competent Hastings Beds is characterised by long strike-faults which determine the pattern of outcrop. Some faults bound horsts, amongst them the central Purbeck inliers. The surrounding incompetent Weald Clay has deformed internally and independently of harder formations above and below. The pattern of outcrop of the Chalk between the Tertiary basins of London and Hampshire and the older Jurassic formations in the west is largely determined by N-facing, highly asymetric or monoclinal folds, such as the Hog's Back, Pewsey and Vale of Wardour folds. The axial traces of some are sinuous and complicated by reverse faulting; others comprise a string of *en echelon* periclines. Folding of comparable style characterises the classical Weymouth–Lulworth–Purbeck–Isle of Wight coastal terrain (Fig. 17.22), though here the deformation is generally more intense, with overturning, crumpling and crushing (Arkell et al. 1947). Moreover, there is a strong post-Wealden pre-Gault phase of folding, faulting and erosion beautifully demonstrated in the Poxwell and Chaldon periclines (House 1961): intra-Cretaceous faults throw down southwards, whereas post- or ?intra-Eocene reverse faults – some of them reactivated intra-Cretaceous normal faults – throw down northwards. Stoneley (1982) regarded the pre-Aptian normal faults as listric growth-faults with associated roll-over anticlines – of which the Poxwell and Chaldon anticlines are examples – accentuated by post-Cretaceous reversed movement.

The **London Basin** (an arm of the southern North Sea Tertiary basin) is extremely discordant with the underlying basement, the axial portion of the Basin striking NE across the ESE trend of the edge of the East Anglian massif. This NE trend is also seen in the Islip anticline and the main syncline of the Plio–Pleistocene 'Crags' of East Anglia but it is not known whether this reflects basement structures. The **Central Somerset Basin** may also, according to Whittaker (1975b), be a graben structure. It merges west-

**Fig. 17.22.** Block diagram of the Poxwell and Chaldon Periclines and the Purbeck Monocline, after Arkell *et al.* (1947) and House (1961). Note the sub-Albian discordance, with Gault and Greensand reposing on folded, faulted and eroded Wealden and older formations.

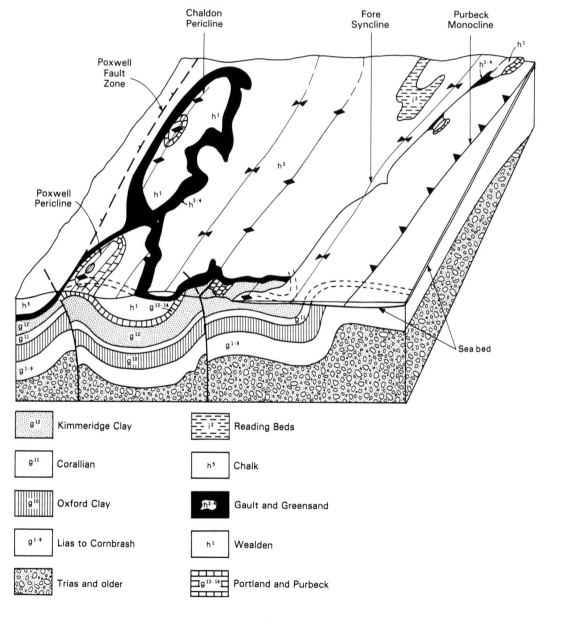

| | | | |
|---|---|---|---|
| $g^{12}$ | Kimmeridge Clay | $i^2$ | Reading Beds |
| $g^{11}$ | Corallian | $h^5$ | Chalk |
| $g^{10}$ | Oxford Clay | $h^{3\text{-}4}$ | Gault and Greensand |
| $g^{1\text{-}9}$ | Lias to Cornbrash | $h^1$ | Wealden |
| | Trias and older | $g^{13\text{-}14}$ | Portland and Purbeck |

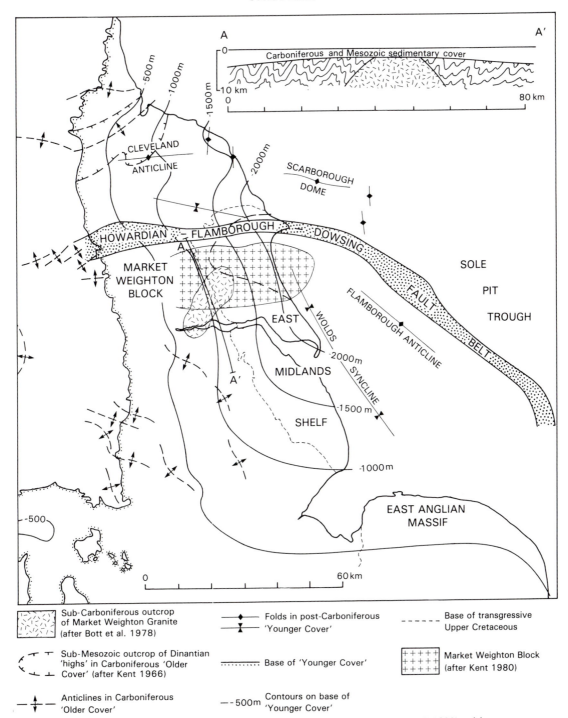

**Fig. 17.23.** Structural map of the North-eastern Basin, after Kent (1966, 1980), with a section of the Market Weighton Block (top) after Bott *et al.* (1978).

552 F. W. DUNNING

wards into the Bristol Channel Basin which does not appear to be fault-bounded (Evans & Thompson 1979). The **Severn** or **Worcester Basin** is a N–S graben bounded on the west by stepped faults along the Malvern line, whilst the eastern boundary fault probably underlies the Vale of Moreton anticline. According to Wills (1978), the basement of the Severn Basin was a mountain tract before Upper Palaeozoic erosion and Triassic downfaulting.

The **Cheshire Basin** is a remarkable structure exactly congruent with the south-east border of the Welsh Caledonides. It is extremely deep: the Prees Borehole reached thin Upper Coal Measures on vertical Ordovician at 4,301 m (Wills 1978, p. 29). It has very steep sides, and even though doubt has been cast on the existence of the Red Rock Fault (Challinor 1978a), faulting must have played a part in its generation. The Basin is broken by many large faults which in the west curve from NNW to WSW to line up with the Bala–Bryneglwys system. Faults of the same group bound the **Manx–Furness Basin** on the east as far north as the Lake District. The Bala Fault also forms the south-east boundary of the **Cardigan Bay Basin**, a 'trapdoor' subsidence whose northern end in Trema-

doc Bay is bounded by the Mochras Fault throwing down the thickest Liassic section in Britain (Woodland 1971). The **Vale of Eden Basin**, connected at its northern end to the Caledonoid **Carlisle–Solway Basin**, is located on the Pennine line, yet another expression of the NNW faulting tendency, along which there have been substantial movements of both early Permian and post-Permian date, accompanied by downwarping of the Permian beds. Bounded on the west by the Pennines and on the south by the East Anglian massif is the **North-eastern Basin**. In its southern half, unconformable Chalk dips gently into the offshore Wolds Syncline above the stable East Midlands Shelf, which terminates in the north in the buoyant Market Weighton Block (Bott *et al.* 1978, Kent 1978, 1980) (Fig. 17.23). North of the Block and forming the northern half of the North-eastern Basin is the Cleveland Jurassic basin, the onshore continuation of the Sole Pit Trough which underwent inversion in early Cretaceous time on its north-eastern flank and in latest Cretaceous time on its south-western flank. Glennie & Boegner (1981) suggested possible causal connections respectively with early Cretaceous opening of the Rockall Trough and late Cretaceous

**Fig. 17.24.** Dissected high-level (500 m) erosion surface in central Wales: accordant summits seen from Bwlch-y-Groes, Gwynedd (F. W. Dunning).

migration of the spreading ridge to the present mid-atlantic ridge.

The main structures in the northern half of the North-eastern Basin, including the Howardian Faults and co-linear Flamborough crush-zone, the Vale of Pickering syncline and Cleveland anticline, trend E by ESE exactly in line with the Hercynian and Caledonian folding in the Askrigg Block and the Austwick inliers.

## Post-Alpine effects

A great variety of erosional landforms, chiefly plateaux and platforms, have been interpreted to imply pulsed epeirogenic uplift of post-Alpine (Neogene) date in southern England and additionally of older Palaeogene date in Wales and northern England. A great deal of controversy surrounds the subject; indeed, the whole edifice of erosion cycles based on these landforms has been attacked by Challinor (1978b). The platforms do nevertheless exist and are indubitably erosion surfaces; the main argument is whether they were formed at about sea level and then rapidly uplifted or were carved by continuous and prolonged subaerial erosion. Notable examples are the high dissected 500 m (1,700′) plateau of central Wales (Fig. 17.24), the remarkably flat 60 m (200′) coastal platform in south-west Dyfed, the 130 m (430′) platform in Cornwall and Devon, and the 180 m (600′) 'Pliocene Platform' of the Weald. The latter is a bench cut into the Chalk with, at Lenham in Kent, solution

pipes containing marine sands (Lenham Beds) of probable late Miocene age (Curry *et al.* 1978). Epeirogenic uplift of at least 180 m is inferred if these beds are *in situ*, but they have also been interpreted as glacially redeposited Neogene from the North Sea (Shepherd-Thorn 1975), with no uplift involved. The amount of uplift that may be inferred from the lower coastal platforms may be slight, since they were probably eroded when the sea level stood appreciably higher in the Neogene and during warm Interglacials. The high inland plateaux, if they are in fact uplifted marine platforms or peneplains, are more likely to be of early Palaeogene or even late Mesozoic age.

Bevan & Hancock (1986) have demonstrated the existence of an extensive system of NW-trending joints and minor faults mesofaults) with throws up to a few metres in Upper Cretaceous and Palaeogene rocks throughout SE England and N France. They are of post-Alpine, Neogene but pre-Red Crag (late Pliocene) age. Bevan and Hancock relate this NW–mesofracture system of the same stress-field responsible for neotectonic movements in the Lower Rhine graben.

Though not caused by orogenesis or epeirogenesis, the many minor structures directly or indirectly due to glacial action might be mentioned: for example, drifts deformed by moving ice in East Anglia; cambers, dip-and-fault structures, gulls and valley bulges in water-logged Jurassic and Lower Cretaceous strata; and, even though its effects are relatively slight in England and Wales, regional uplift due to glacial unloading and isostatic rebound.

## REFERENCES

ALLEN, J. R. L.     1968     Precambrian rocks. C. The Nuneaton District. *In* Sylvester-Bradley, P. C. & Ford, T. D. (Eds.) *The Geology of the East Midlands.* Leicester University Press, Leicester.

ANDERTON, R., BRIDGES, P. H., LEEDER, M. R. & SELLWOOD, B. W.     1979     *A dynamic stratigraphy of the British Isles.* George Allen & Unwin, London, 301 pp.

ANDREWS, J. R., BARKER, A. J. & PAMPLIN, C. F.     1988     A reappraisal of the facing confrontation in north Cornwall: fold- or thrust-dominated tectonics? *J. geol. Soc. London*, **145**, 777–788.

ARKELL, W. J., WRIGHT, C. W. & OSBORNE WHITE, H. J.     1947     The geology of the country around Weymouth, Swanage, Corfe and Lulworth. *Mem. Geol. Surv. G.B.*

ARTHURTON, R. S.     1984     The Ribblesdale fold belt, NW England – a Dinantian-early Namurian dextral shear zone. *In* Hutton, D. H. W. & Sanderson, D. J. (Eds.) Variscan tectonics of the North Atlantic region. *Spec. Pub. geol. Soc. London*, **14**, 131–138.

AUVRAY, B.     1979     *Genèse et évolution de la croûte continentale dans le Nord du Massif Armoricain.* Thèse, University of Rennes, 671 pp.

BADHAM, J. P. N.     1982     Strike-slip orogens – an explanation for the Hercynides. *J. geol. Soc. London*, **139**, 493–504.

BAKER, J. W.     1971     The Proterozoic history of southern Britain. *Proc. Geol. Assoc. London*, **82**, 249–266.

    1973     A marginal late Proterozoic ocean basin in the Welsh region. *Geol. Mag.*, **110**, 447–455.

BAMFORD, D., NUNN, K., PRODEHL, C. & JACOB, B.     1977     LISPB – III. Upper crustal structure of northern Britain. *J. geol. Soc. London*, **133**, 481–488.

BANHAM, P. H., HOPPER, F. M. W. & JACKSON, J. B.     1981     The Gillbrea Nappe in the Skiddaw Group, Cockermouth, Cumbria, England. *Geol. Mag.*, **118**, 509–516.

BARBER, A. J. & MAX, M. D.     1979     A new look at the Mona Complex (Anglesey, North Wales). *J. geol. Soc. London*, **136**, 407–432.

BARBER, A. J., MAX, M. D. & BRÜCK, P. M.     1981     Field Meeting in Anglesey and southeastern Ireland 4–11 June 1977. *Proc. Geol. Assoc. London*, **92**, 269–291.

BARNES, R. P.     1983     The stratigraphy of a sedimentary melange and associated deposits in South Cornwall, England. *Proc. Geol. Assoc. London*, **94**, 217–229.

BARNES, R. P. & ANDREWS, J. R.     1986     Upper Palaeozoic ophiolite generation and obduction in south Cornwall. *J. geol. Soc. London*, **143**, 117–124.

BASSETT, D. A.     1969     Some of the major structures of early Palaeozoic age in Wales and the Welsh Borderland: an historical essay. *In* Wood, A. (Ed.) *The Pre-Cambrian and Lower Palaeozoic rocks of Wales.* University of Wales Press, Cardiff, 67–116.

BEAVON, R. V.     1963     The succession and structure east of the Glaslyn River, North Wales. *Q. J. geol. Soc. London*, **119**, 479–512.

BEVAN, T. G. & HANCOCK, P. L.     1986     A late Cenozoic regional mesofracture system in southern England and northern France. *J. geol. Soc. London*, **143**, 355–362.

BIRPS & ECORS     1986     Deep seismic reflection profiling between England, France and Ireland. *J. geol. Soc. London*, **143**, 45–52.

BOTT, M. H. P., ROBINSON, J. & KOHNSTAMM, M. A.     1978     Granite beneath Market Weighton, east Yorkshire. *J. geol. Soc. London*, **135**, 535–543.

BRANNEY, M. J. & SOPER, N. J.     1988     Ordovician volcano-tectonics in the English Lake District. *J. geol. Soc. London*, **145**, 367–376.

BROMLEY, A. V.     1969     Acid plutonic igneous activity in the Ordovician of North Wales. *In* Wood, A. (Ed.) *The Pre-Cambrian and Lower Palaeozoic rocks of Wales.* University of Wales Press, Cardiff, 387–408.

    1971     Phases of deformation in North Wales. *Geol. Mag.*, **108**, 548–550.

    1976     Granites in mobile belts – the tectonic setting of the Cornubian batholith. *Camborne Sch. Mines J.*, **76**, 40–47.

    1979     Ophiolitic origin of the Lizard complex. *Camborne Sch. Mines J.*, **79**, 25–38.

BROOKS, M.     1970     Pre-Llandovery tectonism and the Malvern structure. *Proc. Geol. Assoc. London*, **81**, 249–268.

BROOKS, M., BAYERLY, M. & LLEWELLYN, D. J.     1977     A new geological model to explain the gravity gradient across Exmoor, north Devon. *J. geol. Soc. London*, **133**, 385–393.

BROOKS, M., DOODY, J. J. & AL-RAWI, F. R. J. 1984 Major crustal reflectors beneath S W England. *J. geol. Soc. London*, **141**, 97–103.

BURGESS, I. C. & WADGE, A. J. 1974 The geology of the Cross Fell area. *Inst. Geol. Sci.* HMSO, London.

BURTON, C. J. & TANNER, P. W. G. 1986 The stratigraphy and structure of the Devonian rocks around Liskeard, east Cornwall, with regional implications. *J. geol. Soc. London*, **143**, 95–105.

BUTCHER, N. E. 1962 The tectonic structure of the Malvern Hills. *Proc. Geol. Assoc. London*, **73**, 103–123.

CAMPBELL, S. D. G., REEDMAN, A. J. & HOWELLS, M. F. 1985 Regional variations in cleavage and fold development in North Wales. *Geol. J.* **20**, 43–52.

CANTRILL, T. C., DIXON, E. E. L., THOMAS, H. H. & JONES, O. T. 1916 The geology of the South Wales coalfield. Part XII. The country around Milford. *Mem. Geol. Surv. England & Wales*.

CHADWICK, R. A., KENOLTY, N. & WHITTAKER, A. 1983 Crustal structure beneath southern England from deep seismic reflection profiles. *J. geol. Soc. London*, **140**, 893–911.

CHALLINOR, J. 1978a The 'Red Rock Fault', Cheshire: a critical review. *Geol. J.*, **13**, 1–10.

1978b A brief review of some aspects of geomorphology in England and Wales. *Mercian Geol.*, **6**, 283–290.

CHAPMAN, T. J., FRY, R. L. & HEAVEY, P. T. 1984 A structural cross-section through SW Devon. *In* Hutton, D. H. W. & Sanderson, D. J. (Eds.) Variscan tectonics of the North Atlantic region. *Spec. Pub. geol. Soc. London*, **14**, 113–118.

COGNÉ, J. & WRIGHT, A. E. 1980 L'orogène cadomien. *In* Geology of Europe from Precambrian to the post-Hercynian sedimentary basins. *Colloquium C6. 26th Int. Geol. Cong.* Paris, 29–55.

COWARD, M. P. & MCCLAY, K. R. 1983 Thrust tectonics of S Devon. *J. geol. Soc. London*, **140**, 215–228.

COWARD, M. P. & SIDDANS, A. W. B. 1979 The tectonic evolution of the Welsh Caledonides. *In* Harris, A. L. Holland, C. H. & Leake, B. E. (Eds.) The Caledonides of the British Isles – reviewed. *Spec. Pub. geol. Soc. London*, **8**, 187–198.

COWARD, M. P. & SMALLWOOD, S. 1984 An interpretation of the Variscan tectonics of SW Britain. *In* Hutton, D. H. W. & Sanderson, D. J. (Eds.) Variscan tectonics of the North Atlantic region. *Spec. Pub. geol. Soc. London*, **14**, 89–102.

COWIE, J. W. & JOHNSON, M. R. W. 1985 Late Precambrian and Cambrian geological time-scale. *In* Snelling, N. J. (Ed.) The Chronology of the Geological Record. *Mem. geol. Soc. London*, **10**, 47–64.

CURRY, D., ADAMS, C. G., BOULTER, M. C., DILLEY, F. C., EAMES, F. E., FUNNELL, B. M. & WELLS, M. K. 1978 A correlation of Tertiary rocks in the British Isles. *Spec. Rept. geol. Soc. London*, **12**.

DAVIES, G. R. 1984 Isotopic evolution of the Lizard Complex. *J. geol. Soc. London*, **141**, 3–14.

DEWEY, J. F. 1969 Structure and sequence in paratectonic British Caledonides. *In* Kay, N. (Ed.) North Atlantic geology and continental drift – a symposium. *Mem. Amer. Assoc. Petrol. Geol.*, **12**, 309–335.

| DEWEY, J. F. | 1982 | Plate tectonics and the evolution of the British Isles. *J. geol. Soc. London*, **139**, 371–412. |
| DODSON, M. H. & REX, D. C. | 1971 | Potassium-argon ages of slates and phyllites from south-west England. *Q. J. geol. Soc. London*, **126**, 465–499. |
| DOODY, J. J. & BROOKS, M. | 1986 | Seismic refraction investigation of the structural setting of the Lizard and Start complexes, SW England. *J. geol. Soc. London*, **143**, 135–140. |
| DUNHAM, K. C. & WILSON, A. A. | 1985 | Geology of the Northern Pennine Orefield: Volume 2, Stainmore to Craven. *Econ. Mem. Br. Geol. Surv.*, Sheets 40, 41, 50 etc. |
| DUNNING, F. W. | 1966 | *Tectonic Map of Great Britain and northern Ireland.* Institute of Geological Sciences. H.M.S.O., London. |
|  | 1977 | Caledonian-Variscan relations in north-west Europe. In *La chaîne varisque d'Europe moyenne et occidentale.* Coll. intern. CNRS, Rennes, No. 243, 165–180. |
| EVANS, A. M. | 1968 | Precambrian rocks. A. Charnwood Forest. *In* Sylvester-Bradley, P. C. & Ford, T. D. (Eds.) *The Geology of the East Midlands.* Leicester University Press, Leicester, 1–12. |
|  | 1979 | The East Midlands aulacogen of Caledonian age. *Mercian Geol.*, **7**, 31–42. |
| EVANS, D. J. & THOMPSON, M. S. | 1979 | The geology of the central Bristol Channel and the Lundy area, South Western Approaches, British Isles. *Proc. Geol. Assoc. London*, **90**, 1–14. |
| FLOYD, P. A. | 1973 | The tectonic environment of south-west England. Reply to contributions to the discussion of a paper by P. A. Floyd. *Proc. Geol. Assoc. London*, **84**, 243–247. |
|  | 1984 | Geochemical characteristics and comparison of the basic rocks of the Lizard Complex and the basaltic lavas within the Hercynian troughs of SW England. *J. geol. Soc. London*, **141**, 61–70. |
| FREEMAN, B., KLEMPERER, S. L. & HOBBS, R. W. | 1988 | The deep structure of northern England and the Iapetus suture zone from BIRPS deep seismic reflection profiles. *J. geol. Soc. London*, **145**, 727–740. |
| FRESHNEY, E. C. | 1977 | The Variscan history of south-west England. In *La chaîne varisque d'Europe moyenne et occidentale.* Coll. intern. CNRS, Rennes, No. 243, 539–546. |
| FYFE, T. B. & WEIR, J. A. | 1976 | The Ettrick Valley Thrust and the upper limit of the Moffat Shales in Craigmichan Scaurs (Dumfries & Galloway region: Annandale & Eskdale district). *Scott. J. Geol.*, **12**, 93–102. |
| GAUSS, G. A. | 1973 | The structure of the Padstow area, north Cornwall. *Proc. Geol. Assoc. London*, **84**, 283–313. |
| GAWTHORPE, R. L. | 1987 | Tectono-sedimentary evolution of the Bowland Basin, N. England, during the Dinantian. *J. geol. Soc. London*, **144**, 59–71. |
| GEORGE, T. N. | 1963 | Palaeozoic growth of the British Caledonides. *In* Johnson, M. R. W. & Stewart, F. H. (Eds.) *The British Caledonides.* Oliver & Boyd, Edinburgh, 1–33. |
|  | 1970 | *South Wales.* British Regional Geology (Third Edition), H.M.S.O. London. |
| GIBBONS, W. | 1983a | The Monian Penmynydd zone of metamorphism in Llŷn, North Wales. *Geol. J.*, **18**, 1–21. |
|  | 1983b | Stratigraphy, subduction and strike-slip faulting in the Mona Complex of North Wales – a review. *Proc. Geol. Assoc. London*, **94**, 147–163. |

| GLENNIE, K. W. & BOEGNER, P. L. E. | 1981 | Sole Pit inversion tectonics. *In* Illing, L. V. & Hobson, G. D. (Eds.) *Petroleum geology of the continental shelf of north-west Europe*. Institute of Petroleum, London, 110–120. |
|---|---|---|
| GREEN, D. H. | 1966 | A re-study and re-interpretation of the geology of the Lizard Peninsula, Cornwall. *In* Hosking, K. F. G. & Shrimpton, G. J. (Eds.) *Present views of some aspects of the geology of Cornwall and Devon*. Truro: Royal Geological Society of Cornwall (Commemorative Volume for 1964), 87–114. |
| GREEN, G. W. & WELCH, F. B. A. | 1965 | Geology of the country around Wells and Cheddar. *Mem. Geol. Surv. Gt. Br.* |
| GREENLY, E. | 1919 | The geology of Anglesey. *Mem. Geol. Surv. Eng. Wales.* 2 vols. |
| GREIG, D. C., WRIGHT, J. E., HAINS, B. A. & MITCHELL, G. H. | 1968 | Geology of the country around Church Stretton, Craven Arms, Wenlock Edge and Brown Clee. *Mem. Geol. Surv. Gt. Br.* |
| HANCOCK, P. L., DUNNE, W. M. & TRINGHAM, M. E. | 1983 | Variscan deformation in southwest Wales. *In* Hancock, P. L. (Ed.) *The Variscan fold belt in the British Isles*. Adam Hilger Ltd., Bristol, 47–73. |
| HARLAND, W. B., COX, A. V., LLEWELLYN, P. G., PICKTON, C. A. G., SMITH, A. G. & WALTERS, R. | 1982 | *A geological time scale*. Cambridge University Press, Cambridge, 131 pp. |
| HAWKINS, T. R. W. & JONES, F. G. | 1981 | Fold and cleavage development within Cambrian meta-sediments of the Vale of Ffestiniog, North Wales. *Geol. J.*, **16**, 65–84. |
| HELM, D. G., ROBERTS, B. & SIMPSON, A. | 1963 | Polyphase folding in the Caledonides south of the Scottish Highlands. *Nature, London*, **200**, 1060–1062. |
| HOBSON, D. M. | 1976 | The structure of the Dartmouth antiform. *Proc. Ussher Soc.*, **3**, 320–332. |
| | 1977 | Polyphase folds from the Start complex. *Proc. Ussher Soc.*, **4**, 102–110. |
| HOBSON, D. M. & SANDERSON, D. J. | 1975 | Major early folds at the southern margin of the Culm synclinorium. *J. geol. Soc. London*, **131**, 337–352. |
| HOLDER, M. T. & LEVERIDGE, B. E. | 1986 | A model for the tectonic evolution of south Cornwall. *J. geol. Soc. London*, **143**, 125–134. |
| HOLLAND, C. H. | 1959 | The Ludlovian and Downtonian rocks of the Knighton district, Radnorshire. *Q. J. geol. Soc. London*, **114**, 449–482. |
| HOUSE, M. R. | 1961 | The structure of the Weymouth Anticline. *Proc. Geol. Assoc. London*, **72**, 221–238. |
| ISAAC, K. P., TURNER, P. J. & STEWART, I. J. | 1982 | The evolution of the Hercynides of central SW England. *J. geol. Soc. London*, **139**, 521–531. |
| JAMES, J. H. | 1956 | The structure and stratigraphy of part of the Pre-Cambrian outcrop between Church Stretton and Linley, Shropshire. *Q. J. geol. Soc. Lond.*, **112**, 315–337. |
| KENT, P. E. | 1949 | A structure contour map of the buried pre-Permian rocks of England and Wales. *Proc. Geol. Assoc. London*, **60**, 87–104. |

| | | |
|---|---|---|
| KENT, P. E. | 1966 | The structure of the concealed Carboniferous rocks of north-eastern England. *Proc. Yorkshire geol. Soc.*, **35**, 323–352. |
| | 1975 | The tectonic development of Great Britain and the surrounding seas. *In* Woodland, A. W. (Ed.) *Petroleum and the continental shelf of north-west Europe*. **I**, *Geology*, Inst. Petrol., London, 3–28. |
| | 1978 | Mesozoic vertical movements in Britain and the surrounding continental shelf. *In* Bowes, D. R. & Leake, B. E. (Eds.) Crustal evolution in northwestern Britain and adjacent regions. *Geol. J. Spec. Issue*, **10**. Seel House Press, Liverpool, 309–324. |
| | 1980 | Eastern England from the Tees to The Wash. *Brit. Reg. Geol.* H.M.S.O., London, 155 pp. |
| KING, W. B. R. & WILCOCKSON, W. H. | 1934 | The Lower Palaeozoic rocks of Austwick and Horton-in-Ribblesdale, Yorkshire. *Q. J. geol. Soc. London*, **90**, 7–31. |
| KIRBY, G. A. | 1979 | The Lizard complex as an ophiolite. *Nature, London*, **282**, 58–61. |
| KLEMPERER, S. L. & MATTHEWS, D. H. | 1987 | Iapetus suture located beneath the North Sea by BIRPS deep seismic profiling. *Geology*, **15**, 195–198. |
| KOHNSTAMM, M. | 1980 | Discussion on 'A new look at the Mona Complex (Anglesey, North Wales)': continuation of discussion in Barber & Max 1979. *J. geol. Soc. London*. **137**, 513–514. |
| KOKELAAR, P. | 1988 | Tectonic controls of Ordovician arc and marginal basin volcanism in Wales. *J. geol. Soc. London*, **145**, 759–775. |
| LAMBERT, R. ST. J. | 1971 | The pre-Pleistocene Phanerozoic time-scale – a review. In *Part 1* of The Phanerozoic Time-scale – a supplement. *Special Publication of the Geological Society, London*, **5**, 9–31. |
| LEAKE, R. C. & STYLES, M. T. | 1984 | Borehole sections through the Traboe hornblende schists, a cumulate complex overlying the Lizard peridotite. *J. geol. Soc. London*, **141**, 41–52. |
| LEEDAL, G. P. & WALKER, G. P. L. | 1950 | A restudy of the Ingletonian Series of Yorkshire. *Geol. Mag.*, **87**, 57–66. |
| LEFORT, J. P. | 1977 | Possible 'Caledonian' subduction under the Domnonian domain – North Armorican area. *Geology*, **5**, 523–526. |
| LE GALL, B. & DARBOUX, J. R. | 1986 | Variscan strain patterns in the Palaeozoic series at the Lizard Front, Southwest England. *Tectonics*, **5**(4), 599–606. |
| LLOYD, G. E. & WHALLEY, J. S. | 1986 | The modification of chevron folds by simple shear: examples from north Cornwall and Devon. *J. geol. Soc. London*, **143**, 89–94. |
| LYNAS, B. T. D. | 1970 | Clarification of the polyphase deformations of North Wales Palaeozoic rocks. *Geol. Mag.*, **107**, 505–510. |
| McKERROW, W. S., LEGGETT, J. K. & EALES, M. H. | 1977 | Imbricate thrust model of the Southern Uplands of Scotland. *Nature, London*, **267**, 237–239. |
| MOSELEY, F. | 1972 | A tectonic history of northwest England. *J. geol. Soc. London*, **128**, 61–598. |
| MUIR, M. D., BLISS, G. M., GRANT, P. R. & FISHER, M. J. | 1979 | Palaeontological evidence for the age of some supposedly Precambrian rocks in Anglesey, North Wales. *J. geol. Soc. London*, **136**, 61–64. |
| NUTT, M. J. C. & SMITH, E. G. | 1981 | Transcurrent faulting and the anomalous position of pre-Carboniferous Anglesey. *Nature, London*, **290**, 492–495. |
| OWEN, T. R., BLOXAM, T. W., JONES, D. G., WALMSLEY, V. G. & WILLIAMS, B. P. | 1971 | Summer (1968) field meeting in Pembrokeshire, South Wales. *Proc. Geol. Assoc. London*, **82**, 17–60. |

| | | |
|---|---|---|
| PENN, J. S. W. & FRENCH, J. | 1971 | *The Malvern Hills.* Guide No. 4. Geol. Assoc. London. |
| PHARAOH, T. C., MERRIMAN, R. J., WEBB, P. C. & BECKINSALE, R. D. | 1987 | The concealed Caledonides of eastern England: preliminary results of a multidisciplinary study. *Proc. Yorkshire geol. Soc.,* **46**, 355–369. |
| PHIPPS, C. B. & REEVE, F. A. E. | 1969 | Structural geology of the Malvern, Abberley and Ledbury Hills. *Q. J. geol. Soc. London,* **125**, 1–37. |
| POWELL, C. M. | 1987 | Inversion tectonics in S.W. Dyfed. *Proc. Geol. Assoc. London,* **98**, 193–203. |
| RAST, N. | 1969 | The relationship between Ordovician structure and volcanicity in Wales. *In* Wood, A. (Ed.) *The Pre-Cambrian and Lower Palaeozoic rocks of Wales.* University of Wales Press, Cardiff, 305–335. |
| RATTEY, P. R. & SANDERSON, D. J. | 1984 | The structure of SW Cornwall and its bearing on the emplacement of the Lizard Complex. *J. geol. Soc. London,* **141**, 86–95. |
| READING, H. G. | 1973 | The tectonic environment of south-west England. Contributions to the discussion of a paper by P. A. Floyd. *Proc. Geol. Assoc. London,* **84**, 239–242. |
| RICHARDSON, S. W. & OXBURGH, E. R. | 1978 | Heat flow, radiogenic heat production and crustal temperatures in England and Wales. *J. geol. Soc. London,* **135**, 323–338. |
| ROBERTS, B. | 1979 | *The geology of Snowdonia and Llŷn: an outline and field guide.* Adam Hilger Ltd., Bristol, 183 pp. |
| ROBERTS, J. L. & SANDERSON, D. J. | 1971 | Polyphase development of slaty cleavage and the confrontation of facing directions in the Devonian rocks of north Cornwall. *Nature, Phys. Sci. London,* **230**, 87–89. |
| SANDERS, L. D. | 1955 | Structural observations on the south-east Lizard. *Geol. Mag.,* **92**, 231–240. |
| SANDERSON, D. J. | 1984 | Structural variation across the northern margin of the Variscides in NW Europe. *In* Hutton, D. H. W. & Sanderson, D. J. (Eds.) *Variscan tectonics of the North Atlantic region. Spec. Pub. geol. Soc. London,* **14**, 149–165. |
| SANDERSON, D. J. & DEARMAN, W. R. | 1973 | Structural zones of the Variscan fold belt in SW England, their location and development. *J. geol. Soc. London,* **129**, 527–536. |
| SEAGO, R. D. & CHAPMAN, T. J. | 1988 | The confrontation of structural styles and the evolution of a foreland basin in central SW England. *J. geol. Soc. London,* **145**, 789–800. |
| SELWOOD, E. B. & THOMAS, J. M. | 1984 | A reinterpretation of the Meldon Anticline in the Belstone area. *Proc. Ussher Soc.,* **6**, 75–81. |
| | 1984 | Structural models of the geology of the north Cornwall coast; a discussion. *Proc. Ussher Soc.,* **6**, 134–136. |
| | 1985 | An alternative model for the structure of north Cornwall – a statement. *Proc. Ussher Soc.,* **6**, 180–182. |
| | 1986a | Upper Palaeozoic successions and nappe structures in north Cornwall. *J. geol. Soc. London,* **143**, 75–82. |
| | 1986b | Observations of the Padstow confrontation, north Cornwall (Abstract). *Proc. Ussher Soc.* **6**(3), 419. |
| SELWOOD, E. B., EDWARDS, R. A. & SIMPSON, S. | 1984 | The geology of the country around Newton Abbot. *Mem. Geol. Surv. Gt. Br.* |
| SELWOOD, E. B., STEWART, I. J. & THOMAS, J. M. | 1985 | Upper Palaeozoic sediments and structure in north Cornwall – a reinterpretation. *Proc. Geol. Assoc., London,* **96**(2), 129–141. |

SHACKLETON, R. M. 1954 The structural evolution of North Wales. *Liverpool Manchester geol. J.*, **1**(3), 261–297.

1969 The Pre-Cambrian of North Wales. *In* Wood, A. (Ed.) *The Pre-Cambrian and Lower Palaeozoic rocks of Wales.* University of Wales, Cardiff, 1–22.

1975 Precambrian rocks of Wales. *In* Harris, A. L. *et al.*, Precambrian. *Spec. Rept. Geol. Soc. London*, **6**, 76–82.

1984 Thin-skinned tectonics, basement control and the Variscan front. *In* Hutton, D. H. W. & Sanderson, D. J. (Eds.) Variscan tectonics of the North Atlantic region. *Spec. Pub. geol. Soc. London*, **14**, 125–129.

SHACKLETON, R. M., 1982 An interpretation of the Variscan structures in SW
RIES, A. C. & England. *J. geol. Soc. London*, **139**, 533–541.
COWARD, M. P.

SHEPHARD-THORN, E. R. 1975 The Quaternary of the Weald – a review. *Proc. Geol. Assoc. London*, **86**, 537–557.

SHIELLS, K. A. G. 1962–63 The geological structure of north-east Northumberland. *Trans. R. Soc. Edinb.*, **65**, 447–481.

SHIELLS, K. A. G. & 1963 Tectonics of the Coldingham Bay area of Berwickshire in
DEARMAN, W. R. the Southern Uplands of Scotland. *Proc. Yorkshire geol. Soc.*, **34**, 209–234.

1966 On the possible occurrence of Dalradian rocks in the Southern Uplands of Scotland. *Scott. J. Geol.*, **2**, 231–242.

SIMPSON, A. 1967 The stratigraphy and tectonics of the Skiddaw Slates and the relationship of the overlying Borrowdale Volcanic Series in part of the Lake District. *Geol. J.*, **5**, 391–418.

SMITH, A. J. & 1975 The structure and geological evolution of the English
CURRY, D. Channel. *Phil. Trans. R. Soc. Lond.*, A, **279**, 3–20.

SMITH, K. & LEAKE, R. C. 1984 Geochemical soil surveys as an aid to mapping and interpretation of the Lizard Complex. *J. geol. Soc. London*, **141**, 71–78.

SMITH, N. J. P. 1985 *Map 2. Contours on the top of the pre-Permian surface of the United Kingdom (South).* Scale 1:1,000,000. British Geological Survey.

SNELLING, N. J. (Ed.) 1985 The chronology of the geological record. *Mem. geol. Soc. London*, **10**, 340 pp.

SOPER, N. J. & 1978 Structure. *In* Moseley, F. (Ed.) *The geology of the Lake
MOSELEY, F. District.* Occasional Pub. No. 3, Yorks. geol. Soc., 45–67.

SOPER, N. J., 1987 Late Caledonian (Acadian) transpression in north-west
WEBB, B. C. & England: timing, geometry and geotectonic significance.
WOODCOCK, N. H. *Proc. Yorkshire geol. Soc.*, **46**, 175–192.

STONELEY, R. 1982 The structural development of the Wessex Basin. *J. geol. Soc. London*, **139**, 543–554.

STRAHAN, A., 1914 The geology of the South Wales coalfield. Part XI. The
CANTRILL, T. C., country around Haverfordwest. *Mem. Geol. Surv. England
DIXON, E. E. L., & Wales.*
THOMAS, H. H. &
JONES, O. T.

STYLES, M. T. & 1984 The Rb-Sr isochron age of the Kennack Gneiss and its
RUNDLE, C. C. bearing on the age of the Lizard Complex, Cornwall. *J. geol. Soc. London*, **141**, 15–19.

THORPE, R. S. 1974 Aspects of magmatism and plate tectonics in the Precambrian of England and Wales. *Geol. J.*, **9**, 115–136.

THORPE, R. S., BECKINSALE, R. D., PATCHETT, P. J., PIPER, J. D. A., DAVIES, G. R. & EVANS, J. A.  1984  Crustal growth and late Precambrian–early Palaeozoic plate tectonic evolution of England and Wales. *J. geol. Soc. London*, **141**, 521–525.

TURNER, J. S.  1936  The structural significance of the Rossendale Anticline. *Trans. Leeds geol. Ass.*, **5**, 157–160.

1949  The deeper structure of central and northern England. *Proc. Yorkshire geol. Soc.*, **27**, 280–297.

TURNER, P. J.  1985  Stratigraphic and structural variations in the Lifton–Marystow area, West Devon, England. *Proc. Geol. Assoc. London*, **96**(4), 323–335.

1986  Stratigraphical and structural variations in central SW England: a critical appraisal. *Proc. Geol. Assoc. London*, **97**, 331–345.

WALLACE, P.  1983  The subsurface Variscides of southern England and their continuation into continental Europe. *In* Hancock, P. L. (Ed.) *The Variscan fold belt in the British Isles*. Adam Hilger Ltd., Bristol, 198–208.

WARREN, P. T.  1964  The stratigraphy and structure of the Silurian (Wenlock) rocks south-east of Hawick, Roxburghshire, Scotland. *Q. J. geol. Soc. London*, **120**, 193–222.

WATSON, J.  1975  The Precambrian rocks of the British Isles – a preliminary review. *In* Harris, A. L. *et al.*, Precambrian. *Spec. Rept. geol. Soc. London*, **6**, 1–10.

WATSON, J. & DUNNING, F. W.  1979  Basement – cover relations in the British Caledonides. *In* Harris, A. L., Holland, C. H. & Leake, B. E. (Eds.) The Caledonides of the British Isles – reviewed. *Sp. Pub. geol. Soc. London*, **8**, 67–91.

WATTS, W. W.  1947  *Geology of the ancient rocks of Charnwood Forest Leicestershire*. Backus, Leicester, 160 pp.

WHITTAKER, A.  1975a  Namurian strata near Cannington Park, Somerset. *Geol. Mag.*, **112**, 325–326.

1975b  A postulated post-Hercynian rift valley system in southern Britain. *Geol. Mag.*, **112**, 137–149.

WILLIAMS, G. D. & CHAPMAN, T. J.  1986  The Bristol–Mendip foreland thrust belt. *J. geol. Soc. London*, **143**, 63–73.

WILLS, L. J.  1978  A palaeogeological map of the Lower Palaeozoic floor below the cover of Upper Devonian, Carboniferous and later formations. *Mem. geol. Soc. London*, **8**, 36 pp.

WOOD, D. S.  1974  Ophiolites, melanges, blueschists, and ignimbrites: Early Caledonian subduction in Wales? *In* Dott, R. H. & Shaver, R. H. (Eds.) Modern and ancient geosynclinal sedimentation. *Soc. Econ. Pal. Min. Spec. Pub.*, **19**, 334–344.

WOODCOCK, N. H.  1984a  Early Palaeozoic sedimentation and tectonics in Wales. *Proc. Geol. Assoc. London*, **94**(4), 323–335.

1984b  The Pontesford Lineament, Welsh Borderland. *J. geol. Soc. London*, **141**, 1001–1014.

WOODLAND, A. W. (Ed.)  1971  The Llanbedr (Mochras Farm) borehole. *Rep. Inst. Geol. Sci.*, **71/18**.

WRIGHT, A. E.  1969  Precambrian rocks of England, Wales and southeast Ireland. *In* Kay, M. (Ed.) North Atlantic – geology and continental drift: a symposium. *Mem. Amer. Assoc. Petrol. Geol.*, **12**, 93–109.

1976  Alternating subduction and the evolution of the Atlantic Caledonides. *Nature, London*, **264**, 156–160.

# 18

# DEEP GEOLOGY

# D. H. Griffiths and G. K. Westbrook

## Introduction

In two small countries with a long and intensive history of geological investigation, it is only comparatively recently that much information has become available on the nature of the crust deeper than 1,000 m. Even so, knowledge of the form, composition and age of Precambrian basement is fragmentary, and the lower crust and mantle are poorly defined despite what must be more than a thousand-fold increase in data over the past two decades. This is, however, a rapidly developing field of investigation, and it is very likely that new information will greatly augment or supersede what is presented in this chapter, over the next few years.

An important component of our knowledge of the deep geology of the country comes from boreholes. Both Wills and Kent were prominent in compiling, interpreting and presenting borehole data in a number of publications (Kent 1949, 1967, 1968, 1974, 1975; Wills 1973, 1978), and more recently, Whittaker (1985) provided a compilation of data on the post Carboniferous sedimentary cover.

While boreholes provide the essential means of establishing the lithologies and ages of rocks at depth, it is through geophysical methods that most of the knowledge of crustal structure has come. England and Wales have good regional gravity and aeromagnetic coverage that is published by the British Geological Survey as companion sheets to the 1:250,000 Universal Transverse Mercator (UTM) geological maps (for current coverage see appendix to Chapter 15), which provide a ready means of comparing geophysical anomalies with surface geology. The British Geological Survey holds, and has used in its compilations, geophysical data acquired by universities and companies in addition to its own, but most of the data from companies is restricted because of its commercial

nature and a substantial body of information exists in the form of unpublished Ph.D. and M.Sc. theses.

Large-scale seismic refraction experiments designed to study the deep structure of the crust are relatively few in number. They provide information on the gross structure of crust in broadly defined layers of different seismic velocities and varying thicknesses. Perhaps the most widely used has been the Lithospheric Seismic Profile in Britain (LISPB) (Bamford et al. 1976, 1977, 1978), upon the results of which geologists have drawn heavily for models of crustal structure and its development. There are, however, ambiguities in the crustal models derived from these experiments, as evidenced by conflict between the interpretations of different seismic experiments crossing regions in common and by reinterpretation of the same data to produce a different model.

There has been a surge of activity in seismic reflection surveys. The overwhelming majority of these surveys have been for hydrocarbon exploration in sedimentary basins, the data from which are not publicly available, but some surveys have been commissioned by the British Geological Survey with sufficiently long recording times to show reflectors in the deep crust and the Moho (Whittaker & Chadwick 1984). The British Institutes Reflection Profiling Syndicate (BIRPS) began in 1982 a series of deep crustal seismic reflection profiles. Profiles from the Irish Sea have been published (Brewer et al. 1983), and others have been obtained around SW England and the Channel and in the North Sea. All the BIRPS profiles to date have been marine, because of the much lower costs of marine versus land surveys, with the recent exception in 1988 of a survey over the Weardale granite in Co. Durham designed to record reflected shear-waves.

Geophysically derived models of geological structures are essentially models of the spatial distribution of rocks with particular physical properties. They must be interpreted within the constraints of known geology and the properties of rock types established by *in situ* and laboratory measurements. This is especially true for gravity and magnetic models which have great inherent ambiguity, and often find their most effective use in discriminating between rival geological hypotheses rather than by providing definitive structural models. Interpretation becomes more difficult when the lower crust and upper mantle are being considered. Qualitative interpretation of magnetic anomalies has to be done with care since the contours of an anomaly can have a very indirect relationship to the contours on the surface of the magnetic body producing the anomaly. Depth estimation techniques, such as those commonly used for deriving depths to magnetic basement, essentially provide values for the maximum depth.

In a short summary such as this it is not possible to include any assessment of the quality or even validity of the information obtained by geophysical methods. Nor has any serious attempt been made to relate deep structure in the crust of England and Wales to the tectonic conditions which produced them. The relationship between the crustal structure and tectonics is not always clear even in currently active regions, and the uncertainty with which much of the structure is known would be compounded in any extrapolation to

**Fig. 18.1.** Summary map of major structural units in England and Wales; based on maps by Kent (1975). Diagonally shaded regions – areas of relative basement uplift during the Upper Palaeozoic. Stippled regions – areas of major subsidence during the Mesozoic. Regions enclosed by dashed lines – areas of relative uplift during Mesozoic and Tertiary. a, Solway Basin; b, Northumberland Basin; c, Vale of Eden; d, Isle of Man, e, Lake District; f, Alston Block; g, Askrigg Block; h, Cheshire Basin; i, Worcester Basin; j, Southern North Sea Basin; k, Manx–Furness basin; l, Cardigan Bay Basin; m, North Celtic Sea Basin; n, Welsh Massif; p, London Brabant Massif; q, Western Approaches Basin; r, Wessex Basin; s, Cornubian Massif.

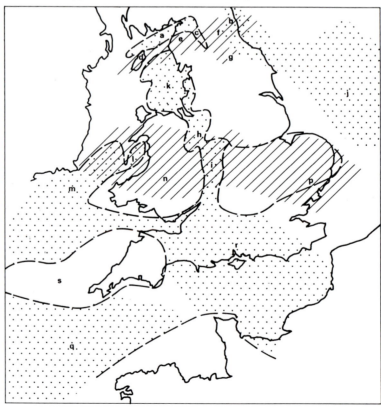

the tectonic environment. Furthermore, adjustments of the crust to accommodate major structures can continue for a long time after their formation, so that they may no longer retain their original form.

Much of the available information on deep geology relates to the depth and configuration of the basement. In this chapter *basement* is used in the geophysical sense as a body of rock having a strong physical contrast, usually seismic velocity, often density, and sometimes magnetisation, with an overlying sedimentary cover. It is usually composed of igneous or metamorphic rocks, but the degree of metamorphism may not necessarily be high, while its age differs from area to area.

The broad tectonic framework of the country is essentially formed by basement ridges or blocks with intervening sedimentary basins of different ages, and is shown in very simplified form in Figure 18.1. The basement feature that dominates the structure of England and Wales is the post-Caledonian Wales–Brabant High which runs eastward across the centre of the country from Wales through the Midlands and East Anglia into the southern North Sea. Outcrops of Precambrian rocks in England occur on this feature. To the north and south lie regions of thick sequences of sedimentary rocks. In the north of England a similar basement ridge runs from the Isle of Man through the Lake District and northern Pennines, County Durham and into the North Sea (Manx–Ringkøbing–Fyn High – see Chapter 15). Four of the principal elements of this feature in the British Isles, the Isle of Man, the Lake District, and Alston and Askrigg Blocks are underlain by granites. Further north lies another sedimentary trough extending from the Solway Firth into Northumberland.

The region south of the Wales–Brabant High is one with a long history of subsidence and sedimentation, complicated by structures formed in Hercynian and later movements. In the south-west lies another positive region, the Cornubian Massif, centred on the Hercynian granite batholiths. With a lower density than the average for crustal rocks the Cornubian granites have clearly been the cause of the relative buoyancy of the region in which they lie, as indeed have the Caledonian granites beneath the northern Pennines and Lake District. This reason for the buoyant behaviour of the basement highs cannot be invoked so easily for the Wales–Brabant high. The acidic intrusives in the east Midlands do not appear to be of any great volume, although it is possible that some negative gravity anomalies with sources in the basement are caused by granitic bodies.

Superimposed on the broad structure outlined above are the series of basins such as the Cheshire and Worcester Basins first formed in the Permo–Triassic. These basins are often grabens or half-grabens,

perhaps genetically related to the North Sea Graben. They generally have a N–S trend and their margins often coincide with major faults of greater age.

The crust beneath the sedimentary cover has been shown by a few large-scale seismic refraction experiments to have a typical thickness of a little over 30 km, with an upper crustal layer of seismic velocity between 5·8 and 6·2 km s$^{-1}$, beneath which at about 15 km depth lies a lower crustal layer in which velocity increases from 6·4 km s$^{-1}$ to about 7 km s$^{-1}$ at the base of the crust.

# Regional Descriptions

## Northern and North-eastern England

The LISPB north–south section across northern England showed that the crust has an average thickness of about 35 km and that it is made up of three layers beneath the Carboniferous and younger sediments (Bamford *et al.* 1978). The uppermost of the three layers, with an average *P*-wave velocity near its surface of 5·8 km s$^{-1}$ increasing to 6·0 km s$^{-1}$ at its base, is thought to be composed predominantly of lower Palaeozoic rocks and thickens towards the north from 7·5 km near Buxton to 14 km near Alston. The shape of the top surface of this layer reflects some of the main structural elements of the northern half of England (Figs. 18.2, 18.3) which are (from north to south): the Northumberland Trough – a Lower Carboniferous sedimentary basin; the Alston Block – a structural high intruded by granite; the Vale of Eden – a Permo–Triassic and Carboniferous sedimentary basin; the Askrigg Block – a structural high intruded by granite; the Craven Basin – a Carboniferous sedimentary basin.

Beneath the 5·8 km s$^{-1}$ layer lies a layer with a seismic velocity of 6·25–6·3 km s$^{-1}$ which was interpreted as a high-grade metamorphic basement of pre-Caledonian age. The quality of the data gave interpretation of the results for the lower crust some ambiguity, but a reflector forming the base of the 6·25–6·3 km s$^{-1}$ layer was discerned beneath the Alston Block and the Vale of Eden (Bamford *et al.* 1978). Also, the data suggested that there is a downward transition from velocities typical of the lower crust (7·3 km s$^{-1}$) to those of the upper mantle (8·0 km s$^{-1}$) over a depth range of 5 km at the base of the crust. The results of the Caledonian Suture Seismic Project along the Northumberland Trough almost at right angles to LISPB, show some important differences from the LISPB model (Bott *et al.* 1985). A layer of average thickness 2 km and seismic velocity 4·4 km s$^{-1}$, Carboniferous, upon a 2 km thick layer of velocity 5·6 km s$^{-1}$, interpreted as Lower Palaeozoic metasediments, overlie a well defined refractor of

6·15 km s⁻¹ that is interpreted to be the top of pre-Caledonian crystalline basement. Beneath the Trough and the western North Sea a lower crustal layer of 6·6 km s⁻¹ occurs at a depth of 16 km. The Moho lies at 30 km beneath a sharp increase in velocity in the lowermost 2 or 3 km of the crust to about 7·2 km s⁻¹ before reaching a mantle velocity of 8·0 km s⁻¹. Beneath the Irish Sea, the lower crustal layer does not have a clearly defined upper boundary

and velocity increases gradually through the crust until a sharp increase just above the Moho at 32 km depth. The conflict with the LISPB model (Fig. 18.2) may not be as great as it appears in the region of lower crust and Moho, because the LISPB model is poorly constrained. Apparently less easy to accommodate is the difference in the upper part of the crust. The pre-Caledonian basement is much closer to the surface than LISPB predicts, even though LISPB does not

**Fig. 18.2.** North–South section of the crust in northern England between the Southern Uplands of Scotland (SU) and Buxton (B) through the Alston Block (AL B) and Askrigg Block (AS B). The seismic P-wave velocity structure shown is taken principally from the LISPB section of Bamford *et al.* (1978) with, for comparison, the orthogonal CSSP section of Bott *et al.* (1985) and refraction experiments in NE England by Bott *et al.* (1984) (dotted boundaries). Part of the difference between the results may be explained by the position of the LISPB line lying west of the Pennines. CSSP also showed the 6·5 km s⁻¹ mid-crustal refractor to be absent beneath the Irish Sea.

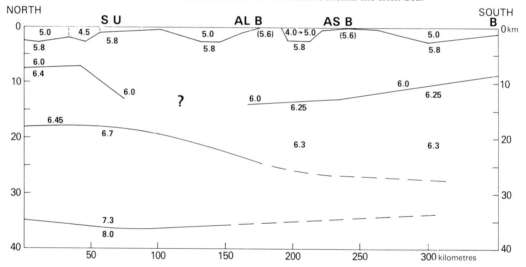

**Fig. 18.3.** Diagrammatic section, N–S, through the northern Pennines, showing stable blocks buoyantly supported by granite batholiths, with intervening Carboniferous basins (after Bott *et al.* 1984).

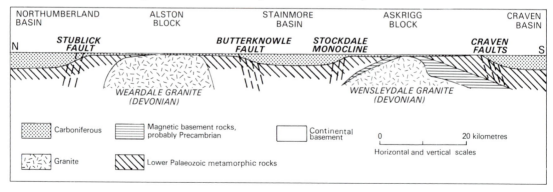

define the upper crust beneath the Northumberland Trough. It is a possibility that the pre-Caledonian basement rises beneath the Trough. Another interpretation presented by Bott *et al.* (1985) essentially discards the LISPB profile and extends the Northumberland Trough structure southward beneath the northern Pennines.

The results of gravity surveys by Bott (1967, 1974), Bott & Masson-Smith (1957), O'Connor (1966), Myers & Wardell (1967) indicated that the Lake District dome, and the Alston and Askrigg blocks are underlain by granite batholiths. The granites are less dense (2·64 Mg m$^{-3}$) than the Lower Palaeozoic (2·75 Mg m$^{-3}$) and Precambrian rocks which they intrude, and give rise to the large negative (Bouguer) gravity anomalies. The presence of the granite beneath the Lake District can be inferred from outcrops (Fig. 16.4) at Eskdale, Ennerdale, Shap and Skiddaw, while drilling at Rookhope on the Alston Block (Dunham *et al.* 1965) and at Taydale on the Askrigg Block (Dunham 1974) cored Devonian granites. (See also Figs. 7.1 and 7.2.)

The line of the LISPB experiment ran along the western margins of the Alston and Askrigg blocks. Seismic refraction experiments directly over them confirm the 5·8 km s$^{-1}$ velocity for the lower Palaeozoic basement and give a velocity of 5·4 km s$^{-1}$ for the granites (Bott *et al.* 1984). A layer with seismic velocity of 6·5 km s$^{-1}$ underlies them at a depth of 12 km, and the Moho is at 28 km depth.

Modelling of the Weardale Granite of the Alston Block showed it to have steep sides and an almost flat top. The Wensleydale Granite of the Askrigg Block has less steeply dipping sides and a narrower top (Fig. 18.3) (Bott 1967). Both granites have an E–W elongation with subsidiary cupolas to east and west. The Lake District granites are considered to form a system of batholiths and interconnecting ridges. The interpretation of the shapes of these granites from the gravity anomalies allows a variety of possible models. The asymmetry of the anomalies over some of the batholiths such as the Eskdale Granite can be explained *either* by lateral changes in composition, and hence density, within an intrusion of simple shape with a floor at constant depth *or* by a more complexly shaped body of uniform composition (Fig. 18.4) Bott

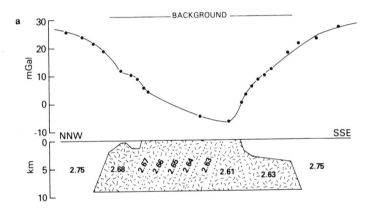

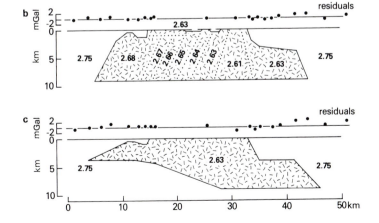

**Fig. 18.4.** Two-dimensional models of the Eskdale granite and Ennerdale granophyre in the western part of the Lake District which produce calculated gravity anomalies that match closely the Bouguer gravity anomaly measured over them. Solid circles are the observed values of the Bouguer gravity anomaly (a). The continuous line is the calculated anomaly. For models (b) and (c), the residuals (observed minus calculated) are shown. The models illustrate variations in the structure and density of the granite that would produce the same gravity anomaly (after Bott 1974).

1974). Without control from boreholes or other geophysical techniques, the 'correct' model cannot be distinguished, but it is known from outcrops that lateral variations in density do exist. Modelling of the Shap Granite produces a batholith of similar form to the Eskdale Granite, except that only a small part of the roof outcrops in the south of the body. Further geophysical work by Locke & Brown (1978) confirmed the overall shape of the Shap Granite. This most easterly outcrop of the Lake District granite complex may be connected with the Weardale Granite by a ridge beneath the Vale of Eden (Bott 1974). The Lake District and Weardale granite complexes lie in an ENE-trending anticlinal structure of Ordovician and Cambrian rocks, which includes the Isle of Man, also intruded by granite (Cornwell 1972). To the north and south of this structure lie Carboniferous rocks in the Northumberland Basin (2–3 km thick) and Stainmore Basin (2·5 km thick) respectively (see Fig. 18.3). Relative movement between Alston Block and these basins took place along the hinge lines of the Stublick

Faults and Ninety Fathom Fault to the north and the Lunedale and Butterknowle Faults to the south. The easterly extensions of these hinge lines in eastern Durham and Northumberland are revealed by the gravity gradients over them and by seismic refraction experiments (Bott *et al.* 1984). The Askrigg Block is bounded to the south by the Craven Basin along the Craven Faults. This basin continues into Lancashire and the Irish Sea, forming the major sedimentary trough to the south of the high of the Lower Palaeozoic rocks (The Manx–Ringkøbing–Fyn High).

A feature of eastern England north of the Wash is a general deepening of the pre-Carboniferous surface eastwards. Northwards this is superimposed upon the ENE-trending features. Of major significance in the basement is a belt of dominantly positive magnetic anomalies running from the Wash to the Askrigg Block, and possibly into the Lake District (Bott 1967). This has been interpreted as a ridge of basement (Wills 1978), although it may be produced by a belt of magnetic rocks within the basement. There is an

**Fig. 18.5.** Simplified contour map of the surface of the pre-Carboniferous basement in the east Midlands (from Kent 1976). Contour interval is 1,000 feet.

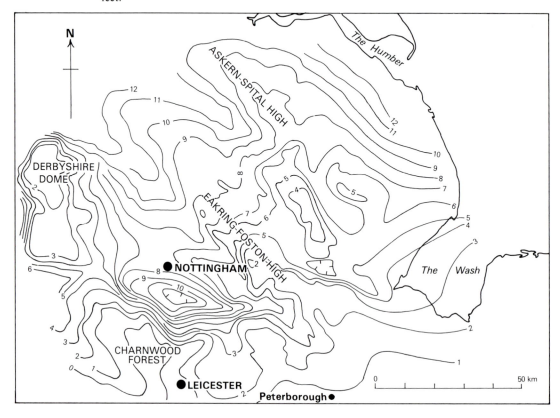

associated linear gravity high between Leeds and the Wash. In Humberside and Lincolnshire the magnetic belt and gravity high correlate with the Askern–Spital High (Fig. 18.5) identified from borehole, seismic reflection, and mining evidence (Kent 1974).

The Wensleydale Granite lies within the magnetic belt and has magnetic anomalies associated with its margins. These magnetic anomalies can be modelled by a weakly magnetised granite (0·25 amp m$^{-1}$) intruded into a more strongly magnetised country rock (0·9 amp m$^{-1}$) (Bott 1961). Some of the rock making up the Askrigg Block, within which the granite lies, has been shown by the Beckermonds Scar borehole to be a magnetite-bearing greywacke of Arenig age (Wilson & Cornwell 1982). It is possible that some magnetisation may be related to the metamorphic aureole of the granite and is caused by pyrrhotite as is the case for magnetic anomalies associated with the Dartmoor Granite (Cornwell 1967), and the Shap Granite (Hughes 1977).

Near Market Weighton, it appears that another granite intrudes the northern flank of the magnetic belt (Bott et al. 1978). The presence of this granite is suggested by a negative gravity anomaly (−20 mGal), elongated in a NE direction, and magnetic anomalies of the type associated with the Wensleydale Granite. The top surfaces of the granite and the magnetic basement are calculated to be at a depth of about 2·5 km. The granite is thought to be of Caledonian age and to have kept the region around it relatively buoyant, producing the Market Weighton uplift during Jurassic and Lower Cretaceous times (see Chapter 11).

Between Market Weighton and the Askrigg Block, the smoother nature of the magnetic anomalies indicates that the basement ridge becomes deeper. A contour map of depth to magnetic basement in the Yorkshire region, compiled by Decca Resources Inc., has been published, with reservations about the depths, by Kent (1974). The contours on this map give values 2 to 3 times greater than those obtained by Bott et al. (1978) and what might be expected from other subsurface data in the south-east of the area. (It has not been possible to ascertain how the contours on the map were derived.) Although the actual values are suspect, the map does indicate a deepening of the basement towards the coast and the presence of the NW-trending basement ridge.

## East Midlands and East Anglia

This area, dominated by a large platform of Pre-cambrian and Lower Palaeozoic rocks with a fairly thin cover of Mesozoic and Tertiary rocks, was called the East Anglian Massif by Kent (1975) and considered to be part of the Wales–Brabant High in the Carboniferous (Wills 1973). The northern part of this platform is made up of Precambrian rocks with Cambrian cover in places. The presence of the Precambrian is well known from boreholes, and it also produces many short-wavelength magnetic anomalies. Early seismic refraction work on the depth to the surface of the pre-Mesozoic platform was carried out by Bullard et al. (1940), and this has been extended and integrated with borehole and gravity data by Chroston & Sola (1982), who showed that the surface of the platform dips northward, but that towards the south is successively composed of Ordovician, Silurian and then Devonian rocks. The long wavelengths of the magnetic anomalies indicate that the magnetised crystalline basement descends southwards beneath these rocks (Brunstrom, in discussion of Stubblefield & Bullerwell (1966), has suggested depths between 6 and 9 km for this basement beneath southern Cambridgeshire and Huntingdonshire). North-east of Norwich, in eastern Norfolk, a thin Carboniferous basin lies on the platform, beneath about 1 km of Mesozoic cover (Allsop & Jones 1981, Allsop 1984).

Short-wavelength magnetic anomalies indicate that shallow Precambrian basement extends northwards beneath the Wash into the ridge structure leading to the Askrigg Block; and NW from the area around Peterborough, through the region of Charnwood Forest, to the Derbyshire Dome (Fig. 18.5 for localities). The crest of the NW basement ridge has been shown from seismic refraction to descend from the surface at Charnwood Forest to a depth of nearly 2 km on the SE flank of the Derbyshire Dome, with small horsts and grabens developed locally along it (Whitcombe & Maguire 1981b). The seismic velocity of 5·64 km s$^{-1}$ of the basement is very similar to the velocities measured in Charnian rocks (Whitcombe & Maguire 1980, Moseley & Ford 1985, see also Maguire 1987), and although the presence of some Lower Palaeozoic rock within the seismic basement cannot be definitely excluded, it is clear that a thickness of about 1·8 km of Lower Palaeozoic rocks, probably Old Red Sandstone and Ordovician mudstones, lies between the Carboniferous limestone of the Derbyshire Dome and the seismic basement (Whitcombe & Maguire 1981b). The Precambrian of the north-eastern flank of this north-westerly ridge is depressed to a depth of at least 3·3 km beneath the Widmerpool Gulf (Chapter 7), south of Nottingham, which is the most southerly part of a Carboniferous basin that extends northwards into the main region of Carboniferous deposition in Lancashire and Yorkshire (Kent 1967), and has a narrow westerly extension between Charnwood Forest and the Derbyshire Dome. Associated with this basin is a large gravity low extending to Doncaster. The basin is complicated by north-westerly trending ridges, the most prominent of which is the Eakring–Foston High (Fig. 18.5).

The pre-Carboniferous basement is shallow beneath the Derbyshire Dome, 272 m in the Woo Dale borehole (Cope 1973). Cope (1949) suggested that the tuffs and altered lavas from the borehole are Precambrian, but this was disputed (Le Bas 1972). Ordovician sediments were found at a depth of 1,803 m in the Eyam borehole to the north-east (Dunham 1973). The overall structure of the basement ridge beneath the Derbyshire Dome has been modelled from gravity data by Maroof (1974) who placed the depth of pre-Carboniferous basement beneath Sheffield at 2·5 km.

Unlike the Alston and Askrigg Blocks, the uplift of the Derbyshire Dome does not appear to be related to the presence of an underlying granite batholith, as there is no large negative gravity anomaly. Intrusive rocks of 'Caledonian' age (the Mountsorrel Granodiorite and the South Leicestershire diorites) do outcrop north and south of Charnwood Forest, and seismic refraction experiments and other geophysical work indicates that there are more concealed intrusives in the same region (Whitcombe & Maguire 1981a, Maguire et al. 1982, Allsop & Arthur 1983). The sizes of these intrusives are all quite small.

An arcuate group of negative gravity anomalies lies between Peterborough and the Wash. The lack of any obvious correlation with the topography of the pre-Carboniferous basement, mainly Precambrian (Fig. 18.5), or to any structure in the overlying sedimentary rocks, implies that the anomalies are caused by structures within the basement (Austin 1979). Magnetic anomalies around the edges of the gravity anomalies indicate that the rocks causing the gravity anomalies are less strongly magnetised than most of the basement rocks. Possible causes are granitic intrusions, perhaps cupolas of a large underlying batholith, or small pre-Carboniferous sedimentary basins.

Allsop (1983) suggested that sedimentary basins are the most probable cause of two anomalies occurring in the Wash and implied that earlier maps of the pre-Carboniferous surface are locally incorrect in this region. This is similar to an earlier interpretation of Sola (1974). In the case of the gravity anomaly immediately NE of Peterborough, a seismic refraction experiment carried out by Durham and Leicester Universities in 1981, showed that the Mesozoic sediments are underlain at depth of 400 m by basement with a seismic velocity of between 5·6 and 5·8 km s$^{-1}$, very similar to that of the Charnian and that consequently a granitic intrusion is the likely cause of the negative gravity anomaly (Arter 1982, Smith 1982, Wilson 1982). The top of the granite probably lies at a depth of between 0·4 and 2·0 km (Rees 1982). If a common origin is invoked for this group of negative gravity anomalies, then it has to be granitic intrusions, presumably related to those of Weardale, Wensleydale

and Market Weighton. The zone of high heat flow running from the Wash to the northern Pennines (Richardson & Oxburgh 1979) may be associated with the granites.

## Cheshire Basin

This major Permo–Triassic sedimentary basin is associated with a large negative gravity anomaly (White 1949, Maroof 1974, IGS UTM Gravity Map). The anomaly is not, however, as great as might be expected from the known thickness of 2·8 km of Permo–Triassic sediments in the Knutsford Borehole (Colter & Barr 1975). The probable reason for this is that the Carboniferous rocks are thinner beneath the basin.

The interpretation of seismic refraction data recorded in the Cheshire Basin and on its margins during the South Irish Sea Seismic Experiment (Ransome & Nunn 1978) shows that the crust beneath the basin is about 35 km thick (Poulter & Westbrook 1979), as is the crust beneath Wales (Nunn 1978) and the crust beneath the Derbyshire Dome (Bamford et al. 1978). The data also indicate that a seismic refractor of velocity 6·4 km s$^{-1}$ rises beneath the basin by about 2 km relative to the margins. In the south of the basin, the Prees Borehole (Colter & Barr 1975) penetrated 3·6 km of Jurassic, Triassic and Permian sediments before passing through 161 m of Upper Carboniferous Coal Measures into Lower Palaeozoic rocks (Ordovician/ Silurian). Although Prees is at the thickest part of the basin, it is on a slight gravity high, and is probably over a basement feature which had positive relief during the Carboniferous.

The eastern margin of the basin is quite complex, particularly in the south-east where magnetic and gravity anomalies indicate that the basement is segmented by a series of en echelon faults. The relationship between a magnetic anomaly and the margin is lost north of Stoke-on-Trent, where the margin is cut by the magnetic anomalies associated with the basement ridge running from the Peak District to the East Anglian Platform. The magnetic anomaly associated with the margin west of Stoke-on-Trent is in some part explained by the presence of volcanics proved in the Apedale Borehole (Giffard 1923), but it is mainly caused by a magnetic basement ridge (Sadler 1977). Gravity modelling of the eastern margin shows that thinning of the Carboniferous beneath the basin is necessary to explain satisfactorily the gravity anomaly (Turnbull 1974, Sadler 1977), and that in the north-east the Permo–Trias/Carboniferous contact must dip at about 45° W, suggesting a faulted margin which finds its surface expression in the Red Rock Fault (see Fig. 8.2). It has been suggested by Challinor (1978) that the fault is an unconformity, and although this is unlikely in the north, it may be so W and SW of Stoke-on-Trent where the contact dips less steeply.

The eastern margin of the Cheshire Basin runs into the Church Stretton Fault system to the south and it has been suggested that this system is part of a major lineament which includes the Dent and Pennine fault systems of northern England. As such it was, presumably, the edge of an English 'microcraton' (Wills 1978). The faults associated with the Permo–Triassic basin system of Cheshire and the Irish Sea (Manx–Furness), however, step to the west in Lancashire,

**Fig. 18.6.** Bouguer gravity anomaly map of England and Wales. Compiled from maps published by Bott & Masson Smith (1957), Maroof (1974), and the Institute of Geological Sciences. Anomalies relative to the 1933 International Gravity Formula. Contours at 5 mGal intervals. The localities of gravity anomalies referred to in the text are as follows: A, Cardigan Bay; B, St George's Channel; C, Central Irish Sea; D, Worcestershire Basin; E, Bristol Channel; F, South-west granites; G, Salisbury–Southampton trough; H, Weardale granite; I, Wensleydale granite; J, Market Weighton anomaly; K, Wash anomalies; L, Cheshire Basin.

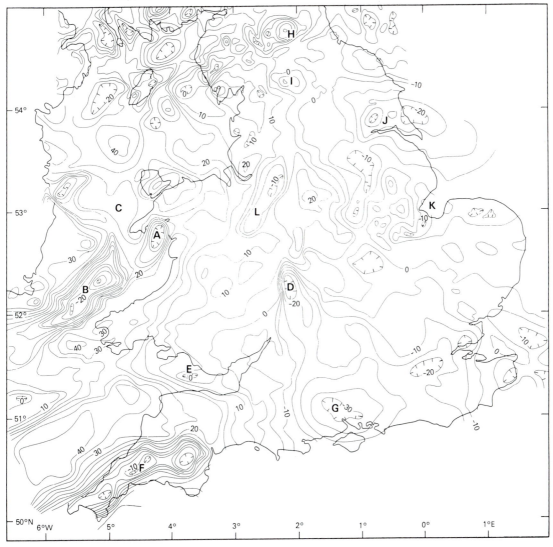

continuing as the Great Haigh Fault and the Lake District Boundary Fault.

There is a disparity between the eastern and western margins of the basin. The western margin has no large magnetic anomalies associated with it and although the gravity anomaly is equally large, the gravity gradient over the western margin is not so steep as that to the east. If it is fault controlled the displacement has probably been distributed across several step faults in the underlying pre-Triassic rocks. Also, the Precambrian rocks exposed on the SE margin of the basin do not occur on the western side. These features indicate a difference between the basement rocks underlying the eastern and western flanks of the basin, and it is probably that this change in basement occurs at the eastern margin of the basin and marks the edge of the English microcraton.

## North Irish Sea

The Cheshire Basin adjoins part of a larger basin system which occupies the most westerly part of Lancashire and the eastern Irish Sea. In the central part of this, the Manx–Furness Basin, boreholes drilled on a structural uplift penetrated a thickness of 3 km of Permo–Trias. The structure of this basin is well known from gravity, seismic reflection and refraction data, and boreholes (Bott 1964, 1968; Bott & Young 1971; Wright et al. 1971; Bacon & McQuillin 1972; Colter & Barr 1975). The Permo–Triassic sediments thin over a prominent structural high which runs from the Isle of Man to the Lake District and thicken again northwards into the Solway (Firth) Basin (Fig. 18.6). Carboniferous sedimentary basins underlie the Permo–Trias to the north and south of the high. On

**Fig. 18.7.** Line drawings of deep seismic reflector sections (Fig. 18.8) through the Irish Sea carried out by BIRPS (Reflection time is the two-way travel time in seconds). The upper part of the sections show sedimentary basins, e.g. the Kish Bank Basin on section E–F. The upper crust beneath the basins lacks reflectors compared with the lower crust, which generally shows many. The Moho is interpreted at the base of the sequence of reflectors in the lower crust (from line drawing provided by M. R. Warner – see Brewer et al. 1983).

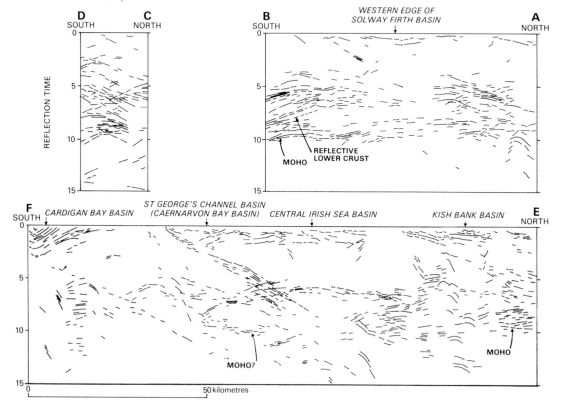

the Isle of Man, gravity lows show the presence of granite masses beneath outcrops at Foxdale and Dhoon (Cornwell 1972).

Once the gravitational effects of sedimentary rocks in the East Irish Sea have been allowed for, the Bouguer gravity anomaly over most of the Irish Sea is about 20 to 30 mGal higher than over the adjacent land areas. This has been interpreted as being caused by thinner or denser crust (Bott 1965, 1968). The Eskdalemuir Seismic Experiment (Agger & Carpenter 1964), with a line of shots between Holyhead and the Solway Firth, showed that the crust does thin beneath the east Irish Sea by about 4 km relative to the crust beneath the Southern Uplands. A minimum thickness obtained of 24 km is based on an assumption of a uniform seismic velocity of $6.12$ km s$^{-1}$ for the crust beneath the sediments, and this must be an underestimate. (The LISPB Experiment put crustal thickness beneath the Southern Uplands at about 33 km, compared with the 28 km estimated by Agger & Carpenter 1964.)

Deep seismic reflection sections obtained along the WINCH line (Western Isles–North Channel Traverse) by BIRPS showed the Moho, apparently at a reflection time of 10 s (two way travel) (Brewer et al. 1983) (Fig. 18.7). If the crust had a uniform seismic velocity of $6.0$ km s$^{-1}$ this would correspond to a depth of 30 km. The sedimentary section at the top of the crust would have lower velocities than this (3–4 km s$^{-1}$), whereas the lower crust would have higher velocity (? $6.5$ km s$^{-1}$). The lower crust, between 5 and 10 s (two way travel), contains many reflectors with gentle to moderate dip, some are curved convex upward probably indicating reflection from out of the plane of the section. The character of the lower crust is quite different from the upper crust (which does not have the same density of reflectors although they are clearly present, particularly in sedimentary section) and from the mantle (where there are very few reflectors). The position of the Moho is identified as the base of the zone of reflectors and is not shown as a separate reflector. In some areas, beneath the Solway Basin and south of the Isle of Man, the Moho cannot be seen on the section.

## Cardigan Bay and South Irish Sea

The first real evidence that Cardigan Bay might contain a Triassic Basin was provided by Powell (1955). He interpreted the steep gravity gradient, parallel to the coast near Harlech, as due to 1,000 m of downfaulted less dense sediments (probably Triassic) to the west. Detailed seismic surveys located the fault and confirmed the presence of 500 m of low velocity sediments (Griffiths et al. 1961). A borehole near the Mochras Fault (Wood & Woodland 1968, 1971) later

proved these to be mainly Tertiary sediments, beneath which occurred 1,300 m of Jurassic strata resting on Trias. From geophysical data Smith and Dabeck (1982) estimated that the fault dipped at 45° and had a throw of about 5 km, this decreasing southward. In Tremadoc Bay Blundell and co-workers (1971) showed that these deposits occupy a sub-basin, the most northerly of a series of basins in the South Irish Sea which strike SW, parallel to the coast of Cardigan Bay, and extend farther south into St Georges Channel and the Celtic Sea. On the north-west the Cardigan Bay–St Georges Channel Basin is separated by the Irish Sea geanticline from the Central Irish Sea Basin (Fig. 18.8). The latter contains a Carboniferous and Permo–Triassic fill (Al Shaikh 1970). A shallow basement ridge extending southward from St Tudwalds peninsula separates the Cardigan Bay Basin from the northward extension of the St Georges Channel Basin. On the south-east the basins are bounded by a complex system of faults, in part probably an extension of the Bala Fault, with a total downthrow comparable with that of the Mochras Fault. From geophysical (Bullerwell & McQuillin 1969, Blundell et al. 1971, Dimitropoulous & Donato 1983) and geological studies (Dobson et al. 1973, Barr et al. 1981, Whittington et al. 1981) it is estimated that the basin is some 7 km or more deep, the fill being mainly sediments of Permo–Triassic, Jurassic and Tertiary age. A prominent salt feature within the basin, the Strumble Head Salt-intrusion, is associated with a major NE-trending fault.

The South Irish Sea Basin is similar in many respects to the Cheshire and Lower Severn basins but is distinguished from the latter by the extent of the Tertiary (intra-Oligocene from seismic evidence) subsidence that occurred – up to 2 km of marine sediments of Palaeogene age being still present in its central part, and 600 m of terrestrial deposits beneath Tremadoc Bay.

## Wales, Welsh Borderland and West Midlands

Because of the lack of deep boreholes and seismic information we can as yet be only very tentative about the deep structure of Wales. Regional gravity shows a marked NW–SE gradient, decreasing from about 45 mGal over Anglesey to 10 mGal at the border. Data from the LISPB–DELTA line across Wales (Nunn 1978) shown in Figure 18.9 indicate that there is a rise in velocity in the crust from $6.1$ to $6.45$ km s$^{-1}$ at depths between 13 km and 18 km. Manchester (1981) showed that depth variations in the high velocity layer accounted satisfactorily for the regional gravity gradient along the line. Although it is not known how far west the mid-crustal interface persists a gradual rise in a NW direction to about 10 km under North

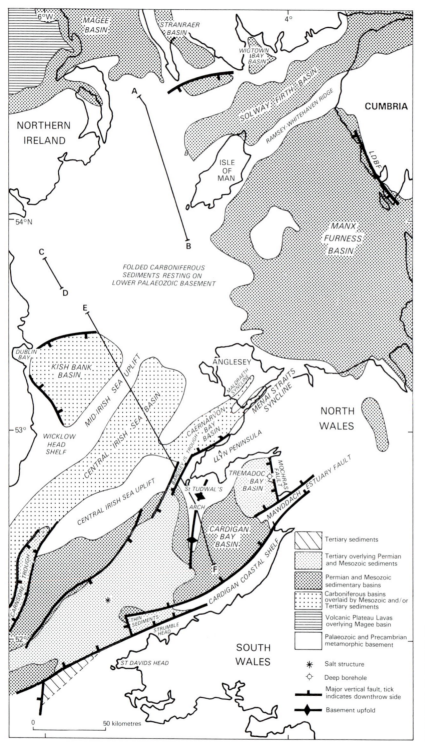

**Fig. 18.8.** Structural map of the Irish Sea (after Naylor & Mounteney 1975). AB, CD, EF are locations of BIRPS seismic reflection lines (Fig. 18.7). LDBF, Lake District Boundary Fault.

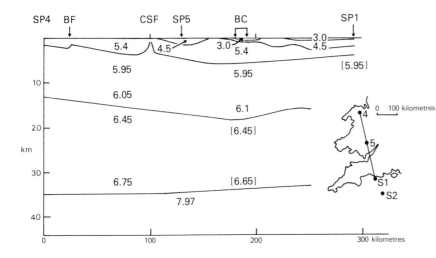

**Fig. 18.9.** Preliminary interpretation of LISPB DELTA seismic refraction line. SP, shot point; BF, Bala Fault; CSF, Church Stretton Fault; BC, Bristol Channel. Velocities in km s$^{-1}$. (By permission of S. K. Nunn.)

Wales is consistent with the gravity field. Overall there is a general correlation between regional anomalies and major tectonic trends, and structures such as the Bala and Dinorwic Faults appear as significant features geophysically, possibly marking the boundaries of crustal blocks. Some residual gravity features can be attributed to basement topography but in general it is not possible to interpret residual gravity in terms of Lower Palaeozoic thickness (Gibb 1961). Locally such a relationship may hold but lateral variations in lithology make it necessary to detemine local density differences.

In mid-Wales results from the LISPB–DELTA line indicate a depth to basement of about 4 km but the interface is probably the base of the bedded sequence and may include rocks of both Lower Palaeozoic and Precambrian age. Cook & Thirlaway (1955), using densities of 2·75 Mg m$^{-3}$ for the Silurian and 2·80 Mg m$^{-3}$ for the Precambrian in the Radnor district showed that there the gravity anomaly can be accounted for by thickness variations in the Silurian. The well-defined linear gravity-high that extends from the exposed Precambrian of the Long Mynd to Old Radnor and beyond appears, on the basis of this and seismic refraction data (Nunn 1978) to be an up-faulted, perhaps in places horst-like, Precambrian ridge. There is a steep drop in gravity on the east side of the ridge south of the Long Mynd to a gravity low west of Ludlow. Calculations based on a density contrast of 0·17 Mg m$^{-3}$ with the Precambrian in this area indicated a thickness of over 2·4 km of Lower Palaeozoic rocks. The ridge is bounded by the Church Stretton fault and the Pontesford Lineament (Woodcock 1984). To the south it swings westward and beyond Llandeilo appears to follow the trend of the Towy anticline into Carmarthenshire (see Chapter 17).

The aeromagnetic map (Fig. 18.10) demonstrates qualitatively the variation in Lower Palaeozoic thickness over Wales. East of the Cambrian rocks of the Harlech Dome and SE of Snowdonia, the magnetic field rapidly becomes smooth. Though Ordovician vulcanicity must contribute to the magnetic anomalies in Snowdonia many of the sources seem to lie within the Precambrian, and are not far below the surface over the crest of the Harlech Dome (Allen & Jackson 1978). It would appear then that the Lower Palaeozoic succession thickens rapidly eastward into mid-Wales, its base rising to the surface again in the Welsh Borderland, this being marked by the appearance of a belt of high amplitude anomalies following the trend of the positive gravity axis (Fig. 18.6). In many places this extends east of the gravity axis, indicating the presence of areas of shallow basement. Eastwards the Precambrian is seen again in the Malvern horst and appears once more in the core of the Lickey axis near Birmingham (Chapter 17). Lying between the two and conspicuous on both the aeromagnetic and gravity maps is the Worcester Basin. Its western edge is revealed by a rapid decrease in gravity east of the Malvern line to a minimum of 30 mGal north of Droitwich. A borehole at Netherton near the axis of the basin proved 1,568 m of Trias underlain by thick andesitic tuff probably of Ordovician age (Wills 1978). Farther south, at Minety (Wills 1978), the Cambrian was met at a depth of 1,448 m beneath the base of the Keuper. Neither Silurian nor Devonian was observed at either locality. The borehole at Stowell Park, Northleach (Green & Melville 1956) ended at 1,501 m in Triassic rocks. No borehole has been drilled on the site of the gravity minimum but it seems possible that the Trias accounts for most of the anomaly. Cook & Thirlaway (1955) assumed the gravity change of over 30 mGal from the Malvern horst at Knightwick into

**Fig. 18.10.** Simplified magnetic anomaly map of England and Wales. Drawn from the 1:625,000 Aeromagnetic Anomaly Maps of Great Britain published by the Institute of Geological Sciences. (nT, gamma; T, magnetic flux density). AA, Welsh Borders magnetic belt; B, Malvern–Newent high; C, Worcestershire Basin low; D, Newbury high; E, Isle of Wight; F, Eastern England magnetic belt; G, Wash basement platform; H, Peterborough–Buxton magnetic belt.

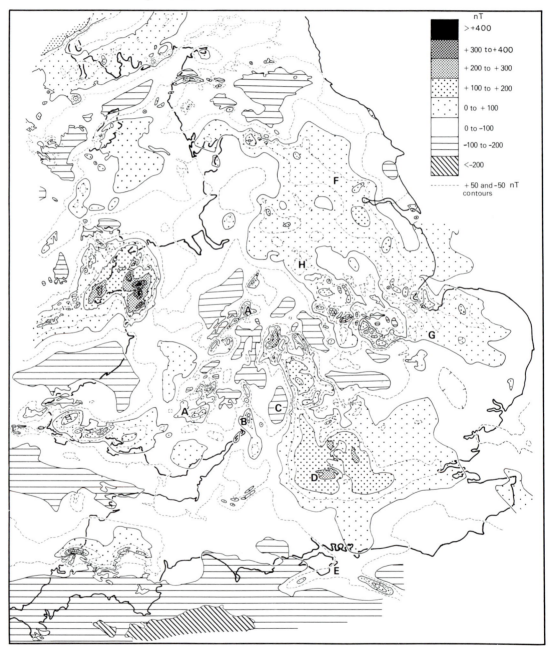

the Worcester Basin to be due entirely to Triassic rocks. They used a density difference of 0·34 Mg m$^{-3}$ and calculated the thickness to be about 2·75 km. Brooks (1968) used a larger density contrast of 0·45 Mg m$^{-3}$ (based on more data) and obtained a thickness of about 2·3 km in the south Malverns. From the gravity profile he calculated the slope of the interface to be at first 45°, decreasing at depth to 20°, the interpretation being given in Figure 18.11. He agreed with the suggestion of Cook & Thirlaway (1955) that the junction is a step-faulted unconformity.

Both the gravity and magnetic fields are high over the exposed rocks of the Malvern Hills, the northerly extension of the anomalies suggesting that the Precambrian axis extends beneath Martley and the Abberley Hills. South of the Malverns, near Newent (13 km NW Gloucester), the gravity field is more obviously indicative of a step-like form for the western boundary of the Severn Basin. Two seismic reflection profiles (Chadwick *et al.* 1983) considerably extend the interpretation based on gravity data. One

of the lines, approximately 7·5 km long, was aligned south-eastwards across the Worcester Basin. The other, 13 km long, extended east–north-east from north of Huntley to a point near Sandhurst. The south-easterly line confirmed the presence of a major fault east of the Malverns with a dip of about 50°, downthrowing the base of the Trias to 2 km. The basement is believed to be Precambrian though some Lower Palaeozoic strata could be present. The north-easterly section appears to indicate a series of easterly dipping faults, downthrowing to the east, the Triassic succession thickening to a fault bounded graben lying between Sandhurst and Hartpurg. Here again the Triassic is some 2 km thick.

To the east of the Worcester Basin lies the London–Brabant Massif. Southwards the basin opens into the Wessex Basin. Westwards there is a connection with the Bristol Channel and South Celtic Sea Basins.

In an early study Falcon & Tarrant (1950) attributed the gravity low in the Oxford area to the presence of Coal Measures beneath the Mesozoic. IGS

**Fig. 18.11.** A two-dimensional interpretation of a gravity profile along a W–E line across the south Malvern Hills. Full line is measured Bouguer anomaly, dots indicate calculated anomaly (from Brooks 1968).

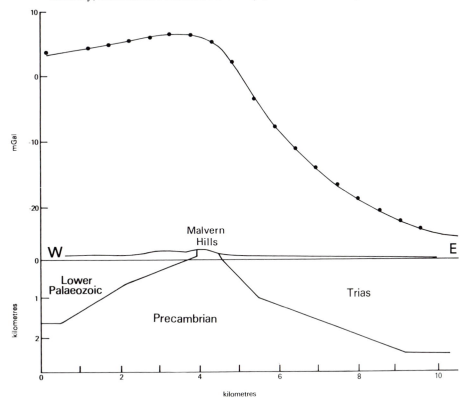

boreholes at Apley Barn, Steeple Aston and Withy-combe Farm have since proved Upper Coal Measures (Dunham & Poole 1978) and it is now believed that the coalfield extends as far north as southern Warwick-shire. The gravity high between Shipston-on-Stour and Banbury is, however, due to lateral density changes beneath the Coal Measures which continue across the area unaffected in thickness.

## South-west England

Our knowledge of the deep structure of SW England owes much to the investigations of Bott and his co-workers (Bott et al. 1958, Bott & Scott 1964). Both the gravity and magnetic fields are dominated by the effect of the Hercynian granites, less dense in general than their host rocks, and surprisingly non-magnetic. The gravity map (Fig. 18.6) shows three main features, a

belt of negative anomalies following the line of the granite outcrops and having an amplitude of 40–50 mGal, a rapid increase southwards over the Start and Lizard peninsulas and a uniform gradient over Exmoor, decreasing to the NE. From a study of cross-profiles and by using simple modelling techniques Bott concluded that the Dartmoor granite has a batholithic form and is bounded by vertical or steeply dipping margins to the north, south and east. The thickness was calculated to be 10 km, probably increasing towards the south. The continuity of the negative belt and the absence of any steep gradients or large changes of anomaly along the axis suggested that individual outcrops were cupolas of one major intrusion, the Cornubian Batholith, extending from west of the Scilly Isles into Devon. A long seismic refraction line (Holder & Bott 1971) placed the base of the low velocity (5·85 km s⁻¹) granite mass along the axis at

**Fig. 18.12.** A two-dimensional interpretation of a gravity profile across the Dartmoor granite batholith. Full line is measured Bouguer anomaly, dots indicate calculated anomaly. Density contrast with country rock in Mg m⁻³ (from Bott & Scott 1964).

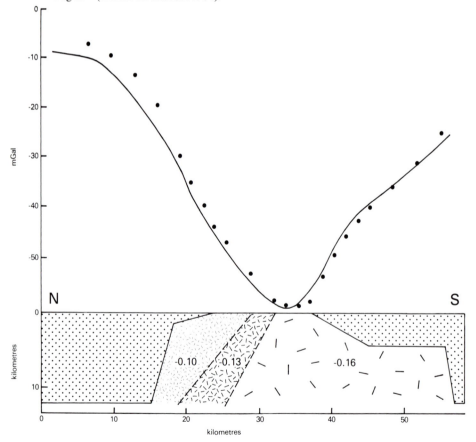

about 10–12 km. Below this velocity increases gradually with depth to 6·9 km s$^{-1}$ at the base of the lower crust at 27 km. Figure 18.12 shows a model of the batholith based on the gravity data. Bott & Scott (1964) obtained a good fit between observed and calculated gravity by combining a reasonable shape for the northern margin with a gradual density increase. Brooks and co-workers (1982), quoting the results of gravity modelling by Al Rawi (1980) conclude that base of the batholith lies at a depth of 9–16 km under Bodmin and 12–22 km west of Dartmoor.

Brooks and co-workers (1984) interpreted a strong near horizontal reflection lying at a depth of between 10 and 15 km, in part beneath the granite and shallowing northwards, as a thrust plane of late Hercynian age postdating granite emplacement. If correct this would indicate a more southerly origin for the intrusion.

Over the Lizard and Start peninsulas the gravity anomalies rise in a southward direction. Bott and co-workers (1958) and Bott & Scott (1964) considered the gradient to be due to a thickening slice of metamorphic and basic rocks thrust from the south (see Chapter 17).

There are difficulties in accepting this interpretation as the gravity gradient commences well north of the Start boundary fault. Magnetic anomalies associated with the Lizard Complex do not extend far beyond the outcrop and the aeromagnetic map shows no clear evidence of a thrust. Brooks and co-workers (1982), discussing the preliminary interpretation of a seismic line between the Lizard and the Start Peninsula commented that the basic/ultrabasic rocks of the Lizard appear to be restricted to the near surface and that a 'Start-type' basement layer can be followed from the Start Peninsula westwards, to pass beneath the Lizard area at about 3–4 km. The Lizard complex may therefore be a thin sheet separated from the basement by possible Devonian metasediments (see also Chapter 17).

Geological and geophysical work in the Bristol Channel (Lloyd *et al.* 1973, Evans & Thompson 1979) established that the southern part of the Channel is occupied by a strong asymmetrical westward-plunging syncline trending ESE, in which is preserved a thick succession of Mesozoic rocks (Bristol Channel Basin). To the east the syncline appears to die out landwards,

**Fig. 18.13.** Geological section across the Bristol Channel Syncline derived from seismic refraction and other data. (a) Swansea Bay to east of Foreland Point. (b) Nash Point to Minehead (from Brooks & Al-Saadi 1977). SL, seismic line. Velocities in km s$^{-1}$.

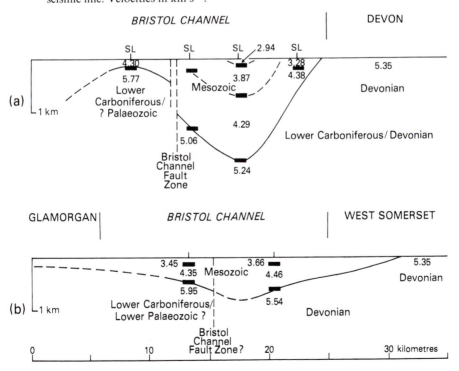

its place possibly taken *en echelon* by the Glastonbury syncline. The two may be separated by an extension of the Cothelestone Fault. North of the Bristol Channel folded Old Red Sandstone and Carboniferous rocks are partly concealed by Mesozoic deposits while to the south are the highly folded Devonian rocks comprising the northern limb of the Cornubian synclinorium. Two geological sections across the Bristol Channel (Fig. 18.13), built up mainly from refraction data (Brooks & Al-Saadi 1977) indicated some 2 km of Mesozoic rocks in the core of the Bristol Channel Basin. The structural nature of the Palaeozoic floor is not altogether certain. However, basing their interpretation on the results of a large body of seismic work in the area Brooks and co-workers (1982) quoted velocity ranges for rocks of various ages and lithologies (Table 18.1). Of these the identifications for layers 1 to 3 are the most firmly established. Layer 5 is virtually continuous beneath the area and its depth has been established by time-term analysis of long refraction lines and quarry blasts (Bayerly & Brooks 1980, Mechie 1980). To the north of the Bristol Channel, under the South Wales Coalfield, basement depth is around 3 km, there being a basement culmination centred under its western half. The basement also deepens northward. South of the culmination it reaches 6 km under the Gower Peninsula, remaining at considerable depth under the Bristol Channel to within a short distance of the North Devon coast. Just off the coast, however, a high velocity layer is met at much shallower depth. Either the basement rises very rapidly near the coast or this layer is a high velocity horizon farther up in the succession. Layer 4 must be at least 3 km thick in the central Channel. Brooks and

co-workers (1982) suggested that earlier identification of Lower Palaeozoic strata immediately beneath the Mesozoic syncline in the Channel may be wrong, the rocks possibly being dolomitised Carboniferous limestones. From Exmoor gravity decreases steadily NE to a long deep trough in the Bristol Channel off Foreland Point, beyond which it rises more gently towards the Glamorgan coastline (Brooks & Thompson 1973). The gravity field is markedly linear with a strike close to that of the Devonian. Bott's original explanation was that the Devonian rocks of Exmoor were overthrust from the south on to a thick sequence of relatively light Carboniferous rocks beneath. In support he advanced the view held by some workers that thrusting was the cause of structures observed in the Cannington inlier. Later Bott & Scott (1964) showed that the gradient could equally well be caused by a northward thickening of low-density rocks of Old Red Sandstone facies unconformably underlying the exposed Devonian. Brooks & Thompson (1973) adhered to the view that thrusting of the Devonian over a basin of Upper Carboniferous sediments was more likely but later seismic and geological evidence led to another explanation (Brooks *et al.* 1977). Assuming a structural culmination under north Devon they interpreted the Exmoor gradient to be a consequence of a southerly dip and increase in thickness of dense Devonian and Culm, with lighter rocks of Lower Palaeozoic or Precambrian age beneath (Fig. 18.14).

In Pembrokeshire the basement deepens southwards from the outcrops of the Pebidian volcanics in the north to a depth of about 4 km in the southeast, rising again to about 3 km in Carmarthen Bay, before deepening again beneath the Bristol Channel, all the evidence being suggestive of the existence of EW-trending highs. However, the Precambrian and Silurian rocks of the Johnston Benton fault block appear to be an isolated slice thrust from the south, the crystalline basement here lying at a depth of several kilometres (Brooks *et al.* 1982).

## Southern England (Wessex Basin)

Seismic reflection profiling and drilling have of late greatly improved knowledge of this area but much of the information is still confidential. However, it has been used, in conjunction with all other available geological information, in the production of the British Geological Survey's atlas of onshore sedimentary basins (Whittaker (ed.) 1985). A thick and complete Jurassic succession with in places 1,500 m or so of Permo–Triassic rocks beneath, has been confirmed while some holes reached the Palaeozoic. Boreholes at Wytch Farm, Dorset, towards the western margin of the basin proved *c.* 1,200 m of Mesozoic strata and *c.* 1,400 m of Permo–Triassic rocks

**Table 18.1.**

| Seismic layer | Refractor velocity range (km s$^{-1}$) | Geological interpretation |
|---|---|---|
| 1 | 4·3–4·7 | Upper Carboniferous |
| 2 | 5·1–5·3 | Carboniferous Limestone |
| 2a | (5·6–5·7) | (dolomitised) |
| 3 | 4·6–4·8 | Old Red Sandstone |
| 4 | 5·5–5·7 | Various layers within pre-Upper Palaeozoic supra-basement |
| 5 | 6·0–6·3 | Precambrian igneous/metamorphic basement |

Suggested interpretation of velocity ranges beneath Bristol Channel (Brooks *et al.* 1982).

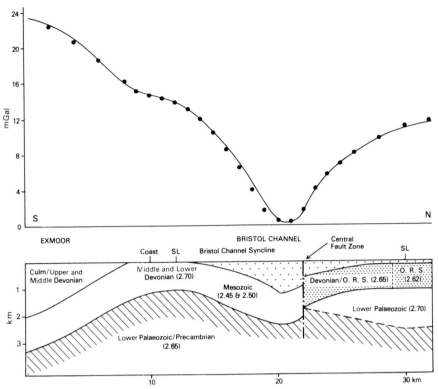

**Fig. 18.14.** Reinterpretation of Bouguer gravity anomaly along longitude 4° 00'W in the Bristol Channel. Full line is measured anomaly, dots indicate calculated anomaly (from Brooks *et al.* 1977). SL, seismic line.

above metamorphosed Devonian basement (Colter & Havard 1981), a somewhat similar succession being obtained at Southampton (Downing *et al.* 1982). Deposition appears to have been in a single basin extending from Dorset through Hampshire and Sussex to East Kent and across the Channel (the Wessex–Paris Basin) bounded on the NE by the London–Brabant Massif and on the west by the Cornubian Massif. Gravity and aeromagnetic maps provide regional coverage, reflecting in the main the post-Palaeozoic history but in places yielding some clues to the lithology and structure of the early formations. In a very general way the Bouguer gravity anomalies (Fig. 18.6) reveal variations in the depth of the Mesozoic basin, gravity decreasing steadily from positive values over the older rocks in Devon to a deep trough in the Salisbury–Southampton area, beyond which it rises steadily eastward to the east Kent platform, though still remaining negative. The anomaly pattern of the aeromagnetic map (Fig. 18.10) both reinforces and supplements information provided by gravity. The magnetic field is smooth over the

Tertiary Hampshire basin and there are no apparent trends. Terris & Bullerwell (1965) quote depths to magnetic basement (calculated from anomalies) in excess of 6 km in some areas, maximum values being obtained in the Southampton–Winchester area. Clearly the magnetic anomalies arise from rocks older and deeper than those forming the floor of the Mesozoic basin. Westwards the zone of maximum depth swings parallel to the outcrop of the Tertiary deposits, passes through Wimbourne and Weymouth and west into Lyme Bay before shallowing. This is also an area of low gravity, interpreted as a basin of Permo–Triassic sediments approximately a kilometre thick (Bacon 1975).

Stoneley (1982) suggested that the evolution of the Wessex Basin was primarily a consequence of Permo–Triassic rifting and related to the formation of the North Sea and other basins in the British Isles. A more detailed interpretation of the tectonics based on the study of deep reflections profiles has been advanced by Whittaker & Chadwick (1984) and is also discussed by Chadwick (1985). They suggested that the continental crust can be divided into three zones. The upper zone shows near horizontal, relatively continuous reflections and in the main comprises the sedimentary section. A middle zone of ill-defined, incoherent low amplitude reflections (sometimes cut by dipping events) is considered to consist of folded and cleaved basement rocks cut by planar thrusts. The nature of the lower zone, the upper part of which is characterised by strong sub-horizontal events of limited lateral extent, is not yet understood.

As a result of crustal extension listric growth faults formed, probably due to reactivation (in a normal sense) of a major thrust at depth. Associated rollover anticlines and normal faults antithetic to the main listric faults then developed. In the Cenozoic, isostatic response to compression brought about inversion, with reverse movement on the normal faults.

The main fault-zones, the Pewsey–London Platform faults, the Wardour–Portsdown faults the Purbeck–Wight faults, trend EW. They are associated with monoclinal flexuring and divide the main basin into sub-basins where dips tend to be small.

Information about the concealed Variscan Front comes from a north to south deep seismic reflection line commencing in the vicinity of Cirencester (Chadwick et al. 1983). Interpretation of the shallow reflections in terms of the geology was based partly on information from boreholes, regional geological and tectonic data being used as an aid to identification of the deeper events. A section is given in Figure 17.18. Mesozoic rocks, possibly down to and including the Permian, conceal older rocks along the length of the section. They are strongly affected by faulting, gentle fold structures are developed and very marked thickness changes occur locally. To the south, beneath the Hercynian unconformity strongly deformed Upper Palaeozoic rocks occur. To the north the rocks are considerably less disturbed, a sharp transition apparently marking the boundary between the more stable region and the mobile belt to the south.

Both Lower Palaeozoic and Precambrian rocks were also tentatively identified, the latter possibly consisting of a thick series of Proterozoic sediments and volcanics (thought to be equivalents to the Longmyndian/Uriconian and Charnian succession of the Midland microcration), overlying and overstepping a crystalline basement of marked topography.

Two prominent planar reflectors on the sections, one dipping at 27° with an azimuth of 155° and the other at 22° with an azimuth of 165° were identified as major thrust zones, the latter apparently levelling off at a depth of about 11 km. In the southern part of the section the thrust zone does not appear to cut the crystalline basement and the authors suggest that it may translate along its upper surface. The more steeply dipping thrust zone is associated with the Variscan Front and appears to extend northwards on to the margin of the Midland craton. A somewhat similar situation has been described by Smalley & Westbrook (1982) in the vicinity of the Hogs Back, west of Guildford, Surrey. Here borehole and geophysical data indicate an underlying crustal discontinuity, a favoured interpretation involving overthrusting to the north on to the stable London Platform, the structures thus forming part of the Variscan Front.

## REFERENCES

| | | |
|---|---|---|
| AGGER, H. E. & CARPENTER, E. W. | 1964 | A crustal study in the vicinity of the Eskdalemuir seismological array station. *Geophys. J. R. astr. Soc.*, **9**, 69–83. |
| ALLEN, P. M. & JACKSON, A. A. | 1978 | Bryn-teg Borehole, North Wales. *Bull. geol. Surv. G.B.*, **61**. |
| ALLSOP, J. M. | 1983 | Geophysical appraisal of two gravity minima in the Wash district. *Rep. Inst. Geol. Sci.*, **83/1**, 28–31. |
| | 1984 | Geophysical appraisal of a Carboniferous basin in north-east Norfolk, England. *Proc. Geol. Assoc. London*, **95**, 175–180. |
| ALLSOP, J. M. & ARTHUR, M. J | 1983 | A possible extension of the South Leicestershire Diorite complex. *Rep. Inst. geol. Sci.*, **83/10**, 25–30. |
| ALLSOP, J. M. & JONES, C. M. | 1981 | A pre-Permian palaeogeological map of the East Midlands and East Anglia. *Trans. Leicestershire Lit. Philos. Soc.*, **75**, 28–33. |
| AL-RAWI, F. R. J. | 1980 | *A geophysical study of deep structure in south west Britain.* Unpubl. Ph.D. Thesis, Univ. Wales. |
| AL SHAIKH, Z. D. | 1970 | The geological structure of part of the Irish Sea. *Geophys. J. R. astron. Soc.*, **20**, 233–237. |

ARTER, G. 1982 *Geophysical investigations of the deep geology of the east Midlands.* Unpubl. Ph.D. Thesis, Univ. Leicester.

AUSTIN, J. A. 1979 *A study of the gravity anomalies in the area around the Wash.* Unpubl. M.Sc. Thesis, Univ. Durham.

BACON, M. 1975 A gravity survey of the eastern English Channel between Lyme Bay and St Brieve Bay. *Philos. Trans. R. Soc. London,* A**279**, 69–78.

BACON, M. & McQUILLIN, R. 1972 Refraction seismic surveys in the north Irish Sea. *J. geol. Soc. London,* **128**, 613–621.

BAMFORD, D., FABER, S., JACOB, B., KAMINSKI, W., PRODEHL, C., FUCHS, K., KING, R. & WILLMORE, R. 1976 A lithospheric seismic profile in Britain. I. Preliminary results. *Geophys. J. R. astr. Soc.,* **44**, 145–160.

BAMFORD, D., NUNN, K., PRODEHL, C. & JACOB, B. 1977 LISPB. III. Upper crustal structure of Northern Britain. *J. geol. Soc. London,* **133**, 481–488.

1978 LISPB. IV. Crustal structure of Northern Britain. *Geophys. J. R. astr. Soc.,* **54**, 43–60.

BARR, K. W., COLTER, V. S. & YOUNG, R. 1981 The geology of the Cardigan Bay–St George's Channel Basin. *In* Illing, L. V. (Ed.) *Petroleum Geology of the Continental Shelf of North-West Europe.* Heyden & Son Ltd., London, 432–443.

BAYERLY, M. & BROOKS, M. 1980 A seismic study of deep structure in South Wales using quarry blasts. *Geophys. J. R. astron. Soc.,* **60**, 1–19.

BLUNDELL, D. J., DAVEY, F. J. & GRAVES, L. J. 1971 Geophysical surveys over the south Irish Sea and Nymphe Bank. *J. geol. Soc. London,* **127**, 339–375.

BOTT, M. H. P. 1961 Geological interpretation of magnetic anomalies over the Askrigg Block. *Q. J. geol. Soc. London,* **117**, 481–495.

1964 Gravity measurements in the north-eastern part of the Irish Sea. *Q. J. geol. Soc. London,* **120**, 369–396.

1965 The deep structure of the northern Irish Sea – a problem of crustal dynamics. *In* Whittard, W. F. & Bradshaw, R. (Eds.) *Submarine Geology and Geophysics, Colston Pap.,* **17**, 179–204. Butterworths, London.

1967 Geophysical investigations of the northern Pennine basement rocks. *Proc. Yorkshire geol. Soc.,* **36**, 139–168.

1968 The geological structure of the Irish Sea Basin. *In* Donovan, D. T. (Ed.) *Geology of Shelf Seas.* Oliver & Boyd, Edinburgh, 93–115.

1974 The geological interpretation of the gravity survey of English Lake District and the Vale of Eden. *J. geol. Soc. London,* **130**, 309–331.

BOTT, M. H. P. & MASSON-SMITH, D. 1957 The geological interpretation of a gravity survey of the Alston Block and the Durham Coalfield, and interpretation of a vertical magnetic survey in North-East England. *Q. J. geol. Soc. London,* **113**, 93–117.

BOTT, M. H. P. & SCOTT, P. 1964 Recent geophysical studies in south-west England. *In* Hosking, K. F. G. & Shrimpton, G. T. (Eds.) *Present views on some aspects of the geology of Cornwall and Devon.* Blackford Ltd., Truro, Cornwall, 25–44.

BOTT, M. H. P. & YOUNG, D. G. G. 1971 Gravity measurements in the North Irish Sea. *Q. J. geol. Soc. London,* **126**, 413–435.

BOTT, M. H. P., DAY, A. H. & MASSON-SMITH, D. 1958 The geological interpretation of gravity and magnetic surveys in Devon and Cornwall. *Philos. Trans. R. Soc. London,* A**251**, 161–191.

BOTT, M. H. P.,              1978    Granite beneath Market Weighton, east Yorkshire. *J. geol.*
  ROBINSON, J. &                     *Soc. London*, **135**, 535–543.
  KOHNSTAM, M. A.

BOTT, M. H. P.,              1984    Deep structure and origin of the Northumberland and
  SWINBURN, P. M. &                  Stainmore Troughs. *Proc. Yorkshire geol. Soc.*, **44**, 479–
  LONG, R. E.                        495.

BOTT, M. H. P.,              1985    Crustal structure south of the Iapetus suture beneath
  LONG, R. E.,                       northern England. *Nature*, **314**, 724–727.
  GREEN, A. S. P.,
  LEWIS, A. H. J.,
  SINHA, M. C. &
  STEVENSON, D. G.

BREWER, J. A.,              1983    BIRPS deep seismic reflection studies of the British
  MATTHEWS, D. H.,                   Caledonides. *Nature, London*, **305**, 206–210.
  WARNER, M. R.,
  HALL, J., SMYTHE, D. K. &
  WHITTINGTON, R. J.

BROOKS, M.                  1968    The geological results of gravity and magnetic surveys in
                                     the Malvern Hills and adjacent districts. *Geol. J.*, **6**, 13–30.

BROOKS, M. &                1977    Seismic refraction studies of geological structure in the
  AL-SAADI, R. H.                    inner part of the Bristol Channel. *J. geol. Soc. London*, **133**,
                                     433–445.

BROOKS, M.,                 1977    A new geological model to explain the gravity gradient
  BAYERLY, M. &                      across Exmoor, North Devon. *J. geol. Soc. London*,
  LLEWELLYN, D. J.                   **133**, 385–393.

BROOKS, M. &                1973    The geological interpretation of a gravity survey of the
  THOMPSON, M. S.                    Bristol Channel. *J. geol. Soc. London*, **129**, 245–274.

BROOKS, M.,                 1984    Major crustal reflections beneath S.W. England. *J. geol.*
  DOODY, J. S. &                     *Soc. London*, **141**, 97–104.
  AL-RAWI, F. R. J.

BROOKS, M., MECHIE, J. &    1982    Geophysical investigations in the Variscides of southwest
  LLEWELLYN, D. J.                   Britain. *In* Hancock, P. L. (Ed.) *The Variscan Fold Belt in
                                     the British Isles.* Hilger, 186–197.

BULLARD, E. C.,             1940    Seismic investigations on the Palaeozoic floor of East
  GASKELL, T. F.,                    England. *Philos. Trans. R. Soc. London*, A**239**, 29–94.
  HARLAND, W. B. &
  KERR-GRANT, C.

BULLERWELL, W. &            1969    Preliminary report on a seismic reflection survey in the
  McQUILLIN, R.                      southern Irish Sea, July 1968. *Rep. Inst. geol. Sci. London*,
                                     **69/2**, 7 pp.

CHADWICK, R. A.             1985    Permian, Mesozoic and Cenozoic structural evolution of
                                     England and Wales in relation to the principle of inversion
                                     tectonics. *In* Whittaker, A. (Ed.) *Atlas of Onshore Sedimen-
                                     tary Basins in England and Wales.* Blackie, Glasgow.

CHADWICK, R. A.,            1983    Crustal structure beneath southern England from deep
  KENOLTY, N. &                      seismic reflection profiles. *J. geol. Soc. London*, **140**,
  WHITTAKER, A.                      893–911.

CHALLINOR, J.              1978    The 'Red Rock Fault', Cheshire: a critical review. *Geol. J.*,
                                     **13**, 1–10.

CHROSTON, P. N. &          1982    Deep boreholes, seismic refraction lines and the interpreta-
  SOLA, M. A.                        tion of gravity anomalies in Norfolk. *J. geol. Soc. London*,
                                     **139**, 255–264.

COLTER, V. S. &            1975    Recent developments in the Geology of the Irish Sea and
  BARR, K. W.                        Cheshire Basins. *In* Woodland, A. W. (Ed.) *Petroleum and
                                     the Continental Shelf of North-West Europe, Volume 1 –
                                     Geology.* App. Sci. Publ., London, 61–76.

COLTER, V. S. & 1981 The Wytch Farm Oil Field, Dorset. *In* Illing, L. V. &
HAVARD, D. J. Hobson, G. D. (Eds.) *Petroleum Geology of the Continental Shelf of North-west Europe.* Heyden & Son Ltd., London, 494–503.

COOK, A. H. & 1955 The geological results of measurements of gravity in the
THIRLAWAY, H. I. S. Welsh Borders. *Q. J. geol. Soc. London,* **111**, 47–70.

COPE, F. W. 1949 Report on a boring at Woo Dale, Buxton. *Abs. Proc. geol. Soc. London,* No. 1446, 24.

1973 Woo Dale Borehole near Buxton, Derbyshire. *Nature Phys. Sci. London,* **243**, 29–30.

CORNWELL, J. D. 1967 The magnetisation of Lower Carboniferous rocks from the northwest border of the Dartmoor granite, Devonshire. *Geophys. J. R. astr. Soc.,* **12**, 381–403.

1972 A gravity survey of the Isle of Man. *Proc. Yorkshire geol. Soc.,* **39**, 93–106.

DIMITROPOULOUS, K. & 1983 The gravity anomaly of the St. George's Channel Basin,
DONATO, J. A. southern Irish Sea – a possible explanation in terms of salt migration. *J. geol. Soc. London,* **140**, 239–244.

DOBSON, M. R., 1973 The geology of the South Irish Sea. *Rep. Inst. geol. Sci.*
EVANS, W. E. & *London,* **73/11**.
WHITTINGTON, R.

DOWNING, R. A., 1982 The Southampton (Western Esplanade) Geothermal Well
ALLEN, D. J., – a preliminary assessment of the resource. *Investigation of*
BURGESS, W. G., *the Geothermal Potential of the U.K.* Institute of Geological
SMITH, I. F. & Sciences, 49 pp.
EDMUNDS, W. M.

DUNHAM, K. C. 1973 A recent deep borehole near Eyam, Derbyshire. *Nature Phys. Sci. London,* **241**, 84–85.

1974 Granite beneath the Pennines in North Yorkshire. *Proc. Yorkshire geol. Soc.,* **40**, 191–194.

DUNHAM, K. C., 1961 Granite beneath the northern Pennines. *Nature, London,*
BOTT, M. H. P., **190**, 899–900.
JOHNSON, G. A. L. &
HODGER, B. L.

DUNHAM, K. C., 1965 Granite beneath Visean sediments with mineralization at
DUNHAM, A. C., Rookhope, northern Pennines. *Q. J, geol. Soc. London,*
HODGE, B. L. & **121**, 383–417.
JOHNSON, G. A. L.

DUNHAM, K. C. & 1974 The Oxfordshire Coalfield. *J. geol. Soc. London,* **130**,
POOLE, E. G. 387–391.

EVANS, D. J. & 1979 The geology of the central Bristol Channel and Lundy area,
THOMPSON, M. S. South Western Approaches, British Isles. *Proc. Geol. Assoc. London,* **90**, 1–14.

FALCON, N. L. & 1950 The gravitational and magnetic exploration of parts of the
TARRANT, L. H. Mesozoic covered areas of south central England. *Q. J. geol. Soc. London,* **106**, 141–170.

GIBB, R. A. 1961 *Gravity Measurements in Wales.* Unpubl. Ph.D. Thesis, Univ. Birmingham.

GIFFARD, H. P. W. 1923 The search for oil in Great Britain. *Trans. Inst. Min. Eng.,* **65**, 221–247.

GREEN, G. W. & 1956 Stratigraphy and the Stowell Park Borehole 1949–51. *Bull.*
MELVILLE, R. V. *geol. Surv. G.B.,* **11**, 1–66.

GRIFFITHS, D. H. & 1965 Bouguer gravity anomalies in Wales. *Geol. J.,* **4**, 335–342.
GIBB, R. A.

GRIFFITHS, D. H., 1961 Geophysical investigations in Tremadoc Bay, North
KING, R. F. & Wales. *Q. J. geol. Soc. London,* **117**, 171–191.
WILSON, C. D. V.

HOLDER, A. P. &       1971    Crustal structure in the vicinity of south-west England.
    BOTT, M. H. P.              *Geophys. J. R. astron. Soc.*, **23**, 465–489.
HUGHES, S. G.         1977    *Magnetic Survey of the Shap Granite region.* Unpubl. M.Sc.
                               Thesis, Univ. Durham.
KENT, P. E.           1949    A structure contour map of the surface of the buried Pre-
                               Permian rocks of England and Wales. *Proc. Geol. Assoc.
                               London*, **60**, 87–104.
                      1967    Contour map of the sub-Carboniferous Surface in the
                               north east Midlands. Symposium on Carboniferous
                               basement rocks of N.E. England. *Proc. Yorkshire geol.
                               Soc.*, **36**, 127–133.
                      1968    The buried floor of Eastern England. *In* Sylvester-Bradley,
                               P. & Ford, T. D. (Eds.) *The Geology of the East Midlands.*
                               University Press, Leicester, 138–148.
                      1974    Structural history. *In* Rayner, D. H. & Hemingway, J. E.
                               (Eds.) *The geology and mineral resources of Yorkshire.*
                               Yorkshire geol. Soc., Leeds, 13–28.
                      1975    The tectonic development of Great Britain and the
                               surrounding seas. *In* Woodland, A. W. (Ed.) *Petroleum and
                               Continental Shelf of North-West Europe.* **1** – *Geology.* Appl.
                               Sci. Publ., London, 3–28.
LE BAS, M. J.         1972    Caledonian igneous rocks beneath central and eastern
                               England. *Proc. Yorks. geol. Soc.*, **39**, 71–84.
LLOYD, A. J.,         1973    The geology of the British Channel floor. *Philos. Trans. R.
    SAVAGE, R. J. G.,          Soc. London*, A**274**, 595–626.
    STRIDE, A. H. &
    DONOVAN, D. T.
LOCKE, C. A. &        1978    Geophysical constraints on the stucture and emplacement
    BROWN, G. C.               of Shap granite. *Nature, London*, **272**, 526–528.
MAGUIRE, P. K. H.     1987    CHARM II – A deep reflection profile within the central
                               England mircrcraton. *J. geol. Soc. London*, **144**, 661–670.
MAGUIRE, P. K. H.,    1982    A deep seismic reflection profile over a Caledonian granite
    ANDREW, E. M.,             in Central England. *Nature, London*, **297**, 671–673.
    ARTER, G.,
    CHADWICK, R. A.,
    GREENWOOD, P.,
    HILL, I. A.,
    KENOLTY, N. &
    KHAN, M. A.
MANCHESTER, R. J.     1981    A crustal model for Wales from Interpretation of Geophys-
                               ical Surveys. UKGA 5 (abstract). *Geophys. J. R. astron.
                               Soc.*, 65.
MAROOF, S. I.         1974    A Bouguer anomaly map of southern Great Britain and the
                               Irish Sea. *J. geol. Soc. London*, **130**, 471–474.
MECHIE, J.            1980    *Seismic studies of deep structure in the Bristol Channel area.*
                               Unpubl. Ph.D. Thesis, Univ. Wales.
MOSELEY F. &          1985    A stratigraphic revision of the late Precambrian rocks of
    FORD, T. D.                Charnwood Forest, Leicestershire. *Mercian Geologist*, **10**,
                               1–18.
MYERS, J. O. &        1967    The gravity anomalies of the Askrigg Block south of
    WARDELL, J.                Wensleydale. *Proc. Yorks. geol. Soc.*, **36**, 169–173.
NUNN, K. R.           1978    Crustal structure beneath Wales from LISPB Line Delta.
                               Abstract in U.K. Geophysical Assembly Supplement.
                               *Geophys. J. R. astr. Soc.*, **53**, 170.
O'CONNOR, D.          1966    *Gravity and magnetic traverse over the Askrigg Block.*
                               Unpubl. M.Sc. Thesis, Univ. Durham.

| POOLE, E. G. | 1978 | Stratigraphy of the Steeple Aston Borehole, Oxfordshire. *Bull. geol. Surv. G.B.*, 57. |
|---|---|---|
| POULTER, M. J. & WESTBROOK, G. K. | 1979 | The deep structure of the Cheshire Basin. Abstract in U.K. Geophysical Assembly Supplement. *Geophys. J. R. astr. Soc.*, **57**, 255. |
| POWELL, D. W. | 1955 | Gravity and magnetic anomalies in North Wales. *Q. J. geol. Soc. London*, **111**, 375–397. |
| RANSOME, C. R. & NUNN, K. R. | 1978 | South Irish Sea Experiment: Preliminary Results. Abstract in U.K. Geophysical Assembly Supplement. *Geophys. J. R. astr. Soc.*, **53**, 170. |
| REES, N. J. | 1982 | *Three dimensional modelling of the Peterborough gravity anomaly.* Unpubl. B.Sc. Dissertation, Univ. Durham. |
| RICHARDSON, S. W. & OXBURGH, E. P. | 1979 | The heat flow field in mainland U.K. *Nature, London*, **282**, 565–567. |
| SADLER, P. B. | 1977 | *Gravity and magnetic study in the area of the Western Anticline of the North Staffordshire Coalfield.* Unpubl. M.Sc. Thesis, Univ. Durham. |
| SMALLEY, S. & WESTBROOK, G. K. | 1982 | Geophysical evidence concerning the southern boundary of the London Platform beneath the Hog's Back, Surrey. *J. geol. Soc. London*, **139**, 139–146. |
| SMITH, A. M. | 1982 | *A seismic refraction profile north of Peterborough.* Unpubl. B.Sc. Dissertation, Univ. Durham. |
| SMITH, I. F. & DABEK, Z. K. | 1982 | Interpretation of the Mochras Fault, North Wales. *U.K.G.A.*, **6** (abstract). *Geophys. J. R. astron. Soc.*, **69**. |
| SOLA, M. A. | 1974 | *A preliminary geophysical study of the pre-Mesozoic rocks in Norfolk.* Unpubl. M.Phil. Thesis, Univ. East Anglia. |
| STONELEY, R. | 1982 | The structural development of the Wessex Basin. *J. geol. Soc. London*, **139**, 543–554. |
| STUBBLEFIELD, C. J. & BULLERWELL, W. | 1967 | Some results of a recent geological survey boring in Huntingdonshire. *Proc. geol. Soc. London*, **1637**, 35–40. |
| TERRIS, A. P. & BULLERWELL, W. | 1965 | Investigations into the Underground structure of Southern England. *Adv. Sci. London*, 232–252. |
| TURNBULL, D. A. R. | 1974 | *Gravity Investigation of the eastern margin of the Cheshire Basin.* Unpubl. M.Sc. Thesis, Univ. Durham. |
| WHITE, P. H. N. | 1949 | Gravity data obtained in Great Britain by the Anglo-American Oil Company Limited. *Q. J. geol. Soc. London*, **104**, 339–364. |
| WHITTAKER, A. (Ed.) | 1985 | *Atlas of Onshore sedimentary basins in England and Wales – Post-Carboniferous tectonics and stratigraphy.* Blackie, Glasgow. |
| WHITTAKER, A. & CHADWICK, R. A. | 1984 | The large-scale structure of the Earth's crust beneath southern Britain. *Geol. Mag.*, **121**, 621–624. |
| WHITCOMBE, D. N. & MAGUIRE, P. K. H. | 1980 | An analysis of the velocity structure of the Precambrian rocks of the Charnwood Forest. *Geophys. J. R. astron. Soc.*, **63**, 405–416. |
| | 1981a | A seismic refraction investigation of the Charnian basement and granite intrusions flanking Charnwood Forest. *J. geol. Soc. London*, **138**, 643–652. |
| | 1981b | Seismic refraction evidence for a basement ridge between the Derbyshire dome and the W. of Charnwood Forest. *J. geol. Soc. London*, **138**, 653–659. |
| WHITTINGTON, R. J., CROCKER, P. F. & DOBSON, M. R. | 1981 | Aspects of the Geology of the South Irish Sea. *Geol. J.*, **16**, 85–88. |
| WILLS, L. J. | 1973 | A palaeogeological map of the Palaeozoic Floor beneath the Permian and Mesozoic Formations in England and Wales. *Mem. geol. Soc. London*, **7**, 23 pp. |

WILLS, L. J.                  1978    A palaeogeological map of the Lower Palaeozoic floor
                                      below the cover of Upper Devonian, Carboniferous and
                                      later formations. *Mem. geol. Soc. London*, **8**, 36 pp.

WILSON. A. A. &               1982    The Institute of Geological Sciences Borehole at Becker-
  CORNWELL, J. A.                     monds Scar, North Yorkshire. *Proc. Yorkshire geol. Soc.*,
                                      **44**, 59–88.

WILSON, J. L.                 1982    *Seismic refraction study in the region of the gravity low to the*
                                      *north of Peterborough.* Unpubl. B.Sc. Dissertation, Univ.
                                      Durham.

WOOD, A. &                    1968    Borehole at Mochras, West of Llanbedr, Merionethshire.
  WOODLAND, A. W.                     *Nature, London*, **219**, 1352–1354.

                              1971    Introduction to the Llanbedr (Mochras Farm) Borehole. *In*
                                      Woodland, A. W. (Ed.) The Llanbedr (Mochras Farm)
                                      Borehole. *Rep. Inst. geol. Sci. London*, **71/18**, 1–9.

WOODCOCK, N. W.               1984    The Pontesford Lineament, Welsh Borderland. *J. geol. Soc.*
                                      *London*, **141**, 1001–1014.

WRIGHT, J. E.,                1971    Irish Sea investigations 1969-70. *Rep. Inst. geol. Sci.*
  HULL, J. H.,                        *London*, **71/19**, 55 pp.
  MCQUILLIN, R. &
  ARNOLD, S. E.

# 19

# ECONOMIC GEOLOGY
# P. McL. D. Duff

## Introduction

It can justly be claimed that it was Britain's mineral resources that enabled it to take the lead in the 'Industrial Revolution' that was occurring in various parts of Europe in the 18th and 19th centuries. Once coal became the chief source of power for the many great mechanical inventions of this period and took the place of charcoal for the smelting of iron, the knowledge that there was abundant coal in England, Wales and Scotland was exploited to the full. As early as 1800, the annual English production of coal was 10,000,000 tons. By 1847, it was over 30,000,000 tons whereas France, at that time, for example, was only producing 5,000,000 tons annually. There was a similar dramatic increase in the annual production in England and Wales of pig-iron – in 1788, 68,000 tons, in 1839, 1,347,000 tons! Britain was, in fact, a leading world producer of coal, and of iron, tin, copper and lead ores, and largely self-sufficient in minerals until the late 19th century (Highley *et al.* 1980).

Britain's mineral resources had, of course, attracted attention long before the 18th century. Stone Age man made tools and weapons from such diverse materials as flint from the Chalk in the south of England and pitchstones from Arran in Scotland. The Bronze Age saw visitors from the Mediterranean bringing metals to England and Wales and encouraging the export of tin from Cornwall. Celtic immigrants 'introduced' the Iron Age and worked the haematite of Cumberland, while the Roman occupation resulted in a considerable increase in mining, including that of lead from the Pennines and coal in northern England and South Wales. Up to the Industrial Revolution there is a continuous history of mining and quarrying of virtually every type of usable material.

Monastic records first mention coal in the 9th century and those of the 12th century show that by then it was worked as a regular source of fuel and indeed was being shipped from Newcastle in the 13th century. (Because of the smoke and 'poisonous odours' emitted during burning, King Edward I imposed the death penalty for a time on anyone found using coal as a fuel!).

Once bricks made it possible for house chimneys to be made easily (15th century) pollution of the atmosphere was cut down and coal became the popular domestic fuel. Industrially, coal reigned supreme from the 19th until the early 20th century with the maximum output achieved in 1913, but gradually declined in importance as oil took over as the main source of power. Mining of metals in general shows a steady decline from the 19th century onwards as overseas sources (e.g. the USA, Chile, Spain and Malaya) became cheaper, until at the present time only minor amounts of tin, lead and zinc ores are being produced (Table 19.11). Without the discovery of gas and oil in the North Sea in the 1960s, Britain by now would have been dependent for the bulk of its energy needs, as well as for its metals, from abroad. Blunden (1975) provided a most valuable account of the historical, socio-economic and quantitative aspects of the exploitation of Britain's mineral resources. Current statistics and developments are summarised annually in the *B.G.S. United Kingdom Annual Yearbook* and in the *Mining Journal Annual Review*.

## Fossil Fuels

### Coal

The maps showing the location of the exposed Carboniferous coalfields of England and Wales in Figs.

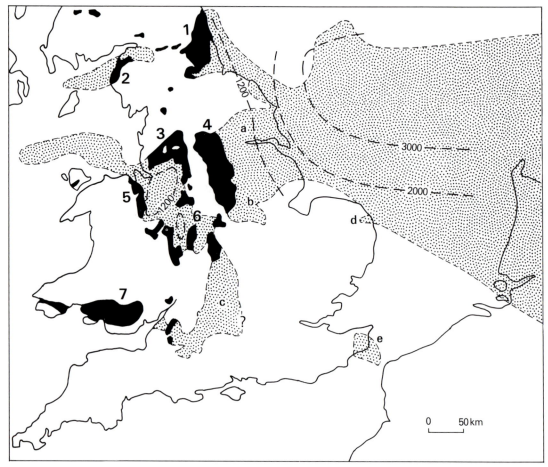

**Fig. 19.2.** Distribution of Westphalian coalfields (1 = Northumberland and Durham; 2 = Cumberland; 3 = Lancashire; 4 = East Pennine; 5 = North Wales; 6 = Midlands; 7 = South Wales). Stippled areas are concealed coalfields (a = Selby; b = Vale of Belvoir; c = Oxfordshire; d = Norfolk; e = Kent). Dashed lines denote depths in metres to top of workable coal measures (adapted from Francis 1979).

19.1a (facing p. 591) and 19.2 could equally be used to show the location of the great industrial areas of England and Wales, where iron- and steel-making, heavy engineering and shipbuilding contributed to the rise of Britain as a great manufacturing and trading nation.

The United Kingdom's annual production of coal for the period 1890–1989/90[1] is shown in Fig. 19.3. (During that period, Scottish production shows a similar pattern (Duff 1983) with c. 40 million tonnes (mt) being produced in 1913 and c. 5 mt in 1989/90.)

From being the world's leading producer of coal in 1913, when the annual output was 292 mt, Britain in

1988–89 was 8th in the world league table with 98·1 mt (of its 103·5 mt annual production) coming from England and Wales. Of the England and Wales total, 13·3 mt were from surface (opencast) mines. The ill-conceived miners' strike of 1984/85 reduced deep mine production in England and Wales to c. 28 mt. After a recovery to 88 mt in 1985/86 production fell to c. 84 mt in 1986/87 with the continuing closure of uneconomic collieries. In 1989/90 underground output was c. 73·6 mt.

At the beginning of 1990, out of a total of 73 underground mines in Britain under the control of the British Coal Corporation (the National Coal

[1] From 1963 onwards, official annual production figures were given from April–March.

Board until 1987) 66 were in England, 6 in South Wales and one in Scotland. [In 1947, when the industry was nationalised, and the National Coal Board (NCB) came into existence, there were 958 underground mines in Great Britain, producing 180 million tonnes of coal.] About 41 mt of deep-mined coal in 1989/90 were produced from the East Pennine (Yorkshire, North Derbyshire, Nottinghamshire) coalfield, 10 mt from Northumberland and Durham, and 3·5 mt from South Wales – the Lancashire. Staffordshire and Leicestershire fields making up the bulk of the remainder of the annual output.

Currently, most of the production in the United Kingdom is now used for power generation (see Table 19.1 which illustrates the changing pattern of use for the last 30 years) and comes entirely from coal-bearing rocks of Carboniferous (Westphalian) age. Inferior coals of Jurassic age were once used locally in Cleveland, Yorkshire, mainly for lime-burning, while Oligocene lignites from Bovey Tracey, Devon (p. 400), were used for fuel in the local pottery industry and to produce montan wax.

The coals grade from bituminous rank to anthracite, with the highest rank coals coming from South Wales and Kent. The NCB had its own coal classification system which is shown in Table 19.2. In general, British coals have a low ash content (2–18 per cent) and are low in Sulphur (0·5–3 per cent).

Wandless (1959) quoted 50 per cent of the then total output (192·5 mt) as having less than 1 per cent S and less than 15 per cent having as much as 2·5 per cent S. Some analyses of coals recently mined in England and Wales are shown in Table 19.3.

There are small areas (e.g. South Staffordshire, SE Durham) where coals abnormally high in Cl content (0·4 wt. per cent) are recorded (see e.g., Caswell *et al.* 1984a, b). The coals affected are exclusively high-volatile bituminous coals and the high Cl values are attributed to ground water concentration levels and the relatively high porosities of coals of that rank.

Thickness of seams extracted (including 'dirt bands') in 1983 averaged 160 cm, with Northumberland and Durham averaging 136 cm, the East Pennine coalfield seams varying between 131 cm and 179 cm, the Lancashire and Staffordshire seams averaging 202 cm and South Wales 172 cm.

The question of the reserves remaining in the coalfields of Britain has been a subject of much debate (see, for example, Francis (1979, p. 21), and Commission on Energy and the Environment (H.M.S.O., 1981, pp. 14–15), the estimates ranging from 4·5 to 45 billion tonnes! The larger figure has been arrived at by including, for example, the coalfields labelled a, b and c in Fig. 19.2 and might be better regarded as one denoting potential 'resources' rather than 'reserves'. While the mining of these 'reserves' is

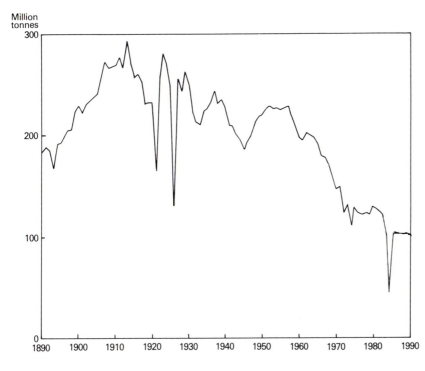

**Fig. 19.3.** Annual production of coal in UK for period 1890–1989/90. (From B.G.S. 1990 and British Coal Corporation Report and Accounts 1989/90. The latter gives summaries of annual output, numbers of mines and employees etc. from 1947–1989/90.)

technically feasible (and British Coal (B.C.C.) is a world leader in technological developments) the capital cost alone involved in their exploitation must rule them out for the immediate future. ('Recoverable' reserves were assessed in 1990 (B.C.C. Annual Report and Accounts 1989/90) as being between 3 and 5 billion tonnes.) Unknown factors include: the continued availability of cheap overseas coal; the future price of oil; the development of nuclear power; privatisation of the electricity industry (B.C.C.'s largest customer) and indeed B.C.C. itself; not to mention the political complexion of future British governments (and their views on the social consequences of closing or maintaining unprofitable mines). All of these combine to make a prediction on the future of coal-mining in Britain difficult.

**Table 19.1.** United Kingdom consumption of coal 1956–1986 (British Geological Survey 1988)

million tonnes

| Year | Collieries | Fuel conversion industries | | | | Direct consumption | | | | | Total (e) inland consumption |
|------|-----------|------------------|------------|---------------|---------------|--------------|----------|------------------|-----------|--------|
| | | Power stations | Gas works | Coke ovens | Others (a) | Industry (b) | Railways | Domestic (c) | Others (d) | |
| 1956 | 8·0 | 47·0 | 28·6 | 30·2 | 2·5 | 44·4 | 12·4 | 39·1 | 8·8 | 221·0 |
| 1957 | 7·3 | 48·0 | 27·1 | 31·5 | 2·9 | 42·3 | 11·7 | 37·2 | 8·3 | 216·3 |
| 1958 | 6·6 | 47·7 | 25·5 | 28·6 | 3·1 | 37·9 | 10·6 | 37·9 | 7·7 | 205·6 |
| 1959 | 5·7 | 47·4 | 23·2 | 26·4 | 2·6 | 35·7 | 9·7 | 35·1 | 6·7 | 192·5 |
| 1960 | 5·1 | 52·7 | 23·0 | 29·3 | 2·3 | 35·5 | 9·0 | 36·1 | 6·9 | 199·9 |
| 1961 | 4·6 | 56·3 | 22·9 | 27·4 | 2·4 | 33·1 | 7·8 | 33·8 | 6·5 | 194·9 |
| 1962 | 4·3 | 62·1 | 22·8 | 24·1 | 2·6 | 31·3 | 6·3 | 34·3 | 6·5 | 194·3 |
| 1963 | 4·0 | 68·6 | 22·8 | 24·1 | 2·7 | 29·6 | 5·1 | 33·6 | 6·7 | 197·1 |
| 1964 | 3·8 | 69·1 | 20·8 | 26·3 | 2·7 | 28·4 | 3·9 | 29·4 | 5·8 | 190·2 |
| 1965 | 3·5 | 71·1 | 18·5 | 26·7 | 2·5 | 27·7 | 2·8 | 28·9 | 5·8 | 187·6 |
| 1966 | 3·1 | 69·7 | 17·2 | 25·2 | 2·7 | 25·6 | 1·7 | 26·9 | 5·3 | 177·5 |
| 1967 | 2·9 | 68·3 | 14·8 | 24·0 | 3·3 | 23·3 | 0·8 | 24·5 | 4·6 | 166·4 |
| 1968 | 2·4 | 74·4 | 10·9 | 25·3 | 3·6 | 23·0 | 0·2 | 23·6 | 3·9 | 167·1 |
| 1969 | 2·0 | 77·1 | 7·0 | 25·7 | 3·9 | 21·7 | 0·2 | 21·9 | 4·1 | 163·7 |
| 1970 | 1·9 | 77·2 | 4·3 | 25·3 | 4·2 | 19·6 | 0·1 | 20·2 | 4·1 | 156·9 |
| 1971 | 1·6 | 72·8 | 1·8 | 23·6 | 4·5 | 15·8 | 0·1 | 17·3 | 3·4 | 140·9 |
| 1972 | 1·4 | 66·7 | 0·6 | 20·5 | 4·5 | 11·7 | 0·1 | 14·6 | 2·9 | 122·9 |
| 1973 | 1·4 | 76·8 | 0·5 | 21·9 | 3·6 | 12·1 | 0·1 | 14·5 | 2·5 | 133·4 |
| 1974 | 1·2 | 67·1 | 0·1 | 18·5 | 3·8 | 11·1 | 0·1 | 13·6 | 2·4 | 117·9 |
| 1975 | 1·2 | 74·6 | 0·0 | 19·1 | 4·1 | 9·7 | 0·1 | 11·6 | 1·9 | 122·2 |
| 1976 | 1·1 | 77·8 | 0·0 | 19·4 | 3·4 | 9·0 | 0·1 | 10·8 | 2·0 | 123·6 |
| 1977 | 1·1 | 80·0 | 0·0 | 17·4 | 3·2 | 9·0 | 0·1 | 11·1 | 2·1 | 124·0 |
| 1978 | 1·0 | 80·6 | 0·0 | 15·0 | 3·1 | 8·6 | 0·1 | 10·2 | 1·9 | 120·5 |
| 1979 | 0·8 | 88·8 | 0·0 | 15·1 | 2·9 | 9·2 | 0·1 | 10·5 | 2·0 | 129·4 |
| 1980 | 0·7 | 89·6 | 0·0 | 11·6 | 3·0 | 7·8 | 0·1 | 8·9 | 1·8 | 123·5 |
| 1981 | 0·6 | 87·2 | 0·0 | 10·8 | 2·5 | 7·0 | 0·1 | 8·4 | 1·8 | 118·4 |
| 1982 | 0·5 | 80·2 | 0·0 | 10·4 | 2·4 | 7·1 | 0·1 | 8·5 | 1·8 | 111·0 |
| 1983 | 0·5 | 81·6 | 0·0 | 10·4 | 2·1 | 7·2 | 0·0 | 7·9 | 1·8 | 111·5 |
| 1984 | 0·2 | 53·4 | 0·0 | 8·2 | 1·2 | 6·0f | 0·0 | 6·4f | 1·7 | 77·3 |
| 1985 | 0·3 | 73·9 | 0·0 | 11·1 | 2·2 | 7·5f | 0·0 | 8·6f | 1·7 | 105·4 |
| 1986 | 0·3 | 82·6g | 0·0 | 11·1g | 2·0 | 8·2f | 0·0 | 8·5f | 1·5 | 114·2 |

a  Low temperature carbonisation and patent fuel plants.
b  From April 1973, the figures relate to disposals.
c  From October 1973, the figures relate to disposals.
d  Agriculture, water transport, public services and miscellaneous.
e  Totals may differ slightly from sum of individual items because of metrication and rounding.
f  Includes estimates of imports of steam coal.
g  Sales from British Coal in 1989/90 were 75·8 and 2·4 respectively (B.C.C. Annual Report and Accounts 1989/90)

Source: Department of Energy

**Table 19.2.** The coal classification system used by the National Coal Board (*Revision of 1964*)

Coal with ash of over 10 per cent must be cleaned before analysis for classification to give maximum yield of coal with ash of 10 per cent or less.

| Coal rank code | | | Volatile matter (d.m.m.f.) (per cent) | Gray-King coke type* | General description |
|---|---|---|---|---|---|
| Main class(es) | Class | Sub-class | | | |
| 100 | | | Under 9·1 | A | |
| | 101† | | Under 6·1 | A | } Anthracites |
| | 102† | | 6·1–9·0 | | |
| 200 | | | 9·1–19·5 | A–G8 | Low-volatile steam coals |
| | 201 | | 9·1–13·5 | A–C | |
| | | 201a | 9·1–11·5 | A–B | } Dry steam coals |
| | | 201b | 11·6–13·5 | B–C | |
| | 202 | | 13·6–15·0 | B–G | |
| | 203 | | 15·1–17·0 | E–G4 | } Coke steam coals |
| | 204 | | 17·1–19·5 | G1–G8 | |
| 300 | | | 19·6–32·0 | A–G9 and over | Medium-volatile coals |
| | 301 | | 19·6–32·0 | G4 and over | |
| | | 301a | 19·6–27·5 | G4 and over | } Prime coking coals |
| | | 301b | 27·6–32·0 | | |
| | 302 | | 19·6–32·0 | G–G3 | Medium-volatile, medium-caking or weakly caking coals |
| | 303 | | 19·6–32·0 | A–F | Medium-volatile, weakly caking to non-caking coals |
| 400 to 900: | | | Over 32·0 | A–G9 and over | High-volatile coals |
| 400 | | | Over 32·0 | G9 and over | |
| | 401 | | 32·1–36·0 | G9 and over | High-volatile, very strongly caking coals |
| | 402 | | Over 36·0 | | |
| 500 | | | Over 32·0 | G5–G8 | |
| | 501 | | 32·1–36·0 | G5–G8 | High-volatile, strongly caking coals |
| | 502 | | Over 36·0 | | |
| 600 | | | Over 32·0 | G1–G4 | |
| | 601 | | 32·1–36·0 | G1–G4 | High volatile, medium-caking coals |
| | 602 | | Over 36·0 | | |
| 700 | | | Over 32·0 | E–G | |
| | 701 | | 32·1–36·0 | E–G | High-volatile, weakly caking coals |
| | 702 | | Over 36·0 | | |
| 800 | | | Over 32·0 | C–D | |
| | 801 | | 32·1–36·0 | C–D | Highly-volatile, very weakly caking coals |
| | 802 | | Over 36·0 | | |
| 900 | | | Over 32·0 | A–B | |
| | 901 | | 32·1–36·0 | A–B | High-volatile, non-caking coals |
| | 902 | | Over 36·0 | | |

\* Coals with volatile matter of under 19·6 per cent are classified by using the parameter of volatile matter alone; the Gray-King coke types quoted for these coals indicate the general ranges found in practice, and are not criteria for classification.

† In order to divide anthracites into two classes, it is sometimes convenient to use a hydrogen content of 3·35 per cent (d.m.m.f.) instead of a volatile matter of 6·0 per cent as the limiting criterion. In the original Coal Survey rank coding system the anthracites were divided into four classes then designated 101, 102, 103 and 104. Although the present division into the two classes satisfied most requirements it may sometimes be necessary to recognise more than two classes.

**NOTES**

1. Coals that have been affected by igneous intrusions ('heat-altered' coals) occur mainly in classes 100, 200 and 300, and when recognised should be distinguished by adding the suffix H to the coal rank code, e.g. 102H, 201bH.

2. Coals that have been oxidised by weathering may occur in any class, and when recognised should be distinguished by adding the suffix W to the coal rank code, e.g. 801W.

**Table 19.3.** Analyses of coals from England and Wales (from Ward 1984)

| Coalfield | Seam | Geological horizon | Moisture (%) | Proximate analysis Volatile matter (%) | Proximate analysis Fixed carbon (%) | Ash (%) | Total sulphur (%) | Specific energy (MJkg⁻¹)* | Crucible swelling number† |
|---|---|---|---|---|---|---|---|---|---|
| Northumberland | Beaumont | Westphalian A | 1–3 | 29–37 | 51–61 | 3–15 | 1·0–2·3 | 29·9–33·7 | 2–8 |
| Durham | Top Busty | Westphalian A | 1–3 | 18–36 | 53–73 | 2–11 | 0·5–1·8 | 30·4–35·5 | 0–9 |
|  | Hutton | Westphalian B | 1–2 | 6–37 | 52–86 | 2–11 | 0·9–3·2 | 28·9–34·8 | 0–9 |
| East Pennine | Parkgate | Westphalian A | 1–6 | 29–37 | 52–66 | 2–11 | 0·7–2·9 | 30·2–34·7 | 1–8 |
|  | Meltonfield | Westphalian B | 3–12 | 31–39 | 47–59 | 2–14 | 0·9–3·7 | 27·2–32·3 | 1–6 |
| Lancashire | Trencherbone | Westphalian A | 2–5 | 31–38 | 50–62 | 2–14 | 1·4–4·2 | 28·3–33·8 | 3–8 |
|  | Crumbouke | Westphalian B | 2–7 | 32–39 | 51–59 | 3–8 | 1·0–3·6 | 29·4–33·1 | 2–7 |
| North Staffs | Banbury | Westphalian A | 1–12 | 32–40 | 50–63 | 2–7 | 0·8–3·3 | 27·9–34·2 | 1–8 |
|  | Moss | Westphalian B | 2–7 | 33–38 | 54–62 | 2–7 | 0·5–1·5 | 29·8–33·8 | 1–9 |
| Warwickshire | Seven Feet | Westphalian A | 6–12 | 36–41 | 43–51 | 3–14 | 1·5–6·9 | 25·6–31·1 | 1–2 |
|  | Two Yard | Westphalian B | 8–15 | 32–38 | 44–57 | 2–10 | 0·6–2·4 | 24·7–30·1 | 0–2 |
| South Wales | Five Feet | Westphalian A | 0·4–2 | 9–33 | 57–83 | 3–10 | 0·6–2·3 | 31·3–35·5 | 0–9 |
|  | Nine Feet | Westphalian B | 0·5–2 | 7–34 | 59–89 | 3–9 | 0·6–1·6 | 32·4–35·3 | 0–9 |
| Kent | Millyard | Westphalian C | 1–2 | 12–31 | 64–74 | 4–13 | 1·0–2·4 | 30·9–34·6 | 7–9 |
| Oxfordshire | Broughton | Westphalian D | 7–13 | 34–37 | 48–52 | 5–10 | 1·9–3·0 | 27·1–28·8 | 1–5 |

* 1·00 MJ kg⁻¹ = 429·923 BTU lb⁻¹.

† Crucible swelling number (or F.S.I.) approximates to Gray-King assay: F.S.I. 0–½ ≃ G. K. Assay A–B; 1–4 ≃ C–G₂; 4½–6 ≃ F–G₄.

## Coalfields

Detailed accounts of the geology of individual coalfields, mostly now only of historical interest, can be found in Trueman (1954) and especially in the Sheet and Coalfield Memoirs of the British Geological Survey. Only the coalfields recently or currently producing or those of possible future interest are dealt with here. The sedimentology of the Coal Measures is dealt with in Chapter 8, but for convenience of reference, vertical sections showing the main seams and market horizons in past and present coalfields are shown in Figs. 19.1a and 19.1b.

Coals worked in England and Wales are from the Westphalian-age Coal Measures (Ch. 8), although the Scremerston Coal Group of Dinantian age in Northumberland contains some 9 or 10 seams which were worked in the past, as were some of similar age in Yorkshire. These older coals, like those of pre-Westphalian age in Scotland, heralded the advance of great deltas from the north which eventually overwhelmed the Dinantian sea (which covered most of northern and central England depositing a variety of limestones) in Namurian and Westphalian times (Ch. 8). Thin coals occur also within the deltaic deposits of Namurian age, but it was during West-phalian times that the vast thicknesses of true Coal Measures were deposited, with some 3,000 m of strata estimated for the middle of the Pennine Province and about 2,400 m in South Wales.

### Northumberland and Durham, Cumberland

In Northumberland and Durham there are some 750 m of Lower and Middle Coal Measures, while in Cumberland, beds of equivalent age total about 450 m. In both coalfields the measures dip east and west respectively under the sea, though in Durham they disappear under a Permian cover, also dipping seawards. In Northumberland and Durham, the measures (c. 250 m) from the Brockwell Seam upwards to the High Main (Fig. 19.1a) contain the best coals. The highest rank coals occur in Durham, particularly in the west, and are of the coking quality required for steel-making. In general, the Coal Measures dip in an easterly direction though there are some gentle folds, while the coalfield is crossed by two main sets of faults, striking WSW–ENE and a few degrees west of north, respectively. Vertical throws of up to 300 m are known for the former set.

Coastal mines, north and south of Tynemouth, had extended their workings under the sea as early as the 18th century, but the last 100 years has seen the greatest development of this practice. By the mid 1980s Ellington Colliery, Northumberland and Westoe Colliery, Co Durham had workings 11 km offshore. Vital information, such as the distance above to unconformable water-bearing Permian beds, was until the 1950s obtained by drilling upwards from the workings. Data on what lay ahead of the workings were provided by drilling ahead or from development tunnels. With increased use of sophisticated mining machinery (and the capital involved) more and better geological information was essential (e.g. occurrence of faults and folds; seam continuity and quality etc). Offshore geophysical surveys, particularly over Permian cover, did not reveal the necessary information.

From 1955 to 1965 from specially constructed boring towers, the first offshore boreholes to prove

**Fig. 19.4a.** 'Big Geordie' – the walking dragline excavator which removes and dumps overburden at the Butterwell opencast site, Northumberland. The bucket has a capacity of 50 m³. (Courtesy of Taylor Woodrow Services Ltd.)

coal under the sea were drilled off the north-east coast of England and in the Firth of Forth, Scotland. Between 1958 and 1965, 18 holes totalling almost 11,000 m were drilled off the Durham coast up to 6·5 km offshore and over an area of some 700 km² (Price & Barnsley 1982, Burn 1986). Both tower and ship-borne drilling rigs were used in further offshore drilling between 1973 and 1985, described by Burn (*op. cit.*) whose comprehensive account includes details of survey methods, costs, results etc. Drilling took place as far as 15 km offshore while the deepest holes went 743 m below seabed (tower rig) and 785 m (ship rig). It appears that the most likely area for any future development is offshore from the Westoe and Wearmouth collieries, south of the Tyne.

While the number of deep mines has decreased considerably during the past decade, opencast min-ing has become increasingly important. Currently the largest opencast mine in Britain (and possibly West-ern Europe) is situated at Butterwell, near Morpeth (Northumberland) (Figs. 19.4a, b), where over 1 mt of power station coal are produced annually by the biggest walking dragline (1550W, 50 m³ bucket) in Western Europe ('Big Geordie'). Fireclay for the refractory and pottery industries is also worked from beneath some of the 14 coal seams present. The basal seam worked (down to a depth of 148 m) is the Bandy ($\simeq$ Brockwell).

Cumberland coals were primarily used for steam-raising and to produce coal, gas and coke. Thin seams and extensive faulting have, however, rendered most of the coalfield uneconomical as far as under-

**Fig. 19.4b.** Butterwell opencast site benches showing four of the thirteen coal seams worked. Note 'Big Geordie' (Fig. 19.4a) with its 80 m long jib. (Courtesy of Taylor Woodrow Services Ltd.)

ground mining is concerned. Opencast workings, however, have proved successful here also.

## Lancashire

The Lancashire coalfield (Figs. 8.2, 19.1a, b) is situated on the northern limb of a large complex syncline centred on Cheshire. The Coal Measures are unconformably overlain by Permian rocks (which, in turn, are succeeded by Triassic rocks) as they dip southwards. A prominent structure, the ENE-trending Rossendale Anticline (bringing up Namurian rocks), separates the main South Lancashire field from the smaller Burnley field (with its prime coking coals) to the north. The field is heavily faulted – the main faults trending NW and having vertical throws of hundreds of metres. Coals are mainly of steam-raising quality and have been mined for centuries, the demand for them in the 18th century resulting in the building of the famous Bridgewater canal from Worsley to Manchester. The field is notable for the depth of working, shafts commonly being of the order of 800–900 m.

## East Pennine

The East Pennine coalfield (Yorkshire, North Derbyshire, Nottinghamshire) (Figs. 8.2, 19.1a, b) is the most important coalfield in Britain, currently producing about 41 mt p.y.[1] of coals of ranks 500–800 (Table 19.2). Like the Northumberland and Durham coalfields, it has a regional dip eastwards, off the Dinantian–Namurian anticlinal ridge of the Pennines. Coal Measures disappear to the east under

[1] p.y. = per year.

the Permian rocks and it is in this concealed portion of the field that future development will take place. Structurally, the field is relatively simple though some prominent NW–SE anticlines and synclines disturb the general dip eastwards, as do the normal faults which tend to strike NW–SE and NE–SW.

Opencast mining is extensive. Underground mines in the exposed area are gradually being worked out and there has been a steady shift eastwards during this century as new shafts have penetrated the Permo-Trias cover to reach the coals below. Coals occur mainly in the Communis to Lower Similis-Pulchra Zones of Westphalian A and B stages (Fig. 19.1a), and of some 80–100 coal-bearing horizons within the Westphalian A and B measures, perhaps 30 have coals (somewhere in the field) of economic value. Some coals are areally extensive with none more so than the famous Barnsley or Top Hard Seam.

Drilling through the Permo-Trias in Yorkshire, near Selby, in the period 1964–67 and subsequently in the 1970s showed that, in this area alone, the Barnsley Seam extended over some 285 km$^2$ with a thickness that varied between 2·0 and 3·25 m at depths ranging from 250 m to 1,100 m. Reserves total some 600 mt and with four other seams the total reserves in the area are estimated to be 2,000 mt. The Selby complex, which includes 6 sites and 10 shafts, is the largest underground mining project in the western world and is planned to produce 10 mt per year for power station use when fully operational. The first stage of the complex, Wistow Mine, commenced production in 1983. It produced over 1·7 mt in 1987/88. Production started from another three (Stillingfleet, Riccall and Whitemoor) of the proposed 5 collieries at the beginning of 1988. In 1988/89 the four mines produced over 5 mt. Wistow attained the European productivity record of 26·7 t per manshift in November 1989. North Selby is scheduled to commence production in 1991.

Future developments are also possible in the concealed portion of the coalfield immediately to the NW of York.

A large-scale project similar to Selby was planned in the north-east Leicestershire coalfield in the Vale of Belvoir, but in 1982 the Department of the Environment rejected the NCB's plans after a lengthy and expensive public enquiry. Recoverable reserves were estimated at 520 mt and a production of about 7 mt per year envisaged. In 1984, a modified scheme was accepted by the Secretary of State for the Environment, whereby a mine would be developed at Asfordby (part of the original larger project). The mine should produce 3 mt p.y. from the 2·6 m thick Deep Main and the Parkgate and Blackshale seams

and ensure employment in the 1990s for much of the workforce freed by pit closures in other parts of Leicestershire.

## Midlands

West and south of the East Pennine coalfield lie the coalfields of the Midlands of England (Figs. 8.2, 19.1a, b). Currently, underground production comes from the North Staffordshire, South Staffordshire and Cannock, Warwickshire, and Leicestershire and South Derbyshire fields. Together, in 1989/90, they produced about 11 mt of coal, mainly for electricity generating stations, though some of the deeper seams from the North Staffordshire field, for example, are suitable for coking purposes.

Geologically, the coalfields of the Midlands are characterised by their variable thicknesses of Coal Measures which can lie, without the intervention of Dinantian and Namurian strata, on Lower Palaeozoic and Devonian rocks (see Ch. 8). Considerable folding and faulting has taken place, hence the original continuous area of deposition is now divided into separate fields, while extensions of the fields under the Triassic cover include valuable untapped reserves. Splitting of seams due to differential subsidence is another feature, e.g. the South Staffordshire Thick Coal (Mitchell & Stubblefield 1945), which, 11 m thick near Dudley, has expanded some 20 km to the north to encompass three workable seams in an aggregate thickness of over 50 m of coal and sediments. Similar splitting takes place in Warwickshire with its Thick Coal (Cope & Jones 1970).

The collieries in the Midlands are noted not only for their depths of working, some shafts being over 1,000 m deep, but also for their high state of mechanisation and their high productivity. Planning proposals have been made to establish a new mine at Hawkhurst Moor, near Coventry, where it is envisaged that 4 mt per year could be produced from the 150 mt reserves calculated to exist in various individual and workable leaves of the 9 m thick Warwickshire Thick Coal.[1]

Historically, mining has taken place in the Midlands since the 13th century, but it was the occurrence of ironstone bands, especially within the Westphalian C Measures of North Staffordshire (Fig. 19.1b), and the use of the coals to smelt them in the latter part of the 18th century that resulted in the Midlands playing such an important part in the Industrial Revolution. (The town of Ironbridge in Shropshire is named after the first (1779) iron bridge in the world.) Coke had started to replace charcoal for the smelting of iron ore in the mid-18th century and from then on the iron industry became based on the coalfields where the sideritic bands and nodules occurring within the Coal

[1] Proposal rejected on environmental grounds in February 1991.

Measures (Ch. 8) became the major source of iron. These could be of the 'Blackband' (Fe 39 per cent; $P_2O_5$ 0·43–1·0 per cent; CaO 0·64–11·0 per cent) or the 'Clayband' (Fe 25–46 per cent; $P_2O_5$ 0·40–2·76 per cent; Ca 1·24–5·07 per cent) type. Clayband ores predominated in South Wales and blackband in North Staffordshire where, in the latter area, unusually thick bands, 1–2 m in thickness, occurred (Dunham *et al.* 1979). Precipitation of ferrous-iron minerals is thought to have taken place during diagenesis within the interstitial pores of the muds in which they occur, the iron having been derived from the weathering of continental rocks surrounding the coal basins and the formation of siderite being favoured by the abundance of $CO_2$ associated with the devolatilisation of underlying peats/coals (see e.g. Curtis & Spears 1968, Dunham *et al.* 1979, Pearson 1979, Young & Taylor 1989).

In addition to the ironstones, the occurrence of suitable clays, particularly in the highest Coal Measure beds, led to the establishment of the famous pottery industries around Stoke-on-Trent.

## South Wales

In South Wales (Figs. 19.1a, b), coal is known to have been used in Glamorgan for cremation during the Bronze Age (3,000–4,000 years ago). The Romans are believed to have been the first to have worked it at out-crop. In the Middle Ages, smelting of tin and lead from Cornwall and silver from North Wales took place, marking the start of the metallurgical industry of the Swansea area. Then, as in other coalfields, the iron smelting of the Industrial Revolution saw the greatest mining developments taking place, with additional demands for coal for steam locomotive, ships, heating, etc. By the end of the 19th century South Wales was the chief coal-exporting region of the world, because of the quality and variety of its coals and the proximity of the mines to tidal waters. From an industry employing 250,000 men, however, a decline set in after the First World War due to the development of coal-mining abroad and the increasing use of fuel oil. Annual output of some 40 mt in the 1920s fell to 21 mt by 1947. In 1989/90 only 6 underground mines remained, producing *c.* 3·5 mt. The famous Rhondda Valley which in its heyday had 60 mines producing 10 mt p.y. saw its last coal hoisted in 1990.

Coals in South Wales (Adams 1967) include anthracites, coking coals for steel and foundry work, steam-raising coals for power generation, and domestic fuels. There is a pronounced increase of rank from the south-east of the coalfield to the anthracite field in the north-west (Fig. 8.9). Essentially, the field is a syncline extending for almost 100 km along its east–west axis, but in detail it is structurally very complex (see, e.g., Owen 1974) with, apart from a variety of fold structures, large NW–SE faults ('cross-faults') and, particularly in the South Crop, strike-thrust faults. There are also two extensive NE–SW fault belts, the Neath and Swansea valley 'disturbances', crossing the field. All faults seem to have histories of intermittent movements, frequently of different directions and over prolonged periods of time.

Possible new developments include a 1·2 mt p.y. coking-coal mine at Margam, south of Port Talbot. Meanwhile opencast mining continues successfully and a 3·3 mt p.y. output is planned by 1995 from Ffostas, near Trimsaran where 11 seams of West-phalian A age occur, including the 'Big' seam which varies between 2 and 5 m in thickness. The structural complexities of the site (with severe faulting and folding, and in places with the stratigraphic succession being completely inverted) exemplify the difficulties of economic underground mining in South Wales.

The obvious pattern of rank variation has been the subject of intense debate and controversy over the years (e.g., see Trotter 1948 and ensuing discussions, 1954). Increase in heat, rather than pressure, is nowadays regarded in most of the world's coalfields as the main cause of 'anthracitisation' despite the almost universal occurrence of anthracites in severely structurally deformed zones. In South Wales, efforts to account for the restriction of the anthracite zone to the north-west corner of the field have ranged from suggestions that this area was: the deepest and hence the hottest part of the basin (some postulated 6,000 m of cover having been subsequently – and conveniently! – eroded); the part of the basin closest to an underlying shallow-dipping thrust-fault (frictional movement along which produced the heat) whose surface expression is seen along the Llandyfaelog, Careg Cennen, Church Stretton line; heated up by an underlying igneous body (for which there was no geophysical evidence).

Davies and Bloxam (1974) noted an increase in trace-metal content of coals which coincided with the increase in rank towards the north-west of the coal-field and suggested the possibility that there had been a source of heat at comparatively shallow depth which caused a high geothermal gradient in the area. Gill *et al.* (1979) confirmed by iso-reflectance studies the increase in rank to the north-west, noted that spores were progressively carbonised but could not account for the high geothermal gradient necessary for the anthracitisation. Geophysical, geochemical and geothermal investigations, however, listed by Bloxam and Owen (1985) do suggest that magmatic heat, perhaps from an igneous mass at a depth of no more than 3·5 km, could have caused the anthracitisation.

## North Wales

The Flintshire Coalfield (Calver & Smith 1974) (Fig. 19.1a) extends south-eastwards from the Point of Ayr, on the Extreme north-west of the Dee estuary to Cairgwrle, some 5 km south of Mold (Figs. 8.2, 19.1a). Mining took place by opencast methods from as early as 1600. Since then, both opencast and underground mines have operated though there is a history of continual closures due to flooding and difficult geological conditions. The Buckley Fireclay Formation (Westphalian 'C' age) was extensively worked at one time for refractory brick manufacture and for pottery clay, and the Cannel (or Yard) seam distilled for oil. Currently, only one pit, at Point of Ayr, exists. Methane is present in considerable quantities in this colliery and is utilised. Originally, seven seams each over 2 m in thickness were worked but the present two workable seams, the Two Yard and Three Yard, are each only about 1 m. Offshore boreholes in the Dee Estuary, combined with 200 km of marine seismic surveys, have, however, proved extensive reserves which it is intended will be worked from an 800 m surface drift from the colliery.

The Denbighshire field to the south (Figs. 8.2, 19.1a and 19.1b) is separated from the Flintshire field by a narrow zone of Namurian and Dinantian rocks, brought to the surface by the Bala Fault and associated anticline, and extends southwards into Shropshire. Mentioned in records as early as 1410, mining has virtually ceased though the eastward dip of measures and various boreholes indicate continuity at depth with the coalfields of Lancashire (Fig. 19.2). Currently, only Bersham Colliery near Wrexham remains active, working the Two Yard seam.

## Kent

The Kent coalfield (Figs. 19.1a, b) is notable because it is the only completely concealed coalfield to be exploited so far in Britain. Predicted to exist under a Tertiary and Mesozoic cover by Godwin-Austen (1855), its reality was proved in a borehole sunk in 1895/96 during test drilling near Dover for a projected Channel tunnel (Hull 1897)! In 1988/89 the sole remaining colliery, Betteshanger, produced 0·44 mt of coking coal. It was planned to increase this amount to c. 0·75 mt p.y. but the miners' rejection of a new shift system resulted in closure later in 1989.

## Oxfordshire

Coal was first proved by drilling in Oxfordshire in 1877 beneath Jurassic rocks at depths of about 360 m, but it was comparatively recently (e.g. Poole 1971, Dunham & Poole 1974) before a summary of the numerous boreholes drilled by the I.G.S. from 1949 onwards indicated the extent of what is now called the Oxfordshire coalfield. While the coals are thin and of moderate quality, it appears there are resources of some 10 billion tonnes under an area of about 1,100 km². In 1982, Allsop et al. on the basis of borehole information and seismic data inferred that Westphalian rocks continued southwards as far as Newbury and Reading though coals were said to be thin.

## Norfolk

A borehole at Somerton, NE Norfolk, through the Chalk, proved 68 m of Carboniferous Coal Measures resting disconformably on Carboniferous Limestone. The southern North Sea borehole 52/5–11 proved Coal Measures at depth also. Geophysical evidence (Allsop 1984) suggests the presence of a small concealed coalfield in the area to the east and north of Norwich (Fig. 19.3).

## Cannel coals and oil shales

While the coals so far discussed are regarded as having been formed from plants accumulated in situ as a forest peat, associated with them, usually as lenticular bands at the top of the seams but sometimes at the base, are found cannel coals – coals formed from the accumulation in ponds of the forest swamps of a fine organic mud. Spores, algal remains and macerated resistant plant debris make up the organic content in varying proportions. ('Cannel' is thought to be derived from 'candle' because slivers of cannel can be lit and will burn like a candle.) Most cannel coals are thin and very impersistent laterally, but occasionally in the past some were worked for their illuminating gas content and in North Wales (Flint, Fig. 19.1a), near Leeswood and Bagillt, the Cannel (or Yard) Seam was at one time distilled for oil.

A cannel coal with increase in clay mineral content can pass transitionally into an oil-shale and the shales immediately above the Cannel Seam were also exploited for distillation. Unlike Scotland, there are no Carboniferous oil-shales or cannel coals in England or Wales in the quantities which gave rise to an industry such as existed in the Lothians region in Scotland (Duff 1983). Oil shales of potential interest do occur, however, in Jurassic-age rocks. The Kimmeridge Clay (Ch. 11) is the most important and contains groups of oil-shale bands concentrated at five different horizons which can be traced intermittently at outcrop from Dorset to Yorkshire (Gallois 1978). Drilling has proved their presence at depth and in fact the Kimmeridge Clay formation with oil shales is recognisable at outcrop in Scotland and in the North Sea (see later). At the present time, they are not considered economic, few of them yielding the 25 litres of oil per tonne of rock (6 US gallons/short ton) necessary to equal the energy to equate with the heat used to distil them, let alone the 42 l/tonne (10 US gallons/short ton) yield necessary currently to break even financially. Their importance is that they are probably the main source rock for the oil found in the northern North Sea (p. 609).

## Hydrocarbons

### History

The discovery in the 1960s of oil and gas in the rocks below the North Sea and the subsequent successful exploitation tends to obscure the fact that, as with coal, natural hydrocarbons were known, and indeed used, in England from earliest recorded times. Pitch, for example, was used by the Romans from seepages from Carboniferous rocks in their town of Wroxeter, near Shrewsbury. Much later, in the 17th century, 'pitch and tar' from a Coal Measures breccia at Pitchford and from Coalbrookdale, both also in Shropshire, were used for waterproofing purposes, in oil lamps and medicinally. ('Betton's British Oil' was used as an inhalant and exported to continental Europe!) In fact, Shrewsbury in the early 18th century

**Fig. 19.5.** Oil and gas wells onshore and offshore (based on Department of Energy 1988).

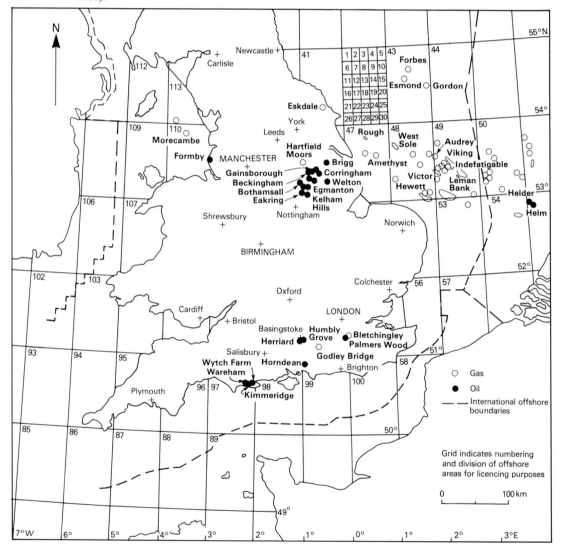

was Britain's first oil-refining centre (Cranfield 1980). From seeps from coal mines in the area (e.g. Hanwood Colliery) sometimes up to 70–80 gallons per day were collected and distilled to make solvents for india rubber, the products being used for the waterproofing cloth required by early balloonists for their gas-bags.

On the other side of the country, in Derbyshire, Riddings Colliery, Alfreton, produced about 100 tons of oil per year in the first half of the 19th century. It attracted the attention of James (later 'Paraffin') Young, the Scots chemist, who set up a distillation plant to produce kerosene, paraffin wax and lamp oil. (He experimented on producing oil from coal and subsequently opened an 'oil-from-coal' plant at Bathgate in Scotland in 1851. His raw material was Lower Carboniferous cannel coal (Boghead coal or Torbanite), and when this seam was exhausted he turned to the nearby Lower Carboniferous oil shales in West Lothian. Thus the Scottish oil shale industry, which lasted until 1962, was born.)

Methane has, of course, long been known to occur in the coal mines. Drainage from collieries is a necessary safety precaution in some areas and the gas is utilised for heating purposes either in the collieries themselves or by industry. Production in 1986 yielded 52 million therms (B.G.S. 1988).

## Onshore exploration

Much information on the history of oil exploration in the United Kingdom is recorded in e.g. Lees & Cox (1937), Lees & Taitt (1946), Falcon & Kent (1960) and Kent (1985). (Localities of present-day oil wells are shown in Fig. 19.5.)

### (a) 1918–1922

Near the end of the First World War, the British government, because of the submarine danger to overseas oil supplies, organised the first real search for oil in Britain and naturally were drawn to areas of known occurrences. Some 11 wells were drilled between 1918 and 1922, particularly on anticlines of Carboniferous rocks flanking the Pennines. Only one well, at Hardstoft (Fig. 19.6), Derbyshire, was successful and this produced oil from the Carboniferous Limestone from 1919 until the 1930s.

### (b) 1934–1939

After this initial period of prospecting, the availability of cheap oil from abroad, coupled with the problems of establishing mineral rights and other legal barriers, discouraged further search. In 1934, however, the Petroleum Production Act vested in the Crown ownership of all mineral oil not discovered up to that date and enabled companies to acquire exploration licences over large areas under attractive operating conditions.

Geological and geophysical (seismic refraction) surveys led to prospecting from 1935 onwards to be concentrated once more on structures in areas of known oil occurrences, Carboniferous rocks were the targets in Nottinghamshire and Lancashire, Mesozoic rocks in Sussex, Hampshire and Dorset (gas had been found when drilling for water near Heathfield in Sussex in 1896; oil seepages were known from shore outcrops on the Dorset coast). In general, results were disappointing, particularly in the south of England, though Permian potash salts (p. 299) were discovered in 1937 above methane-bearing Magnesian Limestone while drilling on the Eskdale anticline (Fig. 19.5) in Yorkshire. In 1939, however, oil in commercial quantities (still being produced in the 1980s) was discovered in Nottinghamshire (Eakring, Figs. 19.5, 19.6, 19.7) in sandstones of Lower Westphalian and Upper Namurian age and, later, boreholes nearby intersected oil-bearing Carboniferous Limestone. In Lancashire (Formby, Fig. 19.5), also in 1939, oil which had migrated upwards from Carboniferous beds was found in Triassic rocks, sealed under 30 m of Pleistocene boulder clay, and production continued in that area until 1965.

**Fig. 19.6.** Schematic cross-section through East Midlands Oilfields (based on Warman *et al.* 1956).

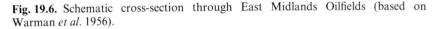

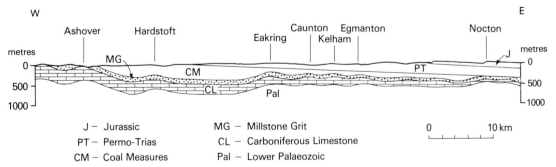

P. McL. D. DUFF

| | 1964 | 1965 | 1966 | 1967 | 1968 | 1969 | 1970 | 1971 | 1972 | 1973 | 1974 | 1975 | 1976 | 1977 | 1978 | 1979 | 1980 |
|---|---|---|---|---|---|---|---|---|---|---|---|---|---|---|---|---|---|
| **(a) Offshore Exploration** | | | | | | | | | | | | | | | | | |
| East of England | 1 | 10 | 20 | 35 | 30 | 34 | 12 | 7 | 8 | 7 | 4 | 4 | 2 | 3 | 5 | | |
| W of Eng./Wales | | | | | | 2 | | | | | | | 4 | 4 | 3 | | |
| Channel & SW Approaches | | | | | | | | | | | | | | | 2 | 4 | |
| **Appraisal Wells** | | | | | | | | | | | | | | | | | |
| East of England | | | 8 | 16 | 7 | 8 | 2 | | 4 | 6 | 2 | 1 | 1 | | 2 | | |
| W of Eng./Wales | | | | | | | | | | | 1 | 1 | | 3 | | | |
| **Development Wells** | | | | | | | | | | | | | | | | | |
| East of England | | | 3 | 13 | 36 | 27 | 28 | 34 | 36 | 21 | 20 | 13 | 7 | 7 | 7 | 2 | |
| W of Eng./Wales | | | | | | | | | | | | | | | | | |
| **(b) Onshore Exploration*** | | | | | | | | | | | | | | | | | |
| United Kingdom | | 6 | 16 | 17 | 2 | 5 | 6 | 8 | 10 | 14 | 6 | 4 | 4 | 3 | 1 | 2 | 8 |
| **Appraisal/Development** | | | | | | | | | | | | | | | | | |
| United Kingdom | | | 20 | 4 | 5 | 2 | ½ | 1 | ½ | 5 | 4 | 15 | 1 | 3 | 8 | 2 | 13 |

| | 1981 | 1982 | 1983 | 1984 | 1985 | 1986 | 1987 | 1988 | 1989 |
|---|---|---|---|---|---|---|---|---|---|
| **(a) Offshore Exploration** | | | | | | | | | |
| East of England | 1 | 9 | 9 | 24 | 17 | 12 | 20 | 25 | 35 |
| W of Eng./Wales | | 4 | 1 | 1 | 1 | 1 | 2 | 5 | 3 |
| Channel & SW Approaches | 1 | 5 | 9 | 3 | 1 | 3 | 1 | 2 | |
| **Appraisal Wells** | | | | | | | | | |
| East of England | 1 | 8 | 17 | 19 | 24 | 17 | 16 | 21 | 16 |
| W of Eng./Wales | | | 3 | | | | | | |
| **Development Wells** | | | | | | | | | |
| East of England | 4 | 11 | 10 | 18 | 28 | 32 | 37 | 47 | 49 |
| W of Eng./Wales | | | 2 | 9 | 4 | 4 | | | 7 |
| **(b) Onshore Exploration*** | | | | | | | | | |
| United Kingdom | 16 | 10 | 13 | 25 | 36 | 37 | 31 | 16 | 16 |
| **Appraisal/Development** | | | | | | | | | |
| United Kingdom | 10 | 6 | 8 | 16 | 31 | 31 | 9 | 31 | 29 |

**Table 19.4.** Wells drilled for oil and gas in seas around England and Wales and onshore* United Kingdom, 1964–89.

* Mainly England.

(Source: Department of Energy 1973–90.)

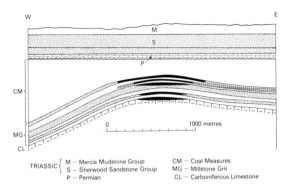

**Fig. 19.7.** The classic anticlinal trap of the Eakring oilfield, showing oil accumulations (black) (Warman *et al.* 1956).

### (c) *1939–1964*

As the discovery at Eakring almost coincided with the outbreak of the Second World War, production rose quickly from 3,634 tonnes in 1939 to 114,800 tonnes in 1942 (Blunden 1975). Continued exploratory drilling in the area through the gently dipping unconformable Permo-Trias cover proved an extension of the Eakring structure at Duke's Wood in 1941 and another Carboniferous age anticline bearing oil at Kelham Hills (Figs. 19.5, 19.6). In 1943, yet another field was found at Caunton (Fig. 19.6). By 1945, these fields produced 410,000 tonnes per year.

After a lull, exploration commenced again in earnest in the 1950s, partly because of the need for gas cheaper than could be produced from coal. During the period 1958–62, commercial oil fields were discovered in Upper Carboniferous sandstones in Nottinghamshire and Lincolnshire, including Corringham, Bothamsall and Gainsborough (the largest) (Fig. 19.5). Gas finds, however, were minimal.

Since the 1937 Eskdale discovery of methane in the Permian Magnesian Limestone, over 40 exploration gas wells have been drilled in the Permian rocks north of the Humber. In 1956, drilling once more on the Eskdale anticlinal structure encountered methane in the Magnesian Limestone and this well, along with that of 1937, was put into production in 1960 to supply Whitby, but lasted only until 1966. At Lockton, near Scarborough, methane gas in commercial quantities was also produced from the Permian dolomites from 1971 until 1974.

Drilling results in the south of England were disappointing for both oil and gas, with the notable exception of Kimmeridge No. 1 (Fig. 19.5) which struck oil in commercial quantities in fractured Jurassic Cornbrash Limestone in 1959. This well assumed full production in 1961. Further drilling in the vicinity in 1964 at Wareham No. 1 (Fig. 19.5) proved oil in fractured Inferior Oolite Limestones, although this well has not been of significant commercial importance.

### (d) *1964–present*

In 1964, the removal, by the Government, of the exemption from duty on indigenous oil production rendered UK oil non-competitive with imports and led to a drastic cutting down (Table 19.4) of exploration (and had led to the closure of the Scottish oil-shale industry).

The discovery of North Sea gas in 1965 (see later), oil off Scotland in 1969–70, plus the 1973 world oil crisis, renewed interest on land and Table 19.4 illustrates how both onshore and offshore drilling have developed since then. Of some 1,500 exploration wells drilled since 1964, 271 have made significant discoveries. The onshore search has concentrated on areas in the north-east of England, the Midlands and the south of England (Fig. 19.5). Geologically, targets continue to be areas of suitable sandstone or carbonate reservoir rocks which structurally are, or were, in close proximity to the currently favourite source rocks – organic-rich marine shales of Carboniferous or Jurassic age. Wytch Farm in Dorset (Fig. 19.5) provides an interesting example.

British Gas Corporation geologists, after a study of the Wessex Basin in 1972 (Colter & Havard 1981), considered that the Bridport Sands (see Fig. 19.8) were a more likely reservoir than the fractured carbonate reservoirs and it was decided to drill a structure in 1973 revealed by BP seismic surveys at Wytch Farm (No. 1). Oil was discovered in the Bridport Sands and further study and geological reasoning led to the belief that movement within the Purbeck–Isle of Wight Disturbance could have resulted in oil having migrated from the inferred Jurassic source upwards, pre the Alpine orogeny, into the stratigraphically older Triassic Sherwood Sandstones Group (see Dranfield *et al.* (1987) for details of the reservoir characteristics). Wytch Farm D5 Well (1977–78) proved the theory correct and Wytch Farm is currently Britain's biggest onshore producer (Table 19.5) and the largest onshore field in Western Europe with reserves in excess of 200 bbl. Production in 1989 is forecast to be 8,220 b/d (3 (M)b/y) of oil, 290,000 cm/y of gas and 185,000 b/y of liquid gas[1] (Shrimpton 1988). (An offshore extension of the field is also probable – p. 610.)

An important discovery in 1980 was at Humbly Grove, near Basingstoke, Hampshire (Fig. 19.5), where oil was found in a mid-Jurassic carbonate reservoir (the Great Oolite Group [Selwood *et al.*

[1] bbl = billion barrels; b/d = barrels per day; (M)b/y = (million) barrels per year; cm/y = cubic metres per year.

South                                                                                  North

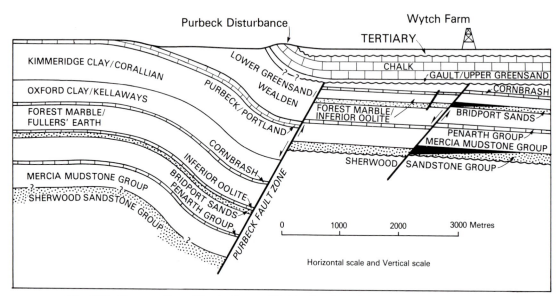

**Fig. 19.8.** Cross-section, Wytch Farm oilfield, showing oil accumulations (black) (Colter & Havard 1981).

1985]). Hancock and Mithen (1987) described the structure as an 'E–W trending horst, fault-closed to the north and south, with a dip closure to east and west' and gave details of the reservoir properties. Lower Lias shales are regarded as the source rock (with some oil migrating into the underlying 'Rhaetic' beds).

Summarising, hydrocarbon exploration to date has been most successful in central eastern England and in southern England (Fig. 19.5) where structural traps have been the main targets. Production figures are given in Table 19.5.

In the first-named region oil (and gas) are being produced in the East Midlands whereas in north

**Table. 19.5.** United Kingdom production of onshore crude petroleum 1975–1989 (from Dept. of Energy 1990)

| | 1975 to end 1984 | 1985 | 1986 | 1987 | 1988 | 1989 | Cumulative total from 1975 | 1989 ('000 bbls) |
|---|---|---|---|---|---|---|---|---|
| Beckingham W | | | 1 | 1 | 3 | 2 | 7 | 12 |
| Crosby Warren | | | 2 | 12 | 13 | 8 | 35 | 65 |
| Farleys Wood | 2 | 5 | 5 | 3 | 2 | 2 | 19 | 11 |
| Herriard | 7 | 3 | 0 | 3 | 4 | 4 | 21 | 28 |
| Horndean | 9 | 3 | 0 | 3 | 9 | 7 | 31 | 52 |
| Humbly Grove | 46 | 13 | 61 | 62 | 48 | 46 | 276 | 348 |
| Nettleham | 12 | 3 | 13 | 17 | 21 | 19 | 85 | 138 |
| Scampton N | | | 2 | 2 | 1 | 15 | 20 | 110 |
| Stainton | | 1 | 1 | 2 | 3 | 2 | 9 | 13 |
| Welton | 23 | 17 | 62 | 112 | 117 | 83 | 414 | 614 |
| Wylch Farm | 960 | 253 | 276 | 292 | 483 | 475 | 2,739 | 3,605 |
| Other Land | 857 | 93 | 96 | 69 | 57 | 59 | 1,231 | 448 |
| Totals | 1,916 | 391 | 519 | 578 | 761 | 722 | 4,887 | 5,444 |

Yorkshire only gas has been discovered so far. Kirby *et al.* (1987) and Scott and Colter (1987) emphasised the importance of the Silesian Edale Shales as source-rocks in the East Midlands with Silesian sandstones as reservoirs. In north Yorkshire, however, methane occurs in Permian dolomites. There, the absence of Westphalian rocks due to pre-Permian erosion, the maturation history of the probable source-rocks of Dinantian/Namurian age shales and the tectonic history's effect on migration pathways are all thought to be factors inhibiting the accumulation of oil.

That Dinantian shales (e.g. Bowland Shales) can be gas-bearing was proved, tragically, in north-west England in 1984. At Abbeystead, in Lancashire, 16 people were killed in a gas explosion in an underground water-installation sited in the Bowland Shales. Selley (1987) compared British Carboniferous shales (and particularly those of Lower Carboniferous age) with economic gas-producing Pennsylvanian Shales in the eastern USA, suggesting that ours too might be considered for small-scale exploitation.

In southern England Brooks and Glennie (1987) included a series of articles (*nos* 10–15) summarising much of the geology of the fields and areas being explored and/or exploited. Reservoir rocks include sandstones and carbonates of Triassic, Jurassic and Cretaceous age with Lower Lias Shales (and possibly the Kimmeridge and Oxford Clays) being the source-rocks.

Apart from the onshore areas already explored the potential of the Variscan Thrust province of south-west England (Taylor 1986), the northern flank of the London–Brabant massif (Tubb *et al.* 1986) the Midland massif (Smith 1987) and the Cambrian of the Welsh Borderland (Parnell 1987) have been reviewed and they add detail to Scott and Colter (1987) and indicate possible sites for future onshore plays.

Finally, with the world-wide interest in Gold's (e.g. 1985) views on the abiogenic origin of methane it is of interest that Ferguson (1988), studying methane in the minerals of the Pennines orefield, and Parnell (1988) examining the occurrence of hydrocarbons in British granites and basement rocks both felt the balance of evidence still favours an organic origin.

## The Continental Shelf

The limited success of onshore hydrocarbon exploration in Britain in the first half of this century, the absence of any international agreement concerning the ownership of minerals on and below the sea bed and the availability of cheap oil from the Middle East, together meant there was little economic incentive to explore the Continental Shelf. The discovery of enormous quantities of natural gas in Permian rocks near Groningen in Holland in 1959, however, and the

knowledge that gas had been encountered in Permian rocks in Yorkshire, led to one of the most rapid and successful hydrocarbons exploration programmes ever mounted and already the literature is considerable. (For sources of information, see particularly, Woodland 1975, Dept. Energy 1975–90, Illing & Hobson 1981, Berry & Wilson 1984, Brennand & Van Hoorn 1986, Brooks *et al.* 1986, Glennie 1986a, b, Brooks & Glennie 1987).

The 1958 Geneva Convention (and subsequent annexes) on the Territorial Sea and Contiguous Zone laid down general rules applicable to all continental-shelf seas and to continental-shelf areas flanked by different countries. Britain finally accepted the principles for dividing up the North Sea in 1964 and offshore production and exploration licences were first issued that year (the 'First Round') under the Petroleum (Production) Act 1934 as extended offshore by the Continental Shelf Act 1964.

At that time, little was known about what lay below the North Sea. Some sea-bed sampling and geophysical work had been carried out, mainly by universities in the late 1950s and early 1960s, but it was not until 1962 that oil companies commenced systematic regional seismic surveys. The Groningen discovery and the presence of gas in rocks of similar age in Yorkshire meant that attention was mainly centred on the southern part of the North Sea. The indications were that the sea covered a thick sedimentary basin of Tertiary and older rocks. The Permian rocks in particular were the prime target.

When applications were invited for the UK Second Round of licences early in 1965, only five (unsuccessful) wells had been drilled in British waters. Off Germany, too, results were discouraging. The West Sole gas field, however, 75 km off the Humber Estuary, was discovered in December 1965 and led to the discovery of other gas fields in the southern North Sea (Fig. 19.5). (It was not, incidentally, until 1969 when over 50 exploration wells had been drilled in northern waters (i.e. British, Norwegian and Danish) that the first commercial *oil* field (Ekofisk, Fig. 19.5) in the North Sea was discovered 350 km due east of Dundee, but just in the Norwegian sector.)

Production figures for the southern North Sea Gas Fields are given in Table 19.6.

## Southern North Sea

While the general geology of the North Sea is covered in Chapter 15 the literature on the southern North Sea gas fields is now so extensive only a brief summary can be given here. Glennie's (1986b) description of the geological development of the southern Permian gas basin of north-west Europe outlines the main features.

Structurally the North Sea is divided into two

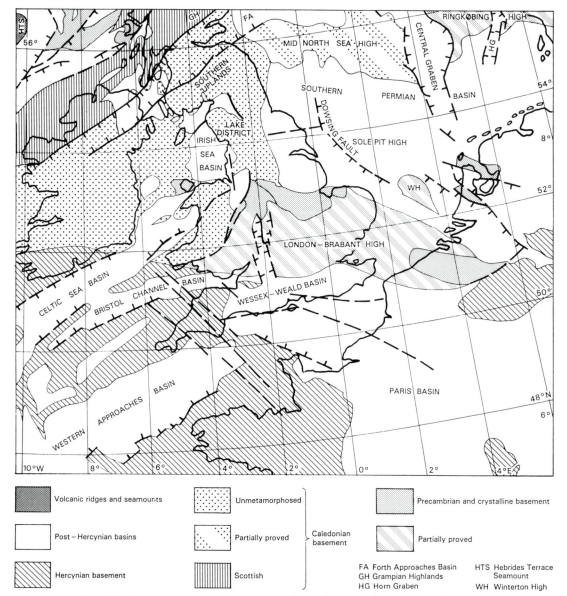

**Fig. 19.9.** Generalised geology and terminology of structural features and sedimentary basins in and offshore England and Wales (based, with permission, on Brooks & Glennie 1987 by kind permission of Brooks Associates).

different hydrocarbon domains by the east–west trending Mid-North Sea High (Figs. 15.4, 19.9). It in turn is cut, at approximately right angles, by the Central Graben in which thick successions of Permian to Triassic sediments (and in its northerly part volcanics) were laid down. Much of the oil in the northern North Sea is associated with this graben.

Basically, methane (the predominant gas) occurs mainly in Permian Rotliegend aeolian sandstone reservoirs. Source rocks are Westphalian coals (Eames 1975, Cornford 1986) and probably also, to a lesser extent Carboniferous black shales. Zechstein halites (Chapter 9) overlying the Rotliegend sandstones form effective cap-rocks. A generalised succession of the rocks proved in the southern North Sea is given in Fig. 19.10.

**Table 19.6.** Gas Production in Southern North Sea and Irish Sea (1968–1988) (Department of Energy (1976–1990) Million cubic metres

| Field (Discovered) | 1968 | 1969 | 1970 | 1971 | 1972 | 1973 | 1974 | 1975 | 1976 | 1977 | 1978 | 1979 | 1980 | 1981 | 1982 | 1983 | 1984 | 1985 | 1986 | 1987 | 1988 | 1989 | Cumulative Total (1989) |
|---|---|---|---|---|---|---|---|---|---|---|---|---|---|---|---|---|---|---|---|---|---|---|---|
| West Sole[1] (1965) | 1,300 | 1,580 | 1,180 | 1,860 | 2,280 | 1,890 | 1,830 | 1,830 | 2,010 | 1,950 | 1,533 | 1,365 | 1,455 | 1,445 | 1,512 | 1,719 | 1,899 | 1,771 | 1,695 | 1,576 | 1,520 | 1,482 | 37,126 |
| Leman Bank (1966) | 750 | 2,920 | 7,970 | 12,890 | 13,130 | 13,100 | 15,610 | 15,010 | 15,370 | 15,580 | 14,719 | 13,831 | 9,482 | 13,207 | 11,675 | 11,985 | 9,376 | 10,943 | 10,029 | 10,903 | 9,506 | 7,241 | 245,204 |
| Hewett Area (1966) | | 560 | 1,950 | 3,380 | 5,150 | 5,710 | 7,060 | 7,640 | 8,110 | 7,850 | 6,392 | 6,288 | 6,568 | 5,048 | 4,108 | 3,851 | 3,631 | 3,413 | 3,336 | 2,720 | 2,176 | 1,908 | 98,856 |
| Indefatigable (1966) | | | | 160 | 4,510 | 4,560 | 5,550 | 6,250 | 6,350 | 6,780 | 6,450 | 6,006 | 6,878 | 5,613 | 5,720 | 4,700 | 5,590 | 5,323 | 6,186 | 6,430 | 6,448 | 4,239 | 101,739 |
| Viking Area (1968) | | | | | 1,390 | 3,590 | 4,770 | 5,510 | 6,050 | 6,330 | 5,238 | 4,397 | 4,689 | 3,307 | 4,381 | 3,413 | 3,197 | 3,450 | 3,712 | 3,166 | 2,599 | 2,068 | 71,258 |
| Rough[2] (1968) | | | | | | | | 10 | 510 | 1,060 | 931 | 1,005 | 467 | 99 | 101 | 27 | 55 | 92 | 7 | 0 | 0 | 0 | 4,370 |
| Victor (1972) | | | | | | | | | | | | | | | | | 472 | 1,485 | 1,600 | 1,578 | 1,436 | 1,414 | 7,985 |
| Esmond (1969) | | | | | | | | | | | | | | | | | | 591 | 970 | 1,202 | 1,049 | 1,317 | 5,129 |
| Forbes (1970) | | | | | | | | | | | | | | | | | | 107 | 305 | 467 | 294 | 161 | 1,334 |
| Gordon (1969) | | | | | | | | | | | | | | | | | | 177 | 528 | 484 | 532 | 571 | 2,292 |
| Thames (1973) | | | | | | | | | | | | | | | | | | | 351 | 1,106 | 860 | 926 | 3,243 |
| North & South Sean (1969/70) | | | | | | | | | | | | | | | | | | | 202 | 391 | 321 | 344 | 1,258 |
| Yare (1969) | | | | | | | | | | | | | | | | | | | | 210 | 264 | 269 | 743 |
| Bure (1983) | | | | | | | | | | | | | | | | | | | | 270 | 291 | 269 | 830 |
| Audrey (1976) | | | | | | | | | | | | | | | | | | | | | 656 | 1,865 | 2,525 |
| Cleeton (1983) | | | | | | | | | | | | | | | | | | | | | 278 | 1,097 | 1,375 |
| North Valiant (1970) | | | | | | | | | | | | | | | | | | | | | 121 | 469 | 590 |
| South Valiant (1970) | | | | | | | | | | | | | | | | | | | | | 110 | 507 | 617 |
| Vanguard (1982) | | | | | | | | | | | | | | | | | | | | | 60 | 436 | 496 |
| Vulcan (1983) | | | | | | | | | | | | | | | | | | | | | 428 | 1,941 | 2,369 |
| Camelot (1987) | | | | | | | | | | | | | | | | | | | | | | 242 | 242 |
| Ravenspurn South (1983) | | | | | | | | | | | | | | | | | | | | | | 92 | 92 |
| South Morecambe (1974) | | | | | | | | | | | | | | | | | | 90 | 604 | 1,441 | 975 | 1,138 | 4,248 |
| | | | | | | | | | | | | | | | | | | | | | | | 593,921 |

[1] Produced 5 Mcm in 1967.  [2] In 1985 converted to off-peak storage unit.

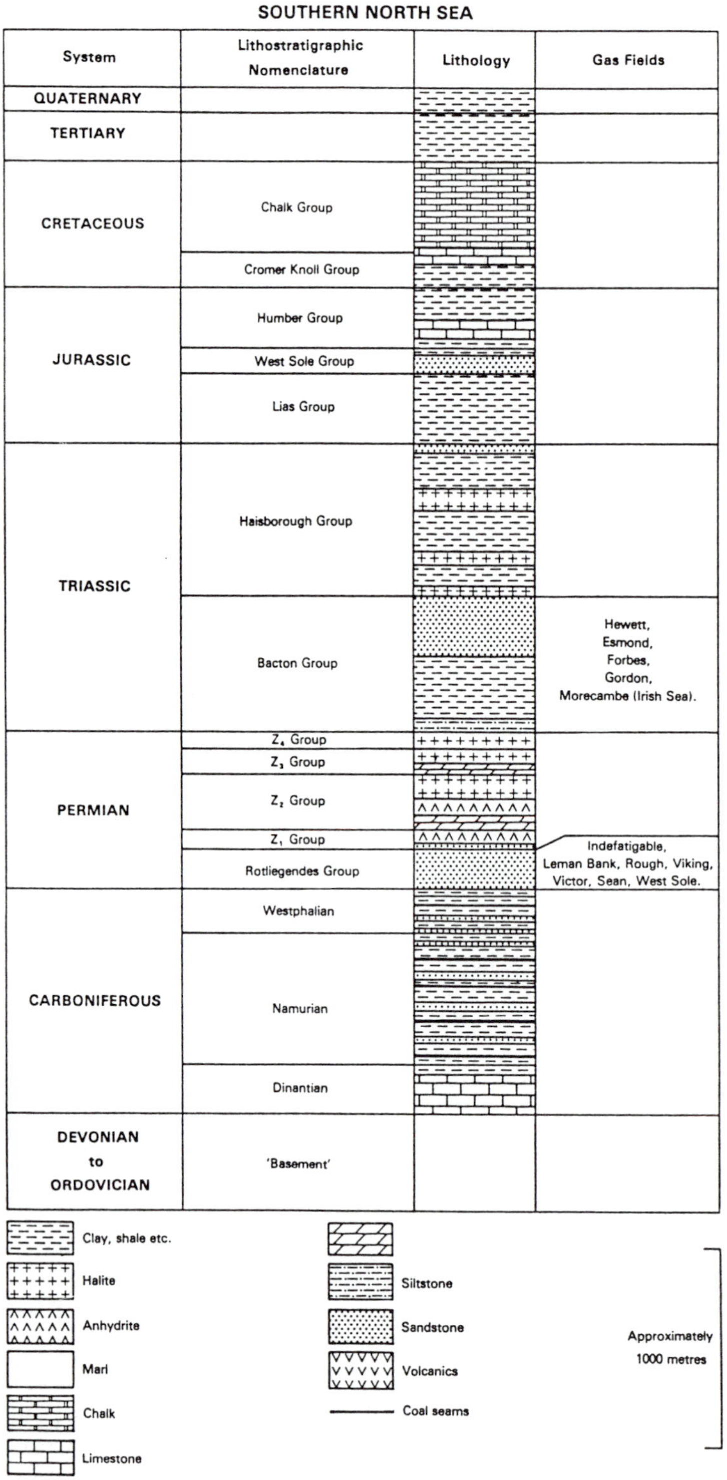

SOUTHERN NORTH SEA

| System | Lithostratigraphic Nomenclature | Lithology | Gas Fields |
|---|---|---|---|
| QUATERNARY | | | |
| TERTIARY | | | |
| CRETACEOUS | Chalk Group | | |
| | Cromer Knoll Group | | |
| JURASSIC | Humber Group | | |
| | West Sole Group | | |
| | Lias Group | | |
| TRIASSIC | Haisborough Group | | |
| | Bacton Group | | Hewett, Esmond, Forbes, Gordon, Morecambe (Irish Sea). |
| PERMIAN | Z₄ Group | | |
| | Z₃ Group | | |
| | Z₂ Group | | |
| | Z₁ Group | | Indefatigable, Leman Bank, Rough, Viking, Victor, Sean, West Sole. |
| | Rotliegendes Group | | |
| CARBONIFEROUS | Westphalian | | |
| | Namurian | | |
| | Dinantian | | |
| DEVONIAN to ORDOVICIAN | 'Basement' | | |

Legend:
- Clay, shale etc.
- Halite
- Anhydrite
- Marl
- Chalk
- Limestone
- Siltstone
- Sandstone
- Volcanics
- Coal seams

Approximately 1000 metres

**Fig. 19.10.** Stratigraphic succession, Southern North Sea (based on Department of Energy, 1985).

The main gasfields to date (Fig. 19.5) lie on a belt containing the Sole Pit High (Figs. 15.12, 15.16 and 19.9) north-east of East Anglia. The Sole Pit Field (Glennie & Boegner 1981) is a classic example of an area of thick basinal deposition which, due to a change of tectonic regime, became an area of uplift, forming a structural high. (The phenomenon has become known as 'inversion', an unfortunate term which puzzled all but the oilfield *cognoscenti* when introduced into the North Sea literature!)

A further striking feature is the evidence of post-depositional lateral and upward flowage of halites by halokinesis and/or halotectonism, demonstrated in text-book style (see e.g. Jenyon 1986) in the North Sea (Figs. 15.12, 15.17). The flowage of the Zechstein halites not only affects the structures but allows them to seal effectively faults, for instance, along which hydrocarbons might escape.

Brooks *et al.* (1986) contains descriptions of many fields by Bifani, R. (Esmond Gas Complex), Conway, S. (Victor), Goodchild, M. W. and Bryant, P. (Rough), ten Have, A. and Hiller, A. (Sean, Block 49/

25a, 15 km SE of Indefatigable). All these authors emphasised that while the Rotliegend sandstones are dominantly aeolian, they display facies changes and interdune sediments (e.g. wadi, sabkha, lacustrine) the distribution of which affects porosity/permeability values. Berry and Wilson (1984) gave a useful summary of typical reservoir properties.

The Hewett Field contains two gas-bearing sandstones in the Triassic Bacton Group (Fig. 19.9), the lowermost sand (Hewett) being a facies of the so-called Bunter Shale. The higher sand (Bunter) is capped by shales and evaporites of the Hailsborough Group. According to Berry & Wilson (1984), no accumulations of gas have been found in post-Triassic rocks. The Carboniferous sandstones are possible reservoirs but their great depth may have rendered their porosity and permeability unsuitable. Nevertheless, deepening of some of the existing wells may yet yield more methane from underlying coals.

Oil is notable for its absence, most probably because the Kimmeridge Clay and other Jurassic source-rocks of the northern oilfields are not buried deeply enough

**Fig. 19.11.** Production Platform 'B', West Sole Gasfield, diving support vessel 'Ragno Due' alongside. (Photograph by British Petroleum.)

to reach maturation level. Conversely the absence of dry gas north of the Mid North Sea High can be accounted for by the absence of the Westphalian coal measures under most of that region.

Figure 19.11 shows a typical North Sea gas production platform.

## Other areas of Continental Shelf

As can be seen from Table 19.4, exploration has also taken (and is taking) place off the west coast of England and Wales and in the English Channel. Penn *et al.* (1987) discussed English Channel prospects. A significant oil discovery offshore, some 10 km east of Wytch Farm was made by B.P. in 1987. Two further holes offshore and an appraisal well drilled in 1988 also produced encouraging results.

So far, the only area where a gas field is under development is the Irish Sea Basin (Fig. 19.5) (see Jackson *et al.* 1987) where the South Morecambe Field (Fig. 19.5) was discovered in 1974 about 40 km west of Blackpool. The reservoir rocks (Bushell 1986) are Triassic sandstones and are overlain by argillaceous and saliferous beds of the Mercia Mudstone group which form the seal (Ebbern 1981). The field is about 25 km long, approximately N–S, and is about 9·5 km E–W wide at its widest point and is the second largest so far discovered on the UK Continental Shelf. It is essentially a faulted anticline, the faults trending N–S, NW–SE and NE–SW. Production of up to 34 M cm/d is forecast.

Drilling has also taken place in the Cardigan Bay (or St George's Channel) Basin (Figs. 15.1, 15.4) and in the Western Approaches Basin (Fig. 19.9) but so far no commercial discoveries have been announced.

## Reserves

Cumulative gas production at the end of 1989 for the UK continental shelf reached 706 bcm. Remaining recoverable reserves are estimated as being between 585–1,770 bcm (Dept. of Energy 1990). Recoverable reserves of gas, as yet undiscovered, in the Southern North Sea and Irish Sea basins could add another 155–755 bcm (Dept. of Energy 1990). Estimates of reserves of recoverable onshore oil in England and Wales are not readily available but undiscovered reserves in central and southern England are estimated as being between 15–75 mt.

# Mineral Deposits

## Ores

The history of metal mining in England and Wales is a long one, stretching back to pre-Roman times. Until the late 19th century, the two countries were virtually

self-sufficient and indeed had been leading world producers in iron, tin, copper and lead ores. The picture is very different today (see p. 626, Table 19.11).

During the past decade and a half, however, interest in the geology of Britain's ore deposits has increased dramatically with an exponential rise in the number of meetings, conferences and publications devoted to the subject after many years of neglect (especially, with a few notable exceptions, by the universities). The British Geological Survey has played an important part in stimulating this interest, particularly through its Mineral Reconnaissance Programme financed by the Department of Trade and Industry commencing in 1973, and subsequently also by the EEC in 1979. The Mineral Exploration and Investment Grants Act (1972) encouraged industry to prospect by allowing 35 per cent subsidies of exploration programmes. Not as much exploration has been undertaken as was hoped, however, due to a combination of the problem of establishing mineral rights, the inevitable protests from the environmentalist lobby, and the world recession and consequent glut of most mineral commodities. Only about £6·05 million of the £50 million allocated to the scheme had been used by 1986/87, for instance, when the last grants were paid.

Comprehensive accounts of the economic minerals of England and Wales are to be found in the *Special Reports on the Mineral Resources of Great Britain, Reports and Mineral Reconnaissance Programme Reports, Economic and District Memoirs* of the British Geological Survey and other publications of that body (see especially Colman 1990). Other important publications include those by the Institution of Mining and Metallurgy (e.g. I.M.M. 1959 and their Transactions), Thomas (1961), Lewis (1967), Dunham *et al.* (1979), and Highley *et al.* (1980). Because of space limitations, however, only selected areas and deposits (Fig. 19.12) will be discussed here.

## South-west England

### Cornwall and Devon

Cornwall and Devon contain the most discussed ore deposits of England and Wales. Cassiterite, from alluvial deposits, is said to have been exploited, along with copper, since the Bronze Age. The Romans are thought to have introduced underground mining of some of the vein minerals and, much later, silver was produced from galena in the 13th century. Tin output increased in the 17th century and the Industrial Revolution saw haematite-bearing veins being mined for the South Wales foundries. An estimate of the total output of major metals from the area was made by Dunham *et al.* (1979) and is given in Table 19.7.

Most of the tin and copper were produced from around Camborne, Redruth, St Just-in-Penwith, St

Austell, between St Ives and Heston, north of Callington and around Tavistock (Fig. 19.13). By the mid-19th century, over 400 mines were active (Edmonds *et al.* 1969) with the annual output of copper reaching its peak around 1860 at over 15,000 tonnes of metal and tin its peak in 1871 with over 10,900 tonnes of metal. In both cases, the output represented half of that of the world's. From then on, however, overseas competition meant a decline in production which has continued until the present day with pauses during the two World Wars and during the spectacular rise in the price of tin through the 1970s until 1985. In October of that year

**Fig. 19.12.** Locality map of mineral occurrences mentioned in text.

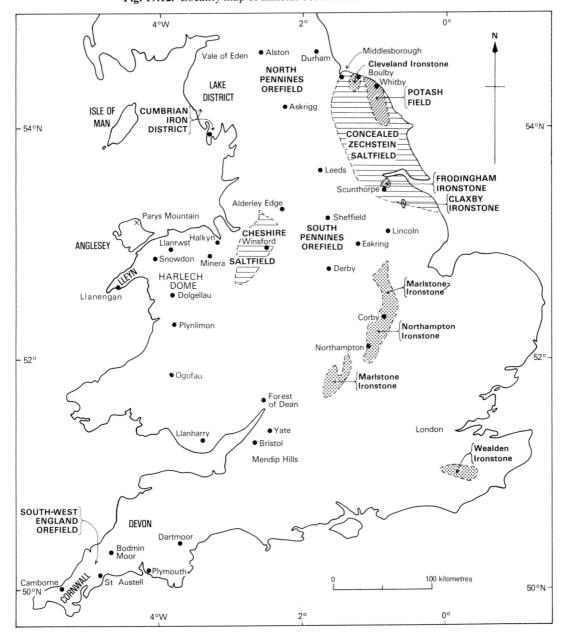

**Table 19.7.** Estimated output of metals and ores from
SW England to date (from Dunham *et al.* 1979)

|  | tonnes |
|---|---|
| Tin metal | 2,500,000 |
| Copper metal | 2,000,000 |
| Lead metal | 250,000 |
| Zinc metal | 25,000 |
| Arsenic (As$_2$O$_3$) | 250,000 |
| Tungsten (WO$_3$) | 5,600 |
| Uranium ore | 2,000 |
| Silver (from lead) | 233 |
| Silver ore | 2,000 |
| Iron ore (haematite & siderite) | 2,000,000 |
| Manganese ore | 100,000 |
| Baryte | 500,000 |
| Fluorite | 10,000 |
| Antimony ores | 1,000 |
| Pyrite | 150,000 |
| Cobalt & nickel ores | 500 |

the International Tin Council could no longer afford
to continue its policy of intervention purchasing and
the price of tin halved, virtually overnight. The
Cornish mines once more became uneconomic and
only Government aid in 1986 enabled two mines to
continue (Wheal Jane, near Truro and South Crofty,
between Camborne and Redruth (see Fig. 19.13)).

The geological literature of the area is vast (see Halls
*et al.* 1985) for a bibliography. Dines (1956) and
Hosking and Shrimpton (1966) are essential starting
points for an introduction to the history and geology
of the mines and ore deposits. Advances in radiometric
age-dating and analytical geochemistry in the past
twenty years have changed views on the metallogenesis
radically (see e.g. Dunham *et al.* 1979, Beer &
Scrivenor 1986, Thorne & Edwards 1985, Stone &
Exley 1986). Embrey and Symes (1987) provided a
most useful introduction to a complex subject.

Mineralisation in the area is spatially and (?)gen-
etically related to the intrusion of a large Hercynian
granite batholith (p. 507), the surface expression of
which is well shown by the outcrop of five major
cupolas (Fig. 19.13). The ore deposits take various
forms, sometimes occurring within the granite,
sometimes in the country rocks. These latter are folded
and faulted Lower Devonian to Upper Carboniferous
slates, quartzites and minor volcanics, the slates being
known in the mining areas as *Killas*. Acid dykes and
sills, known as *elvans*, cut both granite and country
rock.

*Lodes* (Cornish term for veins), which cut the
elvans, generally occupy the planes of normal faults
which have often affected both granite and country
rock and which have a history of repeated movements.
Strikes vary regionally, are sometimes parallel to

**Fig. 19.13.** Distribution of mineral deposits associated with Cornubian granites (after
Moore 1982).

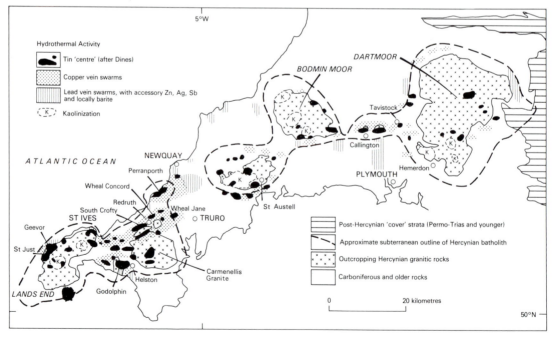

granite joints, and with certain notable exceptions dips are usually steep (over 60°) but can reverse orientation when traced downwards. The ore can fill fault fissures or replace and re-cement fault-breccias. Minerals can include cassiterite, wolframite, pyrite, chalcopyrite, arsenopyrite, galena, sphalerite, stannite, haematite and pitchblende which occur in different quantities and combinations, as do the gangue minerals which include quartz, mica, carbonates, fluorite, tourmaline, chlorite and kaolinite. Altered wall rock [sericitised, greisenised (changed to quartz and muscovite), kaolinised, tourmalinised] is common.

*Stockworks* and greisen-bordered vein swarms involve considerable areas of mineralised rock. Intricate networks of quartz veinlets with accompanying mineralisation may occur in the country rocks above granite or, occasionally, in the granite cupola itself, forming stockworks. Sub-parallel quartz veinlets following local lode directions and bordered by greisen can affect large areas of granite.

*Pipes* of ore have been recorded, for example, from the Carnmenellis granite, one in particular (at East Wheal Lovell mine) some 4 m in diameter having a vertical dimension of 130 m. It consisted of altered granite, quartz, muscovite, cassiterite with sulphides and fluorite. *Carbonas* are similar to pipes, but smaller and less regular.

### Mining of tin and copper

*Wheal Jane*, about 7 km NE of the outcrop of the Carnmenellis granite (see Fig. 19.13), was reopened in 1971 'after a prospecting programme said to have involved as many legal advisors to sort out the mineral rights as it did geologists to prove up the ore reserves' (*Mining Magazine*, November, 1982). The Wheal Jane/Mount Wellington lode system lies within the killas surrounding the Carnmenellis granite and in the main strikes WSW–ENE. The 'B' lode lies immediately below a 16 m thick 'elvan' dyke dipping about 35° WNW then splits beneath 7 and 8 levels (200 m below the surface). Two separate lodes, an upper North Lode (dip 22°) follows the base of the elvan while a lower South Lode (dip 58°) continues downward separately. By 1982, mining was being carried out at a depth of 468 m. In 1988 (Shrimpton 1989) 197,293 t of ore yielded concentrates of Sn (1,278·0 t), Zn (5,502·4 t), Cu (732·1 t) and Ag (1,429 kg). The low price of tin and the cessation of a government subsidy resulted in closure in February 1991.

*South Crofty* lies between Camborne and Redruth and actually contains along a strike length of 4½ km twelve old copper mines which reached the tin zone at depth. (Dolcoath, opened in 1720, had reached a depth of 1,000 m when it closed in 1920.) The workings lie within the Carn Brea granite in ENE trending veins which may dip northwards or southwards at 50° or more. The upper portions of the veins carry chalcopyrite, galena and sphalerite while those currently worked contain cassiterite, arsenopyrite and wolframite along with quartz chlorite and tourmaline. The South Crofty mine in 1988 treated 159,414 t of ore to produce 1,675·6 t of Sn concentrates (Shrimpton *op. cit.*). Mining was continuing in 1991.

*Geevor Mine* near St Just, about 12 km north of Land's End, works lodes carrying cassiterite, some sulphides, quartz, tourmaline and siliceous haematite, which cut both the granite and country rocks. Over thirty lodes have been exploited since 1791 (and probably earlier), their strikes varying between NW–SE and WNW–ESE, their dips generally to the NW or SE being steep and sometimes changing from one direction to the other when followed downwards. Workings from the main Victory Shaft (480 m deep) extend deeper below the sea. The mine, not granted Government aid, closed for a time but re-opened at the end of 1986 hoisting broken ore from the stopes and, during part of 1987, also from pillars, producing 204 t of Sn concentrates. Normal mining was resumed in 1988 on a small scale producing 334 t of tin concentrates by October 1988 (Shrimpton (1989)) but ceased in 1990.

### Other mined minerals

*Iron*, in the form of limonite and haematite and frequently derived from siderite, was produced for the South Wales foundries in the 19th century from mines in north Somerset (where the veins tend to follow the WNW structural grain of the country rocks) and from the Great Perran Lode, an E–W vein west of the St Austell granite near Perranporth, north Cornwall.

*Tungsten*, was formerly worked from veins carrying cassiterite (e.g. at Castle-an-Dinas) and sometimes arsenic (e.g. Agar and Cligga). At Hemerdon, near Plymouth, south Devon, a portion of a granite cupola has been turned into a stockwork carrying wolframite, cassiterite, arsenopyrite and copper and iron sulphides. (Reserves of 42 million tonnes of ore averaging 0·1 per cent $WO_3$ and 0·029 Sn have been proved, but environmentalists' objections to an open pit operation have delayed exploitation.)

Beer and Ball (1986) considered that the enrichment of tin and tungsten in granite aureoles was evidence of their early introduction before the main fissure-filling mineralisation.

*Uranium* minerals (pitchblende, uraninite and various secondary minerals such as autunite and torbernite) were known in SW England early in the 19th century and are generally found in 'cross-courses' along with iron, cobalt, nickel and bismuth minerals, though on occasions (e.g. at Geevor and South Crofty) they occur with tin ore. Uranium ore has been produced from Wheal Trenwith and South Terras mines but despite intensive prospecting (see e.g. Bowie *et al.* 1973) potential economic discoveries have been few. Recent work by the British Geological Survey, however, has discovered a mineralised body NW of the St Austell granite containing 600–1,000 tonnes of uranium at 0·2 per cent U (Institute of Geological Sciences 1982, p. 31).

Finally, of great economic importance, there occur in the Cornubian batholith large areas of kaolinised granite (Fig. 19.12) which are worked for china clay, (see p. 625).

### Origin of ores

During the development of underground mining in and around the granites, it was realised that differences in mineralogy occurred when lodes were traced laterally and vertically and thus Cornwall was an early example of the concept of mineral zoning (which, in essence, envisages minerals precipitating out of solution according to their solubilities, the solutions cooling while moving away from their hot source and thus carrying the lower-temperature minerals farther). In its simplest, earliest form, the concept received support, for example, from the fact that lodes worked for copper minerals in or near the granites tended to have economic tin values at depth. If traced away from the granite, copper-bearing lodes tended to change into predominantly lead-zinc carriers. The zonation is not, however, distributed uniformly around existing granite outcrops or indeed the batholith, because of the area's complex structural and mineralising history and the fact that relative concentrations, changes of pressure, reactions with country rock, etc. play a part in determining order of precipitation of minerals. The faulting and displacement of lodes by later ones (e.g. the *crosscourses* which cut the lodes of the dominant regional direction approximately at right angles) point to episodic injections of mineralising solutions and this is confirmed by age-dating.

**Table 19.8.** Generalised paragenetic sequence of ore and gangue minerals in south-west England (reproduced, with permission of the I.M.M., from Stone & Exley 1986)

| | GREISEN VEINS | HYPOTHERMAL | MESOTHERMAL | EPITHERMAL |
|---|---|---|---|---|
| Temperature (°C) of fluid inclusions | | 500–250 | 350–150 | <150> |
| Ore minerals | | Arsenopyrite ——— Wolframite ——— Cassiterite ——— Molybdenite — Hematite ——— Scheelite — Stannite — Sphalerite — | Chalcopyrite — Pyrite — Pitchblende/Uraninite — Nickeline — Smaltite — Cobaltite — Bismuthinite — Argentite — | Galena — Tetrahedrite — Bournonite — Jamesonite — Stibnite — Siderite — Hematite — Marcasite — |
| Gangue minerals | Quartz — Feldspar — Muscovite — Tourmaline — Chlorite — | Fluorite — Hematite — | Chalcedony — Baryte, dolomite, calcite — | |
| Economically important elements | Arsenic — Tungsten — Tin — | Copper — | Uranium — Nickel — Cobalt — Bismuth — | Zinc — Silver — Lead — Iron — Antimony — |
| Typical emplacement type | Sheeted veins, stockworks, fault-related fractures | Main lodes, caunter lodes, fault-related veins, breccias, stockworks and carbonas | Lodes, caunter lodes, faults, cross-courses | Mainly cross-courses and faults |
| Alteration | Greisenization — Tourmalinization — Silicification — | Feldspathization — Chloritization — Hematization — | | |

Dines (1956) suggested that there must have been many (more than 60) 'emanative centres' scattered throughout the batholith (probably related to high points in an irregular roof) from which hydrothermal ore-fluids moved outwards. Hosking (1966, 1969) described the paragenetic sequence of ore and gangue minerals and Stone & Exley (1985) developed this further (see Table 19.8). Hosking (*op. cit.*) suggested that the presence of different emanative centres within the batholith could be related to the structural trends of the region, which would have determined the orientation of the roof rocks of the batholith both before and after intrusion.

While there is no dispute that the mineralisation was produced from hydrothermal solutions, arguments continue regarding the possible source of the solutions and their relationship to the granites (see e.g. Jackson *et al.* 1982, Darbyshire & Shepherd 1985 for discussions). Opinion seems to have hardened around 280–290 Ma for the period during which granites were intruded but emphasis has been laid on the different ages of the phases of mineralisation, whether associated with an individual granite or from granite to granite (see e.g. Halliday 1980), and the phen-

omenon of 'overprinting' (Thorne & Edwards 1985). Dates of uranium mineralisation range, for example, from 300 Ma to 45 Ma (Dunham *et al.* 1979); the main phase of mineralisation has been calculated as 270–278 Ma (Stone 1982); 220 Ma, 165 Ma and 75 Ma are cited for post-main phase episodes (Jackson *et al.* 1982); and 296 Ma for mineralisation in the Exmoor area (Ineson *et al.* 1977). Darnley (1986) emphasised the importance of taking account of the fact that mineralisation occurred over a period of 200 My.

Jackson *et al.* (1982) in their study of the NW flank of the Dartmoor granite concluded that the granite was produced by anatexis of pelitic-rich crustal rocks richer than normal in certain trace elements. Darbyshire & Shepherd (1985) agreed but emphasised that the various granites had differing chemical characteristics as did Stone and Exley (*op. cit.*) and Thorne and Edwards (*op. cit.*), implying the source rocks were not always identical nor was their chemical evolution. Moore (1982) emphasised that any zonation discerned in particular areas was a local phenomenon of individual vein swarms and due to episodic and prolonged mineralisation which differed in time and type from area to area.

**Fig. 19.14.** Possible convection systems to account for mineralisation at (A) SW England (1 = Sn, W, Cu, Zn, Pb 'emanative centres' of Dines 1956; 2 = Pb, Ag, Ba, U 'cross-course' mineralisation; 3 = Kaolin deposits; PTLS = Permo-Triassic land surface; MLS = modern land surface). (B) Northern England (1 = W, Cu, Pb, Zn, Ba mineralisation in granites and Lower Palaeozoic rocks; 2 = F, Zn, Pb, Ba, Fe (haematite) in Carboniferous rocks. CLS = Caledonian land surface; MLS = modern land surface) (from Moore 1982).

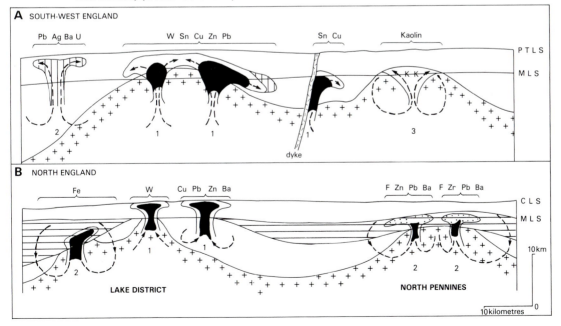

Current views favour convective circulation of mineralised hydrothermal fluids. Their origin is partly meteoric, downward flows scavenging metals from the country rocks and partly juvenile, upward flows receiving a contribution from magmatic metal-bearing waters (Fig. 19.14). Both magmatic and radiogenic heat from the batholith would produce the convective cells that developed from time to time in various parts of the batholith, depending on pressure/temperature changes brought about by tectonic processes and/or erosion of roof-rocks (see, especially, Evans (1982), I.M.M. (1985), Darnley (1986), Stone & Exley (1986), Shepherd & Scrivener (1987), Sams & Thomas-Betts (1988)).

*Marine and alluvial deposits*

The possibility of dredging of the sea-bed sands off the Cornish coasts for cassiterite attracted attention in the late 1960s. Pilot scale operations, involving the processing of 130,000 t of sand between 1979 and 1982, revealed the presence 0·8 km offshore of a sand-layer up to 1·5 m thick in a band 0·8 km wide running parallel to the coast in the St Ives area. The cassiterite present is considered to be derived from long-shore drifting of material both from rivers eroding old mine dumps and from erosion of mineral veins, etc. in the cliffs. Feasibility studies on some 20 mt of potentially economic sand indicate the possibility of a 20 per cent tin concentrate at a grade of 0·15 per cent Sn. Installation of a floating mooring, a 1,000 m floating pipeline and a concentrator near Hayle commenced in 1984.

While marine dredging of cassiterite is new, it should be remembered that alluvial gravels bearing tin in the streams and rivers draining Cornwall and Devon have attracted attention from the Bronze Age onwards. Most of the alluvial tin had been won by the 15th century from the Marazion, Carnon, Par and Pentewan rivers and from marshes such as Goss Moor west of Bodmin. Collins (1975) considered that *gold* in the Cornish tin streams was probably known to the Romans and perhaps even to the Phoenicians and the Iberians. In mediaeval times, Devon was the most important area of gold production in England, some 300–400 miners being employed at Combe Martin (north Devon) during the 13th–14th centuries where the silver in galena from veins in the local slates was also sought, as it was in veins in other areas (e.g. in the Teign Valley, east of Dartmoor). The source of the alluvial gold is considered to be the iron-rich 'gossans' of leached residual material left near the surface after weathering of, say, copper-lead-zinc lodes. Sometimes the gossans themselves were washed for gold as, for example, at the Poltimore mine in the North Molton area of north Devon where copper-bearing veins occurred.

## Somerset and Gloucester

Haematite ores were worked sporadically from Roman times until the beginning of this century in the Forest of Dean (see e.g. Trotter 1942, Kellaway & Welch 1948, Dunham *et al.* 1979). They occur as irregular pockets and veins as replacement deposits in Dinantian-age carbonates (Crease Limestone and Lower Dolomite) and to a lesser extent sandstone (Drybrook Sandstone). Replacement of these rocks by iron-bearing solutions derived from the iron carbonates and sulphides present in overlying Coal Measures shales, intensively weathered during Permo-Trias times, is thought to be responsible for these ores.

Carboniferous Limestone in the Mendip Hills is also the host rock, along with Trias basal conglomerates, for NW to WNW veins in wrench-faults carrying galena, calamine ($ZnCO_3$), calcite and occasionally baryte (Green 1958, Green & Welch 1965, Ford 1976, Ixer 1986). Again, lead mining took place in Roman times though its peak was during the 17th century (the practice of making lead shot by dropping molten lead from a height into water was invented in Bristol). Zinc mining did not start until the 16th century and ceased at the end of the 19th century. As Ixer (*op. cit.*) pointed out there is a paucity of *in situ* evidence to enable much to be said as to the origin of the ores. The field has Mississippi-Valley type characteristics though fluorite is virtually absent.

At Yate in Somerset (Avon) there occurs an important deposit of celestite ($SrSO_4$) which, originally used in pyrotechnics, is now important in the manufacture of colour TV tubes and ferrite magnets. It occurs as stratiform lenses or in nodular beds in the Triassic Mercia Mudstone Group (formerly Keuper Marls). The deposits of celestine are secondary and thought to be redistributed material formally precipitated as an evaporite along with gypsum (see Thomas 1973, and Dunham *et al.* 1979). In 1988, 25,553 t of celestite was produced, Britain being the world's major producer.

## Pennines

Along with copper and tin, the mining of lead ores in England and Wales was also at its peak in the 19th century. From 1850–77 annual lead production, mainly from the Pennine fields (and North Wales) averaged 60,000 t p.y. and sometimes reached 75,000 t (Highley *et al.* 1980).

Lead mining was carried out by the Romans and records indicate virtually continuous mining from the 12th century to the present day, though recently only gangue minerals (fluorite and baryte) have been worked. (See Table 19.11 and B.G.S. (1990) for production details.)

Divided into the Northern and Southern Pennine Orefields (Fig. 19.12) the northern field comprises two distinct tectonic units, traditionally known as the Alston and Askrigg Blocks (see Ch. 7, Fig. 7.7). Here epigenetic Pb-Zn-F-Ba mineralisation occurs in sandstones and limestones of Carboniferous (Asbian–Pendleian) age. The Southern Pennine Orefield contains similar deposits, mainly in the Dinantian limestones of the Derbyshire 'Dome'.

The ore-shoots are mainly in near-vertical veins but can also occur as pipes and 'flats' (i.e. replacements of limestone beds). The minerals include galena, sphalerite, with minor pyrite and marcasite and, occasionally, chalcopyrite. Gangue minerals consist, in varying proportions, of fluorite, baryte, calcite, dolomite and quartz. Ecton in Derbyshire, famous because German miners in 1636 introduced drilling and blasting to British mining, is also notable for having had a copper mine working a brecciated pipe-like body in limestone, containing galena and sphalerite near surface which gave way to chalcopyrite and bornite at depth.

## Northern Pennines

Background information on the geology and mining is to be found in I.M.M. (1959), Dunham (1948,

1952) and Dunham and Wilson (1985). The near-vertical normal faults containing the mineral veins have distinct orientations, those of the Alston block having trends averaging either N65°E (and are galena-rich) or N80°W (fluorite-rich), while in the Askrigg block the directions are mainly either WNW or W–E. There is a distinct zonal pattern of the minerals, the veins in the central part of the Alston block, for instance, containing fluorite as a predominant gangue mineral. Outer concentric zones contain instead baryte, witherite or calcite in its place (Fig. 19.15). Fluid inclusion studies (e.g. Sawkins 1966, Shepherd et al. 1982) indicated the zoning was temperature-controlled, falling from 220°C in the centre of the fluorite zone to perhaps as low as 60°C on the margins. In the Askrigg block, again fluorite is mainly confined to inner areas, but reappears on the southern edge of the orefield.

The zoning in the northern Pennines and the occasional presence of copper minerals at depth led to the suggestion that the source of the fluids was a granite (cf. Cornwall) at depth. Geophysical evidence supported the likely presence of a granite beneath the Alston Block and a borehole at Rookhope (County Durham) between 1960 and 1961 proved its presence.

**Fig. 19.15.** Orefields of Northern England (Moore 1982).

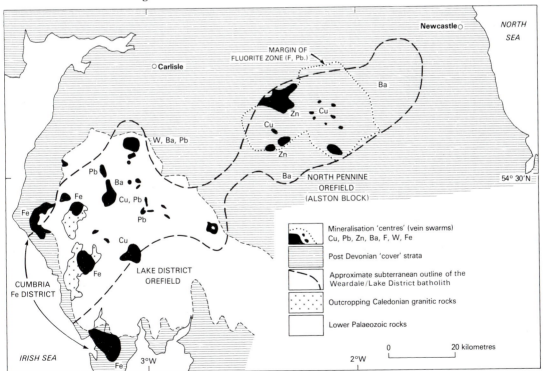

Surprisingly, however, the granite, the Weardale Granite (Fig. 19.15), was *pre-* not *post*-Carboniferous, the contact between it and the Carboniferous Limestone being unconformable (Dunham *et al.* 1965). It appears now that a large granite batholith underlies the Alston–Weardale area and extends westwards below the Lake District (Firman & Lee 1986) where outcrops of granite exist (Fig. 19.14). In 1973, again on the basis of geophysical evidence, drilling at Rayhope, in the Askrigg Block, also revealed a granite unconformably below the beds containing the ore deposits (Dunham 1974). Although the granites are slightly different geochemically and petrologically they both give an age of *c.* 390 ± Ma. Mineralisation cutting the Whin Sill (which is dated at *c.* 296 Ma – see Ch. 16) confirms the granite/mineralisation age relationships.

The origin of the ore-forming solutions has therefore caused considerable debate and much time and effort has been spent on radiometric-dating, fluid-inclusion studies, trace-element analyses etc. Dunham (1983, 1986, 1988), Ixer (1986) and Brown *et al.* (1987–88) contain extensive bibliographies and detailed discussions and are therefore excellent introductions to a continually growing literature.

Opinion favours an origin of the ore-fluids from brines (forced out of the over-pressurised shale basins flanking the Blocks), which collected metals as they moved. Hercynian tectonism provided the faults which formed the main channelways for the fluids and assisted in initiating the convective flow of the fluids (?seismic pumping) towards the batholiths. The granites were areas of high heat flow due to their high thermal conductivity and their radiogenic heat. Metals were collected both from the sediments through which the fluids passed and from the granites themselves, geochemical differences in the basins and batholiths and pressure/temperature differences in the two blocks accounting for variations in mineralogy in and between the two areas.

Differences in detail remain (see Brown *et al.* 1987 and subsequent discussions 1987–88) but in view of past analytical ambiguities that occur (for example in age-dating and fluid-inclusion studies) Moore's (1982) conclusion is perhaps still apposite: 'Current theory suggests a composite origin for the ore minerals is most likely, involving barium, lead, zinc, copper, iron and fluorine derived from Carboniferous sediments and Lower Palaeozoic slates and granites, perhaps in the manner suggested by Russell (1978). Sulphur may have originated from sea water, perhaps through evaporites.'

## Southern Pennines

Mining in Derbyshire also has a long history, dating from Roman times (see e.g. O'Neal 1956 and Varvill

1959). Over 4,000 disused mines have been recorded in the area (Blunden 1975), which has a unique set of mining laws (O'Neal *op. cit.*). The origin of the ores has been discussed at length, both with those of the Northern Pennines and separately (Dunham 1983, 1986; Smith *et al.* 1985; Mostaghel 1985 and Mostaghel & Ford 1986 summarise views and are excellent sources for an extensive literature). Mineralisation is similar both in content and form of the deposits to that of the Northern Pennines, the veins ('rakes') striking generally ENE or NW, the host rocks being Carboniferous limestones of Brigantian and Asbian age. Interbedded basalt lavas ('toadstones') near the top of the succession and the overlying Edale Shales can act as barriers to the rising hydrothermal solutions producing flats. Zoning on a general scale is present in that fluorite is abundant in the east while baryte-calcite is more prominent westwards. [Fluorite has been the main mineral exploited (Table 19.11) for some time, from old dumps, from pipes, flats and limestone replacement bodies (Dunham 1952, Smith *et al.* 1967, Butcher and Hedges 1987).] Galena is also more abundant in the east and mineralisation in the Carboniferous Limestone anticlines of Crich and Ashover, which lie to the east of the main Pennine outcrop, coupled with the presence of fluorite in limestones encountered in oil drillings farther east of Eakring (Fig. 19.6), have been taken to indicate an easterly source of the mineralising fluids.

The orefield differs from the Northern Pennines in that no granite has been proved beneath the field. Smith *et al.* (1985) calculated that Derbyshire Dinantian beds would be under a cover of 2 km of Silesian sediments by the end of the Carboniferous. They suggested that compaction of the marine Edale Shales (Silesian) would expel metalliferous saline formation waters, which, moving to areas of low pressure, perhaps on contact with evaporite sulphates associated with the limestones would eventually precipitate metallic sulphides, etc. Mostaghel and Ford (1986) favoured similar processes emphasising that neighbouring Carboniferous sedimentary basins, in general on compaction, would expel heated saline connate brines which would derive metals from the country rocks *en route* to areas of lowest pressure such as the Derbyshire Dome. (See also Plant and Jones 1989.)

## North-west England

NW England contains numerous steeply dipping veins of mainly non-ferrous metals within Lower Palaeozoic rocks of the Lake District proper and also haematite deposits which replace Carboniferous Limestone in West Cumbria (Fig. 19.15). Eastwood (1959), Ineson (1977), Firman (1978), and Dunham *et al.* (1979) summarised much of the wealth of data on the history and occurrence of the deposits that were once mined in the area. During the 16th century, it was Britain's leading copper producer and in the 19th century the main source of haematite ore. Other metals produced included tungsten, lead, zinc, silver, manganese, antimony, arsenic, cobalt and nickel.

## Lake District

There are two major types of mineralisation in the Lake District, according to Stanley & Vaughan (1982). The first, mainly chalcopyrite-pyrite-arsenopyrite, in faults associated with the Caledonian orogeny, shows a clear relationship to the Lake District granite batholith (Fig. 19.15), is of Lower Devonian (Caledonian) age and, in the vicinity of outcrops of the granite, also displays tungsten or molybdenum mineralisation. The Carrock Fell tungsten deposit is of particular interest because it is the only deposit outside of Cornwall/Devon that has ever produced tungsten

commercially. It lies on the northern margin of the Skiddaw granite where late-stage metasomatism has 'greisenised' the granite. A series of N–S quartz veins in the zone cutting both granite and country rock contain wolframite and scheelite along with a variety of other minerals. Shepherd et al. (1976) from isotopic evidence suggested the tungsten ores were formed soon after the consolidation of the granite from saline mixed magmatic and circulating ground waters. Ball et al. (1985) added further details on the geochemistry.

The second type of mineralisation, mainly galena and sphalerite, is dated as being of Lower Carboniferous age, with early Carboniferous tensional tectonic activity having produced NE–SW or N–S fault-fissures. A later minor mineralisation of Upper Carboniferous-Permian age produced baryte. Moore (1982) pointed out that there were areas of intensive mineralisation with unusually high numbers of vein 'swarms' (Figs. 19.15, 19.16) and that hydrothermal activity was episodic. Ineson & Mitchell (1974) recog-

**Fig. 19.16.** Provinces of high heat flow (greater than 60 mWm$^{-2}$) in England and important mineralised areas (adapted from Brown et al. 1980).

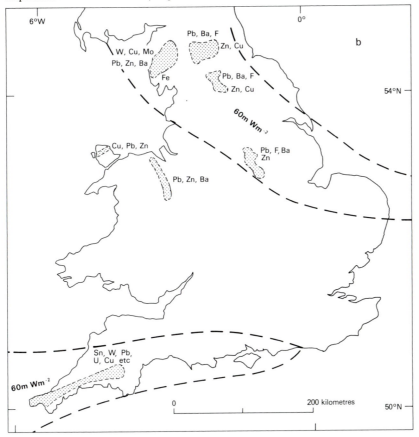

nised from K-Ar studies a series of (hydro)thermal events ranging from 390 Ma (Caledonian) to 180 Ma (Mesozoic).

Stanley & Vaughan (1982) considered, from the evidence of fluid-inclusion studies and experimental data on mineral assemblages, that the source of the mineralising fluids forming the chalcopyrite-pyrite-arsenopyrite suite could have been the Borrowdale Volcanic Group, with some leaching of the granites also having taken place. The high-salinity of the fluids associated with the galena-sphalerite mineralisation suggested that sea water (?Carboniferous) could have been involved in the leaching of the metals from the basement rock. Heat associated with the batholith presumably provided the necessary convective system for both mineral suites (Fig. 19.14).

O'Brien et al. (1985) found that their geochemical studies showed little relationship between the composition, alteration and mineralisation of the various granite outcrops, but also emphasised the importance of the high heat-flow produced by the deeper Lake District batholith in driving the hydrothermal fluids responsible for sporadic and episodic mineralisation. Studies of the Crummock Water aureole, an ENE-elongated zone of bleached, metasomatised Skiddaw Group sediments indicated leaching of metals which was attributed to a hidden portion of the Lake District batholith and therefore pointed to the Group as a likely source of the Lake District ores (Cooper et al. 1988).

## West Cumbria

References to the mining of haematite at Egremont in Cumbria go back to the 12th century (Postlethwaite 1913). In 1988 this was the one remaining area producing iron ore in the United Kingdom. Mining of the Cumbrian ore, most of which occurs as a replacement deposit in Carboniferous Limestone, was greatest during the 18th and 19th centuries, rising to a peak of nearly 3 mt per annum in the 1880s (Firman 1978). In 1988, 2,100 tonnes of haematite (54 per cent Fe) was produced (Shrimpton 1989).

The haematite occurs in the form of veins (occasionally in the Skiddaw Slates) or vein-like bodies in faults, but mainly as tabular bodies ('flats') in limestone beds, or sometimes as *sops* – inverted conical bodies with irregular margins. A replacement origin (Rose & Dunham 1977, Dunham et al. 1979, Firman 1978, Moore 1982) is favoured with hypersaline brines having leached iron from Permo-Trias and/or Lower Palaeozoic rocks. While originally overlying Permo-Trias sediments were favoured as source rocks, it is now considered that the brines could have travelled up dip from buried Permo-Trias in the Irish Sea, for example, or because of the anomalous amount of arsenic in the ores from Lower Palaeozoic slates or

granite (Dunham 1984). Some of the sop deposits contain foundered Triassic blocks which favour the mineralisation being post-Triassic in age, though Evans and El-Nikhely (1982), on the basis of palaeomagnetic measurements on the haematite, considered a Permian date more likely.

## Isle of Man

Gaciri & Ineson (1981) gave a brief resumé of the literature on Isle of Man ore deposits. Veins (mainly fault-fillings) carrying Pb, Zn and Cu sulphides with quartz, calcite, fluorite, baryte and dolomite occur within the Lower Palaeozoic Manx Slates (p. 123). They have been worked at Foxdale, Laxey, Ballacorkish and Kirk Michael. Small granitic intrusions occur at Foxdale and Laxey but no obvious genetic connections can be inferred from this. Mining dates back to the 13th century (Dunham et al. 1979), the lead ores being notable for their high silver content, but little mining has taken place since before the Second World War.

## Central and South-east England

The area of England south of the Pennines and north-east of the Mendips is notably lacking in ore deposits other than the bedded ironstones of Carboniferous, Jurassic and Cretaceous age. Only in Cheshire and Staffordshire, where the Triassic sandstones (e.g. Alderley Edge, Cheshire) contain minor amounts of Cu, Co, Mn, Pb, Ni, V and Ba (see Holmes et al. 1983), and in Shropshire, where Pb-Zn veins with Ba and F occur in pre-Cambrian and Ordovician rocks (Dines in Anon. 1959, Ineson & Mitchell 1975), have any deposits of significance other than iron been discovered. Carboniferous ironstones (see p. 597), however, were worked intermittently in the coalfields of the Midlands from the 12th century onwards, exploitation of them petering out this century; the last area where they were worked (until 1949) being Staffordshire.

The recorded history of iron-ore mining in Britain is a long one (see e.g. various Memoirs and Economic Memoirs of the Geological Survey of Great Britain) and is characterised until the 16th century by workings wherever local ore and charcoal were available. From the 14th century onwards, however, the production of iron did tend to be centred on certain areas, particularly in the Weald of Sussex in the south of England, where Lower Cretaceous (Neocomian) sideritic mudstones (Ch. 12) were exploited from the Wealden Sands.

More important of late, however, have been the 'minette'-type ores occurring in the Jurassic and Cretaceous rocks (Fig. 19.12). They consist essentially of oolitic carbonages and silicates. Table 19.9 sum-

**Table 19.9.** Summary of main Jurassic-Cretaceous Ironstones

| Period | Stage | Name | Main area of workings | Principal ore | Thickness | Grade | Notes |
|---|---|---|---|---|---|---|---|
| Cretaceous | Hauterivian | Claxby Ironstone | N Lincolnshire | Oolitic ironstone | Up to 5m (av. 2m) | Up to 31% Fe | Ceased working 1969 |
| | Valanginian | Wadhurst Clay | Kent, Sussex | Siderite nodules | 3m–4.5m | 31–38% Fe, 2.0% $P_2O_5$ | Developed by Romans, max. production in 17th century |
| | Aalenian | Northampton Sand | Northamptonshire, S Lincolnshire | Sideritic chamositic oolite, limonitic oolite | 3m–4.5m | 31–38% Fe, 2.0% $P_2O_5$ | Developed 1852 onwards. Production in 1942 10.5 m.t.; 1967 – 7 m.t.; 1970s 3 m.t p.a. Reserves 1 bn. t. |
| Jurassic | Domerian (substage of Pliensbachian) | Marlstone Rock Bed ← stratigraphic equivalents → | S Lincolnshire, Leicestershire, and Oxfordshire | Calcitic siderite, chamositic oolites, sideritic limestone | 2.7m–5m 8m, Oxfordshire | 23–28% Fe | Commenced production in 1860s, ceased 1961. 1967 – 530,000 t. Reserves, several hundred m.t. |
| | | Cleveland Ironstone | NE Yorkshire | Chamositic and sideritic oolites & mudstones (magnetite-bearing oolites at Rosedale in S Yorks) | Main seam 3.5m in 24m zone with 4 'iron stones' | 30% Fe (low Ca content, therefore not self-fluxing) | 6 m.t. p.a. 1876–1913 1939 – 2 m.t. Ceased working 1964 |
| | Sinemurian | Frodingham | Lincolnshire | Chamosite and sideritic oolites | 9.2m (interbedded ore, marls and mudstones) | 20–25% (high Ca content – self-fluxing) | Mining dates back to Roman times. Serious exploitation commenced 1859. Underground mining up to 20% of prod'n in 1970s. Ceased working 1988. |

621

marises their stratigraphic occurrence, mineralogy and other characteristics. Their origin is considered to be marine (see Ch. 11) and they have been discussed at length by e.g. Taylor (1949), Hollingworth & Taylor (1951), Whitehead *et al.* (1952), Hemmingway (1974), Hallam (1975), Hallam & Bradshaw (1979), Dixon (1979a), Dunham *et al.* (1979) and Young & Taylor (1989). Since the Second World War, most of the output of indigenous iron ore has been produced from the Jurassic deposits. Only the Frodingham and Northampton ores have been exploited recently.

Iron ore production in England from the Jurassic/Cretaceous ores was at a peak in 1942 (20·2 m tonnes) but has steadily declined – 260,000 tonnes being produced by the British Steel Corporation from only the Frodingham ore at Scunthorpe in 1983, the mine finally closing in 1988.

# North Wales

## Anglesey

Parys Mountain, in the late 18th and early 19th centuries, was Europe's largest copper mine and for a time (1768–88) was the world's greatest producer (3,000–5,000 tonnes of metal per year). It is thought to have been worked in Roman times (because of the discovery of 'pigs' of copper of Roman type) and probably much earlier during the Bronze Age. Little evidence has been found to indicate that working took place between the Roman period and 1768 (Manning 1959). From then on, mining continued intermittently until 1920. In 1985 exploratory drilling commenced and the B.G.S. (1990) quoted mineable reserves as 4·794 mt grading 6·04% Zn, 1·49% Cu, 3·03% Pb, 0·57 g/t Ag and 0·013 oz/t Au. It is planned to sink a 450 m shaft and by 1992 to produce 400,000 t p.y. of ore (25,000 t Zn, 5,840 t Cu, 12,120 t Ph, 810,000 oz Ag and 5,250 oz Au). This would make the mine the largest non-ferrous metal producer ever developed in Britain.

The ores (see e.g. Manning 1959, Pointon & Ixer 1980) occur within Ordovician volcanics and sediments. The Parys Mountain Volcanics are mainly rhyolites and dacites (though Southward [1984] emphasised the importance of basalts in the succession) with associated tuffs and volcaniclastics. Below are Lower Ordovician shales with occasional interbedded breccias and greywackes, while above are cherts and shales which are succeeded by Silurian shales. The ore consists of pyrite, chalcopyrite, sphalerite, galena (with minor amounts of other lead and of bismuth minerals) occurring in more or less strata-bound form either associated with siliceous sinter at the base of the volcanics or at the top of the volcanics as lenses along with chert and shale. Rhyolitic vents in the vicinity and the sinter (?fumarole activity) support an exhalative volcanic-sedimentary origin. Nutt *et al.* (1979), however, favoured a Caledonian epigenetic origin on the basis of K-Ar dates on clay minerals in the country rocks but this view was challenged (Rundle 1981) on the grounds that the K-Ar dates can be interpreted in different ways.

## Llanrwst

Here veins occur in the Upper Ordovician Crafnant Volcanic group (Archer 1959). There are three vein trends – an early ENE-WSW group which are cut by a N-S group, with both cut by a WNW-ESE group. Galena, sphalerite and marcasite in a calcite and quartz gangue are common vein-fillings. The galena has been dated at 340 ± 70 Ma (Moorbath 1962) while K-Ar ages on clay minerals from the vein gouge and altered wall rocks range from 339 ± 6 Ma to 240 ± 3 Ma from a series of specimens collected in the area (Ineson & Mitchell 1975).

At Cae Coch, Trefriw, a stratiform mass of quartzose pyritic ore occurs. Ball & Bland (1985) likened it, because of its occurrence at the contact between the basic tuffs and lavas of the Ordovician Dolgarrog Formation and the succeeding pyritic black mudstones of the Llanrychwyn Slates, to a Kuroko-type deposit. Minor molybdenite and traces of Ba, Ca, Zn, Ni, Co and other metals are present. The ore body is considered to have formed on the surface of the Dolgarrog Volcanics in a brine pool during deposition of euxinic muds.

## Halkyn-Minera

In these districts of Clwyd, E-W steeply dipping veins in normal faults cut Carboniferous Limestone (Schnellman 1959, Hughes 1959a) in the main, though occasionally Namurian sandstones also. Sphalerite and argentiferous galena, in a gangue of calcite and quartz, are common. In cross-courses (N-S strike faults), chalcopyrite and galena occur. Smithsonite, baryte and fluorite have been recorded from both vein sets and from the occasional flat or pipe. The Clwyd area was in fact the largest zinc producer in Britain. In addition, haematite deposits in the limestone have been worked on a small scale as have manganese oxide replacement bodies (Dunham *et al.* 1979).

## Llŷn Peninsula

Bedded manganese ores of Ordovician age were mined at Llanengan (St. Tudwal's Peninsula) from 1894–1927 and from 1941–45 (Dunham *et al.* 1979). At Nant the ore (mainly rhodocrosite in a volcaniclastic host) contained 20–30% Mn while at Benallt the ore (pennantite and other Mn silicates) yielded 30–36% Mn. Traces of oolitic structure suggested an origin as a chamositic mudstone (Dunham *et al.* 1979).

A Mn-enriched zone within Cambrian pelites was described by Bennett (1987) from the area and compared with the Hafotty Formation of Harlech (see later).

Ordovician ironstones, with oolitic structures, consisting of magnetite in a ferruginous sandstone were also worked near Llanengan, though they are better developed near Bangor and other areas north of Snowdon (Trythall *et al.* 1987; see also Young & Taylor (1989).

An E–W lead–zinc copper vein cutting Ordovician rocks near Llanengan was worked during the latter part of the 19th century and carried 158 g Ag/t of lead (Archer 1959, Dunham *et al.* 1979).

## Snowdonia

Copper (and some lead and zinc) sulphide ores were mined in the Snowdon area from quartz-pyrite veins from the 18th century until early this century. Reedman *et al.* (1985) emphasised the association of the north-westerly trending veins with a 14 km-diameter caldera formed during the eruption of the Lower Rhyolitic Tuff Formation of the Ordovician Snowdon Volcanic Group (p. 76 and p. 498).

Mineralising fluids are considered to have been produced from leaching of underlying rocks by hot juvenile and marine waters, the fluids being convectively driven upwards by a sub-caldera heat-source to precipitate out metals in the fractures produced during the caldera formation. The main Snowdonia copper mine, at Drws y coed, was actually outside the caldera and worked E–W veins cutting Cambrian and Ordovician rocks, but it seems likely the origin of the ores was similar.

## Harlech Dome

Of particular interest are the gold deposits of the so-called 'Dolgellau gold-belt' at the southern side of the Harlech Dome. Gold (and silver) ores, unlike other metals, belong to the Crown. While the laws differ slightly for Scotland, Northern Ireland and the Isle of Man, those for England (and Wales, since 1535) emphasise that 'gold and silver mines remain the exclusive property of the Crown, by the royal prerogative, and are mines royal . . . are not regarded recognised as part of the land in which they are found; *a fortiori* they are not part of the minerals of the land' (Collins 1975, p. 54). Gold has been worked intermittently in England and Wales from pre-Roman times. Production reached its peak at the turn of the 19th/20th centuries, as can be seen from Table 19.10. The bulk of gold production has come from the Dolgellau gold-belt in Gwynedd, though the Ogofau (or Roman

Deep Mine), SE of Lampeter, in Dyfed, was evidently worked both opencast and underground during Roman times. The country rocks of the Dolgellau gold-belt which extends around the southern and eastern flanks of the Harlech Dome consist of Middle Cambrian Menevian Beds (or Clogau Shales) and Upper Cambrian Ffestiniog and Maentwrog Shales (Ch. 3). A total thickness of 1,100 m of these Cambrian rocks has been cut by steeply dipping gold-bearing quartz veins. These strike more or less E–W and are spatially associated with intrusive (Ordovician), mainly basic, igneous rocks, though they are clearly post-Caledonian folding in age. Ore-shoots occur when the veins cut the Clogau Shales and may contain, in addition to quartz and gold, pyrite, pyrrhotite, chalcopyrite and minor galena, sphalerite, arsenopyrite and occasionally cobalt, nickel and bismuth sulphides.

The St David's mine (combined Colgau and Vigra mines) was the most important and was originally worked for copper, the main vein being up to 2·7 m wide. Second in importance was the Gwynfynydd mine, north of Dolgellau, which produced gold from the latter part of the 19th century until 1938. Other mines in the area also worked gold from time to time and/or extracted silver from galena. Gold licences obtained from the Crown to explore further the Dolgellau gold-belt resulted in drilling and mapping recommencing in the Gwynfynydd area in 1981.

Alluvial gold exists in the sediments of the River Mawddach and has been worked in the past. Proposals to dredge in the estuary of this river near Dolgellau in 1972 were, however, refused on environmental grounds.

Rice & Sharp (1976) discussed an interesting discovery at Coed-y-Brenin, near Dolgellau, of a 'porphyry-type' copper deposit. Dioritic intrusions, probably co-magmatic with the Rhobell Volcanic

**Table 19.10.** Gold production UK[1], 1860–1939 (Collins 1979)

| Decade | Ore mined (tonnes) | Gold content (kg) |
|---|---|---|
| 1860–69 | 19,000[2] | 516[3] |
| 1870–79 | 140[2] | 85[3] |
| 1880–89 | 11,152[2] | 404[3] |
| 1890–99 | 61,036 | 834[3] |
| 1900–09 | 180,791 | 1,733 |
| 1910–19 | 15,801 | 149 |
| 1920–29 | 234 | 5 |
| 1930–39 | 21,168 | 86 |

[1] 90% production was from Gwynedd and 10% from Dyfed and Scotland.
[2] From incomplete records.
[3] Includes varying amounts of silver.

Group (p. 81), occur within the Cambrian Festiniog Beds. In the 19th century, peat, impregnated with copper carbonate and native copper, had been dug and burnt at the Turf Copper mine and the ash sent to Swansea for smelting. Geochemical and geophysical exploration of the area in the 1970s identified the source and two alteration zones within the diorite carrying chalcopyrite were found to contain about 200 million tonnes of ore averaging 0·3 per cent Cu. Apart from possible environmental objections, the deposit is currently not economic.

Shepherd & Allen (1985) considered, in the light of fluid-inclusion studies, mainly of quartzes, that there were two distinct mineralising episodes in the Harlech Dome area. The first (where the inclusions are richer in $CO_2$) gave rise to the Coed-y-Brenin porphyry deposit and is associated with the emplacement of late-Cambrian diorites. The second, which produces the gold-belt veins, was considered to be due to metal-carrying fluids released from Cambrian and pre-Cambrian sediments during hydraulic fracturing in the closing stages of the Caledonian orogeny.

Bottrell et al. (1988) after further fluid-inclusion studies of the gold-quartz ore of the Clogau–St David's Mine, emphasised the importance of graphite in the shales of the Clogau Formation in promoting gold deposition from incoming aqueous fluids.

## Central Wales

Mining in the North Dyfed/Powys region of Central Wales certainly took place in Tudor times and probably dates back to pre-Roman times (Hughes 1959b). Raybould (1974) described the Cu-Pb-Zn sulphide deposits as being mainly in ENE–WSW trending veins in normal faults which cut across the regional NNE Caledonoid folds. Country rocks are mainly Silurian grits and shales though Ordovician mudstones and grits appear in the core of the major anticlines (e.g. the Plynlimon Dome).

Drilling north-east of Ogofau (Annels 1982) proved a sheared, mineralised zone in black shales and silstones of probable Lower Llandovery age containing Fe, Cu, Pb, Zn sulphides and gold of polyphase origin. Siliceous metalliferous hydrothermal fluids are considered to be derived from underlying Ordovician volcanics or older basement rocks.

Phillips (1972, 1986) discussed the importance of hydraulic fracturing by hydrothermal solutions in producing the faults in which the veins occur and Raybould (1974, 1975) emphasised that fracturing and mineralisation took place simultaneously. As no igneous source is apparent, it was postulated that the hydrothermal fluids were formation waters that had collected metals from underlying rocks. Phillips (1983) suggested that pore fluids were released at depth from

Lower Palaeozoic rocks of low metaporphic grade during a period of regional NNW–SSE extension. Of interest is that K-Ar dates of clay minerals from vein gouges and wall-rocks from Parys Mountain, Shropshire and Plynlimon give ages of $356 \pm 7$ Ma, $355 \pm 8$ Ma and $359 \pm 7$ respectively (Ineson & Mitchell 1975). This has been taken as evidence of a metasomatic event related to the time of mineralisation (i.e. at about the Devonian–Carboniferous boundary).

## South Wales

Iron in the form of siderite in bands and nodules in the Coal Measures (p. 598) was important in the development of the South Wales iron and steel industry. Also historically important were the haematite deposits which occur in dolomites of Carboniferous Limestone age along a 13 km belt between Cardiff and Bridgend on the SE margin of the South Wales coalfield. Irregularly shaped lenses of haematite and goethite both replaced limestone and filled cavities with their upward limits being determined by overlying Namurian shales. Rankin & Criddle (1985), after fluid-inclusion studies, proposed an origin for the ores involving the mixing of downward percolating ground waters, saline and iron-rich from passage through overlying Triassic red beds, with episodic pulses of warmer waters driven from the adjoining Coal Measures by spasmodic tectonic activity. Bedded ores also occur in Lower Limestone Shales and consist essentially of fossiliferous limestone layers (up to 30 cm or so in thickness) of haematised crinoid and bryozoan debris, though these have never proved economic.

The haematite ore needed for the South Wales iron foundries commonly contained about 45–50 per cent Fe and mining took place from the beginning of the 19th century onwards, though statistics of production before 1859 are not readily available. From then until 1925, about 2,600,000 tonnes of ore were produced, the Llanharry mine being the most productive this century. From 1940 until closure in 1975, the haematite mines of South Wales produced about 5 million tonnes of ore.

## Metallogenesis

As can be seen from the various modes of origin suggested for the epigenetic ore deposits of England and Wales, it is no longer accepted that metal-carrying hydrothermal solutions are necessarily late-stage differentiates of a cooling magma. While the spatial relationship between the Hercynian batholith and the ores of Cornubia is obvious and it is agreed there is a Caledonian batholith beneath the ore deposits of the Lake District and the Northern Pennines, a genetic connection is currently doubted. The results of radiogenic age-dating, stable-isotope studies of sulphur, oxygen and hydrogen and rare earth element analysis of ore, gangue minerals and wall-rock minerals and where present their fluid inclusions, together with phase-equilibrium studies, have placed restrictions on anyone attempting to provide definitive models of ore-deposition in England and Wales. Dixon (1979b) gave an excellent account of changing views on metallogenesis over the centuries, and in the last few decades in particular. Studies on British ore deposits encapsulate much of the debates taking place the world over. Disagreement exists concerning the ambiguity, significance and meaning of many analytical results. Ore fluids can be attributed to meteoric water, connate brines, or juvenile water or to mixtures of two or three of these. Age-dates have been particularly susceptible to changes of interpretation. A consensus of opinion over the past decade would seem, however, to have developed concerning the thesis that convective circulation of water could result in scavenging of metals distributed sparsely through all rocks during downward movement. After upwards movement, precipitation of some of these metals where pressure, temperature and solubility conditions were suitable would take place in relatively concentrated amounts (see e.g. Russell 1978, Russell et al. 1981, Moore 1982, Dunham 1988, Russell & Hall 1988).

Exploration for geothermal energy in England (Brown et al. 1980) has revealed high heat flow provinces which coincide with many of the ore deposits (Fig. 19.16). The juxtaposition of granites with unusually high levels of radioactive heat-producing elements and these belts was noted. It is possible 'by virtue of their heat production and relatively high thermal conductivity' long after the period of their consolidation' these granites could produce the heat necessary to induce convection (Brown et al. 1980).

Suffice to say that the theories will be modified and challenged in the future. Metallogenesis, along with many other branches of geology, is the slave of fashion!

## Industrial minerals

The term 'industrial minerals' embraces not only those non-metallic minerals which may be valued for their physical rather than their chemical properties but also rocks, sands, gravels and clays (Table 19.11). Highley et al. (1980) summarised their origin and uses while Blunden (1975) gave a comprehensive account of them (along with the fuels and metallic minerals) with particular reference to the problems of exploitation, environmental constraints and planning laws, etc. The total value of industrial minerals produced in the

United Kingdom in 1988 amounted to over £1·8 billion, with common sand and gravel production being valued at £543,000,000 and limestone and dolomite £498,000,000. Next in importance was china clay (£224,000,000), one of the few minerals (other than fuels) exported from the United Kingdom (B.G.S. 1990).

The sands and gravels are predominantly of fluvial (alluvial and river-terrace) and fluvio-glacial origin and since 1975 comprehensive accounts of their distribution have appeared in the Institute of Geological Sciences Mineral Assessment Reports (the pre-1975 reports appeared as a sub-series of the I.G.S. Report Series).

Most limestone quarried is of Carboniferous age and most of it is used as a source of aggregate in the building and road-making industries or for cement manufacture or agricultural purposes (the important Derbyshire deposits are described in Mineral Assessment Reports Nos. 26, 47, 77, 79, 98, 129 and 144). Permian 'Magnesian Limestone' is used in the iron and steel industries and for the manufacture of refractories and glass. Cretaceous Chalk is used for cement manufacture and as a filler in paper-making.

China clay is the purest of the kaolinite-rich clays which include ball clays, fuller's earth and fireclay. The origin and uses of these clays have been well summarised by Highley (1984, 1975, 1972 and 1982 respectively) and discussed by Vincent & Nicholas (1982).

China clay (consisting, as quarried, of kaolinite, quartz and mica [sometimes lithium-rich]) is produced from kaolinised portions of the SW England granites (Figs. 19.13, 19.14). As the name implies, it was (after beneficiation) of great interest in the ceramics industry, though nowadays most of the production is taken as a 'filler' in the paper industry. It is Britain's second most valuable mineral export after crude oil.

The origin of china clay, like the metals associated with the granites (pp. 612–616), is a matter of debate. Recent fluid-inclusion studies on quartzes from the St Austell granite led to Alderton & Rankin (1983) reviewing various views. They noted the strong association of kaolinised granite with dilute, low salinity, low temperature (70°–170°C) fluids. They felt the temperature of the weathering process (30°C) suggested by Sheppard (1977) for the formation of the inclusions in the quartzes was too low and concluded (op. cit., p. 308) 'that kaolinisation is a "hydrothermal" process bordering on the supergene fluid regime'. Durrance and Bristow (1986) considered convective cells involving groundwaters, brought about by radiogenic heat were part of a two-stage alteration process. Zones of alteration occurred in areas of convective drawdown, the cells being associated with NW–SE transcurrent faults. The kaolinisation process resulted in a mass loss of 20%

**Table 19.11.** United Kingdom production of minerals by countries 1988 (from British Geological Survey 1990)

Thousand tonnes

| Mineral | England | Wales | Scotland | Great Britain | Northern Ireland | United Kingdom | Isle of Man |
|---|---|---|---|---|---|---|---|
| Coal: | | | | | | | |
|   Deep-mined | 75,830 | 5,519 | 2,112 | 83,461 | — | 83,461 | — |
|   Opencast | 12,213 | 1,715 | 3,996 | 17,924 | — | 17,924 | — |
|   Other (a) | 1,130 | 754 | 521 | 2,405 | — | 2,405 | — |
| Natural gas and oil: | | | | | | | |
|   Methane (b) | | | | | | | |
|     Colliery | ... | ... | ... | 204 | — | 204 | — |
|     North Sea | ... | ... | ... | 66,522 | — | 66,522 | — |
|     Other | ... | ... | ... | 42 | — | 42 | — |
|   Crude Oil | | | | | | | |
|     Inland | — | — | — | 761 | — | 761 | — |
|     North Sea | ... | ... | ... | 108,641 | — | 108,641 | — |
|   Condensates and other (c) | ... | ... | ... | 5,004 | — | 5,004 | — |
| Iron ore | 224 | — | — | 224 | — | 224 | — |
| Non-ferrous ores (Metal content): | | | | | | | |
|   Tin | 3·4 | — | — | 3·4 | — | 3·4 | — |
|   Tungsten | — | — | — | — | — | — | — |
|   Lead | 1·2 | — | — | 1·2 | — | 1·2 | — |
|   Copper (e) | 0·7 | — | — | 0·7 | — | 0·7 | — |
|   Zinc (e) | 5·5 | — | — | 5·5 | — | 5·5 | — |
|   Silver (f) | 2·1 | — | — | 2·1 | — | 2·1 | — |
|   Gold | — | 0 | — | 0 | — | 0 | — |
| China stone | 6 | — | — | 6 | — | 6 | — |
| China clay (sales) | 3,277 | — | — | 3,277 | — | 3,277 | — |
| Ball clay (sales) | 716 | — | — | 716 | — | 716 | — |
| Potter's clay | ... | — | — | ... | — | ... | — |
| Fireclay | 791 | ... | ... | 1,057 | ... | ... | — |
| Fuller's earth (sales) (h) | 213 | — | — | 213 | ... | 213 | — |
| Common clay and shale | 16,550 | 437 | 1,547 | 18,534 | 365 | 18,899 | — |
| Slate (i) | ... | ... | ... | 708 | — | 708 | 2 |
| Calcspar | 23 | — | — | 23 | — | 23 | — |
| Limestone (j) | 99,150 | 19,005 | 1,644 | 119,799 | 2,409 | 122,208 | 102 |
| Dolomite (k) | 16,931 | 2,930 | | 19,861 | — | 19,861 | — |
| Chalk | 14,516 | — | — | 14,516 | ... | ... | — |
| Chert and flint | 11 | — | — | 11 | — | 11 | — |
| Silica stone and ganister | 11 | | — | 11 | — | 11 | — |
| Sandstone (l) | 12,035 | 2,296 | 1,700 | 16,031 | 2,870 | 18,901 | — |
| Silica sands | 3,802 | 18 | 520 | 4,340 | — | 4,340 | — |
| Common sand and gravel: | | | | | | | |
|   Land | 97,217 | — | 10,753 | 110,516 | 3,871 | 114,387 | 123 |
|   Marine (m) | 19,829 | 2,189 | — | 22,018 | — | 22,018 | — |
| Igneous rock (n) | 24,698 | 3,719 | 16,219 | 44,636 | 7,419(o) | 51,960 | 91 |
| Gypsum | ... | — | — | ... | — | ... | — |
| Anhydrite | ... | — | — | ... | — | ... | — |
| Rock salt | 877 | — | — | 877 | ... | ... | — |
| Salt from brine | 1,426 | — | — | 1,426 | — | 1,426 | — |
| Salt in brine (p) | 3,827 | — | — | 3,827 | — | 3,827 | — |
| Fluorspar | 104 | — | — | 104 | — | 104 | — |
| Barytes | ... | — | ... | 76 | — | 76 | — |
| Celestite | 26 | — | — | 26 | — | 26 | — |
| Diatomite (q) | — | — | — | — | 0·3 | 0·3 | — |
| Talc | — | — | 14 | 14 | — | 14 | — |
| Lignite | 18 | — | — | 18 | — | 18 | — |
| Honestone | — | 0 | — | 0 | — | 0 | — |
| Potash (r) | 767 | — | — | 767 | — | 767 | — |

a  Slurry etc. recovered and disposed of, other than the British Coal Corporation, from dumps, ponds, rivers etc.

b  Approximate coal equivalent: converted from original data at 250 therms = 1 tonne.

c  Includes ethane, propane and butane, in addition to condensates.

e  Content of mixed concentrate.

f  Silver content of copper-zinc and lead-zinc concentrates. See also pp. 5–19.

h  B.G.S. estimates based on data from producing companies.

i  Slate figures include waste used for constructional fill and powder and granules used in industry.

j  Includes 16,389,000 tonnes of dolomite for constructional uses.

k  Includes dolomite and magnesian limestone used as aggregate and for agricultural purposes as well as for refractory, chemical and other purposes specifically dependent on the high magnesium content.

l  Including grit and conglomerate.

m  Including marine-dredged landings at foreign ports (exports).

n  In addition 237,000 tonnes were produced in Guernsey.

o  Including 96,000 tonnes of granite.

p  Used for purposes other than salt making.

q  Dry weight: estimated by the B.G.S.

r  Marketable product (KCl).

with resulting compensatory isostatic uplift (still taking place today).

*Ball clay* (so called because it was originally obtained in Dorset and Devon by cutting it on the floor of open pits into cubes or 'balls' [Scott 1929]) occur in the Tertiary rocks of Bovey and Petrockstowe in Devon (p. 401) and Wareham in Dorset. The clays are interbedded with sands, other clays and lenticular beds of lignite of Eocene-Oligocene age and presumed to be transported kaolinised material. Most of the output is used in the ceramics industry.

*Fuller's earth* (Chs. 11, 12) is now recognised as an altered volcanic ash and consists mainly of Ca-montmorillonite. By cation exchange, it can be changed to Na-montmorillonite or bentonite. Important deposits occur in the Bathonian-age Fuller's Earth Formation near Bath and in Aptian-age Lower Greensand beds in Surrey, Bedfordshire and Oxfordshire. The name 'fuller's earth' comes from its use in 'fulling' or cleansing woollen fabrics (see Robertson 1986).

The *fireclays* are mainly varieties of the seatearths which occur in Westphalian coal-bearing sequences (Ch. 8) and are generally accepted as the fossil 'soils' of the coal-forming plants. They consist essentially of disordered kaolinite, mica and quartz and are used in the ceramics industry.

Other industrial minerals of interest include the evaporite minerals such as gypsum, anhydrite, rock salt and potash salts. Their geology is dealt with in Chapters 9 and 10 while summaries of their occurrence and uses have been given by Notholt and Highley (1975, 1973) and Notholt (1976) respectively.

Britain's last surviving rock-salt mine (which produced 1·8 Mt in 1988) is at Meadowbank, Winsford, Cheshire, and works, by room and pillar, Triassic salt. New development is planned at Wheatcroft to the north.

Permian potash salts are worked at Boulby, Yorkshire, where seams up to 24 m thick are mined at a depth of 1,100 m and 1 km offshore. In 1988 2·54 Mt of ore was mined producing 767,000 t of potash and 230,000 t of rock salt (Shrimpton 1989).

## Water resources

Because of the distribution of rainfall in England and Wales, surface water supplies are insufficient to meet the needs of the population and industry to the south-east of a line running from about the north Yorkshire coast to the Exe estuary. South-east of that line, luckily, occur the main aquifers which can be tapped for underground water. These are the Cretaceous Chalk, the Permo-Triassic sandstones and the Jurassic limestones, and perhaps some 40 per cent of the population rely on underground water for their needs.

Until recently, problems were encountered in the south-east part of the country because of over-abstraction from the Chalk, but artificial recharge and increased use of water from the River Thames has more or less managed to cope with the demand. Fissures play a large part in making the Chalk permeable, as indeed they do also in the Permo-Triassic sandstones which are the main source of supply in the Midlands of England. Aquifers also exist in the Dinantian limestones, Namurian grits, the Permian Magnesian Limestone, Cretaceous Upper Greensand and Pliocene Crag, and are tapped for relatively local uses. The British Geological Survey has an extensive range of Water Supply papers, Research Reports, Hydrogeological Reports and Maps of the areas where underground water is required.

## Geothermal energy

With the dramatic rise in the price of oil in the 1970s, interest grew throughout the industrial world regarding possibilities of using forms of energy other than fossil fuels. A programme to assess the potential of geothermal energy in the United Kingdom was therefore embarked upon by the British Geological Survey. A series of BGS publications under the title 'Investigation of the Geothermal Potential of the UK' appeared between 1978 and 1985 and a comprehensive account of the resources and possibilities of geothermal energy in the United Kingdom was given by Downing and Gray (1986).

In 1975, there were only 32 observations of surface heat-flow in the UK. By 1984, there were 188 heat-flow observations collected under strictly controlled conditions by the BGS and (under sub-contracts) by the University of Oxford, Imperial College (University of London) and the University of Bath. It has been found that while the area-weighted mean heat flow is about 52 mW/m$^2$ (range $<40->120$ mW/m$^2$), above-average values are particularly associated with the granite batholiths in Cornwall, Northern England and the Grampian Highlands of Scotland and with certain deep Mesozoic basins in eastern and southern England. The contrasting geological environments thus provide two potential sources of geothermal energy (a) 'hot dry rocks' where temperatures of more than 100°C obtain in presumed impermeable rocks at depths of 3 to 4 km (perhaps 200°C about 5·5 km down) and (b) 'low enthalpy' groundwater resources in deep sedimentary basins, where extensive and prolonged circulation has raised the temperature of the waters (Downing and Gray (eds) 1986, Downing & Gray 1986).

The hot dry rocks (HDR) of the granite batholiths owe their heat to the presence of minerals within them containing the heat-producing radioactive elements uranium, thorium and potassium. The granite batholith of Cornwall, long known from mining to have a high geothermal gradient, is being intensively studied by researchers at the Camborne School of Mines (Batchelor 1984), who have already sunk 3 deep (2·15–2·7 kms) boreholes. In the HDR concept, it is proposed that the heat be tapped by fracturing the rock between two closely-spaced boreholes, pumping water down one borehole and then retrieving heated water, which has passed through the fractured zone, up the second borehole (Ledingham 1986).

In northern England, the Lake District and Weardale granites are also HDR possibilities. So far the data are limited and do not indicate particularly high temperatures at depth for either the Shap (148°C at 6 km) or Skiddaw (160°C at 6 km) granites. The concealed Weardale batholith [discovered by the Rookhope Borehole (p. 617)] has, at the borehole, however, the possibility of a temperature of 200°C existing at 6 km depth, which makes it the more promising of the two areas with the present state of knowledge.

The most promising low enthalpy groundwater sources of geothermal heat are the deep Mesozoic basins of east Yorkshire and Lincolnshire, Wessex, Worcester and Cheshire. The suitable aquifers are of Permo-Triassic age and hot groundwaters with temperatures greater than 60°C exist in parts of the aquifers at depths of 2 km. If heat pumps are incorporated into a 'doublet' system of an abstraction and a re-injection well, waters at temperatures as low as 20°C could be used. This temperature can be attained at depths of 400 m, while 40°C can be reached at 1 to 1·5 km. To date, the most promising results have been at Cleethorpes, Lincolnshire, and in and near Southampton where two geothermal wells in the Triassic Sherwood Sandstone Group have been extensively tested. Heat from the 1,700 m deep Southampton well is being used to heat buildings in the new Civic Centre. In the East Yorkshire and Lincolnshire Basin the Triassic Sherwood Sandstone is again the most promising source, though the Basal Permian Sands are another possibility.

Commercial development of the UK's geothermal resources is still in the future. Apart from the unknown legal status of geothermal energy, the best possibilities do not coincide with centres of population and natural gas and oil are still cheap and plentiful. Furthermore, the technical difficulties are still formidable and development is capital-intensive at the beginning of a scheme. The Southampton scheme will provide an all-important comprehensive field-trial for low enthalpy resources. The experiments in Cornwall, along with similar trials in the USA, will indicate the feasibility or otherwise of the HDR concept in its present form (see also Russell & Hall 1988).

## REFERENCES

| | | |
|---|---|---|
| ADAMS, H. F. | 1967 | The seams of the South Wales coalfield. *Instn. Ming. Eng. London*, 38 pp. |
| ALDERTON, D. H. M. & RANKIN, A. H. | 1983 | The character and evolution of hydrothermal fluids associated with the kaolinised St Austell granite, Southwest England. *J. geol. Soc. London*, **140**, 297–309 |
| ALLEN, P. M., COOPER, D. C. & SMITH, F. | 1979 | Mineral exploration in the Harlech Dome, North Wales. *Mineral Reconn. Prog. Rep. Inst. geol. Sci.*, **29**. |
| ALLSOP, J. M. | 1984 | Geophysical appraisal of a Carboniferous basin in north-east Norfolk, England. *Proc. geol. Assoc. London*, **95**, 175–180. |
| ALLSOP, J. M. AND SIX OTHERS | 1982 | Palaeogeological maps of the floors beneath two major unconformities in the Oxford–Newbury–Reading area. *Rep. Inst. geol. Sci. London*, **81/1**, 48–51. |
| ANNELS, A. E. | 1982 | Gold mineralisation at Ogofan, Dyfed, Central Wales. *J. geol. Soc. London* (*Proc.*), **139**, 662. |
| ANON | 1984 | New Hope for Parys Mountain. *Ming. J., London*, **5**, 236. |
| | 1988 | Mining to recommence at Parys Mountain, Wales. *Ming. Mag.* (July), 6. |
| ARCHER, A. A. | 1959 | The Distribution of Non-Ferrous Ores in the Lower Palaeozoic Rocks of North Wales. *In* I.M.M. (1959) *q.v.*, 259–276. |

| | | |
|---|---|---|
| ATKINSON, P. | 1983 | Fluorite mineralisation in the Southern Pennine Orefield. *J. geol. Soc. London (Proc.)*, **140**, 980. |
| BALL, T. K. & BLAND, D. J. | 1985 | The Cae Coch volcanogenic massive sulphide deposit, Trefriw, North Wales. *J. geol. Soc. London*, **142**, 889–898. |
| BALL, T. K., FORTEY, N. J. & SHEPHERD, T. J. | 1985 | Mineralisation at the Carrock Fell Tungsten Mine, N. England; Paragenetic fluid inclusion and geochemical study. *Mineralium Deposita*, **20**, 57–65. |
| BATCHELOR, A. S. | 1984 | Hot dry rock geothermal exploitation in the United Kingdom. *Modern Geology*, **9**, 1–41. |
| BEER, K. E. & BALL, T. K. | 1986 | Tin and tungsten in pelitic rocks from S.W. England and their behaviour in contact zone of granite and mineralised areas. *Proc. Ussher Soc.*, **6** (3), 330–337. |
| BEER, K. E. & SCRIVENER, R. C. | 1982 | *Metalliferous Mineralisation. In* Durrance, E. M. & Laming, D. J. C. (*q.v.*), 117–147. |
| BENNET, M. A. | 1987 | Genesis and diagenesis of the Cambrian manganese deposits, Harlech, North Wales. *Geol. J.*, **22** (Thematic Issue), 7–18. |
| BERRY, J. T. & WILSON, D. C. | 1984 | Geology and reserves (the Southern North Sea). *Trans. Instn. Ming. Metall.*, **93**, B134–B146. |
| B.G.S. (BRITISH GEOLOGICAL SURVEY) | 1990 | *United Kingdom Mineral Yearbook 1989*.[1] British Geological Survey (London, Her Majesty's Stationery Office). |
| BLOXAM. T. W. & OWEN, T. R. | 1985 | Anthracitization of coals in the South Wales coalfield. *Int. J. Coal Geol.*, **4**, 299–307. |
| BLUNDEN, J. | 1975 | *The Mineral Resources of Britain.* Hutchison, London, 545 pp. |
| BOTT, M. H. P. | 1974 | The geological interpretation of a gravity survey of the English Lake District and the Vale of Eden. *J. geol. Soc. London*, **130**, 309–331. |
| BOTTRELL, S. H. & SPIRO, B. | 1988 | A stable isotope study of a black shale-hosted gold mineralization in the Dolgellau Gold Belt, North Wales. *J. geol. Soc. London*, **145**, 941–949. |
| BOTTRELL, S. H., SHEPHERD, T. J., YARDLEY, B. W. D. & DUBESSEY, J. | 1988 | A fluid-inclusion model for the genesis of the ores of the Dolgellau Gold Belt, North Wales. *J. geol. Soc. London*, **145**, 139–145. |
| BRENNAND, T. P. & VAN HOORN, B. | 1986 | Historical Review of North Sea Exploration. *In* Glennie (ed.) 1986 *op. cit.*, 1–24. |
| BROOKS, J. & GLENNIE, K. W. (Eds.) | 1987 | *Petroleum Geology of North West Europe.* Graham & Trotman, London, 2 vols, 1219 pp. |
| BROOKS, J., GOFF, J. C. & VAN HOORN, B. (Eds.) | 1986 | Habitat of Palaeozoic gas in N.W. Europe. *Spec. Publ. geol. Soc. London*, **23**, Scottish Academic Press, Edinburgh, 276 pp. |
| BROWN, G. C. AND FIVE OTHERS | 1980 | Basement heat flow and metalliferous mineralization in England and Wales. *Nature, London*, **288**, 657–659. |
| BROWN, G. C., IXER, R. A., PLANT, J. A. & WEBB, P. C. | 1987–88 | Geochemistry of granites beneath the North Pennines and their role in orefield mineralisation. *Trans. Instn. Ming. Metall. (B)*, **96**, B65–76, B229–232, **97**, B47–49. |
| BURN, H. | 1986 | Aspects of safety and economics in Offshore Boring. *Ming. Engineer*, **146**, 345–356. |
| BUSHELL, T. P. | 1986 | Reservoir Geology of the Morecambe Field. *In* Brooks *et al.* (eds.) 1986 (*q.v.*), 189–208. |

[1] *United Kingdom Mineral Statistics* from 1973–1988.

BUTCHER, N. J. D. &      1987    Exploration and extraction of structurally and litho-
  HEDGES, J. D.                    stratigraphically controlled fluorite deposits in Castleton–
                                   Bradwell area of Southern Pennines Orefield, England.
                                   *Trans. Instn. Ming. Metall.* (**B**), **96**, B149–155.

CASWELL, S. A.,         1984a   Water-soluble chlorine and associated major cations from
  HOLMES, I. F. &                 the coal and mudrocks of the Cannock and North Stafford-
  SPEARS, D. A.                   shire coalfields. *Fuel*, **63**, 774–781.
                        1984b   Total chlorine in coal seam profiles from the South
                                   Staffordshire (Cannock) coalfield. *Fuel*, **63**, 782–787.

COLLINS, R. S.          1975    Gold. *Miner. Dossier, Miner. Resour. Consult. Comm.*, **14**,
                                   66 pp. HMSO, London.

COLTER, C. S. &         1981    The Wytch Farm Oil Field, Dorset. *In* Illing & Hobson,
  HAVARD, D. J.                   1981 (*q.v.*), 494–503.

COLMAN, T. B.           1990    Exploitation for metalliferous and related minerals in
                                   Britain: a guide. (Keyworth, Nottingham: British Geological
                                   Survey.)

COOPER, D. C., LEE, M. K.,  1988  The Crummock Water aureole: a zone of metasomatism
  FORTEY, N. J.,                   and source of ore metals in the English Lake District. *J.
  COOPER, A. B.,                   geol. Soc. London*, **145**, 523–540.
  RUNDLE, C. C.,
  WEBB, R. C. &
  ALLEN, P. M.

COPE, K. G. &           1970    The Warwickshire Thick Coal and its Mining Environ-
  JONES, A. R. L.                 ment. *6me. Cong. int. Strat. Geol. Carb. Sheffield 1967*,
                                   585–598.

CORNFORD, C.            1986    Source Rocks and Hydrocarbons of the North Sea. *In*
                                   Glennie (ed.), (*q.v.*), 197–236.

CRANFIELD, J.           1980    *Britain's on-shore oilfields.* British Petroleum Co Ltd,
                                   London, 11 pp.

CURTISS, C. D. &        1968    The formation of sedimentary iron minerals. *Econ. Geol.*,
  SPEARS, D. A.                   **63**, 257–270.

DARBYSHIRE, D. P. F. &  1985    Chronology of granite magmatism and associated min-
  SHEPHERD, T. J.                 eralization, SW England. *J. geol. Soc. London*, **142**, 1159–
                                   1177.

DARNLEY, A. G.          1986    High heat production (HHP) granites, hydrothermal cir-
                                   culation and ore genesis. (Commentary and discussions of
                                   conference held at St. Austell, Cornwall, 22–25 Sept. 1985.)
                                   *Trans. Instn. Ming. Metall.* (**B**), **95**, B46–49.

DAVIES, M. M. &         1974    The geochemistry of some South Wales Coals. *In* Owen,
  BLOXAM, T. W.                   T. R. (Ed.) 1974b, (*q.v.*).

DEPARTMENT OF ENERGY    1976–   *Development of the oil and gas resources of the United
                        1990    Kingdom.* HMSO, London. (Annual publication).

DINES, H. G.            1956    The Metalliferous Mining Region of South West England.
                                   *Mem. geol. Surv. G.B.* 2 vols. 795 pp.

DIXON, C. J.            1979    *Atlas of Economic Mineral Deposits.* Chapman & Hall,
                                   London, 143 pp.
                        1979a   The Iron Deposits of the Northampton District – UK. *In*
                                   Dixon, C. J. 1979 (*q.v.*), 34–35.
                        1979b   Introduction. *In* Dixon, C. J. 1979 (*q.v.*), 5–11.

DOWNING, R. A. &        1986    *Geothermal Energy – the potential in the United Kingdom.*
  GRAY, D. A. (Eds.)              British Geological Survey, HMSO, London, 200 pp.

DOWNING, R. A. &        1986    Geothermal resources of the United Kingdom. *J. geol. Soc.
  GRAY, D. A.                     London*, **143**, 499–507.

DRANFIELD, P.,          1987    Wytch Farm Oilfield: reservoir characterization of the
  BEGG, S. H. &                  Triassic Sherwood Sandstone for input to reservoir simula-
  CARTER, R. R.                  tion studies. *In* Brooks & Glennie (*q.v.*), 149–160.

| DUNHAM, K. C. | 1948 | Geology of the Northern Pennine Orefield, **1**, Tyne to Stainmore. *Mem. geol. Surv. G.B.* 357 pp. |
| | 1952 | Fluorspar (4th Edn.). *Mem. geol. Surv. G.B.: Spec. Repts. Min. Res. G.B.*, 357 pp. |
| | 1974 | Granite beneath the Pennines in North Yorkshire. *Proc. Yorkshire geol. Soc.*, **40**, 191–194. |
| | 1983 | Ore genesis in the English Pennines: a fluorite subtype. In *Proc. Intnl. Conf. Mississippi Valley type Lead–Zinc Deposits*, Univ. Missouri, Rolla, Missouri, 86–112. |
| | 1984 | Genesis of the Cumbrian hematite deposits. *Proc. Yorkshire geol. Soc.*, **45**, 130. |
| | 1986 | Geology in the real world. *In* Nesbitt, R. W. & Nichol, I. (Eds.). 1986 (*q.v.*), 1–10. |
| | 1988 | Pennine mineralisation in depth. *Proc. Yorkshire geol. Soc.*, **47**, 1–12. |
| DUNHAM, K. C. & POOLE, E. G. | 1974 | The Oxfordshire Coalfield. *J. geol. Soc. London*, **130**, 387–392. |
| DUNHAM, K. C. & WILSON, A. A. | 1985 | Geology of the Northern Pennine Orefield, **2**, Stainmore to Craven. *Econ. Mem. Brit. geol. Surv.* 250 pp. |
| DUNHAM, K. C., DUNHAM, A. C., HODGE, B. L. & JOHNSON, G. A. L. | 1965 | Granite beneath Visean sediments with mineralization at Rookhope, northern Pennines. *Q. J. geol. Soc. London*, **121**, 383–417. |
| DUNHAM, K. C., BEER, K. E., ELLIS, R. A., GALLAGHER, M. J., NUTT, M. J. C. & WEBB, B. C. | 1979 | United Kingdom. *In* Bowie, S. H. U., Kvalheim, A. & Haslam, H. W. (Eds.), *Mineral Deposits of Europe* **1**. Instn. Ming. Metall. and Mineralog. Soc. London, 263–317. |
| DURRANCE, E. M. & BRISTOW, C. M. | 1986 | Kaolinisation and isostatic readjustment in south-west England. *Proc. Ussher. Soc.,* **6** (3), 318–322. |
| DURRANCE, E. M. & LAMING, D. J. (Eds.) | 1982 | *The Geology of Devon.* University of Exeter, 344 pp. |
| EAMES, T. D. | 1975 | Coal rank and gas source relationships – Rotliegendes reservoirs. *In* Woodland, A. W. (Ed.) 1975 (*q.v.*), 191–204. |
| EASTWOOD, T. | 1959 | The Lake District Mining Field. *In* I.M.M. 1959, 149–174. |
| EBBERN, J. | 1981 | The Geology of the Morecambe Gas Field. *In* Illing, L. V. & Hobson, G. D. I. (Eds.) (*q.v.*), 485–493. |
| EMBREY, P. G. & SYMES, R. F. | 1987 | *Minerals of Cornwall & Devon*, Nat. Hist. Mus. Publications, London, 154 pp. |
| EVANS, A. M. (Ed.) | 1982 | *Metallization associated with Acid Magmatism.* John Wiley & Sons, Chichester, N.Y., 385 pp. |
| EVANS, A. M. & Ei-NIKHELY, A. | 1982 | Some palaeomagnetic dates from the West Cumbria hematite deposits, England. *Trans. Instn. Ming. Metall.*, **B**, **91**, B41–43. |
| FALCON, N. L. & KENT, P. E. | 1960 | Geological Results of Petroleum Exploration in Britain 1945–1957. *Mem. geol. Soc. London*, **2**. |
| FERGUSON, J. | 1988 | The nature and origin of light hydrocarbon gases associated with mineralization in the Northern Pennines. *Mar. Petrol. Geol.*, **5**, 378–384. |
| FIRMAN, R. J. | 1978 | Epigenetic mineralization. *In* Moseley, F. (*q.v.*), 226–240. |
| FIRMAN, R. J. & LEE, M. K. | 1986 | Age and structure of the concealed Lake District batholith and its probable influence on subsequent sedimentation, tectonics and mineralization. *In* Nesbitt & Nichol (*q.v.*), 117–127. |
| FRANCIS, E. H. | 1979 | British Coalfields. *Sci. Prog. Oxford*, **66**, 1–23. |
| GACIRI, S. J. & INESON, P. R. | 1981 | Hydrogeochemical lead–zinc–copper anomalies on the Isle of Man. *Trans. Instn. Ming. Metall.*, **B**, **90**, B120–125. |

| GALLOIS, R. W. | 1978 | A pilot study of oil shale occurrences in the Kimmeridge Clay. *Rep. Inst. geol. Sci. London*, **78/13**, 26 pp. |
|---|---|---|
| GARNISH, J. D. | 1976 | Geothermal energy: the case for research in the United Kingdom. *Energy Paper No.* **9**, HMSO, London, 66 pp. |
| GILL, W. D., KHALAF, F. I. & MASSOUD, M. S. | 1979 | Organic matter as indicator of the degree of metamorphism of the Carboniferous rocks in the South Wales Coalfields. *J. Petroleum. Geol.*, **1**, 39–62. |
| GLENNIE, K. W. | 1986 | Development of N.W. Europe's Southern Permian Gas Basin. *In* Brooks *et al.* (*q.v.*), 3–22. |
| GLENNIE, K. W. (Ed.) | 1986 | *Introduction to the Petroleum Geology of the North Sea* (2nd Edn.). Blackwell, Oxford, 278 pp. |
| GLENNIE, K. W. & BOEGNER, P. L. E. | 1981 | Sole Pit inversion tectonics. *In* Illing, L. V. & Hobson, G. D. (Eds.) 1981 (*q.v.*), 110–120. |
| GODWIN-AUSTEN, R. | 1856 | On the possible extension of Coal Measures beneath the South-Eastern part of England. *Q. J. geol. Soc. London*, **12**, 38–73. |
| GOLD, T. | 1985 | The origin of natural gas and petroleum and the prognosis for future supplies. *Ann. Rev. Energy*, **10**, 53–77. |
| GREEN, G. W. | 1958 | The Central Mendip lead–zinc orefield. *Bull. geol. Surv. G.B.*, **14**, 70–90. |
| GREEN, G. W. & WELCH, F. B. | 1965 | Geology of the country around Wells and Cheddar. *Mem. geol. Surv. G.B.*, 225 pp. |
| HALLAM, A. | 1975 | *Jurassic Environments.* Cambridge University Press, 269 pp. |
| HALLAM, A. & BRADSHAW, M. J. | 1979 | Bituminous shales and oolitic ironstones as indicators of transgressions and regressions. *J. geol. Soc. London*, **136**, 157–164. |
| HALLS, C. | 1987 | A mechanistic approach to the paragenetic interpretation of mineral lodes of Cornwall. *Proc. Ussher. Soc.* **6** (4), 548–554. |
| HALLS, C., EXLEY, C. S. & BRUNTON, E. V. | 1985 | A bibliography of magmatism and mineralization in south-west England. *Instn. Ming. Metall. Occ. Pap.*, **5**. |
| HANCOCK, F. R. P. & MITHEN, D. P. | 1987 | The geology of the Humbly Grove Oilfield, Hampshire, U.K. *In* Brooks & Glennie (*q.v.*), 161–170. |
| HEMINGWAY, J. E. | 1974 | Ironstones. *In* Rayner, D. H. & Hemingway, I. E. (Eds.), *The geology and mineral resources of Yorkshire.* Yorkshire Geological Society, Leeds, 329–335. |
| HIGHLEY, D. E. | 1972 | Fuller's Earth. *Miner. Dossier Miner. Resour. consult. Comm.* **3**. HMSO, London, 26 pp. |
| | 1975 | Ball Clay. *Miner. Dossier Miner. Resour. consult. Comm.* **11**. HMSO, London, 32 pp. |
| | 1982 | Geological distribution, exploration and utilisation of fireclays in the United Kingdom. *Trans. Instn. Ming. Metall.* (**B**), **91**, B11–B16. |
| | 1984 | China Clay. *Miner. Dossier Miner. Resour. consult. Comm.* **26**. HMSO, London, 65 pp. |
| HIGHLEY, D. E., CROCKETT, R. N. & DAY, J. B. W. | 1980 | Natural Resources in the United Kingdom. *In Geology of the European Countries.* Graham & Trotman Ltd, London, 422–432. |
| HOLLINGWORTH. S. E. & TAYLOR. J. H. | 1951 | The Northampton Sand Ironstone: stratigraphy, structure and reserves. *Mem. geol. Surv. G.B.*, 211 pp. |
| HOLMES, I., CHAMBERS, A. D., IXER, R. A., TURNER, P. & VAUGHAN, D. J. | 1983 | Diagenetic Processes and the Mineralization in the Triassic of Central England. *Mineralium Deposita*, **18**, 365–377. |

HOSKING, K. F. G.      1966      Permo-Carboniferous and later primary mineralisation of Cornwall and south-west Devon. *In* Hosking, K. F. G. & Shrimpton, G. J. (Eds.) 1966 (*q.v.*).

     1969      The Nature of the Primary Tin Ores of the South-west of England. *In* Fox, W. (Ed.) *2nd Tech. Conf. on Tin, Bangkok.* Inst. Tin. Council London, 1157–1244.

HOSKING, K. F. G. &      1966      *Present views of some aspects of the geology of Cornwall &*
SHRIMPTON, G. J. (Eds.)      *Devon.* Blackford (for Roy. geol. Soc. Cornwall), Truro, 201–245.

HUGHES, W. J.      1959a      The Lead-Zinc possibilities of the Minera district, Denbighshire. *In* I.M.M. (1959) (*q.v.*), 247–250.

     1959b      The non-ferrous mining possibilities of Central Wales. *In* I.M.M. (1959) (*q.v.*), 277–311.

HULL, E.      1897      *Our Coal Resources.* E. & F. N. Spon, London, 157 pp.

ILLING, L. V. &      1981      *Petroleum Geology of the Continental Shelf of North-west*
HOBSON, G. D. (Eds.)      *Europe.* Heyden & Son Ltd., London, 521 pp.

I.M.M.      1959      *The future of non-ferrous mining in Great Britain and Ireland.* Instn. Ming. Metall., London, 614 pp.

     1985      *High heat production (HHP) granites, hydrothermal circulation and ore genesis.* Instn. Ming. Metall., London, 593 pp.

INESON, P. R.      1977      Ores of the Northern Pennines, the Lake District and North Wales. *In* Wolf, K. M. (Ed.) *Handbook of Stratabound and stratiform ore deposits.* Elsevier, Amsterdam.

INESON, P. R. &      1974      K–Ar isotopic age determination from some Lake District
MITCHELL, J. G.      localities. *Geol. Mag.*, **111**, 521–527.

     1975      K–Ar isotopic age determination from some Welsh mineral localities. *Trans. Instn. Ming. Metall.* **(B), 84**, B7–16.

INESON, P. R.,      1977      Potassium-Argon isotopic age determination from some
MITCHELL, J. G. &      North Devon mineral deposits. *Proc. Ussher. Soc.*, **4**, 12–
ROTTENBURY, F. J.      23.

INSTITUTE OF GEOLOGICAL      1982      *Annual Report for 1980–81.* HMSO, London, 31 pp.
SCIENCES

IXER, R. A.      1986      The ore mineralogy and paragenesis of the lead–zinc–fluorite–baryte orefields of the English Pennines and Mendip Hills. *In* Craig, J. R. (Ed.), *Mineral Paragenesis,* Theophrastus Publications, Athens, 179–210.

JACKSON, D. I.,      1987      The geological framework of the East Irish Sea Basin. *In*
MULHOLLAND, P.,      Brooks & Glennie (Eds.) (*q.v.*), 191–203.
JONES, S. M. & WARRINGTON, G.

JACKSON, N. J.,      1982      Hydrothermal activity in the St Just Mining District,
HALLIDAY, A. N.,      Cornwall, England. *In* Evans, A. M. (Ed.) 1982 (*q.v.*), 137–
SHEPPARD, S. M. F. &      179.
MITCHELL, J. G.

JENYON, M. K.      1986      *Salt Tectonics,* Elsevier Applied Science Publishers, London, 191 pp.

KELLAWAY, G. A. &      1948      *Bristol and Gloucester District,* 2nd Edit., British Regional
WELCH, F. B. A.      Geology. HMSO, London, p. 85.

KENT, P. E.      1985      UK onshore oil exploration, 1930–1964. *Mar. Petrol. Geol.*, **2**, 56–64.

KIRBY, G. A.,      1987      Oil and gas generation in eastern England. *In* Brooks &
SMITH, K.,      Glennie (Eds.) (*q.v.*), 171–180.
SMITH, N. J. P. & SWALLOW, P. N.

LAWRENCE, S. R.      1987      Structural development and petroleum potential of the
COSTER, P. W. &      northern flanks of the Bowland Basin (Carboniferous),
IRELAND. R. J.      North-west England. *In* Brooks & Glennie (Eds.) (*q.v.*), 225–233.

LEDINGHAM, P.                1986    Heat Mining. *Ming. Mag.* (March), 241–251.

LEES, G. M. & COX, P. T.     1937    Geological basis of the present search for oil in Great Britain by the D'Arcy Exploration Co. Ltd. *Q. J. geol. Soc. London*, **93**, 156–194.

LEES, G. M. &                1946    The geological results of the search for oilfields in Great
  TAITT, A.                          Britain. *Q. J. geol. Soc. London*, **101**, 255–317.

LEWIS, W. J.                 1967    *Lead mining in Wales.* University of Wales Press, Cardiff, 415 pp.

MANNING, W.                  1959    The Parys and Mona mines in Anglesey. *In* I.M.M. (*q.v.*), 313–325.

MITCHELL, G. H. &            1945    The geology of the northern part of the South Staffordshire
  STUBBLEFIELD, C. J.                Coalfield (Cannock Chase Region), *Geol. Surv. Wartime Pamphlet. Geol. surv. G.B.*, **43**.

MOORE, J. McM.               1982    Mineral Zonation near the Granitic Batholiths of South-west and Northern England and some geothermal analogues. *In* Evans, A. M. (Ed.) 1982 (*q.v.*), 229–241.

MOORBATH, S.                 1962    Lead isotope abundance studies on mineral occurrences in the British Isles and their geological significance. *Philos. Trans. R. Soc. London*, **A254**, 295–360.

MOSELEY, F. (Ed.)            1978    *The Geology of the Lake District.* Yorkshire Geological Society, Leeds, 284 pp.

MOSTAGHEL, M. A.             1985    Classification of the South Pennine Orefield. *Mercian Geologist*, **10**, 27–38.

MOSTAGHEL, M. A. &           1986    A sedimentary basin evolution model for ore genesis in the
  FORD, T. D.                        South Pennine Orefield. *Mercian Geologist*, **10** (3), 209–224.

NESBITT, R. W. &             1986    *Geology in the real world – the Kingsley Dunham Volume.*
  NICHOL, I. (Eds.)                  Instn. Ming. Metall. London, 493 pp.

NOTHOLT, A. J. G. &          1973    Salt. *Miner. Dossier Miner. Resour. Consult. Comm.* **7**.
  HIGHLEY, D. E.                     HMSO, London, 36 pp.

                             1975    Gypsum and Anhydrite. *Miner. Dossier Miner. Resour. Consult. Comm.* **13**, HMSO, London, 38 pp.

NOTHOLT, A. J. G.            1976    Potash. *Miner Dossier Miner. Resour. Consult. Comm.* **16**. HMSO, London, 33 pp.

NUTT, M. J. C.,              1979    The age of mineralization at Parys Mountain, Anglesey. *In*
  INESON, P. R. &                    Harris, A. L., Holland, C. H. & Leake, B. E. (Eds.), *The*
  MITCHELL, J. G.                    *Caledonides of the British Isles – reviewed.* Spec. Pub. 8, Geol. Soc. London, 619–627.

O'BRIEN, C., PLANT, J. A.,   1985    The geochemistry, metasomatism and petrogenesis of the
  SIMPSON, P. R. &                   granites of the English Lake District. *J. geol. Soc. London*,
  TARNEY, J.                         **142**, 1139–1157.

O'NEAL, R. H.                1956    *Derbyshire lead and lead mining – A Bibliography.* Derbyshire County Library, Derby, 59 pp.

OWEN, T. R.                  1974a   The Variscan orogeny in Wales. *In* Owen, T. R. (Ed.) 1974, (*q.v.*).

OWEN, T. R. (Ed.)            1974    *The Upper Palaeozoic and Post Palaeozoic rocks of Wales.* University of Wales Press, Cardiff, 285–294.

PARNELL, J.                  1987    The occurrence of hydrocarbons in Cambrian sandstones of the Welsh Borderland. *Geol. J.*, **22** (Thematic Issue), 173–190.

                             1988    Migration of biogenic hydrocarbons into granites: a review of hydrocarbon in British plutons. *Mar. Petrol. Geol.*, **5**, 385–386.

PEARSON, M. J.               1979    Geochemistry of the Hepworth Carboniferous sedimentary sequence and origin of the diagenetic iron minerals and concretions. *Geochim. cosmochim. Acta.*, **43**, 927–941.

PENN, I. E.    AND FIVE OTHERS    1987    Principal features of the hydrocarbon prospectivity of the Wessex–Channel Basin. *In* Brooks & Glennie (Eds.) (*q.v.*), 109–118.

PHILLIPS, W. J.    1972    Hydraulic fracturing and mineralisation. *J. geol. Soc. London*, **128**, 337–359.

1983    Discussion. *In Trans. Instn. Ming. Metall.* (**B**), **92**, B102–103.

1986    Hydraulic fracturing effects in the formation of mineral deposits. *Trans. Instn. Ming. Metall.* (**B**), **95**, B17–24.

PLANT, J. A. & JONES, D. G. (Eds.)    1989    *Metallogenic models and exploration criteria for buried carbonate hosted ore deposits – a multidisciplinary study in eastern England*. Institution of Mining and Metallurgy and British Geological Survey, London and Keyworth, 161 pp.

POINTON, C. R. & IXER, R. A.    1980    Parys Mountain mineral deposit, Anglesey, Wales: geology and ore mineralogy. *Trans. Instn. Ming. Metall.* (**B**), **89**, B143–155.

POOLE, E. G.    1971    The Oxfordshire Coalfield. *Nature*, London, **232**, 394–395.

POSTLETHWAITE, J.    1913    *Mines and Mining in the (English) Lake District* (3rd Edition). W. H. Moss, Whitehaven, 164 pp.

PRICE, R. J. & BARNSLEY, G. B.    1982    Undersea Coal Exploration. *Coal Energy Qtrly.*, No. **35** (Winter), N.C.B.

RAMSBOTTOM, W. H. C. AND SIX OTHERS    1978    A correlation of Silesian rocks in the British Isles. *Spec. Rep. geol. Soc. London*, **10**, 81 pp.

RANKIN, A. H. & CRIDDLE, A. J.    1985    Mineralizing fluids and metastable low-temperature inclusion brines at Llanharry iron deposit, South Wales. *Trans. Instn. Ming. Metall.* (**B**), **94**, B126–B132.

RAYBOULD, J. G.    1974    Ore textures, paragenesis and zoning in the lead–zinc veins of Mid-Wales. *Trans. Instn. Ming. Metall.* (**B**), **83**, B112–B119.

1975    Discussion on Raybould (1974). *Trans. Instn. Ming. Metall.* (**B**), **84**, B67–69 and **84**, B154–155.

1977    Late Proterozoic stratiform sulphides in southern Caledonides of the British Isles. *Trans. Instn. Ming. Metall.* (**B**), **86**, B99–101.

REEDMAN, A. J., COLMAN, T. B., CAMPBELL, S. D. G. & HOWELLS, M. F.    1985    Volcanogenic mineralization related to the Snowdon Volcanic Group (Ordovician), Gwynedd, North Wales. *J. geol. Soc. London*, **142**, 875–888.

RICE, R. & SHARP, G. J.    1976    Copper mineralization in the forest of Coed-y-Brenin, North Wales. *Trans. Inst. Ming. Metall.* (**B**), **85**, B1–13.

ROBERTSON, R. H. S.    1986    *Fuller's Earth: A History of Calcium Montmorillonite*. Mineralogical Society Occasional Publications, Volturna Press, Hythe, Kent, UK, 421 pp.

ROBINSON, B. W. & INESON, P. R.    1979    Sulphur, oxygen and carbon isotope investigations of lead–zinc–barite–fluorite–calcite mineralization, Derbyshire, England. *Trans. Instn. Ming. Metall.* (**B**), **88**, B107–117.

ROSE, W. C. C. & DUNHAM, K. C.    1977    Geology and hematite deposits of South Cumbria. *Econ. Mem. geol. Surv. G.B.*, 170 pp.

RUNDLE, C. C.    1981    Discussion on the age of mineralization at Parys Mountain, Anglesey. *J. geol. Soc. London*, **138**, 755–756.

RUSSELL, M. J.    1978    Downward excavating hydrothermal cells and Irish-type ore deposits: importance of an underlying thick Caledonian prism. *Trans. Instn. Ming. Metall.* (**B**), **87**, B168–171.

RUSSELL, M. J. & HALL, J.    1988    Mechanics of downward percolation of water in crystalline rocks, with application to problems of geothermal energy extraction. *Trans. Instn. Ming. Metall.* (**B**), **97**, B51–56.

RUSSELL, M. J.,         1981   The genesis of sediment-hosted exhalative zinc and lead
    SOLOMON, M. &              deposits. *Mineralium Deposita*, **16**, 113–127.
    WALSHE, J. L.

SAMS, M. S. &          1988   Models of convective flow and mineralization in south-
    THOMAS-BETTS, A.          west England. *J. geol. Soc. London*, **145**, 809–817.

SAWKINS, F. J.          1966   Ore genesis in the North Pennine orefield in the light of
                              fluid inclusion studies. *Econ. Geol.*, **61**, 385–401.

SCHNELLMAN, G. A.       1959   Lead–zinc mining in the Carboniferous limestone of North
                              Wales. *In* Anon 1959 (*q.v.*), 235–246.

SCOTT, A.               1929   Ball Clays. *Spec. Rep. Min. Resour. G.B.*, **XXXI**, 73 pp.

SCOTT, J. &             1987   Geological aspects of current onshore Great Britain
    COLTER, U. S.             exploration plays. *In* Brooks & Glennie (Eds.) (*q.v.*), 95–
                              107.

SELLEY, R. C.           1987   British shale gas potential scrutinized. *Oil & Gas J.*, June,
                              62–64.

SELLEY, R. C. &         1987   Petroleum habitat in South Dorset. *In* Brooks & Glennie
    STONELY, R.               (Eds.) (*q.v.*), 139–148.

SELWOOD, B. W.,         1985   Stratigraphy and sedimentology of the Great Oolite Group
    SCOTT, J.,                in the Humbly Grove Oilfield, Hampshire. *Mar. Petrol.
    MIKKELSEN, P. &           Geol.*, **2**, 44–55.
    AKRYOD, P.

SHEPHERD, T. J. &       1985   Metallogenesis in the Harlech Dome, North Wales: A fluid
    ALLEN, P. M.              inclusion interpretation. *Mineralium Deposita*, **20**, 159–
                              168.

SHEPHERD, T. J. S. &    1987   Role of basinal brines in the genesis of polymetallic vein
    SCRIVENER, R. C.          deposits, Kit Hill–Gunnislake area, S.W. England. *Proc.
                              Ussher. Soc.* **6** (4), 491–497.

SHEPHERD, T. J.,        1976   Genesis of Carrock Fell tungsten deposits, Cumbria: fluid
    BECKINSALE, R. D.,        inclusion and isotopic study. *Trans. Instn. Ming. Metall.*
    RUNDLE, C. C. &           (**B**), **85**, B63–73.
    DURHAM, J.

SHEPHERD, T. J.,        1982   Rare earth element and isotopic geochemistry of the North
    DARBYSHIRE, D. P. F.,     Pennine ore deposits. *Bull. Bur. Rech. geol. min. Paris* (**2**),
    MOORE, G. R. &            section II, No. 4, pp. 371–377.
    GREENWOOD, D. A.

SHEPPARD, S. M. F.      1977   The Cornubian Batholith South West England: D/H and
                              $^{18}O/^{16}O$ studies of kaolinite and other alteration minerals. *J.
                              geol. Soc. London*, **133**, 573–591.

SHRIMPTON, G. J.        1989   United Kingdom. *Ming. J. Annual Review*, A175–A178.

SMITH, N. J. P.         1987   The deep geology of Central England: the prospectivity of
                              the Palaeozoic rocks. *In* Brooks & Glennie (Eds.) (*q.v.*),
                              217–224.

SMITH, E. G.,           1967   Geology of the country around Chesterfield, Matlock and
    RHYS, G. H. &             Mansfield. *Mem. geol. Surv. G.B.*, 430 pp.
    EDEN, R. A.

SMITH, K.,              1985   The deep structure of Derbyshire. *Geol. J.*, **20**, 215–225.
    SMITH, N. J. P. &
    HOLLIDAY, D. W.

SOUTHWOOD, M.           1984   Basaltic lavas at Parys Mountain, Anglesey: trace-element
                              geochemistry, tectonic setting and exploration implica-
                              tions. *Trans. Instn. Ming. Metall.* (**B**), **93**, B51–54.

STANLEY, C. J. &        1982   Copper, lead, zinc and cobalt mineralization in the English
    VAUGHAN, D. J.            Lake District: classification, conditions of formation and
                              genesis. *J. geol. Soc. London*, **139**, 569–579.

STONE, M.               1982   The behaviour of tin and some other trace elements during
                              granite differentiation, West Cornwall, England. *In* Evans,
                              A. M. (Ed.), 1982 (*q.v.*), 339–355.

| STONE, M. & EXLEY, C. S. | 1986 | High heat production granites of southwest England and their associated mineralization: a review. *Trans. Instn. Ming. Metall.* (**B**), **95**, B25–B36. |
| TAYLOR, J. C. M. | 1986 | Gas prospects in the Variscan Thrust Province of Southern England. *In* Brooks *et al.* (*q.v.*), 37–54. |
| TAYLOR, J. H. | 1949 | Petrology of the Northampton Sand Ironstone formation. *Mem. geol. Surv. G.B.*, 111 pp. |
| THOMAS, I. A. | 1973 | Celestite. *Miner. Dossier Miner. Resour. Consult. Comm.*, **6**. HMSO, London, 26 pp. |
| THOMAS, T. M. | 1961 | *The Mineral Wealth of Wales and its Exploitation.* Oliver & Boyd, Edinburgh, 248 pp. |
| THORNE, M. G. & EDWARDS, R. P. | 1985 | Recent advances in concepts of ore genesis in South West England. *Trans. Roy. Geol. Soc. Cornwall*, **XXI**, 113–152. |
| TROTTER, F. M. | 1942 | Geology of the Forest of Dean Coal and Ironore Field. *Mem. geol. Surv. Eng. & Wales*, 95 pp. |
| | 1948 | The devolatilisation of coal seams in South Wales. *Q. J. geol. Soc. London*, **104**, 389–437. |
| | 1954 | The genesis of the high-rank coals. *Proc. Yorkshire geol. Soc.*, **29**, 267. |
| TRUEMAN, A. E. (Ed.) | 1954 | *The coalfields of Great Britain.* Edward Arnold, London, 396 pp. |
| TRYTHALL, R. J. B., ECCLES, C., MOLYNEUX, S. G. & TAYLOR, W. E. G. | 1987 | Age of controls of ironstone deposition (Ordovician), North Wales. *Geol. J.*, **22** (Thematic Issue), 34–43. |
| TUBB, S. R., SOULBY, A. & LAWRENCE, S. R. | 1986 | Palaeozoic prospects on the Northern flank of the London–Brabant Massif. *In* Brooks *et al.* (*q.v.*), 55–72. |
| VARVILL, W. W. | 1959 | The Future of Lead–Zinc and Fluorspar mining in Derbyshire. *In* Anon, 1959 (*q.v.*), 175–203. |
| VINCENT, A. & NICHOLAS, C. | 1982 | Industrial Minerals. *In* Durrance & Laming (*q.v.*), 291–305. |
| WANDLESS, A. M. | 1959 | Occurrence of Sulphur in British Coals. *J. Inst. Fuel*, **32**, 258–266. |
| WARD, C. R. | 1984 | *Coal Geology and Coal Technology.* Blackwell Scientific Publications, Oxford, 345 pp. |
| WHITEHEAD, T. H., ANDERSON, W., WILSON, V., WRAY, D. A. & DUNHAM, K. C. | 1952 | The Liassic Ironstones. *Mem. geol. Surv. G.B.*, 211 pp. |
| WOODLAND, A. W. | 1956 | The manganese deposits of Great Britain. *In* Reyna, J. G. (Ed.), *Symposium sobre yacimentos de manganeso.* XX Int. geol Congr. (Mexico), **5**, 196–218. |
| WOODLAND, A. W. (Ed.) | 1975 | *Petroleum and the Continental Shelf of Northwest Europe*, **1**, *Geology.* Applied Science Publishers Ltd., Barking, Essex, 501 pp. |
| YOUNG, T. P. & TAYLOR, W. E. G. | 1989 | Phanerozoic Ironstones. *Geol. Soc. Spec. Publ.*, **46**, London, 251 pp. |

# INDEX

Compiled by
HILARY DAVIES

639

*Italic* type refers to figures; **bold** type to tables